S, S	Poynting vector, W m^{-2}
S	flux density, W m^{-2} Hz^{-1}
S, s	distance, m; also surface area, m^2
s	second (of time)
sr	steradian = square radian = rad^2
T	tesla = Wb m^{-2}
T	tera = 10^{12} (prefix)
t	time, s
U	radiation intensity, W sr^{-1}
V	volt
V	voltage (also emf), V
$\mathscr{V}$	emf (electromotive force), V
v	velocity, m s^{-1}
W	watt
Wb	weber
w	energy density, J m^{-3}
X	reactance, Ω
X	reactance/unit length, Ω m^{-1}
$\hat{\mathbf{x}}$	unit vector in x direction
x	coordinate direction
Y	admittance, ℧
Y	admittance/unit length, ℧ m^{-1}
$\hat{\mathbf{y}}$	unit vector in y direction
y	coordinate direction
Z	impedance, Ω
Z	impedance/unit length, Ω m^{-1}
Z_c	intrinsic impedance, conductor, Ω per square
Z_d	intrinsic impedance, dielectric, Ω per square
Z_L	load impedance, Ω
Z_{yz}	transverse impedance, rectangular waveguide, Ω
$Z_{r\phi}$	transverse impedance, cylindrical waveguide, Ω
Z_0	intrinsic impedance, space, Ω per square
Z_0	characteristic impedance, transmission line, Ω
$\hat{\mathbf{z}}$	unit vector in z direction
z	coordinate direction, also red shift
α	(alpha) angle, deg or rad
α	attenuation constant, nep m^{-1}
β	(beta) angle, deg or rad; also phase constant = $2\pi/\lambda$
γ	(gamma) angle, deg or rad
δ	(delta) angle
ε	(epsilon) per constant), .
ε_{ap}	aperture efficiency
ε_M	beam efficiency
ε_m	stray factor
ε_r	relative permittivity
ε_0	permittivity of vacuum, F m^{-1}
η	(eta)
θ	(theta) angle, deg or rad
$\hat{\boldsymbol{\theta}}$	(theta) unit vector in θ direction
κ	(kappa) constant
λ	(lambda) wavelength, m
λ_0	free-space wavelength
μ	(mu) permeability, H m^{-1}
μ_r	relative permeability
μ_0	permeability of vacuum, H m^{-1}
ν	(nu)
ξ	(xi)
π	(pi) = 3.1416
ρ	(rho) electric charge density, C m^{-3}: also mass density, kg m^{-3}
ρ	reflection coefficient, dimensionless
ρ_s	surface charge density, C m^{-2}
ρ_L	linear charge density, C m^{-1}
σ	(sigma) conductivity, ℧ m^{-1}
σ	radar cross section
τ	(tau) tilt angle, polarization ellipse, deg or rad
τ	transmission coefficient
ϕ	(phi) angle, deg or rad
$\hat{\boldsymbol{\phi}}$	(phi) unit vector in ϕ direction
χ	(chi) susceptibility, dimensionless
ψ	(psi) angle, deg or rad
ψ_m	magnetic flux, Wb
Ω	(capital omega) ohm
Ω	(capital omega) solid angle, sr or deg^2
Ω_A	beam area
Ω_M	main beam area
Ω_m	minor lobe area
℧	(upsidedown capital omega) mho (℧ = $1/\Omega$ = **S**, siemens)
ω	(omega) angular frequency (= $2\pi f$), rad s^{-1}

ANTENNAS

Other books by John D. Kraus:

ELECTROMAGNETICS, 3rd ed. (McGraw-Hill)
RADIO ASTRONOMY, 2nd ed. (Cygnus-Quasar)
BIG EAR (Cygnus-Quasar)
OUR COSMIC UNIVERSE (Cygnus-Quasar)

Cover: Geostationary relay satellite with helical antennas. (See Fig. 7-6.)

To Heinrich Hertz, who invented the first antennas . . .

. . . and Guglielmo Marconi, who pioneered in their practical application.

McGraw-Hill Series in Electrical Engineering

Consulting Editor
Stephen W. Director, Carnegie-Mellon University

Circuits and Systems
Communications and Signal Processing
Control Theory
Electronics and Electronic Circuits
Power and Energy
Electromagnetics
Computer Engineering
Introductory
Radar and Antennas
VLSI

Previous Consulting Editors

Ronald N. Bracewell, Colin Cherry, James F. Gibbons, Willis W. Harman, Hubert Heffner, Edward W. Herold, John G. Linvill, Simon Ramo, Ronald A. Rohrer, Anthony E. Siegman, Charles Susskind, Frederick E. Terman, John G. Truxal, Ernst Weber and John R. Whinnery

ANTENNAS

Second Edition

John D. Kraus

Director, Radio Observatory
Taine G. McDougal Professor Emeritus of
Electrical Engineering and Astronomy
The Ohio State University

with sections on
Frequency-Sensitive Surfaces by Benedikt A. Munk
Radar Scattering by Robert G. Kouyoumjian
and
Moment Method by Edward H. Newman
all of the Ohio State University

McGraw-Hill, Inc.
New York St. Louis San Francisco Auckland Bogotá
Caracas Lisbon London Madrid Mexico City Milan
Montreal New Delhi San Juan Singapore
Sydney Tokyo Toronto

This book was set in Times Roman.
The editor was Alar E. Elken;
the designer was Joan E. O'Connor; the production supervisor was
Friederich Schulte. New drawings were done by Robert Davis and
Kristine Hall.
Project supervision was done by Santype International, Ltd.

ANTENNAS

6 7 8 9 10 11 12 13 14 BKMBKM 9 9 8 7 6 5 4

ISBN 0-07-035422-7

Library of Congress Cataloging-in-Publication Data

Kraus, John Daniel. (date).
Antennas.

(McGraw-Hill series in electrical engineering.
Electronics and electronic circuits)
Includes index.
1. Antennas (Electronics) I. Title.
TK7871.6.K74 1988 621.38'028'3 87-15913
ISBN 0-07-035422-7

Radar and Antennas

Consulting Editor
Stephen W. Director, Carnegie-Mellon University

Collin: *Antennas and Radiowave Propagation*
Kraus: *Antennas*
Skolnik: *Introduction to Radar Systems*
Weeks: *Antenna Engineering*

ABOUT THE AUTHOR

John D. Kraus was born in Ann Arbor, Michigan, in 1910 and received his Ph.D. degree in physics from the University of Michigan in 1933. He then did research in nuclear physics with Michigan's newly completed 100-ton cyclotron until World War II when he worked on the degaussing of ships for the U.S. Navy and on radar countermeasures at Harvard University. After the War he came to the Ohio State University where he is now Director of the Radio Observatory and McDougal Professor (Emeritus) of Electrical Engineering and Astronomy.

Dr. Kraus is the inventor of the helical antenna, the workhorse of space communication, the corner reflector, used by the millions for television reception, and many other types of antennas. He designed and built the giant Ohio radio telescope known as "Big Ear." He is the holder of many patents and has published hundreds of scientific and technical articles. He is also the author of the widely used classic textbooks *Antennas* (McGraw-Hill, 1950), considered to be the "Antenna Bible," *Electromagnetics* (McGraw-Hill, 1953, second edition 1973, third edition, 1984), and *Radio Astronomy* (McGraw-Hill, 1966, second edition Cygnus Quasar, 1986). In addition, Dr. Kraus has written two popular books *Big Ear* (1976) and *Our Cosmic Universe* (1980).

Dr. Kraus received the U.S. Navy Meritorious Civilian Service Award in 1946. He was made a Fellow of the Institute of Electrical and Electronic Engineers (IEEE) in 1954 and was elected to the National Academy of Engineering in 1972. He received the Sullivant Medal, Ohio State University's top award, in 1970; the Outstanding Achievement Award of the University of Michigan in 1981; the prestigious Edison Medal of the IEEE in 1985; and the Distinguished Achievement Award of the Antennas and Propagation Society of the IEEE in the same year.

Currently, Dr. Kraus is serving as antenna consultant to government and industry.

CONTENTS

5 The Electric Dipole and Thin Linear Antennas 200

6 The Loop Antenna 238

7 The Helical Antenna 265

18 Antenna Measurements 805

Appendix A Tables for Reference 845

PREFACE

Although there has been an explosion in antenna technology in the years since *Antennas* was published, the basic principles and theory remain unchanged. My aim in this new edition is to blend a central core of basics from the first edition with a representative selection of important new developments and advances resulting in a much enlarged, updated book. It is appropriate that it is appearing just 100 years from the date on which the first antennas were invented by Heinrich Hertz to whom, along with Guglielmo Marconi, this new edition is dedicated.

As with the first edition, physical concepts are emphasized which aid in the visualization and understanding of the radiation phenomenon. More worked examples are given to illustrate the steps and thought processes required in going from a fundamental equation to a useful answer. The new edition stresses practical approaches to real-world situations and much information of value is made available in the form of many simple drawings, graphs and equations.

As with the first edition my purpose is to give a unified treatment of antennas from the electromagnetic theory point of view while paying attention to important applications. Following a brief history of antennas in the first chapter to set the stage, the next three chapters deal with basic concepts and the theory of point sources. These are followed by chapters on the linear, loop, helical, biconical and cylindrical antennas.

Then come chapters on antenna arrays, reflectors, slot, horn, complementary and lens antennas. The last four chapters discuss broadband and frequency-independent antennas, antennas for special applications including electrically small and physically small antennas, temperature, remote sensing, radar, scattering and measurements. The Appendix has many useful tables and references.

The book has over 1000 drawings and illustrations, many of which are unique, providing physical insights into the process of radiation from antennas.

The book is an outgrowth of lectures for antenna courses I have given at

Ohio State University and at Ohio University. The material is suitable for use at late undergraduate or early graduate level and is more than adequate for a one-semester course. The problem sets at the end of each chapter illustrate and extend the material covered in the text. In many cases they include important results on topics listed in the index. There are over 500 problems and worked examples.

Antennas has been written to serve not only as a textbook but also as a reference book for the practicing engineer and scientist. As an aid to those seeking additional information on a particular subject, the book is well documented with references both in footnotes and at the ends of chapters.

A few years ago it was customary to devote many pages of a textbook to computer programs, some with hundreds of steps. Now with many conveniently packaged programs and codes readily available this is no longer necessary. Extensive listings of such programs and codes, particularly those using moment methods, are given in Chapter 9 and in the Appendix. Nevertheless, some relatively short programs are included with the problem sets and in the Appendix.

From my IEEE Antennas and Propagation Society Centennial address (1984) I quote,

> With mankind's activities expanding into space, the need for antennas will grow to an unprecedented degree. Antennas will provide the vital links to and from everything out there. The future of antennas reaches to the stars.

Robert G. Kouyoumjian, Benedikt A. Munk and Edward H. Newman of the Ohio State University have contributed sections on scattering, frequency-sensitive surfaces and moment method respectively. I have edited these contributions to make symbols and terminology consistent with the rest of the book and any errors are my responsibility.

In addition, I gratefully acknowledge the assistance, comments and data from many others on the topics listed:

Walter D. Burnside, Ohio State University, compact ranges
Robert S. Dixon, Ohio State University, phased-arrays
Von R. Eshleman, Stanford University, gravity lenses
Paul E. Mayes, University of Illinois, frequency-independent antennas
Robert E. Munson, Ball Aerospace, microstrip antennas
Leon Peters, Jr., Ohio State University, dipole antennas
David M. Pozar, University of Massachusetts, moment method
Jack H. Richmond, Ohio State University, moment method
Helmut E. Schrank, Westinghouse, low-sidelobe antennas
Chen-To Tai, University of Michigan, dipole antennas

Throughout the preparation of this edition, I have had the expert editorial assistance of Dr. Erich Pacht.

Illustration and manuscript preparation have been handled by Robert Davis, Kristine Hall and William Taylor. McGraw-Hill editors were Sanjeev Rao, Alar Elken and John Morriss.

Although great care has been exercised, some errors or omissions in the text, tables, lists or figures will inevitably occur. Anyone finding them will do me a great service by writing to me so that they can be corrected in subsequent printings.

I also appreciate the very helpful comments of Ronald N. Bracewell, Stanford University, who reviewed the manuscript for McGraw-Hill.

Finally, I thank my wife, Alice, for her patience, encouragement and dedication through all the years of work it has taken.

John D. Kraus
Ohio State University

ANTENNAS

CHAPTER 1

INTRODUCTION

1-1 INTRODUCTION. Since Hertz and Marconi, antennas have become increasingly important to our society until now they are indispensable. They are everywhere: at our homes and workplaces, on our cars and aircraft, while our ships, satellites and spacecraft bristle with them. Even as pedestrians, we carry them.

Although antennas may seem to have a bewildering, almost infinite variety, they all operate according to the same basic principles of electromagnetics. The aim of this book is to explain these principles in the simplest possible terms and illustrate them with many practical examples. In some situations intuitive approaches will suffice while in others complete rigor is needed. The book provides a blend of both with selected examples illustrating when to use one or the other.

This chapter provides an historical background while Chap. 2 gives an introduction to basic concepts. The chapters that follow develop the subject in more detail.

1-2 THE ORIGINS OF ELECTROMAGNETIC THEORY AND THE FIRST ANTENNAS.[1] Six hundred years before Christ, a Greek mathematician, astronomer and philosopher, Thales of Miletus, noted that when amber is rubbed with silk it produces sparks and has a seemingly magical power to

[1] J. D. Kraus, "Antennas Since Hertz and Marconi," *IEEE Trans. Ants. Prop.*, **AP-33**, 131–137, 1985. See also references at end of chapter.

attract particles of fluff and straw. The Greek word for amber is *elektron* and from this we get our words *electricity, electron* and *electronics.* Thales also noted the attractive power between pieces of a natural magnetic rock called loadstone, found at a place called *Magnesia,* from which is derived the words *magnet* and *magnetism.* Thales was a pioneer in both electricity and magnetism but his interest, like that of others of his time, was philosophical rather than practical, and it was 22 centuries before these phenomena were investigated in a serious experimental way.

It remained for William Gilbert of England in about A.D. 1600 to perform the first systematic experiments of electric and magnetic phenomena, describing his experiments in his celebrated book, *De Magnete.* Gilbert invented the electroscope for measuring electrostatic effects. He was also the first to recognize that the earth itself is a huge magnet, thus providing new insights into the principles of the compass and dip needle.

In experiments with electricity made about 1750 that led to his invention of the lightning rod, Benjamin Franklin, the American scientist-statesman, established the law of conservation of charge and determined that there are both *positive* and *negative* charges. Later, Charles Augustin de Coulomb of France measured electric and magnetic forces with a delicate torsion balance he invented. During this period Karl Friedrich Gauss, a German mathematician and astronomer, formulated his famous divergence theorem relating a volume and its surface.

By 1800 Alessandro Volta of Italy had invented the voltaic cell and, connecting several in series, the electric battery. With batteries, electric currents could be produced, and in 1819 the Danish professor of physics Hans Christian Oersted found that a current-carrying wire caused a nearby compass needle to deflect, thus discovering that *electricity could produce magnetism.* Before Oersted, electricity and magnetism were considered as entirely independent phenomena.

The following year, André Marie Ampère, a French physicist, extended Oersted's observations. He invented the solenoidal coil for producing magnetic fields and theorized correctly that the atoms in a magnet are magnetized by tiny electric currents circulating in them. About this time Georg Simon Ohm of Germany published his now-famous law relating current, voltage and resistance. However, it initially met with ridicule and a decade passed before scientists began to recognize its truth and importance.

Then in 1831, Michael Faraday of London demonstrated that a changing magnetic field could produce an electric current. Whereas Oersted found that electricity could produce magnetism, Faraday discovered that *magnetism could produce electricity.* At about the same time, Joseph Henry of Albany, New York, observed the effect independently. Henry also invented the electric telegraph and relay.

Faraday's extensive experimental investigations enabled James Clerk Maxwell, a professor at Cambridge University, England, to establish in a profound and elegant manner the interdependence of electricity and magnetism. In his classic treatise of 1873, he published the first unified theory of electricity and magnetism and founded the science of electromagnetics. He postulated that light was electromagnetic in nature and that electromagnetic radiation of other wavelengths should be possible.

Maxwell unified electromagnetics in the same way that Isaac Newton unified mechanics two centuries earlier with his famous Law of Universal Gravitation governing the motion of *all* bodies both terrestrial *and* celestial.

Although Maxwell's equations are of great importance and, with boundary, continuity and other auxiliary relations, form the basic tenets of modern electromagnetics, many scientists of Maxwell's time were skeptical of his theories. It was more than a decade before his theories were vindicated by Heinrich Rudolph Hertz.

Early in the 1880s the Berlin Academy of Science had offered a prize for research on the relation between electromagnetic forces and dielectric polarization. Heinrich Hertz considered whether the problem could be solved with oscillations using Leyden jars or open induction coils. Although he did not pursue this problem, his interest in oscillations had been kindled and in 1886 as professor at the Technical Institute in Karlsruhe he assembled apparatus we would now describe as a complete radio system with an end-loaded dipole as transmitting antenna and a resonant square loop antenna as receiver.[1] When sparks were produced at a gap at the center of the dipole, sparking also occurred at a gap in the nearby loop. During the next 2 years, Hertz extended his experiments and demonstrated reflection, refraction and polarization, showing that except for their much greater length, radio waves were one with light. Hertz turned the tide against Maxwell around.

Hertz's initial experiments were conducted at wavelengths of about 8 meters while his later work was at shorter wavelengths, around 30 centimeters. Figure 1-1 shows Hertz's earliest 8-meter system and Fig. 1-2 a display of his apparatus, including the cylindrical parabolic reflector he used at 30 centimeters.

Although Hertz was the father of radio, his invention remained a laboratory curiosity for nearly a decade until 20-year-old Guglielmo Marconi, on a summer vacation in the Alps, chanced upon a magazine which described Hertz's experiments. Young Guglielmo wondered if these Hertzian waves could be used to send messages. He became obsessed with the idea, cut short his vacation and rushed home to test it.

In spacious rooms on an upper floor of the Marconi mansion in Bologna, Marconi repeated Hertz's experiments. His first success late one night so elated him he could not wait until morning to break the news, so he woke his mother and demonstrated his radio system to her.

Marconi quickly went on to add tuning, big antenna and ground systems for longer wavelengths and was able to signal over large distances. In mid-December 1901, he startled the world by announcing that he had received radio

[1] His dipole was called a *Hertzian dipole* and the radio waves *Hertzian waves.*

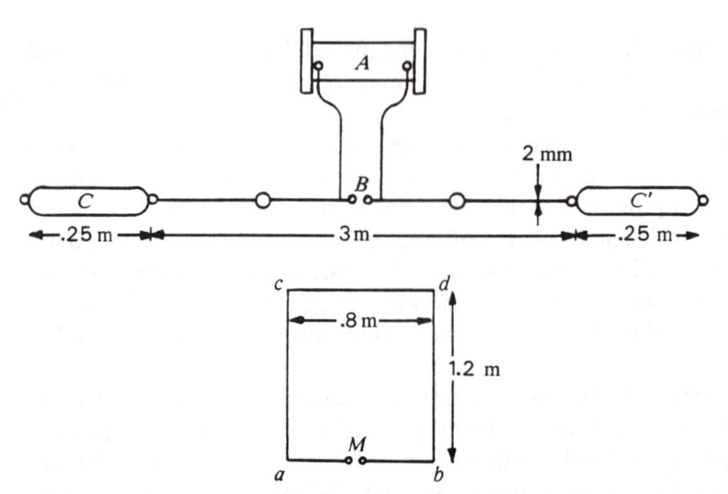

Figure 1-1 Heinrich Hertz's complete radio system of 1886 with end-loaded dipole transmitting antenna (CC') and resonant loop receiving antenna ($abcd$) for $\lambda \simeq 8$ m. With induction coil (A) turned on, sparks at gap B induced sparks at M in the loop receiving antenna. (*From Heinrich Hertz's book Electric Waves, Macmillan, 1893; redrawn with dimensions added.*)

Figure 1-2 Hertz's sphere-loaded $\lambda/2$ dipole and spark gap (resting on floor in foreground) and cylindrical parabolic reflector for 30 centimeters (standing at left). Dipole with spark gap is on the parabola focal axis. (*Photograph by Edward C. Jordan.*)

signals at St. John's, Newfoundland, which had been sent across the Atlantic from a station he had built at Poldhu in Cornwall, England. The scientific establishment did not believe his claim because in its view radio waves, like light, should travel in straight lines and could not bend around the earth from England to Newfoundland. However, the Cable Company believed Marconi and served him with a writ to cease and desist because it had a monopoly on transatlantic communication. The Cable Company's stock had plummeted following Marconi's announcement and it threatened to sue him for any loss of revenue if he persisted. However, persist he did, and a legal battle developed that continued for 27 years until finally the cable and wireless groups merged.

One month after Marconi's announcement, the American Institute of Electrical Engineers (AIEE) held a banquet at New York's Waldorf-Astoria to celebrate the event. Charles Protius Steinmetz, President of the AIEE, was there, as was Alexander Graham Bell, but many prominent scientists boycotted the banquet. Their theories had been challenged and they wanted no part of it.

Not long after the banquet, Marconi provided irrefutable evidence that radio waves could bend around the earth. He recorded Morse signals, inked automatically on tape, as received from England across almost all of the Atlantic while steaming aboard the *SS Philadelphia* from Cherbourg to New York. The ship's captain, the first officer and many passengers were witnesses.

A year later, in 1903, Marconi began a regular transatlantic message service between Poldhu, England, and stations he built near Glace Bay, Nova Scotia, and South Wellfleet on Cape Cod.

In 1901, the Poldhu station had a fan aerial supported by two 60-meter guyed wooden poles and as receiving antenna for his first transatlantic signals at St. John's, Marconi pulled up a 200-meter wire with a kite, working it against an array of wires on the ground. A later antenna at Poldhu, typical of antennas at other Marconi stations, consisted of a conical wire cage. This was held up by four massive self-supporting 70-meter wooden towers (Fig. 1-3). With inputs of 50 kilowatts, antenna wires crackled and glowed with corona at night. Local residents were sure that such fireworks in the sky would alter the weather.

Rarely has an invention captured the public imagination like Marconi's wireless did at the turn of the century. We now call it radio but then it was wireless: Marconi's wireless. After its value at sea had been dramatized by the *SS Republic* and *SS Titanic* disasters, Marconi was regarded with a universal awe and admiration seldom matched. Before wireless, complete isolation enshrouded a ship at sea. Disaster could strike without anyone on the shore or nearby ships being aware that anything had happened. Marconi changed all that. Marconi became the Wizard of Wireless.

Although Hertz had used 30-centimeter wavelengths and Jagadis Chandra Bose and others even shorter wavelengths involving horns and hollow waveguides, the distance these waves could be detected was limited by the technology of the period so these centimeter waves found little use until much later. Radio developed at long wavelengths with very long waves favored for long distances. A popular "rule-of-thumb" of the period was that the range which could be

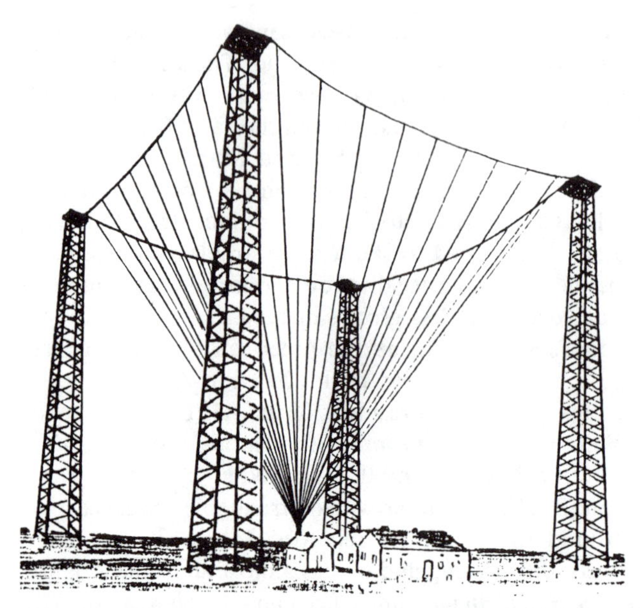

Figure 1-3 Square-cone antenna at Marconi's Poldhu, England, station in 1905. The 70-meter wooden towers support a network of wires which converge to a point just above the transmitting and receiving buildings between the towers.

achieved with adequate power was equal to 500 times the wavelength. Thus, for a range of 5000 kilometers, one required a wavelength of 10 000 meters.

At typical wavelengths of 2000 to 20 000 meters, the antennas were a small fraction of a wavelength in height and their radiation resistances only an ohm or less. Losses in heat and corona reduced efficiencies but with the brute power of many kilowatts, significant amounts were radiated. Although many authorities favored very long wavelengths, Marconi may have appreciated the importance of radiation resistance and was in the vanguard of those advocating shorter wavelengths, such as 600 meters. At this wavelength an antenna could have 100 times its radiation resistance at 6000 meters.

In 1912 the Wireless Institute and the Society of Radio Engineers merged to form the Institute of Radio Engineers.[1] In the first issue of the Institute's *Proceedings*, which appeared in January 1913, it is interesting that the first article was on antennas and in particular on radiation resistance. Another *Proceedings* article noted the youthfulness of commercial wireless operators. Most were in their late teens with practically none over the age of 25. Wireless was definitely a young man's profession.

The era before World War I was one of long waves, of spark, arc and alternators for transmission; and of coherers, Fleming valves and De Forest

[1] In 1963, the Institute of Radio Engineers and the American Institute of Electrical Engineers merged to form the Institute of Electrical and Electronic Engineers (IEEE).

audions for reception. Following the war, vacuum tubes became available for transmission; continuous waves replaced spark and radio broadcasting began in the 200 to 600-meter range.

Wavelengths less than 200 meters were considered of little value and were relegated to the amateurs. In 1921, the American Radio Relay League sent Paul Godley to Europe to try and receive a Greenwich, Connecticut, amateur station operating on 200 meters. Major Edwin H. Armstrong, inventor of the superheterodyne receiver and later of FM, constructed the transmitter with the help of several other amateurs. Godley set up his receiving station near the Firth of Clyde in Scotland. He had two receivers, one a 10-tube superheterodyne, and a Beverage antenna. On December 12, 1921, just 20 years to the day after Marconi received his first transatlantic signals on a very long wavelength, Godley received messages from the Connecticut station and went on to log over 30 other U.S. amateurs. It was a breakthrough, and in the years that followed, wavelengths from 200 meters down began to be used for long-distance communication.

Atmospherics were the bane of the long waves, especially in the summer. They were less on the short waves but still enough of a problem in 1930 for the Bell Telephone Laboratories to have Karl G. Jansky study whether they came from certain predominant directions. Antennas for telephone service with Europe might then be designed with nulls in these directions.

Jansky constructed a rotating 8-element Bruce curtain with a reflector operating at 14 meters (Fig. 1-4). Although he obtained the desired data on atmospherics from thunderstorms, he noted that in the absence of all such static there was always present a very faint hisslike noise or static which moved completely around the compass in 24 hours. After many months of observations, Jansky

Figure 1-4 Karl Guthe Jansky and his rotating Bruce curtain antenna with which he discovered radio emission from our galaxy. (*Courtesy Bell Telephone Laboratories; Jansky inset courtesy Mary Jansky Striffler.*)

concluded that it was coming from beyond the earth and beyond the sun. It was a cosmic static coming from our galaxy with the maximum from the galactic center. Jansky's serendipitous discovery of extraterrestrial radio waves opened a new window on the universe. Jansky became the father of radio astronomy.

Jansky recognized that this cosmic noise from our galaxy set a limit to the sensitivity that could be achieved with a short-wave receiving system. At 14 meters this sky noise has an equivalent temperature of 20 000 kelvins. At centimeter wavelengths it is less, but never less than 3 kelvins. This is the residual sky background level of the primordial fireball that created the universe as measured four decades later by radio astronomers Arno Penzias and Robert Wilson of the Bell Telephone Laboratories at a site not far from the one used by Jansky.

For many years, or until after World War II, only one person, Grote Reber, followed up Jansky's discovery in a significant way. Reber constructed a 9-meter parabolic reflector antenna (Fig. 1-5) operating at a wavelength of about 2 meters which is the prototype of the modern parabolic dish antenna. With it he made the first radio maps of the sky. Reber also recognized that his antenna-receiver constituted a radiometer, i.e., a temperature-measuring device in which his receiver response was related to the temperature of distant regions of space coupled to his antenna via its radiation resistance.

With the advent of radar during World War II, centimeter waves, which had been abandoned at the turn of the century, finally came into their own and the entire radio spectrum opened up to wide usage. Hundreds of stationary communication satellites operating at centimeter wavelengths now ring the earth as though mounted on towers 36 000 kilometers high. Our probes are exploring the solar system to Uranus and beyond, responding to our commands and sending back pictures and data at centimeter wavelengths even though it takes more than an hour for the radio waves to travel the distance one way. Our radio telescopes operating at millimeter to kilometer wavelengths receive signals from objects so distant that the waves have been traveling for more than 10 billion years.

With mankind's activities expanding into space, the need for antennas will grow to an unprecedented degree. Antennas will provide the vital links to and from everything out there. The future of antennas reaches to the stars.

1-3 ELECTROMAGNETIC SPECTRUM. Continuous wave energy radiated by antennas oscillates at radio frequencies. The associated free-space waves range in length from thousands of meters at the long-wave extreme to fractions of a millimeter at the short-wave extreme. The relation of radio waves to the entire electromagnetic spectrum is presented in Fig. 1-6. Short radio waves and long infrared waves overlap into a twilight zone that may be regarded as belonging to both.

The wavelength λ of a wave is related to the frequency f and velocity v of the wave by

$$\boxed{\lambda = \frac{v}{f}} \qquad (1)$$

Figure 1-5 Grote Reber and his parabolic reflector antenna with which he made the first radio maps of the sky. This antenna, which he built in 1938, is the prototype of the modern dish antenna. (*Reber inset courtesy Arthur C. Clarke.*)

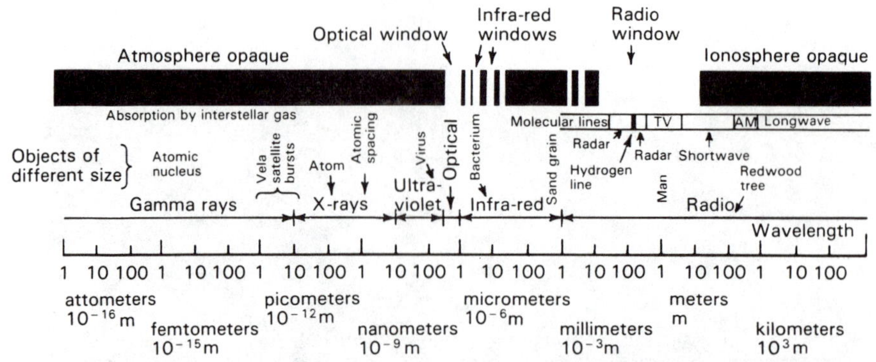

Figure 1-6 The electromagnetic spectrum with wavelength on a logarithmic scale from the shortest gamma rays to the longest radio waves. The atmospheric-ionospheric opacity is shown at the top with the optical and radio windows in evidence.

Thus, the wavelength depends on the velocity v which depends on the medium. In this sense, frequency is a more fundamental quantity since it is independent of the medium. When the medium is free space (vacuum)

$$v = c = 3 \times 10^8 \text{ m s}^{-1} \tag{2}$$

Figure 1-7 shows the relation of wavelength to frequency for $v = c$ (free space). Many of the uses of the spectrum are indicated along the right-hand edge of the figure. A more detailed frequency use listing is given in Table 1-1.

Table 1-1 Radio-frequency band designations

Frequency	Wavelength	Band designation
30–300 Hz	10–1 Mm	ELF (extremely low frequency)
300–3000 Hz	1 Mm–100 km	
3–30 kHz	100–10 km	VLF (very low frequency)
30–300 kHz	10–1 km	LF (low frequency)
300–3000 kHz	1 km–100 m	MF (medium frequency)
3–30 MHz	100–10 m	HF (high frequency)
30–300 MHz	10–1 m	VHF (very high frequency)
300–3000 MHz	1 m–10 cm	UHF (ultra high frequency)
3–30 GHz	10–1 cm	SHF (super high frequency)
30–300 GHz	1 cm–1 mm	EHF (extremely high frequency)
300–3000 GHz	1 mm–100 μm	
Frequency	**Wavelength**	**IEEE Radar Band designation**
1–2 GHz	30–15 cm	L
2–4 GHz	15–7.5 cm	S
4–8 GHz	7.5–3.75 cm	C
8–12 GHz	3.75–2.50 cm	X
12–18 GHz	2.50–1.67 cm	Ku
18–27 GHz	1.67–1.11 cm	K
27–40 GHz	1.11 cm–7.5 mm	Ka
40–300 GHz	7.5–1.0 mm	mm

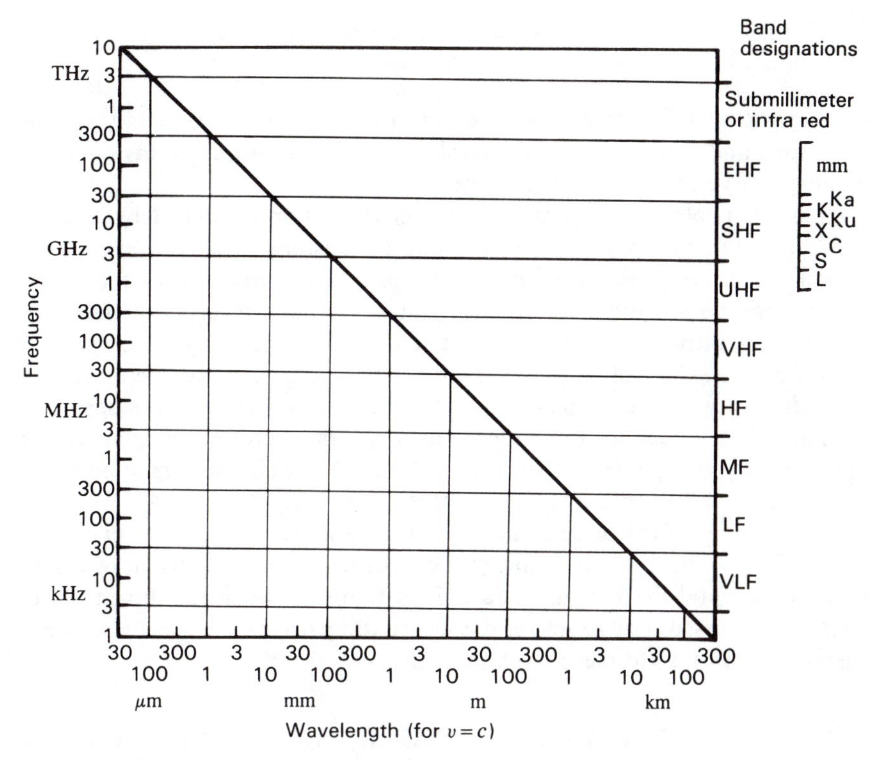

Figure 1-7 Wavelength versus frequency for $v = c$.

Example of wavelength for a given frequency. For a frequency of 300 MHz the corresponding wavelength is given by

$$\lambda = \frac{c}{f} = \frac{3 \times 10^8 \text{ m s}^{-1}}{300 \times 10^6 \text{ Hz}} = 1 \text{ m} \tag{3}$$

In a lossless nonmagnetic dielectric medium with relative permittivity $\varepsilon_r = 2$, the same wave has a velocity

$$v = \frac{c}{\sqrt{\varepsilon_r}} = \frac{3 \times 10^8}{\sqrt{2}} = 2.12 \times 10^8 \text{ m s}^{-1} \tag{4}$$

and

$$\lambda = \frac{v}{f} = \frac{2.12 \times 10^8}{300 \times 10^6} = 0.707 \text{ m} = 707 \text{ mm} \tag{5}$$

1-4 DIMENSIONS AND UNITS.

Lord Kelvin is reported to have said:

When you can measure what you are speaking about and express it in numbers you know something about it; but when you cannot measure it, when you cannot express it in numbers your knowledge is of a meagre and unsatisfactory kind; it may be the beginning of knowledge but you have scarcely progressed in your thoughts to the stage of science whatever the matter may be.

To this it might be added that before we can measure something, we must define its dimensions and provide some standard, or reference unit, in terms of which the quantity can be expressed numerically.

A *dimension* defines some physical characteristic. For example, length, mass, time, velocity and force are dimensions. The dimensions of *length, mass, time, electric current, temperature* and *luminous intensity* are considered as the *fundamental dimensions* since other dimensions can be defined in terms of these six. This choice is arbitrary but convenient. Let the letters L, M, T, I, $\mathscr{T}$ and $\mathscr{I}$ represent the dimensions of length, mass, time, electric current, temperature and luminous intensity. Other dimensions are then secondary dimensions. For example, area is a secondary dimension which can be expressed in terms of the fundamental dimension of length squared (L^2). As other examples, the fundamental dimensions of velocity are L/T and of force are ML/T^2.

A *unit* is a standard or reference by which a dimension can be expressed numerically. Thus, the *meter* is a unit in terms of which the dimension of length can be expressed and the *kilogram* is a unit in terms of which the dimension of mass can be expressed. For example, the length (dimension) of a steel rod might be 2 meters and its mass (dimension) 5 kilograms.

1-5 FUNDAMENTAL AND SECONDARY UNITS. The units for the fundamental dimensions are called the *fundamental* or *base units.* In this book the International System of Units, abbreviated SI, is used.[1] In this system the *meter, kilogram, second, ampere, kelvin* and *candela* are the base units for the six fundamental dimensions of length, mass, time, electric current, temperature and luminous intensity. The definitions for these fundamental units are:

Meter (m). *Length* equal to 1 650 763.73 wavelengths in vacuum corresponding to the $2p_{10}$–$5d_5$ transition of krypton-86.

Kilogram (kg). Equal to *mass* of international prototype kilogram, a platinum-iridium mass preserved at Sèvres, France. This standard kilogram is the only artifact among the SI base units.

Second (s). Equal to *time* duration of 9 192 631 770 periods of radiation corresponding to the transition between two hyperfine levels of the ground state of cesium-133. The second was formerly defined as 1/86 400 part of a mean solar day. The earth's rotation rate is gradually slowing down, but the atomic (cesium-133) transition is

[1] The International System of Units is the modernized version of the metric system. The abbreviation SI is from the French name *Système Internationale d'Unités.* For the complete official description of the system see U.S. Natl. Bur. Stand. Spec. Pub. 330, 1971.

much more constant and is now the standard. The two standards differ by about 1 second per year.

Ampere (A). *Electric current* which if flowing in two infinitely long parallel wires in vacuum separated by 1 meter produces a force of 200 nanonewtons per meter of length (200 nN $m^{-1} = 2 \times 10^{-7}$ N m^{-1}).

Kelvin (K). *Temperature* equal to 1/273.16 of the triple point of water (or triple point of water equals 273.16 kelvins).[1]

Candela (cd). *Luminous intensity* equal to that of 1/600 000 square meter of a perfect radiator at the temperature of freezing platinum.

The units for other dimensions are called *secondary* or *derived* units and are based on these fundamental units.

The material in this book deals principally with the four fundamental dimensions *length, mass, time* and *electric current* (dimensional symbols L, M, T and I). The four fundamental units for these dimensions are the basis of what was formerly called the meter-kilogram-second-ampere (mksa) system, now a subsystem of the SI. The book also includes discussions of temperature but no references to luminous intensity.

The complete SI involves not only units but also other recommendations, one of which is that multiples and submultiples of the SI units be stated in steps of 10^3 or 10^{-3}. Thus, the kilometer (1 km = 10^3 m) and the millimeter (1 mm = 10^{-3} m) are preferred units of length, but the centimeter (= 10^{-2} m) is not. For example, the proper SI designation for the width of motion-picture film is 35 mm, not 3.5 cm.

In this book *rationalized* SI units are used. The rationalized system has the advantage that the factor 4π does not appear in Maxwell's equations (App. A), although it does appear in certain other relations. A complete table of units in this system is given in the Appendix of *Electromagnetics*, 3rd ed., by J. D. Kraus (McGraw-Hill, 1984).

1-6 HOW TO READ THE SYMBOLS AND NOTATION.

In this book *quantities*, or *dimensions*, which are scalars, like charge Q, mass M or resistance R, are always in italics. Quantities which may be vectors *or* scalars are boldface as vectors and italics as scalars, e.g., electric field $\mathbf{E}$ (vector) or E (scalar). Unit vectors are always boldface with a hat (circumflex) over the letter, e.g., $\hat{\mathbf{x}}$ or $\hat{\mathbf{r}}$.[2]

---

[1] Note that the symbol for degrees is not used with kelvins. Thus, the boiling temperature of water (100°C) is 373 kelvins (373 K), *not* 373°K. However, the degree sign is retained with degrees Celsius.

[2] In longhand notation a vector may be indicated by a bar over the letter and hat (ˆ) over the unit vector.

Units are in roman type, i.e., *not* italic; for example, H for henry, s for second, or A for ampere.[1] The abbreviation for a unit is capitalized if the unit is derived from a proper name; otherwise it is lowercase (small letter). Thus, we have C for coulomb but m for meter. Note that when the unit is written out, it is always lowercase even though derived from a proper name. *Prefixes* for units are also roman, like n in nC for nanocoulomb or M in MW for megawatt.

Example 1. $\mathbf{D} = \hat{\mathbf{x}}\ 200\ \mathrm{pC\ m^{-2}}$

means that the electric flux density **D** is a vector in the positive x direction with a magnitude of 200 picocoulombs per square meter ($=2 \times 10^{-10}$ coulomb per square meter).

Example 2. $V = 10\ \mathrm{V}$

means that the voltage V equals 10 volts. Distinguish carefully between V (italics) for voltage, V (roman) for volts, **v** (lowercase, boldface) for velocity and v (lowercase, italics) for volume.

Example 3. $S = 4\ \mathrm{W\ m^{-2}\ Hz^{-1}}$

means that the flux density S (a scalar) equals 4 watts per square meter per hertz. This can also be written $S = 4\ \mathrm{W/m^2/Hz}$ or $4\ \mathrm{W/(m^2\ Hz)}$, but the form $\mathrm{W\ m^{-2}\ Hz^{-1}}$ is more direct and less ambiguous.

Note that for conciseness, prefixes are used where appropriate instead of exponents. Thus, a velocity would be expressed in prefix form as $\mathbf{v} = 215\ \mathrm{Mm\ s^{-1}}$ (215 megameters per second) *not* in the exponential form $2.15 \times 10^8\ \mathrm{m\ s^{-1}}$. However, in solving a problem the exponential would be used although the final answer might be put in the prefix form ($215\ \mathrm{Mm\ s^{-1}}$).

The modernized metric (SI) units and the conventions used herein combine to give a concise, exact and unambiguous notation, and if one is attentive to the details, it will be seen to possess both elegance and beauty.

1-7 EQUATION NUMBERING.

Important equations and those referred to in the text are numbered consecutively beginning with each section. When reference is made to an equation in a different section, its number is preceded by the chapter and section number. Thus, (14-15-3) refers to Chap. 14, Sec. 15, Eq. (3). A reference to this same equation within Sec. 15 of Chap. 14 would read simply (3). Note that chapter and section numbers are printed at the top of each page.

---

[1] In longhand notation no distinction is usually made between quantities (italics) and units (roman). However, it can be done by placing a bar under the letter to indicate italics or writing the letter with a distinct slant.

1-8 DIMENSIONAL ANALYSIS. It is a necessary condition for correctness that every equation be balanced dimensionally. For example, consider the hypothetical formula

$$\frac{M}{L} = DA$$

where M = mass
L = length
D = density (mass per unit volume)
A = area

The dimensional symbols for the left side are M/L, the same as those used. The dimensional symbols for the right side are

$$\frac{M}{L^3} L^2 = \frac{M}{L}$$

Therefore, both sides of this equation have the dimensions of mass per length, and the equation is balanced dimensionally. This is not a guarantee that the equation is correct; i.e., it is not a *sufficient* condition for correctness. It is, however, a *necessary* condition for correctness, and it is frequently helpful to analyze equations in this way to determine whether or not they are dimensionally balanced.

Such *dimensional analysis* is also useful for determining what the dimensions of a quantity are. For example, to find the dimensions of force, we make use of Newton's second law that

$$\text{Force} = \text{mass} \times \text{acceleration}$$

Since acceleration has the dimensions of length per time squared, the dimensions of force are

$$\frac{\text{Mass} \times \text{length}}{\text{Time}^2}$$

or in dimensional symbols

$$\text{Force} = \frac{ML}{T^2}$$

REFERENCES

Bose, Jagadis Chandra: *Collected Physical Papers*, Longmans, Green, 1927.

Bose, Jagadis Chandra: "On a Complete Apparatus for the Study of the Properties of Electric Waves," *Elect. Engr.* (*Lond.*), October 1896.

Brown, George H.: "Marconi," *Cosmic Search*, **2**, 5–8, Spring 1980.

Dunlap, Orrin E.: *Marconi—The Man and His Wireless*, Macmillan, 1937.

Faraday, Michael: *Experimental Researches in Electricity*, B. Quaritch, London, 1855.

Gundlach, Friedrich Wilhelm: "Die Technik der kürzesten elecktromagnetischen Wellen seit Heinrich Hertz," *Elektrotech. Zeit.* (*ETZ*), **7**, 246, 1957.

Hertz, Heinrich Rudolph: "Über Strahlen elecktrischer Kraft," *Wiedemanns Ann. Phys.*, **36**, 769–783, 1889.

Hertz, Heinrich Rudolph: *Electric Waves*, Macmillan, London, 1893; Dover, 1962.

Hertz, Heinrich Rudolph: *Collected Works*, Barth Verlag, 1895.

Hertz, Heinrich Rudolph: "The Work of Hertz and His Successors—Signalling through Space without Wires," Electrician Publications, 1894, 1898, 1900, 1908.

Hertz, Johanna: *Heinrich Hertz*, San Francisco Press, 1977 (memoirs, letters and diaries of Hertz).

Kraus, John D.: *Big Ear*, Cygnus-Quasar, 1976.

Kraus, John D.: "Karl Jansky and His Discovery of Radio Waves from Our Galaxy," *Cosmic Search*, **3**, no. 4, 8–12, 1981.

Kraus, John. D.: "Grote Reber and the First Radio Maps of the Sky," *Cosmic Search*, **4**, no. 1, 14–18, 1982.

Kraus, John D.: "Karl Guthe Jansky's Serendipity, Its Impact on Astronomy and Its Lessons for the Future," in K. Kellermann and B. Sheets (eds.), *Serendipitous Discoveries in Radio Astronomy*, National Radio Astronomy Observatory, 1983.

Kraus, John D.: *Electromagnetics*, 3rd ed., McGraw-Hill, 1984.

Kraus, John D.: "Antennas Since Hertz and Marconi," *IEEE Trans. Ants. Prop.*, **AP-33**, 131–137, February 1985 (Centennial Plenary Session Paper).

Kraus, John D.: *Radio Astronomy*, 2nd ed., Cygnus-Quasar, 1986; Sec. 1–2 on Jansky, Reber and early history.

Kraus, John D.: "Heinrich Hertz—Theorist and Experimenter," *IEEE Trans. Microwave Theory Tech. Hertz Centennial Issue*, **MTT-36**, May 1988.

Lodge, Oliver J.: "Signalling through Space without Wires," Electrician Publications, 1898.

Marconi, Degna: *My Father Marconi*, McGraw-Hill, 1962.

Maxwell, James Clerk: *A Treatise on Electricity and Magnetism*, Oxford, 1873, 1904.

Newton, Isaac: *Principia*, Cambridge, 1687.

Poincaré, Henri, and F. K. Vreeland: *Maxwell's Theory and Wireless Telegraphy*, Constable, London, 1905.

Ramsey, John F.: "Microwave Antenna and Waveguide Techniques before 1900," *Proc. IRE*, **46**, 405–415, February 1958.

Rayleigh, Lord: "On the Passage of Electric Waves through Tubes or the Vibrations of Dielectric Cylinders," *Phil. Mag.*, **43**, 125–132, February 1897.

Righi, A.: "L'Ottica della Oscillazioni Elettriche," Zanichelli, Bologna, 1897.

Rothe, Horst: "Heinrich Hertz, der Entdecker der elektromagnetischen Wellen," *Elektrotech. Zeit. (ETZ)*, **7**, 247–251, 1957.

Wolf, Franz: "Heinrich Hertz, Leben and Werk," *Elektrotech. Zeit (ETZ)*, **7**, 242–246, 1957.

CHAPTER 2

BASIC ANTENNA CONCEPTS

2-1 INTRODUCTION. The purpose of this chapter is to provide introductory insights into antennas and their characteristics. Following a section on definitions, the basic parameters of *radiation resistance*, *temperature*, *pattern*, *directivity*, *gain*, *beam area* and *aperture* are introduced. From the aperture concept it is only a few steps to the important Friis transmission formula. This is followed by a discussion of sources of radiation, field zones around an antenna and the effect of shape on impedance. The sources of radiation are illustrated for both transient (pulse) and continuous waves. The chapter concludes with a discussion of polarization and cross-field.

2-2 DEFINITIONS. A *radio antenna*[1] may be defined as the structure associated with the region of transition between a guided wave and a free-space wave, or vice versa.

In connection with this definition it is also useful to consider what is meant by the terms *transmission line* and *resonator*.

A *transmission line* is a device for transmitting or guiding radio-frequency energy from one point to another. Usually it is desirable to transmit the energy

[1] In its zoological sense, an antenna is the feeler, or organ of touch, of an insect. According to usage in the United States the plural of "insect antenna" is "antennae," but the plural of "radio antenna" is "antennas."

with a minimum of attenuation, heat and radiation losses being as small as possible. This means that while the energy is being conveyed from one point to another it is confined to the transmission line or is bound closely to it. Thus, the wave transmitted along the line is 1-dimensional in that it does not spread out into space but follows along the line. From this general point of view one may extend the term transmission line (or transmission system) to include not only coaxial and 2-wire transmission lines but also hollow pipes, or *waveguides*.

A generator connected to an infinite, lossless transmission line produces a uniform traveling wave along the line. If the line is short-circuited, the outgoing traveling wave is reflected, producing a standing wave on the line due to the interference between the outgoing and reflected waves. A standing wave has associated with it local concentrations of energy. If the reflected wave is equal to the outgoing wave, we have a pure standing wave. The energy concentrations in such a wave oscillate from entirely electric to entirely magnetic and back twice per cycle. Such energy behavior is characteristic of a resonant circuit, or *resonator*. Although the term resonator, in its most general sense, may be applied to any device with standing waves, the term is usually reserved for devices with stored energy concentrations that are large compared with the net flow of energy per cycle.[1] Where there is only an outer conductor, as in a short-circuited section of waveguide, the device is called a *cavity resonator*.

Thus, antennas *radiate (or receive) energy*, transmission lines *guide energy*, while resonators *store energy*.

A guided wave traveling along a transmission line which opens out, as in Fig. 2-1, will radiate as a free-space wave. The guided wave is a plane wave while the free-space wave is a spherically expanding wave. Along the uniform part of the line, energy is guided as a plane wave with little loss, provided the spacing between the wires is a small fraction of a wavelength. At the right, as the transmission line separation approaches a wavelength or more, the wave tends to be radiated so that the opened-out line acts like an antenna which launches a free-space wave. The currents on the transmission line flow out on the transmission line and end there, but the fields associated with them keep on going. To be more explicit, the region of transition between the guided wave and the free-space wave may be defined as an *antenna*.

We have described the antenna as a transmitting device. As a receiving device the definition is turned around, and an antenna is the region of transition between a free-space wave and a guided wave. Thus, *an antenna is a transition device, or transducer, between a guided wave and a free-space wave, or vice versa.*[2]

While transmission lines (or waveguides) are usually made so as to mini-

[1] The ratio of the energy stored to that lost per cycle is proportional to the Q, or sharpness of resonance of the resonator (see Sec. 6-12).

[2] We note that antenna parameters, such as impedance or gain, require that the antenna terminals be specified.

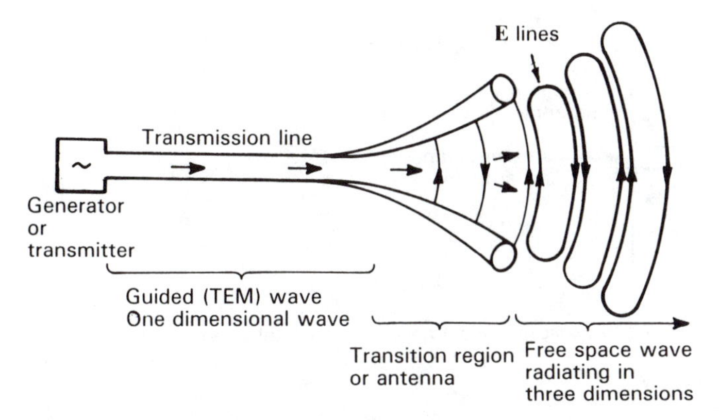

Figure 2-1 The antenna is a region of transition between a wave guided by a transmission line and a free-space wave. The transmission line conductor separation is a small fraction of a wavelength while the separation at the open end of the transition region or antenna may be many wavelengths. More generally, *an antenna interfaces between electrons on conductors and photons in space.* The eye is another such device.

mize radiation, antennas are designed to radiate (or receive) energy as effectively as possible.

The antenna, like the eye, is a transformation device converting electromagnetic photons into circuit currents; but, unlike the eye, the antenna can also convert energy from a circuit into photons radiated into space.[1] In simplest terms *an antenna converts photons to currents or vice versa.*

Consider a transmission line connected to a dipole[2] antenna as in Fig. 2-2. The dipole acts as an antenna because it launches a free-space wave. However, it may also be regarded as a section of an open-ended transmission line. In addition, it exhibits many of the characteristics of a resonator, since energy reflected from the ends of the dipole gives rise to a standing wave and energy storage near the antenna. Thus, a single device, in this case the dipole, exhibits simultaneously properties characteristic of an *antenna*, a *transmission line* and a *resonator*.

2-3 BASIC ANTENNA PARAMETERS.

Referring to Fig. 2-2, the antenna appears from the transmission line as a 2-terminal circuit element having an *impedance* Z with a resistive component called the *radiation resistance* R_r,

---

[1] A *photon* is the quantum unit of electromagnetic energy equal to hf, where h = Planck's constant ($= 6.63 \times 10^{-34}$ J s) and f = frequency (Hz).

[2] A positive electric charge q separated a distance from an equal but negative charge constitutes an *electric dipole*. If the separation is l, then ql is the *dipole moment*. A linear conductor which, at a given instant, has a positive charge at one end and an equal but negative charge at the other end may act as a *dipole antenna*. (A loop may be considered to be a *magnetic dipole antenna* of moment IA, where I = loop current and A = loop area.)

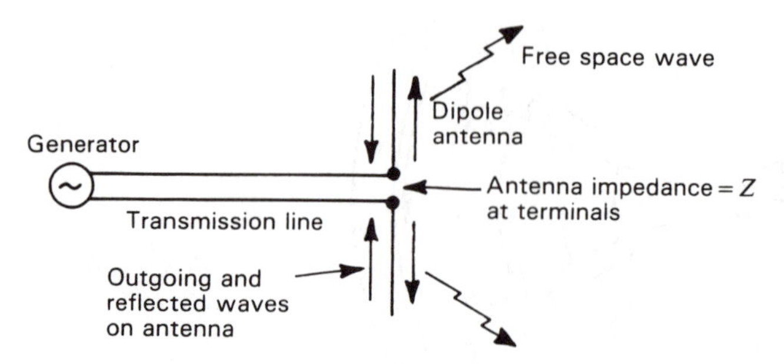

Figure 2-2 The antenna launches a free-space wave but appears as a circuit impedance to the transmission line.

while from space, the antenna is characterized by its *radiation pattern* or patterns involving field quantities.[1]

The radiation resistance R_r is not associated with any resistance in the antenna proper but is a resistance coupled from the antenna and its environment to the antenna terminals. Radiation resistance is discussed in Secs. 2-13 and 2-14 and further in Chap. 5.

Associated with the radiation resistance is also an *antenna temperature* T_A. For a lossless antenna this temperature has nothing to do with the physical temperature of the antenna proper but is related to the temperature of distant regions of space (and nearer surroundings) coupled to the antenna via its radiation resistance. Actually, the antenna temperature is not so much an inherent property of the antenna as it is a parameter that depends on the temperature of the regions the antenna is "looking at." In this sense, a receiving antenna may be regarded as a remote-sensing, temperature-measuring device (see Chap. 17).

Both the radiation resistance R_r and the antenna temperature T_A are single-valued scalar quantities. The radiation patterns, on the other hand, involve the variation of field or power (proportional to the field squared) as a function of the two spherical coordinates θ and ϕ.

2-4 PATTERNS.

Figure 2-3*a* shows a field pattern where r is proportional to the field intensity at a certain distance from the antenna in the direction θ, ϕ. The pattern has its *main-lobe* maximum in the z direction ($\theta = 0$) with *minor lobes* (side and back) in other directions. Between the lobes are nulls in the directions of zero or minimum radiation.

---

[1] *Fields and radiation.* An electromagnetic wave consists of electric and magnetic fields propagating through space, a *field* being a region where electric or magnetic forces act. The electric and magnetic fields in a free-space wave traveling outward at a large distance from an antenna convey energy called *radiation*.

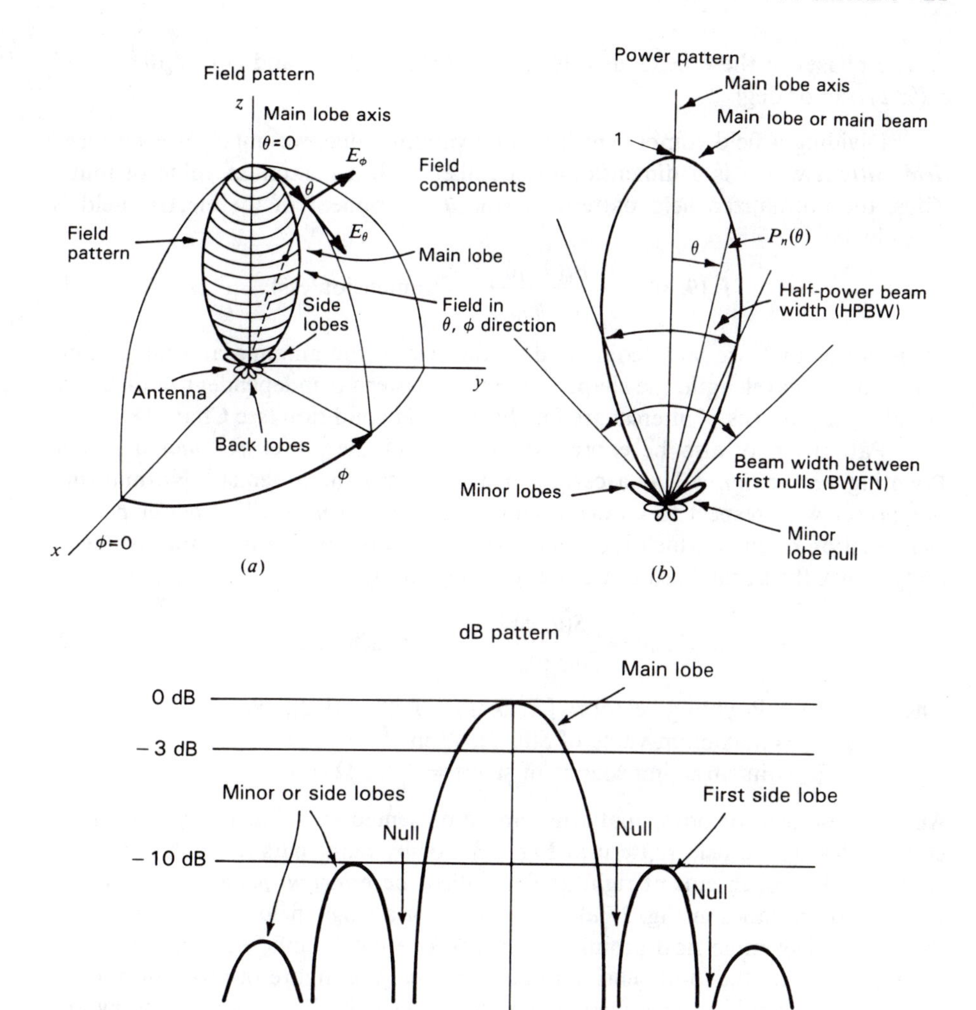

Figure 2-3 (*a*) Antenna field pattern with coordinate system. (*b*) Antenna power pattern in polar coordinates (linear scale). (*c*) Antenna pattern in rectangular coordinates and decibel (logarithmic) scale. Patterns (*b*) and (*c*) are the same.

To completely specify the radiation pattern with respect to field intensity and polarization requires three patterns:

1. The θ component of the electric field as a function of the angles θ and ϕ or $E_\theta(\theta, \phi)$ (V m^{-1})
2. The ϕ component of the electric field as a function of the angles θ and ϕ or $E_\phi(\theta, \phi)$ (V m^{-1})

3. The phases of these fields as a function of the angles θ and ϕ or $\delta_\theta(\theta, \phi)$ and $\delta_\phi(\theta, \phi)$ (rad or deg)

Dividing a field component by its maximum value, we obtain a *normalized field pattern* which is a dimensionless number with a maximum value of unity. Thus, the normalized field pattern for the θ component of the electric field is given by

$$E_\theta(\theta, \phi)_n = \frac{E_\theta(\theta, \phi)}{E_\theta(\theta, \phi)_{\max}} \qquad \text{(dimensionless)} \tag{1}$$

At distances that are large compared to the size of the antenna and large compared to the wavelength, the shape of the field pattern is independent of distance. Usually the patterns of interest are for this *far-field* condition (see Chap. 18).

Patterns may also be expressed in terms of the *power per unit area* [or Poynting vector $S(\theta, \phi)$] at a certain distance from the antenna.[1] Normalizing this power with respect to its maximum value yields a *normalized power pattern* as a function of angle which is a dimensionless number with a maximum value of unity. Thus, the normalized power pattern is given by

$$P_n(\theta, \phi) = \frac{S(\theta, \phi)}{S(\theta, \phi)_{\max}} \qquad \text{(dimensionless)} \tag{2}$$

where $S(\theta, \phi)$ = Poynting vector = $[E_\theta^2(\theta, \phi) + E_\phi^2(\theta, \phi)]/Z_0$, W m^{-2}
$S(\theta, \phi)_{\max}$ = maximum value of $S(\theta, \phi)$, W m^{-2}
Z_0 = intrinsic impedance of space = 376.7 Ω

Any of these field or power patterns can be presented in 3-dimensional spherical coordinates, as the field pattern in Fig. 2-3*a*, or by plane cuts through the main-lobe axis. Two such cuts at right angles, called the *principal plane patterns* (as in the *xz* and *yz* planes in Fig. 2-3*a*), may suffice for a single field component, and if the pattern is symmetrical around the *z* axis, one cut is sufficient. Figure 2-3*b* is such a pattern, the 3-dimensional pattern being a figure-of-revolution of it around the main-lobe axis (similar to the pattern in Fig. 2-3*a*). To show the minor lobes in more detail, the same pattern is presented in Fig. 2-3*c* in rectangular coordinates on a decibel scale, as given by

$$\text{dB} = 10 \log_{10} P_n(\theta, \phi) \tag{3}$$

Although the radiation characteristics of an antenna involve 3-dimensional patterns, many important radiation characteristics can be expressed in terms of simple single-valued scalar quantities. These include:

Beam widths, beam area, main-lobe beam area and *beam efficiency;*
Directivity and *gain;*
Effective aperture, scattering aperture, aperture efficiency and *effective height.*

---

[1] Although the Poynting vector, as the name implies, is a vector (with magnitude and direction), we use here its magnitude, its direction in the far field being radially outward.

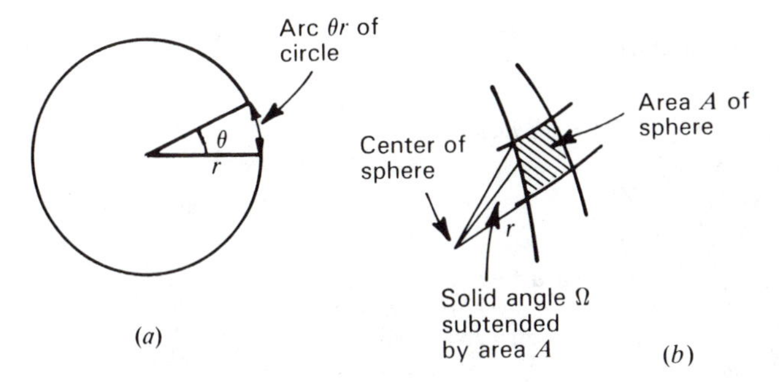

Figure 2-4 (*a*) Arc length $r\theta$ of circle of radius r subtends an angle θ. (*b*) The area A of a sphere of radius r subtends a solid angle Ω.

2-5 BEAM AREA (OR BEAM SOLID ANGLE). The arc of a circle as seen from the center of the circle subtends an angle. Thus, referring to Fig. 2-4*a*, the arc length θr subtends the angle θ. The total angle in the circle is 2π rad (or 360°) and the total arc length is $2\pi r$ (= circumference).

An area A of the surface of a sphere as seen from the center of the sphere subtends a *solid angle* Ω (Fig. 2-4*b*). The total solid angle subtended by the sphere is 4π steradians (or square radians), abbreviated sr.

Let us discuss solid angle in more detail with the aid of Fig. 2-5. Here the incremental area dA of the surface of a sphere is given by

$$dA = (r \sin\theta \, d\phi)(r \, d\theta) = r^2 \sin\theta \, d\theta \, d\phi = r^2 \, d\Omega \tag{1}$$

where $d\Omega$ = solid angle subtended by the area dA

The area of the strip of width $r\,d\theta$ extending around the sphere at a constant angle θ is given by $(2\pi r \sin\theta)(r\,d\theta)$. Integrating this for θ values from 0 to π yields the area of the sphere. Thus,

$$\text{Area of sphere} = 2\pi r^2 \int_0^{\pi} \sin\theta \, d\theta = 2\pi r^2[-\cos\theta]_0^{\pi} = 4\pi r^2 \tag{2}$$

where 4π = solid angle subtended by a sphere, sr

Thus,

$$1 \text{ steradian} = 1 \text{ sr} = (\text{solid angle of sphere})/(4\pi)$$

$$= 1 \text{ rad}^2 = \left(\frac{180}{\pi}\right)^2 (\text{deg}^2) = 3282.8064 \text{ square degrees} \tag{3}$$

Therefore,

$$4\pi \text{ steradians} = 3282.8064 \times 4\pi = 41\,252.96 \simeq 41\,253 \text{ square degrees}$$

$$= \text{solid angle in a sphere} \tag{4}$$

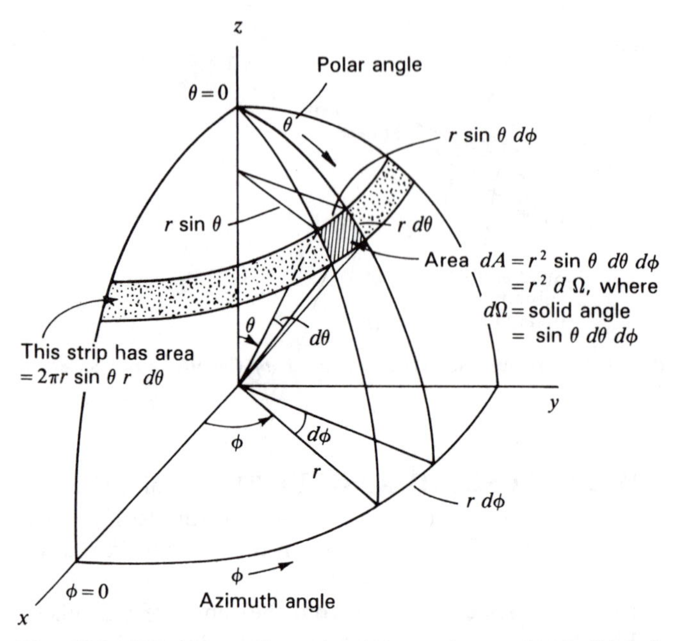

Figure 2-5 Spherical coordinates in relation to the area dA of solid angle $d\Omega = \sin\theta\ d\theta\ d\phi$.

Now the beam area (or *beam solid angle*) Ω_A for an antenna is given by the integral of the normalized power pattern over a sphere (4π sr) or

$$\Omega_A = \int_0^{2\pi}\int_0^{\pi} P_n(\theta, \phi)\ d\Omega \qquad \text{(sr)} \tag{5}$$

where $d\Omega = \sin\theta\ d\theta\ d\phi$

Referring to Fig. 2-6, the beam area Ω_A of an actual pattern is equivalent to the same solid angle subtended by the spherical cap of the cone-shaped (triangular cross-section) pattern.

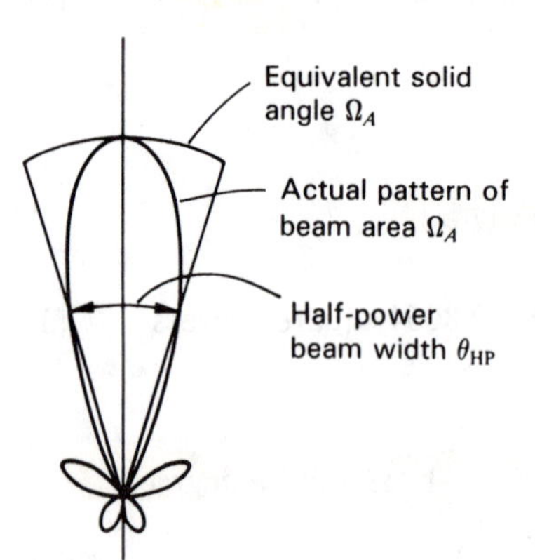

Figure 2-6 Cross section of symmetrical power pattern of antenna showing equivalent solid angle for a cone-shaped (triangular) pattern.

This solid angle can often be described *approximately* in terms of the angles subtended by the *half-power points* of the main lobe in the two principal planes as given by

$$\Omega_A \simeq \theta_{HP} \phi_{HP} \quad \text{(sr)} \tag{6}$$

where θ_{HP} and ϕ_{HP} are the *half-power beam widths* (HPBW) in the two principal planes, minor lobes being neglected.

2-6 RADIATION INTENSITY.

The power radiated from an antenna per unit solid angle is called the *radiation intensity* U (watts per steradian or per square degree). The normalized power pattern of the previous section can also be expressed in terms of this parameter as the ratio of the radiation intensity $U(\theta, \phi)$, as a function of angle, to its maximum value. Thus,

$$P_n(\theta, \phi) = \frac{U(\theta, \phi)}{U(\theta, \phi)_{max}} = \frac{S(\theta, \phi)}{S(\theta, \phi)_{max}} \tag{1}$$

Whereas the Poynting vector S depends on the distance from the antenna (varying inversely as the square of the distance), the radiation intensity U is independent of the distance, assuming in both cases that we are in the far field of the antenna (see Sec. 2-35).

2-7 BEAM EFFICIENCY.

The (total) *beam area* Ω_A (or *beam solid angle*) consists of the main beam area (or solid angle) Ω_M plus the minor-lobe area (or solid angle) Ω_m.[1] Thus,

$$\Omega_A = \Omega_M + \Omega_m \tag{1}$$

The ratio of the main beam area to the (total) beam area is called the (main) *beam efficiency* ε_M. Thus,

$$\varepsilon_M = \frac{\Omega_M}{\Omega_A} = \text{beam efficiency} \tag{2}$$

The ratio of the minor-lobe area (Ω_m) to the (total) beam area is called the *stray factor*. Thus,

$$\varepsilon_m = \frac{\Omega_m}{\Omega_A} = \text{stray factor} \tag{3}$$

It follows that

$$\varepsilon_M + \varepsilon_m = 1 \tag{4}$$

[1] If the main beam is not bounded by a deep null its extent becomes an arbitrary act of judgment.

2-8 DIRECTIVITY. The *directivity* D of an antenna is given by the ratio of the maximum radiation intensity (power per unit solid angle) $U(\theta, \phi)_{\text{max}}$ to the average radiation intensity U_{av} (averaged over a sphere). *Or*, at a certain distance from the antenna the directivity may be expressed as the ratio of the maximum to the average Poynting vector. Thus,

$$D = \frac{U(\theta, \phi)_{\text{max}}}{U_{\text{av}}} = \frac{S(\theta, \phi)_{\text{max}}}{S_{\text{av}}} \qquad \text{(dimensionless)} \tag{1}$$

Both radiation intensity and Poynting vector values should be measured in the far field of the antenna (see Sec. 2-35).

Now the average Poynting vector over a sphere is given by

$$S(\theta, \phi)_{\text{av}} = \frac{1}{4\pi} \int_0^{2\pi} \int_0^{\pi} S(\theta, \phi)\, d\Omega \qquad (\text{W m}^{-2}) \tag{2}$$

Thus, the directivity

$$D = \frac{1}{\dfrac{1}{4\pi} \displaystyle\iint \frac{S(\theta, \phi)}{S(\theta, \phi)_{\text{max}}}\, d\Omega} = \frac{1}{\dfrac{1}{4\pi} \displaystyle\iint P_n(\theta, \phi)\, d\Omega} \tag{3}$$

or

$$D = \frac{4\pi}{\Omega_A} \tag{4}$$

The smaller the beam solid angle, the greater the directivity.

2-9 EXAMPLES OF DIRECTIVITY. If an antenna could be *isotropic* (radiate the same in all directions)

$$P_n(\theta, \phi) = 1 \qquad (\text{for all } \theta \text{ and } \phi) \tag{1}$$

then

$$\Omega_A = 4\pi \tag{2}$$

and

$$D = 1 \tag{3}$$

This is the smallest directivity an antenna can have. Thus, Ω_A must always be equal to or less than 4π, while the directivity D must always be equal to or greater than unity.

Neglecting the effect of minor lobes, we have from (2-8-3) and (2-5-6) the simple *approximation*[1]

$$D \simeq \frac{4\pi}{\theta_{\text{HP}}\,\phi_{\text{HP}}} \simeq \frac{41\,000}{\theta^\circ_{\text{HP}}\,\phi^\circ_{\text{HP}}} \tag{4}$$

[1] 4π sr = 41 253 square degrees. Since (4) is an approximation 41 253 is rounded off to 41 000.

where θ_{HP} = half-power beam width in θ plane, rad
ϕ_{HP} = half-power beam width in ϕ plane, rad
$\theta^{\circ}_{\text{HP}}$ = half-power beam width in θ plane, deg
ϕ°_{HP} = half-power beam width in ϕ plane, deg

Equation (4) is an *approximation* and should be used in this context. To avoid inappropriate usage, see the discussion following Eq. (17) of Sec. 3-13.

If an antenna has a main lobe with both half-power beam widths (HPBWs) = 20°, its directivity from (2-8-4) and (2-5-6) is *approximately*

$$D = \frac{4\pi(\text{sr})}{\Omega_A(\text{sr})} \simeq \frac{41\,000\ (\text{deg}^2)}{\theta^{\circ}_{\text{HP}}\,\phi^{\circ}_{\text{HP}}} = \frac{41\,000\ (\text{deg}^2)}{20^{\circ} \times 20^{\circ}}$$

$$\simeq 103 \simeq 20\ \text{dBi (dB above isotropic)} \qquad (5)$$

which means that the antenna radiates a power in the direction of the main-lobe maximum which is about 100 times as much as would be radiated by a nondirectional (isotropic) antenna for the same power input.

2-10 DIRECTIVITY AND GAIN.

The *gain* of an antenna (referred to a lossless isotropic source) depends on both its directivity and its efficiency.[1] If the efficiency is not 100 percent, the gain is less than the directivity. Thus, the gain

$$G = kD \qquad \text{(dimensionless)} \qquad (1)$$

where k = efficiency factor of antenna ($0 \leq k \leq 1$), dimensionless

This efficiency has to do only with ohmic losses in the antenna. In transmitting, these losses involve power fed to the antenna which is not radiated but heats the antenna structure.

2-11 DIRECTIVITY AND RESOLUTION.

The resolution of an antenna may be defined as equal to half the beam width between first nulls (BWFN/2).[2] For example, an antenna whose pattern BWFN = 2° has a resolution of 1° and, accordingly, should be able to distinguish between transmitters on two adjacent satellites in the Clarke geostationary orbit separated by 1°. Thus, when the antenna beam maximum is aligned with one satellite, the first null coincides with the other satellite.

---

[1] When *gain* is used as a single-valued quantity (like directivity) its maximum nose-on main-beam value is implied in the same way that the power rating of an engine implies its maximum value. Multiplying the gain G by the normalized power pattern $P_n(\theta, \phi)$ gives the gain as a function of angle.

[2] Often called the *Rayleigh resolution*. See Sec. 11-23 and also J. D. Kraus, *Radio Astronomy*, 2nd ed., Cygnus-Quasar, 1986, pp. 6–19.

Half the beam width between first nulls is approximately equal to the half-power beam width (HPBW) or

$$\frac{\text{BWFN}}{2} \simeq \text{HPBW} \tag{1}$$

so from (2-5-6) the product of the BWFN/2 in the two principal planes of the antenna pattern is a measure of the antenna beam area.[1] Thus,

$$\Omega_A = \left(\frac{\text{BWFN}}{2}\right)_\theta \left(\frac{\text{BWFN}}{2}\right)_\phi \tag{2}$$

It then follows that the number N of radio transmitters or point sources of radiation distributed uniformly over the sky which an antenna can resolve is given approximately by

$$N = \frac{4\pi}{\Omega_A} \tag{3}$$

where Ω_A = beam area, sr

However, from (2-8-4),

$$D = \frac{4\pi}{\Omega_A} \tag{4}$$

and we may conclude that *ideally* the number of point sources an antenna can resolve is numerically equal to the directivity of the antenna or

$$D = N \tag{5}$$

Equation (4) states that the *directivity* is equal to the number of beam areas into which the antenna pattern can subdivide the sky and (5) gives the added significance that the *directivity is equal to the number of point sources in the sky that the antenna can resolve* under the assumed ideal conditions of a uniform source distribution.[2]

2-12 APERTURE CONCEPT. The concept of aperture is most simply introduced by considering a receiving antenna. Suppose that the receiving antenna is an electromagnetic horn immersed in the field of a uniform plane wave as suggested in Fig. 2-7. Let the Poynting vector, or power density, of the plane wave be S watts per square meter and the area of the mouth of the horn be A

---

[1] Usually BWFN/2 is slightly greater than HPBW and from (3-13-18) we may conclude that (2) is actually a better approximation to Ω_A than $\Omega_A = \theta_{HP}\phi_{HP}$ as given by (2-5-6).

[2] A strictly regular distribution of points on a sphere is only possible for 4, 6, 8, 12 and 20 points corresponding to the vertices of a tetrahedron, cube, octahedron, isoahedron and dodecahedron.

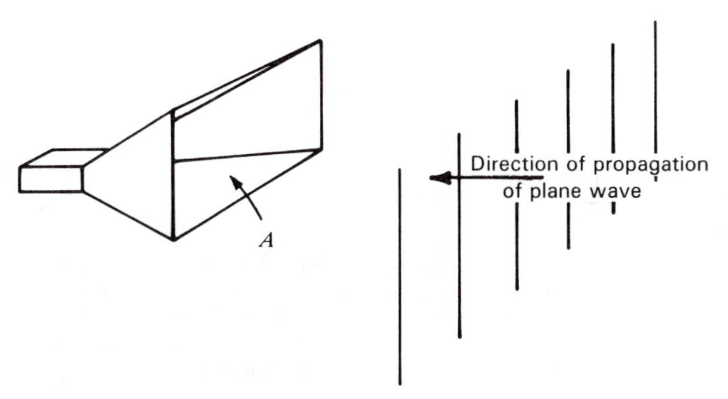

Figure 2-7 Plane wave incident on electromagnetic horn of mouth aperture A.

square meters. If the horn extracts all the power from the wave over its entire area A, then the total power P absorbed from the wave is

$$P = SA \qquad \text{(W)} \tag{1}$$

Thus, the electromagnetic horn may be regarded as an aperture, the total power it extracts from a passing wave being proportional to the aperture or area of its mouth.

It will be convenient to distinguish between several types of apertures, namely, *effective aperture*, *scattering aperture*, *loss aperture*, *collecting aperture* and *physical aperture*. These different types of apertures are defined and discussed in the following sections.

In the following discussion it is assumed, unless otherwise stated, that the antenna has the same polarization as the incident wave and is oriented for maximum response.

2-13 EFFECTIVE APERTURE.

Consider a dipole receiving antenna ($\lambda/2$ or less) situated in the field of a passing electromagnetic wave as suggested in Fig. 2-8*a*. The antenna collects power from the wave and delivers it to the terminating or load impedance Z_T connected to its terminals. The Poynting vector, or power density of the wave, is S watts per square meter. Referring to the equivalent circuit of Fig. 2-8*b*, the antenna may be replaced by an equivalent or Thévenin generator having an equivalent voltage V and internal or equivalent antenna impedance Z_A. The voltage V is induced by the passing wave and produces a current I through the terminating impedance Z_T given by

$$I = \frac{V}{Z_T + Z_A} \tag{1}$$

where I and V are rms or effective values.

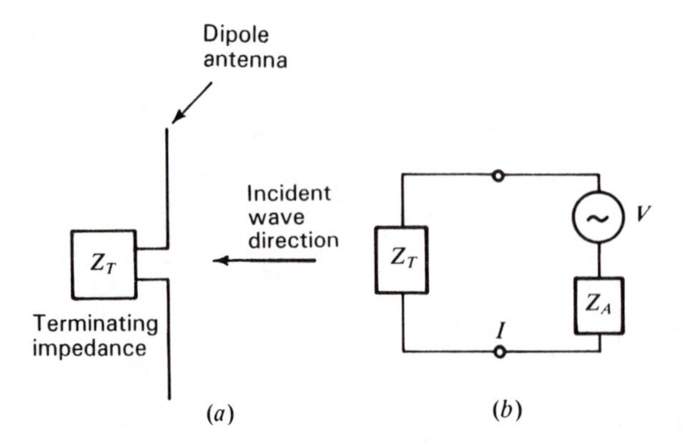

Figure 2-8 Schematic diagram of dipole antenna terminated in impedance Z_T with plane wave incident on antenna (*a*) and equivalent circuit (*b*).

In general, the terminating and antenna impedances are complex; thus

$$Z_T = R_T + jX_T \tag{2}$$

and

$$Z_A = R_A + jX_A \tag{3}$$

The antenna resistance may be divided into two parts, a *radiation resistance* R_r and a nonradiative or *loss resistance* R_L, that is,

$$R_A = R_r + R_L \tag{4}$$

Let the power delivered by the antenna to the terminating impedance be P. Then

$$P = I^2 R_T \tag{5}$$

From (1), (2) and (3), the current magnitude

$$I = \frac{V}{\sqrt{(R_r + R_L + R_T)^2 + (X_A + X_T)^2}} \tag{6}$$

Substituting (6) into (5) gives

$$P = \frac{V^2 R_T}{(R_r + R_L + R_T)^2 + (X_A + X_T)^2} \tag{7}$$

The ratio of the power P in the terminating impedance to the power density of the incident wave is an area A. Thus,

$$\frac{P}{S} = A \tag{8}$$

where P = power in termination, W
S = power density of incident wave, W m^{-2}
A = area, m^2

If S is in watts per square wavelength (W λ^{-2}) then A is in square wavelengths (λ^2), which is often a convenient unit of measurement for areas.

Substituting (7) into (8) we have

$$A = \frac{V^2 R_T}{S[(R_r + R_L + R_T)^2 + (X_A + X_T)^2]} \tag{9}$$

Unless otherwise specified, it is assumed that V is the induced voltage when the antenna is oriented for maximum response and the incident wave has the same polarization as the antenna. The value of A as indicated by (9) takes into account any antenna losses as given by R_L and any mismatch between the antenna and its terminating impedance.

Let us consider now the situation where the terminating impedance is the complex conjugate of the antenna impedance (terminal or load impedance *matched* to antenna) so that maximum power is transferred. Thus,

$$X_T = -X_A \tag{10}$$

and

$$R_T = R_r + R_L \tag{11}$$

Introducing (10) and (11) in (9) yields the *effective aperture* A_e of the antenna. Thus,

$$A_e = \frac{V^2}{4S(R_r + R_L)} \qquad (\text{m}^2 \text{ or } \lambda^2) \tag{12}$$

If the antenna is lossless ($R_L = 0$) we obtain the *maximum effective aperture* A_{em} of the antenna. Thus,

$$A_{em} = \frac{V^2}{4SR_r} \qquad (\text{m}^2 \text{ or } \lambda^2) \tag{13}$$

The aperture A_{em} given by (13) represents the area over which power is extracted from the incident wave and delivered to the load.

Sometimes the terminating impedance is not located physically at the antenna terminals as suggested in Fig. 2-8. Rather, it is in a receiver which is connected to the antenna by a length of transmission line. In this case Z_T is the equivalent impedance which appears across the antenna terminals. If the transmission line is lossless, the power delivered to the receiver is the same as that delivered to the equivalent terminating impedance Z_T. If the transmission line has attenuation, the power delivered to the receiver is less than that delivered to the equivalent terminating impedance by the amount lost in the line.

2-14 SCATTERING APERTURE. In the preceding section we discussed the effective area from which power is absorbed. Referring to Fig. 2-8*b*, the voltage induced in the antenna produces a current through both the antenna impedance Z_A and the terminal or load impedance Z_T. The power P absorbed by the terminal impedance is, as we have seen, the square of this current times the real part of the load impedance. Thus, as given in (2-13-5), $P = I^2 R_T$. Let us now inquire into the power appearing in the antenna impedance Z_A. The real part of this impedance R_A has two parts, the radiation resistance R_r and the loss resistance R_L ($R_A = R_r + R_L$). Therefore, some of the power that is received will be dissipated as heat in the antenna, as given by

$$P' = I^2 R_L \tag{1}$$

The remainder is "dissipated" in the radiation resistance, in other words, is reradiated from the antenna. This reradiated power is

$$P'' = I^2 R_r \tag{2}$$

This *reradiated or scattered power* is analogous to the power that is dissipated in a generator in order that power be delivered to a load. Under conditions of maximum power transfer, as much power is dissipated in the generator as is delivered to the load.

This reradiated power may be related to a *scattering aperture* or scattering cross section. This aperture A_s may be defined as the ratio of the reradiated power to the power density of the incident wave. Thus,

$$A_s = \text{scattering aperture} = \frac{P''}{S} \tag{3}$$

where
$$P'' = I^2 R_r = \frac{V^2 R_r}{(R_r + R_L + R_T)^2 + (X_A + X_T)^2} \tag{4}$$

When $R_L = 0$, and $R_T = R_r$ and $X_T = -X_A$ for maximum power transfer,[1] then

$$A_s = \frac{V^2}{4SR_r} \tag{5}$$

or the scattering aperture equals the maximum effective aperture, that is,

$$A_s = A_{em} \tag{6}$$

Thus, under conditions for which maximum power is delivered to the terminal impedance, *an equal power is reradiated from the receiving antenna.*[2]

Now suppose that the load resistance is zero and $X_T = -X_A$ (antenna resonant). This zero-load-resistance condition may be referred to as a *resonant short-circuit* (RSC) condition. Then for RSC the reradiated power is

$$P'' = \frac{V^2}{R_r} \tag{7}$$

---

[1] Antenna matched.

[2] Referring to Fig. 2-8*a*, note that if the direction of the incident wave changes, the scattered power could increase while V decreases. However, Z_A remains the same.

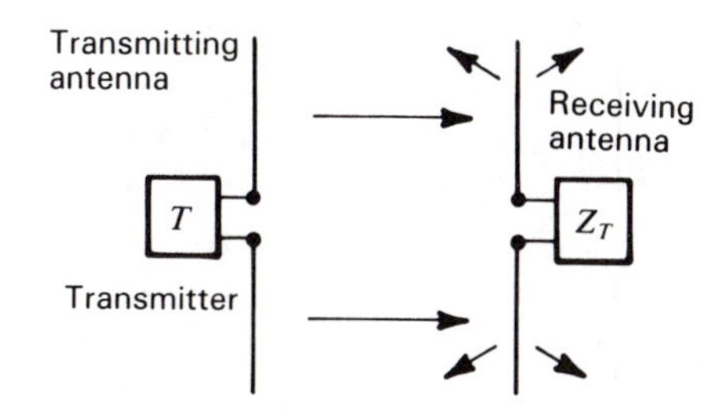

Figure 2-9 $\lambda/2$ dipole antenna receiving (and reradiating) power from $\lambda/2$ dipole transmitting antenna.

and the scattering aperture becomes

$$A_s = \frac{V^2}{SR_r} \tag{8}$$

or

$$A_s = 4A_{em} \tag{9}$$

Thus, for the RSC condition, *the scattering aperture of the antenna is 4 times as great as its maximum effective aperture.*

Figure 2-9 shows two $\lambda/2$ dipoles, one transmitting and the other receiving. Let the receiving antenna be lossless ($R_L = 0$). Consider now three conditions of the receiving antenna:

1. Antenna matched
2. Resonant short circuit
3. Antenna open-circuited ($Z_T = \infty$)

For condition 1 (antenna matched), $A_s = A_{em}$, but for condition 2 (resonant short circuit), $A_s = 4A_{em}$ and 4 times as much power is scattered or reradiated as under condition 1.

Under condition 2 (resonant short circuit), the "receiving" antenna acts like a scatterer and, if close to the transmitting antenna, may absorb and reradiate sufficient power to significantly alter the transmitting antenna radiation pattern. Under these conditions one may refer to the "receiving" antenna as a *parasitic element.* Depending on the phase of the current in the parasitic element, it may act either as a *director* or a *reflector* (see Sec. 11-9*a*). To control its phase, it may be operated off-resonance ($X_T \neq -X_A$), although this also reduces its scattering aperture.

For condition 3 (antenna open-circuited), $I = 0$, $A_e = 0$ and $A_s = 0$.[1]

[1] This is an idealization. Although the scattering may be small it is not zero. See Table 17-2 for scattering from short wires.

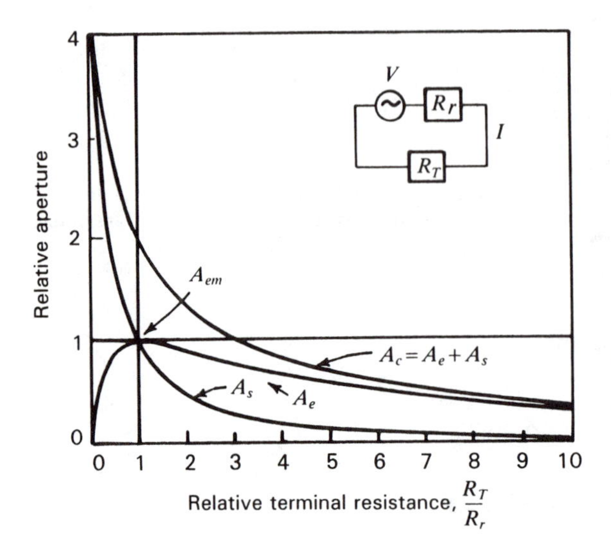

Figure 2-10 Variation of effective aperture A_e, scattering aperture A_s and collecting aperture A_c as a function of the relative terminal resistance R_T/R_r of a small antenna. It is assumed that $R_L = X_A = X_T = 0$.

To summarize:

Condition 1, antenna matched: $A_s = A_{em}$
Condition 2, resonant short circuit: $A_s = 4A_{em}$
Condition 3, antenna open-circuited: $A_s = A_e = 0$

The ratio A_s/A_{em} as a function of the relative terminal resistance R_T/R_r is shown in Fig. 2-10. For $R_T/R_r = 0$, $A_s/A_{em} = 4$, while as R_T/R_r approaches infinity (open circuit), A_s/A_{em} approaches zero.

The ratio of the scattering aperture to the effective aperture may be called the *scattering ratio* β, that is,

$$\text{Scattering ratio} = \frac{A_s}{A_e} = \beta \qquad \text{(dimensionless)} \tag{10}$$

The scattering ratio may assume values between zero and infinity ($0 \le \beta \le \infty$).

For conditions of maximum power transfer and zero antenna losses, the scattering ratio is unity. If the terminal resistance is increased, both the scattering aperture and the effective aperture decrease, but the scattering aperture decreases more rapidly so that the scattering ratio becomes smaller. By increasing the terminal resistance, the ratio of the scattered power to power in the load can be made as small as we please, although by so doing the power in the load is also reduced.

The reradiated or scattered field of an absorbing antenna may be considered as interfering with the incident field so that a shadow may be cast behind the antenna as illustrated in Fig. 2-11-1.

Although the above discussion of scattering aperture is applicable to a single dipole ($\lambda/2$ or shorter), it does not apply in general. (See Sec. 2-18. See also Sec. 17-5.)

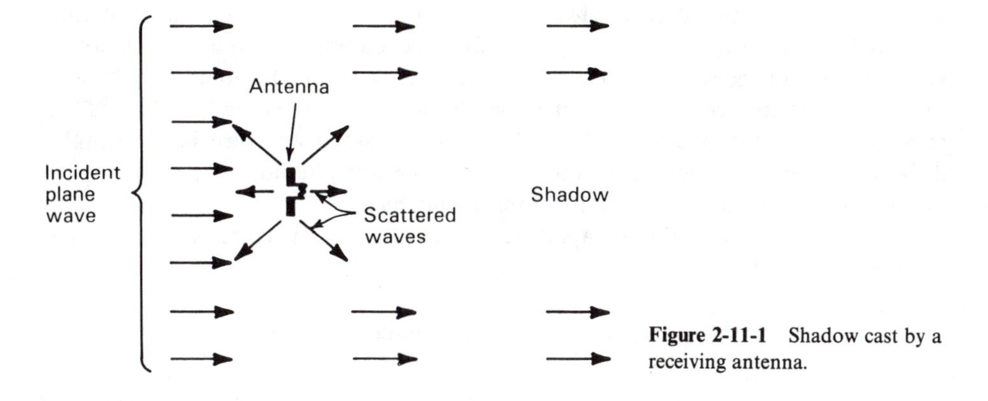

Figure 2-11-1 Shadow cast by a receiving antenna.

2-15 LOSS APERTURE. If R_L is not zero [$k \neq 1$ in (2-10-1)], some power is dissipated as heat in the antenna. This may be related to a *loss aperture* A_L which is given by

$$A_L = \frac{I^2 R_L}{S} = \frac{V^2 R_L}{S[(R_r + R_L + R_T)^2 + (X_A + X_T)^2]} \tag{1}$$

2-16 COLLECTING APERTURE. Three types of apertures have now been discussed: effective, scattering and loss. These three apertures are related to three ways in which power collected by the antenna may be divided: into power in the terminal resistance (effective aperture); into heat in the antenna (loss aperture); or into reradiated power (scattering aperture). By conservation of energy the total power collected is the sum of these three powers. Thus, adding these three apertures together yields what may be called the *collecting aperture* as given by

$$A_c = \frac{V^2(R_r + R_L + R_T)}{S[(R_r + R_L + R_T)^2 + (X_A + X_T)^2]} = A_s + A_L + A_e \tag{1}$$

The variation of A_c with R_T/R_r for the case of $A_L = 0$ is shown in Fig. 2-10.

2-17 PHYSICAL APERTURE AND APERTURE EFFICIENCY. It is often convenient to speak of a fifth type of aperture called the physical aperture A_p. This aperture is a measure of the physical size of the antenna. The manner in which it is defined is entirely arbitrary. For example, it may be defined as the physical cross section (in square meters or square wavelengths) perpendicular to the direction of propagation of the incident wave with the antenna oriented for maximum response. This is a practical definition in the case of many antennas. For example, the physical aperture of an electromagnetic horn is the area of its mouth, while the physical aperture of a linear cylindrical dipole is the cross-sectional area of the dipole. However, in the case of a short stub antenna mounted on an airplane, the physical aperture could be taken as the cross-sectional area of the stub or, since currents associated with the antenna may flow over the entire surface of the airplane, the physical aperture could be taken as the cross-sectional area of the airplane. Thus, the physical aperture has a simple, definite meaning only for some antennas. On the other hand, the *effective aperture has a definite, simply defined value for all antennas.*

The ratio of the effective aperture to the physical aperture is the *aperture efficiency* ε_{ap}, that is,

$$\varepsilon_{ap} = \frac{A_e}{A_p} \qquad \text{(dimensionless)} \qquad (1)$$

Although aperture efficiency may assume values between zero and infinity, it cannot exceed unity for large (in terms of wavelength) broadside apertures.

2-18 SCATTERING BY LARGE APERTURES.

In Sec. 2-14 it was shown that the scattering aperture of a single dipole was equal to the (maximum) effective aperture for the condition of a (conjugate) match and 4 times as much for a resonant short circuit. For a *large broadside aperture A* (dimensions $\gg \lambda$) matched to a uniform wave, all power incident on the aperture can be absorbed over the area A, while an equal power is forward-scattered. Thus, the total *collecting aperture* is $2A$. If the large aperture is a nonabsorbing perfectly conducting flat sheet the power incident on the area A is backscattered while an equal power is forward-scattered, yielding a scattering (and collecting) aperture $2A$. In this case the scattering aperture may be appropriately called a *total scattering cross section* (σ_t), as done in Sec. 17-5. The absorbing and scattering conditions for a large aperture are now discussed in more detail.

The *intrinsic impedance* Z_0 of free (empty) space is 377 Ω ($=\sqrt{\mu_0/\varepsilon_0}$).[1] It is a pure resistance R_0 ($Z_0 = R_0 + j0$). This *intrinsic resistance* takes on more physical significance when we consider the properties of a resistive sheet with a resistance of 377 Ω *per square*.[2] Sheets of this kind (carbon-impregnated paper or cloth) are often called *space paper*, *space cloth* or Salisbury sheets or screens.[3] A square piece of the sheet measures 377 Ω between perfectly conducting bars clamped along opposite edges as in Fig. 2-11-2. For this measurement the size of the sheet makes no difference provided only that it is square. Although the term *ohms per square* is appropriate, the quantity is dimensionally that of resistance (ohms), *not* ohms per square meter.

---

[1] More precisely, $\sqrt{\mu_0/\varepsilon_0} = \mu_0 c = 376.7304$ Ω, where $\mu_0 = 4\pi \times 10^{-7}$ H m^{-1} (by definition) and c = velocity of light.

[2] J. D. Kraus, *Electromagnetics*, 3rd ed., McGraw-Hill, 1984, p. 459.

[3] See also further discussion in Sec. 18-3*c*.

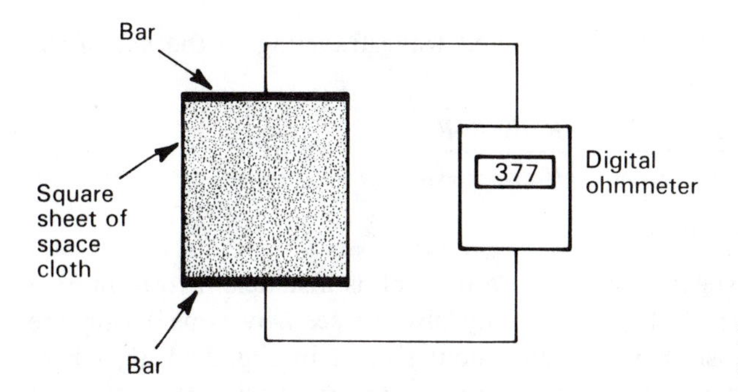

Figure 2-11-2 Space cloth has a resistance of 377 Ω per square.

Consider now what happens when a plane wave is incident normally on an infinite sheet of space cloth (Fig. 2-11-3*a*). Taking the electric field intensity of the incident wave $E_i = 1 \text{ V m}^{-1}$, the field intensity of the transmitted wave continuing to the right of the sheet is

$$E_t = \tau E_i = \frac{2Z_L}{Z_L + Z_0} = \frac{2R_0/2}{(R_0/2) + R_0} = \frac{2}{3} \text{ V m}^{-1} \tag{1}$$

where R_0 = intrinsic resistance of space = 377 Ω

Z_L = load impedance = space cloth in parallel with space behind it $= R_0/2$

τ = transmission coefficient $= \frac{2}{3}$

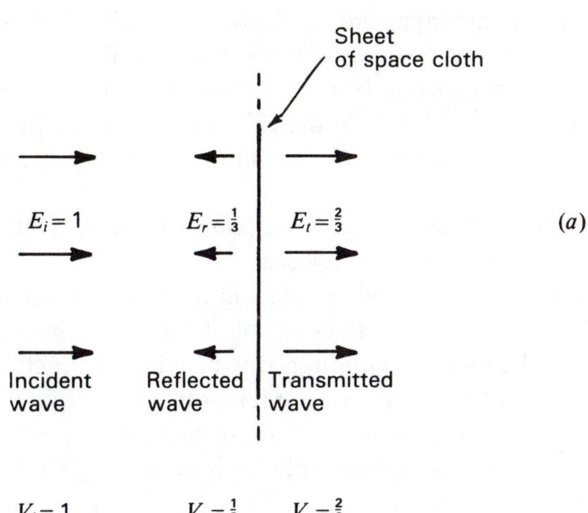

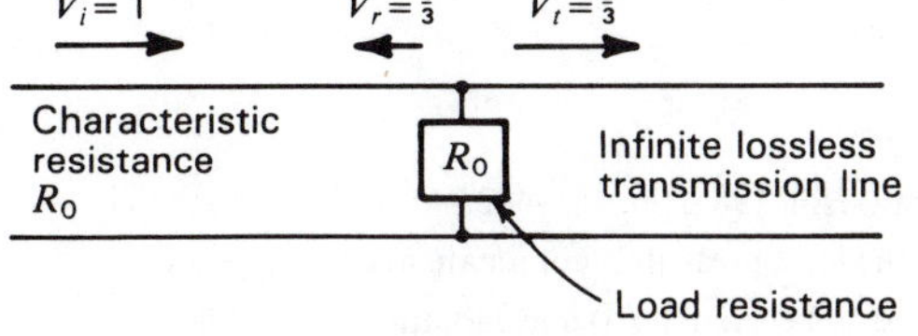

Figure 2-11-3 (*a*) A plane wave traveling to the right incident normally on an infinite sheet of space cloth is partially reflected, partially absorbed and partially transmitted. (*b*) Analogous transmission-line arrangement.

The electric field intensity of the reflected wave traveling to the left of the sheet is

$$E_r = \rho E_i = \frac{Z_L - Z_0}{Z_L + Z_0} = \frac{(R_0/2) - R_0}{(R_0/2) + R_0} = -\frac{1}{3} \text{ V m}^{-1} \tag{2}$$

where ρ = reflection coefficient $= -\frac{1}{3}$

It is apparent that a sheet of space cloth by itself is insufficient to terminate an incident wave without reflection. This may also be seen by considering the analogous lossless transmission line arrangement shown in Fig. 2-11-3*b*, where the load resistance R_0 is in parallel with the line to the right with characteristic resistance R_0.

For both space wave and transmission line, $\frac{1}{9}$ $[=(\frac{1}{3})^2)]$ of the incident power is reflected or scattered back, $\frac{4}{9}$ $[=(\frac{2}{3})^2]$ of the incident power is transmitted or forward-scattered and the remaining $\frac{4}{9}$ absorbed in the space cloth or load. If the area of the space cloth equals A, then the effective aperture $A_e = \frac{4}{9}A$ and the scattering aperture $A_s = \frac{5}{9}A$.

In order to completely absorb the incident wave without reflection or transmission, let an infinite perfectly conducting sheet or reflector be placed parallel to the space cloth and $\lambda/4$ behind it, as portrayed in Fig. 2-11-4*a*. Now the impedance presented to the incident wave at the sheet of space cloth is 377 Ω, being the impedance of the sheet in parallel with an infinite impedance. As a consequence, this arrangement results in the total absorption of the wave by the space cloth.[1] There is, however, a standing wave and energy circulation between the cloth and the conducting sheet and a shadow behind the reflector.

The analogous transmission-line arrangement is illustrated in Fig. 2-11-4*b*, the $\lambda/4$ section (stub) presenting an infinite impedance across the load R_0.

In the case of the plane wave, the perfectly conducting sheet or reflector effectively isolates the region of space behind it from the effects of the wave. In an analogous manner the shorting bar on the transmission line reduces the waves beyond it to a small value.

When the space cloth is backed by the reflector the *wave is matched*. In a similar way, the *line is matched* by the load R_0 with $\lambda/4$ stub.[2]

A transmission line may also be terminated by placing a resistance across the line which is equal to the characteristic resistance of the line, as in Fig. 2-11-3*b*, and disconnecting the line beyond it. Although this provides a practical method of terminating a transmission line, there is no analogous counterpart in the case of a space wave because it is not possible to "disconnect" the space to the right of the termination. A region of space may only be isolated or shielded, as by a perfectly conducting sheet.[3]

---

[1] J. D. Kraus, *Electromagnetics*, 3rd ed., McGraw-Hill, 1984, pp. 461–462.

[2] The stub length can differ from $\lambda/4$ provided the load presents a conjugate match.

[3] The spacing of the transmission line is assumed to be small ($\ll \lambda$) and radiation negligible.

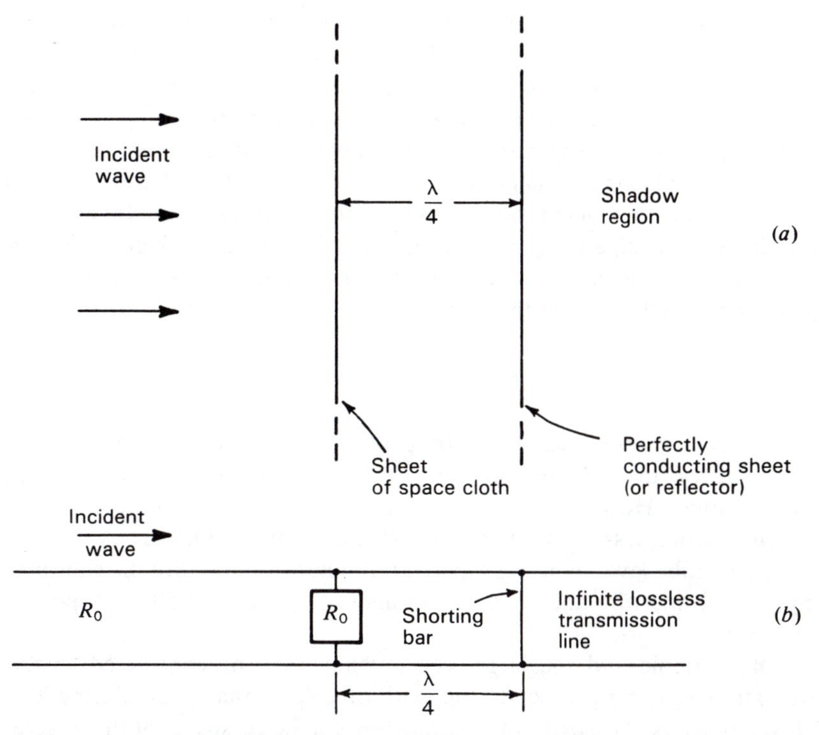

Figure 2-11-4 (*a*) A plane wave traveling to the right incident normally on an infinite sheet of space cloth backed by an infinite perfectly conducting reflecting sheet, as shown, is completely absorbed without reflection. (*b*) Analogous transmission-line arrangement in which a wave traveling to the right is completely absorbed in the load without reflection.

If the space cloth reflector area A is large (dimensions $\gg \lambda$) but *not* infinite in extent the power incident on A is absorbed (as in the infinite case) but there is now scattering of an equal power so that the total collecting aperture A_c is *twice* A or

$$A_c = A_e + A_s = 2A$$

where A_e = effective aperture = A, m^2
A_s = scattering aperture = A, m^2

Thus, as much power is scattered as is absorbed (maximum power transfer condition) ($A_s = A_e$).

If only the flat perfectly conducting reflector of area A is present (no space cloth), the wave incident on the reflector is backscattered instead of absorbed and the wave is totally scattered (half back, half forward) so that the collecting aperture is all scattering aperture and equal to $2A$ ($A_s = 2A = \sigma_t$, see Table 17-1, last row, column 3). In both cases (with and without space cloth) *the incident wave front is disturbed and the energy flow redirected over an area twice the area A.*

Absorption is also possible by methods other than the single space cloth technique as, for example, using thick (multiple space cloth) or other absorbing structures as discussed in Sec. 18-3*c*. These structures, as well as a single space cloth, constitute a *distributed load.* The above conclusions regarding large, but not infinite, apertures also apply to a *large uniform broadside array* of area A connected to a *lumped load* or a *uniformly illuminated parabolic reflector* of area A with power brought to a focus and delivered to a *lumped load.* In all cases (distributed load, broadside array and parabolic reflector), the effective aperture $A_e = A$ (= physical aperture A_p) and the scattering aperture A_s also equal A (= A_p). The aperture efficiency in these cases is given by

$$\varepsilon_{\text{ap}} = \frac{A_e}{A_p} = 1$$

which is the maximum possible value (100 percent efficiency) for large broadside antennas. In theory, the 100 percent limit might be exceeded slightly by using supergain techniques. However, as shown by Rhodes,[1] the practical obstacles are enormous. In practice, less than 100 percent efficiency may be necessary in order to reduce the sidelobe level by using tapered (nonuniform) aperture distributions. Accordingly, large aperture antennas are commonly operated at 50 to 70 percent aperture efficiency.

The single dipole and the large-area antenna may be considered to represent two extremes as regards scattering, with other antenna types intermediate. Table 2-1 summarizes the scattering parameters for large space cloth or array apertures, for transmission lines and for a single dipole ($\lambda/2$ or shorter).

2-19 EFFECTIVE HEIGHT.

The *effective height* h (meters) of an antenna is another parameter related to the aperture. Multiplying the effective height by the incident field E (volts per meter) of the same polarization gives the voltage V induced. Thus,

$$V = hE \tag{1}$$

Accordingly, the effective height may be defined as the ratio of the induced voltage to the incident field or

$$h = \frac{V}{E} \qquad \text{(m)} \tag{2}$$

Consider, for example, a vertical dipole of length $l = \lambda/2$ immersed in an incident field E, as in Fig. 2-12*a*. If the current distribution of the dipole were uniform its effective height would be l. The actual current distribution, however, is nearly sinusoidal with an *average value* $2/\pi = 0.64$ (of the maximum) so that its

---

[1] D. R. Rhodes, "On an Optimum Line Source for Maximum Directivity," *IEEE Trans. Ants. Prop.*, **AP-19**, 485–492, 1971.

Table 2-1 Scattering parameters†

Condition	Space cloth (or array)	Transmission line	Dipole (Fig. 2-8)
Matched (Fig. 2-11-4)	Space cloth (or array) with reflector, area A $A_s = A_e = A$ $A_c = 2A$	Load R_0 with $\lambda/4$ stub No power reflected All power into load	$A_s = A_{em}$
Short circuit	Reflector only, area A $A_s = 2A = \sigma_t$ $A_e = 0$ $A_c = 2A$	All power reflected No power in termination	$A_s = 4A_{em}$ (resonant)
Load only (Fig. 2-11-3)	Space cloth only, area infinite 11% power backscattered 44% power forward-scattered 44% absorbed	Load R_0 only 11% power reflected 44% power transmitted 44% power in load	No dipole
Open circuit	No cloth or reflector $A_s = 0$ $A_e = 0$	No power reflected All power transmitted	$A_s \simeq 0$‡ $A_e = 0$

† Aperture values assume orientation for maximum response, polarization-matched and load conjugate-matched. Uniform aperture response or distribution is assumed for the large areas A (dimensions $\gg \lambda$). The transmission line is assumed to be lossless, the spacing small ($\ll \lambda$) and radiation effects negligible.
‡ Scattering small but not zero.

effective height $h = 0.64\ l$. It is assumed that the antenna is oriented for maximum response.

If the same dipole is used at a longer wavelength so that it is only 0.1λ long, the current tapers almost linearly from the central feed point to zero at the ends in a triangular distribution, as in Fig. 2-12*b*. The average current is $\frac{1}{2}$ of the maximum so that the effective height is $0.5\ l$.

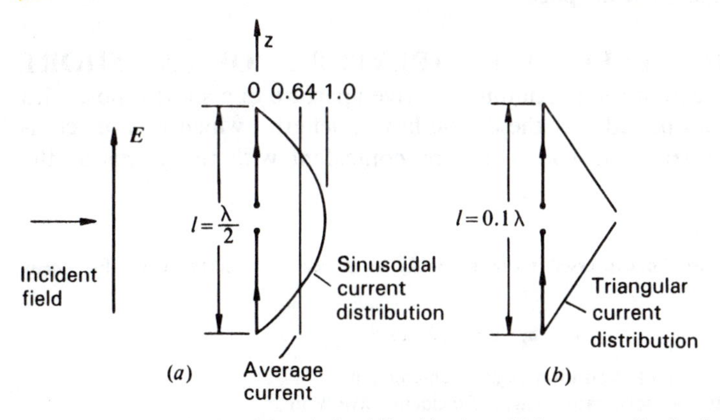

Figure 2-12 (*a*) Dipole of length $l = \lambda/2$ with sinusoidal current distribution. (*b*) Dipole of length $l = 0.1\lambda$ with triangular current distribution.

Thus, another way of defining effective height is to consider the transmitting case and equate the effective height to the physical height (or length l) multiplied by the (normalized) average current or

$$h_e = \frac{1}{I_0} \int_0^{h_p} I(z)\, dz = \frac{I_{av}}{I_0} h_p \qquad \text{(m)} \tag{3}$$

where h_e = effective height, m
h_p = physical height, m
I_{av} = average current, A

It is apparent that *effective height* is a useful parameter for transmitting tower-type antennas.[1] It also has an application for small antennas. The parameter *effective aperture* has more general application to all types of antennas. The two have a simple relation, as will be shown.

For an antenna of radiation resistance R_r matched to its load, the power delivered to the load is equal to

$$P = \frac{1}{4}\frac{V^2}{R_r} = \frac{h^2E^2}{4R_r} \qquad \text{(W)} \tag{4}$$

In terms of the effective aperture the same power is given by

$$P = SA_e = \frac{E^2 A_e}{Z_0} \qquad \text{(W)} \tag{5}$$

where Z_0 = intrinsic impedance of space ($= 377\ \Omega$)

Equating (4) and (5) we obtain

$$h_e = 2\sqrt{\frac{R_r A_e}{Z_0}} \quad \text{(m)} \qquad \text{and} \qquad A_e = \frac{h_e^2 Z_0}{4R_r} \quad (\text{m}^2) \tag{6}$$

Thus, effective height and effective aperture are related via radiation resistance and the intrinsic impedance of space.

2-20 MAXIMUM EFFECTIVE APERTURE OF A SHORT DIPOLE. In this section the maximum effective aperture of a short dipole with uniform current is calculated. Let the dipole have a length l which is short compared with the wavelength ($l \ll \lambda$). Let it be coincident with the y axis at the

---

[1] Effective height can also be expressed more generally as a vector quantity. Thus (for linear polarization) we can write

$$V = \mathbf{h}_e \cdot \mathbf{E} = h_e E \cos\theta$$

where $\mathbf{h}_e$ = effective height and polarization angle of antenna, m
$\mathbf{E}$ = field intensity and polarization angle of incident wave, V m^{-1}
θ = angle between polarization angles of antenna and wave, deg

In a still more general expression (for any polarization state), θ is the angle between polarization states on the Poincaré sphere (see Sec. 2-36).

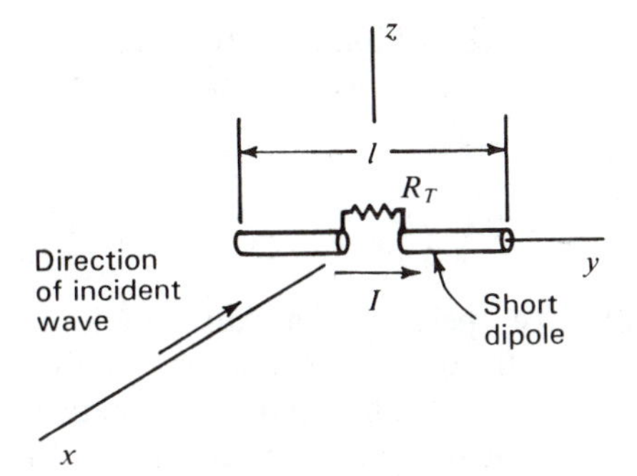

Figure 2-13 Short dipole with uniform current induced by incident wave.

origin as shown in Fig. 2-13, with a plane wave traveling in the negative x direction incident on the dipole. The wave is assumed to be linearly polarized with E in the y direction. The current on the dipole is assumed constant and in the same phase over its entire length, and the terminating resistance R_T is assumed equal to the dipole radiation resistance R_r. The antenna loss resistance R_L is assumed equal to zero.

The maximum effective aperture of an antenna is obtained from (2-13-13) as

$$A_{em} = \frac{V^2}{4SR_r} \tag{1}$$

where the effective value of the induced voltage V is here given by the product of the effective electric field intensity at the dipole and its length, that is,

$$V = El \tag{2}$$

The radiation resistance R_r of a short dipole of length l with uniform current will be shown later (in Sec. 5-3) to be[1]

$$R_r = \frac{80\pi^2 l^2}{\lambda^2}\left(\frac{I_{\text{av}}}{I_0}\right)^2 = 790\left(\frac{I_{\text{av}}}{I_0}\right)^2\left(\frac{l}{\lambda}\right)^2 \quad (\Omega) \tag{3}$$

where λ = wavelength
I_{av} = average current
I_0 = terminal current

The power density, or Poynting vector, of the incident wave at the dipole is related to the field intensity by

$$S = \frac{E^2}{Z} \tag{4}$$

where Z = intrinsic impedance of the medium

---

[1] This relation for the radiation resistance of a short dipole was worked out by Max Abraham in 1904 and R. Rudenberg in 1908. It is very clearly set forth in Jonathan Zenneck's textbook editions of 1905 and 1908 and its English translation, *Wireless Technology*, McGraw-Hill, 1915.

In the present case, the medium is free space so that $Z = 120\pi\ \Omega$. Now substituting (2), (3) and (4) into (1), we obtain for the maximum effective aperture of a short dipole (for $I_{av} = I_0$)

$$A_{em} = \frac{120\pi E^2 l^2 \lambda^2}{320\pi^2 E^2 l^2} = \frac{3}{8\pi} \lambda^2 = 0.119\lambda^2 \tag{5}$$

Equation (5) indicates that the maximum effective aperture of a short dipole is somewhat more than $\frac{1}{10}$ of the square wavelength and is independent of the length of the dipole provided only that it is small ($l \ll \lambda$). The maximum effective aperture neglects the effect of any losses, which probably would be considerable for an actual short dipole antenna. If we assume that the terminating impedance is matched to the antenna impedance but that the antenna has a loss resistance equal to its radiation resistance, the effective aperture from (2-13-12) is $\frac{1}{2}$ the maximum effective aperture obtained in (5).

2-21 MAXIMUM EFFECTIVE APERTURE OF A LINEAR λ/2 ANTENNA. As a further illustration, the maximum effective aperture of a linear $\lambda/2$ antenna will be calculated. It is assumed that the current has a sinusoidal distribution and is in phase along the entire length of the antenna. It is further assumed that $R_L = 0$. Referring to Fig. 2-14*a*, the current I at any point y is then

$$I = I_0 \cos \frac{2\pi y}{\lambda} \tag{1}$$

A plane wave incident on the antenna is traveling in the negative x direction. The wave is linearly polarized with E in the y direction. The equivalent circuit is shown in Fig. 2-14*b*. The antenna has been replaced by an equivalent or Thévenin generator. The infinitesimal voltage dV of this generator due to the voltage

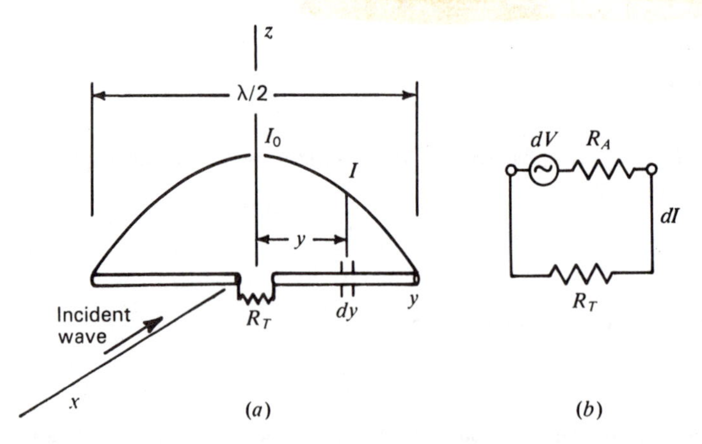

Figure 2-14 Linear $\lambda/2$ antenna in field of electromagnetic wave (*a*) and equivalent circuit (*b*).

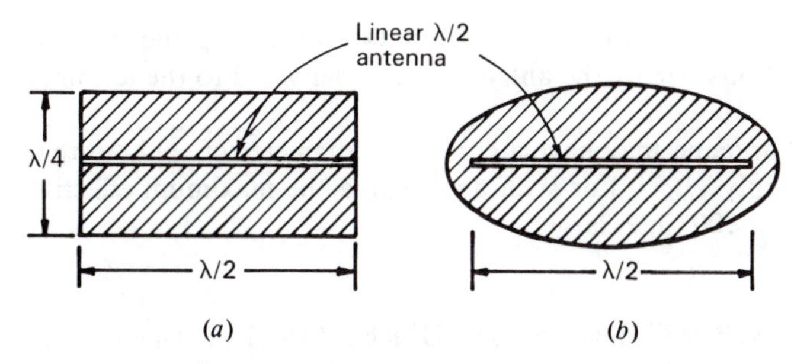

Figure 2-15 (*a*) Maximum effective aperture of linear $\lambda/2$ antenna is approximately represented by rectangle $\frac{1}{2}$ by $\frac{1}{4}\lambda$ on a side. (*b*) Maximum effective aperture of linear $\lambda/2$ antenna represented by elliptical area of $0.13\lambda^2$.

induced by the incident wave in an infinitesimal element of length dy of the antenna is

$$dV = E\,dy\cos\frac{2\pi y}{\lambda} \tag{2}$$

It is assumed that the infinitesimal induced voltage is proportional to the current at the infinitesimal element as given by the current distribution (1).

The total induced voltage V is given by integrating (2) over the length of the antenna. This may be written as

$$V = 2\int_0^{\lambda/4} E\cos\frac{2\pi y}{\lambda}\,dy \tag{3}$$

Performing the integration in (3) we have

$$V = \frac{E\lambda}{\pi} \tag{4}$$

The value of the radiation resistance R_r of the linear $\lambda/2$ antenna will be taken as 73 Ω.[1] The terminating resistance R_T is assumed equal to R_r. The power density at the antenna is as given by (2-20-4). Substituting (4), (2-20-4) and $R_r = 73$ into (2-13-13), we obtain, for the maximum effective aperture of a linear $\lambda/2$ antenna,

$$A_{em} = \frac{120\pi E^2\lambda^2}{4\pi^2E^2 \times 73} = \frac{30}{73\pi}\lambda^2 = 0.13\,\lambda^2 \tag{5}$$

Comparing (5) with (2-20-5), the maximum effective aperture of the linear $\lambda/2$ antenna is about 10 percent greater than that of the short dipole.

The maximum effective aperture of the $\lambda/2$ antenna is approximately the same as an area $\frac{1}{2}$ by $\frac{1}{4}\lambda$ on a side, as illustrated in Fig. 2-15*a*. This area is $\frac{1}{8}\lambda^2$. An elliptically shaped aperture of $0.13\lambda^2$ is shown in Fig. 2-15*b*. The physical

[1] The derivation of this value is given in Sec. 5-6.

significance of these apertures is that power from the incident plane wave is absorbed over an area of this size by the antenna and is delivered to the terminating resistance.

A typical thin $\lambda/2$ antenna may have a conductor diameter of $\frac{1}{400}\lambda$, so that its physical aperture is only $\frac{1}{800}\lambda^2$. For such an antenna the maximum effective aperture of $0.13\lambda^2$ is about 100 times larger.

2-22 EFFECTIVE APERTURE AND DIRECTIVITY.

There is an important relation between effective aperture and directivity of all antennas as will now be shown.

Consider the electric field E_r at a large distance in a direction broadside to a radiating aperture as in Fig. 2-16. If the field intensity in the aperture is constant and equal to E_a (volts per meter), the radiated power is given by

$$P = \frac{|E_a|^2}{Z} A \tag{1}$$

where A = antenna aperture, m^2
Z = intrinsic impedance of the medium, Ω

The power radiated may also be expressed in terms of the field intensity E_r (volts per meter) at a distance r by

$$P = \frac{|E_r|^2}{Z} r^2 \Omega_A \tag{2}$$

where Ω_A = beam solid angle of antenna, sr

It may be shown (Sec. 11-21) that the field intensities E_r and E_a are related by

$$|E_r| = \frac{|E_a| A}{r\lambda} \tag{3}$$

where λ = wavelength, m

Substituting (3) in (2) and equating (1) and (2) yields

$$\lambda^2 = A\Omega_A \tag{4}$$

where λ = wavelength, m
A = antenna aperture, m^2
Ω_A = beam solid angle, sr

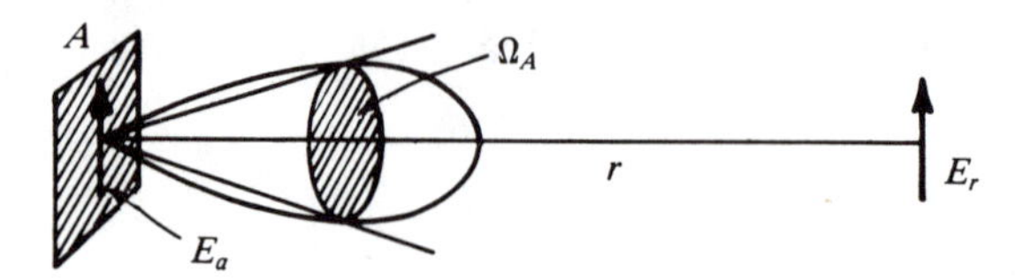

Figure 2-16 Radiation from aperture A with uniform field E_a.

In (4) the aperture A is the physical aperture A_p if the field is uniform over the aperture, as assumed, but in general A is the *maximum effective aperture* A_{em} (losses equal zero). Thus,

$$\boxed{\lambda^2 = A_{em}\Omega_A} \tag{5}$$

We note that A_{em} is determined entirely by the antenna pattern of beam area Ω_A. According to this important relation, the product of the maximum effective aperture of the antenna and the antenna beam solid angle is equal to the wavelength squared. Equation (5) applies to all antennas. From (5) and (2-8-4) we have that

$$D = \frac{4\pi}{\lambda^2} A_{em} \tag{6}$$

When, for simplicity, A_e is substituted for A_{em} in (5) or (6), zero losses are assumed.

Three expressions have now been given for the directivity D of an antenna. They are

$$\boxed{D = \frac{U(\theta, \phi)_{\max}}{U_{\text{av}}} = \frac{S(\theta, \phi)_{\max}}{S_{\text{av}}}} \tag{7}$$

$$\boxed{D = \frac{4\pi}{\Omega_A}} \tag{8}$$

$$\boxed{D = \frac{4\pi}{\lambda^2} A_{em}} \tag{9}$$

From (2-10-1), the gain $G = kD$, where $k = A_e/A_{em}$, so

$$\boxed{G = \frac{4\pi}{\lambda^2} A_e} \tag{10}$$

2-23 BEAM SOLID ANGLE AS A FRACTION OF A SPHERE.

A short dipole with directivity $D = \frac{3}{2}$ has a beam solid angle

$$\Omega_A = \frac{4\pi}{D} = \frac{2}{3} 4\pi \tag{1}$$

Putting $\Omega_{\text{sph}} = 4\pi$ = solid angle of a sphere, the dipole beam solid angle

$$\Omega_A = \tfrac{2}{3}\Omega_{\text{sph}} \tag{2}$$

Thus, the dipole radiation pattern may be said to fill $\frac{2}{3}$ of a sphere. The larger the directivity of an antenna, the smaller is the fraction of a sphere filled by its radiation pattern. At the other extreme, a nondirectional (isotropic) antenna with $D = 1$ completely fills a sphere. This concept, emphasized by Harold A. Wheeler (1964), provides an interesting way of looking at directivity and beam area.

2-24 TABLE OF EFFECTIVE APERTURE, DIRECTIVITY, EFFECTIVE HEIGHT AND OTHER PARAMETERS FOR DIPOLES AND LOOPS

Antenna	Radiation resistance§ R_r, Ω	Maximum effective aperture A_{em}, λ^2	Effective height, maximum value, h, m	Sphere filling factor	Directivity D	Directivity D(dBi)
Isotropic		$\frac{1}{4\pi} = 0.079$		1	1	0
Short dipole,† length l	$80\left(\frac{\pi l I_{av}}{\lambda I_0}\right)^2$	$\frac{3}{8\pi} = 0.119$	$\frac{l I_{av}}{I_0}$	$\frac{2}{3}$	$\frac{3}{2}$	1.76
Short dipole,† $l = \lambda/10$ ($I_{av} = I_0$)	7.9	0.119	$\lambda/10$	$\frac{2}{3}$	$\frac{3}{2}$	1.76
Short dipole,† $l = \lambda/10$ ($I_{av} = \frac{1}{2}I_0$)	1.98	0.119	$\lambda/20$	$\frac{2}{3}$	$\frac{3}{2}$	1.76
Linear, $\lambda/2$ dipole (sinusoidal current distribution)	73	$\frac{30}{73\pi} = 0.13$	$\frac{\lambda}{\pi} = \frac{2l}{\pi}$	0.61	1.64	2.15
Small loop‡ (single turn), any shape	$31\,200\left(\frac{A}{\lambda^2}\right)^2$	$\frac{3}{8\pi} = 0.119$	$2\pi \frac{A}{\lambda}$	$\frac{2}{3}$	$\frac{3}{2}$	1.76
Small square loop‡ (single turn), side length $= l$ Area $A = l^2 = (\lambda/10)^2$	3.12	$\frac{3}{8\pi} = 0.119$	$\frac{2\pi\lambda}{100}$	$\frac{2}{3}$	$\frac{3}{2}$	1.76

§ See Chaps. 5 and 6.
† Length $l \leq \lambda/10$.
‡ Area $A \leq \lambda^2/100$, see Sec. 6-8. For n-turn loop, multiply R_r by n^2 and h by n.

Although the radiation resistance, effective aperture, effective height and directivity are the same for both receiving and transmitting, the current distribution is, in general, not the same. Thus, a plane wave incident on a receiving antenna excites a different current distribution than a localized voltage applied to a pair of terminals for transmitting.

2-25 FRIIS TRANSMISSION FORMULA. The usefulness of the aperture concept will now be illustrated by using it to derive the important *Friis transmission formula* published in 1946 by Harald T. Friis of the Bell Telephone Laboratories.[1]

Referring to Fig. 2-17, this formula gives the power received over a radio communication circuit. Let the transmitter T feed a power P_t to a transmitting

[1] H. T. Friis, "A Note on a Simple Transmission Formula," *Proc. IRE*, **34**, 254–256, 1946.

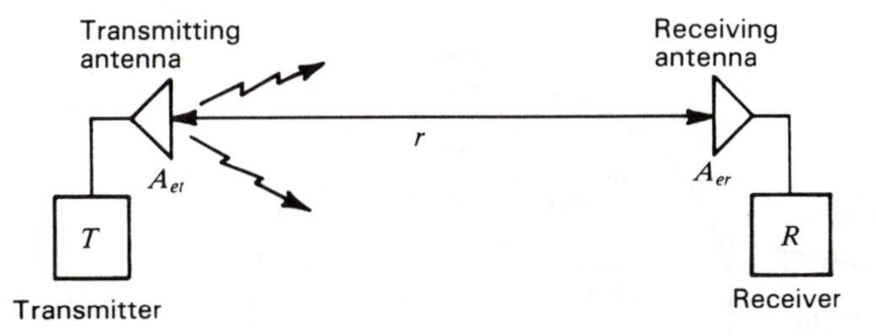

Figure 2-17 Communication circuit with waves from transmitting antenna arriving at the receiving antenna by a direct path of length r.

antenna of effective aperture A_{et}. At a distance r a receiving antenna of effective aperture A_{er} intercepts some of the power radiated by the transmitting antenna and delivers it to the receiver R. Assuming for the moment that the transmitting antenna is isotropic, the power per unit area at the receiving antenna is

$$S_r = \frac{P_t}{4\pi r^2} \quad \text{(W)} \tag{1}$$

If the antenna has gain G_t, the power per unit area at the receiving antenna will be increased in proportion as given by

$$S_r = \frac{P_t G_t}{4\pi r^2} \quad \text{(W)} \tag{2}$$

Now the power collected by the receiving antenna of effective aperture A_{er} is

$$P_r = S_r A_{er} = \frac{P_t G_t A_{er}}{4\pi r^2} \quad \text{(W)} \tag{3}$$

From (2-22-10) the gain of the transmitting antenna can be expressed as

$$G_t = \frac{4\pi A_{et}}{\lambda^2} \tag{4}$$

Substituting this in (3) yields the *Friis transmission formula*

$$\boxed{P_r = P_t \frac{A_{et} A_{er}}{r^2 \lambda^2} \quad \text{(W)}} \tag{5}$$

where P_r = received power (antenna matched), W
P_t = power into transmitting antenna, W
A_{et} = effective aperture of transmitting antenna, m^2
A_{er} = effective aperture of receiving antenna, m^2
r = distance between antennas, m
λ = wavelength, m

It is assumed that each antenna is in the far field of the other.

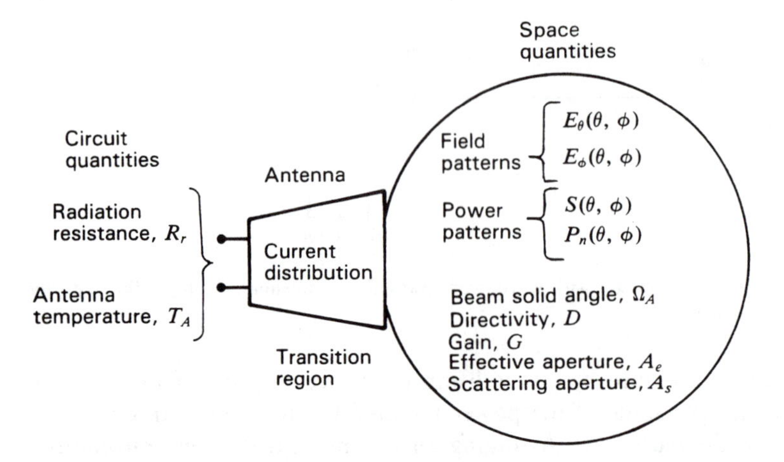

Figure 2-18 Schematic diagram of basic antenna parameters, illustrating the duality of an antenna: a circuit device (with a resistance and temperature) on the one hand and a space device (with radiation patterns, beam angles, directivity, gain and aperture) on the other.

2-26 DUALITY OF ANTENNAS. The duality of an antenna, as a circuit device on the one hand and a space device on the other, is illustrated schematically in Fig. 2-18.

2-27 SOURCES OF RADIATION: RADIATION RESULTS FROM ACCELERATED CHARGES. A stationary electric charge does not radiate (Fig. 2-19*a*) and neither does an electric charge moving at uniform velocity along a straight wire (Fig. 2-19*b*).[1] However, if the charge is accelerated, i.e., its velocity changes with time, it radiates. Thus, as in Fig. 2-19*c*, a charge reversing direction on reflection from the end of a wire radiates. The shorter the pulse for a given charge, the greater the acceleration and the greater the power radiated, or, as in Fig. 2-19*d*, a charge moving at uniform velocity along a curved or bent wire is accelerated and radiates.

Consider a pulse of electric charge moving along a straight conductor in the x direction, as in Fig. 2-20. This moving charge constitutes a momentary electric current I as given by

$$I = q_L \frac{dx}{dt} \quad \text{(A)} \tag{1}$$

where q_L = charge per unit length, C m^{-1}

[1] This can be seen from relativistic considerations, since, for an observer in a reference frame moving with the charge, it will appear stationary.

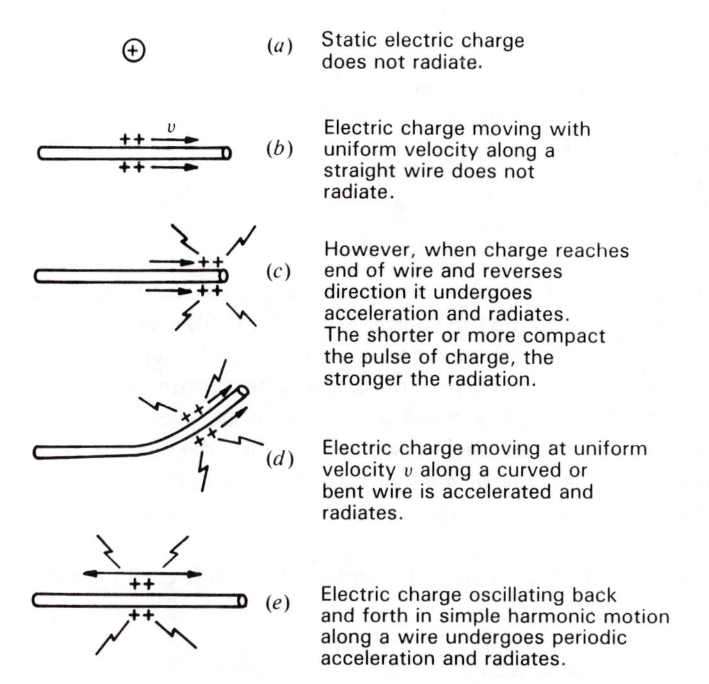

(a) Static electric charge does not radiate.
v
(b) Electric charge moving with uniform velocity along a straight wire does not radiate.
(c) However, when charge reaches end of wire and reverses direction it undergoes acceleration and radiates. The shorter or more compact the pulse of charge, the stronger the radiation.
(d) Electric charge moving at uniform velocity v along a curved or bent wire is accelerated and radiates.
(e) Electric charge oscillating back and forth in simple harmonic motion along a wire undergoes periodic acceleration and radiates.

Figure 2-19 A static electric charge or a charge moving with uniform velocity in a straight line does not radiate. An accelerated charge, however, does radiate.

Multiplying by the length l of the pulse as measured along the conductor gives

$$Il = q_L l \frac{dx}{dt} = qv \qquad \text{(A m)} \tag{2}$$

where $q_L l = q$ = total charge of pulse, C
v = velocity, m s^{-1}

Taking the time derivative

$$\frac{dI}{dt} l = q \frac{d^2x}{dt^2} = q\dot{v} \qquad (\text{A m s}^{-1}) \tag{3}$$

where $\dot{v}$ = acceleration of charge, m s^{-2}

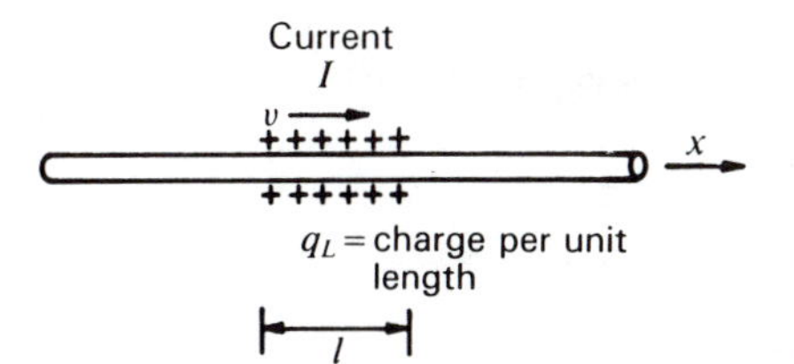

Figure 2-20 Charge pulse of uniform charge density q_L (per unit length) moving with velocity v constitutes an electric current I.

More concisely,

$$\boxed{\dot{I}l = q\dot{v} \qquad (\text{A m s}^{-1})} \tag{4}$$

where $\dot{I}$ = time-changing current, A s^{-1}
l = length of current element, m
q = electric charge, C
$\dot{v}$ = acceleration, m s^{-2}

This is the *basic continuity relation between current and charge for electromagnetic radiation.* Since accelerated[1] charge ($q\dot{v}$) produces radiation, it follows from this equation that *time-changing current* ($\dot{I}$) *produces radiation.* (Fig. 2-19*e*). For transients and pulses we usually focus on charge. For steady-state harmonic variation we usually focus on current. Whereas a pulse radiates a broad spectrum (wide bandwidth) of radiation (the shorter the pulse, the broader the spectrum), a smooth sinusoidal variation of charge or current results in a narrow bandwidth of radiation (theoretically zero at the frequency of the sinusoid if it continues indefinitely).

It may be shown[2] that an accelerated charge radiates a power P as given by[3]

$$P = \frac{\mu^2 q^2 \dot{v}^2}{6\pi Z} \qquad (\text{W}) \tag{5}$$

where μ = permeability of medium, H m^{-1}
q = charge, C
$\dot{v}$ = acceleration, m s^{-2}
Z = impedance of medium

2-28 PULSED OPENED-OUT TWIN-LINE ANTENNAS. The antenna of Fig. 2-1, shown again in Fig. 2-21*a*, has two conductors each resembling an Alpine-type horn used by Swiss mountaineers. The uniform transmission-line section at the left opens out until the conductor separation is a wavelength or more with radiation from the curved region forming a beam to the right. The conductor spacing-diameter ratio is constant, making the characteristic impedance constant over a wide bandwidth. Since radiation occurs from narrower regions at shorter wavelengths, the radiation pattern tends to be relatively

---

[1] Or decelerated.

[2] L. Landau and E. Lifshitz, *The Classical Theory of Fields*, Addison-Wesley, 1951.

[3] Equivalent expressions are

$$\frac{q^2 \dot{v}^2}{6\pi\varepsilon c^3} = \frac{\mu q^2 \dot{v}^2}{6\pi c} \qquad (\text{W}) \tag{6}$$

where ε = permittivity (F m^{-1}) and c = velocity of light (m s^{-1}).

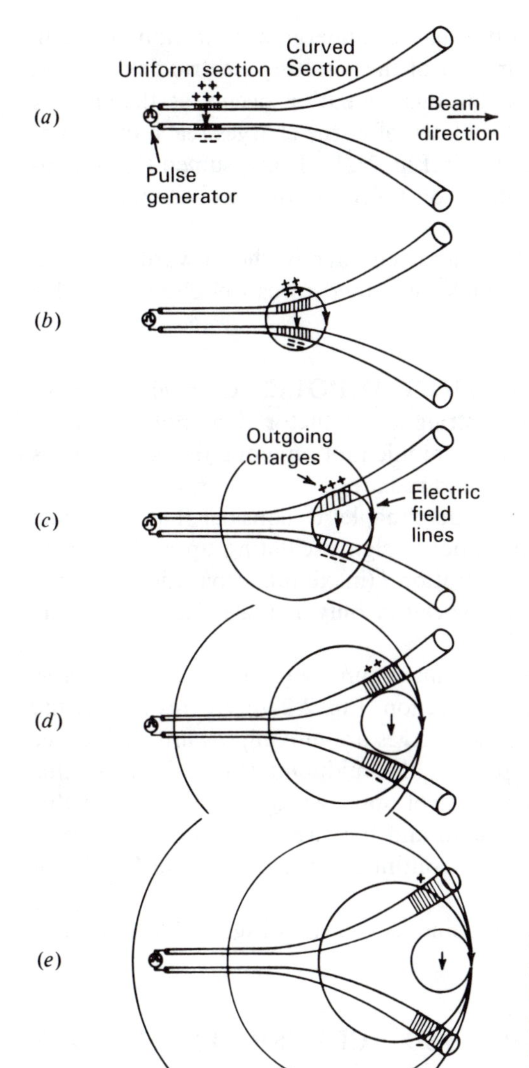

Figure 2-21 Pulsed twin-line antenna. No radiation occurs along the uniform section. However, radiation occurs along the curved portion and is maximum to the right, as suggested by the reinforcement of the electric fields.

constant.[1] These properties make the twin horn a basic broadband antenna.

Let us analyze the process of radiation from this antenna by considering what happens when it is excited by a single short pulse which starts electric charges moving to the right along the uniform transmission-line section at light speed. There is no radiation as the charges travel along the uniform section at the

[1] However, the phase center moves to the right with decrease in frequency.

left but on reaching the curved region the charges undergo acceleration and radiation occurs (Fig. 2-21*b*). The configuration of the resulting electric field a short time later is suggested in Fig. 2-21*c*. Additional radiation occurs as the charges travel further, as shown in Fig. 2-21*d*. Still later, after the charges reach the end of the conductors, the fields are as suggested in Fig. 2-21*e*. It is assumed that due to prior radiation, negligible radiation occurs on reflection of the charges from the open end.[1]

We note that the fields are additive and reinforce in the forward direction (to the right) between the conductors while they tend to cancel elsewhere. This tendency is apparent in Fig. 2-21*c*.

2-29 FIELDS FROM OSCILLATING DIPOLE. Although a charge moving with uniform velocity along a straight conductor does not radiate, a charge moving back and forth in simple harmonic motion along the conductor is subject to acceleration (and deceleration) and radiates.

To illustrate radiation from a dipole antenna, let us consider that the dipole of Fig. 2-22 has two equal charges of opposite sign oscillating up and down in harmonic motion with instantaneous separation l (maximum separation l_0) while focusing attention on the electric field. For clarity only a single electric field line is shown.

At time $t = 0$ the charges are at maximum separation and undergo maximum acceleration $\dot{v}$ as they reverse direction (Fig. 2-22*a*). At this instant the current I is zero. At an $\frac{1}{8}$-period later, the charges are moving toward each other (Fig. 2-22*b*) and at a $\frac{1}{4}$-period they pass at the midpoint (Fig. 2-22*c*). As this happens, the field lines detach and new ones of opposite sign are formed. At this time the equivalent current I is a maximum and the charge acceleration is zero. As time progresses to a $\frac{1}{2}$-period, the fields continue to move out as in Fig. 2-22*d* and *e*.

An oscillating dipole with more field lines is shown in Fig. 2-23 at 4 instants of time.

2-30 RADIATION FROM PULSED CENTER-FED DIPOLE ANTENNAS. Five stages of radiation from a dipole antenna are shown in Fig. 2-24 resulting from a single short voltage pulse applied by a generator at the center of the dipole (positive charge to left, negative charge to right). The pulse length is short compared to the time of propagation along the dipole.

At the first stage [(*a*) top] the pulse has been applied and the charges are moving outward. The electric field lines between the charges expand like a soap bubble with velocity $v = c$ in free space. The charges are assumed to move with

---

[1] With radiation from the curved section, the energy of the pulse decreases as energy is lost to radiation according to (2-27-5). Thus, stated another way, it is assumed that due to prior radiation losses, negligible charge reaches the open end, being absorbed in radiation resistance. Energy lost in radiation resistance is energy radiated.

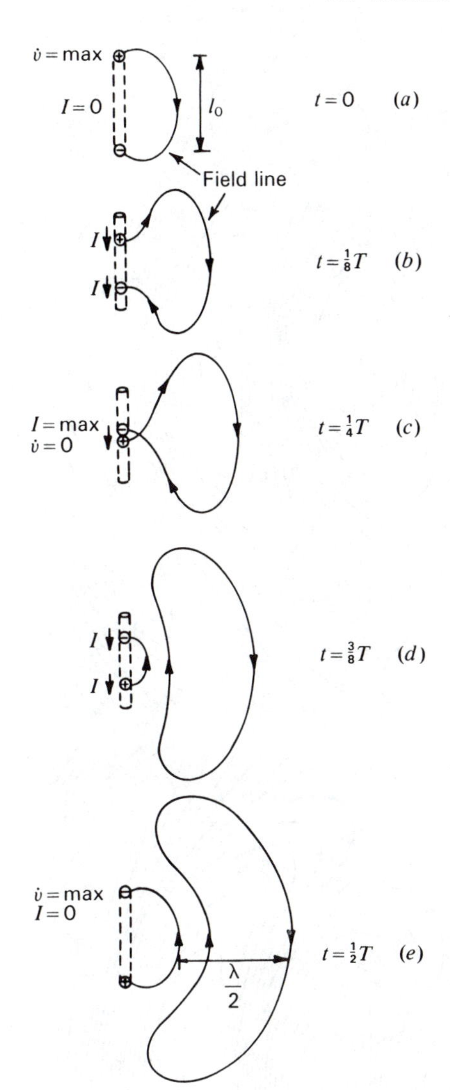

Figure 2-22 Oscillating electric dipole consisting of two electric charges in simple harmonic motion, showing propagation of an electric field line and its detachment (radiation) from the dipole. Arrows next to the dipole indicate current (I) direction.

velocity $v = c$ along the dipole. At the next stage [(*a*) middle] the charges reach the ends of the dipole, are reflected (bounce back) and move inward toward the generator [(*a*) bottom]. If the generator is an impedance match, the pulses are absorbed at the generator but the field lines join, initiating a new pulse from the center of the dipole with the pulse fields somewhat later, as shown in (*b*).

Maximum radiation is broadside to the dipole and zero on axis as with a harmonically excited dipole. Broadside to the dipole ($\theta = 90°$) there is a symmetrical pulse triplet, but, at an angle such as 30° from broadside ($\theta = 60°$), the middle pulse of the triplet splits into two pulses so that the triplet becomes a quadruplet as shown in (*b*). Thus, the pulse pattern is a function of angle. The

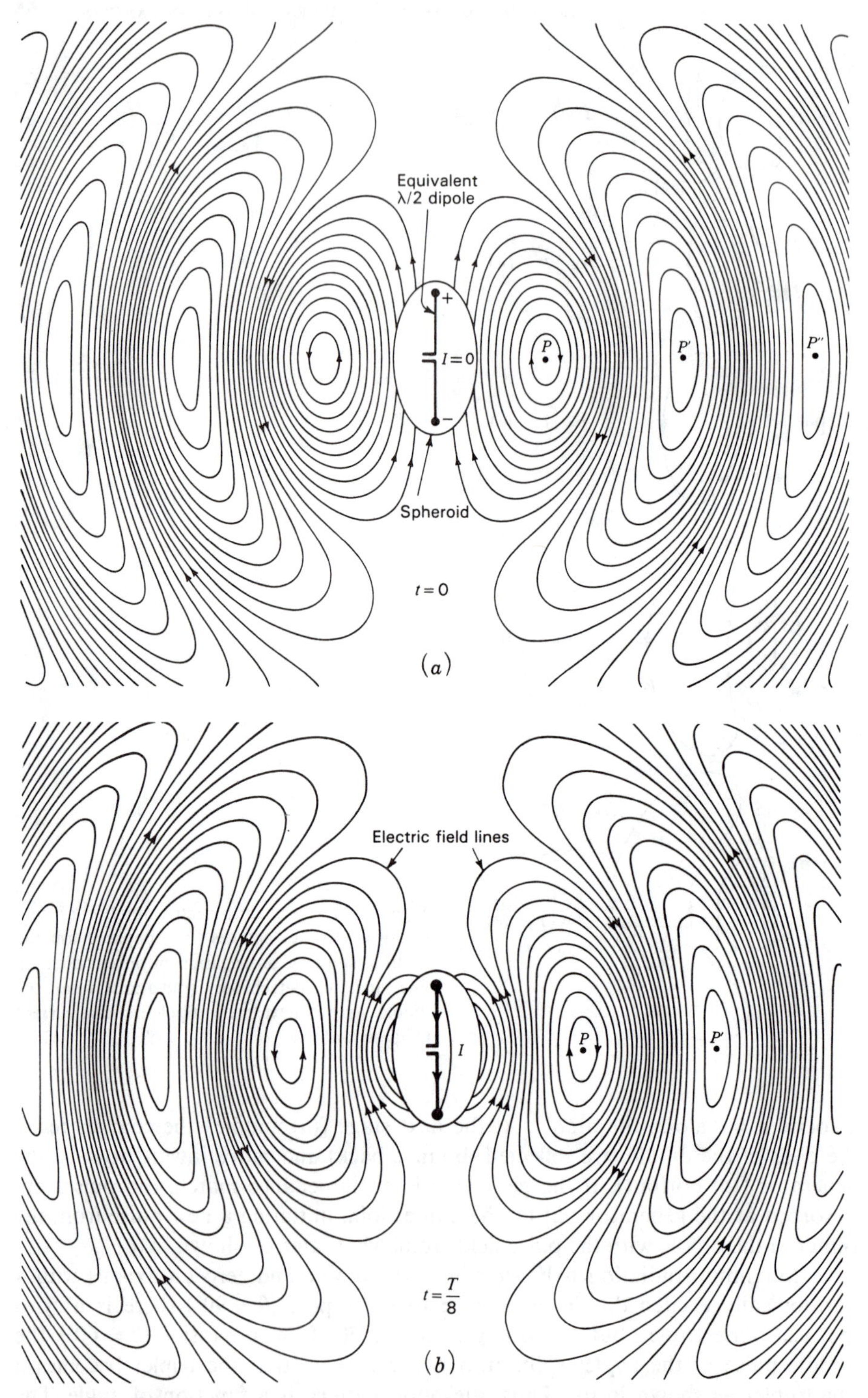

Figure 2-23 Electric field configuration for a $\lambda/2$ antenna at four instants of time: (*a*) $t = 0$, (*b*) $t = T/8$, (*c*) $t = T/4$ and (*d*) $t = \frac{3}{8}T$, where T = period. Note outward movement of the constant-phase points P, P' and P'' as time advances. These points move with a velocity $v = c$ remote from the

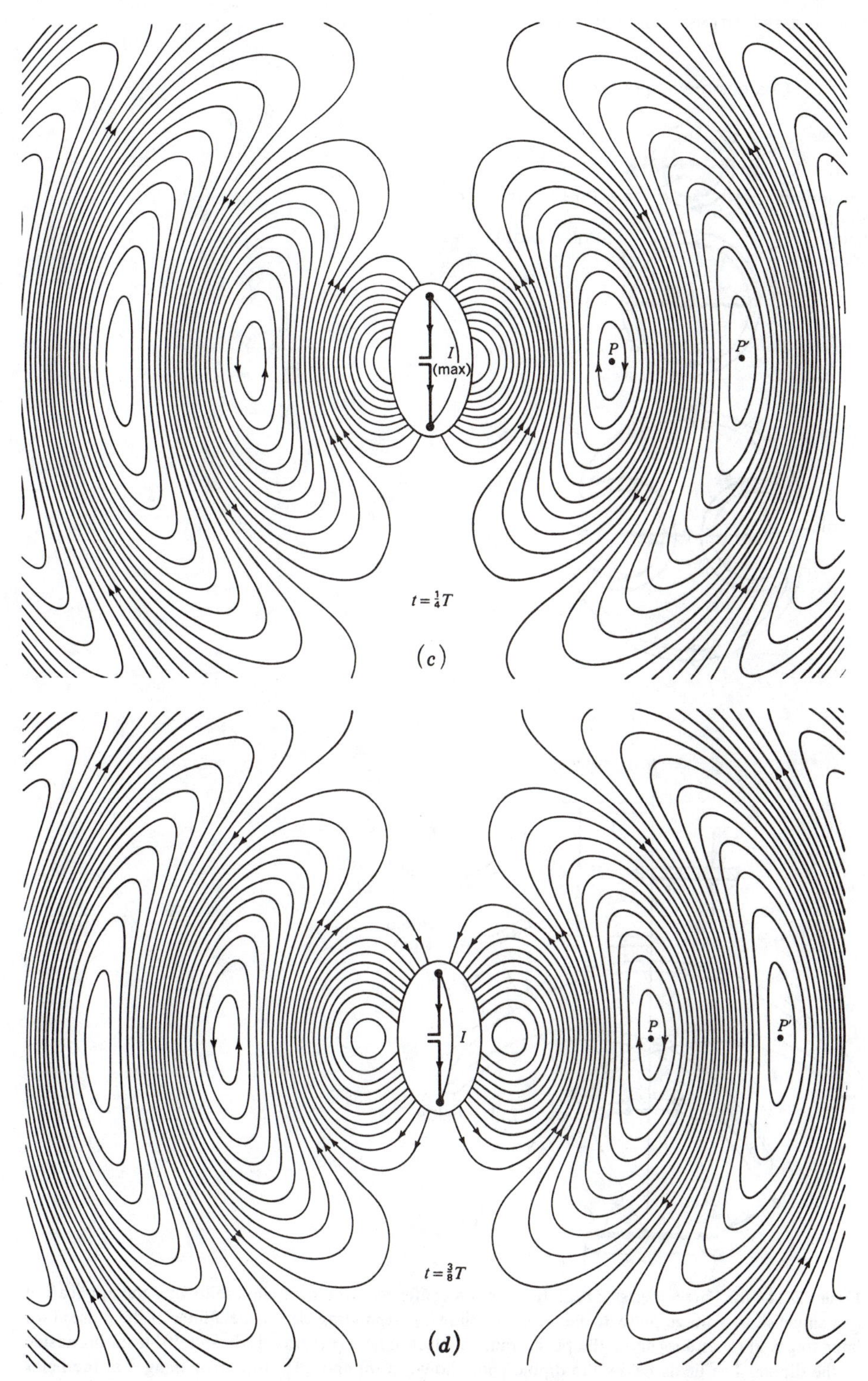

dipole but with $v > c$ in the near field, as may be noted by measuring the distances between successive points. The strength of the electric field is proportional to the density of the lines. (*Produced by Edward M. Kennaugh, courtesy of John D. Cowan, Jr.*)

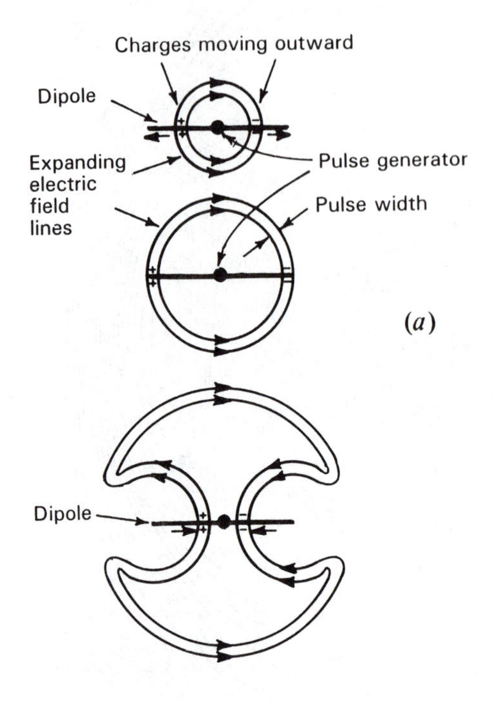

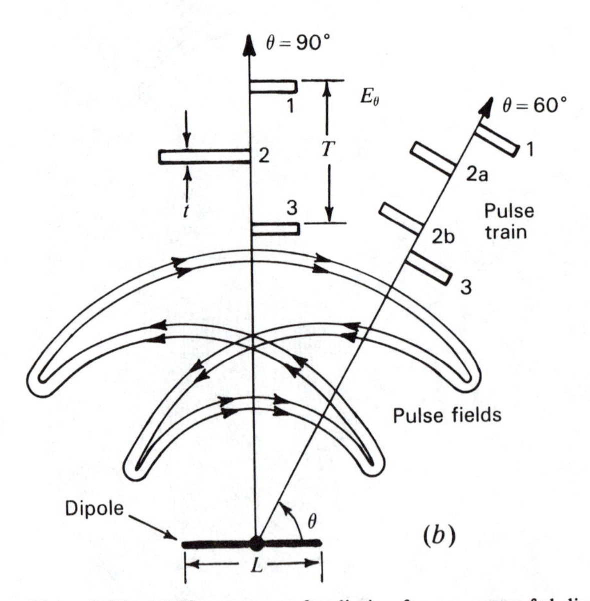

Figure 2-24 (*a*) Three stages of radiation from a center-fed dipole antenna following the application of a single short voltage pulse at the terminals showing expanding electric field lines of the pulse. Two later stages at (*b*) and (*c*) show the pulse trains of electric field (E_θ) broadside and 30° from broadside to the dipole. The fields below the dipole [not shown in (*b*) and (*c*)] are mirror images of the fields above.

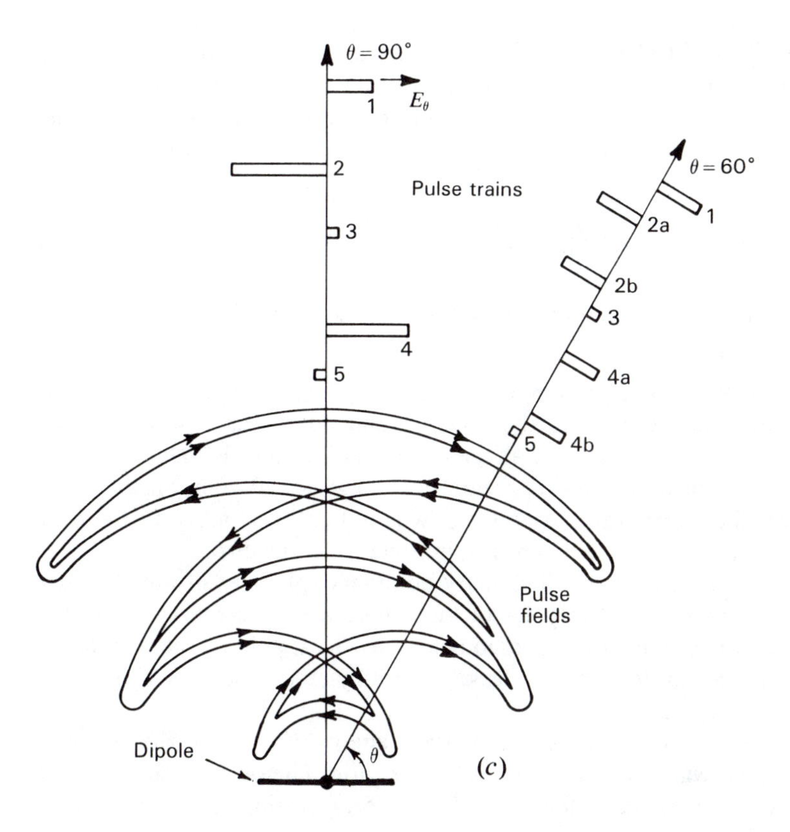

time T between pulses 1 and 3 is the time required for the charges to travel out and back from the generator. This is the same as the travel time over the length L of the dipole, or

$$T = \frac{L}{c} \tag{1}$$

Thus, the dipole length determines the pulse spacing T while the pulse length t determines the much shorter wavelength of the pulse radiation.

If the generator is not an impedance match, then the charges will continue to bounce back and forth on the dipole, resulting in a longer pulse train as suggested in (*c*) for the case of a poor match (some pulse energy is absorbed in the generator but more is reflected). The first 5 pulses broadside (7 at 30° from broadside) of an indefinitely long (gradually damping) pulse train are indicated.

It is evident from Fig. 2-24 that radiation occurs from the points where charge is accelerated, i.e., at the center or feed point and at the ends of the dipole but not along the dipole itself.[1]

[1] G. Franceschetti and C. H. Papas, "Pulsed Antennas," Sensor and Simulation Note 203, Cal. Tech., 1973.

2-31 ANTENNA FIELD ZONES. The fields around an antenna may be divided into two principal regions, one near the antenna called the *near field* or *Fresnel zone* and one at a large distance called the *far field* or *Fraunhofer zone*. Referring to Fig. 2-25, the boundary between the two may be arbitrarily taken to be at a radius

$$R = \frac{2L^2}{\lambda} \quad \text{(m)} \tag{1}$$

where L = maximum dimension of the antenna, m
λ = wavelength, m

In the far or Fraunhofer region, the measurable field components are transverse to the radial direction from the antenna and all power flow is directed radially outward. In the far field the shape of the field pattern is independent of the distance. In the near or Fresnel region, the longitudinal component of the electric field may be significant and power flow is not entirely radial. In the near field, the shape of the field pattern depends, in general, on the distance.

Enclosing the antenna in an imaginary boundary sphere as in Fig. 2-26*a*, it is as though the region near the poles of the sphere acts as a reflector. On the other hand, the waves expanding perpendicular to the dipole in the equatorial region of the sphere result in power leakage through the sphere as if partially transparent in this region.

This results in reciprocating (oscillating) energy flow near the antenna accompanied by outward flow in the equatorial region. The outflow accounts for the power radiated from the antenna, while the reciprocating energy represents reactive power that is trapped near the antenna like in a resonator. This over-simplified discussion accounts in a qualitative way for the field pattern of the $\lambda/2$

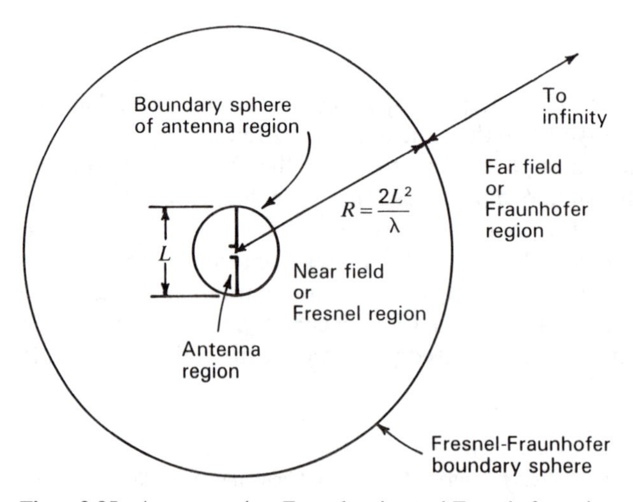

Figure 2-25 Antenna region, Fresnel region and Fraunhofer region.

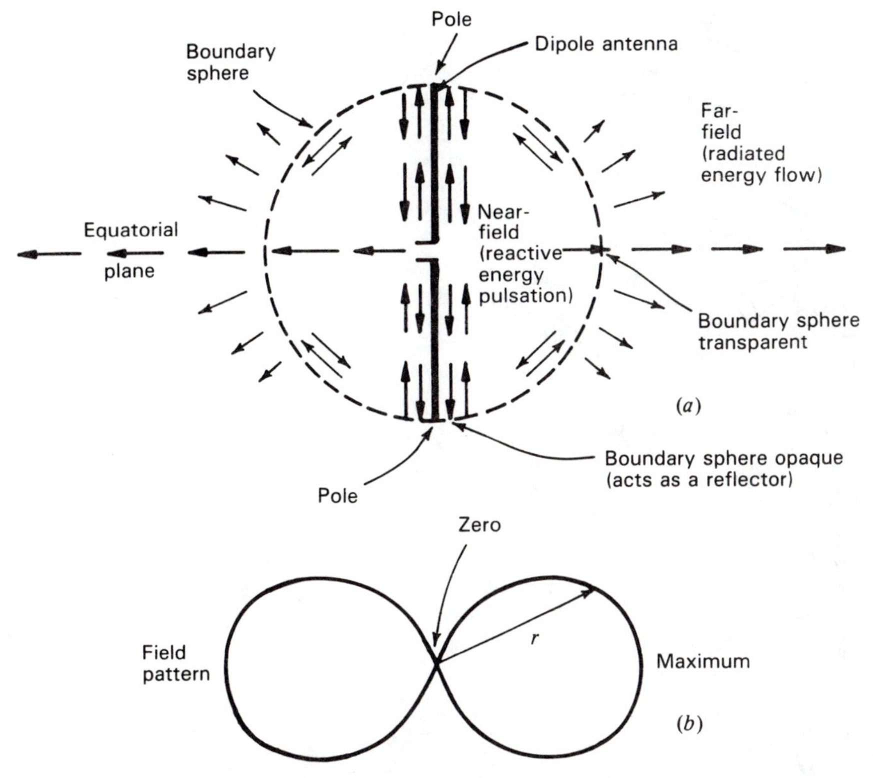

Figure 2-26 Energy flow near a dipole antenna (*a*) and radiation field pattern (*b*). The radius vector *r* is proportional to the field radiated in that direction.

dipole antenna as shown in Fig. 2-26*b*. The energy picture is discussed in more detail in Sec. 5-2 and displayed in Fig. 5-7.

For a $\lambda/2$ dipole antenna, the energy is stored at one instant of time in the electric field, mainly near the ends of the antenna or maximum charge regions, while a $\frac{1}{2}$-period later the energy is stored in the magnetic field mainly near the center of the antenna or maximum current region.

Note that although the term *power flow* is sometimes used, it is actually *energy* which flows, power being the time rate of energy flow. A similar loose usage occurs when we say we pay a power bill, when, in fact, we are actually paying for electric energy.

2-32 SHAPE-IMPEDANCE CONSIDERATIONS. It is possible in many cases to deduce the qualitative behavior of an antenna from its shape. This may be illustrated with the aid of Fig. 2-27. Starting with the opened-out two-conductor transmission line of Fig. 2-27*a*, we find that, if extended far enough, a nearly constant impedance will be provided at the input (left) end for $d \ll \lambda$ and $D \geq \lambda$.

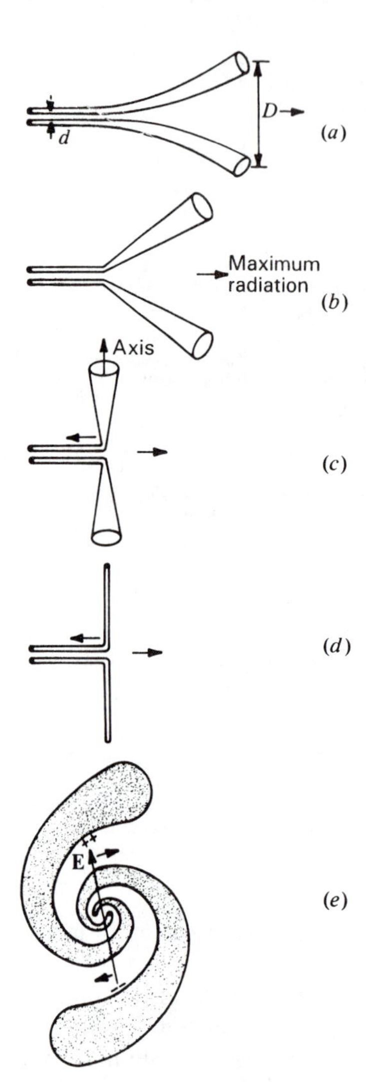

Figure 2-27 Evolution of a thin cylindrical antenna (*d*) from an opened-out twin line (*a*). Curving the conductors as in (*e*) results in the spiral antenna.

In Fig. 2-27*b*, the curved conductors are straightened into regular cones and in Fig. 2-27*c* the cones are aligned colinearly forming a biconical antenna. In Fig. 2-27*d* the cones degenerate into straight wires. In going from Fig. 2-27*a* to *d*, the bandwidth of relatively constant impedance tends to decrease. Another difference is that the antennas of Fig. 2-27*a* and *b* are unidirectional with beams to the right, while the antennas of Fig. 2-27*c* and *d* are omnidirectional in the horizontal plane (perpendicular to the wire or cone axes).

A different modification is shown in Fig. 2-27*e*. Here the two conductors are curved more sharply and in opposite directions, resulting in a spiral antenna with maximum radiation broadside (perpendicular to the page) and with polariz-

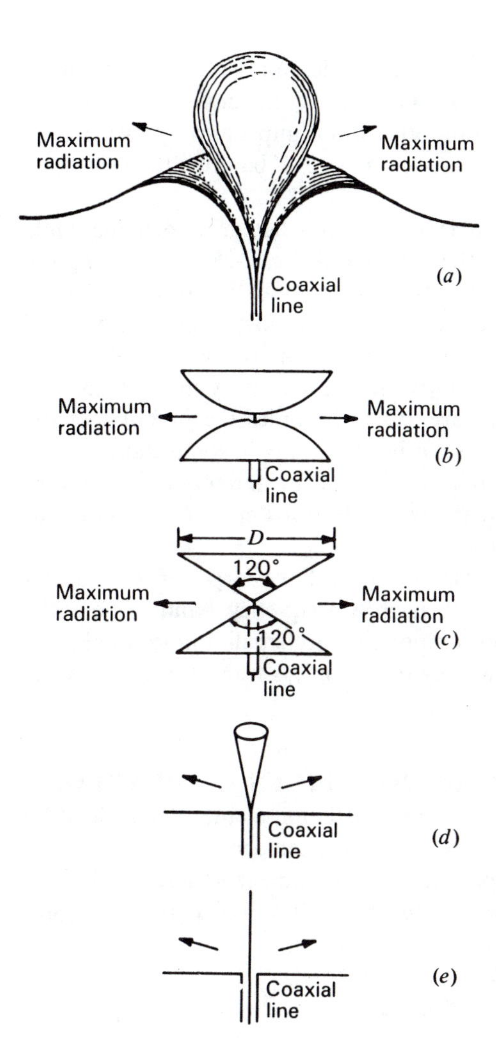

Figure 2-28 Evolution of stub (monopole) antenna (*e*) from volcano-smoke antenna (*a*).

ation which rotates clockwise. This antenna, like the one in Fig. 2-27*a*, exhibits very broadband characteristics (see Chap. 15).

The dipole antennas of Fig. 2-27 are balanced, i.e., they are fed by two-conductor (balanced) transmission lines. Figure 2-28 illustrates a similar evolution of monopole antennas, i.e., antennas fed from coaxial (unbalanced) transmission lines.

By gradually tapering the inner and outer conductors of a coaxial transmission line, a very wide band antenna with an appearance reminiscent of a volcanic crater and puff of smoke is obtained, as suggested in the cutaway view of Fig. 2-28*a*.

In Fig. 2-28*b* the volcano form is modified into a double dish and in Fig. 2-28*c* into two wide-angle cones. All of these antennas are omnidirectional in a plane perpendicular to their axes and all have a wide bandwidth. For example, an actual biconical antenna, as in Fig. 2-28*c*, with a full cone angle of 120° has an omnidirectional pattern and nearly constant 50-Ω input impedance (power reflection less than 1 percent or VSWR < 1.2) over a 6 to 1 bandwidth with cone diameter $D = \lambda$ at the lowest frequency.[1]

Increasing the lower cone angle to 180° or into a flat ground plane while reducing the upper cone angle results in the antenna of Fig. 2-28*d*. Collapsing the upper cone into a thin stub we arrive at the extreme modification of Fig. 2-28*e*. If the antenna of Fig. 2-28*a* is regarded as the most basic form, the stub type of Fig. 2-28*e* is the most degenerate form, with a relatively narrow bandwidth.

As we depart further from the basic type, the discontinuity in the transmission line becomes more abrupt at what eventually becomes the junction of the ground plane and the coaxial line. This discontinuity results in some energy being reflected back into the line. The reflection at the end of the antenna also increases for thinner antennas. At some frequency the two reflections may compensate, but the bandwidth of compensation is narrow.

Antennas with large and abrupt discontinuities have large reflections and act as reflectionless transducers only over narrow frequency bands where the reflections cancel. Antennas with discontinuities that are small and gradual have small reflections and are, in general, relatively reflectionless transducers over wide frequency bands.

2-33 ANTENNAS AND TRANSMISSION LINES COMPARED. A uniform transmission line has a constant characteristic impedance determined by the geometry of its cross section. Thus, the space between the conductors of the coaxial line of Fig. 2-29*a* can be mapped into $5\frac{1}{2}$ curvilinear squares with each square having the characteristic resistance of space ($= 376.7 \simeq 377\ \Omega$).[2] Therefore, this line has a characteristic resistance of [3]

$$Z_0 = \frac{377}{5.5} = 68.5\ \Omega \tag{1}$$

If the line is cut and terminated in a load of 68.5 Ω there would be, ideally, no reflection (line matched). Put another way, if the line is cut and terminated in a load of 5 resistors of 377 Ω, one for each full square, and two 377 Ω resistors in

[1] D. A. McNamara, D. E. Baker and L. Botha, "Some Design Considerations for Biconical Antennas," *IEEE Ants. Prop. Int. Symp. Digest*, 173, 1984.

[2] J. D. Kraus, *Electromagnetics*, 3rd ed., McGraw-Hill, 1984, sec. 3-19.

[3] The analytical (boundary-value) solution (medium air or vacuum) is

$$Z_0 = 138 \log \frac{b}{a}\ (\Omega) = 68.5\ \Omega$$

where b = inner diameter of the outer conductor = 62.7 mm
a = outer diameter of the inner conductor = 20.0 mm

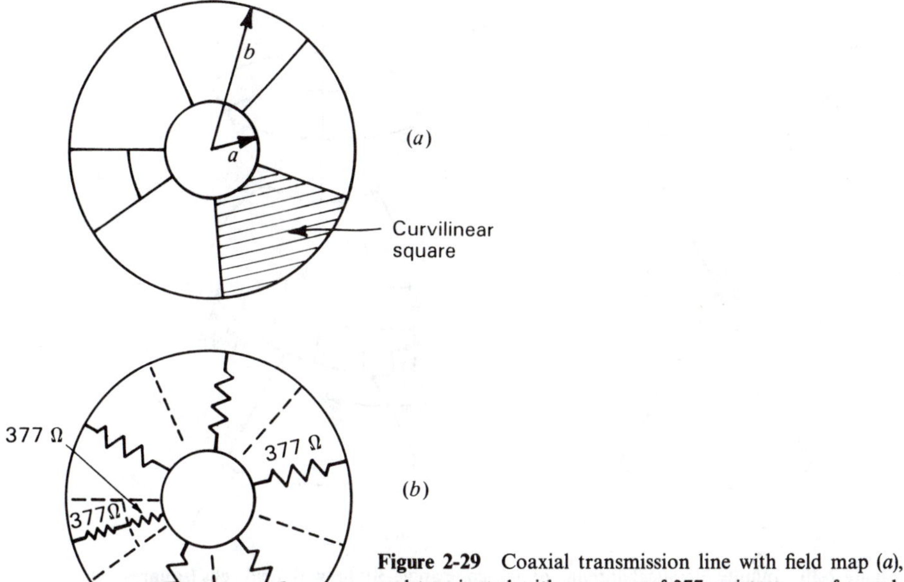

Figure 2-29 Coaxial transmission line with field map (*a*), and terminated with an array of 377 resistors, one for each curvilinear square (*b*). The total resistance of the array is 68.5 Ω.

series for the $\frac{1}{2}$ square, as in Fig. 2-29*b*, the line would be matched. A sheet of space cloth (resistance 377 Ω per square) connected across the end of the line would also behave as a matched load.

Now consider an infinite biconical antenna, as in Fig. 2-30*a*. With a simple graphical field map drawn on a spherical surface, as in Fig. 2-30*b*, the characteristic impedance of the biconical antenna can be obtained and also shown to be a constant, i.e., independent of the distance from the terminals.

Let spherical space be divided into 15° sectors in azimuth (ϕ). Now considering a half-cone angle $\theta = 30°$, the map will require about 5 squares in series from equator to cone and 24 squares in parallel. The map below the equator (not shown) is a mirror image of the one above. Thus, the characteristic resistance of this infinite biconical antenna is

$$Z_0 = \frac{N_s}{N_p} 377 \simeq \frac{2 \times 5}{24} 377 = 157 \ \Omega \tag{2}$$

where N_s = number of squares in series (from cone to cone)
N_p = number of squares in parallel

Schelkunoff's formula (8-2-20) gives a characteristic resistance of 158.0 Ω for a half-cone angle of 30°.

From spherical geometry it follows that the characteristic impedance of an infinite biconical antenna is a (radiation) resistance of constant value since the

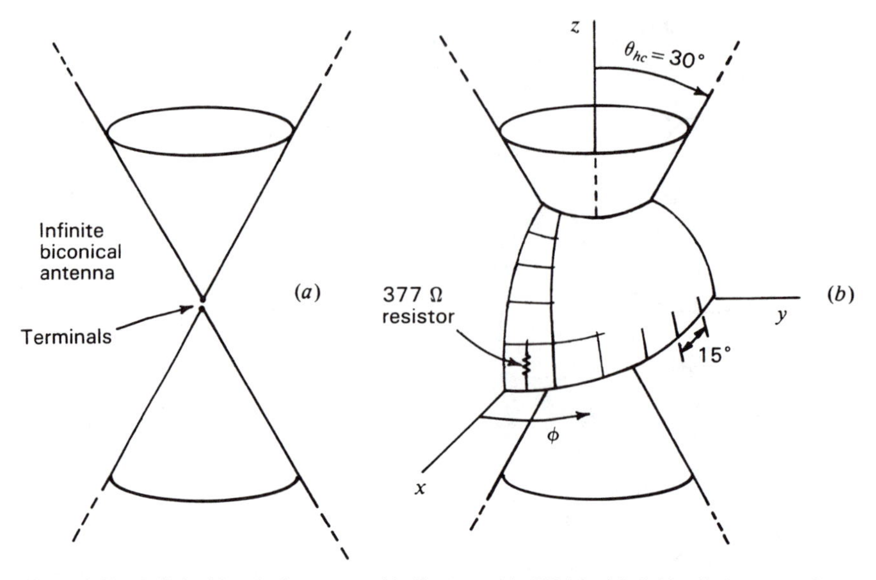

Figure 2-30 Infinite biconical antenna of half-cone angle 30° (*a*) with field cells (squares) (*b*).

ratio of the curvilinear squares in series to the number in parallel is constant, the solid angle subtended by each square being independent of radius.

Figure 2-31 illustrates some of the similarities and differences of antennas and transmission lines. In Fig. 2-31*a* an infinite biconical antenna with $\theta_{hc} = 30°$ and characteristic resistance 160 Ω is compared with an infinite uniform two-conductor transmission line (characteristic resistance 300 Ω). Waves traveling out on both are entirely of the Transverse ElectroMagnetic (TEM) type and the VSWR = 1 for both.

In Fig. 2-31*b* both antenna and line have load resistors equal to their characteristic resistance connected as shown.[1] Beyond the load resistors the VSWR = 1, but between the feed points and the loads the VSWR = 2.

In Fig. 2-31*c* the cones and the line are truncated at the load. On the line the VSWR = 1 (line matched) but the biconical antenna is not matched (VSWR ≠ 1 on the cones). If $d \ll \lambda$ there are no significant waves (radiation negligible) beyond the load at the end of the transmission line, but the biconical antenna is an opened-out radiating system, and higher-order mode waves exist beyond the load in the outer region.

---

[1] Instead of connecting a single 160-Ω resistance between the cones as in Fig. 2-31*b*, a grid of distributed resistors would provide a better arrangement. For example, a 10 × 24 grid of 240 resistors could be employed, one for each curvilinear square of Fig. 2-30*b* with the resistor for each square equal to 377 Ω for a total resistance of 157 Ω. The spherical surface could also be covered with a continuous sheet of space cloth (377 Ω per square).

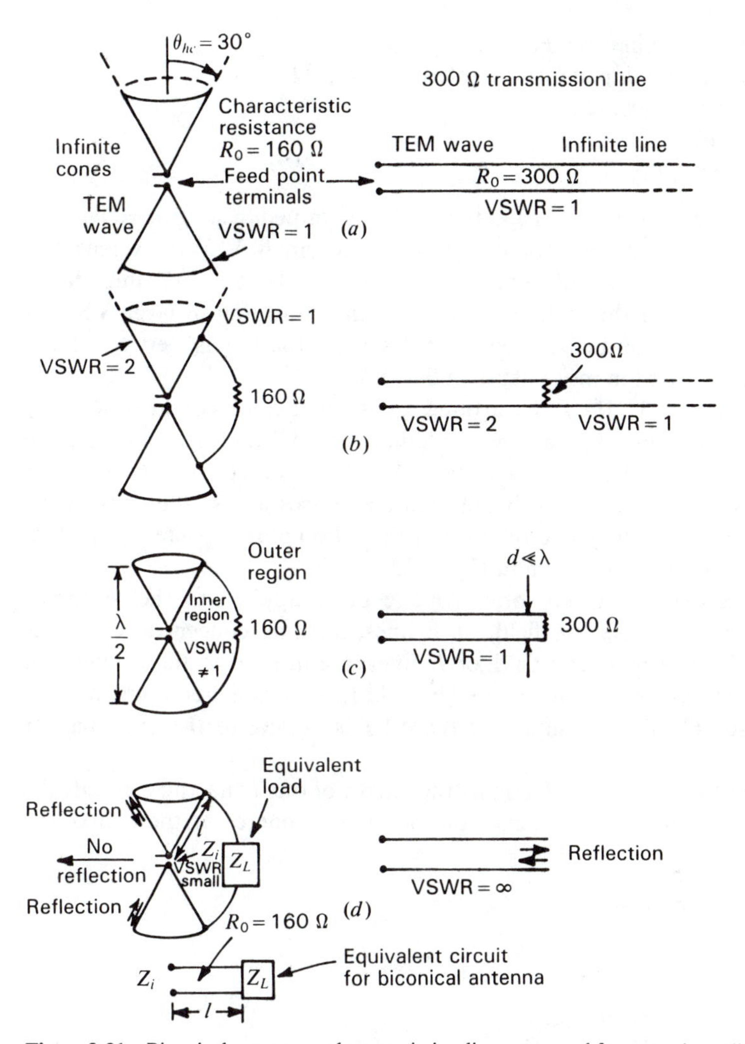

Figure 2-31 Biconical antenna and transmission line compared for several conditions.

In Fig. 2-31*d* the load resistors are removed, resulting in an infinite VSWR on the transmission line but a finite VSWR on the cones of the antenna, because the radiation field of the antenna acts like an equivalent load Z_L. This load impedance transforms to an input (or terminal) impedance

$$Z_i = Z_0 \frac{Z_L + jZ_0 \tan \beta l}{Z_0 + jZ_L \tan \beta l} \quad (\Omega) \tag{3}$$

as though the cones were a transmission line of length l. For the truncated cones there is reflection at the ends of the cones but no reflection of waves radiating in the equatorial plane.

In (3), Z_i = input or terminal impedance, Ω
Z_0 = characteristic impedance of antenna = R_0, Ω
Z_L = load impedance, Ω
$\beta = 2\pi/\lambda$, rad m^{-1}
l = line length, m

For large cone angles ($\theta > 10°$), the terminal impedance approaches the characteristic resistance of the antenna resulting in small VSWRs versus frequency on the cones. For small cone angles ($\theta < 1°$), the terminal impedance departs significantly from the characteristic resistance resulting in large VSWRs versus frequency. For intermediate cone angles the situation is between these extremes. This is discussed in more detail in Sec. 8-4.

Referring to Fig. 2-32, the space around an antenna can be divided into two regions: one next to the antenna, the "antenna region," and one outside, the "outer region." The boundary between the two regions is a sphere whose center is at the middle of the antenna and whose surface passes across the ends of the antenna. (See also Fig. 2-26*a*.) The relation of this "boundary sphere" to a symmetrical, biconical antenna is shown in Fig. 2-32.

The wave caused by a very brief voltage pulse applied to the terminals travels outward with the electric field, or *E* lines, forming concentric circles as shown in Fig. 2-32. The magnetic field, or *H* lines, are normal to the *E* lines and are concentric with the axis of the cones. The field has no radial component. It is strictly transverse (TEM). It is said that these fields belong to the principal, or zero-order, mode.

After a time $t = L/c$, where L equals the length of one cone and c equals the velocity of light, the pulse field reaches the boundary sphere. At the end of the

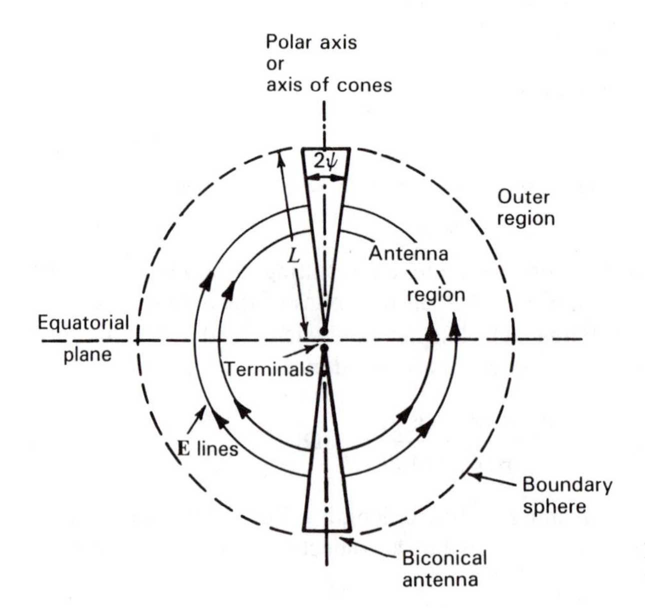

Figure 2-32 Biconical antenna with boundary sphere.

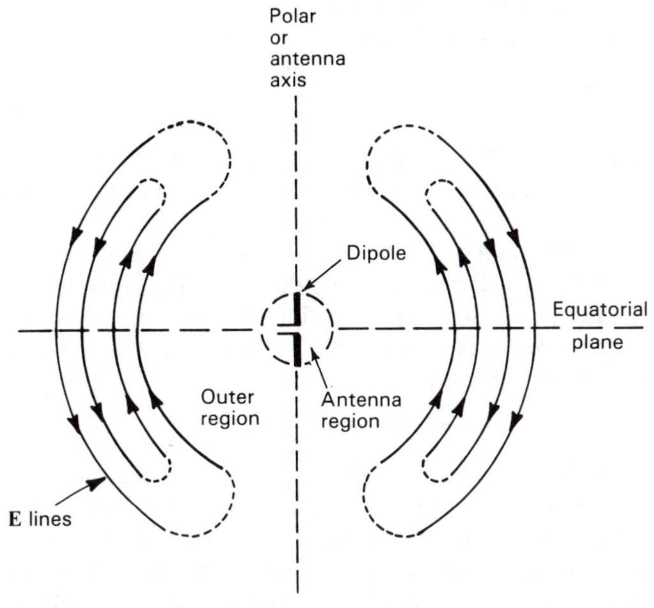

Figure 2-33 Field configuration near dipole antenna.

cones there is an abrupt discontinuity, while at the equator there is none. Hence, there is a large reflection at the end of the cones, and little energy is radiated in this direction. On the other hand, at the equator the energy continues into the outer region, and radiation is a maximum in this direction.

The E lines of principal-mode fields must end on conductors and, hence, cannot exist in free space. The waves which can exist and propagate in free space are higher mode forms in which the E lines form closed loops. The principal-mode wave is called a zero-order wave, and higher-order waves are of order 1 and greater. The configuration of the E lines of a first-order wave in the outer region is illustrated in Fig. 2-33. This wave has been radiated from a short dipole antenna. The wave started on the antenna as a principal-mode wave, has passed through the boundary sphere and has been transformed.[1] The field has a radial component which is largest near the polar axis. At the equatorial plane the radial component is zero and the E lines at this plane travel through the boundary sphere without change. Since the radial components of the field attenuate more rapidly than the transverse components, the radial field becomes negligible in comparison with the transverse field at a large distance from the antenna. Although the field at a large distance from the antenna is of a higher-order type,

[1] Some first-order mode is also present inside the antenna boundary sphere as a reflected wave. This and higher-order modes may exist both inside and outside of the boundary sphere in such a way that there is continuity of the fields at the boundary sphere.

the measurable components are only of the transverse type. To suggest the fact that the radial field components are weak and become negligible at large distances, the E lines in the polar region in Fig. 2-33 are dashed.

2-34 WAVE POLARIZATION.

An important property of an electromagnetic wave is its *polarization*, a quantity describing the orientation of the electric field **E**.[1]

Consider a plane wave traveling out of the page (positive z direction), as in Fig. 2-34*a*, with the electric field at all times in the y direction. This wave is said to be *linearly polarized* (in the y direction). As a function of time and position the electric field of a linearly polarized wave (as in Fig. 2-34*a*) traveling in the positive z direction (out of the page) is given by

$$E_y = E_2 \sin(\omega t - \beta z) \tag{1}$$

In general, the electric field of a wave traveling in the z direction may have both a y component and an x component, as suggested in Fig. 2-34*b*. In this more general situation the wave is said to be *elliptically polarized*. At a fixed value of z the electric vector **E** rotates as a function of time, the tip of the vector describing an ellipse called the *polarization ellipse*. The ratio of the major to minor axes of the polarization ellipse is called the *axial ratio* (AR). Thus, for the wave in Fig. 2-34*b*, AR $= E_2/E_1$. Two extreme cases of elliptical polarization correspond to *circular polarization*, as in Fig. 2-34*c*, and *linear polarization*, as in Fig. 2-34*a*. For circular polarization $E_1 = E_2$ and AR $= 1$, while for linear polarization $E_1 = 0$ and AR $= \infty$.

In the most general case of elliptical polarization the polarization ellipse may have any orientation, as suggested in Fig. 2-35. This elliptically polarized wave may be expressed in terms of two linearly polarized components, one in the x direction and one in the y direction. Thus, if the wave is travelling in the positive z direction (out of the page), the electric field components in the x and y directions are

$$E_x = E_1 \sin(\omega t - \beta z) \tag{2}$$

$$E_y = E_2 \sin(\omega t - \beta z + \delta) \tag{3}$$

where E_1 = amplitude of wave linearly polarized in x direction
E_2 = amplitude of wave linearly polarized in y direction
δ = time-phase angle by which E_y leads E_x

Combining (2) and (3) gives the instantaneous total vector field **E**:

$$\mathbf{E} = \hat{\mathbf{x}} E_1 \sin(\omega t - \beta z) + \hat{\mathbf{y}} E_2 \sin(\omega t - \beta z + \delta) \tag{4}$$

---

[1] Thus, a linearly polarized wave with **E** vertical is called a *vertically polarized* wave, the accompanying magnetic field **H** being horizontal.

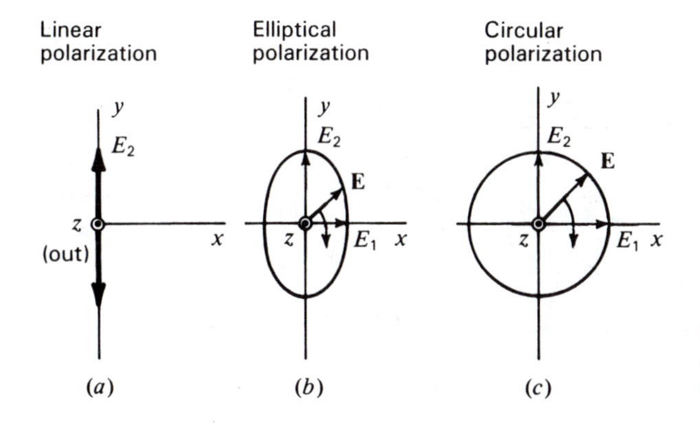

Figure 2-34 Linear, elliptical and circular polarization for wave propagation out of page.

At $z = 0$, $E_x = E_1 \sin \omega t$ and $E_y = E_2 \sin (\omega t + \delta)$. Expanding E_y yields

$$E_y = E_2 (\sin \omega t \cos \delta + \cos \omega t \sin \delta) \tag{5}$$

From the relation for E_x we have $\sin \omega t = E_x/E_1$ and $\cos \omega t = \sqrt{1 - (E_x/E_1)^2}$. Introducing these in (5) eliminates ωt, and on rearranging we obtain

$$\frac{E_x^2}{E_1^2} - \frac{2E_x E_y \cos \delta}{E_1 E_2} + \frac{E_y^2}{E_2^2} = \sin^2 \delta \tag{6}$$

or

$$aE_x^2 - bE_x E_y + cE_y^2 = 1 \tag{7}$$

where

$$a = \frac{1}{E_1^2 \sin^2 \delta} \qquad b = \frac{2 \cos \delta}{E_1 E_2 \sin^2 \delta} \qquad c = \frac{1}{E_2^2 \sin^2 \delta}$$

Equation (7) describes a (polarization) ellipse, as in Fig. 2-35. The line segment OA is the semimajor axis and the line segment OB is the semiminor axis. The tilt angle of the ellipse is τ. The axial ratio is

$$\text{AR} = \frac{OA}{OB} \qquad (1 \leq \text{AR} \leq \infty) \tag{8}$$

For $E_1 = 0$, the wave is linearly polarized in the y direction. For $E_2 = 0$, the wave is linearly polarized in the x direction. If $\delta = 0$ and $E_1 = E_2$, the wave is also linearly polarized but in a plane at an angle of 45° with respect to the x axis ($\tau = 45°$).

For $E_1 = E_2$ and $\delta = \pm 90°$, the wave is circularly polarized. When $\delta = +90°$, the wave is *left-circularly polarized*, and when $\delta = -90°$, the wave is *right-circularly polarized*. For the case $\delta = +90°$ and for $z = 0$ and $t = 0$ we have from (2) and (3) that $\mathbf{E} = \hat{\mathbf{y}} E_2$, as in Fig. 2-36*a*. One-quarter cycle later ($\omega t = 90°$) $\mathbf{E} = \hat{\mathbf{x}} E_1$, as in Fig. 2-36*b*. Thus, at a fixed position ($z = 0$) the electric field vector rotates clockwise (viewing the wave approaching). According to the IEEE

CW = LCP

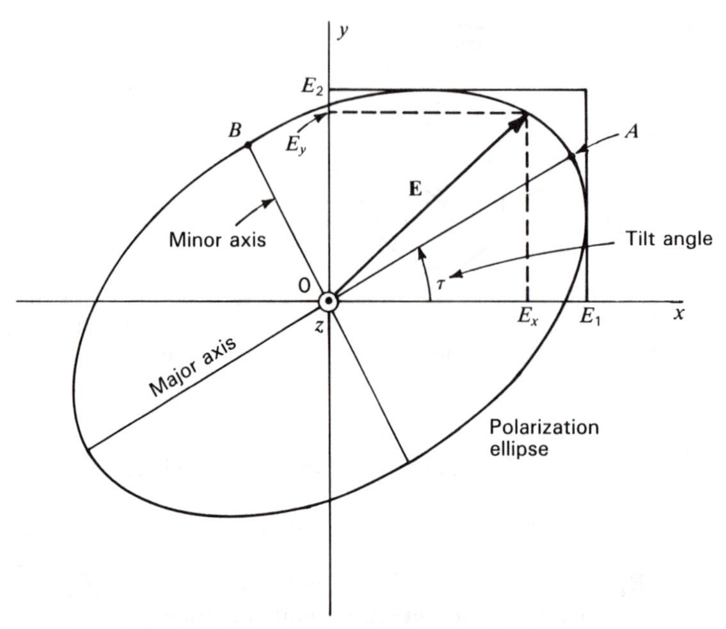

Figure 2-35 Polarization ellipse at tilt angle τ showing instantaneous components E_x and E_y and amplitudes (or peak values) E_1 and E_2.

definition, this corresponds to left-circular polarization.[1] The opposite rotation direction ($\delta = -90°$) corresponds to right-circular polarization. Polarization ellipses, as a function of the ratio E_2/E_1 and phase angle δ (wave approaching), are shown in Fig. 2-37. In special cases, the ellipses become straight lines (linear polarization) or circles (circular polarization).

If the wave is viewed receding (from negative z axis in Fig. 2-36), the electric vector appears to rotate in the opposite direction. Hence, clockwise rotation of **E** with the wave approaching is the same as counterclockwise rotation with the wave receding. Thus, unless the wave direction is specified, there is a possibility of ambiguity as to whether the wave is left- or right-handed. This can be avoided by defining the polarization with the aid of helical antennas (see Chap. 7). Thus, a right-handed monofilar axial-mode helical antenna radiates (or receives) right-circular (IEEE) polarization.[2] A right-handed helix, like a right-handed screw, is right-handed regardless of the position from which it is viewed. There is no possibility here of ambiguity.

The concept of polarization extends to antennas. Thus, an antenna which radiates a linearly polarized wave can be described as a *linearly polarized*

---

[1] This IEEE definition is opposite to the classical optics definition.

[2] A left-handed monofilar axial-mode helical antenna radiates (or receives) left-circular (IEEE) polarization.

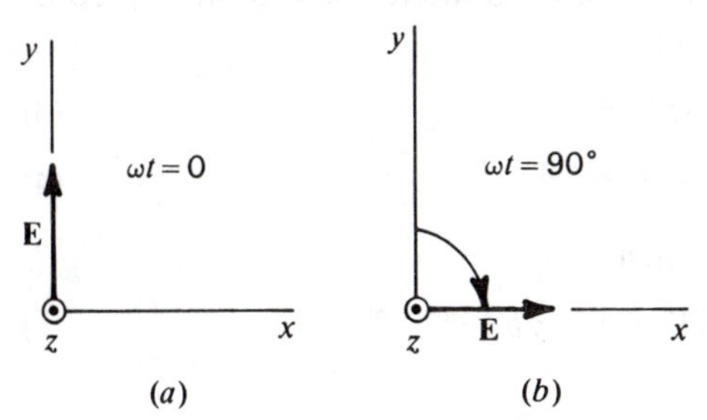

Figure 2-36 Instantaneous orientation of electric field vector E at two instants of time for a left-circularly polarized wave which is approaching (out of page).

antenna, as, for example, a dipole antenna. Likewise, an antenna which radiates a circularly polarized wave can be called a *circularly polarized antenna*, as, for example, a monofilar axial-mode helical antenna (see Chap. 7).

2-35 WAVE POLARIZATION AND THE POYNTING VECTOR.

In complex notation the Poynting vector is

$$\mathbf{S} = \tfrac{1}{2}\mathbf{E} \times \mathbf{H}^* \tag{1}$$

The average Poynting vector is the real part of (1), or

$$\mathbf{S}_{av} = \text{Re}\,\mathbf{S} = \tfrac{1}{2}\text{Re}\,\mathbf{E} \times \mathbf{H}^* \tag{2}$$

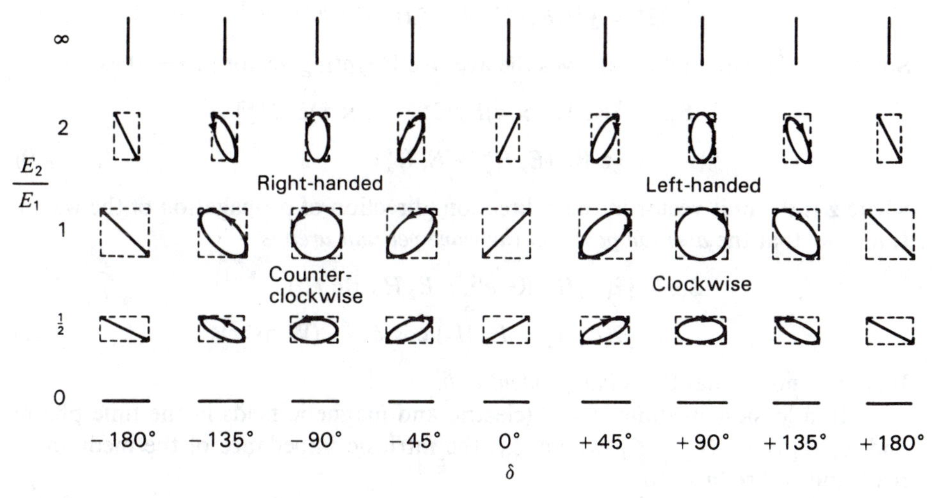

Figure 2-37 Polarization ellipses as a function of the ratio E_2/E_1 and phase angle S with wave approaching. Clockwise rotation of the resultant **E** corresponds to left-handed polarization (IEEE definition) while counterclockwise corresponds to right-handed polarization.

Referring to Fig. 2-35, let the elliptically polarized wave have x and y components with a phase difference δ as given by

$$E_x = E_1 e^{j(\omega t - \beta z)} \tag{3}$$

$$E_y = E_2 e^{j(\omega t - \beta z + \delta)} \tag{4}$$

At $z = 0$ the total electric field (vector) is then

$$\mathbf{E} = \hat{\mathbf{x}} E_x + \hat{\mathbf{y}} E_y = \hat{\mathbf{x}} E_1 e^{j\omega t} + \hat{\mathbf{y}} E_2 e^{j(\omega t + \delta)} \tag{5}$$

where $\hat{\mathbf{x}}$ = unit vector in x direction
$\hat{\mathbf{y}}$ = unit vector in y direction

Note that $\mathbf{E}$ has two components each involving both a space vector and a time-phase factor (phasor, with ωt explicit).

The **H**-field component associated with E_x is

$$H_y = H_1 e^{j(\omega t - \beta z - \xi)} \tag{6}$$

where ξ is the phase lag of H_y with respect to E_x. The **H**-field component associated with E_y is

$$H_x = -H_2 e^{j(\omega t - \beta z + \delta - \xi)} \tag{7}$$

The total **H** field (vector) at $z = 0$ for a wave traveling in the positive z direction is then

$$\mathbf{H} = \hat{\mathbf{y}} H_y - \hat{\mathbf{x}} H_x = \hat{\mathbf{y}} H_1 e^{j(\omega t - \xi)} - \hat{\mathbf{x}} H_2 e^{j(\omega t + \delta - \xi)} \tag{8}$$

The complex conjugate of **H** is equal to (8) except for the sign of the exponents; that is,

$$\mathbf{H}^* = \hat{\mathbf{y}} H_1 e^{-j(\omega t - \xi)} - \hat{\mathbf{x}} H_2 e^{-j(\omega t + \delta - \xi)} \tag{9}$$

Substituting (5) and (9) in (2) gives the average Poynting vector at $z = 0$ as

$$\begin{aligned} \mathbf{S}_{av} &= \tfrac{1}{2}\text{Re}\,[(\hat{\mathbf{x}} \times \hat{\mathbf{y}}) E_x H_y^* - (\hat{\mathbf{y}} \times \hat{\mathbf{x}}) E_y H_x^*] \\ &= \tfrac{1}{2}\hat{\mathbf{z}}\ \text{Re}\,(E_x H_y^* + E_y H_x^*) \end{aligned} \tag{10}$$

where $\hat{\mathbf{z}}$ is the unit vector in the z direction (direction of propagation of the wave). It follows that the *average power of the wave per unit area is*

$$\begin{aligned} \mathbf{S}_{av} &= \tfrac{1}{2}\hat{\mathbf{z}}(E_1 H_1\ \text{Re}\ e^{j\xi} + E_2 H_2\ \text{Re}\ e^{j\xi}) \\ &= \tfrac{1}{2}\hat{\mathbf{z}}(E_1 H_1 + E_2 H_2) \cos \xi \qquad (\text{W m}^{-2}) \end{aligned} \tag{11}$$

It is to be noted that $\mathbf{S}_{av}$ *is independent of* δ.

In a lossless medium, $\xi = 0$ (electric and magnetic fields in the time phase) and $E_1/H_1 = E_2/H_2 = Z_0$ (where Z_0, the intrinsic impedance of the medium, is real), and (11) reduces to

$$\begin{aligned} \mathbf{S}_{av} &= \tfrac{1}{2}\hat{\mathbf{z}}(E_1 H_1 + E_2 H_2) \\ &= \tfrac{1}{2}\hat{\mathbf{z}}(H_1^2 + H_2^2) Z_0 = \tfrac{1}{2}\hat{\mathbf{z}} H^2 Z_0 \qquad (\text{W m}^{-2}) \end{aligned} \tag{12}$$

where $H = \sqrt{H_1^2 + H_2^2}$ is the amplitude of the total **H** field. We can also write

$$\mathbf{S}_{av} = \tfrac{1}{2}\hat{\mathbf{z}}\,\frac{E_1^2 + E_2^2}{Z_0} = \tfrac{1}{2}\hat{\mathbf{z}}\,\frac{E^2}{Z_0} \qquad (\text{W m}^{-2}) \tag{13}$$

where $E = \sqrt{E_1^2 + E_2^2}$ is the amplitude of the total **E** field.

Example. An elliptically polarized wave travelling in the positive z direction in air has x and y components

$$E_x = 3 \sin(\omega t - \beta z) \qquad (\text{V m}^{-1})$$

$$E_y = 6 \sin(\omega t - \beta z + 75°) \quad (\text{V m}^{-1})$$

Find the average power per unit area conveyed by the wave.

Solution. The average power per unit area is equal to the average Poynting vector, which from (13) has a magnitude

$$S_{av} = \frac{1}{2}\frac{E^2}{Z_0} = \frac{1}{2}\frac{E_1^2 + E_2^2}{Z_0}$$

From the stated conditions, the amplitude $E_1 = 3$ V m^{-1} and the amplitude $E_2 = 6$ V m^{-1}. Also, for air, $Z_0 = 376.7\ \Omega$. Hence,

$$S_{av} = \frac{1}{2}\frac{3^2 + 6^2}{376.7} = \frac{1}{2}\frac{45}{376.7} \simeq 60 \text{ mW m}^{-2}$$

2-36 WAVE POLARIZATION AND THE POINCARÉ SPHERE.

In the Poincaré sphere[1] representation of wave polarization, the *polarization state* is described by a point on a sphere where the longitude and latitude of the point are related to parameters of the polarization ellipse (see Fig. 2-38) as follows:

$$\text{Longitude} = 2\tau$$

$$\text{Latitude} = 2\varepsilon \tag{1}$$

where τ = tilt angle, $0° \leq \tau \leq 180°$
$\varepsilon = \cot^{-1}(\mp \text{AR})$, $-45° \leq \varepsilon \leq +45°$

The axial ratio (AR) and angle ε are negative for right-handed and positive for left-handed (IEEE) polarization.

The polarization state described by a point on a sphere can also be expressed in terms of the angle subtended by the great circle drawn from a

[1] H. Poincaré, *Theorie Mathematique de la Luminière*, G. Carré, Paris, 1892.
G. A. Deschamps, "Geometrical Representation of the Polarization of a Plane Electromagnetic Wave," *Proc. IRE*, **39**, 540, May 1951.

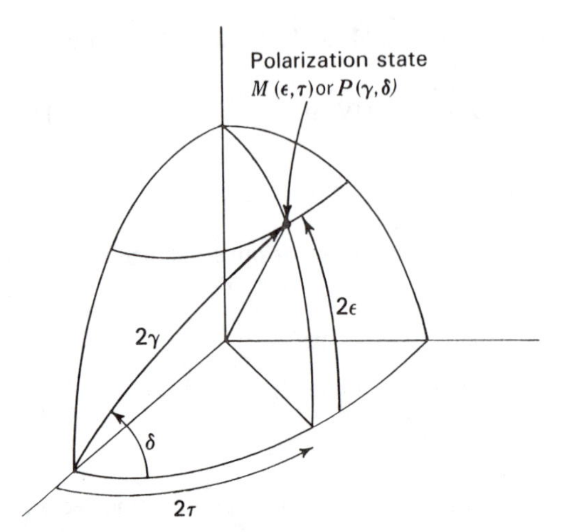

Figure 2-38 Poincaré sphere showing relation of angles ε, τ, γ and δ.

reference point on the equator and the angle between the great circle and the equator (see Fig. 2-38) as follows:

$$\text{Great-circle angle} = 2\gamma$$

$$\text{Equator-to-great-circle angle} = \delta \tag{2}$$

where $\gamma = \tan^{-1}(E_2/E_1)$, $0° \leq \gamma \leq 90°$
δ = phase difference between E_y and E_x, $-180° \leq \delta \leq +180°$

The geometric relation of τ, ε and γ to the polarization ellipse is illustrated in Fig. 2-39. The spherical trigonometric interrelations of τ, ε, γ and δ are as follows:

$$\begin{aligned} \cos 2\gamma &= \cos 2\varepsilon \cos 2\tau \\ \tan \delta &= \frac{\tan 2\varepsilon}{\sin 2\tau} \\ \tan 2\tau &= \tan 2\gamma \cos \delta \\ \sin 2\varepsilon &= \sin 2\gamma \sin \delta \end{aligned} \tag{3}$$

Knowing ε and τ, one can determine γ and δ or vice versa. It is convenient to describe the *polarization state* by either of the two sets of angles (ε, τ) or (γ, δ) which describe a point on the *Poincaré sphere* (Fig. 2-38). Let the polarization state as a function of ε and τ be designated by $M(\varepsilon, \tau)$ or simply M and the polarization state as a function of γ and δ be designated by $P(\gamma, \delta)$ or simply P, as in Fig. 2-38. Two special cases are of interest.

Case 1. For $\delta = 0$ or $\delta = \pm 180°$, E_x and E_y are exactly in phase or out of phase, so that any point on the equator represents *a state of linear polarization.* At the origin ($\varepsilon = \tau = 0$) the polarization is linear and in the x direction ($\tau = 0$), as suggested in Fig. 2-40*a*. On the equator 90° to the right, the polarization is linear with a tilt angle

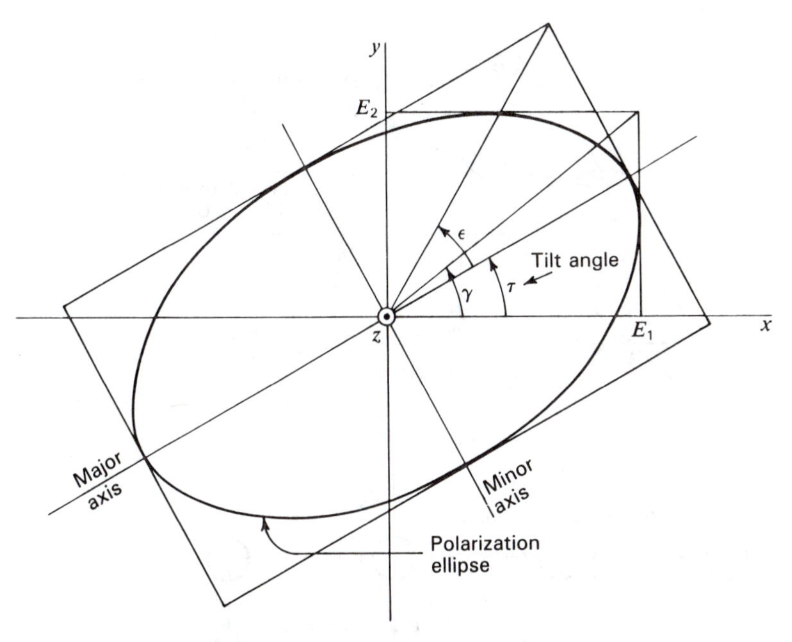

Figure 2-39 Polarization ellipse showing relation of angles ε, γ and τ.

$\tau = 45°$, while 180° from the origin the polarization is linear and in the y direction ($\tau = 90°$). See Fig. 2-40*a* and *b*. One octant of the Poincaré sphere is shown in Fig. 2-40*a*, the full sphere being shown in Fig. 2-40*b* in rectangular projection.

Case 2. For $\delta = \pm 90°$ and $E_2 = E_1$ ($2\gamma = 90°$ and $2\varepsilon = \pm 90°$) E_x and E_y have equal amplitudes but are in phase quadrature, which is the condition for circular polarization. Thus, the poles represent *a state of circular polarization*, the upper pole representing left-circular polarization and the lower pole right-circular (IEEE) polarization, as suggested in Fig. 2-40*a* and *b*.

Cases 1 and 2 represent limiting conditions. In the general case, any point on the upper hemisphere describes a *left-elliptically polarized wave* ranging from pure left circular at the pole to linear at the equator. Likewise, any point on the lower hemisphere describes a *right-elliptically polarized wave* ranging from pure right circular at the pole to linear at the equator. Several elliptical states of polarization are shown by ellipses with appropriate tilt angles τ and axial ratios AR at points on the Poincaré sphere in Fig. 2-40*a* and *b*.

As an application of the Poincaré sphere representation, it may be shown that the voltage response V of an antenna to a wave of arbitrary polarization is given by[1]

$$V = k \cos \frac{MM_a}{2} \tag{4}$$

[1] G. Sinclair, "The Transmission and Reception of Elliptically Polarized Waves," *Proc. IRE*, **38**, 151, 1950.

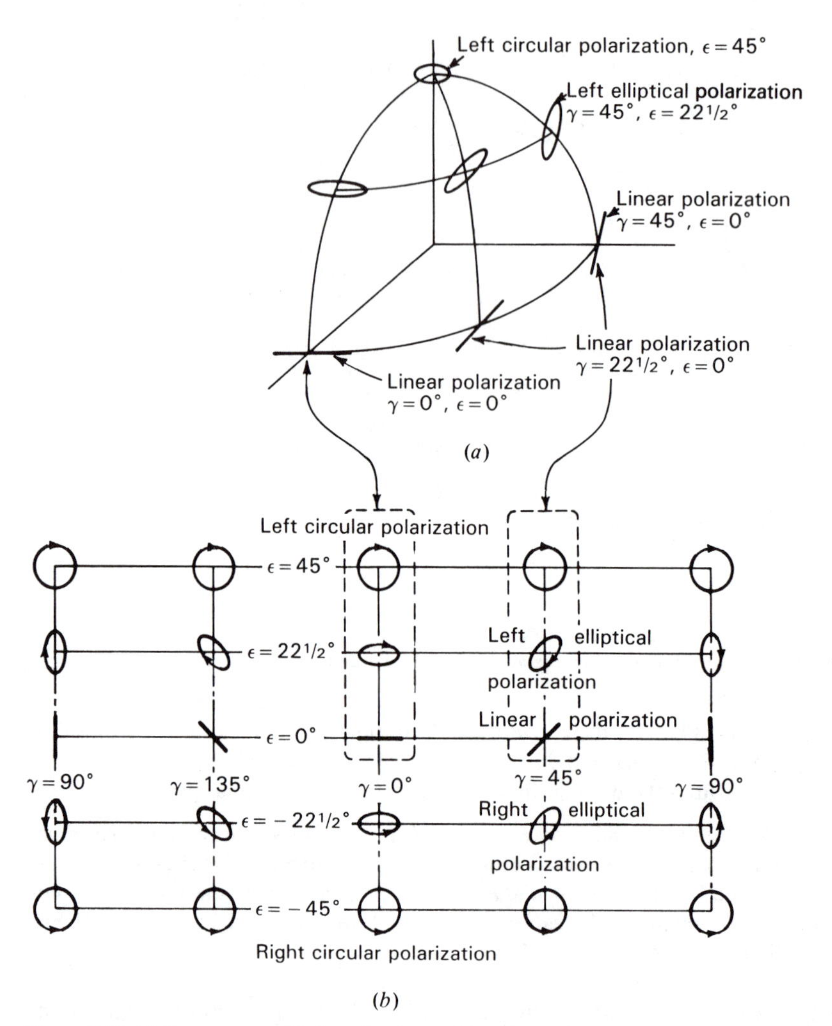

Figure 2-40 (*a*) One octant of Poincaré sphere with polarization states. (*b*) Rectangular projection of Poincaré sphere showing full range of polarization states.

where MM_a = angle subtended by great-circle line from polarization state M to M_a
M = polarization state of wave
M_a = polarization state of antenna
k = constant

The polarization state of the antenna is defined as the polarization state of the wave radiated by the antenna when it is transmitting. The factor k in (4) involves the field strength of the wave and the size of the antenna. An important result to note is that if $MM_a = 0°$, the antenna is matched to the wave (polarization state of wave same as for antenna) and the response is maximized. However, if

$MM_a = 180°$, the response is zero. This can occur, for example, if the wave is linearly polarized in the y direction while the antenna is linearly polarized in the x direction; or if the wave is left-circularly polarized while the antenna is right-circularly polarized. More generally we may say that *an antenna is blind to a wave of opposite (or antipodal) polarization state.*

A more complete discussion of polarization including *unpolarized waves, partial polarization, Stokes parameters* and the *wave-to-antenna coupling factor* is given by Kraus.[1]

2-37 CROSS-FIELD.

For elliptical and circular polarization, the electric field vector **E** at a fixed point rotates with time in a plane *perpendicular* to the direction of wave propagation. There are situations, however, where **E** rotates in a plane *parallel* to the direction of wave propagation. This condition is called *cross-field.*[2] This situation can occur if there is a component of **E** in the direction of propagation. This condition never exists in the case of a single plane wave in free space since such a wave has no field component in the direction of propagation. However, in the near field of an antenna there are field components in both the direction of propagation and normal to this direction so that cross-field is present. This is the case, for example, in the near field of a dipole antenna with field components E_r and E_θ as suggested in Fig. 2-41*a*.

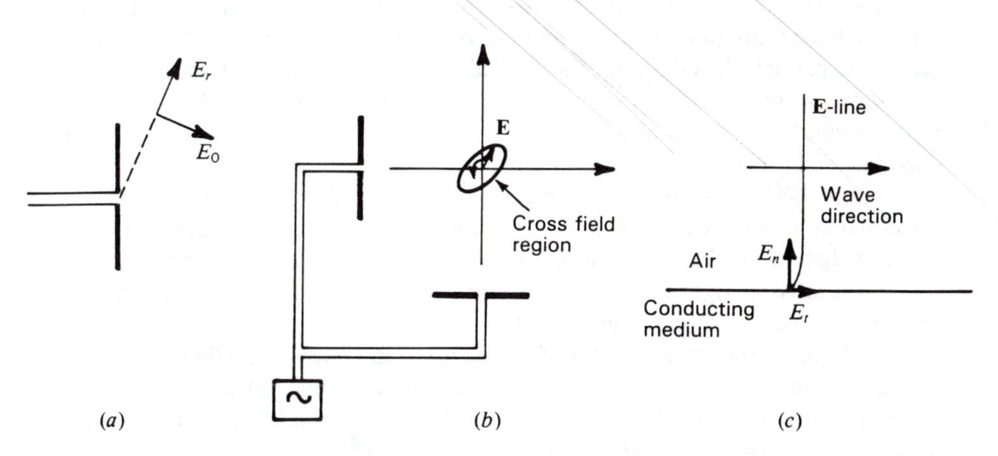

Figure 2-41 Three situations in which cross-field is present: (*a*) in the near field of a dipole antenna and (*b*) in the region exposed to radiation from two dipole antennas. At (*c*) cross-field is present near the surface of a conducting medium along which a plane wave is traveling.

[1] J. D. Kraus, *Radio Astronomy*, 2nd ed., Cygnus-Quasar, 1986, chap. 4; J. D. Kraus, *Electromagnetics*, 3rd ed., McGraw-Hill, 1984, sec. 11-5.

[2] A. Alford, J. D. Kraus and E. C. Barkofsky, *Very High Frequency Techniques*, McGraw-Hill, 1947, chap. 9, p. 200.

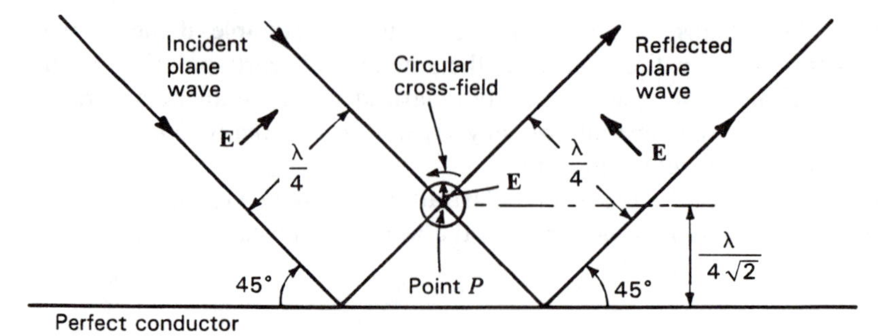

Figure 2-42 Circular cross-field above a perfectly conducting plane at a point where the incident and reflected linearly polarized waves are in time-phase quadrature.

Cross-field may also be present where two waves of the same frequency and traveling in different directions cross. Thus, in the region exposed to radiation from two dipole antennas, as in Fig. 2-41*b*, there is cross-field. The region may be in the far fields of the antennas. Both antennas are linearly polarized in the plane of the page and both radiate at the same frequency (both connected to the same generator). In general, the tip of the **E** vector describes a locus that is an ellipse in a manner similar to that in elliptical polarization except that **E** is confined to the plane of the antennas (plane of the page).

Another situation in which cross-field is present is near the surface of a conducting medium along which a plane wave is traveling (Fig. 2-41*c*). Unless the medium is perfectly conducting, the **E** field is tilted forward near the surface of the medium so that **E** has components both normal to the surface (E_n) and parallel or tangential to the surface (E_t). Since, in general, these components are not in time-phase, elliptical cross-field is present.[1]

Figure 2-42 illustrates the presence of cross-field at a point P above a perfect flat conductor with a linearly polarized plane wave incident at an angle of 45°. At a distance 0.177 wavelength above the conductor the reflected and incident waves cross at right angles and are in time-phase quadrature. This results in circular cross-field at P as suggested in the figure.

With pure cross-field, the tip of **E** describes, in general, an ellipse (elliptical cross-field) which in special cases may become a straight line (linear cross-field) or a circle (circular cross-field). This may be demonstrated analytically in a manner similar to that used in Sec. 2-34 for wave polarization.[2]

Finally, consider the situation at a point where two *circularly polarized* waves cross at right angles as shown in Fig. 2-43. If the waves are of equal amplitude and the same frequency, the loci of **E** at P are ellipses (with $\text{AR} = \sqrt{2}$) which project in the plane of the page as a line AA' (ellipse seen edge-on) or as a

[1] J. D. Kraus, *Electromagnetics*, 3rd ed., McGraw-Hill, 1984, p. 585.

[2] J. D. Kraus, *Electromagnetics*, 1st ed., McGraw-Hill, 1953, p. 385.

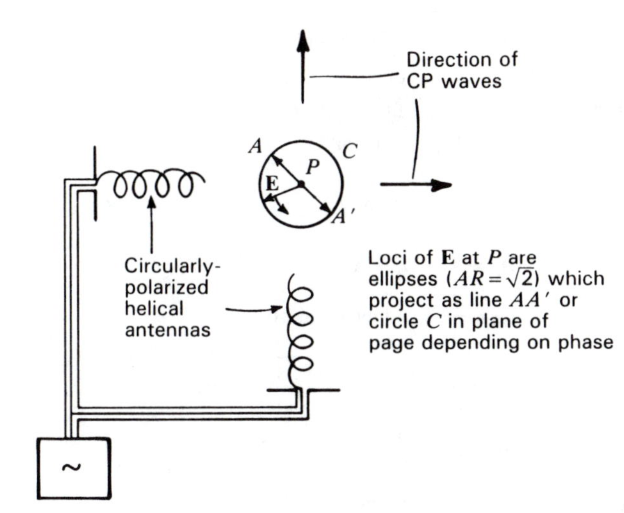

Figure 2-43 Two orthogonal monofilar axial-mode helical antennas producing equal circularly polarized fields result in loci of **E** at P that are ellipses which project as line AA' or as circle C in plane of page depending on phase.

circle C (ellipse seen at a slant angle) depending on the time phase. These loci represent neither pure cross-field nor pure polarization but a combination or hybrid situation; however, provided the two waves are of the same frequency, the locus of the tip of **E** always lies in a plane and, in general, describes an ellipse.

2-38 TABLE SUMMARIZING IMPORTANT RELATIONS OF CHAPTER 2.

Wavelength-frequency	$\lambda = \frac{v}{f}$ (m)
Beam area	$\Omega_A = \iint P_n(\theta, \phi)\, d\Omega$ (sr or deg²)
Beam area (approx.)	$\Omega_A \simeq \theta_{HP}\, \phi_{HP}$ (sr or deg²)
Beam efficiency	$\varepsilon_M = \frac{\Omega_M}{\Omega_A}$ (dimensionless)
Directivity	$D = \frac{U(\theta, \phi)_{max}}{U_{av}} = \frac{S(\theta, \phi)_{max}}{S_{av}}$ (dimensionless)
Directivity	$D = \frac{4\pi}{\Omega_A}$ (dimensionless)
Directivity	$D = \frac{4\pi A_e}{\lambda^2}$ (dimensionless)
Directivity (approx.)	$D \simeq \frac{4\pi}{\theta_{HP}\, \phi_{HP}} \simeq \frac{41\,000}{\theta^\circ_{HP}\, \phi^\circ_{HP}}$ (dimensionless)
Directivity (better approx.)	$D \simeq \frac{41\,000 \varepsilon_M}{k_p\, \theta_{HP}\, \phi_{HP}}$† (dimensionless)

† See (3-13-18).

Gain	$G = kD$	(dimensionless)
Effective aperture and beam area	$A_e \Omega_A = \lambda^2$	(m^2)
Scattering aperture (antenna matched)	$A_s = A_e$	(m^2 or λ^2)
Scattering aperture (antenna short-circuited)	$A_s = 4A_e$	(m^2 or λ^2)
Scattering aperture (antenna open-circuited, small antenna)	$A_s \simeq 0$	(m^2 or λ^2)
Collecting aperture	$A_c = A_e + A_s + A_L$	(m^2 or λ^2)
Aperture efficiency	$\varepsilon_{ap} = \dfrac{A_e}{A_p}$	(dimensionless)
Effective height	$h_e = \dfrac{V}{E} = h_p \dfrac{I_{av}}{I_0} = \sqrt{\dfrac{2R_r A_e}{Z_0}}$	(m)
Friis transmission formula	$P_r = P_t \dfrac{A_{et} A_{er}}{r^2 \lambda^2}$	(W)
Current-charge continuity relation	$\dot{I} l = q \dot{v}$	(A m s^{-1})
Radiation power	$P = \dfrac{\mu^2 q^2 \dot{v}^2}{6\pi Z}$	(W)
Near-field–far-field boundary	$R = \dfrac{2L^2}{\lambda}$	(m)
Average power per unit area of elliptically polarized wave in air	$\mathbf{S}_{av} = \frac{1}{2}\hat{\mathbf{z}} \dfrac{E_1^2 + E_2^2}{Z_0}$	(W m^{-2})

PROBLEMS[1]

2-1 Directivity. Show that the directivity D of an antenna may be written

$$D = \frac{\dfrac{E(\theta, \phi)_{\max} E^*(\theta, \phi)_{\max}}{Z} r^2}{\dfrac{1}{4\pi} \displaystyle\iint_{4\pi} \dfrac{E(\theta, \phi) E^*(\theta, \phi)}{Z} r^2 \, d\Omega}$$

2-2 Directivity. Show that the directivity of an antenna may be expressed as

$$D = \frac{4\pi}{\lambda^2} \frac{\displaystyle\iint_{A_p} E(x, y)\, dx\, dy \iint_{A_p} E^*(x, y)\, dx\, dy}{\displaystyle\iint_{A_p} E(x, y) E^*(x, y)\, dx\, dy}$$

where $E(x, y)$ is the aperture field distribution.

2-3 Effective aperture. What is the maximum effective aperture (approximately) for a beam antenna having half-power widths of 30° and 35° in perpendicular planes intersecting in the beam axis? Minor lobes are small and may be neglected.

[1] Answers to starred (*) problems are given in App. D.

★2-4 Effective aperture. What is the maximum effective aperture of a microwave antenna with a directivity of 900?

2-5 Received power. What is the maximum power received at a distance of 0.5 km over a free-space 1-GHz circuit consisting of a transmitting antenna with a 25-dB gain and a receiving antenna with a 20-dB gain? The gain is with respect to a lossless isotropic source. The transmitting antenna input is 150 W.

★2-6 Spacecraft link over 100 Mm. Two spacecraft are separated by 100 Mm. Each has an antenna with $D = 1000$ operating at 2.5 GHz. If craft A's receiver requires 20 dB over 1 pW, what transmitter power is required on craft B to achieve this signal level?

2-7 Spacecraft link over 3 Mm. Two spacecraft are separated by 3 Mm. Each has an antenna with $D = 200$ operating at 2 GHz. If craft A's receiver requires 20 dB over 1 pW, what transmitter power is required on craft B to achieve this signal level?

2-8 Mars link. (*a*) Design a two-way radio link to operate over earth-Mars distances for data and picture transmission with a Mars probe at 2.5 GHz with a 5-MHz bandwidth. A power of 10^{-19} W Hz^{-1} is to be delivered to the earth receiver and 10^{-17} W Hz^{-1} to the Mars receiver. The Mars antenna must be no larger than 3 m in diameter. Specify effective aperture of Mars and earth antennas and transmitter power (total over entire bandwidth) at each end. Take earth-Mars distance as 6 light-minutes. (*b*) Repeat (*a*) for an earth-Jupiter link. Take the earth-Jupiter distance as 40 light-minutes.

★2-9 Moon link. A radio link from the moon to the earth has a moon-based 5λ-long right-handed monofilar axial-mode helical antenna and a 2-W transmitter operating at 1.5 GHz. What should the polarization state and effective aperture be for the earth-based antenna in order to deliver 10^{-14} W to the receiver? Take the earth-moon distance as 1.27 light-seconds.

2-10 Crossed dipoles for CP and other states. Two $\lambda/2$ dipoles are crossed at 90°. If the two dipoles are fed with equal currents, what is the polarization of the radiation perpendicular to the plane of the dipoles if the currents are (*a*) in phase, (*b*) phase quadrature (90° difference in phase) and (*c*) phase octature (45° difference in phase)?

★2-11 Two LP waves. A wave traveling normally out of the page (toward the reader) has two linearly polarized components

$$E_x = 2 \cos \omega t$$

$$E_y = 3 \cos (\omega t + 90^\circ)$$

(*a*) What is the axial ratio of the resultant wave?
(*b*) What is the tilt angle τ of the major axis of the polarization ellipse?
(*c*) Does **E** rotate clockwise or counterclockwise?

2-12 Two EP waves. A wave traveling normally outward from the page (toward the reader) is the resultant of two elliptically polarized waves, one with components of **E** given by

$$E'_y = 2 \cos \omega t$$

$$E'_x = 6 \cos \left(\omega t + \frac{\pi}{2}\right)$$

and the other with components given by

$$E_y'' = 1 \cos \omega t$$

$$E_x'' = 3 \cos \left(\omega t - \frac{\pi}{2} \right)$$

(*a*) What is the axial ratio of the resultant wave?
(*b*) Does **E** rotate clockwise or counterclockwise?

***2-13 Two LP components.** An elliptically polarized plane wave traveling normally out of the page (toward the reader) has linearly polarized components E_x and E_y. Given that $E_x = E_y = 1$ V m^{-1} and that E_y leads E_x by 72°,
(*a*) Calculate and sketch the polarization ellipse.
(*b*) What is the axial ratio?
(*c*) What is the angle τ between the major axis and the x axis?

2-14 Two LP components. Answer the same questions as in Prob. 2-13 for the case where E_y leads E_x by 72° as before but $E_x = 2$ V m^{-1} and $E_y = 1$ V m^{-1}.

***2-15 Two CP waves.** Two circularly polarized waves intersect at the origin. One (y wave) is traveling in the positive y direction with **E** rotating clockwise as observed from a point on the positive y axis. The other (x wave) is traveling in the positive x direction with **E** rotating clockwise as observed from a point on the positive x axis. At the origin, **E** for the y wave is in the positive z direction at the same instant that **E** for the x wave is in the negative z direction. What is the locus of the resultant **E** vector at the origin?

2-16 Tilt angle. Show that the tilt angle τ can be expressed as

$$\tau = \tfrac{1}{2} \tan^{-1} \frac{2E_1E_2 \cos \delta}{E_1^2 - E_2^2}$$

2-17 Spaceship near moon. A spaceship at lunar distance from the earth transmits 2-GHz waves. If a power of 10 W is radiated isotropically, find (*a*) the average Poynting vector at the earth, (*b*) the rms electric field **E** at the earth and (*c*) the time it takes for the radio waves to travel from the spaceship to the earth. (Take the earth-moon distance as 380 Mm.) (*d*) How many photons per unit area per second fall on the earth from the spaceship transmitter?

***2-18 CP waves.** A wave traveling normally out of the page is the resultant of two circularly polarized components $E_{\text{right}} = 5e^{j\omega t}$ and $E_{\text{left}} = 2e^{-j(\omega t + 90°)}$ (V m^{-1}). Find (*a*) the axial ratio AR, (*b*) the tilt angle τ and (*c*) the hand of rotation (left or right).

2-19 EP wave. A wave traveling normally out of the page (toward the reader) is the resultant of two linearly polarized components $E_x = 3 \cos \omega t$ and $E_y = 2 \cos (\omega t + 90°)$. For the resultant wave find (*a*) the axial ratio AR, (*b*) the tilt angle τ and (*c*) the hand of rotation (left or right).

***2-20 CP waves.** Two circularly polarized waves traveling normally out of the page have fields given by $E_{\text{left}} = 2e^{-j\omega t}$ and $E_{\text{right}} = 3e^{j\omega t}$(V m^{-1}) (rms). For the resultant wave find (*a*) AR, (*b*) the hand of rotation and (*c*) the Poynting vector.

2-21 EP waves. A wave traveling normally out of the page is the resultant of two elliptically polarized (EP) waves, one with components $E_x = 5 \cos \omega t$ and $E_y = 3 \sin \omega t$ and another with components $E_r = 3e^{j\omega t}$ and $E_l = 4e^{-j\omega t}$. For the resultant wave, find (*a*) AR, (*b*) τ and (*c*) the hand of rotation.

*2-22 **CP waves.** A wave traveling normally out of the page is the resultant of two circularly polarized components $E_r = 2e^{j\omega t}$ and $E_l = 4e^{-j(\omega t + 45°)}$. For the resultant wave, find (*a*) AR, (*b*) τ and (*c*) the hand of rotation.

2-23 **More power with CP.** Show that the average Poynting vector of a circularly polarized wave is twice that of a linearly polarized wave if the maximum electric field **E** is the same for both waves. This means that a medium can handle twice as much power before breakdown with circular polarization (CP) than with linear polarization (LP).

2-24 **PV constant for CP.** Show that the instantaneous Poynting vector (PV) of a plane circularly polarized traveling wave is a constant.

*2-25 **EP wave power.** An elliptically polarized wave in a medium with constants $\sigma = 0$, $\mu_r = 2$, $\varepsilon_r = 5$ has H-field components (normal to the direction of propagation and normal to each other) of amplitudes 3 and 4 A m^{-1}. Find the average power conveyed through an area of 5 m^2 normal to the direction of propagation.

2-26 **Circular-depolarization ratio.** If the axial ratio of a wave is AR, show that the circular-depolarization ratio of the wave is given by

$$R = \frac{\text{AR} - 1}{\text{AR} + 1}$$

Thus, for pure circular polarization AR = 1 and $R = 0$ (no depolarization) but for linear polarization AR = ∞ and $R = 1$.

2-27 **Superluminal phase velocity near dipole.** (*a*) By measuring the distances between P, P' and P'' determine the amount of superluminal ($v > c$) phase velocity of the waves near the dipole in Fig. 2-23. (*b*) Under what other conditions are superluminal velocities encountered?

CHAPTER 3

POINT SOURCES

3-1 INTRODUCTION. POINT SOURCE DEFINED. Let us consider an antenna contained within a volume of radius b as in Fig. 3-1a. Confining our attention only to the far field of the antenna, we may make observations of the fields along an observation circle of large radius R. At this distance the measurable fields are entirely transverse, and the power flow, or Poynting vector, is entirely radial. It is convenient in many analyses to assume that the fields of the antenna are everywhere of this type. In fact, we may assume, by extrapolating inward along the radii of the circle, that the waves originate at a fictitious volumeless emitter, or *point source*, at the center O of the observation circle. The actual field variation near the antenna, or "near field," is ignored, and we describe the source of the waves only in terms of the "far field" it produces. Provided that our observations are made at a sufficient distance, any antenna, regardless of its size or complexity, can be represented in this way by a single point source.

Instead of making field measurements around the observation circle with the antenna fixed, the equivalent effect may be obtained by making the measurements at a fixed point Q on the circle and rotating the antenna around the center O. This is usually the more convenient procedure if the antenna is small.

In Fig. 3-1a the center O of the antenna coincides with the center of the observation circle. If the center of the antenna is displaced from O, even to the extent that O lies outside the antenna as in Fig. 3-1b, the distance d between the two centers has a negligible effect on the field patterns at the observation circle

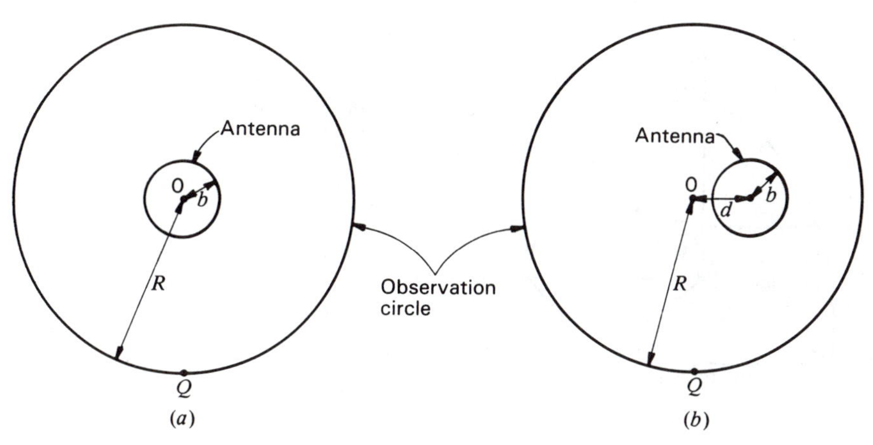

Figure 3-1 Antenna and observation circle.

provided $R \gg d$, $R \gg b$, and $R \gg \lambda$. However, the phase patterns[1] will generally differ, depending on d. If $d = 0$, the phase shift around the observation circle is usually a minimum. As d is increased, the observed phase shift becomes larger.

As discussed in Sec. 2-4, a complete description of the far field of a source requires three patterns: two patterns of orthogonal field components as a function of angle [$E_\theta(\theta, \phi)$ and $E_\phi(\theta, \phi)$] and one pattern of the phase difference of these fields as a function of angle [$\delta(\theta, \phi)$]. For many purposes, however, such a complete knowledge is not necessary. It may suffice to specify only the variation with angle of the power density or Poynting vector magnitude (power per unit area) from the antenna [$S_r(\theta, \phi)$]. In this case the vector nature of the field is disregarded, and the radiation is treated as a scalar quantity. This is done in Sec. 3-2. The vector nature of the field is recognized later in the discussion on the magnitude of the field components in Sec. 3-16. Although the cases considered as examples in this chapter are hypothetical, they could be approximated by actual antennas.

3-2 POWER PATTERNS.

Let a transmitting antenna in free space be represented by a point-source radiator located at the origin of the coordinates in Fig. 3-2 (see also Fig. 2-5). The radiated energy streams from the source in radial lines. The time rate of energy flow per unit area is the *Poynting vector*, or *power density* (watts per square meter). For a point source (or in the far field of any antenna), the Poynting vector **S** has only a radial component S_r with no components in either the θ or ϕ directions ($S_\theta = S_\phi = 0$). Thus, the magnitude of the Poynting vector, or power density, is equal to the radial component ($|\mathbf{S}| = S_r$).

[1] Phase variation around the observation circle.

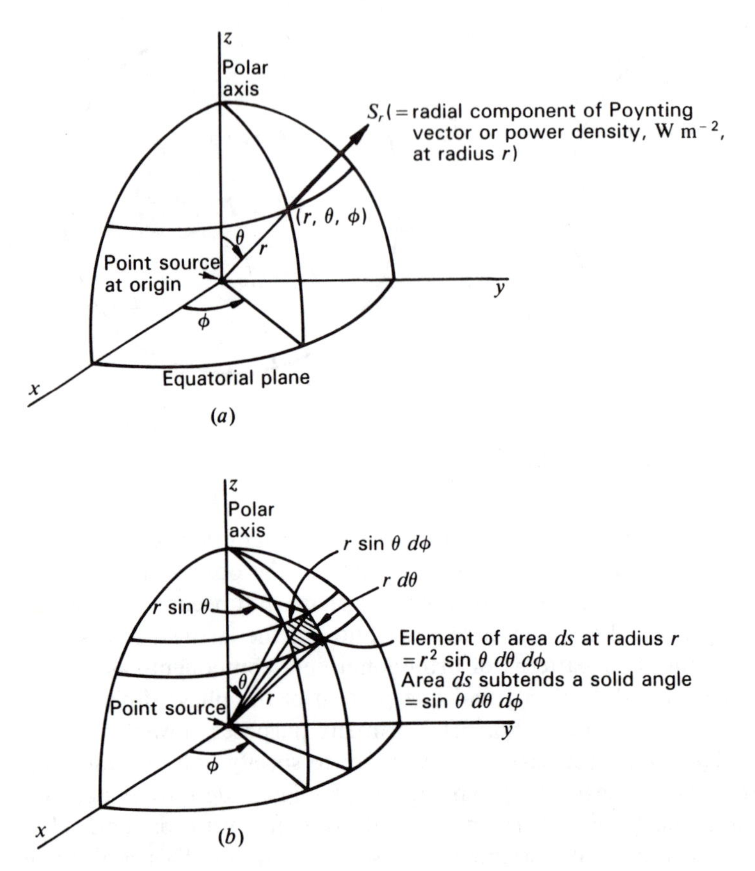

Figure 3-2 Spherical coordinates for a point source of radiation in free space.

A source that radiates energy uniformly in all directions is an *isotropic source*. For such a source the radial component S_r of the Poynting vector is independent of θ and ϕ. A graph of S_r at a constant radius as a function of angle is a Poynting vector, or power-density, pattern, but is usually called a *power pattern*. The 3-dimensional power pattern for an isotropic source is a sphere. In 2 dimensions the pattern is a circle (a cross section through the sphere), as suggested in Fig. 3-3.

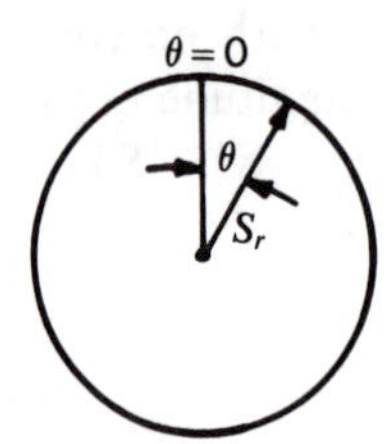

Figure 3-3 Polar power pattern of isotropic source.

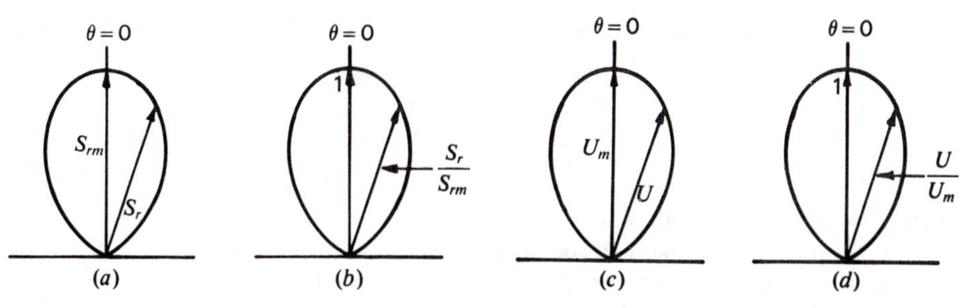

Figure 3-4 Power pattern (*a*), relative power pattern (*b*), radiation-intensity pattern (*c*) and relative radiation-intensity pattern (*d*) for the same directional or anisotropic source. All patterns have the same shape. The relative power and radiation-intensity patterns (*b* and *d*) also have the same magnitude and, hence, are identical.

Although the isotropic source is convenient in theory, it is not a physically realizable type. Even the simplest antennas have directional properties, i.e., they radiate more energy in some directions than in others. In contrast to the isotropic source, they might be called *anisotropic* sources. As an example, the power pattern of such a source is shown in Fig. 3-4*a* where S_{rm} is the maximum value of S_r.

If S_r is expressed in watts per square meter, the graph is an *absolute power pattern.* On the other hand, if S_r is expressed in terms of its value in some reference direction, the graph is a *relative power pattern.* It is customary to take the reference direction such that S_r is a maximum. Thus, the pattern radius for relative power is S_r/S_{rm} where S_{rm} is the maximum value of S_r. The maximum value of the relative power pattern is unity, as shown in Fig. 3-4*b*. A pattern with a maximum of unity is also called a *normalized pattern.*

3-3 A POWER THEOREM[1] AND ITS APPLICATION TO AN ISOTROPIC SOURCE.

If the Poynting vector is known at all points on a sphere of radius r from a point source in a lossless medium, *the total power*

[1] This theorem is a special case of a more general relation for the complex power flow through any closed surface as given by

$$P = \frac{1}{2} \oiint (\mathbf{E} \times \mathbf{H}^*) \cdot \mathbf{ds} \tag{1}$$

where P is the total complex power flow and **E** and $\mathbf{H}^*$ are complex vectors representing the electric and magnetic fields, $\mathbf{H}^*$ being the complex conjugate of **H**. The average Poynting vector is

$$\mathbf{S} = \frac{1}{2} \operatorname{Re} (\mathbf{E} \times \mathbf{H}^*) \tag{2}$$

Now the power flow in the far field is entirely real; hence, taking the real part of (1) and substituting (2), we obtain the special case of (3).

radiated by the source is the integral over the surface of the sphere of the radial component S_r of the average Poynting vector. Thus,

$$P = \oiint \mathbf{S} \cdot \mathbf{ds} = \oiint S_r \, ds \tag{3}$$

where P = power radiated, W
S_r = radial component of average Poynting vector, W m^{-2}
ds = infinitesimal element of area of sphere (see Fig. 3-2*b*)
$= r^2 \sin\theta \, d\theta \, d\phi$, m^2

For an isotropic source, S_r is independent of θ and ϕ so

$$P = S_r \oiint ds = S_r \times 4\pi r^2 \qquad \text{(W)} \tag{4}$$

and

$$S_r = \frac{P}{4\pi r^2} \qquad (\text{W m}^{-2}) \tag{5}$$

Equation (5) indicates that the magnitude of the Poynting vector varies inversely as the square of the distance from a point-source radiator. This is a statement of the well-known law for the variation of power per unit area as a function of the distance.

3-4 RADIATION INTENSITY. Multiplying the power density S_r by the square of the radius r at which it is measured, we obtain the *power per unit solid angle or radiation intensity U*. Thus,

$$r^2 S_r = U = \text{radiation intensity} \tag{1}$$

Whereas the power density S_r is expressed in watts per square meter, the radiation intensity U is expressed in watts per unit solid angle (watts per square radian or steradian). The radiation intensity is independent of the radius. Dimensionally, the radiation intensity is simply power since steradians are dimensionless. Numerically, U is equal to S_r at unit radius.

Substituting (1) into (3-3-3), the power theorem assumes the form

$$P = \iint U \sin\theta \, d\theta \, d\phi = \iint U \, d\Omega \qquad \text{(W)} \tag{2}$$

where $d\Omega = \sin\theta \, d\theta \, d\phi$ = element of solid angle, sr

Thus, the power theorem may be restated as follows:

The total power radiated is given by the integral of the radiation intensity U over a solid angle of 4π.

As already mentioned in Sec. 2-6, power patterns can be expressed in terms of either the Poynting vector (power density) or the radiation intensity. A power pattern in terms of U is shown in Fig. 3-4*c*. The maximum value U_m is in the direction $\theta = 0$. A relative U/U_m pattern has a maximum value of unity as shown in Fig. 3-4*d*. Relative power-density and radiation-intensity patterns are identical.

Applying (2) to an isotropic source gives

$$P = 4\pi U_0 \qquad \text{(W)} \tag{3}$$

where U_0 = radiation intensity of isotropic source, W sr^{-1}.

3-5 SOURCE WITH HEMISPHERIC POWER PATTERN.

As a further example let us consider a source with a power pattern which is a hemisphere; i.e., the radiation intensity equals a constant U_m in the upper hemisphere and is zero in the lower hemisphere, as illustrated by the 3-dimensional diagram of Fig. 3-5*a* and its 2-dimensional cross section of Fig. 3-5*b*. Then, the total power radiated is the radiation intensity integrated over a hemisphere, or

$$P = \iint U \, d\Omega = \int_0^{2\pi} \int_0^{\pi/2} U_m \sin\theta \, d\theta \, d\phi = 2\pi U_m \tag{1}$$

Assuming that the total power P radiated by the hemispheric source is the same as the total power radiated by an isotropic source taken as a reference, (1) and (3-4-3) can be equated, yielding

$$2\pi U_m = 4\pi U_0 \tag{2}$$

or

$$\frac{U_m}{U_0} = 2 = \text{directivity} \tag{3}$$

The ratio of U_m to U_0 in (3) equals the *directivity* of the hemispheric source. The directivity of a source is equal to the ratio of its maximum radiation intensity to its average radiation intensity. The directivity of a source may also be stated as the ratio of its maximum radiation intensity to the radiation intensity of an isotropic source radiating the same total power. By (3), the directivity of the hemispheric source is 2; that is to say, the power per unit solid angle U_m in one

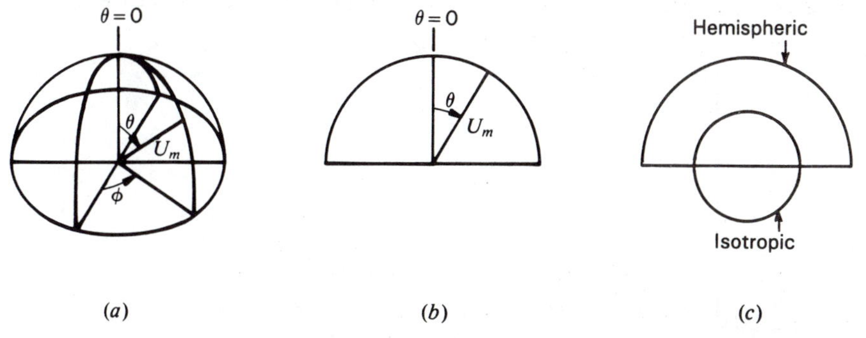

Figure 3-5 Hemispheric power patterns, (*a*) and (*b*), and comparison with isotropic pattern (*c*).

hemisphere from the hemispheric source is twice the power per unit solid angle U_0 from an isotropic source radiating the same total power. This we would expect, since a power P radiated uniformly over one hemisphere will give twice the power per unit solid angle as when radiated uniformly over both hemispheres. The power patterns of a hemispheric source and an isotropic source are compared in Fig. 3-5*c* for the same power radiated by both.

3-6 SOURCE WITH UNIDIRECTIONAL COSINE POWER PATTERN. Let us consider next a source with a cosine radiation-intensity pattern, that is,

$$U = U_m \cos\theta \tag{1}$$

where U_m = maximum radiation intensity

The radiation intensity U has a value only in the upper hemisphere ($0 \leq \theta \leq \pi/2$ and $0 \leq \phi \leq 2\pi$) and is zero in the lower hemisphere. The radiation intensity is a maximum at $\theta = 0$. The pattern is shown in Fig. 3-6. The space pattern is a figure of revolution of this circle around the polar axis.

To find the total power radiated by the cosine source, we apply (3-4-2) and integrate only over the upper hemisphere. Thus

$$P = \int_0^{2\pi}\int_0^{\pi/2} U_m \cos\theta \sin\theta \, d\theta \, d\phi = \pi U_m \tag{2}$$

If the power radiated by the unidirectional cosine source is the same as for an isotropic source, then (2) and (3-4-3) may be set equal, yielding

$$\pi U_m = 4\pi U_0$$

or

$$\text{Directivity} = \frac{U_m}{U_0} = 4 \tag{3}$$

Thus, the maximum radiation intensity U_m of the unidirectional cosine source (in the direction $\theta = 0$) is 4 times the radiation intensity U_0 from an isotropic source radiating the same total power. The power patterns for the two sources are compared in Fig. 3-7 for the same total power radiated by each.

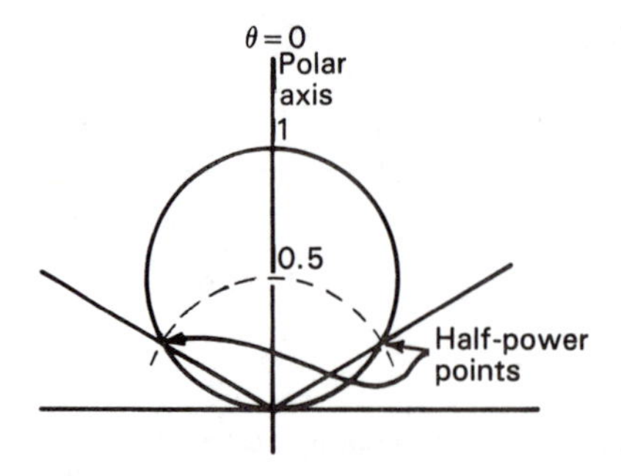

Figure 3-6 Unidirectional cosine power pattern.

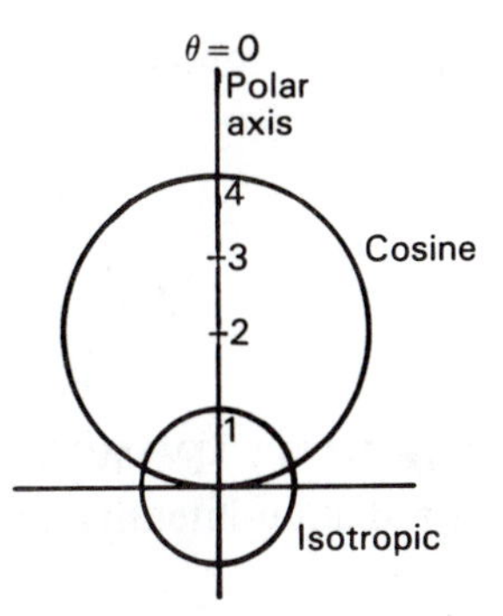

Figure 3-7 Power patterns of unidirectional cosine source compared with isotropic source for same power radiated by both.

3-7 SOURCE WITH BIDIRECTIONAL COSINE POWER PATTERN. Let us assume that the source has a cosine pattern as in the preceding example but that the radiation intensity has a value in both hemispheres, instead of only in the upper one. The pattern is then as indicated by Fig. 3-8. It follows that P is twice its value for the unidirectional cosine power pattern, and hence the directivity is 2 instead of 4.

3-8 SOURCE WITH SINE (DOUGHNUT) POWER PATTERN. Consider next a source having a radiation-intensity pattern given by

$$U = U_m \sin\theta \tag{1}$$

The pattern is shown in Fig. 3-9. The space pattern is a figure-of-revolution of this pattern around the polar axis and has the form of a doughnut. Applying (3-4-2), the total power radiated is

$$P = U_m \int_0^{2\pi} \int_0^{\pi} \sin^2\theta \, d\theta \, d\phi = \pi^2 U_m \tag{2}$$

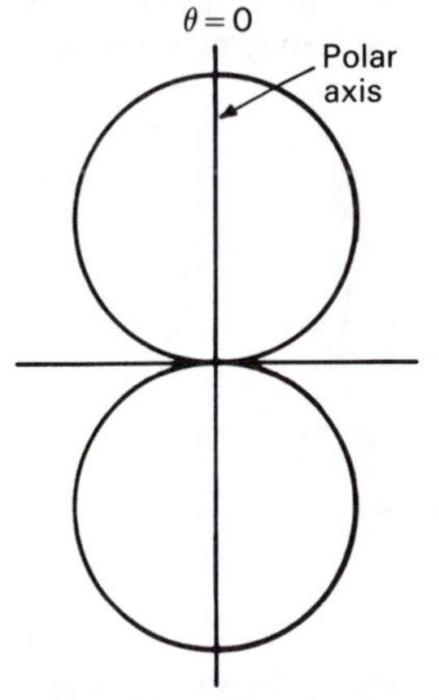

Figure 3-8 Bidirectional cosine power pattern.

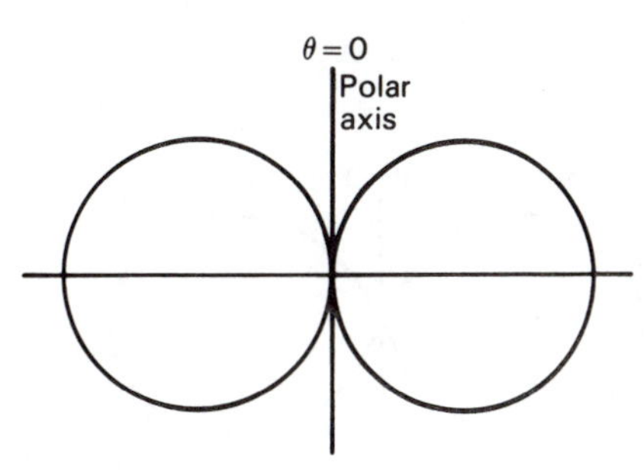

Figure 3-9 Sine power pattern.

If the power radiated by this source is the same as for an isotropic source taken as reference, we have

$$\pi^2 U_m = 4\pi U_0 \tag{3}$$

and

$$\text{Directivity} = \frac{U_m}{U_0} = \frac{4}{\pi} = 1.27 \tag{4}$$

3-9 SOURCE WITH SINE-SQUARED (DOUGHNUT) POWER PATTERN. Next consider a source with a sine-squared radiation-intensity or power pattern. The radiation-intensity pattern is given by

$$U = U_m \sin^2 \theta \tag{1}$$

The power pattern is shown in Fig. 3-10. This type of pattern is of considerable interest because it is the pattern produced by a short dipole coincident with the polar ($\theta = 0$) axis in Fig. 3-10. Applying (3-4-2), the total power radiated is

$$P = U_m \int_0^{2\pi} \int_0^{\pi} \sin^3 \theta \, d\theta \, d\phi = \tfrac{8}{3}\pi U_m \tag{2}$$

If P is the same as for the isotropic source,

$$\tfrac{8}{3}\pi U_m = 4\pi U_0$$

and

$$\text{Directivity} = \frac{U_m}{U_0} = \frac{3}{2} = 1.5 \tag{3}$$

3-10 SOURCE WITH UNIDIRECTIONAL COSINE-SQUARED POWER PATTERN. Let us consider next the case of a source with a unidirectional cosine-squared radiation-intensity pattern as given by

$$U = U_m \cos^2 \theta \tag{1}$$

with the radiation intensity having a value only in the upper hemisphere. The pattern is shown in Fig. 3-11. The 3-dimensional or space pattern is a figure-of-

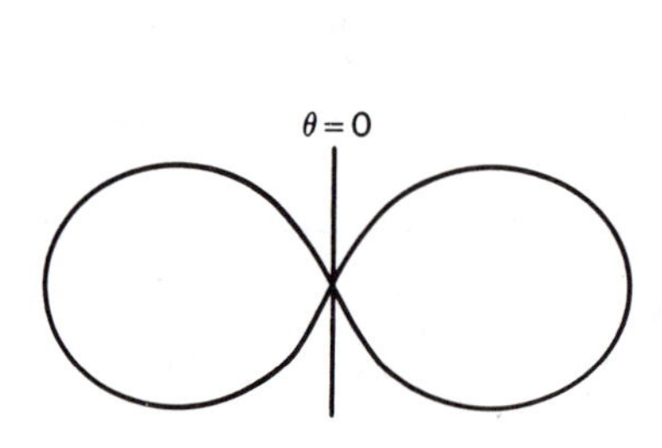

Figure 3-10 Sine-squared power pattern

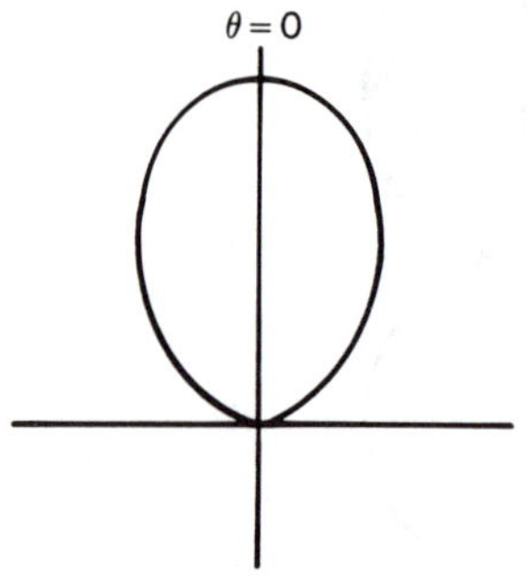

Figure 3-11 Unidirectional cosine-squared power pattern.

revolution of this pattern around the polar ($\theta = 0$) axis and has the form of a prolate spheroid (football shape). The total power radiated is

$$P = U_m \int_0^{2\pi} \int_0^{\pi/2} \cos^2 \theta \sin \theta \, d\theta \, d\phi = \tfrac{2}{3}\pi U_m \tag{2}$$

If P is the same as radiated by an isotropic source,

$$\tfrac{2}{3}\pi U_m = 4\pi U_0$$

and

$$\text{Directivity} = \frac{U_m}{U_0} = 6 \tag{3}$$

Thus, the maximum power per unit solid angle (at $\theta = 0$) from the source with the cosine-squared power pattern is six times the power per unit solid angle from an isotropic source radiating the same power.

3-11 SOURCE WITH UNIDIRECTIONAL COSINE^n POWER PATTERN. A more general case for a unidirectional radiation-intensity pattern which is symmetrical around the polar ($\theta = 0$) axis is given by

$$U = U_m \cos^n \theta \tag{1}$$

where n is any real number. In Fig. 3-12, relative radiation-intensity or power patterns plotted to the same maximum value are shown for the cases where $n = 0, \frac{1}{2}, 1, 2, 3$ and 4. The case for $n = 0$ is the same as the source with the hemispheric power pattern discussed in Sec. 3-5. The cases for $n = 1$ and $n = 2$ were treated in Secs. 3-6 and 3-10. When $n = \frac{1}{2}$, 3 and 4, the directivity is 3, 8 and 10, respectively.[1] These calculations are left to the reader as an exercise. A graph of the directivity of a unidirectional source as a function of n is presented in Fig. 3-13.

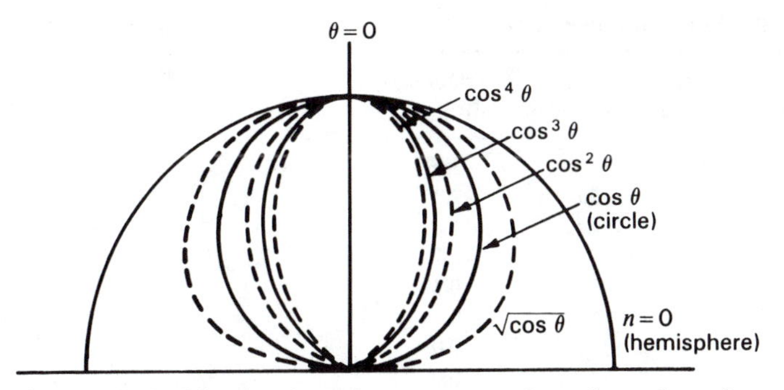

Figure 3-12 Unidirectional $\cos^n \theta$ power patterns for various values of n.

[1] It may be shown that the directivity of sources with power patterns of the type given by (1) can be reduced to the simple expression, directivity $= 2(n + 1)$. The proof is left to the reader as an exercise.

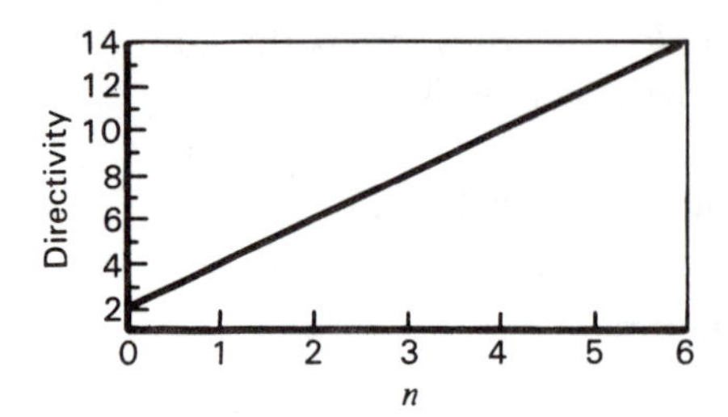

Figure 3-13 Directivity versus n for unidirectional sources with $\cos^n \theta$ power patterns.

The half-power beam widths and exact directivity $[D = 2(n + 1)]$ for $\cos^n \theta$ patterns are listed in Table 3-1 for n values between 0 and 100. The directivity is also given in dBi (dB over isotropic) for both exact and approximate values, where the approximate value is based on the HPBW^2 from (2-9-4). The difference between the exact and approximate values is tabulated in the last column. The approximation is within ± 0.2 dB of the exact value for n between 4 and 10. A further discussion of the (2-9-4) approximation is given following (17) of Sec. 3-13.

3-12 SOURCE WITH UNIDIRECTIONAL POWER PATTERN THAT IS NOT SYMMETRICAL. All the patterns considered thus far have been symmetrical around the polar axis; i.e., the space pattern could be constructed as a figure-of-revolution about the polar axis. Let us now consider a more general case in which the pattern is unidirectional but is unsymmetrical around its major axis. In discussing this type of pattern it will be convenient to shift the direction of the major axis or direction of maximum radiation from the polar ($\theta = 0$) axis to a direction in the equatorial plane as shown in Fig. 3-14 ($\theta = 90°$, $\phi = 90°$). The $\theta = 90°$ plane coincides with the xy plane and the

Table 3-1 Half-power beam widths and directivity of sources with unidirectional $\cos^n \theta$ power patterns.

n	HPBW, deg	D	Exact D, dBi	Approx. D, dBi	Difference, dB
0.0	180.0	2	3.0	1.1	−1.9
0.5	151.0	3	4.8	2.6	−2.2
1	120.0	4	6.0	4.6	−1.4
2	90.0	6	7.8	7.1	−0.7
3	74.9	8	9.0	8.7	−0.3
4	65.5	10	10.0	9.8	−0.2
5	59.0	12	10.8	10.8	0.0
6	54.0	14	11.5	11.5	0.0
8	47.0	18	12.6	12.7	0.1
10	42.2	22	13.4	13.6	0.2
20	30.0	42	16.2	16.6	0.4
50	19.0	102	20.1	20.6	0.5
100	13.5	202	23.1	23.5	0.4

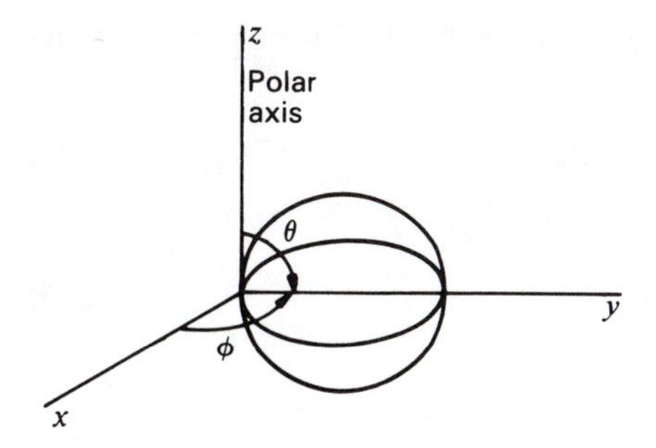

Figure 3-14 Unidirectional source radiating maximum power in the direction $\theta = 90°$, $\phi = 90°$ or along the y axis.

$\phi = 90°$ plane with the yz plane. A rather general expression for the radiation intensity with its maximum at $\theta = 90°$ and $\phi = 90°$ is then given by

$$U = U_m \sin^n \theta \sin^m \phi \tag{1}$$

where n = any real number
m = any real number

and the radiation intensity U has a value only in the right-hand hemisphere (Fig. 3-14) ($0 \le \theta \le \pi$; $0 \le \phi \le \pi$). When $m = n$, (1) becomes the equation for a symmetrical power pattern of the same form as considered in Sec. 3-11. When m and n are not the same, (1) represents the general case in which the pattern has different shapes in the $\theta = 90°$ and $\phi = 90°$ planes. The total power radiated in this general case is

$$P = U_m \int_0^{\pi} \int_0^{\pi} \sin^{n+1} \theta \sin^m \phi \, d\theta \, d\phi \tag{2}$$

3-13 DIRECTIVITY.

The concept of directivity, treated above in some special cases and also introduced in Sec. 2-8, will now be reviewed and developed in more detail.

In Sec. 3-5 directivity was given as the ratio of U_m to U_0 where U_m is the maximum radiation intensity or watts per square radian from the source under consideration and U_0 is the radiation intensity from an isotropic source radiating the same power (or U_0 is the *average* radiation intensity from the source under consideration). Thus,

$$D = \frac{U_m}{U_0} = \frac{\text{maximum radiation intensity}}{\text{average radiation intensity}} \tag{1}$$

where D = directivity

Multiplying numerator and denominator of (1) by 4π gives

$$D = \frac{4\pi U_m}{4\pi U_0} = \frac{4\pi U_m}{P} = \frac{4\pi\ (\text{maximum radiation intensity})}{\text{total power radiated}} \tag{2}$$

Let us now develop a more general expression for the directivity. Let the radiation-intensity pattern be expressed as

$$U = U_a f(\theta, \phi) \tag{3}$$

and its maximum value by

$$U_m = U_a f(\theta, \phi)_{\max} \tag{4}$$

where U_a = a constant

For the special case where

$$f(\theta, \phi)_{\max} = 1 \tag{5}$$

then $U_m = U_a$ and (3) can be written

$$U = U_m f(\theta, \phi) \tag{6}$$

The average radiation intensity is

$$U_0 = \frac{P}{4\pi} = \frac{\iint U_a f(\theta, \phi)\, d\Omega}{4\pi} \tag{7}$$

where P = total power radiated
$d\Omega = \sin\theta\, d\theta\, d\phi$ = element of solid angle

The directivity D is then given by

$$D = \frac{U_m}{U_0} = \frac{U_a f(\theta, \phi)_{\max}}{\dfrac{\iint U_a f(\theta, \phi)\, d\Omega}{4\pi}} = \frac{4\pi f(\theta, \phi)_{\max}}{\iint f(\theta, \phi)\, d\Omega} \tag{8}$$

Equation (8) can be reexpressed as

$$D = \frac{4\pi}{\dfrac{\iint f(\theta, \phi)\, d\Omega}{f(\theta, \phi)_{\max}}} = \frac{4\pi}{\Omega_A} \tag{9}$$

where Ω_A is defined as the *beam area*, or beam solid angle. It is given by

$$\Omega_A = \frac{\iint f(\theta, \phi)\, d\Omega}{f(\theta, \phi)_{\max}} \tag{10}$$

From (1) and (9),

$$D = \frac{U_m}{U_0} = \frac{4\pi}{\Omega_A} \tag{11}$$

and

$$4\pi U_0 = U_m \Omega_A \tag{12}$$

Since $U_0 = P/4\pi$,

$$P = U_m \Omega_A \tag{13}$$

where P = total power radiated

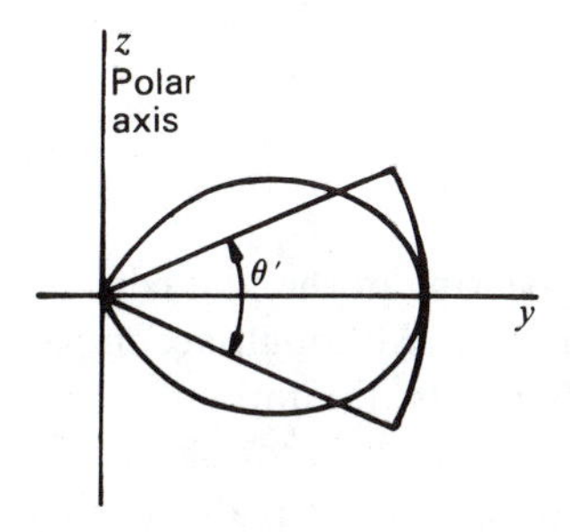

Figure 3-15 Unidirectional power pattern in cross section with included angle θ' of the beam area. The space patterns are figures-of-revolution around the y axis.

Therefore, *the beam area Ω_A is the solid angle through which all the power radiated would stream if the power per unit solid angle equaled the maximum value U_m over the beam area.*

$$\Omega_A = \frac{4\pi U_0}{U_m} \text{ square radians} = 41\,253 \frac{U_0}{U_m} \text{ square degrees} \tag{14}$$

Consider the unidirectional power pattern shown in Fig. 3-15. The pattern is a figure-of-revolution around the y axis. The included angle θ' of the corresponding beam area is also shown. If the power per unit solid angle over the beam area equals the maximum value U_m of the directional source, the power through the beam area equals that radiated by the source.

From this it is only a step to a very simple approximate method of calculating the directivity for a single-lobed pattern, based on an estimate of the beam area from the half-power beam widths of the patterns in two planes at right angles. Thus, let θ_{HP} equal the half-power beam width in the θ plane and ϕ_{HP} the half-power beam width in the ϕ plane. Then, neglecting the effect of minor lobes, we have *approximately*

$$\Omega_A \simeq \theta_{HP}\,\phi_{HP} \tag{15}$$

and, as already presented in (2-9-4),

$$D = \frac{4\pi}{\Omega_A} \simeq \frac{4\pi}{\theta_{HP}\,\phi_{HP}} \simeq \frac{41\,000}{\theta^\circ_{HP}\,\phi^\circ_{HP}} \qquad (16)^1$$

where θ_{HP} = half-power beam width in θ plane, rad
ϕ_{HP} = half-power beam width in ϕ plane, rad
θ°_{HP} = half-power beam width in θ plane, deg
ϕ°_{HP} = half-power beam width in ϕ plane, deg

[1] 4π steradians $= 4\pi(180/\pi)^2$ square degrees $= 41\,253$ square degrees, i.e., there are 41 253 square degrees in a sphere. For the approximate relation (16) the value is rounded off to 41 000.

For a pencil beam where $\theta_{HP} = \phi_{HP}$,

$$D \simeq \frac{4\pi}{\theta_{HP}^2} \simeq \frac{41\,000}{(\theta_{HP}^\circ)^2} \tag{17}$$

Equation (16) has been used very widely since appearing in the first edition, in most cases appropriately but sometimes inappropriately. Although accompanied by a three-paragraph footnote cautioning about its limitations, some authors have ignored these guidelines and argued that the numerator should be 26 358, 32 750 or 34 250 instead of 41 253 square degrees, without realizing that the four cases involve four different classes of antennas.

My original rationale was that formulas like (16) and (17) are useful in the context that if we know that an antenna has a 1° pencil beam and small minor lobes, its gain is approximately 41 000 or 46 dB, give or take a decibel or two, but not that the result is accurate to $\frac{1}{10}$ of a decibel or better.

Some of the limitations of (16) are that: (1) the effect of minor lobes is neglected, (2) the angle product ($\theta_{HP}\,\phi_{HP}$) may not be rigorously related to the true solid angle of the main beam and (3) the angle product relation to the true solid angle varies according to the type of antenna pattern involved.

It is pointed out in the first edition footnote that by introducing a correction factor the result can be improved. It is further pointed out that the value of this factor depends on the antenna type and may be relatively constant for a certain class of antennas.

Instead of introducing one correction factor, let us introduce two,[1] one involving the beam efficiency to correct for the minor lobes and the other to correct for pattern shape so that

$$D \simeq \frac{41\,000\varepsilon_M}{k_p\,\theta_{HP}^\circ\,\phi_{HP}^\circ} \tag{18}$$

where $\varepsilon_M = \dfrac{\Omega_M}{\Omega_A}$ = beam efficiency = 0.75 ± 0.15 for most large antennas

k_p = pattern factor = 1.0 for a uniform field distribution across the antenna aperture

For "ball-park" values (16) may suffice but (18) should be used for closer approximations. The effect of pattern shape is well illustrated by the comparison summarized in Table 3-1, Sec. 3-11.

For a Gaussian distribution (and beam)[2]

$$D \simeq \frac{41\,000 \times 0.88}{\theta_{HP}^\circ\,\phi_{HP}^\circ} \tag{19}$$

---

[1] J. D. Kraus, *Radio Astronomy*, 1st ed., McGraw-Hill, 1966, p. 158; 2nd ed., Cygnus-Quasar, 1986, p. 6-6.

[2] See Prob. 11-20.

3-14 SOURCE WITH PATTERN OF ARBITRARY SHAPE.

Example. Consider a pattern of arbitrary shape as shown in Fig. 3-16 in both polar and rectangular coordinates. The pattern has a pencil beam (symmetrical around the $\theta = 0°$ axis) with a main-lobe HPBW of approximately 22° and four minor lobes. Find the directivity.

Solution. The directivity is given by

$$D = \frac{4\pi}{\int_0^{2\pi} \int_0^{\pi} P_n(\theta) \sin\theta \, d\theta \, d\phi} \tag{1}$$

where the denominator equals the total beam area Ω_A.

Since the pattern is symmetrical (no variation with ϕ), the integral with respect to ϕ yields 2π and (1) reduces to

$$D = \frac{4\pi}{2\pi \int_0^{\pi} P_n(\theta) \sin\theta \, d\theta} \tag{2}$$

We have only the pattern graph available (no analytical expression) so let us divide the pattern (Fig. 3-16b) into 36 steps of 5° each. The approximate value of the integral in the first ($m = 1$) 5° section ($= \pi/36$ rad) is given by

$$\frac{\pi}{36} P_n(\theta_1)_{\text{av}} \sin\theta_1 = \frac{\pi}{36} \frac{1.0 + 0.93}{2} \sin 2.5° \tag{3}$$

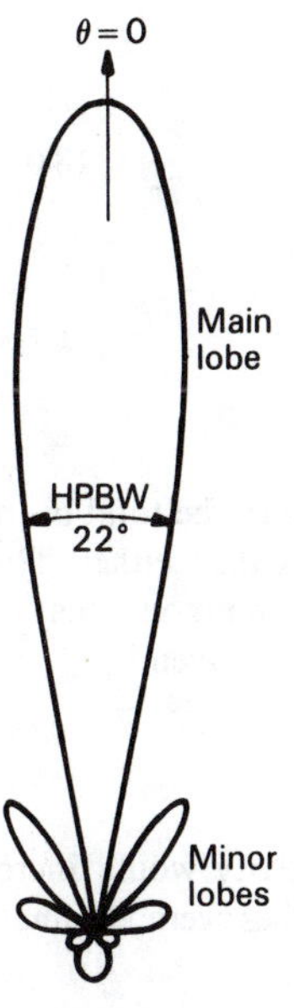

Figure 3-16a Power pattern with main lobe and several minor lobes for worked example calculation of directivity. The pattern is symmetrical around the $\theta = 0$ axis (vertical) with the 3-dimensional pattern a figure-of-revolution around this axis. The same pattern $P_n(\theta)$ is shown in rectangular coordinates in Fig. 3-16b.

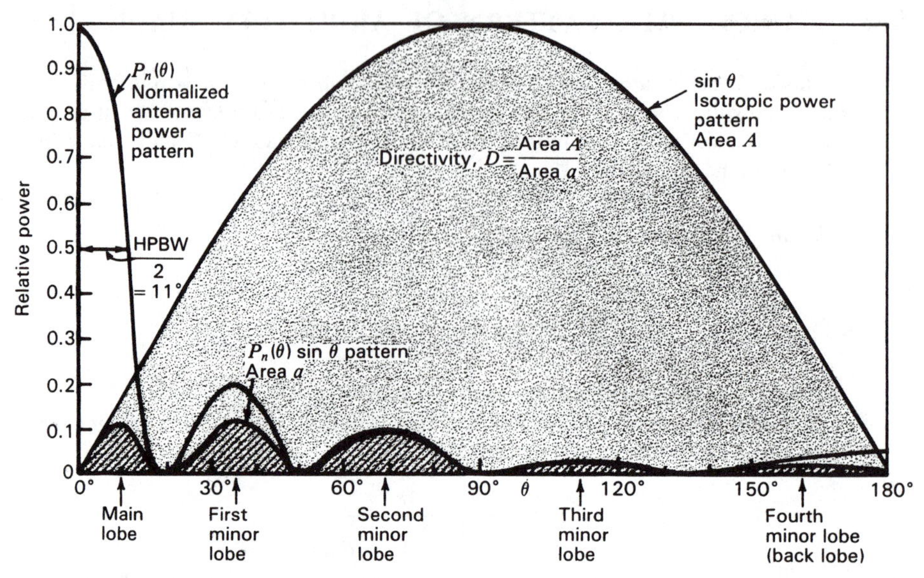

Figure 3-16b Power pattern with main lobe and several minor lobes for worked example calculation of directivity. D equals the ratio of the dot-filled to cross-hatched areas. The same pattern in polar coordinates is shown in Fig. 3-16a.

and the approximate directivity is then given by the summation of all 36 sections or by

$$D \simeq \frac{4\pi}{2\pi(\pi/36) \sum_{m=1}^{m=36} P_n(\theta_m)_{\text{av}} \sin\theta_m} \tag{4}$$

Completing the summation we obtain

$$D = \frac{4\pi}{\Omega_A} \simeq \frac{4\pi}{2\pi(\pi/36)(0.25 + 0.37 + 0.46 + 0.12 + 0.07)} = \frac{72}{1.27\pi} = 18.0 \tag{5}$$

Main lobe	First minor lobe	Second minor lobe	Third minor lobe	Fourth minor lobe (back lobe)

or $D \simeq 12.6$ dBi

It is noteworthy that the second minor lobe contributes most to the total beam area, the first minor lobe almost as much and the main lobe less than either. Thus, the directivity is greatly affected by the minor lobes, which is a common situation with actual antennas. For this antenna pattern the beam efficiency is given by

$$\varepsilon_M = \frac{0.25}{1.27} = 0.20 \tag{6}$$

If the second minor lobe were eliminated, the directivity would increase to 14.5 dBi (up 1.9 dB) and if both first and second minor lobes were eliminated the directivity would increase to 17.1 dBi (up 4.5 dB).

The directivity obtained in the above worked example is approximate. By sufficiently reducing the step size (5° in the example), the summation can be made as precise as the available data will allow. Computation of this numerical integration can be facilitated by using a computer.

The half-power beam width of the pattern in the example is about 22°. Taking $k_p = 1$ and ε_M as in (6), the approximate directivity is then

$$D \simeq \frac{41\,000\,\varepsilon_M}{k_p \times \text{HPBW}^2} = \frac{41\,000 \times 0.2}{(22°)^2} = 16.9 \text{ or } 12.3 \text{ dBi} \tag{7}$$

which is 0.3 dB less than obtained by the 36-step summation.

The beam area of an isotropic source equals 4π steradians. In Fig. 3-16b this corresponds to the area A under the $\sin\theta$ curve. The beam area of the source in the worked example corresponds to the area a under the $P_n(\theta)\sin\theta$ curve. Thus, the directivity is simply A/a or the ratio of the area of the isotropic source to the area of the source being measured. Hence,

$$D = \frac{4\pi}{\Omega_A} = \frac{A}{a} \tag{8}$$

If the areas A and a are cut from a lead sheet of uniform thickness, the directivity equals the ratio of the weight of A to the weight of a.

3-15 GAIN. The definition of directivity in the preceding section is based entirely on the shape of the radiated power pattern. Antenna efficiency is not involved. The *gain* parameter does involve antenna efficiency. The gain[1] of an antenna is defined as

$$G = \frac{\text{maximum radiation intensity}}{\begin{array}{c}\text{maximum radiation intensity from a}\\ \text{reference antenna with same power input}\end{array}} \tag{1}$$

Any type of antenna may be taken as the reference. Often the reference is a linear $\lambda/2$ antenna. Gain includes the effect of losses both in the antenna under consideration (subject antenna) and in the reference antenna.

In many situations it is convenient to assume that the reference antenna is an isotropic source of 100 percent efficiency. The gain so defined for the subject antenna is called the *gain with respect to an isotropic source* or

$$G = \frac{\text{maximum radiation intensity from subject antenna}}{\begin{array}{c}\text{radiation intensity from (lossless) isotropic}\\ \text{source with same power input}\end{array}} \tag{2}$$

---

[1] The gain G as here defined is sometimes called *power gain.* This quantity is equal to the square of the *gain in field intensity* G_f. Thus, if E_1 is the maximum electric field intensity from the antenna at a large distance R and E_0 is the maximum electric field intensity from the reference antenna with the same power input at the same distance R, then the power gain G is given by $G = (E_1/E_0)^2 = G_f^2$.

As given in (2-10-1), the gain with respect to the directivity is given by

$$G = kD \tag{3}$$

where k = efficiency factor of antenna ($0 \leq k \leq 1$)
D = directivity

Thus, the gain of an antenna over a lossless isotropic source equals the directivity if the antenna is 100 percent efficient ($k = 1$) but is less than the directivity if any losses are present in the antenna ($k < 1$). In decibels the gain over an isotropic source as in (2) is expressed as dBi. Directivity is always dBi.

3-16 FIELD PATTERNS. The discussion in the preceding sections is based on considerations of power. This has afforded a simplicity of analysis, since the power flow from a point source has only a radial component which can be considered as a scalar quantity. To describe the field of a point source more completely, we need to consider the electric field **E** and/or the magnetic field **H** (both vectors). For point sources we deal entirely with *far fields* so **E** and **H** are both entirely transverse to the wave direction, are perpendicular to each other, are in-phase and are related in magnitude by the intrinsic impedance of the medium ($E/H = Z = 377\ \Omega$ for free space). For our purposes it suffices to consider only one field vector and we arbitrarily choose the electric field **E**.

Since the Poynting vector around a point source is everywhere radial, it follows that the electric field is entirely transverse, having only E_θ and E_ϕ components. The relation of the radial component S_r of the Poynting vector and the electric field components is illustrated by the spherical coordinate diagram of Fig. 3-17a. The conditions characterizing the far field are then:

1. Poynting vector radial (S_r component only)
2. Electric field transverse (E_θ and E_ϕ components only)

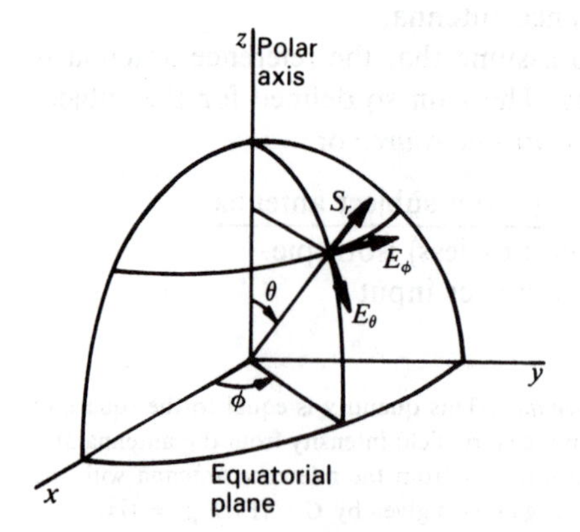

Figure 3-17a Relation of the Poynting vector and the electric field components of the far field.

The Poynting vector and the electric field at a point of the far field are related in the same manner as they are in a plane wave, since, if r is sufficiently large, a small section of the spherical wave front may be considered as a plane.

The relation between the average Poynting vector and the electric field at a point of the far field is

$$S_r = \frac{1}{2}\frac{E^2}{Z} \tag{1}$$

where Z_0 = intrinsic impedance of medium and

$$E = \sqrt{E_\theta^2 + E_\phi^2} \tag{2}$$

where E = amplitude of total electric field intensity
E_θ = amplitude of θ component
E_ϕ = amplitude of ϕ component

The field may be elliptically, linearly or circularly polarized.

If the field components are rms values, rather than amplitudes, the Poynting vector is twice that given in (1).

A pattern showing the variation of the electric field intensity at a constant radius r as a function of angle (θ, ϕ) is called a *field pattern*. In presenting information concerning the far field of an antenna, it is customary to give the field patterns for the two components, E_θ and E_ϕ, of the electric field since the total electric field E can be obtained from the components by (2), but the components cannot be obtained from a knowledge of only E.

When the field intensity is expressed in volts per meter, it is an *absolute field pattern*.[1] On the other hand, if the field intensity is expressed in units relative to its value in some reference direction, it is a *relative field pattern*. The reference direction is usually taken in the direction of maximum field intensity. The relative pattern of the E_θ component is then given by

$$\frac{E_\theta}{E_{\theta m}} \tag{3}$$

and the relative pattern of the E_ϕ component is given by

$$\frac{E_\phi}{E_{\phi m}} \tag{4}$$

where $E_{\theta m}$ = maximum value of E_θ
$E_{\phi m}$ = maximum value of E_ϕ

[1] The magnitude depends on the radius, varying inversely as the distance ($E \propto 1/r$).

The magnitudes of both the electric field components, E_θ and E_ϕ, of the far field vary inversely as the distance from the source. However, they may be different functions, F_1 and F_2, of the angular coordinates, θ and ϕ. Thus, in general,

$$E_\theta = \frac{1}{r} F_1(\theta, \phi) \tag{5}$$

$$E_\phi = \frac{1}{r} F_2(\theta, \phi) \tag{6}$$

Since $S_{rm} = E_m^2/2Z$, where E_m is the maximum value of E, it follows on dividing this into (1) that the relative total power pattern is equal to the square of the relative total field pattern. Thus,

$$P_n = \frac{S_r}{S_{rm}} = \frac{U}{U_m} = \left(\frac{E}{E_m}\right)^2 \tag{7}$$

Example 1. Consider first the case of an antenna whose far field has only an E_ϕ component in the equatorial plane, the E_θ component being zero in this plane. Suppose that the relative equatorial-plane pattern of the E_ϕ component (that is, E_ϕ as a function of ϕ for $\theta = 90°$) is given by

$$\frac{E_\phi}{E_{\phi m}} = \cos\phi \tag{8}$$

This pattern is illustrated at the left of Fig. 3-17b. The length of the radius vector in the diagram is proportional to E_ϕ. A pattern of this form could be produced by a short dipole coincident with the y axis.

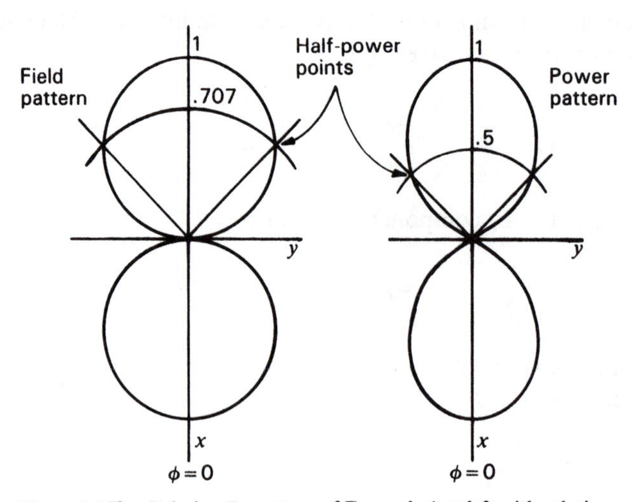

Figure 3-17b Relative E_ϕ pattern of Example 1 at left with relative power pattern at right.

The relative (normalized) power pattern in the equatorial plane is equal to the square of the relative field pattern. Thus

$$P_n = \frac{S_r}{S_{rm}} = \frac{U}{U_m} = \left(\frac{E_\phi}{E_{\phi m}}\right)^2 \tag{9}$$

and substituting (8) into (9) we have

$$P_n = \cos^2 \phi$$

This pattern is illustrated at the right of Fig. 3-17b.

Example 2. Consider next the case of an antenna with a far field that has only an E_θ component in the equatorial plane, the E_ϕ component being zero in this plane. Assume that the relative equatorial-plane pattern of the E_θ component (that is, E_θ as a function of ϕ for $\theta = 90°$) for this antenna is given by

$$\frac{E_\theta}{E_{\theta m}} = \sin \phi \tag{10}$$

This pattern is illustrated by Fig. 3-18*a* and could be produced by a small loop antenna, the axis of the loop coincident with the x axis.

The relative (normalized) power pattern in the equatorial plane is

$$P_n = \sin^2 \phi \tag{11}$$

This pattern is shown by Fig. 3-18*b*.

Example 3. Let us consider finally an antenna whose far field has both E_θ and E_ϕ components in the equatorial plane ($\theta = 90°$). Suppose that this antenna is a composite of the two antennas we have just considered in Examples 1 and 2 and that equal power is radiated by each antenna. If both patterns are of identical shape in 3 dimensions as well as in the xy plane, as from a short dipole and a small loop, it then follows that at a radius r from the composite antenna, $E_{\theta m} = E_{\phi m}$. The individual patterns for the E_θ and E_ϕ components as given by (10) and (8) may then be shown to the same scale by one diagram, as in Fig. 3-19*a*. The relative pattern of the total field E is

$$\frac{E}{E_m} = \sqrt{\sin^2 \phi + \cos^2 \phi} = 1 \tag{12}$$

which is a circle as indicated by the dashed line in Fig. 3-19*a*.

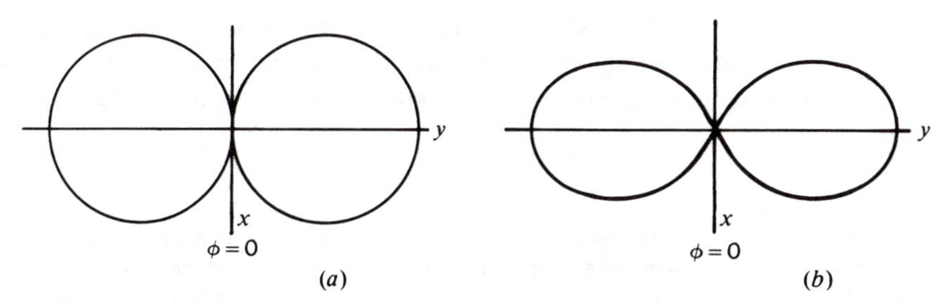

Figure 3-18 Relative E_θ pattern of Example 2 at (*a*) with relative power pattern at (*b*).

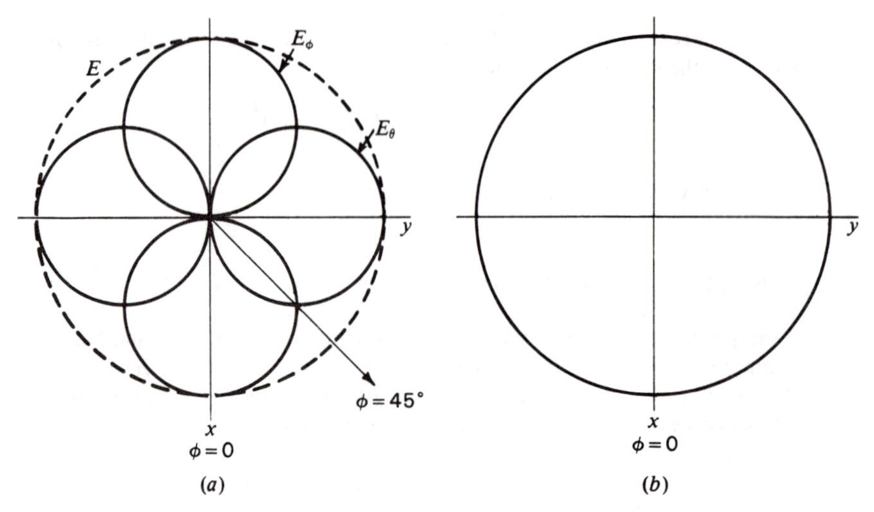

Figure 3-19 (*a*) Relative patterns of E_θ and E_ϕ components of the electric field and the total field E for antenna of Example 3. (*b*) Relative total power pattern.

For this antenna, we may speak of two types of power patterns. One type shows the power variation for one component of the electric field. Thus, the power in the E_θ component of the field is as shown by Fig. 3-18*b* and the power in the E_ϕ component by Fig. 3-17*b*. The second type of power pattern shows the variation of the total power. This is proportional to the square of the total electric field intensity. Accordingly, the relative total power pattern for the composite antenna is

$$P_n = \left(\frac{E}{E_m}\right)^2 = 1$$

The relative pattern in the equatorial plane for the total power is, therefore, a circle of radius unity as illustrated by Fig. 3-19*b*.

We note in Fig. 3-19*a* that at $\phi = 45°$ the magnitudes of the two field components, E_θ and E_ϕ, are equal. Depending on the time phase between E_θ and E_ϕ, the field in this direction could be plane, elliptically or circularly polarized, but regardless of phase the power is the same. To determine the type of polarization requires that the phase angle between E_θ and E_ϕ be known. This is discussed in the next section.

3-17 PHASE PATTERNS.

Assuming that the field varies harmonically with time and that the frequency is known, the far field in all directions from a source may be completely specified by a knowledge of the following four quantities:[1]

[1] In general, for the near or far field, six quantities are required. These are E_θ, E_ϕ, δ, and η, each as a function of r, θ, ϕ and in addition the amplitude of the radial component of the electric field E_r and its phase lag behind E_θ, both as a function of r, θ, ϕ. Since $E_r = 0$ in the far field, only the four quantities given are needed to describe completely the field in the Fraunhofer region.

1. Amplitude of the polar component E_θ of the electric field as a function of r, θ and ϕ
2. Amplitude of the azimuthal component E_ϕ of the electric field as a function of r, θ and ϕ
3. Phase lag δ of E_ϕ behind E_θ as a function of θ and ϕ
4. Phase lag η of either field component behind its value at a reference point as a function of r, θ and ϕ

Since we regard the field of a point source as a far field everywhere, the above four quantities can be considered as those required for a complete knowledge of the field of a point source.

If the amplitudes of the field components are known at a particular radius from a point source in free space, their amplitudes at all distances are known from the inverse-distance law. Thus, it is usually sufficient to specify E_θ and E_ϕ as a function only of θ and ϕ as, for example, by a set of field patterns.

As shown in the preceding sections, the amplitudes of the field components give us directly or indirectly a knowledge of the peak and effective values of the total field and Poynting vector. However, if both field components have a value, the polarization is indeterminate without a knowledge of the phase angle δ between the field components. Focusing our attention on one field component, the phase angle η with respect to the phase at some reference point is a function of the radius and may also be a function of θ and ϕ. A knowledge of η as a function of θ and ϕ is essential when the fields of two or more point sources are to be added.

We now proceed to a discussion of the phase angles, δ and η, and of phase patterns for showing their variation. Let us consider three examples.

Example 1. Consider first a point source that radiates uniformly in the equatorial plane and has only an E_ϕ component of the electric field. Then at a distance r from the source, the instantaneous field $E_{\phi i}$ in the equatorial plane is

$$E_{\phi i} = \frac{\sqrt{2}\,E_\phi}{r}\sin(\omega t - \beta r) \tag{1}$$

where E_ϕ = rms value of ϕ component of electric field intensity at unit radius from the source
ω = $2\pi f$, where f = frequency, Hz
β = $2\pi/\lambda$, where λ = wavelength, m

The relation given by (1) is the equation for the field of a spherical wave traveling radially outward from the source. The equation gives the *instantaneous* value of the field as a function of time and distance. The amplitude or peak value of the field is $\sqrt{2}\,E_\phi/r$. The amplitude is independent of space angle (θ and ϕ) but varies inversely with the distance r. The variation of the instantaneous field with distance for this example is illustrated by the upper graph in Fig. 3-20 in which the amplitude is taken as unity at a distance r. When $r = 0$, the variation of the instantaneous field varies as $\sin \omega t$. It is often convenient to take this variation as a refer-

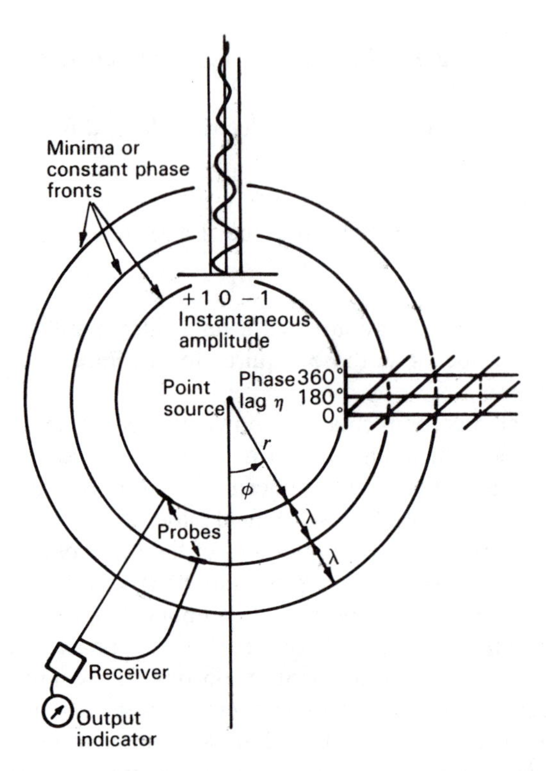

Figure 3-20 Illustration for Example 1. Phase of E_ϕ of point source radiating uniformly in ϕ plane is a function of r but is independent of ϕ. Phase lag η increases linearly with distance r.

ence for the phase, designating it as the phase of the generator or source. The fact that the amplitude at $r = 0$ is infinite need not detract from using the phase at $r = 0$ as a reference. The phase at a distance r is then retarded behind that at the source by the angle βr. A *phase retardation* or lag of E_ϕ with respect to a reference point will, in general, be designated as η. In the present case the reference point is the source;[1] hence,

$$\eta = \beta r = \frac{2\pi r}{\lambda} \quad \text{(rad)} \tag{2}$$

Thus, the phase lag η increases linearly with the distance r from the source. This is illustrated by the chart of phase lag versus distance in Fig. 3-20.

The phase lag η in this example is assumed to be independent of ϕ. To demonstrate experimentally that η depends on r but is independent of ϕ, the arrangement shown at the lower left in Fig. 3-20 could be used. The outputs of two probes or small antennas are combined in a receiver. With both probes at or very near the same point, the receiver output is reduced to a minimum by adjusting the length of one of the probe cables. The voltages from the probes at the receiver are then in phase opposition. With one probe fixed in position, the other is then moved

[1] If the phase is referred to some point at a distance r_1 from the source, then (1) becomes $E_{\phi i} = (\sqrt{2}\, E_\phi/r) \sin(\omega t - \beta d)$, where $d = r - r_1$.

in such a way as to maintain a minimum output. The locus of points for minimum output constitutes a contour (or front) of constant phase. For the point source under consideration, each contour is a circle of constant radius with a separation of λ between contours. The radius of the contours is then given by $r_1 \pm n\lambda$, where r_1 is the radius to the reference probe and n is any integer.

We may define a *phase front* as a (3-dimensional) surface of constant or uniform phase. If our observation circle coincides with a phase front, then we have constant phase along it.

Example 2. Consider next the case of a point source that has only an E_ϕ component and that radiates *nonuniformly* in the equatorial or ϕ plane. Let the instantaneous value in the equatorial plane be given by

$$E_{\phi i} = \frac{\sqrt{2}\,E_{\phi m}}{r} \cos \phi \sin (\omega t - \beta r) \tag{3}$$

where $E_{\phi m}$ = rms value of E_ϕ component at unit radius in the direction of maximum field intensity

Let a point at unit radius and in the direction $\phi = 0$ be taken as the reference for phase. Then at this radius,

$$E_{\phi i} = \sqrt{2}\,E_{\phi m} \cos \phi \sin \omega t \tag{4}$$

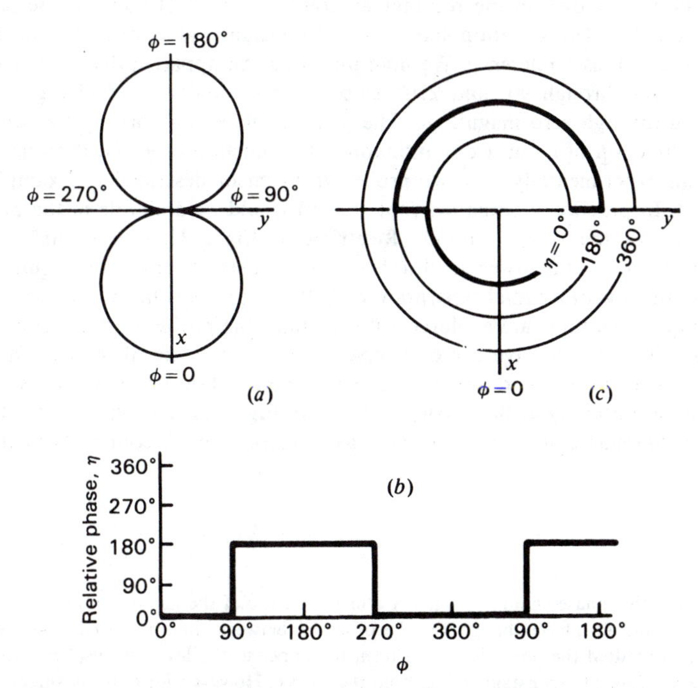

Figure 3-21 Illustration for Example 2. Field pattern is shown at (*a*), the phase pattern in rectangular coordinates at (*b*) and in polar coordinates at (*c*).

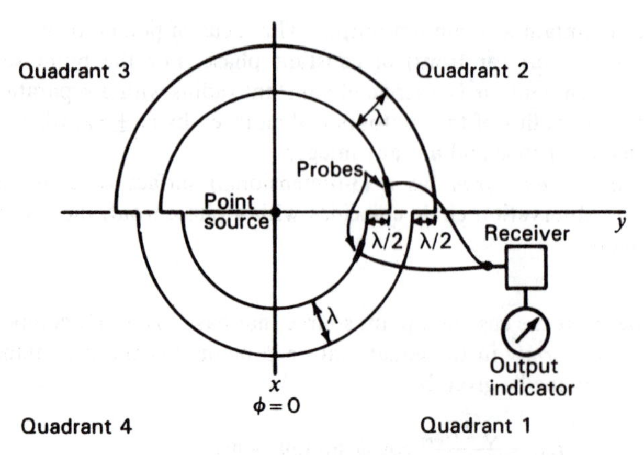

Figure 3-22 Constant-phase contours for source of Example 2.

Setting $\sin \omega t = 1$, the relative field pattern of the E_ϕ component as a function of ϕ is, therefore,

$$E_\phi = \cos \phi \tag{5}$$

as illustrated in Fig. 3-21*a*. A pattern of this type could be obtained by a short dipole coincident with the y axis at the origin. The phase lag η as a function of ϕ is a step function, as shown in the rectangular graph of Fig. 3-21*b* and in the polar graph of Fig. 3-21*c*. The variation shown is at a constant radius with the phase in the direction $\phi = 0$ as a reference. We note that η has an apparent discontinuity of 180° as ϕ passes through 90° and 270°, since at these angles $\cos \phi$ changes sign while passing through zero magnitude. The phase angle η is accordingly a continuous, linear function of r but a discontinuous, step function of ϕ. To demonstrate this variation experimentally, the two-probe arrangement described in Example 1 may be used. In practice, attenuators (not shown) would be desirable in the probe leads to equalize the probe outputs. Referring to Fig. 3-22, if both fixed and movable probes are in the lower quadrants (1 and 4), a set of constant or equiphase circles is obtained with a radial separation of λ. If one probe is fixed in quadrant 1 while the upper quadrants are explored with the movable probe, a set of equiphase circles is obtained which have a radial separation of λ but are displaced radially from the set in the lower quadrants by $\lambda/2$. Thus, the constant-phase contours have an apparent discontinuity at the y axis, as shown in Fig. 3-22. The phase of the field of any linear antenna coincident with the y axis exhibits this discontinuity at the y axis.[1]

[1] It is to be noted that this phase change is actually a characteristic of the method of measurement, since by a second method no phase change may be observed between the upper and lower hemispheres. In the second method the probe is moved from the upper to the lower hemisphere along a circular path in the xz plane at a constant radius from the source. However, for a linear antenna the second method is trivial since it is equivalent to rotating the antenna on its own axis with the probe at a fixed position.

Example 3. Consider lastly a point source which radiates a field with both E_θ and E_ϕ components in the equatorial plane, the instantaneous values being given by

$$E_{\theta i} = \frac{\sqrt{2}\,E_{\theta m}}{r} \sin\phi \sin(\omega t - \beta r) \tag{6}$$

and

$$E_{\phi i} = \frac{\sqrt{2}\,E_{\phi m}}{r} \cos\phi \sin\left(\omega t - \beta r - \frac{\pi}{2}\right) \tag{7}$$

Referring to Fig. 3-23, a field of the form of the E_θ component in the equatorial plane could be produced by a small loop at the origin oriented parallel to the *yz* plane. A field of the form of the E_ϕ component in the equatorial plane could be produced by a short dipole at the origin coincident with the *y* axis. Let a point at unit radius in the first quadrant be taken as the reference for phase. Assuming that the loop and dipole radiate equal power,

$$E_{\theta m} = E_{\phi m} \tag{8}$$

Then at unit radius the relative patterns as a function of ϕ and t are given by

$$E_{\theta i} = \sin\phi \sin\omega t \tag{9}$$

and

$$E_{\phi i} = \cos\phi \sin\left(\omega t - \frac{\pi}{2}\right)$$

$$= -\cos\phi \cos\omega t \tag{10}$$

The relative field patterns in the equatorial plane are shown in Fig. 3-23. With the loop and dipole fed in-phase, their field components are in phase quadrature ($\delta = \pi/2$). In quadrants 1 and 3, E_ϕ lags E_θ by 90°, while in quadrants 2 and 4, E_ϕ leads E_θ by 90°. The phase patterns in the equatorial plane for E_θ and E_ϕ are shown in polar form by Fig. 3-24 and in rectangular form by Fig. 3-25*a*.

Since E_θ, E_ϕ and δ are known, the polarization ellipses may be determined. These polarization ellipses for different directions in the equatorial plane are shown in Fig. 3-25*b*. It is to be noted that in quadrants 1 and 3, where E_ϕ lags E_θ, the **E**

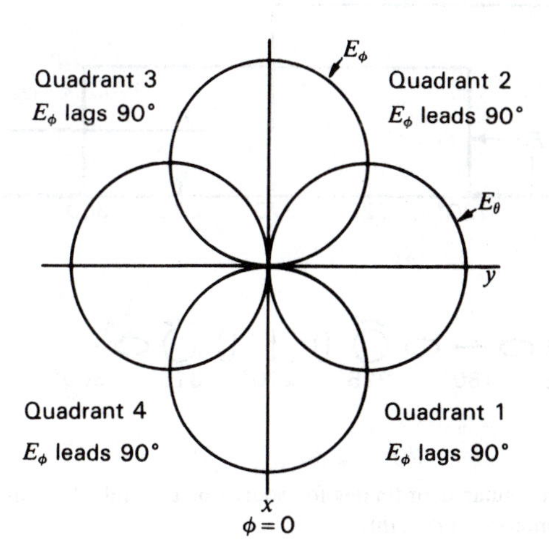

Figure 3-23 Field patterns for source of Example 3.

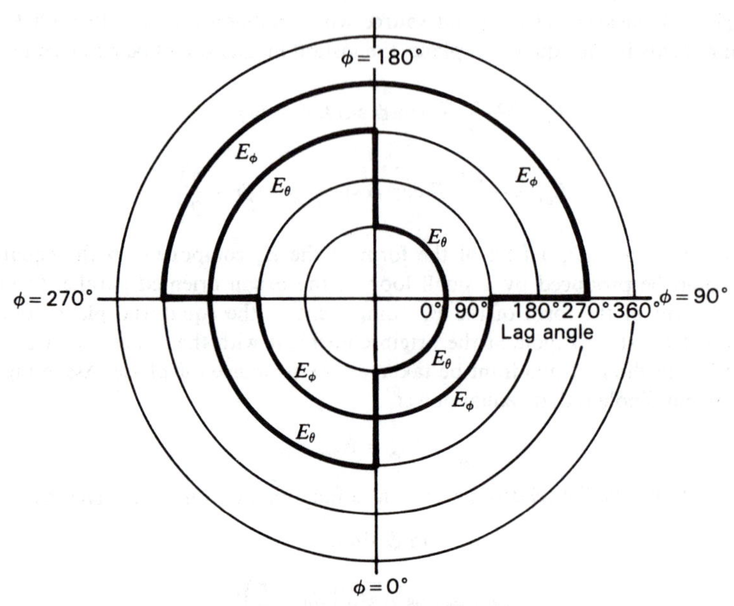

Figure 3-24 Phase lag as a function of ϕ for field components of source of Example 3.

vector rotates counterclockwise, while in quadrants 2 and 4, where E_ϕ leads E_θ, the rotation is clockwise.

At four angles the polarization is circular, **E** rotating counterclockwise at $\phi = 45°$ and 225° and rotating clockwise at $\phi = 135°$ and 315°. The polarization is linear at four angles, being horizontally polarized at 0° and 180° and vertically polarized at 90° and 270°. At all other angles the polarization is elliptical, but the power is constant as a function of ϕ (regardless of the polarization).

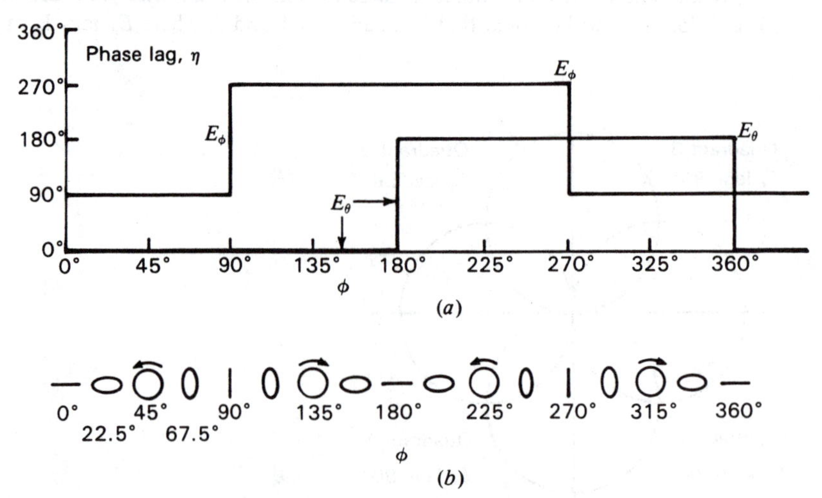

Figure 3-25 Phase patterns in rectangular coordinates for source of Example 3 at (*a*) with polarization ellipses for every 22.5° interval of ϕ at (*b*).

3-18 GENERAL EQUATION FOR THE FIELD OF A POINT SOURCE. Both components of the far field of a point source in free space vary inversely with the distance. Therefore, in general, the two electric field components may be expressed as

$$E_\theta = \frac{E_{\theta m}}{r} f_1(\theta, \phi) \tag{1}$$

and

$$E_\phi = \frac{E_{\phi m}}{r} f_2(\theta, \phi) \tag{2}$$

where $E_{\theta m}$ = rms value of E_θ component at unit radius in the direction of maximum field

$E_{\phi m}$ = rms value of E_ϕ component at unit radius in the direction of maximum field

f_1 and f_2 are, in general, different functions of θ and ϕ but of maximum value unity

The instantaneous values of the field components vary harmonically with time and are given by (1) and (2) multiplied, in general, by different functions of the time. Thus, for the instantaneous field components

$$E_{\theta i} = \frac{\sqrt{2}\,E_{\theta m}}{r} f_1(\theta, \phi) \sin(\omega t - \eta) \tag{3}$$

and

$$E_{\phi i} = \frac{\sqrt{2}\,E_{\phi m}}{r} f_2(\theta, \phi) \sin(\omega t - \eta - \delta) \tag{4}$$

where $\eta = \beta(r - r_1) + f_3(\theta, \phi)$

$\delta = f_4(\theta, \phi)$

r = radius to field point (r, θ, ϕ)

r_1 = radius of point to which phase is referred

f_3 and f_4 are, in general, different functions of θ and ϕ

The instantaneous value of the total electric field at a point (r, θ, ϕ) due to a point source is the vector sum of the instantaneous values of the two components. That is,

$$\mathbf{E}_i = \mathbf{a}_\theta E_{\theta i} + \mathbf{a}_\phi E_{\phi i} \tag{5}$$

where $\mathbf{a}_\theta$ = unit vector in θ direction

$\mathbf{a}_\phi$ = unit vector in ϕ direction

Substituting (3) and (4) into (5) then gives a general equation for the electric field of a point source at any point (r, θ, ϕ) as follows:

$$\mathbf{E}_i = \mathbf{a}_\theta \frac{\sqrt{2}\, E_{\theta m}}{r} f_1(\theta, \phi) \sin(\omega t - \eta) + \mathbf{a}_\phi \frac{\sqrt{2}\, E_{\phi m}}{r} f_2(\theta, \phi) \sin(\omega t - \eta - \delta) \tag{6}$$

In this equation the instantaneous total electric field vector $\mathbf{E}_i$ is a function of both space and time; thus

$$\mathbf{E}_i = f(r, \theta, \phi, t) \tag{7}$$

The far field is entirely specified by (6). When f_1 and f_2 are complicated expressions, it is often convenient to describe $\mathbf{E}_i$ by means of graphs for the four quantities E_θ, E_ϕ, η and δ, as has been discussed. It is assumed that the field varies harmonically with time and that the frequency is known.

PROBLEMS[1]

★3-1 Directivity.

(*a*) Calculate the exact directivity for three unidirectional sources having the following power patterns:

$$U = U_m \sin\theta \sin^2\phi$$
$$U = U_m \sin\theta \sin^3\phi$$
$$U = U_m \sin^2\theta \sin^3\phi$$

U has a value only for $0 \le \theta \le \pi$ and $0 \le \phi \le \pi$ and is zero elsewhere.

(*b*) Calculate the approximate directivity from the product of the half-power beam widths for each of the sources.

(*c*) Tabulate the results for comparison.

3-2 Directivity. Show that the directivity for a source with a unidirectional power pattern given by $U = U_m \cos^n\theta$ can be expressed as $D = 2(n + 1)$. U has a value only for $0 \le \theta \le \pi/2$ and $0 \le \phi \le 2\pi$ and is zero elsewhere.

★3-3 Solar power. The earth receives from the sun 2.2 g cal min^{-1} cm^{-2}.

(*a*) What is the corresponding Poynting vector in watts per square meter?

(*b*) What is the power output of the sun, assuming that it is an isotropic source?

(c) What is the rms field intensity at the earth due to the sun's radiation, assuming all the sun's energy is at a single frequency?

Note: 1 watt = 14.3 g cal min^{-1}
Distance earth to sun = 149×10^6 km

3-4 Directivity and minor lobes. Prove the following theorem: if the minor lobes of a radiation pattern remain constant as the beam width of the main lobe approaches zero, then the directivity of the antenna approaches a constant value as the beam width of the main lobe approaches zero.

---

[1] Answers to starred (*) problems are given in App. D.

3-5 Directivity by integration.

(*a*) Calculate by graphical integration or numerical methods the directivity of a source with a unidirectional power pattern given by $U = \cos\theta$. Compare this directivity value with the exact value. U has a value only for $0 \le \theta \le \pi/2$ and $0 \le \phi \le 2\pi$ and is zero elsewhere.

(*b*) Repeat for a unidirectional power pattern given by $U = \cos^2\theta$.

(*c*) Repeat for a unidirectional power pattern given by $U = \cos^3\theta$.

3-6 Directivity. Calculate the directivity for a source with relative field pattern $E = \cos 2\theta \cos\theta$.

CHAPTER 4

ARRAYS OF POINT SOURCES

4-1 INTRODUCTION. In Chap. 2 an antenna was treated as an aperture. In Chap. 3 an antenna was considered as a single point source. In this chapter we continue with the point-source concept, but extend it to a consideration of arrays of point sources. This approach is of great value since the pattern of any antenna can be regarded as produced by an array of point sources. Much of the discussion will concern arrays of isotropic point sources which may represent many different kinds of antennas. Arrays of nonisotropic but similar point sources are also treated, leading to the principle of pattern multiplication. From arrays of discrete point sources we proceed to continuous arrays of point sources and Huygens' principle.

4-2 ARRAYS OF TWO ISOTROPIC POINT SOURCES. Let us introduce the subject of arrays of point sources by considering the simplest situation, namely, that of two isotropic point sources. As illustrations, five cases involving two isotropic point sources will be discussed.

4-2a Case 1. Two Isotropic Point Sources of Same Amplitude and Phase. The first case we shall analyze is that of two isotropic point sources having equal amplitudes and oscillating in the same phase. Let the two point sources, 1 and 2, be separated by a distance d and located symmetrically with respect to the origin of the coordinates as shown in Fig. 4-1*a*. The angle ϕ is measured counterclockwise from the positive x axis. The origin of the coordi-

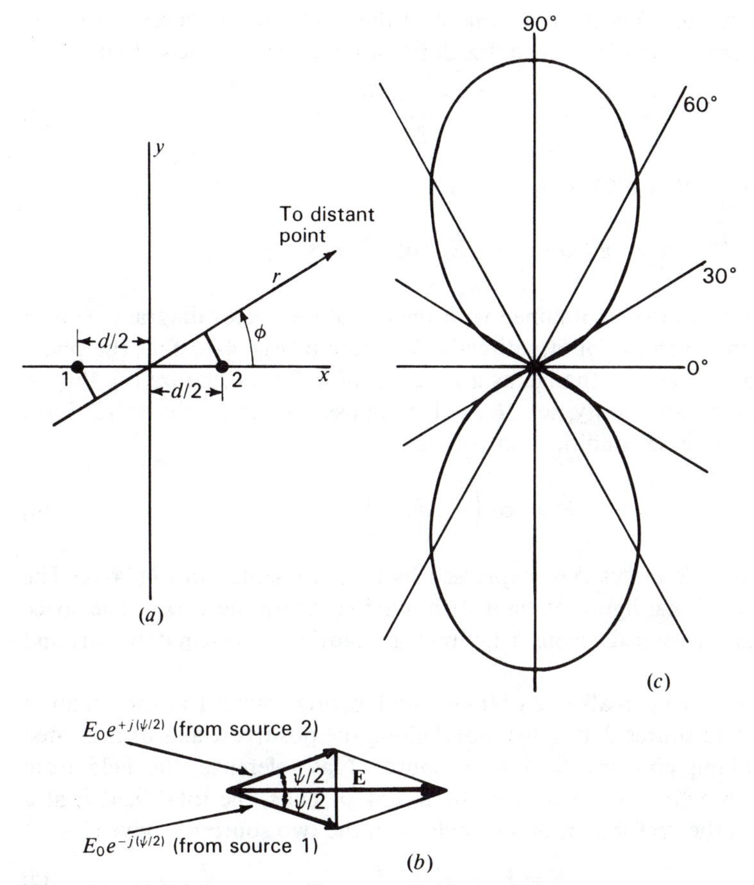

Figure 4-1 (*a*) Relation to coordinate system of 2 isotropic point sources separated by a distance *d*. (*b*) Vector addition of the fields from two isotropic point sources of equal amplitude and same phase located as in (*a*). (*c*) Field pattern of 2 isotropic point sources of equal amplitude and same phase located as in (*a*) for the case where the separation *d* is $\lambda/2$.

nates is taken as the reference for phase. Then at a distant point in the direction ϕ the field from source 1 is retarded by $\frac{1}{2}d_r \cos \phi$, while the field from source 2 is advanced by $\frac{1}{2}d_r \cos \phi$, where d_r is the distance between the sources expressed in radians; that is,

$$d_r = \frac{2\pi d}{\lambda} = \beta d$$

The total field at a large distance r in the direction ϕ is then

$$E = E_0 e^{-j\psi/2} + E_0 e^{+j\psi/2} \tag{1}$$

where $\psi = d_r \cos \phi$ and the amplitude of the field components at the distance r is given by E_0.

The first term in (1) is the component of the field due to source 1 and the second term the component due to source 2. Equation (1) may be rewritten

$$E = 2E_0 \frac{e^{+j\psi/2} + e^{-j\psi/2}}{2} \tag{2}$$

which by a trigonometric identity is

$$E = 2E_0 \cos\frac{\psi}{2} = 2E_0 \cos\left(\frac{d_r}{2}\cos\phi\right) \tag{3}$$

This result may also be obtained with the aid of the vector diagram[1] shown in Fig. 4-1*b*, from which (3) follows directly. We note in Fig. 4-1*b* that the phase of the total field E does not change as a function of ψ. To normalize (3), that is, make its maximum value unity, set $2E_0 = 1$. Suppose further that d is $\lambda/2$. Then $d_r = \pi$. Introducing these conditions into (3) gives

$$E = \cos\left(\frac{\pi}{2}\cos\phi\right) \tag{4}$$

The field pattern of E versus ϕ as expressed by (4) is presented in Fig. 4-1*c*. The pattern is a bidirectional figure-of-eight with maxima along the y axis. The space pattern is doughnut-shaped, being a figure-of-revolution of this pattern around the x axis.

The same pattern can also be obtained by locating source 1 at the origin of the coordinates and source 2 at a distance d along the positive x axis as indicated in Fig. 4-2*a*. Taking now the field from source 1 as reference, the field from source 2 in the direction ϕ is advanced by $d_r \cos\phi$. Thus, the total field E at a large distance r is the vector sum of the fields from the two sources as given by

$$E = E_0 + E_0 e^{+j\psi} \tag{5}$$

where $\psi = d_r \cos\phi$

The relation of these fields is indicated by the vector diagram of Fig. 4-2*b*. From the vector diagram the magnitude of the total field is

$$E = 2E_0 \cos\frac{\psi}{2} = 2E_0 \cos\frac{d_r\cos\phi}{2} \tag{6}$$

as obtained before in (3). The phase of the total field E is, however, not constant in this case but is $\psi/2$, as also shown by rewriting (5) as

$$E = E_0(1 + e^{j\psi}) = 2E_0 e^{j\psi/2}\left(\frac{e^{j\psi/2} + e^{-j\psi/2}}{2}\right) = 2E_0 e^{j\psi/2}\cos\frac{\psi}{2} \tag{7}$$

[1] It is to be noted that the quantities represented here by vectors are not true space vectors but merely vector representations of the time phase (i.e., phasors).

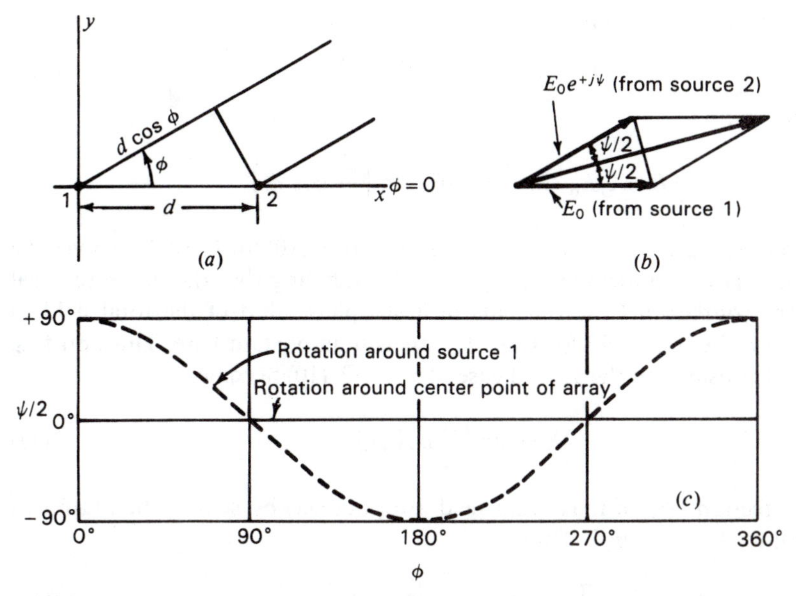

Figure 4-2 (*a*) Two isotropic point sources with the origin of the coordinate system coincident with one of the sources. (*b*) Vector addition of the fields from 2 isotropic point sources of equal amplitude and same phase located as in (*a*). (*c*) Phase of total field as a function of ϕ for 2 isotropic point sources of same amplitude and phase spaced $\lambda/2$ apart. The phase change is zero when referred to the center point of the array but is $\psi/2$ as shown by the dashed curve when referred to source 1.

Normalizing by setting $2E_0 = 1$, (7) becomes

$$E = e^{j\psi/2} \cos \frac{\psi}{2} = \cos \frac{\psi}{2} \underline{/\psi/2} \tag{8}$$

In (8) the cosine factor gives the amplitude variation of E, and the exponential or angle factor gives the phase variation *with respect to source* 1 as the reference. The phase variation for the case of $\lambda/2$ spacing ($d_r = \pi$) is shown by the dashed line in Fig. 4-2*c*. Here the phase angle with respect to the phase of source 1 is given by $\psi/2 = (\pi/2) \cos \phi$. The magnitude variation for this case has already been presented in Fig. 4-1*c*. When the phase is referred to the point midway between the sources (Fig. 4-1*a*), there is no phase change around the array as shown by the solid line in Fig. 4-2*c*. Thus, an observer at a fixed distance observes no phase change when the array is rotated (with respect to ϕ) around its midpoint, but a phase change (dashed curve of Fig. 4-2*c*) is observed if the array is rotated with source 1 as the center of rotation.

4-2b Case 2. Two Isotropic Point Sources of Same Amplitude but Opposite Phase. This case is identical with the one we have just considered except that the two sources are in opposite phase instead of in the same phase. Let the sources be located as in Fig. 4-1*a*. Then the total field in the direction ϕ

at a large distance r is given by

$$E = E_0 e^{+j\psi/2} - E_0 e^{-j\psi/2} \tag{9}$$

from which

$$E = 2jE_0 \sin \frac{\psi}{2} = 2jE_0 \sin \left(\frac{d_r}{2} \cos \phi\right) \tag{10}$$

Whereas in Case 1 (3) involves the cosine of $\psi/2$, (10) for Case 2 involves the sine. Equation (10) also includes an operator j, indicating that the phase reversal of one of the sources in Case 2 results in a 90° phase shift of the total field as compared with the total field for Case 1. This is unimportant here. Thus, putting $2jE_0 = 1$ and considering the special case of $d = \lambda/2$, (10) becomes

$$E = \sin \left(\frac{\pi}{2} \cos \phi\right) \tag{11}$$

The directions ϕ_m of maximum field are obtained by setting the argument of (11) equal to $\pm(2k+1)\pi/2$. Thus,

$$\frac{\pi}{2} \cos \phi_m = \pm(2k+1)\frac{\pi}{2} \tag{11a}$$

where $k = 0, 1, 2, 3, \ldots$. For $k = 0$, $\cos \phi_m = \pm 1$ and $\phi_m = 0°$ and $180°$.

The null directions ϕ_0 are given by

$$\frac{\pi}{2} \cos \phi_0 = \pm k\pi \tag{11b}$$

For $k = 0$, $\phi_0 = \pm 90°$.

The half-power directions are given by

$$\frac{\pi}{2} \cos \phi = \pm(2k+1)\frac{\pi}{4} \tag{11c}$$

For $k = 0$, $\phi = \pm 60°, \pm 120°$.

The field pattern given by (11) is shown in Fig. 4-3. The pattern is a relatively broad figure-of-eight with the maximum field in the same direction as the line joining the sources (x axis). The space pattern is a figure-of-revolution of this pattern around the x axis. The two sources, in this case, may be described as a simple type of "end-fire" array. In contrast to this pattern, the in-phase point sources produce a pattern with the maximum field normal to the line joining the sources, as shown in Fig. 4-1*c*. The two sources for this case may be described as a simple "broadside" type of array.

4-2c Case 3. Two Isotropic Point Sources of the Same Amplitude and in Phase Quadrature. Let the two point sources be located as in Fig. 4-1*a*. Taking the origin of the coordinates as the reference for phase, let source 1 be retarded

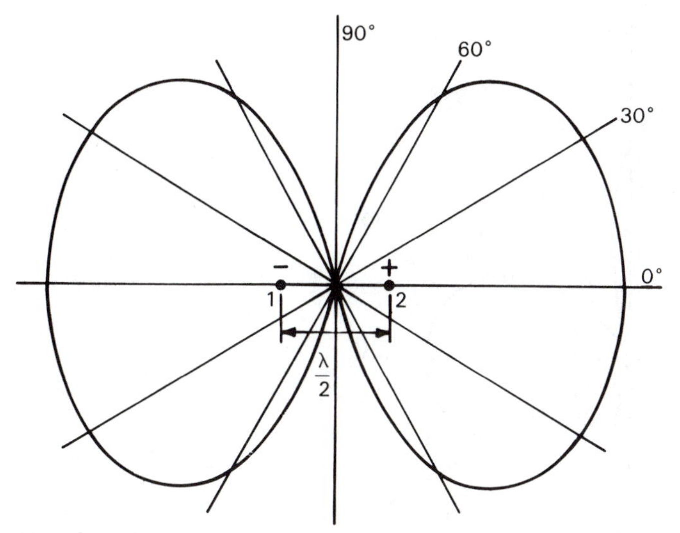

Figure 4-3 Relative field pattern for 2 isotropic point sources of the same amplitude but opposite phase, spaced $\lambda/2$ apart.

by 45° and source 2 advanced by 45°. Then the total field in the direction ϕ at a large distance r is given by

$$E = E_0 \exp\left[+j\left(\frac{d_r \cos\phi}{2} + \frac{\pi}{4}\right)\right] + E_0 \exp\left[-j\left(\frac{d_r \cos\phi}{2} + \frac{\pi}{4}\right)\right] \tag{12}$$

From (12) we obtain

$$E = 2E_0 \cos\left(\frac{\pi}{4} + \frac{d_r}{2}\cos\phi\right) \tag{13}$$

Letting $2E_0 = 1$ and $d = \lambda/2$, (13) becomes

$$E = \cos\left(\frac{\pi}{4} + \frac{\pi}{2}\cos\phi\right) \tag{14}$$

The field pattern given by (14) is presented in Fig. 4-4. The space pattern is a figure-of-revolution of this pattern around the x axis. Most of the radiation is in the second and third quadrants. It is interesting to note that the field in the direction $\phi = 0°$ is the same as in the direction $\phi = 180°$. The directions ϕ_m of maximum field are obtained by setting the argument of (14) equal to $k\pi$, where $k = 0, 1, 2, 3\ldots.$ In this way we obtain

$$\frac{\pi}{4} + \frac{\pi}{2}\cos\phi_m = k\pi \tag{15}$$

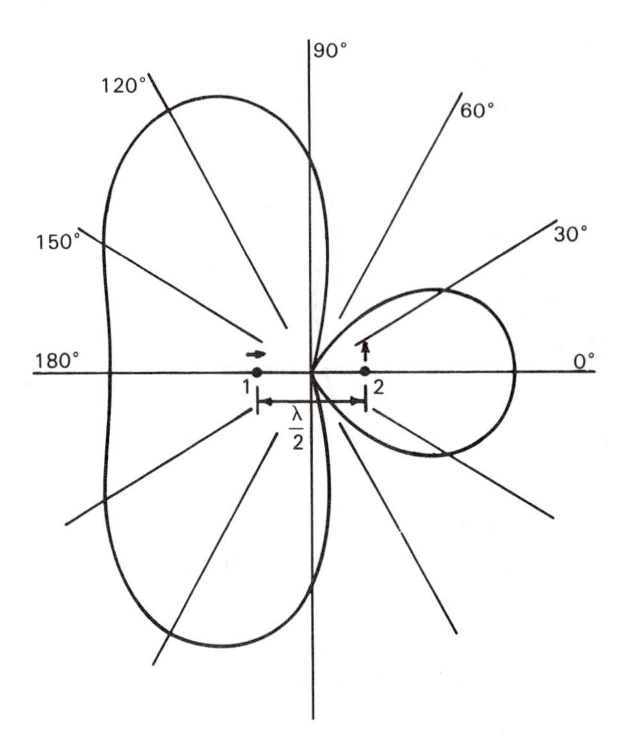

Figure 4-4 Relative field pattern of 2 isotropic point sources of the same amplitude and in phase quadrature for a spacing of $\lambda/2$. The source to the right leads that to the left by 90°.

For $k = 0$,

$$\frac{\pi}{2}\cos\phi_m = -\frac{\pi}{4} \tag{16}$$

and

$$\phi_m = 120° \text{ and } 240° \tag{17}$$

If the spacing between the sources is reduced to $\lambda/4$, (13) becomes

$$E = \cos\left(\frac{\pi}{4} + \frac{\pi}{4}\cos\phi\right) \tag{18}$$

The field pattern for this case is illustrated by Fig. 4-5*a*. It is a cardioid-shaped, unidirectional pattern with maximum field in the negative x direction. The space pattern is a figure-of-revolution of this pattern around the x axis.

A simple method of determining the direction of maximum field is illustrated by Fig. 4-5*b*. As indicated by the vectors, the phase of source 2 is 0° (vector to right) and the phase of source 1 is 270° (vector down). Thus, source 2 leads source 1 by 90°.

To find the field radiated to the left, imagine that we start at source 2 (phase 0°) and travel to the left, riding with the wave (phase 0°) like a surfer rides a breaker. The phase of the wave we are riding is 0° and does not change but by the time we have traveled $\lambda/4$ and arrived at source 1, a $\frac{1}{4}$-period has elapsed so the current in source 1 will have advanced 90° (vector rotated ccw) from 270° to

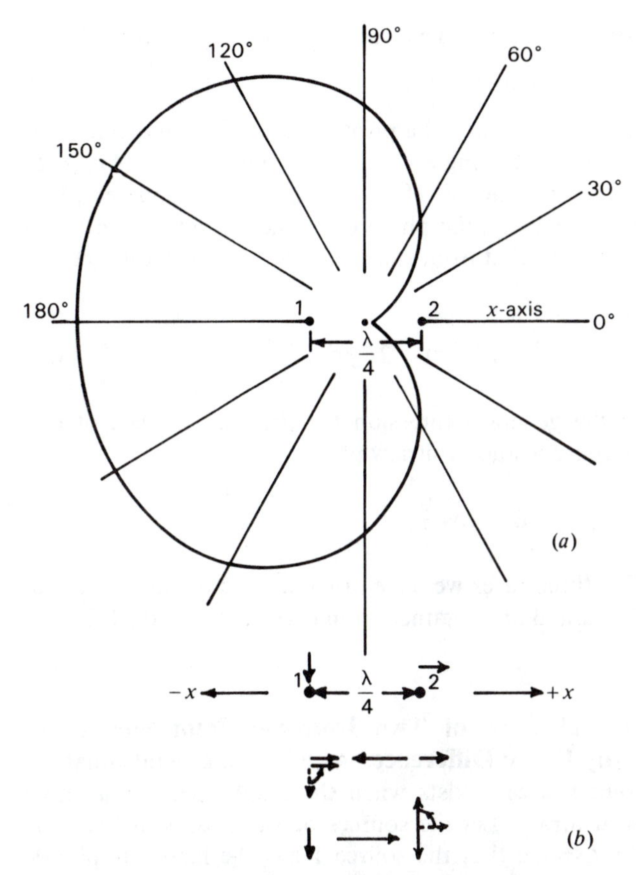

Figure 4-5 (*a*) Relative field pattern of 2 isotropic sources of same amplitude and in phase quadrature for a spacing of $\lambda/4$. Source 2 leads source 1 by 90°. (*b*) Vector diagrams illustrating field reinforcement in the $-x$ direction and field cancellation in the $+x$ direction.

0°, making its phase the same as that of the wave we are riding, as in the middle diagram of Fig. 4-5*b*. Thus, the field of the wave from source 2 reinforces that of the field of source 1, and the two fields travel to the left together in phase producing a maximum field to the left which is twice the field of either source alone.

Now imagine that we start at source 1 with phase 270° (vector down) and travel to the right. By the time we arrive at source 2 the phase of its field has advanced from 0 to 90° so it is in phase opposition and cancels the field of the wave we are riding, as in the bottom diagram in Fig. 4-5*b*, resulting in zero radiation to the right.

4-2d Case 4. General Case of Two Isotropic Point Sources of Equal Amplitude and Any Phase Difference.

Proceeding now to a more general situation, let us consider the case of two isotropic point sources of equal amplitude but of any phase difference δ. The total phase difference ψ between the fields from source 2 and source 1 at a distant point in the direction ϕ (see Fig. 4-2*a*) is then

$$\psi = d_r \cos \phi + \delta \tag{19}$$

Taking source 1 as the reference for phase, the positive sign in (19) indicates that source 2 is advanced in phase by the angle δ. A minus sign would be used to indicate a phase retardation. If, instead of referring the phase to source 1, it is referred to the centerpoint of the array, the phase of the field from source 1 at a distant point is given by $-\psi/2$ and that from source 2 by $+\psi/2$. The total field is then

$$E = E_0(e^{j\psi/2} + e^{-j\psi/2}) = 2E_0 \cos \frac{\psi}{2} \tag{20}$$

Normalizing (20), we have the general expression for the field pattern of two isotropic sources of equal amplitude and arbitrary phase,

$$E = \cos \frac{\psi}{2} \tag{21}$$

where ψ is given by (19). The three cases we have discussed are obviously special cases of (21). Thus, Cases 1, 2 and 3 are obtained from (21) when $\delta = 0°$, $180°$ and $90°$ respectively.

4-2e Case 5. Most General Case of Two Isotropic Point Sources of Unequal Amplitude and Any Phase Difference. A still more general situation, involving two isotropic point sources, exists when the amplitudes are unequal and the phase difference is arbitrary. Let the sources be situated as in Fig. 4-6*a* with source 1 at the origin. Assume that the source 1 has the larger amplitude and that its field at a large distance r has an amplitude of E_0. Let the field from source 2 be of amplitude aE_0 $(0 \le a \le 1)$ at the distance r. Then, referring to Fig. 4-6*b*, the magnitude and phase angle of the total field E is given by

$$E = E_0\sqrt{(1 + a \cos \psi)^2 + a^2 \sin^2 \psi} \;\underline{/\arctan [a \sin \psi/(1 + a \cos \psi)]} \tag{22}$$

where $\psi = d_r \cos \phi + \delta$ and the phase angle ($\angle$) is referred to source 1. This is the phase angle ξ shown in Fig. 4-6*b*.

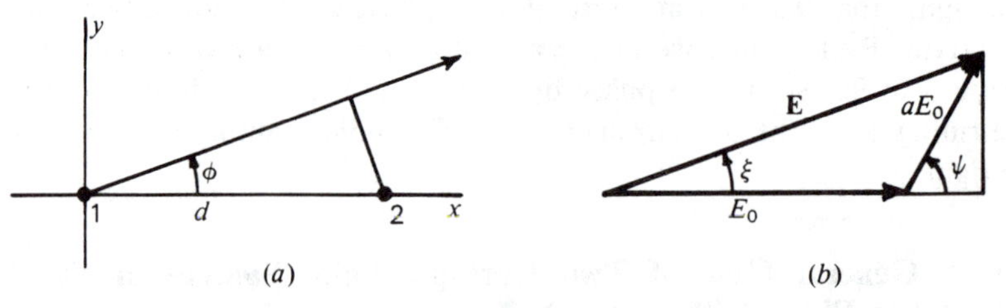

Figure 4-6 (*a*) Two isotropic point sources of unequal amplitude and arbitrary phase with respect to the coordinate system. (*b*) Vector addition of fields from unequal sources arranged as in (*a*). The amplitude of source 2 is assumed to be smaller than that of source 1 by the factor *a*.

4-3 NONISOTROPIC BUT SIMILAR POINT SOURCES AND THE PRINCIPLE OF PATTERN MULTIPLICATION.

The cases considered in the preceding section all involve *isotropic* point sources. These can readily be extended to a more general situation in which the sources are *non-isotropic but similar.*

The word *similar* is here used to indicate that the variation with absolute angle ϕ of both the amplitude and phase of the field is the same.[1] The maximum amplitudes of the individual sources may be unequal. If, however, they are also equal, the sources are not only similar but are *identical.*

As an example, let us reconsider Case 4 of Sec. 4-2d in which the sources are identical, with the modification that both sources 1 and 2 have field patterns given by

$$E_0 = E_0' \sin \phi \tag{1}$$

Patterns of this type might be produced by short dipoles oriented parallel to the x axis as suggested by Fig. 4-7. Substituting (1) in (4-2-20) and normalizing by setting $2E_0' = 1$ gives the field pattern of the array as

$$E = \sin \phi \cos \frac{\psi}{2} \tag{2}$$

where $\psi = d_r \cos \phi + \delta$

This result is the same as obtained by multiplying the pattern of the individual source ($\sin \phi$) by the pattern of two isotropic point sources ($\cos \psi/2$).

If the similar but unequal point sources of Case 5 (Sec. 4-2e) have patterns as given by (1), the total normalized pattern is

$$E = \sin \phi \sqrt{(1 + a \cos \psi)^2 + a^2 \sin^2 \psi} \tag{3}$$

Here again, the result is the same as that obtained by multiplying the pattern of the individual source by the pattern of an array of isotropic point sources.

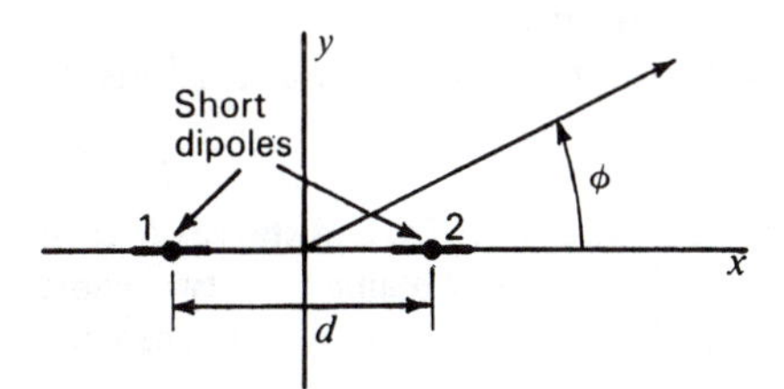

Figure 4-7 Two nonisotropic sources with respect to the coordinate system.

[1] The patterns not only must be of the same shape but also must be oriented in the same direction to be called "similar."

These are examples illustrating the *principle of pattern multiplication*, which may be expressed as follows:

The field pattern of an array of nonisotropic but similar point sources is the product of the pattern of the individual source and the pattern of an array of isotropic point sources, having the same locations, relative amplitudes and phases as the nonisotropic point sources.

This principle may be applied to arrays of any number of sources provided only that they are similar. The individual nonisotropic source or antenna may be of finite size but can be considered as a point source situated at the point in the antenna to which phase is referred. This point is said to be the "phase center."

The above discussion of pattern multiplication has been concerned only with the field pattern or magnitude of the field. If the field of the nonisotropic source and the array of isotropic sources vary in phase with space angle, i.e., have a phase pattern which is not a constant, the statement of the principle of pattern multiplication may be extended to include this more general case as follows:

The total field pattern of an array of nonisotropic but similar sources is the product of the individual source pattern and the pattern of an array of isotropic point sources each located at the phase center of the individual source and having the same relative amplitude and phase, while the total phase pattern is the sum of the phase patterns of the individual source and the array of isotropic point sources.

The total phase pattern is referred to the phase center of the array. In symbols, the total field E is then

$$E = \underbrace{f(\theta, \phi)F(\theta, \phi)}_{\text{Field pattern}} \, \underline{/ \; \underbrace{f_p(\theta, \phi) + F_p(\theta, \phi)}_{\text{Phase pattern}}}$$

where $f(\theta, \phi)$ = field pattern of individual source
$f_p(\theta, \phi)$ = phase pattern of individual source
$F(\theta, \phi)$ = field pattern of array of isotropic sources
$F_p(\theta, \phi)$ = phase pattern of array of isotropic sources

The patterns are expressed in (4) as a function of both polar angles to indicate that the principle of pattern multiplication applies to space patterns as well as to the two-dimensional cases we have been considering.

To illustrate the principle, let us apply to it two special modifications of Case 1 (Sec. 4-2a).

Example 1. Assume two identical point sources separated by a distance d, each source having the field pattern given by (1) as might be obtained by two short dipoles arranged as in Fig. 4-7. Let $d = \lambda/2$ and the phase angle $\delta = 0$. Then the total field pattern is

$$E = \sin \phi \cos \left(\frac{\pi}{2} \cos \phi \right) \tag{5}$$

This pattern is illustrated by Fig. 4-8*c* as the product of the individual source pattern ($\sin \phi$) shown at (*a*) and the array pattern $\{\cos [(\pi/2) \cos \phi]\}$ as shown at

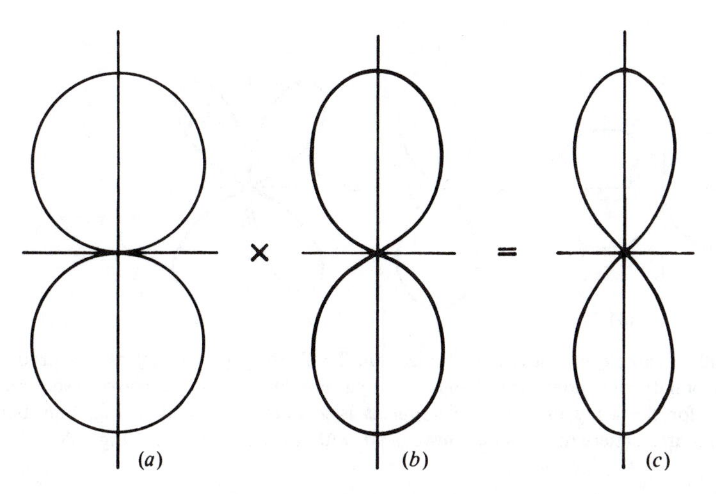

Figure 4-8 Example of pattern multiplication. Two nonisotropic but identical point sources of the same amplitude and phase, spaced $\lambda/2$ apart and arranged as in Fig. 4-7, produce the pattern shown at (*c*). The individual source has the pattern shown at (*a*), which, when multiplied by the pattern of an array of 2 isotropic point sources (of the same amplitude and phase) as shown at (*b*), yields the total array pattern of (*c*).

(*b*). The pattern is sharper than it was in Case 1 (Sec. 4-2a) for the isotropic sources. In this instance, the maximum field of the individual source is in the direction $\phi = 90°$, which coincides with the direction of the maximum field for the array of two isotropic sources.

Example 2. Let us consider next the situation in which $d = \lambda/2$ and $\delta = 0$ as in Example 1 but with individual source patterns given by

$$E_0 = E_0' \cos \phi \tag{6}$$

This type of pattern might be produced by short dipoles oriented parallel to the y axis as in Fig. 4-9. Here the maximum field of the individual source is in the direction ($\phi = 0$) of a null from the array, while the individual source has a null in the direction ($\phi = 90°$) of the pattern maximum of the array. By the principle of pattern multiplication the total normalized field is

$$E = \cos \phi \cos \left(\frac{\pi}{2} \cos \phi \right) \tag{7}$$

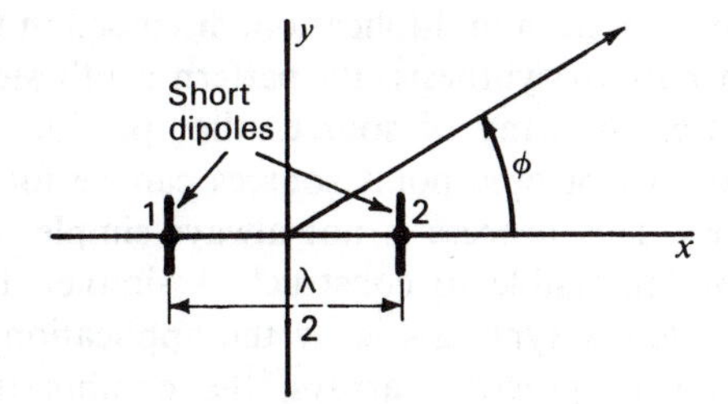

Figure 4-9 Array of 2 nonisotropic sources with respect to the coordinate system.

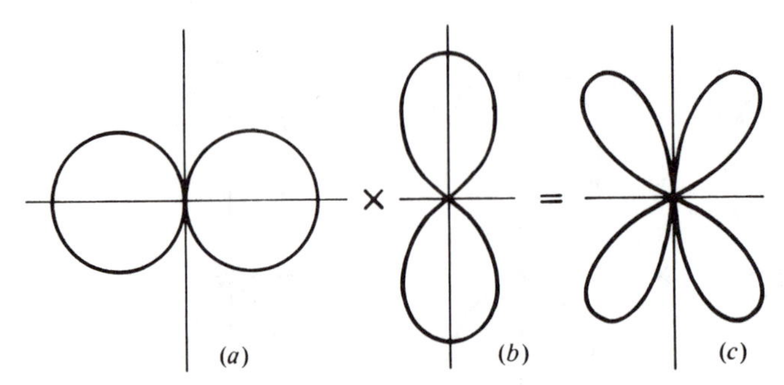

Figure 4-10 Example of pattern multiplication. Total array pattern (*c*) as the product of pattern (*a*) of individual nonisotropic source and pattern (*b*) of array of 2 isotropic sources. The pattern (*b*) for the array of 2 isotropic sources is identical with that of Fig. 4-8*b*, but the individual source pattern (*a*) is rotated through 90° with respect to the one in Fig. 4-8*a*.

The total array pattern in the xy plane as given by (7) is illustrated in Fig. 4-10*c* as the product of the individual source pattern ($\cos \phi$) shown at (*a*) and the array pattern $\{\cos [(\pi/2) \cos \phi]\}$ shown at (*b*). The total array pattern in the xy plane has four lobes with nulls at the x and y axes.

The above examples illustrate two applications of the principle of pattern multiplication to arrays in which the source has a simple pattern. However, in the more general case the individual source may represent an antenna of any complexity provided that the amplitude and phase of its field can be expressed as a function of angle, that is to say, provided that the field pattern and the phase pattern with respect to the phase center are known. If only the total field pattern is desired, phase patterns need not be known provided that the individual sources are identical.

If the arrays in the above examples are parts of still larger arrays, the smaller arrays may be regarded as nonisotropic point sources in the larger array—another application of the principle of pattern multiplication yielding the complete pattern. In this way the principle of pattern multiplication can be applied n times to find the patterns of arrays of arrays of arrays.

4-4 EXAMPLE OF PATTERN SYNTHESIS BY PATTERN MULTIPLICATION. The principle of pattern multiplication, discussed in the preceding section, is of great value in pattern synthesis. By pattern synthesis is meant the process of finding the source or array of sources that produces a desired pattern. Theoretically an array of isotropic point sources can be found that will produce any arbitrary pattern. This process is not always simple and may yield an array that is difficult or impossible to construct. A simpler, less elegant approach to the problem of antenna synthesis is by the application of pattern multiplication to combinations of practical arrays, the combination

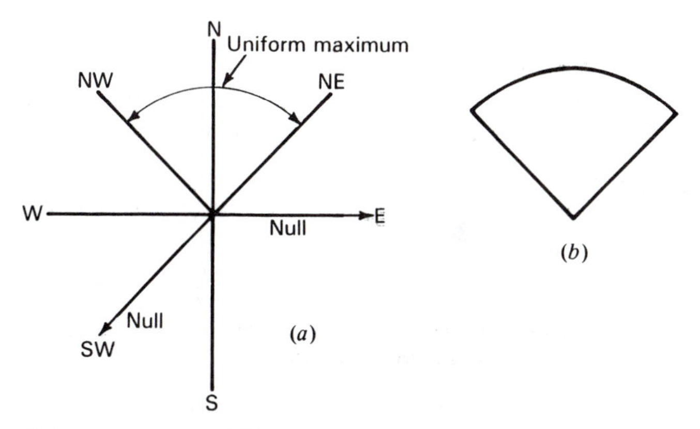

Figure 4-11 (*a*) Requirements for pattern of broadcast station and (*b*) idealized pattern fulfilling them.

which best approximates the desired pattern being arrived at by a trial-and-error process.

To illustrate this application of pattern multiplication, let us consider the following hypothetical problem. A broadcasting station (in the 500- to 1500-kHz frequency band) requires a pattern in the horizontal plane fulfilling the conditions indicated in Fig. 4-11*a*. The maximum field intensity, with as little variation as possible, is to be radiated in the 90° sector between northwest and northeast. No nulls in the pattern can occur in this sector. However, nulls may occur in any direction in the complementary 270° sector, but, as an additional requirement, nulls must be present in the due east and the due southwest directions in order to prevent interference with other stations in these directions. An idealized sector-shaped pattern fulfilling those requirements is illustrated in Fig. 4-11*b*. The antenna producing this pattern is to consist of an array of four vertical towers. The currents in all towers are to be equal in magnitude, but the phase may be adjusted to any relationship. There is also no restriction on the spacing or geometrical arrangement of the towers.

Since we are interested only in the horizontal plane pattern, each tower may be considered as an isotropic point source. The problem then becomes one of finding a space and phase relation of four isotropic point sources located in the horizontal plane which fulfills the above requirements.

The principle of pattern multiplication will be applied to the solution of this problem by seeking the patterns of two pairs of isotropic sources which yield the desired pattern when multiplied together. First let us find a pair of isotropic sources whose pattern fulfills the requirements of a broad lobe of radiation with maximum north and a null southwest. This will be called the "primary" pattern.

Two isotropic sources phased as an end-fire array can produce a pattern with a broader major lobe than when phased as a broadside array (for example, compare Figs. 4-1*c* and 4-5). Since a broad lobe to the north is desired, an end-fire arrangement of two isotropic sources as shown in Fig. 4-12 will be tried.

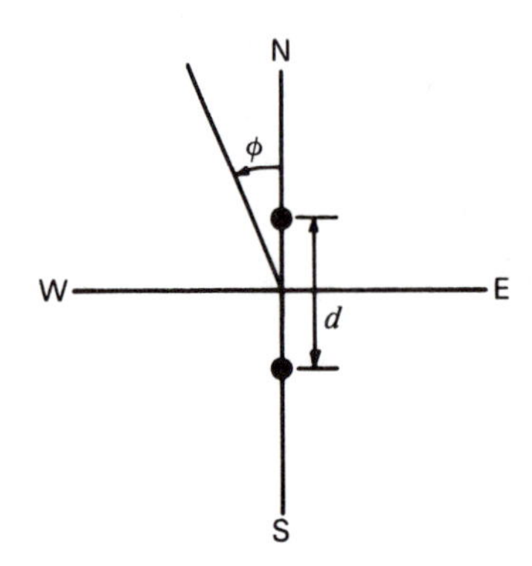

Figure 4-12 Arrangement of 2 isotropic point sources for both primary and secondary arrays.

From a consideration of pattern shapes as a function of separation and phase,[1] a spacing between $\lambda/4$ and $3\lambda/8$ appears suitable (see Fig. 11-11). Accordingly, let $d = 0.3\lambda$. Then the field pattern for the array is

$$E = \cos \frac{\psi}{2} \tag{1}$$

where
$$\psi = 0.6\pi \cos \phi + \delta \tag{2}$$

For there to be a null in the pattern of (1) at $\phi = 135°$ it is necessary that[2]

$$\psi = (2k + 1)\pi \tag{3}$$

where $k = 0, 1, 2, 3, \ldots$

Equating (2) and (3) then gives

$$-0.6\pi \frac{1}{\sqrt{2}} + \delta = (2k + 1)\pi \tag{4}$$

or
$$\delta = (2k + 1)\pi + 0.425\pi \tag{5}$$

For $k = 0$, $\delta = -104°$. The pattern for this case ($d = 0.3\lambda$ and $\delta = -104°$) is illustrated by Fig. 4-13*a*.

Next, let us find the array of two isotropic point sources that will produce a pattern that fulfills the requirements of a null at $\phi = 270°$ and that also has a broad lobe to the north. This will be called the "secondary" pattern. This pattern

---

[1] See, for example, G. H. Brown, "Directional Antennas," *Proc. IRE*, **25**, January 1937; F. E. Terman, *Radio Engineers' Handbook*, McGraw-Hill, New York, 1943, p. 804; C. E. Smith, *Directional Antennas*, Cleveland Institute of Radio Electronics, Cleveland, Ohio, 1946.

[2] The azimuth angle ϕ (Fig. 4-12) is measured counterclockwise (ccw) from the north. This is consistent with the engineering practice of measuring positive angles in a counterclockwise sense. However, it should be noted that the *geodetic azimuth angle* of a point is measured in the opposite, or clockwise (cw), sense from the reference direction, which is sometimes taken as south and sometimes as north.

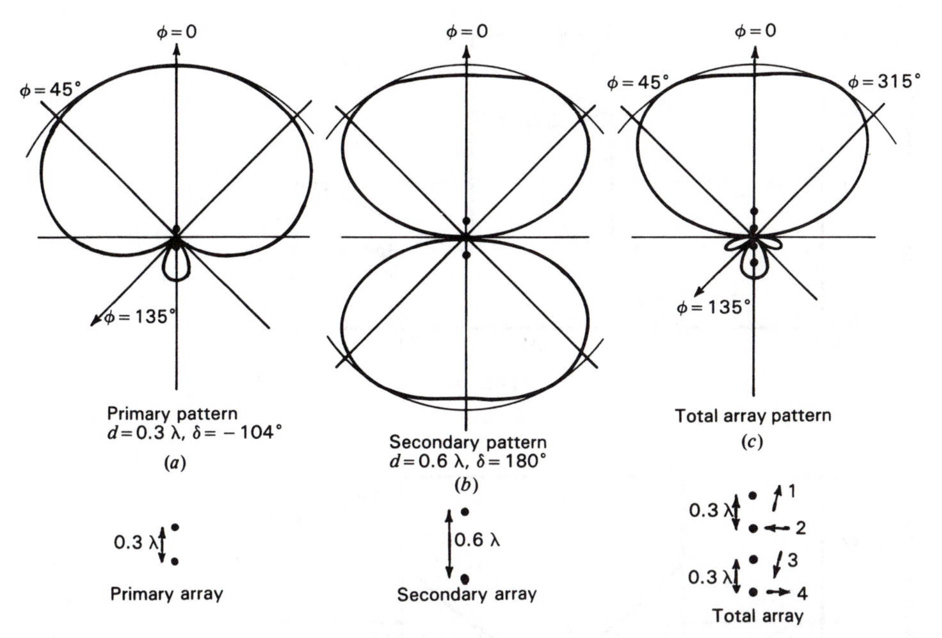

Figure 4-13 Field patterns of primary and secondary arrays of 2 isotropic sources which multiplied together give pattern of total array of 4 isotropic sources.

multiplied by the primary array pattern will then yield the total array pattern. If the secondary isotropic sources are also arranged as in Fig. 4-12 and have a phase difference of 180°, there is a null at $\phi = 270°$. Let the spacing $d = 0.6\lambda$. Then the secondary pattern is given by (1) where

$$\psi = 1.2\pi \cos \phi + \pi \tag{6}$$

The pattern is illustrated by Fig. 4-13*b*. By the principle of pattern multiplication, the total array pattern is the product of this pattern and the primary array pattern, or

$$E = \cos (54° \cos \phi - 52°) \cos (108° \cos \phi + 90°) \tag{7}$$

This pattern, which is illustrated by Fig. 4-13*c*, satisfies the pattern requirements. The complete array is obtained by replacing each of the isotropic sources of the secondary pattern by the two-source array producing the primary pattern. The midpoint of each primary array is its phase center, so this point is placed at the location of a secondary source. The complete antenna is then a linear array of four isotropic point sources as shown in the lower part of Fig. 4-13, where now each source represents a single vertical tower. All towers carry the same current. The current of tower 2 leads tower 1 and the current of tower 4 leads tower 3 by 104°, while the current in towers 1 and 3 and 2 and 4 are in phase opposition. The relative phase of the current is illustrated by the vectors in the lower part of Fig. 4-13*c*.

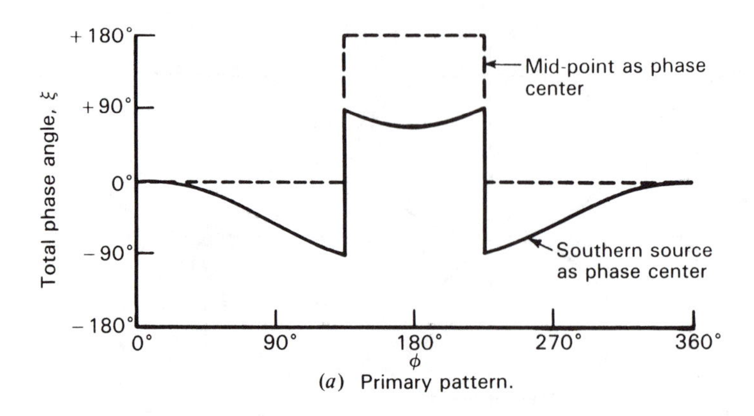

(a) Primary pattern.

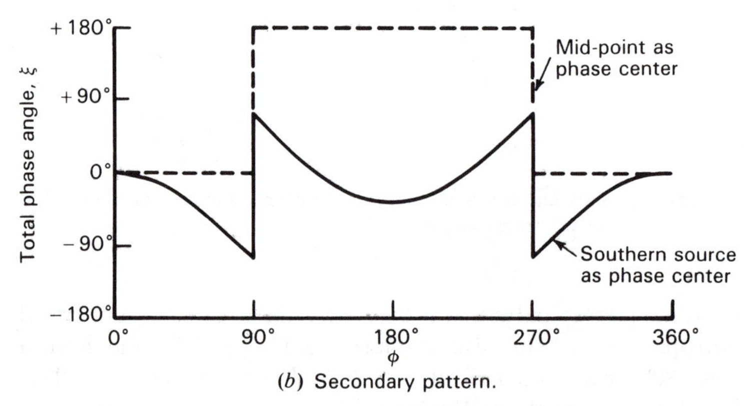

(b) Secondary pattern.

Figure 4-14 Phase patterns of primary, secondary and total arrays having the field patterns shown in Fig. 4-13. Phase patterns are given for the phase center at the midpoint of the array and at the southernmost source, the arrangement of the arrays and the phase centers being shown at (*d*). The phase angle ξ is adjusted to zero at $\phi = 0$ in all cases.

The solution obtained is only one of an infinite number of possible solutions involving four towers. It is, however, a satisfactory and practical solution to the problem.

The phase variation ξ around the primary, secondary and total arrays is shown in Fig. 4-14*a*, *b* and *c* with the phase center at the centerpoint of each array and also at the southernmost source. The arrangement of the arrays with their phase centers is illustrated in Fig. 4-14*d* for both cases.

4-5 NONISOTROPIC AND DISSIMILAR POINT SOURCES.

In Sec. 4-3 nonisotropic but similar point sources were discussed, and it was shown that the principle of pattern multiplication could be applied. However, if the sources are dissimilar, this principle is no longer applicable and the fields of the sources must be added at each angle ϕ for which the total field is calculated.

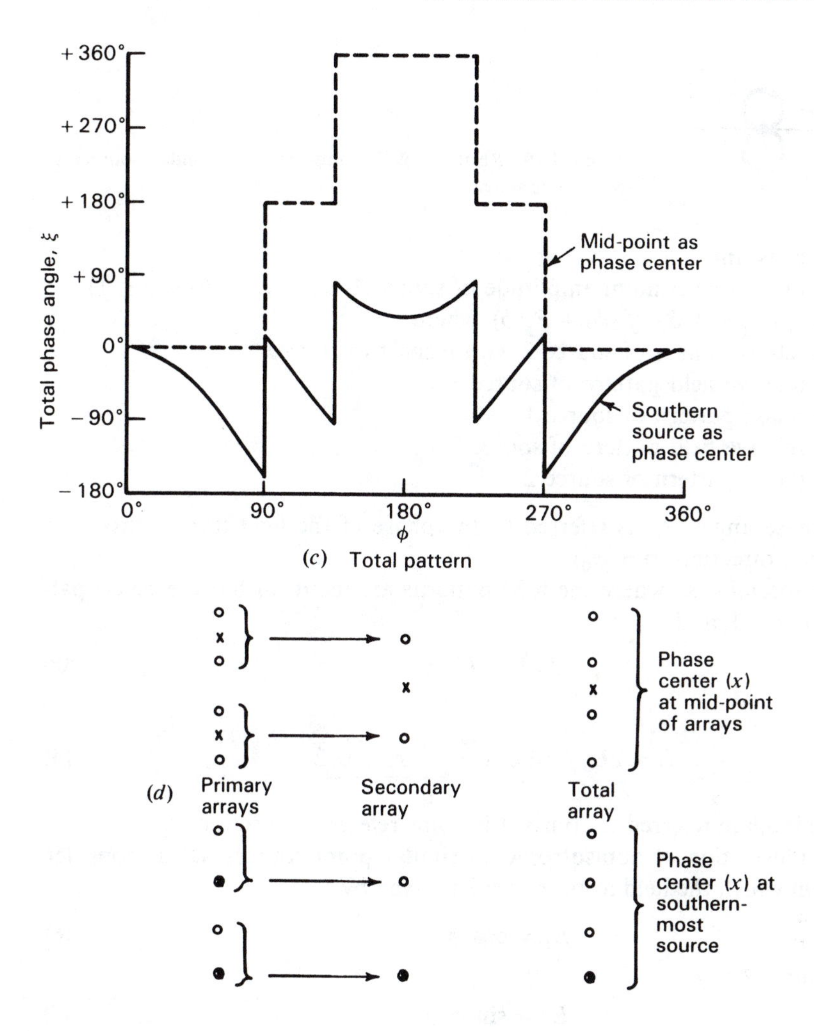

(c) Total pattern

(d)

Thus, for two dissimilar sources 1 and 2 situated on the x axis with source 1 at the origin and the sources separated by a distance d (same geometry as Fig. 4-6) the total field is in general

$$E = E_1 + E_2 = E_0\sqrt{[f(\phi) + aF(\phi)\cos\psi]^2 + [aF(\phi)\sin\psi]^2}$$

$$\underline{/f_p(\phi) + \arctan\left[aF(\phi)\sin\psi/(f(\phi) + aF(\phi)\cos\psi)\right]} \quad (1)$$

where the field from source 1 is taken as

$$E_1 = E_0\, f(\phi)\, \underline{/f_p(\phi)} \quad (2)$$

and from source 2 as

$$E_2 = aE_0\, F(\phi)\, \underline{/F_p(\phi) + d_r\cos\phi + \delta} \quad (3)$$

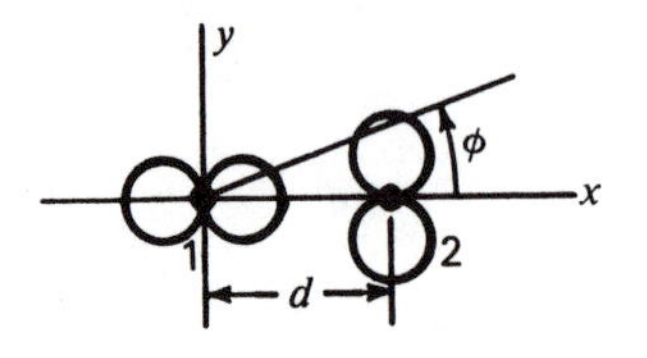

Figure 4-15 Relation of 2 nonisotropic dissimilar sources to coordinate system.

where E_0 = constant
a = ratio of maximum amplitude of source 2 to source 1 ($0 \le a \le 1$)
$\psi = d_r \cos \phi + \delta - f_p(\phi) + F_p(\phi)$, where
δ = relative phase of source 2 with respect to source 1
$f(\phi)$ = relative field pattern of source 1
$f_p(\phi)$ = phase pattern of source 1
$F(\phi)$ = relative field pattern of source 2
$F_p(\phi)$ = phase pattern of source 2

In (1) the phase angle ($\angle$) is referred to the phase of the field from source 1 in some reference direction ($\phi = \phi_0$).

In the special case where the field patterns are identical but the phase patterns are not, $a = 1$, and

$$f(\phi) = F(\phi) \tag{4}$$

from which

$$E = 2E_0 f(\phi) \cos \frac{\psi}{2} \underline{/f_p(\phi) + \psi/2} \tag{5}$$

where phase is again referred to source 1 in some reference direction ϕ_0.

As an illustration of nonisotropic, dissimilar point sources, let us consider an example in which the field from source 1 is given by

$$E_1 = \cos \phi \; \underline{/0} \tag{6}$$

and from source 2 by

$$E_2 = \sin \phi \; \underline{/\psi} \tag{7}$$

where $\psi = d_r \cos \phi + \delta$

The relation of the two sources to the coordinate system and the individual field patterns is shown in Fig. 4-15. Source 1 is located at the origin. The total field E is then the vector sum of E_1 and E_2, or

$$E = \cos \phi + \sin \phi \; \underline{/\psi} \tag{8}$$

Let us consider the case for $\lambda/4$ spacing ($d = \lambda/4$) and phase quadrature of the sources ($\delta = \pi/2$). Then

$$\psi = \frac{\pi}{2}(\cos \phi + 1) \tag{9}$$

The calculation for this case is easily carried out by graphical vector addition. The resulting field pattern for the total field E of the array is presented in Fig.

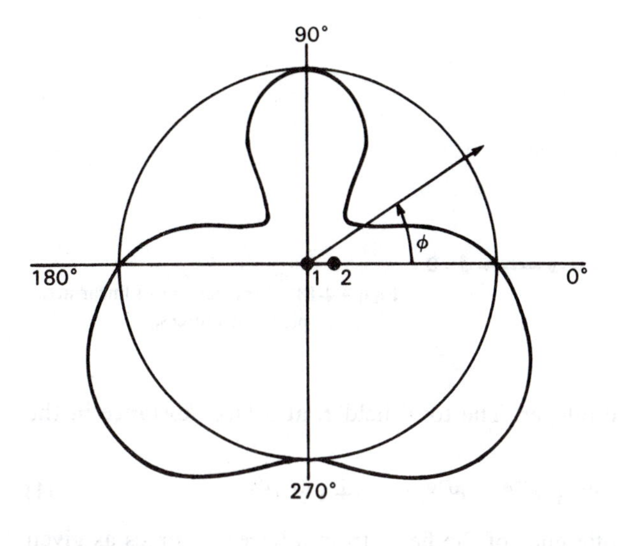

Figure 4-16 Field pattern of array of 2 nonisotropic dissimilar sources of Fig. 4-15 for $d = \lambda/4$ and $\delta = 90°$.

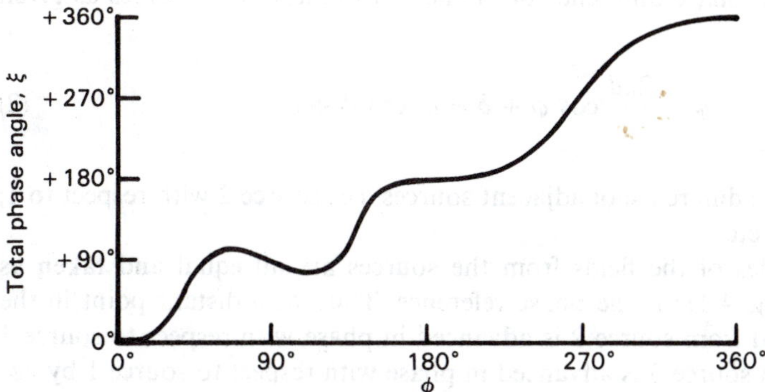

Figure 4-17 Phase pattern of array having field pattern of Fig. 4-16. The phase angle ξ is with respect to source 1 as phase center.

4-16, and the resulting phase pattern for the angle ξ is given in Fig. 4-17. The angle ξ is the phase angle between the total field and the field of source 1 in the direction $\phi = 0$.

4-6 LINEAR ARRAYS OF n ISOTROPIC POINT SOURCES OF EQUAL AMPLITUDE AND SPACING.[1]

4-6a Introduction. Let us now proceed to the case of n isotropic point sources of equal amplitude and spacing arranged as a linear array, as indicated in Fig.

[1] S. A. Schelkunoff, *Electromagnetic Waves*, Van Nostrand, New York, 1943, p. 342.
J. A. Stratton, *Electromagnetic Theory*, McGraw-Hill, New York, 1941, p. 451.

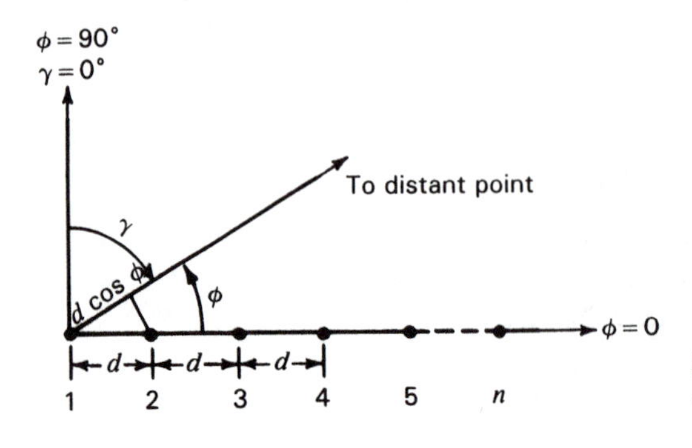

Figure 4-18 Arrangement of linear array of n isotropic point sources.

4-18, where n is any positive integer. The total field E at a large distance in the direction ϕ is given by

$$E = 1 + e^{j\psi} + e^{j2\psi} + e^{j3\psi} + \cdots + e^{j(n-1)\psi} \tag{1}$$

where ψ is the total phase difference of the fields from adjacent sources as given by

$$\psi = \frac{2\pi d}{\lambda} \cos\phi + \delta = d_r \cos\phi + \delta \tag{2}$$

where δ is the phase difference of adjacent sources, i.e., source 2 with respect to 1, 3 with respect to 2, etc.

The amplitudes of the fields from the sources are all equal and taken as unity. Source 1 (Fig. 4-18) is the phase reference. Thus, at a distant point in the direction ϕ the field from source 2 is advanced in phase with respect to source 1 by ψ, the field from source 3 is advanced in phase with respect to source 1 by 2ψ, etc.

Equation (1) is a geometric series. Each term represents a phasor, and the amplitude of the total field E and its phase angle ξ can be obtained by phasor (vector) addition as in Fig. 4-19. Analytically, E can be expressed in a simple trigonometric form which we now develop as follows:
Multiply (1) by $e^{j\psi}$, giving

$$Ee^{j\psi} = e^{j\psi} + e^{j2\psi} + e^{j3\psi} + \cdots + e^{jn\psi} \tag{3}$$

Now subtract (3) from (1) and divide by $1 - e^{j\psi}$, yielding

$$E = \frac{1 - e^{jn\psi}}{1 - e^{j\psi}} \tag{4}$$

Equation (4) may be rewritten as

$$E = \frac{e^{jn\psi/2}}{e^{j\psi/2}} \left(\frac{e^{jn\psi/2} - e^{-jn\psi/2}}{e^{j\psi/2} - e^{-j\psi/2}} \right) \tag{5}$$

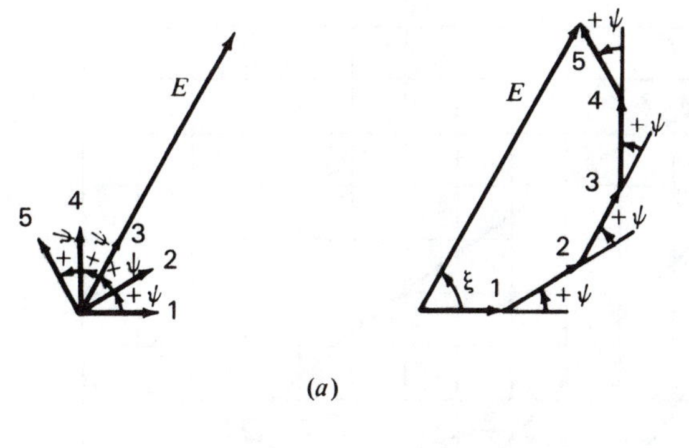

(a)

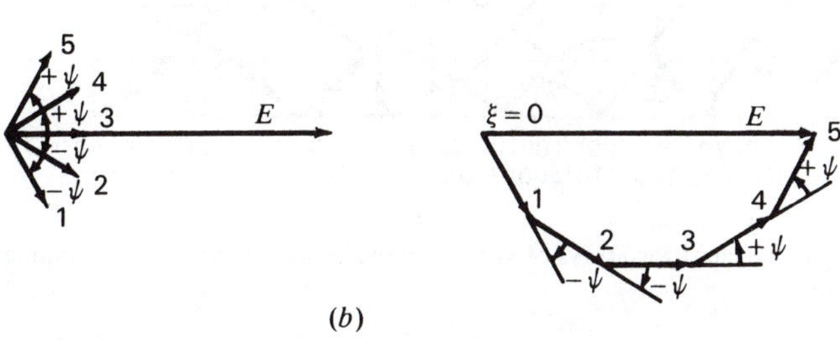

(b)

Figure 4-19 (a) Vector addition of fields at a large distance from the linear array of 5 isotropic point sources of equal amplitude with source 1 as the phase center (reference for phase). (b) Same, but with midpoint of array (source 3) as phase center.

from which

$$E = e^{j\xi} \frac{\sin (n\psi/2)}{\sin (\psi/2)} = \frac{\sin (n\psi/2)}{\sin (\psi/2)} \angle \xi \tag{6}$$

where ξ is referred to the field from source 1. The value of ξ is given by

$$\xi = \frac{n-1}{2}\psi \tag{7}$$

If the phase is referred to the centerpoint of the array, (6) becomes

$$E = \frac{\sin (n\psi/2)}{\sin (\psi/2)} \tag{8}$$

In this case the phase pattern is a step function as given by the sign of (8). The phase of the field is constant wherever E has a value but changes sign when E goes through zero.

When $\psi = 0$, (6) or (8) is indeterminate so that for this case E must be obtained as the limit of (8) as ψ approaches zero. Thus, for $\psi = 0$ we have the

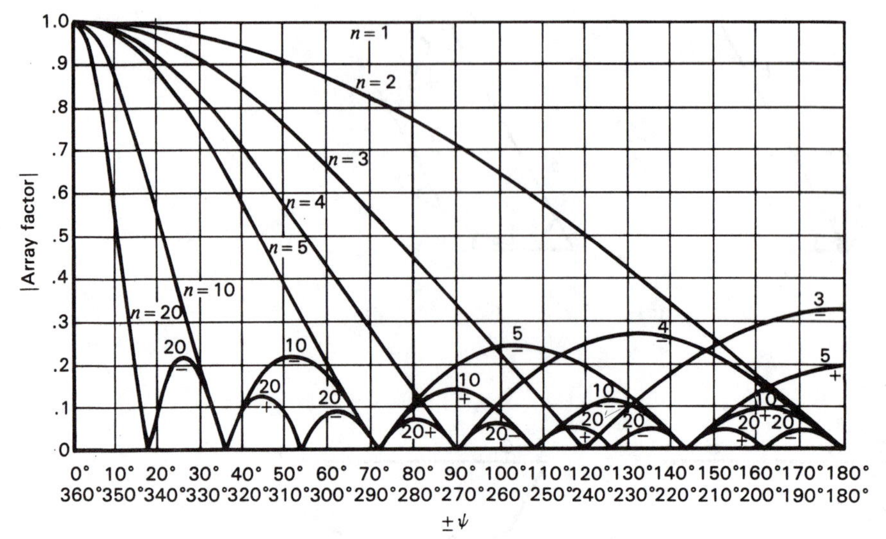

Figure 4-20 Universal field-pattern chart for arrays of various numbers n of isotropic point sources of equal amplitude and spacing.

relation that

$$E = n \tag{8a}$$

This is the maximum value that E can attain. Hence, the normalized value of the total field for $E_{max} = n$ is

$$E = \frac{1}{n} \frac{\sin (n\psi/2)}{\sin (\psi/2)} \tag{9}$$

The field as given by (9) will be referred to as the "array factor." Values of the array factor as obtained from (9) for various numbers of sources are presented in Fig. 4-20. If ψ is known as a function of ϕ, then the field pattern can be obtained directly from Fig. 4-20.

We may conclude from the above discussion that the field from the array will be a maximum in any direction ϕ for which $\psi = 0$. Stated in another way, the fields from the sources all arrive at a distant point in the same phase when $\psi = 0$. In special cases, ψ may not be zero for any value of ϕ, and in this case the field is usually a maximum at the minimum value of ψ.

To illustrate some of the properties of linear arrays (9) will now be applied to several special cases. See BASIC programs in App. B for calculating patterns involving these different cases. See also Probs. 4-35 and 4-40.

4-6b Case 1. Broadside Array (Sources in Phase). The first case is a linear array of n isotropic sources of the same amplitude and phase. Therefore, $\delta = 0$

and

$$\psi = d_r \cos \phi \tag{10}$$

To make $\psi = 0$ requires that $\phi = (2k + 1)(\pi/2)$, where $k = 0, 1, 2, 3, \ldots$. The field is, therefore, a maximum when

$$\phi = \frac{\pi}{2} \quad \text{and} \quad \frac{3\pi}{2} \tag{10a}$$

That is, the maximum field is in a direction normal to the array. Hence, this condition, which is characterized by in-phase sources ($\delta = 0$), results in a "broadside" type of array.

As an example, the pattern of a broadside array of four in-phase isotropic point sources of equal amplitude is shown in Fig. 4-21*a*. The spacing between sources is $\lambda/2$.[1] The field pattern in rectangular coordinates and the phase patterns for this array are presented in Fig. 4-21*b*.

4-6c Case 2. Ordinary End-Fire Array. Let us now find the phase angle between adjacent sources that is required to make the field a maximum in the direction of the array ($\phi = 0$). An array of this type may be called an "end-fire" array. For this we substitute the conditions $\psi = 0$ and $\phi = 0$ into (2), from which

$$\delta = -d_r \tag{11}$$

Hence, for an end-fire array, the phase between sources is retarded progressively by the same amount as the spacing between sources in radians. Thus, if the spacing is $\lambda/4$, source 2 in Fig. 4-18 should lag source 1 by 90°, source 3 should lag source 2 by 90°, etc.

As an example, the field pattern of an end-fire array of four isotropic point sources is presented in Fig. 4-22*a*. The spacing between sources is $\lambda/2$ and $\delta = -\pi$. The field pattern in rectangular coordinates and the phase patterns are shown in Fig. 4-22*b*. The same shape of field pattern is obtained in this case if $\delta = +\pi$ since, with $d = \lambda/2$, the pattern is bidirectional. However, if the spacing is less than $\lambda/2$, the maximum radiation is in the direction $\phi = 0$ when $\delta = -d_r$ and in the direction $\phi = 180°$ when $\delta = +d_r$.

4-6d Case 3. End-Fire Array with Increased Directivity. The situation discussed in Case 2, namely, for $\delta = -d_r$, produces a maximum field in the direction $\phi = 0$ but does not give the maximum directivity. It has been shown by

[1] If the spacing between elements exceeds λ, sidelobes appear which are equal in amplitude to the main (center) lobe. These are called *grating lobes* (see Sec. 11-26).

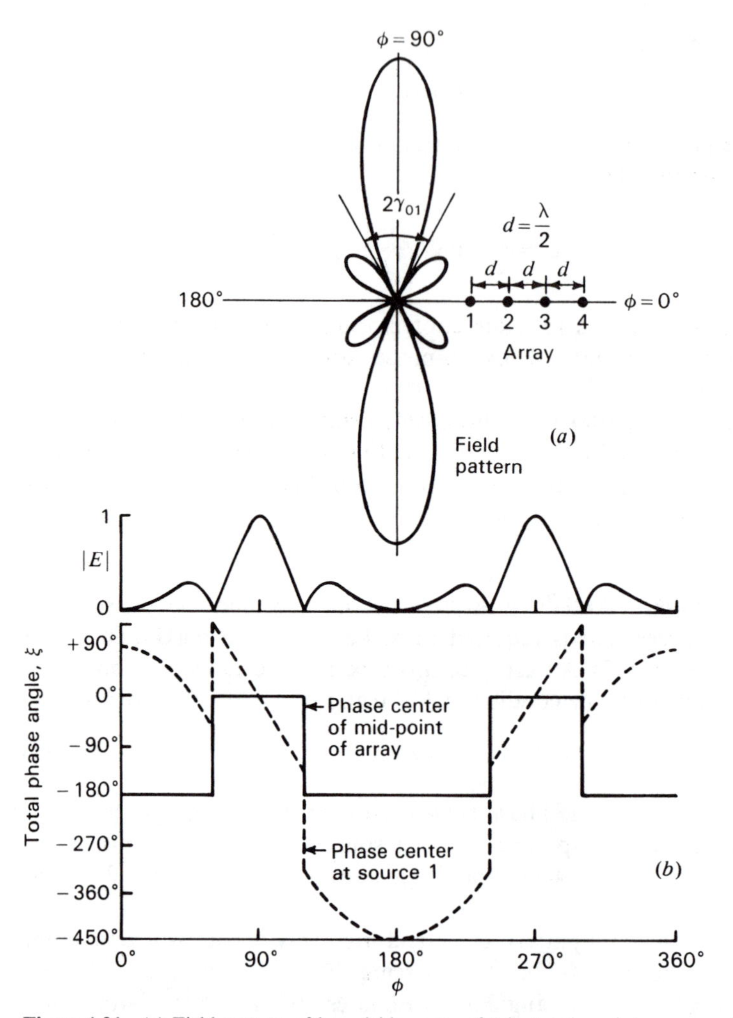

Figure 4-21 (*a*) Field pattern of broadside array of 4 isotropic point sources of the same amplitude and phase. The spacing between sources is $\lambda/2$. (*b*) Field pattern in rectangular coordinates and phase patterns of same array with phase center at midpoint and at source 1. The reference direction for phase is at $\phi = 90°$.

Hansen and Woodyard[1] that a larger directivity is obtained by increasing the phase change between sources so that

$$\delta = -\left(d_r + \frac{\pi}{n}\right) \tag{12}$$

This condition will be referred to as the condition for "increased directivity."

[1] W. W. Hansen and J. R. Woodyard, "A New Principle in Directional Antenna Design," *Proc. IRE*, **26**, 333–345, March 1938.

Thus for the phase difference of the fields at a large distance we have

$$\psi = d_r(\cos\phi - 1) - \frac{\pi}{n} \tag{13}$$

As an example, the field pattern of an end-fire array of four isotropic point sources for this case is illustrated in Fig. 4-23. The spacing between sources is $\lambda/2$, and therefore $\delta = -(5\pi/4)$. Hence, the conditions are the same as for the array with the pattern of Fig. 4-22, except that the phase difference between sources is increased by $\pi/4$. Comparing the field patterns of Figs. 4-22*a* and 4-23, it is apparent that the additional phase difference yields a considerably sharper main

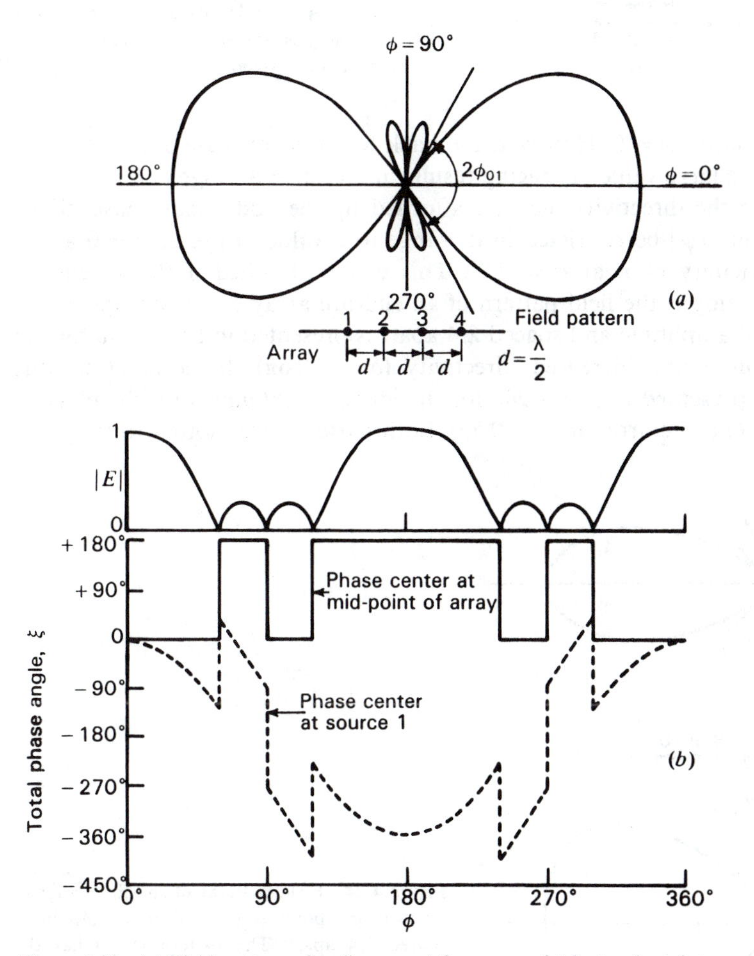

Figure 4-22 (*a*) Field pattern of ordinary end-fire array of 4 isotropic point sources of same amplitude. Spacing is $\lambda/2$ and the phase angle $\delta = -\pi$. (*b*) Field pattern in rectangular coordinates and phase patterns of same array with phase center at midpoint and at source 1. The reference direction for phase is at $\phi = 0$.

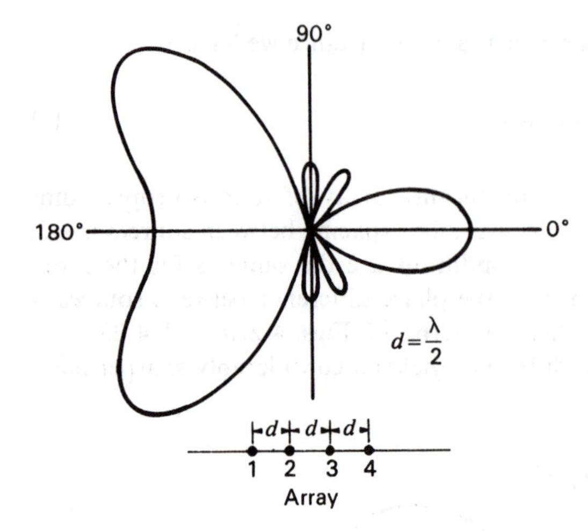

Figure 4-23 Field pattern of end-fire array of 4 isotropic point sources of equal amplitude spaced $\lambda/2$ apart. The phasing is adjusted for increased directivity ($\delta = -\frac{5}{4}\pi$).

lobe in the direction $\phi = 0$. However, the back lobes in this case are excessively large because the large value of spacing results in too great a range in ψ.

To realize the directivity increase afforded by the additional phase difference requires that $|\psi|$ be restricted in its range to a value of π/n at $\phi = 0$ and a value in the vicinity of π at $\phi = 180°$. This can be fulfilled if the spacing is reduced. For example, the field pattern of an end-fire array of 10 isotropic point sources of equal amplitude and spaced $\lambda/4$ apart is presented in Fig. 4-24*a* for the phase condition giving increased directivity ($\delta = -0.6\pi$). In contrast to this pattern, one is presented in Fig. 4-24*b* for the identical antenna with the phasing of an ordinary end-fire array ($\delta = -0.5\pi$). Both patterns are plotted to the same

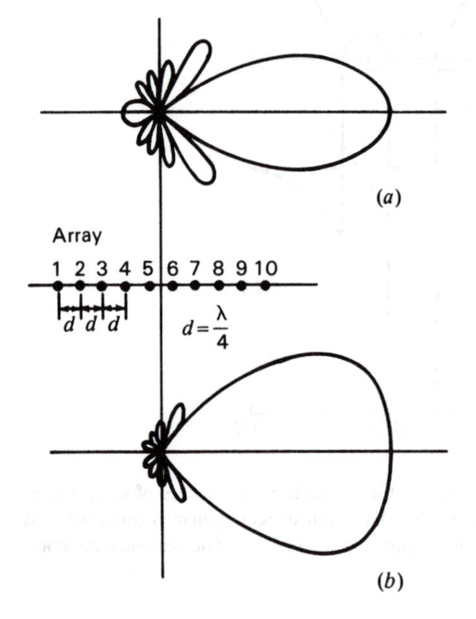

Figure 4-24 Field patterns of end-fire arrays of 10 isotropic point sources of equal amplitude spaced $\lambda/4$ apart. The pattern at (*a*) has the phase adjusted for increased directivity ($\delta = -0.6\pi$), while the pattern at (*b*) has the phasing of an ordinary end-fire array ($\delta = -0.5\pi$).

Table 4-1

	Ordinary end-fire array	End-fire array with increased directivity
Beam width between half-power points	69°	38°
Beam width between first nulls	106°	74°
Directivity	11	19

maximum. The increased directivity is apparent from the greater sharpness of the upper pattern. Integrating the pattern, including the minor lobes, the directivity of the upper pattern is found to be about 19 and of the lower pattern about 11. The beam widths and directivities for the two patterns are compared in Table 4-1.

The maximum of the field pattern of Fig. 4-24a occurs at $\phi = 0$ and $\psi = -\pi/n$. In general, any increased directivity end-fire array, with maximum at $\psi = -\pi/n$, has a normalized field pattern given by

$$E = \sin\left(\frac{\pi}{2n}\right)\frac{\sin(n\psi/2)}{\sin(\psi/2)} \tag{14}$$

4-6e Case 4. Array with Maximum Field in an Arbitrary Direction. Scanning Array. Let us consider the case of an array with a field pattern having a maximum in some arbitrary direction ϕ_1 not equal to $k\pi/2$ where $k = 0, 1, 2$ or 3. Then (2) becomes

$$0 = d_r \cos\phi_1 + \delta \tag{15}$$

By specifying the spacing d_r, the required phase difference δ is then determined by (15). Conversely, by changing δ the beam direction ϕ_1 can be shifted or scanned.

As an example, suppose that $n = 4$, $d = \lambda/2$ and that we wish to have a maximum field in the direction of $\phi = 60°$. Then $\delta = -\pi/2$, yielding the field pattern shown in Fig. 4-25.

4-7 NULL DIRECTIONS FOR ARRAYS OF n ISOTROPIC POINT SOURCES OF EQUAL AMPLITUDE AND SPACING.

In this section simple methods are discussed for finding the directions of the pattern nulls of the arrays considered in Sec. 4-6.

Following the procedure given by Schelkunoff,[1] the null directions for an array of n isotropic point sources of equal amplitude and spacing occur when

[1] S. A. Schelkunoff, *Electromagnetic Waves*, Van Nostrand, New York, 1943, p. 343.
S. A. Schelkunoff, "A Mathematical Theory of Arrays," *Bell System Tech. J.*, **22**, 80–107, January 1943.

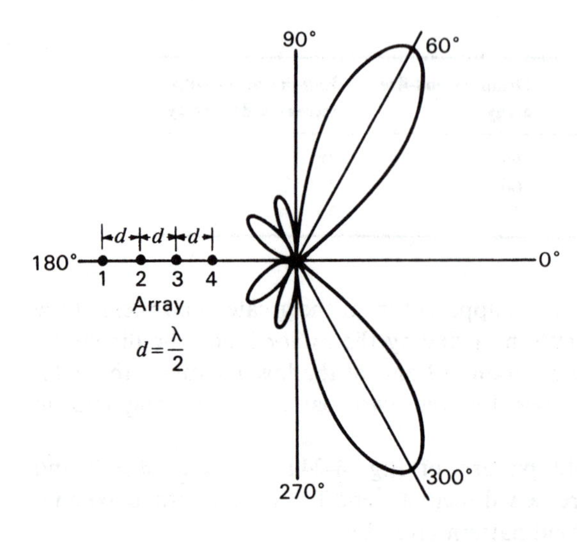

Figure 4-25 Field pattern of array of 4 isotropic point sources of equal amplitude with phasing adjusted to give the maximum at $\phi = 60°$. The spacing is $\lambda/2$.

$E = 0$ or, provided that the denominator of (4-6-4) is not zero, when

$$e^{jn\psi} = 1 \tag{1}$$

Equation (1) requires that

$$n\psi = \pm 2K\pi \tag{2}$$

where $K = 1, 2, 3, \ldots$

Equating the value of ψ in (2) to its value in (4-6-2) gives

$$\psi = d_r \cos \phi_0 + \delta = \pm \frac{2K\pi}{n} \tag{3}$$

Thus,

$$\phi_0 = \arccos \left[\left(\pm \frac{2K\pi}{n} - \delta \right) \frac{1}{d_r} \right] \tag{4}$$

where ϕ_0 gives the direction of the pattern nulls. Note that values of K must be excluded for which $K = mn$, where $m = 1, 2, 3, \ldots$. Thus, if $K = mn$, (2) reduces to $\psi = \pm 2m\pi$ and the denominator of (4-6-4) equals zero so that the null condition of (1), that the numerator of (4-6-4) be zero, is insufficient.

In a broadside array $\delta = 0$, so that for this case (4) becomes

$$\phi_0 = \arccos \left(\pm \frac{2K\pi}{nd_r} \right) = \arccos \left(\pm \frac{K\lambda}{nd} \right) \tag{5}$$

As an example, the field pattern of Fig. 4-21 ($n = 4$, $d = \lambda/2$, $\delta = 0$) has the null directions

$$\phi_0 = \arccos \left(\pm \frac{K}{2} \right) \tag{6}$$

For $K = 1$, $\phi_0 = \pm 60°$ and $\pm 120°$, and for $K = 2$, $\phi_0 = 0°$ and $180°$. These are the six null directions for this array.

If ϕ_0 in (3) is replaced by its complementary angle γ_0 (see Fig. 4-18), then (5) becomes

$$\gamma_0 = \arcsin\left(\pm\frac{K\lambda}{nd}\right) \tag{7}$$

If the array is long, so that $nd \gg K\lambda$,

$$\gamma_0 \simeq \pm\frac{K\lambda}{nd} \tag{8}$$

The first nulls either side of the maximum occur for $K = 1$. These angles will be designated γ_{01}. Thus,

$$\gamma_{01} \simeq \pm\frac{\lambda}{nd} \tag{9}$$

and the total beam width of the main lobe between first nulls *for a long broadside array* is then

$$2\gamma_{01} \simeq \frac{2\lambda}{nd} \tag{10}$$

For the field pattern in Fig. 4-21 this width is exactly 60°, while as given by (10) it is 1 rad, or 57.3°. This pattern is for an array 2λ long. The agreement would be better with longer arrays.

Turning next to *end-fire arrays*, the condition for an ordinary end-fire array is that $\delta = -d_r$. Thus, for this case (3) becomes

$$\cos\phi_0 - 1 = \pm\frac{2K\pi}{nd_r} \tag{11}$$

from which we obtain

$$\frac{\phi_0}{2} = \arcsin\left(\pm\sqrt{\frac{K\pi}{nd_r}}\right) \tag{12}$$

or

$$\phi_0 = 2\arcsin\left(\pm\sqrt{\frac{K\lambda}{2nd}}\right) \tag{13}$$

As an example, the field pattern of Fig. 4-22 ($n = 4$, $d = \lambda/2$, $\delta = -\pi$) has the null directions

$$\phi_0 = 2\arcsin\left(\pm\sqrt{\frac{K}{4}}\right) \tag{14}$$

For $K = 1$, $\phi_0 = \pm 60°$; for $K = 2$, $\phi_0 = \pm 90°$, etc.

If the array is long, so that $nd \gg K\lambda$, (13) becomes

$$\phi_0 \simeq \pm\sqrt{\frac{2K\lambda}{nd}} \tag{15}$$

The first nulls either side of the main lobe occur for $K = 1$. These angles will be designated ϕ_{01}. Thus,

$$\phi_{01} \simeq \pm\sqrt{\frac{2\lambda}{nd}} \tag{16}$$

and the total beam width of the main lobe between first nulls *for a long ordinary end-fire array* is then

$$2\phi_{01} \simeq 2\sqrt{\frac{2\lambda}{nd}} \tag{17}$$

For the field pattern in Fig. 4-22 this width is exactly 120°, while as given by (17) it is 2 rad, or 115°.

For end-fire arrays with increased directivity as proposed by Hansen and Woodyard, the condition is that $\delta = -(d_r + \pi/n)$. Thus, for this case (3) becomes

$$d_r(\cos\phi_0 - 1) - \frac{\pi}{n} = \pm 2\frac{K\pi}{n} \tag{18}$$

from which

$$\frac{\phi_0}{2} = \arcsin\left[\pm\sqrt{\frac{\pi}{2nd_r}(2K-1)}\right] \tag{19}$$

or

$$\phi_0 = 2\arcsin\left[\pm\sqrt{\frac{\lambda}{4nd}(2K-1)}\right] \tag{20}$$

If the array is long, so that $nd \gg K\lambda$, (20) becomes

$$\phi_0 \simeq \pm\sqrt{\frac{\lambda}{nd}(2K-1)} \tag{21}$$

The first nulls either side of the main lobe, ϕ_{01}, occur for $K = 1$. Thus,

$$\phi_{01} \simeq \pm\sqrt{\frac{\lambda}{nd}} \tag{22}$$

and the total beam width of the main lobe between first nulls *for a long end-fire array with increased directivity* is then

$$2\phi_{01} \simeq 2\sqrt{\frac{\lambda}{nd}} \tag{23}$$

This width is $1/\sqrt{2}$, or 71 percent, of the width of the ordinary end-fire array. As an example, the ordinary end-fire array pattern of Fig. 4-24*b* has a beam width between first nulls of 106°. The width of the pattern in Fig. 4-24*a* for the array with increased directivity is 74°, or 70 percent as much.

Table 4-2 lists the formulas for null directions and beam widths for the different arrays considered above. The null directions in column 2 apply to arrays of any length. The formulas in the third and fourth columns are approximate and apply only to long arrays.

The formulas in Table 4-2 have been used to calculate the curves presented in Fig. 4-26. These curves show the beam width between first nulls as a function of nd_λ for three types of arrays: broadside, ordinary end-fire and end-fire with increased directivity. The quantity nd_λ $(=nd/\lambda)$ is approximately equal to the length of the array in wavelengths for long arrays. The exact value of the array length is $(n-1)d_\lambda$.

The beam width of long broadside arrays is inversely proportional to the array length, whereas the beam width of long end-fire types is inversely proportional to the square root of the array length. Hence, the beam width in the plane of a long linear broadside array is much smaller than for end-fire types of the same length as shown by Fig. 4-26. It should be noted, however, that the broadside array has a disc-shaped pattern with a narrow beam width in a plane

Table 4-2 Null directions and beam widths between first nulls for linear arrays of n isotropic point sources of equal amplitude and spacing.

(For $n \geq 2$. The angles in columns 3 and 4 are expressed in radians. To convert to degrees, multiply by 57.3.)

Type of array	Null directions (array any length)	Null directions (long array)	Beam width between first nulls (long array)
General case	$\phi_0 = \arccos\left[\left(\frac{\pm 2K\pi}{n} - \delta\right)\frac{1}{d_r}\right]$		
Broadside	$\gamma_0 = \arcsin\left(\pm\frac{K\lambda}{nd}\right)$	$\gamma_0 \simeq \pm\frac{K\lambda}{nd}$	$2\gamma_{01} \simeq \frac{2\lambda}{nd}$
Ordinary end-fire	$\phi_0 = 2\arcsin\left(\pm\sqrt{\frac{K\lambda}{2nd}}\right)$	$\phi_0 \simeq \pm\sqrt{\frac{2K\lambda}{nd}}$	$2\phi_{01} \simeq 2\sqrt{\frac{2\lambda}{nd}}$
End-fire with increased directivity (Hansen and Woodyard)	$\phi_0 = 2\arcsin\left[\pm\sqrt{\frac{\lambda}{4nd}(2K-1)}\right]$	$\phi_0 \simeq \pm\sqrt{\frac{\lambda}{nd}(2K-1)}$	$2\phi_{01} \simeq 2\sqrt{\frac{\lambda}{nd}}$

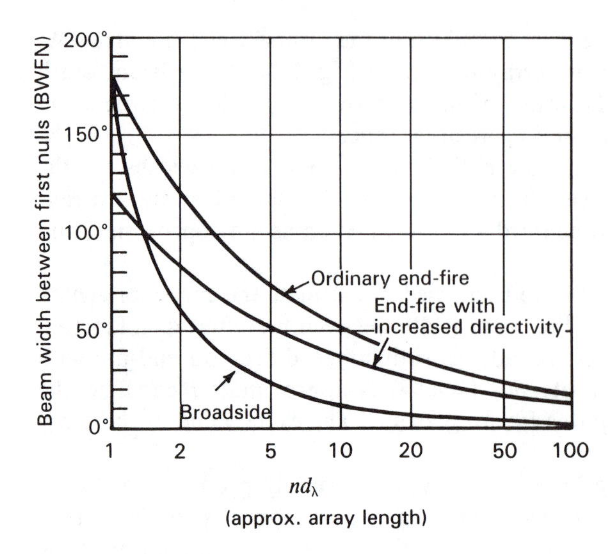

Figure 4-26 Beam width between first nulls as a function of nd_λ for arrays of n isotropic point sources of equal amplitude. For long arrays, nd_λ is approximately equal to the array length.

through the array axis but a circular pattern (360° beam width) in the plane normal to the array axis. On the other hand, the end-fire array has a cigar-shaped pattern with the same beam width in all planes through the array axis.

4-8 BROADSIDE VERSUS END-FIRE ARRAYS. TURNS VERSUS DIPOLES AND 3-DIMENSIONAL ARRAYS. Assuming that the half-power beam width (HPBW) is $\frac{1}{2}$ the beam width between first nulls (BWFN) (an approximation), we have from (2-9-4) and (4-7-10) that the directivity of a *linear broadside array* is given approximately by

$$D \simeq \frac{4\pi}{\theta_{\mathrm{HP}}\,\phi_{\mathrm{HP}}} = \frac{4\pi nd}{2\pi\lambda} = 2L_\lambda \tag{1}$$

where n = number of sources
d = spacing between sources, m
λ = wavelength, m
$L_\lambda = nd/\lambda = L/\lambda$ = length of array in wavelengths, dimensionless

The pattern is disc-shaped so $\theta = 360°$, or 2π radians. It is assumed that $\phi_{\mathrm{HP}} = \mathrm{BWFN}/2 = \lambda/nd$; also that $L_\lambda \gg 1$ so that $L \simeq nd$.

From (4-7-17), the directivity of an *ordinary end-fire array* is approximately

$$D \simeq 4\pi \frac{nd}{2\lambda} = 2\pi L_\lambda \tag{2}$$

while from (4-7-23) the directivity of an *end-fire array with increased directivity* is approximately

$$D \simeq 4\pi \frac{nd}{\lambda} = 4\pi L_\lambda \tag{3}$$

A general expression for directivity, as given by (2-22-9), is

$$D = \frac{4\pi A_e}{\lambda^2} \tag{4}$$

where A_e = effective aperture, m^2

For a *square unidirectional broadside array or aperture* with 100 percent aperture efficiency (uniform aperture distribution) we have

$$A_e = A_p = L^2 \qquad (\mathrm{m}^2)$$

and

$$D = 4\pi L_\lambda^2 \tag{5}$$

For 100 percent aperture efficiency (uniform aperture distribution) the directivity of a *circular unidirectional broadside array or aperture* is

$$D = \pi^2 d_\lambda^2 \tag{6}$$

where $d_\lambda = d/\lambda$ = diameter of array or aperture in wavelengths, dimensionless

These directivity relations are summarized in Table 4-3, which also gives half-power beam widths and numerical directivities for array lengths (or diameters) of 1 to 1000 as measured in wavelengths.

For array dimensions of the order of λ, the increased directivity end-fire array and square broadside array have comparable directivities. We note, however, that the end-fire directivity is proportional to the length L_λ while the directivity of the broadside square array is proportional to the *square* of the side length L_λ. Hence, for a high directivity an end-fire array must have a much greater dimension than a square broadside array. For example, a square array with 1000λ on a side has a directivity (see Table 4-3) of 12.6 million. To equal this directivity an increased-directivity end-fire array must be 1 million λ long and an ordinary end-fire array 2 million λ long. Even if all the sources or elements of such a long end-fire array could be fed with equal amplitude and in proper phase, the great length of the array is a severe disadvantage. Thus, it is apparent why broadside arrays or apertures are invariably used for high-directivity (high-gain) applications. The broadside aperture may consist of an array of $\lambda/2$ dipoles or it may be the aperture of a parabolic reflector antenna with single feed point or, as discussed next, a broadside array of intermediate-length end-fire antennas.

With uniform amplitude of all $\lambda/2$ dipole elements across a large aperture (and dipole spacing no more than $\lambda/2$), there is nothing to be gained by replacing each $\lambda/2$ dipole by a more directional end-fire antenna. However, if each end-fire

Table 4-3 Directivities and beamwidths of arrays and apertures†

Array (or aperture)‡	Directivity formula	Directivity for L_λ or d_λ equal to 1	10	100	1000	Half-power beam widths
Linear broadside array of length L_λ	$2L_\lambda$	2	20	200	2000	$\frac{50.8°}{L_\lambda} \times 360°$
Ordinary end-fire array of length L_λ	$2\pi L_\lambda$	6.3	63	630	6300	$\frac{108°}{\sqrt{L_\lambda}}$
Increased-directivity end-fire array of length L_λ	$4\pi L_\lambda$	12.6	126	1 260	12 600	$\frac{52°}{\sqrt{L_\lambda}}$
Square broadside aperture with side length L_λ	$4\pi L_\lambda^2$	12.6	1260	126 000	1.26×10^7	$\frac{50.8°}{L_\lambda} \times \frac{50.8°}{L_\lambda}$
Circular broadside aperture with diameter d_λ	$\pi^2 d_\lambda^2$	9.9	990	99 000	9.9×10^6	$\frac{58°}{d_\lambda}$
Flat array (length L_λ) of ordinary end-fire antennas (length $L_{\lambda E}$)§	$\pi L_\lambda \sqrt{8L_{\lambda E}}$					
Same but square ($L_{\lambda E} = L_\lambda$)§	$\pi\sqrt{8L_\lambda^3}$	8.9	281	8900	281 000	
Flat array (length L_λ) of increased-directivity end-fire antennas (length $L_{\lambda E'}$)§	$4\pi L_\lambda \sqrt{L_{\lambda E'}}$					
Same but square ($L_{\lambda E'} = L_\lambda$)§	$4\pi\sqrt{L_\lambda^3}$	12.6	398	12 600	398 000	

† The directivities for the arrays (broadside and end-fire) are approximate while for the apertures (square and circular) the directivities are exact. Note that if the directivity of the square or circular aperture is calculated using the approximation $41\,000/\theta_{HP}^{\circ}\phi_{HP}^{\circ}$ the result is larger than the directivity given in the table. See discussion regarding this approximation in connection with (3-13-16) and (3-13-18).

Arrays or apertures are assumed to be large compared to the wavelength and to have uniform amplitude distribution. Specifically, the square and circular apertures are assumed to have 100 percent aperture efficiency. See text for other assumptions involved.

The directivities for the square and circular apertures are identical in form to the more general exact relation given by (4), which applies to an aperture of any shape.

‡ L_λ = array length, wavelengths
d_λ = array diameter, wavelengths
$L_{\lambda E}$ = ordinary end-fire array length, wavelengths
$L_{\lambda E'}$ = increased-directivity end-fire array length, wavelengths

§ Beam off edge, *not* perpendicular to flat side.

antenna is used to *replace several* $\lambda/2$ dipoles then some benefits may accrue as described in the following example involving a broadside array of end-fire antennas. An array of this type may be called a *volume or 3-dimensional array.*

Figure 4-27*a* is an edge view of a 12 × 12 broadside array of 144 point sources with $\lambda/2$ spacing. The figure shows 12 elements as seen from one edge. Let each source consist of a $\lambda/2$ dipole element as suggested in partial broadside view in Fig. 4-27*e*. The array of Fig. 4-27*a* is bidirectional but if backed at $\lambda/4$ spacing by an identical array fed with equal amplitude currents and 90° phase difference, each pair of dipoles has a unidirectional pattern (as in Fig. 4-5) and

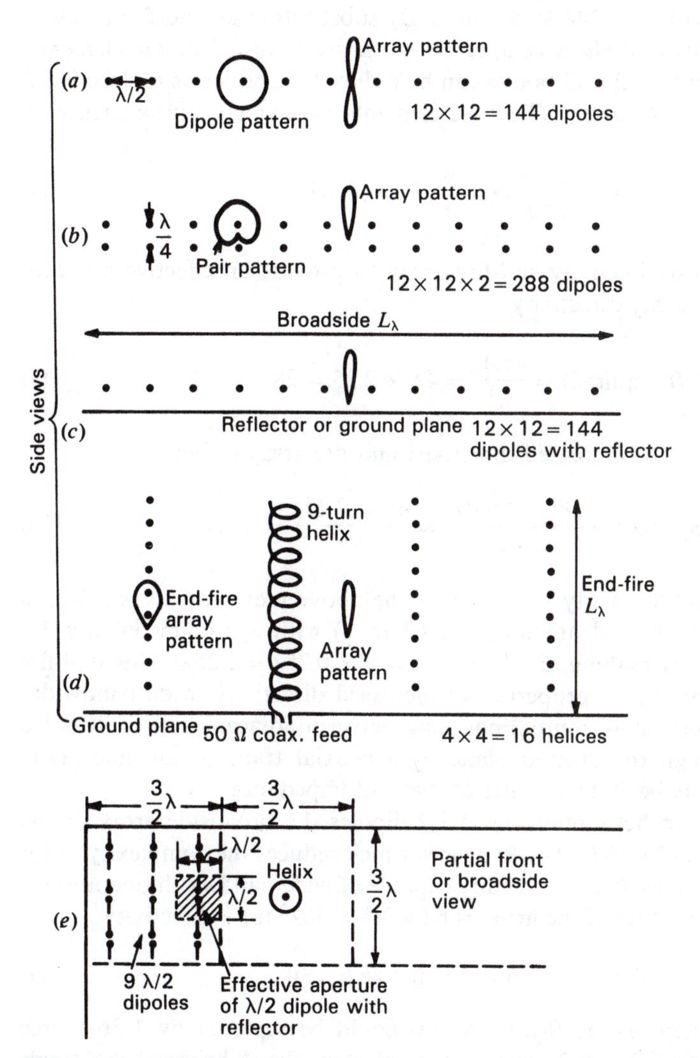

Figure 4-27 Equivalence of broadside array of $\lambda/2$ dipoles and 3-dimensional broadside-end-fire array. Parts *a*, *b*, *c* and *d* are edge or side views while *e* is a front or broadside view.

the entire $12 \times 12 \times 2$ array is unidirectional as suggested in Fig. 4-27*b*. An alternative, simpler arrangement is to replace the second array with a conducting flat-sheet reflector or *ground plane* at an appropriate spacing, as in Fig. 4-27*c*.

A $\lambda/2$ dipole has a directivity $D = 1.64$, and taking the directivity of dipole with reflector as twice this value or 3.28 ($= 2 \times 1.64$), the equivalent effective aperture is

$$A_e = \frac{D\lambda^2}{4\pi} = \frac{3.28}{4\pi}\lambda^2 = 0.26\lambda^2 \tag{7}$$

or approximately $\lambda^2/4$, as suggested in Fig. 4-27*e*.

The 12×12 array has 144 feed points. By substituting an end-fire array of appropriate directivity and effective aperture for a group of $\lambda/2$ dipole elements, the number of elements and feed points can be reduced. Thus, let us replace 9 $\lambda/2$ dipoles by a single end-fire array. The 9 $\lambda/2$ elements have an effective aperture of

$$A_e \simeq \left(\frac{3}{2}\right)^2 \lambda^2 = \frac{9}{4}\lambda^2 = 2.25\lambda^2 \tag{8}$$

For an increased-directivity end-fire array to provide an effective aperture of $2.25\lambda^2$ requires an array directivity

$$D(\text{required}) = \frac{4\pi A_e}{\lambda^2} = 4\pi \times 2.25 = 28.3 \tag{9}$$

The required length of the increased-directivity end-fire array is then

$$L(\text{required}) = \frac{D(\text{required})\lambda}{4\pi} = \frac{4\pi \times 2.25\lambda}{4\pi} = 2.25\lambda \tag{10}$$

An effective end-fire array which meets the above requirements is a 9-turn monofilar axial-mode helical antenna (see Chap. 7) with λ/π diameter and $\lambda/4$ spacing between turns making the length $L = 9 \times 0.25\lambda = 2.25\lambda$. This end-fire antenna has the remarkable properties of increased directivity, wide bandwidth (over 2 to 1) and very small mutual coupling between adjacent helices. It is fed from one end through the ground plane by a coaxial transmission line (as in Fig. 4-27*d*) which may be 50 Ω or other convenient impedance.

With each 9-turn helix replacing 9 $\lambda/2$ dipoles the broadside array of 144 $\lambda/2$ dipoles is replaced by 16 ($= 144/9$) helices which reduces the complexity of the feed system and provides an array readily capable of wide bandwidth operation.

The effective aperture of the array is $6\lambda \times 6\lambda = 36\lambda^2$ for a directivity

$$D = 4\pi \times 36 = 452 \text{ (or 26.6 dBi)} \tag{11}$$

As a further step, the 16 9-turn helices could be replaced by 4 36-source end-fire arrays (36-turn helices), and as a final step, the 4 helices by a single 144-source end-fire array (144-turn helix) 36λ long. Such a long helix is not practical but, even if it were, its great length is a disadvantage as compared to a more compact array with a number of shorter helices.

In the above example each turn of a helix (as an end-fire source) has a directivity equal to a $\lambda/2$ dipole with reflector (as a broadside source), but polarizations differ (helix circularly polarized and dipole linearly polarized). In summary, we considered the following cases:

1. 12×12 array of reflector-backed $\lambda/2$ dipoles: $12 \times 12 = 144$ sources
2. 4×4 array of 9-turn helical end-fire antennas: $4 \times 4 \times 9 = 144$ sources
3. 2×2 array of 36-turn helical end-fire antennas: $2 \times 2 \times 36 = 144$ sources
4. Single 144-turn helical end-fire antenna: $1 \times 144 = 144$ sources

Thus, for constant directivity and effective aperture, the *number of sources is a constant* whether the sources are arranged in a flat broadside array or in 3-dimensional broadside-end-fire configurations, all sources having uniform amplitude. Although the constant directivity and effective aperture in the above example may not apply fully for all 3-dimensional arrays, the example illustrates the principle that for a given number of sources, various configurations may produce (ideally) similar, if not identical, directivities and effective apertures.

Another broadside-end-fire combination is a *flat* (*or planar*) *array* consisting of a linear array with only a single row of end-fire antennas (helices) as in Fig. 4-27*d* (no other arrays stacked perpendicular to the page). The beam width in one plane is determined by the broadside length L_λ and the beam width in the other plane by the end-fire length $L_{\lambda E}$ (for ordinary end-fire) or length $L_{\lambda E'}$ (for increased directivity).

The directivity for the ordinary end-fire case is

$$D(\text{ordinary}) = \pi L_\lambda \sqrt{8 L_{\lambda E}} \tag{12}$$

and if $L_\lambda = L_{\lambda E}$ (square flat array),

$$D(\text{ordinary}) = \pi \sqrt{8 L_\lambda^3} \tag{13}$$

The directivity for the increased directivity case is

$$D(\text{increased directivity}) = 4\pi L_\lambda \sqrt{L_{\lambda E'}} \tag{14}$$

and if $L_\lambda = L_{\lambda E'}$ (square flat array),

$$D(\text{increased directivity}) = 4\pi \sqrt{L_\lambda^3} \tag{15}$$

These relations are summarized in Table 4-3.

An early application of end-fire antennas in a large 3-dimensional array, which I designed and built in 1951, is shown in Fig. 7-4. It has 96 helices of 11 turns each in a 4×24 configuration. Equivalent to an array of 1056 ($= 11 \times 96$)$\lambda/2$ dipoles with reflectors, it has a wide bandwidth (over 2 to 1) and is simple to feed.

4-9 DIRECTIONS OF MAXIMA FOR ARRAYS OF n ISOTROPIC POINT SOURCES OF EQUAL AMPLITUDE AND SPACING. Let us now proceed to a discussion of the methods for locating the positions of the pattern maxima. The major-lobe maximum usually occurs when $\psi = 0$. This is the case for the broadside or ordinary end-fire array. The main lobes of the broadside array are then at $\phi = 90°$ and $270°$, while for the ordinary end-fire array the main lobe is at $0°$ or $180°$ or both. For the end-fire array with increased directivity the main-lobe maximum occurs at a value of $\psi = \pm\pi/n$ with the main lobe at $0°$ or $180°$. Referring to Fig. 4-24*a*, the main-lobe maximum (first maximum) for this case occurs at the first maximum of the numerator of (4-6-8).

The maxima of the minor lobes are situated between the first- and higher-order nulls. It has been pointed out by Schelkunoff that these maxima occur approximately whenever the numerator of (4-6-8) is a maximum, i.e., when

$$\sin \frac{n\psi}{2} = 1 \tag{1}$$

Referring to Fig. 4-28, we note that the numerator of (4-6-8) varies as a function of ψ more rapidly than the denominator $\sin(\psi/2)$. This is especially true when n is large. Thus, although the nulls occur exactly where $\sin(n\psi/2) = 0$, the maxima occur approximately where $\sin(n\psi/2) = 1$. This condition requires that

$$\frac{n\psi}{2} = \pm(2K + 1)\frac{\pi}{2} \tag{2}$$

where $K = 1, 2, 3, \ldots$

Substituting the value of ψ from (2) into (4-6-2) gives

$$d_r \cos \phi_m + \delta = \frac{\pm(2K + 1)\pi}{n} \tag{3}$$

Therefore

$$\phi_m \simeq \arccos \left\{\left[\frac{\pm(2K + 1)\pi}{n} - \delta\right]\frac{1}{d_r}\right\} \tag{4}$$

where ϕ_m = direction of the minor-lobe maxima

For a *broadside array*, $\delta = 0$ so that (4) becomes

$$\phi_m \simeq \arccos \frac{\pm(2K + 1)\lambda}{2nd} \tag{5}$$

As an example, the field pattern of Fig. 4-21 ($n = 4$, $d = \lambda/2$, $\delta = 0$) has the minor-lobe maxima at

$$\phi_m \simeq \arccos \frac{\pm(2K + 1)}{4} \tag{6}$$

For $K = 1$, $\phi_m = \pm 41.4°$ and $\pm 138.6°$. These are the approximate directions for the maxima of the four minor lobes of this pattern.

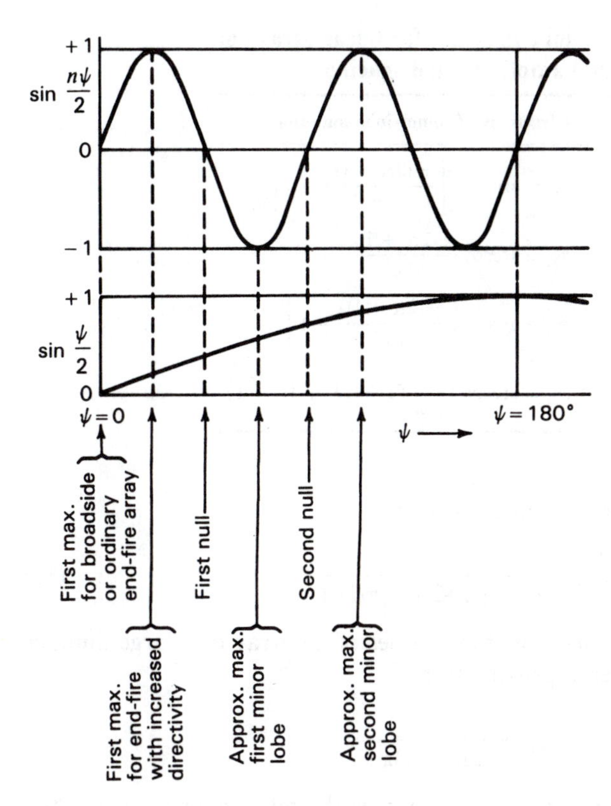

Figure 4-28 Graphs of the numerator ($\sin n\psi/2$) and denominator ($\sin \psi/2$) of the array factor as functions of ψ, showing the values of ψ corresponding to maxima and nulls of a field pattern for the case $n = 8$.

For an *ordinary end-fire array*, $\delta = -d_r$ so that (4) becomes

$$\phi_m \simeq \arccos \left[\frac{\pm(2K+1)\lambda}{2nd} + 1 \right] \tag{7}$$

while, for an *end-fire array with increased directivity*, $\delta = -(d_r + \pi/n)$ and

$$\phi_m \simeq \arccos \left\{ \frac{\lambda}{2nd} [1 \pm (2K+1)] + 1 \right\} \tag{8}$$

The above formulas for the approximate location of the minor-lobe maxima are listed in Table 4-4 ($K = 1$ for first minor lobe, $K = 2$ for second minor lobe, etc.).

The amplitudes of the field at the minor-lobe maxima are also of interest. It has been shown by Schelkunoff that since the numerator of (4-6-9) is approximately unity at the maximum of a minor lobe, the relative amplitude of a minor-lobe maximum E_{ML} is given by

$$E_{ML} \simeq \frac{1}{n \sin (\psi/2)} \tag{9}$$

Table 4-4 Directions of minor-lobe maxima for linear arrays of n isotropic point sources of equal amplitude and spacing

Type of array	Directions of minor-lobe maxima
General	$\phi_m \simeq \arccos \left\{ \left[\frac{\pm(2K+1)\pi}{n} - \delta \right] \frac{1}{d_r} \right\}$
Broadside	$\phi_m \simeq \arccos \frac{\pm(2K+1)\lambda}{2nd}$
Ordinary end-fire	$\phi_m \simeq \arccos \left[\frac{\pm(2K+1)\lambda}{2nd} + 1 \right]$
End-fire with increased directivity (Hansen and Woodyard)	$\phi_m \simeq \arccos \left\{ \frac{\lambda}{2nd} [1 \pm (2K+1)] + 1 \right\}$

Introducing the value of ψ from (2) into (9) yields

$$E_{\text{ML}} \simeq \frac{1}{n \sin [(2K+1)\pi/2n]} \tag{10}$$

When $n \gg K$, that is, for the first few minor lobes of an array of a large number of sources, we have the further approximation

$$E_{\text{ML}} \simeq \frac{2}{(2K+1)\pi} \tag{11}$$

Thus, for arrays of a large number of sources the relative amplitude of the first few minor lobes is given by (11) for $K = 1, 2, 3$, etc. In a broadside or ordinary end-fire array, the major-lobe maximum is unity so that the relative amplitudes of the maximum and first five minor lobes for arrays of these types and many sources are 1, 0.21, 0.13, 0.09, 0.07 and 0.06. From the curve for $n = 20$ in Fig. 4-20 we have the corresponding relative amplitudes given by 1, 0.22, 0.13, 0.09, 0.07 and 0.06. For an end-fire array with increased directivity the maximum for $\phi = 0$ and $n = 20$ occurs at $\psi = \pi/20 = 9°$. At this value of ψ the array factor is 0.63. Putting the maximum equal to unity then makes the relative amplitudes 1, 0.35, 0.21, 0.14, 0.11 and 0.09. It is interesting to note in (10) that the maximum amplitude of the smallest minor lobe occurs for $2K + 1 = n$. Then

$$\sin \left[\frac{(2K+1)\pi}{2n} \right] = 1 \tag{12}$$

and

$$E_{\text{ML}} \simeq \frac{1}{n} \tag{13}$$

The condition $2K + 1 = n$ is exactly fulfilled when n is odd for the minor-lobe maximum at $\psi = 180°$ (see Fig. 4-20). When n is even, the condition is approximately fulfilled by the minor lobes nearest $\psi = 180°$. Thus, the maximum amplitude of the smallest minor lobe of the field pattern of any array of n isotropic point sources of equal amplitude and spacing will never be less than $1/n$ of the major-lobe maximum. An exception to this is where the range of ψ ends after a null in the array factor has been passed but before the next maximum has been reached. In this case the maximum of the smallest minor lobe may be arbitrarily small.

4-10 LINEAR BROADSIDE ARRAYS WITH NONUNIFORM AMPLITUDE DISTRIBUTIONS. GENERAL CONSIDERATIONS.

In the preceding section, our discussion was limited to linear arrays of n isotropic sources of *equal* amplitude. This discussion will now be extended to the more general case where the amplitude distribution may be nonuniform. In introducing this subject, it is instructive to compare field patterns of four types of amplitude distributions, namely, uniform, binomial, edge and optimum. To be specific, let us consider a linear array of five isotropic point sources with $\lambda/2$ spacing. If the sources are in phase and all equal in amplitude, we may calculate the pattern as discussed in Sec. 4-6, the result being as shown in Fig. 4-29 by the pattern designated *uniform*. A uniform distribution yields the maximum directivity. The pattern has a half-power beam width of 23°, but the minor lobes are relatively large. The amplitude of the first minor lobe is 24 percent of the major-lobe maximum (see Fig. 4-20, $n = 5$). In some applications this minor-lobe amplitude may be undesirably large.

To reduce the sidelobe level of linear in-phase broadside arrays, John Stone Stone[1] proposed that the sources have amplitudes proportional to the coefficients of a binomial series of the form

$$(a+b)^{n-1} = a^{n-1} + (n-1)a^{n-2}b + \frac{(n-1)(n-2)}{2!}a^{(n-3)}b^2 + \cdots \qquad (1)$$

where n is the number of sources. Thus, for arrays of three to six sources the relative amplitudes are given by

n	Relative amplitudes **(Pascal's triangle)**
3	1 2 1
4	1 3 3 1
5	1 4 6 4 1
6	1 5 10 10 5 1

where the amplitudes are arranged as in *Pascal's triangle* (any inside number is equal to the sum of the adjacent numbers in the row above).

Applying the binomial distribution to the array of five sources spaced $\lambda/2$ apart, the sources have the relative amplitudes 1, 4, 6, 4, 1. The resulting pattern,

[1] John Stone Stone, U.S. Patents 1,643,323 and 1,715,433.

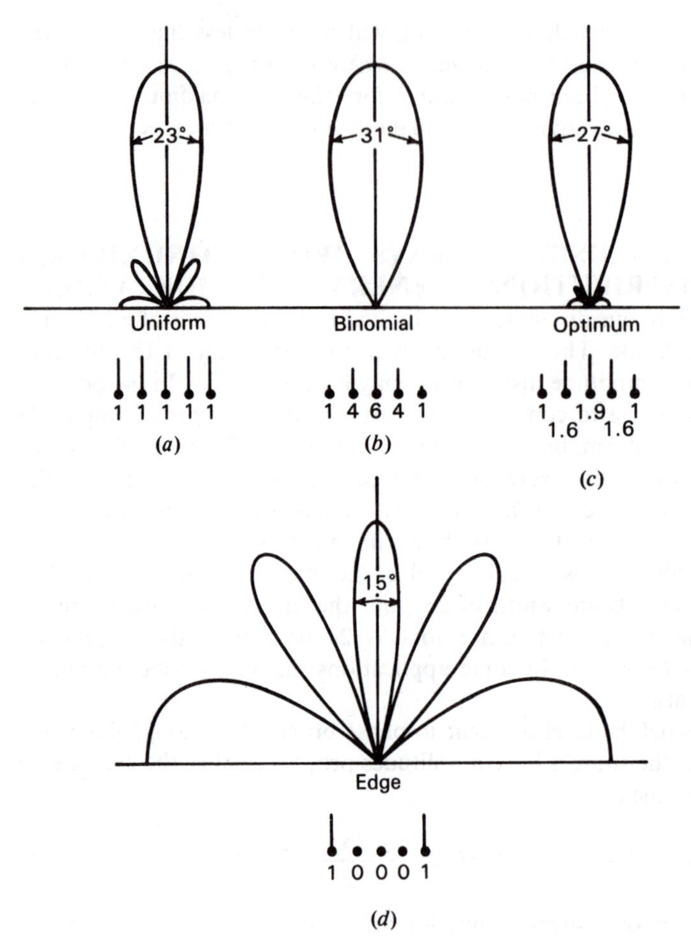

Figure 4-29 Normalized field patterns of broadside arrays of 5 isotropic point sources spaced $\lambda/2$ apart. All sources are in the same phase, but the relative amplitudes have four different distributions: uniform, binomial, optimum and edge. Only the upper half of the pattern is shown. The relative amplitudes of the 5 sources are indicated in each case by the array below the pattern, the height of the line at each source being proportional to its amplitude. All patterns are adjusted to the same maximum amplitude.

designated *binomial*, is shown in Fig. 4-29. Methods of calculating such patterns are discussed in the next section. The pattern has no minor lobes, but this has been achieved at the expense of an increased beam width (31°). For spacings of $\lambda/2$ or less between elements, the minor lobes are eliminated by Stone's binomial distribution. However, the increased beam width and the large ratio of current amplitudes required in large arrays are disadvantages.

At the other extreme from the binomial distribution, we might try an edge distribution in which only the end sources of the array are supplied with power, the three central sources being either omitted or inactive. The relative amplitudes of the five-source array are, accordingly, 1, 0, 0, 0, 1. The array has, therefore, degenerated to two sources 2λ apart and has the field pattern designated as *edge* in Fig. 4-29. The beam width between half-power points of the "main" lobe (normal to the array) is 15°, but "minor" lobes are the same amplitude as the "main" lobe.

Comparing the binomial and edge distributions for the five-source array with $\lambda/2$ spacing, we have

Type of distribution	Half-power beam width	Minor-lobe amplitude (% of major lobe)
Binomial	31°	0
Edge	15°	100

Although for most applications it would be desirable to combine the 15° beam width of the edge distribution with the zero minor-lobe level of the binomial distribution, this combination is not possible. However, if the distribution is between the binomial and the edge type, a compromise between the beam width and the sidelobe level can be made; i.e., the sidelobe level will not be zero, but the beam width will be less than for the binomial distribution. An amplitude distribution of this nature for linear in-phase broadside arrays has been proposed by Dolph[1] which has the further property of optimizing the relation between beam width and sidelobe level; i.e., if the sidelobe level is specified, the beam width between first nulls is minimized, or, conversely, if the beam width between first nulls is specified, the sidelobe level is minimized. Dolph's distribution is based on the properties of the Tchebyscheff polynomials and accordingly will be referred to as the Dolph-Tchebyscheff or optimum distribution.

Applying the Dolph-Tchebyscheff distribution to our array of five sources with $\lambda/2$ spacing, let us specify a sidelobe level 20 dB below the main lobe, i.e., a minor-lobe amplitude 10 percent of the main lobe. The relative amplitude distribution for this sidelobe level is 1, 1.6, 1.9, 1.6, 1 and yields the pattern designated *optimum* in Fig. 4-29. Methods of calculating the distribution and pattern are discussed in the next section. The beam width between half-power points is 27°, which is less than for the binomial distribution. Smaller beam widths can be obtained only by raising the sidelobe level. The Dolph-Tchebyscheff distribution includes all distributions between the binomial and the edge. In fact, the binomial and edge distributions are special cases of the Dolph-Tchebyscheff distribution, the binomial distribution corresponding to an infinite ratio between main- and sidelobe levels and the edge distribution to a ratio of unity. The uniform distribution is, however, not a special case of the Dolph-Tchebyscheff distribution.

---

[1] C. L. Dolph, "A Current Distribution for Broadside Arrays Which Optimizes the Relationship between Beam Width and Side-Lobe Level," *Proc. IRE*, **34**, no. 6, 335–348, June 1946.
H. J. Riblet, "Discussion on Dolph's Paper," *Proc. IRE*, **35**, no. 5, 489–492, May 1947.

Referring to Fig. 4-29, we may draw a number of general conclusions regarding the relation between patterns and amplitude distributions. We note that if the amplitude tapers to a small value at the edge of the array (binomial distribution), minor lobes can be eliminated. On the other hand, if the distribution has an inverse taper with maximum amplitude at the edges and none at the center of the array (edge distribution), the minor lobes are accentuated, being in fact equal to the "main" lobe. From this we may quite properly conclude that the minor-lobe level is closely related to the abruptness with which the amplitude distribution ends at the edge of the array. An abrupt discontinuity in the distribution results in large minor lobes, while a gradually tapered distribution approaching zero at the edge minimizes the discontinuity and the minor lobe amplitude. In the next section, we shall see that the abrupt discontinuity produces large higher "harmonic" terms in the Fourier series representing the pattern. On the other hand, these higher harmonic terms are small when the distribution tapers gradually to a small value at the edge. There is an analogy between this situation and the Fourier analysis of wave shapes. Thus, a square wave has relatively large higher harmonics, whereas a pure sine wave has none, the square wave being analogous to the uniform array distribution while the pure sine wave is analogous to the binomial distribution.

The preceding discussion has been concerned with arrays of discrete sources separated by finite distances. However, the general conclusions concerning amplitude distributions which we have drawn can be extended to large arrays of continuous distributions of an infinite number of point sources, such as might exist in the case of a continuous current distribution on a metal sheet or in the case of a continuous field distribution across the mouth of an electromagnetic horn. If the amplitude distribution follows a Gaussian error curve, which is similar to a binomial distribution for discrete sources, then minor lobes are absent but the beam width is relatively large. An increase of amplitude at the edge reduces the beam width but results in minor lobes, as we have seen. Thus, in the case of a high-gain parabolic reflector type of antenna, the illumination of the reflector by the primary antenna is usually arranged to taper toward the edge of the parabola. However, a compromise is generally made between beam width and side-lobe level so that the illumination is not zero at the edge but has an appreciable value as in a Dolph-Tchebyscheff distribution.

4-11 LINEAR ARRAYS WITH NONUNIFORM AMPLITUDE DISTRIBUTIONS. THE DOLPH-TCHEBYSCHEFF OPTIMUM DISTRIBUTION.

In this section linear in-phase arrays with nonuniform amplitude distributions are analyzed, and the development and application of the Dolph-Tchebyscheff distribution are discussed.

Let us consider a linear array of an even number n_e of isotropic point sources of uniform spacing d arranged as in Fig. 4-30*a*. All sources are in the same phase. The direction $\theta = 0$ is taken normal to the array with the origin at the center of the array as shown. The individual sources have the amplitudes A_0,

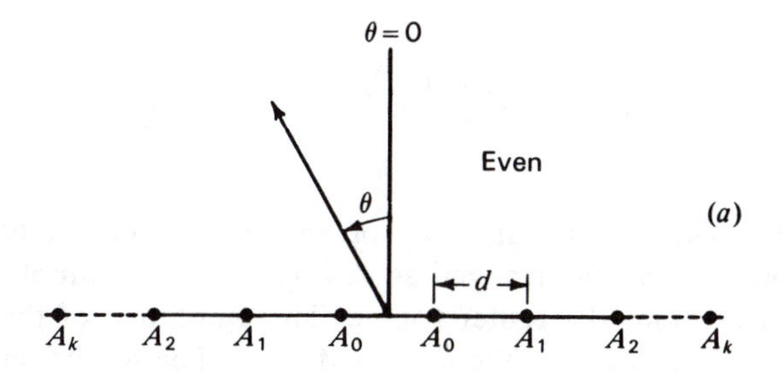

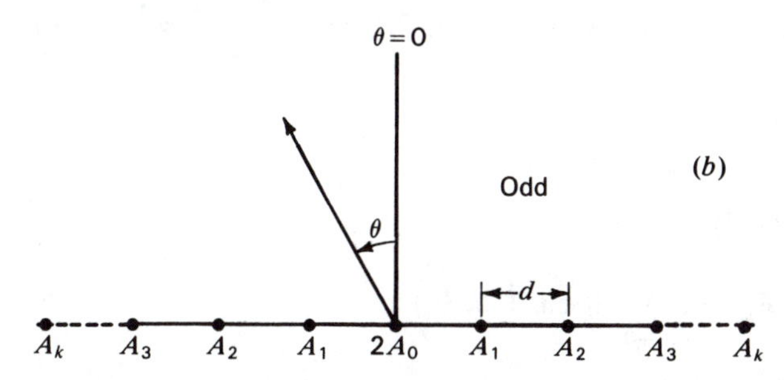

Figure 4-30 Linear broadside arrays of n isotropic sources with uniform spacing for n even (a) and n odd (b).

A_1, A_2, etc., as indicated, the amplitude distribution being symmetrical about the center of the array. The total field E_{n_e} from the even number of sources at a large distance in a direction θ is then the sum of the fields of the symmetrical pairs of sources, or

$$E_{n_e} = 2A_0 \cos \frac{\psi}{2} + 2A_1 \cos \frac{3\psi}{2} + \cdots + 2A_k \cos \left(\frac{n_e - 1}{2} \psi \right) \tag{1}$$

where
$$\psi = \frac{2\pi d}{\lambda} \sin \theta = d_r \sin \theta \tag{2}$$

Each term in (1) represents the field due to a symmetrically disposed pair of the sources.

Now let

$$2(k + 1) = n_e$$

where $k = 0, 1, 2, 3, \ldots$ so that

$$\frac{n_e - 1}{2} = \frac{2k + 1}{2}$$

Then (1) becomes

$$E_{n_e} = 2 \sum_{k=0}^{k=N-1} A_k \cos\left(\frac{2k+1}{2}\psi\right) \tag{3}$$

where $N = n_e/2$

Next let us consider the case of a linear array of an odd number n_o of isotropic point sources of uniform spacing arranged as in Fig. 4-30*b*. The amplitude distribution is symmetrical about the center source. The amplitude of the center source is taken as $2A_0$, the next as A_1, the next as A_2, etc. The total field E_{n_o} from the odd number of sources at a large distance in a direction θ is then

$$E_{n_o} = 2A_0 + 2A_1 \cos\psi + 2A_2 \cos 2\psi + \cdots + 2A_k \cos\left(\frac{n_o - 1}{2}\psi\right) \tag{4}$$

Now for this case let

$$2k + 1 = n_o$$

where $k = 0, 1, 2, 3, \ldots$. Then (4) becomes

$$E_{n_o} = 2 \sum_{k=0}^{k=N} A_k \cos\left(2k\frac{\psi}{2}\right) \tag{5}$$

where $N = (n_o - 1)/2$

The series expressed by (4) or by (5) may be recognized as a finite Fourier series of N terms.[1] For $k = 0$ we have a constant term $2A_0$ representing the contribution of the center source. For $k = 1$ we have the term $2A_1 \cos\psi$ representing the contribution of the first pair of sources on either side of the center source. For each higher value of k we have a higher harmonic term which in each case represents the contribution of a pair of symmetrically disposed sources. Thus, the total field pattern is simply the sum of a series of terms of increasing order in the same way that the waveform of an alternating current can be represented as a Fourier series involving, in general, a constant term, a fundamental term and higher harmonic terms. The field pattern of an even number of sources as given by (1) or (3) is also a finite Fourier series but one which has no constant term and only odd harmonics. The coefficients $A_0, A_1, \ldots$ in both series are arbitrary and express the amplitude distribution.

To illustrate the Fourier nature of the field-pattern expression, let us consider the simple example of an array of 9 isotropic point sources spaced $\lambda/2$ apart, having the same amplitude and phase. Hence, the coefficients are related as follows: $2A_0 = A_1 = A_2 = A_3 = A_4 = \frac{1}{2}$. The number of sources is odd; hence the expression for the field pattern is then given by (5) as

$$E_9 = \tfrac{1}{2} + \cos\psi + \cos 2\psi + \cos 3\psi + \cos 4\psi \tag{6}$$

[1] Irving Wolff, "Determination of the Radiating System Which Will Produce a Specified Directional Characteristic," *Proc. IRE*, **25**, 630–643, May 1937.

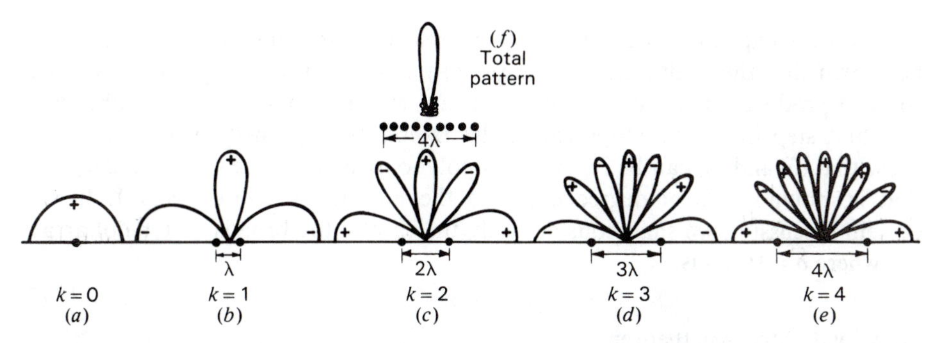

Figure 4-31 Resolution of total pattern of array of 9 isotropic sources into Fourier components due to center source and pairs of symmetrically disposed sources. The relative field pattern of the entire array is shown by (*f*). The lower halves of patterns are not shown. (Note that the end-fire lobes are wider than the broadside lobes.)

The first term ($k = 0$) is a constant so that the field pattern is a circle of amplitude $\frac{1}{2}$ as shown in Fig. 4-31*a*. The second term ($k = 1$) may be regarded as the fundamental term of the Fourier series and gives the pattern of the two sources (A_1 in Fig. 4-30*b*) on either side of the center. This pattern has 4 lobes of maximum amplitude of unity, as illustrated in Fig. 4-31*b*. The next term ($k = 2$) may be regarded as the second harmonic term and gives the pattern of the next pair of sources (A_2 in Fig. 4-30*b*). This pattern has 8 lobes as shown by Fig. 4-31*c*. The last two terms represent the third and fourth harmonics, and the patterns have 12 and 16 lobes, respectively, as indicated by Fig. 4-31*d* and *e*. The above relations may be summarized as in Table 4-5.

The algebraic sum of the patterns given by the five terms is the total far-field pattern of the array which is presented in Fig. 4-31*f*. If the middle source of the array has zero amplitude or is omitted, the total pattern is then the sum of the four terms for which $k = 1, 2, 3$ and 4. If in addition the pair of sources A_1 is omitted, the total pattern is the sum of three terms for which $k = 2, 3$ and 4. Since these are higher harmonic terms, we may properly expect that in this case the minor lobes of the total pattern will be accentuated. It is apparent from the above discussion that the field pattern of any symmetrical amplitude distribution can be expresed as a series of the form of (3) or (5).

Table 4-5

k	Sources	Spacing	Fourier term	Pattern
0	1	0	Constant	Circle
1	2	1λ	Fundamental	4 lobes
2	2	2λ	Second harmonic	8 lobes
3	2	3λ	Third harmonic	12 lobes
4	2	4λ	Fourth harmonic	16 lobes

Proceeding now to the Dolph-Tchebyscheff amplitude distribution, it will be shown that the coefficients of the pattern series[1] can be uniquely determined so as to produce a pattern of minimum beam width for a specified sidelobe level. The first step in the development of the Dolph-Tchebyscheff distribution is to show that (3) and (5) can be regarded as polynomials of degree $n_e - 1$ and $n_o - 1$, that is, polynomials of degree equal to the number of sources less 1. In the present discussion we shall consider only the case of the broadside type of array, i.e., where $\delta = 0$. Thus,

$$\psi = d_r \sin\theta \tag{7}$$

Now by de Moivre's theorem,

$$e^{jm\psi/2} = \cos m\frac{\psi}{2} + j\sin m\frac{\psi}{2} = \left(\cos\frac{\psi}{2} + j\sin\frac{\psi}{2}\right)^m \tag{8}$$

On taking real parts of (8) we have

$$\cos m\frac{\psi}{2} = \mathrm{Re}\left(\cos\frac{\psi}{2} + j\sin\frac{\psi}{2}\right)^m \tag{9}$$

Expanding (9) as a binomial series gives

$$\cos m\frac{\psi}{2} = \cos^m\frac{\psi}{2} - \frac{m(m-1)}{2!}\cos^{m-2}\frac{\psi}{2}\sin^2\frac{\psi}{2} + \frac{m(m-1)(m-2)(m-3)}{4!}\cos^{m-4}\frac{\psi}{2}\sin^4\frac{\psi}{2} - \cdots \tag{10}$$

Putting $\sin^2(\psi/2) = 1 - \cos^2(\psi/2)$, and substituting particular values of m, (10) then reduces to the following:

$$\left.\begin{aligned} m &= 0, & \cos m\frac{\psi}{2} &= 1 \\ m &= 1, & \cos m\frac{\psi}{2} &= \cos\frac{\psi}{2} \\ m &= 2, & \cos m\frac{\psi}{2} &= 2\cos^2\frac{\psi}{2} - 1 \\ m &= 3, & \cos m\frac{\psi}{2} &= 4\cos^3\frac{\psi}{2} - 3\cos\frac{\psi}{2} \\ m &= 4, & \cos m\frac{\psi}{2} &= 8\cos^4\frac{\psi}{2} - 8\cos^2\frac{\psi}{2} + 1 \\ &\text{etc.} \end{aligned}\right\} \tag{11}$$

[1] Equations (1), (3), (4) and (5).

Now let

$$x = \cos\frac{\psi}{2} \tag{12}$$

whereupon the equations of (11) become

$$\left.\begin{aligned} &\cos m\frac{\psi}{2} = 1, && \text{when } m = 0 \\ &\cos m\frac{\psi}{2} = x, && \text{when } m = 1 \\ &\cos m\frac{\psi}{2} = 2x^2 - 1, && \text{when } m = 2 \\ &\text{etc.} \end{aligned}\right\} \tag{13}$$

The polynomials of (13) are called Tchebyscheff polynomials, which may be designated in general by

$$T_m(x) = \cos m\frac{\psi}{2} \tag{14}$$

For particular values of m, the first eight Tchebyscheff polynomials are

$$\left.\begin{aligned} T_0(x) &= 1 \\ T_1(x) &= x \\ T_2(x) &= 2x^2 - 1 \\ T_3(x) &= 4x^3 - 3x \\ T_4(x) &= 8x^4 - 8x^2 + 1 \\ T_5(x) &= 16x^5 - 20x^3 + 5x \\ T_6(x) &= 32x^6 - 48x^4 + 18x^2 - 1 \\ T_7(x) &= 64x^7 - 112x^5 + 56x^3 - 7x \end{aligned}\right\} \tag{15}$$

We note in (15) that the degree of the polynomial is the same as the value of m.

The roots of the polynomials occur when $\cos m(\psi/2) = 0$ or when

$$m\frac{\psi}{2} = (2k - 1)\frac{\pi}{2} \tag{16}$$

where $k = 1, 2, 3, \ldots$

The roots of x, designated x', are thus

$$x' = \cos\left[(2k - 1)\frac{\pi}{2m}\right] \tag{17}$$

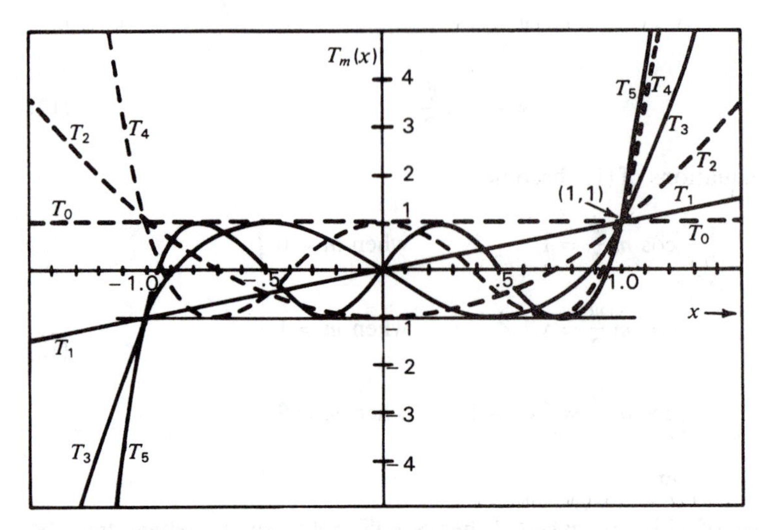

Figure 4-32 Tchebyscheff polynomials of degree $m = 0$ through $m = 5$.

We have shown that cos $m(\psi/2)$ can be expressed as a polynomial of degree m. Thus, (3) and (5) are expressible as polynomials of degree $2k + 1$ and $2k$ respectively, since each is the sum of cosine polynomials of the form cos $m(\psi/2)$. For an even number n_e of sources $2k + 1 = n_e - 1$, while for an odd number n_o, $2k = n_o - 1$. Therefore, (3) and (5), which express the field pattern of a symmetric in-phase equispaced linear array of n isotropic point sources, are polynomials of degree equal to the number of sources less 1. If we now set the array polynomial as given by (3) or (5) equal to the Tchebyscheff polynomial of like degree ($m = n - 1$) and equate the array coefficients to the coefficients of the Tchebyscheff polynomial, then the amplitude distribution given by these coefficients is a Tchebyscheff distribution and the field pattern of the array corresponds to the Tchebyscheff polynomial of degree $n - 1$.

The Tchebyscheff polynomials of degree $m = 0$ through $m = 5$ are presented in Fig. 4-32. Referring to Fig. 4-32, the following properties of the polynomials are worthy of note:

1. All pass through the point (1, 1).
2. For values of x in the range $-1 \leq x \leq +1$, the polynomials all lie between ordinate values of $+1$ and -1. All roots occur between $-1 \leq x \leq +1$, and all maximum values in this range are ± 1.

We may now describe Dolph's method of applying the Tchebyscheff polynomial to obtain an optimum pattern. Suppose that we have an array of 6 sources. The field pattern is then a polynomial of degree 5. If this polynomial is equated to the Tchebyscheff polynomial of degree 5, shown in Fig. 4-33, then the optimum pattern may be derived as follows. Let the ratio of the main-lobe

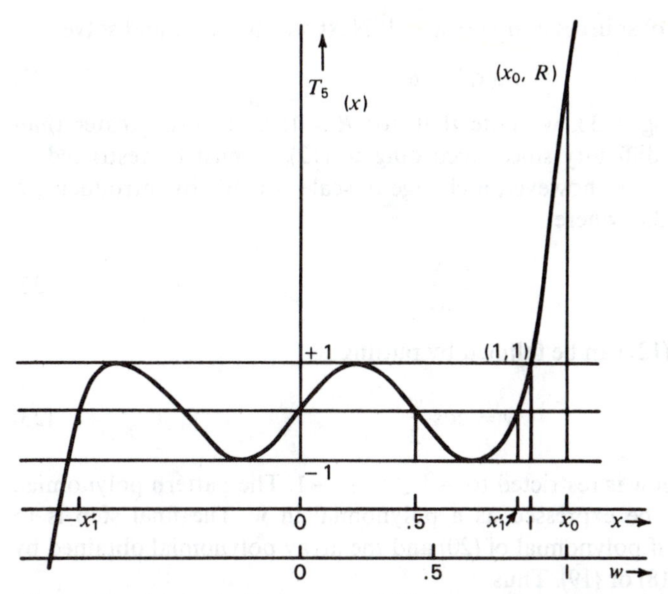

Figure 4-33 Tchebyscheff polynomial of fifth degree with relation to coordinate scales.

maximum to the minor-lobe level be specified as R; that is,

$$R = \frac{\text{main-lobe maximum}}{\text{sidelobe level}}$$

The point (x_0, R) on the $T_5(x)$ polynomial curve then corresponds to the main-lobe maximum, while the minor lobes are confined to a maximum value of unity. The roots of the polynomial correspond to the nulls of the field pattern. The important property of the Tchebyscheff polynomial is that *if the ratio R is specified, the beam width to the first null* ($x = x_1'$) *is minimized.* The corollary also holds that *if the beam width is specified, the ratio R is maximized* (sidelobe level minimized).

The procedure will now be summarized. Let us write (3) and (5) again. It is to be noted that they are functions of $\psi/2$. Thus,

$$E_{n_e} = 2 \sum_{k=0}^{k=N-1} A_k \cos\left[(2k+1)\frac{\psi}{2}\right] \qquad (n \text{ even}) \tag{18}$$

and

$$E_{n_o} = 2 \sum_{k=0}^{k=N} A_k \cos\left(2k\frac{\psi}{2}\right) \qquad (n \text{ odd}) \tag{19}$$

Since we are usually interested only in the relative field pattern, the factor 2 before the summation sign in (18) and (19) may be dropped.

For an array of n sources, the first step is to select the Tchebyscheff polynomial of the same degree as the array polynomial, (3) or (5). This is given by

$$T_{n-1}(x) \tag{20}$$

where n is the number of sources and $m = n - 1$. Next we choose R and solve

$$T_m(x_0) = R \tag{21}$$

for x_0. Referring to Fig. 4-33, we note that, for $R > 1$, x_0 is also greater than unity. This presents a difficulty since, according to (12), x must be restricted to the range $-1 \leq x \leq +1$. If, however, a change of scale is made by introducing a new abscissa w (Fig. 4-33), where

$$w = \frac{x}{x_0} \tag{22}$$

then the restriction of (12) can be fulfilled by putting

$$w = \cos\frac{\psi}{2} \tag{23}$$

where now the range of w is restricted to $-1 \leq w \leq +1$. The pattern polynomial, (18) or (19), may now be expressed as a polynomial in w. The final step is to equate the Tchebyscheff polynomial of (20) and the array polynomial obtained by substituting (23) into (18) or (19). Thus,

$$T_{n-1}(x) = E_n \tag{24}$$

The coefficients of the array polynomial are then obtained from (24), yielding the Dolph-Tchebyscheff amplitude distribution which is an optimum for the sidelobe level specified.

As a proof of the optimum property of the Tchebyscheff polynomial, let us consider any other polynomial $P(x)$ of degree 5 which passes through (x_0, R) in Fig. 4-33 and the highest root x_1' and for all smaller values of x lies between $+1$ and -1. If the range in ordinate of $P(x)$ is less than ± 1, then this polynomial would give a smaller sidelobe level for this same beam width, and $T_5(x)$ would not be optimum. Since $P(x)$ lies between ± 1 in the range $-x_1' \leq x \leq +x_1'$ it must intersect the curve $T_5(x)$ in at least $m + 1 = 6$ points, including (x_0, R). Two polynomials of the same degree m which intersect in $m + 1$ points must be the same polynomial,[1] so that

$$P(x) = T_5(x)$$

and the $T_5(x)$ polynomial is, therefore, the optimum.

If the spacing between sources exceeds $\lambda/2$, it should be noted that as the spacing approaches λ a large lobe develops at $\theta = \pm 90°$ which equals the main lobe when $d = \lambda$. However, if the individual sources of the array are nonisotropic, i.e., are directional with the maximum at $\theta = 0$ and with little or no radiation

---

[1] This follows from the fact that a polynomial of degree m has $m + 1$ arbitrary constants. Further, if $m + 1$ points on the polynomial's curve are specified, $m + 1$ independent equations with $m + 1$ unknowns can be written and the $m + 1$ constants thereby determined.

at $\theta = \pm 90°$, then by pattern multiplication the lobes of the total pattern at $\theta = \pm 90°$ can be made small.

4-12 EXAMPLE OF DOLPH-TCHEBYSCHEFF DISTRIBUTION FOR AN ARRAY OF 8 SOURCES.

To illustrate the method for finding the Dolph-Tchebyscheff distribution, let us work the following problem.

An array of $n = 8$ in-phase isotropic sources, spaced $\lambda/2$ apart, is to have a sidelobe level 26 dB below the main-lobe maximum. Find the amplitude distribution fulfilling this requirement that produces the minimum beam width between first nulls, and plot the field pattern.

Since

$$\text{Sidelobe level in dB below main-lobe maximum} = 20 \log_{10} R \tag{1}$$

it follows that

$$R = 20 \tag{2}$$

The Tchebyscheff polynomial of degree $n - 1$ is $T_7(x)$. Thus, we set

$$T_7(x_0) = 20 \tag{3}$$

The value of x_0 may be determined by trial and error from the $T_7(x)$ expansion as given in (4-11-15) or x_0 may be calculated from

$$x_0 = \tfrac{1}{2}[(R + \sqrt{R^2 - 1})^{1/m} + (R - \sqrt{R^2 - 1})^{1/m}] \tag{4}$$

Substituting $R = 20$ and $m = 7$ in (4) yields

$$x_0 = 1.15 \tag{5}$$

Now substituting (4-11-23) in (4-11-18) and dropping the factor 2, we have

$$\begin{aligned} E_8 = {} & A_0 w + A_1(4w^3 - 3w) + A_2(16w^5 - 20w^3 + 5w) \\ & + A_3(64w^7 - 112w^5 + 56w^3 - 7w) \end{aligned} \tag{6}$$

But $w = x/x_0$, so making this substitution in (6) and grouping terms of like degree,

$$\begin{aligned} E_8 = {} & \frac{64A_3}{x_0^7} x^7 + \frac{16A_2 - 112A_3}{x_0^5} x^5 + \frac{4A_1 - 20A_2 + 56A_3}{x_0^3} x^3 \\ & + \frac{A_0 - 3A_1 + 5A_2 - 7A_3}{x_0} x \end{aligned} \tag{7}$$

The Tchebyscheff polynomial of like degree is

$$T_7(x) = 64x^7 - 112x^5 + 56x^3 - 7x \tag{8}$$

Now equating (7) and (8),

$$E_8 = T_7(x) \tag{9}$$

For (9) to be true requires that the coefficients of (7) equal the coefficients of the terms of like degree in (8). Therefore,

$$\frac{64A_3}{x_0^7} = 64 \tag{10}$$

or

$$A_3 = x_0^7 = 1.15^7 = 2.66 \tag{11}$$

In a similar way we find that

$$\left.\begin{aligned} A_2 &= 4.56 \\ A_1 &= 6.82 \\ A_0 &= 8.25 \end{aligned}\right\} \tag{12}$$

The relative amplitudes of the 8 sources are then

1, 1.7, 2.6, 3.1, 3.1, 2.6, 1.7, 1

To obtain the field pattern given by the Dolph-Tchebyscheff distribution, we recall that $\psi/2 = (d_r \sin\theta)/2$, $\cos(\psi/2) = w$, and $w = x/x_0$, from which

$$x = x_0 \cos\frac{d_r \sin\theta}{2} \tag{13}$$

The value of x corresponding to a given value of θ, as obtained from (13), is then introduced in the appropriate Tchebyscheff polynomial, in this case $T_7(x)$, or scaled from a graph of this polynomial, as shown in Fig. 4-34. The value of the polynomial for this value of x is then the relative field strength in the direction θ. In general, as θ ranges from $-\pi/2$ to $+\pi/2$, the variables $\psi/2$, w and x range as indicated by Table 4-6. Thus, in general, as θ ranges from $-\pi/2$ to 0 to $+\pi/2$, x ranges from some point, such as a in Fig. 4-34, to x_0 and back again to a, the ordinate value giving the relative field intensity.

In our problem, $d_r = \pi$ and $x_0 = 1.15$, so that the range of x is as shown in Table 4-7. Hence, at $\theta = -90°$ we start at the origin in Fig. 4-34 (point b), and as

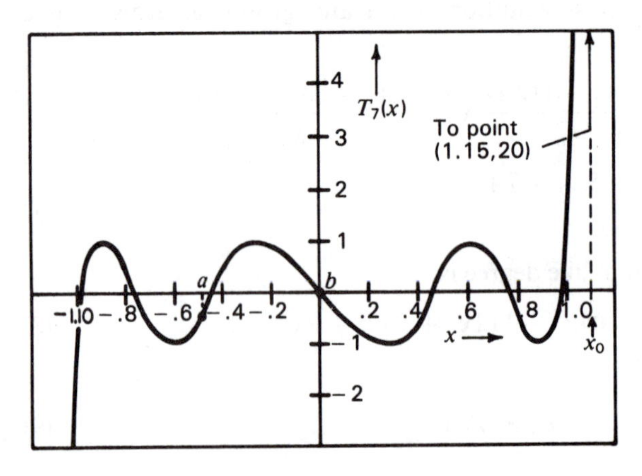

Figure 4-34 Tchebyscheff polynomial of the seventh degree.

Table 4-6

Variable	Range		
θ	$-\frac{\pi}{2}$	0	$+\frac{\pi}{2}$
$\frac{\psi}{2}$	$-\frac{d_r}{2}$	0	$+\frac{d_r}{2}$
w	$\cos \frac{d_r}{2}$	1	$\cos \frac{d_r}{2}$
x	$x_0 \cos \frac{d_r}{2}$	x_0	$x_0 \cos \frac{d_r}{2}$

θ approaches 0° we proceed to the right along the polynomial curve, reaching the point (x_0, $R = 1.15$, 20) when $\theta = 0°$. As θ continues to increase, we retrace the polynomial curve, reaching the origin when $\theta = 90°$. Thus, the pattern is symmetrical about the $\theta = 0°$ direction.

As a preliminary step to plotting the field pattern, it is usually helpful to make a plot of x versus θ from (13). Then, knowing the values of x for the nulls and maxima of the $T_m(x)$ curve, the corresponding values of θ may be determined. As many intermediate points as are needed may also be obtained in the same manner. Following this procedure, the field pattern for our problem of the 8-source array is presented in Fig. 4-35*a* in rectangular coordinates and in Fig. 4-35*b* in polar coordinates.

4-13 COMPARISON OF AMPLITUDE DISTRIBUTIONS FOR 8-SOURCE ARRAYS. In the problem worked in the preceding section, the sidelobe level was 26 dB below the maximum of the main beam ($R = 20$). It is of interest to compare the amplitude for this case with the distributions for other sidelobe levels. This is done in Fig. 4-36, in which the relative amplitude distributions are shown for 8-source arrays with sidelobe levels ranging from 0 dB to an infinite number of decibels below the main beam maximum. The infinite decibel case corresponds to $R = \infty$ (zero sidelobe level) and is identical with Stone's binomial distribution. The relative amplitudes for this case are 1, 7, 21, 35, 35, 21,

Table 4-7

Variable	Range		
θ	$-\frac{\pi}{2}$	0	$+\frac{\pi}{2}$
x	0	1.15	0

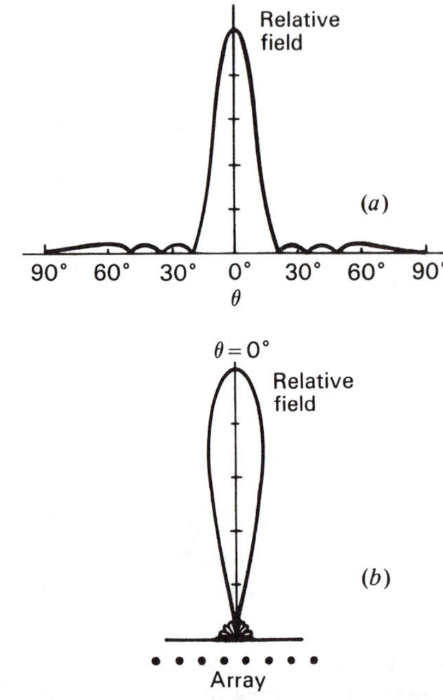

Figure 4-35 Relative field pattern of broadside array of 8 isotropic sources spaced $\lambda/2$ apart. The amplitude distribution gives a minimum beam width for a sidelobe level $\frac{1}{20}$ of the main lobe. The pattern is shown in rectangular coordinates at (*a*) and in polar at (*b*). Both diagrams show the pattern only from $-90°$ to $+90°$, the other half of the pattern being identical.

7, 1.[1] The ratio of amplitudes of the center sources to the edge sources is 35 to 1. Such a large ratio would be very difficult to achieve in practice. As the sidelobe level increases (R decreases), the amplitude distribution becomes more uniform, the ratio of the center to edge amplitudes being only about 3 to 1 for the 26-dB ($R = 20$) case. The 20-dB case ($R = 10$) is more uniform, with an amplitude ratio of only 1.7 to 1. The 14-dB case ($R = 5$) exhibits a still more uniform distribution but shows an inversion, the maximum amplitude having shifted to the outermost sources (1 and 8). The uniform distribution is not a special case of the Dolph-Tchebyscheff distribution, an inversion occurring before the uniform case is reached. As the sidelobe level is raised still further, the distribution tends more towards an edge type, the amplitude of the inner sources decreasing still further. In the extreme case, where the sidelobes are equal to the main-lobe level (0 dB, or $R = 1$), the amplitudes of all of the inner sources are zero, and the distribution is of the edge type discussed in connection with Fig. 4-29*d*. Thus, both the binomial and edge distributions are special cases of the Dolph-Tchebyscheff distribution, but the uniform amplitude distribution is not. The point of nearest approach to the uniform distribution is for an R value between 5 and 10. Referring to Fig. 4-20 and interpolating for $n = 8$ between the curves for $n = 10$ and $n = 5$, it is interesting to note that the ratio of the main-lobe maximum to the minor-lobe

[1] As may be noted by extending Pascal's triangle (Sec. 4-10) to $n = 8$.

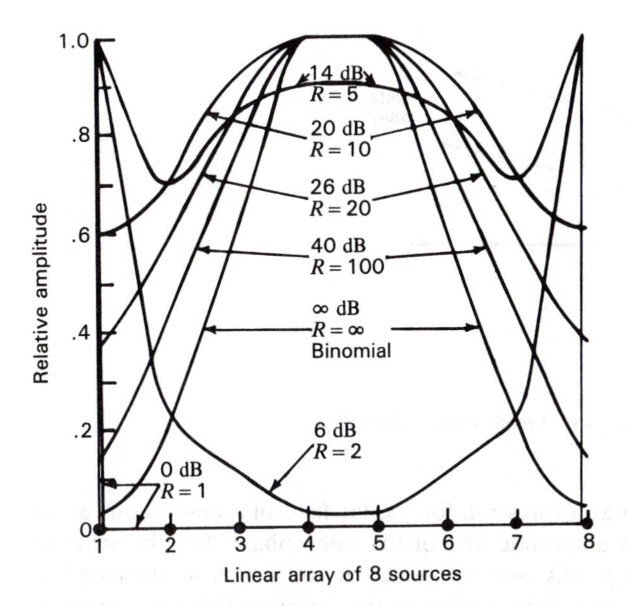

Figure 4-36 Comparison of Dolph-Tchebyscheff amplitude distribution envelopes for various sidelobe levels.

maxima ranges from about 4.3 to 8 for an array of eight sources of uniform amplitude.

The Dolph-Tchebyscheff optimum amplitude distribution, as discussed in the preceding sections, is optimum only if $d \geq \lambda/2$, which covers the cases of most interest for broadside arrays. By a generalization of the method, however, cases with smaller spacings can also be optimized.[1]

In conclusion, it should be pointed out that the properties of the Tchebyscheff polynomials may be applied not only to antenna patterns as discussed above but also to other situations. It is necessary, however, that the function to be optimized be expressible as a polynomial.

4-14 CONTINUOUS ARRAYS. In the preceding sections, the discussion has been restricted to arrays of discrete point sources, i.e., to arrays of a finite number of sources separated by finite distances. We now proceed to a consideration of *continuous* arrays of point sources, i.e., arrays of an infinite number of sources separated by infinitesimal distances. By Huygens' principle, a continuous array of point sources is equivalent to a continuous field distribution. In this way, our discussion of continuous arrays can be extended to include the radiation patterns of field distributions across apertures, as, for example, the pattern of an electromagnetic horn where the field distribution across the mouth of the horn is known.

---

[1] H. J. Riblet, *Proc. IRE*, **35**, no. 5, 489–492, May 1947.

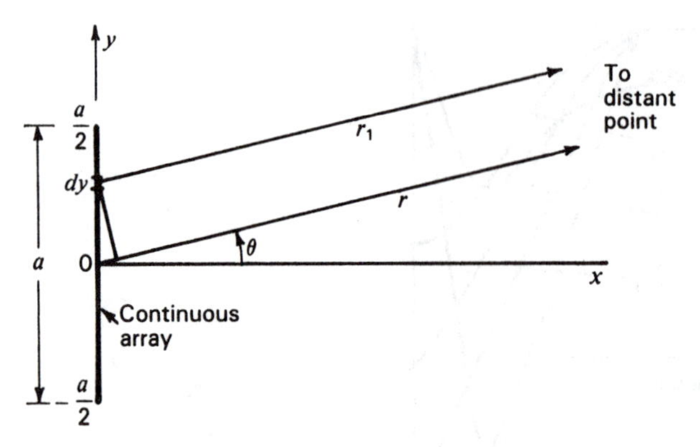

Figure 4-37 Continuous broadside array of point sources of length a.

We shall now develop an expression for the far field of a continuous array of point sources of uniform amplitude and of the same phase. Let the array of length a be parallel to the y axis with its center at the origin as indicated in Fig. 4-37. Then the field dE at a distant point in the direction θ due to the point sources in the infinitesimal length dy at a distance y from the origin is

$$dE = \frac{A}{r_1} e^{j\omega[t-(r_1/c)]}\, dy = \frac{A}{r_1} e^{j(\omega t - \beta r_1)}\, dy \tag{1}$$

where $\beta = \omega/c = 2\pi/\lambda$ and A is a constant involving amplitude. The total field E at the distant point is then the integrated value of (1) over the array of length a as given by

$$E = \int_{-a/2}^{a/2} \frac{A}{r_1} e^{j(\omega t - \beta r_1)}\, dy \tag{2}$$

Both A and the time factor may be taken outside the integral, and r_1 may also be if $r_1 \gg a$. Thus,

$$E = \frac{Ae^{j\omega t}}{r_1} \int_{-a/2}^{a/2} e^{-j\beta r_1}\, dy \tag{3}$$

However, referring to Fig. 4-37,

$$r_1 = r - y \sin\theta \tag{4}$$

Substituting (4) in (3) and taking the constant factor $e^{-j\beta r}$ outside the integral, we have

$$E = A' \int_{-a/2}^{a/2} e^{j\beta y \sin\theta}\, dy \tag{5}$$

where

$$A' = \frac{Ae^{j(\omega t - \beta r)}}{r_1} \tag{6}$$

Integrating (5) yields

$$E = \frac{2A'}{\beta \sin \theta} \frac{e^{j(\beta a/2) \sin \theta} - e^{-j(\beta a/2) \sin \theta}}{2j} \tag{7}$$

which may be written as

$$E = \frac{2A'}{\beta \sin \theta} \sin \left(\frac{\beta a}{2} \sin \theta \right) \tag{8}$$

Let

$$\psi' = \beta a \sin \theta = a_r \sin \theta \tag{9}$$

where $a_r = \beta a = 2\pi a/\lambda$ = array length, rad

Then

$$E = \frac{2A'}{\beta \sin \theta} \sin \frac{\psi'}{2} \tag{10}$$

However, from (9),

$$\beta \sin \theta = \frac{\psi'}{a}$$

so that (10) becomes

$$E = aA' \frac{\sin (\psi'/2)}{\psi'/2} \tag{11}$$

Normalizing (11) gives finally

$$E = \frac{\sin (\psi'/2)}{\psi'/2} \tag{12}$$

Equation (12) expresses the far field, or Fraunhofer diffraction pattern, of a continuous broadside array of length a having uniform amplitude and phase. For n discrete, equally spaced sources, it was previously shown by (4-6-9) that the normalized value of the total field is

$$E = \frac{\sin (n\psi/2)}{n \sin (\psi/2)} \tag{13}$$

where $\psi = d \cos \phi + \delta$

For in-phase sources, $\delta = 0$. Comparing Figs. 4-18 and 4-37 we note that $\phi = \theta + \pi/2$, so that

$$\psi = -d_r \sin \theta = -\beta d \sin \theta \tag{14}$$

For small values of ψ, which occur for small values of θ, d or both, (13) can be expressed as

$$E = \frac{\sin (n\psi/2)}{n\psi/2} = \frac{\sin [(\beta n d/2) \sin \theta]}{(\beta n d/2) \sin \theta} \tag{15}$$

The length a of the array of discrete sources is

$$a = d(n - 1) \tag{16}$$

where n = number of sources
d = spacing

If $n \gg 1$, $a \simeq nd$ and (15) becomes

$$E = \frac{\sin [(\beta a/2) \sin \theta]}{(\beta a/2) \sin \theta} = \frac{\sin [(a_r/2) \sin \theta]}{(a_r/2) \sin \theta} \tag{17}$$

where $a_r = \beta a = 2\pi a/\lambda$

By (9) this can now be expressed as

$$E = \frac{\sin (\psi'/2)}{\psi'/2} \tag{18}$$

which is identical with the value obtained in (12) for the continuous array. Thus, the field pattern for an array of many discrete sources ($n \gg 1$) and for small values of ψ is the same as the pattern of a continuous array of the same length. If the array is long, that is, if $nd \gg \lambda$, the main beam and the first minor lobes are confined to small values of θ. It therefore follows that the main features of the pattern of a large array are the same, whether the array has many discrete sources or is a continuous distribution of sources. Many of the conclusions derived in previous sections concerning amplitude distributions for arrays of discrete sources can also be applied to continuous arrays provided that the arrays are large.

The null directions θ_0 of the continuous array pattern are given by

$$\frac{\psi'}{2} = \pm K\pi \tag{19}$$

where $K = 1, 2, 3, \ldots$

Thus,

$$\theta_0 = \arcsin \left(\pm \frac{K\lambda}{a} \right) \tag{20}$$

For a long array (20) can be expressed as

$$\theta_0 \simeq \pm \frac{K}{a_\lambda} \text{ (rad)} \simeq \pm \frac{57.3K}{a_\lambda} \text{ (deg)} \tag{21}$$

where $a_\lambda = a/\lambda$

The beam width between first nulls ($K = 1$) for a long array is then

$$2\theta_{01} \simeq \frac{2}{a_\lambda} \text{ (rad)} \simeq \frac{115}{a_\lambda} \text{ (deg)} \tag{22}$$

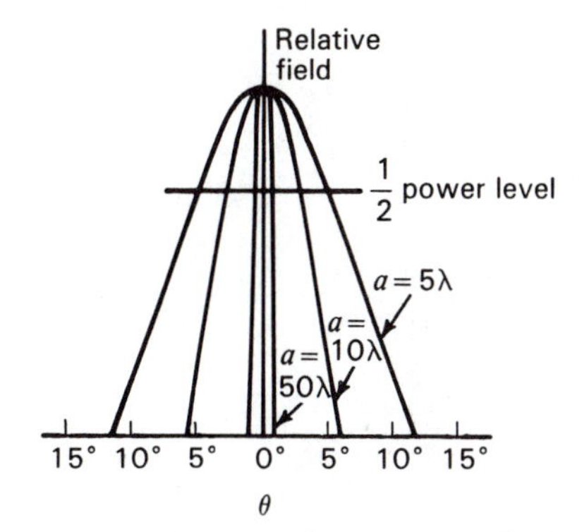

Figure 4-38 Main-lobe field patterns of continuous uniform broadside arrays 5, 10 and 50λ long.

It is to be noted that (20), (21) and (22) are identical with the expressions given for the broadside array of discrete sources, if nd is replaced by a (see Table 4-2, Sec. 4-7). Therefore, the null locations for long arrays of either discrete or continuous sources are the same provided only that $n \gg 1$.

The field patterns of the main beam of continuous arrays of point sources 5, 10 and 50λ long are compared in Fig. 4-38. It may be noted that the beam width between half-power points, θ_{HP}, of a long, uniform broadside array is given approximately by

$$\theta_{HP} = 0.9\theta_{01} = \frac{0.9}{a_\lambda} \qquad \text{(rad)} \tag{23}$$

or

$$\theta_{HP} = \frac{51}{a_\lambda} \qquad \text{(deg)} \tag{24}$$

4-15 HUYGENS' PRINCIPLE.[1]

The principle proposed by Christian Huygens (1629–1695) has been of fundamental importance to the development of wave theory. Huygens' principle states that *each point on a primary wave front can be considered to be a new source of a secondary spherical wave and that a secondary wave front can be constructed as the envelope of these secondary waves*, as suggested in Fig. 4-39. Thus, a spherical wave from a single point source propagates as a spherical wave as indicated in Fig. 4-39*a*, while an infinite plane wave continues as a plane wave as suggested in Fig. 4-39*b*. This principle of physical optics can be used to explain the apparent bending of electromagnetic waves around obstacles, i.e., the diffraction of waves, a *diffracted* ray being one that follows a path that cannot be interpreted as either reflection or refraction.

[1] C. Huygens, *Traité de la Lumière*, Leyden, 1690.
Max Born, *Optik*, Springer-Verlag, 1933.
Arnold Sommerfeld, "Theorie der Beugung," in Frank and von Mises (eds.), *Differential und Integralgleichungen der Mechanik und Physik*, Vieweg, 1935.

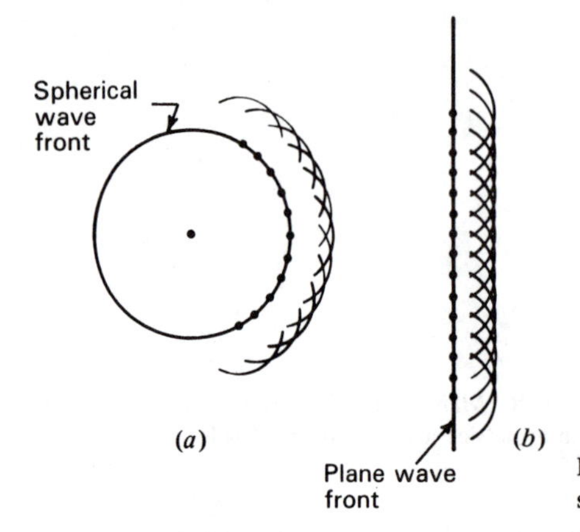

Figure 4-39 Spherical and plane wave fronts with secondary waves of Huygens.

Let us consider the situation shown in Fig. 4-40*a* in which an infinite plane electromagnetic wave is incident on an infinite flat sheet which is opaque to the waves. The sheet has a slot of width a and of infinite length in the direction normal to the page. The field everywhere to the right of the sheet is the result of the section of the wave that passes through the slot. If a is many wavelengths, the field distribution across the slot may be assumed, in the first approximation, to be uniform, as shown in Fig. 4-40*b*. By Huygens' principle the field everywhere to

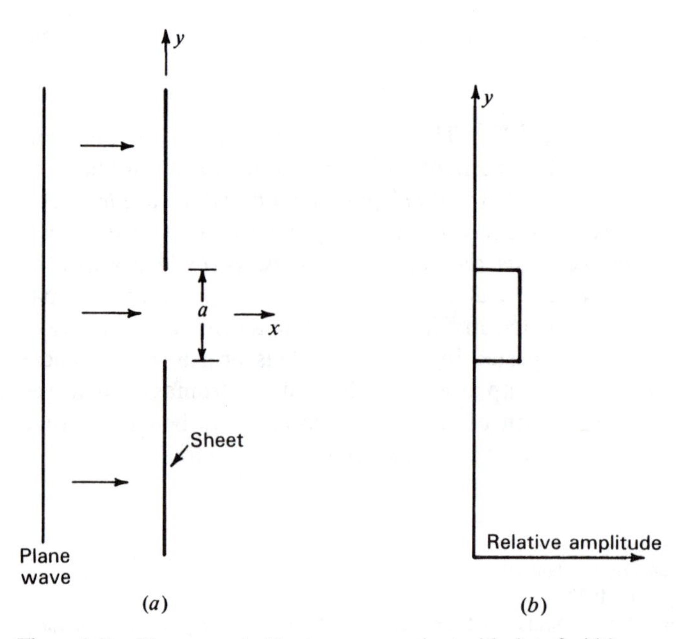

Figure 4-40 Plane wave incident on opaque sheet with slot of width a.

the right of the sheet is the same as though each point in the plane of the slot is the source of a new spherical wave. Each of these point sources is of equal amplitude and phase. Thus, by Huygens' principle the slotted sheet with a uniform field across the opening can be replaced by a continuous array of point sources which just fills the opening. The field pattern in the xy plane (Fig. 4-40*a*) is then calculated in the same way as for a continuous linear array of point sources of length a oriented parallel to the y axis.

The far field, or Fraunhofer diffraction pattern, of such an array was shown in the preceding section to be given by

$$E = \frac{\sin(\psi'/2)}{\psi'/2} \tag{1}$$

where $\psi' = (2\pi a/\lambda)\sin\theta$ and where θ is in the xy plane (Fig. 4-37). This pattern, in the xy plane, is independent of the extent of the array in the z direction (normal to the page).

In deriving (1), that is, (4-14-12) of Sec. 4-14, the total field at a point was obtained by integrating the contributions from a continuous array of sources distributed over a length a. For points at a great distance from the array the integral can be simplified, and the integration is straightforward, as demonstrated in the preceding section. For points near to the array, however, the integral does not simplify in this way but can be reduced to the form of Fresnel's integrals. The field variation near the slot as obtained in this way is commonly called a Fresnel diffraction pattern. Along a straight line parallel to the slot and a short distance from it, the field variation is as suggested at (*a*) in Fig. 4-41, the variation approximating the uniform distribution of field at the slot as shown in Fig. 4-40*b*. As the distance x from the slot is increased, the Fresnel patterns change through a series of transitional forms, such as suggested at (*b*) in Fig. 4-41, until at large distances we enter the Fraunhofer region and the pattern assumes a form as suggested by (*c*) in Fig. 4-41. Ordinarily the Fraunhofer pattern is obtained by rotating the slot around its center so that the field is observed at a constant radius rather than at a constant distance x. The resulting field pattern in polar coordinates is then as suggested at (*d*) in Fig. 4-41. Once we have entered the Fraunhofer region, this pattern is the same at all greater distances. For a point to be in the Fraunhofer region, it must be at a sufficient distance from the slot so that we can make the assumption that lines extending from the edges of the slot to the point are parallel. This is commonly assumed to be the case when the point is at a distance r from the slot given by

$$r \geq \frac{2a^2}{\lambda} \tag{2}$$

where a is the width or aperture of the slot, which is assumed to be large. Thus, the larger the aperture or the shorter the wavelength, the greater must be the distance at which the pattern is measured if we wish to avoid the effects of Fresnel diffraction. This is discussed further in Sec. 18-3a.

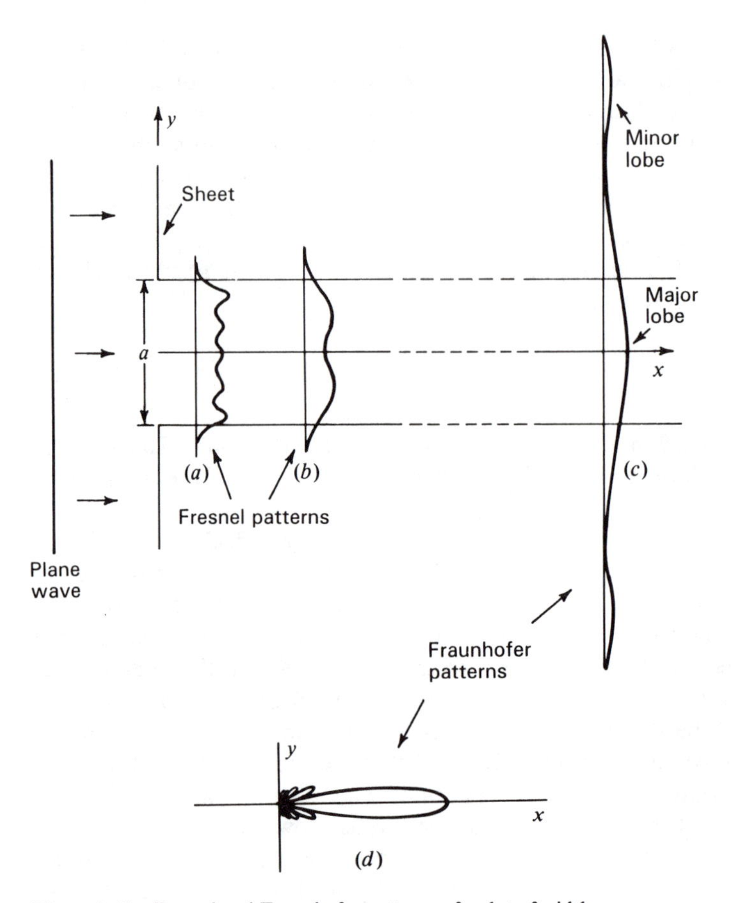

Figure 4-41 Fresnel and Fraunhofer patterns of a slot of width a.

A nearly uniform type of field distribution across an aperture such as discussed above in connection with Figs. 4-40 and 4-41 occurs in optics when a beam of light is incident on a slit. It also may be realized by the field distribution across the mouth of a long electromagnetic horn antenna as in Fig. 4-42*a*. Since the pattern of a uniform field distribution is the same as the pattern of a uniform distribution of point sources of equal extent, another form of antenna equivalent to the optical slit or electromagnetic horn is a uniform current sheet. This can be approximated by a "billboard" type of array, as in Fig. 4-42*b*, having many dipole antennas carrying equal currents. The expressions which have been developed can thus be applied to a calculation of the Fraunhofer diffraction pattern of an optical slit or the far field of a horn or uniform current sheet. If the field or current distribution across the slit or antenna aperture is not uniform, the form factor for the distribution will appear in the integral for the field expression. If the aperture is large, the relations developed for amplitude distributions of arrays of discrete sources can be applied to the case of continuous arrays of sources.

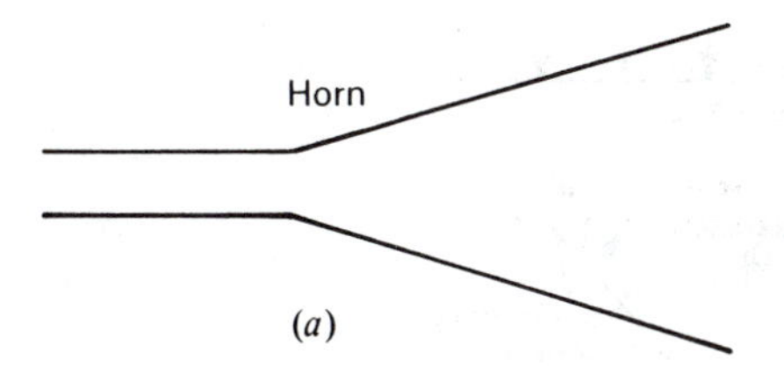

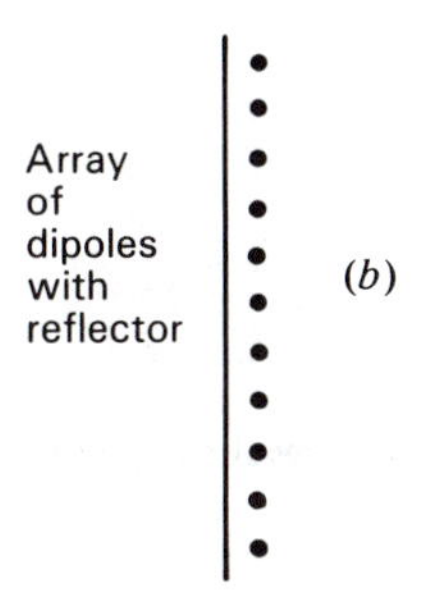

Figure 4-42 Electromagnetic horn antenna and array of dipoles with reflector.

Huygens' principle is not without its limitations. Thus, it neglects the vector nature of the electromagnetic field. It also neglects the effect of currents which flow at the edge of the slot, as in Figs. 4-40 and 4-41, or at the edge of the horn, as in Fig. 4-42*a*. However, if the aperture is sufficiently large and we confine our attention to directions roughly normal to aperture, the scalar theory of Huygens' principle gives satisfactory results.

4-16 HUYGENS' PRINCIPLE APPLIED TO THE DIFFRACTION OF A PLANE WAVE INCIDENT ON A FLAT SHEET. PHYSICAL OPTICS. Consider a uniform plane wave incident on a perfectly conducting half-plane, as in Fig. 4-43*a*.[1] We want to calculate the electric field at point P at a distance r behind the plane. By Huygens' principle,

$$E = \int_{\substack{\text{over} \\ x \text{ axis}}} dE \tag{1}$$

where dE is the electric field at P due to a point source at a distance x from the origin, as in Fig. 4-43*b*. Thus,

$$dE = \frac{E_0}{r} e^{-j\beta(r+\delta)} \, dx \tag{2}$$

[1] J. D. Kraus, *Radio Astronomy*, 2nd ed., Cygnus-Quasar Books, 1986.

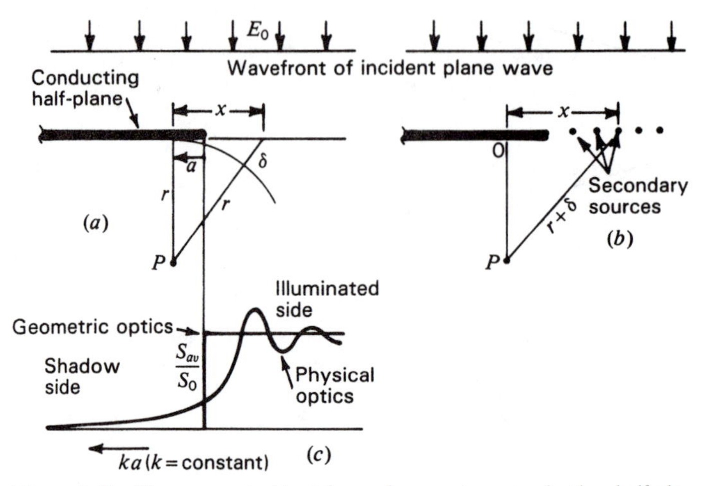

Figure 4-43 Plane wave incident from above onto a conducting half-plane with resultant power-density variation below the plane as obtained by physical optics.

so that

$$E = \frac{E_0}{r} e^{-j\beta r} \int_a^\infty e^{-j\beta\delta} \, dx \tag{3}$$

If $\delta \ll r$, it follows that

$$\delta = \frac{x^2}{2r} \tag{4}$$

When we let $2/r\lambda = k^2$ and $kx = u$, (3) becomes

$$E = \frac{E_0}{kr} e^{-j\beta r} \int_{ka}^\infty e^{-j\pi u^2/2} \, du \tag{5}$$

which can be rewritten as

$$E = \frac{E_0}{kr} e^{-j\beta r} \int_0^\infty e^{-j\pi u^2/2} \, du - \int_0^{ka} e^{-j\pi u^2/2} \, du \tag{6}$$

The integrals in (6) have the form of Fresnel integrals so (6) can be written

$$E = \frac{E_0}{kr} e^{-j\beta r} \{\tfrac{1}{2} + \tfrac{1}{2}j - [C(ka) + jS(ka)]\} \tag{7}$$

where

$$C(ka) = \int_0^{ka} \cos \frac{\pi u^2}{2} \, du = \text{Fresnel cosine integral} \tag{8}$$

$$S(ka) = \int_0^{ka} \sin \frac{\pi u^2}{2} \, du = \text{Fresnel sine integral} \tag{9}$$

where $ka = \sqrt{\frac{2}{r\lambda}}\, a$, dimensionless.

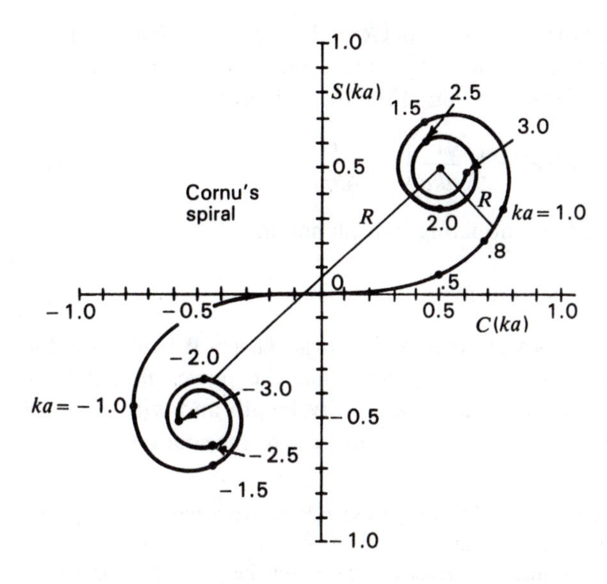

Figure 4-44 Cornu spiral showing $C(ka)$ and $S(ka)$ as a function of ka values along the spiral. For example, when $ka = 1.0$, $C(ka) = 0.780$ and $S(ka) = 0.338$. When $ka = \infty$, $C(ka) = S(ka) = \frac{1}{2}$.

A graph of $C(ka)$ and $S(ka)$ yields the Cornu spiral (Fig. 4-44). Since $C(-ka) = -C(ka)$ and $S(-ka) = -S(ka)$, the spiral for negative values of ka is in the third quadrant and is symmetrical with respect to the origin for the spiral in the first quadrant.

The power density as a function of ka is then

$$S_{av} = \frac{EE^*}{Z} = S_0 \tfrac{1}{2}\{[\tfrac{1}{2} - C(ka)]^2 + [\tfrac{1}{2} - S(ka)]^2\} \qquad (\text{W m}^{-2}) \tag{10}$$

where $$S_0 = \frac{E_0^2 \lambda}{2Zr} \qquad (\text{W m}^{-2}) \tag{11}$$

The power density variation of (10) as a function of ka (with r, λ and k constant) is shown in Fig. 4-43*c*. Assuming that the plane wave originates from a distant source we have

1. For no obscuration, $ka = -\infty$ and $S_{av} = S_0$.
2. For source, observer and edge of obscuring plane in line, $ka = 0$ and $S_{av} = \frac{1}{4}S_0$.
3. For complete obscuration, $ka = +\infty$ and $S_{av} = 0$.

Thus, the power density does not go to zero abruptly as the point of observation goes from the illuminated side ($ka < 0$) to the shadow side ($ka > 0$); rather, there are fluctuations followed by a gradual decrease in power density.

From (10) and (11) the relative power density as a function of ka is

$$S_{av}(\text{relative}) = \frac{S_{av}}{S_0} = \tfrac{1}{2}\{[\tfrac{1}{2} - C(ka)]^2 + [\tfrac{1}{2} - S(ka)]^2\} \tag{12}$$

The relative power density (12) is equal to $\frac{1}{2}R^2$, where R is the distance from a ka value on the Cornu spiral to the point $(\frac{1}{2}, \frac{1}{2})$ (see Fig. 4-44). For large positive values of ka, R approaches $1/\pi ka$, so that (12) reduces approximately to

$$S_{av}(\text{relative}) = \frac{1}{2}\left(\frac{1}{\pi ka}\right)^2 = \frac{r\lambda}{4\pi^2 a^2} \tag{13}$$

where r = distance from obstacle (conducting half-plane), m
λ = wavelength, m
a = distance into shadow region, m

Equation (13) gives the relative power density for large ka (>3) (well into the shadow region). For this condition it is apparent that the power flux density (Poynting vector) due to diffraction increases with wavelength and with distance (from edge) but decreases as the square of the distance a into the shadow region.

Example. A vertical conducting wall 25 m high extends above a flat ground plane. A λ = 10-cm transmitter is situated 25 m above the ground plane at a large distance to one side of the vertical wall and a receiver is located on the ground plane 100 m to the other side of the wall. Find the signal level at the receiver due to diffraction over the wall as compared to the direct path signal without the wall.

Solution. The constant $k = \sqrt{2/r\lambda} = \sqrt{2/100 \times 0.1} = 0.44$ and $a = 25$ m, so $ka = 11$ which is greater than 3. Thus, (13) is applicable and

$$S_{av}(\text{relative}) = \frac{r\lambda}{4\pi^2 a^2} = \frac{100 \times 0.1}{4\pi^2 \times 25^2} \simeq \frac{1}{2500} \text{ or } -34 \text{ dB}$$

Thus, the vertical wall causes 34 dB of attenuation as compared to a direct path signal.

If the half-plane in Fig. 4-43 is replaced by a strip of width D and length $\gg D$, diffraction occurs from both edges, scattering radiation into the shadow region behind the strip. On the centerline of the strip, diffraction fields from both edges are equal in magnitude and of the same phase since the path lengths from both edges are equal. Thus, the diffracted field has a maximum or central peak on the centerline.

If the strip is replaced by a disk of diameter D, there is diffraction around its entire edge and all diffracted fields arrive in phase on the centerline behind the disk producing a larger central peak. In optics this peak is called the *axial bright spot*. In a similar way, the diffracted fields from the feed system at the focus of a parabolic dish reflector can produce a back lobe on the axis of the parabola. See additional discussions on diffraction in Secs. 2-18, 12-2, 13-3, 17-5 and 18-3d.

4-17 RECTANGULAR-AREA BROADSIDE ARRAYS.

The method of obtaining the field patterns of linear arrays discussed in the preceding sections

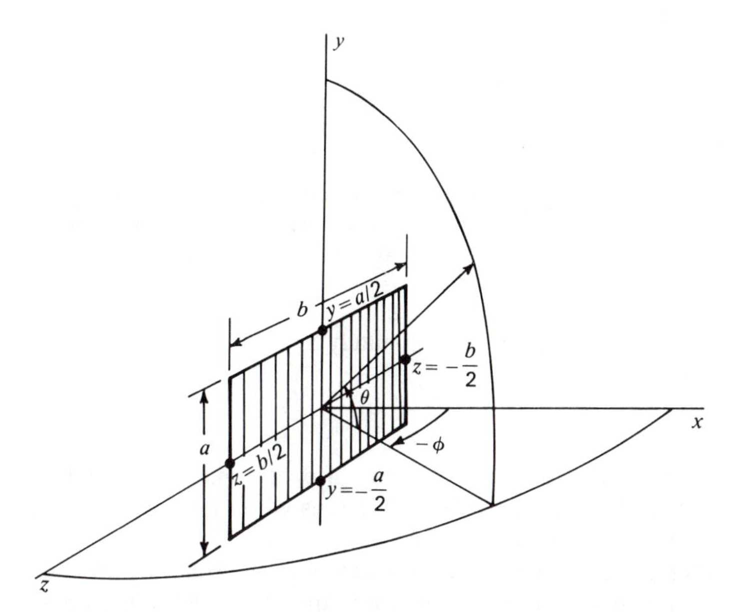

Figure 4-45 Rectangular broadside array of height a and length b with relation to coordinates.

can be easily extended to the case of rectangular broadside arrays, i.e., arrays of sources which occupy a flat area of rectangular shape, as in Fig. 4-45. For such a rectangular array, the field pattern in the xy plane (as a function of θ) depends only on the y-dimension a of the array, while the field pattern in the xz plane (as a function of ϕ) depends only on the z-dimension b of the array. The assumption is made that the field or current distribution across the array in the y direction is the same for any values of z between $\pm b/2$. Likewise, it is assumed that the amplitude distribution across the array in the z direction is the same for all values of y between $\pm a/2$. Therefore, the field pattern in the xy plane is calculated as though the array consists only of a single linear array of height a coincident with the y axis (y array). In the same way, the pattern in the xz plane is obtained by calculating the pattern of a single linear array of length b coincident with the z axis (z array). If the array also has depth in the x direction, i.e., has end-fire directivity, then the pattern in the xy plane is the product of the patterns of the single linear x and y arrays, while the pattern in the xz plane is the product of the patterns of the x and z arrays.

If the area occupied by the array is not rectangular in shape, the above principles do not hold. However, the approximate field patterns may be obtained in the case of an array of elliptical area, for example, by assuming that it is a rectangular area as in Fig. 4-46*a* or in the case of a circular area by assuming that it is square as in Fig. 4-46*b*.

From the field patterns in two planes (xy and xz) of a rectangular array the beam widths between half-power points can be obtained. If the minor lobes are

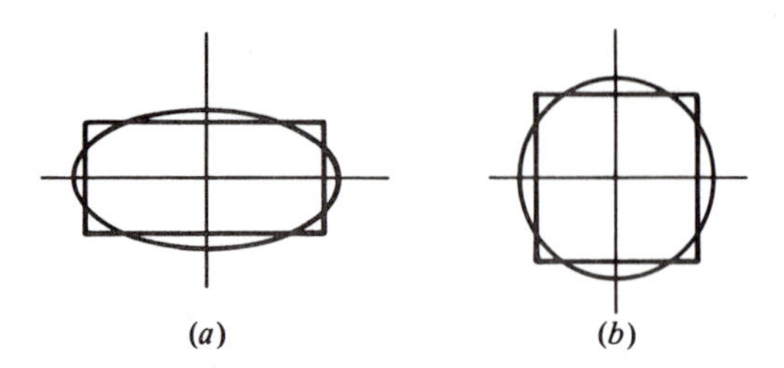

Figure 4-46 Elliptical array with equivalent rectangular array (*a*) and circular array with equivalent square array (*b*).

not large, the directivity D is then given approximately by

$$D = \frac{41\,000}{\theta_1^\circ \phi_1^\circ} \tag{1}$$

where θ_1° and ϕ_1° are the half-power beam widths in degrees in the xy and xz planes respectively. The limitations of (1) are discussed following (3-13-16), leading to (3-13-18) which includes correction factors.

An expression for the directivity of a large rectangular broadside array of height a and width b (Fig. 4-45) and with a uniform amplitude distribution may also be derived rigorously as follows. By (3-13-8) the directivity of an antenna is given by

$$D = \frac{4\pi f(\theta, \phi)_{\max}}{\iint f(\theta, \phi) \sin\theta \, d\theta \, d\phi} \tag{2}$$

where $f(\theta, \phi)$ is the space power pattern, which varies as the square of the space field pattern. From (4-14-17) the space field pattern of a large rectangular array is

$$E(\theta, \phi) = \frac{\sin\left[(a_r \sin\theta)/2\right]}{(a_r \sin\theta)/2} \frac{\sin\left[(b_r \sin\phi)/2\right]}{(b_r \sin\phi)/2} \tag{3}$$

where $a_r = 2\pi a/\lambda$
$b_r = 2\pi b/\lambda$

The main beam maximum is in the direction $\theta = \phi = 0$ in Fig. 4-45. In (3), $\theta = 0$ at the equator, while in (2), $\theta = 0$ at the zenith. For large arrays and relatively sharp beams we can therefore replace $\sin\theta$ and $\sin\phi$ in (3) by the angles, while $\sin\theta$ in (2) can be set equal to unity. Assuming that the array is unidirectional (no field in the $-x$ direction), the integral in the denominator of (2) then becomes

$$\int_{-\pi/2}^{\pi/2} \int_{-\pi/2}^{\pi/2} \frac{\sin^2(\pi a\theta/\lambda)}{(\pi a\theta/\lambda)^2} \frac{\sin^2(\pi b\phi/\lambda)}{(\pi b\phi/\lambda)^2} \, d\theta \, d\phi \tag{4}$$

Making the limits of integration $-\infty$ to $+\infty$ instead of $-\pi/2$ to $+\pi/2$, (4) may be evaluated as λ^2/ab. Therefore, the approximate directivity D of a large unidirectional rectangular broadside array with a uniform amplitude distribution is

$$D = \frac{4\pi ab}{\lambda^2} = 12.6 \frac{ab}{\lambda^2} \tag{5}$$

As an example, the directivity of a broadside array of height $a = 10\lambda$ and length $b = 20\lambda$ is, from (5), equal to 2520, or 34 dB.

4-18 ARRAYS WITH MISSING SOURCES AND RANDOM ARRAYS. A linear array of 5 isotropic point sources with $\lambda/2$ spacing is discussed in Sec. 4-10 for several amplitude distributions including uniform, binomial and Dolph-Tchebyscheff. Let us consider this 5-source array again with all sources of equal amplitude with pattern as in Fig. 4-47*a* (same as Fig. 4-29*a*) and

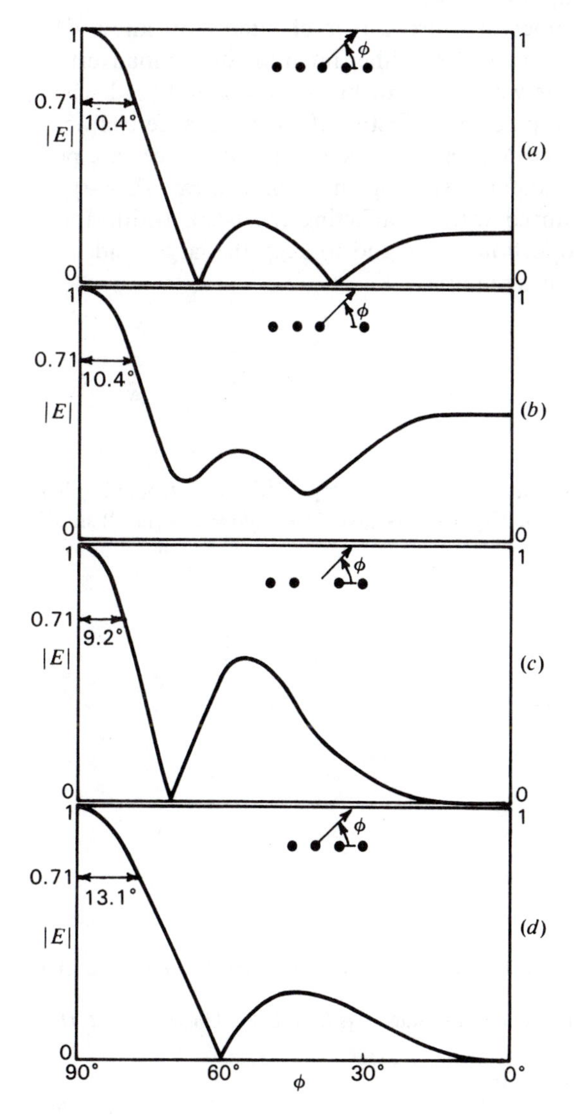

Figure 4-47 Field patterns of linear array of 5 isotropic point sources of equal amplitude and $\lambda/2$ spacing: (*a*) all 5 sources on, (*b*) one source (next to the edge) off, (*c*) one source (at the center) off and (*d*) one source (at the edge) off.

note what happens to the pattern if one of the sources is turned off (amplitude reduced to zero). If the off source is next to the edge source, the beam width is essentially unchanged but the minor-lobe level is up and nulls are filled as shown in Fig. 4-47*b*. When the center source is off, the beam width is reduced but the minor-lobe level is higher, as indicated in Fig. 4-47*c*. If an edge source is off, the array is identical with a uniform array of 4 sources and, as shown in Fig. 4-47*d*, has a larger beam width with the minor-lobe level slightly higher than for the uniform source array of Fig. 4-47*a*. (Compare curves for $n = 4$ and 5 in Fig. 4-20.)

If the amplitude distribution of the array is tapered (binomial or Dolph-Tchebyscheff) as in Fig. 4-29*b* and *c*, turning off a source at the edge will have less effect than in the uniform-amplitude case.

In an array of a large number of sources, it is of interest to know what happens if one or more sources are turned off either intentionally or inadvertently. Also to reduce cost, a designer would like to know how many (and which) sources can be omitted without appreciably affecting the performance characteristics. It has been shown by Lo and Maher and Cheng[1] that if sources are positioned with random instead of uniform spacing in a large array ($L \gg \lambda$) the number n of sources can be reduced without affecting the beam width appreciably. The gain, however, is proportional to n and to keep the largest sidelobes below a certain level, a minimum n is required.

PROBLEMS[2]

★4-1 Two point sources.

(*a*) Show that the relative $E(\phi)$ pattern of an array of 2 identical isotropic in-phase point sources arranged as in Fig. P4-1 is given by $E(\phi) = \cos\,[(d_r/2)\sin\phi]$, where $d_r = 2\pi d/\lambda$.

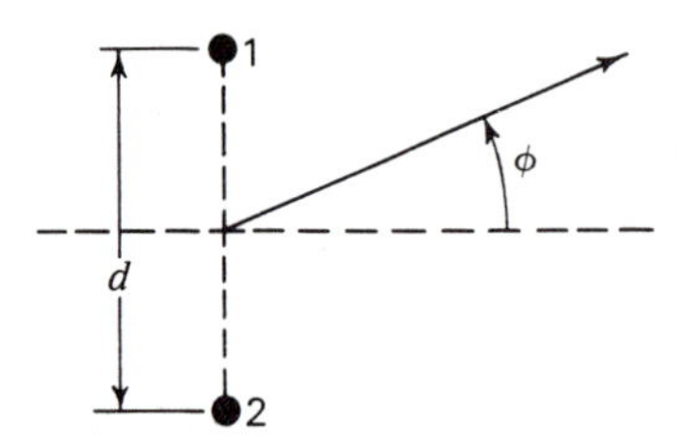

Figure P4-1 Two point sources.

[1] Y. T. Lo, "A Probabilistic Approach to the Design of Large Antenna Arrays," *IEEE Trans. Ants. Prop.*, **AP-11**, 95–96, 1963.

T. M. Maher and D. K. Cheng, "Random Removal of Radiators from Large Linear Arrays," *IEEE Trans. Ants. Prop.*, **AP-11**, 106–112, 1963.

[2] Answers to starred (*) problems are given in App. D.

(*b*) Show that the maxima, nulls and half-power points of the pattern are given by the following relations:

Maxima: $$\phi = \arcsin\left(\pm\frac{k\lambda}{d}\right)$$

Nulls: $$\phi = \arcsin\left[\pm\frac{(2k+1)\lambda}{2d}\right]$$

Half-power points: $$\phi = \arcsin\left[\pm\frac{(2k+1)\lambda}{4d}\right]$$

where $k = 0, 1, 2, 3, \ldots$.

(*c*) For $d = \lambda$ find the maxima, nulls and half-power points, and from these points and any additional points that may be needed plot the $E(\phi)$ pattern for $0° \leq \phi \leq 360°$. There are four maxima, four nulls and eight half-power points.

(*d*) Repeat for $d = 3\lambda/2$.

(*e*) Repeat for $d = 4\lambda$.

(*f*) Repeat for $d = \lambda/4$. Note that this pattern has two maxima and two half-power points but no nulls. The half-power points are minima.

4-2 Four sources in square array.

(*a*) Derive an expression for $E(\phi)$ for an array of 4 identical isotropic point sources arranged as in Fig. P4-2. The spacing d between each source and the center point of the array is $3\lambda/8$. Sources 1 and 2 are in phase, and sources 3 and 4 in opposite phase with respect to 1 and 2.

(*b*) Plot, approximately, the normalized field pattern.

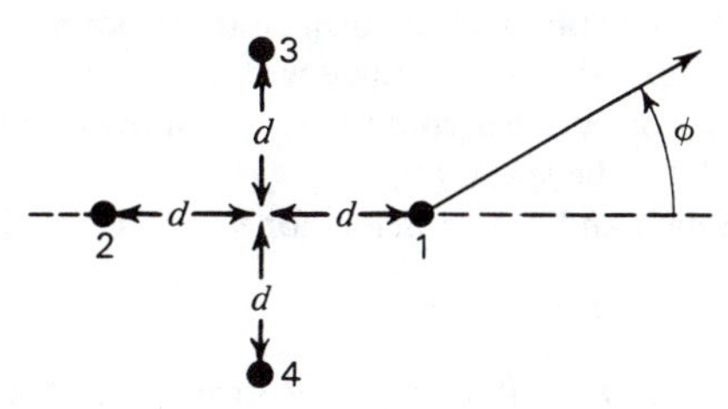

Figure P4-2 Four sources in square array.

4-3 Two point sources.

(*a*) What is the expression for $E(\phi)$ for an array of 2 point sources arranged as in the figure for Prob. 4-1. The spacing d is $3\lambda/8$. The amplitude of source 1 in the ϕ plane is given by $|\cos\phi|$ and the phase by ϕ. The amplitude of the field of source 2 is given by $|\cos(\phi - 45°)|$ and the phase of the field by $\phi - 45°$.

(*b*) Plot the normalized amplitude and the phase of $E(\phi)$ referring the phase to the centerpoint of the array.

★4-4 Four sources in broadside array.

(*a*) Derive an expression for $E(\phi)$ for a linear in-phase broadside array of 4 identical isotropic point sources. Take $\phi = 0$ in the broadside direction. The spacing between sources is $5\lambda/8$.

(*b*) Plot, approximately, the normalized field pattern ($0° \leq \phi \leq 360°$).

(*c*) Repeat parts (*a*) and (*b*) with the changed condition that the amplitudes of the 4 sources are proportional to the coefficients of the binomial series for $(a + b)^{n-1}$.

4-5 Tchebyscheff $T_3(x)$ and $T_6(x)$.

(*a*) Calculate and plot $\cos\theta$ as x and $\cos 3\theta$ as y, for $-1 \leq x \leq +1$. Compare with the curve for $T_3(x)$.

(*b*) Calculate and plot $\cos\theta$ as x and $\cos 6\theta$ as y, for $-1 \leq x \leq +1$. Compare with the curve for $T_6(x)$.

★4-6 Five source Dolph-Tchebyscheff (D-T) distribution.

(*a*) Find the Dolph-Tchebyscheff current distribution for the minimum beam width of a linear in-phase broadside array of five isotropic point sources. The spacing between sources is $\lambda/2$ and the sidelobe level is to be 20 dB down. Take $\phi = 0$ in the broadside direction.

(*b*) Locate the nulls and maxima of the minor lobes.

(*c*) Plot, approximately, the normalized field pattern $(0° \leq \phi \leq 360°)$.

(*d*) What is the half-power beam width?

4-7 Eight-source D-T distribution.

(*a*) Find the Dolph-Tchebyscheff current distribution for the minimum beam width of a linear in-phase broadside array of 8 isotropic sources. The spacing between elements is $3\lambda/4$ and the sidelobe level is to be 40 dB down. Take $\phi = 0$ in the broadside direction.

(*b*) Locate the nulls and the maxima of the minor lobes.

(*c*) Plot, approximately, the normalized field pattern $(0° \leq \phi \leq 360°)$.

(*d*) What is the half-power beam width?

4-8 *n*-source array.

(*a*) Derive an expression for $E(\psi)$ for an array of n identical isotropic point sources where $\psi = f(\phi, d, \delta)$. ϕ is the azimuthal position angle with $\phi = 0$ in the direction of the array. δ is the phase lag between sources as one moves along the array in the $\phi = 0°$ direction and d is the spacing.

(*b*) Plot the normalized field as ordinate and ψ as abscissa for $n = 2, 4, 6, 8, 10$ and 12 for $0° \leq \psi \leq 180°$.

4-9 Ten-source end-fire array.

(*a*) Plot $E(\phi)$ for an end-fire array of $n = 10$ identical isotropic point sources spaced $3\lambda/8$ apart with $\delta = -3\pi/4$.

(*b*) Repeat with $\delta = -\pi[(3/4) + (1/n)]$.

★4-10 Two-source broadside array.

(*a*) Calculate the directivity of a broadside array of two identical isotropic in-phase point sources spaced $\lambda/2$ apart along the polar axis, the field pattern being given by

$$E = \cos\left(\frac{\pi}{2}\cos\theta\right)$$

where θ is the polar angle.

(*b*) Show that the directivity for a broadside array of two identical isotropic in-phase point sources spaced a distance d is given by

$$D = \frac{2}{1 + (\lambda/2\pi d)\sin(2\pi d/\lambda)}$$

4-11 Two-source end-fire array.

(*a*) Calculate the directivity of an end-fire array of two identical isotropic point sources in phase opposition, spaced $\lambda/2$ apart along the polar axis, the relative field pattern being given by

$$E = \sin\left(\frac{\pi}{2}\cos\theta\right)$$

where θ is the polar angle.

(*b*) Show that the directivity of an ordinary end-fire array of two identical isotropic point sources spaced a distance d is given by

$$D = \frac{2}{1 + (\lambda/4\pi d)\sin(4\pi d/\lambda)}$$

4-12 Four-tower BC array. A broadcasting station requires the horizontal plane pattern indicated by Fig. P4-12. The maximum field intensity is to be radiated northeast with as little decrease as possible in field intensity in the 90° sector between north and east. No nulls are permitted in this sector. Nulls may occur in any direction in the complementary 270° sector. However, it is required that nulls must be present for the directions of due west and due southwest, in order to prevent interference with other stations in these directions.

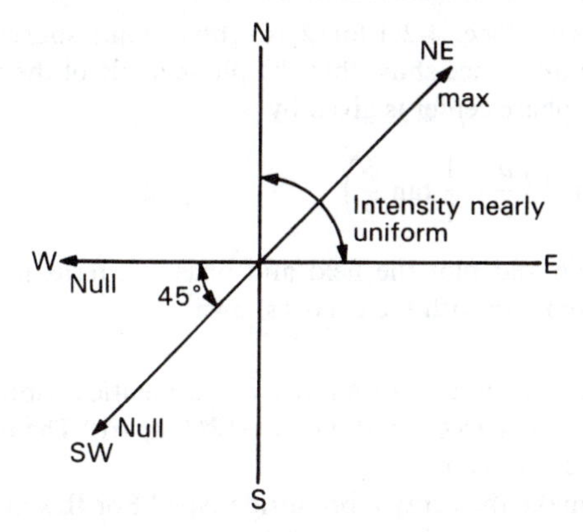

Figure P4-12 Four-tower BC array pattern requirements.

Design a four-vertical-tower array to fulfill these requirements. The currents are to be equal in magnitude in all towers, but the phase may be adjusted to any relationship. There is also no restriction on the spacing or geometrical arrangements of the towers. Plot the field pattern.

4-13 Two-source patterns. Calculate and plot the field and phase patterns for an array of two isotropic sources of the same amplitude and phase, for two cases:

(*a*) $d = \frac{3}{4}\lambda$

(*b*) $d = \frac{3}{2}\lambda$

Plot the field pattern in polar coordinates and phase pattern in rectangular coordinates with:

1. Phase center at source 1
2. Phase center at midpoint

4-14 Field and phase patterns. Calculate and plot the field and phase patterns of an array of 2 nonisotropic dissimilar sources for which the total field is given by

$$E = \cos \phi + \sin \phi \underline{/\psi}$$

where $\psi = d \cos \phi + \delta = \frac{\pi}{2} (\cos \phi + 1)$

Take source 1 as the reference for phase. See Fig. P4-14.

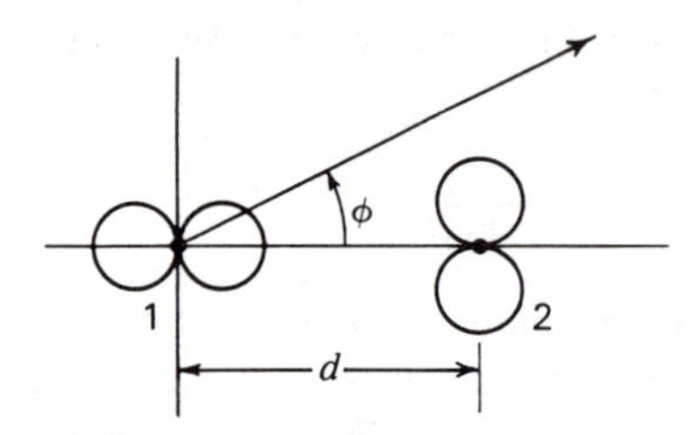

Figure P4-14 Field and phase patterns.

★**4-15 DT 6-source array.** Calculate the Dolph-Tchebyscheff distribution of a six-source broadside array for $R = 5$, 7 and 10. Explain the variation.

4-16 Two-unequal-source array. In Case 5 (Sec. 4-2e) for 2 isotropic point sources of unequal amplitude and any phase difference show that the phase angle of the total field with midpoint of the array as phase center is given by

$$\tan^{-1}\left(\frac{a-1}{a+1} \tan \frac{\psi}{2}\right)$$

4-17 Field and phase patterns. Calculate and plot the field and phase patterns for the cases of Figs. 4-21 and 4-22 and compare with the curves shown.

★**4-18 Five-source array.**

(*a*) What is an expression for the field pattern of an array of 5 identical isotropic point sources arranged in line and spaced a distance d ($\leq \lambda/2$) apart? The phase lead of source 2 over 1, 3 over 2, etc., is δ.

(*b*) What value should δ have to make the array a broadside type? For this broadside case, what are the relative current magnitudes of the sources for:

1. Maximum directivity
2. No sidelobes
3. Sidelobes equal in magnitude to "main" lobe

4-19 Two-tower BC array. A broadcast array of 2 vertical towers with equal currents is to have a horizontal plane pattern with a broad maximum of field intensity to the north and a null at an azimuth angle of 131° measured counterclockwise from the north. Specify the arrangement of the towers, their spacing, and phasing. Calculate and plot the field pattern in the horizontal plane.

4-20 Three-tower BC array. A broadcast array with 3 vertical towers arranged in a straight horizontal line is to have a horizontal-plane pattern with a broad maximum of field intensity to the north and nulls at azimuth angles of 105°, 147° and 213° measured counterclockwise from the north. The towers need not have equal currents. For the purpose of analysis the center tower (2) may be regarded as

2 towers, $2a$ and $2b$, $2a$ belonging to an array of itself and tower 1 and $2b$ to an array of itself and tower 3. Specify the arrangement of towers, their spacing, currents and phasing. Calculate and plot the field pattern in the horizontal plane.

4-21 Four-tower BC array. A broadcast array of 4 vertical towers with equal currents is to have a symmetrical 4-lobed pattern in the horizontal plane with maximum field intensity to the north, east, south and west and a reduced field intensity to the northeast, southeast, southwest and northwest equal to $\frac{1}{2}$ the maximum. Specify the array arrangement, orientation, spacing and phasing. Calculate and plot the field pattern in the horizontal plane.

★**4-22 Eight-source end-fire array.**

(*a*) Calculate and plot the field pattern of a linear array of 8 isotropic point sources of equal amplitude spaced 0.2λ apart for the ordinary end-fire condition.

(*b*) Repeat, assuming that the phasing satisfies the Hansen and Woodyard increased-directivity condition.

(*c*) Calculate the directivity in both cases by graphical or numerical integration of the entire pattern.

★**4-23 Rectangular current sheet.** Calculate and plot the patterns in both planes perpendicular to a rectangular sheet carrying a current of uniform density and everywhere of the same direction and phase if the sheet measures 10 by 20λ. What is the approximate directivity?

4-24 Twelve-source end-fire array.

(*a*) Calculate and plot the field pattern of a linear end-fire array of 12 isotropic point sources of equal amplitude spaced $\lambda/4$ apart for the ordinary end-fire condition.

(*b*) Calculate the directivity by graphical or numerical integration of the entire pattern. Note that it is the power pattern (square of field pattern) which is to be integrated. It is most convenient to make the array axis coincide with the polar or z axis of Fig. 3-2 so that the pattern is a function only of θ.

(*c*) Calculate the directivity by the approximate half-power beam-width method and compare with that obtained in (*b*).

★**4-25 Twelve-source broadside array.**

(*a*) Calculate and plot the pattern of a linear broadside array of 12 isotropic point sources of equal amplitude spaced $\lambda/4$ apart with all sources in the same phase.

(*b*) Calculate the directivity by graphical or numerical integration of the entire pattern and compare with the directivity obtained in Prob. 4-24 for the same size array operating end-fire.

(*c*) Calculate the directivity by the approximate half-power beam-width method and compare with that obtained in (*b*).

4-26 Twelve-source end-fire array with increased directivity.

(*a*) Calculate and plot the pattern of a linear end-fire array of 12 isotropic point sources of equal amplitude spaced $\lambda/4$ apart and phased to fulfill the Hansen and Woodyard increased-directivity condition.

(*b*) Calculate the directivity by graphical or numerical integration of the entire pattern and compare with the directivity obtained in Probs. 4-24 and 4-25.

(*c*) Calculate the directivity by the approximate half-power beam-width method and compare with that obtained in (*b*).

4-27 *n*-source array. Variable phase velocity. Referring to Fig. 4-18, assume that the uniform array of n isotropic point sources is connected by a transmission system extending along the array with the feed point at source 1 so that the phase of source 2 lags 1 by $\omega d/v$, 3 lags 1 by $2\omega d/v$, etc., where v is the phase velocity to the right along the transmission system. Show that the far field is given by (4-6-8) where $\psi = d_r[\cos\phi - (1/p)]$, where p is the relative phase velocity, that is, $p = v/c$ where c is the velocity of light. Show also that $p = \infty$ for the broadside case, $p = 2$ for the maximum field at $\phi = 60°$, $p = 1$ for the ordinary end-fire case and $p = 1/[1 + (1/2nd_\lambda)]$ for the increased-directivity end-fire case.

4-28 Continuous array. Variable phase velocity. Consider that the array of discrete sources in Fig. 4-18 is replaced by a continuous array of length L and assume that it is energized like the array of Prob. 4-27. Show that the far field for the general case of any phase lag δ' per unit distance along the continuous array is given by (4-14-18) where $\psi' = L_r \cos\phi - \delta' L = L_r[\cos\phi - (1/p)]$, where $p = v/c$ as in Prob. 4-27. Show also that for the four cases considered in Prob. 4-27 the p values are the same except for the increased-directivity end-fire case where $p = 1/[1 + (1/2L_\lambda)]$.

4-29 Binomial distribution. Use the principle of pattern multiplication to show that a linear array with binomial amplitude distribution has a pattern with no minor lobes.

4-30 Two-source array. Show that for a 2-source array the field patterns

$$E = \frac{\sin(n\psi/2)}{\sin(\psi/2)} \qquad \text{and} \qquad E = 2\cos\frac{\psi}{2}$$

are equivalent.

4-31 Directivity of ordinary end-fire array. Show that the directivity of an ordinary end-fire array may be expressed as

$$D = \frac{n}{1 + (\lambda/2\pi nd)\sum_{k=1}^{n-1}[(n-k)/k]\sin(4\pi kd/\lambda)}$$

Note that

$$\left[\frac{\sin(n\psi/2)}{\sin(\psi/2)}\right]^2 = n + \sum_{k=1}^{n-1} 2(n-k)\cos 2k\frac{\psi}{2}$$

4-32 Directivity of broadside array. Show that the directivity of a broadside array may be expressed as

$$D = \frac{n}{1 + (\lambda/\pi nd)\sum_{k=1}^{n-1}[(n-k)/k]\sin(2\pi kd/\lambda)}$$

4-33 Phase center. Show that the phase center of a uniform array is at its centerpoint.

★4-34 Three-source array. The center source of a 3-source array has a (current) amplitude of unity. For a sidelobe level 0.1 of the main-lobe maximum field, find the Dolph-Tchebyscheff value for the amplitude of the end sources. The source spacing $d = \lambda/2$.

4-35 End-fire arrays. $\lambda/2$ spacing. The following BASIC program provides antenna field-pattern graphs in polar and rectangular coordinates for an array of 4 sources as illustrated in Fig. P4-35. Using modifications of this program, produce graphs of the patterns for a larger number of sources spaced $\lambda/2$ apart with end-fire phasing, such as (*a*) 5 sources, (*b*) 6 sources, (*c*) 8 sources and (*d*) 12 sources.

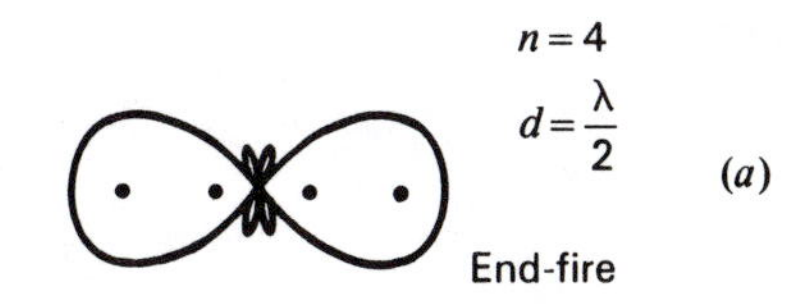

(*a*)

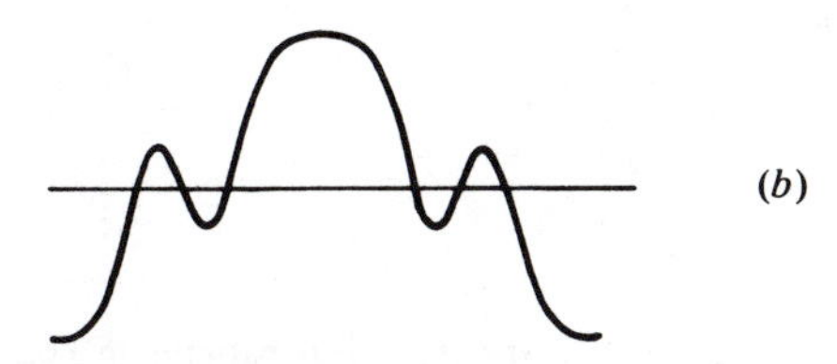

(*b*)

Figure P4-35 End-fire array of 4 sources with $\lambda/2$ spacing. Field patterns in polar and rectangular coordinates.

```
END-FIRE ARRAY N = 4  d = λ/2
POLAR PLOT
10 HOME
20 HGR
30 HCOLOR = 3
40 FOR A = .02 TO 3.12 STEP .01
50 R = 15 * SIN(6.28 * (COS(A) - 1))/SIN(1.57 * (COS(A) - 1))
60 HPLOT 138 + R * COS(A),79 + R * SIN(A)
61 HPLOT 138 + R * (-COS(A)),79 + R * SIN(A)
70 NEXT A
RECTANGULAR PLOT (POLAR PLOT STEPS 10 60 61 70 OMITTED)
60 HPLOT A *· 30,R + 75
61 HPLOT (A + 3.16) * 30,-R + 75
70 NEXT A
```

See also App. B.

4-36 End-fire arrays. $\lambda/4$ spacing. Repeat Prob. 4-35 for the case where the spacing is $\lambda/4$ instead of $\lambda/2$. The patterns in this case will be unidirectional instead of bidirectional as with $\lambda/2$ spacing.

4-37 Two-element interferometer. Using a computer as in the above problems, produce graphs of the field patterns of 2 isotropic in-phase sources with spacings of (*a*) 8λ, (*b*) 16λ and (*c*) 32λ.

★4-38 Two sources in phase. Two isotropic point sources of equal amplitude and same phase are spaced 2λ apart. (*a*) Plot a graph of the field pattern. (*b*) Tabulate the angles for maxima and nulls.

4-39 Two sources in opposite phase. Two isotropic sources of equal amplitude and opposite phase have 1.5λ spacing. Find the angles for all maxima and nulls.

4-40 Broadside arrays. $\lambda/2$ spacing. The following BASIC program provides antenna field-pattern graphs in polar and rectangular coordinates for an array of 4 sources as illustrated in Fig. P4-40. Using modifications of this program, produce graphs of the patterns for a larger number of in-phase sources spaced $\lambda/2$ apart, such as (*a*) 6 sources, (*b*) 8 sources and (*c*) 12 sources.

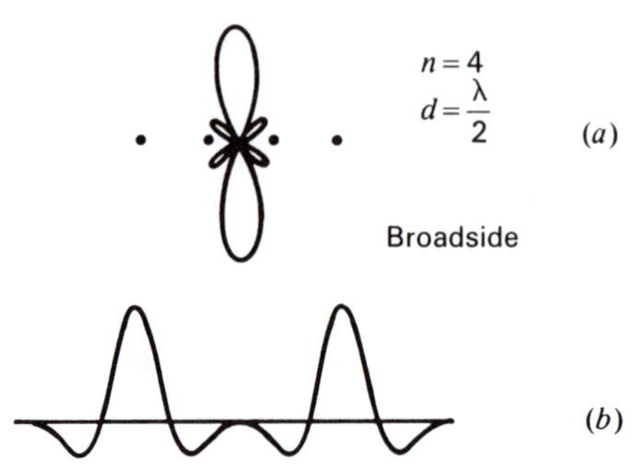

Figure P4-40 Broadside array of 4 sources with $\lambda/2$ spacing. Field patterns in polar and rectangular coordinates.

```
BROADSIDE ARRAY N = 4     d = λ/2
POLAR PLOT
10 HOME
20 HGR
30 HCOLOR = 3
40 FOR A = .02 TO 6.26 STEP .01
50 R = 15 * SIN(6.28 * COS(A))/SIN(1.57 * COS(A))
60 HPLOT 138 + R * COS(A),79 + R * SIN(A)
70 NEXT A
RECTANGULAR PLOT (POLAR PLOT STEPS 10 60 70 OMITTED)
60 HPLOT A * 30,R + 75
70 NEXT A
```

See also App. B.

4-41 Three unequal sources. Three isotropic in-line sources have $\lambda/4$ spacing. The middle source has 3 times the current of the end sources. If the phase of the middle source is 0°, the phase of one end source +90° and the phase of the other end source −90°, make a graph of the normalized field pattern.

4-42 Long broadside array. Show that the HPBW of a long uniform broadside array is given (without approximation) by $50.8°/L_\lambda$, where $L_\lambda = L/\lambda$ = length of array in wavelengths.

4-43 Phase center of 2-source array. An array consists of 2 isotropic point sources, one at the origin and one at a distance of $\lambda/2$ in the x direction. If the source at the origin has twice the amplitude (field) of the other source, find the position of the phase center of the array.

★**4-44 24-source end-fire array.** A uniform linear array has 24 isotropic point sources with a spacing of $\lambda/2$. If the phase difference $\delta = -\pi/2$ (ordinary end-fire condition), calculate exactly (*a*) the HPBW, (*b*) the first sidelobe level, (*c*) the beam solid angle, (*d*) the beam efficiency, (*e*) the directivity and (*f*) the effective aperture.

4-45 Stray factor and directive gain. The ratio of the main beam solid angle Ω_M to the (total) beam solid angle Ω_A is called the *main beam efficiency*. The ratio of the minor-lobe solid angle Ω_m to the (total) beam solid angle Ω_A is called the *stray factor*. It follows that $\Omega_M/\Omega_A + \Omega_m/\Omega_A = 1$. Show that the *average directive gain* over the minor lobes of a highly directive antenna is nearly equal to the stray factor. The directive gain is equal to the directivity multiplied by the normalized power pattern $[=DP_n(\theta, \phi)]$, making it a function of angle with the maximum value equal to D.

4-46 Power patterns. Write and run power-pattern programs for Probs. 4-35 and 4-40.

4-47 End-fire array with increased gain. Write and run normalized field- and power-pattern programs for end-fire arrays with $d = \lambda/4$, $\delta = -(2\pi d/\lambda) - (\pi/n)$ for $n = 4$, 8, 12 and 16.

4-48 Grating lobe pattern. Write and run field-pattern programs for broadside arrays with $d = \lambda$ for $n = 4$, 8, 12 and 16. With $d \geq \lambda$, grating lobes appear.

CHAPTER 5

THE ELECTRIC DIPOLE AND THIN LINEAR ANTENNAS

5-1 THE SHORT ELECTRIC DIPOLE. Since any linear antenna may be considered as consisting of a large number of very short conductors connected in series, it is of interest to examine first the radiation properties of short conductors. From a knowledge of the properties of short conductors, we can then proceed to a study of long linear conductors such as are commonly employed in practice.

A short linear conductor is often called a short *dipole*. In the following discussion, a short dipole is always of finite length even though it may be very short. If the dipole is vanishingly short, it is an infinitesimal dipole.

Let us consider a short dipole such as shown in Fig. 5-1*a*. The length L is very short compared to the wavelength ($L \ll \lambda$). Plates at the ends of the dipole provide capacitive loading. The short length and the presence of these plates result in a uniform current I along the entire length L of the dipole. The dipole may be energized by a balanced transmission line, as shown. It is assumed that the transmission line does not radiate and, therefore, its presence will be disregarded. Radiation from the end plates is also considered to be negligible. The diameter d of the dipole is small compared to its length ($d \ll L$). Thus, for purposes of analysis we may consider that the short dipole appears as in Fig. 5-1*b*. Here it consists simply of a thin conductor of length L with a uniform current I

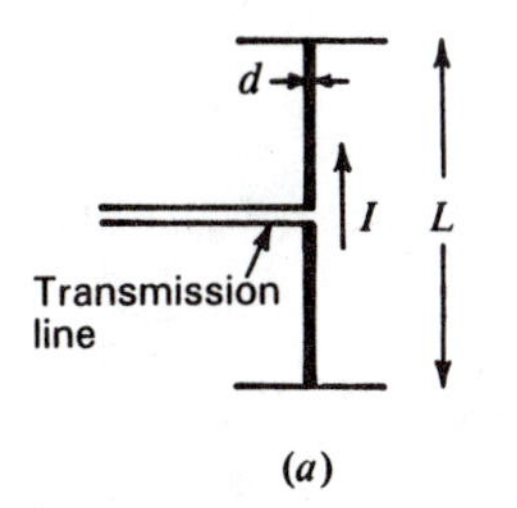

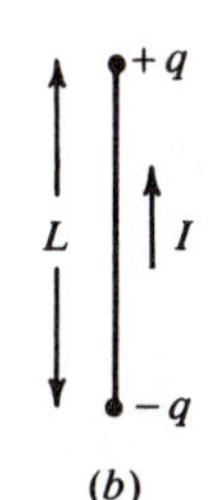

Figure 5-1 A short dipole antenna (*a*) and its equivalent (*b*).

and point charges q at the ends. The current and charge are related by

$$\frac{dq}{dt} = I \tag{1}$$

5-2 THE FIELDS OF A SHORT DIPOLE. Let us now proceed to find the fields everywhere around a short dipole. Let the dipole of length L be placed coincident with the z axis and with its center at the origin as in Fig. 5-2. The relation of the electric field components, E_r, E_θ and E_ϕ, is then as shown. It is assumed that the medium surrounding the dipole is air or vacuum.

In dealing with antennas or radiating systems, the propagation time is a matter of great importance. Thus, if a current is flowing in the short dipole of Fig. 5-3, the effect of the current is not felt instantaneously at the point P, but only after an interval equal to the time required for the disturbance to propagate over the distance r. We have already recognized this in Chap. 4 in connection with the pattern of arrays of point sources, but here we are more explicit and describe it as a *retardation* effect.

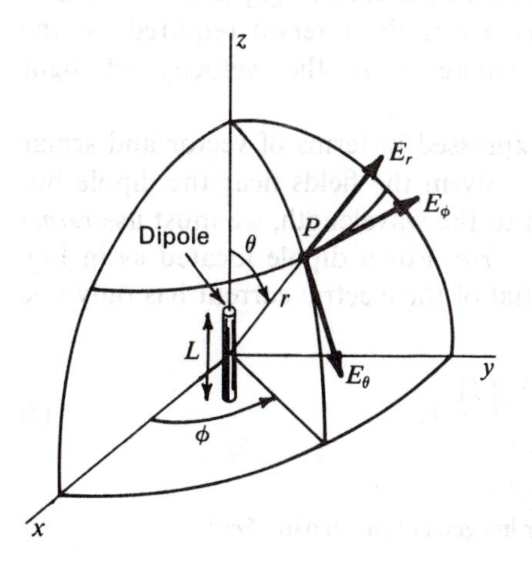

Figure 5-2 Relation of dipole to coordinates.

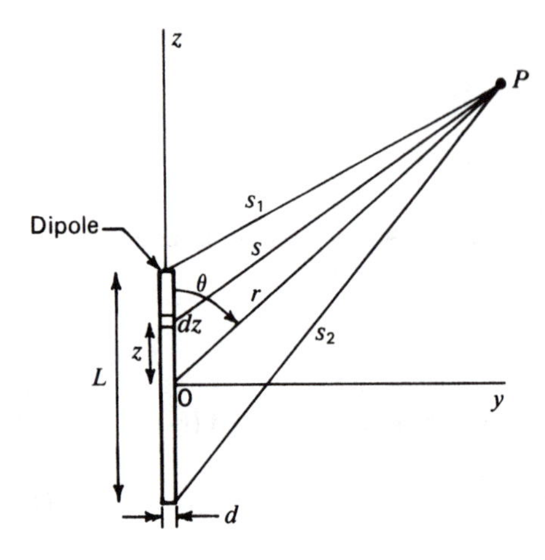

Figure 5-3a Geometry for short dipole.

Accordingly, instead of writing the current I as[1]

$$I = I_0 e^{j\omega t} \tag{1}$$

which implies instantaneous propagation of the effect of the current, we introduce the propagation (or retardation) time as done by Lorentz and write

$$[I] = I_0 e^{j\omega[t-(r/c)]} \tag{2}$$

where $[I]$ is called the *retarded current.* Specifically, the retardation time r/c results in a *phase retardation* $\omega r/c = 2\pi fr/c$ radians $= 360° \, fr/c = 360° \, t/T$, where $T = 1/f =$ time of one period or cycle (seconds) and $f =$ frequency (hertz, Hz = cycles per second). The brackets may be added as in (2) to indicate explicitly that the effect of the current is retarded.

Equation (2) is a statement of the fact that the disturbance at a time t and at a distance r from a current element is caused by a current $[I]$ that occurred at an earlier time $t - r/c$. The time difference r/c is the interval required for the disturbance to travel the distance r, where c is the velocity of light ($=300$ Mm s^{-1}).

Electric and magnetic fields can be expressed in terms of vector and scalar potentials. Since we will be interested not only in the fields near the dipole but also at distances which are large compared to the wavelength, we must use *retarded potentials*, i.e., expressions involving $t - r/c$. For a dipole located as in Fig. 5-2 or Fig. 5-3a, the retarded vector potential of the electric current has only one component, namely, A_z. Its value is

$$A_z = \frac{\mu_0}{4\pi} \int_{-L/2}^{L/2} \frac{[I]}{s} \, dz \tag{3}$$

[1] It is assumed that we take either the real ($\cos \omega t$) or imaginary ($\sin \omega t$) part of $e^{j\omega t}$.

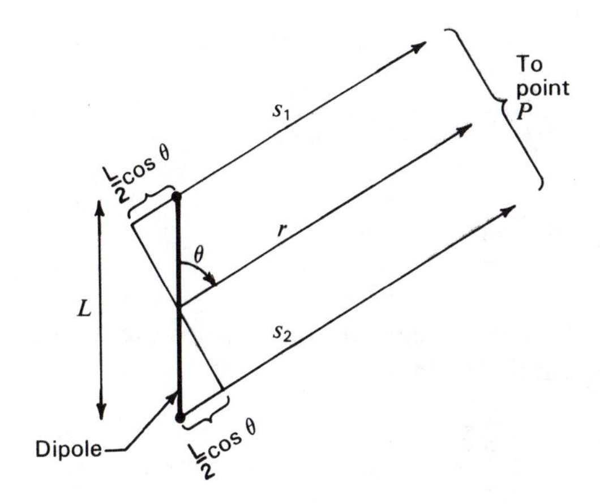

Figure 5-3b Relations for short dipole when $r \gg L$.

where $[I]$ is the retarded current given by

$$[I] = I_0 e^{j\omega[t-(s/c)]} \tag{3a}$$

In (3) and (3a),

z = distance to a point on the conductor
I_0 = peak value in time of current (uniform along dipole)
μ_0 = permeability of free space = $4\pi \times 10^{-7}$ H m^{-1}

If the distance from the dipole is large compared to its length ($r \gg L$) and if the wavelength is large compared to the length ($\lambda \gg L$), we can put $s = r$ and neglect the phase differences of the field contributions from different parts of the wire. The integrand in (3) can then be regarded as a constant, so that (3) becomes

$$A_z = \frac{\mu_0 L I_0 e^{j\omega[t-(r/c)]}}{4\pi r} \tag{4}$$

The retarded scalar potential V of a charge distribution is

$$V = \frac{1}{4\pi\varepsilon_0} \int_V \frac{[\rho]}{s} \, d\tau \tag{5}$$

where $[\rho]$ is the retarded charge density given by

$$[\rho] = \rho_0 e^{j\omega[t-(s/c)]} \tag{6}$$

and $d\tau$ = infinitesimal volume element
ε_0 = permittivity or dielectric constant of free space = 8.85×10^{-12} F m^{-1}

Since the region of charge in the case of the dipole being considered is confined to the points at the ends as in Fig. 5-1b, (5) reduces to

$$V = \frac{1}{4\pi\varepsilon_0} \left\{ \frac{[q]}{s_1} - \frac{[q]}{s_2} \right\} \tag{7}$$

From (5-1-1) and (3*a*),

$$[q] = \int [I]\,dt = I_0 \int e^{j\omega[t-(s/c)]}\,dt = \frac{[I]}{j\omega} \tag{8}$$

Substituting (8) into (7),

$$V = \frac{I_0}{4\pi\varepsilon_0 j\omega}\left[\frac{e^{j\omega[t-(s_1/c)]}}{s_1} - \frac{e^{j\omega[t-(s_2/c)]}}{s_2}\right] \tag{9}$$

Referring to Fig. 5-3b, when $r \gg L$, the lines connecting the ends of the dipole and the point P may be considered as parallel so that

$$s_1 = r - \frac{L}{2}\cos\theta \tag{10}$$

and

$$s_2 = r + \frac{L}{2}\cos\theta \tag{11}$$

Substituting (10) and (11) into (9) and clearing fractions, we have

$$V = \frac{I_0\, e^{j\omega[t-(r/c)]}}{4\pi\varepsilon_0 j\omega} \times \frac{e^{j(\omega L/2c)\cos\theta}[r + (L/2)\cos\theta] - e^{-j(\omega L/2c)\cos\theta}[r - (L/2)\cos\theta]}{r^2} \tag{12}$$

where the term $L^2 \cos^2\theta/4$ in the denominator has been neglected in comparison with r^2 by assuming that $r \gg L$. By de Moivre's theorem (12) becomes

$$V = \frac{I_0\, e^{j\omega[t-(r/c)]}}{4\pi\varepsilon_0 j\omega r^2}\left[\left(\cos\frac{\omega L\cos\theta}{2c} + j\sin\frac{\omega L\cos\theta}{2c}\right)\left(r + \frac{L}{2}\cos\theta\right) - \left(\cos\frac{\omega L\cos\theta}{2c} - j\sin\frac{\omega L\cos\theta}{2c}\right)\left(r - \frac{L}{2}\cos\theta\right)\right] \tag{13}$$

If the wavelength is much greater than the length of the dipole ($\lambda \gg L$), then

$$\cos\frac{\omega L\cos\theta}{2c} = \cos\frac{\pi L\cos\theta}{\lambda} \simeq 1 \tag{14}$$

and

$$\sin\frac{\omega L\cos\theta}{2c} \simeq \frac{\omega L\cos\theta}{2c} \tag{15}$$

Introducing (14) and (15) into (13), the expression for the scalar potential then reduces to

$$V = \frac{I_0 L\cos\theta\, e^{j\omega[t-(r/c)]}}{4\pi\varepsilon_0 c}\left(\frac{1}{r} + \frac{c}{j\omega}\frac{1}{r^2}\right) \tag{16}$$

Equations (4) and (16) express the vector and scalar potentials everywhere due to a short dipole. The only restrictions are that $r \gg L$ and $\lambda \gg L$. These equations give the vector and scalar potentials at a point P in terms of the distance r to the

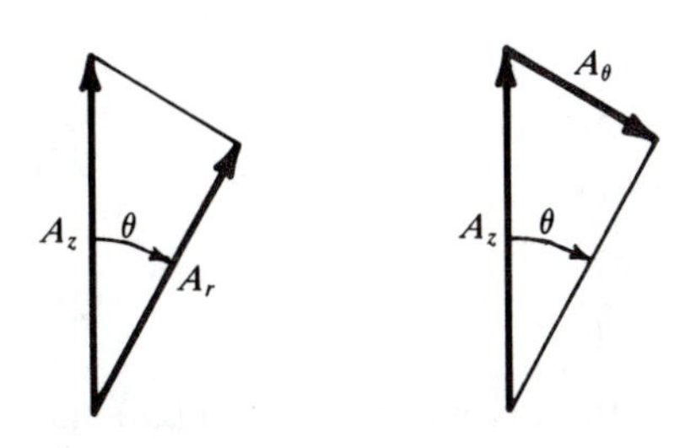

Figure 5-4 Resolution of vector potential into A_r and A_θ components.

point from the center of the dipole, the angle θ, the length of the dipole L, the current on the dipole and some constants.

Knowing the vector potential **A** and the scalar potential V, the electric and magnetic fields may then be obtained from the relations

$$\mathbf{E} = -j\omega\mathbf{A} - \nabla V \tag{17}$$

and

$$\mathbf{H} = \frac{1}{\mu} \nabla \times \mathbf{A} \tag{18}$$

It will be desirable to obtain **E** and **H** in polar coordinates. The polar coordinate components for the vector potential are

$$\mathbf{A} = \hat{\mathbf{r}} A_r + \hat{\boldsymbol{\theta}} A_\theta + \hat{\boldsymbol{\phi}} A_\phi \tag{19}$$

Since the vector potential for the dipole has only a z component, $A_\phi = 0$, and A_r and A_θ are given by (see Fig. 5-4)

$$A_r = A_z \cos\theta \tag{20}$$

$$A_\theta = -A_z \sin\theta \tag{21}$$

where A_z is as given by (4). In polar coordinates the gradient of V is

$$\nabla V = \mathbf{a}_r \frac{\partial V}{\partial r} + \mathbf{a}_\theta \frac{1}{r}\frac{\partial V}{\partial \theta} + \mathbf{a}_\phi \frac{1}{r\sin\theta}\frac{\partial V}{\partial \phi} \tag{22}$$

Calculating now the electric field **E** from (17), let us first express **E** in its polar coordinate components. Thus,

$$\mathbf{E} = \mathbf{a}_r E_r + \mathbf{a}_\theta E_\theta + \mathbf{a}_\phi E_\phi \tag{23}$$

From (17), (19) and (22) the three components of **E** are then

$$E_r = -j\omega A_r - \frac{\partial V}{\partial r} \tag{24}$$

$$E_\theta = -j\omega A_\phi - \frac{1}{r}\frac{\partial V}{\partial \theta} \tag{25}$$

$$E_\phi = -j\omega A_\phi - \frac{1}{r\sin\theta}\frac{\partial V}{\partial \phi} \tag{26}$$

In (26), $A_\phi = 0$. The second term is also zero since V in (16) is independent of ϕ so that $\partial V/\partial \phi = 0$. Therefore, $E_\phi = 0$. Substituting (20) into (24) and (21) into

(25), we have

$$E_r = -j\omega A_z \cos\theta - \frac{\partial V}{\partial r} \tag{27}$$

and

$$E_\theta = j\omega A_z \sin\theta - \frac{1}{r}\frac{\partial V}{\partial \theta} \tag{28}$$

Introducing now the values of A_z from (4) and V from (16) into (27) and (28) and performing the indicated operations, we obtain

$$E_r = \frac{I_0 L \cos\theta\, e^{j\omega[t-(r/c)]}}{2\pi\varepsilon_0}\left(\frac{1}{cr^2} + \frac{1}{j\omega r^3}\right) \tag{29}$$

and

$$E_\theta = \frac{I_0 L \sin\theta\, e^{j\omega[t-(r/c)]}}{4\pi\varepsilon_0}\left(\frac{j\omega}{c^2 r} + \frac{1}{cr^2} + \frac{1}{j\omega r^3}\right) \tag{30}$$

In obtaining (29) and (30) the relation was used that $\mu_0 \varepsilon_0 = 1/c^2$, where c = velocity of light.

Turning our attention now to the magnetic field, this may be calculated by (18). In polar coordinates the curl of **A** is

$$\nabla \times \mathbf{A} = \frac{\hat{\mathbf{r}}}{r\sin\theta}\left[\frac{\partial(\sin\theta)A_\phi}{\partial\theta} - \frac{\partial(A_\theta)}{\partial\phi}\right]$$

$$+ \frac{\hat{\boldsymbol{\theta}}}{r\sin\theta}\left[\frac{\partial A_r}{\partial\phi} - \frac{\partial(r\sin\theta)A_\phi}{\partial r}\right] + \frac{\hat{\boldsymbol{\phi}}}{r}\left[\frac{\partial(rA_\theta)}{\partial r} - \frac{\partial A_r}{\partial\theta}\right] \tag{31}$$

Since $A_\phi = 0$, the first and fourth terms of (31) are zero. From (4) and (20) and (21) we note that A_r and A_θ are independent of ϕ, so that the second and third terms of (31) are also zero. Thus, only the last two terms in (31) contribute, so that $\nabla \times \mathbf{A}$, and hence also **H**, have only a ϕ component. Introducing (20) and (21) into (31), performing the indicated operations and substituting this result into (18), we have

$$|\mathbf{H}| = H_\phi = \frac{I_0 L \sin\theta\, e^{j\omega[t-(r/c)]}}{4\pi}\left(\frac{j\omega}{cr} + \frac{1}{r^2}\right) \tag{32}$$

and

$$H_r = H_\theta = 0 \tag{33}$$

Thus, the fields from the dipole have only three components E_r, E_θ and H_ϕ. The components E_ϕ, H_r and H_θ are everywhere zero.

When r is very large, the terms in $1/r^2$ and $1/r^3$ in (29), (30) and (32) can be neglected in favor of the terms in $1/r$. Thus, in the *far field* E_r is negligible, and we have effectively only two field components, E_θ and H_ϕ, given by

$$E_\theta = \frac{j\omega I_0 L \sin\theta\, e^{j\omega[t-(r/c)]}}{4\pi\varepsilon_0 c^2 r} = j\frac{I_0 \beta L}{4\pi\varepsilon_0 cr}\sin\theta\, e^{j\omega[t-(r/c)]} \tag{34}$$

and

$$H_\phi = \frac{j\omega I_0 L \sin\theta\, e^{j\omega[t-(r/c)]}}{4\pi cr} = j\frac{I_0 \beta L}{4\pi r}\sin\theta\, e^{j\omega[t-(r/c)]} \tag{35}$$

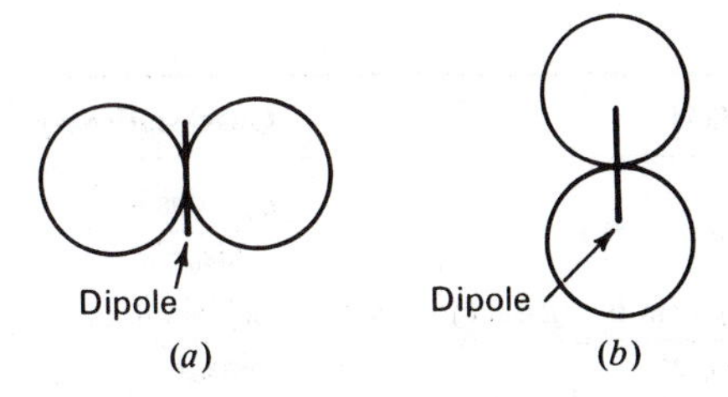

Figure 5-5 Near- and far-field patterns of E_θ and H_ϕ components for short dipole (*a*) and near-field pattern of E_r component (*b*).

Taking the ratio of E_θ to H_ϕ as given by (34) and (35), we obtain

$$\frac{E_\theta}{H_\phi} = \frac{1}{\varepsilon_0 c} = \sqrt{\frac{\mu_0}{\varepsilon_0}} = 377\ \Omega \tag{36}$$

This is the *intrinsic impedance of free space* (a pure resistance).

Comparing (34) and (35) we note that E_θ and H_ϕ are in time phase in the far field. We note also that the field patterns of both are proportional to sin θ. The pattern is independent of ϕ, so that the space pattern is doughnut-shaped, being a figure-of-revolution of the pattern in Fig. 5-5*a* about the axis of the dipole. Referring to the near-field expressions given by (29), (30) and (32), we note that for a small r the electric field has two components E_r and E_θ, which are both in time-phase quadrature with the magnetic field, as in a resonator. At intermediate distances, E_θ and E_r can approach time-phase quadrature so that the total electric field vector rotates in a plane parallel to the direction of propagation, thus exhibiting the phenomenon of *cross-field*. For the E_θ and H_ϕ components, the near-field patterns are the same as the far-field patterns, being proportional to sin θ (Fig. 5-5*a*). However, the near-field pattern for E_r is proportional to cos θ as indicated by Fig. 5-5*b*. The space pattern for E_r is a figure-of-revolution of this pattern around the dipole axis.

Let us now consider the situation at very low frequencies. This will be referred to as the *quasi-stationary*, or dc, case. Since from (8),

$$[I] = I_0 e^{j\omega[t-(r/c)]} = j\omega[q] \tag{37}$$

(29) and (30) can be rewritten as

$$E_r = \frac{[q]L \cos\theta}{2\pi\varepsilon_0}\left(\frac{j\omega}{cr^2} + \frac{1}{r^3}\right) \tag{38}$$

and

$$E_\theta = \frac{[q]L \sin\theta}{4\pi\varepsilon_0}\left(-\frac{\omega^2}{c^2 r} + \frac{j\omega}{cr^2} + \frac{1}{r^3}\right) \tag{39}$$

The magnetic field is given by (32) as

$$H_\phi = \frac{[I]L \sin\theta}{4\pi}\left(\frac{j\omega}{cr} + \frac{1}{r^2}\right) \tag{40}$$

Table 5-1 Fields of a short electric dipole†

Component	General expression	Far field	Quasi-stationary
E_r	$\dfrac{[I]L\cos\theta}{2\pi\varepsilon_0}\left(\dfrac{1}{cr^2}+\dfrac{1}{j\omega r^3}\right)$	0	$\dfrac{q_0 L\cos\theta}{2\pi\varepsilon_0 r^3}$
E_θ	$\dfrac{[I]L\sin\theta}{4\pi\varepsilon_0}\left(\dfrac{j\omega}{c^2 r}+\dfrac{1}{cr^2}+\dfrac{1}{j\omega r^3}\right)$	$\dfrac{[I]Lj\omega\sin\theta}{4\pi\varepsilon_0 c^2 r}=\dfrac{j60\pi[I]\sin\theta}{r}\dfrac{L}{\lambda}$	$\dfrac{q_0 L\sin\theta}{4\pi\varepsilon_0 r^3}$
H_ϕ	$\dfrac{[I]L\sin\theta}{4\pi}\left(\dfrac{j\omega}{cr}+\dfrac{1}{r^2}\right)$	$\dfrac{[I]Lj\omega\sin\theta}{4\pi cr}=\dfrac{j[I]\sin\theta}{2r}\dfrac{L}{\lambda}$	$\dfrac{I_0 L\sin\theta}{4\pi r^2}$

† The restriction applies that $r \gg L$ and $\lambda \gg L$. The quantities in the table are in SI units, that is, E in volts per meter, H in amperes per meter, I in amperes, r in meters, etc. $[I]$ is as given by (37). Three of the field components of an electric dipole are everywhere zero, that is,

$$E_\phi = H_r = H_\theta = 0$$

At low frequencies, ω approaches zero so that the terms with ω in the numerator can be neglected. As $\omega \to 0$, we also have

$$[q] = q_0 e^{j\omega[t-(r/c)]} = q_0 \tag{41}$$

and

$$[I] = I_0 \tag{42}$$

Thus, for the quasi-stationary, or dc, case, the field components become, from (38), (39) and (40),

$$E_r = \frac{q_0 L\cos\theta}{2\pi\varepsilon_0 r^3} \tag{43}$$

$$E_\theta = \frac{q_0 L\sin\theta}{4\pi\varepsilon_0 r^3} \tag{44}$$

$$H_\phi = \frac{I_0 L\sin\theta}{4\pi r^2} \tag{45}$$

The restriction that $r \gg L$ still applies.

The expressions for the electric field, (43) and (44), are identical to those obtained in electrostatics for the field of two point charges, $+q_0$ and $-q_0$, separated by a distance L. The relation for the magnetic field, (45), may be recognized as the Biot-Savart relation for the magnetic field of a short element carrying a steady or slowly varying current. Since in the expressions for the quasi-stationary case the fields decrease as $1/r^2$ or $1/r^3$, the fields are confined to the vicinity of the dipole and there is negligible radiation. In the general expressions for the fields, (38), (39) and (40), it is the $1/r$ terms which are important in the far field and hence take into account the radiation.

The expressions for the fields from a short dipole developed above are summarized in Table 5-1.

If we had been interested only in the far field, the development beginning with (5) could have been much simplified. The scalar potential V does not contribute to the far field, so that both **E** and **H** may be determined from **A** alone.

Thus, from (17), **E** and **H** of the far field may be obtained very simply from

$$|\mathbf{E}| = E_\theta = -j\omega A_\theta \tag{45a}$$

and

$$|\mathbf{H}| = H_\phi = \frac{E_\theta}{Z} = -\frac{j\omega}{Z} A_\theta \tag{45b}$$

where $Z = \sqrt{\mu_0/\varepsilon_0} = 377\ \Omega$

H_ϕ may also be obtained as before from (18) and E_θ from H_ϕ. Thus,

$$H_\phi = |\mathbf{H}| = \frac{1}{\mu_0} |\nabla \times \mathbf{A}| \tag{45c}$$

and neglecting terms in $1/r^2$,

$$E_\theta = ZH_\phi = \frac{Z}{\mu_0} |\nabla \times \mathbf{A}| \tag{45d}$$

Equation (30) for the θ component of the electric field may be reexpressed as

$$E_\theta = \frac{I_0 e^{j\omega t} L_\lambda Z \sin\theta}{\lambda} \left[\frac{1}{2r_\lambda} \underline{/-360^\circ\ r_\lambda + 90^\circ} + \frac{1}{4\pi r_\lambda^2} \underline{/-360^\circ\ r_\lambda + 0^\circ} + \frac{1}{8\pi^2 r_\lambda^3} \underline{/-360^\circ\ r_\lambda - 90^\circ} \right] \quad (\text{V m}^{-1}) \tag{46}$$

where $L_\lambda = L/\lambda$
$r_\lambda = r/\lambda$
$Z = 377\ \Omega$

The restrictions apply that

$$\lambda \gg L \qquad (L_\lambda \ll 1)$$
$$r \gg L \qquad (r_\lambda \gg L_\lambda) \tag{47}$$

If we let

$$A = \frac{1}{2r_\lambda} \underline{/-360^\circ\ r_\lambda + 90^\circ}$$

$$B = \frac{1}{4\pi r_\lambda^2} \underline{/-360^\circ\ r_\lambda + 0^\circ}$$

$$C = \frac{1}{8\pi^2 r_\lambda^3} \underline{/-360^\circ\ r_\lambda - 90^\circ}$$

(46) becomes

$$E_\theta = \frac{I_0 e^{j\omega t} L_\lambda Z \sin\theta}{\lambda} (A + B + C) \tag{48}$$

The magnitudes of the components A, B and C are shown in Fig. 5-6 as a function of r_λ (distance in wavelengths). For r_λ greater than the radian distance

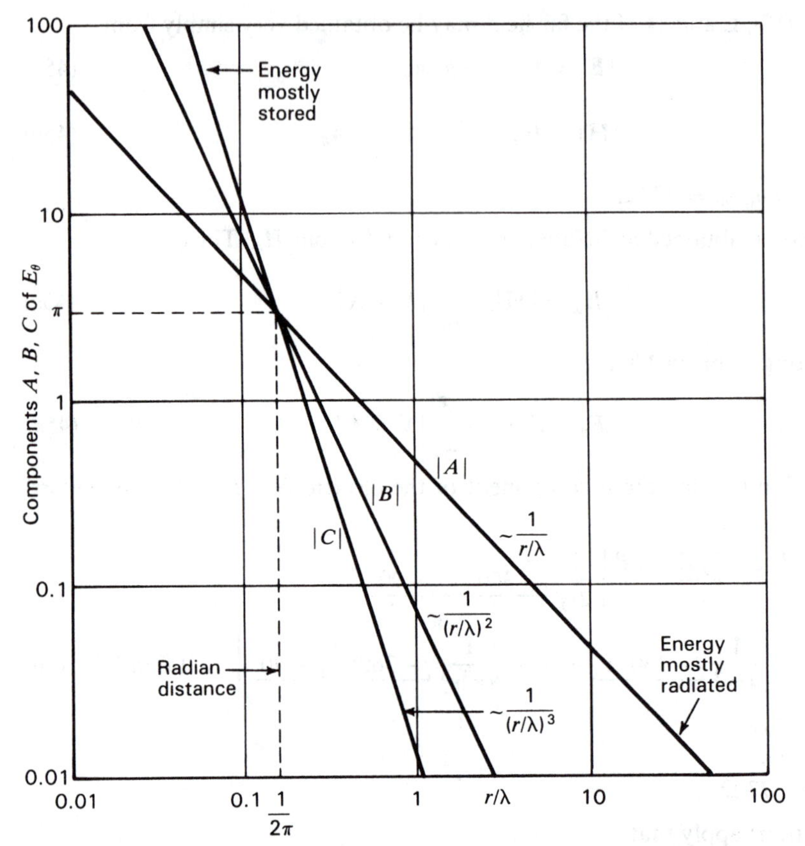

Figure 5-6 Variation of the magnitudes of the components of E_θ of a short electric dipole as a function of distance (r/λ). The magnitudes of all components equal π at the radian distance $1/(2\pi)$. At larger distances energy is mostly radiated, at smaller distances mostly stored.

$[1/(2\pi)]$, component A of the electric field is dominant, for r_λ less than the radian distance component C of the electric field is dominant, while *at* the radian distance only B contributes $(=\pi)$ because although $|A| = |B| = |C| = \pi$, A and C are in phase opposition and cancel.

Equation (32) for the ϕ component (only component) of the magnetic field may be reexpressed as

$$H_\phi = \frac{I_0 e^{j\omega t} L_\lambda \sin\theta}{\lambda}\left[\frac{1}{2r_\lambda}\underline{/-360^\circ\, r_\lambda + 90^\circ} + \frac{1}{4\pi r_\lambda^2}\underline{/-360^\circ\, r_\lambda + 0^\circ}\right] \tag{49}$$

The restrictions of (47) apply.

For r_λ greater than the radian distance the first component of the magnetic field (in brackets) is dominant, for r_λ less than the radian distance the second component of the magnetic field is dominant, while *at* the radian distance both components are equal $(=\pi)$ and in phase quadrature.

Note that the ratio of E_θ to H_ϕ, as given by the ratio of (46) to (49), is an impedance whose value becomes exactly equal to Z, the intrinsic impedance of space ($\simeq 377\ \Omega$), for $r_\lambda \gg 1/(2\pi)$.

For the special case where $\theta = 90°$ (perpendicular to the dipole in the xy plane of Fig. 5-2) and at $r_\lambda \gg 1/(2\pi)$,

$$|H_\phi| = \frac{I_0 L_\lambda}{2r} \qquad (\text{A m}^{-1}) \tag{50}$$

while at $r_\lambda \ll 1/(2\pi)$,

$$|H_\phi| = \frac{I_0 L}{4\pi r^2} \tag{50a}$$

which is identical to the relation for the magnetic field perpendicular to a short linear conductor carrying direct current as given by (45).

The magnetic field *at any distance r* from an infinite linear conductor with direct current is given by

$$H_\phi = \frac{I_0}{2\pi r} \tag{50b}$$

which is *Ampere's law.*[1]

Remarkably, the magnitude of the magnetic field in the equatorial plane ($\theta = 90°$) *in the far field of an oscillating $\lambda/2$ dipole* is identical to (50*b*) (Ampere's law). It is assumed that the current distribution on the $\lambda/2$ dipole is sinusoidal. This is discussed in more detail in Sec. 5-5. The above magnetic field relations are summarized in Table 5-2.

Rearranging the three field components of Table 5-1 for a short electric dipole, we have

$$E_r = \frac{[I]L_\lambda Z \cos\theta}{\lambda}\left[\frac{1}{2\pi r_\lambda^2} - j\frac{1}{4\pi^2 r_\lambda^3}\right] \tag{51}$$

$$E_\theta = \frac{[I]L_\lambda Z \sin\theta}{\lambda}\left[j\frac{1}{2r_\lambda} + \frac{1}{4\pi r_\lambda^2} - j\frac{1}{8\pi^2 r_\lambda^3}\right] \tag{52}$$

$$H_\phi = \frac{[I]L_\lambda \sin\theta}{\lambda}\left[j\frac{1}{2r_\lambda} + \frac{1}{4\pi r_\lambda^2}\right] \tag{53}$$

We note that the constant factor in each of the terms in brackets differs from the factors of adjacent terms by a factor of 2π, as do the constant factors in A, B and C of (48).

---

[1] See, for example, J. D. Kraus, *Electromagnetics*, 3rd ed., McGraw-Hill, 1984, p. 170.

Table 5-2 Magnetic fields from dipoles and linear conductors†

Short oscillating dipole at $r_\lambda \gg 1/(2\pi)$, $r \gg L$, $\lambda \gg L$	$\lvert H_\phi \rvert = \dfrac{I_0 L_\lambda}{2r}$	(A m^{-1})
Short oscillating dipole at $r_\lambda \ll 1/(2\pi)$, $r \gg L$, $\lambda \gg L$	$\lvert H_\phi \rvert = \dfrac{I_0 L}{4\pi r^2}$	(A m^{-1})
Short linear conductor with direct current $r \gg L$	$H_\phi = \dfrac{I_0 L}{4\pi r^2}$	(A m^{-1})
$\lambda/2$ oscillating dipole, far field	$\lvert H_\phi \rvert = \dfrac{I_0}{2\pi r}$	(A m^{-1})
Infinite linear conductor with direct current at any r	$H_\phi = \dfrac{I_0}{2\pi r}$ (Ampere's law)	(A m^{-1})

† Magnetic field at distance r from dipoles and linear conductors (in direction perpendicular to dipole, in xy plane of Fig. 5-2) with current I_0.

At the radian distance $[r_\lambda = 1/(2\pi)]$, (51), (52) and (53) reduce to

$$E_r = \frac{2\sqrt{2}\,\pi[I]L_\lambda Z \cos\theta}{\lambda} \angle{-45^\circ} \tag{54}$$

$$E_\theta = \frac{\pi[I]L_\lambda Z \sin\theta}{\lambda} \tag{55}$$

$$H_\phi = \frac{\sqrt{2}\,\pi[I]L_\lambda \sin\theta}{\lambda} \angle{45^\circ} \tag{56}$$

The magnitude of the average power flux or Poynting vector in the θ direction is given by

$$S_\theta = \tfrac{1}{2}\,\text{Re}\, E_r H_\phi^* = \tfrac{1}{2} E_r H_\phi \,\text{Re}\, 1\angle{-90^\circ} = \tfrac{1}{2} E_r H_\phi \cos(-90^\circ) = 0 \tag{57}$$

indicating that no power is transmitted. However, the product $E_r H_\phi$ represents imaginary or reactive energy that oscillates back and forth from electric to magnetic energy twice per cycle.

In like manner the magnitude of the power flux or Poynting vector in the r direction is given by

$$S_r = \tfrac{1}{2} E_\theta H_\phi \cos(-45^\circ) = \frac{1}{2\sqrt{2}} E_\theta H_\phi \tag{58}$$

indicating energy flow in the r direction.

Much closer to the dipole $[r_\lambda \ll 1/(2\pi)]$, but with the restrictions of (47) still applying, (51), (52) and (53) reduce approximately to

$$E_r = -j\frac{[I]L_\lambda Z \cos\theta}{4\pi^2\lambda r_\lambda^3} \tag{59}$$

$$E_\theta = -j\frac{[I]L_\lambda Z \sin\theta}{8\pi^2\lambda r_\lambda^3} \tag{60}$$

$$H_\phi = \frac{[I]L_\lambda \sin\theta}{4\pi\lambda r_\lambda^2} \tag{61}$$

From these equations it is apparent that $S_r = S_\theta = 0$. However, the products $E_r H_\phi$ and $E_\theta H_\phi$ represent imaginary or reactive energy oscillating back and forth but not going anywhere. Thus, close to the dipole there is a region of almost complete energy storage.

Remote from the dipole $[r_\lambda \gg 1/(2\pi)]$, (51), (52) and (53) reduce approximately to

$$E_r = 0 \tag{62}$$

$$E_\theta = j\frac{[I]L_\lambda Z \sin\theta}{2\lambda r_\lambda} \tag{63}$$

$$H_\phi = j\frac{[I]L_\lambda \sin\theta}{2\lambda r_\lambda} \tag{64}$$

Since $E_r = 0$, there is no energy flow in the θ direction ($S_\theta = 0$). However, since E_θ and H_ϕ are in time phase, their product represents real power flow in the outward radial direction. This power is radiated.

The Poynting vector or power flux around a short dipole antenna is shown by means of vectors in Fig. 5-7a. The length of the vectors is proportional to the Poynting vector magnitude. Double-ended vectors indicate imaginary or reactive power (vars per square meter) while single-ended vectors represent real power flow (watts per square meter) in the direction indicated.

The region near the dipole is one of stored energy (reactive power) while regions remote from the dipole are ones of radiation. The radian sphere at $r_\lambda = 1/(2\pi)$ marks a zone of transition from one region to the other with a nearly equal division of the imaginary and real (radiated) power.

The region close to the dipole may be likened to a spherical resonator within which pulsating energy is trapped, but with some leakage which is radiated. There is no exact boundary to this resonator region, but if we arbitrarily put it at the radian distance a qualitative picture may be sketched as in Fig. 5-7b.

5-3 RADIATION RESISTANCE OF SHORT ELECTRIC DIPOLE. Let us now calculate the radiation resistance of the short dipole of Fig. 5-1*b*. This may be done as follows. The Poynting vector of the far field is integrated over a large sphere to obtain the total power radiated. This power is

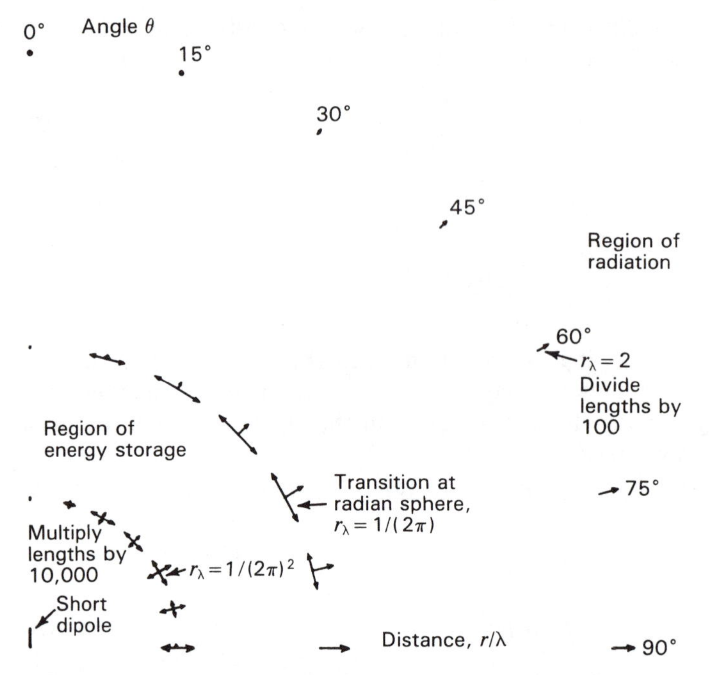

Figure 5-7a Power flux vectors at three distances near a short dipole antenna. Double-ended vectors indicate reactive power (vars per square meter) while single-ended vectors represent real power (watts per square meter). At the innermost distance $[r_\lambda = 1/(2\pi)^2]$, the power is almost entirely imaginary (reactive) with stored energy oscillating from electric to magnetic twice per cycle. At the outermost distance, the power is almost entirely real and flowing radially outward as radiation. At the radian sphere $[r_\lambda = 1/(2\pi)]$, the condition is in transition with energy pulsating in the θ direction and also radiating in the radial direction. Some stored energy (not shown) is also pulsating in the radial direction. Note that for proper scale, vectors at the innermost distance should be 10 000 times larger while at the outermost distance they should be 100 times smaller. The three radial distances are not to scale. The other quadrants are mirror images.

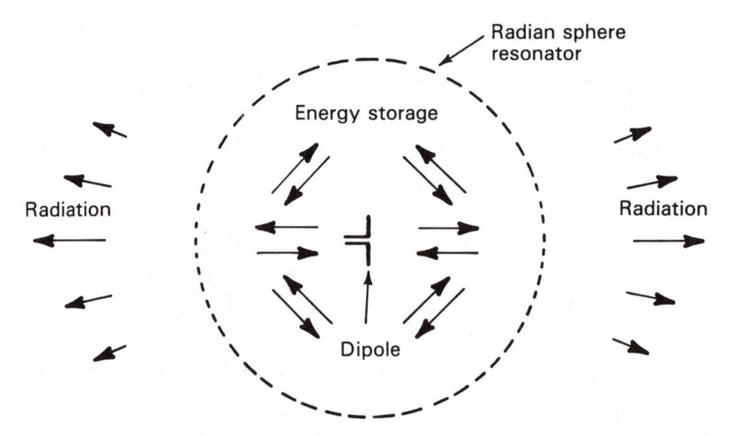

Figure 5-7b Sketch suggesting that within the radian sphere the situation is like that inside a resonator with high-density pulsating energy accompanied by leakage which is radiated.

then equated to I^2R where I is the rms current on the dipole and R is a resistance, called the radiation resistance of the dipole.

The *average* Poynting vector is given by

$$\mathbf{S} = \tfrac{1}{2}\,\text{Re}\,(\mathbf{E} \times \mathbf{H}^*) \tag{1}$$

The far-field components are E_θ and H_ϕ so that the radial component of the Poynting vector is

$$S_r = \tfrac{1}{2}\,\text{Re}\,E_\theta H_\phi^* \tag{2}$$

where E_θ and H_ϕ^* are complex.

The far-field components are related by the intrinsic impedance of the medium. Hence,

$$E_\theta = H_\phi Z = H_\phi \sqrt{\frac{\mu}{\varepsilon}} \tag{3}$$

Thus, (2) becomes

$$S_r = \tfrac{1}{2}\,\text{Re}\,ZH_\phi H_\phi^* = \tfrac{1}{2}|H_\phi|^2\,\text{Re}\,Z = \tfrac{1}{2}|H_\phi|^2\sqrt{\frac{\mu}{\varepsilon}} \tag{4}$$

The total power P radiated is then

$$P = \iint S_r\,ds = \frac{1}{2}\sqrt{\frac{\mu}{\varepsilon}}\int_0^{2\pi}\int_0^{\pi}|H_\phi|^2 r^2 \sin\theta\,d\theta\,d\phi \tag{5}$$

where the angles are as shown in Fig. 5-2 and $|H_\phi|$ is the absolute value of the magnetic field, which from (5-2-35) is

$$|H_\phi| = \frac{\omega I_0 L \sin\theta}{4\pi c r} \tag{6}$$

Substituting this into (5) we have

$$P = \frac{1}{32}\sqrt{\frac{\mu}{\varepsilon}}\frac{\beta^2 I_0^2 L^2}{\pi^2}\int_0^{2\pi}\int_0^{\pi}\sin^3\theta\,d\theta\,d\phi \tag{7}$$

The double integral equals $8\pi/3$ and (7) becomes

$$P = \sqrt{\frac{\mu}{\varepsilon}}\frac{\beta^2 I_0^2 L^2}{12\pi} \tag{8}$$

This is the *average* power or rate at which energy is streaming out of a sphere surrounding the dipole. Hence, it is equal to the power radiated. Assuming no losses, it is also equal to the power delivered to the dipole. Therefore, P must be equal to the square of the rms current I flowing on the dipole times a resistance R_r called the *radiation resistance* of the dipole. Thus,

$$\sqrt{\frac{\mu}{\varepsilon}}\frac{\beta^2 I_0^2 L^2}{12\pi} = \left(\frac{I_0}{\sqrt{2}}\right)^2 R_r \tag{9}$$

Solving for R_r,

$$R_r = \sqrt{\frac{\mu}{\varepsilon}} \frac{\beta^2 L^2}{6\pi} \tag{10}$$

For air or vacuum $\sqrt{\mu/\varepsilon} = \sqrt{\mu_0/\varepsilon_0} = 377 = 120\pi\ \Omega$ so that (10) becomes[1]

$$R_r = 80\pi^2\left(\frac{L}{\lambda}\right)^2 = 80\pi^2 L_\lambda^2 = 790 L_\lambda^2 \qquad (\Omega) \tag{11}$$

As an example suppose that $L_\lambda = \frac{1}{10}$. Then $R_r = 7.9\ \Omega$. If $L_\lambda = 0.01$, then $R_r = 0.08\ \Omega$. Thus, the radiation resistance of a short dipole is small.

In developing the field expressions for the short dipole, which were used in obtaining (11), the restriction was made that $\lambda \gg L$. This made it possible to neglect the phase difference of field contributions from different parts of the dipole. If $L_\lambda = \frac{1}{2}$ we violate this assumption, but, as a matter of interest, let us find what the radiation resistance of a $\lambda/2$ dipole is, when calculated in this way. Then for $L_\lambda = \frac{1}{2}$, we obtain $R_r = 197\ \Omega$. The correct value is $168\ \Omega$ (see Prob. 5-3), which indicates the magnitude of the error introduced by violating the restriction that $\lambda \gg L$ to the extent of taking $L = \lambda/2$.

It has been assumed that with end-loading (see Fig. 5-1*a*) the dipole current is uniform. However, with no end-loading the current must be zero at the ends and, if the dipole is short, the current tapers almost linearly from a maximum at the center to zero at the ends, as in Fig. 2-12*b*, with an average value of $\frac{1}{2}$ of the maximum. Modifying (8) for the general case where the current is not uniform on the dipole, the *radiated power* is

$$P = \sqrt{\frac{\mu}{\varepsilon}} \frac{\beta^2 I_{av}^2 L^2}{12\pi} \qquad (\mathrm{W}) \tag{12}$$

where I_{av} = amplitude of *average current* on dipole (peak value in time)

The *power delivered* to the dipole is, as before,

$$P = \tfrac{1}{2} I_0^2 R_r \qquad (\mathrm{W}) \tag{13}$$

where I_0 = amplitude of *terminal current* of center-fed dipole (peak value in time)

Equating the *power radiated* (12) to the *power delivered* (13) yields, for free space ($\mu = \mu_0$ and $\varepsilon = \varepsilon_0$), a radiation resistance

$$R_r = 790\left(\frac{I_{av}}{I_0}\right)^2 L_\lambda^2 \qquad (\Omega) \tag{14}$$ [2]

---

[1] $\sqrt{\mu_0/\varepsilon_0} = 376.73\ \Omega$. 377 and 120π are convenient approximations.

[2] As already given by (2-20-3). See also footnote accompanying (2-20-3).

For a short dipole without end-loading, we have $I_{av} = \frac{1}{2}I_0$, as noted above, and (14) becomes

$$R_r = 197L_\lambda^2 \qquad (\Omega) \tag{15}$$

5-4 THE FIELDS OF A SHORT DIPOLE BY THE HERTZ VECTOR METHOD. In Sec. 5-2 the fields of a short dipole were obtained by a method involving the use of vector and scalar potentials. Another equivalent method which is sometimes employed makes use of the Hertz vector. Since this method is frequently found in the literature, it will be of interest to use it to find the fields of a short electric dipole. The fields so obtained are identical with those found by the vector-scalar potential method, indicating the equivalence of the two procedures.

The retarded vector potential of any electric-current distribution is given by

$$\mathbf{A} = \frac{\mu}{4\pi}\int_V \frac{[\mathbf{J}]}{r}\,d\tau \tag{1}$$

where the retarded current density $[\mathbf{J}]$ is given by

$$[\mathbf{J}] = \mathbf{J}_0\, e^{j\omega[t-(r/c)]} \tag{2}$$

Multiplying numerator and denominator by ε, (1) may be written as

$$\mathbf{A} = \mu\varepsilon\,\frac{\partial\mathbf{\Pi}}{\partial t} \tag{3}$$

where

$$\frac{\partial\mathbf{\Pi}}{\partial t} = \frac{1}{4\pi\varepsilon}\int_V \frac{[\mathbf{J}]}{r}\,d\tau \tag{4}$$

where t represents time and τ volume. The quantity $\mathbf{\Pi}$ is the retarded Hertz vector or retarded Hertzian potential. Since $[\mathbf{J}]$ is the only time-dependent quantity on the right-hand side of (4), we have for the retarded Hertz vector

$$\mathbf{\Pi} = \frac{1}{4\pi\varepsilon}\int_V \frac{\int |\mathbf{J}|\,dt}{r}\,d\tau = \frac{1}{4\pi\varepsilon j\omega}\int_V \frac{[\mathbf{J}]}{r}\,d\tau \tag{5}$$

Since

$$\mathbf{\Pi} = \mathbf{\Pi}_0\, e^{j\omega[t-(r/c)]}$$

we obtain from (3)

$$\mathbf{A} = j\omega\mu\varepsilon\mathbf{\Pi} \tag{6}$$

and

$$\mathbf{\Pi} = -\frac{j}{\omega\mu\varepsilon}\mathbf{A} \tag{7}$$

If the retarded Hertz vector is known, both **E** and **H** everywhere can be calculated from the relations

$$\mathbf{E} = \omega^2 \mu\varepsilon \mathbf{\Pi} + \nabla(\nabla \cdot \mathbf{\Pi}) \tag{8}$$

$$\mathbf{H} = j\omega\varepsilon \nabla \times \mathbf{\Pi} \tag{9}$$

Thus, **E** and **H** are derivable from a single potential function, **Π**. Substituting (7) into (8) and (9), these relations may be also reexpressed in terms of **A** alone. Thus,

$$\mathbf{E} = -j\omega\mathbf{A} - \frac{j}{\omega\mu\varepsilon} \nabla(\nabla \cdot \mathbf{A}) \tag{10}$$

$$\mathbf{H} = \frac{1}{\mu} \nabla \times \mathbf{A} \tag{11}$$

Let us now find the retarded Hertz vector for a short electric dipole. The vector potential for the dipole has only a z component as given by (5-2-4). Therefore, from (7) the Hertz vector has only a z component given by

$$\Pi_z = -\frac{jI_0 L e^{j\omega[t-(r/c)]}}{4\pi r\omega\varepsilon} \tag{12}$$

In polar coordinates **Π** has two components, obtained in the same way as the components of **A** in (5-2-20) and (5-2-21). Thus,

$$\mathbf{\Pi} = \hat{\mathbf{r}}\Pi_z \cos\theta - \hat{\boldsymbol{\theta}}\Pi_z \sin\theta \tag{13}$$

Substituting (12) into (13), and this in turn in (9) and performing the indicated operations, yields the result that

$$H_\phi = \frac{[I]L \sin\theta}{4\pi}\left(\frac{j\omega}{cr} + \frac{1}{r^2}\right) \tag{14}$$

This result is identical with that obtained previously in (5-2-32). We could have anticipated this result since substituting (7) into (9) gives (11), from which (5-2-32) was obtained.

Substituting (12) into (13) and this in turn in (8) then gives the electric field **E** everywhere. The expressions for the two components, E_r and E_θ, so obtained are identical with those arrived at in (5-2-29) and (5-2-30) by the use of vector and scalar potentials.

5-5 THE THIN LINEAR ANTENNA. In this section expressions for the far-field patterns of thin linear antennas will be developed. It is assumed that the antennas are symmetrically fed at the center by a balanced two-wire transmission line. The antennas may be of any length, but it is assumed that the current distribution is sinusoidal. Current-distribution measurements indicate that this is a good assumption provided that the antenna is thin, i.e., when the conductor diameter is less than, say, $\lambda/100$. Thus, the sinusoidal current distribution approximates the natural distribution on thin antennas. Examples of the approximate

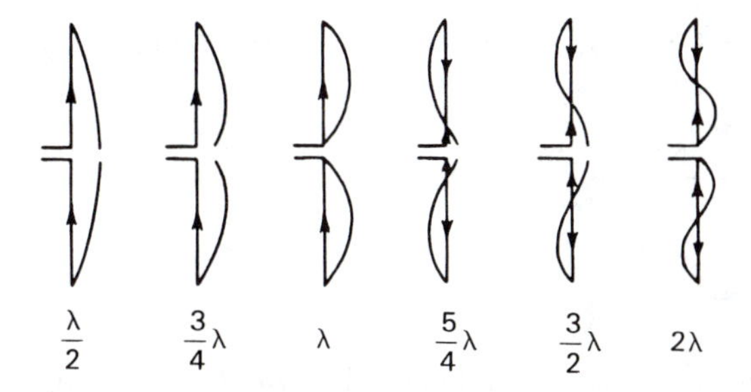

Figure 5-8 Approximate natural-current distribution for thin, linear, center-fed antennas of various lengths.

natural-current distributions on a number of thin, linear center-fed antennas of different length are illustrated in Fig. 5-8. The currents are in phase over each $\lambda/2$ section and in opposite phase over the next.

Referring to Fig. 5-9, let us now proceed to develop the far-field equations for a symmetrical, thin, linear, center-fed antenna of length L. The retarded value of the current at any point z on the antenna referred to a point at a distance s is

$$[I] = I_0 \sin\left[\frac{2\pi}{\lambda}\left(\frac{L}{2} \pm z\right)\right] e^{j\omega[t-(r/c)]} \tag{1}$$

In (1) the function

$$\sin\left[\frac{2\pi}{\lambda}\left(\frac{L}{2} \pm z\right)\right]$$

is the form factor for the current on the antenna. The expression $(L/2) + z$ is used when $z < 0$ and $(L/2) - z$ is used when $z > 0$. By regarding the antenna as made up of a series of infinitesimal dipoles of length dz, the field of the entire antenna may then be obtained by integrating the fields from all of the dipoles making up

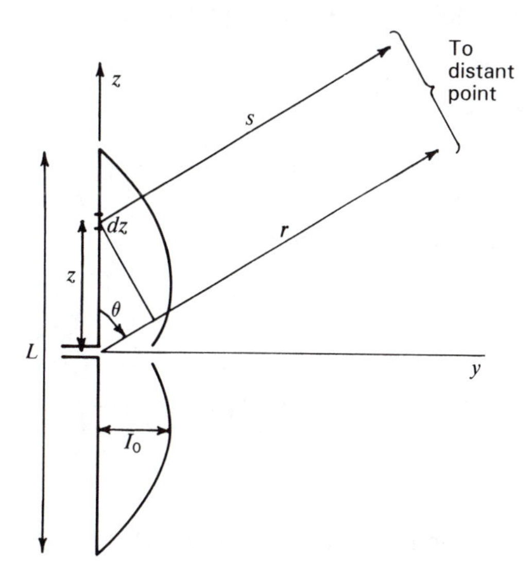

Figure 5-9 Relations for symmetrical, thin, linear, center-fed antenna of length L.

the antenna. The far fields dE_θ and dH_ϕ at a distance s from the infinitesimal dipole dz are (see Table 5-1)

$$dE_\theta = \frac{j60\pi[I] \sin \theta \, dz}{s\lambda} \tag{2}$$

$$dH_\phi = \frac{j[I] \sin \theta \, dz}{2s\lambda} \tag{3}$$

Since $E_\theta = ZH_\phi = 120\pi H_\phi$, it will suffice to calculate H_ϕ. The value of the magnetic field H_ϕ for the entire antenna is the integral of (3) over the length of the antenna. Thus,

$$H_\phi = \int_{-L/2}^{L/2} dH_\phi \tag{4}$$

Now introducing the value of $[I]$ from (1) into (3) and substituting this into (4) we have

$$H_\phi = \frac{jI_0 \sin \theta \, e^{j\omega t}}{2\lambda} \left\{ \int_{-L/2}^{0} \frac{1}{s} \sin \left[\frac{2\pi}{\lambda} \left(\frac{L}{2} + z \right) \right] e^{-j\omega s/c} \, dz + \int_{0}^{L/2} \frac{1}{s} \sin \left[\frac{2\pi}{\lambda} \left(\frac{L}{2} - z \right) \right] e^{-j\omega s/c} \, dz \right\} \tag{5}$$

In (5), $1/s$ affects only the amplitude, and hence at a large distance it may be regarded as a constant. Also at a large distance, the difference between s and r can be neglected in its effect on the amplitude although its effect on the phase must be considered. Further, from Fig. 5-9,

$$s = r - z \cos \theta \tag{6}$$

Substituting (6) into (5) and also r for s in the amplitude factor, (5) becomes

$$H_\phi = \frac{jI_0 \sin \theta \, e^{j\omega[t-(r/c)]}}{2\lambda r} \left\{ \int_{-L/2}^{0} \sin \left[\frac{2\pi}{\lambda} \left(\frac{L}{2} + z \right) \right] e^{j(\omega \cos \theta) z/c} \, dz + \int_{0}^{L/2} \sin \left[\frac{2\pi}{\lambda} \left(\frac{L}{2} - z \right) \right] e^{j(\omega \cos \theta) z/c} \, dz \right\} \tag{7}$$

Since $\beta = \omega/c = 2\pi/\lambda$ and $\beta/4\pi = 1/(2\lambda)$, (7) may be rewritten as

$$H_\phi = \frac{j\beta I_0 \sin \theta \, e^{j\omega[t-(r/c)]}}{4\pi r} \left\{ \int_{-L/2}^{0} e^{j\beta z \cos \theta} \sin \left[\beta \left(\frac{L}{2} + z \right) \right] dz + \int_{0}^{L/2} e^{j\beta z \cos \theta} \sin \left[\beta \left(\frac{L}{2} - z \right) \right] dz \right\} \tag{8}$$

The integrals are of the form

$$\int e^{ax} \sin (c + bx) \, dx = \frac{e^{ax}}{a^2 + b^2} [a \sin (c + bx) - b \cos (c + bx)] \tag{9}$$

where for the first integral

$$a = j\beta \cos \theta$$

$$b = \beta$$

$$c = \beta L/2$$

For the second integral a and c are the same as in the first integral, but $b = -\beta$. Carrying through the two integrations, adding the results and simplifying yields

$$H_\phi = \frac{j[I_0]}{2\pi r} \left[\frac{\cos [(\beta L \cos \theta)/2] - \cos (\beta L/2)}{\sin \theta} \right] \tag{10}$$

Multiplying H_ϕ by $Z = 120\pi$ gives E_θ as

$$E_\theta = \frac{j60[I_0]}{r} \left[\frac{\cos [(\beta L \cos \theta)/2] - \cos (\beta L/2)}{\sin \theta} \right] \tag{11}$$

where $[I_0] = I_0 e^{j\omega[t-(r/c)]}$

Equations (10) and (11) are the expressions for the far fields, H_ϕ and E_θ, of a symmetrical, center-fed, thin linear antenna of length L. The shape of the far-field pattern is given by the factor in the brackets. The factors preceding the brackets in (10) and (11) give the instantaneous magnitude of the fields as functions of the antenna current and the distance r. To obtain the rms value of the field, we let $[I_0]$ equal the rms current at the location of the current maximum. There is no factor involving phase in (10) or (11), since the center of the antenna is taken as the phase center. Hence any phase change of the fields as a function of θ will be a jump of 180° when the pattern factor changes sign.

As examples of the far-field patterns of linear center-fed antennas, three antennas of different lengths will be considered. Since the amplitude factor is independent of the length, only the relative field patterns as given by the pattern factor will be compared.

5-5a Case 1. $\lambda/2$ Antenna. When $L = \lambda/2$, the pattern factor becomes

$$E = \frac{\cos [(\pi/2) \cos \theta]}{\sin \theta} \tag{12}$$

This pattern is shown in Fig. 5-10a. It is only slightly more directional than the pattern of an infinitesimal or short dipole which is given by $\sin \theta$. The beam width between half-power points of the $\lambda/2$ antenna is 78° as compared to 90° for the short dipole.

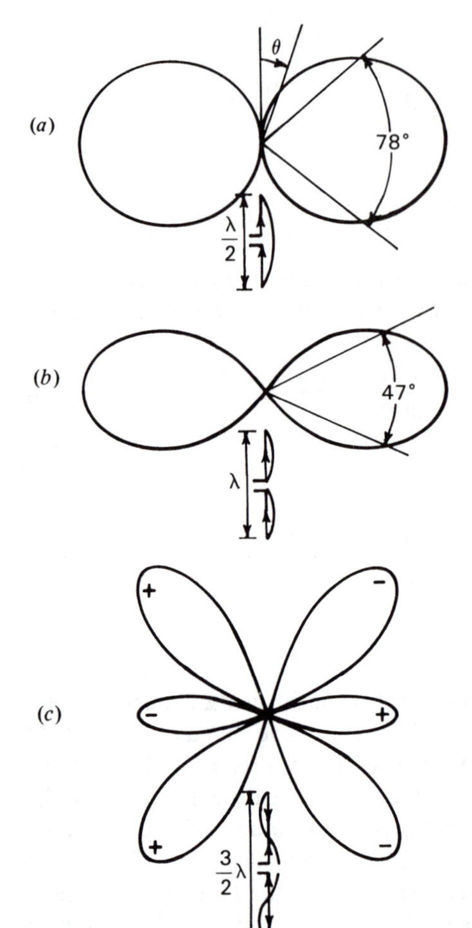

Figure 5-10 Far-field patterns of $\lambda/2$, full-wave and $3\lambda/2$ antennas. The antennas are center-fed and the current distribution is assumed to be sinusoidal.

5-5b Case 2. Full-Wave Antenna. When $L = \lambda$, the pattern factor becomes

$$E = \frac{\cos(\pi \cos\theta) + 1}{\sin\theta} \tag{13}$$

This pattern is shown in Fig. 5-10*b*. The half-power beam width is 47°.

5-5c Case 3. $3\lambda/2$ Antenna. When $L = 3\lambda/2$, the pattern factor is

$$E = \frac{\cos(\frac{3}{2}\pi \cos\theta)}{\sin\theta} \tag{14}$$

The pattern for this case is presented in Fig. 5-10*c*. With the midpoint of the antenna as phase center, the phase shifts 180° at each null, the relative phase of the lobes being indicated by the + and − signs. In all three cases, (*a*), (*b*) and (*c*), the space pattern is a figure-of-revolution of pattern shown around the axis of the antenna.

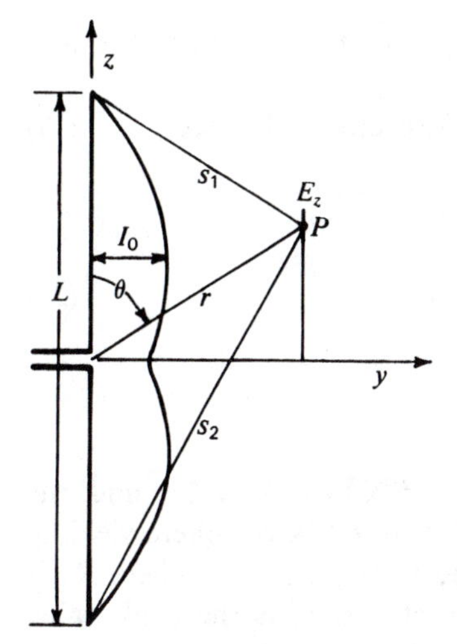

Figure 5-11 Symmetrical center-fed dipole with sinusoidal current distribution. The field component E_z at any distance can be expressed as the sum of 3 components radiating from the ends and the center of the dipole.

5-5d Field at any Distance from Center-Fed Dipole. The geometry for the field at the point P from a symmetrical center-fed dipole of length L with sinusoidal current distribution is presented in Fig. 5-11. The maximum current is I_0. It may be shown that the z component of the electric field at the point P is given by

$$E_z = \frac{-jI_0 Z}{4\pi}\left[\frac{e^{-j\beta s_1}}{s_1} + \frac{e^{-j\beta s_2}}{s_2} - 2\cos\frac{\beta L}{2}\frac{e^{-j\beta r}}{r}\right] \tag{15}$$

The ϕ component of the magnetic field at the point P (Fig. 5-11) is given by

$$H_\phi = \frac{jI_0}{4\pi r \sin\theta}\left(e^{-j\beta s_1} + e^{-j\beta s_2} - 2\cos\frac{\beta L}{2}e^{-j\beta r}\right) \tag{16}$$

Whereas the other field equations for oscillating dipoles given in this chapter apply only with the restrictions of (5-2-47), (15) and (16) apply without distance restrictions. Equations (15) and (16) are reminiscent of the pulsed center-fed dipole of Fig. 2-24 in that the field at P is made up of 3 field components, one from each end of the dipole and one from the center.

If P lies on the y axis ($\theta = 90°$) and the dipole is $\lambda/2$ long, (15) becomes

$$E_z = \frac{I_0 Z}{2\pi\lambda\sqrt{\frac{1}{16} + r_\lambda^2}} \underline{/-360°\sqrt{\tfrac{1}{16} + r_\lambda^2} - 90°} \qquad (\text{V m}^{-1}) \tag{17}$$

and (16) becomes

$$H_\phi = \frac{I_0}{2\pi r} \underline{/-360°\sqrt{\tfrac{1}{16} + r_\lambda^2} + 90°} \qquad (\text{A m}^{-1}) \tag{18}$$

where $r_\lambda = r/\lambda$
$Z = 377\ \Omega$
I_0 = maximum current = terminal current

At a large distance the ratio of E_z as given by (17) to H_ϕ as given by (18) is

$$\frac{E_z}{H_\phi} = Z = 377\ \Omega = \text{intrinsic impedance (resistance) of space} \tag{19}$$

The magnitude of H_ϕ is

$$|H_\phi| = \frac{I_0}{2\pi r} \qquad (\text{A m}^{-1}) \tag{20}$$

as given in Table 5-2.

5-6 RADIATION RESISTANCE OF $\lambda/2$ ANTENNA.

To find the radiation resistance, the Poynting vector is integrated over a large sphere yielding the power radiated, and this power is then equated to $(I_0/\sqrt{2})^2 R_0$, where R_0 is the radiation resistance at a current maximum point and I_0 is the peak value in time of the current at this point. The total power P radiated was given in (5-3-5)[1] in terms of H_ϕ for a short dipole. In (5-3-5), $|H_\phi|$ is the absolute value. Hence, the corresponding value of H_ϕ for a linear antenna is obtained from (5-5-10) by putting $|j[I_0]| = I_0$. Substituting this into (5-3-5), we obtain

$$P = \frac{15 I_0^2}{\pi} \int_0^{2\pi} \int_0^{\pi} \frac{\{\cos[(\beta L/2)\cos\theta] - \cos(\beta L/2)\}^2}{\sin\theta}\, d\theta\, d\phi \tag{1}$$

$$= 30 I_0^2 \int_0^{\pi} \frac{\{\cos[(\beta L/2)\cos\theta] - \cos(\beta L/2)\}^2}{\sin\theta}\, d\theta \tag{2}$$

Equating the radiated power as given by (2) to $I_0^2 R_0/2$ we have

$$P = \frac{I_0^2 R_0}{2} \tag{3}$$

and

$$R_0 = 60 \int_0^{\pi} \frac{\{\cos[(\beta L/2)\cos\theta] - \cos(\beta L/2)\}^2}{\sin\theta}\, d\theta \tag{4}$$

where the radiation resistance R_0 is referred to the current maximum. In the case of a $\lambda/2$ antenna this is at the center of the antenna or at the terminals of the transmission line (see Fig. 5-8).

Proceeding now to evaluate (4), let

$$u = \cos\theta \qquad \text{and} \qquad du = -\sin\theta\, d\theta \tag{5}$$

by which (4) is transformed to

$$R_0 = 60 \int_{-1}^{+1} \frac{[\cos(\beta L u/2) - \cos(\beta L/2)]^2}{1 - u^2}\, du \tag{6}$$

---

[1] $P = \iint \mathbf{S} \cdot d\mathbf{s} = \frac{1}{2}\sqrt{\mu/\varepsilon} \iint |H_\phi|^2\, ds$

However,

$$\frac{1}{1-u^2} = \frac{1}{(1+u)(1-u)} = \frac{1}{2}\left(\frac{1}{1+u} + \frac{1}{1-u}\right) \tag{7}$$

Also putting $k = \beta L/2$, (6) becomes

$$R_0 = 30 \int_{-1}^{+1} \left[\frac{(\cos ku - \cos k)^2}{1+u} + \frac{(\cos ku - \cos k)^2}{1-u}\right] du \tag{8}$$

This integral gives the radiation resistance for a thin linear antenna of any length L. For the special case being considered where $L = \lambda/2$, we have $k = \pi/2$. Thus, in the case of a thin $\lambda/2$ antenna, (8) reduces to

$$R_0 = 30 \int_{-1}^{+1} \left[\frac{\cos^2(\pi u/2)}{1+u} + \frac{\cos^2(\pi u/2)}{1-u}\right] du \tag{9}$$

Now in the first term let

$$1 + u = \frac{v}{\pi} \qquad \text{and} \qquad du = \frac{dv}{\pi} \tag{10}$$

and in the second term let

$$1 - u = \frac{v'}{\pi} \qquad \text{and} \qquad du = -\frac{dv'}{\pi} \tag{11}$$

Noting also that $(v - \pi)/2 = (\pi - v')/2$, Eq. (9) becomes

$$R_0 = 60 \int_0^{2\pi} \frac{\cos^2[(v-\pi)/2]}{v} dv \tag{12}$$

But $\cos^2(x/2) = \frac{1}{2}(1 + \cos x)$ so that

$$R_0 = 30 \int_0^{2\pi} \frac{1 + \cos(v - \pi)}{v} dv = 30 \int_0^{2\pi} \frac{1 - \cos v}{v} dv \tag{13}$$

The last integral in (13) is often designated as Cin (x). Thus,

$$\text{Cin}(x) = \int_0^x \frac{1 - \cos v}{v} dv = \ln \gamma x - \text{Ci}(x)$$

$$= 0.577 + \ln x - \text{Ci}(x) \tag{14}$$

where $\gamma = e^c = 1.781$ or $\ln \gamma = c = 0.577 =$ Euler's constant

The part of this integral given by

$$\text{Ci}(x) = \ln \gamma x - \text{Cin}(x) \tag{15}$$

is called the cosine integral. The value of this integral is given by

$$\text{Ci}(x) = \int_\infty^x \frac{\cos v}{v} dv = \ln \gamma x - \frac{x^2}{2!2} + \frac{x^4}{4!4} - \frac{x^6}{6!6} + \cdots \tag{16}$$

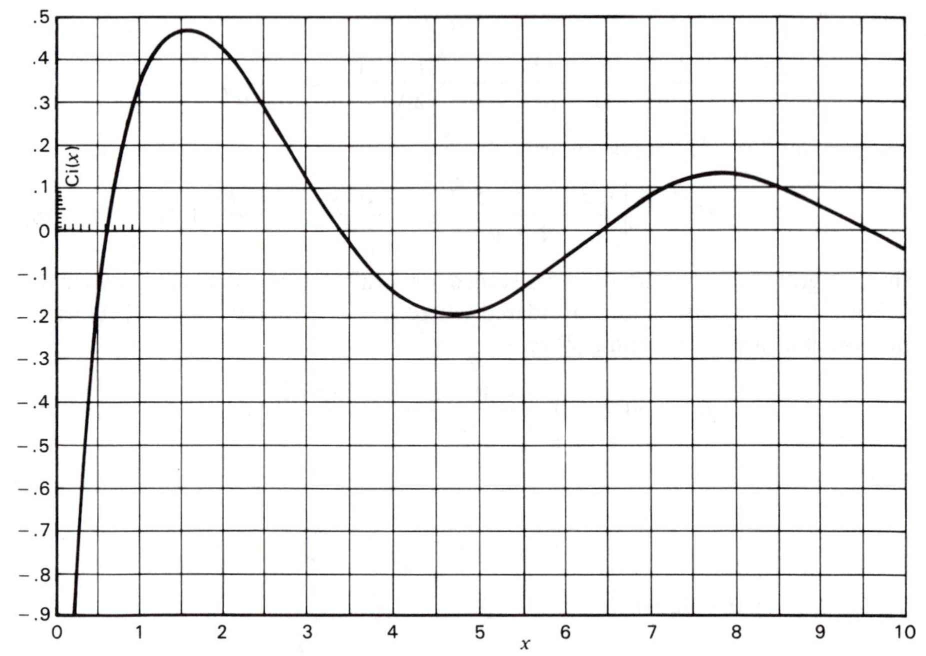

Figure 5-12a Cosine integral.

When x is small ($x < 0.2$),

$$\text{Ci}(x) \simeq \ln \gamma x = 0.577 + \ln x \tag{17}$$

When x is large ($x \gg 1$),

$$\text{Ci}(x) = \frac{\sin x}{x} \tag{18}$$

A curve of the cosine integral as a function of x is presented in Fig. 5-12a. It is to be noted that Ci (x) converges around zero at large values of x. From (16) and (14) we obtain Cin (x) as an infinite series,

$$\text{Cin}(x) = \frac{x^2}{2!2} - \frac{x^4}{4!4} + \frac{x^6}{6!6} - \cdots \tag{19}$$

While discussing Cin (x) and Ci (x), mention may be made of another integral which commonly occurs in impedance calculations. This is the sine integral, Si (x), given by

$$\text{Si}(x) = \int_0^x \frac{\sin v}{v}\, dv = x - \frac{x^3}{3!3} + \frac{x^5}{5!5} - \cdots \tag{20}$$

When x is small ($x < 0.5$),

$$\text{Si}(x) \simeq x \tag{21}$$

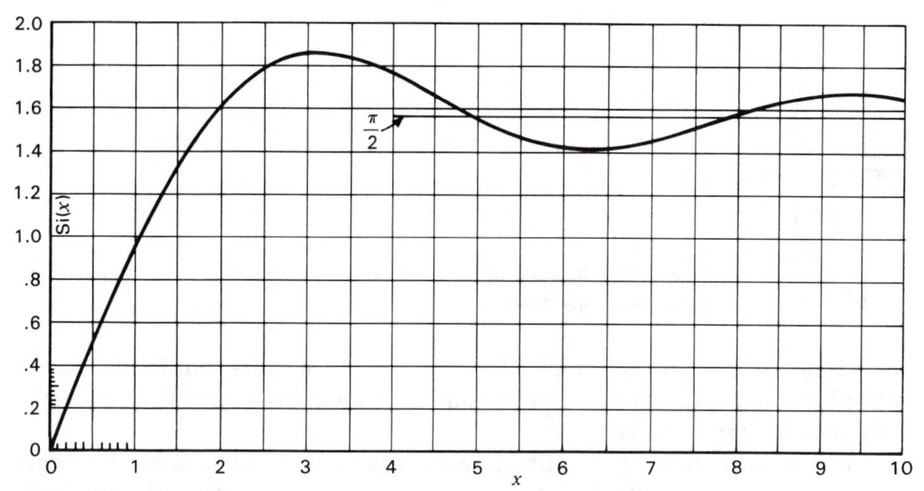

Figure 5-12b Sine integral.

When x is large ($x \gg 1$),

$$\text{Si}\,(x) \simeq \frac{\pi}{2} - \frac{\cos x}{x} \tag{22}$$

A curve of the sine integral as a function of x is presented in Fig. 5-12b. It is to be noted that Si (x) converges around $\pi/2$ at large values of x.

Returning now to (13), this can be written as

$$R_0 = 30\ \text{Cin}\,(2\pi) = 30 \times 2.44 = 73\ \Omega \tag{23}$$

This is the well-known value for the radiation resistance of a thin, linear, center-fed, $\lambda/2$ antenna with sinusoidal current distribution. The terminal impedance also includes some inductive reactance in series with R_0 (see Chap. 10). To make the reactance zero, i.e., to make the antenna resonant, requires that the antenna be a few per cent less than $\lambda/2$. This shortening also results in a reduction in the value of the radiation resistance.

5-7 RADIATION RESISTANCE AT A POINT WHICH IS NOT A CURRENT MAXIMUM. If we calculate, for example, the radiation resistance of a $3\lambda/4$ antenna (see Fig. 5-8) by the above method, we obtain its value at a current maximum. This is not the point at which the transmission line is connected. Neglecting antenna losses, the value of radiation resistance so obtained is the resistance R_0 which would appear at the terminals of a transmission line connected at a current maximum in the antenna, *provided* that the current distribution on the antenna is the same as when it is center-fed as in Fig. 5-8. Since a change of the feed point from the center of the antenna may change the current distribution, the radiation resistance R_0 is not the value which would be measured on a $3\lambda/4$ antenna or on any symmetrical antenna whose

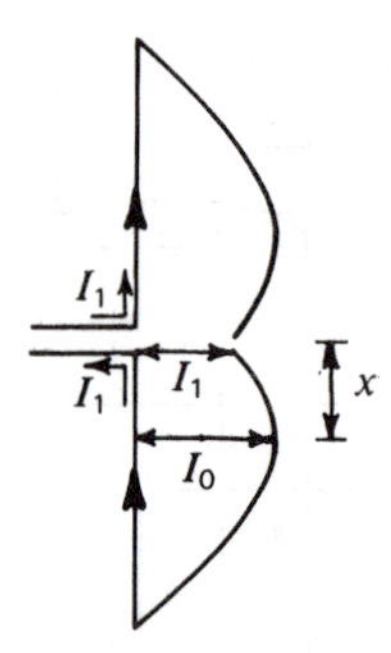

Figure 5-13 Relation of current I_1 at transmission-line terminals to current I_0 at current maximum.

length is not an odd number of $\lambda/2$. However, R_0 can be easily transformed to the value which would appear across the terminals of the transmission line connected at the center of the antenna.

This may be done by equating (5-6-3) to the power supplied by the transmission line, given by $I_1^2R_1/2$, where I_1 is the current amplitude at the terminals and R_1 is the radiation resistance at this point (see Fig. 5-13). Thus,

$$\frac{I_1^2}{2} R_1 = \frac{I_0^2}{2} R_0 \tag{1}$$

where R_0 is the radiation resistance calculated at the current maximum. Thus, the radiation resistance appearing at the terminals is

$$R_1 = \left(\frac{I_0}{I_1}\right)^2 R_0 \tag{2}$$

The current I_1 at a distance x from the nearest current maximum, as shown in Fig. 5-13, is given by

$$I_1 = I_0 \cos \beta x \tag{3}$$

where I_1 = terminal current
I_0 = maximum current

Therefore, (2) can be expressed as

$$R_1 = \frac{R_0}{\cos^2 \beta x} \tag{4}$$

When $x = 0$, $R_1 = R_0$; but when $x = \lambda/4$, $R_1 = \infty$ if $R_0 \neq 0$. However, the radiation resistance measured at a current minimum ($x = \lambda/4$) is not infinite as would be calculated from (4), since an actual antenna is not infinitesimally thin and the current at a minimum point is not zero. Nevertheless, the radiation resistance at a current minimum may in practice be very large, i.e., thousands of ohms.

5-8 FIELDS OF A THIN LINEAR ANTENNA WITH A UNIFORM TRAVELING WAVE. The foregoing discussion has been

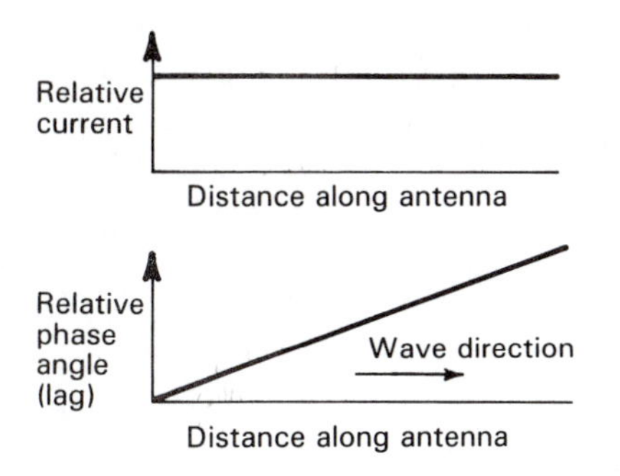

Figure 5-14 Current amplitude and phase relations along an antenna carrying a single uniform traveling wave.

confined to the case of antennas with sinusoidal current distributions. This current distribution may be regarded as the standing wave produced by two uniform (unattenuated) traveling waves of equal amplitude moving in opposite directions along the antenna. If, however, only one such wave is present on the antenna, the current distribution is uniform. The amplitude is a constant along the antenna, and the phase changes linearly with distance as suggested by Fig. 5-14.

The condition of a uniform traveling wave on an antenna is one of considerable importance, as this condition may be approximated in a number of antenna systems. For example, a single-wire antenna terminated in its characteristic impedance, as in Fig. 5-15*a*, may have essentially a uniform traveling wave.[1] This type of antenna is often referred to as a Beverage or wave antenna. A terminated rhombic antenna (Fig. 5-15*b*) may also have essentially a single traveling wave. The Beverage and rhombic antennas are discussed further in Chap. 16. Other types of antennas that have, in the first approximation, a single outgoing traveling wave, are a long monofilar axial-mode helical antenna and a long, thick linear antenna as illustrated in Fig. 5-15*c* and *d*. These antennas have no terminating impedance but behave in a similar way to terminated antennas. Thus, the thick linear conductor has a current distribution similar to a thin terminated linear conductor, and the patterns are similar if the conductor diameter is not too large. The results for a traveling wave on a linear conductor can be applied to a helix, as shown in Chap. 7, by considering that the helix consists of a number of short linear segments. On the linear antennas, the phase velocity of the traveling wave is substantially equal to the velocity of light. However, the phase velocity along the conductor of a monofilar axial-mode helical antenna may differ appreciably from the velocity of light. Hence, to make the results applicable to any of the antenna types shown in Fig. 5-15, the fields from an antenna with a traveling

[1] Since the fields of an antenna are not confined to the immediate vicinity of the antenna, it is not possible to provide a nonreflecting termination with a lumped impedance. However, a lumped impedance may greatly reduce reflections at the termination.

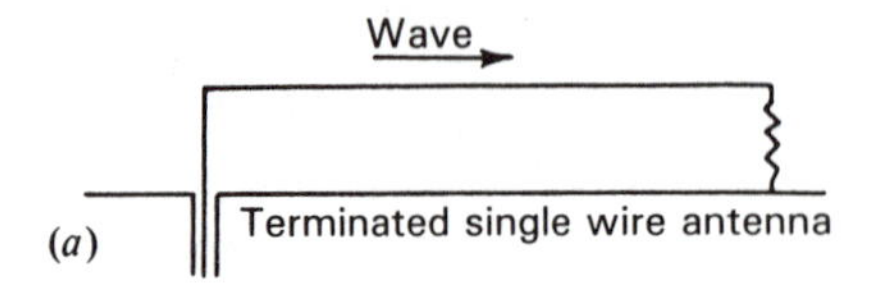

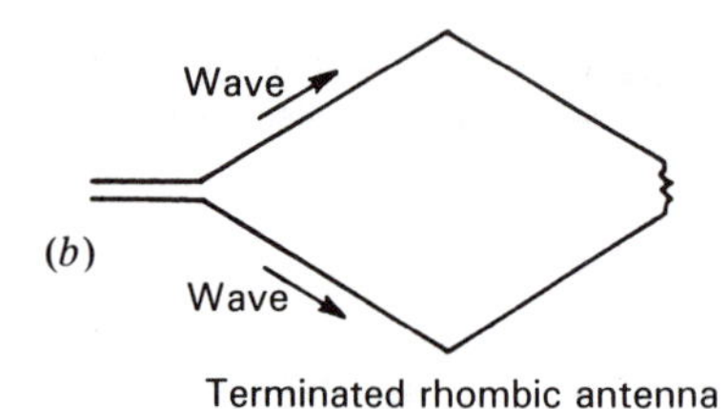

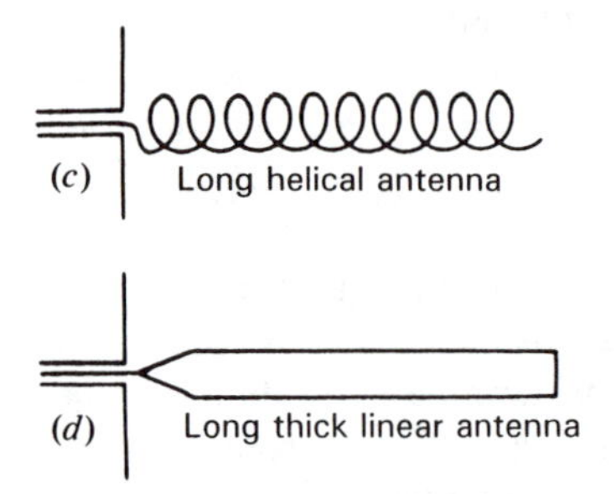

Figure 5-15 Various antennas having essentially a single traveling wave.

wave will be developed for the general case where the phase velocity v of the wave along the conductor may have any arbitrary value.[1]

Proceeding now to find the field radiated by a traveling wave on a thin linear conductor, let us consider a conductor of length b coincident with the z axis and with one end at the origin of a cylindrical coordinate system (ρ, ξ, z) as in Fig. 5-16. It is assumed that a single, uniform traveling wave is moving to the right along the conductor.

Since the current is entirely in the z direction, the magnetic field has but one component H_ξ. The ξ direction is normal to the page at P in Fig. 5-16, and its positive sense is outward from the page. The magnetic field H_ξ can be obtained from the Hertz vector $\mathbf{\Pi}$. Since the current is entirely in the z direction, the Hertz

[1] A. Alford, "A Discussion of Methods Employed in Calculations of Electromagnetic Fields of Radiating Conductors," *Elec. Commun.*, **15**, 70–88, July 1936. Treats case where velocity is equal to light.
J. D. Kraus and J. C. Williamson, "Characteristics of Helical Antennas Radiating in the Axial Mode," *J. App. Phys.*, **19**, 87–96, January 1948. Treats general case.
J. Grosskopf, "Über die Verwendung zweier Lösungsansätze der Maxwellschen Gleichungen bei der Berechnung der electromagnetischen Felder strahlender Leiter," *Hochfrequenztechnik und Electroakustik*, **49**, 205–211, June 1937. Treats case where velocity is equal to light.

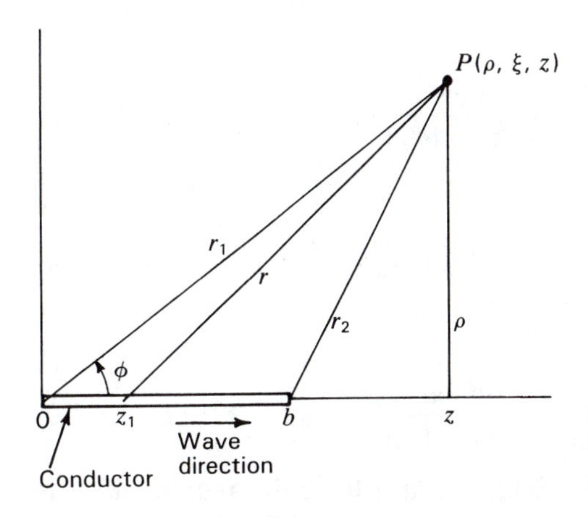

Figure 5-16 Relation of conductor of length b with single traveling wave to cylindrical coordinate system.

vector has only a z component. Thus,

$$H_\xi = j\omega\varepsilon(\nabla \times \boldsymbol{\Pi})_\xi = -j\omega\varepsilon \frac{\partial \Pi_z}{\partial \rho} \tag{1}$$

where Π_z is the z component of the retarded Hertz vector at the point P, as given by

$$\Pi_z = \frac{1}{4\pi j\omega\varepsilon} \int_0^b \frac{[I]}{r}\, dz_1 \tag{2}$$

where

$$[I] = I_0 \sin \omega\left(t - \frac{r}{c} - \frac{z_1}{v}\right) \tag{3}$$

where z_1 = a point on the conductor

and

$$v = pc \qquad \text{or} \qquad p = \frac{v}{c} \tag{4}$$

In (4), p is the ratio of the velocity along the conductor v to the velocity of light c. This ratio will be called the *relative phase velocity*.

All the conditions required for calculating the magnetic field due to a single traveling wave on the linear conductor are contained in the relations (1) through (4). That is, if $[I]$ in (3) is substituted into (2) and Π_z from this equation into (1) and the indicated operations performed, we obtain the field H_ξ. Let us now proceed to carry through this calculation. To do this, let

$$u = t - \frac{r}{c} - \frac{z_1}{v} \tag{5}$$

Now since

$$r = [(z - z_1)^2 + \rho^2]^{1/2} \tag{6}$$

we have

$$\frac{du}{dz_1} = \frac{z - z_1}{rc} - \frac{1}{pc} \tag{7}$$

Equation (2) now becomes

$$\Pi_z = \frac{I_0 c}{4\pi j\omega\varepsilon} \int_{u_1}^{u_2} \frac{\sin \omega u}{z - z_1 - (r/p)} \, du \tag{8}$$

where the new limits are

$$u_1 = t - \frac{r_1}{c} \qquad \text{and} \qquad u_2 = t - \frac{r_2}{c} - \frac{b}{v} \tag{9}$$

Introducing (8) into (1) we have

$$H_\xi = \frac{I_0 c}{4\pi} \frac{\partial}{\partial \rho} \int_{u_1}^{u_2} \frac{\sin \omega u}{z - z_1 - (r/p)} \, du \tag{10}$$

Confining our attention now to the far field, i.e., at a large distance r, which is very much larger than b, the quantity z_1 can be neglected and the denominator of the integrand considered to be a constant $z - (r/p)$. Therefore (10) becomes

$$H_\xi = -\frac{I_0 c}{4\pi\omega} \frac{\partial}{\partial \rho} \left[\frac{-\cos \omega u_2 + \cos \omega u_1}{z - (r/p)} \right] \tag{11}$$

Performing the differentiation with respect to ρ, (11) becomes

$$H_\xi = \frac{I_0 \rho}{4\pi r} \left\{ \frac{[z - (r/p)](\sin \omega u_2 - \sin \omega u_1) + [\lambda/(2\pi p)](\cos \omega u_2 - \cos \omega u_1)}{[z - (r/p)]^2} \right\} \tag{12}$$

At arbitrarily large distances, i.e., where

$$\left| z - \frac{r}{p} \right| \gg \frac{\lambda}{2\pi p}$$

and for the case where

$$\sin \omega u_2 - \sin \omega u_1 \neq 0$$

(12) reduces to

$$H_\xi = \frac{I_0}{4\pi r} \frac{\sin \phi}{\cos \phi - (1/p)} (\sin \omega u_2 - \sin \omega u_1) \tag{13}$$

where the relations have been introduced for $r \gg b$ that

$$\frac{z}{r} = \cos \phi \qquad \text{and} \qquad \frac{\rho}{r} = \sin \phi \tag{14}$$

Introducing the values of u_1 and u_2 into (13) from (9) and by trigonometric manipulation, (13) can be put in the form

$$H_\xi = \frac{I_0 p}{2\pi r_1} \left\{ \frac{\sin \phi}{1 - p \cos \phi} \sin \left[\frac{\omega b}{2pc} (1 - p \cos \phi) \right] \right\} \bigg/ \left[\omega \left(t - \frac{r_1}{c} \right) - \frac{\omega b}{2pc} (1 - p \cos \phi) \right] \tag{15}$$

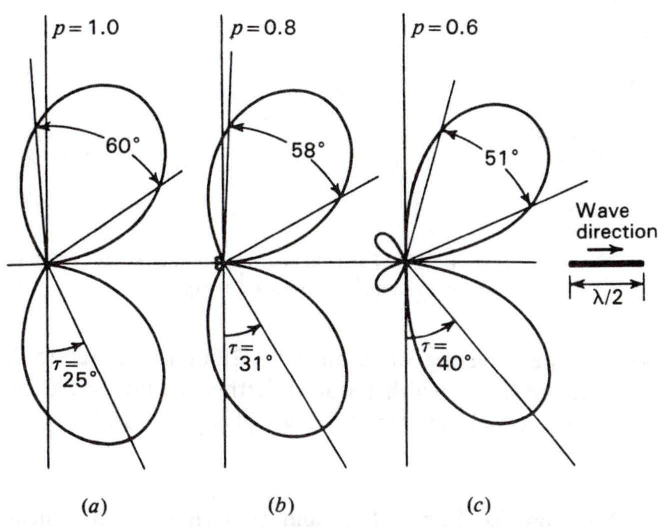

Figure 5-17 Far-field patterns of linear $\lambda/2$ antenna carrying a uniform traveling wave (to right) for three conditions of relative phase velocity ($p = 1.0$, 0.8 and 0.6). The tilt angle τ and the half-power beam widths are indicated for each pattern.

Equation (15) gives the instantaneous magnetic field at large distances from the linear antenna carrying a single traveling wave of amplitude I_0, in terms of the distance r_1, direction angle ϕ, relative phase velocity p, radian frequency ω, conductor length b, time t and velocity of light c. The distant or far electric field E_ϕ is obtained from H_ξ by $E_\phi = H_\xi Z$, where $Z = 377\ \Omega$.

In (15) the shape of the field pattern is given by the expression in the braces { }. The expression indicated as an angle $\angle$ gives the phase of the field referred to the origin of the coordinates (see Fig. 5-16) as the phase center. The relative phase pattern at a constant distance is given by the right-hand term, $[\omega b/(2pc)]$ $(1 - p \cos \phi)$.

Several examples will now be considered to illustrate the nature of the field patterns obtained on linear conductors carrying a uniform traveling wave.

5-8a Case 1. Linear $\lambda/2$ Antenna. Let us consider a linear antenna, $\lambda/2$ long as measured in free-space wavelengths. Thus, assuming that $p = 1$, the phase velocity along the antenna is equal to that of light and the pattern calculated from (15) is as shown by Fig. 5-17*a*. The difference between this pattern and that for a linear $\lambda/2$ antenna with a sinusoidal current distribution or standing wave (Fig. 5-10*a*) is striking. The lobes are sharper and also tilted forward in the case of the traveling wave antenna (Fig. 5-17*a*). The tilt is in the direction of the traveling wave. The tilt angle τ of the direction of maximum radiation is 25° and the beam width between half-power points is about 60°. This is in contrast to $\tau = 0$ and a beam width of 78° for the $\lambda/2$ antenna with a sinusoidal current distribution or standing wave.

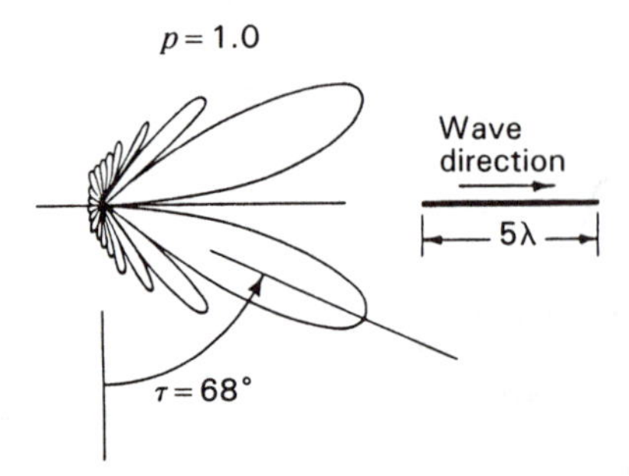

Figure 5-18 Far-field pattern of linear 5λ antenna carrying a uniform traveling wave ($p = 1$).

As the phase velocity of the traveling wave on the $\lambda/2$ antenna is reduced, the tilt angle is increased and the beam width reduced further, as illustrated by the patterns of Fig. 5-17*b* and *c*, which are for the cases of $p = 0.8$ and $p = 0.6$ respectively.

5-8b Case 2. Linear Antenna 5λ Long. The field pattern for a 5λ linear antenna with a single traveling wave is presented in Fig. 5-18 for the case where $p = 1$ (that is, $v = c$). This pattern is typical of those for long, terminated antennas, the radiation being beamed forward in a cone having the antenna as its axis. The tilt angle for this antenna is about 68°.

5-8c Case 3. Linear Antennas $\lambda/2$ to 25λ Long. As the length of the antenna is increased the tilt angle increases further, reaching about 78° (12° from antenna) when the length is 20λ for $p = 1$. The variation of the angle of the conical beam from the antenna is shown in Fig. 5-19 as a function of the antenna length for a wave traveling at the velocity of light ($p = 1$). Note that α in Fig. 5-19 is the complement of the tilt angle τ, that is, $\alpha = 90° - \tau$.

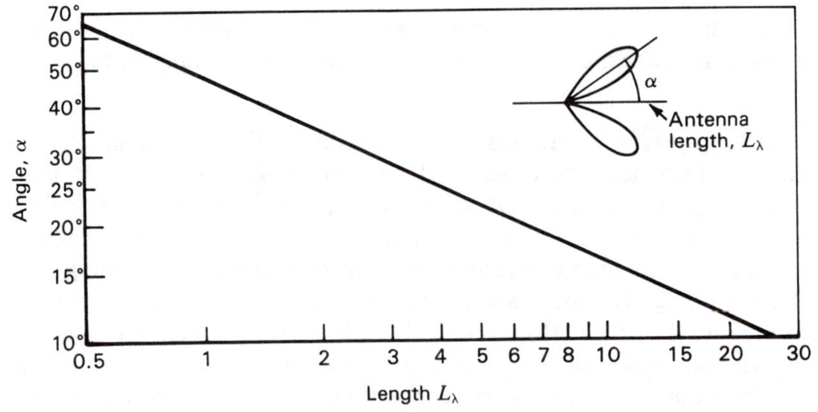

Figure 5-19 Angle α of main beam maximum from a linear traveling wave antenna as a function of antenna length in wavelengths (L_λ) with $v = c$ ($p = 1$) for antennas $\lambda/2$ to 25λ long.

PROBLEMS[1]

★5-1 Electric dipole.

(*a*) Two equal static electric charges of opposite sign separated by a distance L constitute a static electric dipole. Show that the electric potential at a distance r from such a dipole is given by

$$V = \frac{QL \cos \theta}{4\pi\varepsilon r^2}$$

where Q is the magnitude of each charge and θ is the angle between the radius r and the line joining the charges (axis of dipole). It is assumed that r is very large compared to L.

(*b*) Find the vector value of the electric field **E** at a large distance from a static electric dipole by taking the gradient of the potential expression in part (*a*).

5-2 2λ antenna. The instantaneous current distribution of a thin linear center-fed antenna 2λ long is sinusoidal as shown in Fig. P5-2.

(*a*) Calculate and plot the pattern of the far field.
(*b*) What is the radiation resistance referred to a current loop?
(*c*) What is the radiation resistance at the transmission-line terminals as shown?
(*d*) What is the radiation resistance $\lambda/8$ from a current loop?

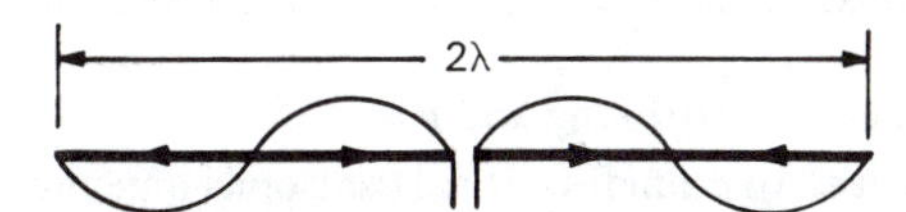

Figure P5-2 2λ antenna.

★5-3 λ/2 antenna. Assume that the current is of uniform magnitude and in phase along the entire length of a $\lambda/2$ thin linear element.

(*a*) Calculate and plot the pattern of the far field.
(*b*) What is the radiation resistance?
(*c*) Tabulate for comparison:

1. Radiation resistance of part (*b*) above
2. Radiation resistance at the current loop of a $\lambda/2$ thin linear element with sinusoidal in-phase current distribution
3. Radiation resistance of a $\lambda/2$ dipole calculated by means of the short dipole formula

(*d*) Discuss the three results tabulated in part (c) and give reasons for the differences.

5-4 λ/2 antennas in echelon. Calculate and plot the radiation-field pattern in the plane of two thin linear $\lambda/2$ antennas with equal in-phase currents and the spacing relationship shown in Fig. P5-4. Assume sinusoidal current distributions.

---

[1] Answers to starred (★) problems are given in App. D.

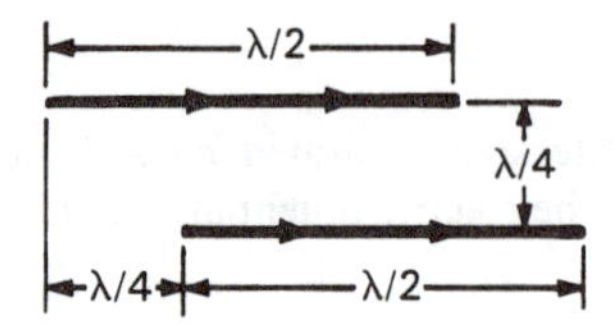

Figure P5-4 $\lambda/2$ antennas in echelon.

5-5 **1λ antenna with standing wave.** Calculate the field pattern in the plane of the 1λ antenna shown in Fig. P5-5. Assume that the current distribution on each wire is sinusoidal and that all currents are in phase. Plot the pattern.

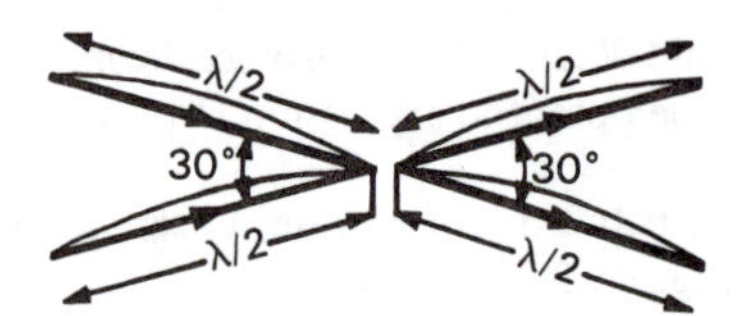

Figure P5-5 1λ antenna with standing wave.

★5-6 **1λ and 10λ antennas with traveling waves.**
(*a*) Calculate and plot the far-field pattern in the plane of a thin linear element 1λ long, carrying a single uniform traveling wave for 2 cases of the relative phase velocity $p = 1$ and 0.5.
(*b*) Repeat for the single case of an element 10λ long and $p = 1$.

★5-7 **Isotropic antenna. Radiation resistance.** An omnidirectional (isotropic) antenna has a field pattern given by $E = 10I/r$ (V m^{-1}), where I = terminal current (A) and r = distance (m). Find the radiation resistance.

5-8 **Short dipole.** For a thin center-fed dipole $\lambda/15$ long find (*a*) directivity D, (*b*) gain G, (*c*) effective aperture A_e, (*d*) beam solid angle Ω_A and (*e*) radiation resistance R_r. The antenna current tapers linearly from its value at the terminals to zero at its ends. The loss resistance is 1 Ω.

★5-9 **Conical pattern.** An antenna has a conical field pattern with uniform field for zenith angles (θ) from 0 to 60° and zero field from 60 to 180°. Find exactly (*a*) the beam solid angle and (*b*) directivity. The pattern is independent of the azimuth angle (ϕ).

5-10 **Conical pattern.** An antenna has a conical field pattern with uniform field for zenith angles (θ) from 0 to 45° and zero field from 45 to 180°. Find exactly (*a*) the beam solid angle, (*b*) directivity and (*c*) effective aperture. (*d*) Find the radiation resistance if the field $E = 5$ V m^{-1} at a distance of 50 m for a terminal current $I = 2$ A (rms). The pattern is independent of the azimuth angle (ϕ).

★5-11 **Directional pattern in θ and ϕ.** An antenna has a uniform field pattern for zenith angles (θ) between 45 and 90° and for azimuth (ϕ) angles between 0 and 120°. If $E = 3$ V m^{-1} at a distance of 500 m from the antenna and the terminal current is 5 A, find the radiation resistance of the antenna. $E = 0$ except within the angles given above.

5-12 **Directional pattern in θ and ϕ.** An antenna has a uniform field $E = 2$ V m^{-1} (rms) at a distance of 100 m for zenith angles between 30 and 60° and azimuth angles ϕ between 0 and 90° with $E = 0$ elsewhere. The antenna terminal current is 3 A (rms). Find (*a*) directivity, (*b*) effective aperture and (*c*) radiation resistance.

★**5-13 Directional pattern with back lobe.** The field pattern of an antenna varies with zenith angle (θ) as follows: E_n ($= E_{\text{normalized}}$) $= 1$ between $\theta = 0°$ and $\theta = 30°$ (main lobe), $E_n = 0$ between $\theta = 30°$ and $\theta = 90°$ and $E_n = \frac{1}{3}$ between $\theta = 90°$ and $\theta = 180°$ (back lobe). The pattern is independent of azimuth angle (ϕ). (*a*) Find the exact directivity. (*b*) If the field equals 8 V m^{-1} (rms) for $\theta = 0°$ at a distance of 200 m with a terminal current $I = 4$ A (rms), find the radiation resistance.

5-14 Short dipole. The radiated field of a short-dipole antenna with uniform current is given by $|E| = 30\beta l(I/r)\sin\theta$, where l = length, I = current, r = distance and θ = pattern angle. Find the radiation resistance.

5-15 Equivalence of pattern factors. Show that the field pattern of an ordinary end-fire array of a large number of colinear short dipoles as given by (4-6-8), multiplied by the dipole pattern $\sin\phi$, is equivalent to (5-8-15) for a long linear conductor with traveling wave for $p = 1$.

5-16 Relation of radiation resistance to beam area. Show that the radiation resistance of an antenna is a function of its beam area Ω_A as given by

$$R_r = \frac{Sr^2}{I^2}\,\Omega_A$$

where S = Poynting vector at distance r in direction of pattern maximum
I = terminal current

5-17 Cross-field. Find the locations in the field of a short dipole where circular cross-field exists.

CHAPTER 6

THE LOOP ANTENNA

This chapter is devoted to the loop antenna. First, the field pattern of a small loop is derived very simply by considering that the loop is square and consists of four short linear dipoles. The same field equations are then developed by a somewhat longer method based on the assumption that the small loop is equivalent to a short magnetic dipole. Finally, the general case of the loop antenna with uniform current is treated for loops of any size. Although most of the development concerns circular loops, square loops are also discussed, and it is shown that the far fields of circular and square loops of the same area are the same when they are small but different when they are large in terms of wavelength.

6-1 THE SMALL LOOP. A very simple method of finding the field pattern of a small loop is treated in this section. Consider a circular loop of radius a with a uniform in-phase current as suggested by Fig. 6-1a. The radius a is very small compared to the wavelength ($a \ll \lambda$). Suppose now that the circular loop is represented by a square loop of side length d, also with a uniform in-phase current, as shown in Fig. 6-1b. In this way, the loop can be treated as four short linear dipoles, whose properties we have already investigated in Chap. 5. Let d be chosen so that the area of the square loop is the same as the area of the circular loop; that is,

$$d^2 = \pi a^2 \tag{1}$$

If the loop is oriented as in Fig. 6-2, its far electric field has only an E_ϕ component. To find the far-field pattern in the yz plane, it is only necessary to

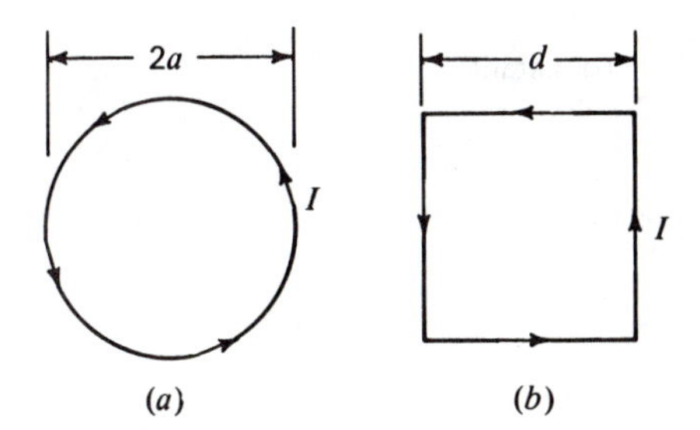

Figure 6-1 Circular loop (*a*) and square loop (*b*).

consider two of the four small linear dipoles (2 and 4). A cross section through the loop in the *yz* plane is presented in Fig. 6-3. Since the individual small dipoles 2 and 4 are nondirectional in the *yz* plane, the field pattern of the loop in this plane is the same as that for two isotropic point sources as treated in Sec. 4-2. Thus,

$$E_\phi = -E_{\phi 0}\, e^{j\psi/2} + E_{\phi 0}\, e^{-j\psi/2} \tag{2}$$

where $E_{\phi 0}$ = electric field from individual dipole and

$$\psi = \frac{2\pi d}{\lambda} \sin\theta = d_r \sin\theta \tag{3}$$

It follows that

$$E_\phi = -2jE_{\phi 0} \sin\left(\frac{d_r}{2}\sin\theta\right) \tag{4}$$

The factor j in (4) indicates that the total field E_ϕ is in phase quadrature with the field $E_{\phi 0}$ of the individual dipole. This may be readily seen by a vector construction of the type of Fig. 4-1*b* of Chap. 4. Now if $d \ll \lambda$, (4) can be written

$$E_\phi = -jE_{\phi 0}\, d_r \sin\theta \tag{5}$$

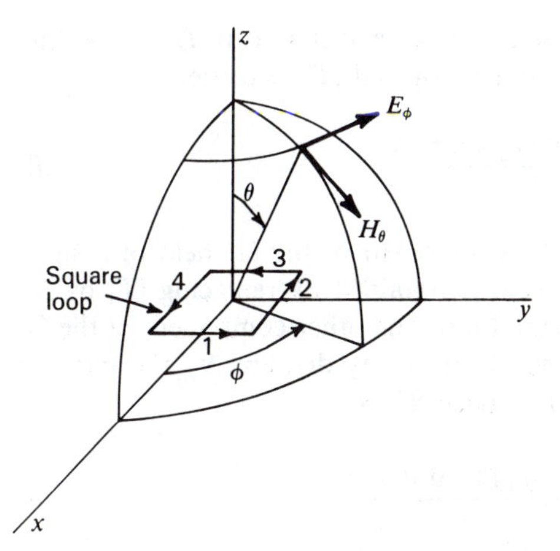

Figure 6-2 Relation of square loop to coordinates.

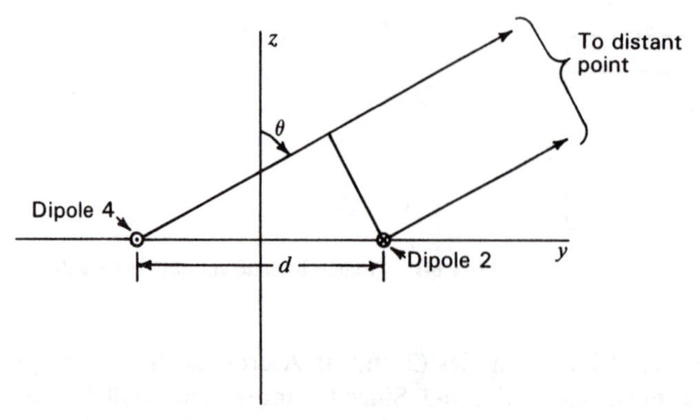

Figure 6-3 Construction for finding far field of dipoles 2 and 4 of square loop.

The far field of the individual dipole was developed in Chap. 5, being given in Table 5-1. In developing the dipole formula, the dipole was in the z direction, whereas in the present case it is in the x direction (see Figs. 6-2 and 6-3). The angle θ in the dipole formula is measured from the dipole axis and is 90° in the present case. The angle θ in (5) is a different angle with respect to the dipole, being as shown in Figs. 6-2 and 6-3. Thus, we have for the far field $E_{\phi 0}$ of the individual dipole

$$E_{\phi 0} = \frac{j60\pi[I]L}{r\lambda} \tag{6}$$

where $[I]$ is the retarded current on the dipole and r is the distance from the dipole. Substituting (6) in (5) then gives

$$E_{\phi} = \frac{60\pi[I]Ld_r \sin\theta}{r\lambda} \tag{7}$$

However, the length L of the short dipole is the same as d, that is, $L = d$. Noting also that $d_r = 2\pi d/\lambda$ and that the area A of the loop is d^2, (7) becomes

$$E_{\phi} = \frac{120\pi^2[I] \sin\theta}{r} \frac{A}{\lambda^2} \tag{8}$$

This is the instantaneous value of the E_{ϕ} component of the far field of a small loop of area A. The peak value of the field is obtained by replacing $[I]$ by I_0, where I_0 is the peak current in time on the loop. The other component of the far field of the loop is H_{θ}, which is obtained from (8) by dividing by the intrinsic impedance of the medium, in this case, free space. Thus,

$$H_{\theta} = \frac{E_{\phi}}{120\pi} = \frac{\pi[I] \sin\theta}{r} \frac{A}{\lambda^2} \tag{9}$$

6-2 THE SHORT MAGNETIC DIPOLE. EQUIVALENCE TO A LOOP. Another method of treating the small loop is by making use of its equivalence to a short magnetic dipole. Thus, a small loop of area A and carrying a uniform in-phase electric current I is replaced by an equivalent magnetic dipole of length l as shown in Fig. 6-4*a*. The magnetic dipole is assumed to carry a fictitious magnetic current I_m.

The relation between the loop and its equivalent magnetic dipole will now be developed. The moment of the magnetic dipole is $q_m l$ where q_m is the pole strength at each end as in Fig. 6-4*b*. The magnetic current is related to this pole strength by

$$I_m = -\mu \frac{dq_m}{dt} \tag{1}$$

where $I_m = I_{m0}\, e^{j\omega t}$

Integrating (1) with respect to time,

$$q_m = -\frac{I_m}{j\omega\mu} \tag{2}$$

The magnetic moment of the loop is IA. Equating this to the moment of the magnetic dipole, we have

$$q_m l = IA \tag{3}$$

Substituting (2) in (3),

$$\frac{I_m l}{j\omega\mu} = -IA \tag{4}$$

This may be reexpressed as

$$I_m l = -j\omega\mu IA = -j2\pi f \frac{\lambda}{\lambda} \mu IA = -j2\pi \frac{Z_0}{\lambda} IA \tag{5}$$

or

$$I_m l = -j240\pi^2 I \frac{A}{\lambda} \tag{6}$$

In retarded form (6) is

$$[I_m]l = -j240\pi^2[I] \frac{A}{\lambda} \tag{7}$$

where $[I_m] = I_{m0}\, e^{j\omega[t-(r/c)]}$

$[I] = I_0\, e^{j\omega[t-(r/c)]}$

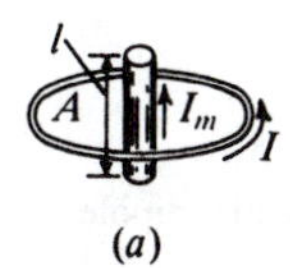

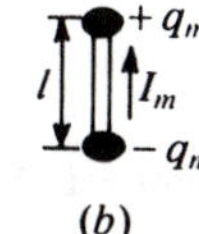

Figure 6-4 (*a*) Relation of small loop of area A to short magnetic dipole of length l. (*b*) Short magnetic dipole.

Equations (6) and (7) relate a loop of area A and carrying a current I to its equivalent magnetic dipole of length l carrying a fictitious magnetic current I_m.

6-3 THE SHORT MAGNETIC DIPOLE. FAR FIELDS. In this section the far fields of a short magnetic dipole will be calculated. Then applying the equivalence relation between a loop and magnetic dipole developed in Sec. 6-2, we obtain the far field of a small circular loop.

The method of finding the fields of a short magnetic dipole is formally the same as that employed in Sec. 5-2 to find the far field of a short electric dipole. The only difference is that electric current I is replaced by a fictitious magnetic current I_m and that **E** is replaced by **H**. Then with the magnetic dipole oriented as in Fig. 6-5, the retarded vector potential **F** of the *magnetic* current is

$$\mathbf{F} = \frac{\mu}{4\pi} \iiint \frac{[\mathbf{J}_m]}{r}\, dv = \hat{\mathbf{z}} \frac{\mu}{4\pi} \int_{-l/2}^{+l/2} \frac{[I_m]}{r}\, dz \qquad (\mathrm{V^2\,s\,A^{-1}\,m^{-1}}) \tag{1}$$

The vector potential **F** has only a z component F_z. Introducing the value of the retarded current

$$F_z = \frac{\mu I_{m0}}{4\pi} \int_{-l/2}^{+l/2} \frac{e^{j\omega[t-(r/c)]}}{r}\, dz \tag{2}$$

If $r \gg l$ and $\lambda \gg l$, the phase difference of the contributions of the various current elements of length dz along the magnetic dipole can be neglected. Hence, the integrand in (2) may be regarded as a constant, and (2) becomes

$$F_z = \frac{\mu I_{m0}\, l e^{j\omega[t-(r/c)]}}{4\pi r} \tag{3}$$

The electric field **E** is obtained from **F** by the relation

$$\mathbf{E} = \frac{1}{\mu} \nabla \times \mathbf{F} \tag{4}$$

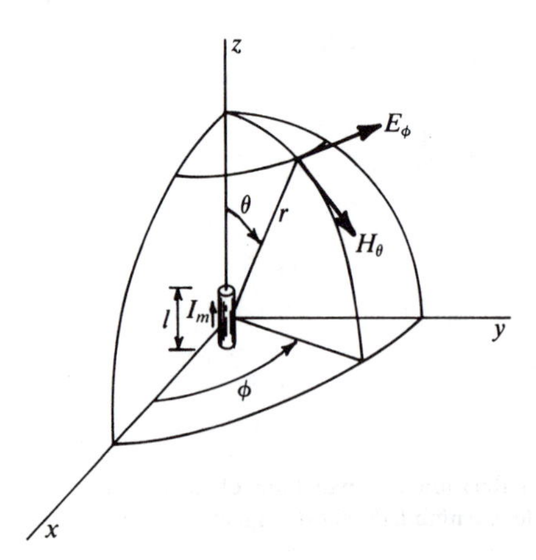

Figure 6-5 Relation of short magnetic dipole to coordinates.

Resolving F_z into its spherical or polar coordinate components F_θ and F_r and taking the curl of **F** as in (4), the E_ϕ component of the electric field is found to be

$$E_\phi = \frac{[I_m]l \sin\theta}{4\pi}\left(\frac{j\omega}{cr} + \frac{1}{r^2}\right) \tag{5}$$

This is the only component of the electric field produced by a magnetic dipole oriented as in Fig. 6-5. It is interesting to note that (5) is identical with the expression for H_ϕ developed for a short electric dipole, provided that E in (5) is replaced by H and I_m by I (see Table 5-1).

The relation of (5) applies at any distance from the magnetic dipole, provided only that $r \gg l$ and $\lambda \gg l$. At a large distance r the second term of (5) can be neglected, and (5) becomes

$$E_\phi = \frac{j[I_m]\omega l \sin\theta}{4\pi cr} = \frac{j[I_m]\sin\theta}{2r}\frac{l}{\lambda} \tag{6}$$

This is the far electric field from a short magnetic dipole of length l and carrying a fictitious magnetic current I_m. The far magnetic field H_θ of the magnetic dipole is related to E_ϕ by the intrinsic impedance of the medium, in this case, free space. Hence

$$H_\theta = \frac{j[I_m]\sin\theta}{240\pi r}\frac{l}{\lambda} \tag{7}$$

Substituting (6-2-7) for the moment $[I_m]l$ in (6) and (7), we obtain

$$E_\phi = \frac{120\pi^2[I]\sin\theta}{r}\frac{A}{\lambda^2} \tag{8}$$

and

$$H_\theta = \frac{\pi[I]\sin\theta}{r}\frac{A}{\lambda^2} \tag{9}$$

These are then the far-field equations in a plane perpendicular to a small loop of area A carrying a current I. They are identical with (6-1-8) and (6-1-9) developed in Sec. 6-1 by the method using a square loop of four short linear electric dipoles. The field pattern in the plane of a circular loop with uniform current is by symmetry a circle. The far-field pattern in the plane of a small square loop with uniform current may also be shown to be a circle (Prob. 6-6). Thus, it appears that the far fields of *small* circular and square loops are identical *provided that both have the same area.*

Both E_ϕ and H_θ vary as the sine of the angle θ measured from the polar axis as illustrated in Fig. 6-6. The fields are independent of ϕ. Hence, the space patterns are figures-of-revolution of the pattern of Fig. 6-6 around the polar axis, the

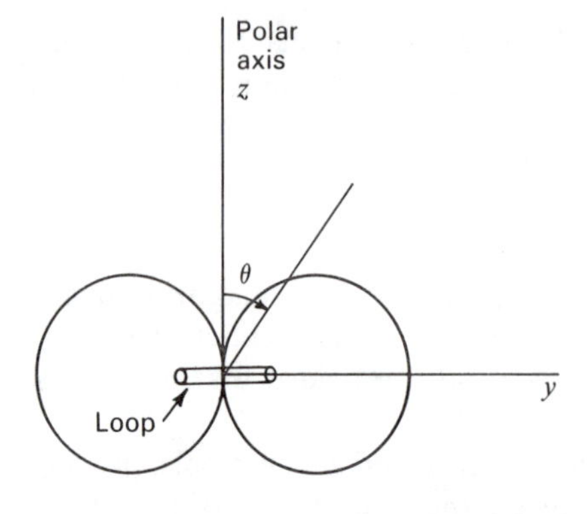

Figure 6-6 Far-field pattern for a small loop.

form being that of a doughnut. This pattern is identical in shape to that of a short electric dipole oriented parallel to the polar or z axis.

6-4 COMPARISON OF FAR FIELDS OF SMALL LOOP AND SHORT DIPOLE. It is of interest to compare the far-field expressions for a small loop with those for a short electric dipole. The comparison is made in Table 6-1. The presence of the operator j in the dipole expressions and its absence in the loop equations indicate that the fields of the electric dipole and of the loop are in time-phase quadrature, the current I being in the same phase in both the dipole and loop. This quadrature relationship is a fundamental difference between the fields of loops and dipoles. See Prob. 6-9.

The formulas in Table 6-1 apply to a loop oriented as in Fig. 6-2 and a dipole oriented parallel to the polar or z axis. The formulas are exact only for vanishingly small loops and dipoles. However, they are good approximations for loops up to $\lambda/10$ in diameter and dipoles up to $\lambda/10$ long.

6-5 THE LOOP ANTENNA. GENERAL CASE. The general case of a loop antenna with uniform, in-phase current will now be discussed. The size of the loop is not restricted to a small value compared to the wavelength as in the

Table 6-1 Far fields of small electric dipoles and loops

Field	Electric dipole	Loop
Electric	$E_\theta = \dfrac{j60\pi[I] \sin\theta}{r} \dfrac{L}{\lambda}$	$E_\phi = \dfrac{120\pi^2[I] \sin\theta}{r} \dfrac{A}{\lambda^2}$
Magnetic	$H_\phi = \dfrac{j[I] \sin\theta}{2r} \dfrac{L}{\lambda}$	$H_\theta = \dfrac{\pi[I] \sin\theta}{r} \dfrac{A}{\lambda^2}$

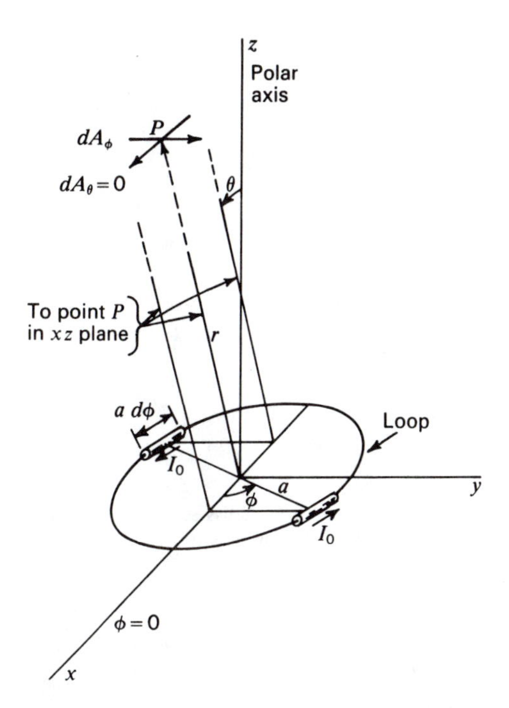

Figure 6-7 Loop of any radius a with relation to coordinates.

preceding sections but may assume any value. The method of treatment follows that given by Foster.[1]

Let the loop of radius a be located with its center at the origin of the coordinates as in Fig. 6-7. The current I is uniform and in phase around the loop. Although this condition is readily obtained when the loop is small, it is not a natural condition for large loops energized at a point. For loops with perimeters of about $\lambda/4$ or larger, phase shifters of some type must be introduced at intervals around the periphery in order to approximate a uniform, in-phase current on the loop. Assuming that the current is uniform and in phase, the far-field expressions will be derived with the aid of the vector potential of the electric current. The vector potential will first be developed for a pair of short, diametrically opposed electric dipoles of length $a\,d\phi$, as in Fig. 6-7. Then integrating over the loop, the total vector potential is obtained, and from this the far-field components are derived.

Since the current is confined to the loop, the only component of the vector potential having a value is A_ϕ. The other components are zero: $A_\theta = A_r = 0$. The infinitesimal value at the point P of the ϕ component of A from two diametrically

[1] Donald Foster, "Loop Antennas with Uniform Current," *Proc. IRE*, **32**, 603–607, October 1944. A discussion of circular loops of circumference less than $\lambda/2$ ($C_\lambda < \frac{1}{2}$) with nonuniform current distribution is given by G. Glinski, "Note on Circular Loop Antennas with Nonuniform Current Distribution," *J. Appl. Phys.*, **18**, 638–644, July 1947.

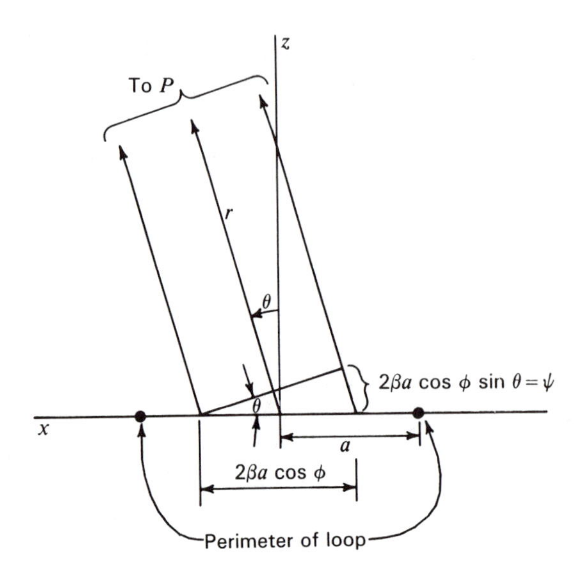

Figure 6-8 Cross section in *xz* plane through loop of Fig. 6-7.

opposed infinitesimal dipoles is

$$dA_\phi = \frac{\mu \, dM}{4\pi r} \tag{1}$$

where dM is the current moment due to one pair of diametrically opposed infinitesimal dipoles of length, $a \, d\phi$. In the $\phi = 0$ plane (Fig. 6-7) the ϕ component of the retarded current moment due to one dipole is

$$[I]a \, d\phi \cos \phi \tag{2}$$

where $[I] = I_0 e^{j\omega[t-(r/c)]}$ and I_0 is the peak current in time on the loop.

Figure 6-8 is a cross section through the loop in the xz plane of Fig. 6-7. Referring now to Fig. 6-8, the resultant moment dM at a large distance due to a pair of diametrically opposed dipoles is

$$dM = 2j[I]a \, d\phi \cos \phi \sin \frac{\psi}{2} \tag{3}$$

where $\psi = 2\beta a \cos \phi \sin \theta$ radians

Introducing this value for ψ into (3) we have

$$dM = 2j[I]a \cos \phi \, [\sin (\beta a \cos \phi \sin \theta)] \, d\phi \tag{4}$$

Now substituting (4) into (1) and integrating,

$$A_\phi = \frac{j\mu[I]a}{2\pi r} \int_0^\pi \sin (\beta a \cos \phi \sin \theta) \cos \phi \, d\phi \tag{5}$$

or

$$A_\phi = \frac{j\mu[I]a}{2r} J_1(\beta a \sin \theta) \tag{6}$$

where J_1 is a Bessel function of the first order and of argument ($\beta a \sin \theta$). The integration of (5) is performed on equivalent dipoles which are all situated at the origin but have different orientations with respect to ϕ. The retarded current $[I]$ is referred to the origin and, hence, is constant in the integration.

The far electric field of the loop has only a ϕ component given by

$$E_\phi = -j\omega A_\phi \tag{7}$$

Substituting the value of A_ϕ from (6) into (7) yields

$$E_\phi = \frac{\mu\omega[I]a}{2r} J_1(\beta a \sin \theta) \tag{8}$$

or

$$E_\phi = \frac{60\pi\beta a[I]}{r} J_1(\beta a \sin \theta) \tag{9}$$

This expression gives the instantaneous electric field at a large distance r from a loop of any radius a. The peak value of E_ϕ is obtained by putting $[I] = I_0$, where I_0 is the peak value (in time) of the current on the loop. The magnetic field H_θ at a large distance is related to E_ϕ by the intrinsic impedance of the medium, in this case, free space. Thus,

$$H_\theta = \frac{\beta a[I]}{2r} J_1(\beta a \sin \theta) \tag{10}$$

This expression gives the instantaneous magnetic field at a large distance r from a loop of any radius a.

6-6 FAR-FIELD PATTERNS OF CIRCULAR LOOP ANTENNAS WITH UNIFORM CURRENT.

The far-field patterns for a loop of any size are given by (6-5-9) and (6-5-10). For a loop of a given size, βa is constant and the shape of the far-field pattern is given as a function of θ by

$$J_1(C_\lambda \sin \theta) \tag{1}$$

where C_λ is the circumference of the loop in wavelengths. That is,

$$C_\lambda = \frac{2\pi a}{\lambda} = \beta a \tag{2}$$

The value of $\sin \theta$ as a function of θ ranges in magnitude between zero and unity. When $\theta = 90°$, the relative field is $J_1(C_\lambda)$, and as θ decreases to zero, the values of the relative field vary in accordance with the J_1 curve from $J_1(C_\lambda)$ to zero. This is illustrated by Fig. 6-9 in which a rectified first-order Bessel curve is shown as a function of $C_\lambda \sin \theta$.

As an example, let us find the pattern for a loop 1λ in diameter ($C_\lambda = \pi = 3.14$). The relative field in the direction $\theta = 90°$ is then 0.285. As θ decreases, the field intensity rises, reaching a maximum of 0.582 at angle θ of about 36°. As θ decreases further, the field intensity also decreases, reaching zero at $\theta = 0°$. The

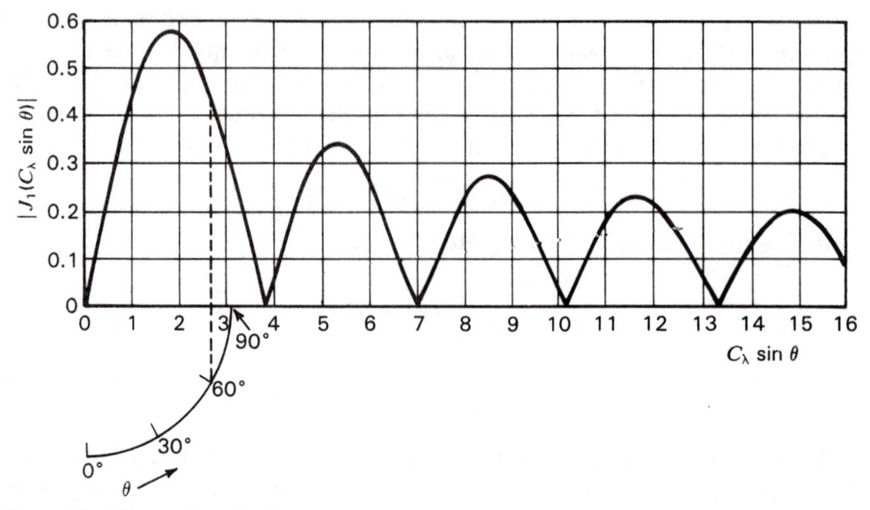

Figure 6-9 Pattern chart for loops with uniform current as given by first-order Bessel curve as a function of $C_\lambda \sin \theta$.

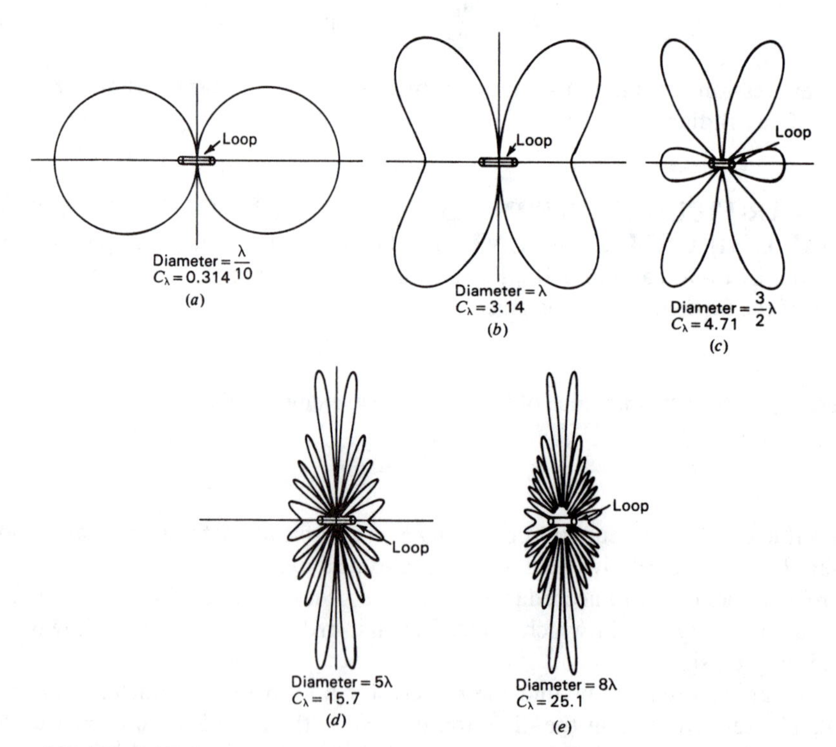

Figure 6-10 Far-field patterns of loops of 0.1, 1, 1.5, 5 and 8λ diameter. Uniform in-phase current is assumed on the loops.

pattern in the other four quadrants is symmetrical, the complete pattern being as presented in Fig. 6-10*b*.

It is possible to obtain the pattern by a graphical construction. This is illustrated for the case we have just considered of $C_\lambda = \pi$ by the auxiliary circle quadrant in Fig. 6-9. The angle θ is laid off around the arc of the circle. The radius of the circle is equal to $C_\lambda \sin 90° = C_\lambda$, which in this case is π. The field in the direction $\theta = 60°$, for instance, is then given by drawing a perpendicular to the axis of the abscissa and continuing this perpendicular until it intersects the J_1 curve, giving a value of relative field, in this case, of 0.443, as shown in Fig. 6-9.

Turning now to a consideration of loops of other size, it is to be noted from Fig. 6-9 that the maximum field is in the direction $\theta = 90°$ for all loops which are less than 1.84λ in circumference (less than 0.585λ in diameter). As an example, the pattern for a loop $\lambda/10$ in diameter is presented in Fig. 6-10*a*. The pattern is practically a sine pattern as would be obtained with a very small loop.

By way of contrast, the pattern for a loop 5λ in diameter is shown in Fig. 6-10*d*. In this case, which is typical for large circular loops with uniform current, the maximum field is in a direction nearly normal to the plane of the loop, while the field in the direction of the plane of the loop is small.

All patterns in Fig. 6-10 are adjusted to the same maximum. The space patterns for the five cases in Fig. 6-10 are figures-of-revolution of the patterns around the polar axis. It is to be noted that the field exactly normal to the loop is always zero, regardless of the size of the loop.

6-7 THE SMALL LOOP AS A SPECIAL CASE.

The relations of (6-5-9) and (6-5-10) apply to loops of any size. It will now be shown that for the special case of a small loop, these expressions reduce to the ones obtained previously.

For small arguments of the first-order Bessel function, the following approximate relation can be used:[1]

$$J_1(x) = \frac{x}{2} \tag{1}$$

where x is any variable. When $x = \frac{1}{3}$, the approximation of (1) is about 1 percent in error. The relation becomes exact as x approaches zero. Thus, if the perimeter of the loop is $\lambda/3$ or less ($C_\lambda < \frac{1}{3}$), (1) may be applied to (6-5-9) and (6-5-10) with an error which is about 1 percent or less. Equations (6-5-9) and (6-5-10) then

[1] For small arguments, the J_1 curve is nearly linear (see Fig. 6-9). The general relation for a Bessel function of any order n is $J_n(x) \simeq x^n/n!\,2^n$, where $|x| \ll 1$.

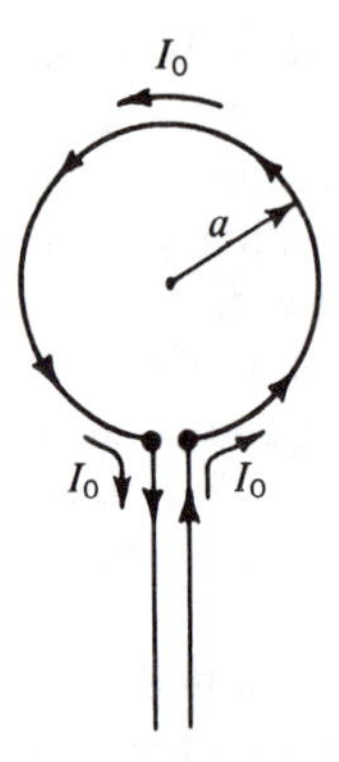

Figure 6-11 Loop and transmission line.

become

$$E_\phi = \frac{60\pi\beta a[I]\beta a \sin\theta}{2r} = \frac{120\pi^2[I] \sin\theta}{r}\frac{A}{\lambda^2} \tag{2}$$

$$H_\theta = \frac{\beta a[I]\beta a \sin\theta}{4r} = \frac{\pi[I] \sin\theta}{r}\frac{A}{\lambda^2} \tag{3}$$

These far-field equations for a small loop are identical with those obtained in earlier sections (see Table 6-1).

6-8 RADIATION RESISTANCE OF LOOPS.[1] To find the radiation resistance of a loop antenna, the Poynting vector is integrated over a large sphere yielding the total power P radiated. This power is then equated to the square of the effective current on the loop times the radiation resistance R_r:

$$P = \frac{I_0^2}{2} R_r \tag{1}$$

where I_0 = peak current in time on the loop. The radiation resistance so obtained is the value which would appear at the loop terminals connected to the transmission line, as shown in Fig. 6-11. The situation shown in Fig. 6-11 occurs naturally only on small loops. However, it will be assumed that the current is uniform and in phase for any radius a, this condition being obtained by means of phase shifters, multiple feeds or other devices (see Fig. 16-19). The average Poynting vector of a far field is given by

$$S_r = \tfrac{1}{2}|H|^2 \operatorname{Re} Z \tag{2}$$

where $|H|$ is the absolute value of the magnetic field and Z is the intrinsic imped-

[1] The procedure follows that given by Donald Foster, "Loop Antennas with Uniform Current," *Proc. IRE*, **32**, 603–607, October 1944.

ance of the medium, which in this case is free space. Substituting the absolute value of H_θ from (6-5-10) for $|H|$ in (2) yields

$$S_r = \frac{15\pi(\beta a I_0)^2}{r^2} J_1^2(\beta a \sin\theta) \tag{3}$$

The total power radiated P is the integral of S_r over a large sphere; that is,

$$P = \iint S_r \, ds = 15\pi(\beta a I_0)^2 \int_0^{2\pi} \int_0^{\pi} J_1^2(\beta a \sin\theta) \sin\theta \, d\theta \, d\phi \tag{4}$$

or

$$P = 30\pi^2(\beta a I_0)^2 \int_0^{\pi} J_1^2(\beta a \sin\theta) \sin\theta \, d\theta \tag{5}$$

In the case of a loop that is small in terms of wavelengths, the approximation of (6-7-1) can be applied. Thus (5) reduces to

$$P = \frac{15}{2} \pi^2(\beta a)^4 I_0^2 \int_0^{\pi} \sin^3\theta \, d\theta = 10\pi^2\beta^4 a^4 I_0^2 \tag{6}$$

Since the area $A = \pi a^2$, (6) becomes

$$P = 10\beta^4 A^2 I_0^2 \tag{7}$$

Assuming no antenna losses, this power equals the power delivered to the loop terminals as given by (1). Therefore,

$$R_r \frac{I_0^2}{2} = 10\beta^4 A^2 I_0^2 \tag{8}$$

and

$$R_r = 31\,171\left(\frac{A}{\lambda^2}\right)^2 = 197 C_\lambda^4 \qquad (\Omega) \tag{9}$$

or

$$R_r \simeq 31\,200\left(\frac{A}{\lambda^2}\right)^2 \qquad (\Omega) \tag{10}$$

This is the radiation resistance of a small single-turn loop antenna, circular or square, with uniform in-phase current. The relation is about 2 percent in error when the loop perimeter is $\lambda/3$. A circular loop of this perimeter has a diameter of about $\lambda/10$. Its radiation resistance by (10) is nearly 2.5 Ω.

The radiation resistance of a small loop consisting of one or more turns is given by[1]

$$R_r = 31\,200\left(n \frac{A}{\lambda^2}\right)^2 \qquad (\Omega)$$

where n = number of turns

[1] A. Alford and A. G. Kandoian, "Ultrahigh-Frequency Loop Antennas," *Trans. AIEE*, **59**, 843–848, 1940.

Let us now proceed to find the radiation resistance of a circular loop of any radius a. To do this we must integrate (5). However, the integral of (5) may be reexpressed. Thus, in general,[1]

$$\int_0^\pi J_1^2(x \sin \theta) \sin \theta \, d\theta = \frac{1}{x} \int_0^{2x} J_2(y) \, dy \tag{11}$$

where y is any function

Applying (11) to (5) we obtain

$$P = 30\pi^2 \beta a I_0^2 \int_0^{2\beta a} J_2(y) \, dy \tag{12}$$

Equating (12) and (1) and putting $\beta a = C_\lambda$ yields

$$R_r = 60\pi^2 C_\lambda \int_0^{2C_\lambda} J_2(y) \, dy \qquad (\Omega) \tag{13}$$

This is the radiation resistance as given by Foster for a single-turn circular loop with uniform in-phase current and of any circumference C_λ.

When the loop is large ($C_\lambda \geq 5$), we can use the approximation

$$\int_0^{2C_\lambda} J_2(y) \, dy \simeq 1 \tag{14}$$

so that (13) reduces to

$$R_r = 60\pi^2 C_\lambda = 592 C_\lambda = 3720 \frac{a}{\lambda} \tag{15}$$

For a loop of 10λ perimeter, the radiation resistance by (15) is nearly 6000 Ω.

For values of C_λ between $\frac{1}{3}$ and 5 the integral in (13) can be evaluated using the transformation

$$\int_0^{2C_\lambda} J_2(y) \, dy = \int_0^{2C_\lambda} J_0(y) \, dy - 2J_1(2C_\lambda) \tag{16}$$

where the expressions on the right of (16) are tabulated functions.[2]

For perimeters of over 5λ ($C_\lambda > 5$) one can also use the asymptotic development,

$$\int_0^{2x} J_2(y) \, dy \simeq 1 - \frac{1}{\sqrt{\pi x}} \left[\sin\left(2x - \frac{\pi}{4}\right) + \frac{11}{16x} \cos\left(2x - \frac{\pi}{4}\right) \right] \tag{17}$$

where $x = \beta a = C_\lambda$

[1] G. N. Watson, *A Treatise on the Theory of Bessel Functions*, Cambridge University Press, London, 1922.

[2] The integral involving J_0 for the interval $0 \leq x \leq 5$ (where $x = C_\lambda$) is given by A. N. Lowan and M. Abramowitz, *J. Math. Phys.*, **22**, 2–12, May 1943; and also by Natl. Bur. Standards Tech. Memo 20.

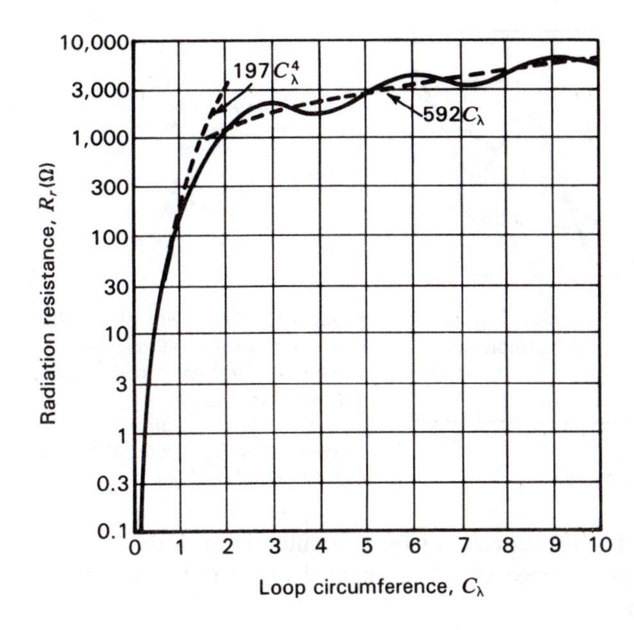

Figure 6-12 Radiation resistance of single-turn circular loop with uniform, in-phase current as a function of the loop circumference in wavelengths, C_λ.

For small values of x, one can use a series obtained by integrating the ascending power series for J_2. Thus,

$$\int_0^{2x} J_2(y)\,dy = \frac{x^3}{3}\left(1 - \frac{x^2}{5} + \frac{x^4}{56} - \frac{x^6}{1080} + \frac{x^8}{31\,680} - \cdots\right) \tag{18}$$

When $x = C_\lambda = 2$ (perimeter of 2λ), the result with four terms is about 2 percent in error. This same percentage error is obtained with one term when the perimeter is about $\lambda/3$.

A graph showing the radiation resistance of single-turn loops with uniform current as a function of the circumference in wavelengths is presented in Fig. 6-12. The data for the curve are based on Foster's formulas as given above. Curves for the approximate formulas of small and large loops are shown by the dashed lines.

6-9 DIRECTIVITY OF CIRCULAR LOOP ANTENNAS WITH UNIFORM CURRENT. The directivity D of an antenna was defined in (2-8-1) as the ratio of maximum radiation intensity to the average radiation intensity. The maximum radiation intensity for a loop antenna is given by r^2 times (6-8-3). The average radiation intensity is given by (6-8-5) divided by 4π. Thus, the directivity of a loop is

$$D = \frac{2C_\lambda[J_1^2(C_\lambda \sin\theta)]_{\max}}{\int_0^{2C_\lambda} J_2(y)\,dy} \tag{1}$$

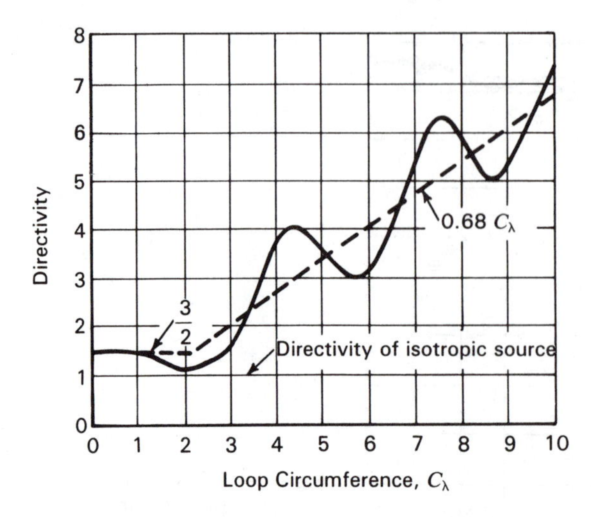

Figure 6-13 Directivity of circular loop antenna with uniform, in-phase current as a function of loop circumference in wavelengths, C_λ. (*After D. Foster, "Loop Antennas with Uniform Current,"* Proc. IRE, **32**, *603–607, October 1944.*)

This is Foster's expression for the directivity of a circular loop with uniform in-phase current of any circumference C_λ. The angle θ in (1) is the value for which the field is a maximum.

For a small loop ($C_\lambda < \frac{1}{3}$), the directivity expression reduces to

$$D = \tfrac{3}{2}\sin^2\theta = \tfrac{3}{2} \tag{2}$$

since the field is a maximum at $\theta = 90°$. The value of $\frac{3}{2}$ is the same as for a short electric dipole. This is to be expected since the pattern of a short dipole is the same as for a small loop.

For a large loop ($C_\lambda > 5$), (1) reduces to

$$D = 2C_\lambda[J_1^2(C_\lambda \sin\theta)]_{\max} \tag{3}$$

From Fig. 6-9 we note that for any loop with $C_\lambda \geq 1.84$, the maximum value of $J_1(C_\lambda \sin\theta)$ is 0.582. Thus, the directivity expression of (3) for a large loop becomes

$$D = 0.68C_\lambda \tag{4}$$

The directivity of a loop antenna as a function of the loop circumference C_λ is presented in Fig. 6-13. Curves based on the approximate relations of (2) and (4) for small and large loops are indicated by dashed lines.

6-10 TABLE OF LOOP FORMULAS. The relations developed in the preceding sections are summarized in Table 6-2. The general and large loop formulas are based on Foster's results.

6-11 SQUARE LOOPS. It was shown in Sec. 6-3 that the far-field patterns of square and circular loops of the same area are identical when the loops are small ($A < \lambda^2/100$). As a generalization, we may say that the properties depend

Table 6-2 Formulas for circular loops with uniform current

Quantity	General expression (any size loop)	Small loop† $A < \lambda^2/100$ $C_\lambda < \frac{1}{3}$	Large loop $C_\lambda > 5$
Far E_ϕ	$\frac{60\pi[I]C_\lambda J_1(C_\lambda \sin\theta)}{r}$	$\frac{120\pi^2[I]\sin\theta}{r}\frac{A}{\lambda^2}$	Same as general
Far H_θ	$\frac{[I]C_\lambda J_1(C_\lambda \sin\theta)}{2r}$	$\frac{\pi[I]\sin\theta}{r}\frac{A}{\lambda^2}$	Same as general
Radiation resistance, Ω	$60\pi^2 C_\lambda \int_0^{2C_\lambda} J_2(y)\,dy$	$31\,200\left(\frac{A}{\lambda^2}\right)^2 = 197C_\lambda^4$	$3720\frac{a}{\lambda} = 592C_\lambda$
Directivity	$\frac{2C_\lambda[J_1^2(C_\lambda \sin\theta)]_{\max}}{\int_0^{2C_\lambda} J_2(y)\,dy}$	$\frac{3}{2}$	$4.25\frac{a}{\lambda} = 0.68C_\lambda$

A = area of loop; C_λ = circumference of circular loop, wavelengths.

† The small loop formulas apply not only to circular loops but also to square loops of area A and in fact to small loops of any shape having an area A. The formula involving C_λ applies, of course, only to a circular loop. See Sec. 16-9 about simulating uniform current on a large loop.

only on the area and that the shape of the loop has no effect when the loop is small. However, this is not the case when the loop is large. The pattern of a circular loop of any size is independent of the angle ϕ but is a function of θ (see Fig. 6-2). On the other hand, the pattern of a large square loop is a function of both θ and ϕ. Referring to Fig. 6-14, the pattern in a plane normal to the plane of the loop and parallel to two sides (1 and 3), as indicated by the line AA', is simply the pattern of two point sources representing sides 2 and 4 of the loop. The pattern in a plane normal to the plane of the loop and passing through diagonal corners, as indicated by the line BB', is different. The complete range in the

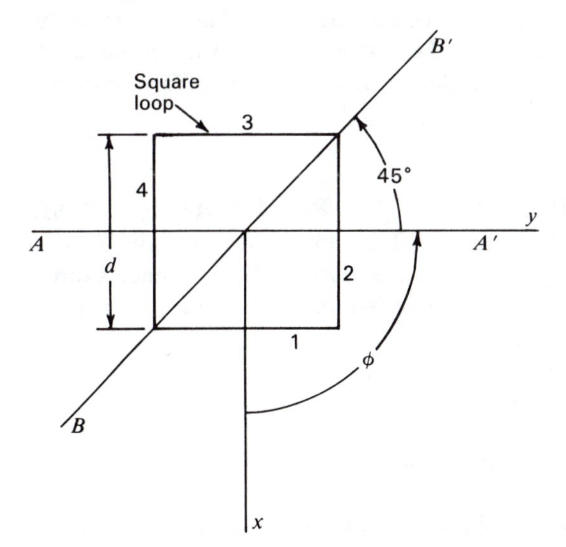

Figure 6-14 Large square loop.

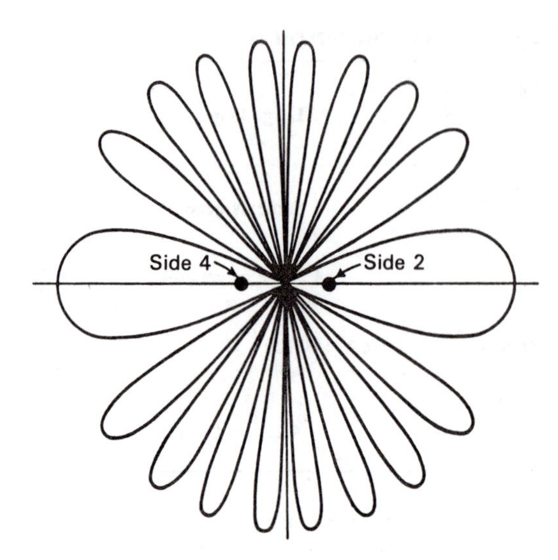

Figure 6-15 Pattern of square loop with uniform, in-phase current. The loop is 4.44λ on a side. The pattern is in a plane normal to the plane of the loop and through the line AA' of Fig. 6-14.

pattern variation as a function of ϕ is contained in this 45° interval between AA' and BB' in Fig. 6-14.

An additional difference of large circular and square loops is in the θ patterns. For instance, Fig. 6-10*d* shows the pattern as a function of θ for a circular loop 5λ in diameter. By way of comparison, the pattern for a square loop of the same area is presented in Fig. 6-15. The square loop is 4.44λ on a side. The pattern is in a plane perpendicular to the plane of the loop and parallel to the sides (plane contains AA' in Fig. 6-14). Comparing Figs. 6-10*d* and 6-15, we note that the pattern lobes of the circular loop decrease in magnitude as θ approaches 90° while the lobes of the square loop are of equal magnitude. This illustrates the difference of the Bessel function pattern of the circular loop and the trigonometric function pattern of the square loop. In the above discussion, uniform in-phase currents are assumed.

6-12 RADIATION EFFICIENCY, Q, BANDWIDTH AND SIGNAL-TO-NOISE RATIO. In Sec. 2-10 we noted that the gain G of an antenna with respect to an isotropic source is identical with the antenna's directivity D provided no losses other than radiation are present. For the more general case we write as in (2-10-1) that

$$G = kD \tag{1}$$

where k = efficiency factor ($0 \le k \le 1$), dimensionless

For a lossless antenna, $k = 1$, but with ohmic losses k is less than 1.

If an antenna has a radiation resistance R_r and a loss resistance R_L, then its (radiation) efficiency factor

$$k = \frac{R_r}{R_r + R_L} \tag{2}$$

and the gain

$$G = \frac{R_r}{R_r + R_L} \frac{4\pi A_{em}}{\lambda^2} = \frac{4\pi A_{em} k}{\lambda^2} \tag{3}$$

For antennas which are small compared to the wavelength, the radiation resistance R_r is small and, if ohmic losses R_L are significant, radiation efficiency is reduced. Thus, short dipoles and small loops may be inefficient radiators when losses are present. For example, when $R_L = R_r$ the radiation efficiency is 50 percent; only half of the power input to the antenna is radiated, the other half being dissipated as heat in the antenna structure.

An rf wave entering a conductor attenuates to $1/e$ of its surface value in a distance δ given by[1]

$$\delta = \frac{1}{\sqrt{f\pi\mu\sigma}} \tag{4}$$

where f = frequency, Hz
μ = permeability of medium, H m^{-1}
σ = conductivity of medium, ℧ m^{-1}

It is assumed that $\sigma \gg \omega\varepsilon$. The induced current density in the conductor also attenuates in the same way. This means that the current density associated with a wave traveling along a conductor is greatest close to the surface, the so-called *skin effect*. The quantity δ is referred to as the *1/e depth of penetration*. It follows that the rf resistance of a round wire or solid cylindrical conductor is equivalent to the dc resistance of a hollow tube of the same material of wall thickness δ. It is assumed that the wire or conductor diameter is much larger than δ. Thus, assuming that the perimeter or circumference L is much smaller than the wavelength so that the current is essentially uniform around the loop, the *ohmic* (or loss resistance) of a small loop antenna is given by

$$R_L = \frac{L}{\sigma\pi d\delta} = \frac{L}{d}\sqrt{\frac{f\mu_0}{\pi\sigma}} \quad (\Omega) \tag{5}$$

where L = loop length (perimeter or circumference), m
d = wire or conductor diameter, m

[1] J. D. Kraus, *Electromagnetics*, 3rd ed., McGraw-Hill, 1984, pp. 447–451.

From (6-8-10) the radiation resistance of a small loop is

$$R_r \simeq 31\,200\left(\frac{A}{\lambda^2}\right)^2 \simeq 197C_\lambda^4 \tag{6}$$

where A = loop area (square or circular), m^2
$C_\lambda = C/\lambda$, where C = circumference of circular loop

Assuming that the loop's inductive reactance is balanced by a capacitor, the terminal impedance will be resistive and equal to

$$R_T = R_r + R_L \tag{7}$$

and the radiation efficiency, or ratio of power radiated to input power, will be

$$k = \frac{1}{1 + (R_L/R_r)} \tag{8}$$

For a 1-turn copper-conductor circular loop (perimeter $L = C$) in air ($\sigma = 5.7 \times 10^7$ ℧ m^{-1}, $\mu_0 = 4\pi \times 10^{-7}$ H m^{-1}),

$$\frac{R_L}{R_r} = \frac{3430}{C^3 f_{\text{MHz}}^{3.5}\, d} \tag{9}$$

where C = circumference of loop, m
f_{MHz} = frequency, MHz
d = wire (or conductor) diameter, m

For small square loops of side length l ($L = 4l$), we may take $C = 3.5l$.

Example. Find the radiation efficiency of a 1-m diameter loop ($C = \pi$ m) of 10-mm diameter copper wire at (*a*) 1 MHz and (*b*) at 10 MHz.

Solution
(*a*) From (9),

$$\frac{R_L}{R_r} = \frac{3430}{\pi^3 \times 1 \times 10^{-2}} = 11\,000 \tag{10}$$

and the radiation efficiency

$$k = \frac{1}{1 + 11\,000} = 9 \times 10^{-5} \text{ (or } -40.5 \text{ dB)} \tag{10a}$$

(*b*) At 10 MHz we have

$$k = 0.22 \text{ (or } -6.6 \text{ dB)} \tag{11}$$

The radiation efficiency as a function of frequency for a small single-turn copper loop in air is shown in Fig. 6-16. It is assumed that the loop is small compared to the wavelength ($C \ll \lambda$) and that the wire or conductor diameter is small compared to the loop circumference ($d \ll C$). Dielectric losses are neglected.

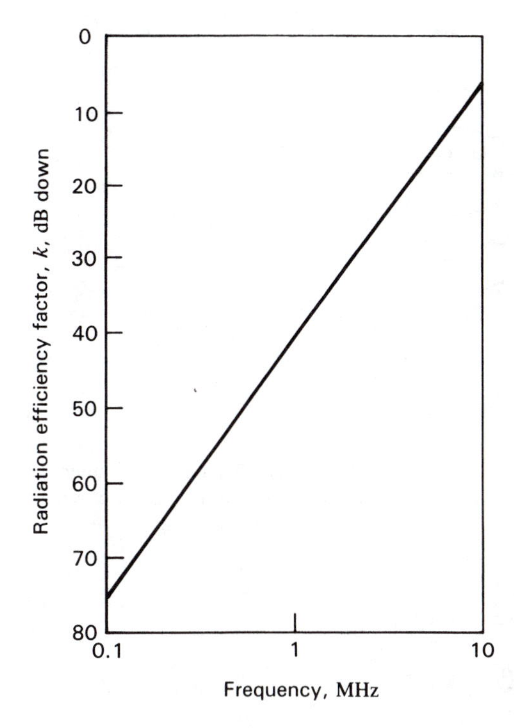

Figure 6-16 Radiation efficiency factor as a function of frequency for a 1-m diameter single-turn copper loop in air ($C = \pi$ m, $d = 10$ mm).

In spite of the low efficiency of a small loop, there are many applications where such loops are useful in receiving applications provided the received signal-to-noise ratio is acceptable as discussed later in this section [see (19)].

For loops with n turns, R_r increases in proportion to n^2 while R_L increases in proportion to n. Hence, for multiturn loops (9) becomes

$$\frac{R_L}{R_r} = \frac{3430}{C^3 f_{\text{MHz}}^{3.5} nd} \tag{12}$$

and the radiation efficiency k is increased by a factor which approaches n if R_L/R_r is large. In (12) the effect of capacitance between turns has been neglected but if the turns are spaced sufficiently and are few in number, (12) can be a useful approximation.

The radiation efficiency of a multiturn loop or coil antenna can be increased by introducing a *ferrite rod* into the coil as in Fig. 6-17. Here the coil (horizontal to receive vertical polarization) serves the function of both an antenna and also (with a series capacitor) of the resonant circuit for the first (mixer) stage of a broadcast receiver (500 to 1600 kHz).

The radiation resistance of a *ferrite loaded loop or coil* is given by

$$R_r = 31\,200 \mu_{er}^2 n^2 \left(\frac{A}{\lambda^2} \right)^2 = 197 \mu_{er}^2 n^2 C_\lambda^4 \qquad (\Omega) \tag{13}$$

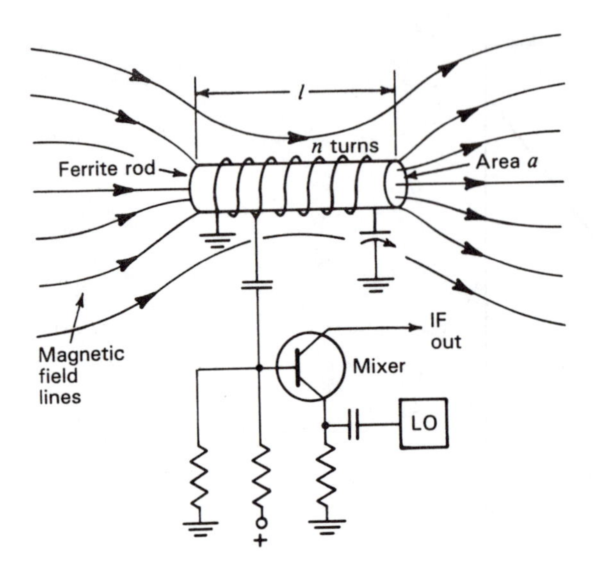

Figure 6-17 Ferrite rod antenna and associated tuned circuit of receiver front-end mixer stage.

and the loss resistance (due to the ferrite rod) by

$$R_f = 2\pi f \mu_{er} \frac{\mu_r''}{\mu_r'} \mu_0 n^2 \frac{a}{l} \qquad (\Omega) \tag{14}$$

where f = frequency, Hz
μ_{er} = effective relative permeability of ferrite rod, dimensionless
μ_r' = real part of relative permeability of ferrite material, dimensionless
μ_r'' = imaginary part of relative permeability of ferrite material, dimensionless
μ_0 = $4\pi \times 10^{-7}$, H m^{-1}
n = number of turns
a = ferrite rod cross-sectional area, m^2
l = length of ferrite rod, m

Because of its open geometry (as contrasted to a closed core or ring) a ferrite rod with a relative permeability μ_r will have a smaller *effective relative permeability* μ_{er} (due to demagnetization effect). Typically for a rod with $\mu_r = 250$ and a length-diameter ratio of 10, the effective relative permeability is about 50.

The ohmic loss resistance R_L of the coil is as given by (5). The radiation efficiency factor for the ferrite rod coil antenna is then

$$k = \frac{R_r}{R_r + R_L + R_f} = \frac{1}{1 + [(R_L + R_f)/R_r]} \tag{15}$$

Dielectric loss is neglected.

Knowing the total resistance $(R_r + R_L + R_f)$ of the ferrite rod antenna, one can calculate the Q and bandwidth of the tuned circuit of which it is a part. The

Q (ratio of energy stored to energy lost per cycle) is given by

$$Q = \frac{2\pi f_0 L}{R_r + R_L + R_f} = \frac{f_0}{\Delta f_{HP}} \tag{16}$$

where f_0 = center frequency, Hz
$L = \mu_{er} n^2 a \mu_0 / l$ = inductance, H[1]
Δf_{HP} = bandwidth at half-power, Hz

Example. The multiturn ferrite rod antenna of a broadcast receiver has 10 turns of 1 mm diameter enameled copper wire wound on a ferrite rod 1 cm in diameter and 10 cm long. The ferrite rod $\mu_r = \mu_r' - \mu_r'' = 250 - j2.5$. Take $\mu_{er} = 50$. At 1 MHz find (*a*) the radiation efficiency, (*b*) the Q and (*c*) the half-power bandwidth.

Solution

(*a*) From (13), $R_r = 1.91 \times 10^{-4}\ \Omega$. From (14), $R_f = 0.31\ \Omega$. From (4), $\delta = 7 \times 10^{-5}$ m. Thus, the ratio $d/\delta = 14.3$ so we can use (5) (times n), which makes $R_L = 0.026\ \Omega$. Accordingly, $(R_L + R_f)/R_r = 1790$ and $k = 1/1790 = 5.6 \times 10^{-4}$. (Dielectric losses are neglected.)

(*b*) From (16), $Q = 162$.

(*c*) From (16), $\Delta f_{HP} = 6.170$ kHz.

Although 6.17 kHz is adequate front-end *selectivity* for the 10-kHz channel spacing of the broadcast band, the low aperture efficiency of less than 0.06 percent makes it uncertain whether the *sensitivity* is adequate. To determine this, a calculation of the *signal-to-noise* ratio for a typical application is required. From the Friis transmission formula (2-25-5), the power received from a transmitter of power P_t at a distance r is

$$P_r = \frac{P_t A_{et} A_{er}}{r^2 \lambda^2} \qquad \text{(W)} \tag{17}$$

where A_{et} = effective aperture of transmitting antenna, m^2
A_{er} = effective aperture of receiving antenna, m^2

For a small loop receiving antenna, $D = \frac{3}{2}$ so

$$A_{er} = \frac{1.5\lambda^2 k}{4\pi} \qquad (m^2) \tag{18}$$

where k = radiation efficiency factor

The signal-to-noise ratio (S/N) is given by

$$\frac{S}{N} = \frac{P_r}{N} \qquad \text{(dimensionless)} \tag{19}$$

---

[1] Distinguish between L for inductance in (16) and L for length in (5).

where P_r = received power, W
$N = kT_{sys}\,\Delta f$ [from (17-3-8)]

At 1 MHz, T_{sys} is dominated by the sky (antenna) temperature T_s, so taking $T_{sys} = T_s$ we have for this case

$$N = kT_s\,\Delta f \tag{20}$$

where k = Boltzmann's constant = 1.38×10^{-23} J K^{-1}
T_s = sky background temperature, K
Δf = bandwidth, Hz

Distinguish between k in (20) for Boltzmann's constant and k in (18) for the radiation efficiency factor.

Example. Find the S/N ratio for a receiver with the ferrite rod antenna of the above example at a distance of 100 km from a 1-MHz 10-kW broadcast station with an omnidirectional antenna. Take the receiver band width as $\Delta f = 10^4$ Hz.

Solution. Assuming that the radiation pattern of the transmitting antenna fills a half-sphere, its directivity is 2 and, hence,

$$A_{et} = \frac{2\lambda^2}{4\pi} \qquad (\text{m}^2)$$

From the previous example, $k = 5.6 \times 10^{-4}$ so from (17) and (18)

$$P_r = 10^{-6}\ \text{W}$$

Taking the 1-MHz sky background temperature $T_s \simeq 10^{10}$ K,

$$N = 1.38 \times 10^{-23} \times 10^{10} \times 10^4 = 1.4 \times 10^{-9}\ \text{W}$$

Therefore, the signal-to-noise ratio at the receiver input is

$$\frac{\text{S}}{\text{N}} = \frac{10^{-6}}{1.4 \times 10^{-9}} = 714 \text{ or } 28.5 \text{ dB}$$

which is adequate for AM reception.

Thus, an antenna which is less than 0.06 percent efficient is adequate for AM reception under the circumstances of the above example. Even smaller, less-efficient ferrite rod antennas may be satisfactory and are used in popular low-cost pocket-size radios. However, the very low efficiency makes the antenna unsuitable for transmission except under the special circumstances of low power and short range.

Formulas for radiation resistance, loss resistance and radiation efficiency of small loop antennas and also formulas for Q, bandwidth and signal-to-noise ratio are summarized in Table 6-3.

Table 6-3 Radiation resistance, loss resistance and radiation efficiency of small loop antennas and also formulas for Q, bandwidth and signal-to-noise ratio†

Quantity	Formula	Reference equation for units
Radiation resistance, single turn	$R_r = 31\,200\left(\frac{A}{\lambda^2}\right)^2$ (Ω)	(6-8-10)
Radiation resistance, single turn	$R_r = 197 C_\lambda^4$ (Ω)	(6-8-9)
Loss resistance, n-turn	$R_L = \frac{nL}{d}\sqrt{\frac{f\mu_0}{\pi\sigma}}$ (Ω)	(6-12-5)
Radiation efficiency, n-turn	$k = \frac{1}{1 + (R_L/R_r)}$	(6-12-8)
R_L/R_r ratio, n-turn copper conductor	$\frac{R_L}{R_r} = \frac{3430}{C^3 f_{\text{MHz}}^{3.5} nd}$	(6-12-12)
Loss resistance, n-turn ferrite rod antenna	$R_f = 2\pi f \mu_{er} \frac{\mu_r''}{\mu_r'} \mu_0 n^2 \frac{a}{l}$ (Ω)	(6-12-14)
Radiation efficiency, n-turn ferrite rod antenna	$k = \frac{1}{1 + [(R_L + R_f)/R_r]}$	(6-12-15)
Q	$Q = \frac{2\pi f_0 L}{R_r + R_L + R_f}$	(6-12-16)
Bandwidth	$\Delta f_{\text{HP}} = \frac{f_0}{Q}$	(6-12-16)
Signal-to-noise ratio	$\frac{\text{S}}{\text{N}} = \frac{P_r}{N} = \frac{P_t A_{et} A_{er}}{r^2 \lambda^2 k T_s \Delta f_{\text{HP}}}$	(6-12-17), (6-12-19) and (6-12-20)

† L in third row for loss resistance is perimeter length whereas L in eighth row for Q is inductance [see (6-12-16)].

PROBLEMS[1]

6-1 **The $3\lambda/4$ diameter loop.** Calculate and plot the far-field pattern normal to the plane of a circular loop $3\lambda/4$ in diameter with a uniform in-phase current distribution.

***6-2** **The 1λ square loop.** Calculate and plot the far-field pattern in a plane normal to the plane of a square loop and parallel to one side. The loop is 1λ on a side. Assume uniform in-phase currents.

[1] Answers to starred (*) problems are given in App. D.

6-3 The $\lambda/10$ diameter loop. What is the maximum effective aperture of a thin loop antenna 0.1λ in diameter with a uniform in-phase current distribution?

***6-4 Radiation resistance of loop.** What is the radiation resistance of the loop of Prob. 6-1?

6-5 Pattern, radiation resistance and directivity of loops. A circular loop antenna with uniform in-phase current has a diameter D. What is (*a*) the far-field pattern (calculate and plot), (*b*) the radiation resistance and (*c*) the directivity for each of three cases where (1) $D = \lambda/4$, (2) $D = 1.5\lambda$ and (3) $D = 8\lambda$?

6-6 Small square loop. Resolving the small square loop with uniform current into four short dipoles, show that the far-field pattern in the plane of the loop is a circle.

***6-7 Circular loop.** A circular loop antenna with uniform in-phase current has a diameter D. Find (*a*) the far-field pattern (calculate and plot), (*b*) the radiation resistance and (*c*) the directivity for the following three cases: (1) $D = \lambda/3$, (2) $D = 0.75\lambda$ and (3) $D = 2\lambda$.

6-8 Small-loop resistance. (*a*) Using a Poynting vector integration, show that the radiation resistance of a small loop is equal to $320\pi^4(A/\lambda^2)^2$ Ω where A = area of loop (m^2). (*b*) Show that the effective aperture of an isotropic antenna equals $\lambda^2/4\pi$.

6-9 Loop and dipole for circular polarization. If a short electric dipole antenna is mounted inside a small loop antenna (on polar axis, Fig. 6-6) and both dipole and loop are fed in phase with equal power, show that the radiation is everywhere circularly polarized with a pattern as in Fig. 6-6.

CHAPTER 7

THE HELICAL ANTENNA

7-1 INTRODUCTION. In 1946, a few months after joining the faculty at Ohio State University, I attended an afternoon lecture on traveling-wave tubes by a famous scientist who was visiting the campus. In these tubes an electron beam is fired down the inside of a long wire helix for amplification of waves traveling along the helix. The helix is only a small fraction of a wavelength in diameter and acts as a guiding structure. After the lecture I asked the visitor if he thought a helix could be used as an antenna. "No," he replied, "I've tried it and it doesn't work." The finality of his answer set me thinking. If the helix were larger in diameter than in a traveling-wave tube, I felt that it would have to radiate in some way, but how, I did not know. I determined to find out.

That evening in the basement of my home I wound a 7-turn helical coil of wire 1λ in circumference and fed it via coaxial line and ground plane from my 12-cm oscillator (Fig. 7-1). I was thrilled to find that it produced a sharp beam of circularly polarized radiation off its open end.

Next I wound other helices with larger and smaller diameters, noting little change in behavior. Adding more turns, however, resulted in sharper beams. Although my invention/discovery had come quickly, I realized then that much work would be required to understand this remarkable antenna. Actually it took years of extensive measurements and calculations. I published many articles, a few with students to whom I had assigned studies of specific properties of the

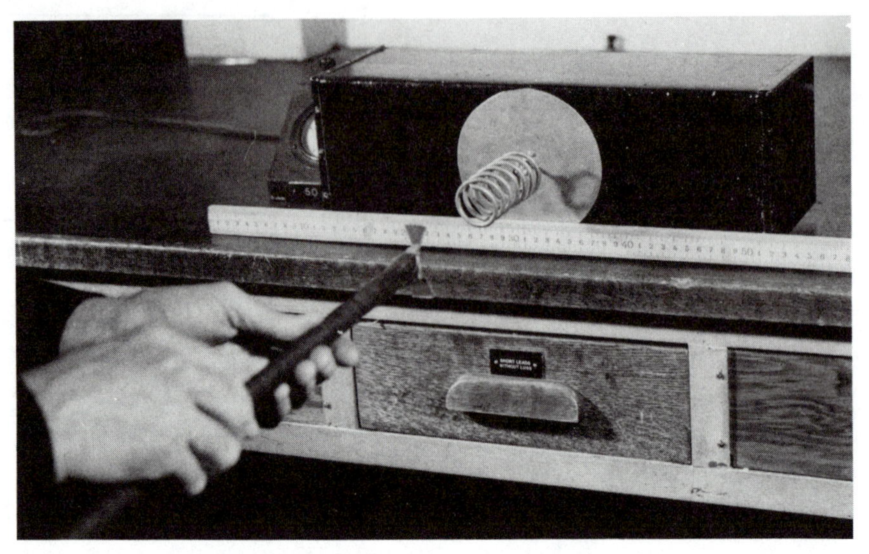

Figure 7-1 The first axial-mode helical antenna (1946). When I rotated the hand-held dipole there was no change in response, indicating circular polarization.

antenna.[1] I also derived equations suitable for engineering design purposes and summarized them in Chap. 7 of the first edition of *Antennas*.

The steps taken to unravel the mystery of the helix went something like this. The input impedance was measured and found to be essentially resistive and constant over a wide bandwidth. This suggested that the helix behaved like a terminated (matched) transmission line. This was hard to understand because the open end of the helix was completely unterminated. New insights came when we measured the current distribution along the helix. This we did by rotating a helix and its ground plane while holding a small loop (current probe) under the helical conductor (Fig. 7-2). At a low frequency (helix circumference about $\lambda/2$) there was an almost pure standing wave (VSWR $\rightarrow \infty$) all along the helix (outgoing and reflected waves nearly equal) (Fig. 7-3*a*), but as the frequency increased, the dis-

---

[1] J. D. Kraus, "Helical Beam Antenna," *Electronics*, **20**, 109–111, April 1947.

J. D. Kraus and J. C. Williamson, "Characteristics of Helical Antennas Radiating in the Axial Mode," *J. Appl. Phys.*, **19**, 87–96, January 1948.

O. J. Glasser and J. D. Kraus, "Measured Impedances of Helical Beam Antennas," *J. Appl. Phys.*, **19**, 193–197, February 1948.

J. D. Kraus, "Helical Beam Antennas for Wide-Band Applications," *Proc. IRE*, **36**, 1236–1242, October 1948.

J. D. Kraus, "The Helical Antenna," *Proc. IRE*, **37**, 263–272, March 1949.

J. D. Kraus, "Helical Beam Antenna Design Techniques," *Communications*, **29**, 6–9, 34–35, September 1949.

T. E. Tice and J. D. Kraus, "The Influence of Conductor Size on the Properties of Helical Beam Antennas," *Proc. IRE*, **37**, 1296, November 1949.

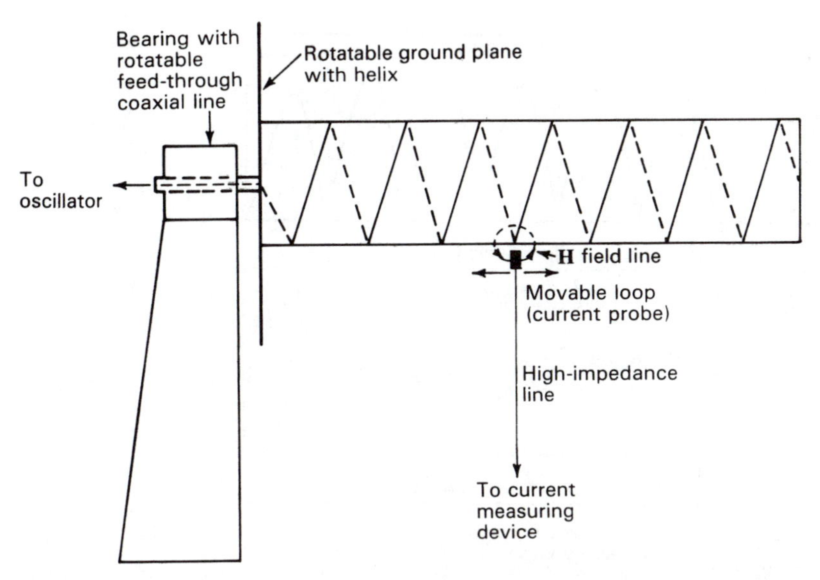

Figure 7-2 Helix and ground plane mounted to rotate on the helix axis for current distribution measurements along the helical conductor using a loop probe. (*After Kraus and Williamson, "Characteristics of Helical Antennas Radiating in the Axial Mode,"* J. Appl. Phys., **19**, *87–96, January 1948.*) As the helix was rotated, the probe was moved horizontally to follow the helical conductor.

tribution changed dramatically. For a helix circumference of about 1λ three regions appeared: near the input end the current decayed exponentially, near the open end there was a standing wave over a short distance, while between the ends there was a relatively uniform current amplitude (small VSWR) which extended over most of the helix (Fig. 7-3*b*). The decay at the input end could be understood as a transition between a helix-to-ground-plane mode and a pure helix mode. The reflection of the outgoing wave at the open end also decayed exponentially to a much smaller reflected wave, leaving the outgoing wave dominant over most of the helix (VSWR small). The small VSWR ripple was sufficient, however, to measure the relative phase velocity ($=\lambda_h/\lambda_0$) along the helix, which was useful for an understanding of the radiation patterns. The current distribution resolved into outgoing and reflected waves is shown in Fig. 7-3*c*.

Our extensive pattern measurements showed that the end-fire beam mode persists over a frequency range of about 2 to 1 centered on the frequency for which the circumference is 1λ. Thus, the diameter I had chosen for the first helix I tried was optimum!

Although the helix is continuous, it can also be regarded as a periodic structure. Thus, assuming that an n-turn helix is an end-fire array of n sources, I calculated the pattern using the formula (4-6-9) for the ordinary end-fire condition. Surprisingly, the measured helix patterns were much sharper. Could the helix be operating in the increased-directivity condition? I calculated patterns for this condition using the formula (4-6-14) and obtained good agreement with the

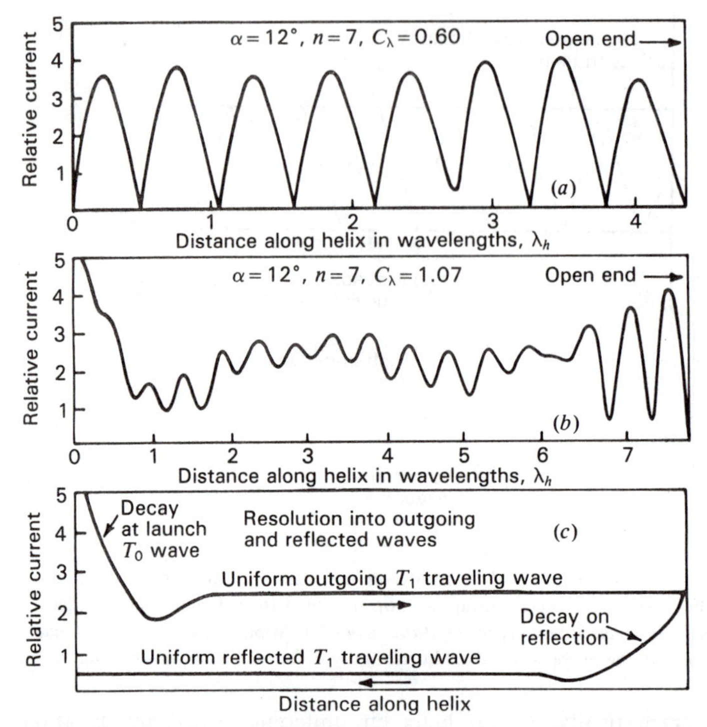

Figure 7-3 Typical measured current distribution (*a*) at a frequency below the axial mode of operation and (*b*) at a frequency near the center of the axial mode region. (*c*) Resolution of currents into outgoing and reflected waves. (*From Kraus and Williamson, "Characteristics of Helical Antennas Radiating in the Axial Mode,"* J. Appl. Phys., **19**, *87–96, January 1948.*) Compare with distribution on the long, thick cylindrical conductor in Fig. 9-18.

measured patterns. Furthermore, this condition persisted over a wide bandwidth, indicating that the phase velocity on the helix changes by just the right amount to maintain the increased-directivity condition. The phase velocity measurements we had made also confirmed this. Thus, the helix locks onto the increased-directivity condition and automatically stays locked over the full bandwidth. Not only does the helix have a nearly uniform resistive input over a wide bandwidth but it also operates as a supergain end-fire array over the same bandwidth! Furthermore, it is noncritical to an unprecedented degree and is easy to use in arrays because of an almost negligible mutual impedance.

The helix immediately found wide application. I employed it in an array of 96 11-turn helices in a radio telescope I designed and built with my students in 1951 (Fig. 7-4). Operating at frequencies of 200 to 300 MHz, the array measured 50 m in length and had a gain of 35 dB. With it we produced some of the first and most extensive maps of the radio sky.[1] Others employed the helix over a wide range of frequencies, some at frequencies as low as 10 MHz (Fig. 7-5).

---

[1] J. D. Kraus, *Radio Astronomy*, 2nd ed., Cygnus-Quasar, 1986, p. 8-2.

Figure 7-4 Radio telescope at the Ohio State University Radio Observatory with array of 96 11-turn monofilar axial-mode helical antennas mounted on a tiltable ground plane 50 m long. This array was used to make some of the first and most extensive maps of the radio sky. See also Sec. 7-12.

Figure 7-5 Rotatable (in azimuth) 6-turn helical antenna about 45 m long for operation at frequencies around 10 MHz ($\lambda = 30$ m). Note workmen on arch at far end for scale. (*Courtesy Electro-Physics Laboratory.*)

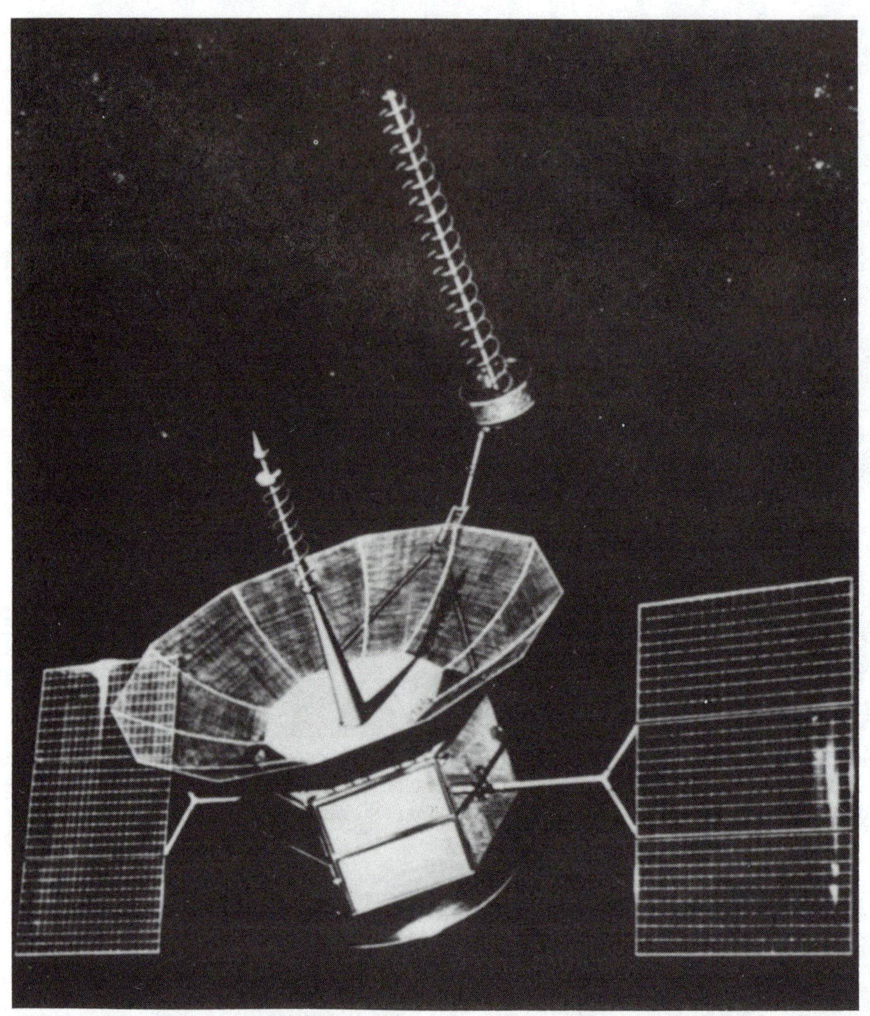

Figure 7-6 Fleetsatcom geostationary relay satellite with monofilar axial-mode helical antennas for transmission and reception with one as the feed for a dish. These satellites provide global communications for the U.S. military forces. (*Courtesy TRW Corp.; H. E. King and J. L. Wong, "Antenna System for the FleetSatCom Satellites,"* IEEE International Symposium on Antennas and Propagation, *pp. 349–352, 1977.*)

Following Sputnik the helical antenna became the workhorse of space communications for telephone, television and data, being employed both on satellites and at ground stations. Many U.S. satellites, including its weather satellites, Comsat, Fleetsatcom (Fig. 7-6), GOES (global environmental satellites), Leasat, Navstar-GPS (global position satellites) (Fig. 7-7), Westar and tracking and data-relay satellites, all have helical antennas, the latter with arrays of 30. Russian satellites also have helical antennas, each of the Ekran class satellites being

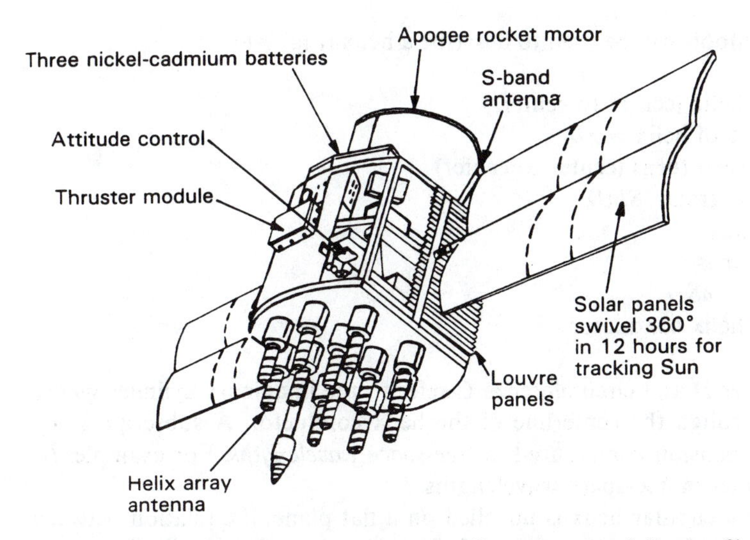

Figure 7-7 Navstar GPS (global position) satellite with array of 10 monofilar axial-mode helical antennas. Eighteen of these satellites are in elliptical orbits around the earth. From them one can determine one's absolute position anywhere on the earth (latitude, longitude and altitude), at any time and in any weather, to a precision of a few centimeters and relative position to a few millimeters.

equipped with an array of 96 helicals. The helical antenna has been carried to the Moon and Mars. It is also on many other probes of planets and comets, being used alone, in arrays or as feeds for parabolic reflectors, its circular polarization, high gain and simplicity making it especially attractive for space applications.[1]

This short account provides a brief introduction to the helix in which some of the experimental and analytical steps taken to understand its behavior are outlined. Specifically the helix discussed can be described as a *monofilar* (*one-wire*) *axial-mode helical antenna.* Its operation is explained in more detail in the following sections along with treatments of many variants and related forms of helical antennas including, later in the chapter, helices with 2 or more wires.

7-2 HELICAL GEOMETRY.

The helix is a basic 3-dimensional geometric form. A helical wire on a uniform cylinder becomes a straight wire when unwound by rolling the cylinder on a flat surface. Viewed end-on, a helix projects as a circle. Thus, a helix combines the geometric forms of a straight line, a circle and a cylinder. In addition a helix has handedness; it can be either left- or right-handed.

[1] A more detailed personal account of my early work on the helical antenna is given in my book *Big Ear*, Cygnus-Quasar, 1976.

The following symbols will be used to describe a helix (Fig. 7-8):

D = diameter of helix (center to center)
C = circumference of helix = πD
S = spacing between turns (center to center)
α = pitch angle = arctan $S/\pi D$
L = length of 1 turn
n = number of turns
A = axial length = nS
d = diameter of helix conductor

The diameter D and circumference C refer to the imaginary cylinder whose surface passes through the centerline of the helix conductor. A subscript λ signifies that the dimension is measured in *free-space wavelengths*. For example, D_λ is the helix diameter in free-space wavelengths.

If 1 turn of a circular helix is unrolled on a flat plane, the relation between the spacing S, circumference C, turn length L and pitch angle α are as illustrated by the triangle in Fig. 7-9.

The dimensions of a helix are conveniently represented by a diameter-spacing chart or, as in Fig. 7-10, by a circumference-spacing chart. On this chart the dimensions of a helix may be expressed either in rectangular coordinates by the spacing S_λ and circumference C_λ or in polar coordinates by the length of 1 turn L_λ and the pitch angle α. When the spacing is zero, $\alpha = 0$ and the helix becomes a loop. On the other hand, when the diameter is zero, $\alpha = 90°$ and the helix becomes a linear conductor. Thus, in Fig. 7-10 the ordinate axis represents loops while the abscissa axis represents linear conductors. The entire area between the two axes represents the general case of the helix.

Suppose that we have a 1-turn helix with a turn length of 1λ ($L_\lambda = 1$). When $\alpha = 0$, the helix is a loop of 1λ circumference or of diameter equal to $1\lambda/\pi$. As the pitch angle α increases, the circumference decreases and the dimensions of the helix move along the $L_\lambda = 1$ curve in Fig. 7-10 until, when $\alpha = 90°$, the "helix" is a straight conductor 1λ long.

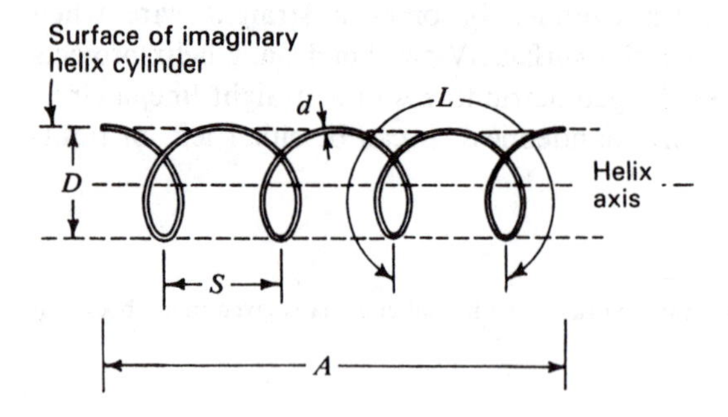

Figure 7-8 Helix and associated dimensions.

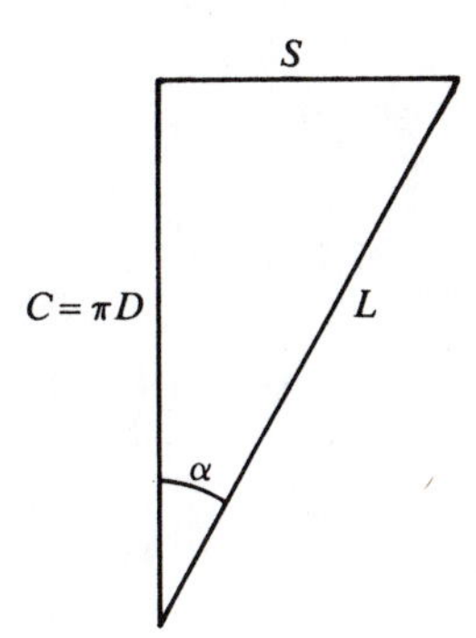

Figure 7-9 Relation between circumference, spacing, turn length and pitch angle of a helix.

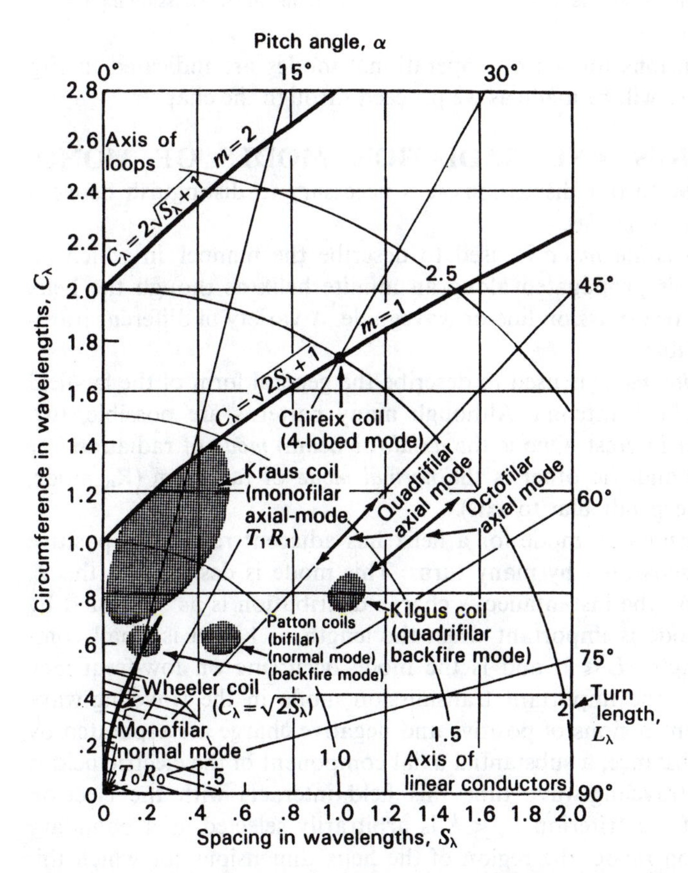

Figure 7-10 Helix chart showing the location of different modes of operation as a function of the helix dimensions (diameter, spacing and pitch angle). As a function of frequency the helix moves along a line of constant pitch angle. Along the vertical axis the helices become loops and along the horizontal axis they become linear conductors.

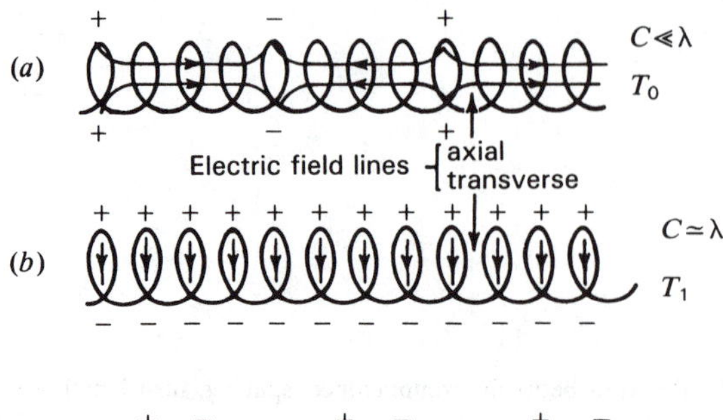

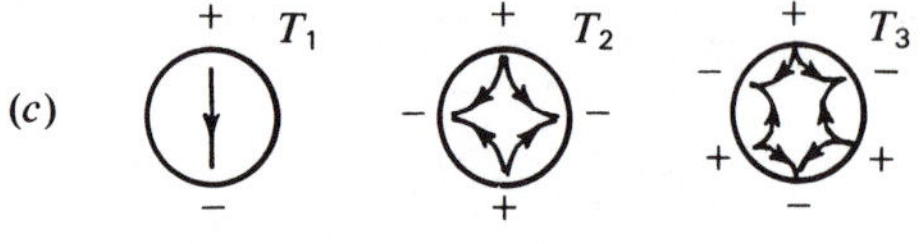

End view of helices

Figure 7-11 Instantaneous charge and field configurations on helices for different transmission modes.

The helix dimensions for various operational modes are indicated in Fig. 7-10. Reference to these will be made as we proceed through the chapter.

7-3 TRANSMISSION AND RADIATION MODES OF MONOFILAR[1] HELICES. In our discussions it is necessary to distinguish between *transmission* and *radiation modes.*

The term *transmission mode* is used to describe the manner in which an electromagnetic wave is propagated along an infinite helix as though the helix constitutes an infinite transmission line or waveguide. A variety of different transmission modes is possible.

The term *radiation mode* is used to describe the general form of the far-field pattern of a finite helical antenna. Although many patterns are possible, two kinds are of particular interest. One is the *axial* (or beam) *mode* of radiation (R_1 mode, beam on axis) and the other is the *normal mode* of radiation (R_0 mode, radiation maximum perpendicular to axis).

The lowest transmission mode for a helix has adjacent regions of positive and negative charge separated by many turns. This mode is designed as the T_0 transmission mode and the instantaneous charge distribution is as suggested by Fig. 7-11*a*. The T_0 mode is important when the length of 1 turn is small compared to the wavelength ($L \ll \lambda$) and is the mode occurring on low-frequency inductances. It is also the important transmission mode in the traveling-wave tube. Since the adjacent regions of positive and negative charge are separated by an appreciable axial distance, a substantial axial component of the electric field is present, and in the traveling-wave tube this field interacts with the electron stream (Fig. 7-11*a*). If the criterion $L_\lambda < \frac{1}{3}$ is arbitrarily selected as a boundary for the T_0 transmission mode, the region of the helix dimensions for which this mode is important is within the $T_0 R_0$ area in Fig. 7-10.

---

[1] *Monofilar* = unifilar = one wire = single wire = single conductor (terms used to distinguish the one-wire helix from helices with 2 or more wires).

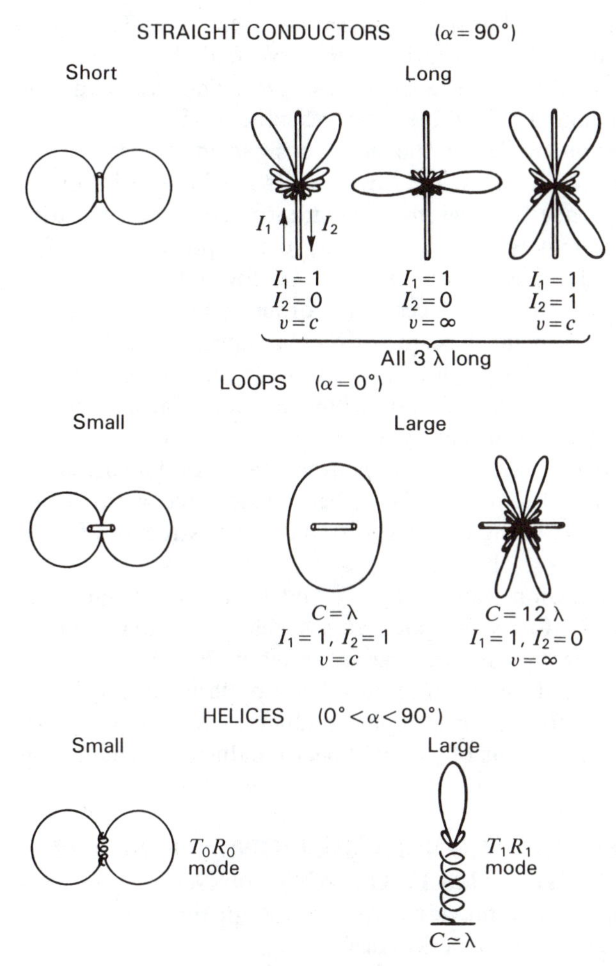

Figure 7-12 Patterns of straight conductor, loop and helix compared. I_1 and I_2 represent current magnitudes of waves traveling in opposite directions on antennas. If $I_2 = I_1$ there is a pure standing wave. If $I_2 = 0$, only a pure traveling wave is present. (v = velocity of wave along antenna, c = velocity of light, C = circumference.)

A helix excited in the T_0 transmission mode may radiate. Let us consider the case when the helix is very short ($nL \ll \lambda$) and the current is assumed to be of uniform magnitude and in phase along the entire helix. It is theoretically possible to approximate this condition on a small, end-loaded helix. However, its radiation resistance is small. The maximum field from the helix is normal to the helix axis for all helix dimensions provided only that $nL \ll \lambda$. Thus, this condition is called a "normal radiation mode" (R_0). Any component of the field has a sine variation with θ as shown in Fig. 7-12 (lower left). The space pattern is a figure-of-revolution of the pattern shown around the vertical axis. The field is, in general, elliptically polarized but for certain helix dimensions may be circularly polarized and for other dimensions, linearly polarized (see Sec. 7-19). The transmission mode and radiation mode appropriate for very small helices can be described by combining the T_0 and R_0 designations as $T_0 R_0$. This designation is applied in Fig. 7-10 to the region of helix dimensions near the origin.

A first-order transmission mode on the helix, designated T_1, is possible when the helix circumference C in free-space wavelengths is of the order of 1λ. For small pitch angles, this mode has regions of adjacent positive and negative charge separated by approximately $\frac{1}{2}$-turn (or near the opposite ends of a diameter) as shown in Fig. 7-11*b* and also in end view by Fig. 7-11*c*. It is found that radiation from helices with circumferences of the order of 1λ ($C_\lambda \sim 1$) and a number of turns ($n > 1$) is a well-defined beam with a maximum in the direction of the helix axis. Hence, this type of operation is called the "axial (or beam) mode of radiation" with designation R_1. The radiation from this monofilar axial mode helical antenna is circularly polarized or nearly so.

The monofilar axial mode of radiation occurs over a range of dimensions as indicated by the shaded ($T_1 R_1$) area in Fig. 7-10; being associated with the T_1 transmission mode, the combined designation appropriate to this region of helix dimensions is $T_1 R_1$.

Still higher-order transmission modes, T_2, T_3 and so forth, become permissible for larger values of C_λ. For small pitch angles, the approximate charge distribution around the helix for these modes is as suggested by Fig. 7-11*c*.

The axial ($T_1 R_1$) and normal radiation mode ($T_0 R_0$) patterns of a helix are compared with the radiation patterns for straight conductors and loops in Fig. 7-12. It is to be noted that the patterns of a short linear conductor, a small loop and a small helix are the same.

7-4 PRACTICAL DESIGN CONSIDERATIONS FOR THE MONOFILAR AXIAL-MODE HELICAL ANTENNA.[1] Before analyzing the many facets of the antenna individually, an overall picture will be given by describing the performance of some practical designs.

The monofilar axial-mode helical antenna is very noncritical and one of the easiest of all antennas to build. Nevertheless, attention to details can maximize its performance.

The important parameters are:

1. Beam width
2. Gain
3. Impedance
4. Axial ratio

[1] When I first described the *helical beam antenna* in 1946, there was little chance of ambiguity but now there are many variations of the basic type making more explicit names desirable—hence the term, *monofilar axial-mode helical antenna. Uniform 1-wire end-fire helix* or *single-conductor helical beam antenna* are also possible designations while some call it a *Kraus-type helix* or *Kraus coil.*

Gain and beam width, which are interdependent [$G \propto (1/\text{HPBW}^2)$], and the other parameters are all functions of the number of turns, the turn spacing (or pitch angle) and the frequency. For a given number of turns the behavior of the beam width, gain, impedance and axial ratio determines the useful *bandwidth*. The nominal center frequency of this bandwidth corresponds to a helix circumference of about 1λ ($C_\lambda = 1$). For a given bandwidth to be completely useful, all 4 parameters must be satisfactory over the entire bandwidth.

The parameters are also functions of the ground plane size and shape, the helical conductor diameter, the helix support structure and the feed arrangement. The ground plane may be flat (either circular or square) with a diameter or side dimension of at least $3\lambda/4$ or the ground plane (launching structure) may be cup-shaped forming a shallow cavity (Fig. 7-13*b*).

A 2-turn flush-mounted design described by Bystrom and Bernsten[1] for aircraft applications is shown in Fig. 7-13*c*. These authors found that 2 turns are required to obtain satisfactory pattern and impedance characteristics but that no significant improvement is obtained with a deeper cavity and a larger number of turns since the size of the aperture opening remains the same (like an open-ended cylindrical waveguide).

The deep conical arrangement of Fig. 7-13*d* is effective in reducing the side- and back-lobe radiation.[2]

Launching a wave on the helix may also be done without a ground plane, producing a back-fire beam (see Fig. 7-56). By slicing the outer conductor of the coaxial cable longitudinally and separating it gradually from the inner conductor over the first turn or two of the helix, Munk and Peters devised a ground-planeless feed that produces a forward end-fire beam.[3]

Conductor size is not critical[4] and may range from 0.005λ or less to 0.05λ or more (Fig. 7-14). The helix may be supported by a few radial insulators mounted on an axial dielectric or metal rod or tube whose diameter is a few hundredths of a wavelength, by one or more longitudinal dielectric rods mounted peripherally (secured directly to the helical conductor) or by a thin-wall dielectric tube on which the helix is wound. With the latter arrangement the operating bandwidth is shifted to lower frequencies so that for a given frequency the antenna is smaller. Several of these mounting arrangements are illustrated in Fig. 7-15.

---

[1] A. Bystrom, Jr., and D. G. Bernsten, "An Experimental Investigation of Cavity-Mounted Helical Antennas," *IRE Trans. Ants. Prop.*, **AP-4**, 53–58, January 1956.

[2] K. R. Carver, "The Helicone: A Circularly Polarized Antenna with Low Side Lobe Level," *Proc. IEEE*, **55**, 559, April 1967.

K. R. Carver and B. M. Potts, "Some Characteristics of the Helicone Antenna," *Antennas and Propagation International Symposium*, 1970, pp. 142–150.

[3] B. A. Munk and L. Peters, "A Helical Launcher for the Helical Antenna," *IEEE Trans. Ants. Prop.*, **AP-16**, 362–363, May 1968.

[4] T. E. Tice and J. D. Kraus, "The Influence of Conductor Size on the Properties of Helical Beam Antennas," *Proc. IRE*, **37**, 1296, November 1949.

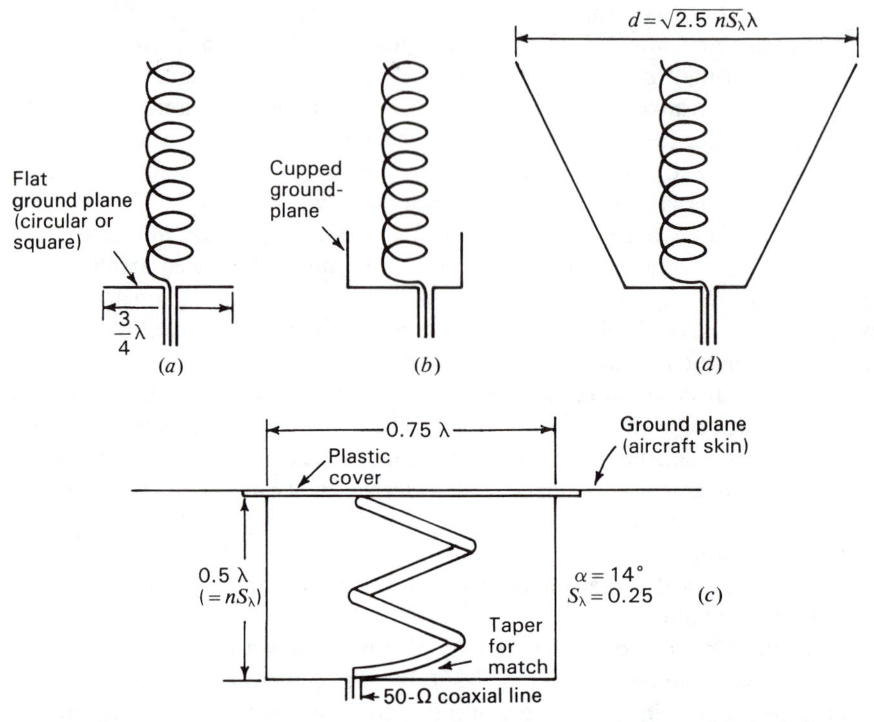

Figure 7-13 (*a*) Monofilar axial-mode helical antenna on flat ground plane and (*b*) in shallow cupped ground plane (see also Fig. 7-16c). (*c*) General-purpose flush-mounted 2-turn monofilar axial-mode helical antenna with taper feed for matching to a 50-Ω coaxial line (*after Bystrom and Bernsten, ref. 1, p. 277*) (see also Fig. 7-16*a* and *b*). (*d*) Deep conical ground-plane enclosure for reducing side and back lobes. (*After K. R. Carver, ref. 2, p. 277*).

The helix may be fed axially, peripherally or from any convenient location on the ground-plane launching structure with the inner conductor of the coaxial line connected to the helix and the outer conductor bonded to the ground plane.

With axial feed the terminal impedance (resistive) is given within 20 percent by

$$R = 140C_\lambda \qquad (\Omega) \tag{1}$$

while with peripheral feed Baker[1] gives its value within 10 percent as

$$R = \frac{150}{\sqrt{C_\lambda}} \qquad (\Omega) \tag{2}$$

[1] D. E. Baker, "Design of a Broadband Impedance Matching Section for Peripherally Fed Helical Antennas," *Antenna Applications Symposium*, University of Illinois, September 1980.

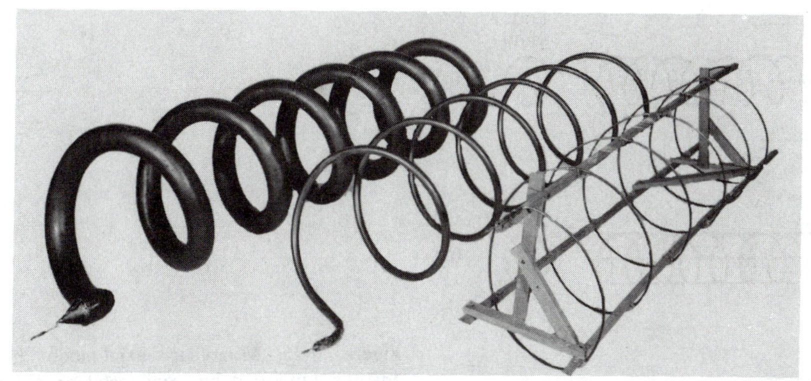

Figure 7-14 Peripherally fed monofilar axial-mode helical antennas with helix conductors of 0.055, 0.017 and 0.0042λ diameter at center frequency of 400 MHz for determining effect of conductor diameter on helix performance. Only minor differences were measured. (*After T. E. Tice and J. D. Kraus, "The Influence of Conductor Size on the Properties of Helical Beam Antennas,"* Proc. IRE, **37**, *1296, November 1949.*) The 0.055λ diameter tubing (4.1 cm diameter) is about the largest size which could be bent to the radius of 11 cm ($=\lambda/2\pi$).

These relations have the restrictions that $0.8 \leq C_\lambda \leq 1.2$, $12° \leq \alpha \leq 14°$ and $n \geq 4$.

With a suitable matching section the terminal impedance (resistive) can be made any desired value from less than 50 Ω to more than 150 Ω. Thus, by bringing the last $\frac{1}{4}$-turn of the helix parallel to the ground plane in a gradual manner, a tapered transition between the 140- or 150-Ω helix impedance and a 50-Ω coaxial line can be readily accomplished.[1] This can be done with either axially or peripherally fed helices but is more convenient with a peripheral feed. Details of a suitable arrangement are shown in Fig. 7-16*a* and *b*.

As the helix tubing is brought close to the ground plane, it is gradually flattened until it is completely flat at the termination, where it is spaced from the ground plane by a dielectric sheet (or slab). The appropriate height *h* (or thickness of the sheet) is given by[2]

$$h = \frac{w}{[377/(\sqrt{\varepsilon_r}Z_0)] - 2} \tag{3}$$

where w = width of conductor at termination

h = height of conductor above ground plane (or thickness of dielectric sheet) in same units as w

ε_r = relative permittivity of dielectric sheet

Z_0 = characteristic impedance of coaxial line

---

[1] J. D. Kraus, "A 50-ohm Input Impedance for Helical Beam Antennas," *IEEE Trans. Ants. Prop.*, **AP-25**, 913, November 1977.

[2] J. D. Kraus, *Electromagnetics*, 3rd ed., McGraw-Hill, 1984, pp. 397–398.

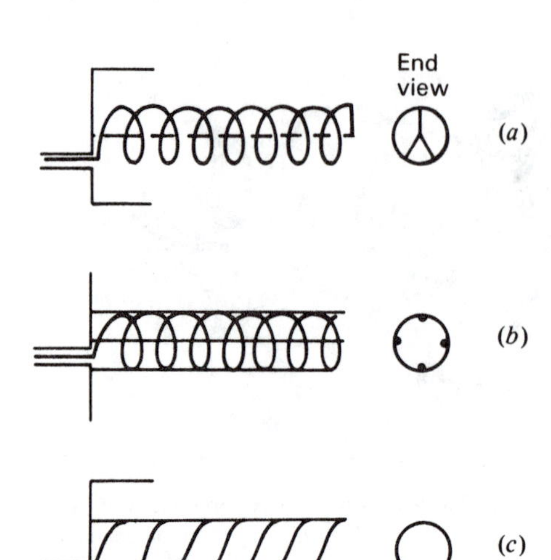

Figure 7-15 Monofilar axial-mode helical antenna supported by axial metal or dielectric rod (or tube) with radial insulators (*a*), by four peripheral dielectric rods secured to the helix (*b*) and by a dielectric tube on which the helix is wound (*c*).

Example. If the flattened tubing width is 5 mm, find the required thickness of a polystyrene sheet ($\varepsilon_r = 2.7$) for matching to a 50-Ω coaxial transmission line.

Solution. From (3),

$$h = \frac{5}{[377/(\sqrt{2.7} \times 50)] - 2} = 1.9 \text{ mm}$$

A typical peripherally fed monofilar axial-mode helical antenna with cup ground-plane launcher matched to a 50-Ω line, as in Fig. 7-16*a* and *b*, is shown in

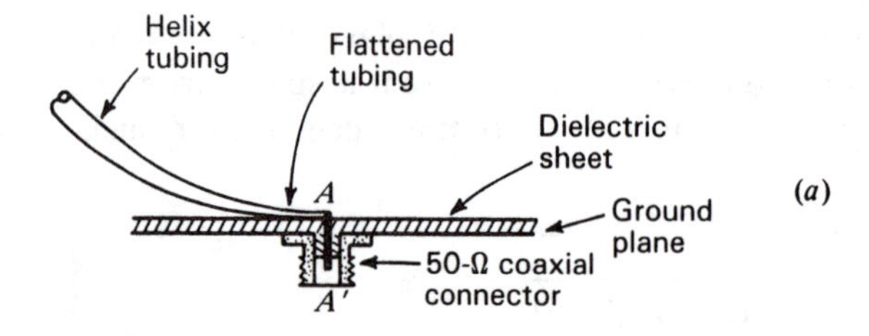

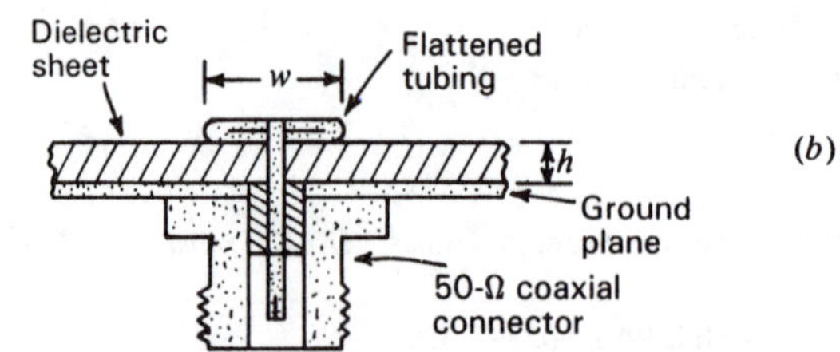

Figure 7-16 (*a*) Gradually tapered transition from helix to coaxial line with detailed cross section at (*b*).

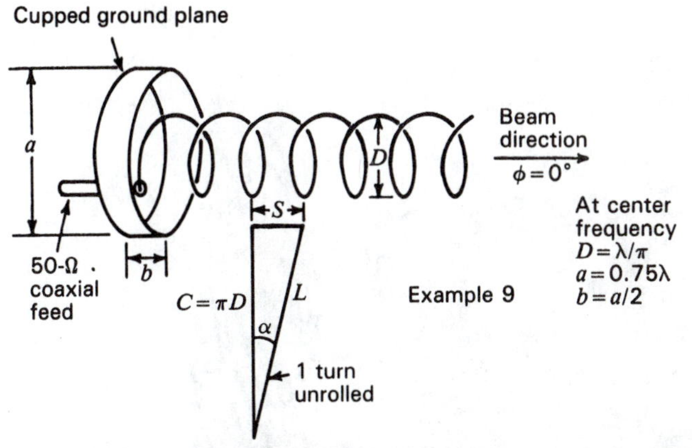

Figure 7-16c Typical peripherally fed monofilar axial-mode helical antenna with cupped ground plane matched to a 50-Ω coaxial transmission line as in Fig. 7-16*a* and *b*. The turn spacing $S = 0.225\lambda$ and the circumference $C = \lambda$ at the center frequency. The relative phase velocity p changes automatically by just the right amount to lock onto end-fire ($\phi = 0°$) with supergain over a frequency range of about one octave. Typical dimensions of the cupped ground plane are $a = 0.75\lambda$ and $b = a/2$ at the center frequency.

Fig. 7-16c with dimensions given in wavelengths at the center frequency for which $C_\lambda = 1$. Support may be an axial rod with radial insulators or one or more peripheral rods (Fig. 7-15*a* and *b*).

A monofilar axial-mode helical antenna with flat circular ground plane and supported by dielectric members is shown in Fig. 7-17 and one with cupped ground plane supported by a dielectric cylinder is illustrated in Fig. 7-18.

Measured patterns of a 6-turn helix as a function of frequency are presented in Fig. 7-19 and patterns at the center frequency ($C_\lambda = 1$) as a function of length (number of turns) are shown in Fig. 7-20.

Based on a large number of such pattern measurements which we made during 1948 and 1949, the beam widths were found to be given by the following quasi-empirical relations.

$$\text{HPBW (half-power beam width)} \simeq \frac{52}{C_\lambda\sqrt{nS_\lambda}} \quad \text{(deg)} \tag{4}$$

$$\text{BWFN (beam width between first nulls)} \simeq \frac{115}{C_\lambda\sqrt{nS_\lambda}} \quad \text{(deg)} \tag{5}$$

The HPBW as given by (4) is shown graphically in Fig. 7-21. Dividing the square of (4) into the number of square degrees in a sphere (41 253) yields an

Figure 7-17 Dielectric-member-supported monofilar axial-mode helical antenna with flat circular ground plane. The pitch angle is 12.5°. The axial feed is directly from a 150-Ω coaxial cable (no matching section required). The conductor tubing is 0.02λ in diameter. Note that the open grid of the ground plane has both circular and radial conductors. Both are essential. (Built by the author.)

Figure 7-18 Thin-wall plastic-cylinder-supported $6\frac{1}{2}$-turn monofilar axial-mode helical antenna with solid metal cupped ground plane. Feeding is via a matching transition from a 50-Ω cable connected through a fitting mounted on the back of the cup ground plane at a point between the plastic cylinder and the lip of the cup. The helix is a flat metal strip bonded to the plastic cylinder. The strip width is 0.03λ (equivalent to a 0.015λ diameter round conductor). The pitch angle is 12.8°. Built by the author for UHF TV band operation with VSWR < 2 from channel 25 to 83 (524 to 890 MHz) and less than 1.2 from channel 27 to 75 (548 to 842 MHz).

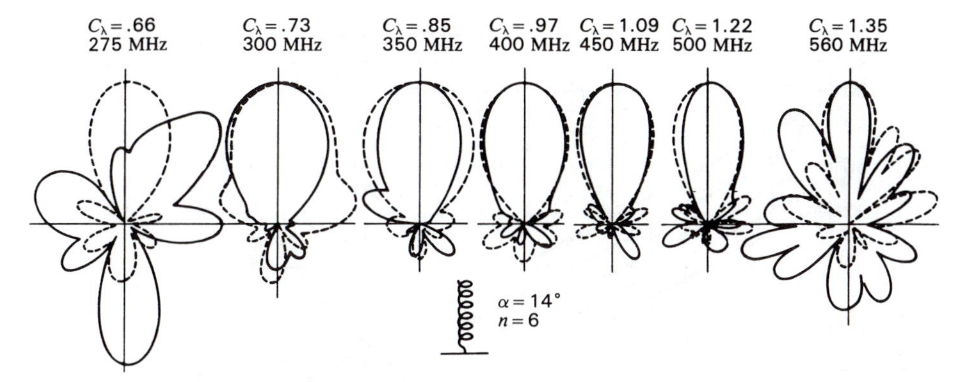

Figure 7-19 Measured field patterns of monofilar axial-mode helical antenna of 6 turns and 14° pitch angle. Patterns are characteristic of the axial mode of radiation over a range of circumferences from about 0.73 to 1.22λ. Both the circumference and the frequency (in megahertz) are indicated. The solid patterns are for the horizontally polarized field component (E_ϕ) and the dashed for the vertically polarized (E_θ). Both are adjusted to the same maximum. (*After Kraus.*)

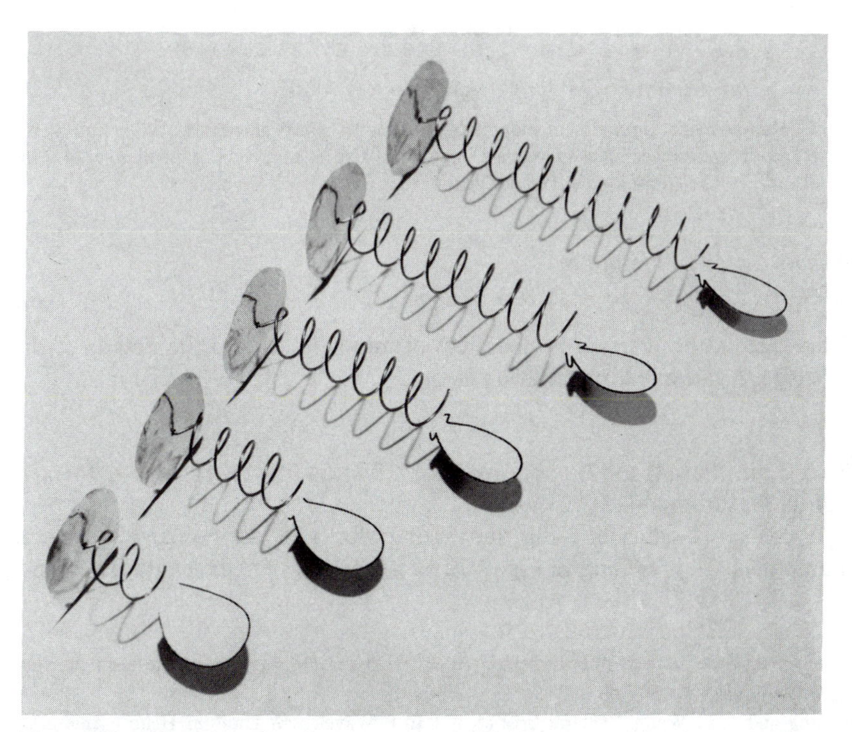

Figure 7-20 Effect of number of turns on measured field patterns. Helices have 12.2° pitch angle and 2, 4, 6, 8, 10 turns. Patterns shown are average of measured E_θ and E_ϕ patterns. (*After Kraus.*)

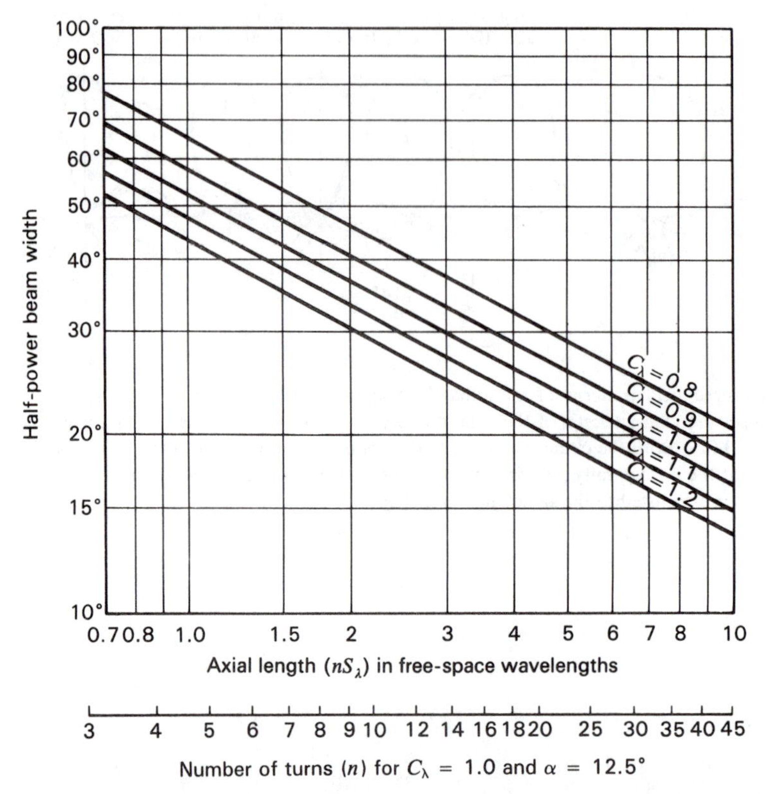

Figure 7-21 Half-power beam width of monofilar axial-mode helical antenna as a function of the axial length and circumference in free-space wavelengths and also as a function of the number of turns for $C_\lambda = 1.0$ and $\alpha = 12.5°$ (lower scale). (*After Kraus.*)

approximate directivity relation:[1]

$$D \simeq 15C_\lambda^2 nS_\lambda \tag{6}$$

This calculation disregards the effect of minor lobes and the details of the pattern shape. A more realistic relation is

$$D \simeq 12C_\lambda^2 nS_\lambda \tag{7}$$

Restrictions are that (4) to (7) apply only for $0.8 < C_\lambda < 1.15$, $12° < \alpha < 14°$ and $n > 3$. Note that from (4-8-3), $D = 4\pi nS_\lambda$.

The measured gains of King and Wong[2] for 12.8° monofilar axial-mode helical antennas are presented in Fig. 7-22 as a function of helix length ($L_\lambda = nS_\lambda$)

---

[1] It is assumed that the patterns of both field components are of the same shape and are figures-of-revolution around the helix axis.

[2] H. E. King and J. L. Wong, "Characteristics of 1 to 8 Wavelength Uniform Helical Antennas," *IEEE Trans. Ants. Prop.*, **AP-28**, 291, March 1980.

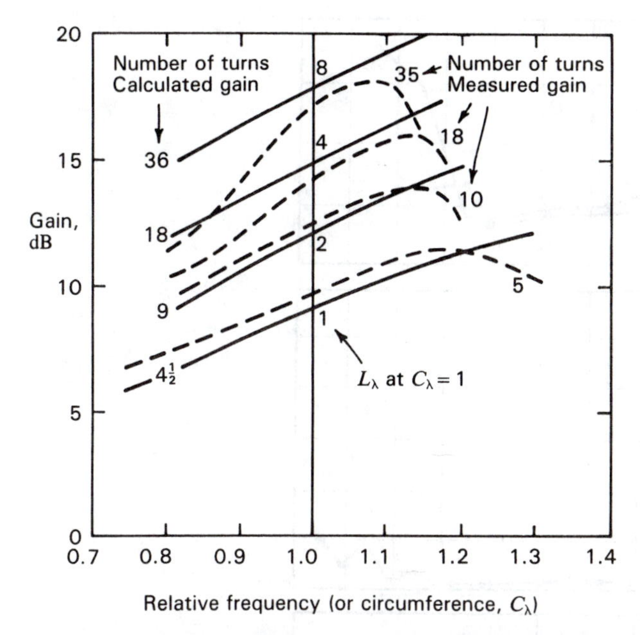

Figure 7-22 Measured (dashed) gain curves of monofilar axial-mode helical antennas as a function of relative frequency for different numbers of turns for a pitch angle of $\alpha = 12.8°$. (*After H. E. King and J. L. Wong, "Characteristics of 1 to 8 Wavelength Uniform Helical Antennas,"* IEEE Trans. Ants. Prop., **AP-28**, *291, March 1980.*) Calculated (solid) gain curves are also shown for different numbers of turns.

and frequency. Although higher gains are obtained by an increased number of turns, the bandwidth tends to become smaller. Highest gains occur at 10 to 20 percent above the center frequency for which $C_\lambda = 1$. Although the gains in Fig. 7-22 tend to be less than calculated from (7), they were measured on helices with 0.08λ diameter axial metal tubes.

Although pitch angles as small as 2°, as noted by MacLean and Kouyoumjian,[1] and as large as 25°, as noted by Kraus, can be used, angles of 12° to 14° (corresponding to turn spacings of 0.21 to 0.25λ at $C_\lambda = 1$) are optimum. King and Wong found that on helices with metal axial tubes, smaller pitch angles (near 12°) resulted in a slightly higher (1 dB) gain but a narrower bandwidth than larger angles (near 14°).

Turning to other parameters, the pattern, axial ratio and impedance (VSWR) performance as a function of frequency for a 6-turn, 14° pitch angle monofilar axial-mode helical antenna are summarized in Fig. 7-23. This is the same antenna for which the patterns are shown in Fig. 7-19. The half-power beam width is taken between half-power points regardless of whether these occur

[1] T. S. M. MacLean and R. G. Kouyoumjian, "The Bandwidth of Helical Antennas," *IRE Trans. Ants. Prop.*, **AP-7**, S379–386, December 1959.

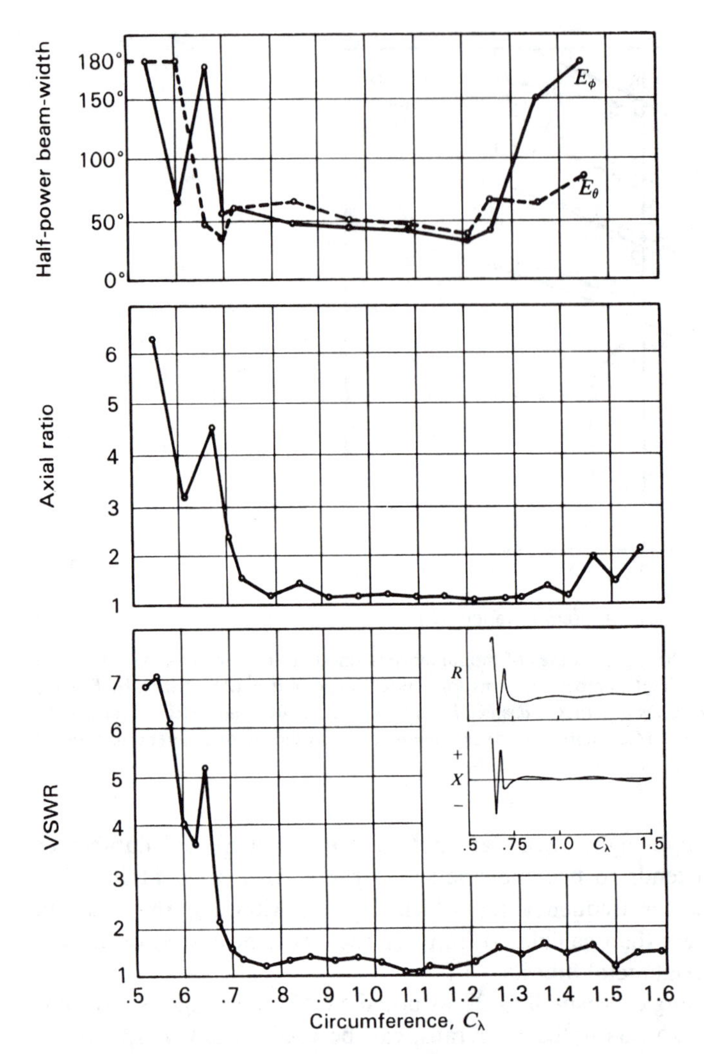

Figure 7-23 Summary of measured performance of 6-turn, 14° monofilar axial-mode helical antenna. The curves show the HPBW for both field components, the axial ratio and the VSWR on a 53-Ω line as a function of the relative frequency (or circumference C_λ). Trends of (relative) resistance R and reactance X are shown in the VSWR inset. Note the relatively constant R and small X for $C_\lambda > 0.7$. (*After Kraus.*)

on the major lobe or on minor lobes. This definition is arbitrary but is convenient to take into account a splitting up of the pattern into many lobes of large amplitude at frequencies outside the beam mode. Beam widths of 180° or more are arbitrarily plotted as 180°. The axial ratio is the value measured in the direction of the helix axis. The standing-wave ratio is the value measured on a 53-Ω coaxial line. A transformer section $\lambda/4$ long at the center frequency is located at the helix terminals to transform the terminal resistance of approximately 130 to 53 Ω. Considered altogether, these pattern, polarization and impedance characteristics represent remarkably good performance over a wide frequency range for a circularly polarized beam antenna.

The onset of the axial mode at a relative frequency of about 0.7 is very evident with axial-mode operation, extending from this frequency over at least an octave for VSWR and axial ratio and almost an octave for pattern.

Several arrangements have been proposed to reduce the axial ratio and VSWR to even lower values. These include a conical end-taper section by Wong and King,[1] Donn,[2] Angelakos and Kajfez[3] and Jamwal and Vakil,[4] and a flat

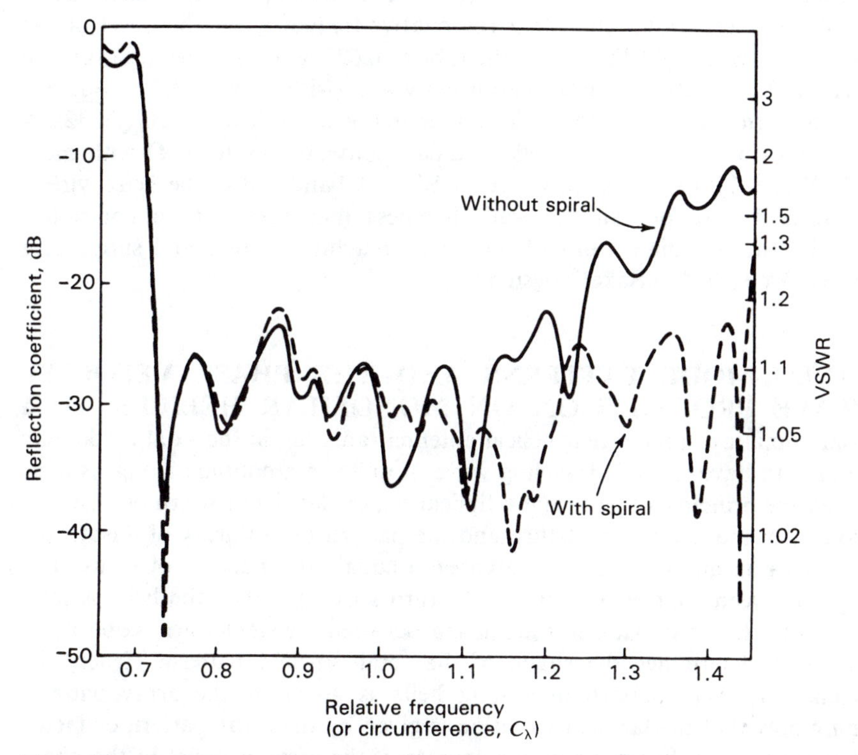

Figure 7-24 Reflection coefficient and VSWR for an impedance-matched peripherally fed 10-turn, 13.8° monofilar axial-mode helical antenna as a function of relative frequency (or circumference C_λ) without spiral termination (solid) and with it (dashed). (*After D. E. Baker, "Design of a Broadband Impedance Matching Section for Peripherally Fed Helical Antennas,"* Antenna Applications Symposium, *University of Illinois, September 1980.*)

---

[1] J. L. Wong and H. E. King, "Broadband Quasi-Taper Helical Antennas," *IEEE Trans. Ants. Prop.*, **AP-27**, 72–78, January 1979.

[2] C. Donn, "A New Helical Antenna Design for Better On-and-Off Boresight Axial Ratio Performance," *IEEE Trans. Ants. Prop.*, **AP-28**, 264–267, March 1980.

[3] D. J. Angelakos and D. Kajfez, "Modifications on the Axial-Mode Helical Antenna," *Proc. IEEE*, **55**, 558–559, April 1967.

[4] K. K. S. Jamwal and R. Vakil, "Design Analysis of Gain-Optimized Helix Antennas for X-band Frequencies," *Microwave J.*, 177–183, September 1985.

spiral termination by Baker.[1] The flat spiral adds no axial length to the helix. The reflection coefficient (or VSWR) as measured by Baker for a 10-turn peripherally fed monofilar axial-mode helical antenna with and without the spiral termination is presented in Fig. 7-24. The improvement occurs at relative frequencies above 1.1 which, however, is a region where the gain is decreasing.

A dielectric tube supporting a helical conductor may significantly affect performance. The magnitude of the effect depends on the dielectric's properties and its geometry, especially the thickness of the tubing wall. For a peripherally fed helix supported on a polyvinyl chloride (PVC) tube with $\varepsilon_r = 2.70$, Baker,[1] using the VSWR as a criterion, found that the relative frequency for the onset of the axial mode shifted from 0.72 without the tube to 0.625 with the tube for a ratio of 1.15. Thus, the effective relative permittivity ε_{eff} (with tube) $= 1.32$ $(=1.15^2)$, making the terminal resistance at the center frequency 130 Ω $(=150/\sqrt{1.32})$. A precision matching section designed by Baker converts this to 50 Ω with measured VSWR < 1.2 ($\rho_v < -20$ dB) over a 1.7 to 1 bandwidth. The helix wire is wound in a groove of half the wall thickness machined with a computer-controlled lathe. All dimensions of the helix, matching section and supporting structure are specified in Baker's design.

7-5 AXIAL-MODE PATTERNS AND THE PHASE VELOCITY OF WAVE PROPAGATION ON MONOFILAR HELICES.[2]

As a first approximation, a monofilar helical antenna radiating in the axial mode may be assumed to have a single traveling wave of uniform amplitude along its conductor. By the principle of pattern multiplication, the far-field pattern of a helix is the product of the pattern for 1 turn and the pattern for an array of n isotropic point sources as in Fig. 7-25. The number n equals the number of turns. The spacing S between sources is equal to the turn spacing. When the helix is long (say, $nS_\lambda > 1$), the array pattern is much sharper than the single-turn pattern and hence largely determines the shape of the total far-field pattern. Hence, the approximate far-field pattern of a long helix is given by the array pattern. Assuming now that the far-field variation is given by the array pattern or factor and that the phase difference between sources of the array is equal to the phase shift over 1 turn length L_λ for a single traveling wave, it is possible to obtain a simple, approximate expression for the phase velocity required to produce axial-mode radiation. This value of phase velocity is then used in pattern calculations.

The array pattern or array factor E for an array of n isotropic point sources arranged as in Fig. 7-25 is given by (4-6-8). Thus,

$$E = \frac{\sin (n\psi/2)}{\sin (\psi/2)} \tag{1}$$

---

[1] D. E. Baker, "Design of a Broadband Impedance Matching Section for Peripherally Fed Helical Antennas," *Antenna Applications Symposium*, University of Illinois, September 1980.

[2] J. D. Kraus, "The Helical Antenna," *Proc. IRE*, **37**, 263–272, March 1949.

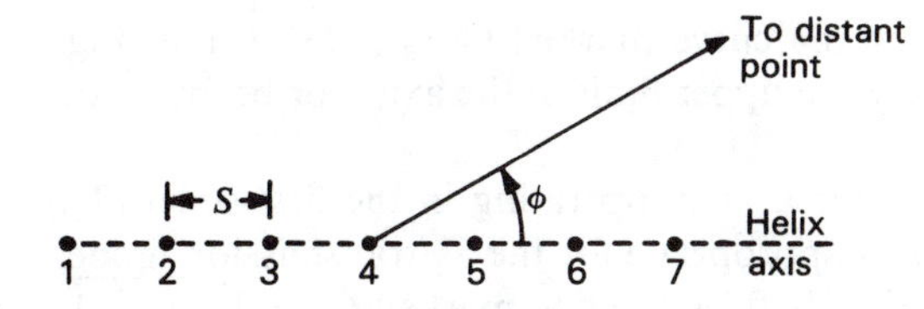

Figure 7-25 Array of isotropic sources, each source representing 1 turn of the helix.

where n = number of sources and

$$\psi = S_r \cos \phi + \delta \tag{2}$$

where $S_r = 2\pi S/\lambda$

In the present case, (2) becomes

$$\psi = 2\pi\left(S_\lambda \cos \phi - \frac{L_\lambda}{p}\right) \tag{3}$$

where $p = v/c$ = relative phase velocity of wave propagation along the helical conductor, v being the phase velocity along the helical conductor and c being the velocity of light in free space.

If the fields from all sources are in phase at a point on the helix axis ($\phi = 0$), the radiation will be in the axial mode. For the fields to be in phase (ordinary end-fire condition) requires that

$$\psi = -2\pi m \tag{4}$$

where $m = 0, 1, 2, 3, \ldots$

The minus sign in (4) results from the fact that the phase of source 2 is retarded by $2\pi L_\lambda/p$ with respect to source 1. Source 3 is similarly retarded with respect to source 2, etc.

Now putting $\phi = 0$ and equating (3) and (4), we have

$$\frac{L_\lambda}{p} = S_\lambda + m \tag{5}$$

When $m = 1$ and $p = 1$, we have the relation

$$L_\lambda - S_\lambda = 1 \qquad \text{or} \qquad L - S = \lambda \tag{6}$$

This is an approximate relation between the turn length and spacing required for a helix radiating in the axial mode. Since for a helix $L^2 = \pi^2 D^2 + S^2$, (6) can be rewritten as

$$D_\lambda = \frac{\sqrt{2S_\lambda + 1}}{\pi} \qquad \text{or} \qquad C_\lambda = \sqrt{2S_\lambda + 1} \tag{7}$$

Equation (7) is shown graphically by the curve marked $C_\lambda = \sqrt{2S_\lambda + 1}$ in Fig. 7-10. The curve defines approximately the upper limit of the axial- or beam-mode region.

When $m = 1$, (5) is appropriate for a helix operating in the first-order (T_1) transmission mode. When $m = 2$, (5) is appropriate for the T_2 transmission mode, etc. A curve for $m = 2$ is shown in Fig. 7-10 by the line marked $C_\lambda = 2\sqrt{S_\lambda + 1}$. Hence, m corresponds to the order of the transmission mode on a helix radiating a maximum field in the axial direction. The case of particular interest here is where $m = 1$.

The case where $m = 0$ does not represent a realizable condition, unless p exceeds unity, since when $m = 0$ and $p = 1$ in (5) we have $L = S$. This is the condition for an end-fire array of isotropic sources excited by a straight wire connecting them ($\alpha = 90°$). However, the field in the axial direction of a straight wire is zero so that there can be no axial mode of radiation in this case.

Returning now to a consideration of the case where $m = 1$ and solving (5) for p, we have

$$p = \frac{L_\lambda}{S_\lambda + 1} \tag{8}$$

From the triangle of Fig. 7-9, (8) can also be expressed as

$$p = \frac{1}{\sin\alpha + [(\cos\alpha)/C_\lambda]} \tag{9}$$

Equation (9) gives the required variation in the relative phase velocity p as a function of the circumference C_λ for in-phase fields in the axial direction. The variation for helices of different pitch angles is illustrated in Fig. 7-26. These curves indicate that when a helix is radiating in the axial mode ($\frac{3}{4} < C_\lambda < \frac{4}{3}$) the value of p may be considerably less than unity. This is borne out by direct measurements of the phase velocity. In fact, the observed phase velocity is found to be slightly less than called for by (8) or (9). Calculating the array pattern for a 7-turn helix using values of p from (8) and (9) yields patterns much broader than observed. The p value of (8) or (9) corresponds to the ordinary end-fire condition discussed in Chap. 4. If the increased directivity condition of Hansen and Woodyard is presumed to exist, (4) becomes

$$\psi = -\left(2\pi m + \frac{\pi}{n}\right) \tag{10}$$

Now equating (10) and (3), putting $\phi = 0$ and solving for p we have

$$p = \frac{L_\lambda}{S_\lambda + m + (1/2n)} \tag{11}$$

For the case of interest $m = 1$ and

$$p = \frac{L_\lambda}{S_\lambda + [(2n + 1)/2n]} \tag{12}$$

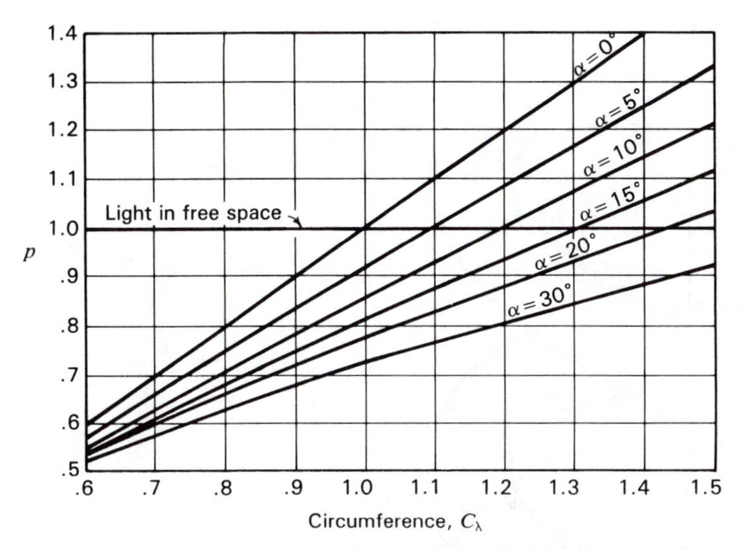

Figure 7-26 Relative phase velocity p for different pitch angles as a function of the helix circumference C_λ for the condition of in-phase fields in the axial direction.

For large values of n, (12) reduces to (8). Equation (12) can also be expressed as[1]

$$p = \frac{1}{\sin\alpha + [(2n+1)/2n][(\cos\alpha)/C_\lambda]} \tag{13}$$

Using p as obtained from (12) or (13) to calculate the array factor yields patterns in good agreement with measured patterns. The p value from (12) or (13) also is in closer agreement with measured values of the relative phase velocity. Hence, it appears that the increased directivity condition is approximated as a natural condition on helices radiating in the axial mode.[2]

Another method of finding the relative phase velocity p on helical antennas radiating in the axial mode is by measuring the angle ϕ_0 at which the first minimum or null occurs in the far-field pattern. This corresponds to the first null in the array factor, which is at ψ_0 (see Fig. 4-20). Then in this case (4) becomes

$$\psi = -(2\pi m + \psi_0) \tag{14}$$

Now equating (14) and (3) and putting $m = 1$ and solving for p, we have

$$p = \frac{L_\lambda}{S_\lambda \cos\phi_0 + 1 + (\psi_0/2\pi)} \tag{15}$$

[1] It is to be noted that, as n becomes large, this relation (13) for increased directivity reduces to (9).

[2] The axial mode region is shown by the shaded (T_1R_1) area in Fig. 7-10. Helices with dimensions in this region radiate in the axial mode, and (9), or more properly (13), applies. Outside this region these equations generally do not apply.

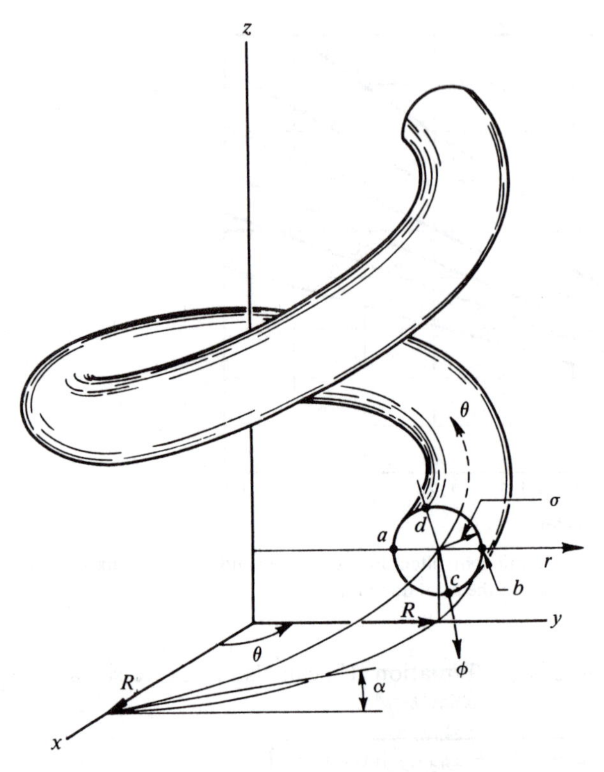

Figure 7-27 Helix showing points c and d at the conductor surface.

Three relations for the relative phase velocity p have been discussed for helices radiating in the axial mode with transmission in the T_1 mode. These are given by (9), (13) and (15).

A fourth relation for p appropriate to the T_1 and higher-order transmission modes on infinite helices has been obtained by Bagby[1] by applying boundary conditions approximating a helical conductor to a solution of the general wave equation expressed in a new coordinate system he called "helicoidal cylindrical coordinates." Bagby's solution is obtained by applying boundary conditions to the two points c and d in Fig. 7-27. His value of the relative phase velocity is given by

$$p = \frac{C_\lambda}{m \cos \alpha + hR \sin \alpha} \tag{16}$$

where

$$hR = \tan \alpha \frac{m\, J_m^2(kR)}{J_{m-1}(kR)\, J_{m+1}(kR)} \tag{17}$$

[1] C. K. Bagby, "A Theoretical Investigation of Electro-magnetic Wave Propagation on the Helical Beam Antenna," Master's thesis, Electrical Engineering Department, Ohio State University, 1948.

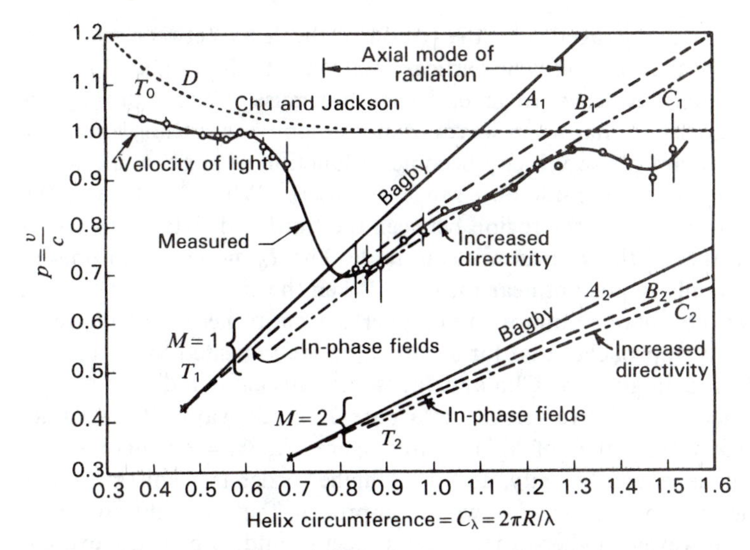

Figure 7-28 Relative phase velocity p as a function of the helix circumference C_λ for 13° helices. The solid curve is measured on a 13°, 7-turn helix. Curves A_1 and A_2 are as calculated by Bagby for T_1 and T_2 transmission modes on an infinite 13° helix. Curves B_1 and B_2 are calculated for in-phase fields and curves C_1 and C_2 for increased directivity for T_1 and T_2 transmission modes. Curve D is from data by Chu and Jackson as calculated for the T_0 transmission mode. (*After Kraus.*)

where m = order of transmission mode ($=1, 2, 3, \ldots$) ($m \neq 0$)
R = radius of helix cylinder
$kR = \sqrt{C_\lambda^2 - (hR)^2}$
h = constant
J = Bessel function of argument kR

The variation of p as a function of C_λ for a 13° helix as calculated by (16) and (17) for the case $m = 1$ is illustrated by the curve A_1 in Fig. 7-28. A curve for the T_1 transmission mode ($m = 1$) as calculated for the in-phase condition from (9) is shown by B_1. A curve for the increased directivity condition on a 13°, 7-turn helix, with $m = 1$, is presented by C_1.

Curves for the T_2 transmission mode for each of the three cases considered above are also presented in Fig. 7-28. In addition, a curve of the measured relative phase velocity on a 13°, 7-turn helix is shown for circumferences between about 0.4 and 1.5λ. It is to be noted that in the circumference range where the helix is radiating in the axial mode ($\frac{3}{4} < C_\lambda < \frac{4}{3}$), the increased directivity curve, of the three calculated curves, lies closest to the measured curve.[1] The measured curve gives the value of the total or resultant phase velocity owing to all modes

[1] The increased directivity curve is the only curve calculated for a helix of 7 turns. The in-phase field curve refers to no specific length while Bagby's curve is for an infinite helix.

present (T_0, T_1, etc.) as averaged over the region of the helix between the third and sixth turns from the feed end. The vertical lines indicate the spread, if any, in values observed at one frequency. In general, each transmission mode propagates with a different velocity so that when waves of more than one transmission mode are present the resultant phase velocity becomes a function of position along the helix and may vary over a considerable range of values.[1] When $\frac{3}{4} < C_\lambda < \frac{4}{3}$ the phase velocity as measured in the region between the third and sixth turns corresponds closely to that of the T_1 transmission mode. The T_0 mode is also present on the helix but is only important near the ends. When the circumference $C_\lambda < \frac{2}{3}$, the T_0 mode may be obtained almost alone over the entire helix and the measured phase velocity approaches that for a pure T_0 mode indicated by curve D in Fig. 7-28, based on data given by Chu and Jackson.[2] This curve indicates that at small circumferences the relative velocity of a pure T_0 mode wave attains values considerably greater than that of light in free space. At $C_\lambda = \frac{2}{3}$, curve D has decreased to a value of nearly unity, and if no higher-order transmission mode were permissible, the phase velocity would approach that of light for large circumferences. However, higher-order modes occur, and, when C_λ exceeds about $\frac{2}{3}$, the resultant velocity drops abruptly, as shown by the measured curve in Fig. 7-28. This change corresponds to a transition from the T_0 to the T_1 transmission mode. For a circumference in the transition region, such as 0.7λ, both T_0 and T_1 modes are of about equal importance.

When C_λ is about $\frac{3}{4}$ or somewhat more, the measured phase velocity approaches a value associated with the T_1 mode. As C_λ increases further, the relative phase velocity increases in an approximately linear fashion, agreeing most closely with the theoretical curve for the increased directivity condition (curve C_1). When C_λ reaches about $\frac{4}{3}$, a still higher order transmission mode (T_2) appears to become partially effective, causing further dips in the measured curve. However, the radiation may no longer be in the axial mode.

The formulas given for helical antennas operating in the first-order transmission mode ($m = 1$) are summarized in Table 7-1.

As mentioned above, the approximate far-field pattern of a monofilar helix radiating in the axial mode is given by the array factor for n isotropic point sources, each source replacing a single turn of the helix (see Fig. 7-25).

The normalized array factor is

$$E = \sin \frac{\pi}{2n} \frac{\sin (n\psi/2)}{\sin (\psi/2)} \tag{18}$$

where $\psi = 2\pi[S_\lambda \cos \phi - (L_\lambda/p)]$

[1] J. A. Marsh, "Measured Current Distributions on Helical Antennas," *Proc. IRE*, **39**, 668–675, June 1951.

[2] L. J. Chu and J. D. Jackson, "Field Theory of Traveling Wave Tubes," *Proc. IRE*, **36**, 853–863, July 1948.

Table 7-1 Relative phase velocities for first-order transmission mode on helical antennas

Condition	Relative phase velocity
In-phase fields (ordinary end-fire)	$p = \dfrac{L_\lambda}{S_\lambda + 1} = \dfrac{1}{\sin\alpha + [(\cos\alpha)/C_\lambda]}$
Increased directivity	$p = \dfrac{L_\lambda}{S_\lambda + [(2n+1)/2n]}$ $= \dfrac{1}{\sin a + [(2n+1)/2n][(\cos\alpha)/C_\lambda]}$
From first null of measured field pattern	$p = \dfrac{L_\lambda}{S_\lambda \cos\phi_0 + (\psi_0/2\pi) + 1}$
Helicoidal cylindrical coordinate solution	$p = \dfrac{C_\lambda}{\cos\alpha + hR\sin\alpha}$ where hR is as given by (17)

The normalizing factor is $\sin(\pi/2n)$ instead of $1/n$ since the increased directivity end-fire condition is assumed to exist (see Sec. 4-6d, Case 3). For a given helix, S_λ and L_λ are known and p can be calculated from (12) or (13). ψ is then obtained as a function of ϕ. From (18), these values of ψ give the field pattern.

As an illustration, the calculated array factor patterns for a 7-turn, 12° helix with $C_\lambda = 0.95$ are shown in Fig. 7-29 for p values corresponding to increased directivity and also in-phase fields and for $p = 1$, 0.9 and 0.725. A measured curve (average of E_ϕ and E_θ) is shown for comparison. It is apparent that the pattern calculated for the increased-directivity condition ($p = 0.76$) agrees most closely with the measured pattern. The measured pattern was taken on a helix mounted on a ground plane 0.88λ in diameter. The calculated patterns neglect the effect of a ground plane. This effect is small if the back lobe is small compared to the front lobe, as it is for $p = 0.802$ and $p = 0.76$.

The sensitivity of the pattern to the phase velocity is very apparent from Fig. 7-29. In particular, we note that as little as a 5 percent difference in phase velocity from that required for the increased directivity condition ($p = 0.76$) produces marked changes as shown by the patterns for $p = 0.802$ (5 percent high) and $p = 0.725$ (5 percent low).

7-6 MONOFILAR AXIAL-MODE SINGLE-TURN PATTERNS.

In this section expressions will be developed for the far-field patterns from a single turn of a monofilar helix radiating in the axial mode. It is assumed that the single turn has a uniform traveling wave along its entire length. The product of the single-turn pattern and the array factor then gives the total helix pattern.

A circular helix may be treated approximately by assuming that it is of square cross section. The total field from a single turn is then the resultant of the

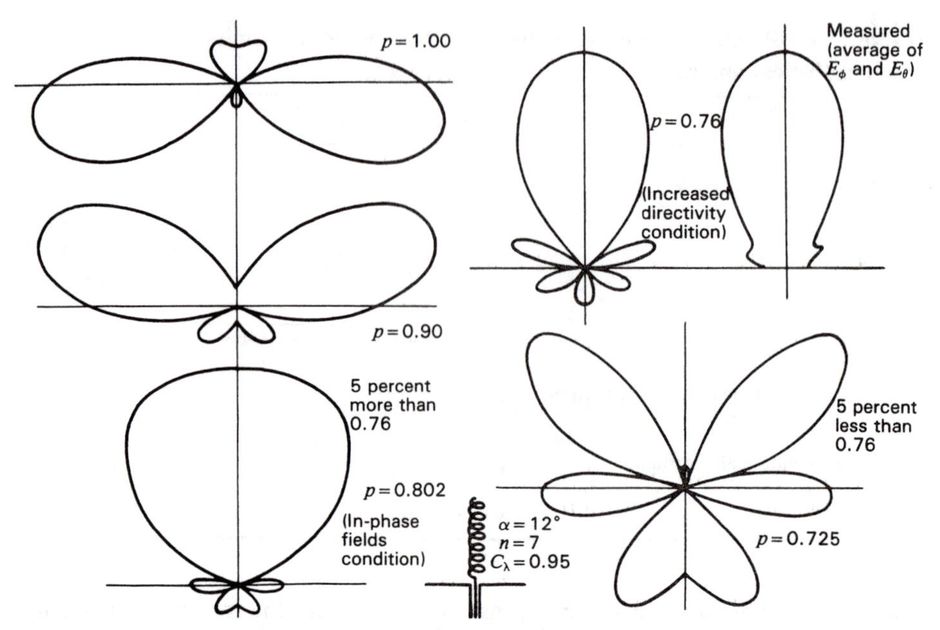

Figure 7-29 Array factor patterns for 12°, 7-turn helix with $C_\lambda = 0.95$. Patterns are shown for $p = 1$, 0.9, 0.802 (in-phase fields or ordinary end-fire condition), 0.76 (increased directivity) and 0.725. A measured curve is also presented. All patterns are adjusted to the same maximum. The sensitivity of the pattern to phase velocity is evident. A change of as little as 5 percent produces a drastic change in pattern, as may be noted by comparing the pattern for $p = 0.802$ (5 percent high) and the one for $p = 0.725$ (5 percent low) with the one for $p = 0.76$ which matches the measured pattern.

fields of four short, linear antennas as shown in Fig. 7-30*a*. A helix of square cross section can, of course, be treated exactly by this method. Measurements indicate that the difference between helices of circular and square cross section is small.

Referring to Fig. 7-31, the far electric field components, $E_{\phi T}$ and $E_{\theta T}$, in the xz plane will be calculated as a function of ϕ for a single-turn helix.

Let the area of the square helix be equal to that of the circular helix so that

$$g = \frac{\sqrt{\pi}\, D}{2} \tag{1}$$

where g is as shown in Fig. 7-30*a*.

The far magnetic field for a linear element with a uniform traveling wave is given by (5-8-15). Multiplying (5-8-15) by the intrinsic impedance Z of free space, putting $\gamma = (3\pi/2) + \alpha + \phi$, $t = 0$ and $b = g/\cos\alpha$, we obtain the expression for the ϕ component $E_{\phi 1}$ of the far field in the xz plane due to element 1 of the square helix as follows:

$$E_{\phi 1} = k\,\frac{\sin\gamma}{A}\sin BA \,\underline{\left/\left(-\frac{\omega r_1}{c} - BA\right)\right.} \tag{2}$$

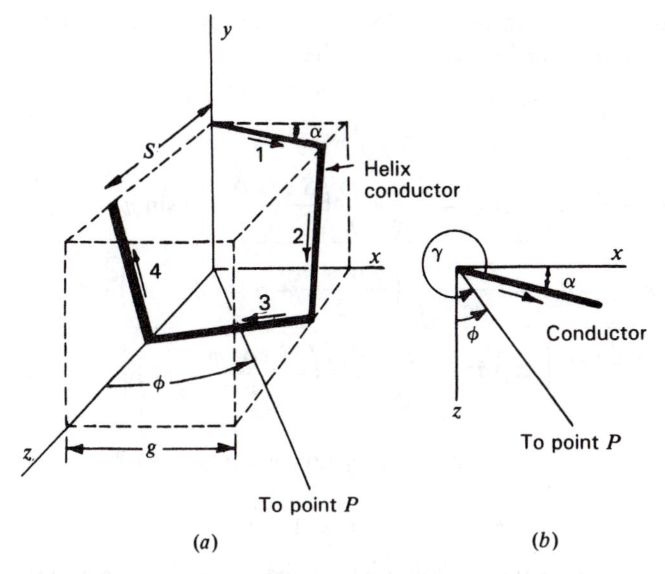

Figure 7-30 Square helix used in calculating single-turn pattern.

where $k = \dfrac{I_0 pZ}{2\pi r_1}$

$$A = 1 - p \cos \gamma$$

$$B = \frac{\omega g}{2pc \cos \alpha}$$

The expressions for $E_{\phi 2}$, $E_{\phi 3}$, etc., due to elements 2, 3 and 4 of the square turn are obtained in a similar way. Since the elements are all dissimilar sources, the total ϕ component, $E_{\phi T}$, from a single square turn is obtained by adding the

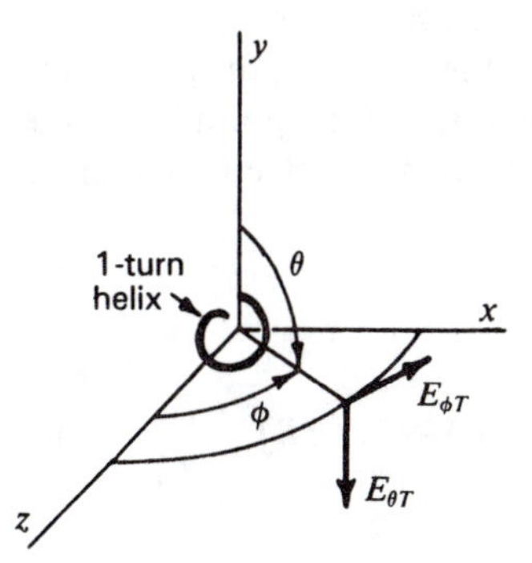

Figure 7-31 Field components with relation to single-turn helix.

fields from the four elements at each angle ϕ for which the total field is calculated. The sum of the fields from the four elements is then

$$
\begin{aligned}
E_{\phi T} = {} & k \frac{\sin \gamma}{A} \sin BA \, \underline{/ \left(-BA - \frac{\omega r_1}{c} \right)} \\
& + k \frac{\sin BA'' \sin \alpha \sin \phi}{A''} \underline{/ \left[-BA'' - \frac{L\omega}{4pc} + \frac{\omega}{c} \left(\frac{S \cos \phi}{4} + g \sin \phi - r_1 \right) \right]} \\
& + k \frac{\sin \gamma'}{A'} \sin BA' \underline{/ \left[-BA' - \frac{L\omega}{2pc} + \frac{\omega}{c} \left(\frac{S \cos \phi}{2} + g \sin \phi - r_1 \right) \right]} \\
& + k \frac{\sin BA'' \sin \alpha \sin \phi}{A''} \underline{/ \left[-BA'' - \frac{3L\omega}{4pc} + \frac{\omega}{c} \left(\frac{3S \cos \phi}{4} - r_1 \right) \right]} \qquad (3)
\end{aligned}
$$

where $\gamma = \dfrac{3\pi}{2} + \alpha + \phi, \qquad \gamma' = \dfrac{\pi}{2} - \alpha + \phi, \qquad \gamma'' = \arccos(\sin \alpha \cos \phi)$

$$A = 1 - p \cos \gamma, \qquad A' = 1 - p \cos \gamma', \qquad A'' = 1 - p \cos \gamma''$$

When a helix of circular cross section is being calculated, $L = \pi D / \cos \alpha$ in (3), while for a helix of square cross section $L = 4b$.

If the contributions of elements 2 and 4 are neglected, which is a good approximation when both α and ϕ are small, the expression for $E_{\phi T}$ is considerably simplified. Making this approximation, letting $k = 1$ and $r_1 =$ constant, we obtain

$$
\begin{aligned}
E_{\phi T} = {} & \frac{\sin \gamma}{A} \sin BA \, \underline{/(-BA)} \\
& + \frac{\sin \gamma'}{A'} \sin BA' \, \underline{/[-BA' - 2\sqrt{\pi} B + \pi(S_\lambda \cos \phi + \sqrt{\pi} D_\lambda \sin \phi)]} \qquad (4)
\end{aligned}
$$

Equation (4) applies specifically to helices of circular cross section, so that B in (4) is

$$B = \frac{D_\lambda \pi^{3/2}}{2p \cos \alpha} \qquad (5)$$

Equation (4) gives the approximate pattern of the ϕ component of the far field in the xz plane from a single-turn helix of circular cross section.

In the case of the θ component of the far field in the xz plane, only elements 2 and 4 of the square turn contribute. Putting $k = 1$, the magnitude of the approximate θ pattern of the far field of a single-turn helix of circular cross section can be shown to be

$$
\begin{aligned}
|E_{\theta T}| = {} & 2 \frac{\sin \gamma'' \sin BA'' \cos \alpha}{A''(1 - \sin^2 \alpha \cos^2 \phi)^{1/2}} \\
& \times \sin \tfrac{1}{2}[\pi(S_\lambda \cos \phi - \sqrt{\pi} D_\lambda \sin \phi) - 2\sqrt{\pi} B] \qquad (6)
\end{aligned}
$$

where B is as given by (5) and γ'' and A'' are as in (3).

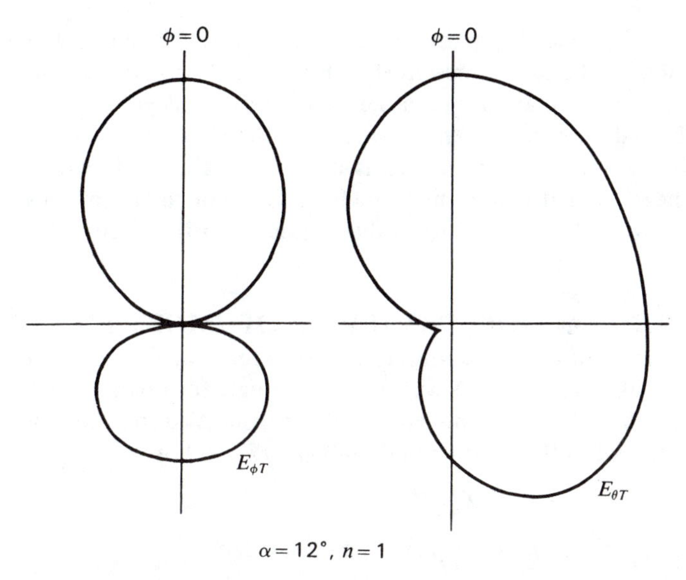

Figure 7-32 Calculated patterns for $E_{\phi T}$ and $E_{\theta T}$ fields of single turn of a 12° helix.

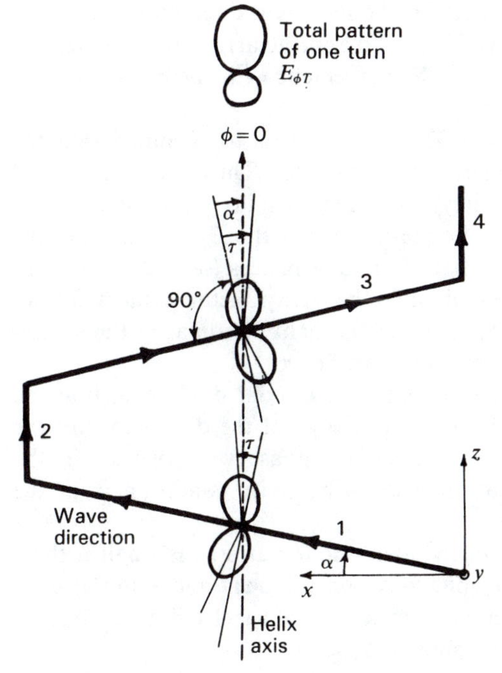

Figure 7-33 Individual E_ϕ patterns of elements 1 and 3 and total pattern of single turn, $E_{\phi T}$. The single turn is shown in plan view (in the xz plane of Fig. 7-30). The single turn and coordinate axes have been rotated around the y axis so that the z direction ($\phi = 0$) is toward the top of the page.

As an example, the $E_{\phi T}$ and $E_{\theta T}$ patterns for a single-turn 12° helix with $C_\lambda = 1.07$ have been calculated and are presented in Fig. 7-32. Although the two patterns are of different form, both are broad in the axial direction ($\phi = 0$).

The individual E_ϕ patterns of elements 1 and 3 of the single turn are as suggested in Fig. 7-33. One lobe of each pattern is nearly in the axial direction, the tilt angle τ being nearly equal to the pitch angle α. The individual patterns add to give the $E_{\phi T}$ pattern for the single turn as shown (see also Fig. 7-32).

7-7 COMPLETE AXIAL-MODE PATTERNS OF MONOFILAR HELICES. By the principle of pattern multiplication, the total far-field pattern of a helix radiating in the axial mode is the product of the single-turn pattern and the array factor E. Thus, the total ϕ component E_ϕ of the distant electric field of a helix of circular cross section is the product of (7-6-4) and (7-5-18) or

$$E_\phi = E_{\phi T} E \tag{1}$$

The total θ component E_θ is the product of (7-6-6) and (7-5-18) or

$$E_\theta = E_{\theta T} E \tag{2}$$

As examples, the approximate E_ϕ and E_θ patterns, as calculated by the above procedure, for a 12°, 7-turn uniform helix of circular cross section with $C_\lambda = 1.07$ are presented in Fig. 7-34 at (*a*) and (*c*). The helix is shown at (*e*), with E_ϕ in the plane of the page and E_θ normal to the page. The array factor is shown at (*b*). The single-turn patterns are as presented in Fig. 7-32. The value of p used in these calculations is approximately that for the increased directivity condition. The product of the single-turn patterns (Fig. 7-32) and the array factor pattern at (*b*) yields the total patterns at (*a*) and (*c*). The agreement with the measured patterns shown at (*d*) and (*f*) is satisfactory.

Comparing the patterns of Figs. 7-32 and 7-34, it is to be noted that the array factor is much sharper than the single-turn patterns. Thus, the total E_ϕ and E_θ patterns (*a*) and (*c*) (Fig. 7-34) are nearly the same, in spite of the difference in the single-turn patterns. Furthermore, the main lobes of the E_ϕ and E_θ patterns are very similar to the array factor pattern. For long helices (say, $nS_\lambda > 1$) it is, therefore, apparent that a calculation of only the array factor suffices for an approximate pattern of any field component of the helix. Ordinarily the single-turn pattern need not be calculated except for short helices.

The far-field patterns of a helix radiating in the axial mode can, thus, be calculated to a good approximation from a knowledge of the dimensions of the helix and the wavelength. The value of the relative phase velocity used in the calculations may be computed for the increased-directivity condition from the helix dimensions and number of turns.

The effect of the ground plane on the axial-mode patterns is small if there are at least a few turns, since the returning wave on the helix and also the back lobe of the outgoing wave are both small. Hence, the effect of the ground plane may be neglected unless the helix is very short ($nS_\lambda < \frac{1}{2}$).

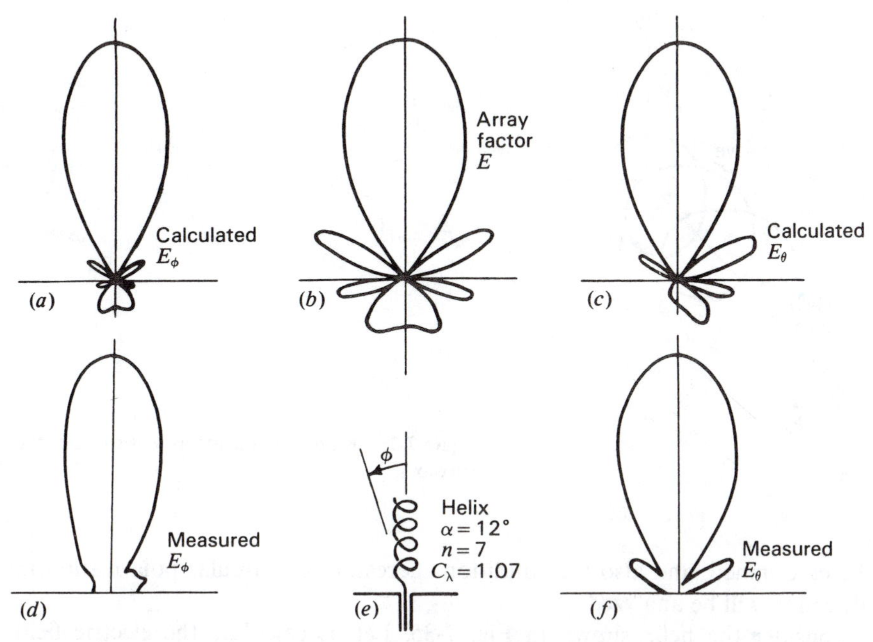

Figure 7-34 Comparison of complete calculated patterns (product of single-turn pattern and array factor) with measured patterns for a 12°, 7-turn helix with $C_\lambda = 1.07$ radiating in the axial mode. Agreement is satisfactory.

The approximate pattern of an axial-mode helix can be calculated very simply, while including the approximate effect of the single-turn pattern, by assuming that the single-turn pattern is given by cos ϕ. Then the normalized total radiation pattern is expressed by

$$E = \left(\sin \frac{90°}{n}\right) \frac{\sin (n\psi/2)}{\sin (\psi/2)} \cos \phi \tag{3}$$

where n = number of turns and

$$\psi = 360°[S_\lambda(1 - \cos \phi) + (1/2n)] \tag{4}$$

The value of ψ in (4) is for the increased-directivity condition and is obtained by substituting (7-5-12) in (7-5-3) and simplifying. The first factor in (3) is a normalizing factor, i.e., makes the maximum value of E unity.

7-8 AXIAL RATIO AND CONDITIONS FOR CIRCULAR POLARIZATION OF MONOFILAR AXIAL-MODE HELICAL ANTENNAS.[1]

In this section the axial ratio in the direction of the helix axis

---

[1] For a general discussion of elliptical and circular polarization see Secs. 2-34 through 2-37; also see J. D. Kraus, *Radio Astronomy*, 2nd ed., Cygnus-Quasar, 1986, chap. 4.

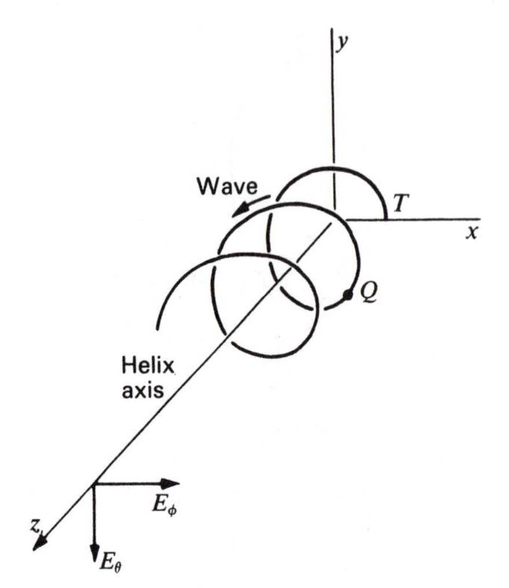

Figure 7-35 Field components as viewed from the helix axis.

will be determined, and also the conditions necessary for circular polarization in this direction will be analyzed.

Consider the helix shown in Fig. 7-35. Let us calculate the electric field components E_ϕ and E_θ, as shown, at a large distance from the helix in the z direction. The helix is assumed to have a single uniform traveling wave as indicated. The relative phase velocity is p. The diameter of the helix is D and the spacing between turns is S. Unrolling the helix in the xz plane, the relations are as shown in Fig. 7-36. The helix as viewed from a point on the z axis is as indicated in Fig. 7-37. The angle ξ is measured from the xz plane. The coordinates of a point Q on the helix can be specified as r, ξ, z. The point Q is at a distance l from the terminal point T as measured along the helix. From the geometry of Figs. 7-36 and 7-37, we can write

$$\left.\begin{aligned} h &= l \sin \alpha \\ z_p - h &= z_p - l \sin \alpha \\ \alpha &= \arctan \frac{S}{\pi D} = \arccos \frac{r\xi}{l} \\ r\xi &= l \cos \alpha \end{aligned}\right\} \tag{1}$$

where z_p is the distance from the origin to the distant point P on the z axis.

At the point P the ϕ component E_ϕ of the electric field for a helix of an integral number of turns n is

$$E_\phi = E_0 \int_0^{2\pi n} \sin \xi \exp \left[j\omega \left(t - \frac{z_p}{c} + \frac{l \sin \alpha}{c} - \frac{l}{pc} \right) \right] d\xi \tag{2}$$

where E_0 is a constant involving the current magnitude on the helix.

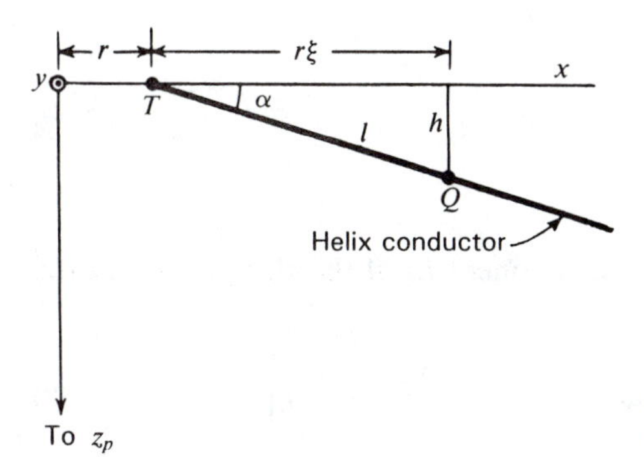

Figure 7-36 Geometry for calculating fields in the z direction.

From (1) the last two terms of the exponent in (2) may be rewritten. Thus,

$$\frac{l \sin \alpha}{c} - \frac{l}{pc} = \frac{r\xi}{c}\left(\tan \alpha - \frac{1}{p \cos \alpha}\right) = \frac{r\xi q}{c} \tag{3}$$

where

$$q = \tan \alpha - \frac{1}{p \cos \alpha} \tag{4}$$

When $\alpha = 0$, the helix becomes a loop and $q = -1/p$. The relation being obtained is, thus, a general one, applying not only to helices but also to loops as a special case. Equation (2) now reduces to

$$E_\phi = E_0 \, e^{j(\omega t - \beta s_p)} \int_0^{2\pi n} \sin \xi \; e^{jk\xi} \, d\xi \tag{5}$$

where quantities independent of ξ have been taken outside the integral and where

$$\beta = \frac{\omega}{c} = \frac{2\pi}{\lambda}$$

and

$$k = \beta r q = L_\lambda\left(\sin \alpha - \frac{1}{p}\right) \tag{6}$$

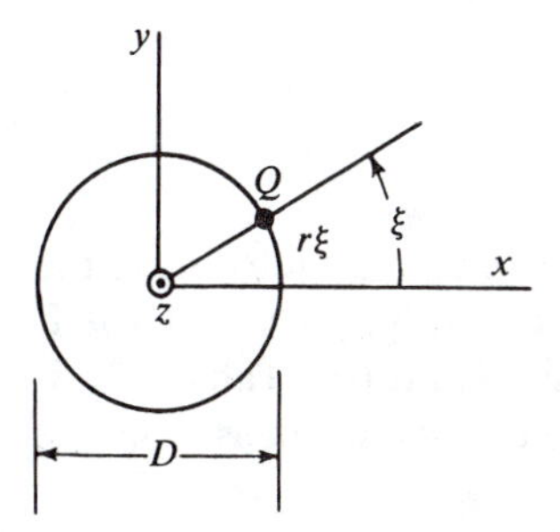

Figure 7-37 Helix of Fig. 7-35 as viewed from the positive z axis.

On integration (5) becomes

$$E_\phi = \frac{E_1}{k^2 - 1}(e^{j2\pi nk} - 1) \tag{7}$$

where $E_1 = E_0\, e^{j(\omega t - \beta z_p)}$

In a similar way we have for the θ component E_θ of the electric field at the point P,

$$E_\theta = E_0 \int_0^{2\pi n} \cos \xi \exp\left[j\omega\left(t - \frac{z_p}{c} + \frac{l \sin \alpha}{c} - \frac{l}{pc}\right)\right] d\xi \tag{8}$$

Making the same substitutions as in (2), we obtain from (8)

$$E_\theta = \frac{jE_1 k}{k^2 - 1}(e^{j2\pi nk} - 1) \tag{9}$$

The condition for circular polarization in the direction of the z axis is

$$\frac{E_\phi}{E_\theta} = \pm j \tag{10}$$

The ratio of (7) to (9) gives

$$\frac{E_\phi}{E_\theta} = \frac{1}{jk} = -\frac{j}{k} \tag{11}$$

Accordingly, for circular polarization in the axial direction of a helix of an integral number of turns, k must equal ± 1.

Equation (11) indicates that E_ϕ and E_θ are in time-phase quadrature. Therefore, the axial ratio AR is given by the magnitude of (11) or

$$\text{AR} = \frac{|E_\phi|}{|E_\theta|} = \left|\frac{1}{jk}\right| = \frac{1}{k} \tag{12}$$

The axial ratio will be restricted here to values between unity and infinity. Hence, if (12) is less than unity, its reciprocal is taken.

Substituting the value of k from (6) into (12) yields

$$\text{AR} = \frac{1}{|L_\lambda[\sin \alpha - (1/p)]|} \tag{13}$$

or

$$\text{AR} = \left|L_\lambda\left(\sin \alpha - \frac{1}{p}\right)\right| \tag{14}$$

Either (13) or (14) is used so that $1 \leq \text{AR} \leq \infty$.

From (13) and (14), it appears that the axial ratio can be calculated from the turn length L_λ and pitch angle α of the helix, and the relative phase velocity p. If we introduce the value of p for the condition of in-phase fields (see Table 7-1), it is found that $\text{AR} = 1$. In other words, the in-phase field condition is also the condition for circular polarization in the axial direction.

This may also be shown by noting that (11) satisfies the condition for circular polarization when $k = -1$, or

$$L_\lambda\left(\sin \alpha - \frac{1}{p}\right) = -1 \tag{15}$$

Solving (15) for p, we obtain

$$p = \frac{L_\lambda}{S_\lambda + 1} \tag{16}$$

which is identical with the relation for in-phase fields (ordinary end-fire condition).

Our previous discussion on phase velocity indicated that p followed more closely the relation for increased directivity than the relation for in-phase fields. Thus, introducing p in (14) for the condition of increased directivity, we obtain

$$\text{AR (on axis)} = \frac{2n + 1}{2n} \tag{17}$$

where n is the number of turns of the helix. If n is large the axial ratio approaches unity and the polarization is nearly circular.[1]

When I first derived (17) in 1947, it came as a pleasant surprise that the axial ratio could be given by such a simple expression.

As an example, let us consider the axial ratio in the direction of the helix axis for a 13°, 7-turn helix. The axial ratio is unity if the relative velocity for the condition of in-phase fields is used. By (17) the axial ratio for the condition of increased directivity is $15/14 = 1.07$. This axial ratio is independent of the frequency or circumference C_λ as shown by the dashed line in Fig. 7-38. In this figure, the axial ratio is presented as a function of the helix circumference C_λ in free-space wavelengths.

If the axial ratio is calculated from (13) or (14), using the measured value of p shown in Fig. 7-28, an axial ratio variation is obtained as indicated by the solid curve in Fig. 7-38. This type of axial ratio versus circumference curve is typical of ones measured on helical beam antennas. Usually, however, the measured axial ratio increases more sharply as C_λ decreases to values less than about $\frac{3}{4}$. This difference results from the fact that the calculation of axial ratio by (13) or (14) neglects the effect of the back wave on the helix. This is usually small when the helix is radiating in the axial mode but at lower frequencies or smaller circumferences ($C_\lambda < \frac{3}{4}$) the back wave is important. The back wave on the helix produces a wave reflected from the ground plane having the opposite direction of

[1] With circularly polarized feed an AR = 1 can be obtained on the axis for a helix of any length according to R. G. Vaughan and J. B. Andersen ("Polarization Properties of the Axial Mode Helix Antenna," *IEEE Trans. Ants. Prop.*, **AP-33**, 10–20, January 1985). They also deduce the axial ratio as a function of the off-axis angle.

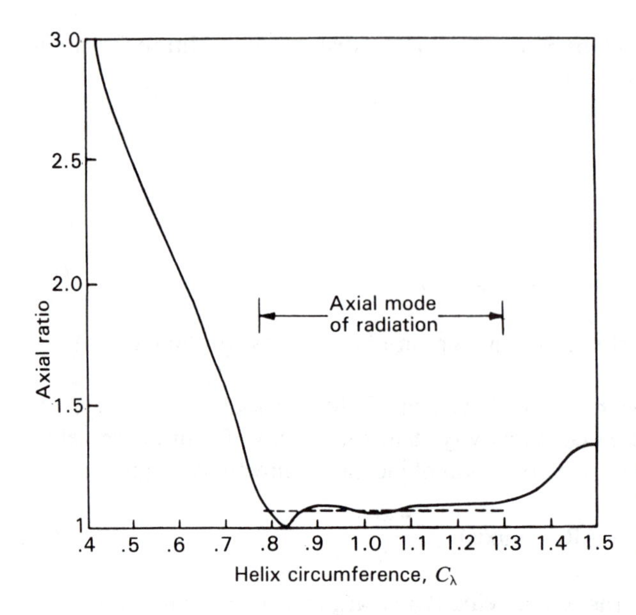

Figure 7-38 Axial ratio as a function of helix circumference C_λ for a 13°, 7-turn monofilar axial-mode helical antenna. The dashed curve is from (17). (*After Kraus.*)

field rotation to that produced by the outgoing traveling wave on the helix. This causes the axial ratio to increase more rapidly than indicated in Fig. 7-38.

The foregoing discussion applies to helices of an integral number of turns. Let us now consider a long helix where the number of turns may assume non-integral values. Hence, the length of the helical conductor will be specified as ξ_1 instead of $2\pi n$. It is further assumed that k is nearly unity. Thus, (5) becomes

$$E_\phi = \frac{E_1}{2j} \int_0^{\xi_1} (e^{j(k+1)\xi} - e^{j(k-1)\xi})\, d\xi \tag{18}$$

Since $k \simeq -1$, $k + 1 \simeq 0$, and it follows that

$$e^{j(k+1)\xi_1} \simeq 1 + j(k+1)\xi_1 \tag{19}$$

Now integrating (18) and introducing the condition that k is nearly equal to -1 and the approximation of (19), we have

$$E_\phi = -\frac{E_1}{2}\left(j\xi_1 - \frac{e^{j(k-1)\xi_1} - 1}{k-1}\right) \tag{20}$$

Similarly the θ component E_θ of the electric field is

$$E_\theta = +\frac{E_1}{2j}\left(j\xi_1 + \frac{e^{j(k-1)\xi_1} - 1}{k-1}\right) \tag{21}$$

When the helix is very long

$$\xi_1 \gg 1$$

and (20) and (21) reduce to

$$E_\phi = -j\frac{E_1\xi_1}{2} \quad \text{and} \quad E_\theta = +\frac{E_1\xi_1}{2} \tag{22}$$

Taking the ratio of E_ϕ to E_θ,

$$\frac{E_\phi}{E_\theta} = -j \tag{23}$$

which fulfills the condition for circular polarization.

Still another condition resulting in circular polarization is obtained when $(k \pm 1)\xi_1 = 2\pi m$, where m is an integer. This condition is satisfied when either the positive or negative sign in $k \pm 1$ is chosen but not for both.

To summarize, the important conditions for circular polarization are as follows:

1. The radiation in the axial direction from a helical antenna of any pitch angle and of an integral number of 1 or more turns will be circularly polarized if $k = -1$ (in-phase fields or ordinary end-fire condition).
2. The radiation in the axial direction from a helical antenna of any pitch angle and a large number of turns, which are not necessarily an integral number, is nearly circularly polarized if k is nearly -1.

7-9 WIDEBAND CHARACTERISTICS OF MONOFILAR HELICAL ANTENNAS RADIATING IN THE AXIAL MODE.

The helical beam antenna[1] has inherent broadband properties, possessing desirable pattern, impedance and polarization characteristics over a relatively wide frequency range. The natural adjustment of the phase velocity so that the fields from each turn add nearly in phase in the axial direction accounts for the persistence of the axial mode of radiation over a nearly 2 to 1 range in frequency. If the phase velocity were constant as a function of frequency, the axial-mode patterns would be obtained only over a narrow frequency range. The terminal impedance is relatively constant over the same frequency range because of the large attenuation of the wave reflected from the open end of the helix. The polarization is nearly circular over the same range in frequency because the condition of fields in phase is also the condition for circular polarization.

As shown in Fig. 7-39*a*, the dimensions of a helix in free-space wavelengths move along a constant pitch-angle line as a function of frequency. If F_1 is the lower frequency limit of the axial mode of radiation and F_2 the upper frequency limit of this mode, then the range in dimensions for a 10° helix would be as suggested by the heavy line on the diameter-spacing chart of Fig. 7-39*a*. The center frequency F_0 is arbitrarily defined as $F_0 = (F_1 + F_2)/2$.

[1] Or monofilar axial-mode helical antenna.

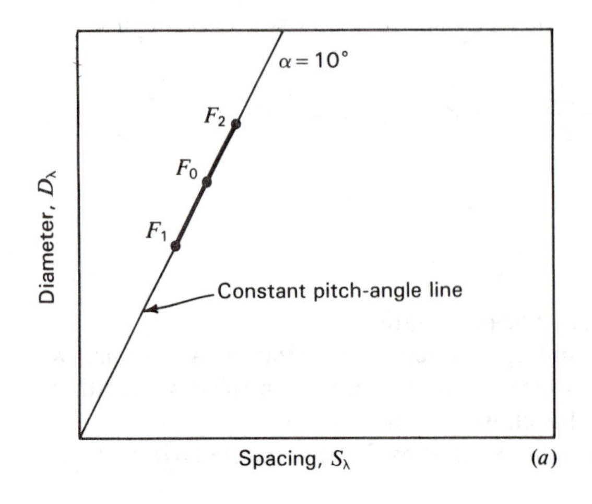

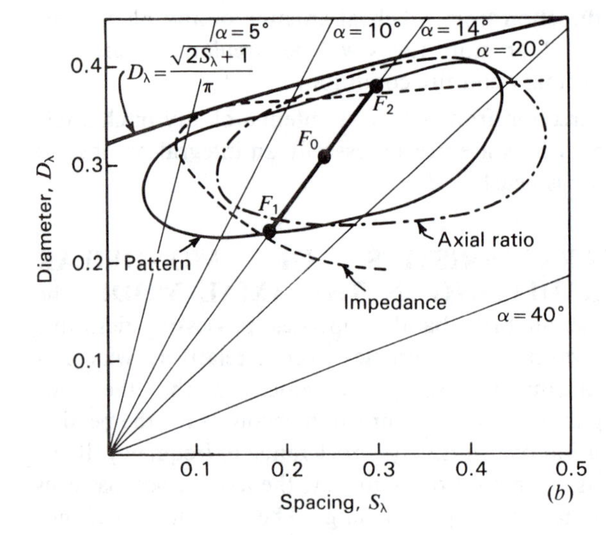

Figure 7-39 Diameter-spacing charts for monofilar helices with measured performance contours (*b*) for the axial mode of radiation.

The properties of a helical beam antenna are a function of the pitch angle. The angle resulting in a maximum frequency range $F_2 - F_1$ of the axial mode of radiation is said to be an "optimum" pitch angle. To determine an optimum angle, the pattern, impedance and polarization characteristics of helical antennas may be compared on a diameter-spacing chart as in Fig. 7-39*b*. The three contours indicate the region of satisfactory pattern, impedance and polarization values as determined by measurements on helices of various pitch angle as a function of frequency. The axial length of the helices tested is about 1.6λ at the center frequency. The pattern contour in Fig. 7-39*b* indicates the approximate region of satisfactory patterns. A satisfactory pattern is considered to be one with a major lobe in the axial direction and with relatively small minor lobes. Inside the pattern contour, the patterns are of this form and have half-power beam widths of less than 60° and as small as 30°. Inside the impedance contour in Fig. 7-39*b* the terminal impedance is relatively constant and is nearly a pure resistance of 100 to 150 Ω. Inside the axial ratio contour, the axial ratio in the direction of the helix axis is less than 1.25. Note that all contours lie below the line for which $D_\lambda = \sqrt{2S_\lambda + 1/\pi}$. This line may be regarded as an upper limit for the beam mode. It is apparent that the frequency range $F_2 - F_1$ is small if the pitch angle is either too small or too large. A pitch angle of about 12 or 14° would appear to be "optimum" for helices about 1.6λ long at the center frequency. Since the properties of the helix change slowly in the vicinity of the optimum angle, there is nothing critical about this value. The contours are arbitrary but are suitable for a general-purpose beam antenna of moderate directivity. The exact values of the frequency limits, F_1 and F_2, are also arbitrary but are relatively well defined by the close bunching of the contours near the frequency limits.

Based on the above conclusions, I constructed a 14°, 6-turn helix and measured its properties. The helix has a diameter of 0.31λ at the center frequency (400 MHz). The diameter of the conductor is about 0.02λ. Conductor diameters of 0.005 to 0.05λ can be used with little difference in the properties of this helix in the frequency range of the beam mode.[1]

The measured patterns between 275 and 560 MHz are presented in Fig. 7-19. It is apparent that the patterns are satisfactory over a frequency range from 300 MHz ($C_\lambda = 0.73$) to 500 MHz ($C_\lambda = 1.22$).

A summary of the characteristics of this antenna are given in Fig. 7-23 in which the half-power beam width, axial ratio and standing-wave ratio are shown as a function of the helix circumference.

7-10 TABLE OF PATTERN, BEAM WIDTH, GAIN, IMPEDANCE AND AXIAL RATIO FORMULAS. Expressions developed in the preceding sections for calculating the pattern, beam width, directivity, terminal resistance and axial ratio for axial-mode helical antennas are summarized in Table 7-2. These relations apply to helices for $12° < \alpha < 15°$, $\frac{3}{4} < C_\lambda < \frac{4}{3}$ and $n > 3$ or to the more specific restrictions listed in the footnote to the table.

7-11 RADIATION FROM LINEAR PERIODIC STRUCTURES WITH TRAVELING WAVES WITH PARTICULAR REFERENCE TO THE HELIX AS A PERIODIC STRUCTURE ANTENNA. Radiation from continuous linear antennas carrying a traveling wave was discussed in Sec. 5-8. Although the helical beam antenna consists of a continuous

---

[1] Design data for a 12.5° helix are given by J. D. Kraus, "Helical Beam Antenna Design Techniques," *Communications*, **29**, September 1949.

Table 7-2 Formulas for monofilar axial-mode helical antennas

Pattern	$E = \left(\sin \frac{90^\circ}{n}\right) \frac{\sin (n\psi/2)}{\sin (\psi/2)} \cos \phi$ where $\psi = 360^\circ \left[S_\lambda(1 - \cos \phi) + \frac{1}{2n} \right]$
Beam width (half-power) See restrictions	$\text{HPBW} = \frac{52^\circ}{C_\lambda \sqrt{nS_\lambda}}$
Beam width (first nulls) See restrictions	$\text{BWFN} = \frac{115^\circ}{C_\lambda \sqrt{nS_\lambda}}$
Directivity (or gain†) See restrictions	$D = 12C_\lambda^2 nS_\lambda$
Terminal resistance See restrictions	$R = 140C_\lambda\ \Omega$ (axial feed) $R = 150/\sqrt{C_\lambda}\ \Omega$ (peripheral feed)
Axial ratio (on axis)	$\text{AR} = \frac{2n+1}{2n}$ (increased directivity)
Axial ratio (on axis)	$\text{AR} = \left| L_\lambda \left(\sin \alpha - \frac{1}{p} \right) \right|$ (p unrestricted)

n = number of turns of helix
C_λ = circumference in free-space wavelengths
S_λ = spacing between turns in free-space wavelengths
L_λ = turn length in free-space wavelengths
α = pitch angle
p = relative phase velocity
ϕ = angle with respect to helix axis
Restriction for beam width and directivity: $0.8 < C_\lambda < 1.15$; $12^\circ < \alpha < 14^\circ$; $n > 3$.
Restriction for terminal resistance: $0.8 < C_\lambda < 1.2$; $12^\circ < \alpha < 14^\circ$; $n \geq 4$.
† Assuming no losses.

conductor carrying a traveling wave, it is also a periodic structure with period equal to the turn spacing as considered in Sec. 7-5.

Now let us develop the periodic structure approach in a more general way which illustrates the relation of helical antennas to other periodic-structure (dipole) antennas.[1]

A linear array of n isotropic point sources of equal amplitude and spacing is shown in Fig. 7-40 representing a linear periodic structure carrying a traveling wave. As discussed previously the phase difference of the fields from adjacent sources as observed at a distance point is given by

$$\psi = \frac{2\pi}{\lambda_0} S \cos \phi - \delta \tag{1}$$

[1] Although the helix is a periodic structure, it is continuous. The dipole arrays are discontinuous.

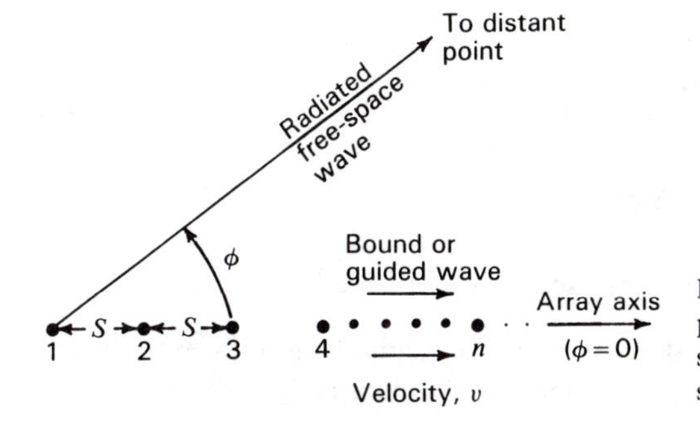

Figure 7-40 Linear array of n isotropic point sources of equal amplitude and spacing, S, representing a linear periodic structure carrying a traveling wave.

where S = spacing between sources, m
λ_0 = free-space wavelength, m
ϕ = angle between array axis and direction of distant point, rad or deg
δ = phase difference of source 2 with respect to source 1, 3 with respect to 2, etc., rad or deg

We assume that the array is fed by a wave traveling along it from left to right via a guiding structure which may, for example, be an open-wire transmission line, a waveguide or a helix.[1] The *phase constant* of the traveling wave is given by

$$\beta = \frac{2\pi}{\lambda_0 p} \text{ (rad m}^{-1}) = \frac{360^\circ}{\lambda_0 p} \text{ (deg m}^{-1}) \tag{2}$$

where λ_0 = free-space wavelength, m
$p = v/c$ = relative phase velocity, dimensionless
v = wave velocity, m s^{-1}
c = velocity of light, m s^{-1}

The *phase difference* between sources is given by

$$\delta = \frac{2\pi}{\lambda_0 p} S \text{ (rad)} = \frac{360^\circ}{\lambda_0 p} S \text{ (deg)} \tag{3}$$

where S = spacing between sources, m

In general, for the fields from the n sources to be in-phase at a distant point requires that

$$\psi = 2\pi m \qquad \text{(rad or deg)} \tag{4}$$

where m = mode number = 0, ± 1, ± 2, etc.

---

[1] We assume a uniform traveling wave. Although such a wave is approximated with a monofilar axial-mode helix, it is not necessarily realized with the dipole arrays presented in this section without the addition of suitable impedance matching networks (not shown). See Sec. 11-10 on Phased Arrays and Sec. 16-21 on Leaky Wave Antennas.

Introducing (3) and (4) into (1) yields

$$2\pi m = \frac{2\pi}{\lambda_0} S \cos \phi - \frac{2\pi}{\lambda_0 p} S \tag{5}$$

or

$$2\pi m = \beta_0 S \cos \phi - \beta S \tag{6}$$

where $\beta_0 = \dfrac{2\pi}{\lambda_0}$ = phase constant of free-space wave, rad m^{-1}

For *mode number* $m = 0$, the phase difference of the fields from adjacent sources at a distant point is zero; for $m = 1$, the phase difference is 2π; for $m = 2$, the phase difference is 4π; etc.

$\beta = \dfrac{2\pi}{\lambda_0 p}$ = phase constant of guided wave, rad m^{-1}

$\beta_0 S$ = electrical distance between sources for a free-space wave, rad

βS = electrical distance between sources for the guided wave, rad

$\beta_0 S \cos \phi$ = electrical distance between sources for a free-space wave in direction of distant point

From (6) we have

$$\beta_0 \cos \phi = \beta + \frac{2\pi m}{S} \tag{7}$$

or

$$\cos \phi = \frac{\beta}{\beta_0} + \frac{2\pi m}{\beta_0 S} = \frac{1}{p} + \frac{m}{S/\lambda_0} \tag{8}$$

Let us now consider several examples.

Example 1 Mode number $m = 0$. For different relative phase velocities p, the beam angles ϕ, as given by (8), are as tabulated:

Relative phase velocity, p	$\cos \phi$	ϕ	**Beam direction**
1 ($v = c$)	1	0°	End-fire
∞	0	90°	Broadside
-1 ($v = c$ with wave right-to-left)	-1	180°	Back-fire
<1	>1	Imaginary	No beam

For the last entry ($p < 1$), ϕ is imaginary. This implies that all of the wave energy is bound to the array (guided along it) and that there is no radiation (no beam).

Summary. For p values from $+1$ to $+\infty$ and $-\infty$ to -1, the beam swings from end-fire ($\phi = 0°$) through broadside ($\phi = 90°$) to back-fire ($\phi = 180°$). For p values between -1 and $+1$ ($-1 < p < 1$) ϕ is imaginary (no radiation). We note that these results are independent of the spacing S.

Example 2 Mode number $m = -1$. Relative phase velocity $p = 1$ ($v = c$). Fields from adjacent sources have 2π ($=360°$) phase difference at a distant point in the direction of the beam maximum. For different spacings S, the beam angles ϕ, as given by (8), are as tabulated:

Spacing S	$\cos\phi$	ϕ	Beam direction
λ_0	0	90°	Broadside†
$\lambda_0/2$	−1	180°	Back-fire‡

† For this case, there are also equal beams end-fire ($\phi = 0°$) and back-fire ($\phi = 180°$).

‡ For this case, there is an equal end-fire lobe ($\phi = 0°$).

Summary. For spacings between $\lambda_0/2$ and λ_0 the beam swings between 90 and 180°. Larger spacings are required to swing the beam to angles less than 90° but other lobes also appear.

Example 3 Mode number $m = -1$. Relative phase velocity $p = \frac{1}{2}$ (slow wave). For different spacings S, the beam angles ϕ, as given by (8), are as tabulated:

Spacing S	$\cos\phi$	ϕ	Beam direction
λ_0	1	0°, 90°, 180°	End-fire, broadside and backfire
$\lambda_0/2$	0	90°	Broadside
$\lambda_0/3$	−1	180°	Back-fire

Summary. For spacings between $\lambda_0/3$ and λ_0 the beam swings from back-fire (180°) through broadside (90°) to end-fire (0°), but for $S = \lambda_0$ broadside and back-fire lobes also appear. For spacings greater than λ_0 or less than $\lambda_0/3$, ϕ is imaginary (no beam).

Example 4 Mode number $m = -1$. Relative phase velocity $p = \frac{1}{5}$ (slow wave). For different spacings S, the beam angles ϕ, as given by (8), are as tabulated:

Spacing S	$\cos\phi$	ϕ	Beam direction
$\lambda_0/4$	1	0°	End-fire
$\lambda_0/5$	0	90°	Broadside
$\lambda_0/6$	−1	180°	Back-fire

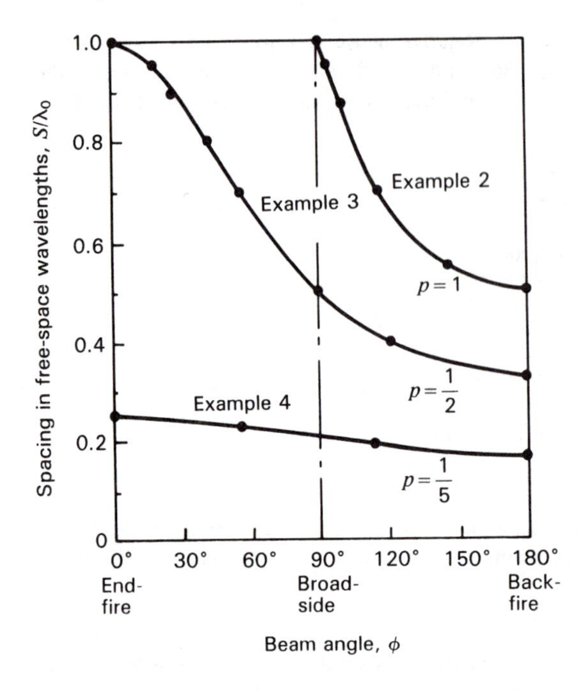

Figure 7-41 Relation of spacing and beam angle for linear arrays with traveling waves of relative phase velocities $p = 1$, $\frac{1}{2}$ and $\frac{1}{5}$ (Examples 2, 3 and 4) all with mode number $m = -1$.

Summary. For spacings between $\lambda_0/4$ and $\lambda_0/6$ the beam swings from end-fire, through broadside to back-fire. For spacings less than $\lambda_0/6$ or more than $\lambda_0/4$, ϕ is imaginary (no beam).

The results of Examples 2, 3 and 4 are shown graphically in Fig. 7-41.

Another way of analyzing an array, and periodic structures in general, is to plot the electrical spacing $\beta_0 S$ of the free-space wave (as ordinate) versus the electrical spacing βS of the guided wave traveling along the array (as abscissa). Dividing both coordinates by 2π, we obtain S/λ_0 as ordinate and $S/(p\lambda_0) = S/\lambda$ as abscissa (where $\lambda = p\lambda_0$ = wavelength on the array). This type of S-S diagram[1] is presented in Fig. 7-42, illustrating the three arrays of Examples 2, 3 and 4 (shown also in the S-ϕ diagram of Fig. 7-41).

For a relative phase velocity $p = 1$, $S/\lambda = S/\lambda_0$, and the array operates along the $p = 1$ line (at 45° to the axes). Back-fire occurs at $S/\lambda_0 = \frac{1}{2}$ and broadside at $S/\lambda_0 = 1$ (Example 2).

For a relative phase velocity $p = \frac{1}{2}$, the array operates along the $p = \frac{1}{2}$ line with back-fire at $S/\lambda_0 = \frac{1}{3}$, broadside at $S/\lambda_0 = \frac{1}{2}$ and end-fire at $S/\lambda_0 = 1$ at right edge of diagram (Example 3).

[1] Some authors use k for β_0 and refer to the graph as a k-β diagram, also called a *Brillouin diagram* after Leon Brillouin, *Wave Propagation in Periodic Structures*, McGraw-Hill, 1946.

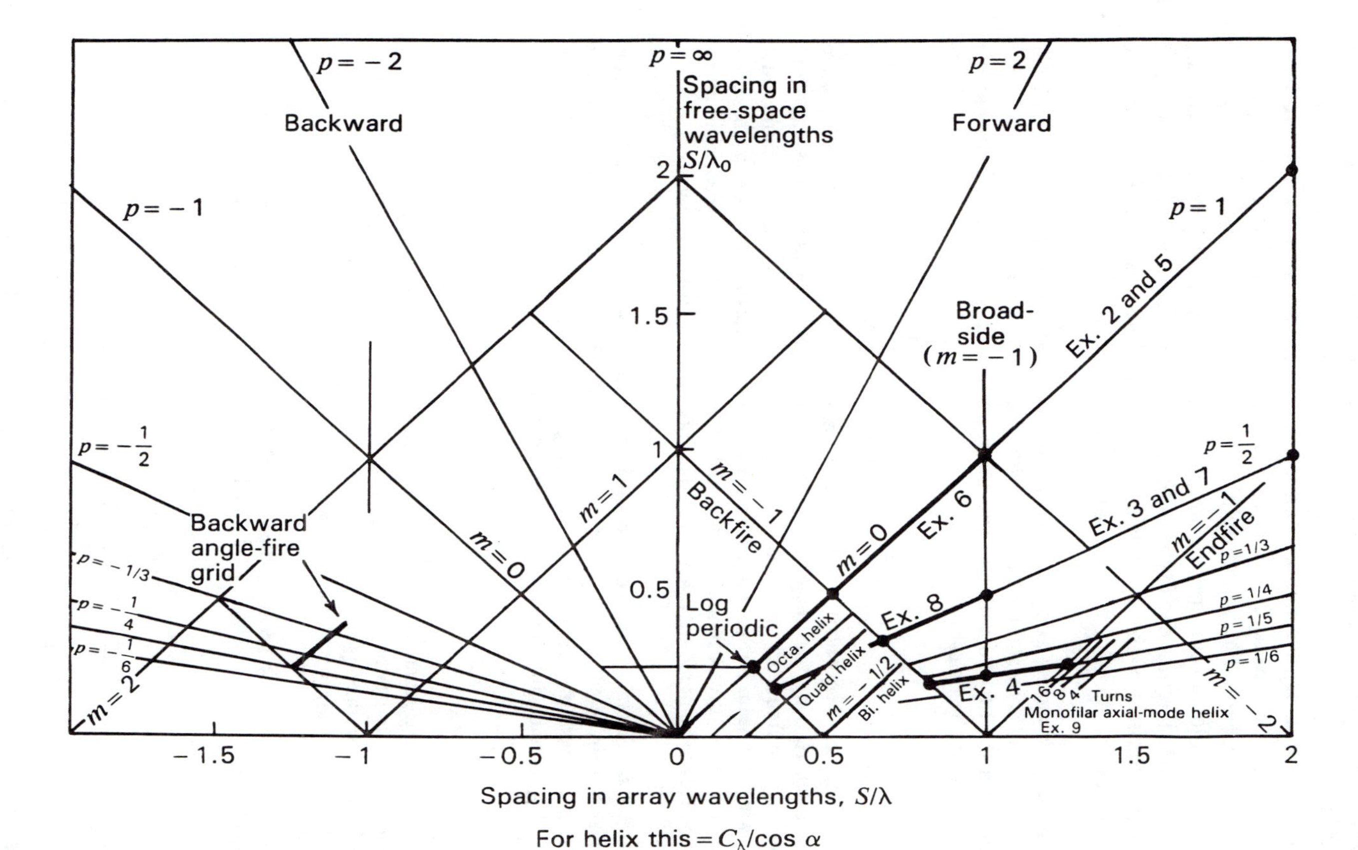

Figure 7-42 *S-S* diagram for linear traveling-wave antennas showing regions of operation for different types of arrays including Examples 2 through 9. Note that for a helix, S/λ here corresponds to the turn length L_λ.

For a relative phase velocity $p = \frac{1}{5}$, the array operates along the $p = \frac{1}{5}$ line with back-fire at $S/\lambda_0 = \frac{1}{6}$, broadside at $S/\lambda_0 = \frac{1}{5}$ and end-fire at $S/\lambda_0 = \frac{1}{4}$ (Example 4). For a higher mode number $m = -2$, the array again produces beams swinging from back-fire at $S/\lambda_0 = \frac{1}{3}$, through broadside at $S/\lambda_0 = \frac{2}{5}$ to end-fire at $S/\lambda_0 = \frac{1}{2}$ (off diagram at right).

Let us consider several more examples of traveling-wave periodic-structure arrays. On an S-S diagram each type of array occupies a unique location or niche which is characteristic of the antenna's behavior. Differences between arrays are clearly evident from their locations on the diagram.

Example 5 Scanning array of dipoles with $S = \lambda_0$ at center frequency (Fig. 7-43). Beam scanning is by shifting frequency. Physical element spacing is constant. Array is fed from the left end with a 2-wire transmission line ($p = 1$). The mode number $m = -1$, so (8) becomes

$$\cos \phi = \frac{1}{p} - \frac{1}{S/\lambda_0} = 1 - \frac{1}{S/\lambda_0} \tag{9}$$

At the center frequency $S = \lambda_0$ and from (9) $\cos \phi = 0$ and $\phi = 90°$ (beam broadside). Halving the frequency makes $S = \lambda_0/2$ and $\phi = 180°$ (beam back-fire). Doubling the frequency makes $S = 2\lambda_0$ and $\phi = 60°$ (only 30° beyond broadside). To swing the beam further toward end-fire requires a further increase in frequency. The position of this scanning array is shown in Fig. 7-42 (line labeled Examples 2 and 5). It is to be noted that there are other lobes present not given by (9).

Example 6 Scanning array of alternately reversed dipoles with $S = \lambda_0/2$ at center frequency (Fig. 7-44). Beam scanning is by shifting frequency. Physical element spacing is constant. Array is fed from the left end with a 2-wire transmission line ($p = 1$). The mode number $m = -\frac{1}{2}$, so (8) becomes

$$\cos \phi = 1 - \frac{1}{2S/\lambda_0} \tag{10}$$

At the center frequency $S = \lambda_0/2$ and from (10) $\cos \phi = 0$ and $\phi = 90°$ (beam broadside). Halving the frequency makes $S = \lambda_0/4$ and $\phi = 180°$ (beam back-fire). Doubling the frequency makes $S = \lambda_0$, swinging the beam to $\phi = 60°$. The position of this scanning array is shown in Fig. 7-42.

Example 7 Scanning array of dipoles with $S = \lambda_0/2$ at center frequency with slow wave ($p = \frac{1}{2}$) (Fig. 7-45). Beam scanning is by shifting frequency. Physical element

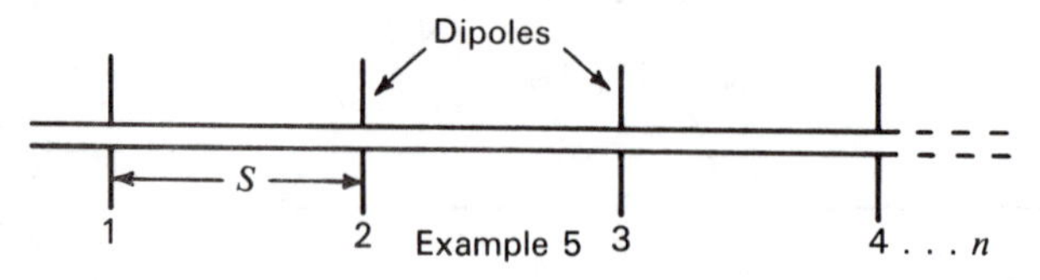

Figure 7-43 Scanning array of dipoles with $S = \lambda_0$ at center frequency, relative phase velocity $p = 1$ ($v = c$) and mode number $m = -1$. The beam angle ϕ (with respect to dipole 1) is outward from the page. (Example 5.)

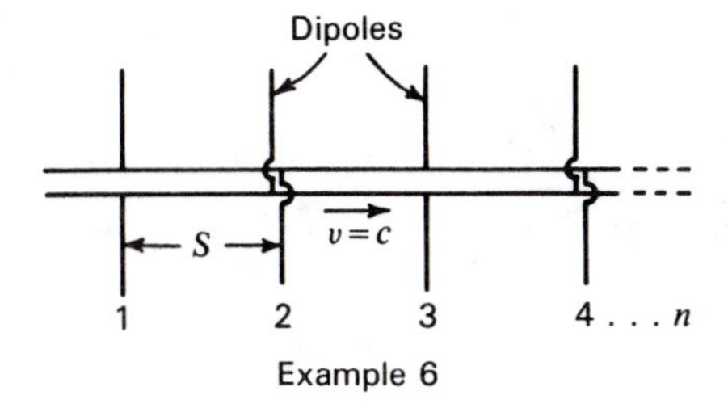

Figure 7-44 Scanning array of alternately reversed dipoles with $S = \lambda_0/2$ at center frequency, relative phase velocity $p = 1$ and mode number $m = -\frac{1}{2}$. The beam angle ϕ (with respect to dipole 1) is outward from the page. (Example 6.)

spacing is constant. Array is fed from the left end by a 2-wire transmission line with line length $2S$ between dipoles so that $p = \frac{1}{2}$. The mode number $m = -1$, so (8) becomes

$$\cos \phi = 2 - \frac{1}{S/\lambda_0} \tag{11}$$

At the center frequency $S = \lambda_0/2$ and from (11) $\phi = 90°$ (beam broadside). Reducing the frequency so $S = \lambda_0/3$ makes $\phi = 180°$ (beam back-fire). Doubling the frequency makes $S = \lambda_0$ and $\phi = 0°$ (beam end-fire). The position of this scanning array is shown in Fig. 7-42 (line labeled Examples 3 and 7).

Example 8 Scanning array of alternately reversed dipoles with $S = \lambda_0/4$ at center frequency with slow wave ($p = \frac{1}{2}$) (Fig. 7-46). Beam scanning is by shifting frequency. Physical element spacing is constant. Array is fed from the left end by a 2-wire transmission line with line length $2S$ between dipoles so that $p = \frac{1}{2}$. The mode number $m = -\frac{1}{2}$, so (8) becomes

$$\cos \phi = 2 - \frac{1}{2S/\lambda_0} \tag{12}$$

At the center frequency $S = \lambda_0/4$ and from (12) $\phi = 90°$ (beam broadside). Decreasing the frequency to $\frac{1}{6}$ makes $S = \lambda_0/6$ and $\phi = 180°$ (beam back-fire). Doubling the frequency makes $S = \lambda_0/2$ and $\phi = 0°$ (beam end-fire). The position of this scanning array is also shown in Fig. 7-42.

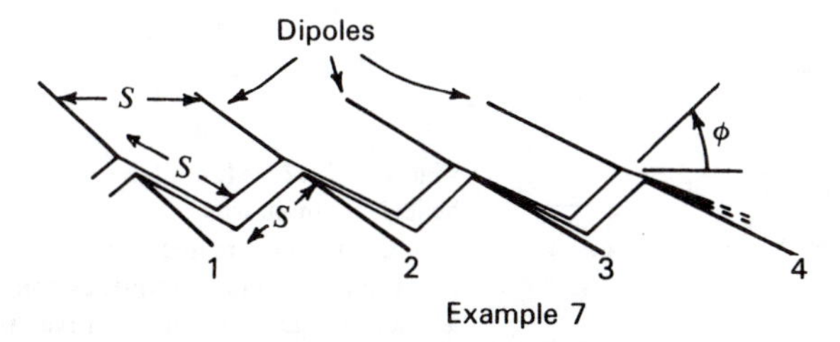

Figure 7-45 Scanning array of dipoles with $S = \lambda_0/2$ at center frequency, relative phase velocity $p = \frac{1}{2}$ (slow wave) and mode number $m = -1$. (Example 7.)

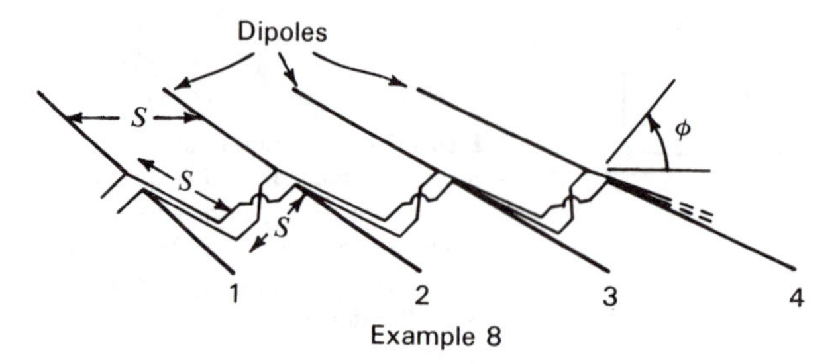

Figure 7-46 Scanning array of alternately reversed dipoles with $S = \lambda_0/4$ at center frequency, relative phase velocity $p = \frac{1}{2}$ (slow wave) and mode number $m = -\frac{1}{2}$. (Example 8.)

Now let us consider the monofilar helical antenna in comparison to the above examples of other periodic structure arrays with traveling waves.

Example 9 Monofilar axial-mode helical antenna with turn-spacing $S = \lambda_0/4$ and circumference $C = \lambda_0$ at center frequency (Fig. 7-47). As discussed in Sec. 7-5, the monofilar axial-mode helical antenna operates in the increased-directivity condition resulting in a supergain end-fire beam ($\phi = 0°$). The mode number $m = -1$ and from (5) we have for $\phi = 0°$ that

$$-2\pi - \frac{\pi}{n} = \frac{2\pi}{\lambda_0} S - \frac{2\pi}{\lambda_0} \frac{S}{p} \tag{13}$$

where n = number of turns

$p = (v/c) \sin \alpha$ = relative phase velocity of wave *in direction of helix axis* (not along the helical conductor as in Sec. 7-5)

α = pitch angle

Rearranging (13) we obtain

$$p = \frac{1}{1 + \dfrac{2n+1}{2n} \dfrac{1}{S/\lambda_0}} \tag{14}$$

For large n,

$$p \simeq \frac{1}{1 + [1/(S/\lambda_0)]} \tag{15}$$

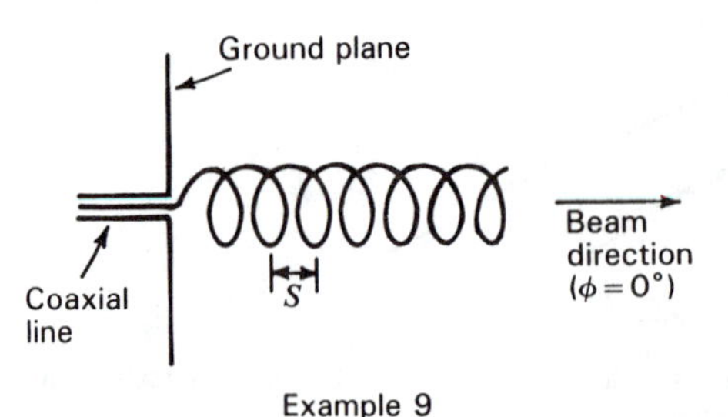

Figure 7-47 Monofilar axial-mode helical antenna which locks on end-fire ($\phi = 0°$) with increased directivity over more than an octave frequency range by an automatic shift in relative phase velocity p (see Fig. 7-42). The mode number $m = -1$.

Thus, end-fire ($\phi = 0°$) with $S = \lambda_0/4$ requires that $p = 0.20$. If the frequency decreases so that $S = \lambda_0/5$ (a 25 percent change in frequency) we have from (5) that

$$\cos \phi = \frac{1}{p} - \frac{1}{S/\lambda_0} = \frac{1}{0.20} - 5 = 0 \tag{16}$$

making $\phi = 90°$ (beam broadside). However, with frequency change the beam does *not* swing broadside but remains locked on end-fire ($\phi = 0°$) because *the phase velocity changes automatically by just the right amount to not only compensate for the frequency change but also to provide increased directivity and supergain.* This is one of the very remarkable properties of the monofilar axial-mode helical antenna.

The increased-directivity condition involves not only the turn spacing S but also the number of turns n. However, for large n the difference in p for the two end-fire conditions is small. Thus, for large n we find from (15) that p values range from about 0.25 for $S = \lambda_0/3$ through 0.20 for $S = \lambda_0/4$ to 0.167 for $S = \lambda_0/5$, locking the beam on end-fire with supergain over a frequency range of about 2 to 1, which is in marked contrast to the beam-swinging of the scanning arrays discussed above. The position of the monofilar helical antenna over a 2 to 1 frequency range is shown in Fig. 7-42 for $n = 4$, 8 and 16 turns. For very large n, the position moves along the end-fire line which intersects $S/\lambda = 1$ on the $S/\lambda_0 = 0$ axis. We note that while other arrays move along constant p lines as the frequency changes, the *monofilar axial-mode helical antenna moves along a constant beam-angle (end-fire) line, cutting across lines of constant p value.*

Another remarkable property and great advantage of the monofilar helical antenna is that the input impedance is an almost constant resistance over an octave bandwidth, the resistance being easily set at any convenient value from 50 to 150 Ω. This is in contrast to the large impedance fluctuations of the above dipole arrays with change in frequency.

7-12 ARRAYS OF MONOFILAR AXIAL-MODE HELICAL ANTENNAS.

With arrays of monofilar axial-mode helical antennas the designer must strike a balance between the number and length of helices needed to achieve a desired gain. As discussed in Sec. 4-8 the choice is between more lower-gain antennas and fewer higher-gain antennas appropriately spaced. As an illustration consider the following problem.

Example. Design a circularly polarized antenna using one or more end-fire elements to produce a gain of 24 dB for operation at a given wavelength λ.

Solution. The highest end-fire gain is obtained with the increased-directivity condition which is automatic with monofilar axial-mode helical antennas. From (7-4-7) for $\alpha = 12.7°$ and $C_\lambda = 1.05$, the required length of a single helix is

$$L = nS = \frac{252}{12 \times 1.05^2} = 19\lambda \tag{1}$$

requiring an 80-turn helix (Fig. 7-48*a*).

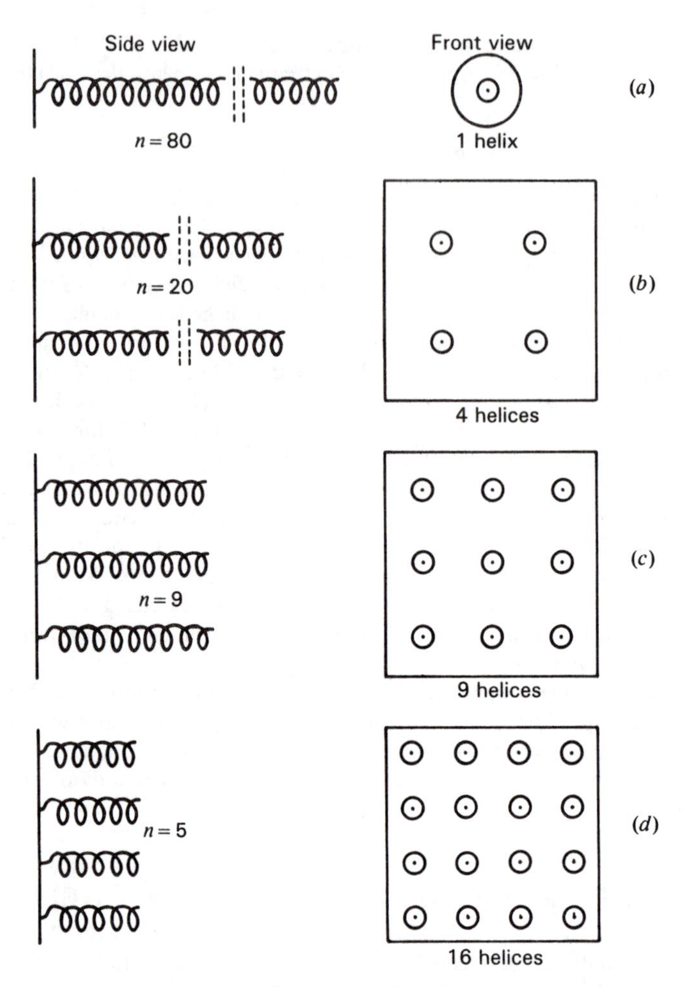

Figure 7-48 Single monofilar axial-mode helix with 80 turns (*a*) compared with an array of four 20-turn helices (*b*), an array of nine 9-turn helices (*c*) and sixteen 5-turn helices (*d*) for worked example. All have 24 dB gain. Note also that the product of the number of helices and number of turns for each array equals 80 (± 1). See Sec. 4-8.

A more compact configuration results if four 20-turn helices are used in a broadside array. Assuming uniform aperture distribution, the effective aperture of each helix is

$$A_e = \frac{D\lambda^2}{4\pi} = \frac{63\lambda^2}{4\pi} = 5.0\lambda^2 \tag{2}$$

Assuming a square aperture, the side length is 2.24λ ($=\sqrt{5.0}$). With each helix placed at the center of its aperture area, the spacing between helices is 2.24λ (Fig. 7-48*b*).

A third configuration results if nine 9-turn helices are used in a broadside array (Fig. 7-48*c*).

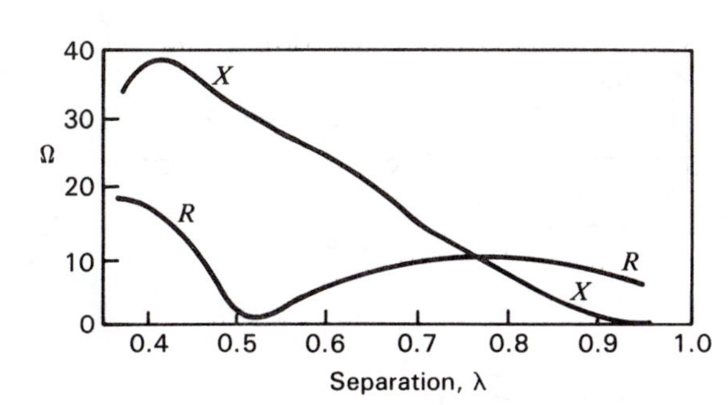

Figure 7-49 Resistive (R) and reactive (X) components of the mutual impedance of a pair of same-handed 8-turn monofilar axial-mode helical antennas of 12° pitch angle as a function of the separation distance (center-to-center) in wavelengths. Helix circumference $C_\lambda = 1$. Conductor diameter is 0.016λ. Self-impedance $Z_{11} \simeq 140 + j0\ \Omega$. (*After E. A. Blasi, "Theory and Application of the Radiation Mutual Coupling Factor," M.S. thesis, Electrical Engineering Department, Ohio State University, 1952.*)

A fourth possible configuration results if sixteen 5-turn helices are used in a broadside array with a spacing of 1.12λ between helices (Fig. 7-48*d*).

Which of the above configurations should be used? The decision will depend on considerations of support structure and feed connections. The single helix has a single feed point and a small ground plane but is very long. The other configurations have larger ground planes but are more compact. The 4- and 16-helix configurations have the advantage that the helices can be fed by a symmetrical corporate structure. Also the 16-helix configuration can be operated as a phased array.

With the multiple-helix arrays the mutual impedance of adjacent helices is a consideration. Figure 7-49 shows the resistive and reactive components of the mutual impedance of a pair of same-handed 8-turn monofilar axial-mode helical antennas as a function of the separation distance in wavelengths ($C_\lambda = 1$, $\alpha = 12°$) as measured by Blasi.[1] At spacings of a wavelength or more, as is typical in helix arrays, the mutual impedance is only a few percent or less of the helix self-impedance (140 Ω resistive). Thus, in designing the feed connections for a helix array the effect of mutual impedance can often be neglected without significant consequences. As examples, let us consider the feed systems for two helix arrays, one with 4 helices and the other with 96 helices.

7-12a Array of 4 Monofilar Axial-Mode Helical Antennas. This array, shown in Fig. 7-50*a*, which I constructed in 1947, has four 6-turn, 14° pitch angle helices mounted with 1.5λ spacing on a $2.5 \times 2.5\lambda$ square ground plane.[2] The helices are fed axially through an insulated fitting in the ground plane. Each feed

[1] E. A. Blasi, "Theory and Application of the Radiation Mutual Coupling Factor," M.S. thesis, Electrical Engineering Department, Ohio State University, 1952.

[2] J. D. Kraus, "Helical Beam Antennas for Wide-Band Applications," *Proc. IRE*, **36**, 1236–1242, October 1948.

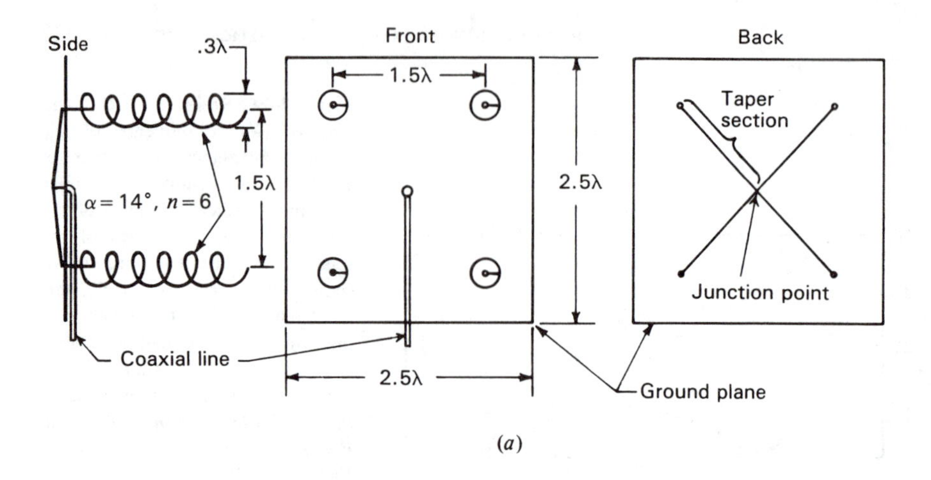

(a)

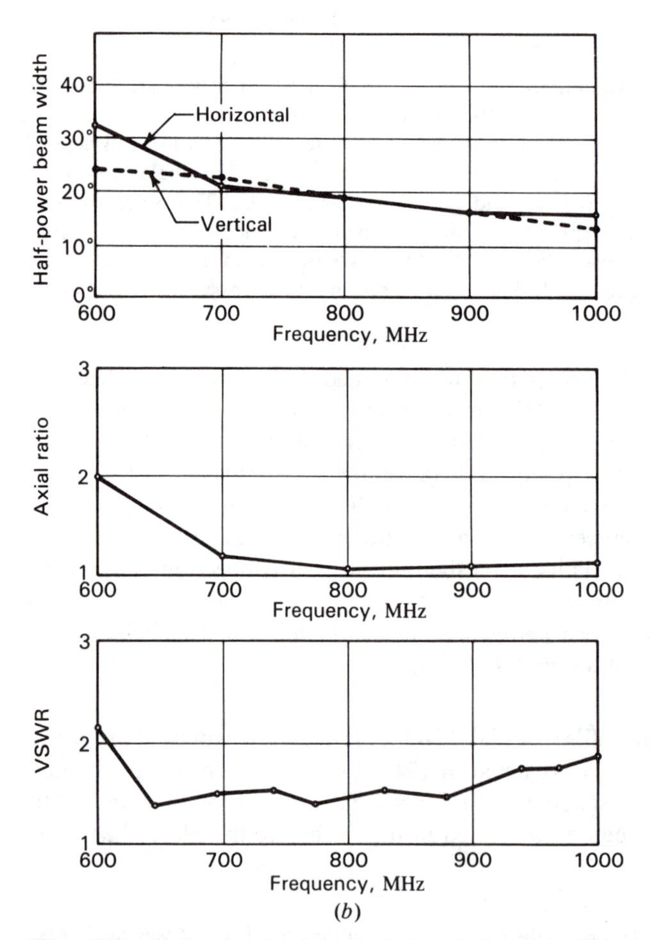

(b)

Figure 7-50 (a) Constructional details for broadside array of four 6-turn, 14° monofilar axial-mode helical antennas. Dimensions are in wavelengths at the center frequency. (b) Measured performance of 4-helix array of (a) showing beam widths, axial ratio and VSWR on a 53-Ω line as a function of frequency. (*After J. D. Kraus, "Helical Beam Antennas for Wide-Band Applications,"* Proc. IRE, **36**, *1236–1242, October 1948.*)

point is connected on the back side of the ground plane to a junction point at the center of the ground plane. Each conductor acts as a single-wire versus ground-plane transmission line with a spacing between wire and ground plane which tapers gradually so that the approximately 140 Ω at each helix is transformed to 200 Ω at the junction. The four 200-Ω lines in parallel yield 50 Ω at the junction which is fed through an insulated connector to a 50-Ω coaxial fitting on the front (helix) side of the ground plane. Since the taper sections are about 1λ long, the arrangement provides a low VSWR on the 50-Ω line connected to the junction over a wide bandwidth. Beam width, axial ratio and VSWR performance of the array are presented in Fig. 7-50*b*. The array gain at the center frequency (800 MHz) is about 18.5 dB and at 1000 MHz about 21.5 dB.

7-12b Array of 96 Monofilar Axial-Mode Helical Antennas. This array, shown in Fig. 7-4, which I designed and constructed in 1951,[1] has 96 11-turn 12.5° pitch angle helices mounted on a tiltable flat ground plane 40λ long by 5λ wide for operation at a center frequency of 250 MHz. Each helix is fed by a 150-Ω coaxial cable. Equal-length cables from each helix of one bay or group of 12 helices are connected in parallel to one end of a 2λ long tapered transition section which transforms the 12.5-Ω (=150/12) resistive impedance to 50 Ω. Equal-length 50-Ω coaxial cables then connect the transition section of each bay to a central location resulting in a uniform in-phase aperture distribution with low VSWR over a wide bandwidth. The array produces a gain of about 35 dB at the center frequency of 250 MHz ($\lambda = 1.2$ m) and increased gain at higher frequencies. At 250 MHz the beam is fan shaped with half-power beam widths of 1 by 8°.

7-13 THE MONOFILAR AXIAL-MODE HELIX AS A PARASITIC ELEMENT (see Fig. 7-51)

Helix-helix (Fig. 7-51*a*). If the conductor of a 6-turn monofilar axial-mode helical antenna is cut at the end of the second turn, the antenna continues to operate with the first 2 turns launching the wave and the remaining 4 turns acting as a parasitic director.

Polyrod-helix (Fig. 7-51*b*). By slipping a parasitic helix of several turns over a linearly polarized polyrod antenna, it becomes a circularly polarized antenna.

Horn-helix (Fig. 7-51*c*). By placing a parasitic helix of several turns in the throat of a linearly polarized pyramidal horn antenna without touching the horn walls, the horn radiation becomes circularly polarized.

---

[1] J. D. Kraus, "The Ohio State Radio Telescope," *Sky and Tel.*, **12**, 157–159, April 1953.

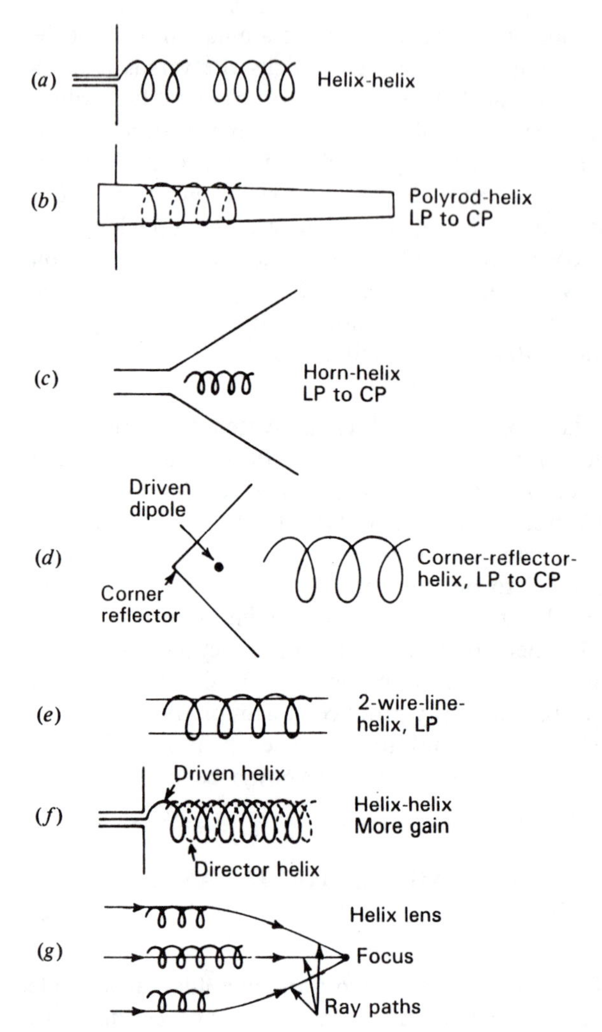

Figure 7-51 The monofilar axial-mode helical antenna in 7 applications as a parasitic element.

Corner-helix (Fig. 7-51*d*). A parasitic helix in front of a corner-reflector antenna results in a circularly polarized antenna.

The 2-wire-line-helix (Fig. 7-51*e*). If a parasitic helix of many turns is slipped over a 2-wire transmission line without touching it (helix diameter slightly greater than line spacing), it is reported that the combination becomes a linearly polarized end-fire antenna with **E** parallel to the plane of the 2-wire line.[1]

[1] G. Broussaud and E. Spitz, "Endfire Antennas," *Proc. IRE*, **49**, 515–516, February 1961.

Helix-helix (Fig. 7-51*f*). If a parasitic helix is wound between the turns of a driven monofilar axial-mode helical antenna without touching it (diameters the same), Nakano *et al.* report that the combination gives an increased gain of about 1 dB without an increase in the axial length of the antenna. The increased gain occurs for helices of any number of turns between 8 and 20. The parasitic helix may be regarded as a director for the driven helix.[1]

Helix lens (Fig. 7-51*g*). A monofilar axial-mode helical antenna (or, for that matter, any end-fire antenna) acts as a lens. An array of parasitic helices of appropriate length arranged in a broadside configuration can operate as a large aperture-lens antenna (see Fig. 14-22).

7-14 THE MONOFILAR AXIAL-MODE HELICAL ANTENNA AS A PHASE AND FREQUENCY SHIFTER.

The monofilar axial-mode helical antenna is a simple, beautiful device for changing phase or frequency. Thus, if the monofilar axial-mode helical antenna in Fig. 7-52 is transmitting at a frequency F, rotating the helix on its axis by 90° will advance the phase of the radiated wave by 90° (or retard it, depending on the direction of rotation). Rotating the helix continuously f times per second results in radiation at a frequency $F \pm f$ depending on the direction of rotation (see also Sec. 18-10).

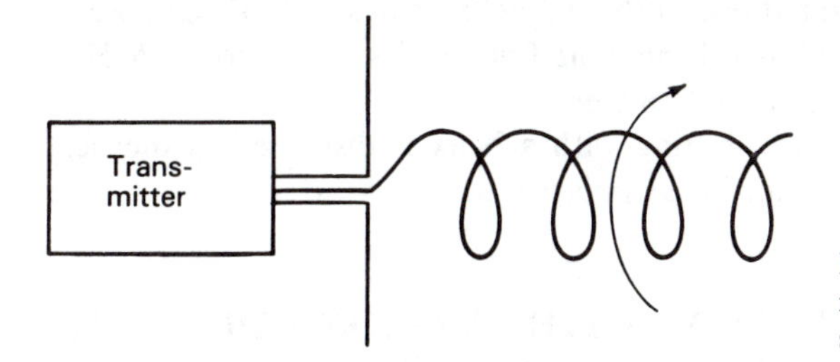

Figure 7-52 The monofilar axial-mode helical antenna as a phase-shifting device.

As an application, consider the 3-helix lobe-sweeping antenna of Fig. 7-53. All helices are of the same hand. By rotating helix 1 clockwise and helix 3 counterclockwise with helix 2 at the center stationary, a continuously swept lobe is obtained as suggested in the figure. In operation a small lobe appears about 30° to the left, then grows in amplitude while sweeping to the right, reaching a maximum at 0° (at right angles to the array). Sweeping further to the right, the lobe decreases to a small amplitude at an angle of about 30° and simultaneously a small lobe appears at 30° to the left and the process is repeated, giving a continuously sweeping lobe (left to right) which crosses the 0° direction n times per

---

[1] H. Nakano, T. Yamane, J. Yamauchi and Y. Samada, "Helical Antenna with Increased Power Gain," *IEEE AP-S Int. Symp.*, **1**, 417–420, 1984.

H. Nakano, Y. Samada and J. Yamauchi, "Axial Mode Helical Antennas," *IEEE Trans. Ants. Prop.*, **AP-34**, 1143–1148, September 1986.

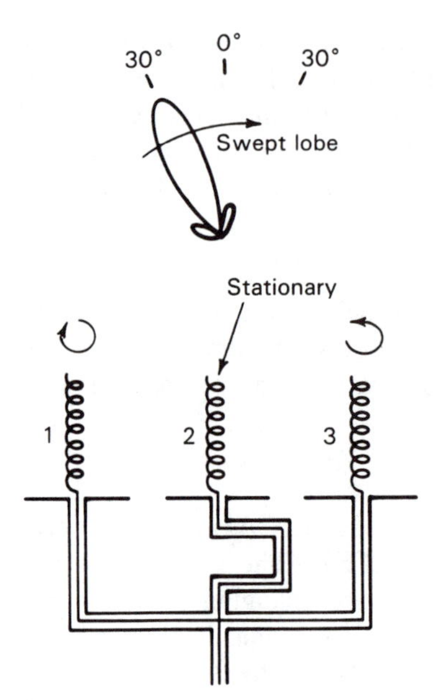

Figure 7-53 Array of 3 right-handed monofilar axial-mode helical antennas with outer 2 rotating in opposite directions to produce a continuously sweeping lobe.

minute for a helix rotation speed of n revolutions per minute. By using more helices, the beam width of the swept lobe can be made arbitrarily small.

I designed, built and operated one of these 3 helix arrays in 1957 for observing radio emissions from the planet Jupiter at frequencies of 25 to 35 MHz.[1] Each helix had 3 turns and was 3 m in diameter.

Another application of phase-shifting with a helix is discussed in the next section in connection with helices for linear polarization.

7-15 LINEAR POLARIZATION WITH MONOFILAR AXIAL-MODE HELICAL ANTENNAS. If two monofilar axial-mode helical antennas are mounted side by side and fed equal power, the radiation on axis will be linearly polarized provided the helices are of opposite hand but otherwise identical (Fig. 7-54*a*). With switch 1 right, switch 2 left and switch 3 up, polarization is linear. Rotating one of the helices on its axis 90° rotates the plane of linear polarization by 45°. Rotating one helix through 180° rotates the plane of linear polarization 90°. With switches 1 and 3 left, as in the figure, the polarization is LCP (left-handed circular polarization). With switches 2 and 3 right, the polarization is RCP (right-handed circular polarization). Thus, the two helices can provide either left or right circular polarization or any plane of linear polarization.

---

[1] J. D. Kraus, "Planetary and Solar Radio Emission at 11 Meters Wavelength," *Proc. IRE*, **46**, 266–274, January 1958.

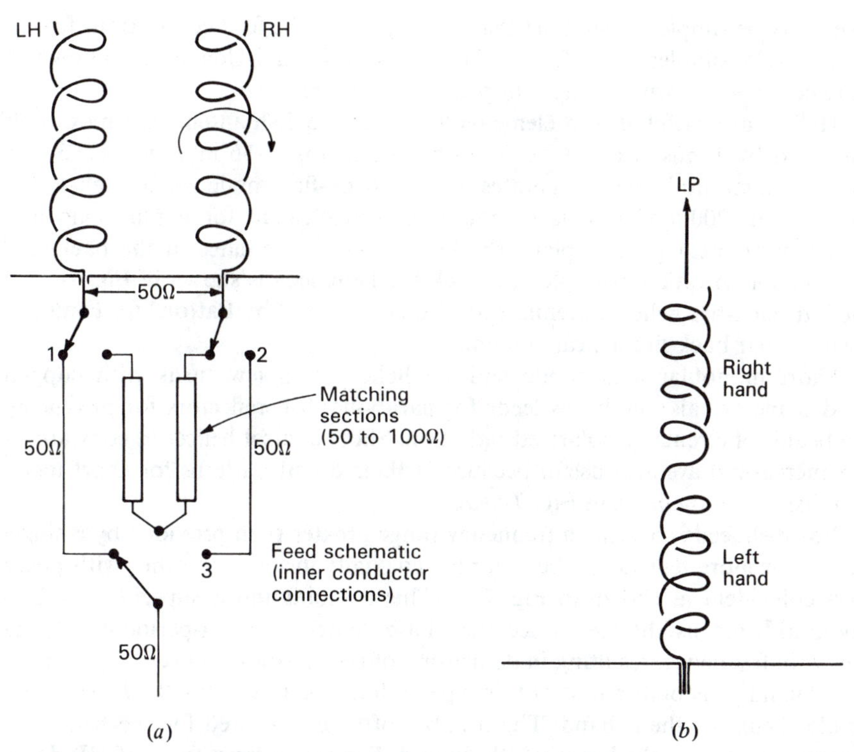

Figure 7-54 (*a*) Arrangement for producing left circular polarization (LCP), right circular polarization (RCP) or any plane of linear polarization (LP). (*b*) Two helices of opposite hand in series for producing linear polarization (LP).

Another method of obtaining linear polarization is to connect a left- and a right-handed helix in series as in Fig. 7-54*b*.

A third method has already been discussed in Sec. 7-13 (2-wire-line-helix).

Elliptical polarization approaching linear polarization can be obtained by flattening a helix so that its cross section is elliptical instead of circular.

7-16 MONOFILAR AXIAL-MODE HELICAL ANTENNAS AS FEEDS. Figure 7-55 shows a driven helix feeding an array of crossed dipoles acting as directors for producing circular polarization. Although this arrangement has less gain and bandwidth than a full helix of the same length the crossed

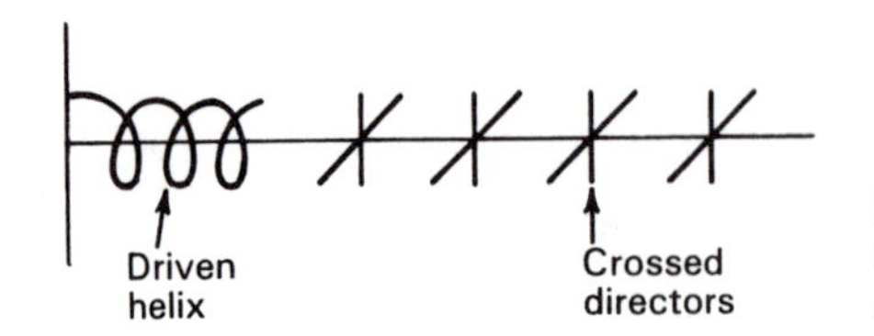

Figure 7-55 Monofilar axial-mode helical antenna as feed for an end-fire array of crossed dipoles.

dipoles may be simpler to support than a long helix. The feed connections for the helix are also simpler than for a pair of crossed Yagi-Uda antennas, which require equal power with voltages in phase quadrature.

Helices are useful as feed elements for parabolic dish antennas. An example constructed by Johnson and Cotton[1] is shown in Fig. 7-56 in which a 3.5-turn monofilar axial-mode helix operates in the back-fire mode as a high-power unpressurized (200 kW) circularly polarized feed element for a parabolic dish reflector. Without a ground plane the helix naturally radiates in the backward axial direction. Another example of a back-fire helix feed is shown in Fig. 7-6.[2] A monofilar back-fire helical antenna was also constructed by Patton[3] for comparison with bifilar back-fire helical antennas.

Short monofilar axial-mode end-fire helices of a few turns with cupped ground plane are also useful as feeds for parabolic dish reflectors for producing sharp beams of circularly polarized radiation. Short conical helices (α constant, D and S increasing) are also useful because of their broad patterns for short focal-length dishes (see the helix in Fig. 7-59*a*).

For dish feeds covering a frequency range greater than provided by a single helix, two or more helices can be mounted coaxially inside each other with phase centers coincident as shown in Fig. 7-57. This combination is superior to a log-periodic antenna as the feed since the phase center of a log-periodic antenna moves with frequency, resulting in defocusing of the parabolic reflector system.

Holland[4] has built a feed of this type with a larger helix for the L band and a smaller helix for the S band. The number of turns required for the helix feed antennas depends on the beam width desired. For the pattern to be 10 dB down at the edge of the parabolic dish reflector, the required number of turns is approximately given by

$$n \simeq \frac{8400}{\phi^2 S_\lambda} \tag{1}$$

[1] R. C. Johnson and R. B. Cotton, "A Backfire Helical Feed," Georgia Institute of Technology, Engineering Experimental Station Report, 1982.

[2] H. E. King and J. L. Wong, "Antenna System for the FleetSatCom Satellites," *IEEE International Symposium on Antennas and Propagation*, pp. 349–352, 1977.

[3] W. T. Patton, "The Backfire Bifilar Helical Antenna," Ph.D. thesis, University of Illinois, 1963.

[4] J. Holland, "Multiple Feed Antenna Covers *L*, *S* and *C*-Band Segments," *Microwave J.*, 82–85, October 1981.

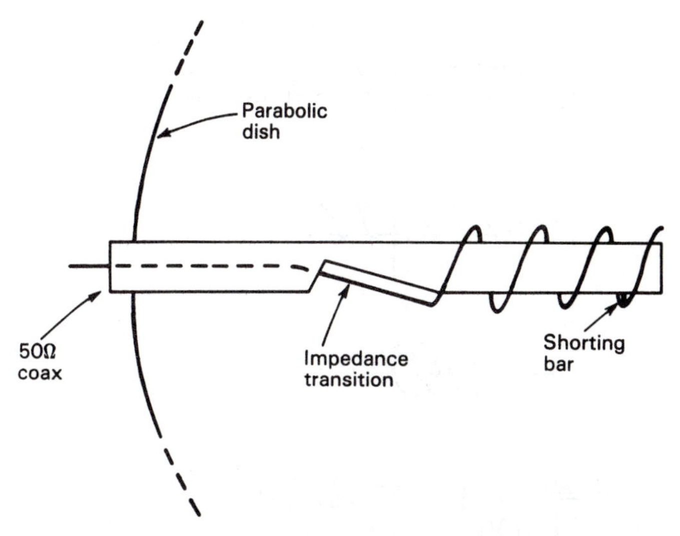

Figure 7-56 Short back-fire monofilar axial-mode helix as high-power (200 kW) circularly polarized feed for a parabolic dish antenna. (*After R. C. Johnson and R. B. Cotton, "A Backfire Helical Feed," Georgia Institute of Technology, Engineering Experimental Station Report, 1982.*)

where ϕ = 10 dB beam width
S_λ = turn spacing in wavelengths

Thus, if $S_\lambda = 0.21$ ($\alpha = 12°$) and the required value of $\phi = 115°$ we have from (1) that $n = 3$.

To reduce mutual coupling of the helices, Holland placed the peripheral feed points of the two helices on opposite sides of the axis, as suggested in Fig. 7-57, obtaining 2-port fixed-phase-center operation over a 5 to 1 bandwidth.

7-17 TAPERED AND OTHER FORMS OF AXIAL-MODE HELICAL ANTENNAS. In this section a number of variants of the uniform (constant diameter, constant pitch) monofilar axial-mode helical antenna are discussed. Some of these forms are shown in Figs. 7-58, 7-59 and 7-60 which are reproduced here without changes from the first edition of *Antennas* (1950). In

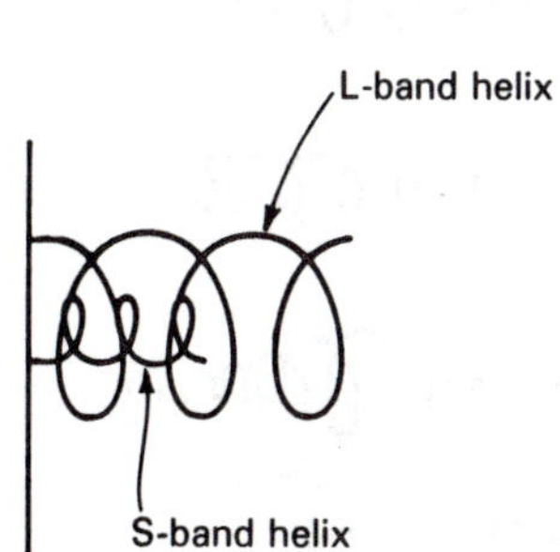

Figure 7-57 Coaxially-mounted peripherally-fed monofilar axial-mode helical antennas of same hand as parabolic dish feeds with same stationary phase centers for covering a 5 to 1 frequency range. (*After J. Holland, "Multiple Feed Antenna Covers L, S and C-Band Segments,"* Microwave J., *82–85, October 1981.*)

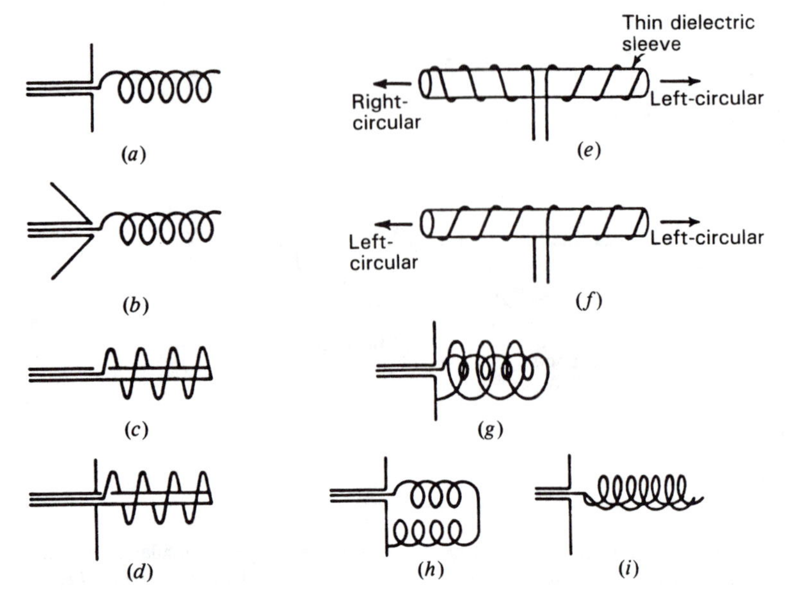

Figure 7-58 Axial-mode helices showing various constructional and feed arrangements. (*After Kraus.*)

Fig. 7-58 we recognize in (*a*) a uniform helix with ground plane and in (*c*) a uniform helix without ground plane (the same configuration as in Fig. 7-56 for a back-fire feed). The double winding in (*i*) is similar to the one of Nakano *et al.* (Fig. 7-51*f*) except that both windings in (*i*) are driven while in Nakano *et al.*'s one is parasitic.

Figure 7-59 shows 9 forms of tapered monofilar axial-mode helical antennas grouped into 3 classes: (1) pitch angle α constant but turn spacing S and diameter D variable, (2) diameter D constant but pitch angle α and turn spacing S

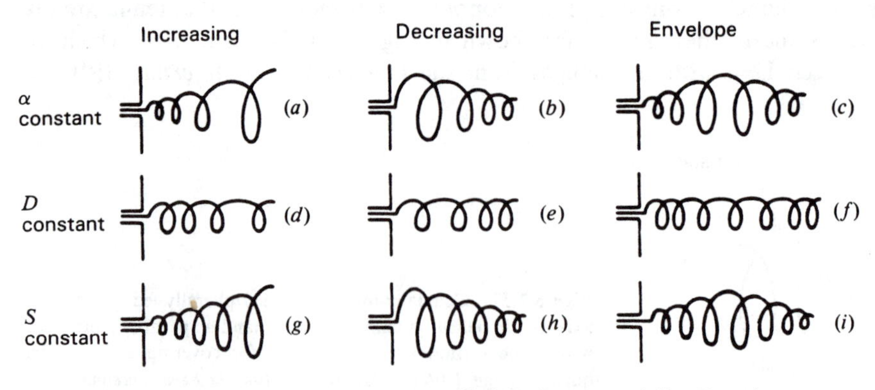

Figure 7-59 Types of tapered monofilar axial-mode helical antennas. (*After Kraus.*)

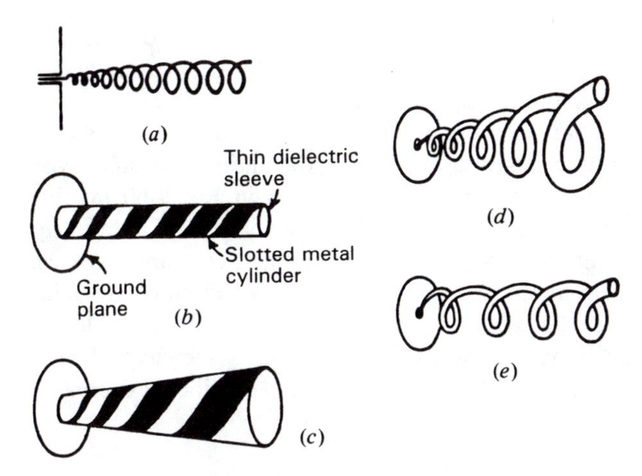

Figure 7-60 Types of tapered monofilar axial-mode helical antennas including ones in which conductor size is tapered. (*After Kraus.*)

variable and (3) turn spacing S constant but pitch angle α and diameter D variable. Many of these forms have been investigated—the class (2) form by Day.[1]

Day measured patterns of monofilar axial-mode helical antennas of 6 turns with the diameter constant but α increasing or decreasing (D constant but α and S variable as in Fig. 7-59*d*, *e* and *f* or the middle row of Fig. 7-59). The helix conductor diameter was 0.02λ. Pitch angles were varied on a given helix from 1 to 20°, 5 to 17° or 9 to 15°, both increasing and decreasing. These were compared with a constant pitch angle of 12.5° at helix circumferences C_λ of 0.6, 0.8, 1.0, 1.2 and 1.4—a total of 35 cases. For pitch angle tapers between 5 and 17° and $0.8 \leq C_\lambda \leq 1.2$, the pattern variations are minor. However, at $C_\lambda = 1.2$ and with the pitch angle decreasing from 17° at the feed end to 5° at the open end, the gain is 1 dB more than for $C_\lambda = 1.2$ and $\alpha = 12.5°$ (constant pitch). This is a significant improvement since the highest gain for a uniform 6-turn helix occurs when α is approximately 12.5° and the circumference C_λ approximately 1.2. Thus, the center helix (D constant, α and S decreasing), of the 9 shown in Fig. 7-59 (i.e., Fig. 7-59*e*), is a useful variant of the uniform helix.

The conical helix in Fig. 7-59*a* for which α is constant and D (or C) and S are increasing has been investigated by Chatterjee,[2] Nakano, Mikawa and Yamauchi[3] and others. With small pitch angles Chatterjee found that very broad

[1] P. C. Day, "Some Characteristics of Tapered Helical Beam Antennas," M.S. thesis, Ohio State University, 1950.

[2] J. S. Chatterjee, "Radiation Field of a Conical Helix," *J. Appl. Phys.*, **24**, 550–559, May 1953; **26**, 331–335, March 1955.

[3] H. Nakano, T. Mikawa and J. Yamauchi, "Numerical Analysis of Monofilar Conical Helix," *IEEE AP-S Int. Symp.*, **1**, 177–180, 1984.

patterns can be obtained over a 5 to 1 bandwidth. According to Nakano, Mikama and Yamauchi, the currents involved are those of the attenuating wave-launching region close to the feed point (see Fig. 7-3*b* and *c*).

Additional tapered types are shown in Fig. 7-60. The one at (*a*) has a tapered and uniform section but reversed in order from the uniform-to-taper type of Wong and King[1] and others. The other designs shown in Fig. 7-60 involve variation in the diameter of the helical conductor d or the width w of a flat strip conductor. Thus, there are 4 quantities which can be varied, α, D, S and d (or w). Since the characteristics of the monofilar axial-mode helical antenna are relatively insensitive to moderate changes in dimensions, the effect of moderate departures from uniformity is, in general, not large. However, some changes may produce significant increases in gain, as discussed above, and significant decreases in axial ratio and VSWR (see Fig. 7-24). See also Sec. 15-4.

7-18 MULTIFILAR AXIAL-MODE (KILGUS COIL AND PATTON COIL) HELICAL ANTENNAS. Four wires, each $\lambda/2$ long and forming $\frac{1}{2}$-turn of a helix as in Fig. 7-61, produce a cardiod-shaped back-fire circularly polarized pattern (HPBW $\simeq$ 120°) when the two pairs are fed in phase quadrature. This Kilgus coil may be described as two $\frac{1}{2}$-turn bifilar helices or one $\frac{1}{2}$-turn quadrifilar helix.[2] The antenna is resonant and the bandwidth is narrow (about 4 percent). The 4 wires can also be $\lambda/4$ or λ long. For these lengths the lower ends are open-circuited instead of short-circuited as for the $\lambda/2$ wires of Fig. 7-61. Each bifilar helix can be balun-fed at the top from a coaxial line extending to the top along the central axis. By increasing the number of turns Kilgus reports that shaped-conical patterns can be obtained which may be more useful for some applications than a cardiod (heart-shaped) pattern.[3]

A bifilar helix end-fed by a balanced 2-wire transmission line produces a back-fire beam when operated above the cutoff frequency of the principal mode of the helical waveguide. The maximum directivity of this Patton coil occurs slightly above the cutoff frequency.[4] The pattern broadens with increasing frequency and at pitch angles of about 45° the back-fire beam splits and scans toward side-fire.

Below the cutoff frequency, there is a standing-wave current distribution along the helical conductor. Above cutoff, the standing wave gives way to a gradually decaying traveling wave. With a further increase in frequency the rate of decay increases and a low-level standing wave appears, indicating the existence of

[1] J. L. Wong and H. E. King, "Broadband Quasi–Taper Helical Antennas," *IEEE Trans. Ants. Prop.*, **AP-27**, 72–78, January 1979.

[2] C. C. Kilgus, "Resonant Quadrifilar Helix," *IEEE Trans. Ants. Prop.*, **AP-17**, 349–351, May 1969.

[3] C. C. Kilgus, "Shaped-Conical Radiation Pattern Performance of the Backfire Quadrifilar Helix," *IEEE Trans. Ants. Prop.*, **AP-23**, 392–397, May 1975.

[4] W. T. Patton, "The Backfire Helical Antenna," Ph.D. thesis, University of Illinois, 1963.

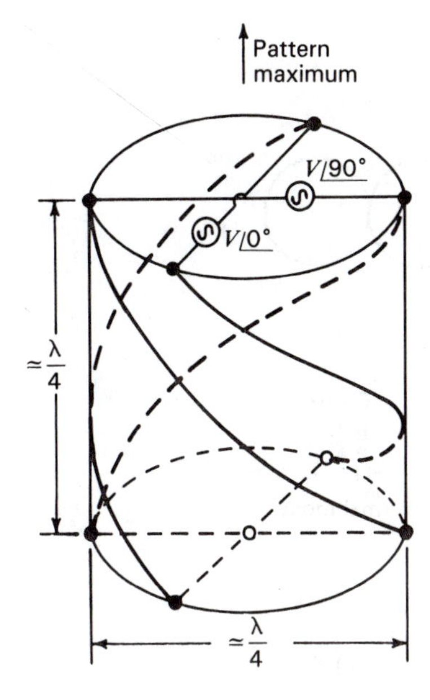

Figure 7-61 Resonant narrowband back-fire quadrifilar Kilgus coil for very broad circularly polarized pattern. (*After C. C. Kilgus, "Resonant Quadrifilar Helix,"* IEEE Trans. Ants. Prop., **AP-17**, *349–351, May 1969.*) Wires are situated in space as though wrapped around a cylinder as suggested in the figure.

a higher-order helical waveguide mode. This establishes the upper frequency limit of the bifilar helix back-fire radiation.

Quadrifilar and octofilar axial-mode forward end-fire circularly polarized helical antennas using large pitch angles (30 to 60°) have been investigated by Gerst and Worden[1] and Adams *et al.*[2]

7-19 MONOFILAR AND MULTIFILAR NORMAL-MODE HELICAL ANTENNAS. THE WHEELER COIL. The previous sections deal with *axial-mode* helical antennas with maximum radiation in the direction of the helix axis. The radiation may be (forward) end-fire or back-fire. In this section the *normal mode* of radiation is discussed, *normal* being used in the sense of *perpendicular to* or at *right angles to* the helix axis. This radiation with its maximum normal to the helix axis may also be described as *side-fire* or *broadside.*

When the helix circumference is approximately a wavelength the axial mode of radiation is dominant, but when the circumference is much smaller the normal mode is dominant. Figure 7-62*a* and *c* shows helices radiating in both modes while Fig. 7-62*b* shows a 4-lobed mode helix (Chireix coil) with the relative sizes for producing the modes being indicated.

---

[1] C. W. Gerst and R. A. Worden, "Helix Antennas Take a Turn for the Better," *Electronics*, 100–110, Aug. 22, 1966.

[2] A. A. Adams, R. K. Greenough, R. F. Wallenberg, A. Mendelovicz and C. Lumjiak, "The Quadrifilar Helix Antenna, *IEEE Trans. Ants. Prop.*, **AP-22**, 173–178, March 1974.

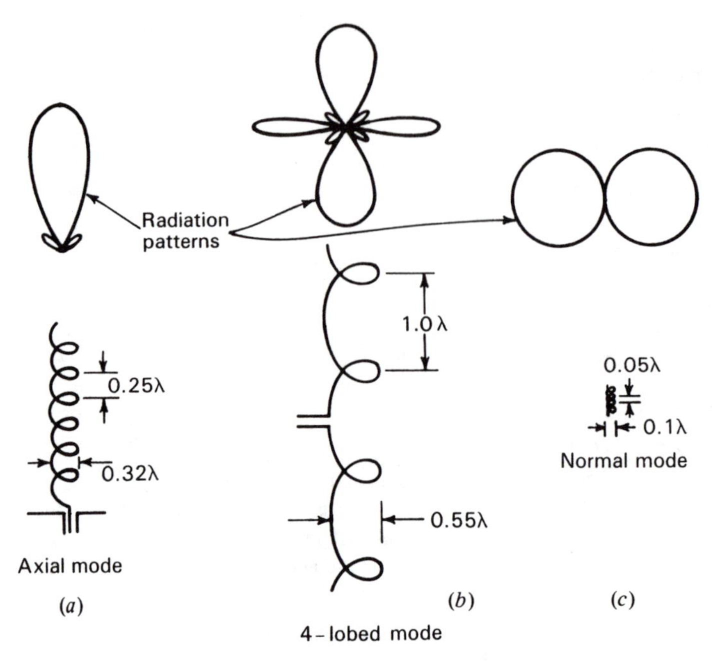

Figure 7-62 Field patterns of axial, 4-lobed and normal radiation modes of helical antennas with relative size indicated.

Now let us examine the requirements for normal-mode radiation in more detail. Consider a helix oriented with axis coincident with the polar or z axis as in Fig. 7-63*a*. *If the dimensions are small* ($nL \ll \lambda$), the maximum radiation is in the xy plane for a helix oriented as in Fig. 7-63*a*, with zero field in the z direction.

When the pitch angle is zero, the helix becomes a loop as in Fig. 7-63*b*. When the pitch angle is 90°, the helix straightens out into a linear antenna as in Fig. 7-63*c*, the loop and straight antenna being limiting cases of the helix.

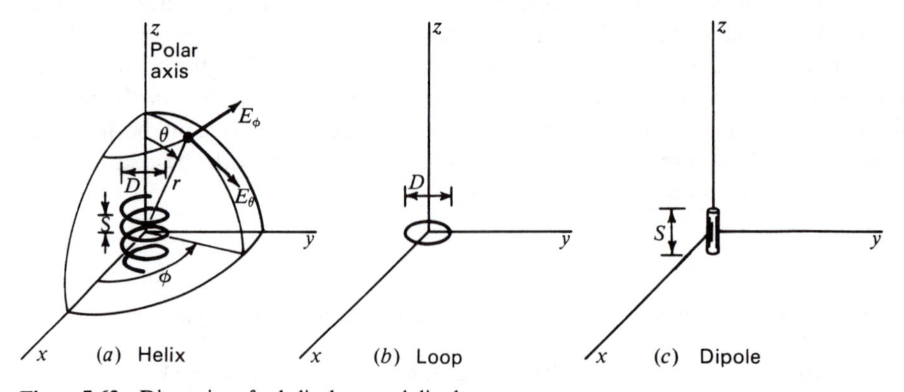

Figure 7-63 Dimensions for helix, loop and dipole.

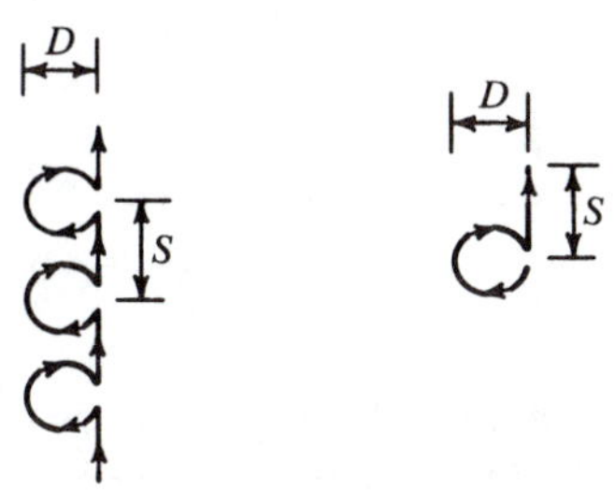

Figure 7-64 Modified helix for normal-mode calculations.

The far field of the helix may be described by two components of the electric field, E_ϕ and E_θ, as shown in Fig. 7-63*a*. Let us now develop expressions for the far-field patterns of these components for a small short helix. The development is facilitated by assuming that the helix consists of a number of small loops and short dipoles connected in series as in Fig. 7-64*a*. The diameter D of the loops is the same as the helix diameter, and the length of the dipoles S is the same as the spacing between turns of the helix. Provided that the helix is small, the modified form of Fig. 7-64*a* is equivalent to the true helix of Fig. 7-63*a*. The current is assumed to be uniform in magnitude and in phase over the entire length of the helix. Since the helix is small, the far-field pattern is independent of the number of turns. Hence, it suffices to calculate the far-field patterns of a single small loop and one short dipole as illustrated in Fig. 7-64*b*.

The far field of the small loop has only an E_ϕ component. Its value is given in Table 6-1 as

$$E_\phi = \frac{120\pi^2[I] \sin\theta}{r} \frac{A}{\lambda^2} \tag{1}$$

where the area of the loop $A = \pi D^2/4$

The far field of the short dipole has only an E_θ component. Its value is given in the same table as

$$E_\theta = j\,\frac{60\pi[I] \sin\theta}{r} \frac{S}{\lambda} \tag{2}$$

where S has been substituted for L as the length of the dipole.

Comparing (1) and (2), the j operator in (2) and its absence in (1) indicates that E_ϕ and E_θ are in phase quadrature. The ratio of the magnitudes of (1) and (2) then gives the axial ratio of the polarization ellipse of the far field. Hence, dividing the magnitude of (2) by (1) we obtain for the axial ratio:

$$\text{AR} = \frac{|E_\theta|}{|E_\phi|} = \frac{S\lambda}{2\pi A} = \frac{2S\lambda}{\pi^2 D^2} = \frac{2S_\lambda}{C_\lambda^2} \tag{3}$$

Three special cases of the polarization ellipse are of interest. (1) When $E_\phi = 0$, the axial ratio is infinite and the polarization ellipse is a vertical line indicating

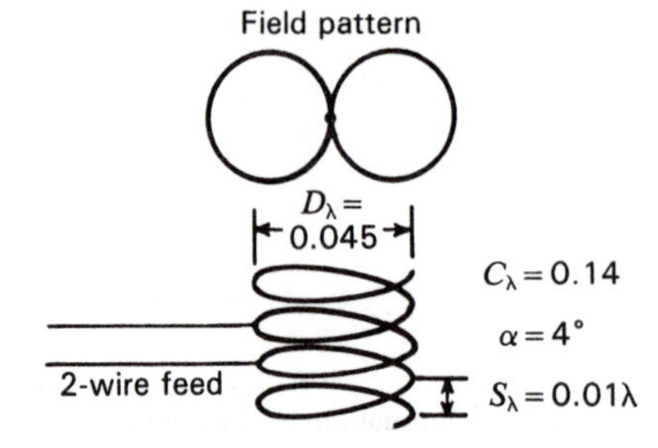

Figure 7-65 Resonant narrowband circularly polarized monofilar normal-mode Wheeler coil. Pattern is that of a short dipole.

linear vertical polarization. The helix in this case is a vertical dipole. (2) When $E_\theta = 0$, the axial ratio is zero[1] and the polarization ellipse is a horizontal line indicating linear horizontal polarization. The helix in this case is a horizontal loop. (3) The third special case of interest occurs when $|E_\theta| = |E_\phi|$. For this case the axial ratio is unity and the polarization ellipse is a circle, indicating circular polarization. Thus, setting (3) equal to unity yields

$$\pi D = \sqrt{2S\lambda} \qquad \text{or} \qquad C_\lambda = \sqrt{2S_\lambda} \tag{4}$$

This relation was first obtained by Wheeler in an equivalent form.[2] The radiation is circularly polarized in all directions in space but with zero field on axis (z direction, Fig. 7-63*a*). A monofilar normal-mode helix or *Wheeler coil* fulfilling condition (4) is shown in Fig. 7-65. It is a resonant, narrowband antenna.

We have considered three special cases of the polarization ellipse involving linear and circular polarization. In the general case, the radiation is elliptically polarized. Therefore, the radiation from a helix of constant turn-length changes progressively through the following forms as the pitch angle is varied. When $\alpha = 0$, we have a loop (Fig. 7-63*b*) and the polarization is linear and horizontal. As α increases, let us consider the helix dimensions as we move along a constant L_λ line (circle with center at origin, Fig. 7-10). As α increases from zero, the polarization becomes elliptical with the major axis of the polarization ellipse horizontal. When α reaches a value such that $C_\lambda = \sqrt{2S_\lambda}$ the polarization is circular. With the aid of Fig. 7-9, this value of α is given by

$$\alpha = \arcsin \frac{-1 + \sqrt{1 + L_\lambda^2}}{L_\lambda} \tag{5}$$

As α increases still further, the polarization again becomes elliptical but with the major axis of the polarization ellipse vertical. Finally, when α reaches 90°, we have a dipole (Fig. 7-63*c*) and the polarization is linear and vertical. Wheeler's

---

[1] The axial ratio is here allowed to range from 0 to ∞, instead of from 1 to ∞ as customarily, in order to distinguish between linear vertical and linear horizontal polarization.

[2] H. A. Wheeler, "A Helical Antenna for Circular Polarization," *Proc. IRE*, **35**, 1484–1488, December 1947.

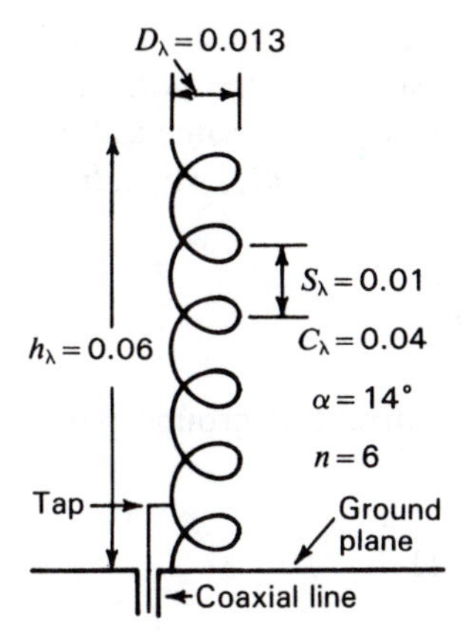

Figure 7-66 Short resonant narrowband monofilar normal-mode helical antenna mounted over a ground plane as substitute for a $\lambda/4$ stub.

relation for circular polarization from a helix radiating in the normal mode as given by (4) or (5) is shown in Fig. 7-10 by the curve marked $C_\lambda = \sqrt{2S_\lambda}$.

In the preceding discussion on the normal mode of radiation, the assumption is made that the current is uniform in magnitude and in phase over the entire length of the helix. This condition could be approximated if the helix is very small ($nL \ll \lambda$) and is end-loaded. However, the bandwidth of such a small helix is very narrow, and the radiation efficiency is low. The bandwidth and radiation efficiency could be increased by increasing the size of the helix, but to approximate the uniform, in-phase current distribution requires that some type of phase shifter be placed at intervals along the helix. This may be inconvenient or impractical. Hence, the production of the normal mode of radiation from a helix has practical limitations.

An antenna having four slanting dipoles that is suggestive of a fractional-turn quadrifilar helix radiating in the normal mode had been built by Brown and Woodward[1] (see Fig. 16-20*f*). Their arrangement is based on a design described by Lindenblad.[2]

Resonant monofilar normal-mode helical antennas are useful as short, essentially vertically polarized, radiators. Referring to Fig. 7-66, the helix mounted on a ground plane with axis vertical acts as a resonant narrowband substitute for a $\lambda/4$ vertical stub or monopole above a ground plane. The helix in Fig. 7-66 is 0.06λ in height or about $\frac{1}{4}$-height of a $\lambda/4$ stub. From (3) the axial ratio of the helix is given by

$$\mathrm{AR} = \frac{2S_\lambda}{C_\lambda^2} = \frac{2 \times 0.01}{(0.04)^2} = 12.5 \tag{6}$$

[1] G. H. Brown and O. M. Woodward, "Circularly Polarized Omnidirectional Antenna," *RCA Rev.*, **8**, 259–269, June 1947.

[2] N. E. Lindenblad, "Antennas and Transmission Lines at the Empire State Television Station," *Communications*, **21**, 10–14, 24–26, April 1941.

with the major axis of the polarization ellipse vertical. The polarization is, thus, essentially linear and vertical with an omnidirectional pattern in the horizontal plane (plane of ground plane). The radiation resistance is nearly the same as for a short monopole of height h_λ above the ground plane where $h_\lambda = nS_\lambda$, which from (2-20-3) or (5-3-14) (for a short dipole) is given by

$$R_s = \tfrac{1}{2} \times 790\left(\frac{I_{av}}{I_0}\right)^2 h_\lambda^2 \quad (\Omega) \tag{7}$$

Assuming a sinusoidal current distribution (maximum current at ground plane, zero at open end),

$$R_s = 395 \times \left(\frac{2}{\pi}\right)^2 \times 0.06^2 = 0.6\ \Omega \tag{8}$$

This is the radiation resistance between the base of the helix and ground. Connection to a coaxial line would require an impedance transformer, but with the shunt feed of Fig. 7-66 the helix can be matched directly to a coaxial line by adjusting the tap point on the helix. With such a small radiation resistance, any loss resistance can reduce efficiency (see Secs. 2-15 and 6-12). The advantage of the helix over a straight stub or monopole is that its inductance can resonate the antenna.

A center-fed monofilar helix ($\alpha = 30°$) with $S_\lambda = 1$, $L_\lambda = 2$ and $C_\lambda = \sqrt{3}$ has a 4-lobed pattern, 1 lobe each way on axis and 2 lobes normal to the axis. Its location is indicated on the $m = 1$ line of Fig. 7-10 where the $L_\lambda = 2$ and $\alpha = 30°$ lines intersect, for which also $C_\lambda = 1.73$ and $S_\lambda = 1$ (Chireix coil; see also Fig. 7-62*b*).

Patton[1] has demonstrated that a bifilar helix end-fed by a balanced 2-wire transmission line can produce circularly polarized omnidirectional side-fire radiation when pitch angles of about 45° are used.

Some other monofilar and multifilar normal mode (side-fire) helices for omnidirectional FM and TV broadcasting are described by King and Wong and by DuHamel.[2]

PROBLEMS[3]

★7-1 An 8-turn helix. A monofilar helical antenna has $\alpha = 12°$, $n = 8$, $D = 225$ mm. (*a*) What is p at 400 MHz for (1) in-phase fields and (2) increased directivity? (*b*) Calculate and plot the field patterns for $p = 1.0$, 0.9 and 0.5 and also for p equal to the value for in-phase fields and increased directivity. Assume each turn is an isotropic point source. (*c*) Repeat (*b*) assuming each turn has a cosine pattern.

---

[1] W. T. Patton, "The Backfire Helical Antenna," Ph.D. thesis, University of Illinois, 1963.

[2] H. E. King and J. L. Wong, pp. 13–18, and R. H. DuHamel, pp. 28–35, in R. C. Johnson and H. Jasik (eds.), *Radio Engineering Handbook*, McGraw-Hill, 1984.

[3] Answers to starred (★) problems are given in App. D.

7-2 A 10-turn helix. A right-handed monofilar helical antenna has 10 turns, 100 mm diameter and 70 mm turn spacing. The frequency is 1 GHz. (*a*) Calculate and plot the far-field pattern. (*b*) What is the HPBW? (*c*) What is the gain? (*d*) What is the polarization state? (*e*) Repeat the problem for a frequency of 300 MHz.

7-3 A 30-turn helix. A right-handed monofilar axial-mode helical antenna has 30 turns, $\lambda/3$ diameter and $\lambda/5$ turn spacing. Find (*a*) HPBW, (*b*) gain and (*c*) polarization state.

Note regarding Probs. 7-1, 7-2 and 7-3: The patterns for monofilar axial-mode helices may be calculated using the BASIC program in App. B-2 where in line 10:

N = number of turns
D = spacing = $2\pi S_\lambda$
S = phase shift between turns = $2\pi S_\lambda + (\pi/N)$
MF = multiplying or normalizing factor = $67 \sin(\pi/2N)$

and to account for the single-turn pattern, line 80 should read:

R = MF*ABS(R)*CA.

7-4 Helices, left and right. Two monofilar axial-mode helical antennas are mounted side-by-side with axes parallel (in the x direction). The antennas are identical except that one is wound left-handed and the other right-handed. What is the polarization state in the x direction if the two antennas are fed (*a*) in phase and (*b*) in opposite phase?

7-5 A 6-turn helix. A monofilar axial-mode helical antenna has 6 turns, 231 mm diameter and 181 mm turn spacing. Neglect the effect of the ground plane. Assume that the relative phase velocity p along the helical conductor satisfies the increased-directivity condition. Calculate and plot the following patterns as a function of ϕ (0 to 360°) in the $\theta = 90°$ plane at 400 MHz. Use the square helix approximation. (*a*) $E_{\phi T}$ for a single turn and E_ϕ for the entire helix. (*b*) Repeat (*a*) neglecting the contribution of sides 2 and 4 of the square turn. (*c*) $E_{\theta T}$ for a single turn and E_θ for the entire helix.

★7-6 Normal-mode helix.

(*a*) What is the approximate relation required between the diameter D and height H of an antenna having the configuration shown in Fig. P7-6, in order to obtain a circularly polarized far field at all points at which the field is not zero. The loop is circular and is horizontal, and the linear conductor of length H is vertical. Assume D and H are small compared to the wavelength, and assume the current is of uniform magnitude and in phase over the system.

(*b*) What is the pattern of the far circularly polarized field?

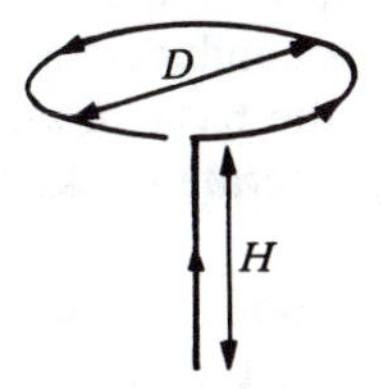

Figure P7-6 Normal-mode helix.

CHAPTER 8

THE BICONICAL ANTENNA AND ITS IMPEDANCE

8-1 INTRODUCTION. Sir Oliver Lodge constructed a biconical antenna in 1897, while the single cone working against ground was popularized by Marconi (Fig. 1-3). The fan (flat triangular) antenna was also used by Marconi and others. The broadband characteristics of monoconical (single-cone) and biconical (double-cone) antennas make them useful for many applications. In this chapter a fundamental analysis is given and both theoretical and experimental results are presented.

In the chapters preceding 7 it is usually assumed that the antenna conductor is thin, in fact, infinitesimally thin. From known or assumed current distributions, the far-field patterns are calculated. The effect of the conductor thickness on the pattern is negligible provided that the diameter of the conductor is a small fraction of a wavelength. Thus, the patterns calculated on the basis of an infinitesimally thin conductor are applicable to conductors of moderate thickness, say for $d < 0.05\lambda$ where d is the conductor diameter.

The radiation resistance of thin linear conductors and loops is calculated in Chaps. 5 and 6. This calculation is based on a knowledge of the pattern and a known or assumed current distribution. The values so obtained apply strictly to an infinitesimally thin conductor. The conductor thickness, up to moderate diam-

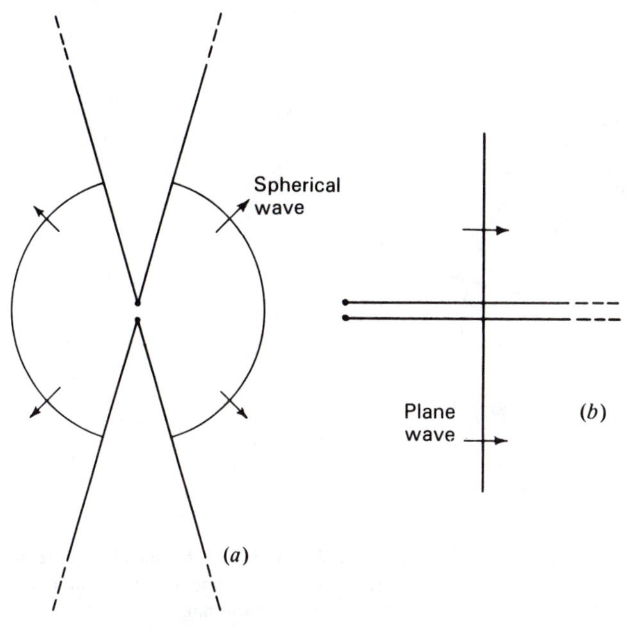

Figure 8-1 An infinite biconical antenna (*a*) is analogous to an infinite uniform transmission line (*b*).

eters, has only a small effect on the resistance at or near a current loop but may have a large effect on the resistance at or near a current minimum.[1]

In this chapter, we consider the problem of finding the input terminal resistance and also the reactance, taking into account the effect of conductor thickness. This problem is most simply approached by Schelkunoff's treatment of the biconical antenna[2] which will be outlined in the following sections. Beginning with the infinite biconical antenna, the analysis proceeds to terminated biconical antennas, i.e., ones of finite length. This method of treatment bears a striking similarity to that usually employed with transmission lines in which the infinite transmission line is discussed first, followed by the terminated line of finite length.

8-2 THE CHARACTERISTIC IMPEDANCE OF THE INFINITE BICONICAL ANTENNA. The infinite biconical antenna is analogous to an infinite uniform transmission line. The biconical antenna acts as a guide for a spherical wave in the same way that a uniform transmission line acts as a guide for a plane wave. The two situations are compared in Fig. 8-1.

The characteristic impedance of a biconical antenna will now be derived and will be shown to be uniform. Let a generator be connected to the terminals

---

[1] This is discussed in more detail in Chap. 9.

[2] S. A. Schelkunoff, *Electromagnetic Waves*, Van Nostrand, New York, 1943, chap. 11, p. 441.

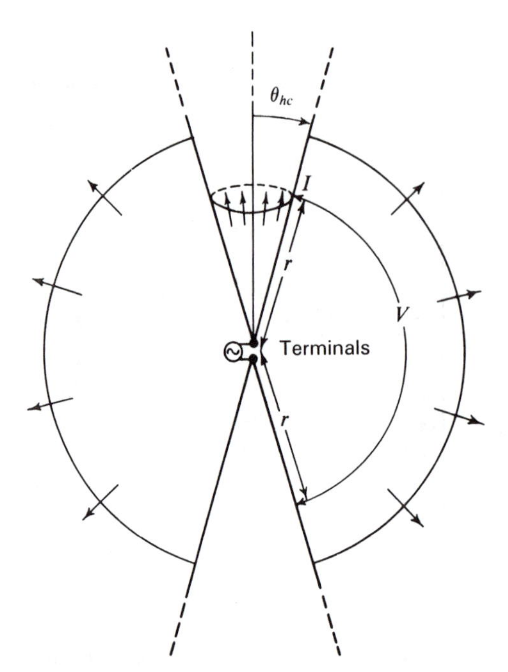

Figure 8-2 Infinite biconical antenna showing voltage V and current I at a distance r from the terminals.

of an infinite biconical antenna as in Fig. 8-2. The generator causes waves with spherical phase fronts to travel radially outward from the terminals as suggested. The waves produce currents on the cones and a voltage between them. Let V be the voltage between points on the upper and lower cones a distance r from the terminals as in Fig. 8-2. Let I be the total current on the surface of one of the cones at a distance r from the terminals. As on an ordinary transmission line, the ratio V/I is the characteristic impedance of the antenna. For the characteristic impedance to be uniform, it is necessary that the ratio V/I be independent of r.

Before V and I can be calculated, we must determine the nature of the electric and magnetic fields existing in the space between the conducting cones. Although the biconical transmission line can support an infinite number of transmission modes, let us assume that only the TEM or principal transmission mode is present. For the TEM mode, both **E** and **H** are entirely transverse, i.e., they have no radial component. The **E** lines are along great circles passing through the polar axis as shown in Fig. 8-3. This satisfies the boundary conditions since **E** is normal to the surface of the cones. The **H** lines are circles lying in planes normal to the polar axis.

Maxwell's equation from Faraday's law for harmonically varying fields is

$$\nabla \times \mathbf{E} = -j\omega\mu\mathbf{H} \tag{1}$$

The biconical antenna is most readily handled in spherical coordinates. Let the spherical coordinates r, θ, ϕ be related to the antenna as in Fig. 8-4. Expanding

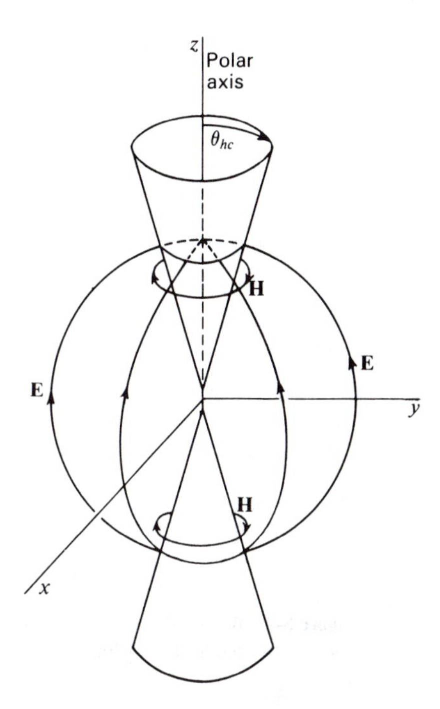

Figure 8-3 E and H lines of outgoing TEM wave on biconical antenna.

the left side of (1) in spherical coordinates, we have

$$\nabla \times \mathbf{E} = \frac{\hat{\mathbf{r}}}{r^2 \sin\theta}\left[\frac{\partial(r \sin\theta\, E_\phi)}{\partial\theta} - \frac{\partial(rE_\theta)}{\partial\phi}\right] + \frac{\hat{\boldsymbol{\theta}}}{r\sin\theta}\left[\frac{\partial E_r}{\partial\phi} - \frac{\partial(r\sin\theta\, E_\phi)}{\partial r}\right] + \frac{\hat{\boldsymbol{\phi}}}{r}\left[\frac{\partial(rE_\theta)}{\partial r} - \frac{\partial E_r}{\partial\theta}\right] \tag{2}$$

Since **E** has only a θ component, which by symmetry is independent of ϕ, only the fifth term of (2) does not vanish. Thus,

$$\nabla \times \mathbf{E} = \frac{\hat{\boldsymbol{\phi}}}{r}\frac{\partial(rE_\theta)}{\partial r} \tag{3}$$

Expanding the right side of (1) in spherical coordinates,

$$-j\omega\mu\mathbf{H} = -j\omega\mu(\hat{\mathbf{r}}H_r + \hat{\boldsymbol{\theta}}H_\theta + \hat{\boldsymbol{\phi}}H_\phi) \tag{4}$$

Since **H** has only a ϕ component, (4) reduces to

$$-j\omega\mu\mathbf{H} = -\hat{\boldsymbol{\phi}}j\omega\mu H_\phi \tag{5}$$

Now equating (3) and (5) we have

$$\frac{1}{r}\frac{\partial(rE_\theta)}{\partial r} = -j\omega\mu H_\phi \tag{6}$$

This is Maxwell's equation (1) reduced to a special form appropriate to a spherical wave.

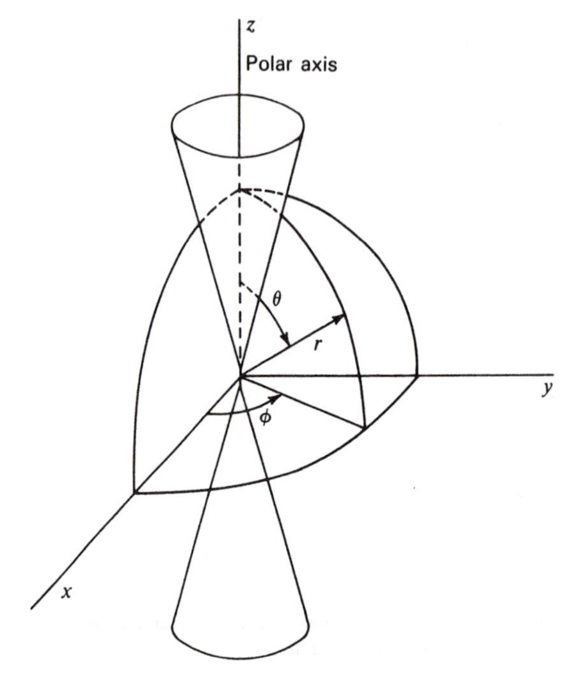

Figure 8-4 Biconical antenna with relation to spherical coordinates r, θ, ϕ.

Maxwell's equation from Ampère's law for harmonically varying fields in a nonconducting medium is

$$\nabla \times \mathbf{H} = j\omega\varepsilon\mathbf{E} \tag{7}$$

H has only a ϕ component and **E** only a θ component. Since $E_r = 0$ it follows that

$$\frac{\partial(\sin\theta\, H_\phi)}{\partial\theta} = 0 \tag{8}$$

Equation (7) can be reduced by a similar procedure as used for (1) to the form

$$\frac{\partial(rH_\phi)}{\partial r} = -j\omega\varepsilon(rE_\theta) \tag{9}$$

Now differentiating (9) with respect to r and introducing (6), we obtain a wave equation in (rH_ϕ). Thus,

$$\frac{\partial^2(rH_\phi)}{\partial r^2} = -\omega^2\mu\varepsilon(rH_\phi) \tag{10}$$

The condition of (8) requires that H_ϕ vary inversely as the sine of θ; that is,

$$H_\phi \propto \frac{1}{\sin\theta} \tag{11}$$

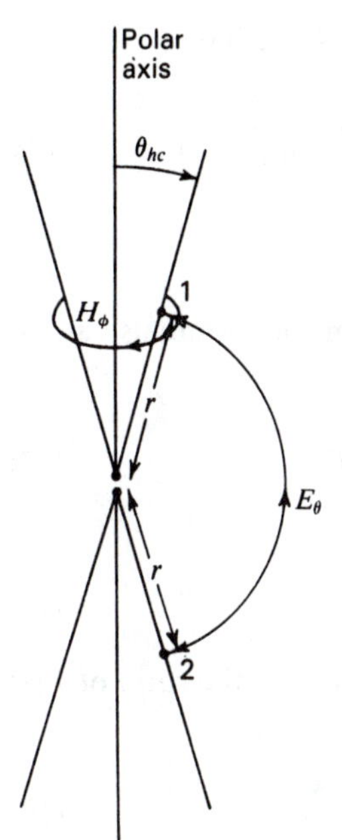

Figure 8-5 E_θ and H_ϕ field components at a distance r from the terminals of a biconical antenna.

Hence, a solution of (10) which also fulfills (11) is

$$H_\phi = \frac{1}{r \sin \theta} H_0 e^{-j\beta r} \tag{12}$$

where $\beta = \omega\sqrt{\mu\varepsilon} = 2\pi/\lambda$

This solution represents an outgoing traveling wave on the antenna. Since the biconical antenna is assumed to be infinitely long, only the outgoing wave need be considered.

The electric and magnetic fields of a TEM wave are related by the intrinsic impedance Z_0 of the medium. Thus, we have

$$E_\theta = Z_0 H_\phi = \frac{Z_0}{r \sin \theta} H_0 e^{-j\beta r} \tag{13}$$

Equations (12) and (13) give the variation of the magnetic and electric fields of a TEM outgoing wave in the space between the cones of a biconical antenna as a function of θ and r. The fields are independent of ϕ.

The voltage $V(r)$ between points 1 and 2 on the cones at a distance r from the terminals (see Fig. 8-5) can now be obtained by taking the line integral of E_θ along a great circle between the two points. Thus,

$$V(r) = \int_{\theta_{hc}}^{\pi - \theta_{hc}} E_\theta r \, d\theta \tag{14}$$

where θ_{hc} is the half-angle of the cone. Substituting (13) in (14) we have

$$V(r) = Z_0 H_0 e^{-j\beta r} \int_{\theta_{hc}}^{\pi - \theta_{hc}} \frac{d\theta}{\sin \theta} = Z_0 H_0 e^{-j\beta r} \ln \frac{\cot (\theta_{hc}/2)}{\tan (\theta_{hc}/2)} \tag{15}$$

or
$$V(r) = 2Z_0 H_0 e^{-j\beta r} \ln \cot \frac{\theta_{hc}}{2} \tag{16}$$

The total current $I(r)$ on the cone at a distance r from the terminals can be obtained by applying Ampère's law. Thus,

$$I(r) = \int_0^{2\pi} H_\phi r \sin \theta \, d\phi = 2\pi r H_\phi \sin \theta \tag{17}$$

Now substituting H_ϕ from (12) in (17) yields

$$I(r) = 2\pi H_0 e^{-j\beta r} \tag{18}$$

The characteristic impedance Z_k of the biconical antenna is the ratio of $V(r)$ to $I(r)$ as given by (16) and (18) or

$$Z_k = \frac{V(r)}{I(r)} = \frac{Z_0}{\pi} \ln \cot \frac{\theta_{hc}}{2} \tag{19}$$

For a medium of free space between the cones, $Z_0 = 120\pi \ \Omega$ so that (19) becomes

$$Z_k = 120 \ln \cot \frac{\theta_{hc}}{2} \quad (\Omega) \tag{20}$$

When θ_{hc} is small ($\theta_{hc} < 20°$), $\cot (\theta_{hc}/2) \simeq 2/\theta_{hc}$ so that

$$Z_k = 120 \ln \frac{2}{\theta_{hc}} \quad (\Omega) \tag{21}$$

Equations (20) and (21) are Schelkunoff's relations for the characteristic impedance of a biconical antenna. Since these equations are independent of r, the biconical antenna has a uniform characteristic impedance.

8-3 INPUT IMPEDANCE OF THE INFINITE BICONICAL ANTENNA. The input impedance of a biconical antenna with TEM waves is given by the ratio $V(r)/I(r)$ as r approaches zero. For an infinite biconical antenna this ratio is independent of r, so that the input impedance of the infinite biconical antenna equals the characteristic impedance. The input impedance depends only on the TEM wave and is unaffected by higher-order waves. Thus

$$Z_i = Z_k \tag{1}$$

where Z_i is the input impedance of the biconical antenna and Z_k is the characteristic impedance as given by (8-2-20) or for small cone angles by (8-2-21). The characteristic and input impedances are pure resistances, a characteristic resist-

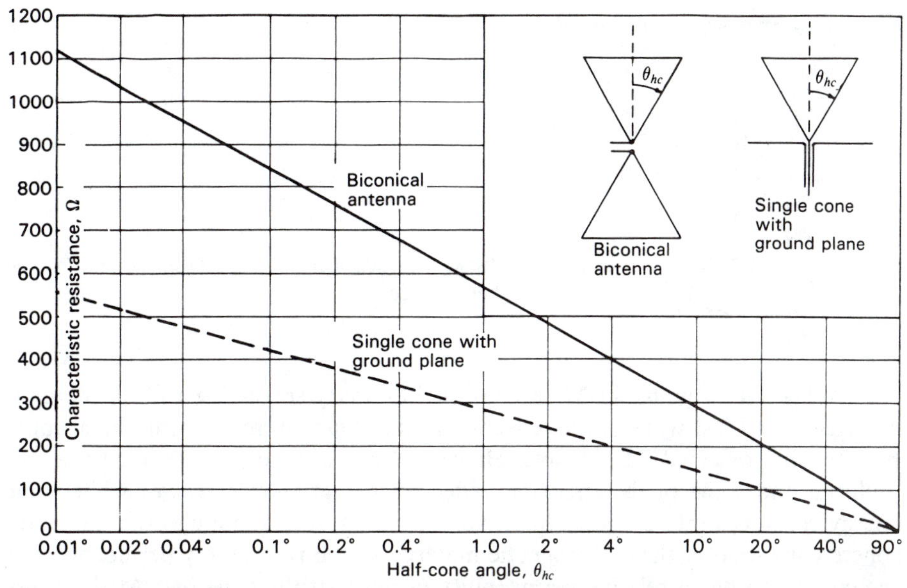

Figure 8-6 Characteristic resistance of biconical antenna and of single cone with ground plane (monoconical antenna) as a function of the half-cone angle in degrees. If the antenna is infinitely long, the terminal impedance is equal to the characteristic resistance as given in the figure.

ance R_k and an input resistance R_i. They are given by

$$R_k = R_i = 120 \ln \cot \frac{\theta_{\text{hc}}}{2} \qquad (\Omega) \tag{2}$$

The variation of this resistance as a function of the half-cone angle θ_{hc} is presented by the solid curve in Fig. 8-6. An infinite biconical antenna of 2° total cone angle ($\theta_{\text{hc}} = 1°$) has a resistance of 568 Ω, while one with a total cone angle of 100° ($\theta_{\text{hc}} = 50°$) has a resistance of 91 Ω. If the lower cone is replaced by a large ground plane (see insert in Fig. 8-6), the resistance is $\frac{1}{2}$ the value given by (2), as shown by the dashed line in Fig. 8-6.

8-4 INPUT IMPEDANCE OF THE FINITE BICONICAL ANTENNA. In this section we will consider the finite biconical antenna. This is analogous to a finite or terminated transmission line.

A TEM-mode wave can exist along the biconical conductors, but in the space beyond the cones transmission can be only in higher-order modes. Schelkunoff has defined the sphere coinciding with the ends of the cones as the *boundary sphere*, as indicated in Fig. 8-7. The radius of the sphere is l, being equal to the length of the cones ($r = l$). Inside this sphere TEM waves can exist, and also higher-order modes may be present, but outside only the higher-order modes can exist.

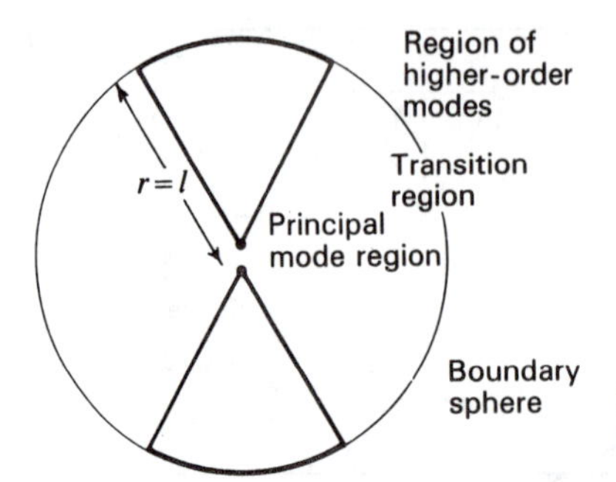

Figure 8-7 Schelkunoff's finite biconical antenna and boundary sphere.

When an outgoing TEM reaches the boundary sphere, part of its energy is reflected as a TEM wave. If the reflection at the sphere were uniform, there would be only this reflected TEM wave. However, the reflection at the sphere is not uniform, and some of the energy is reflected in higher-order waves while some energy continues into space as higher-order waves. It is as though the boundary sphere consists of a shell of magnetic material which is infinitely permeable near the cones and has a relative permeability of unity at the equator.[1] At the cones most of the outgoing TEM waves are reflected, but near the equator most of the energy escapes, as suggested in Fig. 8-8. It is but a step to imagine that, from the impedance viewpoint, the magnetic shell acts like a terminating or load impedance Z_L connected across the open end of the cones as suggested in Fig. 8-9*a*. Neglecting the effect of the end caps of the cones, the finite biconical antenna can now be treated as a transmission line of characteristic impedance Z_k terminated in the load impedance Z_L (see Fig. 8-9*b*). If the impedance Z_L can be found, the impedance Z_i at the input terminals of the biconical antenna is calculable as the impedance Z_L reflected back over a line of characteristic impedance Z_k and length l. Thus (see App. A, Sec. A-2),

$$Z_i = Z_k \frac{Z_L + jZ_k \tan \beta l}{Z_k + jZ_L \tan \beta l} \tag{1}$$

Thus, the problem resolves itself into one of finding Z_L. Reduced to simple terms, Schelkunoff's method of finding Z_L consists first of calculating Z_m at a current maximum on a very thin biconical antenna, a sinusoidal current distribution being assumed. In Fig. 8-10*a* a thin biconical antenna of length l is shown. Z_m is the impedance which appears between the current maximum on one cone and the corresponding point on the other cone. Since this impedance occurs $\lambda/4$ from the open end of the antenna, Z_L is then equal to Z_m transformed over a line $\lambda/4$ long as in Fig. 8-10*b*. Finally, the input impedance Z_i is Z_L transformed over a line of characteristic impedance Z_k and length l as in Fig. 8-9*b*.

---

[1] The shell is assumed to have zero electrical conductivity and a relative permittivity of unity.

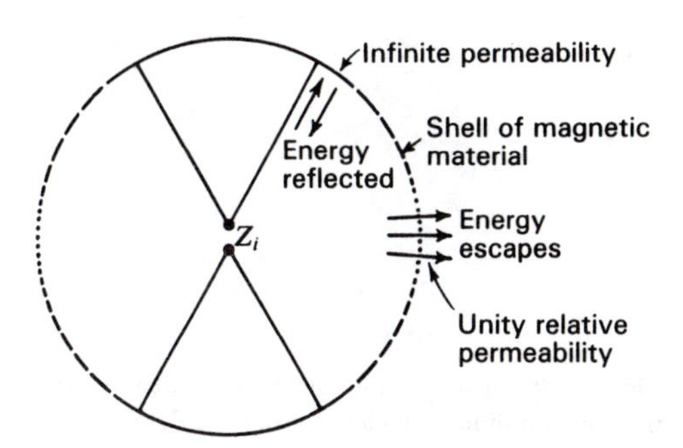

Figure 8-8 Finite biconical antenna with boundary sphere replaced by a shell of magnetic material.

The impedance Z_L is obtained from Z_m by the transmission-line relation (see App. A, Sec. A-2)

$$Z_L = Z_k \frac{Z_m + jZ_k \tan \beta x}{Z_k + jZ_m \tan \beta x} \tag{2}$$

However, the line is $\lambda/4$ long so that $\beta x = \pi/2$ and (2) reduces to

$$Z_L = \frac{Z_k^2}{Z_m} \tag{3}$$

Whereas Z_k is entirely real, Z_m in (3) may have both real and imaginary parts. Thus,

$$Z_m = R_m + jX_m$$

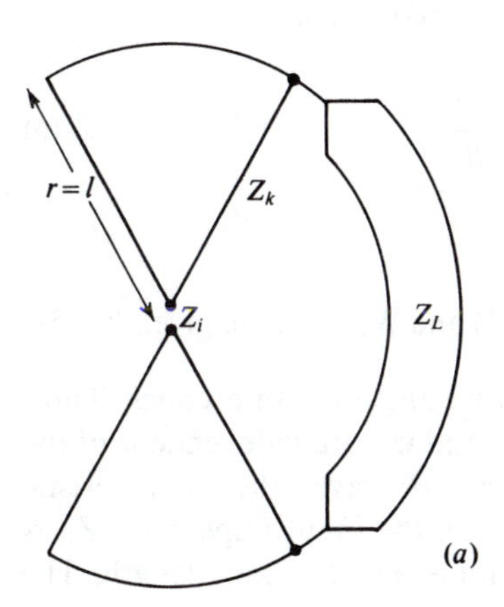

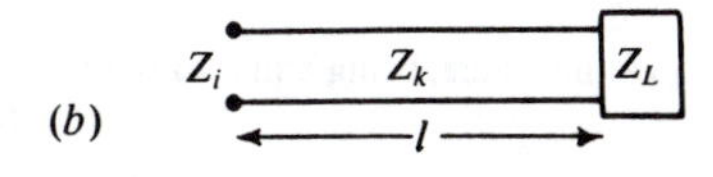

Figure 8-9 Finite or terminated biconical antenna and equivalent transmission line.

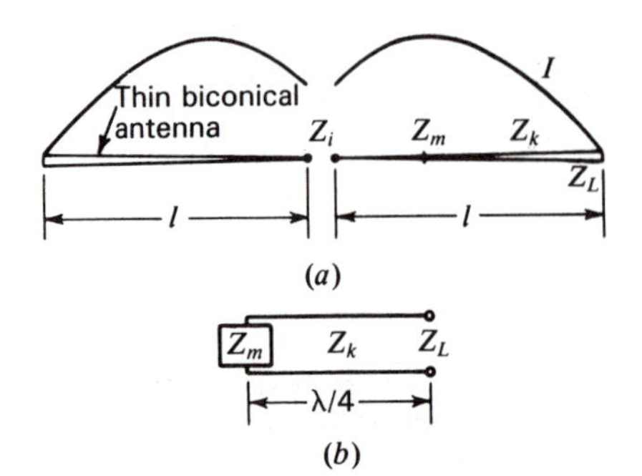

Figure 8-10 Thin finite biconical antenna and transmission-line equivalent for finding Z_L.

The real part R_m is the same as the radiation resistance at a current maximum of a very thin linear antenna. It has been calculated by Schelkunoff as[1]

$$R_m = 60 \text{ Cin } 2\beta l + 30(0.577 + \ln \beta l - 2 \text{ Ci } 2\beta l + \text{Ci } 4\beta l) \cos 2\beta l$$
$$+ 30(\text{Si } 4\beta l - 2 \text{ Si } 2\beta l) \sin 2\beta l \qquad (\Omega) \tag{4}$$

Provided only that the antenna is thin, the radiation resistance R_m is independent of the shape of the antenna (i.e., whether cylindrical or conical). However, the radiation reactance depends on the shape and has been calculated by Schelkunoff for a thin cone as

$$X_m = 60 \text{ Si } 2\beta l + 30(\text{Ci } 4\beta l - \ln \beta l - 0.577) \sin 2\beta l$$
$$- 30(\text{Si } 4\beta l) \cos 2\beta l \qquad (\Omega) \tag{5}$$

Now substituting (3) for Z_L into (1), the input impedance is

$$Z_i = Z_k \frac{Z_k + jZ_m \tan \beta l}{Z_m + jZ_k \tan \beta l} \tag{6}$$

where l = length of one cone
Z_k = value given by (8-2-21)
$Z_m = R_m + jX_m$, where R_m = value given by (4) and X_m = value given by (5)

The value of Z_m becomes independent of cone angle for thin cones. Thus, the real and imaginary parts of Z_m, as given by (4) and (5), are independent of the cone angle, being functions only of the cone length l. However, the characteristic impedance Z_k is a function of the cone angle. Hence, the input impedance Z_i as calculated by (6) is a function of both the cone angle and the cone length. The

[1] l equals half the total length of the antenna. In Chap. 5, L is twice this value, being equal to the total antenna length (that is, $L = 2l$).

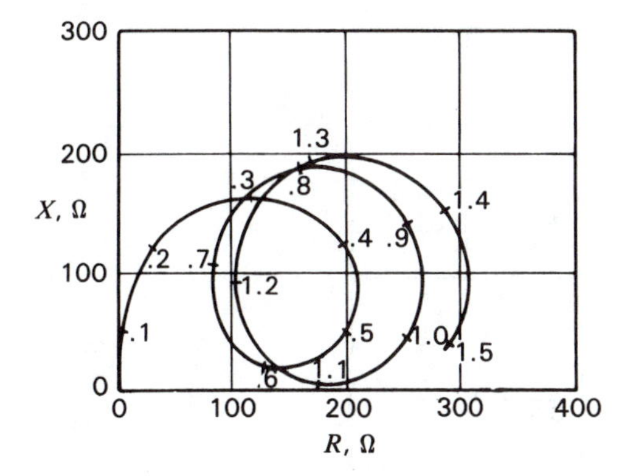

Figure 8-11 Resistance R_m and reactance X_m of radiation impedance Z_m of a biconical antenna as a function of the cone length in wavelengths (l_λ).

limitation in calculating Z_m (that the cone angle be small) also limits the use of (6) to small cone angles, say, half-cone angles of less than about 3°.[1]

The radiation impedance Z_m at the current maximum of Schelkunoff's biconical antenna as given by (4) and (5) is presented in Fig. 8-11. The impedance is given as a function of cone length, l_λ, in wavelengths, where $l_\lambda = l/\lambda$. This impedance applies to small cone angles.

Introducing Z_m into (6), the input impedance can be obtained for cones of different characteristic impedance. As illustrations, the input impedance of a biconical antenna of 1000 Ω characteristic impedance (half-cone angle, $\theta_{\text{hc}} = 0.027°$) and for one of 450 Ω characteristic impedance (half-cone angle, $\theta_{\text{hc}} = 2.7°$) are given in Fig. 8-12,[2] as functions of the cone length in wavelengths (l_λ). If the lower cone is replaced by a large ground plane (see insert in Fig. 8-6), the input impedance is halved.

It is significant that the terminal impedance of the thicker biconical antenna (lower characteristic impedance) is more constant as a function of cone length than the impedance of the thinner antenna. This difference in impedance behavior of thick and thin antennas is typical not only of conical antennas but also of antennas of other shapes, such as cylindrical antennas. We thus conclude that the impedance characteristics of a thick antenna are, in general, more suitable for wideband applications than those of a thin antenna.

The curve in Fig. 8-12 for the 2.7° half-angle biconical antenna spirals inward and would eventually end at the point $R = 450$, $X = 0$, when the length l_λ becomes infinite. Likewise, the curve for the 0.027° antenna spirals into $R = 1000$, $X = 0$, when $l_\lambda = \infty$. The effect of the cone angle is greatest near the second, fourth, or even, resonances ($l_\lambda \simeq \frac{1}{2}$, 1, etc.) and least near the first, third, or odd, resonances ($l_\lambda \simeq \frac{1}{4}, \frac{3}{4}$, etc.).

---

[1] Approximate solutions for wide cone angles are discussed by C. T. Tai, "Application of a Variational Principle to Biconical Antennas," *J. Appl. Phys.*, **20**, 1076–1084, November 1949.
P. D. P. Smith, "The Conical Dipole of Wide Angle," *J. Appl. Phys.*, **19**, 11–23, January 1948.

[2] The curves in Figs. 8-11 and 8-12 are plotted from data given by S. A. Schelkunoff, *Electromagnetic Waves*, Van Nostrand, New York, 1943.

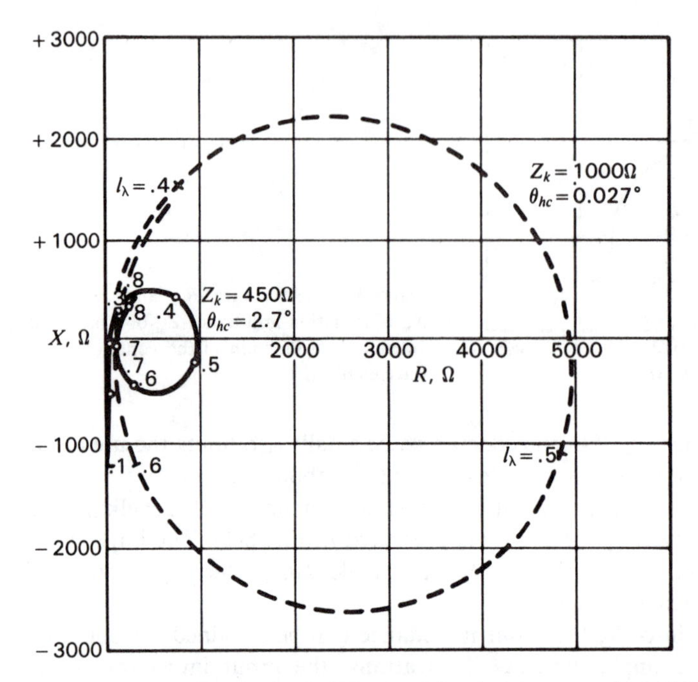

Figure 8-12 Calculated input impedance of biconical antennas with 2.7° half-cone angle (solid curve) and with 0.027° half-cone angle (dashed curve). The resistance R and reactance X of the input impedance Z_i are presented as a function of the length of one cone in wavelengths (l_λ), the length being indicated in 0.1λ intervals.

We note in Fig. 8-12 that the geometric mean resistance R_{12} of the resistance at the first and second resonances is about $\frac{1}{2}$ the characteristic resistance of the biconical antenna. We take $R_{12} = \sqrt{R_1 R_2}$, where R_1 is the resistance at the first resonance ($l_\lambda \simeq \frac{1}{4}$) and R_2 is the resistance at the second resonance ($l_\lambda \simeq \frac{1}{2}$). Thus, for the antenna with 2.7° half-cone angle, $R_{12} = 224$, which is about $\frac{1}{2}$ the characteristic resistance ($R_k = 450$). For the antenna with the 0.027° half-cone angle, $R_{12} = 500$ or $\frac{1}{2}$ the characteristic resistance ($R_k = 1000$). The geometric mean resistance R_{23} of the resistance at the second and third resonances is closer to the characteristic resistance. We take $R_{23} = \sqrt{R_2 R_3}$, where R_3 is the resistance at the third resonance ($l_\lambda \simeq \frac{3}{4}$). Thus, for the antenna with the 2.7° half-cone angle, $R_{23} = 317$ ($R_k = 450$) while for the antenna with the 0.027° half-cone angle, $R_{23} = 710$ ($R_k = 1000$). The geometric mean of successive higher resonant resistances would be expected to approach closer yet to the characteristic resistance around which the impedance spiral converges.

The impedance spirals in Fig. 8-12 are for a biconical antenna. If the lower cone is replaced by a large ground plane, the impedance values are halved. Measured impedances of single cones with ground plane are presented in Fig. 8-13 for cones with half-angles of 5, 10, 20 and 30° and characteristic resistances ($R_k = Z_k$)

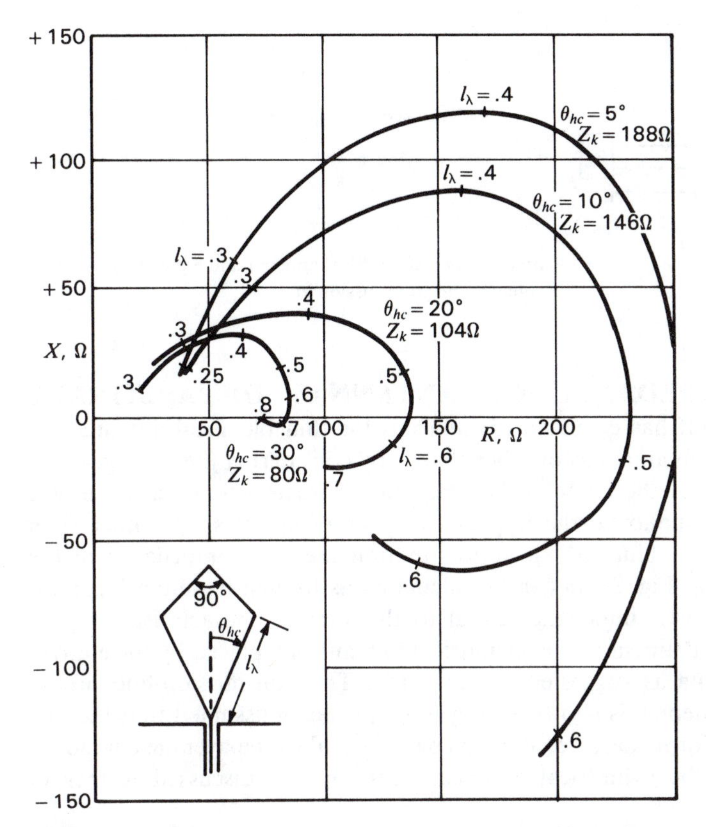

Figure 8-13 Measured input impedance of single cones with top hat as a function of cone length in wavelengths (l_λ). Impedance curves are presented for cones with half-angles of 5, 10, 20 and 30°.

of 188, 146, 104 and 80 Ω respectively.[1] The cones measured had a top hat consisting of an inverted cone of 90° total included angle (see insert in Fig. 8-13). It is to be noted that the trend toward reduced impedance variation with increasing cone angle, as predicted by the calculated curves of Fig. 8-12, is continued for the larger cone angles.

8-5 PATTERN OF BICONICAL ANTENNA.

The far-field pattern of a biconical antenna will be nearly the same as for an infinitesimally thin linear antenna provided that the cone angle is small. It is assumed that the current distribution is sinusoidal. Thus, Eqs. (5-5-10) and (5-5-11) can be used for thin biconical antennas, the substitution being made that $L = 2l$, where l is the length of one cone.

---

[1] The curves in Fig. 8-13 are plotted from data presented by A. Dorne, in *Very High Frequency Techniques*, by Radio Research Laboratory Staff, McGraw-Hill, New York, 1947, chap. 4.

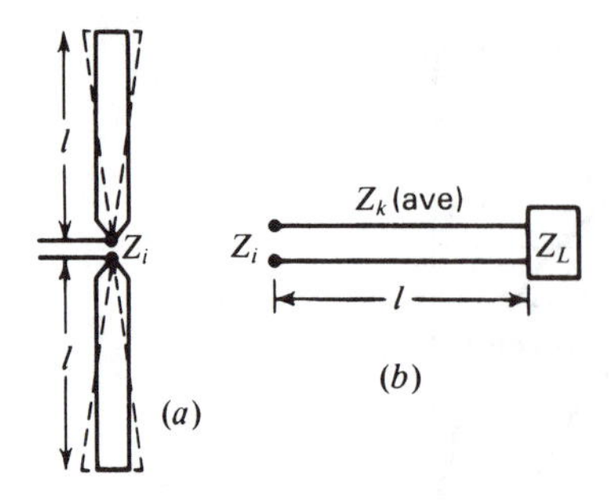

Figure 8-14 Cylindrical antenna and equivalent biconical antenna and transmission line.

8-6 INPUT IMPEDANCE OF ANTENNAS OF ARBITRARY SHAPE. Schelkunoff has extended his analysis for thin biconical antennas, as outlined above, to thin antennas of other shapes by considering the average characteristic impedance of the antenna. Whereas the characteristic impedance of a biconical antenna is uniform, the impedance of antennas of shape other than conical is nonuniform. Thus, as an approximation the input impedance of the cylindrical antenna in Fig. 8-14*a* can be calculated as though it were a biconical antenna of characteristic impedance equal to the *average* characteristic impedance of the cylindrical antenna. The cylindrical antenna is replaced by the equivalent biconical antenna as suggested in Fig. 8-14*a*. The transmission-line circuit, equivalent to the antenna, is shown in Fig. 8-14*b*, it being assumed that the line of length l has a uniform characteristic impedance equal to the average characteristic impedance of the cylindrical antenna. This topic is discussed further in Sec. 9-11.

8-7 MEASUREMENTS OF CONICAL AND TRIANGULAR ANTENNAS. THE BROWN-WOODWARD (BOW-TIE) ANTENNA. In 1945 Brown and Woodward made an extensive set of impedance measurements (published in 1952)[1] of both cones and triangles operating against a ground plane and also patterns of both biconical and triangular (bow-tie) dipoles. The impedance data apply, of course, also to biconical and triangular dipoles by multiplying impedance values by 2.

Figure 8-15 shows Brown and Woodward's results for conical and triangular antennas operating against a ground plane as a function of the length l_λ (or height)[2] for flare angles θ of 30, 60 and 90°. Although the conical measurements were made with open-ended cones, Brown and Woodward found no significant

[1] G. H. Brown and O. M. Woodward, "Experimentally Determined Radiation Characteristics of Conical and Triangular Antennas," *RCA Rev.*, **13**, 425–452, December 1952.

[2] Note that l_λ here is measured perpendicular to the ground plane whereas in Fig. 8-13 it is measured along the side of the cone.

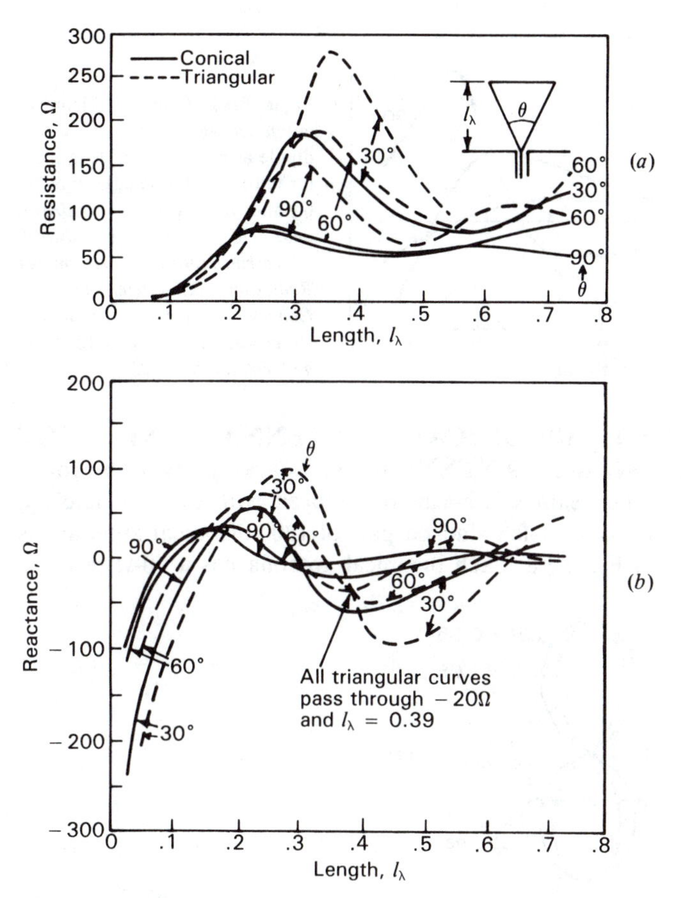

Figure 8-15 Measured resistance (a) and reactance (b) values for monoconical and monotriangular (flat sheet) antennas as a function of length l_λ for flare angles θ of 30, 60 and 90°. (*After G. H. Brown and O. M. Woodward, "Experimentally Determined Radiation Characteristics of Conical and Triangular Antennas,"* RCA Rev., **13**, *425–452, December 1952.*) Brown and Woodward give results in smaller flare angle increments between 0 and 90°.

difference in impedance values for a 60° cone with and without end caps. The gain of conical dipoles of length $2l_\lambda$ with respect to a $\lambda/2$ dipole is shown in Fig. 8-16. The gains are calculated from measured patterns.

Although the conical antennas have a smaller resistance fluctuation with frequency than the triangular antennas, the flat geometry of the triangles is attractive. The measured performance of a Brown-Woodward (bow-tie) antenna 34 cm long connected to a 300-Ω twin line for frequencies between 480 and 900 MHz (UHF TV channels 15 to 83) is presented in Fig. 8-17.

A biconical antenna with a flare angle of 120° is shown in Fig. 2-28c, which has a VSWR < 2 over a 6 to 1 bandwidth with a cone diameter $D = \lambda$ at the lowest frequency.

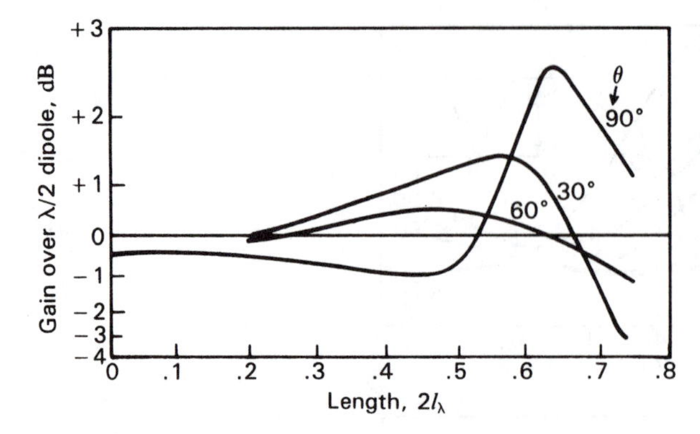

Figure 8-16 Gain of biconical antennas with respect to a $\lambda/2$ dipole as a function of length $2l_\lambda$ for full cone (flare) angles θ of 30, 60 and 90°. (*After G. H. Brown and O. M. Woodward, "Experimentally Determined Radiation Characteristics of Conical and Triangular Antennas,"* RCA Rev., **13**, *425–452, December 1952.*)

8-8 THE STACKED BICONICAL ANTENNA AND THE PHANTOM BICONICAL ANTENNA. By stacking two biconical antennas, the vertical-plane beam width can be reduced with approximately a 3-dB increase in gain. A coaxially-fed stacked pair of 120° biconical antennas is shown in cross section in Fig. 8-18. Each biconical antenna has a 50-Ω resistive

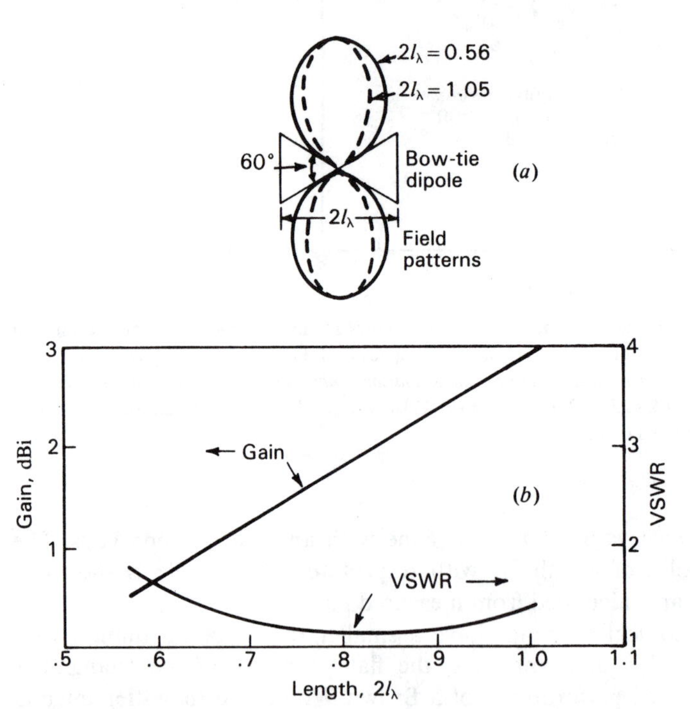

Figure 8-17 Gain in dBi and VSWR of UHF Brown-Woodward (bow-tie) antenna with 60° flare angle as a function of the length $2l_\lambda$. (*After G. H. Brown and O. M. Woodward, "Experimentally Determined Radiation Characteristics of Conical and Triangular Antennas,"* RCA Rev., **13**, *425–452, December 1952.*) The field patterns shown in (*a*) are actually those with the plane of the bow-tie perpendicular to the page (rotated 90° on its axis) instead of with the plane of the bow-tie parallel to the page as drawn.

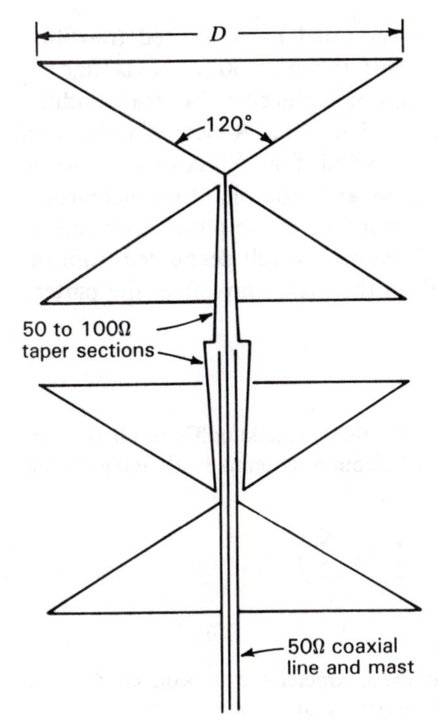

Figure 8-18 Stacked pair of broadband 120° biconical antennas in cross section. $D = \lambda$ at the lowest frequency used.

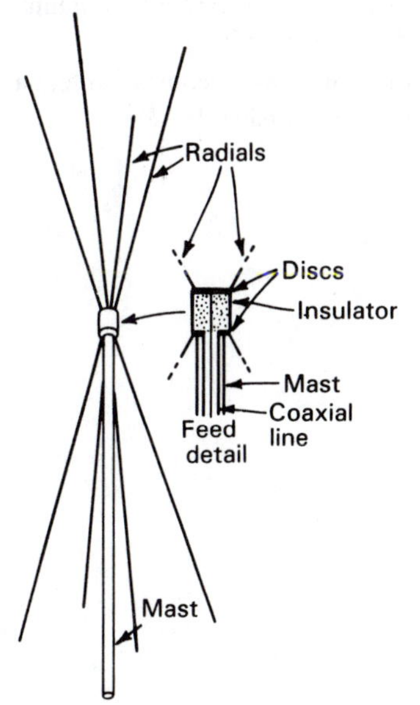

Figure 8-19 Phantom 30° biconical antenna with 4 radial rods replacing each cone.

input impedance (see Fig. 2-28*c*) which is transformed by a tapered transition section of coaxial line to 100 Ω at the junction with the main 50 Ω coaxial line.

For reduced wind resistance each cone can be replaced by several conducting radial rods resulting in a "phantom" biconical antenna. A view of a phantom arrangement for a 30° biconical antenna with 4 radials for each cone is shown in Fig. 8-19. By making the radial rods easily removable, the phantom biconical is well adapted for portable applications. A disadvantage of the phantom biconical antenna is that currents on the lower "cone" are not as well decoupled from the mast as with a solid cone. Currents induced on the mast may affect the pattern and gain adversely.

PROBLEMS[1]

8-1 Biconical antenna with unequal cone angles. Confirm Schelkunoff's result that the characteristic impedance of an unsymmetrical biconical antenna (with unequal cone angles) is

$$Z_k = 60 \ln \left(\cot \frac{\theta'_{hc}}{2} \cot \frac{\theta''_{hc}}{2} \right)$$

where θ'_{hc} = half the upper cone angle
θ''_{hc} = half the lower cone angle

8-2 Single cone and ground plane. Prove that the characteristic impedance Z_k for a single cone and ground plane is half Z_k for a biconical antenna.

★8-3 The 2° cone. Calculate the terminal impedance of a conical antenna of 2° total angle operating against a very large ground plane. The length l of the cone is $3\lambda/8$.

8-4 Bow-tie antenna. What is the gain (dBi) and input impedance of a 1λ Brown-Woodward bow-tie antenna with 60° flare angle?

8-5 Monotriangular antenna. What is the input impedance of a monotriangular antenna 0.39λ long coaxially fed from a ground plane if the flare angle is 90°?

8-6 Monoconical antenna. What is the input impedance of a monoconical antenna 0.39λ long coaxially fed from a ground plane if the full-cone angle is 90°?

[1]Answers to starred (★) problems are given in App. D.

CHAPTER 9

THE CYLINDRICAL ANTENNA; THE MOMENT METHOD (MM)

9-1 INTRODUCTION.[1] In previous chapters, the assumption is made that the current distribution on a finite antenna is sinusoidal. This assumption is a good one provided that the antenna is very thin. In this chapter, a method for calculating the current distribution of a cylindrical center-fed antenna will be discussed, taking into account the thickness of the antenna conductor.

This is a boundary-value problem. The antenna as a boundary-value problem was treated many years ago by Abraham,[2] who obtained an exact solution for a freely oscillating elongated ellipsoid of revolution. However, the earliest treatments of the cylindrical center-driven antenna as a boundary-value problem

[1] In other chapters sufficient steps are given in most analyses that the reader should be able to supply the intermediate ones without undue difficulty. However, this is not the case in this chapter since in most instances a large number of steps is omitted between those given in order to reduce the length of the development.

[2] M. Abraham, "Die electrischen Schwingungen um einen stabförmigen Leiter, behandelt nach der Maxwellschen Theorie," *Ann. Physik*, **66**, 435–472, 1898.

are those of Hallén[1] and King.[2] More recently the problem has been discussed by Synge and Albert.[3] Hallén's method leads to an integral equation, approximate solutions of which yield the current distribution. Knowing the current distribution and the voltage applied at the input terminals, the input impedance is then obtained as the ratio of the voltage to the current at the terminals.

Hallén's integral-equation method will not be presented in detail, but the important steps and results will be discussed in the following sections. Later in the chapter the Moment Method (MM) is applied to the solution of the current distribution, impedance and radar cross section of a short dipole antenna.

9-2 OUTLINE OF THE INTEGRAL-EQUATION METHOD.

Since this method is a long one, an outline of the important steps is given in this section.

The *objective* of the method is twofold:

1. To obtain the current distribution of a cylindrical center-fed antenna in terms of its length and diameter.
2. To obtain the input impedance.

An *outline of the procedure* is given by the following steps. These are treated more fully in the sections which follow.

1. The field **E** inside the conductor is expressed in terms of the current and skin effect resistance.
2. The field **E** outside the conductor is expressed in terms of the vector potential.
3. The tangential components of **E** are equated, obtaining a wave equation in the vector potential **A**.

Steps 1 through 3 are discussed in Sec. 9-3.

4. The wave equation in **A** is solved as the sum of a complementary function and a particular integral.

---

[1] Erik Hallén, "Theoretical Investigations into the Transmitting and Receiving Qualities of Antennae," *Nova Acta Regiae Soc. Sci. Upsaliensis*, ser. IV, **11**, no. 4, 1–44, 1938.

[2] L. V. King, "On the Radiation Field of a Perfectly Conducting Base-Insulated Cylindrical Antenna over a Perfectly Conducting Plane Earth, and the Calculation of the Radiation Resistance and Reactance," *Phil. Trans. R. Soc. (Lond.)*, **236**, 381–422, 1937.

[3] G. E. Albert and J. L. Synge, "The General Problem of Antenna Radiation. I," *Q. Appl. Math.*, **6**, 117–131, July 1948; and J. L. Synge, "The General Problem of Antenna Radiation and the Fundamental Integral Equation, With Application to an Antenna of Revolution. II," *Q. Appl. Math.*, **6**, 133–156, July 1948.

5. The constant C_2 in the solution is evaluated in terms of the conditions at the input terminals.
6. The vector potential **A** is expressed in terms of the antenna current I.
7. The value of C_2 from 5 and of **A** from 6 are inserted in the solution 4, obtaining Hallén's integral equation. This is an integral equation in the current I.

Steps 4 through 7 are discussed in Sec. 9-4.

8. A partial solution for the current I is then obtained by evaluating one of the integrals so that the current is expressed as the sum of several terms, some of which also involve I.
9. Neglecting certain terms in I, an approximate (zero-order) solution is obtained for I.
10. This value of I is substituted back in the current equation, obtaining a first-order approximation for the current. This process of iteration can be continued, yielding second-order and higher-order solutions.
11. The constant C_1 is evaluated and an asymptotic expansion obtained for the current; that is,

$$I_z = \frac{jV_T}{60\Omega'}\left[\frac{\sin \beta(l - |y|) + (b_1/\Omega') + (b_2/\Omega'^2) + \cdots}{\cos \beta l + (d_1/\Omega') + (d_2/\Omega'^2) + \cdots}\right]$$

where $\Omega' = 2 \ln (2l/a)$, where l is the half-length of the antenna and a the radius. The first-order approximation involves terms only as high as b_1/Ω' and d_1/Ω'. A second-order approximation involves b_2/Ω'^2 and d_2/Ω'^2, etc.

Steps 8 through 11 are discussed in Sec. 9-5.

12. The input impedance is then obtained as the ratio of the input terminal voltage V_T to the current at the input terminals I_T. This is discussed in Sec. 9-9.

9-3 THE WAVE EQUATION IN THE VECTOR POTENTIAL A.[1]

Consider the center-fed cylindrical antenna of total length $2l$ and diameter $2a$ as shown in Fig. 9-1. Let us first state the boundary conditions. Since the tangential components of the electric field are equal at a boundary,

$$E'_z = E_z \tag{1}$$

[1] The development in this section and in Sec. 9-4, leading up to Hallén's integral equation, follows the presentation of Ronold King and C. W. Harrison, Jr., "The Distribution of Current along a Symmetrical Center-Driven Antenna," *Proc. IRE*, **31**, 548–567, October 1943.

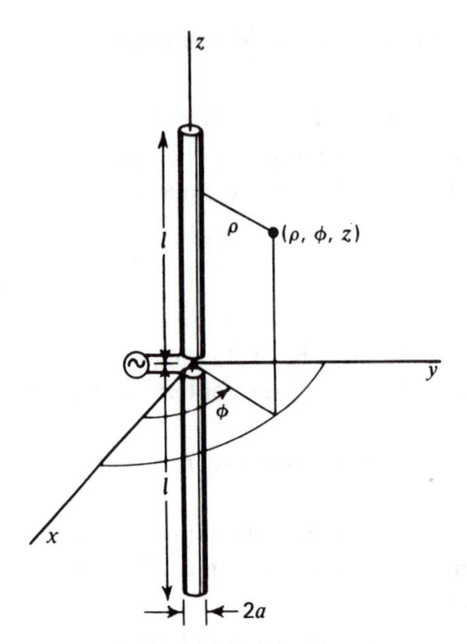

Figure 9-1 Symmetrical center-fed cylindrical antenna with relation to coordinates.

along the cylindrical surface. In (1), E'_z is the field just inside the conductor ($\rho = a - da$) and E_z is the field just outside the surface of the conductor ($\rho = a + da$), as indicated in Fig. 9-2. At the end faces of the antenna we have

$$E'_\rho = E_\rho \tag{2}$$

where E'_ρ = radial field just inside the face ($z = l - dl$)
E_ρ = radial field just outside ($z = l + dl$), as suggested in Fig. 9-2

To simplify the problem, it is assumed that l is much larger than a ($l \gg a$) and that the radius is very small compared to the wavelength ($\beta a \ll 1$). The effect of the end face can then be neglected and the current I_z taken equal to zero at $z = \pm l$. Then

$$E'_z = ZI_z \tag{3}$$

where Z = conductor impedance in ohms per meter length of the conductor due to skin effect
I_z = total current

The electric field **E** outside the conductor is derivable entirely from the vector potential **A**, i.e., as given by (5-4-10),

$$\mathbf{E} = -j\frac{c^2}{\omega}\nabla(\nabla \cdot \mathbf{A}) - j\omega\mathbf{A} \tag{4}$$

Neglecting the end faces, the tangential **E** outside the conductor will have only an E_z component. Since the current is entirely in the z direction, **A** has only

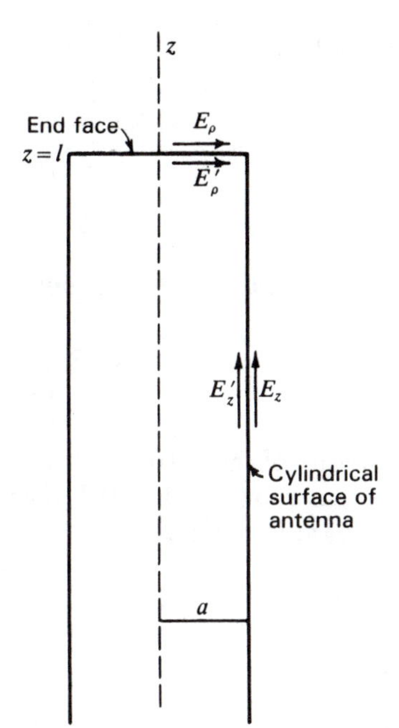

Figure 9-2 The tangential components of the electric field at the surface of the antenna are equal.

a z component. Hence, at the conductor surface (4) becomes

$$E_z = -j\frac{\omega}{\beta^2}\left(\frac{\partial^2 A_z}{\partial z^2} + \beta^2 A_z\right) \tag{5}$$

Now equating (3) and (5) in accordance with the boundary condition of (1), we obtain a wave equation,

$$\frac{\partial^2 A_z}{\partial z^2} + \beta^2 A_z = \frac{j\beta^2}{\omega} ZI_z \tag{6}$$

This completes the first three steps in the outline of Sec. 9-2.

9-4 HALLÉN'S INTEGRAL EQUATION.

We next proceed to obtain a solution of (9-3-6), which is a one-dimensional wave equation in the vector potential A_z. The equation is of the second order and first degree. If the antenna conductivity is infinite, $Z = 0$ and the equation becomes homogeneous. However, when Z is not zero, the equation is not homogeneous and its solution is given as the sum of a complementary function A_c and a particular integral A_r; that is,

$$A_z = A_c + A_r \tag{1}$$

Introducing the values of A_c and A_r, (1) becomes

$$A_z = -\frac{j}{c}(C_1 \cos \beta z + C_2 \sin \beta z) + \frac{jZ}{c}\int_0^z I(s) \sin \beta(z - s)\, ds \tag{2}$$

Assume that the antenna is excited symmetrically by a pair of closely spaced terminals. Then

$$I_z(z) = I_z(-z)$$

and

$$A_z(z) = A_z(-z) \tag{3}$$

The constant C_2 in (2) may be evaluated as equal to $\frac{1}{2}$ the applied terminal voltage V_T. Thus,

$$C_2 = \tfrac{1}{2}V_T \tag{4}$$

Let us now express the vector potential A_z in terms of the current on the antenna. For a conductor of length $z = -l$ to $z = +l$, as shown in Fig. 9-3, the vector potential A_z at any point outside the conductor or at its surface is

$$A_z = \frac{\mu}{4\pi}\int_{-l}^{+l} \frac{[I_{z_1}]}{r}\, dz_1$$

$$= \frac{\mu e^{j\omega t}}{4\pi}\int_{-l}^{+l} \frac{I_{z_1} e^{-j\beta r}}{r}\, dz_1 \tag{5}$$

where $r = [\rho^2 + (z - z_1)^2]^{1/2}$
z_1 = a point on the conductor $(-l \le z_1 \le +l)$

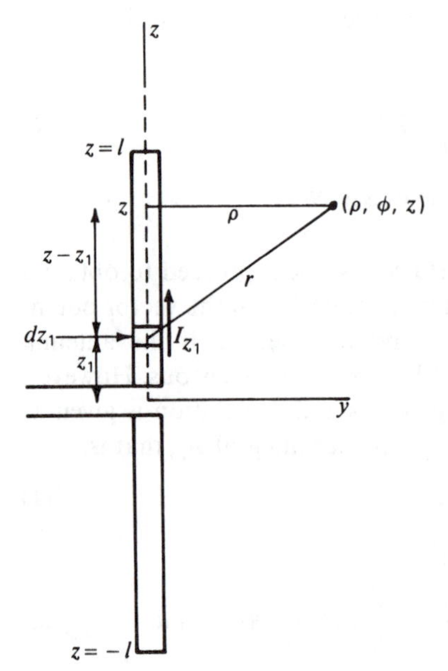

Figure 9-3 Construction for obtaining vector potential A_z.

Inserting the values of C_2 and A_z from (4) and (5) in (2) and rearranging yields Hallén's integral equation,[1]

$$\frac{jc\mu e^{j\omega t}}{4\pi}\int_{-l}^{+l}\frac{I_{z_1}e^{-j\beta r}}{r}\,dz_1 = C_1\cos\beta z + \frac{V_T}{2}\sin\beta|z| - Z\int_0^z I(s)\sin(z-s)\,ds \tag{6}$$

The absolute value sign on z in the second term of the right side of (6) has been introduced because of the symmetry condition of (3). Hallén's equation (6) is an equation in the current I_{z_1} on the conductor. If (6) could be solved for I_{z_1}, the current distribution could be obtained as a function of the antenna dimensions and the conductor impedance.

The term with Z has a negligible effect provided that the antenna is a good conductor. Thus, assuming that $Z = 0$ (conductivity infinite), we can put *Hallén's integral equation* in a simplified form as follows:

$$30j\int_{-l}^{+l}\frac{I_{z_1}e^{-j\beta r}}{r}\,dz_1 = C_1\cos\beta z + \frac{V_T}{2}\sin\beta|z| \tag{7}$$

In (7) we have put $e^{j\omega t} = 1$ and written $c\mu/4\pi = 30$. This completes steps 4 through 7 in the outline of Sec. 9-2.

9-5 FIRST-ORDER SOLUTION OF HALLÉN'S EQUATION.[2]

The problem now is to obtain a solution of (9-4-7) for the antenna current I_z which can be evaluated. As a first step in the solution, let the integral in (9-4-7) be expanded by adding and subtracting I_z; that is,

$$\int_{-l}^{+l}\frac{I_{z_1}e^{-j\beta r}}{r}\,dz_1 = \int_{-l}^{+l}\frac{I_z + I_{z_1}e^{-j\beta r} - I_z}{r}\,dz_1 \tag{1}$$

$$= I_z\int_{-l}^{+l}\frac{dz_1}{r} + \int_{-l}^{+l}\frac{I_{z_1}e^{-j\beta r} - I_z}{r}\,dz_1 \tag{2}$$

[1] An integral equation is an equation in which an unknown function appears under the integral sign. In this case, the unknown function is the antenna current I_{z_1}.

In the integral equation approach to a boundary-value problem, the independent variable ranges over the boundary surface (in this case, the antenna) so that the boundary conditions are incorporated in the integral equation. This is in contrast to the situation with the differential equation approach, in which the independent variable ranges throughout space, with a solution being sought that satisfies the boundary conditions.

[2] The development in this and following sections is similar to that given by Erik Hallén, "Theoretical Investigations into the Transmitting and Receiving Qualities of Antennae," *Nova Acta Regiae Soc. Sci. Upsaliensis*, ser. IV, **11**, no. 4, 1–44, 1938; also by Ronold King and C. W. Harrison, Jr., "The Distribution of Current along a Symmetrical Center-Driven Antenna," *Proc. IRE*, **31**, 548–567, October 1943.

Integrating the first term in (2) and putting $\rho = a$ we have

$$\int_{-l}^{+l} \frac{dz_1}{r} = \Omega' + \ln\left[1 - \left(\frac{z}{l}\right)^2\right] + \delta \tag{3}$$

where

$$\Omega' = 2 \ln \frac{2l}{a} = 2 \ln \frac{\text{total antenna length}}{\text{conductor radius}} \tag{4}$$ [1]

and

$$\delta = \ln \left\{\frac{1}{4}\left[\sqrt{1 + \left(\frac{a}{l - z}\right)^2} + 1\right]\left[\sqrt{1 + \left(\frac{a}{l + z}\right)^2} + 1\right]\right\} \tag{5}$$

Substituting (3) into (2), and this in turn into (9-4-7), yields

$$I_z = \frac{-j}{30\Omega'}\left(C_1 \cos \beta z + \frac{1}{2} V_T \sin \beta |z|\right)$$
$$- \frac{1}{\Omega'}\left\{I_z \ln\left[1 - \left(\frac{z}{l}\right)^2\right] + I_z \delta + \int_{-l}^{+l} \frac{I_{z_1} e^{-j\beta r} - I_z}{r}\, dz_1\right\} \tag{6}$$

At the ends of the antenna the current is zero. Thus, when $z = l$, $I_z = 0$ so that (6) reduces to

$$0 = \frac{-j}{30\Omega'}\left(C_1 \cos \beta l + \frac{1}{2} V_T \sin \beta l\right) + \frac{1}{\Omega'} \int_{-l}^{+l} \frac{I_{z_1} e^{-j\beta r_1}}{r_1}\, dz_1 \tag{7}$$

where $r_1 = \sqrt{(l - z_1)^2 + a^2}$

Now subtracting (7) from (6) as done by Hallén, we have

$$I_z = \frac{-j}{30\Omega'}\left[C_1(\cos \beta z - \cos \beta l) + \frac{1}{2} V_T(\sin \beta |z| - \sin \beta l)\right]$$
$$- \frac{1}{\Omega'}\left\{I_z \ln\left[1 - \left(\frac{z}{l}\right)^2\right] + I_z \delta + \int_{-l}^{+l} \frac{I_{z_1} e^{-j\beta r} - I_z}{r}\, dz_1 - \int_{-l}^{+l} \frac{I_{z_1} e^{-j\beta r_1}}{r_1}\, dz_1\right\} \tag{8}$$

Proceeding with Hallén's solution, the quantity in the braces in (8) is taken as zero so that the current I_z, given by the terms in the brackets, becomes a zero-order approximation, designated I_{z0}. Thus,

$$I_{z0} = -\frac{j}{30\Omega'}\left(C_1 F_{0z} + \frac{1}{2} V_T G_{0z}\right) \tag{9}$$

[1] Distinguish between the coefficient Ω' (with prime) as given by (4) and Ω (without prime) for ohms as given with impedances.

where the following symbols have been introduced

$$F_{0z} = \cos \beta z - \cos \beta l$$
$$G_{0z} = \sin \beta |z| - \sin \beta l \tag{10}$$

Subsituting I_{z0}, as given by (9), for I_z on the right side of (8), a first-order approximation I_{z_1} can be obtained for the current; that is,

$$I_{z1} = -\frac{j}{30\Omega'}\left[C_1\left(F_{0z} + \frac{F_{1z}}{\Omega'}\right) + \frac{1}{2}V_T\left(G_{0z} + \frac{G_{1z}}{\Omega'}\right)\right] \tag{11}$$

where $F_{1z} = F_1(z) - F_1(l)$

$$F_1(z) = -F_{0z} \ln\left[1 - \left(\frac{z}{l}\right)^2\right] + F_{0z}\delta - \int_{-l}^{+l} \frac{F_{0z_1}e^{-j\beta r} - F_{0z}}{r}\, dz_1$$

$$F_1(l) = -\int_{-l}^{+l} \frac{F_{0z_1}e^{-j\beta r_1}}{r_1}\, dz_1$$

$$G_{1z} = G_1(z) - G_1(l)$$

$G_1(z)$ is the same as $F_1(z)$ with G substituted for F and $G_1(l)$ is the same as $F_1(l)$ with G substituted for F.

If (11) is now substituted for I_z on the right side of (8), a second-order approximation for the current can be obtained. Continuing this process yields third-order and higher-order approximations, and the solution for the antenna current I_z takes the form

$$I_z = \frac{-j}{30\Omega'}\left[C_1\left(F_{0z} + \frac{F_{1z}}{\Omega'} + \frac{F_{2z}}{\Omega'^2} + \cdots\right) + \frac{1}{2}V_T\left(G_{0z} + \frac{G_{1z}}{\Omega'} + \frac{G_{2z}}{\Omega'^2} + \cdots\right)\right] \tag{12}$$

Substituting I_z as given by (12) into (7) yields

$$C_1 = -\frac{1}{2}V_T\left[\frac{G_0(l) + (1/\Omega')G_1(l) + \cdots}{F_0(l) + (1/\Omega')F_1(l) + \cdots}\right] \tag{13}$$

Inserting C_1 from (13) in (12) and rearranging, the current is given by the asymptotic expansion,

$$I_z = \frac{jV_T}{60\Omega'}\left[\frac{\sin \beta(l - |z|) + (b_1/\Omega') + (b_2/\Omega'^2) + \cdots}{\cos \beta l + (d_1/\Omega') + (d_2/\Omega'^2) + \cdots}\right] \tag{14}$$

where $b_1 = F_1(z) \sin \beta l - F_1(l) \sin \beta |z| + G_1(l) \cos \beta z - G_1(z) \cos \beta l$
$d_1 = F_1(l)$

Neglecting b_2, d_2 and higher-order terms, the first-order solution for the antenna current is

$$I_z = \frac{jV_T}{60\Omega'}\left[\frac{\sin \beta(l - |z|) + (b_1/\Omega')}{\cos \beta l + (d_1/\Omega')}\right] \tag{15}$$

The quantities b_1 and d_1 have been calculated in terms of real and imaginary functions[1] by King and Harrison for several values of l and curves given.[2] This completes steps 8 through 11 in the outline of Sec. 9-2.

9-6 LENGTH-THICKNESS PARAMETER Ω'. The above development is based on the assumptions that $l \gg a$ and $\beta a \ll l$. The condition that $l \gg a$ will be arbitrarily taken to mean that

$$\frac{l}{a} \geq 60 \tag{1}$$

The ratio l/a equals the ratio of the total length of the cylindrical antenna to the diameter. Thus,

$$\frac{\text{Total length}}{\text{Diameter}} = \frac{2l}{2a} = \frac{l}{a}$$

When $l/a = 60$, the value of Ω' from (9-5-4) is

$$\Omega' = 2 \ln \frac{2l}{a} = 2 \ln 120 \simeq 9.6$$

A graph of Ω' as a function of the ratio of the total length to the conductor diameter is presented in Fig. 9-4.

Another reason for restricting l/a to large values ($l/a \geq 60$) is that for asymptotic convergence of (9-5-14) Ω' must exceed a certain value. If Ω' is too small, the series may diverge.

9-7 EQUIVALENT RADIUS OF ANTENNAS WITH NON-CIRCULAR CROSS SECTION. The above discussion in this and preceding sections deals with uniform cylindrical antennas, i.e., antennas of circular cross section (radius $= a$). According to Hallén,[3] uniform antennas with non-circular cross section can also be treated by taking an equivalent radius. For squares cross sections of side length g (Fig. 9-5), the equivalent radius is

$$a = 0.59g \tag{1}$$

while for thin flat strips of width w the equivalent radius is

$$a = 0.25w \tag{2}$$

---

[1] $b_1 = M_1^{\text{I}} + jM_1^{\text{II}}$ and $d_1 = A_1^{\text{I}} + jA_1^{\text{II}}$.

[2] Ronold King and C. W. Harrison, Jr., "The Distribution of Current along a Symmetrical Center-Driven Antenna," *Proc. IRE*, **31**, 548–567, October 1943.

[3] Erik Hallén, "Theoretical Investigations into the Transmitting and Receiving Qualities of Antennae," *Nova Acta Regiae Soc. Sci. Upsaliensis*, ser. IV, **11**, no. 4, 1–44, 1938.

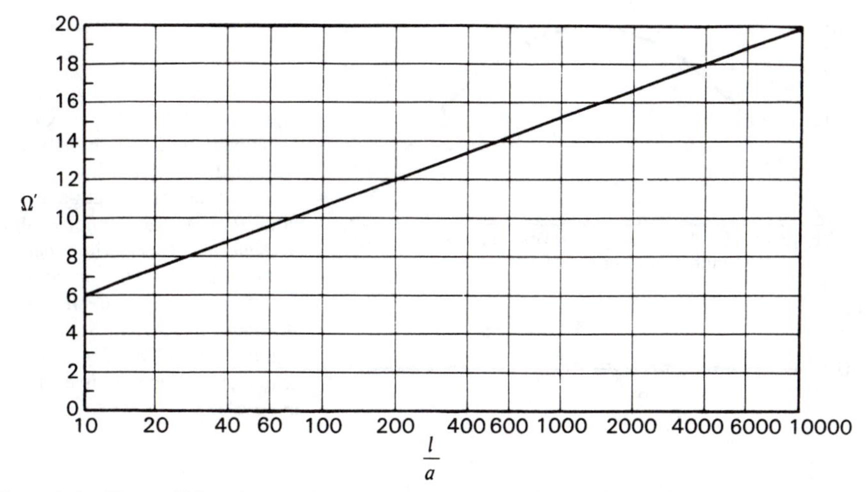

Figure 9-4 The coefficient Ω' as a function of the total length-diameter ratio ($2l/2a$) or length-radius ratio (l/a) of a cylindrical antenna.

For any shape of cross section there exists equivalent radius and hence a value of Ω'. In all cases it is assumed that the cross section is uniform over the entire length of the antenna.

9-8 CURRENT DISTRIBUTIONS.

The amplitude and phase of the current along cylindrical antennas of three lengths and two values of the total length-diameter ratio (l/a) are presented in Figs. 9-6, 9-7 and 9-8. Figure 9-6 is for a $\lambda/2$ antenna ($2l = \lambda/2$), Fig. 9-7 for a full-wavelength antenna ($2l = \lambda$) and Fig. 9-8 for a $5\lambda/4$ antenna ($2l = 5\lambda/4$). For each length the relative amplitude and phase of the current are presented for $\Omega' = 10$ and $\Omega' = \infty$ corresponding to total length-diameter ratios (l/a) of 75 and ∞. The amplitude curves are adjusted to the same maximum value, and all phase curves are adjusted to the same value at the ends of the antenna.

It is generally assumed that the current distribution of an infinitesimally thin antenna ($l/a = \infty$) is sinusoidal, and that the phase is constant over a $\lambda/2$

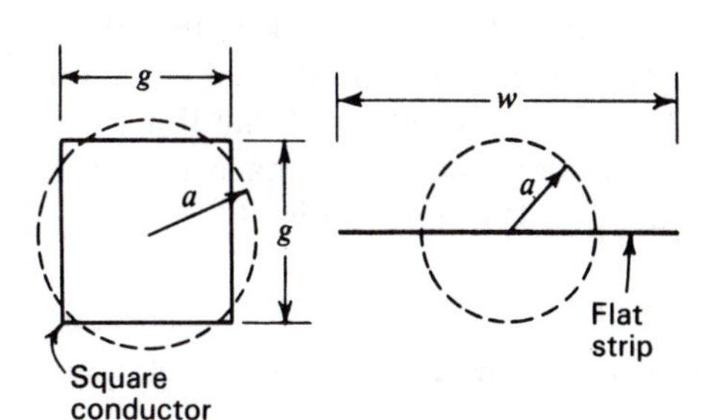

Figure 9-5 Conductors of square and flat cross section with equivalent circular conductors of radius a.

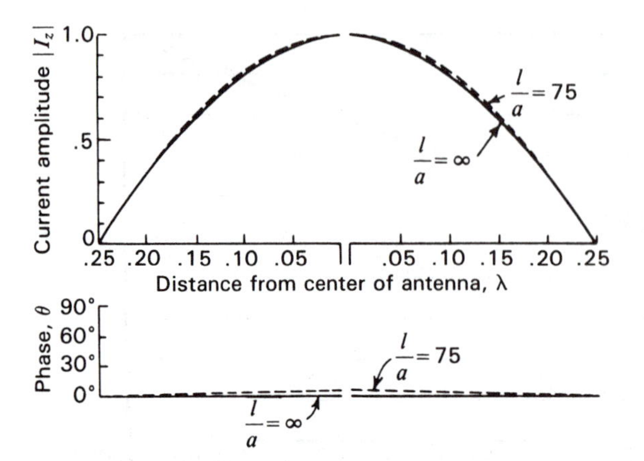

Figure 9-6 Relative current amplitude and phase along a center-fed $\lambda/2$ cylindrical antenna ($2l = \lambda/2$) for total length-diameter ratios (l/a) of 75 and infinity. (*After R. King and C. W. Harrison, Jr., "The Distribution of Current along a Symmetrical Center-Driven Antenna,"* Proc. IRE, **31**, *548–567, October 1943.*) Distance from the center of the antenna is expressed in wavelengths.

interval, changing abruptly by 180° between intervals. This behavior is illustrated by the solid curves in Figs. 9-6, 9-7 and 9-8.

The dashed curves illustrate the current amplitude and phase variation for $l/a = 75$ ($\Omega' = 10$). The difference between these curves and the solid curves ($l/a = \infty$) is not large but is appreciable. The dashed curves ($l/a = 75$) are from the distributions given by King and Harrison[1] as calculated from (9-5-15), the current being expressed in terms of its amplitude and the phase angle relative to a reference point. Thus,

$$I_z = |I_z| \underline{/\theta} \tag{1}$$

The effect of the length-thickness ratio on the current amplitude is well illustrated by Fig. 9-7 for a full-wavelength antenna. When the antenna is infinitesimally thin, the current is zero at the center. As the antenna becomes thicker, the current minimum increases and at the same time shifts slightly toward the end of the antenna. For still thicker antennas ($l/a < 75$), Eq. (9-5-15) is no longer a good approximation for the current, but it might be expected that the above trend would continue.

The effect of the length-thickness ratio on the phase variation is well illustrated by Fig. 9-8 for a $5\lambda/4$ antenna. When the antenna is infinitesimally thin, the phase varies as a step function, being constant over $\lambda/2$ and changing by 180° at the end of the $\lambda/2$ interval (solid line, Fig. 9-8). This type of phase variation is observed in a pure standing wave. As the antenna becomes thicker, the phase shift at the end of the $\lambda/2$ interval tends to become less abrupt (dashed curve for $l/a = 75$). For still thicker antennas ($l/a < 75$), it might be expected that this trend would continue and for very thick antennas would tend to approach that of a pure traveling wave, as indicated by the straight dashed lines in Fig. 9-8.

[1] Ronold King and C. W. Harrison, Jr., "The Distribution of Current along a Symmetrical Center-Driven Antenna," *Proc. IRE*, **31**, 548–567, October 1943.

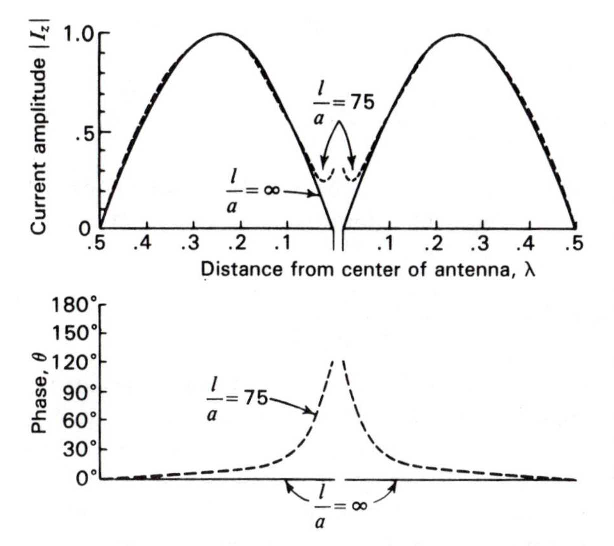

Figure 9-7 Relative current amplitude and phase along a center-fed full-wavelength cylindrical antenna ($2l = \lambda$) for total length-diameter ratios (l/a) of 75 and infinity. (*After R. King and C. W. Harrison, Jr., "The Distribution of Current along a Symmetrical Center-Driven Antenna,"* Proc. IRE, **31**, *548–567, October 1943.*) Distance from the center of the antenna is expressed in wavelengths.

9-9 INPUT IMPEDANCE.

The input impedance Z_T of a center-fed cylindrical antenna is found by taking the ratio of the input or terminal voltage V_T and the current I_T at the input terminals; that is,

$$Z_T = \frac{V_T}{I_T} = R_T + jX_T \tag{1}$$

where $I_T = I_z(0)$
R_T = terminal resistance
X_T = terminal reactance

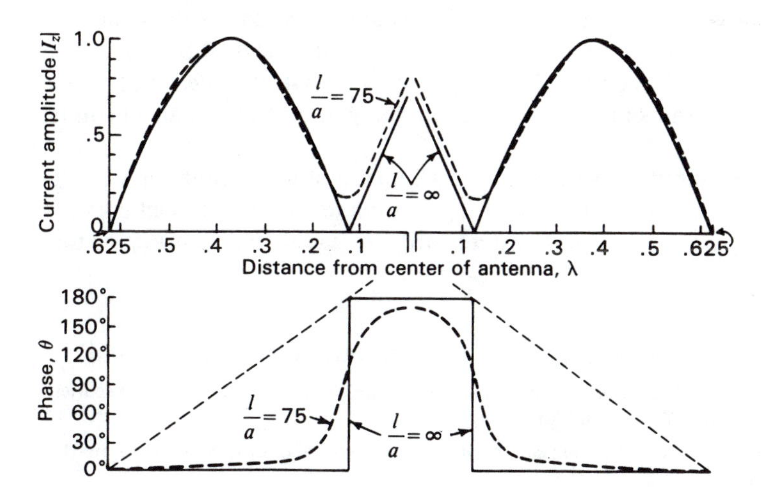

Figure 9-8 Relative current amplitude and phase along a center-fed $5\lambda/4$ cylindrical antenna ($2l = 5\lambda/4$) for total length-diameter ratios (l/a) of 75 and infinity. (*After R. King and C. W. Harrison, Jr., "The Distribution of Current along a Symmetrical Center-Driven Antenna,"* Proc. IRE, **31**, *548–567, October 1943.*) Distance from the center of the antenna is expressed in wavelengths.

Therefore, setting $z = 0$ in (9-5-15) and inserting this value of current in (1) yields Hallén's relation for the input impedance,

$$Z_T = -j60\Omega' \left[\frac{\cos \beta l + (d_1/\Omega')}{\sin \beta l + (b_1/\Omega')} \right] \tag{2}$$

This is a first-order approximation for the input impedance. If the second-order terms are included [see (9-5-14)], Hallén's input-impedance expression has the form

$$Z_T = -j60\Omega' \left[\frac{\cos \beta l + (d_1/\Omega') + (d_2/\Omega'^2)}{\sin \beta l + (b_1/\Omega') + (b_2/\Omega'^2)} \right] \tag{3}$$

This relation has been evaluated by Hallén[1] who has also presented the results in chart form.[2] Impedance spirals based on Hallén's data are presented in Fig. 9-9 for center-fed cylindrical antennas with ratios of total length to diameter (l/a) of 60 and 2000. The half-length l of the antenna is given along the spirals in free-space wavelengths. The impedance variation is that which would be obtained as a function of frequency for an antenna of fixed physical dimensions. The difference in the impedance behavior of the thinner antenna ($l/a = 2000$) and of the thicker antenna ($l/a = 60$) is striking, the variation in impedance with frequency of the thicker antenna being much less than that of the thinner antenna.

The impedance, given by (2) or (3), applies to center-fed cylindrical antennas of total length $2l$ and diameter $2a$. To obtain the impedance of a cylindrical stub antenna of length l and diameter $2a$ operating against a very large perfectly conducting ground plane, (2) and (3) are divided by 2. The impedance curve based on Hallén's calculations for a cylindrical stub antenna with an l/a ratio of 60 is given by the solid spiral in Fig. 9-10. The length l of the stub is indicated in free-space wavelengths along the spiral. The measured impedance variation of the same type of antenna ($l/a = 60$) as given by Dorne[3] is also shown in Fig. 9-10 by the dashed spiral. The agreement is good considering the fact that the measured curve includes the effect of the shunt capacitance at the gap and the small but finite antenna terminals.

The measured input impedance of a cylindrical stub antenna with an l/a ratio of 20 is shown in Fig. 9-11. Comparing this curve with the dashed curve of Fig. 9-10, it is apparent that the trend toward decreased impedance variation

[1] Erik Hallén, "On Antenna Impedances," *Trans. Roy. Inst. Technol.*, Stockholm, no. 13, 1947.

[2] Erik Hallén, "Admittance Diagrams for Antennas and the Relation between Antenna Theories," Cruft Laboratory Tech. Rept. 46, Harvard University, 1948.

[3] A. Dorne, in *Very High Frequency Techniques*, by Radio Research Laboratory Staff, McGraw-Hill, New York, 1947, chap. 4.

See also G. H. Brown and O. M. Woodward, "Experimentally Determined Impedance Characteristics of Cylindrical Antennas," *Proc. IRE*, **33**, 257–262, April 1945; D. D. King, "The Measured Impedance of Cylindrical Dipoles," *J. Appl. Phys.*, **17**, 844–852, October 1946; and C-T. Tai, "Dipoles and Monopoles," in *Antenna Engineering Handbook*, McGraw-Hill, 1984, chap. 4.

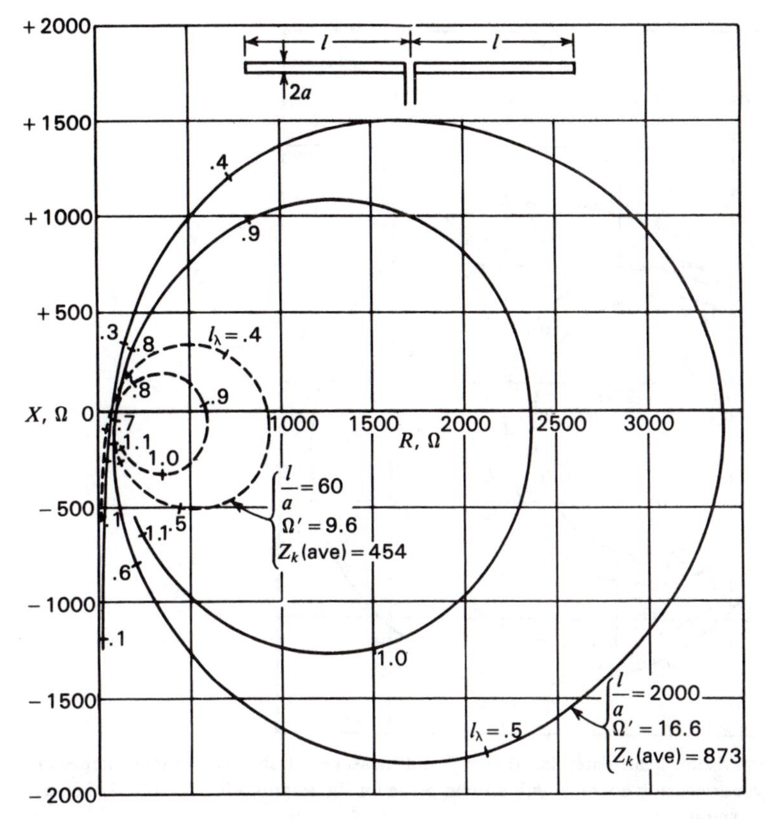

Figure 9-9 Calculated input impedance ($R + jX$, Ω') for cylindrical center-fed antennas with ratios of total length to diameter ($2l/2a$) of 60 and 2000 as a function of l_λ (along spiral). (*After E. Hallén.*)

with smaller l/a ratio (increased thickness) suggested by Fig. 9-9 is continued to smaller l/a ratios. A measured impedance curve for $l/a = 472$ is also included in Fig. 9-11.[1]

An antenna is said to be resonant when the input impedance is a pure resistance. On the impedance diagrams of Figs. 9-9, 9-10 and 9-11 resonance occurs where the spirals cross the $X = 0$ axis. At zero frequency all the impedance spirals start at $R = 0$ and $X = -\infty$. As the frequency increases, the reactance decreases and the resistance also increases, although more slowly. The first resonance occurs when the length l of the antenna is about $\lambda/4$. The resistance at the first resonance is designated R_1. As the frequency is increased, the length of the antenna becomes greater and the second resonance occurs when the length l is about $\lambda/2$. The resistance at the second resonance is designated R_2. At the third resonance (resistance $= R_3$), the antenna length l is about $3\lambda/4$ and at the

[1] The curves in Fig. 9-11 are based on data presented by Dorne (Ref. 3 on p. 372).

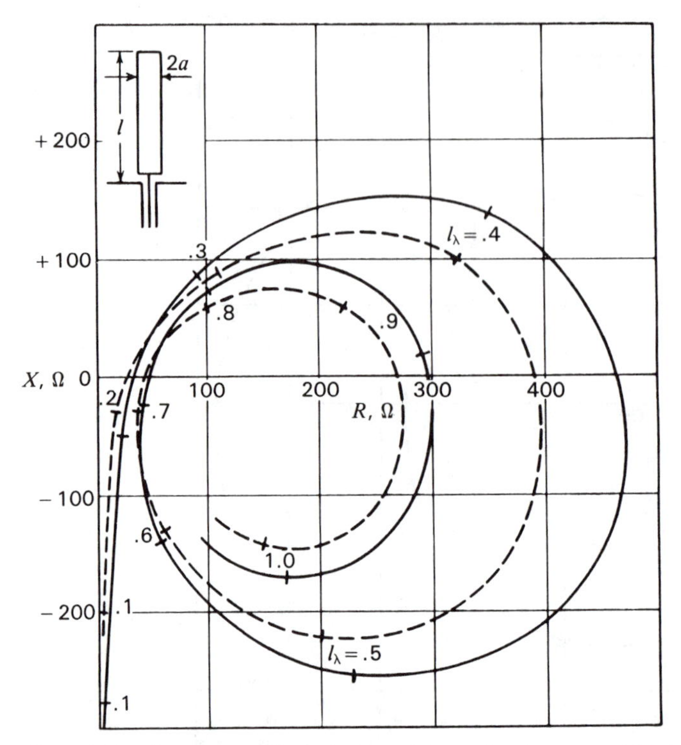

Figure 9-10 Comparison of calculated (solid curve) and measured (dashed curve) input impedance ($R + jX$, Ω′) for cylindrical stub antenna with ground plane for the length-radius ratio (l/a) of 60 as a function of l_λ (along spiral).

fourth resonance (resistance = R_4) l is about 1λ. As the frequency is increased indefinitely, an infinite number of such resonances can be obtained except where the impedance stays reactive.

Since it is common practice to operate antennas at or near resonance, the values of the resonant resistances are of interest. Curves based on Hallén's calculated graphs[1] are presented in Fig. 9-12 for the first four resonances of a cylindrical stub antenna with large ground plane as a function of the length-radius ratio (l/a). The lowest value of l/a for which Hallén gives data is 60, since the accuracy of (3) tends to deteriorate for smaller l/a values. Thus, the solid part of the curves ($l/a > 60$) are according to Hallén's calculated values. The dashed parts of the curves are extrapolations to smaller values of l/a. The extrapolation is without theoretical basis but is probably not much in error. A few measured values of resonant resistance from Dorne's data[2] are shown as points in

[1] Erik Hallén, "Admittance Diagrams for Antennas and the Relation between Antenna Theories," Cruft Laboratory Tech. Rept. 46, Harvard University, 1948.

[2] A. Dorne, in *Very High Frequency Techniques*, by Radio Research Laboratory Staff, McGraw-Hill, New York, 1947, chap. 4.

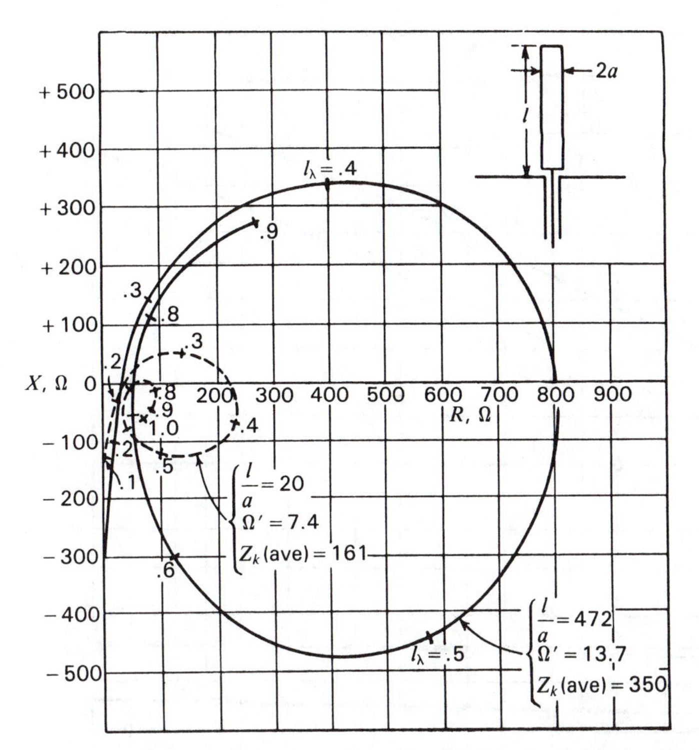

Figure 9-11 Measured input impedance ($R + jX$, Ω') of cylindrical stub antenna with ground plane for the length-radius ratio (l/a) of 20 and 472 as a function of l_λ (along spiral).

Fig. 9-12, the dotted lines indicating to which resonant resistance the points correspond.

Figure 9-12 illustrates the difference in the effect of antenna thickness on the resistance at odd and even resonances. The resistance at odd resonances (R_1, R_3, etc.) is nearly independent of the antenna thickness. The first resonant resistance R_1 is about 35 Ω and the third resonant resistance R_3 is about 50 Ω over a large range of l/a ratios. On the other hand, the antenna thickness has a large effect on the resistance at even resonances (R_2, R_4, etc.). The thicker the antenna, the smaller the resistance. For example, the second resonant resistance R_2 is about 200 Ω when $l/a = 10$ and increases to about 1500 Ω at $l/a = 1000$. The fourth resonant resistance behaves in a similar fashion, the values being somewhat less.

The difference in the resistance behavior at odd and even resonances is related to the current distribution. Thus, at odd resonances the antenna length l is an odd number of $\lambda/4$ (approximately), and a current maximum appears at or near the input terminals. At even resonances the antenna length l is an even number of $\lambda/4$ (approximately), and a current minimum appears at or near the input terminals. As indicated by the current distribution curves of Figs. 9-7 and 9-8, one of the most noticeable effects of an increase in antenna thickness is the

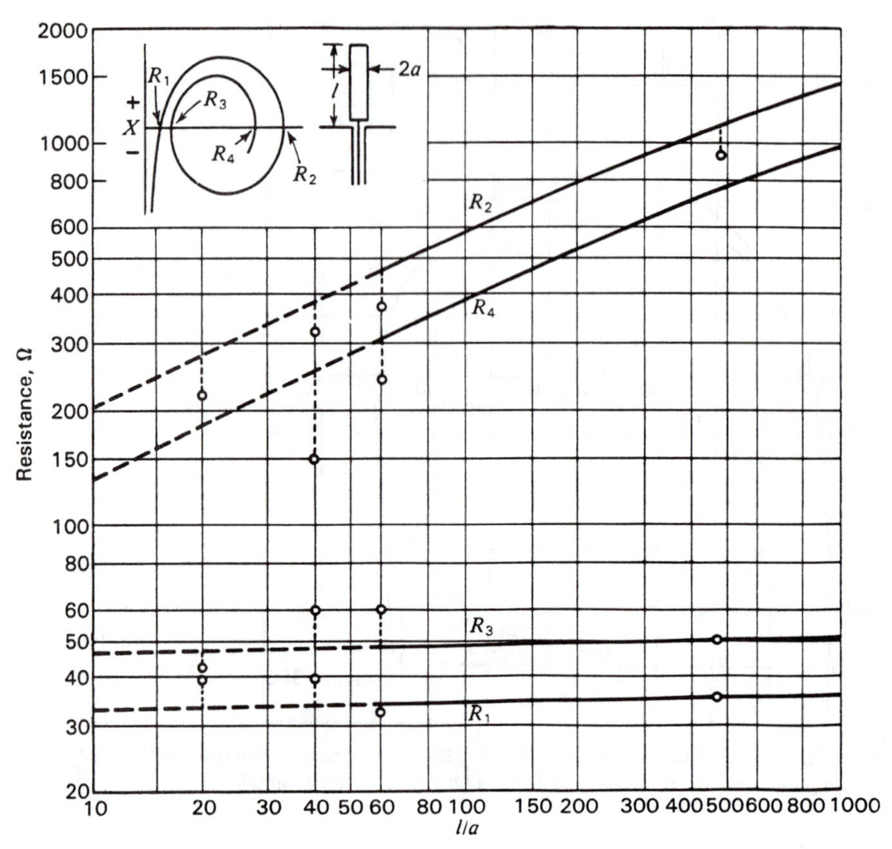

Figure 9-12 Resonant resistance of cylindrical stub antenna with ground plane as a function of the length-radius ratio (l/a). Curves are shown for the first four resonances. For cylindrical center-fed antennas (total length $2l$) multiply the resistance by 2.

increase of the current at current minima. Thus, when a current minimum is at or near the input terminals, an increase in the antenna thickness raises the input current I_T for a constant input voltage V_T so that the resonant resistance given by the ratio V_T/I_T is reduced.

9-10 PATTERNS OF CYLINDRICAL ANTENNAS. Formulas for calculating the far-field patterns of thin linear antennas were developed in Chap. 5. Although these relations apply strictly to infinitesimally thin conductors, they provide a first approximation to the pattern of even a relatively thick cylindrical antenna. This is illustrated by the patterns in Fig. 9-13 for center-fed linear cylindrical antennas of total length $2l$ equal to $\lambda/2$, 1λ, $3\lambda/2$ and 2λ. The calculated patterns for infinitesimally thin antennas are shown in the top row. Three of these patterns were given previously in Fig. 5-10. In the next three rows patterns

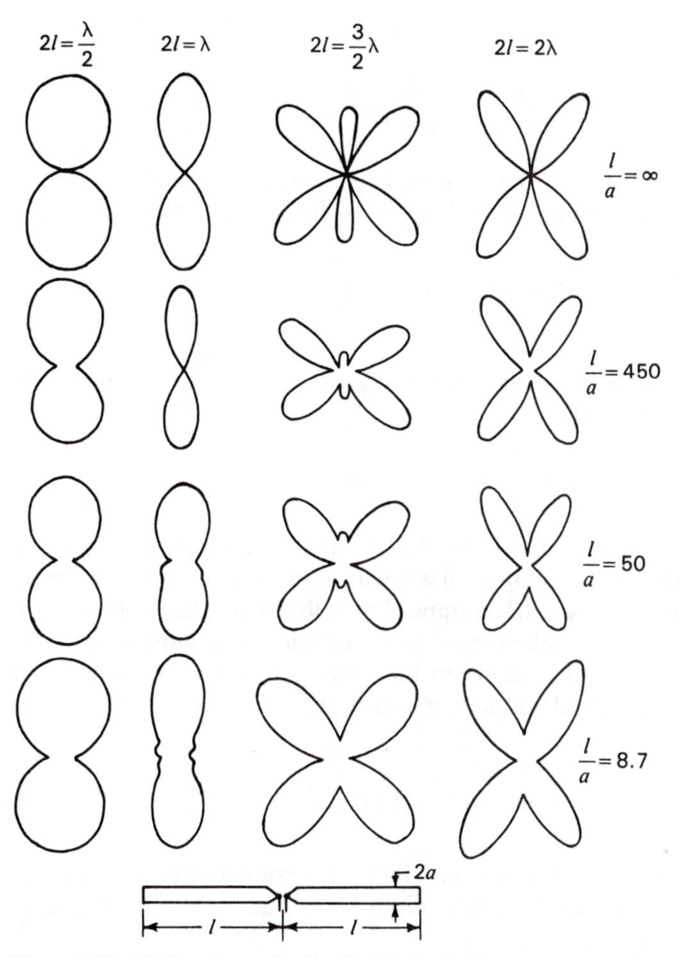

Figure 9-13 Field patterns of cylindrical center-fed linear antennas of total length $2l$ as a function of the total length-diameter ratio (l/a) and also as a function of the total length ($2l$) in wavelengths.

measured by Dorne[1] are given for l/a ratios of 450, 50 and 8.7. The principal effect of increased antenna thickness appears to be that some of the pattern nulls are filled in and that some minor lobes are obliterated (note the patterns in the third column for $2l = 3\lambda/2$).

9-11 THE THIN CYLINDRICAL ANTENNA. If the assumption is made that the cylindrical antenna is infinitesimally thin ($\Omega' \to \infty$), the current

---

[1] A. Dorne, in *Very High Frequency Techniques*, by Radio Research Laboratory Staff, McGraw-Hill, New York, 1947, chap. 4.

distribution given by (9-5-14) or (9-5-15) reduces to

$$I_z = \frac{jV_T}{60\Omega'} \frac{\sin \beta(l - |z|)}{\cos \beta l} \tag{1}$$

Although Ω' approaches infinity, the ratio V_T/Ω' may be maintained constant by also letting V_T approach infinity. According to (1), the shape of the current distribution is sinusoidal; that is,

$$I_z = k \sin \beta(l - |z|) \tag{2}$$

where k = a constant

The input impedance Z_T is the ratio V_T/I_T where I_T is the current at the terminals ($z = 0$). Thus from (1),

$$Z_T = \frac{V_T}{I_T} = -j60\Omega' \cot \beta l \tag{3}$$

In (3) we may regard Ω' as large but finite. The terminal impedance Z_T according to (3) is a pure reactance X_T. Equation (3) is identical to the relation for the input impedance of an open-circuited lossless transmission line of length βl (see App. A, Sec. A-2) provided that $60\Omega'$ is taken equal to the characteristic impedance of the line. If, by analogy, $60\Omega'$ is taken equal to the average characteristic impedance Z_k (ave) of the center-fed cylindrical antenna then, from the value of Ω' given in Sec. 9-6,

$$Z_k \text{ (ave)} = 60\Omega' = 120 \ln \frac{2l}{a} \tag{4}$$

This relation is of the same form as Schelkunoff's expression for the characteristic impedance Z_k of a thin biconical antenna given by (8-2-21) since for small cone angles $\theta_{hc} = a/l$ so that (8-2-21) becomes

$$Z_k = 120 \ln \frac{2l}{a} \tag{5}$$

where a = end radius of the cone as shown in Fig. 9-14.

The average characteristic impedance of a center-fed cylindrical antenna as given by Schelkunoff is

$$Z_k \text{ (ave)} = 120\left(\ln \frac{2l}{a} - 1\right) \tag{6}$$

The average impedance of a cylindrical stub antenna with a large ground plane is $\frac{1}{2}$ the value of (6).

As $l/a \to \infty$, (6) reduces to the form given in (4). However, for finite values of l/a, the average characteristic impedance of a cylindrical antenna is the same as for a biconical antenna of the same length l but with an end radius a which is larger than the radius of the cylindrical conductor. This is suggested in Fig.

Figure 9-14 Biconical antenna of end radius a and length l.

8-14a. For example, a cylindrical antenna with an l/a ratio of 500 has an average characteristic impedance equal to that of a biconical antenna of the same length with an end radius 2.8 times larger than the radius of the cylindrical conductor.

In Fig. 9-9 the calculated input impedance is presented for cylindrical center-fed antennas with total length-diameter ratios ($2l/2a = l/a$) of 60 and 2000. The average characteristic impedance of these antennas by (6) is 454 and 873 Ω respectively. The curve for the l/a ratio of 60 [Z_k (ave) = 454 Ω] has approximately the same form as the calculated impedance spiral in Fig. 8-12 for a 2.7° half-angle biconical antenna ($Z_k = 450$ Ω).

In Fig. 9-11 the measured input impedance is shown for cylindrical stub antennas with l/a ratios of 20 and 472. The average characteristic impedance of these antennas as given by $\frac{1}{2}$ of (6) is 161 and 350 Ω respectively. The curve for $l/a = 20$ [Z_k (ave) = 161 Ω] is of the same general form (although displaced downward), as would be anticipated from Fig. 8-13 since a spiral for Z_k (ave) = 161 Ω should lie between those shown in Fig. 8-13 for $Z_k = 146$ Ω and $Z_k = 188$ Ω.

9-12 CYLINDRICAL ANTENNAS WITH CONICAL INPUT SECTIONS. It is common practice to construct cylindrical antennas with short conical sections at the input terminals as indicated at the bottom of Fig. 9-13. If the cylinders are of large cross section, the conical sections are particularly desirable in order to reduce the shunt capacitance at the gap. Since the measured impedance of an antenna includes the effect of the gap capacitance and the small but finite terminals, the measured impedances will differ more or less from the theoretical values. It is to be expected that measured values will agree better with calculated ones when end cones are used rather than when the ends of the cylinders are butted close together.

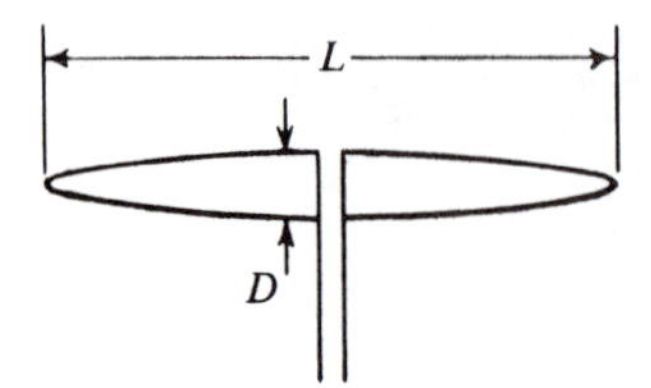

Figure 9-15 Prolate spheroidal antenna.

9-13 ANTENNAS OF OTHER SHAPES. THE SPHEROIDAL ANTENNA. The solution of a boundary-value problem may be facilitated if the boundary can be specified by one coordinate of an appropriate coordinate system. A spherical antenna or one in the shape of an elongated ellipsoid of revolution (prolate spheroid), as in Fig. 9-15, is amenable to such treatment since the surface of the spheroid corresponds to a particular value of one coordinate of a spheroidal coordinate system. By varying the eccentricity of the ellipsoid, one may study the properties of the sphere at the one extreme of eccentricity and of a long thin conductor at the other extreme. This problem has been treated at length by Stratton and Chu[1] and by Page and Adams.[2] Stratton and Chu give admittance and impedance curves for various length-diameter (L/D) ratios (see Fig. 9-15). For long, thin ellipsoids the impedance characteristics are similar to those deduced by other methods. The current distribution for thin $\lambda/2$ spheroids is also found to be nearly sinusoidal.

A point of interest is that for spheroids of the order of $\lambda/2$ long, the impedance variation with frequency decreases with decreasing L/D ratios (thicker spheroids); that is to say, resonance with thick spheroids is broader than with thin ones. This is in agreement with the well-known fact that thick antennas have broader band impedance characteristics than thin ones.

9-14 CURRENT DISTRIBUTIONS ON LONG CYLINDRICAL ANTENNAS. On a matched lossless transmission line an outgoing wave has a uniform current magnitude and a linear phase change with distance (Fig. 9-16). If the line is mismatched and the reflected (returning) wave is $\frac{1}{2}$ the magnitude of the outgoing wave, a standing wave appears on the line with VSWR given by[3]

$$\text{VSWR} = \frac{I_{\text{max}}}{I_{\text{min}}} = \frac{I_0 + I_1}{I_0 - I_1} = \frac{1 + \frac{1}{2}}{1 - \frac{1}{2}} = 3 \tag{1}$$

[1] J. A. Stratton and L. J. Chu, "Steady State Solutions of Electromagnetic Field Problems," *J. Appl. Phys.*, **12**, 230–248, March 1941.

[2] L. Page and N. I. Adams, "The Electrical Oscillations of a Prolate Spheroid," *Phys. Rev.*, **53**, 819–831, 1938.

[3] Properly this should be ISWR for the current standing-wave ratio. However, the VSWR = ISWR, although their standing-wave patterns are displaced in position.

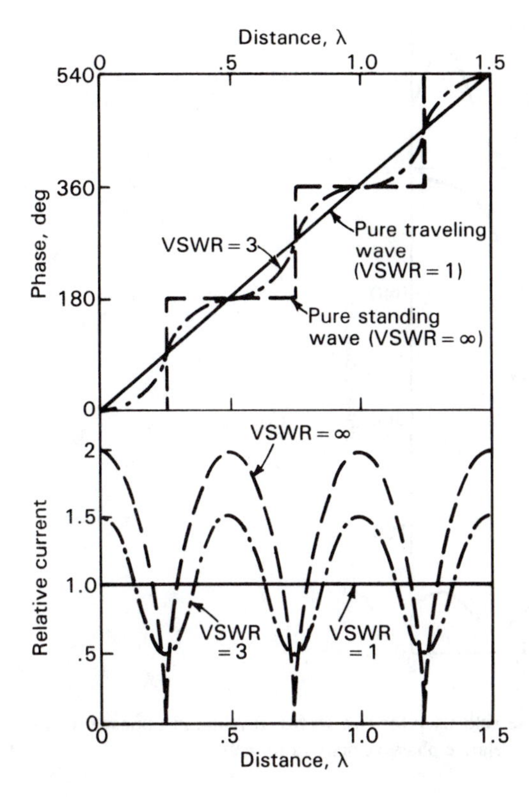

Figure 9-16 Current distribution and phase variation (1) of single uniform traveling wave (solid lines), (2) of two waves traveling in opposite directions with magnitudes 1 and $\frac{1}{2}$ (dash-dot lines) and (3) of 2 waves of equal magnitudes (dashed lines). The last case represents a full (pure) standing wave.

where I_0 = current magnitude of outgoing wave
I_1 = current magnitude of returning wave

The phase change is also nonuniform (fluctuating with distance), as indicated in Fig. 9-16.

When the line is completely mismatched (open- or short-circuited), so that the returning wave equals the outgoing wave, the VSWR = ∞ and the phase changes in a stepwise fashion (Fig. 9-16).

The phase velocity of a wave on the line is given by

$$v = \frac{\omega}{\beta(x)} = \frac{\omega}{d\phi/dx} \tag{2}$$

where $\omega = 2\pi f$, Hz
ϕ = phase, rad or deg
$\beta = 2\pi/\lambda$, rad (or deg) λ^{-1}

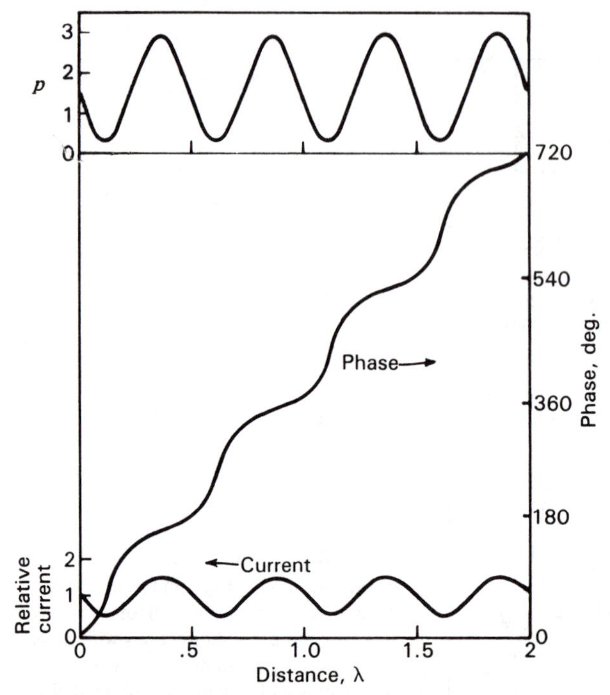

Figure 9-17 Current, phase and relative phase velocity p for two waves traveling in opposite directions with magnitudes 1 and $\frac{1}{2}$ (VSWR = 3 and relative phase velocity ratio = 9).

In general, for two opposite current waves of unequal magnitude and velocity, the phase velocity is[1]

$$v = \omega\{I_0^2 + I_1^2 + 2I_0 I_1 \cos[(\beta_1 + \beta_0)x + \gamma]\} \Bigg/ \Bigg\{ I_0^2 \beta_0 - I_1^2 \beta_1 + I_0 I_1(\beta_0 - \beta_1)\cos[(\beta_1 - \beta_0)x + \gamma] + \sin[(\beta_1 + \beta_0)x + \gamma]\left(I_0 \frac{dI_1}{dx} - I_1 \frac{dI_0}{dx}\right)\Bigg\} \quad (3)$$

where $\beta_0 = 2\pi/\lambda_0$
$\beta_1 = 2\pi/\lambda_1$
λ_0 = wavelength for outgoing wave, m
λ_1 = wavelength for returning wave, m
γ = phase difference of two waves, rad or deg

For the case where $\gamma = 0$, the velocity of both waves is the same ($\beta_1 = \beta_0$), I_0 and I_1 are constant with distance and (3) reduces to

$$v = \frac{\omega(I_0^2 + I_1^2 + 2I_0 I_1 \cos 2\beta x)}{\beta_0(I_0^2 - I_1^2)} \quad (4)$$

[1] J. A. Marsh, "A Study of Phase Velocity on Long Cylindrical Conductors," Ph.D. thesis, Ohio State University, 1949.

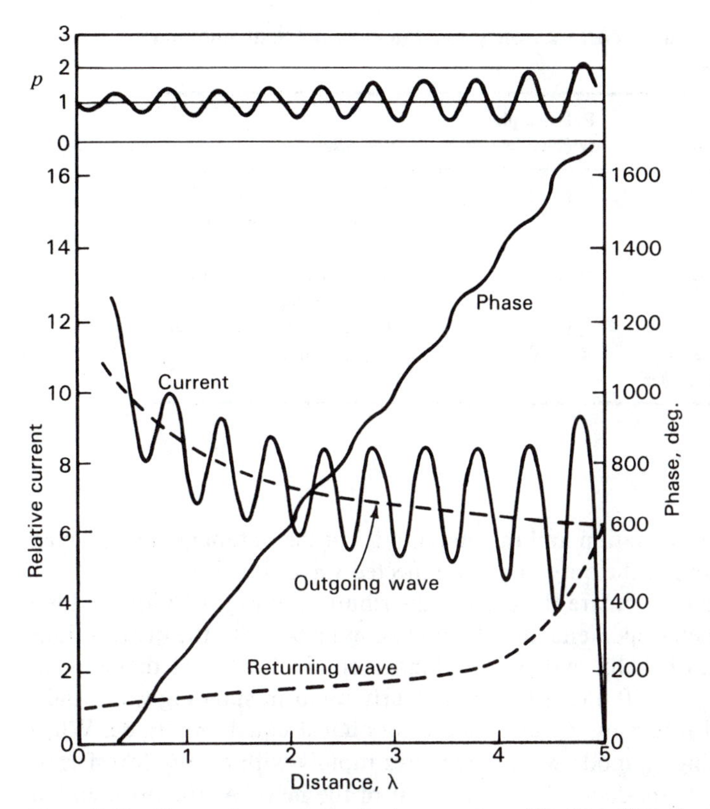

Figure 9-18 Measured current distribution on a long (5λ), thick (0.2λ diameter) cylindrical conductor with attendant phase and relative phase velocity (p) of the total wave. (*After J. A. Marsh, "A Study of Phase Velocity on Long Cylindrical Conductors," Ph.D. thesis, Ohio State University, 1949.*) Resolution into outgoing and returning (reflected) waves is indicated.

Dividing by c $(=\omega/\beta_0)$ yields the *relative phase velocity* p. The *ratio* of the maximum to minimum relative phase velocity is

$$\frac{p_{\max}}{p_{\min}} = \frac{(I_0 + I_1)^2}{(I_0 - I_1)^2} \tag{5}$$

Comparing (5) and (1) we note that

$$\frac{p_{\max}}{p_{\min}} = \frac{(I_0 + I_1)^2}{(I_0 - I_1)^2} = \left(\frac{I_{\max}}{I_{\min}}\right)^2 = \text{VSWR}^2 \tag{6}$$

The variation of p $(=v/c)$, phase (ϕ) and current magnitude ($|I|$) for $I_1 = \frac{1}{2}I_0$ are presented in Fig. 9-17 over a distance of 2λ. We note that the VSWR = 3 and the relative phase velocity ratio equals 9, and also that the relative phase velocity p is a maximum where I is a maximum.

The current and phase measured by Marsh along a 5λ open-ended cylindrical conductor 0.2λ in diameter are shown in Fig. 9-18, as well as the deduced

Table 9-1 Comparison of currents on long, thick cylindrical antenna and on helical antennas

Antenna	Mode	Relative phase velocity, $p(=v/c)$†	Remarks
Cylindrical conductor	All T_0	1	Gradual attenuation of outgoing and returning T_0 waves
Helix, $C_\lambda = 0.6$	All T_0	>1	Uniform equal outgoing and returning T_0 waves
Helix, $C_\lambda = 1.07$ (axial mode)	T_0 at ends, T_1 over remainder	>1 $<1(\sim 0.8)$	T_0 waves attenuate rapidly T_1 waves uniform

† For single traveling wave.

relative phase velocity variation and the magnitudes of the outgoing and reflected waves. The attenuation of the outgoing and reflected waves is evident.

It is interesting to compare the current distribution of Fig. 9-18 for the long (5λ), thick (0.2λ diameter open-ended) cylindrical conductor with the distributions of Fig. 7-3 for a 7-turn helix with much thinner conductor (0.02λ diameter at $C_\lambda = 1.07$). When $C_\lambda = 0.60$, the helix has a nearly uniform standing wave, indicating outgoing and returning T_0 mode waves of almost equal amplitude. When $C_\lambda = 1.07$ the outgoing T_0 mode wave attenuates rapidly with energy transferred to a nearly uniform T_1 mode wave over the rest of the helix. At the open end, a reflected or returning T_0 mode wave is excited which attenuates rapidly while transforming into a small nearly uniform returning T_1 mode wave.

On the cylindrical conductor (Fig. 9-18) a T_0 mode wave attenuates gradually over the length of the conductor and on reflection from the open end excites a gradually attenuating returning wave.

The behavior of the two antennas is summarized in Table 9-1.

9-15 INTEGRAL EQUATIONS AND THE MOMENT METHOD (MM) IN ELECTROSTATICS.

As an introduction to the moment method, let us consider its application to an electrostatics problem.

In calculus we deal with *differential equations* as, for example,

$$\frac{dx}{dt} = v \tag{1}$$

or the rate of change of distance (x) with time (t) equals the velocity (v). It is implied that we know x as a function of t or how x varies with time [$x(t)$].

On the other hand, suppose we know how the velocity varies as a function of time [$v(t)$]. Then the distance is given by the *integral* of the velocity with

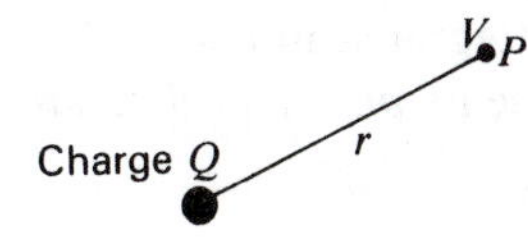

Figure 9-19 Electric potential V at point P is inversely proportional to the distance r from charge Q.

respect to time or

$$x = \int_0^{t_1} v(t)\,dt \tag{2}$$

If v is constant, then (2) becomes

$$x = vt_1 \tag{3}$$

or x is the distance traveled at velocity v in time t_1. Now suppose that x is known at $t = 0$ and $t = t_1$ but v is not known during this period of time. Then (2) is an *integral equation* with the problem being to obtain a solution for the velocity as a function of time $[v(t)]$.

Referring to Fig. 9-19, a basic relation of electrostatics is that the electric potential V at a point P due to a charge Q is given by

$$V = \frac{Q}{4\pi\varepsilon r} \quad \text{(V)} \tag{4}$$

where Q = charge, C
r = distance from P to the charge Q, m
ε = permittivity of medium, F m^{-1}

For a line of charge of density ρ_L (C m^{-1}) as in Fig. 9-20, then V at some observation point P is given by the integral of (4) over the length l of the line or

$$V = \frac{1}{4\pi\varepsilon}\int_0^l \frac{\rho_L(x)}{r}\,dx \tag{5}$$

where $\rho_L(x)$ = charge per unit of length of line as a function of x, C m^{-1}

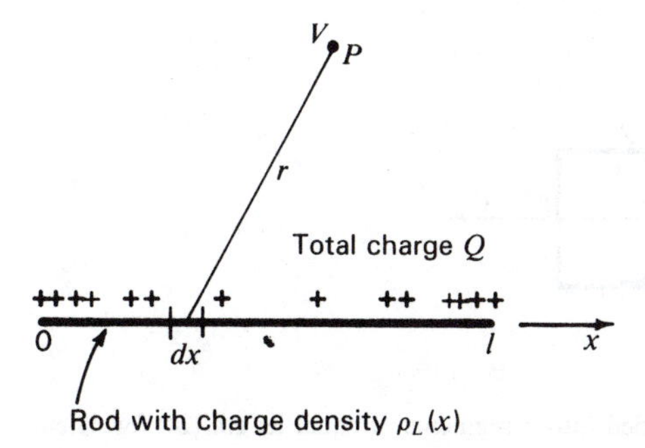

Figure 9-20 Electric potential V at point P due to a rod with charge density $\rho_L(x)$ which is a function of position (x).

If $\rho_L(x)$ is known as a function of x, then (5) can be integrated in a straightforward manner. However, if $\rho_L(x)$ is not known, (5) represents *an integral equation* with the problem being to find a solution for $\rho_L(x)$.

Example. Charge distribution on wire. Let the line be an isolated conducting rod or wire of radius a and length $l = 8a$ on which a total charge $+Q$ has been placed. Since like charges repel, it may be anticipated that the charge will tend to separate and pile up near the ends of the rod, making the charge distribution along the rod nonuniform. The problem is to determine this charge distribution $\rho_L(x)$ using an incremental numerical technique or moment method as an introduction to integral equations.

Solution. First, let us divide the rod of length l into 4 segments or increments with each segment of length $2a$ as in Fig. 9-21 ($l = 8a$). Let the total charge on segment 1 be Q_1 and on segment 2 be Q_2. By symmetry, the charge on segment 3 is the same as on segment 2, or Q_2. Likewise, the charge on segment 4 is equal to Q_1. Specifically, our problem is to find the ratio of Q_1 to Q_2 [a first step or approximation in solving for $\rho_L(x)$].

Let us assume that all of the charge on each segment is concentrated on a circle on the surface of the segment around its midpoint with the observation or test points on the wire axis. Since the distance r from an observation point to any point on the circle of charge is constant, we may consider that all of the charge of a segment is at one point (charge or source point), as in Fig. 9-21. The situation may now be regarded as one with 4 points of charge (source points) in empty space with the potential at observation or test points on the axis to be determined. Thus, from (4) the potential at point P_{12} is given by

$$V(P_{12}) = \frac{1}{4\pi\varepsilon}\left[\frac{Q_1}{\sqrt{a^2 + a^2}} + \frac{Q_2}{\sqrt{a^2 + a^2}} + \frac{Q_2}{\sqrt{a^2 + 9a^2}} + \frac{Q_1}{\sqrt{a^2 + 25a^2}}\right] \tag{6}$$

Likewise at point P_{23} the potential is given by

$$V(P_{23}) = \frac{1}{4\pi\varepsilon}\left[\frac{Q_1}{\sqrt{a^2 + 9a^2}} + \frac{Q_2}{\sqrt{a^2 + a^2}} + \frac{Q_2}{\sqrt{a^2 + a^2}} + \frac{Q_1}{\sqrt{a^2 + 9a^2}}\right] \tag{7}$$

A *boundary condition* is that (even though the charge density varies along the rod) the potential is constant. Therefore, $V(P_{12}) = V(P_{23})$ so that equating (6) and (7) we find that

$$Q_1 = 1.45Q_2 \tag{8}$$

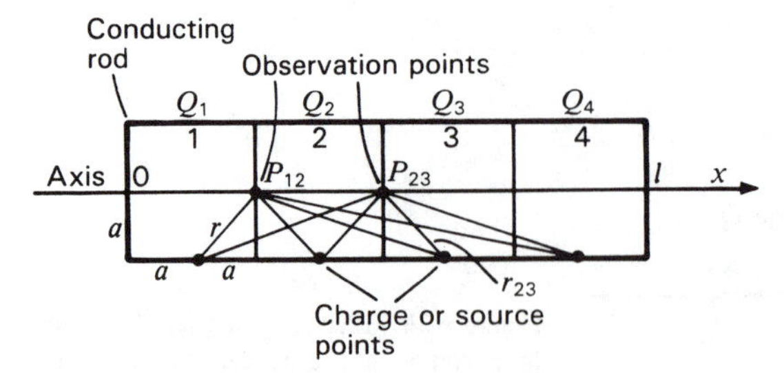

Figure 9-21 Charged rod of radius a divided into 4 segments of equal length ($2a$) for calculation of charge ratios.

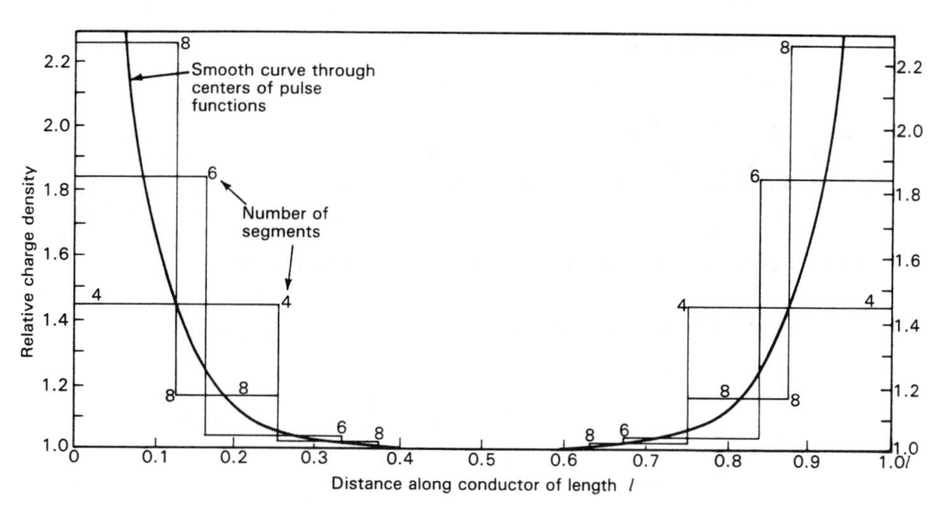

Figure 9-22 Relative charge density along straight conducting rod of radius a and length $8a$ as calculated by the moment method using 4, 6 and 8 segments.

Thus, the charge (or average charge density) for the outer segments is 45 per cent greater than for the inner segments and we can write

$$Q_1 : Q_2 = 1.45 : 1.00 \tag{9}$$

Dividing the rod into 6 segments and proceeding as above results in the ratios

$$Q_1 : Q_2 : Q_3 = 1.84 : 1.03 : 1.00 \tag{10}$$

The charge density distribution along the rod is shown by the step or pulse functions in Fig. 9-22 for the cases of 4, 6 and 8 segments. A smooth curve is also drawn through the centers of the pulse functions.

To simplify the above calculations we neglected the effect of the end surfaces of the rod. For 4 segments the cylindrical area of a segment is 4 times the end of the rod so that the effect of neglecting the ends is not large. However, with more segments the effect becomes greater, especially for the charge on the end segments.

Let us now discuss the problem more formally. From (5),

$$V = \frac{1}{4\pi\varepsilon} \int_0^l \frac{\rho_L(x)}{r}\, dx \tag{11}$$

Referring to Fig. 9-23, let the line be divided into N segments of equal length Δx with average charge density $\overline{\rho_L(x)_n}$ for segment Δx_n. Then the charge on segment n is given by

$$Q_n = \overline{\rho_L(x)_n}\, \Delta x_n, \qquad n = 1, 2, 3, \ldots, N \tag{12}$$

and the total charge on the wire by

$$Q = \sum_{n=1}^{N} Q_n \tag{13}$$

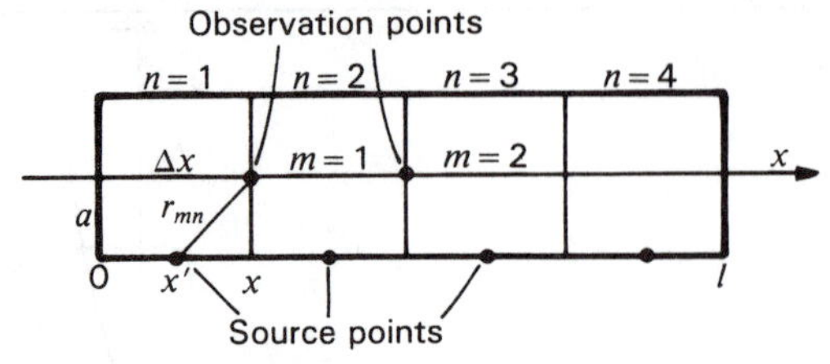

Figure 9-23 Charged rod for calculation of charge density distribution.

Equation (11) can now be written as

$$\sum_{n=1}^{N} l_{mn} Q_n = V_m \tag{14}$$

where

$$l_{mn} = \frac{1}{4\pi\varepsilon r_{mn}}, \qquad m = 1, 2, 3, \ldots, M \tag{15}$$

$$r_{mn} = \sqrt{a^2 + (x - x')^2}$$

a = rod radius

x = axial distance of observation or test point m

x' = axial distance of source point at middle of segment n

In matrix notation (14) is

$$[l_{mn}][Q_n] = [V_m] \tag{16}$$

or

$$\begin{bmatrix} l_{11} & l_{12} & l_{13} & \cdots & l_{1N} \\ l_{21} & l_{22} & l_{23} & \cdots & l_{2N} \\ l_{31} & l_{32} & l_{33} & \cdots & l_{3N} \\ \vdots & \vdots & \vdots & & \vdots \\ l_{M1} & l_{M2} & l_{M3} & \cdots & l_{M4} \end{bmatrix} \begin{bmatrix} Q_1 \\ Q_2 \\ Q_3 \\ \vdots \\ Q_N \end{bmatrix} = \begin{bmatrix} V_1 \\ V_2 \\ V_3 \\ \vdots \\ V_M \end{bmatrix} \tag{17}$$

Thus, the integral equation (11) has been transformed into a set of N simultaneous linear algebraic equations (17) where l_{mn} represents a known function (the inverse distance relation), V_m represents potentials determined by the boundary conditions and Q_n represents the N unknown charges whose values are sought.

In the charged rod example we have from symmetry that $M = N/2$ and from the boundary condition that

$$V_1 = V_2 = V_3 = \cdots \tag{18}$$

For 4 segments ($m = 2$, $n = 4$), (17) then reduces to

$$\begin{bmatrix} l_{11} & l_{12} & l_{13} & l_{14} \\ l_{21} & l_{22} & l_{23} & l_{24} \end{bmatrix} \begin{bmatrix} Q_1 \\ Q_2 \\ Q_2 \\ Q_1 \end{bmatrix} = V \tag{19}$$

Introducing (15) for l_{mn}, (19) is identical in form with (6) and (7) for increments of $\Delta x = 2a$, where m designates the test or observation points (P_{12} for $m = 1$ and P_{23}

for $m = 2$ in Fig. 9-21) and r_{mn} is equal to the distance between the test point m and the source point of segment n. Thus, r_{23} is the distance between test point $m = 2$ (P_{23} in Fig. 9-21) and the center of segment 3.

In the above we have used a pulse or step-function approximation in which the boundary condition (V = constant) is not enforced everywhere along the rod but only at certain observation or test points. In between observation points the boundary condition may not be satisfied. However, as the number of segments and observation points increase, the boundary condition is enforced at more points (the solution converging) and the accuracy of the results should improve. In the sense that the residual discrepancies or moments should vanish with a sufficient number of properly selected pulse functions, the procedure we have discussed may be called a *moment method*.

9-16 THE MOMENT METHOD (MM) AND ITS APPLICATION TO A WIRE ANTENNA. As discussed in the previous section, an integral equation can be transformed into a set of simultaneous linear algebraic equations (or matrix equation) which may then be solved by numerical techniques. Roger Harrington[1] has unified the various procedures into a general moment method (MM) now widely used with powerful computers for solving electromagnetic field problems.

In this section the method will be developed for a wire antenna and applied to an example for a short dipole.

Consider a cylindrical current-carrying conductor (or wire) of radius a isolated in free space (Fig. 9-24). Let its conductivity $\sigma = \infty$ so that we can consider the radio-frequency current to be entirely on the surface ($1/e$ depth = 0). The total current at point z' on the conductor is

$$I(z') = K(z')\, 2\pi a \tag{1}$$

where $K(z')$ = surface current density at z' (A m^{-1})

All of the current is at a distance a from the conductor axis (z axis), and we will consider it as flowing in empty space along an infinitesimally thin filament parallel to the z axis at a distance a, as in Fig. 9-25, with the conductor no longer present.

The electric field of charges and current is given by

$$\mathbf{E} = -j\omega\mu_0 \mathbf{A} - \nabla V \tag{2}$$

where $\mathbf{A}$ = vector potential
V = scalar potential

[1] R. F. Harrington, *Field Computation by Moment Methods*, Macmillan, 1968.

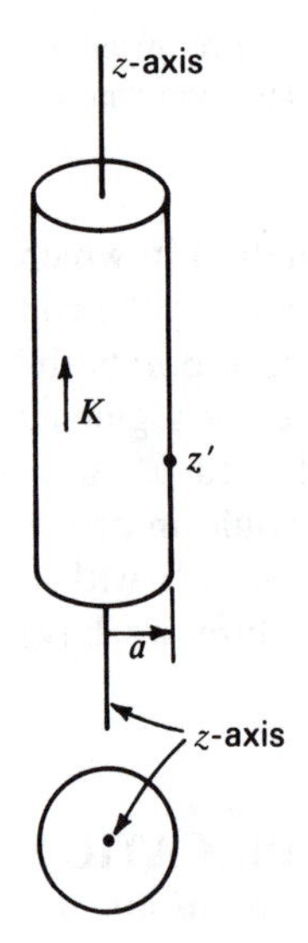

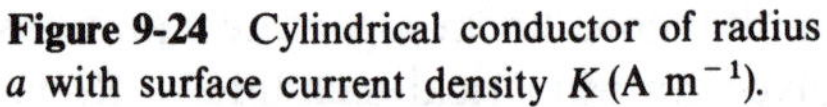

Figure 9-24 Cylindrical conductor of radius a with surface current density K (A m^{-1}).

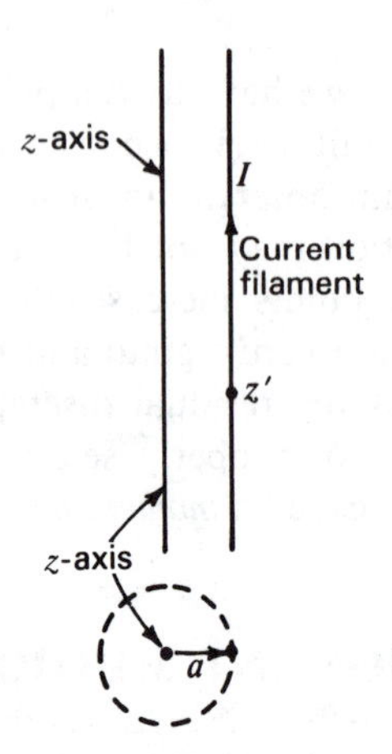

Figure 9-25 Conductor replaced by current filament $I = 2\pi aK$ (A) at distance a from the z-axis.

For current only in the z direction (2) becomes

$$E_z = -j\omega\mu_0 A_z - \frac{\partial V}{\partial z} \tag{3}$$

We also have

$$\frac{\partial A_z}{\partial z} = -j\omega\varepsilon_0 V \tag{4}$$

and its z derivative

$$\frac{\partial^2 A_z}{\partial z^2} = -j\omega\varepsilon_0 \frac{\partial V}{\partial z} \tag{4a}$$

or

$$-\frac{\partial V}{\partial z} = \frac{1}{j\omega\varepsilon_0} \frac{\partial^2 A_z}{\partial z^2} \tag{5}$$

Introducing (5) into (3),

$$E_z = \frac{1}{j\omega\varepsilon_0} \frac{\partial^2 A_z}{\partial z^2} - j\omega\mu_0 A_z \tag{6}$$

$$= \frac{1}{j\omega\varepsilon_0} \left(\frac{\partial^2 A_z}{\partial z^2} + \omega^2 \mu_0 \varepsilon_0 A_z \right) \tag{7}$$

$$= \frac{1}{j\omega\varepsilon_0} \left(\frac{\partial^2 A_z}{\partial z^2} + \beta^2 A_z \right) \tag{8}$$

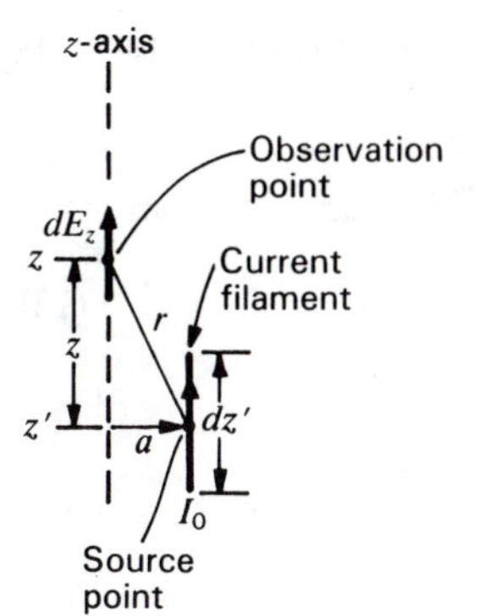

Figure 9-26 Source point on current filament with field dE_z at distance r on the z-axis.

For a current element dz, the vector potential

$$dA_z = \frac{I(z')e^{-j\beta r}\, dz}{4\pi r} \tag{9}$$

where $e^{-j\beta r}/r = G_{zz'}$ = free space Green's function
$r = \sqrt{(z - z')^2 + a^2}$
z = observation point
z' = source point (see Fig. 9-26)

The field from this current element is then

$$dE_z = \frac{I(z')}{j\omega\varepsilon_0 4\pi}\left(\frac{\partial^2}{\partial z^2} G_{zz'} + \beta^2 G_{zz'}\right) dz' \tag{10}$$

For a conductor of length L, the total field is given by the integral of (10), known as *Pocklington's equation:*[1]

$$E_z = \frac{1}{4\pi j\omega\varepsilon_0}\int_{-L/2}^{L/2}\left(\frac{\partial^2}{\partial z^2} G_{zz'} + \beta^2 G_{zz'}\right) I(z')\, dz' \tag{11}$$

E_z is the radiated field due to the current $I(z')$, resulting from an impressed or source field $E_{z'}$, from, for example, a voltage applied at the antenna terminals or from an incident plane wave (scattering case). On (and inside) the conductor the sum of these fields must vanish ($\sigma = \infty$) so

$$E_z = -E_{z'} \tag{12}$$

---

[1] H. C. Pocklington, "Electrical Oscillations in Wires," *Camb. Phil. Soc. Proc.*, **9**, 324–332, 1897.

Richmond[1] has differentiated and rearranged (11) in a more convenient form as follows:

$$-E_{z'} = \frac{\lambda Z_0}{8\pi^2 j} \int_{-L/2}^{L/2} \frac{e^{-j\beta r}}{r^5} [(1 + j\beta r)(2r^2 - 3a^2) + \beta^2 a^2 r^2] I(z')\, dz' \qquad (\text{V m}^{-1}) \tag{13}$$

where r = distance between source and observation points

$= \sqrt{(z' - z)^2 + a^2}$, m

$Z_0 = 377\ \Omega$

With parameters in *dimensionless form*, (13) becomes

$$-V = -\Delta z\ E(z')$$
$$= -j \frac{Z_0 \Delta z_\lambda}{8\pi^2} \int_{-L/2}^{L/2} \frac{e^{-j2\pi r_\lambda}}{r_\lambda^3} \left\{(1 + j2\pi r_\lambda)\left[2 - 3\left(\frac{a}{r}\right)^2\right] + 4\pi^2 a_\lambda^2\right\} I(z')\, dz_\lambda \quad (\text{V}) \tag{14}$$

where $r_\lambda = r/\lambda$, dimensionless

V = voltage developed by $E(z')$ over Δz_λ, V

For brevity let $-E(z')$ in (14) be written as

$$-E(z') = \int_{-L/2}^{L/2} I(z')G(r_{mn})\, dz' \qquad (\text{V m}^{-1}) \tag{15}$$

$$\text{where } G(r_{mn}) = \frac{-jZ_0}{8\pi^2\lambda^2}\left(\frac{e^{-j2\pi r_\lambda}}{r_\lambda^3}\right)\left\{(1 + j2\pi r_\lambda)\left[2 - 3\left(\frac{a}{r}\right)^2\right] + 4\pi^2 a_\lambda^2\right\} \qquad (\Omega\ \text{m}^{-2}) \tag{16}$$

$r = r_{mn}$

m = observation point

n = source point

Approximating the current with a series expansion we let

$$I(z') = \sum_{n=1}^{N} I_n F_n(z') \tag{17}$$

where $F_n(z')$ is a pulse function (equal to zero or unity) for incremental segments $\Delta z'_n$. (Other functions are possible, e.g., overlapping segments, each with a tri-

[1] Jack H. Richmond, "Digital Computer Solutions of the Rigorous Equations for Scattering Problems," *Proc. IEEE*, **53**, 796–804, August 1965.

angular or piecewise sinusoidal current distribution.)[1] For the mth segment we have

$$-E_{z'}(z_m) = \sum_{n=1}^{N} I_n \int_{\Delta z_n'} G(r_{mn})\, dz' \qquad (\text{V m}^{-1}) \tag{18}$$

Putting

$$G_{mn} = \int_{\Delta z_n'} G(r_{mn})\, dz' \simeq G(r_{mn})\, \Delta z_n' \qquad (\Omega\ \text{m}^{-1}) \tag{19}$$

(18) becomes approximately

$$-E_{z'}(z_m) = I_1 G_{m1} + I_2 G_{m2} + \cdots + I_n G_{m3} + \cdots + I_N G_{m4} \tag{20}$$

and the antenna equation now takes the form of a network equation. Writing (20) for each of the N segments ($m = 1, 2, 3, \ldots, N$), we obtain a set of equations:

$$\begin{aligned} I_1 G_{11} + I_2 G_{12} + \cdots + I_N G_{1N} &= -E_{z'}(z_1) \\ I_1 G_{21} + I_2 G_{22} + \cdots + I_N G_{2N} &= -E_{z'}(z_2) \\ \vdots \qquad \vdots \qquad\quad \vdots \quad &\qquad \vdots \\ I_1 G_{N1} + I_2 G_{N2} + \cdots + I_N G_{NN} &= -E_{z'}(z_N) \end{aligned} \tag{21}$$

which may be expressed in matrix form as

$$\begin{bmatrix} G_{11} & G_{12} & \cdots & G_{1N} \\ G_{21} & G_{22} & \cdots & G_{2N} \\ \vdots & \vdots & & \vdots \\ G_{N1} & G_{N2} & \cdots & G_{NN} \end{bmatrix} \begin{bmatrix} I_1 \\ I_2 \\ \vdots \\ I_N \end{bmatrix} = \begin{bmatrix} -E_{z'}(z_1) \\ -E_{z'}(z_2) \\ \vdots \\ -E_{z'}(z_N) \end{bmatrix} \tag{22}$$

and in compact notation by

$$[G_{mn}][I_n] = -[E_m] \qquad (\text{V } \lambda^{-1} \text{ or V m}^{-1}) \tag{23}$$

where $m = 1, 2, 3, \ldots, N$
$n = 1, 2, 3, \ldots, N$

Multiplying both sides of the equation by the distance Δz,

$$\Delta z[G_{mn}][I_n] = -\Delta z[E_m] \qquad (\text{V}) \tag{24}$$

---

[1] The piecewise sinusoidal distribution (section of a sine curve) is assumed to be somewhat more appropriate than a strictly triangular distribution. A piecewise sinusoidal distribution is used in Sec. 9-17. The differences between distributions are most significant for a small number of segments. With a large number of segments the different distribution functions should all give equivalent results.

we have on the left the product of an impedance (Z) and current (I) and on the right a voltage (V) as in an electric circuit equation (Ohm's law):

$$[Z_{mn}][I_n] = -[V_m] \quad \text{(V)} \tag{25}$$

Example. Current distribution and impedance of short dipole. To illustrate the application of the moment method (circuit equation) technique, use (16), (19) and (21) to calculate the current distribution and input impedance of a perfectly conducting center-fed cylindrical dipole 0.1λ long with a radius of 0.001λ. We presume no knowledge of what the impedance or current distribution should be except that the current distribution is symmetrical.

Solution. Referring to Fig. 9-27, the dipole is divided into 3 segments $\Delta z' = 0.033\lambda$ long and each assumed to have a uniform current over each segment (pulse function) given by I_1, I_2 and I_3. Thus, $N = 3$ and (21) becomes

$$\begin{aligned} I_1 G_{11} + I_2 G_{12} + I_3 G_{13} &= -E(z'_1) \\ I_1 G_{21} + I_2 G_{22} + I_3 G_{23} &= -E(z'_2) \\ I_1 G_{31} + I_2 G_{32} + I_3 G_{33} &= -E(z'_3) \end{aligned} \tag{26}$$

The upper and lower halves of the dipole are symmetrical so that $r_{12} = r_{21} = r_{23} = r_{32}$ and $r_{13} = r_{31}$. Also $r_{11} = r_{22} = r_{33}$. Accordingly, we need to evaluate (16) for only 3 distances, r_{11}, r_{12} and r_{13}, between the source and observation points (point matching).

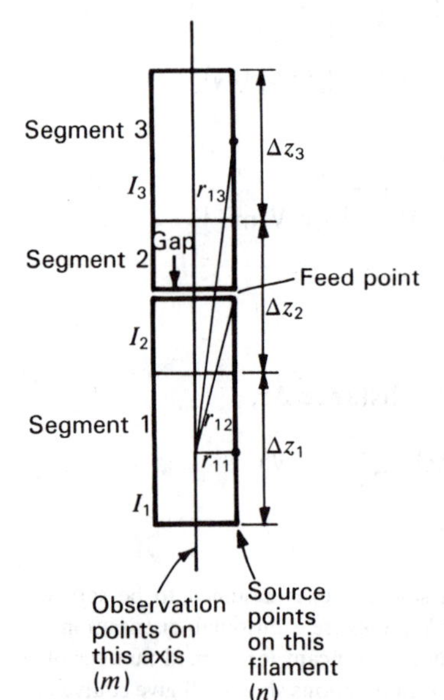

Figure 9-27 Short center-fed dipole 0.1λ long divided into 3 segments $\Delta z_1 = \Delta z_2 = \Delta z_3 = 0.033\lambda$ long.

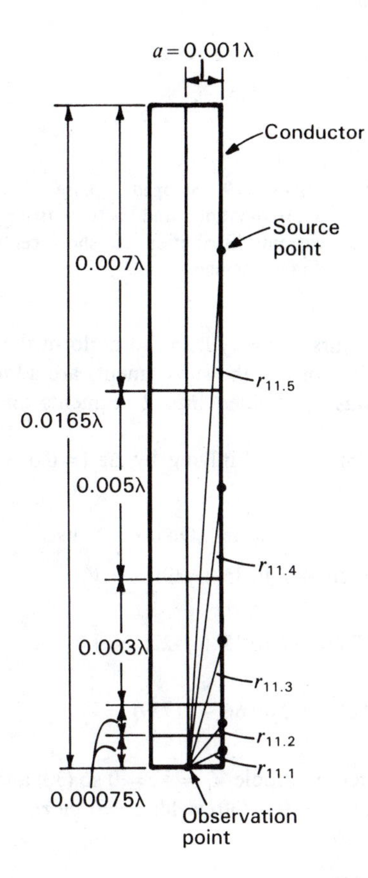

Figure 9-28 Half of one dipole segment divided into 5 subsegments for calculations of G_{11} in the example.

Introducing numerical values for $r_{12} = 0.066\lambda$, $a = 0.001\lambda$ and $\Delta z = 0.033\lambda$ into (16),

$$G_{13} = \frac{-j377 \times 0.033}{8\pi^2 \times 0.066^3} (\cos 2\pi \times 0.066 - j \sin 2\pi \times 0.066)$$

$$\times \left\{(1 + j2\pi \times 0.066)\left[2 - 3\left(\frac{0.001}{0.066}\right)^2\right] + 4\pi^2 \times 0.001^2\right\}$$

$$= -25.8 - j1184 \qquad (\Omega\,\lambda^{-1}) \tag{27}$$

Similarly,

$$G_{11} = -20.0 + j52\,700 \qquad (\Omega\,\lambda^{-1}) \tag{28}$$

$$G_{12} = -25.6 - j12\,800 \qquad (\Omega\,\lambda^{-1}) \tag{29}$$

Since the integrand of (16) is sensitive to small changes in r when r is small (particularly in the region where $r = \sqrt{1.5}\,a$ or less), G_{11} was divided into 5 subsegments, as indicated in Fig. 9-28. The segment with the smallest r contributes negligible resistance to the dipole impedance but a large negative reactance. For larger r's there is a contribution to the resistance and the reactance becomes posi-

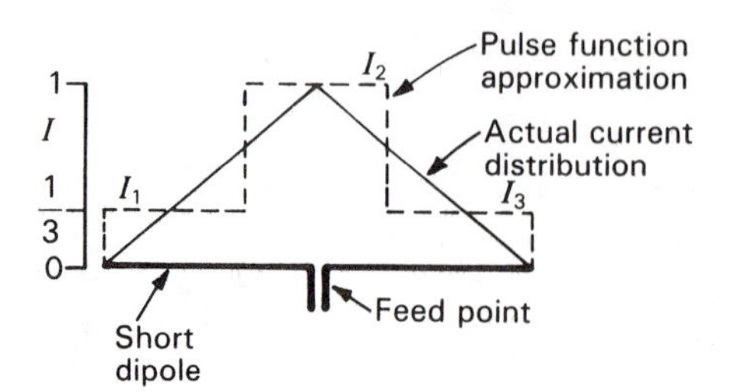

Figure 9-29 Stepped (pulse function) approximation and actual (triangular) current distribution on short center-fed dipole antenna.

tive. The change in sign of the reactance occurs at $r = \sqrt{1.5}\,a$. To perform the integration [by summation of (19)], the contributions of the subsegments are added to obtain the value in (28). Likewise, G_{12} was subdivided into 2 segments and the contributions added to obtain (29).

Introducing (27), (28) and (29) in (26) and multiplying by Δz $(=0.033\lambda)$ we obtain

Amperes Ohms Volts

$$I_1(0.66 - j1739) + I_2(0.85 + j422) + I_3(0.85 + j39) = V_1 \tag{30}$$

$$I_1(0.85 + j422) + I_2(0.66 - j1739) + I_3(0.85 + j422) = V_2 \tag{31}$$

$$I_1(0.85 + j39) + I_2(0.85 + j422) + I_3(0.66 - j1739) = V_3 \tag{32}$$

By symmetry $I_1 = I_3$; also for a center-fed dipole $V_1 = V_3 = 0$ so (30) and (32) are identical. Introducing the condition that $V_1 = 0$ in (30) yields a *current ratio* of

$$\frac{I_1}{I_2} = 0.25 + j0.0002 \qquad (ans.)$$

Putting this ratio in (31), dividing by I_2 and setting $V_2 = 1$ V yields an *input impedance* of

$$Z = \frac{V_2}{I_2} = R + jX = 1.1 - j1528\ \Omega \qquad (ans.) \tag{33}$$

For a short center-fed dipole the current distribution is nearly triangular so that the pulse function current ratio I_1/I_2 should be $\frac{1}{3}$ as suggested in Fig. 9-29. Ideally, the moment method should yield this value. If this ratio ($\frac{1}{3}$) is substituted in (31) we obtain

$$Z = 1.2 - j1458\ \Omega \tag{34}$$

For a dipole of the same length (0.1λ) but only $\frac{1}{10}$ the radius (0.0001λ), Richmond's convergence value of $Z = 1.852 - j1895\ \Omega$ (see Prob. 9-6).

It is not expected with the few segments used and the assumptions made that high accuracy would be obtained. The purpose of the example is to illustrate the moment method with actual numerical values for a simple case which can be solved with a small pocket calculator. It took the author an hour or so using one. However, such calculations if done frequently are most appropriately accomplished with a computer, especially one programmed to accommodate more segments and a range of dipole parameters such as length and diameter. The literature describing programs for doing this is extensive. The book *Antenna Design Using Personal Computers* by David M. Pozar[1] includes a number of programs on wire antennas. One of these called DIPOLE uses a standard moment method solution in a 254-step BASIC program which can handle a wide range of dipole dimensions.

The use of personal computers for antenna problems has been presented by Miller and Burke[2] with discussions of the Numerical Electromagnetics Code (NEC) and its subset MININEC for wire antennas. Many additional references on computer programs are listed at the end of this chapter and in App. B.

The next section includes 3 moment method examples which can be done on a pocket calculator.

9-17 SELF-IMPEDANCE, RADAR CROSS SECTION AND MUTUAL IMPEDANCE OF SHORT DIPOLES BY THE METHOD OF MOMENTS

By Edward H. Newman.[3]

In this section the moment method (MM) with piecewise sinusoidal current modes is used to calculate the current distribution, input impedance and radar cross section of a short dipole. The mutual coupling of two adjacent short dipoles is also calculated. Using a simplification of the mutual impedance equations of Howard E. King[4] the calculations can be done on a scientific hand calculator in an hour or so. Thus, the MM can be illustrated without resorting to a digital computer and taking the time and effort to write an appropriate computer program.

As in (9-16-17), the dipole current can be approximated by the series expansion[5]

$$I(z') = \sum_{n=1}^{N} I_n F_n(z') \tag{1}$$

where $F_n(z')$ is a piecewise sinusoidal mode. For example, the dipole may be divided into 4 equal segments of length $d = l/4$ as shown in Fig. 9-30. Segment n

---

[1] Artech House, 1985.

[2] E. K. Miller and G. J. Burke, "Personal Computer Applications in Electromagnetics," *IEEE Ants. Prop. Soc. Newsletter*, 5–9, August 1983.

[3] ElectroScience Laboratory, Ohio State University.

[4] H. E. King, "Mutual Impedance of Unequal Length Antennas in Echelon," *IEEE Trans. Ants. Prop.*, **AP-5**, 306–313, July 1957.

[5] We assume that the dipole is perfectly conducting and that the surrounding medium is free space.

extends from z_n to z_{n+1}. The piecewise sinusoidal modes are placed on the dipole in an overlapping fashion, with mode n existing on segment n and $n + 1$. Mode n has endpoints z_n and z_{n+2}, and center or terminals at z_{n+1}. F_n is a filament of electric current, located a radius a from the wire centerline (i.e., on the surface of the wire) and with current

$$F_n(z) = \frac{\sin \beta(d - |z - z_{n+1}|)}{\sin \beta d} \qquad \text{(A)} \qquad (2)$$

where $\beta = 2\pi/\lambda$

$F_n(z)$ is zero at its endpoints and rises sinusoidally to a maximum at its center with terminal current of $I_{n0} = F_n(z_{n+1}) = 1$ A. Note that the piecewise sinusoidal modes produce a current which is continuous and also zero at the dipole endpoints. Except at the dipole endpoints, the dipole current of (1) at z_{n+1} is I_n amperes. Equation (1) produces a sinusoidal interpolation of the current values at the $N + 2$ points.

We require that the radiated and impressed fields satisfy (9-16-12). Substituting (1) into (9-16-12) we have

$$-\sum_{n=1}^{N} I_n E_{zn} \simeq E_{z'} \qquad (3)$$

where E_{zn} = free-space z component of the electric field of F_n
$E_{z'}$ = z component of the incident field

E_{zn} is available in terms of simple functions given by Schelkunoff and Friis[1] and King.[2]

The weighting functions in the MM solution are chosen identical to the expansion functions, except that they are located along the centerline of the dipole. This is because we enforce (9-16-12) on the centerline. Then multiplying both sides of (3) by the sequence of N weighting functions, F_m ($m = 1, 2, \ldots, N$), (3) becomes an $N \times N$ system of simultaneous linear algebraic equations which can be written compactly in matrix form as

$$[Z]I = V \qquad (4)$$

Here I is the current column vector whose N components contain the I_n of (1). $[Z]$ is the $N \times N$ impedance matrix whose typical term is

$$Z_{mn} = -\int_m E_{zn} F_m \, dz \qquad (5)$$

In general $[Z]$ is dependent on the geometry and material composition of the scatterer, but not on the incident fields. A typical element of the right-hand-side

[1] S. A. Schelkunoff and H. T. Friis, *Antennas Theory and Practice*, Wiley, 1952.

[2] H. E. King, "Mutual Impedance of Unequal Length Antennas in Echelon," *IEEE Trans. Ants. Prop.*, **AP-5**, 306–313, July 1957.

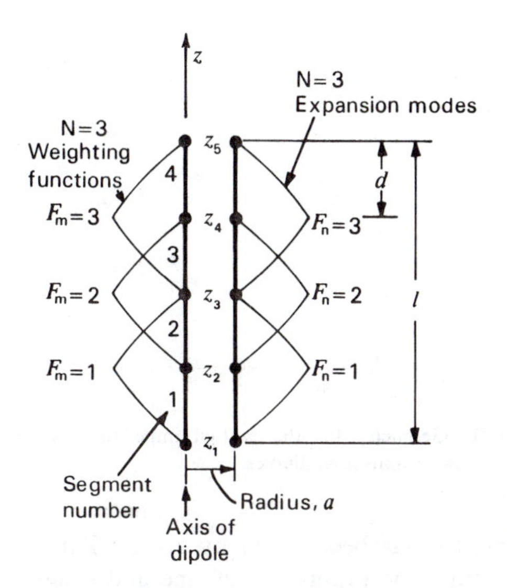

Figure 9-30 Three piecewise sinusoidal dipole modes on a dipole divided into 4 equal segments for Example 1.

or voltage vector V is given by

$$V_m = \int_m E_{z'} F_m \, dz \tag{6}$$

The integration in (5) and (6) is on the dipole centerline, and over the extent of F_m, that is, from $z = z_m$ to z_{m+2}. The dimensions of the elements of $[Z]$ and V are volt-amperes (VA), while the elements of I are dimensionless. If the Z_{mn} were divided by $I_{m0} I_{n0}$, then the Z_{mn} would have dimensions of ohms (Ω). Since in our case the modal terminal currents are $I_{n0} = 1$ A, Z_{mn} can be considered to have the dimensions of ohms. In any case, the $[Z]$ matrix is usually referred to as an impedance matrix and V as a voltage vector since the matrix equation (4) resembles an N-port generalization of Ohm's law.

The major problem in an MM solution is usually the evaluation of the elements in the impedance matrix. Typically this involves numerical integrations and/or the evaluation of special functions. As a result, most MM solutions are done on a digital computer and require a great deal of programming time and effort. For this reason, most MM solutions are not suitable as a simple example problem which can be accomplished in about an hour using only a hand calculator. Relatively simple expressions for the elements in the dipole MM impedance matrix are presented here, thus eliminating the need for a digital computer to carry out the MM solution to the examples given.

For the dipole antenna, the elements in the impedance matrix, as given by (5), are the mutual impedances between parallel piecewise sinusoidal dipole modes. Figure 9-31 shows two parallel piecewise sinusoidal dipole modes of length $2d$. The bottom of weighting mode m is located a distance h above the center of expansion mode n, and the modes are staggered by the distance r. For convenience, the expansion mode has its center at $z = 0$. Exact expressions for

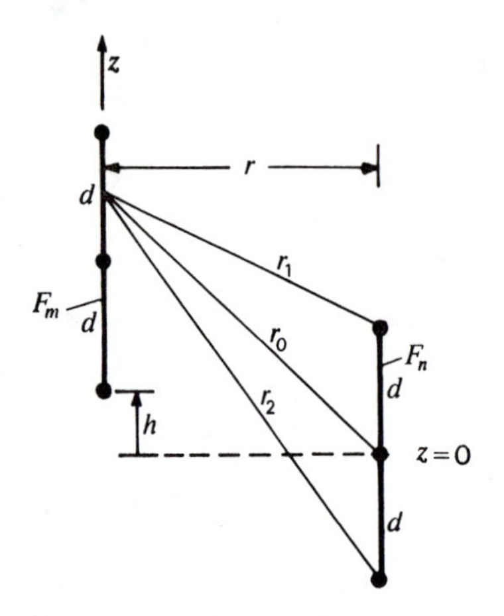

Figure 9-31 Geometry for the mutual impedance of 2 parallel piecewise sinusoidal dipoles.

the mutual impedance between these modes has been given by King.[1] King's expressions are very lengthy and also require the evaluation of sine and cosine integrals. In order to simplify King's expressions we assume that the modes are electrically small and electrically close. Specifically, if we assume that βd, βh and βr are all $\ll 1$, it may be shown that King's expressions for the mutual impedance between modes m and n reduce to[2]

$$Z_{mn} = R_{mn} + jX_{mn} \tag{7}$$

where

$$R_{mn} = 20(\beta d)^2 \tag{8}$$

and

$$\begin{aligned} X_{mn} = -\frac{30}{\beta d^2}[&-4A + 6B - 4C + D + E + 4h \ln(2A + 2h) \\ &- 6(d + h)\ln(2B + 2h + 2d) + 4(2d + h)\ln(2C + 2h + 4d) \\ &+ (d - h)\ln(2D + 2h - 2d) - (3d + h)\ln(2E + 2h + 6d)] \end{aligned} \tag{9}$$

where $A = \sqrt{r^2 + h^2}$
$B = \sqrt{r^2 + (d + h)^2}$
$C = \sqrt{r^2 + (2d + h)^2}$
$D = \sqrt{r^2 + (d - h)^2}$
$E = \sqrt{r^2 + (3d + h)^2}$

[1] H. E. King, "Mutual Impedance of Unequal Length Antennas in Echelon," *IEEE Trans. Ants. Prop.*, **AP-5**, 306–313, July 1957.

[2] I appreciate the assistance of Linda Bingham in obtaining these simplified expressions.

Equations (8) and (9) are suitable for a hand calculator, since they involve no operations more complicated than a logarithm and square root. Note that R_{mn} is the well-known formula for the radiation resistance of a short dipole, and is independent of mode separation. Equation (9) can be further simplified if we assume $r = a \ll d$ (wire radius much less than the segment length) and also consider certain special values of h. For self-impedance terms, $m = n$, and

$$X_{mm}(h = -d) = -\frac{30}{\beta d}\left[-4 + 4 \ln\left(\frac{d}{a}\right)\right] \tag{10}$$

For adjacent modes with one overlapping segment, $|m - n| = 1$ and

$$X_{mn}(h = 0) = -\frac{30}{\beta d}\left[1 + 2 \ln\left(\frac{1.54a}{d}\right)\right] \tag{11}$$

If $|m - n| = 2$, then the modes share a single point and

$$X_{mn}(h = d) = -\frac{30}{\beta d}[-0.68] \tag{12}$$

If $|m - n| \geq 3$, then the modes are not touching and

$$X_{mn}(h \geq 2d) = -\frac{30}{\beta d}\left[\frac{h}{d} \ln \frac{h^4(2d + h)^4}{(d + h)^6(h - d)(3d + h)} + \ln \frac{(2d + h)^8(h - d)}{(d + h)^6(3d + h)^3}\right] \tag{13}$$

Now consider the evaluation of the right-hand-side vector V. As seen in (6), V is dependent upon the z component of the incident electric field.

First, consider the case where the dipole is excited by a *voltage generator*. The simplest, and probably the most commonly used, model for a voltage generator is the so-called *delta-gap model.*[1] A delta-gap generator is one that creates an extremely large, but highly localized, electric field polarized parallel to the wire centerline. A delta-gap generator located at $z = z'$ has the incident field

$$E_{z'} = v\, \delta(z - z') \tag{14}$$

where v = generator voltage
$\delta(z)$ = unit area Dirac delta function

Normally the generators are placed at the center or terminals of the piecewise sinusoidal modes. Thus, referring to Fig. 9-30 for a dipole with $N = 3$ modes, the generator could be placed at $z = z_2$, z_3 or z_4, which would be at the terminals of modes 1, 2 or 3 respectively. Inserting the incident field from (14) into (6) shows that, if a delta-gap generator of v_m volts is placed at the terminals of mode m, then

$$V_m = v_m \tag{15}$$

[1] W. L. Stutzman and G. A. Thiele, *Antenna Theory and Design*, Wiley, 1981, chap. 7.

Element m of V is nonzero only if a nonzero generator is placed at the terminals of mode m.

Now consider the effect of placing a *lumped load* in the wire. A lumped load of Z_{lm} ohms, placed at the terminals of mode m, will produce a voltage of $-I_m Z_{lm}$ volts at these terminals. If we treat this voltage as a dependent delta-gap generator, then according to (15) we should add $-I_m Z_{lm}$ to V_m. However, this is an unknown voltage, since initially I_m is unknown. Since it is conventional to write all unknowns on the left-hand side of the matrix equation, we add $I_m Z_{lm}$ to both sides of row m of the matrix equation. Thus, it can be seen that a lumped load of Z_{lm} ohms placed at the terminals of mode m simply results in Z_{mm} being replaced by $Z_{mm} + Z_{lm}$.

There is no physical break or gap in the wire where a generator or load is placed. Thus, the current is continuous through generators and loads. However, there is a slope discontinuity, or jump in the derivative of the current, at the generator or load. Note that the piecewise sinusoidal modes account for this behavior by enforcing continuity of current on the wire and by allowing a slope discontinuity at their terminals.

Next consider the situation where the wire is excited by a normally incident plane wave. If a z-polarized plane wave is incident from the $+x$ axis with magnitude E_0, then

$$E_{z'} = E_0 e^{j\beta x} \tag{16}$$

Inserting (16) into (6) and integrating yields

$$V_m = \frac{2E_0}{\beta} \tan\left(\frac{\beta d}{2}\right) \tag{17}$$

Example 1 Dipole current distribution and input impedance. Compute the current distribution and input impedance of a center-fed dipole antenna. The dipole length $l = \lambda/10$ and radius $a = \lambda/10\,000$.

Solution. As illustrated in Fig. 9-30 we use $N = 3$ piecewise sinusoidal modes on the dipole and segment the dipole into $N + 1 = 4$ equal segments of length $d = l/4 = 0.025\lambda$. In this case the 3×3 MM matrix (4) can be explicitly written as

$$\begin{bmatrix} Z_{11} & Z_{12} & Z_{13} \\ Z_{21} & Z_{22} & Z_{23} \\ Z_{31} & Z_{32} & Z_{33} \end{bmatrix} \begin{bmatrix} I_1 \\ I_2 \\ I_3 \end{bmatrix} = \begin{bmatrix} V_1 \\ V_2 \\ V_3 \end{bmatrix} \tag{18}$$

Since the current on the center-fed dipole is symmetric, $I_1 = I_3$. In this case, we can add column 3 of the matrix equation to column 1 and reduce the order 3 matrix equation to the order 2 matrix equation

$$\begin{bmatrix} (Z_{11} + Z_{13}) & Z_{12} \\ (Z_{21} + Z_{23}) & Z_{22} \end{bmatrix} \begin{bmatrix} I_1 \\ I_2 \end{bmatrix} = \begin{bmatrix} V_1 \\ V_2 \end{bmatrix} \tag{19}$$

Table 9-2 Elements of the [Z] matrix (VA) for Example 1

Element	Approximate	Exact
Z_{11}	$0.4935 - j3454$	$0.4944 - j3426$
Z_{12}	$0.4935 + j1753$	$0.4945 + j1576$
Z_{13}	$0.4935 + j129.9$	$0.4885 + j132.2$

Reducing the order of the matrix equation from 3 to 2 greatly reduces the effort in the hand calculations required to solve the matrix equation.

Although [Z] in (18) and (19) contain 9 elements, only 3 are distinct, since from the symmetry of the dipole,

$$Z_{11} = Z_{22} = Z_{33} \tag{20}$$

$$Z_{12} = Z_{21} = Z_{23} = Z_{32} \tag{21}$$

$$Z_{13} = Z_{31} \tag{22}$$

The real part of each Z_{mn} is given by (8) as

$$R_{mn} = 0.4935 \text{ VA} \tag{23}$$

The imaginary part of the Z_{mn} can be computed from (9); however, here we choose to use the simpler forms of (10), (11) and (12). Table 9-2 shows the elements in the first row of the [Z] matrix of (18) computed by (7), (8), (10), (11) and (12) and by King's exact expressions. Note that the approximate values of [Z] are within 11 percent of the exact values.

If the approximate values of [Z] from Table 9-2 are substituted into (19) we obtain

$$\begin{bmatrix} 0.9869 - j3324 & 0.4935 + j1753 \\ 0.9869 + j3506 & 0.4935 - j3454 \end{bmatrix} \begin{bmatrix} I_1 \\ I_2 \end{bmatrix} = \begin{bmatrix} 0 \\ 1 \end{bmatrix} \tag{24}$$

where we have set $V_2 = 1$ VA since there is a 1-V generator at the terminals of mode 2. Equation (24) can be easily solved using Cramer's rule. The results for the elements in the dimensionless current vector are

$$I_1 = I_3 = 0.000\,328\,6 \,\underline{/89.892^\circ} \tag{25}$$

$$I_2 = 0.000\,623\,0 \,\underline{/89.926^\circ} \tag{26}$$

Dividing (26) by (25), the current ratio I_2/I_1 is very nearly equal to 1.9, indicating a nearly triangular current distribution on the dipole.

The dipole current in amperes can now be obtained by inserting these coefficients into (1) with $N = 3$. Thus, the dipole input or terminal current is I_2 amperes. The input impedance is given by the ratio of the input voltage to the input current; that is,

$$Z_{\text{in}} = \frac{1}{I_2} = 2.083 - j1605 \ \Omega \qquad (ans.) \tag{27}$$

By contrast, if we were to use the exact values of [Z] from Table 9-2, then the results for the current distribution and input impedance would be

$$I_1 = I_3 = 0.000\,249\,8 \underline{/90.0^\circ} \tag{28}$$

$$I_2 = 0.000\,521\,9 \underline{/89.9^\circ} \tag{29}$$

$$Z_{\text{in}} = 1.892 - j1916\ \Omega \qquad (ans.) \tag{30}$$

Example 2 Scattering from a short dipole. Compute the radar cross section[1] of the same dipole considered in Example 1 for a wave at normal (broadside) incidence with the dipole terminated in a (conjugate) matched load.

Solution. To terminate the dipole in its conjugate matched load, we place a lumped load of Z_{in}^* at the center of the dipole, i.e., at the terminals of mode 2. Using the value of Z_{in} from (27),

$$Z_{l2} = Z_{\text{in}}^* = 2.083 + j1605\ \Omega \tag{31}$$

The impedance matrix for the loaded dipole is identical to that of the unloaded dipole except that we add Z_{l2} to the self-impedance of mode 2 to obtain

$$Z_{22} = (0.4935 - j3454) + (2.083 + j1605) = 2.576 - j1849\ \text{VA} \tag{32}$$

If the incident electric field is a unit amplitude z-polarized plane wave incident from the $+x$ axis, then the elements of the right-hand-side vector are identical and given by (17) with $E_0 = 1$:

$$V_m = 0.025\,05\ \text{VA}, \qquad m = 1, 2, 3 \tag{33}$$

Since the excitation and loading of the antenna are symmetric with respect to the center of the dipole, the current on the dipole remains symmetric. Thus, the current vector can still be computed from the order 2 matrix (19):

$$\begin{bmatrix} 0.9869 - j3324 & 0.4935 + j1753 \\ 0.9869 + j3506 & 2.576 \ \ - j1849 \end{bmatrix} \begin{bmatrix} I_1 \\ I_2 \end{bmatrix} = \begin{bmatrix} 0.025\,05 \\ 0.025\,05 \end{bmatrix} \tag{34}$$

Equation (34) can be solved using Cramer's rule. The results for the elements in the dimensionless current vector are

$$I_1 = I_3 = 0.006\,515 \underline{/0.581^\circ} \tag{35}$$

$$I_2 = 0.012\,35 \underline{/0.548^\circ} \tag{36}$$

The scattered field is given by

$$E_s = \sum_{n=1}^{N} I_n E_{zn} \tag{37}$$

where E_{zn} is the free-space electric field of expansion mode F_n as in (3). For a field point on the $+x$ axis (i.e., in the backscatter direction) and in the far zone of the

[1] See Sec. 17-5.

dipole, the electric field of F_n will be z polarized and given by

$$E_{zn} = j60 \tan\left(\frac{\beta d}{2}\right)\frac{e^{-j\beta x}}{x} \qquad (\text{V m}^{-1}) \tag{38}$$

Using (38) and the above values for the I_n, the far-zone backscattered electric field is

$$E_s = 0.1199 \,\underline{/90.6^\circ}\, \frac{e^{-j\beta x}}{x} \qquad (\text{V m}^{-1}) \tag{39}$$

The *radar cross section* of the dipole is

$$\sigma = 4\pi x^2 \frac{|E_s|^2}{|E_0|^2} = 0.1806\lambda^2 \qquad (\textit{ans.}) \tag{40}$$

By contrast, if we were to repeat this example with the exact value of $[Z]$ from Table 9-2, the results would be

$$I_1 = I_3 = 0.006\,199 \,\underline{/0.0^\circ} \tag{41}$$

$$I_2 = 0.012\,95 \,\underline{/0.0^\circ} \tag{42}$$

$$\sigma = 0.1801\lambda^2 \qquad (\textit{ans.}) \tag{43}$$

From (2-20-5) the maximum effective aperture of a short matched lossless dipole is given by

$$A_{em} = \frac{3}{8\pi}\lambda^2 = 0.119\lambda^2 \tag{44}$$

provided that the dipole length $l \ll \lambda$. The total scattering aperture equals A_{em} and the radar cross section is this value times the short dipole directivity D $(=1.5)$ or

$$\sigma = DA_{em} = \frac{1.5 \times 3}{8\pi}\lambda^2 = 0.179\lambda^2 \tag{45}$$

as compared to 0.1806 and $0.1801\lambda^2$ above.

Example 3 Mutual impedance of 2 short dipoles. Compute the mutual impedance of 2 short side-by-side dipoles separated by $\lambda/100$ as in Fig. 9-32. The dipoles are identical to the ones in Examples 1 and 2.

Solution. To simplify the computations, we place only one piecewise sinusoidal mode on each dipole. Thus, the order $N = 2$ matrix equation for this example is

$$\begin{bmatrix} Z_{11} & Z_{12} \\ Z_{21} & Z_{22} \end{bmatrix}\begin{bmatrix} I_1 \\ I_2 \end{bmatrix} = \begin{bmatrix} V_1 \\ V_2 \end{bmatrix} \tag{46}$$

Only Z_{11} and Z_{12} need be computed, since from the symmetry of the dipoles $Z_{11} = Z_{22}$ and $Z_{12} = Z_{21}$. Z_{11} is evaluated from (7), (8) and (10) with $d = -h = l/2 = 0.05\lambda$ and $r = a = 0.0001\lambda$. Z_{12} is evaluated from the same equations with a replaced by $s = 0.01\lambda$, since $h = -d$. The results for Z_{11} and Z_{12} are shown in

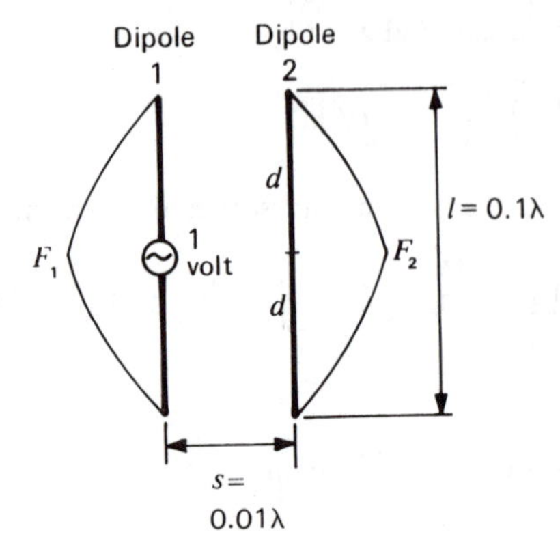

Figure 9-32 Geometry for 2 coupled dipoles of Example 3. The radius of each dipole is 0.0001λ.

Table 9-3 where they are compared with the exact values of King[1] and of Richmond.[2] For this simple one mode per dipole solution, Z_{12} is the *mutual impedance* between the two dipoles.

Inserting the approximate values from Table 9-3 into (46) we obtain

$$\begin{bmatrix} 1.9739 - j1992 & 1.9739 - j232.8 \\ 1.9739 - j232.8 & 1.9739 - j1992 \end{bmatrix} \begin{bmatrix} I_1 \\ I_2 \end{bmatrix} = \begin{bmatrix} 1 \\ 0 \end{bmatrix} \tag{47}$$

In (47) we set $V_1 = 1$, since to compute the input impedance of dipole 1, we place a 1-V generator at the terminals of mode 1. Equation (47) has the solution

$$I_1 = 0.000\,509\,0 \,\angle 89.955^\circ \tag{48}$$

$$I_2 = 0.000\,059\,49 \,\angle -89.616^\circ \tag{49}$$

The input impedance is then

$$Z_{\text{in}} = \frac{1}{I_1} = 1.539 - j1964 \ \Omega \tag{50}$$

The $N = 1$ mode solution for the input impedance of dipole 1 alone is

$$Z_{\text{in}} = Z_{11} = 1.974 - j1992 \ \Omega \tag{51}$$

Table 9-3 Elements of the [Z] matrix (VA) for Example 3

Element	Approximate	Exact
Z_{11}	1.9739 − j1992	2.0000 − j1921
Z_{12}	1.9739 − j232.8	1.9971 − j325.1

[1] H. E. King, "Mutual Impedance of Unequal Length Antennas in Echelon," *IEEE Trans. Ants. Prop.*, **AP-5**, 306–313, July 1957.

[2] J. H. Richmond, "Radiation and Scattering by Thin-Wire Structures in a Homogeneous Conducting Medium," *IEEE Trans. Ants. Prop.*, **AP-22**, 365, March 1974.

Table 9-4 Summary of self-impedance, mutual impedance and radar cross section for short dipoles†

	Modes	Equation
$Z_s = 1.974 - j1992\ \Omega$	$N = 1$ (approx. King)	(51)
$= 2.000 - j1921\ \Omega$	$N = 1$ (exact King)	Z_{11} (exact), Table 9-3
$= 2.083 - j1605\ \Omega$	$N = 3$ (approx. King)	(27)
$= 1.892 - j1916\ \Omega$	$N = 3$ (exact King)	(30)
$= 1.864 - j1905\ \Omega$	$N = 5$ (exact King)	J. H. Richmond,
$= 1.856 - j1899\ \Omega$	$N = 7$ (exact King)	FORTRAN IV program (see References below)
$Z_m = 1.974 - j233\ \Omega$	$N = 1$ (approx. King)	Z_{12} (approx.), Table 9-3
$= 1.997 - j325\ \Omega$	$N = 1$ (exact King)	Z_{12} (exact), Table 9-3
$Z_\Delta = Z_s - Z_m = 0.000 - j1759\ \Omega$	$N = 1$ (approx. King)	(51) $-Z_{12}$ (approx.)
$= 0.003 - j1596\ \Omega$	$N = 1$ (exact King)	Z_{11} (exact) $-Z_{12}$ (exact)
$\sigma = 0.1806\lambda^2$ (approx. King)		(40)
$= 0.1801\lambda^2$ (exact King)		(43)
$= 0.179\lambda^2$ (short dipole theory, Chap. 2)		(45)

† Z_s = self-impedance = input impedance of isolated dipole, Ω
Z_m = mutual-impedance, Ω
σ = radar cross section, λ^2
Dipole length, $l = \lambda/10$
Dipole radius, $a = \lambda/10\,000$
Dipole separation, $s = \lambda/100$ (for Z_m)
Note that Z_s should approach the true value as N increases provided the solution converges, also that the self-mutual difference $Z_\Delta \to 0$ as $s \to 0$ and $Z_\Delta \to Z_s$ as $s \to \infty$ ($Z_m \to 0$).

If the exact values of Table 9-3 are used, the impedance of dipole 1 in the presence of dipole 2 is

$$Z_{\text{in}} = 1.382 - j1822\ \Omega \tag{52}$$

Table 9-4 gives a summary of self-impedance, mutual impedance and radar cross section for short dipoles.

ADDITIONAL REFERENCES FOR CHAP. 9

Andreasen, M. G.: "Scattering from Parallel Metallic Cylinders with Arbitrary Cross Sections," *IEEE Trans. Ants. Prop.*, **AP-12**, 746–754, November 1964.
Andreasen, M. G.: "Scattering from Bodies of Revolution by an Exact Method," *IEEE Trans. Ants. Prop.*, **AP-13**, 303–310, March 1965.
Balanis, C. A.: *Antenna Theory: Analysis and Design*, Harper and Row, 1982.
Burke, G. J.: *The Numerical Electromagnetics Code* (*NEC*), The SCEEE Press, 1981.
Elliott, R. S.: *Antenna Theory and Design*, Prentice-Hall, 1981.
Kennaugh, E. M.: "Multipole Field Expansions and Their Use in Approximate Solutions of Electromagnetic Scattering Problems," Ph.D. dissertation, Ohio State University, 1959.
Mei, K. K., and J. G. Van Bladel: "Scattering by Perfectly Conducting Rectangular Cylinders," *IEEE Trans. Ants. Prop.*, **AP-11**, 185–192, March 1963.
Newman, E. H.: *Simple Examples of the Method of Moments in Electromagnetics*, pending publication.

Richmond, J. H.: "Scattering by a Dielectric Cylinder of Arbitrary Cross Section Shape," *IEEE Trans. Ants. Prop.*, **AP-13**, 334–341, May 1965.

Richmond, J. H.: "Scattering by an Arbitrary Array of Parallel Wires," *IEEE Trans. Microwave Theory and Tech.*, **MTT-13**, 408–412, July 1965.

Richmond, J. H.: "Digital Computer Solutions of the Rigorous Equations for Scattering Problems," *Proc. IEEE*, **53**, 796–804, August 1965.

Richmond, J. H.: "Radiation and Scattering by Thin-Wire Structures in a Homogeneous Conducting Medium," FORTRAN IV program giving current distribution, impedance, gain, patterns, scattering cross section and other parameters for dipoles, loops, plates, spheres, cones, and wire grid models of aircraft and ships. ASIS-NAPS Document NAPS-02223,[1] *IEEE Trans. Ants. Prop.*, **AP-22**, 365, March 1974.

Stutzman, W. L., and G. A. Thiele: *Antenna Theory and Design*, John Wiley, 1981.

Tsai, L. S., and C. E. Smith: "Moment Methods in Electromagnetics for Undergraduates," *IEEE Trans. Educ.*, **E-21**, 14–22, February 1978.

Additional references are in App. B.

PROBLEMS

9-1 Hallén's equation.

(*a*) What is the initial relation used in developing Hallén's integral equation?

(*b*) Indicate the principal steps required to arrive at the current distribution and terminal impedance of a cylindrical antenna by means of Hallén's integral equation.

9-2 Charge distribution. Determine the electrostatic charge distribution on a cylindrical conducting rod with a length-diameter ratio of 6.

9-3 Dipole impedance. Calculate the input impedance of a cylindrical center-fed dipole antenna $\lambda/12$ long with a length-diameter ratio of 25.

9-4 Input impedance of short dipole. Calculate the current distribution and input impedance of a center-fed dipole antenna $\lambda/15$ long with a length-diameter ratio of 200 using MM.

9-5 Mutual impedance of short dipoles. Calculate the mutual impedance of 2 side-by-side dipole antennas separated by $\lambda/25$ with each dipole $\lambda/8$ long and $\lambda/100$ diameter using MM.

9-6 $\lambda/10$ dipole impedance. Show that the convergence or true value of the self-impedance Z_s of the dipole of Table 9-4 is $1.852 - j1895\ \Omega$.

[1] Available from ASIS-NAPS c/o Microfiche Publications, 305 E. 46th St., New York, NY 10017.

CHAPTER 10

SELF AND MUTUAL IMPEDANCES

10-1 INTRODUCTION. The impedance presented by an antenna to a transmission line can be represented by a 2-terminal network. This is illustrated in Fig. 10-1 in which the antenna is replaced by an equivalent impedance Z connected to the terminals of the transmission line. In designing a transmitter and its associated transmission line, it is convenient to consider that the antenna is simply a 2-terminal impedance. This impedance into which the transmission line operates is called the *terminal* or *driving-point impedance.* If the antenna is isolated, i.e., remote from the ground or other objects, and is lossless,[1] its terminal impedance is the same as the *self-impedance* of the antenna. This impedance has a real part called the *self-resistance* (radiation resistance) and an imaginary part called the *self-reactance.* The self-impedance is the same for reception as for transmission.

In case there are nearby objects, say several other antennas, the terminal impedance can still be replaced by a 2-terminal network. However, its value is determined not only by the self-impedance of the antenna but also by the mutual impedances between it and the other antennas and the currents flowing on them. The terminal impedance is the same for both transmission and reception (see next section).

[1] By lossless is meant that there is no Joule heating associated with the antenna. There may, of course, be radiation. If the antenna is not lossless, an equivalent loss resistance appears at the terminals in series with the self-resistance or radiation resistance (see Sec. 2-15).

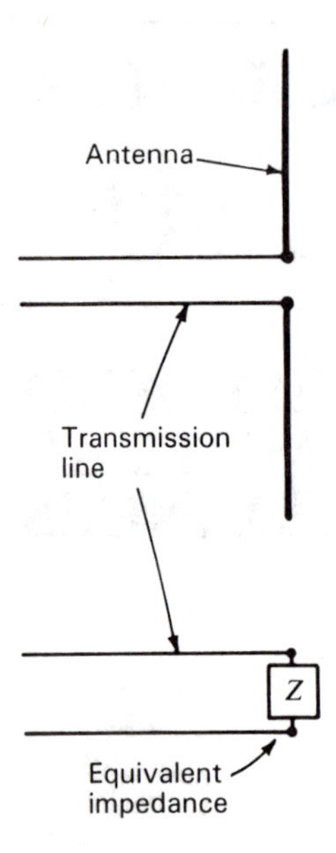

Figure 10-1 Transmission line with antenna and with equivalent impedance.

In Chap. 5 an expression was developed for the radiation resistance (or self-resistance) of thin linear antennas. In the following sections this analysis is extended to yield expressions for both the self-resistance and the self-reactance. In addition, expressions are developed for the *mutual resistance* and *mutual reactance* of two thin linear antennas. These expressions will be used in Chap. 11 to find the driving-point impedance in an array of linear antennas. Even though the impedances apply strictly to infinitesimally thin antennas, they are useful in connection with practical types of cylindrical antennas, provided that the antennas are thin.

In developing the subject of antenna impedance, an important and much-used theorem is that of *reciprocity*. Accordingly, this topic is discussed first and then applied to the impedance problem.

10-2 RECIPROCITY THEOREM FOR ANTENNAS.

The Rayleigh-Helmholtz reciprocity theorem[1] has been generalized by Carson[2] to include continuous media. This theorem as applied to antennas may be stated as follows:

---

[1] Lord Rayleigh, *The Theory of Sound*, The Macmillan Company, New York, vol. 1 (1877, 1937), pp. 98 and 150–157, and vol. 2 (1878, 1929), p. 145.

[2] J. R. Carson, "A Generalization of the Reciprocal Theorem," *Bell System Tech. J.*, **3**, 393–399, July, 1924.

J. R. Carson, "Reciprocal Theorems in Radio Communication," *Proc. IRE*, **17**, 952–956, June 1929.

Stuart Ballantine, "Reciprocity in Electromagnetic, Mechanical, Acoustical, and Interconnected Systems," *Proc. IRE*, **17**, 929–951, June 1929.

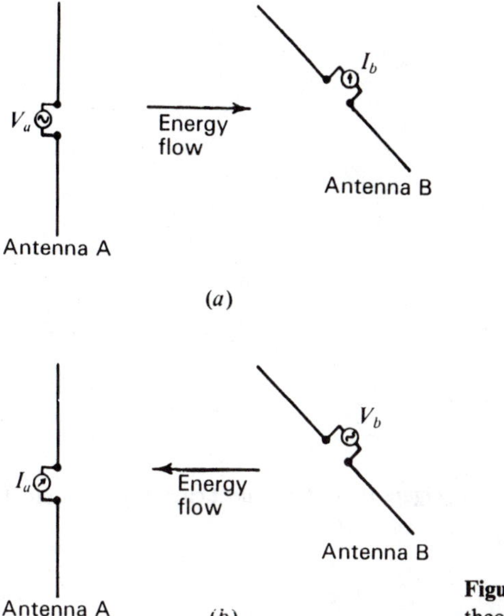

Figure 10-2 Illustrations for reciprocity theorem.

If an emf is applied to the terminals of an antenna A and the current measured at the terminals of another antenna B, then an equal current (in both amplitude and phase) will be obtained at the terminals of antenna A if the same emf is applied to the terminals of antenna B.

It is assumed that the emfs are of the same frequency and that the media are linear, passive and also isotropic. An important consequence of this theorem is the fact that under these conditions the transmitting and receiving patterns of an antenna are the same. Also, for matched impedances, the power flow is the same either way.

As an illustration of the reciprocity theorem for antennas, consider the following two cases.

Case 1. Let an emf V_a be applied to the terminals of antenna A as in Fig. 10-2*a*. This antenna acts as a transmitting antenna, and energy flows from it to antenna B, which may be considered as a receiving antenna, producing a current I_b at its terminals.[1] It is assumed that the generator supplying the emf and the ammeter for measuring the current have zero impedance or, if not zero, that the generator and ammeter impedances are equal.

Case 2. If an emf V_b is applied to the terminals of antenna B, then it acts as a

---

[1] Although the emf V_a and the current I_b are scalar space quantities, they are complex or vector quantities with respect to time phase. The term "phasor" is sometimes used to distinguish such a quantity from a true space vector.

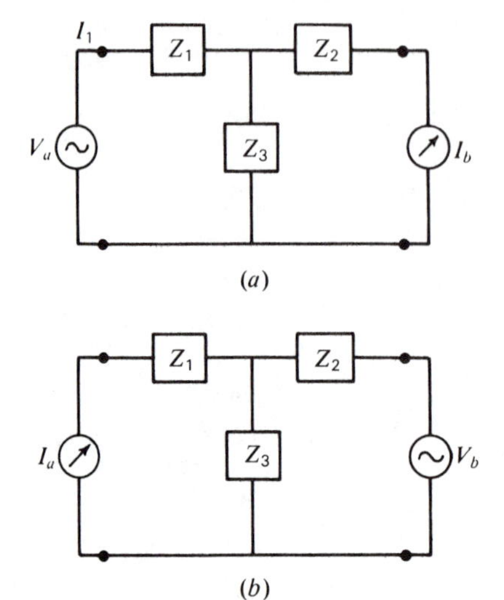

Figure 10-3 Equivalent circuits used in proof of reciprocity theorem.

transmitting antenna and energy flows from it to antenna A as in Fig. 10-2*b*, producing a current I_a at its terminals.

Now if $V_b = V_a$, then by the reciprocity theorem $I_a = I_b$.

The ratio of an emf to a current is an impedance. In Case 1, the ratio of V_a to I_b may be called the *transfer impedance* Z_{ab}, and in Case 2 the ratio V_b to I_a may be called the transfer impedance Z_{ba}. Then, by the reciprocity theorem it follows that these impedances are equal. Thus,

$$\frac{V_a}{I_b} = Z_{ab} = Z_{ba} = \frac{V_b}{I_a} \tag{1}$$

In order to prove the reciprocity theorem for antennas, let the antennas and the space between them be replaced by a network of linear, passive, bilateral impedances. Since any 4-terminal network can be reduced to an equivalent T section,[1] the antenna arrangement of Case 1 (see Fig. 10-2*a*) can be replaced by the network of Fig. 10-3*a*.

The current through the meter is

$$I_b = I_1 \frac{Z_3}{Z_2 + Z_3} \tag{2}$$

where

$$I_1 = \frac{V_a}{Z_1 + [Z_2 Z_3/(Z_2 + Z_3)]} = \frac{V_a(Z_2 + Z_3)}{Z_1 Z_2 + Z_2 Z_3 + Z_3 Z_1} \tag{3}$$

[1] This is true insofar as the amplitude and phase of the input voltage and output current are concerned.

Introducing (3) into (2) yields the current through the meter in terms of the emf V_a and the network impedances. Thus,

$$I_b = \frac{V_a Z_3}{Z_1 Z_2 + Z_2 Z_3 + Z_3 Z_1} \tag{4}$$

If the locations of the emf and current meter are interchanged, as in Fig. 10-3*b*, we obtain

$$I_a = \frac{V_b Z_3}{Z_1 Z_2 + Z_2 Z_3 + Z_3 Z_1} \tag{5}$$

Comparing (4) and (5), it follows that if $V_a = V_b$ then $I_a = I_b$, proving the theorem.

10-3 SELF-IMPEDANCE OF A THIN LINEAR ANTENNA.

In Sec. 9-16 we discussed how the current distribution and self-impedance of a dipole antenna can be obtained using the *moment method.* In this section we consider an *induced emf method* used by Carter[1] in 1932, decades before the general availability of powerful computers. Only the self-impedance can be determined with the induced emf method, the current distribution being assumed at the outset. Since measurements indicate that the current distribution on thin dipoles is nearly sinusoidal, except near current minima, this form of distribution is assumed. It results in satisfactory values for dipoles with length-diameter ratios as small as 100, provided the terminals are at a current maximum.

Consider a thin center-fed dipole antenna with lower end located at the origin of the coordinates as shown in Fig. 10-4. The antenna is situated in air or vacuum and is remote from other objects. Since the antenna is thin, a sinusoidal current distribution will be assumed with the maximum current I_1 at the terminals. Only lengths L which are an odd multiple of $\lambda/2$ will be considered so that the current distribution is symmetrical, with a current maximum at the terminals. The current distribution shown in Fig. 10-4 is for the case where $L = \lambda/2$. The current at a distance z from the origin is designated I_z. Then

$$I_z = I_1 \sin \beta z \tag{1}$$

Suppose that an emf V_{11} applied to the terminals of the antenna of Fig. 10-4 produces a current I_z at a distance z from the lower end. The ratio of V_{11} to I_z

[1] P. S. Carter, "Circuit Relations in Radiating Systems and Applications to Antenna Problems," *Proc. IRE*, **20**, 1004–1041, June 1932.

A. A. Pistolkors, "The Radiation Resistance of Beam Antennas," *Proc. IRE*, **17**, 562–579, March 1929.

R. Bechmann, "Calculation of Electric and Magnetic Field Strengths of Any Oscillating Straight Conductors," *Proc. IRE*, **19**, 461–466, March 1931.

R. Bechmann, "On the Calculation of Radiation Resistance of Antennas and Antenna Combinations," *Proc. IRE*, **19**, 1471–1480, August 1931.

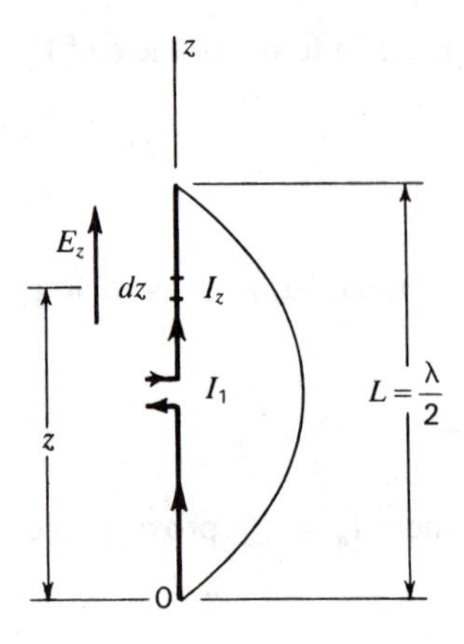

Figure 10-4 Center-fed linear $\lambda/2$ antenna.

may be designated as the transfer impedance Z_{1z}. Thus,

$$Z_{1z} = \frac{V_{11}}{I_z} \tag{2}$$

Next let the *applied field* at the antenna and parallel to it be E_z. This is the field produced by the antenna's own current as though it were flowing in empty space. This field *induces a field* E_{zi} at the conductor such that the boundary conditions are satisfied. For a perfect conductor these conditions are that the total field E_{zt} is zero or that $E_{zt} = E_z + E_{zi} = 0$ and therefore $E_{zi} = -E_z$. The emf dV_z produced by the induced field over a length dz is $-E_z\,dz$ or

$$dV_z = -E_z\,dz \tag{3}$$

If the antenna is short-circuited, this emf will produce a current dI_1 at the terminals. Then the transfer impedance Z_{z1} is given by

$$Z_{z1} = \frac{dV_z}{dI_1} \tag{4}$$

Since the reciprocity theorem (Sec. 10-2) holds not only for 2 separate antennas but also for 2 points on the same antenna, it follows that the transfer impedances of (2) and (4) are equal. Therefore,

$$\frac{V_{11}}{I_z} = Z_{1z} = Z_{z1} = \frac{dV_z}{dI_1} = \frac{-E_z\,dz}{dI_1} \tag{5}$$

and

$$V_{11}\,dI_1 = -I_z E_z\,dz \tag{6}$$

The terminal impedance Z_{11} of the antenna is given by the ratio of V_{11} to the total terminal current I_1. Thus,

$$Z_{11} = \frac{V_{11}}{I_1} \tag{7}$$

The impedance Z_{11} is a constant and is independent of the current amplitude. This follows from the fact that the system is linear. Therefore, Z_{11} can also be expressed as the ratio of an infinitesimal emf dV_{11} at the terminals to an infinitesimal current dI_1 at the terminals, or

$$Z_{11} = \frac{V_{11}}{I_1} = \frac{dV_{11}}{dI_1} \tag{8}$$

from which

$$V_{11}\, dI_1 = I_1\, dV_{11} \tag{9}$$

Substituting (9) into (6),

$$dV_{11} = \frac{-I_z}{I_1} E_z\, dz \tag{10}$$

Integrating (10) over the length of the antenna, we obtain

$$V_{11} = -\frac{1}{I_1} \int_0^L I_z E_z\, dz \tag{11}$$

where V_{11} is the emf which must be applied at the terminals to produce the current I_1 at the terminals. The terminal impedance Z_{11} is then

$$Z_{11} = \frac{V_{11}}{I_1} = -\frac{1}{I_1^2} \int_0^L I_z E_z\, dz \tag{12}$$

Since the antenna is isolated, this impedance is called the *self-impedance*. In (12) E_z is the z component of the electric field *at* the antenna caused by its own current. It will be convenient to indicate explicitly this type of field by the symbol E_{11} in place of E_z. Introducing also the value I_z from (1) into (12), we obtain for the self-impedance

$$Z_{11} = -\frac{1}{I_1} \int_0^L E_{11} \sin \beta z\, dz \tag{13}$$

To evaluate (13), it is first necessary to derive an expression for the field E_{11} along the antenna produced by its own current. Substituting this into (13) and integrating, it is possible to obtain an expression which can be evaluated numerically. The steps in this development are given in the following paragraphs.

If expressions can be written for the retarded scalar potential V due to charges on the antenna and for the retarded vector potential $\mathbf{A}$ due to currents on the antenna, then the electric field everywhere is derivable from the relation

$$\mathbf{E} = -\nabla V - j\omega \mathbf{A} \tag{14}$$

More particularly, the z component of $\mathbf{E}$ is given by

$$E_z = -\frac{\partial V}{\partial z} - j\omega A_z \tag{15}$$

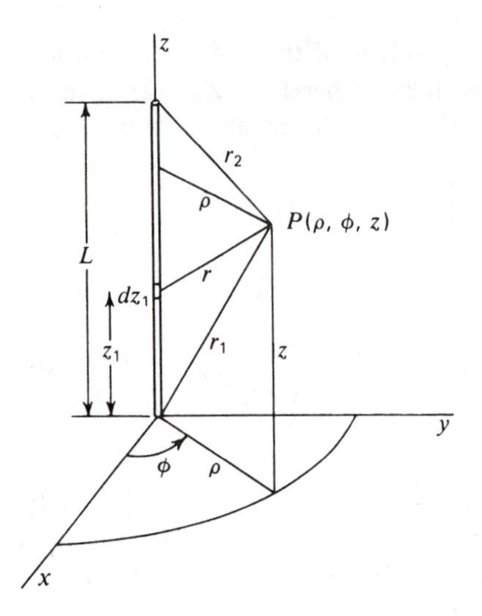

Figure 10-5 Relation of coordinates to antenna.

Referring to Fig. 10-5, let the antenna be coincident with the z axis. A point on the antenna is designated by its distance z_1 from the origin. A point P in space is given in cylindrical coordinates by ρ, ϕ, z. Other distances are as shown. Only lengths L which are an odd multiple of $\lambda/2$ will be considered. Thus,

$$L = \frac{n\lambda}{2}$$

where $n = 1, 3, 5, \ldots$

The scalar potential V at any point is given by

$$V = \frac{1}{4\pi\varepsilon_0} \iiint \frac{\rho}{r} \, d\tau \tag{16}$$

where ρ is the volume charge density, r the distance from the charge element to the point and $d\tau$ is a volume element. From Fig. 10-5,

$$r = \sqrt{\rho^2 + (z - z_1)^2}$$

In the case of a thin wire of length L, (16) reduces to

$$V = \frac{1}{4\pi\varepsilon_0} \int_0^L \frac{\rho_L}{r} \, dz_1 \tag{17}$$

where ρ_L = linear charge density on the wire

The vector potential **A** at any point is given by

$$\mathbf{A} = \frac{\mu_0}{4\pi} \iiint \frac{\mathbf{J}}{r} \, d\tau \tag{18}$$

where **J** = the current density

In the case of a thin wire (18) reduces to

$$A_z = \frac{\mu_0}{4\pi} \int_0^L \frac{I_{z1}}{r} \, dz_1 \tag{19}$$

where I_{z_1} = the current on the wire

By the continuity relation between current and linear charge density

$$\rho_L = -\int \frac{\partial I_{z_1}}{\partial z_1} \, dt \tag{20}$$

The current on the antenna is assumed to have a sinusoidal distribution as given by (1). Introducing the retarded time factor, we have for the retarded current

$$I_{z_1} = I_1 \sin \beta z_1 \, e^{j\omega[t-(r/c)]} \tag{21}$$

Substituting (21) into (20) and performing the indicated operations, the retarded linear charge density is

$$\rho_L = \frac{j\beta L_1}{\omega} \cos \beta z_1 \, e^{j\omega[t-(r/c)]} \tag{22}$$

Introducing (22) into (17) and noting that $\beta/\omega = 1/c$, the retarded scalar potential is

$$V = \frac{jI_1 e^{j\omega t}}{4\pi\varepsilon_0 c} \int_0^L \frac{\cos \beta z_1 e^{-j\beta r}}{r} \, dz_1 \tag{23}$$

Likewise, introducing (21) into (19), the z component of the retarded vector potential is

$$A_z = \frac{\mu_0 I_1 e^{j\omega t}}{4\pi} \int_0^L \frac{\sin \beta z_1 e^{-j\beta r}}{r} \, dz_1 \tag{24}$$

By de Moivre's theorem,

$$\cos \beta z_1 = \tfrac{1}{2}(e^{j\beta z_1} + e^{-j\beta z_1}) \tag{25}$$

and

$$\sin \beta z_1 = \frac{1}{2j}(e^{j\beta z_1} - e^{-j\beta z_1}) \tag{26}$$

Making these substitutions in (23) and (24),

$$V = \frac{jI_1 e^{j\omega t}}{8\pi\varepsilon_0 c} \int_0^L \frac{e^{-j\beta(z_1+r)} + e^{j\beta(z_1-r)}}{r} \, dz_1 \tag{27}$$

and

$$A_z = \frac{j\mu_0 I_1 e^{j\omega t}}{8\pi} \int_0^L \frac{e^{-j\beta(z_1+r)} - e^{j\beta(z_1-r)}}{r} \, dz_1 \tag{28}$$

Equations (27) and (28) give the retarded scalar and vector potentials caused by current on the antenna with the assumed sinusoidal distribution. Substituting these equations into (15) yields an expression for the z component of the electric field everywhere. Thus,

$$E_z = -\frac{jI_1e^{j\omega t}}{8\pi\varepsilon_0 c}\int_0^L \frac{\partial}{\partial z}\left(\frac{e^{-j\beta(z_1+r)} + e^{j\beta(z_1-r)}}{r}\right) dz_1 + \frac{\omega\mu_0 I_1 e^{j\omega t}}{8\pi}\int_0^L \left(\frac{e^{-j\beta(z_1+r)} - e^{j\beta(z_1-r)}}{r}\right) dz_1 \tag{29}$$

$$E_z = -\frac{jI_1e^{j\omega t}}{4\pi\varepsilon_0 c}\left(\frac{e^{-j\beta r_1}}{r_1} + \frac{e^{-j\beta r_2}}{r_2}\right) \tag{30}$$

where

$$r_1 = \sqrt{\rho^2 + z^2} \tag{31}$$

and

$$r_2 = \sqrt{\rho^2 + (L-z)^2} \tag{32}$$

The factor $1/4\pi\varepsilon_0 c \simeq 120\pi/4\pi = 30$. Also putting the time factor equal to its absolute value $e^{j\omega t} = 1$, Eq. (30) becomes

$$E_z = -j30I_1\left(\frac{e^{-j\beta r_1}}{r_1} + \frac{e^{-j\beta r_2}}{r_2}\right) \tag{33}$$

At the antenna (31) and (32) become

$$r_1 = z \tag{34}$$

and

$$r_2 = L - z \tag{35}$$

Substituting these into (33) yields the value of the z component of the electric field E_{11} *at* the antenna due to its own current. Thus,

$$E_{11} = -j30I_1\left(\frac{e^{-j\beta z}}{z} + \frac{e^{-j\beta(L-z)}}{L-z}\right) \tag{36}$$

Introducing (36) into (13) we obtain the self-impedance Z_{11} of a thin linear antenna an odd number of $\lambda/2$ long. Hence,

$$Z_{11} = j30\int_0^L \left(\frac{e^{-j\beta z}}{z} + \frac{e^{-j\beta(L-z)}}{L-z}\right)\sin\beta z\, dz \tag{37}$$

Applying de Moivre's theorem to $\sin\beta z$,

$$Z_{11} = -15\int_0^L \left[\frac{e^{-j2\beta z} - 1}{z} - \frac{e^{-j\beta L}(e^{j2\beta z} - 1)}{L-z}\right] dz \tag{38}$$

For $L = n\lambda/2$ where $n = 1, 3, 5, \ldots$, $e^{-j\beta L} = e^{-j\pi n} = -1$, so that (38) becomes

$$Z_{11} = -15\int_0^L \left(\frac{e^{-j2\beta z} - 1}{z} + \frac{e^{j2\beta z} - 1}{L-z}\right) dz \tag{39}$$

or

$$Z_{11} = 15\int_0^L \frac{1 - e^{-j2\beta z}}{z}\, dz + 15\int_0^L \frac{1 - e^{j2\beta z}}{L-z}\, dz \tag{40}$$

In the first integral let

$$u = 2\beta z \qquad \text{or} \qquad du = 2\beta \, dz$$

The upper limit $z = L$ becomes $u = 2\beta L = 2\pi n$, while the lower limit is unchanged. The first integral then transforms to

$$15 \int_0^{2\pi n} \frac{1 - e^{-ju}}{u} \, du \tag{41}$$

In the second integral let

$$v = 2\beta(L - z) \qquad \text{or} \qquad dv = -2\beta \, dz$$

The upper limit becomes zero while the lower limit becomes $2\pi n$. The second integral then transforms to

$$-15 \int_{2\pi n}^{0} \frac{1 - e^{j(2\pi n - v)}}{v} \, dv = 15 \int_0^{2\pi n} \frac{1 - e^{-jv}}{v} \, dv \tag{42}$$

Equations (41) and (42) are definite integrals of identical form. Since their limits are the same, they are equal. Therefore (40) becomes

$$Z_{11} = 30 \int_0^{2\pi n} \frac{1 - e^{-ju}}{u} \, du \tag{43}$$

If we now put $w = ju$, (43) transforms to

$$Z_{11} = 30 \int_0^{j2\pi n} \frac{1 - e^{-w}}{w} \, dw \tag{44}$$

The integral in (44) is an exponential integral with imaginary argument. It is designated by Ein (jy). Thus,[1]

$$\text{Ein}\,(jy) = \int_0^{jy} \frac{1 - e^{-w}}{w} \, dw \tag{45}$$

In our case $y = 2\pi n$. This integral can be expressed in terms of the sine and cosine integrals discussed in Sec. 5-6. Thus,

$$\text{Ein}\,(jy) = \text{Cin}\,(y) + j\,\text{Si}\,(y) \tag{46}$$

or

$$\text{Ein}\,(jy) = 0.577 + \ln y - \text{Ci}\,(y) + j\,\text{Si}\,(y) \tag{47}$$

Hence, the self-impedance is

$$Z_{11} = R_{11} + jX_{11} = 30[\text{Cin}\,(2\pi n) + j\,\text{Si}\,(2\pi n)] \tag{48}$$

or

$$Z_{11} = 30[0.577 + \ln (2\pi n) - \text{Ci}\,(2\pi n) + j\,\text{Si}\,(2\pi n)] \quad (\Omega) \tag{49}$$

---

[1] See, for example, S. A. Schelkunoff, *Applied Mathematics for Engineers and Scientists*, Van Nostrand, New York, 1948, p. 377.

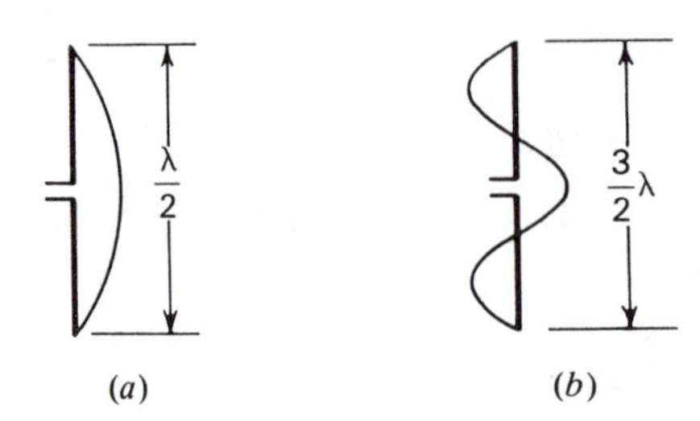

Figure 10-6 $\lambda/2$ and $3\lambda/2$ antennas.

The self-resistance is

$$R_{11} = 30 \text{ Cin } (2\pi n) = 30[0.577 + \ln (2\pi n) - \text{Ci } (2\pi n)] \quad (\Omega) \tag{50}$$

and the self-reactance is

$$X_{11} = 30 \text{ Si } (2\pi n) \quad (\Omega) \tag{51}$$

These equations give the impedance values for a thin linear center-fed antenna that is an odd number (n) of $\lambda/2$ long. The current distribution is assumed to be sinusoidal (Fig. 10-6*a*). The values are those appearing at the terminals at the center of the antenna.

In the case of a $\lambda/2$ *antenna* as shown in Fig. 10-6*a*, $n = 1$, and we have for the self-resistance and self-reactance

$$R_{11} = 30 \text{ Cin } (2\pi) \quad (\Omega) \tag{52}$$

and

$$X_{11} = 30 \text{ Si } (2\pi) \quad (\Omega) \tag{53}$$

The value of (52) is identical with that given for the radiation resistance of a $\lambda/2$ antenna, in Sec. 5-6, Eq. (5-6-23). Evaluating (52) and (53), we obtain for the self-impedance

$$Z_{11} = R_{11} + jX_{11} = 73 + j42.5 \ \Omega \tag{54}$$

Since X_{11} is not zero, an antenna exactly $\lambda/2$ long is not resonant. To obtain a resonant antenna, it is common practice to shorten the antenna a few percent to make $X_{11} = 0$. In this case the self-resistance is somewhat less than 73 Ω.

For a $3\lambda/2$ *antenna* as shown in Fig. 10-6*b*, $n = 3$, and the self-impedance is

$$Z_{11} = 30[\text{Cin } (6\pi) + j \text{ Si } (6\pi)]$$

or

$$Z_{11} = 105.5 + j45.5 \ \Omega \tag{55}$$

It is interesting that the self-reactance of center-fed antennas, an exact odd number of $\lambda/2$ long, is always positive since the sine integral Si $(2\pi n)$ is always positive. For large n the sine integral converges around a value of $\pi/2$ (see Fig. 5-12b) which corresponds to a reactance of 47.1 Ω. It should be noted that for antenna lengths not an exact odd number of $\lambda/2$ the reactance may be positive or negative. However, the foregoing analysis of this section is limited to antennas that are an exact odd number of $\lambda/2$ long.

For large n, the self-resistance expression (50) approaches the value

$$R_{11} = 30[0.577 + \ln (2\pi n)] \tag{56}$$

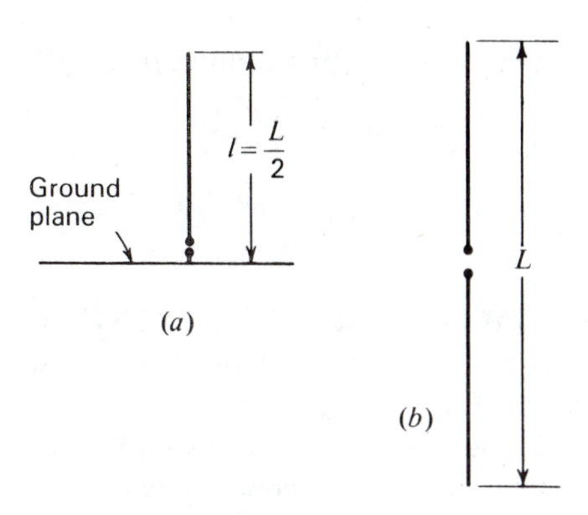

Figure 10-7 Stub antenna of length l at (*a*) and center-fed antenna of length L at (*b*).

since Ci $(2\pi n)$ approaches zero. Thus, the self-resistance continues to increase indefinitely with increasing n but at a logarithmic rate.

The more general situation, where the antenna length L is not restricted to an odd number of $\lambda/2$, has also been treated.[1] The antenna is center-fed, and the current distribution is assumed to be sinusoidal (see Fig. 5-8). The self-resistance for this case is

$$R_{11} = 30\left[\left(1 - \cot^2 \frac{\beta L}{2}\right) \text{Cin } 2\beta L + 4 \cot^2 \frac{\beta L}{2} \text{Cin } \beta L + 2 \cot \frac{\beta L}{2} (\text{Si } 2\beta L - 2 \text{ Si } \beta L)\right] \quad (\Omega) \quad (57)$$

When the length L is small, (57) reduces very nearly to

$$R_{11} = 5(\beta L)^2 \qquad (\Omega) \tag{58}$$

which is the same as (2-20-3) when $I_{av} = (1/2)\, I_0$.

For the special case of $L = n\lambda/2$, where $n = 1, 3, 5, \ldots$, (57) reduces to the relation given previously by (50).

The above discussion of this section applies to balanced center-fed antennas. For a thin linear stub antenna of height l perpendicular to an infinite, perfectly conducting ground plane as in Fig. 10-7*a*, the self-impedance is $\frac{1}{2}$ that for the corresponding balanced type (Fig. 10-7*b*). The general formula (57) for self-resistance can be converted for a stub antenna above a ground plane by changing the factor 30 to 15 and making the substitution $L = 2l$. The formulas (50) and (51) can be converted for a stub antenna with ground plane where the antenna is an odd number n of $\lambda/4$ long by changing the factor 30 to 15. Thus, for

---

[1] G. H. Brown and R. King, "High Frequency Models in Antenna Investigations," *Proc. IRE*, **22**, 457–480, April 1934.
J. Labus, "Recherische Ermittlung der Impedanz von Antennen," *Hochfrequenztechnik und Electro-akustik*, *17*, January 1933.

a $\lambda/4$ antenna perpendicular to an infinite, perfectly conducting ground plane, the self-impedance is

$$Z_{11} = 36.5 + j21 \ \Omega$$

10-4 MUTUAL IMPEDANCE OF TWO PARALLEL LINEAR ANTENNAS. The *mutual impedance* of 2 coupled circuits is defined in circuit theory as the negative of the ratio of the emf V_{21} *induced* in circuit 2 to the current I_1 flowing in circuit 1 with circuit 2 open. Consider, for example, the coupled circuit of Fig. 10-8 consisting of the primary and secondary coils of a transformer. The mutual impedance Z_{21} is then

$$Z_{21} = -\frac{V_{21}}{I_1} \tag{1}$$

where V_{21} is the emf induced across the terminals of the open-circuited secondary by the current I_1 in the primary. The mutual impedance, so defined, is not the same as a transfer impedance such as discussed in connection with the reciprocity theorem in Sec. 10-2. In general, a *transfer impedance* is the ratio of an emf *impressed* in one circuit to the resulting current in another with all circuits closed. For example, if the generator in Fig. 10-8 is removed from the primary and is connected to the secondary terminals, the ratio of the emf V *applied* by this generator to the current I_1 in the closed primary circuit is a transfer impedance Z_T. Thus,

$$\frac{V}{I_1} = Z_T \tag{2}$$

This impedance is not the same as the mutual impedance Z_{21} given in (1).

Instead of the coupled circuit of Fig. 10-8, let us consider now the case of 2 coupled antennas 1 and 2 as shown in Fig. 10-9. Suppose a current I_1 in antenna 1 induces an emf V_{21} at the open terminals of antenna 2. Then the ratio of $-V_{21}$ to I_1 is the mutual impedance Z_{21}. Thus,

$$Z_{21} = \frac{-V_{21}}{I_1} \tag{3}$$

If the generator is moved to the terminals of antenna 2, then by reciprocity the mutual impedance Z_{12} or ratio of $-V_{12}$ to I_2 is the same as before, where V_{12}

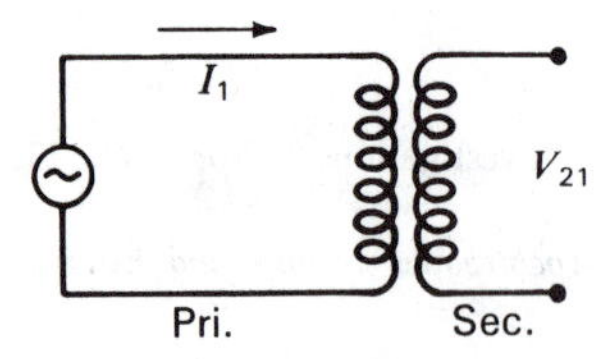

Figure 10-8 Coupled circuit or transformer.

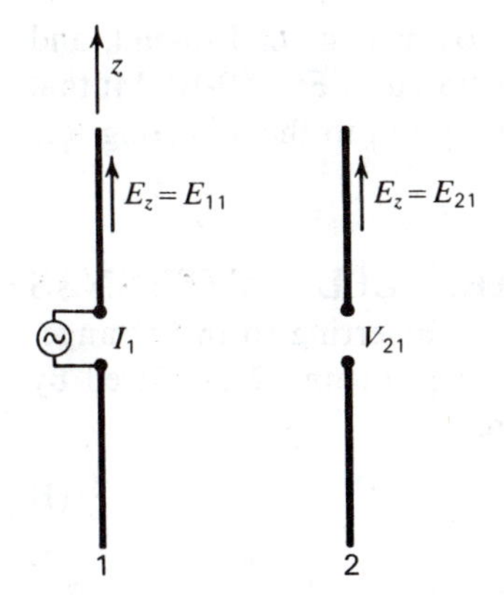

Figure 10-9 Parallel coupled antennas.

is the emf induced at the open terminals of antenna 1 by the current I_2 in antenna 2. Thus,

$$\frac{-V_{21}}{I_1} = Z_{21} = Z_{12} = \frac{-V_{12}}{I_2} \tag{4}$$

To calculate the mutual impedance, we need to know V_{21} and I_1. Let the antennas be in the z direction as shown in Fig. 10-9. The emf $-V_{11}$ induced in an antenna by its own current is indicated by (10-3-11). To obtain the emf V_{21} induced at the open terminals of antenna 2 by the current in antenna 1, we set $E_z = E_{21}$, $V_{11} = -V_{21}$, and $I_1 = I_2$ in (10-3-11). Then

$$V_{21} = \frac{1}{I_2} \int_0^L I_z E_{21}\, dz \tag{5}$$

where I_2 is the maximum current and I_z the value at a distance z from the lower end of antenna 2 with its terminals closed, and where E_{21} is the electric field along antenna 2 produced by the current in antenna 1. Assuming that this current distribution is sinusoidal as given by

$$I_z = I_2 \sin \beta z \tag{6}$$

so that (5) becomes

$$V_{21} = \int_0^L E_{21} \sin \beta z\, dz \tag{7}$$

then

$$Z_{21} = \frac{-V_{21}}{I_1} = -\frac{1}{I_1} \int_0^L E_{21} \sin \beta z\, dz \tag{8}$$

This is the general expression for the mutual impedance of two thin linear, parallel, center-fed antennas with sinusoidal current distribution. We will consider first the situation where both antennas are the same length L, where L is an odd number of $\lambda/2$ long ($L = n\lambda/2$; $n = 1, 3, 5, \ldots$). A case of particular interest is where both antennas are $\lambda/2$ long ($n = 1$). The relative positions of the antennas may be divided into three situations: side-by-side, collinear or end-to-end, and staggered or in echelon. These arrangements are illustrated in Fig. 10-10. Mutual impedance expressions for the three arrangements are given in the following sections.

10-5 MUTUAL IMPEDANCE OF PARALLEL ANTENNAS SIDE-BY-SIDE.

Let d be separation of the antennas. Referring to the arrangement of Fig. 10-10a and Fig. 10-11, the field E_{21} along antenna 2 produced by the current I_1 in antenna 1 is given by (10-3-33) where

$$r_1 = \sqrt{d^2 + z^2} \tag{1}$$

and

$$r_2 = \sqrt{d^2 + (L - z)^2} \tag{2}$$

Substituting this into (10-4-8), the mutual impedance becomes

$$Z_{21} = j30 \int_0^L \left\{ \frac{\exp(-j\beta\sqrt{d^2 + z^2})}{\sqrt{d^2 + z^2}} + \frac{\exp[-j\beta\sqrt{d^2 + (L - z)^2}]}{\sqrt{d^2 + (L - z)^2}} \right\} \sin \beta z \, dz \tag{3}$$

Carter has shown that upon integration of (3),

$$Z_{21} = 30\{2 \text{ Ei}(-j\beta d) - \text{Ei}[-j\beta(\sqrt{d^2 + L^2} + L)] - \text{Ei}[-j\beta(\sqrt{d^2 + L^2} - L)]\} \quad (\Omega) \tag{4}$$

where the exponential integral

$$\text{Ei}(\pm jy) = \text{Ci}(y) \pm j \text{ Si}(y) \tag{5}$$

Thus, the mutual resistance is

$$R_{21} = 30\{2 \text{ Ci}(\beta d) - \text{Ci}[\beta(\sqrt{d^2 + L^2} + L)] - \text{Ci}[\beta(\sqrt{d^2 + L^2} - L)]\} \quad (\Omega) \tag{6}$$

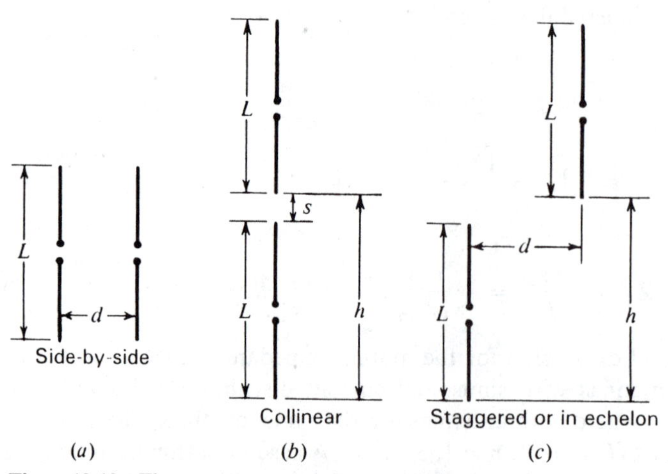

Figure 10-10 Three arrangements of two parallel antennas.

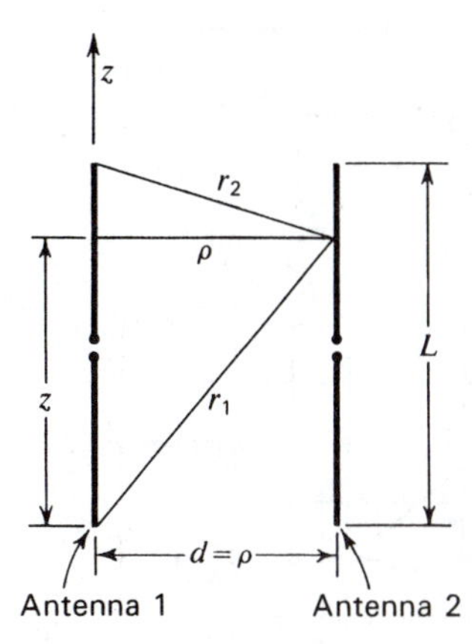

Figure 10-11 Parallel coupled antennas with dimensions.

and the mutual reactance is

$$X_{21} = -30\{2\ \mathrm{Si}\ (\beta d) - \mathrm{Si}\ [\beta(\sqrt{d^2 + L^2} + L)] - \mathrm{Si}\ [\beta(\sqrt{d^2 + L^2} - L)]\} \quad (\Omega) \qquad (7)$$

where

$$R_{21} + jX_{21} = Z_{21} = Z_{12} = R_{12} + jX_{12} \qquad (8)$$

and where $L = n\lambda/2$ for n odd.

The mutual resistance and reactance calculated by (6) and (7) for the case of $\lambda/2$ antennas ($L = \lambda/2$) are presented by the solid curves in Fig. 10-12 as a function of the spacing d. The mutual resistance R_{21} is also listed in Table 10-1.

An integral-equation method for the calculation of the mutual impedance of linear antennas has been presented by King and Harrison[1] and by Tai.[2] The method is related to that discussed in Chap. 9. In this method the diameter of the antenna conductor is a factor. By way of comparison, curves for the mutual resistance and reactance given by Tai are also shown in Fig. 10-12. The dashed curves are for a total length-diameter ratio (L/D) of 11 000 (very thin antenna) and the dotted curves for a ratio of 73.

In Table 10-1 the quantity $R_{11} - R_{21}$, which is important in array calculations, is also tabulated. When d is small, it has been shown by Brown[3] that this quantity is given approximately by the simple relation

$$R_{11} - R_{21} = 60\pi^2\left(\frac{d}{\lambda}\right)^2 = 592.2\left(\frac{d}{\lambda}\right)^2 \quad (\Omega) \qquad (9)$$

where λ = the free-space wavelength

[1] R. King and C. W. Harrison, Jr., "Mutual and Self Impedance for Coupled Antennas," *J. Appl. Phys.*, **15**, 481–495, June 1944.

[2] C. T. Tai, "Coupled Antennas," *Proc. IRE*, **36**, 487–500, April 1948.

[3] G. H. Brown, personal communication to the author, June 16, 1938.

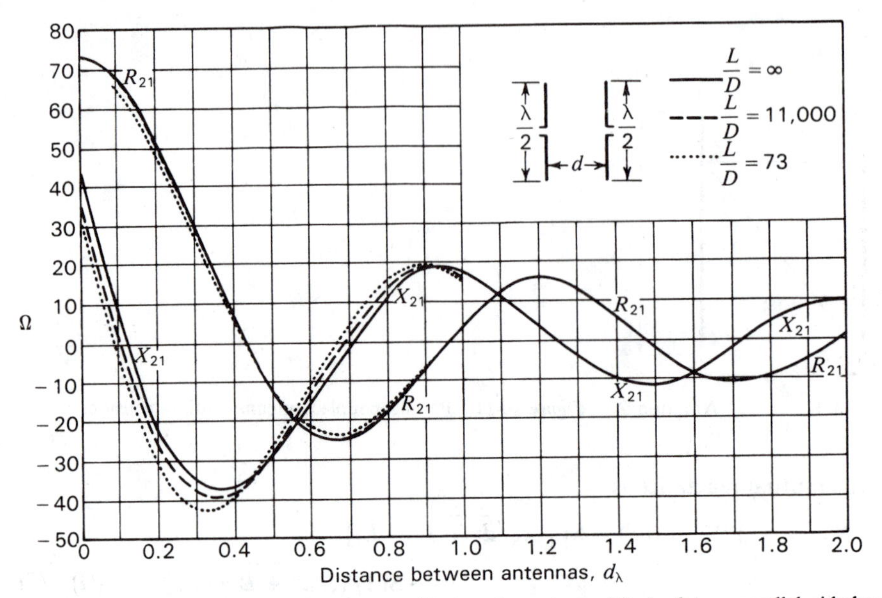

Figure 10-12 Curves of mutual resistance (R_{21}) and reactance (X_{21}) of two parallel side-by-side linear $\lambda/2$ antennas as a function of distance between them. Solid curves are for infinitesimally thin antennas as calculated from Carter's formulas. Dashed and dotted curves between 0 and 1.0λ spacing are from Tai's data for antennas with L/D ratios of 11 000 and 73 respectively.

This relation is accurate to within 1 percent when $d \leq 0.05\lambda$ and to within about 5 percent when $d \leq 0.1\lambda$.

In the more general situation where the antenna length L is not restricted to an odd number of $\lambda/2$, the mutual resistance and reactance are given by Brown and King[1] as

$$R_{21} = 30 \frac{1}{\sin^2 (\beta L/2)} \Bigg\{ 2(2 + \cos \beta L) \text{ Ci } \beta d$$
$$- 4 \cos^2 \frac{\beta L}{2} \left[\text{Ci } \frac{\beta}{2} (\sqrt{4d^2 + L^2} - L) + \text{Ci } \frac{\beta}{2} (\sqrt{4d^2 + L^2} + L) \right]$$
$$+ \cos \beta L [\text{Ci } \beta(\sqrt{d^2 + L^2} - L) + \text{Ci } \beta(\sqrt{d^2 + L^2} + L)]$$
$$+ \sin \beta L \Bigg[\text{Si } \beta(\sqrt{d^2 + L^2} + L) - \text{Si } \beta(\sqrt{d^2 + L^2} - L)$$
$$- 2 \text{ Si } \frac{\beta}{2} (\sqrt{4d^2 + L^2} + L) + 2 \text{ Si } \frac{\beta}{2} (\sqrt{4d^2 + L^2} - L) \Bigg] \Bigg\} \quad (\Omega) \qquad (10)$$

[1] G. H. Brown and R. King, "High Frequency Models in Antenna Investigations," *Proc. IRE*, **22**, 457–480, April 1934.

Table 10-1 Mutual resistance versus spacing for thin center-fed side-by-side $\lambda/2$ antennas ($\beta L = 180°$), with sinusoidal current distribution

Spacing d	Mutual resistance R_{21}, Ω	Self minus mutual resistance $(R_{11} - R_{21})$, Ω
0.00	73.13	0.00
0.01	73.07	0.06
0.05	71.65	1.48
0.10	67.5	5.63
0.125	64.4	8.7
0.15	60.6	12.5
0.20	51.6	21.5
0.25	40.9	32.2
0.3	29.4	43.7
0.4	+6.3	66.8
0.5	−12.7	85.8
0.6	−23.4	96.5
0.7	−24.8	97.9
0.8	−18.6	91.7
0.9	−7.2	80.3
1.0	+3.8	69.3
1.1	+12.1	61.0
1.2	+15.8	57.3
1.3	+12.4	60.7
1.4	+5.8	67.3
1.5	−2.4	75.5
1.6	−8.3	81.4
1.7	−10.7	83.8
1.8	−9.4	82.5
1.9	−4.8	77.9
2.0	+1.1	72.0

and

$$
\begin{aligned}
X_{21} = 30 \frac{1}{\sin^2 (\beta L/2)} \Bigg\{ & -2(2 + \cos \beta L) \operatorname{Si} \beta d \\
& + 4 \cos^2 \frac{\beta L}{2} \left[\operatorname{Si} \frac{\beta}{2} \left(\sqrt{4d^2 + L^2} - L\right) + \operatorname{Si} \frac{\beta}{2} \left(\sqrt{4d^2 + L^2} + L\right) \right] \\
& - \cos \beta L \left[\operatorname{Si} \beta\left(\sqrt{d^2 + L^2} - L\right) + \operatorname{Si} \beta\left(\sqrt{d^2 + L^2} + L\right)\right] \\
& + \sin \beta L \Bigg[\operatorname{Ci} \beta\left(\sqrt{d^2 + L^2} + L\right) - \operatorname{Ci} \beta\left(\sqrt{d^2 + L^2} - L\right) \\
& - 2 \operatorname{Ci} \frac{\beta}{2} \left(\sqrt{4d^2 + L^2} + L\right) + 2 \operatorname{Ci} \frac{\beta}{2} \left(\sqrt{4d^2 + L^2} - L\right) \Bigg] \Bigg\} \quad (\Omega) \qquad (11)
\end{aligned}
$$

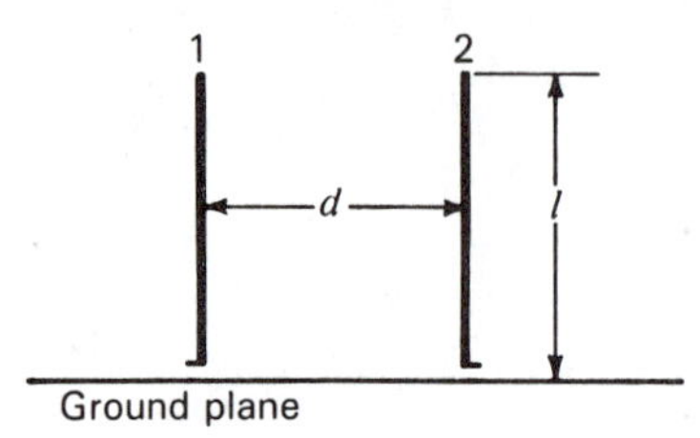

Figure 10-13 Two coupled linear parallel stub antennas.

In the special case of $L = n\lambda/2$, where n is odd, (10) and (11) reduce to the relations given previously by (6) and (7).

The above relations of this section apply to balanced center-fed antennas. The mutual impedance of two stub antennas of height $l = L/2$ above an infinite, perfectly conducting ground plane as in Fig. 10-13 is $\frac{1}{2}$ that given by (6) and (7) or (10) and (11). These relations are converted to the ground-plane case by changing the factor 30 to 15 and making the substitution $L = 2l$.

10-6 MUTUAL IMPEDANCE OF PARALLEL COLLINEAR ANTENNAS.

Let each antenna be an odd number of $\lambda/2$ long and arranged as in Fig. 10-10b. For the case where h is greater than L, Carter[1] gives the mutual resistance and reactance as

$$R_{21} = -15 \cos \beta h \left[-2 \text{ Ci } 2\beta h + \text{Ci } 2\beta(h - L) + \text{Ci } 2\beta(h + L) - \ln \left(\frac{h^2 - L^2}{h^2} \right) \right]$$
$$+ 15 \sin \beta h \, [2 \text{ Si } 2\beta h - \text{Si } 2\beta(h - L) - \text{Si } 2\beta(h + L)] \quad (\Omega) \quad (1)$$

and

$$X_{21} = -15 \cos \beta h [2 \text{ Si } 2\beta h - \text{Si } 2\beta(h - L) - \text{Si } 2\beta(h + L)]$$
$$+ 15 \sin \beta h \left[2 \text{ Ci } 2\beta h - \text{Ci } 2\beta(h - L) - \text{Ci } 2\beta(h + L) - \ln \left(\frac{h^2 - L^2}{h^2} \right) \right] \quad (\Omega) \quad (2)$$

Curves for R_{21} and X_{21} of parallel collinear $\lambda/2$ antennas ($L = \lambda/2$) are presented in Fig. 10-14 as a function of the spacing s where $s = h - L$ (see Fig. 10-10b).

10-7 MUTUAL IMPEDANCE OF PARALLEL ANTENNAS IN ECHELON.

For this case the antennas are staggered or in echelon as in Fig. 10-10c. Each antenna is an odd number of $\lambda/2$ long. The mutual resistance

[1] P. S. Carter, "Circuit Relations in Radiating Systems and Applications to Antenna Problems," *Proc. IRE*, **20**, 1004–1041, June 1932.

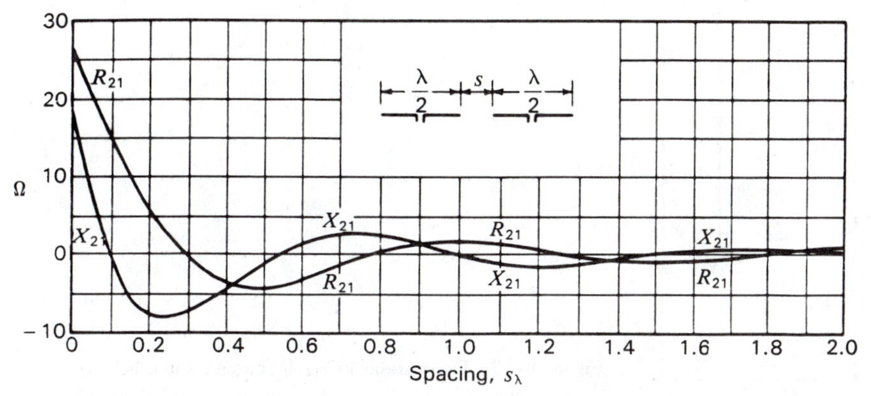

Figure 10-14 Curves of mutual resistance (R_{21}) and reactance (X_{21}) of two parallel collinear infinitesimally thin $\lambda/2$ antennas as a function of the spacing s between adjacent ends.

and reactance of two such antennas are given by Carter[1] as

$$R_{21} = -15 \cos \beta h(-2 \text{ Ci } A - 2 \text{ Ci } A' + \text{Ci } B + \text{Ci } B' + \text{Ci } C + \text{Ci } C')$$
$$+ 15 \sin \beta h(2 \text{ Si } A - 2 \text{ Si } A' - \text{Si } B + \text{Si } B' - \text{Si } C + \text{Si } C') \quad (\Omega) \quad (1)$$

and

$$X_{21} = -15 \cos \beta h(2 \text{ Si } A + 2 \text{ Si } A' - \text{Si } B - \text{Si } B' - \text{Si } C - \text{Si } C')$$
$$+ 15 \sin \beta h(2 \text{ Ci } A - 2 \text{ Ci } A' - \text{Ci } B + \text{Ci } B' - \text{Ci } C + \text{Ci } C') \quad (\Omega) \quad (2)$$

where $A = \beta(\sqrt{d^2 + h^2} + h)$
$A' = \beta(\sqrt{d^2 + h^2} - h)$
$B = \beta[\sqrt{d^2 + (h - L)^2} + (h - L)]$
$B' = \beta[\sqrt{d^2 + (h - L)^2} - (h - L)]$
$C = \beta[\sqrt{d^2 + (h + L)^2} + (h + L)]$
$C' = \beta[\sqrt{d^2 + (h + L)^2} - (h + L)]$

Values of the mutual resistance in ohms as calculated from (1) are listed in Table 10-2[2] as a function of d and h for the case where the antennas are $\lambda/2$ long ($L = \lambda/2$), as indicated in Fig. 10-15.

The staggered or echelon arrangement is the more general situation of which the side-by-side position (Sec. 10-5) and the collinear position (Sec. 10-6) are special cases.

[1] P. S. Carter, "Circuit Relations in Radiating Systems and Applications to Antenna Problems," *Proc. IRE*, **20**, 1004–1041, June 1932. See also H. E. King, "Mutual Impedance of Unequal Length Antennas in Echelon," *IEEE Trans. Ants. Prop.*, **AP-5**, 306–313, July 1957.

[2] All but a few values are from a table by A. A. Pistolkors, "The Radiation Resistance of Beam Antennas," *Proc. IRE*, **17**, 562–579, March 1929.

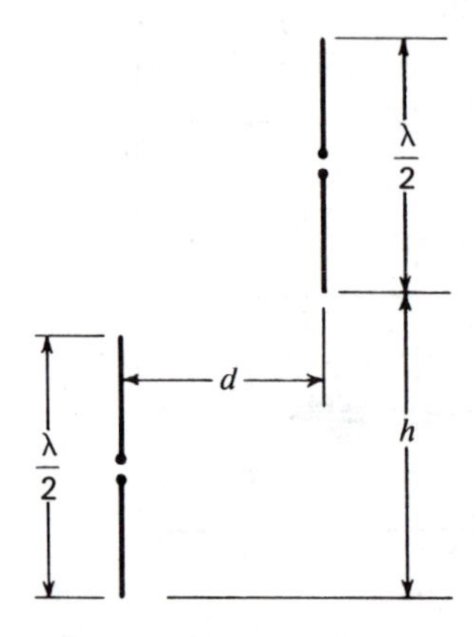

Figure 10-15 Two parallel linear $\lambda/2$ antennas in echelon.

10-8 MUTUAL IMPEDANCE OF OTHER CONFIGURATIONS.

There are many other antenna configurations for which the mutual impedance may be of interest. The variety is enormous, but two will be mentioned and references given which the reader may consult for further information.

1. *Parallel antennas of unequal height.* This case has been treated by Cox.[1] His data apply specifically to stub antennas perpendicular to an infinite, perfectly conducting ground, but can be used with symmetrical center-fed antennas of twice the length by multiplying the resistance and reactance values by 2 (see also Howard E. King[2]).
2. *V or skew antennas.* Some antenna systems involve nonparallel linear radiators. The mutual impedance of such inclined antennas are readily calculated by the moment method, as, for example, by J. H. Richmond's FORTRAN IV program, ASIS-NAPS Document NAPS-02223 (see References at the end of Chap. 9).

For two short dipoles, however, a simple, useful relation can be derived as follows.

Referring to Fig. 10-16, consider the 2 short center-fed dipoles 1 and 2 of length L_λ and L'_λ separated by a distance r_λ with orientation angles θ and θ' as indicated. The mutual impedance Z_{21} is given by the ratio of the voltage V_{21} induced in dipole 2 by the current I_1 flowing in dipole 1. Then for $L_\lambda \ll 1$, $L'_\lambda \ll 1$ and $r_\lambda \gg 1$, we have, from (5-2-34),

$$Z_{21} = \frac{V_{21}}{I_1} = \frac{60\pi L_\lambda}{r_\lambda} \sin\theta \, e^{-j\beta r} L'_\lambda \sin\theta' \tag{1}$$

[1] C. R. Cox, "Mutual Impedance between Vertical Antennas of Unequal Heights," *Proc. IRE*, **35**, 1367–1370, November 1947.

[2] H. E. King, "Mutual Impedance of Unequal Length Antennas in Echelon," *IEEE Trans. Ants. Prop.*, **AP-5**, 306–313, July 1957.

Table 10-2 Mutual resistance as a function of d and h (Fig. 10-15) for thin $\lambda/2$ antennas in echelon

	Spacing h						
Spacing d	0.0λ	0.5λ	1.0λ	1.5λ	2.0λ	2.5λ	3.0λ
0.0λ	+73.1	+26.4	−4.1	+1.8	−1.0	+0.6	−0.4
0.5λ	−12.7	−11.8	−0.8	+0.8	−1.0	+0.5	−0.3
1.0λ	+3.8	+8.8	+3.6	−2.9	+1.1	−0.4	+0.1
1.5λ	−2.4	−5.8	−6.3	+2.0	+0.6	−1.0	+0.9
2.0λ	+1.1	+3.8	+6.1	+0.2	−2.6	+1.6	−0.5
2.5λ	−0.8	−2.8	−5.7	−2.4	+2.7	−0.3	−0.1
3.0λ	+0.4	+1.9	+4.5	+3.2	−2.1	−1.6	+1.7
3.5λ	−0.3	−1.5	−3.9	−3.8	+0.7	+2.7	−1.0
4.0λ	+0.2	+1.1	+3.1	+3.7	+0.5	−2.5	−0.1
4.5λ	−0.2	−0.9	−2.5	−3.4	−1.3	+2.0	+1.1
5.0λ	+0.2	+0.7	+2.1	+3.1	+1.8	−1.4	−1.9
5.5λ	−0.1	−0.6	−1.8	−2.9	−2.2	+0.5	+1.8
6.0λ	+0.1	+0.5	+1.6	+2.6	+2.3	−0.1	−2.0
6.5λ	−0.1	−0.5	−1.2	−2.3	−2.3	−0.5	+1.7
7.0λ	+0.1	+0.4	+1.1	+2.1	+2.3	+0.9	−1.3
7.5λ	0.0	−0.3	−1.0	−1.9	−2.1	−1.0	+0.7

Magnitude Orientation Periodic function

or

$$Z_{21} = \frac{60\pi L_\lambda L'_\lambda}{r_\lambda}(\sin\theta\sin\theta')(\sin 2\pi r_\lambda + j\cos 2\pi r_\lambda) \tag{2}$$

with maximum value

$$Z_{21}(\text{max}) = \frac{60\pi L_\lambda L'_\lambda}{r_\lambda} \quad (\Omega) \tag{3}$$

We note in (2) that there are 3 factors: the first is a *magnitude factor* involving the lengths of the dipoles and their separation, the second involves their mutual *orientation*, while the third factor is a *periodic or complex function* of unit magnitude giving the phase as a function of the separation distance. The mutual impedance for antennas, in general, involves these 3 factors.

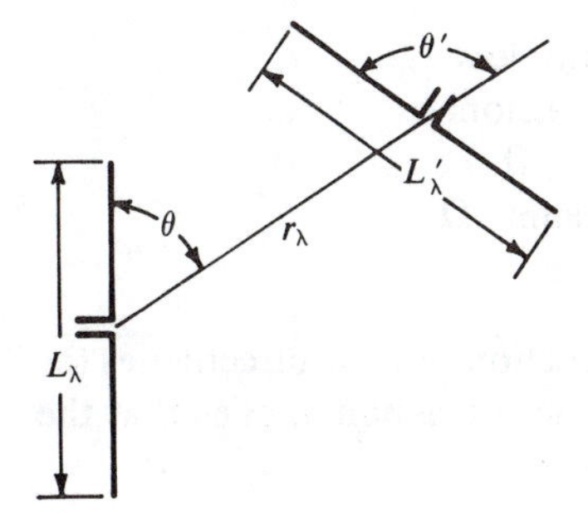

Figure 10-16 Two short center-fed dipoles of length L_λ and L'_λ with separation r_λ and orientation angles θ and θ' for mutual impedance equation.

10-9 MUTUAL IMPEDANCE IN TERMS OF DIRECTIVITY AND RADIATION RESISTANCE. Consider a transmitting antenna and a receiving antenna separated by a distance r. The power delivered to the receiving antenna load under matched conditions is

$$P_r = \frac{(V_r/2)^2}{R_r} \qquad \text{(W)} \tag{1}$$

where V_r = voltage induced at terminals of receiving antenna, V
R_r = radiation resistance of receiving antenna, Ω

The power transmitted is given by

$$P_t = I_t^2 R_t \qquad \text{(W)} \tag{2}$$

where I_t = current at terminals of transmitting antenna, A
R_t = radiation resistance of transmitting antenna, Ω

From (1) and (2), the magnitude of the mutual impedance of the antennas is

$$|Z_m| = \frac{V_r}{I_t} = \frac{2\sqrt{P_r R_r}}{\sqrt{P_t/R_t}} = 2\sqrt{\frac{P_r}{P_t}}\sqrt{R_r R_t} \tag{3}$$

From the Friis transmission formula (2-25-5),

$$\frac{P_r}{P_t} = \frac{A_{er} A_{et}}{r^2 \lambda^2} \tag{4}$$

where A_{er} = effective aperture or receiving antenna, m^2
A_{et} = effective aperture of transmitting antenna, m^2

Introducing (4) in (3), the maximum mutual impedance becomes

$$Z_m(\max) = \frac{2\sqrt{A_{er} A_{et} R_r R_t}}{r\lambda} \tag{5}$$

However, the directivity $D = 4\pi A_e/\lambda^2$, so the *maximum mutual impedance* can be reexpressed as

$$|Z_m|(\max) = \frac{\sqrt{D_r D_t R_r R_t}}{2\pi r_\lambda} \qquad (\Omega) \tag{6}$$

where D_r = directivity of receiving antenna, dimensionless
D_t = directivity of transmitting antenna, dimensionless
R_r = radiation resistance of receiving antenna, Ω
R_t = radiation resistance of transmitting antenna, Ω
r_λ = separation, wavelengths

Thus, the mutual impedance of two antennas is a function of their directivities (or apertures) and radiation resistances. In the above, the restriction applies that the separation distance is large compared to the antenna dimensions and also that $r \gg \lambda$.

Example. Calculate the maximum mutual impedance of 2 center-fed dipole antennas 0.1λ long separated by 10λ. Assume uniform current distribution on the dipoles. The receiving antenna is terminated for maximum power transfer.

Solution. From (6) and (2-20-3) (and also the table of Sec. 2-24),

$$|Z_m|(\max) = \frac{1.5 \times 0.8\pi^2}{2\pi \times 10} = 0.1885\ \Omega$$

From (10-8-3),

$$Z_{21}(\max) = \frac{60\pi \times 0.1^2}{10} = 0.1885\ \Omega$$

Although (6) and (10-8-3) bear little resemblance, except for the r_λ in the denominator, both yield an identical impedance.

ADDITIONAL REFERENCES FOR CHAP. 10

Rhodes, D. R.: "A Reactance Theorem," *Proc. Roy. Soc. Lond. A*, **353**, 1–10, 1977.

Uda, S., and Y. Mushiake: "On the Theory of Antennae with Discontinuous Thickness," *Tech. Rept. Tohuku Univ.*, **14**, no. 2, 1950.

Uda, S., and Y. Mushiake: "Theoretical Calculation of the Input Impedances of Two Parallel Antennae," *Sci. Rept. Res. Inst. Tohoku Univ.*, **B-1**, **2**, no. 1, 1951.

PROBLEMS[1]

★10-1 **A $5\lambda/2$ antenna.** Calculate the self-resistance and self-reactance of a thin, symmetrical center-fed linear antenna $5\lambda/2$ long.

10-2 **Parallel side-by-side $\lambda/2$ antennas.** Calculate the mutual resistance and mutual reactance for two parallel side-by-side thin linear $\lambda/2$ antennas with a separation of 0.15λ.

10-3 **Two $\lambda/2$ antennas in echelon.** Calculate the mutual resistance and reactance of two parallel thin linear $\lambda/2$ antennas in echelon for the case where $d = 0.25\lambda$ and $h = 1.25\lambda$ (see Fig. 10-15).

10-4 **Brown's equation.** Prove Brown's relation $R_{11} - R_{21} = 60\pi^2(d/\lambda)^2$ given in (10-5-9).

★10-5 **Three side-by-side antennas.** Three antennas are arranged as shown in Fig. P10-5. The currents are of the same magnitude in all antennas. The currents are in phase in (*a*) and (*c*), but the current in (*b*) is in antiphase. The self-resistance of each

[1] Answers to starred (★) problems are given in App. D.

antenna is 100 Ω, while the mutual resistances are: $R_{ab} = R_{bc} = 40$ Ω and $R_{ac} = -10$ Ω. What is the radiation resistance of each of the antennas? The resistances are referred to the terminals, which are in the same location in all antennas.

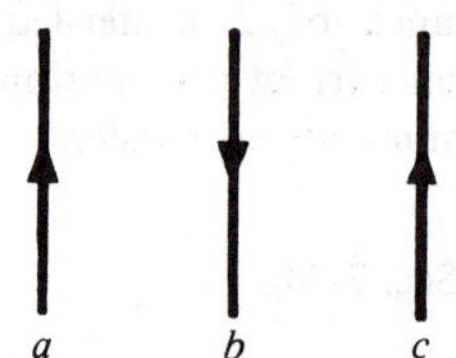

Figure P10-5 Three side-by-side antennas.

10-6 Self-resistance and mutual resistance. Explain why the mutual resistance of two antennas can be both positive and negative but the self-resistance of a single antenna can only be positive.

10-7 Terminal impedance.

(*a*) Show by means of an equivalent network that at the terminals of a receiving antenna, the equivalent, or Thévenin, generator has an impedance $Z_{22} - (Z_m^2/Z_{11})$ and an emf $V_1 Z_m/Z_{11}$, where Z_{11} = self-impedance of transmitting antenna, Z_{22} = self-impedance of receiving antenna, Z_m = mutual impedance and V_1 = emf applied to terminals of transmitting antenna.

(*b*) What load impedance connected to the terminals of the receiving antenna results in the maximum power transfer?

CHAPTER 11

ARRAYS OF DIPOLES AND OF APERTURES

11-1 INTRODUCTION. Essential background for this chapter is covered in Chap. 4 on arrays of point sources, Chap. 5 on linear antennas, Chap. 7 on driven element arrays (Secs. 7-11 and 7-12) and Chap. 10 on self and mutual impedances. The heading for this chapter could appropriately be "Arrays of Antenna Elements" where an *element* refers to the basic unit of which an array is constructed. In the first part of this chapter the "elements" are mostly thin linear *dipoles* while in the latter sections of the chapter the array "elements" are *apertures* in general, which may be helices, horns, big reflectors or arrays of dipoles (arrays of arrays).

The *far-* or *radiation field pattern*, the *driving point impedance* and the *array gain* are first derived in that order for several different arrays of dipoles. The method of analysis is general and is applicable to other dipole arrays, the specific types discussed serving merely as examples. Array gain is calculated by treating the dipoles as circuit elements having self and mutual impedances. Although direct pattern integration could be used to determine the gain, the circuit approach is simpler provided impedance values are available (patterns having been utilized in the impedance calculations). In most of the arrays the dipoles are driven but Sec. 11-9 discusses arrays having parasitic dipoles.

Retro, phased, scanning, adaptive, microstrip, low-sidelobe, long-wire and curtain arrays are topics of Secs. 11-10 through 11-16. The remaining sections

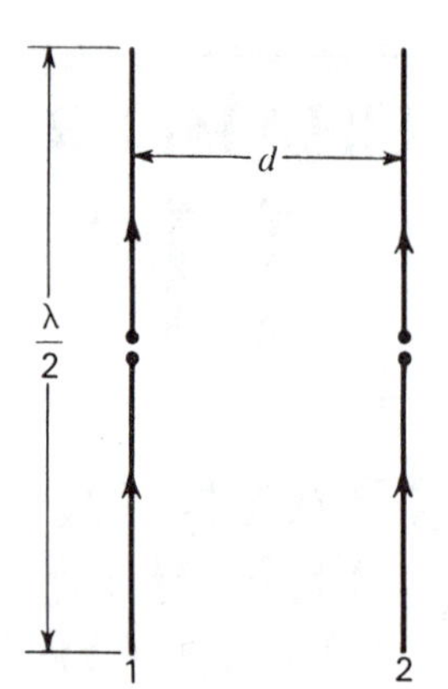

Figure 11-1 Broadside array of 2 in-phase $\lambda/2$ elements.

discuss continuous apertures, Fourier transform relations between far-field patterns and aperture distributions, total power and correlation arrays, interferometers, aperture synthesis, very large arrays and very long baseline arrays.

11-2 ARRAY OF TWO DRIVEN λ/2 ELEMENTS. BROADSIDE CASE.

Consider two center-fed $\lambda/2$ elements (dipoles) arranged side by side with a spacing d as in Fig. 11-1. Two special cases will be considered: the *broadside case*[1] treated in this section, in which the two elements are fed with equal *in-phase* currents, and the *end-fire case*[2] (Sec. 11-3), in which the two elements are fed with equal currents in *opposite phase*. The more general case where the currents are equal in magnitude but in any phase relation is treated in Sec. 11-4.

11-2a Field Patterns.

The first part of the analysis will be to determine the absolute far-field patterns. It is convenient to obtain two pattern expressions, one for the horizontal plane and one for the vertical plane. Ordinarily, the relative patterns would be sufficient. However, the absolute patterns will be needed in gain calculations. Let the elements be vertical as shown in Fig. 11-2*a*. It is assumed that the array is in free space, i.e., at an infinite distance from the ground or other objects. The field intensity $E_1(\phi)$ from a single element as a function of ϕ

---

[1] In the so-called "broadside case" there is always a major lobe of radiation broadside to the array, although at large spacings there may be an end-fire lobe of equal magnitude (as, for example, when the spacing is 1λ).

[2] In the so-called "end-fire case" the pattern always has zero radiation broadside. The maximum radiation is always end-fire if the spacing is $\lambda/2$ or less. However, for greater spacings the maximum radiation is, in general, not end-fire. Since spacings of $\lambda/2$ or less are of principal interest, the array may be referred to as an end-fire type.

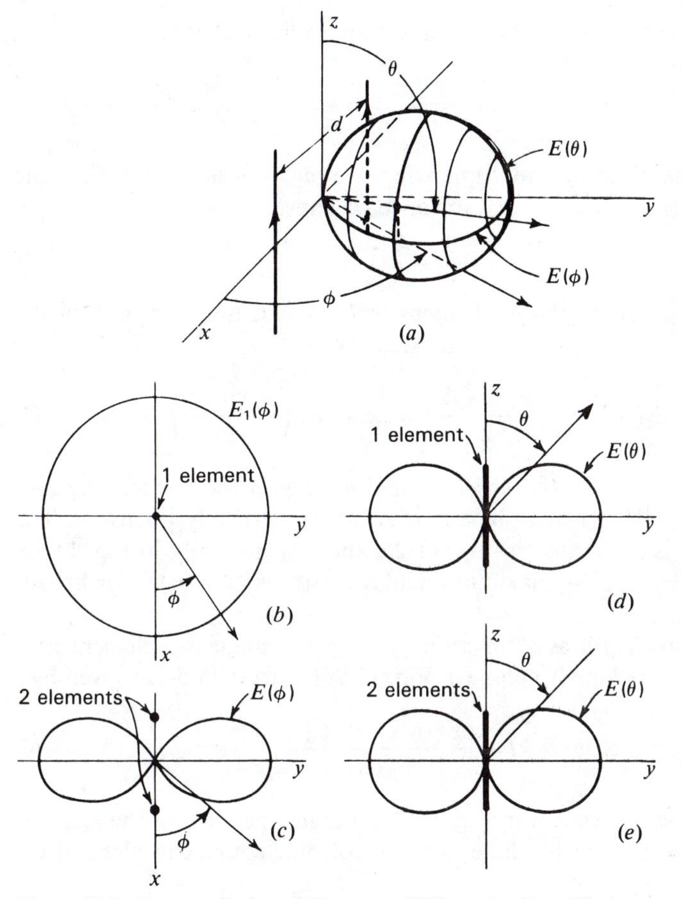

Figure 11-2 Patterns for broadside array of 2 linear in-phase $\lambda/2$ elements with spacing $d = \lambda/2$.

and at a large distance D ($D \gg d$) in a horizontal plane ($\theta = 90°$ or xy plane in Fig. 11-2a) is

$$E_1(\phi) = kI_1 \tag{1}$$

where k is a constant ($\Omega\ \mathrm{m}^{-1}$) involving the distance D and I_1 is the terminal current. Equation (1) is the absolute field pattern in the horizontal plane. It is independent of ϕ so that the relative pattern is a circle as indicated in Fig. 11-2b.

Next let the elements be replaced by isotropic point sources of equal amplitude. The pattern $E_{\mathrm{iso}}(\phi)$ as a function of ϕ in the horizontal plane for two such isotropic in-phase point sources is given by (4-2-6) as

$$E_{\mathrm{iso}}(\phi) = 2E_0 \cos\left(\frac{d_r \cos \phi}{2}\right) \tag{2}$$

where d_r is the distance between sources expressed in radians; that is,

$$d_r = \frac{2\pi d}{\lambda} \tag{3}$$

Applying the principle of pattern multiplication, we may consider that E_0 is the field intensity from a single element at a distance D. Thus,

$$E_0 = E_1(\phi) = kI_1 \tag{4}$$

Introducing (4) into (2) yields the field intensity $E(\phi)$ as a function of ϕ in the horizontal plane at a large distance D from the array, or

$$E(\phi) = E_1(\phi) 2 \cos\left(\frac{d_r \cos\phi}{2}\right) = 2kI_1 \cos\left(\frac{d_r \cos\phi}{2}\right) \tag{5}$$

This expression may be called the *absolute field pattern in the horizontal plane.* The electric field at points in this plane is everywhere vertically polarized. The shape of this pattern is illustrated in Fig. 11-2*c*, and also partially in Fig. 11-2*a*, for the case where $d = \lambda/2$. The maximum field intensity is at $\phi = 90°$ or broadside to the array.

The field intensity $E_1(\theta)$ as a function of θ from a single $\lambda/2$ element at a distance D in the vertical plane (yz plane in Fig. 11-2*a*) is, from (5-5-12), given by

$$E_1(\theta) = kI_1 \frac{\cos[(\pi/2)\cos\theta]}{\sin\theta} \tag{6}$$

The shape of this pattern is shown in Fig. 11-2*d*. It is independent of the angle ϕ. The pattern $E_{iso}(\theta)$ in the vertical plane for two isotropic sources in place of the two elements is

$$E_{iso}(\theta) = 2E_0 \tag{7a}$$

Applying the principle of pattern multiplication, we put

$$E_0 = E_1(\theta) \tag{7b}$$

so that the field intensity $E(\theta)$ in the vertical plane at a distance D from the array is

$$E(\theta) = 2kI_1 \frac{\cos[(\pi/2)\cos\theta]}{\sin\theta} \tag{8}$$

This may be called the *absolute field pattern in the vertical plane.* This pattern has the same shape as the pattern for a single element in the vertical plane and is independent of the spacing. The relative pattern is presented in Fig. 11-2*e* and also partially in Fig. 11-2*a*. The relative 3-dimensional field variation for the case where $d = \lambda/2$ is suggested in Fig. 11-2*a*. This pattern is actually bidirectional, only half being shown.

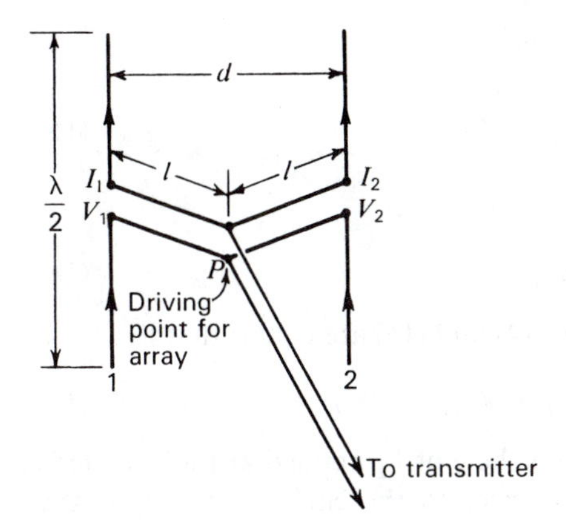

Figure 11-3 Broadside array of 2 linear $\lambda/2$ elements with arrangement for driving elements with equal in-phase currents.

11-2b Driving-Point Impedance. Suppose that the array is energized by the transmission-line arrangement shown in Fig. 11-3. Two transmission lines of equal length l join at P to a third line extending to a transmitter. Let us find the driving-point impedance presented to the third line at the point P.[1] This will be called the driving point for the array.

Let V_1 be the emf applied at the terminals of element 1. Then

$$V_1 = I_1 Z_{11} + I_2 Z_{12} \tag{9}$$

where I_1 is the current in element 1, I_2 the current in element 2, Z_{11} is the self-impedance of element 1 and Z_{12} is the mutual impedance between the two elements. Likewise, if V_2 is the emf applied at the terminals of element 2,

$$V_2 = I_2 Z_{22} + I_1 Z_{12} \tag{10}$$

where Z_{22} = the self-impedance of element 2

The currents are equal and in phase so

$$I_1 = I_2 \tag{11}$$

Therefore, (9) and (10) become

$$V_1 = I_1(Z_{11} + Z_{12}) \tag{12}$$

and

$$V_2 = I_2(Z_{22} + Z_{12}) \tag{13}$$

The terminal impedance Z_1 of element 1 is

$$Z_1 = \frac{V_1}{I_1} = Z_{11} + Z_{12} \tag{14}$$

---

[1] G. H. Brown, "A Critical Study of the Characteristics of Broadcast Antennas as Affected by Antenna Current Distribution," *Proc. IRE*, **24**, 48–81, January 1936.
G. H. Brown, "Directional Antennas," *Proc. IRE*, **25**, 78–145, January 1937.

and the terminal impedance Z_2 for element 2 is

$$Z_2 = \frac{V_2}{I_2} = Z_{22} + Z_{12} \tag{15}$$

Since the elements are identical

$$Z_{22} = Z_{11} \tag{16}$$

Therefore, the terminal impedances given by (14) and (15) are equal; that is,

$$Z_1 = Z_2 = Z_{11} + Z_{12} \tag{17}$$

Since $Z_1 = Z_2$ and $I_1 = I_2$ it is necessary that the emf V_1 applied at the terminals of element 1 be equal and in phase with respect to the emf V_2 applied at the terminals of element 2.

For the case where the spacing d is $\lambda/2$, the terminal impedance Z_1 of each element is

$$\begin{aligned} Z_1 = Z_{11} + Z_{12} &= R_{11} + R_{12} + j(X_{11} + X_{12}) \\ &= 73 - 13 + j(43 - 29) \\ &= 60 + j14 \quad \Omega \end{aligned} \tag{18}$$

Suppose that the reactance of 14 Ω is tuned out at the terminals by a series capacitance.[1] The terminal impedance then becomes a pure resistance of 60 Ω. If the length l of each transmission line between the antenna terminals and P is $\lambda/2$, the driving-point impedance of the array at P is a pure resistance of 30 Ω. This value is independent of the characteristic impedance of the $\lambda/2$ lines. However, a resistance of 30 Ω is too low to be matched readily by an open-wire transmission line. Therefore, a more practical arrangement would be to make l equal to $\lambda/4$. Suppose that we wish to have a driving-point resistance of 600 Ω. To do this, we let the characteristic impedance of each $\lambda/4$ line be $\sqrt{1200 \times 60} = 269$ Ω.[2] Each line transforms the 60 Ω to 1200 Ω and since two such lines are connected in parallel at P, the driving-point impedance for the array is a pure resistance of 600 Ω. This is the impedance presented to the line to the transmitter. For an impedance match this line should have a characteristic impedance of 600 Ω.

11-2c Gain in Field Intensity. As the last part of the analysis of the array, let us determine the gain in field intensity for the array. This could be done by

[1] It is often simpler to resonate the elements by shortening them slightly. This modifies the resistive component of the impedance and also alters the $E(\theta)$ field pattern, but to a first approximation these effects can usually be neglected.

[2] For the special case of a $\lambda/4$ line, the general transmission-line formula reduces to $Z_{in} = Z_0^2/Z_L$ where Z_{in} is the input impedance, Z_0 the characteristic impedance and Z_L the load impedance. Thus, $Z_0 = \sqrt{Z_{in} Z_L}$.

pattern integration as in Chap. 3, but with self- and mutual-impedance values available a shorter method is as follows.

Let the total power input (real power) to the array be P.[1] Assuming no heat losses, the power P_1 in element 1 is

$$P_1 = I_1^2(R_{11} + R_{12}) \tag{19}$$

and the power P_2 in element 2 is

$$P_2 = I_2^2(R_{22} + R_{12}) \tag{20}$$

where I_1 and I_2 are rms currents. However, $R_{22} = R_{11}$ and $I_2 = I_1$. Making these substitutions and adding (19) and (20) to obtain the total power P, we have

$$P = P_1 + P_2 = 2I_1^2(R_{11} + R_{12}) \tag{21}$$

and

$$I_1 = \sqrt{\frac{P}{2(R_{11} + R_{12})}} \tag{22}$$

Suppose that we express the gain with respect to a single $\lambda/2$ element as the reference antenna. Let the same power P be supplied to this antenna. Then assuming no heat losses, the current I_0 at its terminals is

$$I_0 = \sqrt{\frac{P}{R_{00}}} \tag{23}$$

where R_{00} is the self-resistance of the reference antenna $(= R_{11})$.

In general, the *gain in field intensity*[2] of an array over a reference antenna is given by the ratio of the field intensity from the array to the field intensity from the reference antenna when both are supplied with the same power P. The comparison is, of course, made in the same direction from both the array and the reference antenna. In the present case it will be convenient to obtain two gain expressions, one for the horizontal plane and the other for the vertical plane.

In the horizontal plane the field intensity $E_{HW}(\phi)$, as a function of ϕ, at a distance D from a single vertical center-fed $\lambda/2$ reference antenna is of the form of (1). Thus,

$$E_{HW}(\phi) = kI_0 \tag{24}$$

---

[1] It is important that the antenna power P be considered constant. Most transmitters are essentially constant power devices which can be coupled to a wide range of antenna impedance. Until the antenna power was considered constant by G. H. Brown (*Proc. IRE*, January 1937) the advantages of closely spaced elements were not apparent. Prior to this time the antenna current had usually been considered constant.

[2] The *power gain* discussed in Chap. 2 is equal to the square of the *gain in field intensity*. The power gain is the ratio of the radiation intensities (power per unit solid angle) for the array and reference antennas, the radiation intensity being proportional to the square of the field intensity.

where I_0 is the terminal current and "HW" indicates "Half-Wavelength ($\lambda/2$) antenna." Substituting the value of I_0 from (23), we obtain

$$E_{\mathrm{HW}}(\phi) = k\sqrt{\frac{P}{R_{00}}} \tag{25}$$

The field intensity $E(\phi)$ in the horizontal plane at a distance D from the array is given by (5). Introducing the value of the terminal current I_1 from (22) into (5) yields

$$E(\phi) = k\sqrt{\frac{2P}{R_{11} + R_{12}}} \cos\left(\frac{d_r \cos\phi}{2}\right) \tag{26}$$

The ratio of (26) to (25) gives the gain in field intensity of the array (as a function of ϕ in the horizontal plane) with respect to a vertical $\lambda/2$ reference antenna with the same power input. This gain will be designated by the symbol $G_f(\phi)$[A/HW] where the expression in the brackets is by way of explanation that it is the gain in field of the *array* (A) with respect to a *half-wavelength reference antenna* (HW)[1] in the same direction from both array and reference antenna. Thus,

$$G_f(\phi)\left[\frac{\mathrm{A}}{\mathrm{HW}}\right] = \frac{E(\phi)}{E_{\mathrm{HW}}(\phi)} = \sqrt{\frac{2R_{00}}{R_{11} + R_{12}}}\left|\cos\left(\frac{d_r \cos\phi}{2}\right)\right| \tag{27}$$

The absolute value bars || are introduced so that the gain will be confined to positive values (or zero) regardless of the values of d_r and ϕ. A negative gain would merely indicate a phase difference between the fields of the array and the reference antenna.

If the gain is the ratio of the *maximum* field of the array to the *maximum* field of the reference antenna, it is designated by G_f (not a function of angle).

The self-resistances $R_{00} = R_{11} = 73\ \Omega$. For the case where the spacing is $\lambda/2$, $d_r = \pi$ and $R_{12} = -13\ \Omega$ so that (27) becomes

$$G_f(\phi)\left[\frac{\mathrm{A}}{\mathrm{HW}}\right] = 1.56 \cos\left(\frac{\pi}{2}\cos\phi\right) \tag{28}$$

In the broadside direction ($\phi = \pi/2$), the pattern factor becomes unity. The gain is then 1.56. This is the ratio of the maximum field of the array to the maximum field of the reference antenna (see Fig. 11-4). Hence, $G_f = 1.56$.

---

[1] Both the array and the $\lambda/2$ reference antenna are assumed to be in free space. Thus, to be more explicit, the expression $G_f(\phi)$[AFS/HWFS], meaning the gain in field intensity of the *Array in Free Space* (AFS) with respect to a *Half-Wavelength reference antenna in Free Space* (HWFS), might be used. However, to simplify the notation, the letters "FS" will be omitted when *both* antennas are in free space.

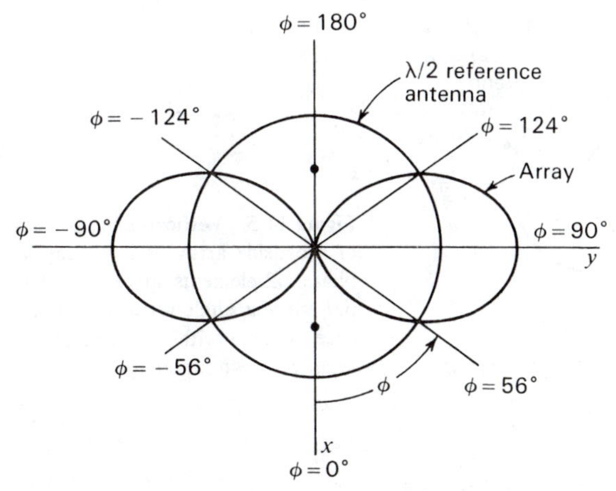

Figure 11-4 Horizontal plane pattern of broadside array of 2 vertical in-phase $\lambda/2$ elements spaced $\lambda/2$. The pattern of a single vertical $\lambda/2$ reference antenna with the same power input is shown for comparison.

It is also of interest to find the angle ϕ_0 for which the gain is unity. For this condition (28) becomes

$$\cos\left(\frac{\pi}{2}\cos\phi_0\right) = 0.64 \tag{29}$$

or $$\phi_0 = \pm 56^\circ \quad \text{or} \quad \pm 124^\circ$$

These angles are shown in Fig. 11-4. The array has a gain of greater than unity in both broadside directions over an angle of 68°.

The gain as a decibel ratio is given by the relation

$$\text{Gain} = 20 \log_{10} G_f \qquad \text{(dB)}$$

where G_f = gain in field intensity

Thus, a field-intensity gain of 1.56 is equal to 3.86 dB.

Turning our attention now to the gain in the vertical plane (yz plane of Fig. 11-2a), the field intensity $E_{\text{HW}}(\theta)$ as a function of θ in this vertical plane at a distance D from a single vertical $\lambda/2$ reference antenna with the same power input is of the form of (6). Thus,

$$E_{\text{HW}}(\theta) = kI_0 \frac{\cos[(\pi/2)\cos\theta]}{\sin\theta} \tag{30}$$

where I_0 = the terminal current

Substituting its value from (23), we get

$$E_{\text{HW}}(\theta) = k\sqrt{\frac{P}{R_{00}}}\frac{\cos[(\pi/2)\cos\theta]}{\sin\theta} \tag{31}$$

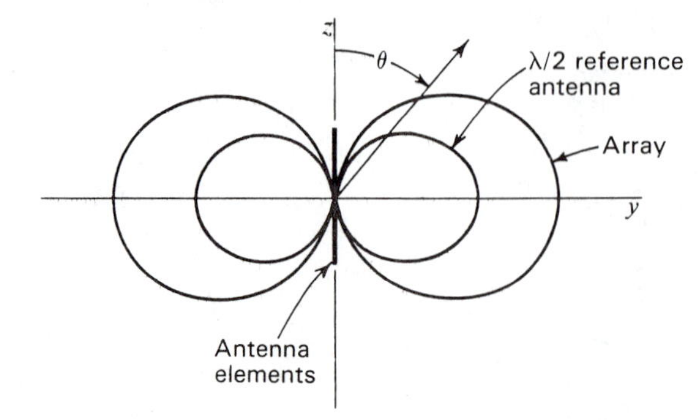

Figure 11-5 Vertical-plane pattern of broadside array of 2 vertical in-phase $\lambda/2$ elements spaced $\lambda/2$. The pattern of a single vertical $\lambda/2$ reference antenna with the same power input is shown for comparison.

The field intensity $E(\theta)$ as a function of θ in the vertical plane at a distance D from the array is given by (8). Introducing the value of the terminal current I_1 from (22) into (8), we have

$$E(\theta) = k\sqrt{\frac{2P}{R_{11} + R_{12}}}\,\frac{\cos\,[(\pi/2)\cos\theta]}{\sin\theta} \tag{32}$$

The ratio of (32) to (31) gives the gain in field intensity, $G_f(\theta)[\mathrm{A/HW}]$, of the array as a function of θ in the vertical plane over a vertical $\lambda/2$ reference antenna with the same power input. Thus,

$$G_f(\theta)\left[\frac{\mathrm{A}}{\mathrm{HW}}\right] = \frac{E(\theta)}{E_{\mathrm{HW}}(\theta)} = \sqrt{\frac{2R_{00}}{R_{11} + R_{12}}} \tag{33}$$

The gain is a constant, being independent of the angle θ. For the case where the spacing is $\lambda/2$, (33) becomes

$$G_f(\theta)\left[\frac{\mathrm{A}}{\mathrm{HW}}\right] = 1.56 \text{ (or 3.86 dB)} \tag{34}$$

The shape of the pattern for the array and for the $\lambda/2$ reference antenna is the same as shown in Fig. 11-5, but the ratio of the radius vectors in a given direction is a constant equal to 1.56.

If the reference antenna is an isotropic source instead of a $\lambda/2$ antenna, the gain in the vertical plane is a function of the angle θ. The maximum gain in field intensity of the array over an isotropic source with the same power input is $\sqrt{1.64}$ times greater than the voltage gain over a $\lambda/2$ reference antenna [$D(\lambda/2) = 1.64$, see Sec. 2-24]. Thus, when the spacing is $\lambda/2$, the *maximum gain in*

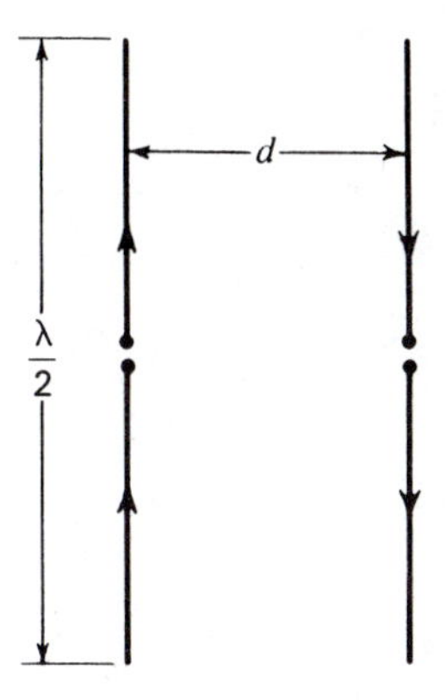

Figure 11-6 End-fire array of 2 linear $\lambda/2$ elements with currents of equal magnitude but opposite phase.

*field intensity of the array with respect to an **isotropic** source* is

$$G_f\left[\frac{\text{A}}{\text{iso}}\right] = 1.56 \times \sqrt{1.64} = 2.0 \text{ (or 6.0 dBi)}^1 \tag{35}$$

This value is in the broadside direction ($\phi = \theta = 90°$).

11-3 ARRAY OF 2 DRIVEN λ/2 ELEMENTS. END-FIRE CASE. Consider an array of 2 center-fed vertical $\lambda/2$ elements (dipoles) in free space arranged side by side with a spacing d and equal currents in opposite phase as in Fig. 11-6. The only difference between this case and the one discussed in Sec. 11-2 is that the currents in the elements are taken to be in the opposite phase instead of in the same phase. As in Sec. 11-2, the analysis will be divided into 3 subsections on the field patterns, driving-point impedance and gain in field intensity.

11-3a Field Patterns. The field intensity $E_1(\phi)$ as a function of ϕ at a distance D in a horizontal plane (xy or ϕ plane in Fig. 11-7*a*) from a single element is

$$E_1(\phi) = kI_1$$

where k = a constant involving the distance D
I_1 = the terminal current

Replacing the elements by isotropic point sources of equal amplitude, the pattern $E_{\text{iso}}(\phi)$ in the horizontal plane for two such isotropic out-of-phase sources is given by (4-2-10) as

$$E_{\text{iso}}(\phi) = 2E_0 \sin\left(\frac{d_r \cos\phi}{2}\right) \tag{1}$$

[1] Distinguish between "dB" for gain with respect to a *reference antenna* ($\lambda/2$ dipole in the present case) and "dBi" for gain with respect to an *isotropic source.*

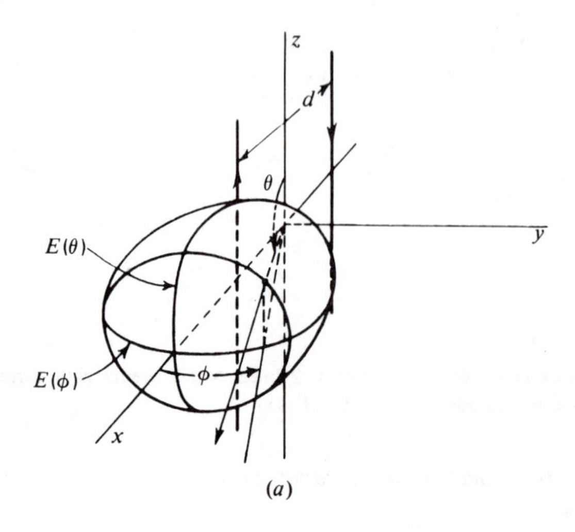

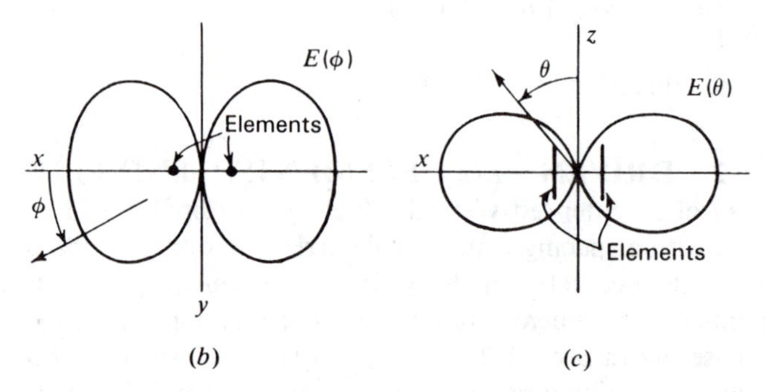

Figure 11-7 Patterns for end-fire array of 2 linear out-of-phase $\lambda/2$ elements with spacing $d = \lambda/2$.

Applying the principle of pattern multiplication, we may consider that E_0 is the field intensity from a single element at a large distance D. Thus

$$E_0 = E_1(\phi) = kI_1 \tag{2}$$

and the field intensity $E(\phi)$ as a function of ϕ in the horizontal plane at a large distance D from the array is

$$E(\phi) = 2kI_1 \sin\left(\frac{d_r \cos\phi}{2}\right) \tag{3}$$

This is the absolute field pattern in the horizontal plane. The electric field at points in this plane is everywhere vertically polarized. The relative pattern for the case where the spacing d is $\lambda/2$ is shown in Fig. 11-7b and also partially in Fig. 11-7a. The maximum field intensity is at $\phi = 0°$ and $\phi = 180°$. Hence, the array is commonly referred to as an "end-fire" type.

The field intensity $E_1(\theta)$ as a function of θ from a single $\lambda/2$ element at a distance D in the vertical plane (xz plane in Fig. 11-7a) is, from (5-5-12), given by

$$E_1(\theta) = kI_1 \frac{\cos [(\pi/2) \cos \theta]}{\sin \theta} \tag{4}$$

The pattern $E_{\text{iso}}(\theta)$ as a function of θ in the vertical plane for two isotropic sources in place of the two elements is, from (4-2-10),

$$E_{\text{iso}}(\theta) = 2E_0 \sin \left(\frac{d_r \sin \theta}{2} \right) \tag{5}$$

Note that θ is complementary to ϕ in (4-2-10), so $\cos \phi = \sin \theta$.

Putting $E_0 = E_1(\theta)$, the field intensity $E(\theta)$ as a function of θ in the vertical plane at a large distance D from the array is

$$E(\theta) = 2kI_1 \frac{\cos [(\pi/2) \cos \theta]}{\sin \theta} \sin \left(\frac{d_r \sin \theta}{2} \right) \tag{6}$$

This is the absolute field pattern in the vertical plane. The relative pattern is illustrated in Fig. 11-7c, and also partially in Fig. 11-7a, for the case where the spacing is $\lambda/2$. The relative 3-dimensional field variation for this case ($d = \lambda/2$) is suggested in Fig. 11-7a. This pattern is actually bidirectional, only half being shown.

11-3b Driving-Point Impedance. Let V_1 be the emf applied to the terminals of element 1. Then

$$V_1 = I_1 Z_{11} + I_2 Z_{12} \tag{7}$$

Likewise, if V_2 is the emf applied to the terminals of element 2,

$$V_2 = I_2 Z_{22} + I_1 Z_{12} \tag{8}$$

The currents are equal in magnitude but opposite in phase so

$$I_2 = -I_1 \tag{9}$$

Therefore, (7) and (8) become

$$V_1 = I_1(Z_{11} - Z_{12}) \tag{10}$$

and

$$V_2 = I_2(Z_{22} - Z_{12}) \tag{11}$$

The terminal impedance Z_1 of element 1 is

$$Z_1 = \frac{V_1}{I_1} = Z_{11} - Z_{12} \tag{12}$$

and the terminal impedance Z_2 of element 2 is

$$Z_2 = \frac{V_2}{I_2} = Z_{22} - Z_{12} \tag{13}$$

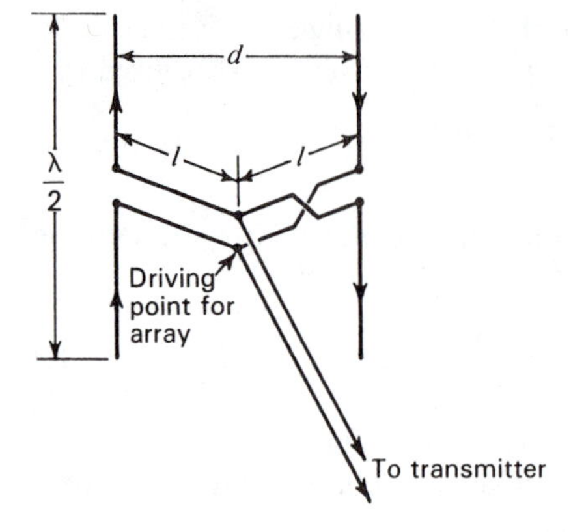

Figure 11-8 End-fire array of 2 linear $\lambda/2$ elements with arrangement for driving elements with currents of equal magnitude but opposite phase.

Therefore,

$$Z_1 = Z_2 = Z_{11} - Z_{12} \tag{14}$$

or

$$\frac{V_1}{I_1} = \frac{V_2}{I_2} \tag{15}$$

Since $I_2 = -I_1$ it follows from (15) that $V_2 = -V_1$. This means that the 2 elements must be energized with emfs which are equal in magnitude and opposite in phase. This may be done by means of a crossover in the transmission line from the driving point P to one of the elements as shown in Fig. 11-8. The length l of each line is the same.

For the case where the spacing between elements is $\lambda/2$, the terminal impedance of each element is

$$Z_1 = R_{11} - R_{12} + j(X_{11} - X_{12})$$

$$= 86 + j72 \quad \Omega \tag{16}$$

Consider that the reactance of 72 Ω is tuned out by a series capacitance at the terminals of each element. The terminal impedance is then a pure resistance of 86 Ω. To obtain a driving-point resistance of 600 Ω, let the length l of the line from P to each element be $\lambda/4$ and let the line impedance be $\sqrt{1200 \times 86} =$ 321 Ω. For an impedance match, the line from the driving point P to the transmitter should have a characteristic impedance of 600 Ω.

11-3c Gain in Field Intensity. Using the same method as in Sec. 11-2*c*, the current I_1 in each element for a power input P to the array is given by

$$I_1 = \sqrt{\frac{P}{2(R_{11} - R_{12})}} \tag{17}$$

It is assumed that there are no heat losses. The current I_0 in a single $\lambda/2$ reference antenna is given by (11-2-23). The gain in field intensity $G_f(\phi)$[A/HW] as a function of ϕ in the horizontal plane with respect to a $\lambda/2$ reference antenna is

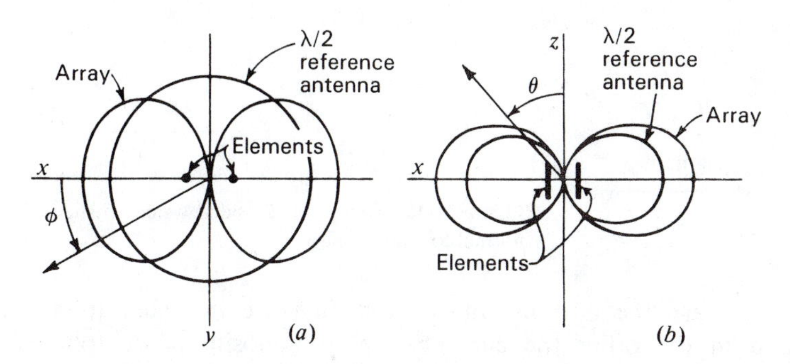

Figure 11-9 Horizontal plane pattern (*a*) and vertical plane pattern (*b*) of end-fire array of 2 vertical $\lambda/2$ elements with $\lambda/2$ spacing. The patterns of a vertical $\lambda/2$ reference antenna with the same power input are shown for comparison.

obtained by substituting (17) in (3) and taking the ratio of this result to (11-2-25). This yields

$$G_f(\phi)\left[\frac{\text{A}}{\text{HW}}\right] = \sqrt{\frac{2R_{00}}{R_{11} - R_{12}}}\left|\sin\left(\frac{d_r \cos\phi}{2}\right)\right| \tag{18}$$

For a spacing of $\lambda/2$, (18) reduces to

$$G_f(\phi)\left[\frac{\text{A}}{\text{HW}}\right] = 1.3\left|\sin\left[\left(\frac{\pi}{2}\right)\cos\phi\right]\right| \tag{19}$$

In the end-fire directions ($\phi = 0°$ and $180°$) the pattern factor becomes unity, and the gain is 1.3 or 2.3 dB. This is the gain G_f (see Fig. 11-9).

The gain in field intensity $G_f(\theta)[\text{A/HW}]$ as a function of θ in the vertical plane (xz plane of Fig. 11-7*a*) with respect to a $\lambda/2$ reference antenna is found by substituting (17) in (6) and taking the ratio of this result to (11-2-31), obtaining

$$G_f(\theta)\left[\frac{\text{A}}{\text{HW}}\right] = \sqrt{\frac{2R_{00}}{R_{11} - R_{12}}}\left|\sin\left(\frac{d_r \sin\theta}{2}\right)\right| \tag{20}$$

which is of the same form as the gain expression (18) for the horizontal plane (note that maximum radiation is in a direction $\theta = 90°$, $\phi = 0°$).

The gain in field intensity G_f of the array over an isotropic source with the same power input is $1.3 \times \sqrt{1.64} = 1.66$ (or 4.4 dBi).

11-4 ARRAY OF 2 DRIVEN λ/2 ELEMENTS. GENERAL CASE WITH EQUAL CURRENTS OF ANY PHASE RELATION.[1] In the preceding sections 2 special cases of an array of two $\lambda/2$ driven elements have

[1] For a more detailed discussion of this case and also of the most general case where the current amplitudes are unequal, see G. H. Brown, "Directional Antennas," *Proc. IRE*, **25**, 78–145, January 1937.

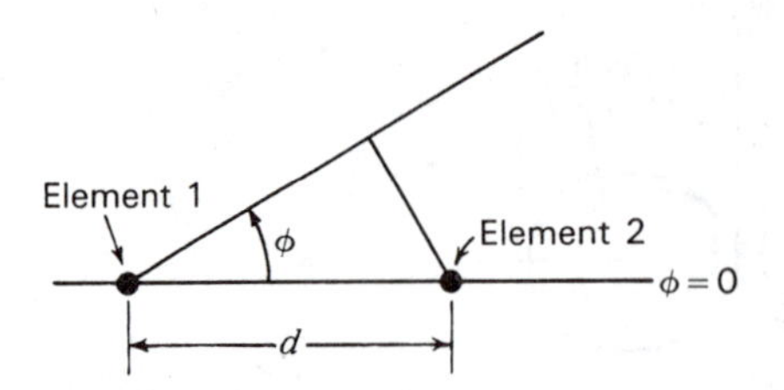

Figure 11-10 Array of 2 side-by-side elements normal to plane of page.

been treated. In one case the currents in the elements are in phase (phase difference = 0°) and in the other the currents are in opposite phase (phase difference = 180°). In this section the more general case is considered where the phase difference may have any value. As in the preceding cases, the two $\lambda/2$ elements are arranged side by side with a spacing d and are driven with currents of equal magnitude.

For the general-phase case the radiation-field pattern in the horizontal plane (xy plane of Fig. 11-7a) is, from (4-2-20), given by

$$E(\phi) = 2kI_1 \cos\frac{\psi}{2} \tag{1}$$

where ψ is the total phase difference between the fields from element 1 and element 2 at a large distance in the direction ϕ (see Fig. 11-10). Thus,

$$\psi = d_r \cos\phi + \delta \tag{2}$$

where δ = the phase difference of the currents in the elements

A positive sign in (2) indicates that the current in element 2 of Fig. 11-10 is advanced in phase by an angle δ with respect to the current in element 1; that is,

$$I_2 = I_1 \underline{/\delta}$$

or

$$I_1 = I_2 \underline{/-\delta} \tag{3}$$

The voltages applied at each element are

$$V_1 = I_1 Z_{11} + I_2 Z_{12} = I_1(Z_{11} + Z_{12} \underline{/\delta}) \tag{4}$$

and

$$V_2 = I_2 Z_{22} + I_1 Z_{12} = I_2(Z_{22} + Z_{12} \underline{/-\delta}) \tag{5}$$

The driving-point impedances of the elements are then

$$Z_1 = \frac{V_1}{I_1} = Z_{11} + Z_{12} \underline{/\delta} \tag{6}$$

and

$$Z_2 = \frac{V_2}{I_2} = Z_{22} + Z_{12} \underline{/-\delta} \tag{7}$$

The real parts of the driving-point resistances are

$$R_1 = R_{11} + |Z_{12}| \cos(\tau + \delta) \tag{8}$$

and

$$R_2 = R_{22} + |Z_{12}| \cos(\tau - \delta) \tag{9}$$

where τ = the phase angle of the mutual impedance Z_{12} (that is, $\tau = \arctan X_{12}/R_{12}$ where $Z_{12} = R_{12} + jX_{12}$)

Therefore, the power P_1 in element 1 is

$$P_1 = |I_1|^2 R_1 = |I_1|^2[R_{11} + |Z_{12}| \cos(\tau + \delta)] \tag{10}$$

and the power P_2 in element 2 is

$$P_2 = |I_2|^2[R_{22} + |Z_{12}| \cos(\tau - \delta)] \tag{11}$$

Since $R_{11} = R_{22}$, the total power P is

$$\begin{aligned} P = P_1 + P_2 &= |I_1|^2\{2R_{11} + |Z_{12}|[\cos(\tau + \delta) + \cos(\tau - \delta)]\} \\ &= 2|I_1|^2(R_{11} + |Z_{12}| \cos \tau \cos \delta) \\ &= 2|I_1|^2(R_{11} + R_{12} \cos \delta) \end{aligned} \tag{12}$$

It follows that the gain in field intensity as a function of ϕ in the horizontal plane[1] of the array over a single $\lambda/2$ element with the same power input is

$$G_f(\phi)\left[\frac{\text{A}}{\text{HW}}\right] = \sqrt{\frac{2R_{11}}{R_{11} + R_{12}\cos\delta}}\left|\cos\left(\frac{d_r \cos\phi + \delta}{2}\right)\right| \tag{13}$$

A polar plot of (13) with respect to the azimuth angle ϕ gives the radiation-field pattern of the array in the horizontal plane, the ratio of the magnitude of the radius vector to a unit radius indicating the gain over a reference $\lambda/2$ antenna. Brown[2] has calculated such patterns as a function of phase difference δ and spacing d_r. Examples of these are shown in Fig. 11-11.

The radiation-field pattern in the vertical plane containing the elements (in the plane of the page of Fig. 11-12) is

$$E(\theta) = 2kI_1 \cos\left(\frac{d_r \sin\theta + \delta}{2}\right)\frac{\cos[(\pi/2)\cos\theta]}{\sin\theta} \tag{14}$$

Thus, the pattern in the vertical plane has the shape of the patterns of Fig. 11-11 multiplied by the pattern of a single $\lambda/2$ antenna. The gain in the vertical plane over a vertical $\lambda/2$ reference antenna with the same power input is then

$$G_f(\theta)\left[\frac{\text{A}}{\text{HW}}\right] = \sqrt{\frac{2R_{11}}{R_{11} + R_{12}\cos\delta}}\left|\cos\left(\frac{d_r \sin\theta + \delta}{2}\right)\right| \tag{15}$$

It is often convenient to refer the gain to an isotropic source with the same power input. Since the power gain of a $\lambda/2$ antenna over an isotropic source is 1.64 [$D(\lambda/2) = 1.64$, see Sec. 2-24], the gain in field intensity as a function of θ in

---

[1] This is the plane of the page in Fig. 11-10.

[2] G. H. Brown, "Directional Antennas," *Proc. IRE*, **25**, 78–145, January 1937.

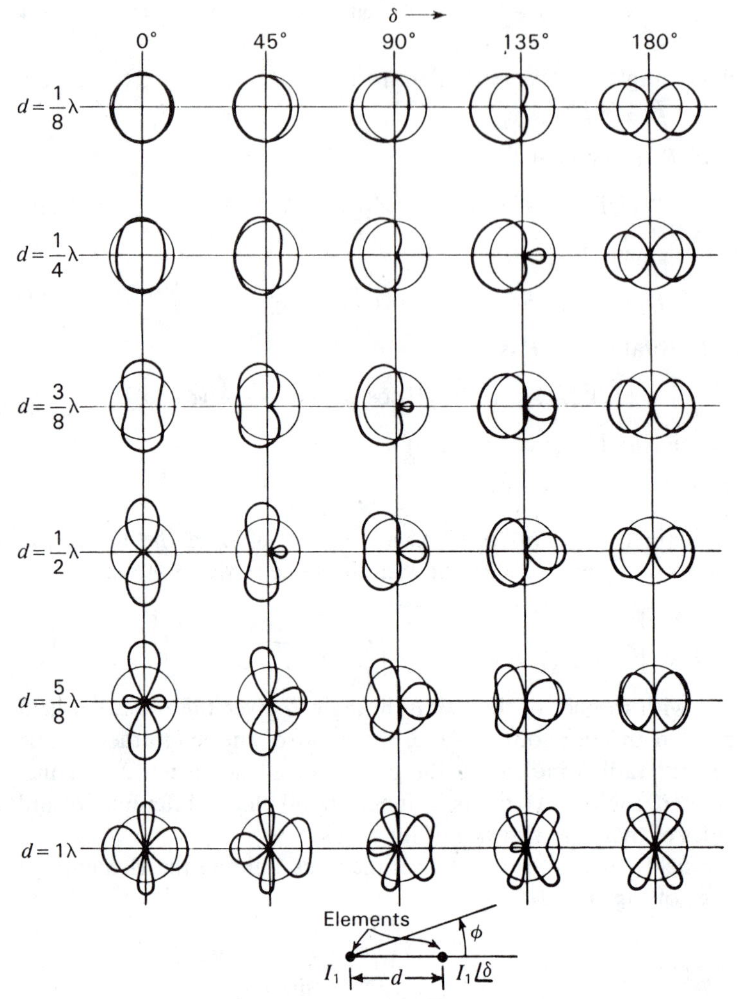

Figure 11-11 Horizontal-plane field patterns of 2 vertical elements as a function of the phase difference δ and spacing d. (*After G. H. Brown, "Directional Antennas,"* Proc. IRE, **25**, *78–145, January 1937.*) Both elements are the same length and have currents of equal magnitude. The circles indicate the field intensity of a single reference element of the same length with the same power input.

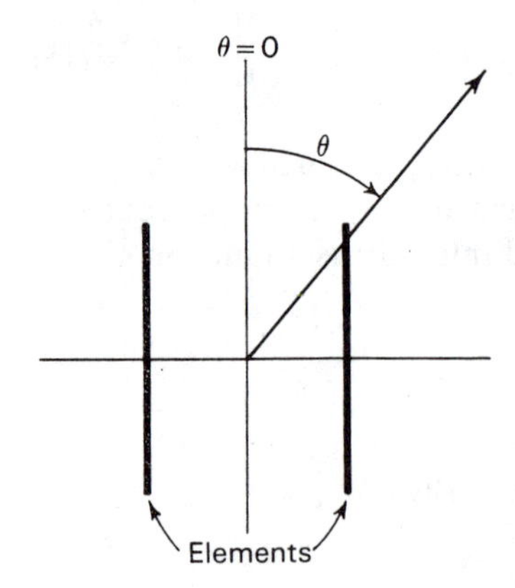

Figure 11-12 Relation of polar angle θ in the plane of the elements.

the vertical plane of a vertical $\lambda/2$ antenna in free space over an isotropic source is

$$G_f(\theta)\left[\frac{\text{HW}}{\text{iso}}\right] = \sqrt{1.64}\left|\frac{\cos\left[(\pi/2)\cos\theta\right]}{\sin\theta}\right| \tag{16}$$

The gain in field intensity in the vertical plane of the array over the isotropic source is then the product of (15) and (16) or

$$G_f(\theta)\left[\frac{\text{A}}{\text{iso}}\right] = G_f(\theta)\left[\frac{\text{A}}{\text{HW}}\right]G_f(\theta)\left[\frac{\text{HW}}{\text{iso}}\right]$$

$$= \sqrt{\frac{3.28R_{11}}{R_{11} + R_{12}\cos\delta}}\left|\cos\left(\frac{d_r\sin\theta + \delta}{2}\right)\frac{\cos\left[(\pi/2)\cos\theta\right]}{\sin\theta}\right| \tag{17}$$

11-5 CLOSELY SPACED ELEMENTS

11-5a Introduction. In 1932 I joined the Institute of Radio Engineers (IRE, now IEEE). The following year I received my Ph.D. degree in physics from the University of Michigan, publishing my dissertation research in the *Proceedings of the IRE*. I read every issue with interest. On opening the January 1937 *Proceedings* I delved into a monumental treatise on "Directional Antennas" by George H. Brown of RCA.[1] Buried deep in the article was, to me, an astonishing calculation which indicated that parallel linear dipoles with spacings of $\lambda/8$ or less had higher gains than the customary larger spacings.

Within one week of the time I received my *Proceedings* I had designed and built an array of 4 close-spaced $\lambda/2$ dipoles at my amateur station W8JK. Operating at a wavelength of 20 m, the array was phenomenally effective. I published the design and in subsequent articles extended it to an entire family of close-spaced arrays. The antennas outperformed all others. I called them "flat-top beams" but everyone else called them W8JK arrays.[2] They were soon in use by the thousands world-wide.

[1] G. H. Brown, "Directional Antennas," *Proc. IRE*, **25**, 78–145, January 1937.

[2] J. D. Kraus, "A Small but Effective Flat Top Beam Antenna," *Radio*, no. 213, 56–58, March 1937 and no. 216, 10–16, June 1937.

J. D. Kraus, "Rotary Flat Top Beam Antennas," *Radio*, no. 222, 11–16, December 1937.

J. Kraus, "Directional Antennas with Closely-Spaced Elements," *QST*, **22**, 21–23, January 1938.

J. D. Kraus, "Characteristics of Antennas with Closely-Spaced Elements," *Radio*, no. 236, 9–19, February 1939.

J. D. Kraus, "Antenna Arrays with Closely-Spaced Elements," *Proc. IRE*, **28**, 76–84, February 1940.

J. Kraus, "W8JK 5-Band Rotary Beam Antenna," *QST*, **54**, 11–14, July 1970.

J. Kraus, "The W8JK Antenna," *QST*, **66**, 11–14, June 1982.

In 1937 close spacing was a new and revolutionary concept. In George Brown's autobiography[1] he states:

> Ironically, the particular portion of my paper which John Kraus used so effectively was a small paper which I submitted to the *Proceedings* in 1932 only to have it rejected by a reviewer who denied its validity. When I prepared "Directional Antennas" I tucked this older material into the middle of this bulky manuscript on the assumption that the reviewer would not notice it.

George Brown's ruse worked and the world finally learned of his idea but only after it had languished in obscurity for five years!

In the next part of this section, the benefits and limitations of close spacing are analyzed and its application to the W8JK array is outlined.

11-5b Closely Spaced Elements and Radiating Efficiency. The W8JK Array. The end-fire array of two side-by-side, out-of-phase $\lambda/2$ dipole elements discussed in Sec. 11-3 can produce substantial gains when the spacing is decreased to small values. As indicated by the $R_L = 0$ curve in the gain-versus-spacing graph of Fig. 11-13*a*, the gain approaches 3.9 dB at small spacings. At $\lambda/2$ spacing the gain is 2.3 dB. This curve is calculated from (11-3-18) for $\phi = 0°$ or (11-3-20) for $\theta = 90°$. As the spacing d approaches zero, the coupling factor becomes infinite, but at the same time the pattern factor approaches zero. The product of the two or gain stays finite, leveling off at a value of about 3.9 dB (6.0 dBi) for small spacings, as illustrated by Fig. 11-13*b*. The fact that increased gain is associated with small spacings makes this arrangement attractive for many applications.

Thus far it has been assumed that there are no heat losses in the antenna system. In many antennas such losses are small and can be neglected. However, in the W8JK antenna such losses may have considerable effect on the gain (Fig. 11-13). Therefore, the question of losses and of radiating efficiency will be treated in this section in connection with a discussion of arrays of 2 closely spaced, out-of-phase elements. The term "closely spaced" will be taken to mean that the elements are spaced $\lambda/4$ or less.

A transmitting antenna is a device for radiating radio-frequency power. Let the *radiating efficiency* be defined as the ratio of the power radiated to the power input of the antenna. The real power delivered to the antenna that is not radiated is dissipated in the *loss resistance* and appears chiefly in the form of heat in the antenna conductor, in the insulators supporting the antenna, etc. An antenna with a total terminal resistance R_{1T} may be considered to have a terminal radiation resistance R_1 and an equivalent terminal loss resistance R_{1L} (see Sec. 2-13)

[1] G. H. Brown, "And Part of Which I Was," Angus Cupar, 117 Hunt Drive, Princeton, NJ 08540, 1982.

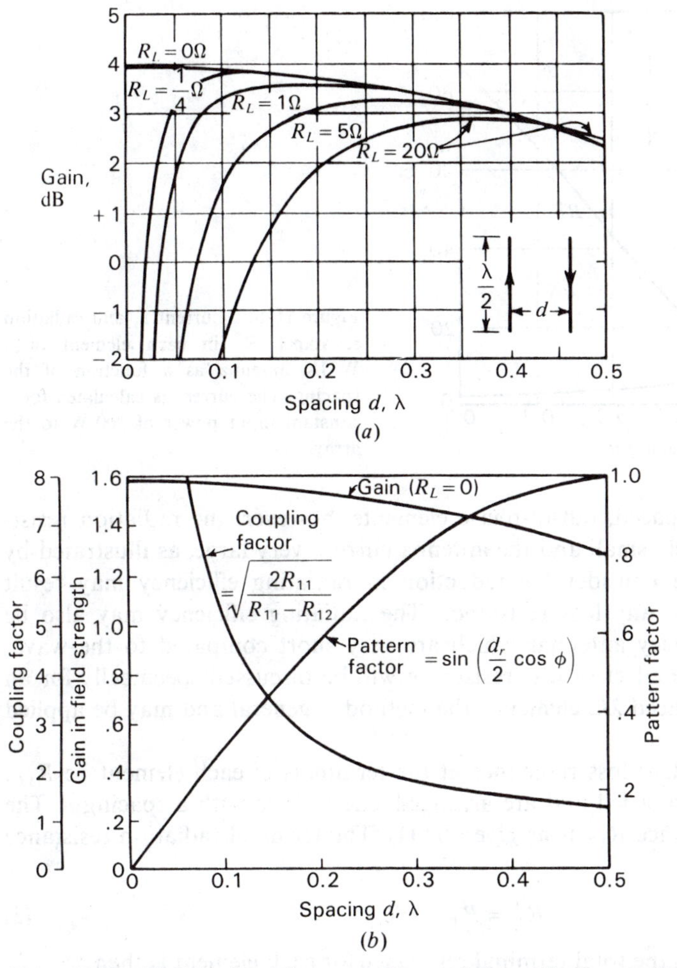

Figure 11-13 (*a*) Gain of end-fire array of 2 out-of-phase $\lambda/2$ elements (W8JK array) with respect to a $\lambda/2$ reference antenna as a function of the spacing for 5 values of the loss resistance R_L. (*b*) Gain curve for $R_L = 0$ with variation of its component factors, the coupling factor and the pattern factor, for $\phi = 0$.

such that

$$R_{1T} = R_1 + R_{1L} \tag{1}$$

It follows that

$$\text{Radiating efficiency (\%)} = \frac{R_1}{R_1 + R_{1L}} \times 100 \tag{2}$$

Since many types of high-frequency antennas have radiation resistances that are large compared to any loss resistance, the efficiencies are high. In an

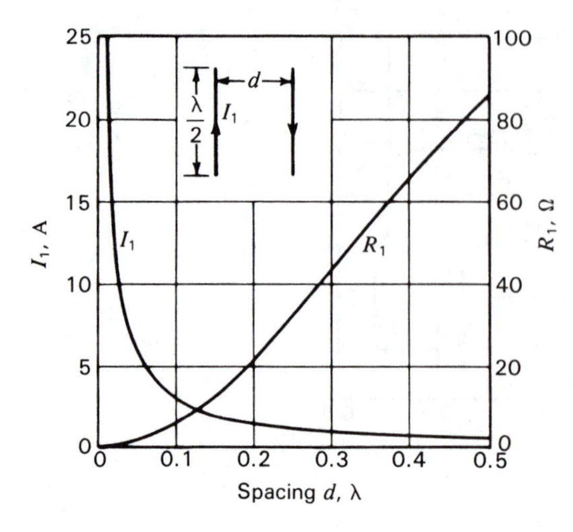

Figure 11-14 Current I_1 and radiation resistance R_1 in each element of a W8JK antenna as a function of the spacing. The current is calculated for a constant input power of 100 W to the array.

array with closely spaced, out-of-phase elements, however, the radiation resistance may be relatively small and the antenna current very large, as illustrated by Fig. 11-14. Hence, a considerable reduction in radiating efficiency may result from the presence of any loss resistance. The radiating efficiency may also be small for low-frequency antennas which are very short compared to the wavelength. Although the effect of loss resistance will be discussed specifically for an array of 2 closely spaced $\lambda/2$ elements, the method is general and may be applied to any type of antenna.

Let the equivalent loss resistance at the terminals of each element be R_{1L}. The elements are center fed and are arranged side by side with a spacing d. The total terminal resistance R_{1T} is as given by (1). The terminal radiation resistance R_1 is given by

$$R_1 = R_{11} - R_{12} \tag{3}$$

Substituting (3) in (1) the total terminal resistance for each element is then

$$R_{1T} = R_{11} + R_{1L} - R_{12} \tag{4}$$

If a power P is supplied to the 2-element array, the current I_1 in each element is

$$I_1 = \sqrt{\frac{P}{2(R_{11} + R_{1L} - R_{12})}} \tag{5}$$

The total terminal resistance R_{0T} of a single, center-fed $\lambda/2$ reference antenna is

$$R_{0T} = R_{00} + R_{0L} \tag{6}$$

where R_{00} is the self-resistance and R_{0L} the loss resistance of the reference antenna. The current I_0 at the terminals of the reference antenna is then

$$I_0 = \sqrt{\frac{P}{R_{00} + R_{0L}}} \tag{7}$$

With the array elements vertical, the gain in field intensity as a function of ϕ in the horizontal plane (xy plane in Fig. 11-7a) is obtained by substituting (5) in (11-3-3), (7) in (11-2-24) and taking the ratio which gives

$$G_f(\phi)\left[\frac{\text{A}}{\text{HW}}\right] = \sqrt{\frac{2(R_{00} + R_{0L})}{R_{11} + R_{1L} - R_{12}}}\left|\sin\left(\frac{d_r \cos\phi}{2}\right)\right| \tag{8}$$

This expression reduces to (11-3-18) if the loss resistances are zero ($R_{0L} = R_{1L} = 0$).

In a similar way the gain in field intensity as a function of θ in the vertical plane (xz plane in Fig. 11-7a) is

$$G_f(\theta)\left[\frac{\text{A}}{\text{HW}}\right] = \sqrt{\frac{2(R_{00} + R_{0L})}{R_{11} + R_{1L} - R_{12}}}\left|\sin\left(\frac{d_r \sin\theta}{2}\right)\right| \tag{9}$$

This reduces to (11-3-20) if the loss resistances are zero.

The effect of loss resistance on the gain of a closely spaced array of 2 out-of-phase $\lambda/2$ elements over a $\lambda/2$ reference antenna is illustrated by the curves in Fig. 11-13a. The gain presented is the maximum gain which occurs in the directions of maximum radiation from the array ($\phi = 0$ and $180°$; $\theta = 90°$). The top curve is for zero loss resistance ($R_{0L} = R_{1L} = 0$). The lower curves are for 4 different values of assumed loss resistance: $\frac{1}{4}$, 1, 5 and 20 Ω. The assumption is made that the loss resistance R_{1L} of each element of the array is the same as the loss resistance R_{0L} of the reference $\lambda/2$ antenna (that is, $R_{1L} = R_{0L}$). It is apparent from the curves that a loss resistance of only 1 Ω seriously limits the gain at spacings of less than $\lambda/10$, and larger loss resistances cause reductions in gain at considerably greater spacings. If the loss resistance is taken to be 1 Ω (a not unlikely value for a typical high-frequency antenna), the gain is almost constant (within 0.1 dB) for spacings between $\lambda/8$ and $\lambda/4$. Smaller spacings result in reduced gain because of decreased efficiency while larger spacings also give reduced gain, not because of decreased efficiency but because of the decrease in the coupling factor. A spacing of $\lambda/8$ has the advantage that the physical size of the antenna is less. However, resonance is sharper for this spacing than for wider spacings. Hence, a spacing of $\lambda/4$ is to be preferred if a wide bandwidth is desired. In some situations an intermediate or compromise spacing is indicated.

The Q of an antenna, like the Q of any resonant circuit, is proportional to the ratio of the energy stored to the energy lost (in heat or radiation) per cycle. For a constant power input to the closely spaced array *the Q is nearly proportional to the square of the current I in each element.* Referring to Fig. 11-14, it is apparent that the current for $\lambda/8$ spacing is about twice the value for $\lambda/4$ spacing. Hence the Q for $\lambda/8$ spacing is about 4 times the Q for $\lambda/4$ spacing. A large Q indicates a large amount of stored energy near the antenna in proportion to the energy radiated per cycle. This also means that the antenna acts like a sharply tuned circuit. Since the bandwidth (if it is narrow) is inversely proportional to the Q, a spacing of $\lambda/4$ provides about 4 times the bandwidth obtained with $\lambda/8$ spacing. Although the efficiency of an array with closely spaced, out-of-phase

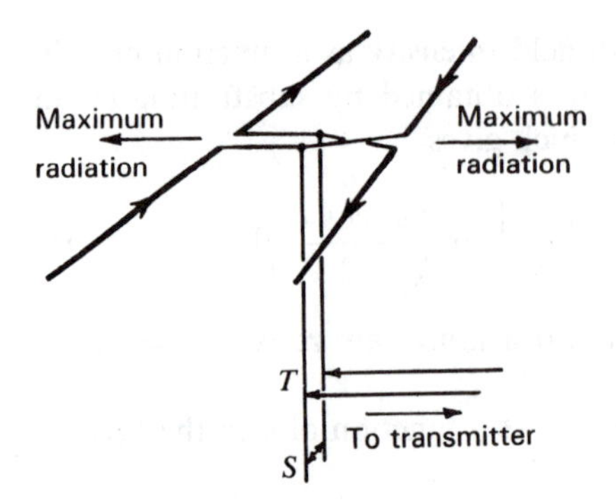

Figure 11-15 Horizontally polarized W8JK array with closely spaced elements carrying equal out-of-phase currents.

elements might be increased, e.g., by using a large diameter conductor for each element, any substantial increase in bandwidth requires an increase in the spacing between the elements. This increase also raises the radiating efficiency.

A single-section horizontally polarized W8JK closely spaced array consists of two side-by-side, out-of-phase $\lambda/2$ elements as indicated in Fig. 11-15. A single-section array is also shown in Fig. 11-16*a*. Five other examples of W8JK antennas are shown in Fig. 11-16 with arrows located at current maxima, indicating the instantaneous current directions. The type at Fig. 11-16*b* has an additional collinear $\lambda/2$ section, the 2 sections being energized from the center. A 4-section center-fed array is illustrated in Fig. 11-16*c*. The additional sections yield a higher gain by virtue of the sharper beam in the plane of the elements. The antennas of Fig. 11-16*d*, *e* and *f* are end-fed types corresponding to the center-fed arrays in the left-hand group. The spacing d is usually between $\lambda/8$ and $\lambda/4$.

Referring to the horizontally polarized W8JK array in Fig. 11-15, the location of the short circuit S on the vertical line is adjusted for resonance (total length from open end of horizontal dipole to S an odd number of $\lambda/4$

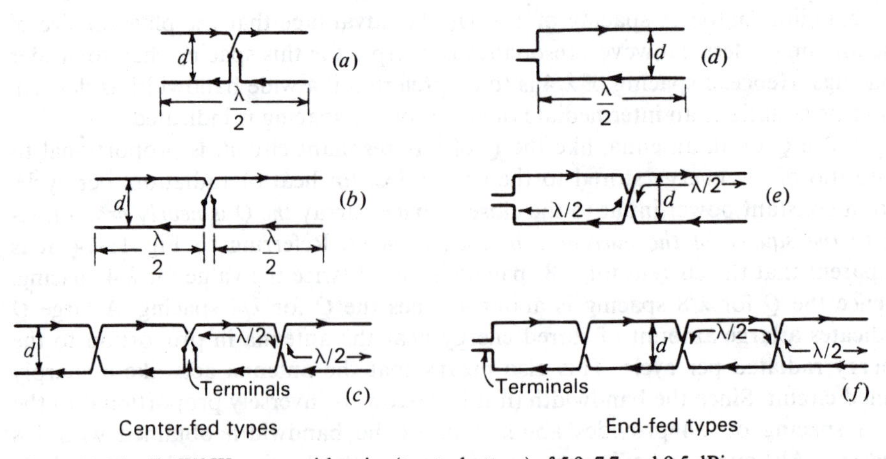

Figure 11-16 Six W8JK arrays with gains (top to bottom) of 5.9, 7.7 and 9.5 dBi.

approximately). The distance of the tap point T above S is then adjusted for minimum VSWR on the line to the transmitter (or receiver). The other types in Fig. 11-16 can also be matched in the same way. An adjustable trombone tuner matching a W8JK array to a coaxial line is shown in Fig. 16-29 (see also accompanying discussion). A W8JK array fed by constant-impedance lines without tuners is shown in Fig. 11-62.

11-6 ARRAY OF n DRIVEN ELEMENTS.

The field pattern of an array of many elements can often be obtained by an application of the principle of pattern multiplication. As an example, consider the volume array of Fig. 11-17 consisting of 16 $\lambda/2$ dipole elements with equal currents. In the y direction the spacing between elements is d, in the x direction the spacing is a and in the z direction the spacing is h. Let the y-direction and z-direction arrays be broadside types and the x-direction array an end-fire type such that the maximum radiation of the entire volume array is in the positive x direction. Let $d = h = \lambda/2$ and $a = \lambda/4$. Consider that the currents in all elements are equal in magnitude and that the currents in the front 8 elements are in phase but retarded by 90° with respect to the currents in the rear 8 elements. By the principle of pattern multiplication the pattern of the array is given by the pattern of a single element multiplied by the pattern of a volume array of point sources, where the point sources have the same space distribution as the elements. In general, the field pattern $E(\theta, \phi)$ of a volume array as a function of θ and ϕ is

$$E(\theta, \phi) = E_s(\theta, \phi)E_x(\theta, \phi)E_y(\theta, \phi)E_z(\theta, \phi) \tag{1}$$

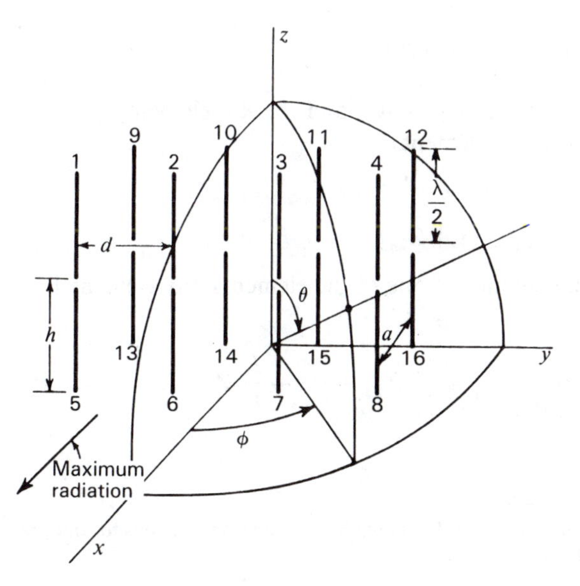

Figure 11-17 Array of 16 $\lambda/2$ dipole elements.

where $E_s(\theta, \phi)$ = pattern of single element
$E_x(\theta, \phi)$ = pattern of linear array of point sources in x direction
$E_y(\theta, \phi)$ = pattern of linear array of point sources in y direction
$E_z(\theta, \phi)$ = pattern of linear array of point sources in z direction

The product of the last 3 terms in (1) is the pattern of a volume array of point sources. If, for instance, we wish to obtain the pattern of the entire array $E(\phi)$ as a function of ϕ in the xy plane ($\theta = 90°$), we introduce the appropriate pattern expression in this plane for each component array in (1). For the example being considered the normalized pattern becomes

$$E(\phi) = \frac{\sin(2\pi \sin \phi)}{4 \sin[(\pi/2) \sin \phi]} \cos\left[\frac{\pi}{4}(1 - \cos \phi)\right] \tag{2}$$

Only the $E_y(\phi)$ broadside pattern and the $E_x(\phi)$ end-fire pattern contribute to the array pattern in the xy plane, since in this plane the $E_s(\phi)$ pattern of a single element and the $E_z(\phi)$ broadside pattern are uniform.

The impedance relations for an array of any number n of identical elements are derived by an extension of the analysis used in the special cases in the preceding sections.[1] Thus, for n driven elements we have

$$\left.\begin{aligned}
V_1 &= I_1 Z_{11} + I_2 Z_{12} + I_3 Z_{13} + \cdots + I_n Z_{1n} \\
V_2 &= I_1 Z_{21} + I_2 Z_{22} + I_3 Z_{23} + \cdots + I_n Z_{2n} \\
V_3 &= I_1 Z_{31} + I_2 Z_{32} + I_3 Z_{33} + \cdots + I_n Z_{3n} \\
&\ \vdots \\
V_n &= I_1 Z_{n1} + I_2 Z_{n2} + I_3 Z_{n3} + \cdots + I_n Z_{nn}
\end{aligned}\right\} \tag{3}$$

where V_n = terminal voltage of the nth element
I_n = terminal current of the nth element
Z_{1n} = mutual impedance between element 1 and the nth element
Z_{nn} = self-impedance of the nth element

In matrix form (3) can be expressed as

$$[V_n] = [I_n][Z_{nn}] \tag{4}$$

The driving-point or terminal impedance of one of the elements, say element 1, is then

$$Z_1 = \frac{V_1}{I_1} = Z_{11} + \frac{I_2}{I_1} Z_{12} + \frac{I_3}{I_1} Z_{13} + \cdots + \frac{I_n}{I_1} Z_{1n} \tag{5}$$

[1] See, for example, P. S. Carter, "Circuit Relations in Radiating Systems and Applications to Antenna Problems," *Proc. IRE*, **20**, 1007, June 1932.

If the currents in the elements and the self and mutual impedances are known, the driving-point impedance Z_1 can be evaluated.

The voltage gain of an array of n elements over a single element can be determined in the same manner as outlined for the special cases considered in the previous sections. For instance, the gain in field intensity as a function of ϕ in the xy plane ($\theta = 90°$) for the array of Fig. 11-17 with respect to a single vertical $\lambda/2$ element with the same power input is

$$G_f(\phi)\left[\frac{\text{A}}{\text{HW}}\right] = \sqrt{\frac{R_{11} + R_{1L}}{R_{11} + R_{1L} + R_{13} + R_{15} + R_{17} + \frac{3}{2}(R_{12} + R_{16}) + \frac{1}{2}(R_{14} + R_{18})}} \times \frac{\sin(2\pi \sin\phi)}{\sin[(\pi/2)\sin\phi]} \cos\left[\frac{\pi}{4}(1 - \cos\phi)\right] \tag{6}$$

where R_{11} = self-resistance of one element
R_{1L} = loss resistance of one element
R_{12} = mutual resistance between element 1 and element 2
R_{13} = mutual resistance between element 1 and element 3, etc.

The numbering of the elements is as indicated in Fig. 11-17. It is assumed that $d = h = \lambda/2$ and $a = \lambda/4$ and that the current magnitudes are equal, the currents in the front 8 elements being all in the same phase but retarded 90° with respect to the currents in the rear 8 elements.

11-7 HORIZONTAL ANTENNAS ABOVE A PLANE GROUND.

In the previous discussions it has been assumed that the antenna is in free space, i.e., infinitely remote from the ground. Although the fields near elevated microwave antennas may closely approximate this idealized situation, the fields of most antennas are affected by the presence of the ground. The change in the pattern from its free-space shape is of primary importance. The impedance relations may also be different than when the array is in free space, especially if the array is very close to the ground. In this section the effect of the ground on horizontal antennas is discussed. In Sec. 11-8 the effect of the ground is analyzed for vertical antennas. A number of special cases are treated in each section, these being limited to single elements or to simple arrays of several elements. Perfectly conducting ground is assumed. The effect of imperfect (ordinary) ground is discussed in Sec. 16-4.

11-7a Horizontal λ/2 Antenna above Ground.

Consider the horizontal $\lambda/2$ antenna shown in Fig. 11-18 at a height h above a plane ground of infinite extent. Owing to the presence of the ground, the field at a distant point P is the resultant of a direct wave and a wave reflected from the ground as in Fig. 11-19. Assuming that the ground is perfectly conducting, the tangential component of the electric

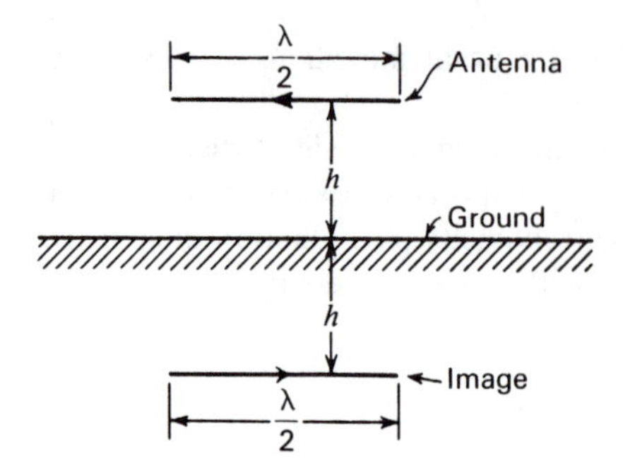

Figure 11-18 Horizontal $\lambda/2$ antenna at height h above ground with image at equal distance below ground.

field must vanish at its surface. To fulfill this boundary condition, the reflected wave must suffer a phase reversal of 180° at the point of reflection.

To obtain the field at a distant point P, it is convenient to transform the problem by the *method of images*. In this method the ground is replaced by an image of the antenna situated a distance h below the ground plane. By taking the current in the image equal in magnitude but reversed in phase by 180° with respect to the antenna current, the condition of zero tangential electric field is met at all points along a plane everywhere equidistant from the antenna and the image. This is the plane of the ground which the image replaces. In this way, the problem of a horizontal antenna above a perfectly conducting ground[1] of infinite extent can be transformed into the problem already treated in Sec. 11-3 of a so-called end-fire array. One point of difference is that in developing the gain expression it is assumed that if a power P is delivered to the antenna, an equal

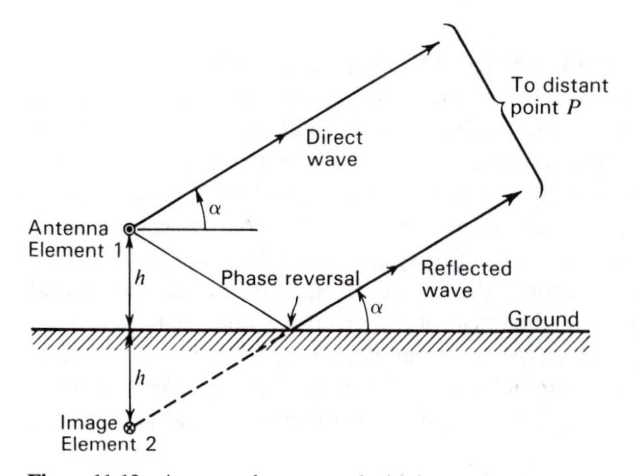

Figure 11-19 Antenna above ground with image showing direct and reflected waves.

[1] It is also possible to apply the method of images to the case of a ground of infinite extent but of finite conductivity σ and of dielectric constant ε by properly adjusting the relative magnitude and phase of the image current with respect to the antenna current.

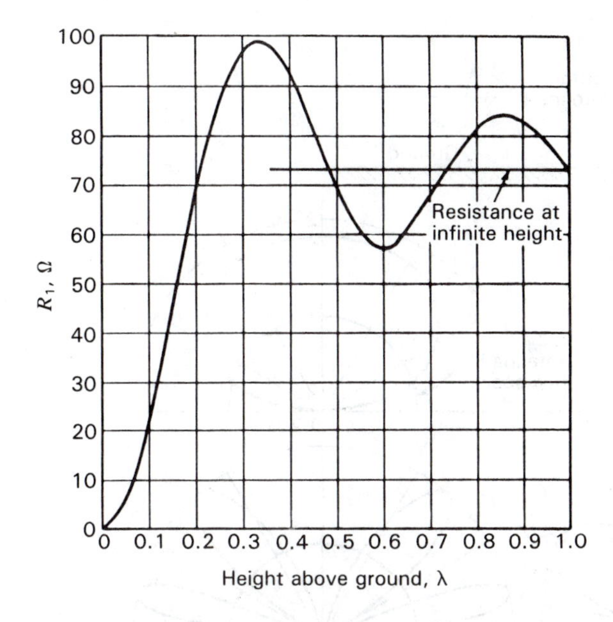

Figure 11-20 Driving- or feed-point resistance R_1 at the center of a horizontal $\lambda/2$ dipole antenna as a function of its height above a perfectly conducting ground.

power is also supplied to the image. Hence, a total power $2P$ is furnished to the "end-fire array" consisting of the antenna and its image.

Owing to the presence of the ground, the driving-point impedance of the antenna is, in general, different from its free-space value. Thus, the applied voltage at the antenna terminals is

$$V_1 = I_1 Z_{11} + I_2 Z_m \tag{1}$$

where I_1 = the antenna current
I_2 = the image current
Z_{11} = the self-impedance of the antenna
Z_m = the mutual impedance of the antenna and its image at a distance of $2h$

Since $I_2 = -I_1$, the driving- or feed-point impedance of the antenna is

$$Z_1 = \frac{V_1}{I_1} = Z_{11} - Z_m \tag{2}$$

The real part of (2) or driving-point radiation resistance is

$$R_1 = R_{11} - R_m \tag{3}$$

The variation of this resistance at the center of the $\lambda/2$ antenna is shown in Fig. 11-20 as a function of the antenna height h above the ground. As the height becomes very large, the effect of the image on the resistance decreases, the radiation resistance approaching its free-space value.

Since the antenna and image have currents of equal magnitude but opposite phase, there is zero radiation in the horizontal plane, i.e., in the direction for

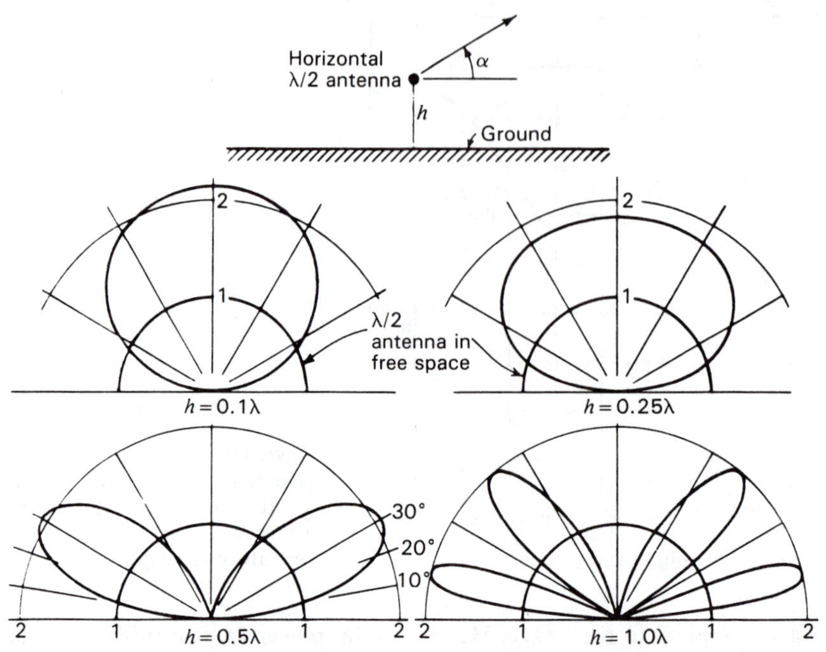

Figure 11-21 Vertical-plane field patterns of a horizontal $\lambda/2$ dipole at various heights h above a perfectly conducting ground as calculated from (4) for $R_{1L} = 0$. Patterns give gain in field intensity over a $\lambda/2$ dipole with the same power input. Note that the presence of the ground increases the field by approximately 6 dB or more in certain directions.

which the elevation angle α is zero (see Fig. 11-19). If the height h is $\lambda/4$ or less, the maximum radiation is always in the vertical direction ($\alpha = 90°$). For larger heights the maximum radiation is, in general, at some elevation angle between 0 and 90°.

It is convenient to compare the horizontal $\lambda/2$ antenna at a height h above ground with respect to a $\lambda/2$ antenna in free space with the same power input. At a large distance the gain in field intensity of the "Half-Wavelength antenna Above Ground" (HWAG) with respect to the "Half-Wavelength antenna in Free Space" (HWFS) is given by

$$G_f(\alpha)\left[\frac{\text{HWAG}}{\text{HWFS}}\right] = \sqrt{\frac{R_{11} + R_{1L}}{R_{11} + R_{1L} - R_m}}\,|2 \sin (h_r \sin \alpha)| \tag{4}$$

where $h_r = (2\pi/\lambda)h$

R_{11} = self-resistance of $\lambda/2$ antenna

R_{1L} = loss resistance of $\lambda/2$ antenna

R_m = mutual resistance of $\lambda/2$ antenna and its image at a distance of $2h$

Equation (4) gives the gain in the vertical plane normal to the antenna as a function of α (see Fig. 11-21).

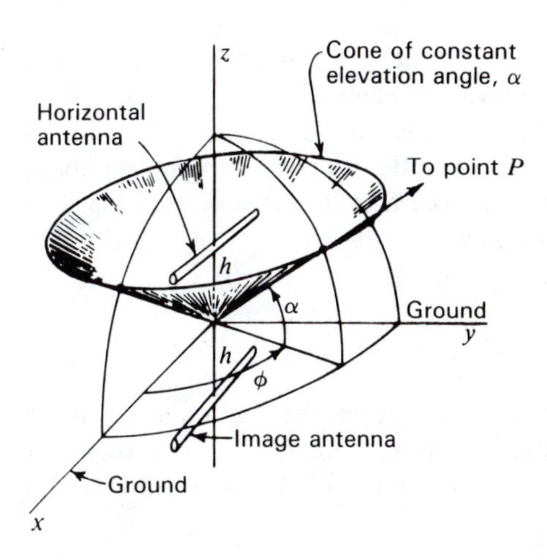

Figure 11-22 Horizontal antenna at height h above ground (xy plane) showing azimuth angle ϕ and elevation angle α for a distant point P.

The vertical-plane patterns of a horizontal $\lambda/2$ antenna are shown in Fig. 11-21 for heights $h = 0.1, 0.25, 0.5$ and 1.0λ. The circular pattern is for a $\lambda/2$ antenna in free space (i.e., with the ground removed) with the same power input. It is assumed that loss resistances are zero.

It is also of interest to calculate the field pattern as a function of the azimuth angle ϕ for a constant elevation angle α. The radius vector to the distant point P then sweeps out a cone as suggested in Fig. 11-22. To find this field pattern, let us first consider the field pattern of a horizontal antenna in free space as in Fig. 11-23. The xy plane is horizontal. The field intensity at a large distance

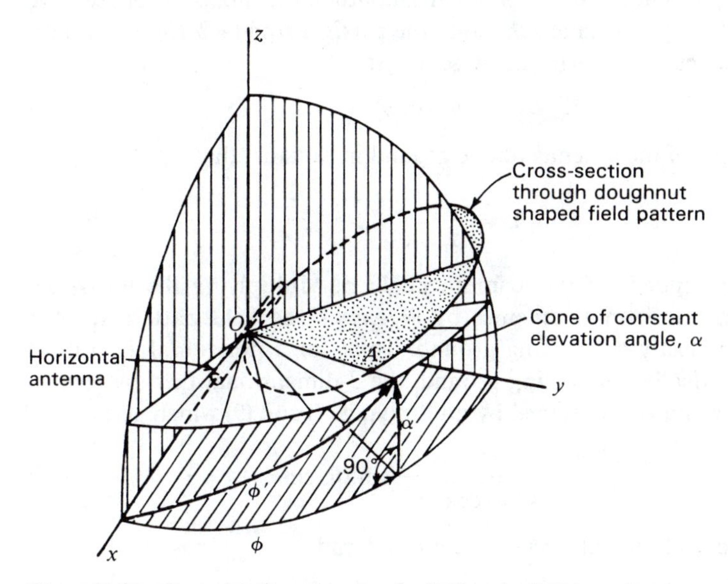

Figure 11-23 Geometrical construction for finding the field intensity at a constant elevation angle α in direction of line OA.

in the direction of α and ϕ is then given by the length OA between the origin and the point of intersection of a cone of elevation angle α and the surface of the 3-dimensional doughnut field pattern of the antenna as suggested in Fig. 11-23. This length is obtained from the field-pattern formula of the antenna in free space by expressing the polar angle ϕ' from the antenna axis in terms of α and ϕ. For the spherical right triangle in Fig. 11-23 we have

$$\cos\phi' = \cos\phi\cos\alpha \tag{5}$$

or

$$\sin\phi' = \sqrt{1-\cos^2\phi\cos^2\alpha} \tag{6}$$

Substituting these relations in the pattern formula, we get the field intensity in the direction α, ϕ. For example, by substituting (5) and (6) into (5-5-12), noting that ϕ' in (5) and (6) equals θ in (5-5-12), we obtain for the field of a $\lambda/2$ horizontal dipole antenna

$$E(\alpha,\phi) = \frac{\cos[(\pi/2)\cos\phi\cos\alpha]}{\sqrt{1-\cos^2\phi\cos^2\alpha}} \tag{7}$$

Then the relative field pattern of the horizontal $\lambda/2$ dipole antenna in free space as a function of ϕ at a fixed elevation angle α_0 is given by

$$E(\phi) = \frac{\cos[(\pi/2)\cos\phi\cos\alpha_0]}{\sqrt{1-\cos^2\phi\cos^2\alpha_0}} \tag{8}$$

To obtain the field pattern of the antenna when situated at a height h above a perfectly conducting ground, we multiply the above free-space relations by the pattern of 2 isotropic point sources of equal amplitude but opposite phase. The sources are separated by a distance $2h$ along the z axis. From (4-2-10) the pattern of the isotropic sources becomes in the present case

$$E_{\text{iso}} = \sin(h_r\sin\alpha) \tag{9}$$

where h_r is the height of the antenna above ground in radians; that is,

$$h_r = \frac{2\pi h}{\lambda}$$

The pattern is independent of the azimuth angle ϕ. Multiplying the free-space field pattern of any horizontal antenna by (9) yields the field pattern for the antenna above a perfectly conducting ground. Thus, for a horizontal $\lambda/2$ dipole antenna above a perfectly conducting ground, the 3-dimensional field pattern as a function of both α and ϕ is obtained by multiplying (7) and (9) which gives

$$E = \frac{\cos[(\pi/2)\cos\phi\cos\alpha]}{\sqrt{1-\cos^2\phi\cos^2\alpha}}\sin(h_r\sin\alpha) \tag{10}$$

where h_r = the height of the antenna above ground, rad

As an example, the field patterns as a function of the azimuth angle ϕ at elevation angles $\alpha = 10$, 20 and 30° are presented in Fig. 11-24 as calculated from

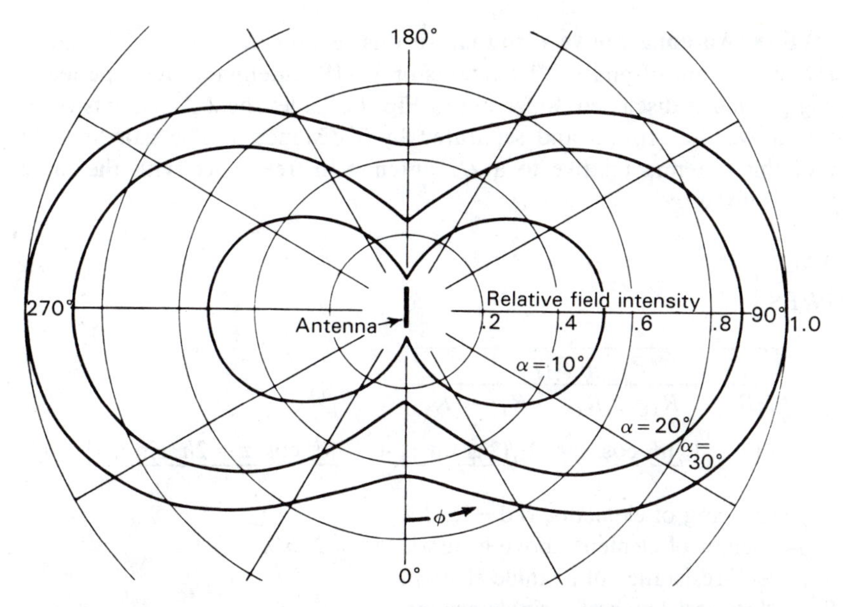

Figure 11-24 Azimuthal field patterns of horizontal $\lambda/2$ antenna $\lambda/2$ above ground at elevation angles $\alpha = 10$, 20 and 30°.

(10) for a horizontal $\lambda/2$ antenna at a height of $\lambda/2$ ($h_r = \pi$) above a perfectly conducting ground of infinite extent. The relative magnitudes of these patterns at $\phi = 90$ or 270° are seen to correspond to the field intensities at $\alpha = 10$, 20 and 30° in the vertical-plane pattern of Fig. 11-21 for $h = 0.5\lambda$. It should be noted that the field is horizontally polarized at $\phi = 90$ or 270° and is vertically polarized at $\phi = 0°$ and $\phi = 180°$. At intermediate azimuth angles the field is linearly polarized at a slant angle.

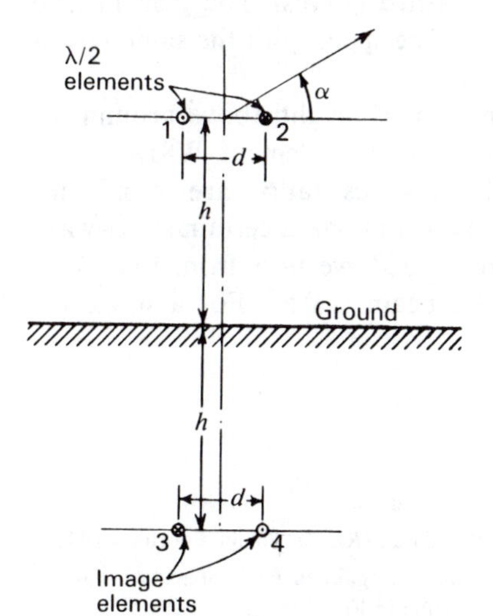

Figure 11-25 W8JK antenna above ground.

11-7b W8JK Antenna above Ground. In this section the case of 2 horizontal, closely spaced, out-of-phase $\lambda/2$ elements or W8JK antenna above a perfectly conducting ground is discussed. Referring to Fig. 11-25, let the $\lambda/2$ elements be at a height h above the ground and separated by a distance d. The gain in field intensity of this antenna relative to a $\lambda/2$ antenna in free space with the same power input is given by[1]

$$G_f(\alpha)\left[\frac{\text{AAG}}{\text{HWFS}}\right]^{\dagger}$$

$$= \sqrt{\frac{R_{11} + R_{1L}}{2(R_{11} + R_{1L} + R_{14} - R_{12} - R_{13})}}$$
$$\times \left| \left[1 - 1 \underline{/(d_r \cos \alpha)} - 1 \underline{/(2h_r \sin \alpha)} + 1 \underline{/(d_r \cos \alpha + 2h_r \sin \alpha)}\right] \right| \quad (11)$$

where d_r = spacing of elements, rad = $2\pi d/\lambda$
h_r = height of element above ground, rad = $2\pi h/\lambda$
R_{11} = self-resistance of a single element
R_{1L} = loss resistance of a single element
R_{12} = mutual resistance of elements 1 and 2
R_{13} = mutual resistance of elements 1 and 3, etc.

where the elements are numbered as in Fig. 11-25. The gain in (11) is expressed as a function of α in the vertical plane normal to the elements.

Polar plots calculated by (11) for the gain in field intensity of a W8JK antenna consisting of two $\lambda/2$ elements spaced $\lambda/8$ apart are presented by the solid curves in Fig. 11-26 for antenna heights of $\lambda/2$ and $3\lambda/4$ above ground. Patterns of a single $\lambda/2$ antenna at the same heights above ground and with the same power input are shown for comparison (dashed curves). The gain in field intensity is expressed relative to a $\lambda/2$ antenna in free space with the same power input.

In Fig. 11-27 the gain is given as a function of height above ground for several elevation angles. Curves are shown for both a 2-element W8JK and a single horizontal $\lambda/2$ antenna. It is assumed that loss resistances are zero. If, for example, the effective elevation angle at a particular time on a certain short-wave circuit (transmission via ionospheric reflections) is 30°, we note from Fig. 11-27 that the optimum height for a 2-element W8JK beam is 0.5λ. For a single $\lambda/2$ antenna the optimum height is about 0.57λ.

---

[1] J. D. Kraus, "Antenna Arrays with Closely Spaced Elements," *Proc. IRE*, **28**, 76–84, February 1940.

† The symbols in the brackets are by way of explanation that the gain in field intensity is for the "Array Above Ground with respect to a Half-Wavelength antenna in Free Space."

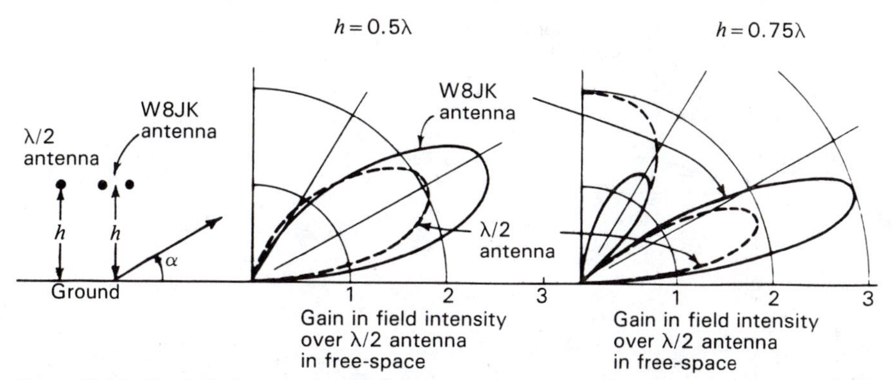

Figure 11-26 Vertical-plane patterns (solid curves) of 2-element W8JK antenna with $\lambda/8$ spacing at heights of $\lambda/2$ and $3\lambda/4$ above ground. The patterns are plotted relative to a $\lambda/2$ antenna in free space with the same power input. The vertical plane patterns of a single $\lambda/2$ antenna at the same heights above ground and with the same power input are shown for comparison by the dashed curves. The left-hand quadrants of the vertical planes are mirror images.

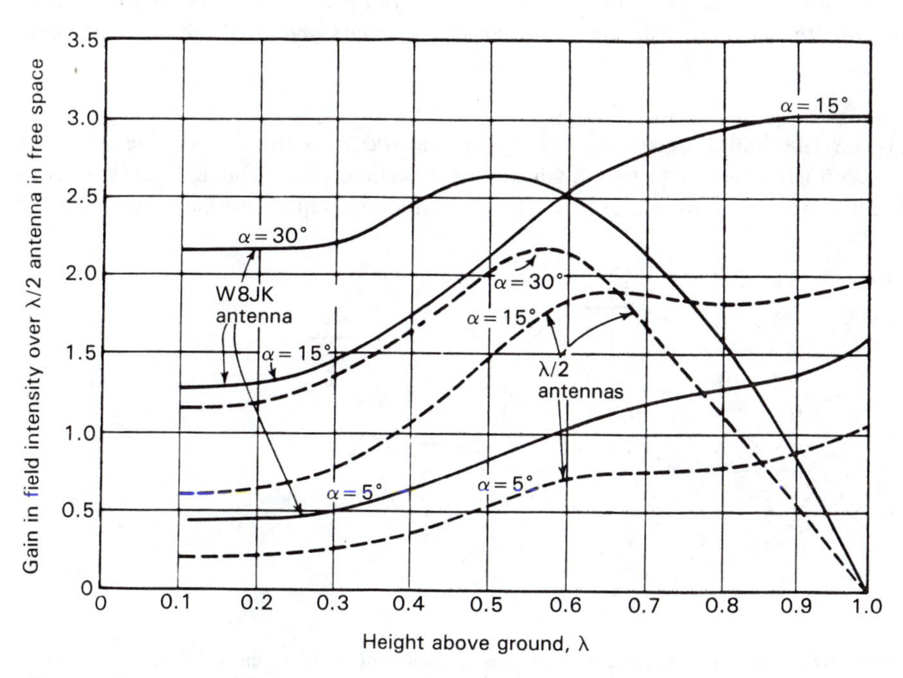

Figure 11-27 Gain in field intensity of 2-element W8JK antenna with $\lambda/8$ spacing (solid curves) and of a single $\lambda/2$ antenna (dashed curves) as a function of the height above a perfectly conducting ground. Gains are relative to a single $\lambda/2$ antenna in free space with the same power input. Curves are given for elevation angles $\alpha = 5$, 15 and 30°. We note that the gain of a 1-section (2-element) W8JK antenna at $\alpha = 15°$ and $h = \lambda$ exceeds 3 (= 11.8 dBi). For a 4-section W8JK antenna at this angle and height the gain is about 16 dBi.

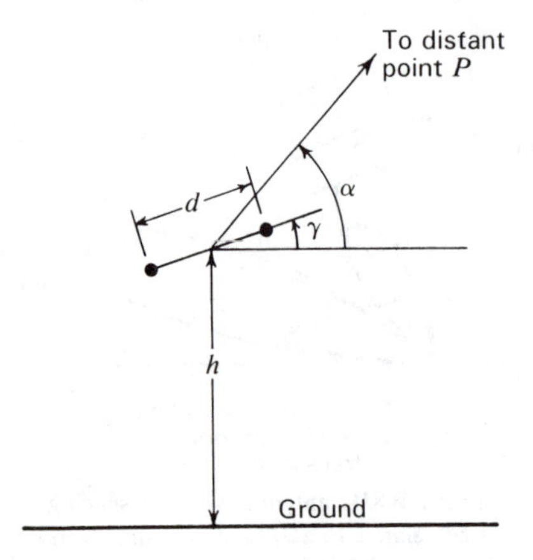

Figure 11-28 Tilted W8JK antenna.

It is interesting to consider the effect of tilting the plane of the W8JK elements by an angle γ as in Fig. 11-28. Results calculated by an extension of the above analysis are illustrated in Figs. 11-29 and 11-30 for 2-element arrays at average heights of $\lambda/2$ and $3\lambda/4$ above a perfectly conducting ground.[1] Patterns are shown for tilt angles $\gamma = 0$, 30, 45 and 90°. *In all cases the effect of the tilt is to increase the field intensity at large elevation angles and to decrease it at small angles.*

11-7c Stacked Horizontal λ/2 Antennas above Ground.

Consider the case of two horizontal $\lambda/2$ elements stacked in a vertical plane above a perfectly conducting ground of infinite extent. The elements have equal in-phase currents. The

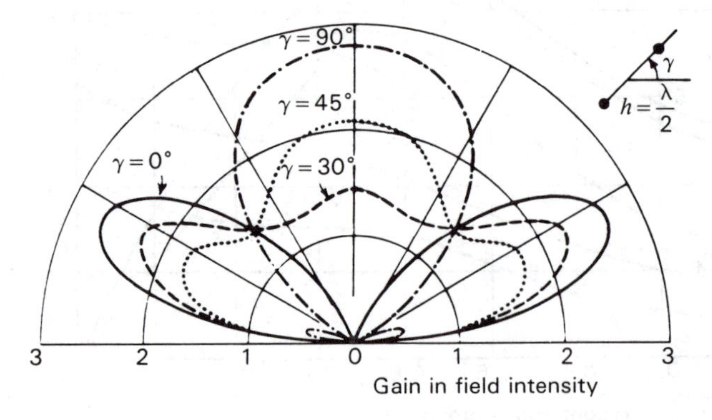

Figure 11-29 Vertical-plane patterns for horizontal 2-element W8JK antenna with $\lambda/8$ spacing at an average height of $\lambda/2$ above ground for tilt angles $\gamma = 0$, 30, 45 and 90°. Patterns give gain in field intensity over a single $\lambda/2$ antenna in free space with the same power input.

[1] J. D. Kraus, "Characteristics of Antennas with Closely Spaced Elements," *Radio*, 9–19, February 1939.

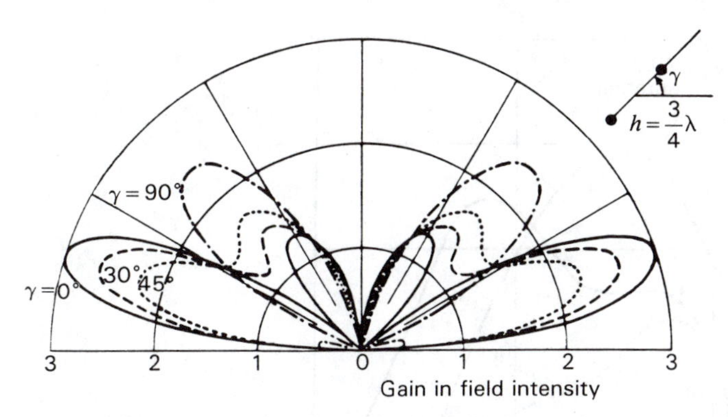

Figure 11-30 Same as for Fig. 11-29 but with array elements at average height of $3\lambda/4$ above ground.

arrangement of the elements and their images is shown in Fig. 11-31. The height of the upper element above ground is h. Let the spacing between elements be $\lambda/2$ so that the height of the lower element above ground is $h - \lambda/2$. The gain in field intensity of this array over a single $\lambda/2$ dipole antenna in free space with the same power input is

$$G_f(\alpha)\left[\frac{\text{AAG}}{\text{HWFS}}\right] = \sqrt{\frac{R_{00} + R_{0L}}{2(R_{11} + R_{1L} + R_{12} - R_{13}) - R_{23} - R_{14}}} \times 2|\{\sin(h_r \sin\alpha) + \sin[(h_r - \pi)\sin\alpha]\}| \tag{12}$$

where R_{12} is the mutual resistance between elements 1 and 2, R_{13} the mutual resistance between elements 1 and 3, etc. The elements are numbered as in Fig. 11-31. This expression gives the gain as a function of h and of the elevation

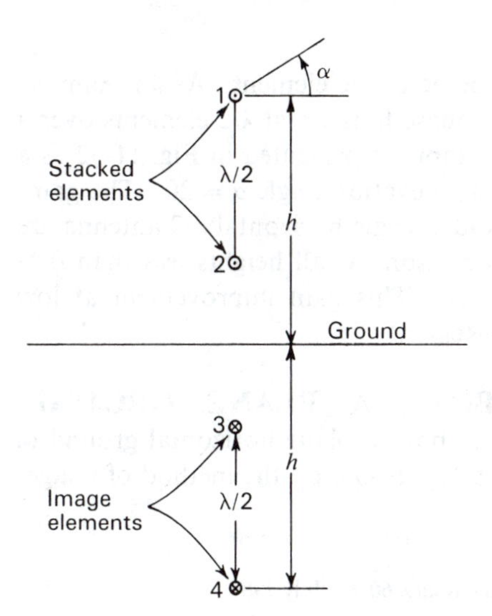

Figure 11-31 Array of stacked horizontal $\lambda/2$ elements.

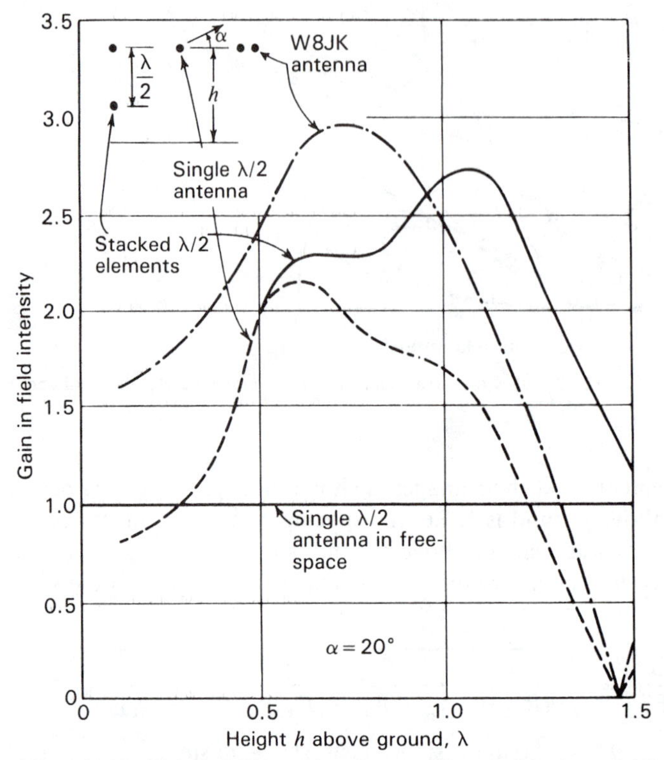

Figure 11-32 Gain in field intensity of array of 2 stacked horizontal $\lambda/2$ elements as a function of the height above ground of the upper element for an elevation angle of 20°. The elements are stacked $\lambda/2$ apart. The gain is relative to a single $\lambda/2$ dipole antenna in free space with the same power input. Gains of a 2-element W8JK antenna and single $\lambda/2$ antenna as a function of the height above ground are also shown for comparison at the same elevation angle. At all heights less than 0.9λ the W8JK antenna produces the highest gain.

angle α in the vertical plane normal to the plane of the elements. As an example, the gain in field intensity for two stacked in-phase horizontal $\lambda/2$ elements over a free-space $\lambda/2$ antenna with the same power input is presented in Fig. 11-32 as a function of the height h above ground, for an elevation angle $\alpha = 20°$. The gains at $\alpha = 20°$ for a 2-element W8JK antenna and a single horizontal $\lambda/2$ antenna are also shown as a function of height for comparison. At all heights less than 0.9λ the W8JK antenna produces the highest gain. This gain improvement at low heights and angles has been emphasized by Regier.[1]

11-8 VERTICAL ANTENNAS ABOVE A PLANE GROUND.

Consider a vertical stub antenna of length l above a plane horizontal ground of infinite extent and perfect conductivity as in Fig. 11-33. By the method of images

[1] F. Regier, "A New Look at the W8JK Antenna," *Ham Radio*, 60–63, July 1981.

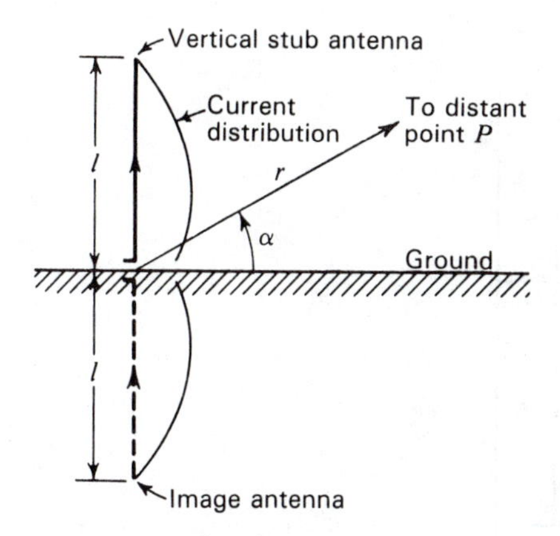

Figure 11-33 Vertical stub antenna above a ground plane.

the ground may be replaced by an image antenna of length l with sinusoidal current distribution and instantaneous current direction as indicated. The problem of the stub antenna above ground then reduces to the problem already treated in Chap. 5 of a linear center-fed antenna with symmetrical current distribution. The electric field intensity as a function of the elevation angle α and distance r may be derived from (5-5-11), obtaining

$$E(\alpha, r) = \frac{60}{r}\sqrt{\frac{P}{R_{11} + R_{1L}}}\,\frac{\cos(l_r \sin\alpha) - \cos l_r}{\cos\alpha} \qquad (\text{V m}^{-1}) \tag{1}$$

where $l_r = \beta l = (2\pi/\lambda)l$, dimensionless
R_{11} = self-resistance of a vertical stub antenna of length l referred to the point of current maximum, Ω
R_{1L} = effective loss resistance of antenna referred to same point, Ω
P = power input, W
r = distance, m

Values of the self-resistance referred to the current loop of a vertical stub antenna above a perfectly conducting ground have been given by Brown[1] and by Labus.[2] These values are presented as a function of antenna length in Fig. 11-34. Using these values of self-resistance, or radiation resistance, the field intensity of a vertical stub antenna of any length l and power input P can be calculated by (1) at any elevation angle α and distance r. Thus, the field intensity by (1) along the ground ($\alpha = 0$) for a $\lambda/4$ vertical antenna ($l_r = \pi/2$) with a power input $P = 1$ W at a distance of 1609 m is 6.5 mV m^{-1}. The value of R_{11} for a $\lambda/4$ stub antenna is 36.5 Ω, and R_{1L} is assumed to be zero.

[1] G. H. Brown, "A Critical Study of the Characteristics of Broadcast Antennas as Affected by Antenna Current Distribution," *Proc. IRE*, **24**, 48–81, January 1936.

[2] J. Labus, "Rechnerische Ermittlung der Impedanz von Antennen," *Hochfrequenztechnik und Electro-akustik*, **41**, 17–23, January 1933.

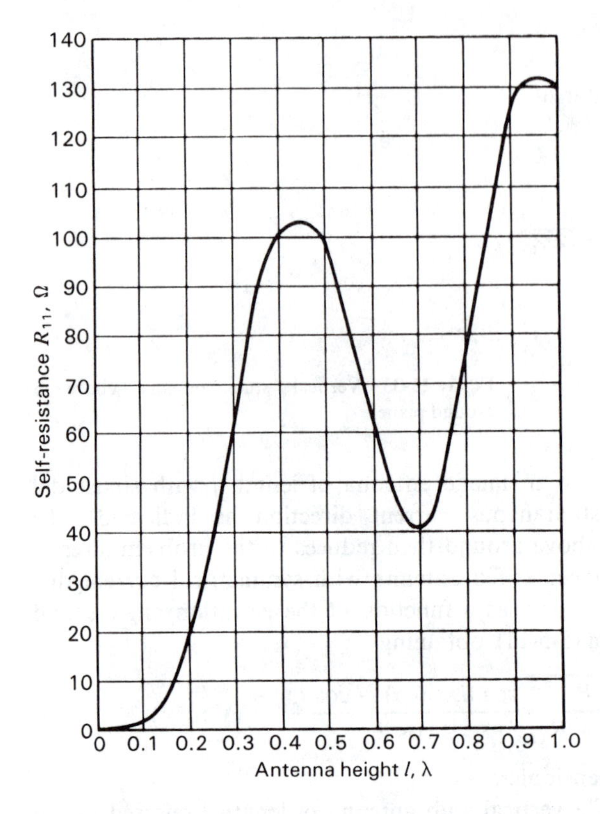

Figure 11-34 Radiation resistance at the current maximum of a thin vertical antenna as a function of the height l of the antenna. (*After G. H. Brown, "A Critical Study of the Characteristics of Broadcast Antennas as Affected by Antenna Current Distribution,"* Proc. IRE, **24**, *48–81, January 1936; and also J. Labus, "Rechnerische Ermittlung der Impedanz von Antennen,"* Hochfrequenztechnik und Electroakustik, **41**, *17–23, January 1933.*)

Vertical stub antennas, singly or in directional arrays, are very widely used for AM broadcasting. In this application the field intensity along the ground ($\alpha = 0$) is of particular interest. It is also customary to compare field intensities at some standard distance, say 1.61 km, and for some standard input such as 1 kW. For this case (1) reduces to

$$E = \frac{1.18(1 - \cos l_r)}{\sqrt{R_{11} + R_{1L}}} \quad (\text{V m}^{-1}) \tag{2}$$

where E is the field intensity along the ground at a distance of 1.61 km for a power input of 1 kW. The variation of E as given by (1) is presented in Fig. 11-35 as a function of antenna length.[1,2] The vertical-plane patterns calculated by (1) as a function of the elevation angle α for vertical antennas of various heights are presented in Fig. 11-36.[1,2] A length of about 0.64λ yields the greatest field intensity along the ground, but as pointed out by Brown[1] the large high-angle radi-

[1] G. H. Brown, "A Critical Study of the Characteristics of Broadcast Antennas as Affected by Antenna Current Distribution," *Proc. IRE*, **25**, 78–145, January 1937.

[2] C. E. Smith, *Directional Antennas*, Cleveland Institute of Radio Electronics, Cleveland, Ohio, 1946.

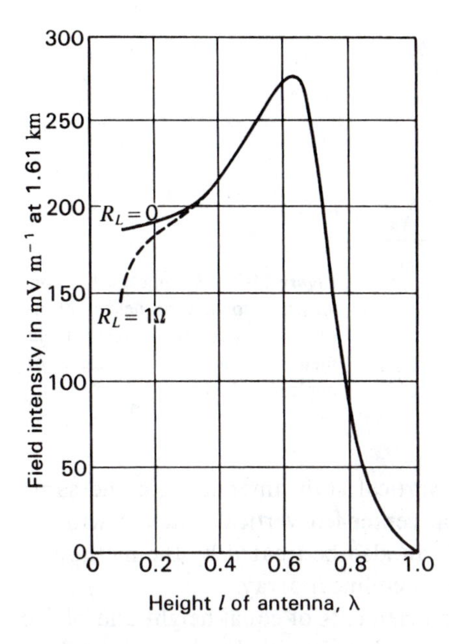

Figure 11-35 Field intensity at the ground (zero elevation angle) at a distance of 1.61 km from a vertical antenna with 1 kW input as a function of its height l. Perfectly conducting ground is assumed. The solid curve is for an assumed loss resistance $R_L = 0$ and the dashed curve for $R_L = 1\ \Omega$.

ation (at $\alpha = 60°$) for this length reduces the nonfading range at broadcast frequencies (500 to 1500 kHz) as compared, for example, with an antenna about $\lambda/2$ long. The nonfading range is largest for an antenna height of 0.528λ. It is assumed that the loss resistance $R_{1L} = 0$, that is, the entire input to the antenna is radiated. The small amount of high-angle radiation, which is an important factor in reducing fading, is apparent for the $l = 0.528\lambda$ antenna as compared to other lengths (see Fig. 11-36*b* and *c*).

The analysis of arrays of several vertical stub antennas can be reduced in a similar fashion to arrays of symmetrical center-fed antennas. Many of these have been treated in previous sections. In this case it is often convenient to compare

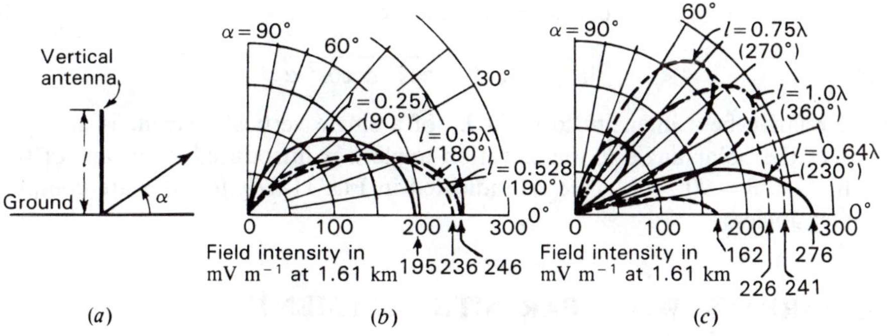

Figure 11-36 Vertical-plane field patterns of vertical antennas for several values of antenna height l. The field intensity is expressed in millivolts per meter at a distance of 1.61 km for 1 kW input. Perfectly conducting ground and zero loss resistance are assumed.

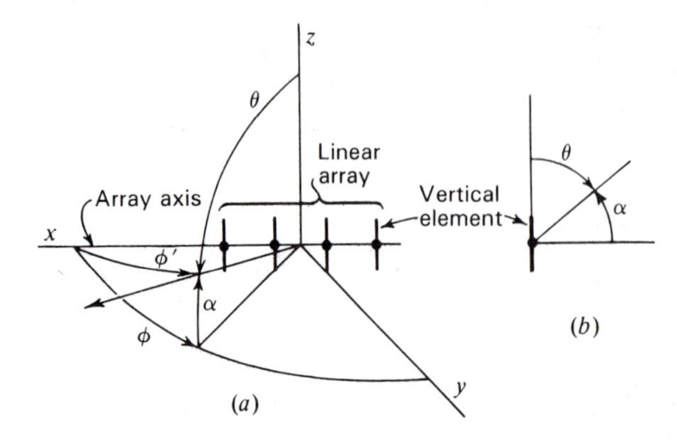

Figure 11-37 Geometrical construction for finding field intensity of a linear array of vertical elements at a constant elevation angle α.

the pattern and refer the gain to a single vertical stub antenna with the same power input. The situation of a symmetrical center-fed vertical antenna with its lower end some distance above the ground can also be treated by the method of images. In this case the antenna is reduced to a collinear array.

For the case of a linear array of vertical elements of equal height and of the same current distribution, the pattern $E(\phi)$ as a function of the azimuth angle ϕ at a constant elevation angle α is given by

$$E(\phi) = E_{\text{iso}}(\phi') \times E_1 \tag{3}$$

where $E_{\text{iso}}(\phi')$ = relative pattern of array of isotropic point sources used to replace elements

E_1 = relative field intensity of a single vertical element at the elevation angle α

The angle ϕ' in the pattern formula of the array of isotropic sources is the angle with respect to the array axis or x axis in Fig. 11-37*a*. Before inserting this formula into (3), it is necessary to express ϕ' in terms of the azimuth angle ϕ and elevation angle α (Fig. 11-37*a*). This is done by the substitutions

$$\cos \phi' = \cos \phi \cos \alpha \tag{4}$$

and

$$\sin \phi' = \sqrt{1 - \cos^2 \phi \cos^2 \alpha} \tag{5}$$

If the relative field intensity formula E_1 of a single vertical element is given in terms of the polar angle θ, the elevation angle α is introduced by means of the substitution $\theta = 90° - \alpha$, since, as indicated in Fig. 11-37*b*, θ and α are complementary angles.

11-9 ARRAYS WITH PARASITIC ELEMENTS

11-9a Introduction. In the above sections it has been assumed that all the array elements are driven, i.e., all are supplied with power by means of a trans-

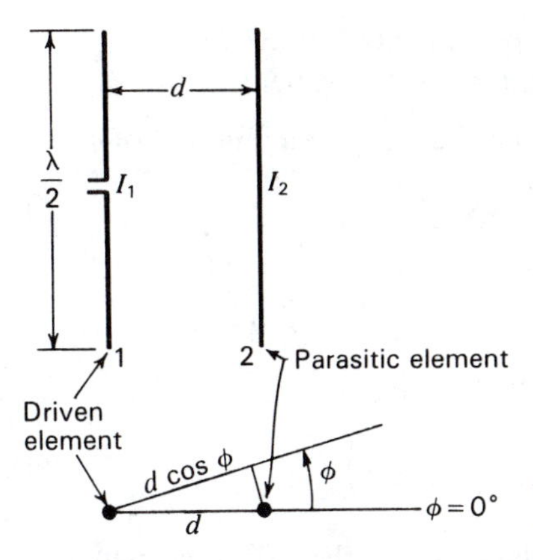

Figure 11-38 Array with one driven dipole element and one parasitic element.

mission line. Directional arrays can also be constructed with the aid of elements in which currents are induced by the fields of a driven element. Such elements have no transmission-line connection to the transmitter and are usually referred to as "parasitic elements."

Let us consider the case of an array in free space consisting of one driven $\lambda/2$ dipole element (element 1) and one parasitic element (element 2), as in Fig. 11-38. The procedure follows that used by Brown.[1] Suppose that both elements are vertical so that the azimuth angle ϕ is as indicated. The circuit relations for the elements are

$$V_1 = I_1 Z_{11} + I_2 Z_{12} \tag{1}$$

$$0 = I_2 Z_{22} + I_1 Z_{12} \tag{2}$$

From (2) the current in element 2 is

$$I_2 = -I_1 \frac{Z_{12}}{Z_{22}} = -I_1 \frac{|Z_{12}| \underline{/\tau_m}}{|Z_{22}| \underline{/\tau_2}} = -I_1 \left| \frac{Z_{12}}{Z_{22}} \right| \underline{/\tau_m - \tau_2} \tag{3}$$

or

$$I_2 = I_1 \left| \frac{Z_{12}}{Z_{22}} \right| \underline{/\xi} \tag{4}$$

where $\xi = \pi + \tau_m - \tau_2$, in which

$$\tau_m = \arctan \frac{X_{12}}{R_{12}}$$

$$\tau_2 = \arctan \frac{X_{22}}{R_{22}}$$

[1] G. H. Brown, "Directional Antennas," *Proc. IRE*, **25**, 78–145, January 1937.

where $R_{12} + jX_{12} = Z_{12}$ = mutual impedance of elements 1 and 2, Ω
$R_{22} + jX_{22} = Z_{22}$ = self-impedance of the parasitic element, Ω

The electric field intensity at a large distance from the array as a function of ϕ is

$$E(\phi) = k(I_1 + I_2 \underline{/d_r \cos \phi}) \tag{5}$$

where $d_r = \beta d = \dfrac{2\pi}{\lambda} d$

Substituting (4) for I_2 in (5),

$$E(\phi) = kI_1\left(1 + \left|\frac{Z_{12}}{Z_{22}}\right| \underline{/\xi + d_r \cos \phi}\right) \tag{6}$$

Solving (1) and (2) for the driving-point impedance Z_1 of the driven element, we get

$$Z_1 = Z_{11} - \frac{Z_{12}^2}{Z_{22}} = Z_{11} - \frac{|Z_{12}^2| \underline{/2\tau_m}}{|Z_{22}| \underline{/\tau_2}} \tag{7}$$

The real part of Z_1 is

$$R_1 = R_{11} - \left|\frac{Z_{12}^2}{Z_{22}}\right| \cos (2\tau_m - \tau_2) \tag{8}$$

Adding a term for the effective loss resistance, if any is present, we have

$$R_1 = R_{11} + R_{1L} - \left|\frac{Z_{12}^2}{Z_{22}}\right| \cos (2\tau_m - \tau_2) \tag{9}$$

For a power input P to the driven element,

$$I_1 = \sqrt{\frac{P}{R_1}} = \sqrt{\frac{P}{R_{11} + R_{1L} - |Z_{12}^2/Z_{22}| \cos (2\tau_m - \tau_2)}} \tag{10}$$

and substituting (10) for I_1 in (6) yields the electric field intensity at a large distance from the array as a function of ϕ. Thus,

$$E(\phi) = k\sqrt{\frac{P}{R_{11} + R_{1L} - |Z_{12}^2/Z_{22}| \cos (2\tau_m - \tau_2)}} \times \left(1 + \left|\frac{Z_{12}}{Z_{22}}\right| \underline{/\xi + d_r \cos \phi}\right) \tag{11}$$

For a power input P to a single vertical $\lambda/2$ element the electric field intensity at the same distance is

$$E_{HW}(\phi) = kI_0 = k\sqrt{\frac{P}{R_{00} + R_{0L}}} \qquad (\text{V m}^{-1}) \tag{12}$$

where R_{00} = self-resistance of single $\lambda/2$ element, Ω
R_{0L} = loss resistance of single $\lambda/2$ element, Ω

The gain in field intensity (as a function of ϕ) of the array with respect to a single $\lambda/2$ antenna with the same power input is the ratio of (11) to (12). Since $R_{00} = R_{11}$ and letting $R_{0L} = R_{1L}$, we have

$$G_f(\phi)\left[\frac{\text{A}}{\text{HW}}\right] = \sqrt{\frac{R_{11} + R_{1L}}{R_{11} + R_{1L} - |Z_{12}^2/Z_{22}| \cos(2\tau_m - \tau_2)}} \times \left(1 + \left|\frac{Z_{12}}{Z_{22}}\right| \underline{/\xi + d_r \cos\phi}\right) \tag{13}$$

If Z_{22} is made very large by detuning the parasitic element (i.e., by making X_{22} large), (13) reduces to unity, that is to say, the field of the array becomes the same as the single $\lambda/2$ dipole comparison antenna.

By means of a relation equivalent to (13), Brown[1] analyzed arrays with a single parasitic element for various values of parasitic element reactance (X_{22}) and was the first to point out that spacings of less than $\lambda/4$ were desirable.

The magnitude of the current in the parasitic element and its phase relation to the current in the driven element depends on its tuning. The parasitic element may have a fixed length of $\lambda/2$, the tuning being accomplished by inserting a lumped reactance in series with the antenna at its center point. Alternatively, the parasitic element may be continuous and the tuning accomplished by adjusting the length. This method is often simpler in practice but is more difficult of analysis. *When the $\lambda/2$ parasitic dipole element is inductive (longer than its resonant length) it acts as a reflector. When it is capacitive (shorter than its resonant length) it acts as a director.*[2]

Arrays may be constructed with both a reflector and a director. A 3-element array of this type is shown in Fig. 11-39, one parasitic element acting as a reflector and the other as a director. The analysis for the 3-element array is more complex than for the 2-element type treated above.

Experimentally measured field patterns of a horizontal 3-element array situated 1λ above a square horizontal ground plane about 13λ on a side are presented in Fig. 11-40. The element lengths and spacings are as indicated. The gain at $\alpha = 15°$ for this array at a height of 1λ is about 5 dB with respect to a single $\lambda/2$

[1] G. H. Brown, "Directional Antennas," *Proc. IRE*, **25**, 78–145, January 1937.

[2] From Fig. 9-9, we note that the reactance of a thin linear element varies rapidly as a function of frequency when its length is about $\lambda/2$, going from positive (inductive) reactance through zero reactance (resonance) to negative (capacitive) reactance values as the length is reduced.

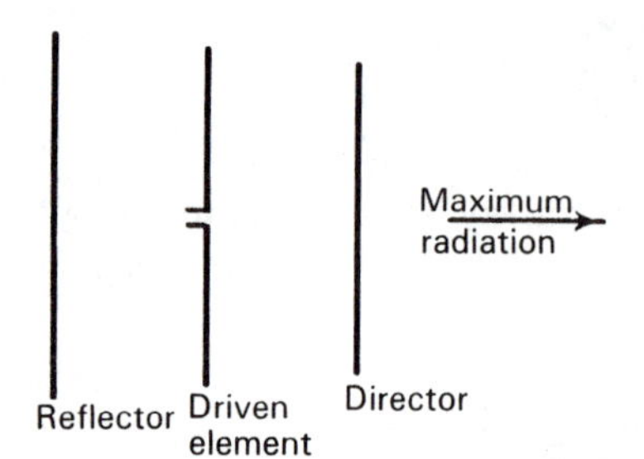

Figure 11-39 Three-element array.

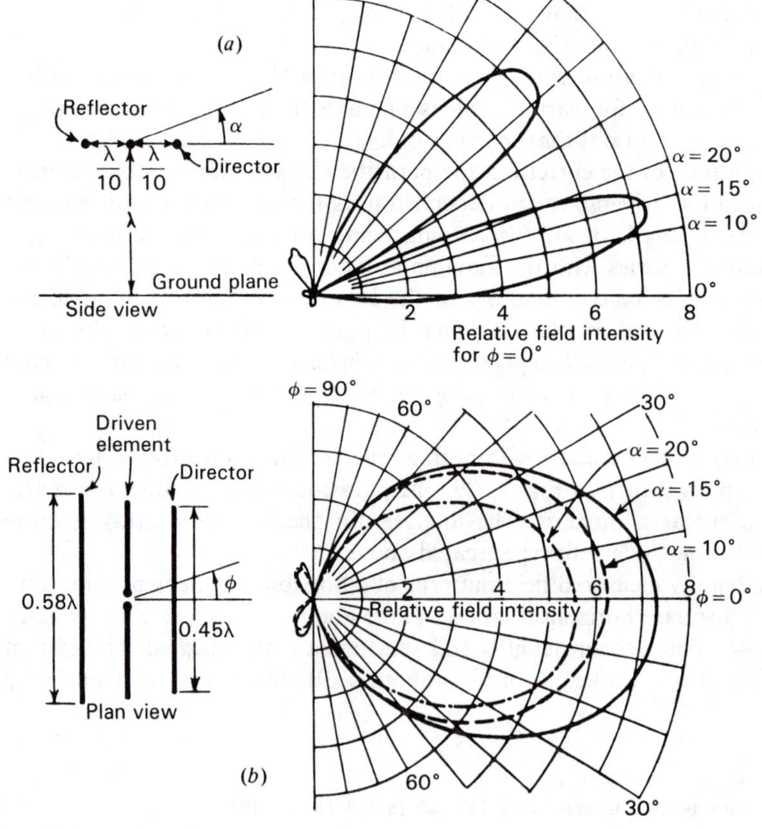

Figure 11-40 Measured vertical plane pattern (*a*) and horizontal plane patterns (*b*) at 3 elevation angles for a 3-element array located 1λ above a large ground plane. (*Patterns by D. C. Cleckner, Ohio State University.*)

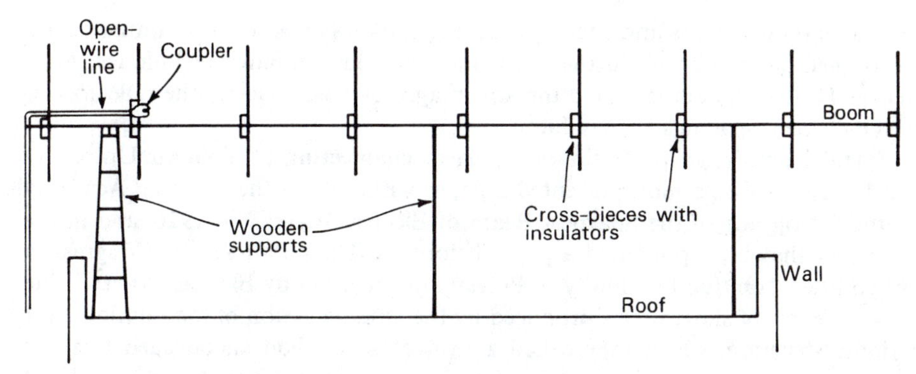

Figure 11-41a Shintaro Uda's experimental antenna with 1 reflector and 7 directors on the roof of his laboratory at Tohoku University for vertically polarized transmission tests during 1927 and 1928 over land and sea paths up to 135 km using a wavelength $\lambda = 4.4$ m. The horizontal wooden boom supporting the array elements is 15 m long.

dipole antenna at the same height.[1] The vertical plane pattern is shown in Fig. 11-40*a*. It is interesting to note that because of the finite size of the ground plane there is radiation at negative elevation angles. This phenomenon is characteristic of antennas with finite ground planes, the radiation at negative angles being largely the result of currents on the edges of the ground plane or beneath it. The azimuthal patterns at elevation angles $\alpha = 10$, 15 and 20° are shown in Fig. 11-40*b*. A parasitic array of this type with closely spaced elements has a small driving-point radiation resistance and a relatively narrow bandwidth.

11-9b The Yagi-Uda Array. Shintaro Uda, an assistant professor at Tohoku University, had not turned 30 when he conducted experiments on the use of parasitic reflector and director elements in 1926. This led to his publication of a series of 11 articles (from March 1926 to July 1929) in the *Journal of the Institute of Electrical Engineers of Japan* titled "On the Wireless Beam of Short Electric Waves."[2] He measured patterns and gains with a single parasitic reflector, a single parasitic director and with a reflector and as many as 30 directors. One of his many experimental arrays is shown in Fig. 11-41a. He found the highest gain with the reflector about $\lambda/2$ in length and spaced about $\lambda/4$ from the driven element, while the best director lengths were about 10 percent less than $\lambda/2$ with optimum spacings about $\lambda/3$. Even though many patterns were measured in the

---

[1] Note that it is necessary to specify both the height and elevation angle at which the comparison is made. In comparing one antenna with another, the gain as a function of elevation angle at a given height (or as a function of height at a given elevation angle) may, in general, range from zero to infinity.

[2] S. Uda, "On the Wireless Beam of Short Electric Waves," *JIEE* (*Japan*), March 1926, pp. 273–282 (I); April 1926 (II); July 1926 (III); January 1927 (IV); January 1927 (V); April 1927 (VI); June 1927 (VII); October 1927 (VIII); November 1927, pp. 1209–1219 (IX); April 1928 (X); July 1929 (XI).

near field, these lengths and spacings agree remarkably well with optimum values determined since then by further experimental and computer techniques. After George H. Brown demonstrated the advantages of close spacing, the reflector-to-driven-element spacings were reduced.

Hidetsugu Yagi, professor of electrical engineering at Tohoku University and 10 years Uda's senior, presented a paper with Uda at the Imperial Academy on the "Projector of the Sharpest Beam of Electric Waves" in 1926, and in the same year they both presented a paper before the Third Pan-Pacific Congress in Tokyo titled "On the Feasibility of Power Transmission by Electric Waves." The narrow beams of short waves produced by the guiding action of the multidirector periodic structure, which they called a "wave canal," had encouraged them to suggest using it for short-wave power transmission, an idea now being considered for beaming solar power to the earth from a space station or from earth to a satellite (but not necessarily with their antennas).

It is reported that Professor Yagi had received a substantial grant from Sendai businessman Saito Zenuemon which supported the antenna research done by Uda with Yagi's collaboration. Then in 1928 Yagi toured the United States presenting talks before Institute of Radio Engineers sections in New York, Washington and Hartford, and in the same year Yagi published his now famous article on "Beam Transmission of Ultra Short Waves" in the *Proceedings of the IRE.*[1] Although Yagi noted that Uda had already published 9 papers on the antenna and acknowledged that Uda's ingenuity was mainly responsible for its successful development, the antenna soon came to be called "a Yagi." In deference to Uda's contributions, I refer to the array as a *Yagi-Uda antenna*, a practice now becoming common. Uda has summarized his researches on the antenna in two data-packed books.[2]

A typical modern-version 6-element Yagi-Uda antenna is shown in Fig. 11-41b. It consists of a driven element (folded $\lambda/2$ dipole) fed by a 300-Ω 2-wire transmission line (twin line), a reflector and 4 directors. Dimensions (lengths and spacings) are indicated on the figure. The antenna provides a gain of about 12 dBi (maximum) with a bandwidth at half-power of 10 percent. By adjusting lengths and spacings appropriately (tweeking), the dimensions can be optimized, producing an increase in gain of another decibel.[3] However, the dimensions are critical.

---

[1] H. Yagi, "Beam Transmission of Ultra Short Waves," *Proc. IRE*, **16**, 715–740, June 1928.

[2] S. Uda and Y. Mushiake, *Yagi-Uda Antenna*, Maruzden, Tokyo, 1954.

S. Uda, *Short Wave Projector; Historical Records of My Studies in the Early Days*, published by Uda, 1974.

[3] C. A. Chen and D. K. Cheng, "Optimum Spacings for Yagi-Uda Arrays," *IEEE Trans. Ants. Prop.*, **AP-21**, 615–623, September 1973.

C. A. Chen and D. K. Cheng, "Optimum Element Lengths for Yagi-Uda Arrays," *IEEE Trans. Ants. Prop.*, **AP-23**, 8–15, 1975.

P. P. Viezbicke, "Yagi Antenna Design," NBS Technical Note 688, December 1968.

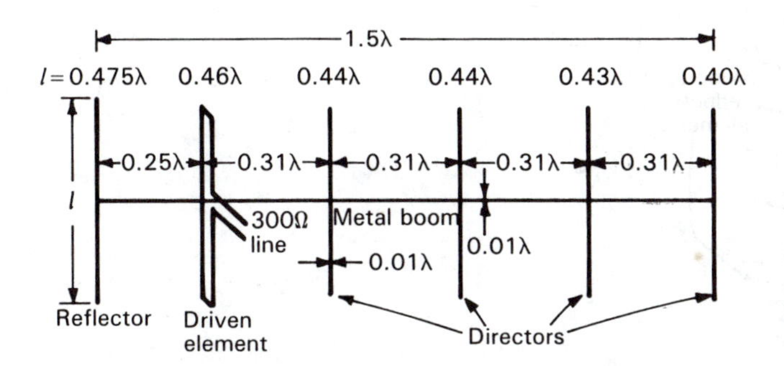

Figure 11-41b Modern-version 6-element Yagi-Uda antenna with dimensions. It has a maximum directivity of about 12 dBi at the center of a bandwidth of 10 percent at half-power.

The inherently narrow bandwidth of the Yagi-Uda antenna can be broadened to 1.5 to 1 by lengthening the reflector to improve operation at low frequencies and shortening the directors to improve high-frequency operation. However, this is accomplished at a sacrifice in gain of as much as 5 dB.

11-9c Square-Corner–Yagi-Uda Hybrid. The corner reflector (Sec. 12-3) is an inherently wideband antenna producing substantial gains. Thus, a square-corner reflector with maximum dimension equal to the 1.5λ length of the Yagi-Uda antenna of Fig. 11-41b produces about 1 dB more gain. Furthermore, the corner reflector can operate over a 2 to 1 bandwidth with 5 or 6 dB more gain than a 1.5 to 1 bandwidth Yagi-Uda antenna of the same size.

Although reducing the length of the reflector elements and reducing their number (by increasing the spacing between them) decreases the gain of a corner reflector antenna, the gain may still be more than that of a wideband Yagi-Uda antenna. However, if the increased spacing is as much as $\lambda/6$ or $\lambda/4$ at the low-frequency end of a 2 to 1 band, the spacing becomes $\lambda/3$ to $\lambda/2$ at the high-frequency end, reducing the effectiveness of the reflector. To compensate for this effect at the high-frequency end a number of directors can be added in front of the driven element of the corner reflector, resulting in a corner reflector with directors or *corner–Yagi-Uda hybrid antenna* as shown in Fig. 11-42. The length and spacing of the directors are made appropriate for the high-frequency end of the band. The thinned corner reflector is effective at the low and mid frequencies while the directors are effective at the high frequencies, resulting in an average gain of 7 or 8 dBi over a bandwidth of nearly 2 to 1, which is adequate for the UHF U.S. TV band. Although less than the gain of a full-element corner reflector antenna, the hybrid has less wind resistance.

In the competitive high-volume UHF TV antenna market, in which millions of corner reflector and Yagi-Uda antennas have been sold, the hybrid offers an alternative. At least half of the reflector elements can be removed from the

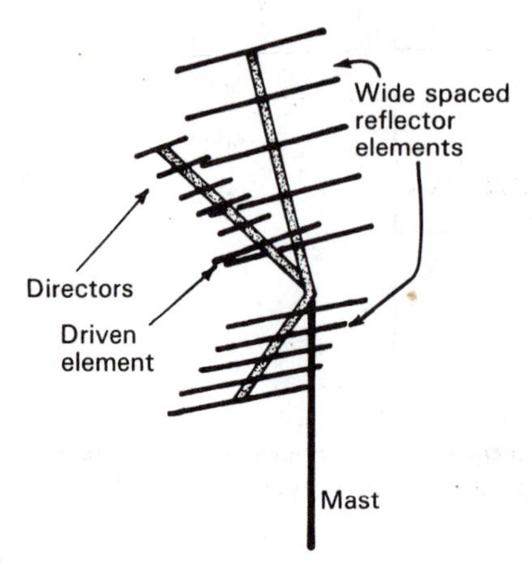

Figure 11-42 Square-corner–Yagi-Uda hybrid antenna. It has an average directivity of about 8 dBi over a 2 to 1 bandwidth.

corner reflector. The directors which are added are much shorter and may be fewer in number, resulting in a net savings in cost.

Compare the wide bandwidth corner reflector of Fig. 12-14 and also the Yagi-Uda–corner–log-periodic array described in Sec. 15-6.

11-9d Circular polarization with a Yagi-Uda Antenna. To produce circular polarization, 2 Yagi-Uda antennas can be crossed (elements at right angles on the same boom) with the driven elements fed in phase quadrature, or both driven elements can be fed in phase but with one array displaced $\lambda/4$ along the boom with respect to the other. Another alternative is to feed the crossed director pairs with a monofilar axial-mode helical antenna (see Fig. 7-55). An advantage of this arrangement is that it can be fed by a single coaxial transmission line.

11-9e The Landsdorfer Shaped-Dipole Array. The gain of a Yagi-Uda antenna can be increased by adding more directors and increasing the length of the array. As with all end-fire arrays, twice the gain (3 dB improvement) requires a 4-fold increase in length. Another method of obtaining a 3 dB improvement is to stack 2 arrays. As yet another alternative, Landsdorfer[1] has demonstrated that higher gain can be obtained by extending and shaping the conductors of a 3-element close-spaced Yagi-Uda antenna.

Consider the center-fed $3\lambda/2$ dipole shown in Fig. 11-43*a*. Assuming a sinusoidal current distribution, the field pattern is as indicated in (*b*) (see also Fig. 5-9). There are small broadside lobes and also large lobes at an angle. If the center $\lambda/2$

[1] F. M. Landsdorfer, "Zur Optimalen Form von Linearantennen," *Frequenz*, **30**, 344–349, 1976.
F. M. Landsdorfer, "A New Type of Directional Antenna," *Ant. Prop. Soc. Int. Symp. Digest*, 169–172, 1976.
F. M. Landsdorfer and R. R. Sacher, *Optimization of Wire Antennas*, Wiley, 1985.

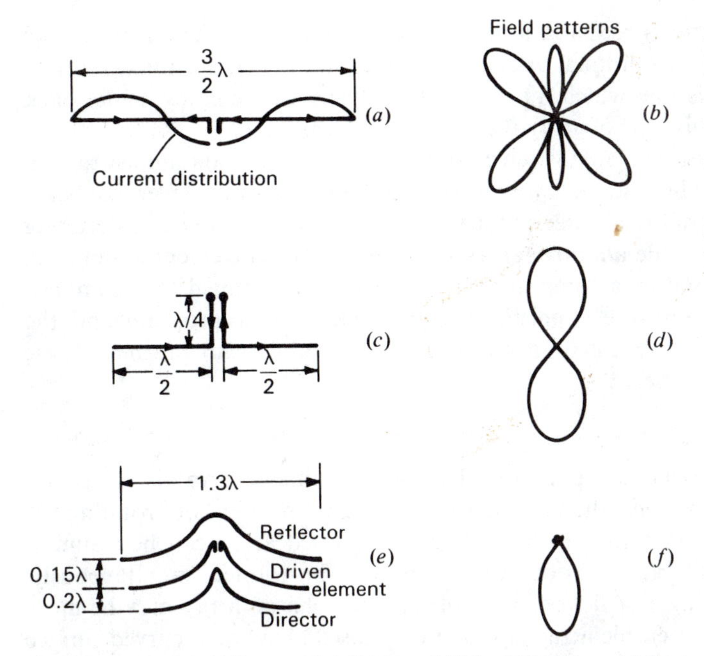

Figure 11-43 (*a*) A $3\lambda/2$ center-fed dipole and (*b*) its field pattern. (*c*) Center section folded into a $\lambda/4$ stub reducing length to 1λ and (*d*) the field pattern. (*e*) Further shaping and addition of shaped director and reflector forms a Landsdorfer array with the field pattern (*f*).

section is folded into a $\lambda/4$ stub as in (*c*) the antenna reduces to an in-phase 1λ dipole with a bidirectional broadside field pattern as in (*d*). Now pulling the stub apart and shaping it and the $\lambda/2$ sections, it can be arranged with a similarly shaped director and reflector as done by Landsdorfer and shown in (*e*) with the unidirectional pattern at (*f*). The overall length is 1.3λ with gain reported to be about 11.5 dBi. This compares to about 8.5 dBi for a close-spaced 3-element array of $\lambda/2$ dipoles.

11-10 PHASED ARRAYS

11-10a Introduction. Although the elements of any antenna array must be phased in some manner, the term *phased array* has come to mean an array of many elements with the phase (and also, in general, the amplitude) of each element being a variable, providing *control of the beam direction* and pattern shape including sidelobes. These arrays are discussed in the second part of this section.

Specialized phased arrays given different names are the frequency scanning array, the retro-array and the adaptive array.

In the scanning array, phase change is accomplished by varying the frequency. These *frequency scanning arrays* are among the simplest phased arrays since no phase control is required at each element. Several of these arrays were discussed in Sec. 7-11. Additional ones are described in the next section (11-11).

A *retro-array* is one which automatically reflects an incoming signal back toward its source. This type of array is considered in Sec. 11-12.

Adaptive arrays have an awareness of their environment and adjust to it in a desired fashion. Thus, an *adaptive array* can automatically steer its beam toward a desired signal while steering a null toward an undesired or interfering signal. In a more versatile adaptive array the output of each element is sampled, digitized and processed by a computer which can be programmed to accomplish tasks limited mainly by the sophistication of the computer program and the available computer power. Such an array may be called a *smart antenna*. These arrays are described in Sec. 11-13.

11-10b Phased Array Designs. An objective of a phased array is to accomplish beam steering without the mechanical and inertial problems of rotating the entire array. In principle, the beam steering of a phased array can be instantaneous and, with suitable networks, all beams can be formed simultaneously. However, the look angle or field of view of a planar phased array may be more restricted than for a steerable array (although a phased array on a curved surface may cover as much solid angle). Also the beam of a rotatable array maintains its shape with change in direction whereas a phased array beam may not.

Another objective of the phased array is to provide beam control at a fixed frequency or at any number of frequencies within a certain bandwidth in a *frequency-independent manner.*

In its most simplistic form, beam steering of a phased array can be done by mechanical switching. Thus, consider the case of the rudimentary 3-element array of Fig. 11-44*a*. Let each element be a $\lambda/2$ dipole (seen end-on in the figure). An incoming wave arriving broadside as in (*a*) will induce voltages in the transmission lines (or cables) in the same phase so that if all cables are of the same length ($l_1 = l_2 = l_3$) the voltages will be in phase at the (dashed) in-phase line. By bringing all 3 cables to a common point as in (*b*), the 3-element array will operate as a broadside array. For an impedance match, the cable to the receiver (or transmitter) should be $\frac{1}{3}$ the impedance of the 3 cables, or a 3 to 1 impedance transformer can be inserted at the common junction point with all cables of the same impedance.

Now consider a wave arriving at an angle of 45° from broadside as in Fig. 11-44*c*. If the wave velocity $v = c$ on the cables, the (dashed) in-phase line is parallel to the wave front of the incoming wave, as suggested in (*c*). However, if $v < c$, the lengths l_2 and l_3 must be increased as suggested in order for all phases to be the same (the in-phase condition). Then, if cables of these lengths are joined as in (*d*) the 3-element array will have its beam 45° from broadside.

By installing a switch at each antenna element and one at the common feed point as in Fig. 11-44*e* and mechanically ganging all switches together, the beam can be shifted from broadside to 45° by operating the ganged switch.

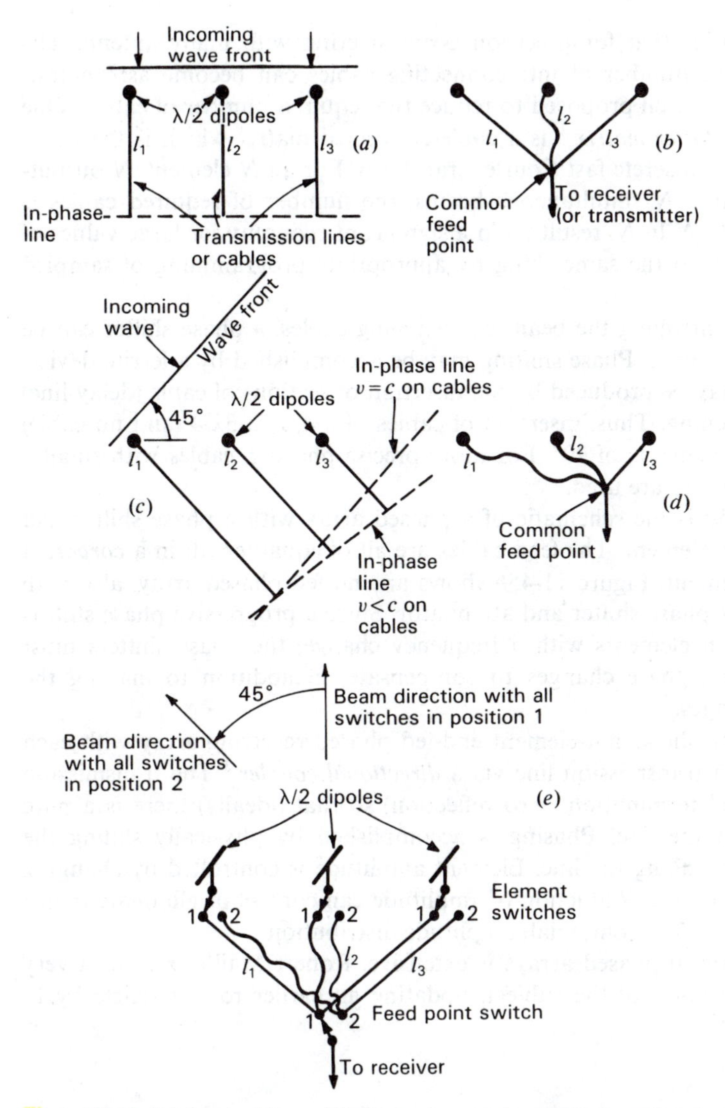

Figure 11-44 (*a*) Array of three $\lambda/2$ dipoles (seen end-on) with incoming wave broadside. (*b*) Equal length cables joined. (*c*) Incoming wave at 45° from broadside. (*d*) Appropriate lengths of cable joined. (*e*) Switches for shifting from broadside to 45° reception.

By adding more switch points and more cables of appropriate length the beam can be steered to an arbitrarily large number of directions. With more elements, narrower beams can be formed. With diodes (PIN type[1]) in place of mechanical switches, control can be electronic. However, even with these modifi-

[1] PIN (Positive Intrinsic Negative): high open-circuit impedance, low short-circuit impedance.

cations, it is obvious that for precision beam steering with many antenna elements, the required number of interconnecting cables can become astronomic. Many schemes have been proposed to reduce the required number of cables. One of these, called a *Butler matrix*,[1] is a cable-connected matrix which is the hard-wire equivalent of a discrete fast Fourier transform. For an N element, N output-port matrix, forming N simultaneous beams, the number of required cables is reduced from N^2 to $N \ln N$, resulting in a significant economy for large values of N. Computers can do the same thing by appropriate programming of sampled signals.

Instead of controlling the beam by switching cables, a phase shifter can be installed at each element. Phase shifting may be accomplished by a ferrite device. The same effect may be produced by the insertion of sections of cable (delay line) by electronic switching. Thus, insertion of cables of $\lambda/4$, $\lambda/2$, $3\lambda/4$ (and no cable) provides phase increments of 90°. For more precise phasing, cables with smaller incremental differences are used.

Figure 11-45*a* is the schematic of a phased array with a phase shifter and attenuator at each element. The feed cables are all of equal length in a *corporate structure*[2] arrangement. Figure 11-45*b* shows an end-fed phased array, also with individual element phase shifter and attenuator. Since a progressive phase shift is introduced between elements with a frequency change, the phase shifters must introduce opposing phase changes to compensate, in addition to making the desired phase changes.

Figure 11-45*c* shows a 4-element end-fed phased receiving array with each element fed from a transmission line via a *directional coupler*.[3] The transmission line has a matched termination (zero reflection) so that (ideally) there is a pure traveling wave on the line. Phasing is accomplished by physically sliding the directional couplers along the line. Element amplitude is controlled by changing the closeness of coupling. Reduction of amplitude can control or eliminate minor lobes as with a 1 : 3 : 3 : 1 (binomial) amplitude distribution.

The literature on phased arrays is extensive. Robert Mailloux gives a very comprehensive overview of the subject, updating an earlier review article by L. Stark.[4]

---

[1] J. Butler and R. Lowe, "Beamforming Matrix Simplifies Design of Electronically Scanned Antennas," *Elect. Des.*, **9**, 170–173, 1961.

J. L. Butler, in R. C. Hansen (ed.), *Microwave Scanning Antennas*, vol. 3, Academic Press, 1966, chap. 3 (includes 80 references).

[2] Named after the organizational *structure of a corporation* with president over 2 vice presidents each over 2 subordinates, etc., the diagram in Fig. 11-45*a* being an upside-down version.

[3] For explanation of directional coupler action, see J. D. Kraus, *Electromagnetics*, 3rd ed., McGraw-Hill, 1984, pp. 419–420.

[4] R. J. Mailloux, "Phased Array Theory and Technology," *Proc. IEEE*, **70**, 246–291, March 1982 (includes 174 references).

L. Stark, "Microwave Theory of Phased Array Antennas—A Review," *Proc. IEEE*, **62**, 1661–1701, 1974.

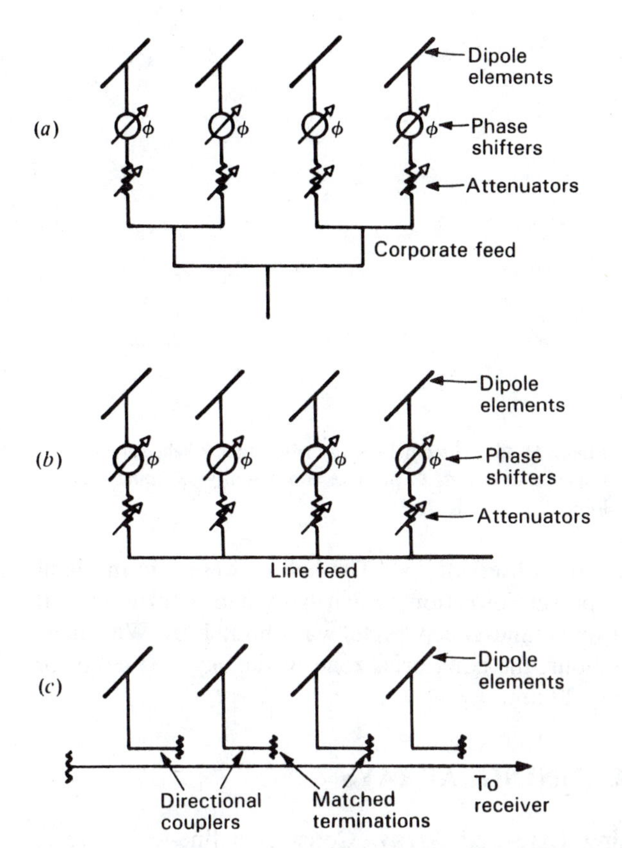

Figure 11-45 (*a*) Schematic of phased array fed by corporate structure and (*b*) end-fed. (*c*) Phased array with each element fed from a matched transmission line via a directional coupler.

11-10c Rotatable Helix Phased Array. With monofilar axial-mode helices as elements of the array, phasing can be accomplished by rotating the helices on their axes, a rotation of 90° providing a 90° shift in phase of the (circularly polarized) wave. For example, with 3 helices of the same hand connected as in Fig. 11-46 the beam direction can be steered by rotation of the outer helices (1 and 3) with helix 2 stationary. Thus, continuously rotating helix 1 clockwise and helix 3 counterclockwise will result in a continuously sweeping beam between two angular extremes, as suggested in the figure. In this type of operation a lobe appears at the left extreme of the sweep angle, grows in amplitude as it sweeps to the right, reaching maximum amplitude at broadside. It then becomes smaller as it sweeps further to the right. After reaching the extreme right of the sweep angle, a new lobe appears at the left extreme and the process repeats. The *angle of sweep* is determined by the pattern of a single helix.

I designed and built a 3-helix beam-sweeping array of this type in 1958 for operation at 25 to 35 MHz at the Ohio State University Radio Observatory for

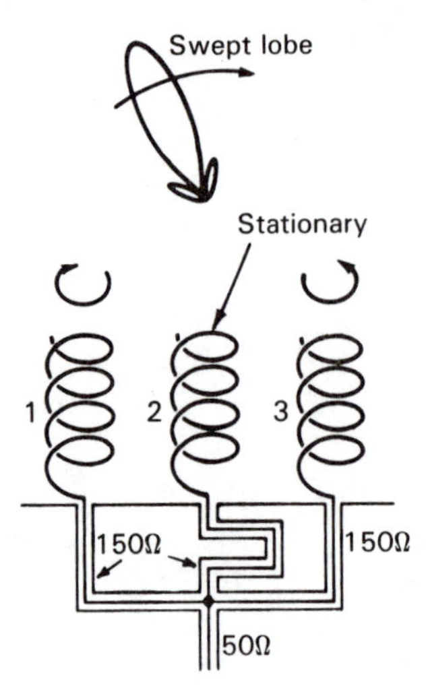

Figure 11-46 Three-helix array with outer 2 helices rotating in opposite directions to produce continuously sweeping beam or lobe.

planetary (Jupiter) and solar radio observations.[1] The helices were 3 m in diameter, the outer 2 rotating in opposite directions. Each helix had 3 turns so that the beam width between first nulls (and sweep angle) was about 130°. With helix rotation at 3 revolutions per hour, the equatorial zone of the sky was swept or scanned from east to west every 20 min.

11-11 FREQUENCY-SCANNING ARRAYS

11-11a Frequency-Scanning Line-Fed Array. Consider a line-fed array of uniformly spaced elements (dipoles) with a receiver connected at the right end of the line as suggested by the schematic of Fig. 11-47. Each element is fed from the transmission line via a directional coupler. This arrangement is similar to that in Fig. 11-45*c* but in the present case the coupler positions are fixed, with beam sweeping or scanning done by changing the frequency. The transmission line is matched to eliminate reflections and insure an essentially pure traveling wave on the line. From (7-11-8),

$$\cos\phi = \frac{1}{p} + \frac{m}{d/\lambda_0} \tag{1}$$

where ϕ = beam angle from array axis, rad or deg
p = phase velocity on transmission line = v/c, dimensionless
m = mode number, dimensionless
d = element spacing, m
λ_0 = free-space wavelength at center frequency of array operation, m

[1] J. D. Kraus, "Planetary and Solar Radio Emission at 11 m Wavelength," *Proc. IRE*, **46**, 266–274, January 1958.
J. D. Kraus, *Radio Astronomy*, 2nd ed., Cygnus-Quasar, 1986, p. 6-40 (1st ed., 1966, p. 193).

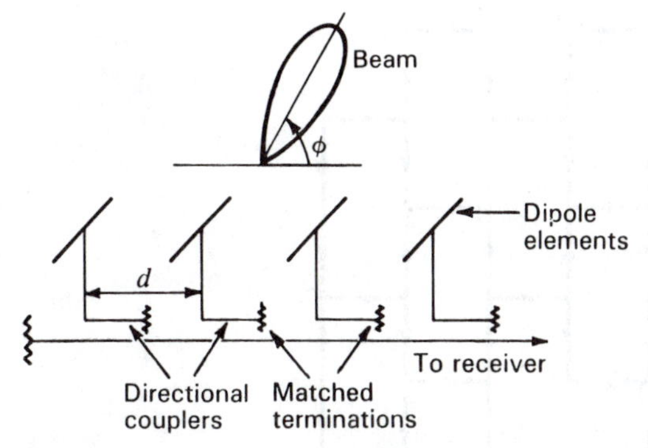

Figure 11-47 Frequency-scanning line-fed array of uniformly spaced elements with tunable receiver at left end. Beam angle ϕ is a function of the frequency.

For $p = 1$ and $m = 0$, $\phi = 0°$ (beam fixed at end-fire independent of λ_0).

Consider now the situation for $p = 1$, $m = -1$, $d = 1$ m and $\lambda_0 = 1$ m, or

$$\cos \phi = 1 - 1 = 0 \tag{2}$$

and $\phi = 90°$ (beam broadside).

Suppose next that the frequency is increased so that the wavelength is $0.9\lambda_0$ or 0.9 m; then

$$\cos \phi = 1 - 0.9 = 0.1 \tag{3}$$

and $\phi = 84.3°$, or 5.7° right of broadside. Shifting to a lower frequency so that the wavelength is $1.1\lambda_0$ or 1.1 m,

$$\cos \phi = 1 - 1.1 = -0.1$$

and $\phi = 95.7°$, or 5.7° left of broadside.

Thus, a $\pm$ 10 percent shift in wavelength (or frequency) swings the beam $\pm 5.7°$ from broadside (total scan 11.4°) with larger frequency shifts resulting in larger scan angles. To eliminate or reduce beams at $\phi = -90°$ (mirror image), and also end-fire and back-fire beams, the $\lambda/2$ dipoles can be replaced by identical unidirectional elements, the scan angle then being restricted to the beam width of the individual element.

This frequency-scanning array has no moving parts, no phase shifters and no switches, making it one of the simplest types of phased arrays.

11-11b Frequency-Scanning Backward Angle-Fire Grid and Chain Arrays.[1]

The current on folded-wire antennas usually assumes a sinusoidal (standing-wave) distribution. In the 1930s I had constructed and used folded-wire

---

[1] Some call the *frequency-scanning backward angle-fire grid array* a *Kraus grid* and the *frequency-scanning backward angle-fire chain array* a *Tiuri chain*.

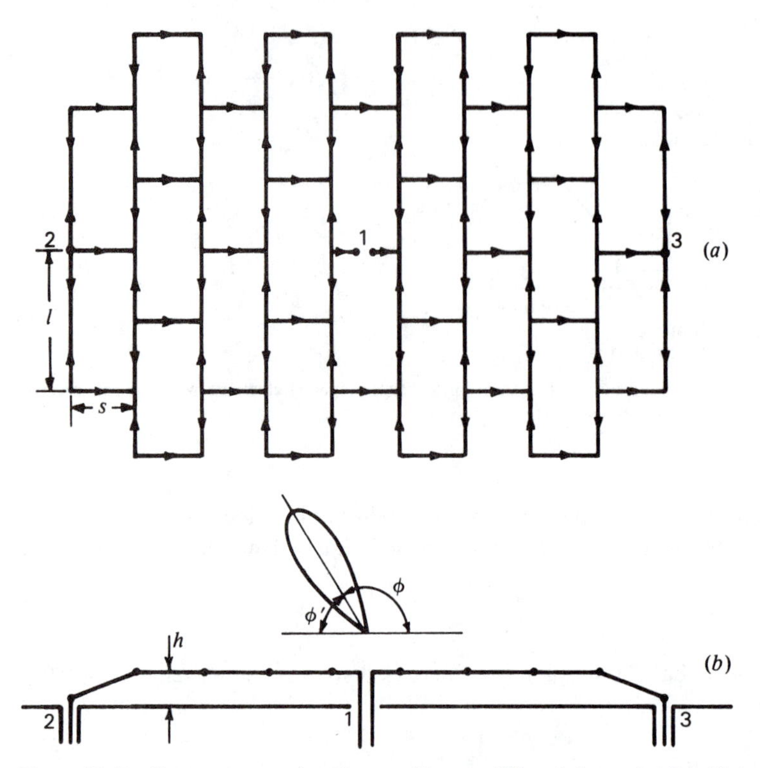

Figure 11-48 Frequency-scanning Kraus grid array. When fed at point 2 (with terminals at point 1 short-circuited and a matched load connected at point 3) the beam is at an angle ϕ which is a function of the frequency. A frequency shift of ± 14 percent swings the beam angle ϕ through about 75°. Switching the feed point to 3 and load to 2 puts the beam in the right-hand quadrant, making the total scan angle 150°. Typically, $l \simeq \lambda$, $s < \lambda/2$, $h < \lambda/4$.

antennas of this kind, such as the Bruce curtain (Fig. 11-58*b*), and became familiar with their operation. A broadside array of these antennas (for increased gain) may require many interconnecting transmission lines to feed them. In thinking about these arrays during the winter of 1961–1962, I wondered if it would be possible to construct a continuous wire grid as a broadside array and feed it at a single centrally located point.

The basic arrangement is illustrated in Fig. 11-48. The dimensions of each mesh of the grid are $l \simeq \lambda$ by $s \simeq \lambda/2$. Assuming standing waves, the instantaneous current distribution would be as indicated by the arrows, one located at each current maximum point. Currents on all of the short sides ($\lambda/2$ long, horizontal in the figure) would be in phase, while on the long sides of the meshes (λ long, vertical in the figure) there are as many current maxima in one direction as in the opposite so that radiation broadside from the long sides should (ideally) be zero. Thus, I reasoned, the array should produce a linearly polarized (horizontal in the figure) beam broadside to the array with a gain proportional to the number of $\lambda/2$ sides (or elements) (31 in the figure).

I constructed an array similar to the one in Fig. 11-48, mounted it approximately $\lambda/4$ from a flat conducting ground plane and fed it with a balanced transmission line at the central point (1 in the figure). To my surprise, I found that the radiation was not in a single broadside beam but was split into 2 equal lobes, one left and one right of broadside.

It was apparent that the current distribution was not that of a resonant standing wave but rather of two traveling waves, one to the left and one to the right from point 1. Accordingly, in order to have only one (left-to-right) traveling wave across the entire grid, I short-circuited the terminals at point 1 and fed the grid with a coaxial line at the left edge (point 2). This resulted in a single beam in the back-fire direction (opposite to the traveling wave) as indicated in Fig. 11-48, with the beam angle ϕ a function of the frequency. Thus, in this mode of operation, the antenna is a nonresonant frequency-scanning array with the long sides of the meshes behaving essentially as transmission-line sections and the short sides as both radiating and transmission-line sections. Although the terminals at the other edge of the array (point 3) may be left open, connecting a matched load reduces any reflected wave which could degrade the desired condition of a single left-to-right traveling wave. Thus, the antenna may be regarded as a traveling-wave-fed array of discrete radiating elements.[1]

Feeding the array at point 2 (terminals at 1 short-circuited and 3 matched), the beam direction ϕ as a function of the grid parameters is given by

$$\frac{2\pi s}{\lambda}\cos\phi - \frac{\pi l}{\lambda p_x} - \frac{2\pi s}{\lambda p_y} = -2\pi m \tag{4}$$

where s = length of short side, m
l = length of long side, m
ϕ = beam angle from array axis
p_x = relative phase velocity along short side = v_x/c
p_y = relative phase velocity along long side = v_y/c
m = mode number = integer

For $m = 1$, $p_x = p_y = 1$, (4) becomes

$$-2\pi = \frac{2\pi}{\lambda}\left(s\cos\phi - \frac{l}{2} - s\right) \tag{5}$$

or typically ($l = 2.75s$)

$$\cos\phi = 2.37 - \frac{1}{s/\lambda} \tag{6}$$

[1] J. D. Kraus, "A Backward Angle-Fire Array Antenna," *IEEE Trans. Ants. Prop.*, **AP-12**, 48–50, January 1964.
J. D. Kraus, "Backward Angle Traveling Wave Wire Mesh Antenna Array," U.S. Patent 3,290,688, applied for June 11, 1962, granted Dec. 6, 1966.

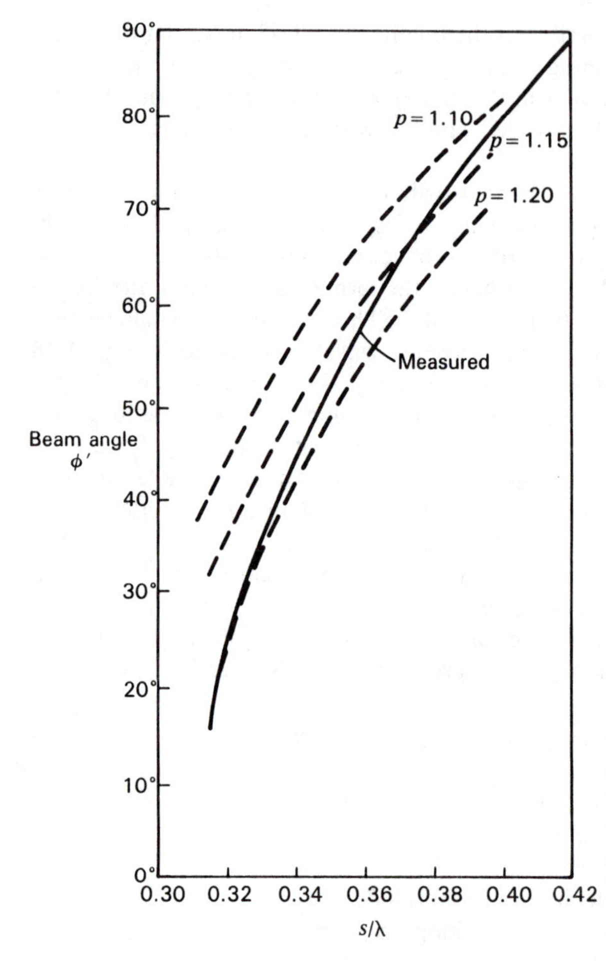

Figure 11-49 Beam angle ϕ' as a function of frequency expressed in terms of s/λ. The dashed curves are calculated for different values of relative phase velocity p. The solid curve is measured, suggesting that p is a (weak) function of the frequency.

The beam direction ϕ' (complement to ϕ in Fig. 11-48) varies from about 15 to 90° for changes in s from 0.3 to 0.42λ. A wavelength (or frequency) change of ± 14 percent swings the beam through a scan angle of 75°. By switching the feed point to 3 and matched load to point 2 the beam can be placed in the right quadrant, increasing the total scan angle to 150°. Although this analysis is oversimplified it illustrates the basic relations. Comparing calculations with measurements indicates that the relative phase velocity p $(=p_x = p_y)$ is a function of the frequency, as suggested by Fig. 11-49, and not a constant $(=1)$ as assumed above. Additional measurements suggest further that $p_x \neq p_y$.[1]

[1] J. D. Kraus and K. R. Carver, "Wave Velocities on the Grid-Structure Backward Angle-Fire Antenna," *IEEE Trans. Ants. Prop.*, **AP-12**, 509–510, July 1964.

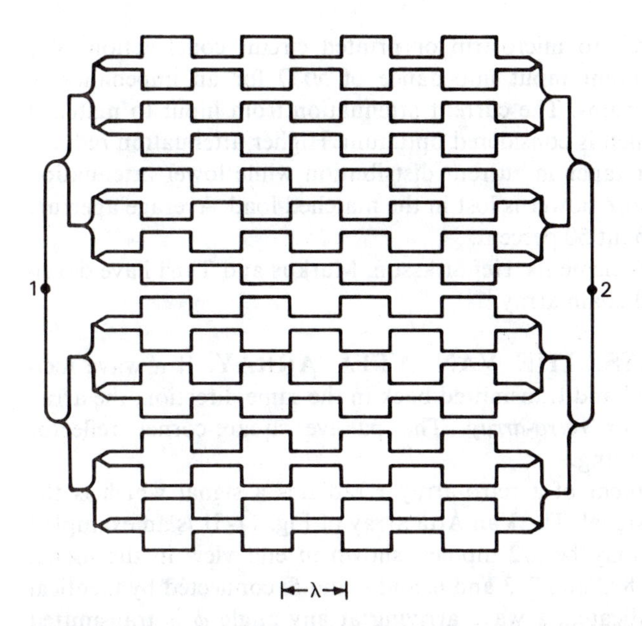

Figure 11-50 Frequency-scanning Tiuri chain array for feeding at one end (point 1) with matched load at point 2, or vice versa. With array mounted about 0.15λ from a flat conducting ground plane, the impedance of a single chain is typically about 300 Ω and the terminal impedance (at points 1 or 2) typically 50 Ω. Polarization is horizontal. (*After S. Tallqvist, "Theory of the Meander Type Chain and Grid Antenna," Lic. thesis, Helsinki University of Technology, 1977.*)

My experimentally determined "best" average value of s $(=0.36\lambda)$ corresponds to an average value for the long sides of $l = 2.75s \simeq \lambda$. Thus, in practice the long sides of the meshes are $\sim\lambda$ long, as envisioned in the initial design, but the short sides are less than $\lambda/2$.

Considering the array as a transmission line, the extra path length $\lambda/2$ between radiating elements reduces the effective phase velocity (left to right) in the ratio $s/[s + (\lambda/2)]$. Typically, $l = 2.75s$ so that the effective phase velocity is about $0.4c$, making the grid a *slow-wave* structure. The average gain of a grid array as in Fig. 11-48 is about 17 dBi.

A microstrip version of the array is described by Conti *et al.*[1]

The *Kraus grid* array principle has been extended by Tiuri, Tallqvist and Urpo[2] and by Tallqvist[3] to an array in which the meshes are divided into parallel matched chains, with a typical configuration as in Fig. 11-50. This *Tiuri chain*

[1] R. Conti, J. Toth, T. Dowling and J. Weiss, "The Wire Grid Microstrip Antenna," *IEEE Trans. Ants. Prop.*, **AP-29**, 157–166, January 1981.

[2] M. Tiuri, S. Tallqvist and S. Urpo, "Chain Antenna," *IEEE Ant. Prop. Soc. Int. Symp. Digest*, June 10, 1974.

[3] S. Tallqvist, "Theory of the Meander Type Chain and Grid Antenna," Lic. thesis, Helsinki University of Technology, 1977.

design is also well adapted to microstrip or printed circuit construction. The array shown has an endpoint input impedance of 50 Ω for an impedance of 300 Ω for the individual chains. The current attenuation from input to matched output is about 10 dB, which is considered optimum. Higher attenuation reduces the gain due to the larger taper in current distribution while lower attenuation lowers the gain because more power is lost in the matched load. Average aperture efficiencies are typically about 50 percent.

By bending the chain elements, Hendriksson, Markus and Tiuri have developed a circularly polarized chain array.[1]

11-12 RETRO-ARRAYS. THE VAN ATTA ARRAY. If a wave incident on an array is received and transmitted back in the same direction, the array acts as a *retro-reflector* or *retro-array*. The passive square-corner reflector (Sec. 12-3*b*) does the same thing.

In general, each element of a retro-array reradiates a signal which is the conjugate of the received signal. The Van Atta array of Fig. 11-51 is an example.[2] The 8 identical elements may be $\lambda/2$ dipoles, shown in end view in the figure. With element pairs (1 and 8, 2 and 7, 3 and 6, and 4 and 5) connected by identical equal-length cables, as indicated, a wave arriving at any angle ϕ is transmitted back in the same direction. The array shown in Fig. 11-51, like the square-corner reflector, is passive. An adaptive (active) array (Sec. 11-13) can also be made retrodirective by using a mixer to produce a conjugate phase shift for each element.[3] An advantage of an active array is that the elements need not be arranged in a line or, in a 2-dimensional case, in a plane. Active retro-arrays can also incorporate amplifiers.[4]

11-13 ADAPTIVE ARRAYS AND SMART ANTENNAS. The antenna elements and their transmission-line interconnections discussed so far produce a beam or beams in predetermined directions. Thus, when receiving, these arrays look in a given direction regardless of whether any signals are arriving from that direction or not. However, by processing the signals from the individual elements, an array can become active and react intelligently to its environment, steering its beam toward a desired signal while simultaneously steering a null toward an undesired, interfering signal and thereby maximizing the signal-to-noise ratio of the desired signal. The term *adaptive array* is applied to this kind of antenna.

---

[1] J. Hendriksson, K. Markus and M. Tiuri, "A Circularly Polarized Traveling-Wave Chain Antenna," *European Microwave Conf.*, Brighton, September 1979.

[2] L. C. Van Atta, "Electromagnetic Reflector," U.S. Patent 2,909,002, Oct. 6, 1959.

[3] C. Y. Pon, "Retrodirective Array Using the Heterodyne Method," *IEEE Trans. Ants. Prop.*, **AP-12,** 176–180, March 1964.

[4] S. N. Andre and D. J. Leonard, "An Active Retrodirective Array for Satellite Communications," *IEEE Trans. Ants. Prop.*, **AP-12,** 181–186, March 1964.

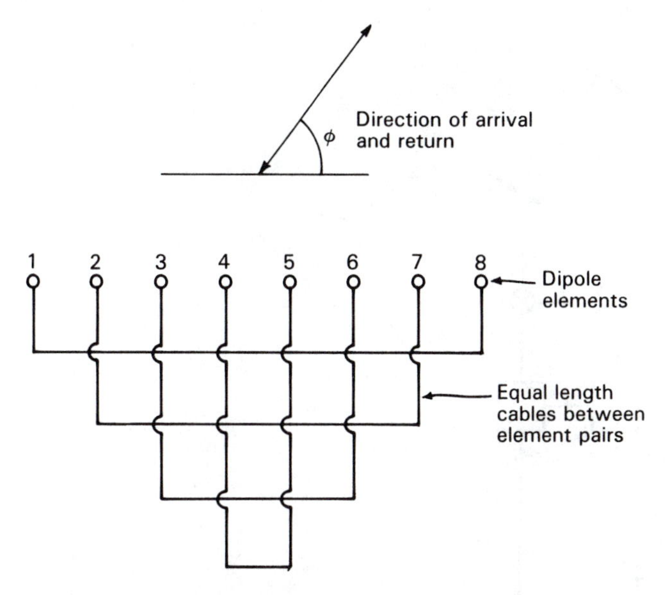

Figure 11-51 Eight-element Van Atta retro-array. Element pairs are connected by equal length lines.

Also, by suitable signal processing, performance may be further enhanced, giving *simulated patterns*[1] of higher resolution and lower sidelobes. In addition, by appropriate sampling and digitizing the signals at the terminals of each element and processing them with a computer, a very intelligent or *smart antenna* can, in principle, be built. For a given number of elements, such an antenna's capabilities are limited, mainly by the ingenuity of the programmer and the available computer power. Thus, for example, multiple beams may be simultaneously directed toward many signals arriving from different directions within the field of view of the antenna (ideally $= 2\pi$ sr for a planar array). These antennas are sometimes called Digital Beam Forming (DBF) antennas.[2]

As a rudimentary example of an adaptive array, a simple 2-element system is shown in Fig. 11-52 with $\lambda/2$ spacing between the elements at the signal frequency f_s. Let each element be a $\lambda/2$ dipole seen end-on in Fig. 11-52 so that the patterns of the elements are uniform in the plane of the page. With elements operating in phase, the beam is broadside (up in the figure).

Consider now the case of a signal at 30° from broadside as suggested in Fig. 11-52 so that the wave arriving at element 2 travels $\lambda/4$ farther than to element 1, thus retarding the phase of the signal by 90° at element 2. Each element is equipped with its own mixer, Voltage-Controlled Oscillator (VCO), intermediate frequency amplifier and phase detector. An oscillator at the interme-

---

[1] Simulated patterns are ones that exist only in the signal-processing domain.

[2] H. Steyskal, "Digital Beamforming Antennas," *Microwave J.*, **30**, 107–124, January 1987.

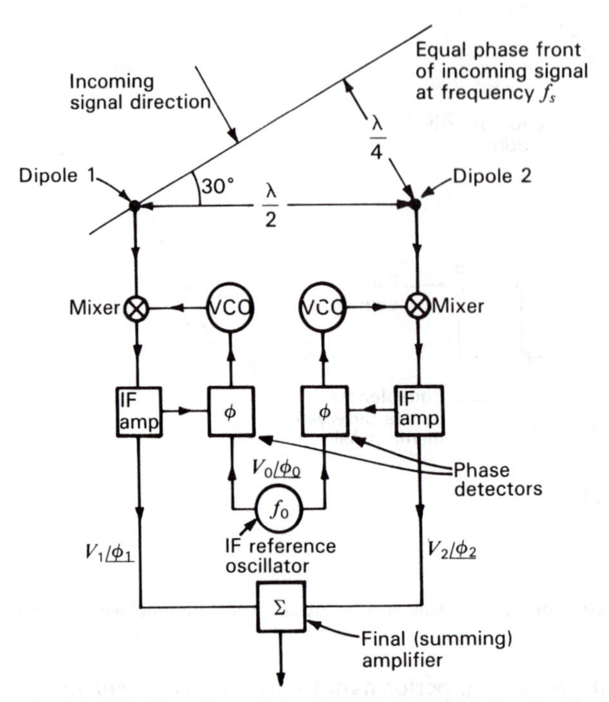

Figure 11-52 Two-element adaptive array with signal-processing circuitry.

diate frequency f_0 is connected to each phase detector as reference. The phase detector compares the *phase of the downshifted signal* with the *phase of the reference oscillator* and produces a voltage proportional to the phase difference. This voltage, in turn, advances or retards the phase of the VCO output so as to reduce the phase difference to zero (phase locking). The voltage for the VCO of element 1 would ideally be equal in magnitude but of opposite sign to the voltage for the VCO of element 2 so that the downshifted signals from both elements are locked in phase, making

$$\phi_1 = \phi_2 = \phi_0 \tag{1}$$

where ϕ_1 = phase of downshifted signal from element 1
ϕ_2 = phase of downshifted signal from element 2
ϕ_0 = phase of reference oscillator

With equal gain from both IF amplifiers the voltages V_1 and V_2 from both elements should be equal so that

$$V_1\underline{/\phi_1} = V_2\underline{/\phi_2} \tag{2}$$

making the voltage from the summing amplifier proportional to $2V_1$ $(=2V_2)$ and maximizing the response of the array to the incoming signal by steering the beam onto the incoming signal. In our example, 45° phase corrections of opposite sign would be required by the VCOs (+for element 1, −for element 2).

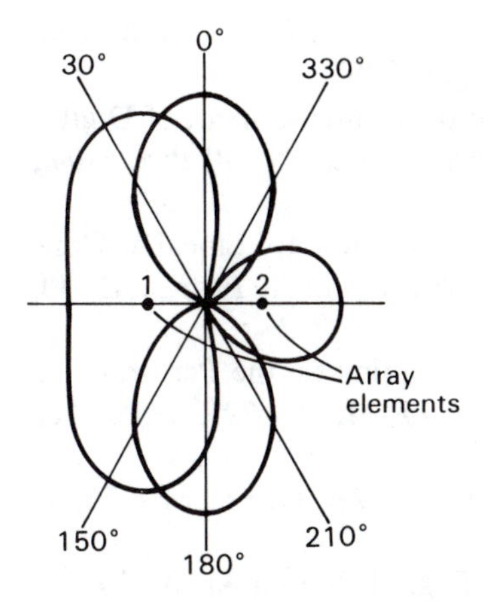

Figure 11-53 Patterns of 2-element adaptive array for signals from 0 and 30° directions. For the 0° signal, nulls are at 90 and 270° while for the 30° signal, nulls are at 210 and 330°. These patterns are identical with those of Figs. 4-1 and 4-4.

In our rudimentary 2-element example, the beam will be in the 0° direction for a signal from the 0° direction and at 30° for a signal from that direction, as shown by the patterns in Fig. 11-53. If interfering signals are arriving from the 210 and 330° directions when the main signal is at 30°, the nulls at 210 and 330° will suppress the interference. However, an interfering signal at 150° would be at a pattern maximum, the same as the desired signal at 30°. To provide more effective adaptation to its environment, an array with more elements and more sophisticated signal processing is required. For example, the main beam may be steered toward the desired signal by changing the progressive phase difference between elements, while, independently, one or more nulls are steered toward interfering signals by modifying the array element amplitudes with digitally controlled attenuators.

11-13a Literature on Adaptive Arrays. There is an extensive literature on adaptive arrays. Three special issues of the *IEEE Transactions on Antennas and Propagation* have been devoted to adaptive arrays. They are: vol. AP-12, March 1964; vol. AP-24, September 1976; and vol. AP-34, March 1986. Additional references are as follows:

Bickmore, R. W.: "Time Versus Space in Antenna Theory," in R. C. Hansen (ed.), *Microwave Scanning Antennas*, vol. 3, Academic Press, 1966, pp. 289–339.

Blank, S.: "An Algorithm for the Empirical Optimization of Antenna Arrays," *IEEE Trans. Ants. Prop.*, **AP-31**, 685–689, July 1983.

Butler, J. L.: "Digital, Matrix, and Intermediate Frequency Scanning," in R. C. Hansen (ed.), *Microwave Scanning Antennas*, vol. 3, Academic Press, 1966, pp. 217–288.

Dinger, R. J.: "A Computer Study of Interference Nulling by Reactively Steered Adaptive Arrays," *Ant. Prop. Soc. Int. Symp. Proc.*, **2**, 807–810, 1984.

Einarsson, O.: Optimization of Planar Arrays, *IEEE Trans. Ants. Prop.*, **AP-27**, 86–92, January 1979.

Ersoy, O.: "Real Discrete Fourier Transform," *IEEE Trans. Acoustics, Speech, and Signal Processing*, **ASSP-33**, 880–882, August 1985.

Fan, H., E. I. El-Masry and W. K. Jenkins: "Resolution Enhancement of Digital Beamformers," *IEEE Trans. Acoustics, Speech, and Signal Processing*, **ASSP-32**, 1041–1052, October 1984.

Griffiths, L. J., and C. W. Jim: "An Alternative Approach to Linearly Constrained Adaptive Beamforming," *IEEE Trans. Ants. Prop.*, **AP-30**, January 1982.

Gupta, I. J., and A. A. Ksienski: "Effect of Mutual Coupling on the Performance of Adaptive Arrays," *IEEE Trans. Ants. Prop.*, **AP-31**, 785–791, September 1983.

Hansen, R. C.: "Gain Limitations of Large Antennas," *IRE Trans. Ants. Prop.*, **AP-8**, 491–495, September 1960.

Hatcher, B. R.: "Granularity of Beam Positions in Digital Phased Arrays," *Proc. IEEE*, **56**, November 1968.

Johnson, H. W., and C. S. Burrus: "The Design of Optimal DFT Algorithms Using Dynamic Programming," *IEEE Trans. Acoustics, Speech, and Signal Processing*, **ASSP-31**, 378–387, April 1983.

Laxpati, S. R.: "Planar Array Synthesis with Prescribed Pattern Nulls," *IEEE Trans. Ants. Prop.*, **AP-30**, 1176–1183, November 1982.

Mucci, R. A.: "A Comparison of Efficient Beamforming Algorithms," *IEEE Trans. Acoustics, Speech, and Signal Processing*, **ASSP-32**, 548–558, June 1984.

Ricardi, L. J.: "Adaptive Antennas," in R. C. Johnson and H. Jasik (eds.), *Antenna Engineering Handbook*, McGraw-Hill, 1984, chap. 22.

Shelton, J. P., and K. S. Kelleher: "Multiple Beams from Linear Arrays," *IRE Trans. Ants. Prop.*, **AP-9**, 154–161, March 1961.

Sorenson, H. V., M. T. Heideman and C. S. Burrus: "On Computing the Split-Radix FFT," *IEEE Trans. Acoustics, Speech, and Signal Processing*, **ASSP-34**, 152–156, February 1986.

Steinberg, B. D.: "Design Approach for a High-Resolution Microwave Imaging Radio Camera," *J. Franklin Inst.*, **296**, 415–432, December 1973.

Volakis, J. L., and J. D. Young: "Phase Linearization of a Broad-Band Antenna Response in Time Domain," *IEEE Trans. Ants. Prop.*, **AP-30**, 309–313, March 1982.

Vu, T. B.: "Null Steering by Controlling Current Amplitudes Only," *Ant. Prop. Soc. Int. Symp. Proc.*, **2**, 811–814, 1984.

Waldron, T. P., S. K. Chin and R. J. Naster: "Distributed Beamsteering Control of Phased Array Radars," *Microwave J.*, **29**, 133–146, September 1986.

Weber, M. E., and R. Heisler: "A Frequency-Domain Beamforming Algorithm for Wideband Coherent Signal Processing," *J. Acoustic Society of America*, **76**, October 1984.

Wheeler, H. A.: "The Grating-Lobe Series for the Impedance Variation in a Planar Phased-Array Antenna," *IEEE Trans. Ants. Prop.*, **AP-14**, 707–714, November 1966.

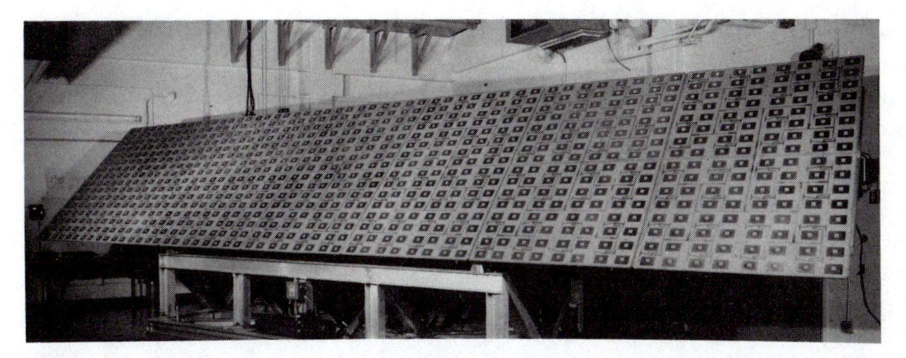

Figure 11-54 An 896-element microstrip array for remote sensing the earth from space. The 9.5 × 2.4 m array has 34 dBi gain at $\lambda \simeq 30$ cm. (*Courtesy Robert Munson, Ball Aerospace, Boulder, Colorado.*)

11-14 MICROSTRIP ARRAYS. Printed circuit and microstrip techniques facilitate the construction of multielement arrays for microwave frequencies.[1] The Kraus grid and Tiuri chain arrays (Sec. 11-11*b*) are examples. The 896-element microstrip antenna for space research shown in Fig. 11-54 is another. All elements are photoetched from one side of a printed circuit board. The corporate structure feed has amplitude taper in the narrow dimension. The 9.5 × 2.4 m array has a 34-dBi gain at $\lambda \simeq 30$ cm.

Although the microstrip element is inherently narrowband, log-periodic patch arrays have achieved bandwidths of 4 to 1.[2] Patch (or microstrip) antennas are discussed further in Chap. 16.

11-15 LOW-SIDELOBE ARRAYS. Ideally it may be desirable for an antenna to have a narrow, well-defined beam with no sidelobes, or at least none above a certain prescribed level. A prime factor affecting the sidelobe level is the aperture distribution (Sec. 11-22). Although some distributions yield (theoretically) zero sidelobes, they may be more difficult of realization on reflector antennas than on phased arrays. The typical large parabolic reflector antenna (Chap. 12) may also have significant sidelobes due to other causes such as aperture blocking and diffraction from its prime focus feed or Cassegrain reflector, from struts of the supporting structure, from feed spillover, from surface irregularities and from the edge of the parabola. The edge diffraction may be reduced by using a rolled edge (see Sec. 18-3*d*). Although offset feeds or integral horn reflectors may have significantly lower minor lobes, accurate construction techniques providing precise amplitude and phase control of all elements of a phased

---

[1] H. G. Oltman and D. A. Huebner, "Electromagnetically Coupled Microstrip Dipoles," *IEEE Trans. Ants. Prop.*, **AP-29**, 151–157, January 1981.

[2] P. S. Hall, "Microstrip Antenna Array with Multi-Octave Bandwidth," *Microwave J.*, **29**, 133–139, March 1986.

array have resulted in phased array designs with sidelobe levels down 50 dB or more. This is as good or better than has been achieved to date with reflector antennas.[1] However, phased arrays are inherently narrower band than reflector antennas.

11-16 LONG-WIRE ANTENNAS. Most of the preceding parts of this chapter deal with arrays of individual, discrete elements (usually $\lambda/2$ long) interconnected by transmission lines. A linear wire antenna, many wavelengths long, may also be regarded as an array of $\lambda/2$ elements but connected in a continuous linear fashion with each element serving as both a radiator and a transmission line. The long-wire antennas discussed in this section are the V, rhombic and Beverage types. The V antenna may be either unterminated (with standing wave) or terminated (with traveling wave). The rhombic and Beverage antennas are almost always terminated (with traveling wave).

11-16a V Antennas.[2] By assuming a sinusoidal (standing-wave) current distribution, the pattern of a long thin wire antenna can be calculated as described in Chap. 5. A typical pattern is shown in Fig. 11-55*a* for a wire 2λ long. The main lobes are at an angle $\beta = 36°$ with respect to the wire. By arranging two such wires in a V with an included angle $\gamma = 72°$ as in Fig. 11-55*b*, a bidirectional pattern can be obtained. This pattern is the sum of the patterns of the individual wires or legs. Although an included angle $\gamma = 2\beta$ results in the alignment of the major lobes at zero elevation angle (wires horizontal) and in free space, it is necessary to make γ somewhat less than 2β in order to obtain alignment at elevation angles greater than zero.[3] This is because the space pattern of a single wire is conical, being obtained by revolving the pattern of Fig. 11-55*a*, for example, with the wire as the axis.

If the legs of the thin-wire V antenna are terminated in their characteristic impedance, as in Fig. 11-55*c*, so that the wires carry only an outgoing traveling wave, the back-radiation is greatly reduced. The patterns of the individual wires can be calculated, assuming a single traveling wave as done in Chap. 5.

A similar effect may be produced without terminations by the use of V conductors of considerable thickness. The reflected wave on such a conductor may be small compared to the outgoing wave, and a condition approaching that of a single traveling (outgoing) wave may result. For example, a V antenna con-

[1] H. E. Schrank, "Low Sidelobe Phased Array Antennas," *IEEE Ant. Prop. Soc. Newsletter*, **25**, 5–9, April 1983.

[2] P. S. Carter, C. W. Hansell and N. E. Lindenblad, "Development of Directive Transmitting Antennas by R. C. A. Communications, Inc.," *Proc. IRE*, **19**, 1773–1842, October 1931.

P. S. Carter, "Circuit Relations in Radiating Systems and Applications to Antenna Problems," *Proc. IRE*, **20**, 1004–1041, June 1932.

[3] *The A.R.R.L. Antenna Book*, American Radio Relay League, West Hartford, Conn., 1984, p. 7-4.

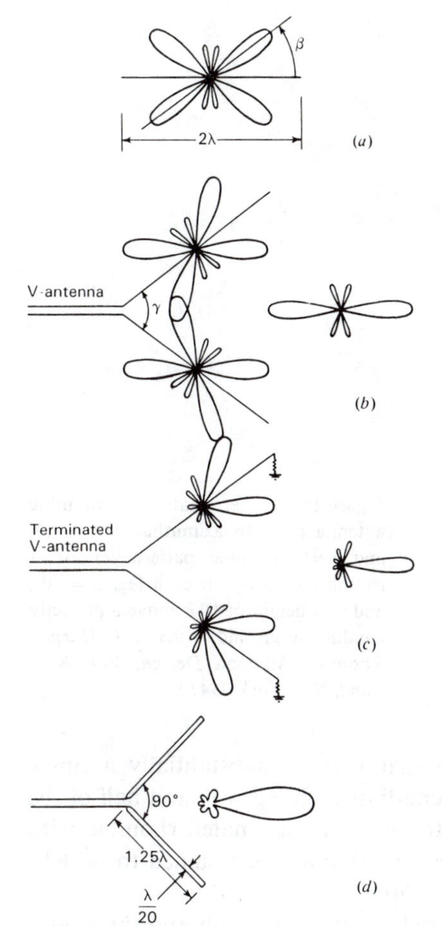

Figure 11-55 (*a*) Calculated pattern of 2λ wire with standing wave, (*b*) V antenna of two such wires, (*c*) terminated V antenna with legs 2λ long and (*d*) V antenna of cylindrical conductors 1.25λ long with measured pattern.

sisting of two cylindrical conductors 1.25λ long and $\lambda/20$ diameter with an included angle $\gamma = 90°$ has the highly unidirectional pattern[1] of Fig. 11-55*d*.

11-16b Rhombic Antennas.[2] A rhombic antenna may be regarded as a double-V type. The wires at the end remote from the feed end are in close proximity, as in Fig. 11-56*a*. A terminating resistance, usually 600 to 800 Ω, can be

---

[1]A. Dorne, in *Very High Frequency Techniques*, Radio Research Laboratory Staff, McGraw-Hill, New York, 1947, chap. 4, p. 115.

[2] E. Bruce, "Development in Short-wave Directive Antennas," *Proc. IRE*, **19**, 1406–1433, August 1931.

E. Bruce, A. C. Beck and L. R. Lowry, "Horizontal Rhombic Antennas," *Proc. IRE*, **23**, 24–46, January 1935.

A. E. Harper, *Rhombic Antenna Design*, Van Nostrand, New York, 1941.

Donald Foster, "Radiation from Rhombic Antennas," *Proc. IRE*, **25**, 1327–1353, October 1937.

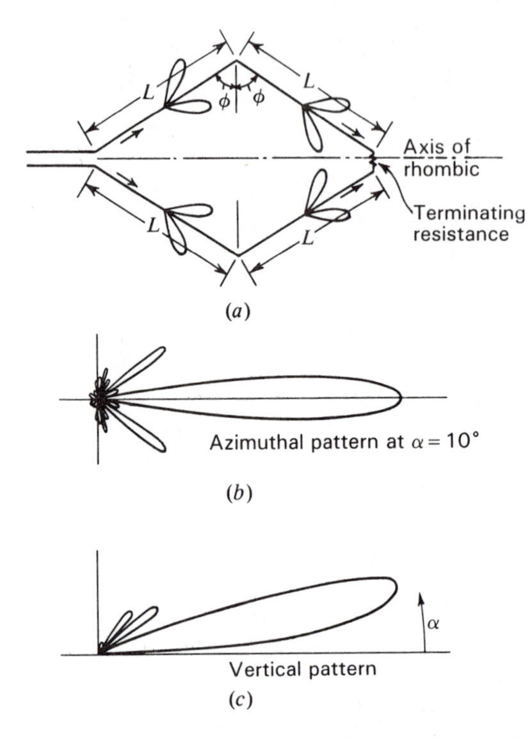

Figure 11-56 Terminated rhombic antenna (*a*) with azimuthal pattern (*b*) and vertical plane pattern (*c*) for a rhombic 6λ long on each leg, $\phi = 70°$, and at a height of 1.1λ above a perfectly conducting ground. (*After A. E. Harper,* Rhombic Antenna Design, *Van Nostrand, New York, 1941.*)

conveniently connected at this location so that there is substantially a single outgoing traveling wave on the wires. The length of each leg is L, and half of the included side angle is ϕ. The calculated patterns[1] of a terminated rhombic with legs 6λ long are shown in Fig. 11-56*b* and *c*. The rhombic is assumed to be 1.1λ above a perfectly conducting ground, and $\phi = 70°$.

In designing a rhombic antenna, the angle ϕ, the leg length and the height above ground may be so chosen that (1) the maximum of the main lobe coincides with the desired elevation angle α (*alignment design*) or (2) the maximum relative field intensity E for a constant antenna current is obtained at the desired elevation angle α (*maximum E design*).[2]

If the height above ground is less than that required for these designs, alignment may be obtained by increasing the leg length. If the height is maintained but the leg length is reduced, alignment may be obtained by changing the angle ϕ. As a third possibility, if both the height and the leg length are reduced, the angle ϕ can be changed to produce alignment. Any of these 3 modifications results in a so-called *compromise design*[2] having reduced gain. If moderate departures from optimum performance are acceptable, a rhombic antenna can be operated without adjustment over a frequency band of the order of 2 to 1.

[1] From A. E. Harper, *Rhombic Antenna Design*, Van Nostrand, New York, 1941.

The pattern of a rhombic antenna may be calculated as the sum of the patterns of four tilted wires each with a single outgoing traveling wave. The effect of a perfectly conducting ground may be introduced by the method of images. For a horizontal rhombic of perfectly conducting wire above a perfectly conducting plane ground, Bruce, Beck and Lowry[2] give the relative field intensity E in the vertical plane coincident with the rhombic axis[3] as a function of α, ϕ, L_λ and H_λ as

$$E = \frac{(\cos\phi)[\sin(H_r \sin\alpha)][\sin(\psi L_r)]^2}{\psi} \tag{1}$$

where α = elevation angle with respect to ground
ϕ = half included side angle of rhombic antenna
$H_\lambda = H/\lambda$ = height of rhombic antenna above ground
$L_\lambda = L/\lambda$ = leg length
$H_r = 2\pi H_\lambda = 2\pi(H/\lambda)$
$L_r = 2\pi L_\lambda = 2\pi L/\lambda$
$\psi = (1 - \sin\phi\cos\alpha)/2$

A uniform antenna current is assumed and mutual coupling is neglected.

Following the procedure of Bruce, Beck and Lowry, the various parameters may be determined as follows. For the maximum E condition, E is maximized with respect to H_λ, that is, we make

$$\frac{\partial E}{\partial H_\lambda} = 0 \tag{2}$$

which yields

$$\cos(2\pi H_\lambda \sin\alpha) = 0$$

which is satisfied when

$$2\pi H_\lambda \sin\alpha = n\frac{\pi}{2}$$

where $n = 1, 3, 5, \ldots$

For the lowest practical height, $n = 1$. Therefore,

$$H_\lambda = \frac{1}{4\sin\alpha} \tag{3}$$

[2] E. Bruce, A. C. Beck and L. R. Lowry, "Horizontal Rhombic Antennas," *Proc. IRE*, **23**, 24–26, January 1935.

[3] The radiation in this plane is horizontally polarized. However, in other planes the polarization is not, in general, horizontal.

Equation (3) gives the height H_λ for the antenna. To find the leg length, E is maximized with respect to L_λ, obtaining

$$L_\lambda = \frac{1}{2(1 - \sin \phi \cos \alpha)} \tag{4}$$

Finally, by maximizing E with respect to ϕ and introducing the condition of (4),

$$\phi = 90° - \alpha \tag{5}$$

Substituting (5) back into (4) yields

$$L_\lambda = \frac{1}{2 \sin^2 \alpha} \tag{6}$$

Equations (3), (5) and (6) then give the height in wavelengths H_λ, the half-side angle ϕ and the leg length in wavelengths L_λ, for *maximum* E at the desired elevation angle α. This is for a uniform antenna current. It does not follow that the field intensity at the desired elevation angle is a maximum for a given power input to the antenna. However, it is probably very close to this condition. It is also of interest that for the maximum E condition the maximum point of the main lobe of radiation is not, in general, aligned with the desired elevation angle.

In the *alignment design* the maximum point of the main lobe of radiation is aligned with the desired elevation angle α. For this condition, E at α is slightly less than for the maximum E condition. Alignment is accomplished by maximizing E with respect to α and introducing the condition of (3). This gives

$$L_\lambda = \frac{0.371}{1 - \sin \phi \cos \alpha} \tag{7}$$

Substituting (7) in (1) and maximizing the resulting relation for the field with respect to ϕ gives

$$\phi = 90° - \alpha \tag{8}$$

as before. Finally substituting (8) in (7) we obtain

$$L_\lambda = \frac{0.371}{\sin^2 \alpha} \tag{9}$$

Equations (3), (8) and (9) then give H_λ, ϕ and L_λ for *alignment* of the maximum point of the main lobe of radiation with the desired elevation angle α. Only the length is different in the alignment design, being $0.371/0.5 = 0.74$ of the value for the maximum E design.

The above design relations are summarized in Table 11-1 together with design formulas for 3 kinds of compromise designs.

An end-to-end receiving array of a number of rhombics may be so connected as to provide an electrically controllable vertical plane pattern which can

Table 11-1 Design formulas for terminated rhombic antennas†

Type of rhombic antenna	Formulas
Maximum E at elevation angle α	$H_\lambda = \dfrac{1}{4 \sin \alpha}$ $\phi = 90° - \alpha$ $L_\lambda = \dfrac{0.5}{\sin^2 \alpha}$
Alignment of major lobe with elevation angle α	$H_\lambda = \dfrac{1}{4 \sin \alpha}$ $\phi = 90° - \alpha$ $L_\lambda = \dfrac{0.371}{\sin^2 \alpha}$
Reduced height H' Compromise design for alignment at elevation angle α	$\phi = 90° - \alpha$ $L_\lambda = \dfrac{\tan [(\pi L_\lambda) \sin^2 \alpha]}{\sin \alpha} \left[\dfrac{1}{2\pi \sin \alpha} - \dfrac{H'_\lambda}{\tan (H'_r \sin \alpha)}\right]$ where $H'_\lambda = \dfrac{H'}{\lambda}$ and $H'_r = 2\pi \dfrac{H'}{\lambda}$
Reduced length L' Compromise design for alignment at elevation angle α	$H_\lambda = \dfrac{1}{4 \sin \alpha}$ $\phi = \arcsin \left[\dfrac{L'_\lambda - 0.371}{L'_\lambda \cos \alpha}\right]$ where $L'_\lambda = L'/\lambda$
Reduced height H' and length L' Compromise design for alignment at elevation angle α	Solve this equation for ϕ: $\dfrac{H'_\lambda}{\sin \phi \tan \alpha \tan (H'_r \sin \alpha)} = \dfrac{1}{4\pi\psi} - \dfrac{L'_\lambda}{\tan (\psi L'_r)}$ where $\psi = \dfrac{1 - \sin \phi \cos \alpha}{2}$ and $L'_r = 2\pi \dfrac{L'}{\lambda}$

† After E. Bruce, A. C. Beck, and L. R. Lowry, "Horizontal Rhombic Antennas," *Proc. IRE*, **23**, 24–26, January 1935.

be adjusted to coincide with the optimum elevation angle of downcoming waves. This Multiple Unit Steerable Antenna,[1] or MUSA, is a vertically steerable system of this kind for long-distance short-wave reception of horizontally polarized downcoming waves.

[1] H. T. Friis and C. B. Feldman, "A Multiple Unit Steerable Antenna for Short-Wave Reception," *Proc. IRE*, **25**, 841–917, July 1937.

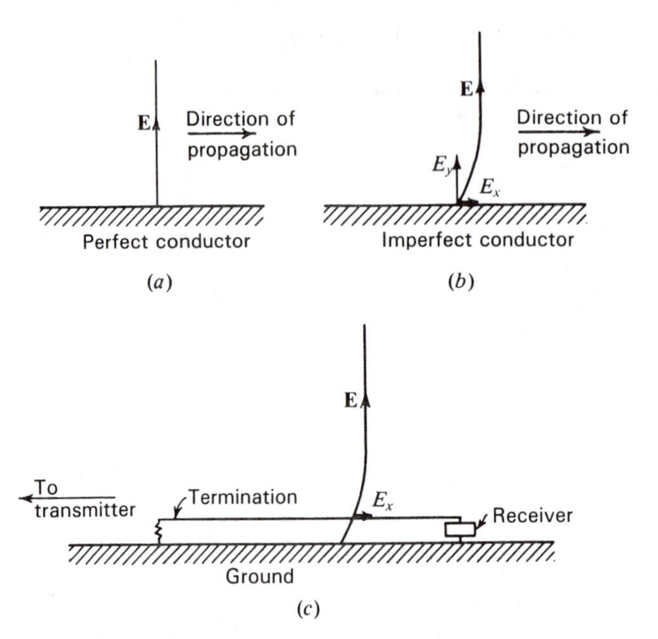

Figure 11-57 (*a*) Wave front over a perfect conductor. (*b*) Wave front over imperfect conductor. (*c*) Beverage antenna.

11-16c Beverage Antennas. The electric field of a wave traveling along a perfectly conducting surface is perpendicular to the surface as in Fig. 11-57*a*. However, if the surface is an imperfect conductor, such as the earth's surface or ground, the electric field lines have a forward tilt near the surface as in Fig. 11-57*b*. Hence, the field at the surface has a vertical component E_y and a horizontal component E_x.[1] The component E_x is associated with that part of the wave that enters the surface and is dissipated as heat. The E_y component continues to travel along the surface.

The fact that a horizontal component E_x exists is applied in the wave antenna of Beverage, Rice and Kellogg for receiving vertically polarized waves.[2] This antenna consists of a long horizontal wire terminated in its characteristic impedance at the end toward the transmitting station as in Fig. 11-57*c*. The ground acts as the imperfect conductor. The emfs induced along the antenna by the E_x component, as the wave travels toward the receiver, all add up in the same phase at the receiver. Energy from a wave arriving from the opposite direction is

[1] Actually the wave exhibits elliptical cross-field, i.e., the electric vector describes an ellipse whose plane is parallel to the direction of propagation. However, the axial ratio of this ellipse is usually very large.

[2] H. H. Beverage, C. W. Rice and E. W. Kellogg, "The Wave Antenna, a New Type of Highly Directive Antenna," *Trans. AIEE*, **42**, 215, 1923.

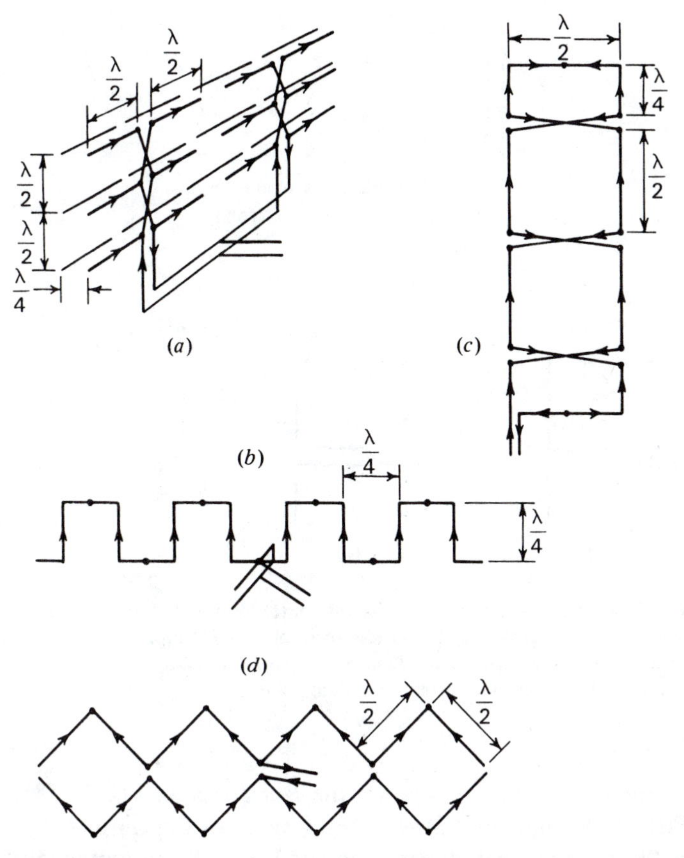

Figure 11-58 (*a*) Array of $\lambda/2$ dipoles with reflectors, (*b*) symmetrical Bruce antenna, (*c*) Sterba curtain array and (*d*) Chireix-Mesny array. Arrows indicate instantaneous current directions and dots indicate current minimum points.

largely absorbed in the termination. Hence, the antenna exhibits a directional pattern in the horizontal plane with maximum response in the direction of the termination (to the left in Fig. 11-57*c*). The Beverage antenna finds application as a receiving antenna in the low- and medium-frequency range.

11-17 CURTAIN ARRAYS. In short-wave communications the curtain type of array finds many applications. As an example, a curtain type is illustrated in Fig. 11-58*a* that consists of an array of $\lambda/2$ dipoles with a similar curtain at a distance of about $\lambda/4$ acting as a reflector.[1] If the array is large in terms of wavelengths, the reflector curtain is nearly equivalent to a large sheet reflector.

[1] H. Brückmann, *Antennen, ihre Theorie und Technik*, S. Hirzel, Leipzig, 1939, p. 300.

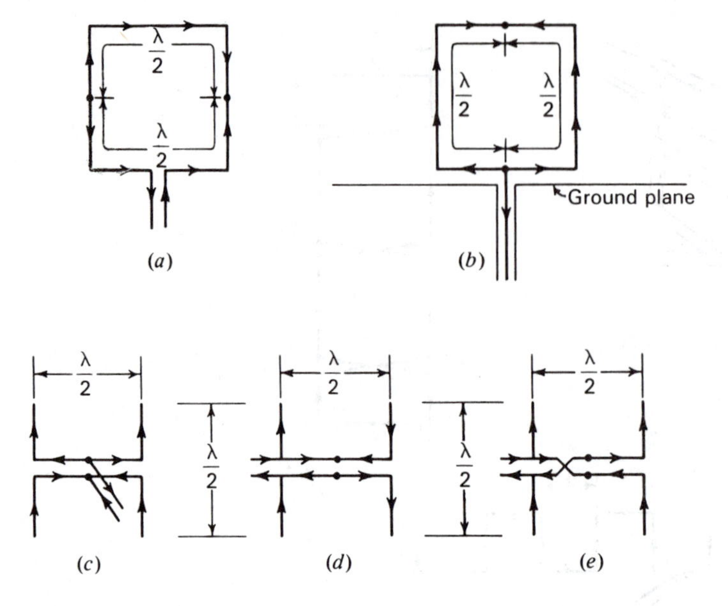

Figure 11-59 (*a*) Loop with 2-wire feed for horizontal polarization, (*b*) loop with 1-wire feed from coaxial line for vertical polarization, (*c*) center-fed broadside array of two $\lambda/2$ dipoles, (*d*) end-fed end-fire array of two $\lambda/2$ dipoles and (*e*) end-fed broadside array of two $\lambda/2$ dipoles. Arrows indicate instantaneous current directions and dots indicate current minimum points.

Several other examples of curtain arrays are the *Bruce type* of Fig. 11-58*b*, the *Sterba type*[1] of Fig. 11-58*c* and the *Chireix-Mesny type*[2] of Fig. 11-58*d*. The arrows are located at or near current maxima and indicate the instantaneous current direction. The small dots indicate the locations of current minima.

11-18 LOCATION AND METHOD OF FEEDING ANTENNAS.

It is interesting to note the effect that the method and location of feeding has on the characteristics of an antenna. As illustrations, let us consider the following cases.

If an antenna is fed with a balanced 2-wire line, equal out-of-phase currents must flow at the feed point. Thus, a square loop 1λ in perimeter and fed at the bottom as in Fig. 11-59*a* must have the current distribution indicated. The

[1] E. J. Sterba, "Theoretical and Practical Aspects of Directional Transmitting Systems," *Proc. IRE*, **19**, 1184–1215, July 1931.

[2] H. Chireix, "French System of Directional Aerials for Transmission on Short Waves," *Exp. Wireless and Wireless Engr.*, **6**, 235, May 1929.

arrows indicate the instantaneous current directions and the dots the locations of current minima. The radiation normal to this loop is horizontally polarized.

Consider now the situation shown in Fig. 11-59*b*. Here the loop is fed at the same location. However, the loop is continuous and is fed at a point by an unbalanced line. In this case, the antenna currents flowing to the feed point are equal and in phase, so that the current distribution on the antenna must be as indicated. The radiation normal to this loop is vertically polarized.

The location at which an antenna is energized also may be important. For example, two $\lambda/2$ elements have in-phase currents when symmetrically fed as in Fig. 11-59*c* but out-of-phase currents when fed from one end as in Fig. 11-59*d*. For the currents to be in phase when the array is fed from one end requires that the line between the elements be transposed as in Fig. 11-59*e*.

11-19 FOLDED DIPOLE ANTENNAS. A simple $\lambda/2$ dipole has a terminal resistance of about 70 Ω so that an impedance transformer is required to match this antenna to a 2-wire line of 300 to 600 Ω characteristic impedance. However, the terminal resistance of the modified $\lambda/2$ dipole shown in Fig. 11-60*a* is nearly 300 Ω so that it can be directly connected to a 2-wire line having a characteristic impedance of the same value. This "ultra close-spaced type of array" is called a *folded dipole*. More specifically the one in Fig. 11-60*a* is a 2-wire folded $\lambda/2$ dipole. The antenna consists of 2 closely spaced $\lambda/2$ elements connected together at the outer ends. The currents in the elements are substantially equal and in phase.

Assuming that both conductors of the dipole have the same diameter, the approximate value of the terminal impedance may be deduced very simply as

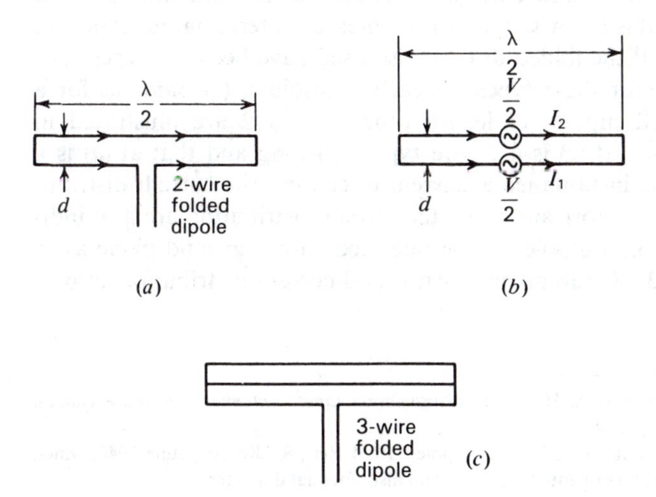

Figure 11-60 Folded dipoles.

follows.[1] Let the emf V applied to the antenna terminals be divided between the 2 dipoles as in Fig. 11-60*b*. Then

$$\frac{V}{2} = I_1 Z_{11} + I_2 Z_{12} \tag{1}$$

where I_1 = current at terminals of dipole 1
I_2 = current at terminals of dipole 2
Z_{11} = self-impedance of dipole 1
Z_{12} = mutual impedance of dipoles 1 and 2

Since $I_1 = I_2$, (1) becomes

$$V = 2I_1(Z_{11} + Z_{12}) \tag{2}$$

Further, since the 2 dipoles are close together, usually d is of the order of $\lambda/100$, $Z_{12} \simeq Z_{11}$. Thus, the terminal impedance Z of the antenna is given by

$$Z = \frac{V}{I_1} \simeq 4Z_{11} \tag{3}$$

Taking $Z_{11} \simeq 70 + j0\ \Omega$ for a $\lambda/2$ dipole, the terminal impedance of the 2-wire folded dipole becomes

$$Z \simeq 280\ \Omega$$

For a 3-wire folded $\lambda/2$ dipole as in Fig. 11-60*c* the terminal resistance calculated in this way is $9 \times 70 = 630\ \Omega$. In general, for a folded $\lambda/2$ dipole of N wires, the terminal resistance is $70N^2\ \Omega$. Equal currents in all wires are assumed.

Several other types of folded-wire antennas[2] are shown in Fig. 11-61. The one at (*a*) is a 3-wire type which differs from the one in Fig. 11-60*c* in that there are no closed loops. The measured terminal resistance of this antenna is about 1200 Ω. The antenna at (*b*) is a 4-wire type with a measured terminal resistance of about 1400 Ω. Thus far, all the folded dipoles discussed have been $\lambda/2$ types. The total current distribution for these types is nearly sinusoidal, the same as for a simple $\lambda/2$ dipole. Folded dipoles of length other than $\lambda/2$ are illustrated in Fig. 11-61*c* and *d*. The one at (*c*) is a 2-wire type $3\lambda/4$ long and that at (*d*) is a 4-wire type $3\lambda/8$ long. The instantaneous current directions, the current distribution on the individual conductors and the total current distribution are also indicated. Half of the 2-wire $3\lambda/4$ dipole can be operated with a ground plane as in Fig. 11-61*e*, yielding the $3\lambda/8$ stub antenna with total current distribution shown.

---

[1] R. W. P. King, H. R. Mimno and A. H. Wing, *Transmission Lines, Antennas and Wave Guides*, McGraw-Hill, New York, 1945, p. 224.
W. V. B. Roberts, "Input Impedance of a Folded Dipole," *RCA Rev.*, **8**, 289–300, June 1947, which treats folded dipoles with conductors of equal diameter and also unequal diameter.

[2] J. D. Kraus, "Multi-wire Dipole Antennas," *Electronics*, **13**, 26–27, January 1940.

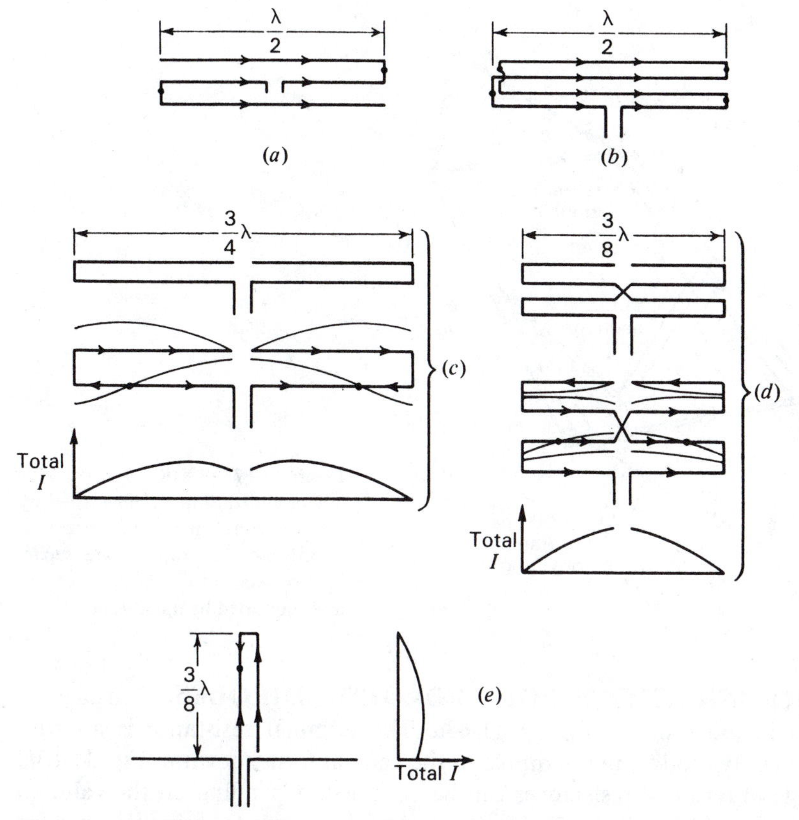

Figure 11-61 (*a*) Three-wire folded $\lambda/2$ dipole, (*b*) 4-wire folded $\lambda/2$ dipole, (*c*) 2-wire $3\lambda/4$ antenna, (*d*) 4-wire $3\lambda/8$ antenna and (*e*) 2-wire $3\lambda/8$ stub antenna. Arrows indicate instantaneous current directions and dots indicate current minimum points. (*After Kraus.*)

The measured terminal resistance of the 2-wire $3\lambda/4$ dipole is about 450 Ω, of the 4-wire $3\lambda/8$ dipole about 225 Ω and of the 2-wire $3\lambda/8$ stub antenna about 225 Ω.

An application of the 3-wire folded $\lambda/2$ dipole of Fig. 11-61*a* to a W8JK array with $\lambda/5$ spacing is shown in Fig. 11-62.[1] The impedance of each folded dipole in free space is about 1200 Ω (resistive) but in the array is reduced to 300 Ω, which transforms via a $\lambda/4$ 600-Ω line to 1200 Ω. At the junction of the two transformers the impedance is half 1200 Ω, or 600 Ω, matching a 600-Ω line to the transmitter or receiver. Thus, the W8JK array is fed entirely by lines of constant impedance (600 Ω) with no resonant stubs or tuning adjustments required.

---

[1] J. D. Kraus, "Twin-Three Flat-Top Beam Antenna," *Radio*, no. 243, 10–16, November 1939.

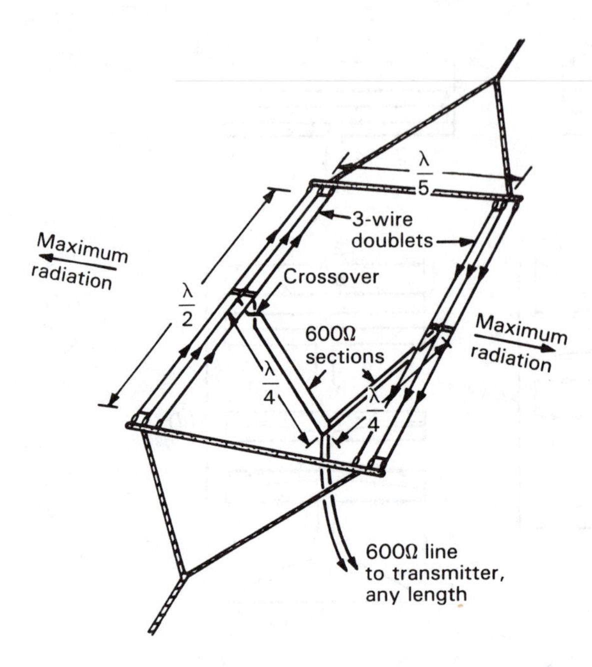

Figure 11-62 W8JK array with 3-wire folded dipole elements fed by transmission lines of constant impedance. The dipoles are separated by wooden or plastic spreaders and supported by nylon rope.

11-20 MODIFICATIONS OF FOLDED DIPOLES. Consider a 2-wire folded dipole shown in Fig. 11-63*a*. The terminal resistance is approximately 300 Ω. By modifying the dipole to the general form shown in Fig. 11-63*b*, a wide range of terminal resistances can be obtained, depending on the value of D. This arrangement is called a *T-match antenna.*[1] Dimensions in wavelengths for providing an impedance match to a 600 Ω line are shown in Fig. 11-63*c*.

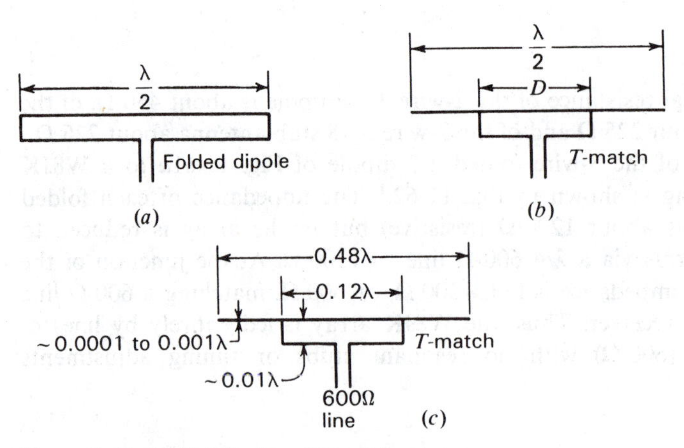

Figure 11-63 Folded dipole and T-match antennas.

[1] J. D. Kraus and S. S. Sturgeon, "The T-Matched Antenna," *QST*, **24**, 24–25, September 1940.

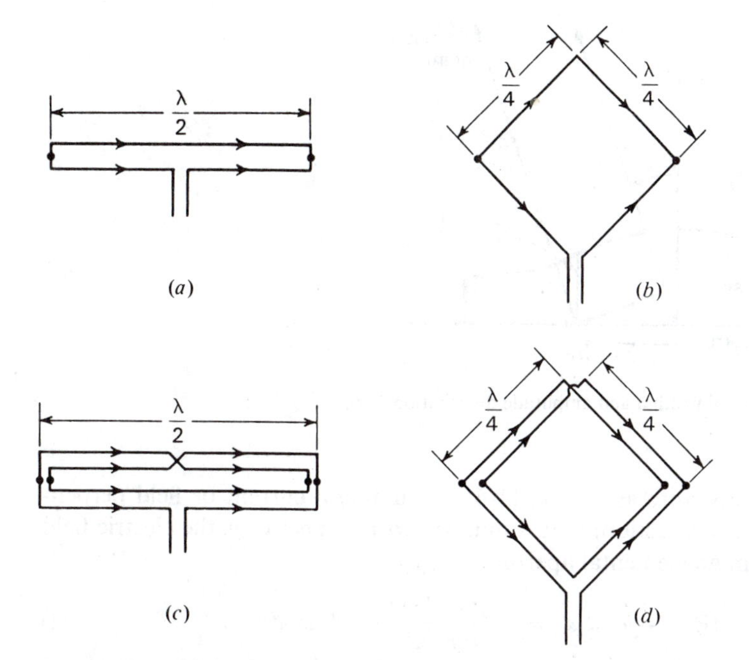

Figure 11-64 (*a*) Two-wire folded dipole and (*b*) as modified to form single-turn loop. (*c*) Four-wire folded dipole and (*d*) as modified to form 2-turn loop.

A 2-wire folded $\lambda/2$ dipole is also shown in Fig. 11-64*a*. The arrows indicate the instantaneous current direction and the small dots indicate the locations of current minima. By pulling the dipole wires apart at the center, the single-turn loop antenna of Fig. 11-64*b* is obtained. The length of each side is $\lambda/4$. The loop has a lower terminal resistance than the folded dipole.

A 4-wire folded $\lambda/2$ dipole is shown in Fig. 11-64*c*. This dipole is the same type as shown in Fig. 11-61*b*. It is, however, sketched in a different manner. By pulling this dipole apart at the center the 2-turn loop or *quad antenna* of Fig. 11-64*d* results.

The directivity of all the types shown in Fig. 11-64 is nearly the same as for a simple $\lambda/2$ dipole, although the types at (*b*) and (*d*) are equivalent to 2 horizontal dipoles stacked $\sim 0.18\lambda$ giving a small increase. With the loop types vertical and the terminals at the lowest corner, the radiation normal to the plane of the loops is horizontally polarized.

11-21 CONTINUOUS APERTURE DISTRIBUTION.[1]

Extending our discussion of Sec. 4-14, consider now a continuous-current sheet or field dis-

[1] The following sections (11-21 through 11-25) are from J. D. Kraus, *Radio Astronomy*, 2nd ed., Cygnus-Quasar, 1986.

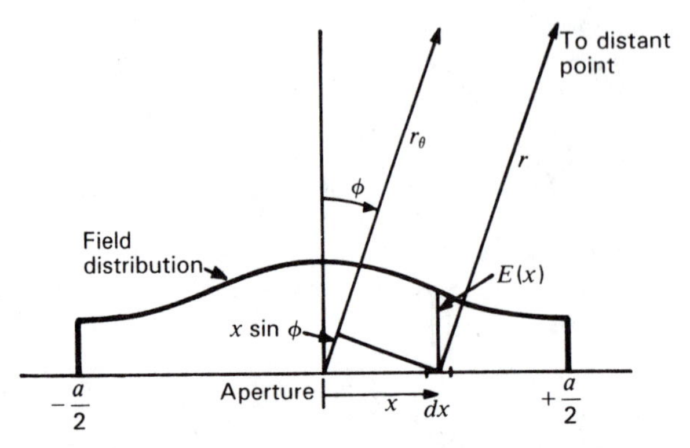

Figure 11-65 Aperture of width a and amplitude distribution $E(x)$.

tribution over an aperture as in Fig. 11-65. Assuming a current or field perpendicular to the page (y direction) that is uniform with respect to y, the electric field at a distance r from an elemental aperture $dx\,dy$ is[1]

$$dE = -j\omega\, d\mathscr{A}_y = -\frac{j\omega\mu}{4\pi r}\frac{E(x)}{Z}e^{-j\beta r}\,dx\,dy \tag{1}$$

where $\mathscr{A}_y$ = vector potential $\left(= \frac{\mu}{4\pi}\iiint \frac{J_y}{r}\,dv, \text{ in general}\right)$, V s m^{-1}

J_y = current density, A m^{-2}

$E(x)$ = aperture electric-field distribution, V m^{-1}

Z = intrinsic impedance of medium, Ω square^{-1}

$\omega = 2\pi f$ (f = frequency), rad s^{-1}

μ = permeability of medium, H m^{-1}

For an aperture with a uniform dimension y_1 perpendicular to the page and with the field distribution over the aperture a function only of x, the electric field as a function of ϕ at a large distance from the aperture ($r \gg a$) is, from (1),

$$E(\phi) = \frac{-j\omega\mu y_1 e^{-j\beta r_0}}{4\pi r_0 Z}\int_{-a/2}^{+a/2} E(x)e^{j\beta x \sin\phi}\,dx \tag{2}$$

The magnitude of $E(\phi)$ is then

$$|E(\phi)| = \frac{y_1}{2r_0\lambda}\int_{-a/2}^{+a/2} E(x)e^{j\beta x \sin\phi}\,dx \tag{3}$$

---

[1] Note that $E(x)/Z = E_y(x)/Z = H_x = J_y z$, where z (up in Fig. 11-65) is the thickness of the current sheet. Also $dE = dE_y$.

where $\beta = 2\pi/\lambda$. For a uniform aperture distribution $[E(x) = E_a]$, (3) reduces to

$$|E(\phi)| = \frac{y_1 E_a}{2r_0 \lambda} \int_{-a/2}^{+a/2} e^{j\beta x \sin\phi}\, dx \tag{4}$$

and on axis ($\phi = 0$) we have

$$|E(\phi)| = \frac{E_a a y_1}{2r_0 \lambda} = \frac{E_a A}{2r_0 \lambda} \tag{5}$$

where A = aperture area $(= ay_1)$
E_a = electric field in aperture plane

For unidirectional radiation from the aperture (in direction $\phi = 0$ but not in direction $\phi = 180°$), $|E(\phi)|$ is twice the value given in (5).

Integration of (4) gives

$$|E(\phi)| = k_0 \frac{\sin[(\beta a/2)\sin\phi]}{(\beta a/2)\sin\phi} \tag{6}$$

where

$$k_0 = \frac{AE_a}{2r_0 \lambda} \tag{7}$$

From (4-14-17) the field of a long array of n discrete sources of spacing d is

$$E = nE_0 \frac{\sin[(\beta a'/2)\sin\phi]}{(\beta a'/2)\sin\phi} \tag{8}$$

where the length of the long array is $a' = (n-1)d \simeq nd$ and E_0 = field of one source. It is also assumed in (8) that ϕ is restricted to small angles. This is not an undue restriction if the array is large and only the main lobe and first sidelobes are of interest. Under these conditions it is clear that the field pattern (8) of the long array of discrete sources is the same as the pattern (6) for the continuous array of the same length ($a = a'$) (as already noted in Sec. 4-14).

11-22 FOURIER TRANSFORM RELATIONS BETWEEN THE FAR-FIELD PATTERN AND THE APERTURE DISTRIBUTION.

According to Booker and Clemmow[1] a 1-dimensional aperture distribution $E(x_\lambda)$ and its far-field distribution $E(\sin\phi)$ are reciprocal Fourier transforms as given

[1] H. G. Booker and P. C. Clemmow, "The Concept of an Angular Spectrum of Plane Waves and Its Relation to That of Polar Diagram and Aperture Distribution," *Proc. Instn. Elec. Engrs. London*, ser. 3, **97**, 11–17, January 1950.
For the more general 2-dimensional case see R. N. Bracewell, "Radio Astronomy Techniques," in S. Flügge (ed.), *Handbuch der Physik*, vol. 54, Springer-Verlag OHG, Berlin, 1962, pp. 42–129.

by

$$E(\sin\phi) = \int_{-\infty}^{\infty} E(x_\lambda)e^{j2\pi x_\lambda \sin\phi}\, dx_\lambda \tag{1}$$

and

$$E(x_\lambda) = \int_{-\infty}^{\infty} E(\sin\phi)e^{-j2\pi x_\lambda \sin\phi}\, d(\sin\phi) \tag{2}$$

where $x_\lambda = x/\lambda$. For real values of ϕ, $|\sin\phi| \leq 1$, the field distribution represents radiated power, while for $|\sin\phi| > 1$ it represents reactive or stored power.[1] The field distribution $E(\sin\phi)$, or angular spectrum, refers to an angular distribution of plane waves. Except for $|\sin\phi| > 1$ the angular spectrum for a finite aperture is the same as the far-field pattern $E(\phi)$ (the far-field condition $r \gg a$ does not hold for an infinite aperture, i.e., where $a = \infty$). Thus, for a finite aperture the Fourier integral representation of (1) may be written

$$E(\phi) = \int_{-a_\lambda/2}^{+a_\lambda/2} E(x_\lambda)e^{j2\pi x_\lambda \sin\phi}\, dx_\lambda \tag{3}$$

This is identical with (11-21-2) except for constant factors. Equation (11-21-2) is an absolute relation, whereas (3) is relative. Examples of the far-field patterns $E(\phi)$ for several aperture distributions $E(x_\lambda)$ of the same extent are presented in Fig. 11-66.

Taking the uniform distribution as reference, the more tapered distributions (triangular and cosine) have larger beam widths and smaller minor lobes, while the most gradually tapered distributions (cosine squared and Gaussian) have still larger beam widths but no minor lobes. On the other hand, an inverse taper (less amplitude at the center than at the edge), such as shown in Fig. 11-66*f*, yields a smaller beam width but larger minor lobes than for the uniform distribution. Such an inverse taper might inadvertently result from aperture blocking due to a feed structure in front of the aperture. Carrying the inverse taper to its extreme limit results in the edge distribution of Fig. 11-66*g*. This distribution is equivalent to that of a 2-element interferometer and has a beam width $\frac{1}{2}$ that of the uniform distribution but sidelobes equal in amplitude to the main lobe.

A uniform line source or rectangular aperture distribution (Fig. 11-66*a*) produces the highest directivity. However, the first sidelobe is only about 13 dB down. Thus, aperture distributions used in practice are a trade-off or compromise between desired directivity (or gain) and sidelobe level, as already discussed in Sec. 4-10 and Sec. 4-11 on the Dolph-Tchebyscheff distribution. A table of beam-width and sidelobe level for various rectangular and circular aperture distributions is given in App. A, Sec. A-10.

[1] D. R. Rhodes, "The Optimum Line Source for the Best Mean Square Approximation to a Given Radiation Pattern," *IEEE Trans. Ants. Prop.*, **AP-11**, 440–446, July 1963.

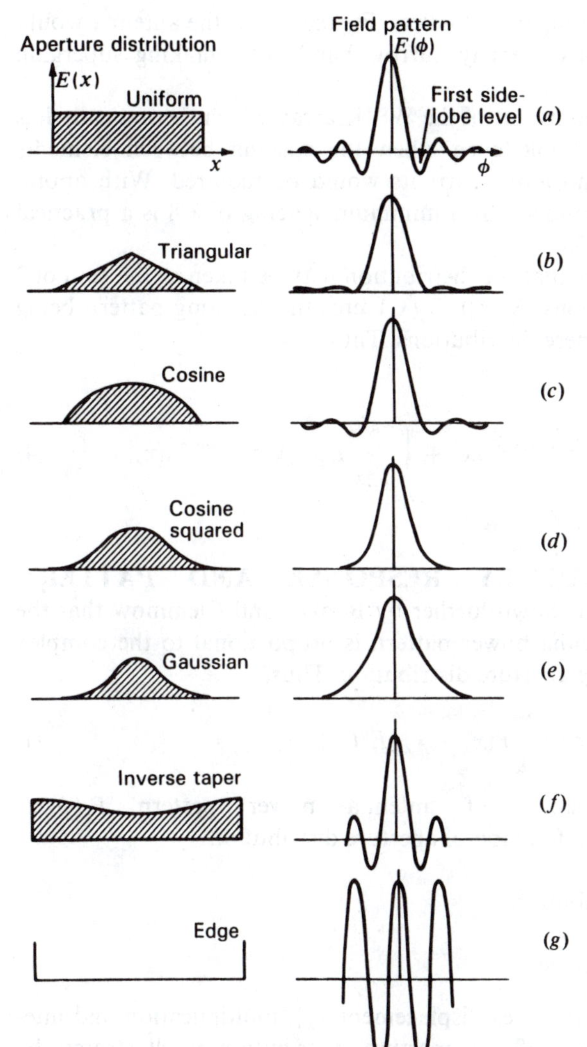

Figure 11-66 Seven different antenna aperture distributions with associated far-field patterns.

Theoretically, the directivity of a uniform aperture distribution can be exceeded (supergain condition) by large field fluctuations near the edges of the aperture. However, according to Rhodes,[1] to obtain a $\frac{1}{2}$ dB increase in directivity from a 5λ aperture requires that the field near the edges be at least 20 times the field for a uniform aperture. The currents to produce these fields would have I^2R losses which offset any gain unless all conductors were perfectly conducting and

[1] D. R. Rhodes, "On an Optimum Line Source for Maximum Directivity," *IEEE Trans. Ants. Prop.*, **AP-19**, 485–492, July 1971.

the surrounding media were completely lossless. Furthermore, the antenna would have an enormous Q and an extremely narrow bandwidth, making supergain attempts impractical.

The situation here is reminiscent of the W8JK array which theoretically has a 4-dB gain over a single dipole even when the spacing between elements approaches zero; however, enormous currents would be required. With appreciable losses and for usable bandwidths a minimum spacing of $\lambda/8$ is a practical limit, as indicated in Fig. 11-13*a*.

A useful property of (3) is that the distribution may be taken as the sum of 2 or more component distributions, $E_1(x_\lambda)$, $E_2(x_\lambda)$, etc., the resulting pattern being the sum of the transforms of these distributions. Thus,

$$E_1(\phi) + E_2(\phi) + \cdots$$

$$= \int_{-a_\lambda/2}^{+a_\lambda/2} E_1(x_\lambda)e^{j2\pi x_\lambda \sin\phi}\, dx_\lambda + \int_{-a_\lambda/2}^{+a_\lambda/2} E_2(x_\lambda)e^{j2\pi x_\lambda \sin\phi}\, dx_\lambda + \cdots \quad (4)$$

11-23 SPATIAL FREQUENCY RESPONSE AND PATTERN SMOOTHING. It has been shown further by Booker and Clemmow that the Fourier transform of the antenna power pattern is proportional to the complex autocorrelation function of the aperture distribution. Thus,

$$\bar{P}(x_{\lambda 0}) \propto \int_{-\infty}^{\infty} E(x_\lambda - x_{\lambda 0})E^*(x_\lambda)\, dx_\lambda \quad (1)$$

where $\bar{P}(x_{\lambda 0})$ = Fourier transform of antenna power pattern $P_n(\phi) \propto$ autocorrelation function of aperture distribution

$E(\phi)$ = field pattern

$E(x_\lambda)$ = aperture distribution

$x_\lambda = x/\lambda$ = distance, λ

$x_{\lambda 0} = x/\lambda_0$ = displacement, λ

The autocorrelation function involves displacement $x_{\lambda 0}$, multiplication and integration. The situation for a uniform aperture distribution is illustrated by Fig. 11-67. The aperture distribution is shown at (*b*) and as displaced by $x_{\lambda 0}$ at (*a*). The autocorrelation function, as shown at (*c*), is proportional to the area under the product curve of the upper 2 distributions or, in this case, to the area of overlap. It is apparent that the autocorrelation function is zero for values of $x_{\lambda 0}$ greater than the aperture width a_λ since $\bar{P}(x_{\lambda 0}) = 0$ for $|x_{\lambda 0}| > a_\lambda$.

If a source moves through the beam of an antenna (or if the source is fixed and the antenna is rotated) the *observed response* of this scanning process is proportional to the convolution of the antenna power pattern and the source brightness distribution. Thus,

$$S(\phi_0) = \int_{-\infty}^{\infty} B(\phi)\tilde{P}_n(\phi_0 - \phi)\, d\phi \quad (2)$$

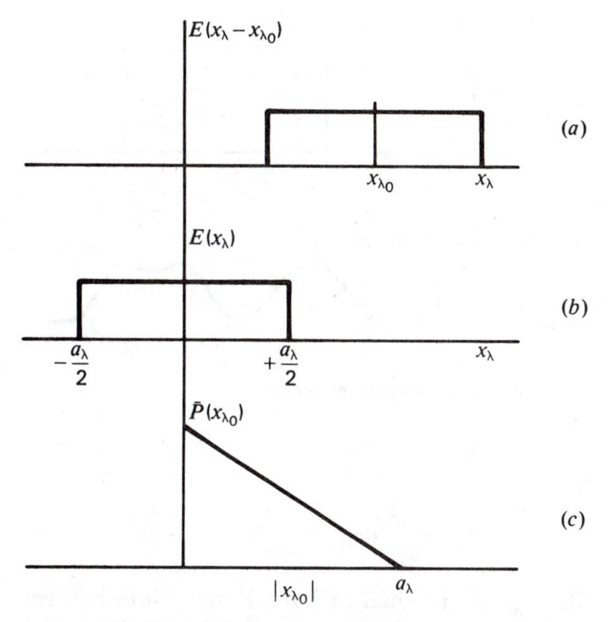

Figure 11-67 The autocorrelation function of the aperture distribution yields the Fourier transform of the antenna pattern.

where $S(\phi_0)$ = observed power distribution, W m^{-2} Hz^{-1}
$B(\phi)$ = true source brightness distribution, W m^{-2} Hz^{-1} sr^{-1}
$\tilde{P}_n(\phi)$ = mirror image of normalized antenna power pattern
ϕ_0 = displacement angle (scan angle)

It follows that

$$\bar{S}(x_\lambda) = \bar{B}(x_\lambda)\bar{P}(x_{\lambda 0}) \tag{3}$$

where the bars mean the Fourier transform. Since $\bar{P}(x_{\lambda 0})$ varies as the autocorrelation function of the aperture distribution, it follows that $\bar{S}(x_\lambda)$ and $S(\phi_0)$ are zero where $\bar{P}(x_{\lambda 0}) = 0$. This means that there is a cutoff for all values of $x_{\lambda 0}$ greater than a_λ.[1] The quantity $x_{\lambda 0}$ is called the *spatial frequency* (wavelengths per aperture) and a_λ its *cutoff value*. Thus,

$$x_{\lambda_c} = a_\lambda \tag{4}$$

where x_{λ_c} = spatial-frequency cutoff

---

[1] R. N. Bracewell and J. A. Roberts, "Aerial Smoothing in Radio Astronomy," *Australian J. Phys.*, 7, 615–640, 1954.

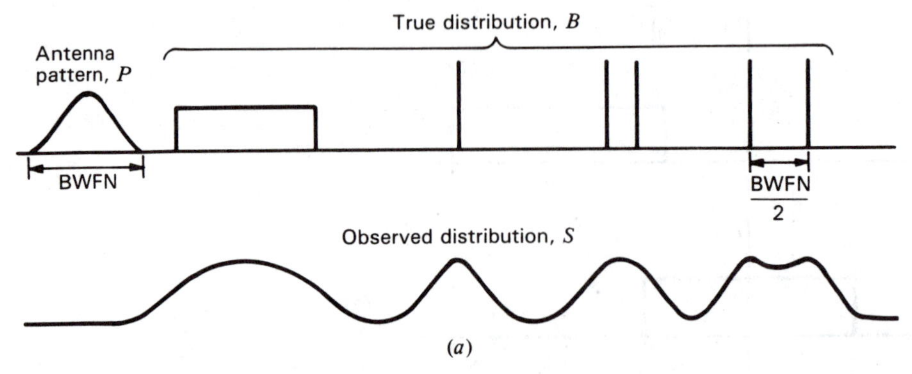

Figure 11-68a Smoothed distribution S observed with antenna pattern P.

The reciprocal of x_{λ_c} gives an angle

$$\phi_c = \frac{1}{a_\lambda} \text{ rad} = \frac{57.3}{a_\lambda} \text{ deg} \tag{5}$$

It follows that this (cutoff) angle ϕ_c is equal to $\frac{1}{2}$ the Beam Width between First Nulls (BWFN) for a uniform aperture distribution (ϕ_c = BWFN/2), and is 12 percent greater than the beam width at half-power (ϕ_c = 1.12 HPBW). The significance of ϕ_c is that structure in the source distribution having a period of less than BWFN/2 will not appear in the observed response. Thus, the antenna tends to smooth the true brightness distribution.[1] This is illustrated in Fig. 11-68a. Half of the beam width between first nulls (BWFN/2) is equal to the *Rayleigh resolution.*[2] Thus, 2 point sources separated by this distance will be just resolved, as indicated at the right in Fig. 11-68a.

The observed half-power width as a function of the source width in half-power beam widths for a large uniform linear-aperture antenna and a uniform 1-dimensional source is shown in Fig. 11-68b. A source of half-power width equal to the antenna half-power beam width produces about 20 percent beam broadening, or an observed width of 1.2 beam widths. For larger source widths the observed width approaches the actual source width. Thus, from the amount of broadening an estimate may be made of the equivalent source extent.

11-24 THE SIMPLE (ADDING) INTERFEROMETER. The resolution of an antenna or of a radio telescope can be improved, for example, by increasing the aperture a. However, this may not be economically feasible. A

[1] R. N. Bracewell and J. A. Roberts, "Aerial Smoothing in Radio Astronomy," *Australian J. Phys.*, **7**, 615–640, 1954.

[2] The *Rayleigh resolution* can, in principle, be improved upon in the signal-processing domain of an adaptive array (see Sec. 11-13).

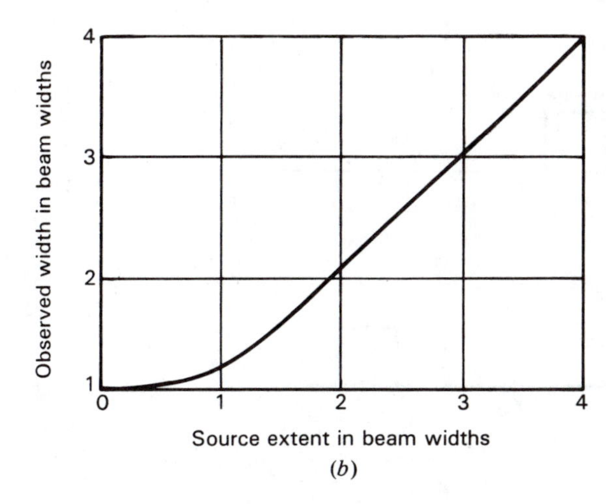

Figure 11-68b Observed half-power width as a function of source width in half-power beam widths for a large uniform linear-aperture antenna and a uniform 1-dimensional source.

common expedient is to use two antennas spaced a distance s apart, as in Fig. 11-69. If each antenna has a uniform aperture distribution of width a, the resulting autocorrelation function is as shown in Fig. 11-70. It is apparent that by making observations with spacings out to s_λ it is possible to obtain higher spatial-frequency components in the observed pattern to a cutoff

$$x_{\lambda_c} = s_\lambda + a_\lambda \tag{1}$$

and a smaller resolution angle

$$\phi_c = \frac{1}{s_\lambda + a_\lambda} \quad \text{(rad)} = \frac{57.3}{s_\lambda + a_\lambda} \quad \text{(deg)} \tag{2}$$

In the following analysis it will be shown that if observations are made to sufficiently large spacings, it is possible, in principle, to deduce the true source distribution.

The normalized far-field pattern of the 2-element array is

$$E(\phi) = E_n(\phi) \cos \frac{\psi}{2} \tag{3}$$

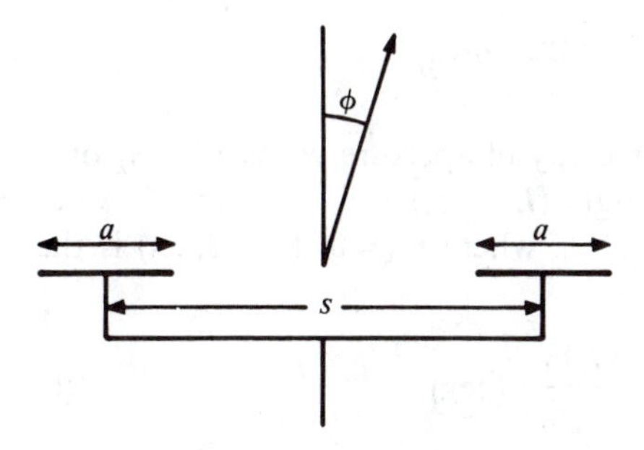

Figure 11-69 Simple 2-element interferometer.

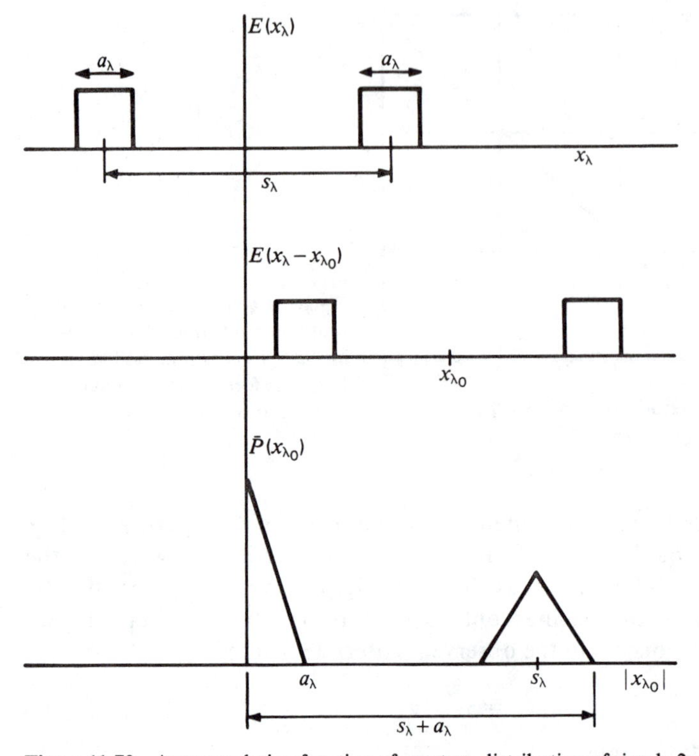

Figure 11-70 Autocorrelation function of aperture distribution of simple 2-element interferometer.

where $E_n(\phi)$ = normalized field pattern of individual array element
$\psi = 2\pi s_\lambda \sin \phi$

The relative power pattern is equal to the square of $|E(\phi)|$, or

$$P(\phi) = |E(\phi)|^2 = |E_n(\phi)|^2 \cos^2 \frac{\psi}{2} = |E_n(\phi)|^2 \frac{1 + \cos \psi}{2} \tag{4}$$

For large spacings the pattern has many lobes, which, in optics, are referred to as fringes. The first null occurs when $\psi = \pi$, from which the beam width between first nulls, or *fringe spacing*, is

$$\text{BWFN} = \frac{1}{s_\lambda} \text{ (rad)} = \frac{57.3}{s_\lambda} \text{ (deg)} \tag{5}$$

This is the BWFN/2 value for a continuous array of aperture width $a_\lambda = s_\lambda$ or a large array of discrete sources of the same length ($L_\lambda = s_\lambda$).

The pattern maxima occur when $\psi = 2\pi n$, where n (=0, 1, 2, 3, ...) is the *fringe order*. Thus,

$$\phi_{\max} = \frac{n}{s_\lambda} \text{ (rad)} = \frac{57.3n}{s_\lambda} \text{ (deg)} \tag{6}$$

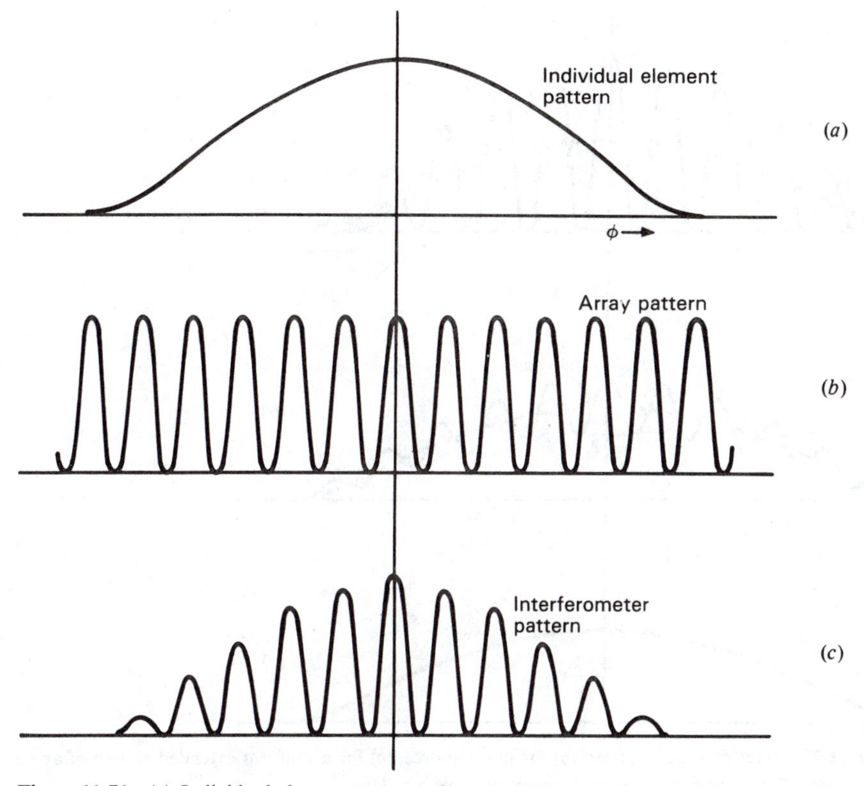

Figure 11-71 (*a*) Individual-element pattern, (*b*) array pattern and (*c*) the resultant interferometer pattern for the case of a point source.

Referring to Fig. 11-71, the first factor in (4) represents the individual-element pattern, as shown in (*a*), and the second factor the pattern of the array of 2 elements, as in (*b*). The product of the 2 factors gives the interferometer pattern, as indicated in (*c*). In these patterns a point source is implied. In the general case, for a source of angular extent α the observed flux density is the convolution of the true source distribution and the antenna power pattern. Assuming that the source extent is small compared to the individual-element pattern, so that $|E_n(\phi)|$ is essentially constant across the source, we have in the 1-dimensional case that

$$S(\phi_0, s_\lambda) = |E_n(\phi)|^2 \int_{-\alpha/2}^{+\alpha/2} B(\phi)\{1 + \cos[2\pi s_\lambda \sin(\phi_0 - \phi)]\}\, d\phi$$

$$= |E_n(\phi)|^2 \left\{\int_{-\alpha/2}^{+\alpha/2} B(\phi)\, d\phi + \int_{-\alpha/2}^{+\alpha/2} B(\phi) \cos[2\pi s_\lambda \sin(\phi_0 - \phi)]\, d\phi\right\}$$

$$= |E_n(\phi)|^2 \left\{S_0 + \int_{-\alpha/2}^{+\alpha/2} B(\phi) \cos[2\pi s_\lambda \sin(\phi_0 - \phi)]\, d\phi\right\} \tag{7}$$

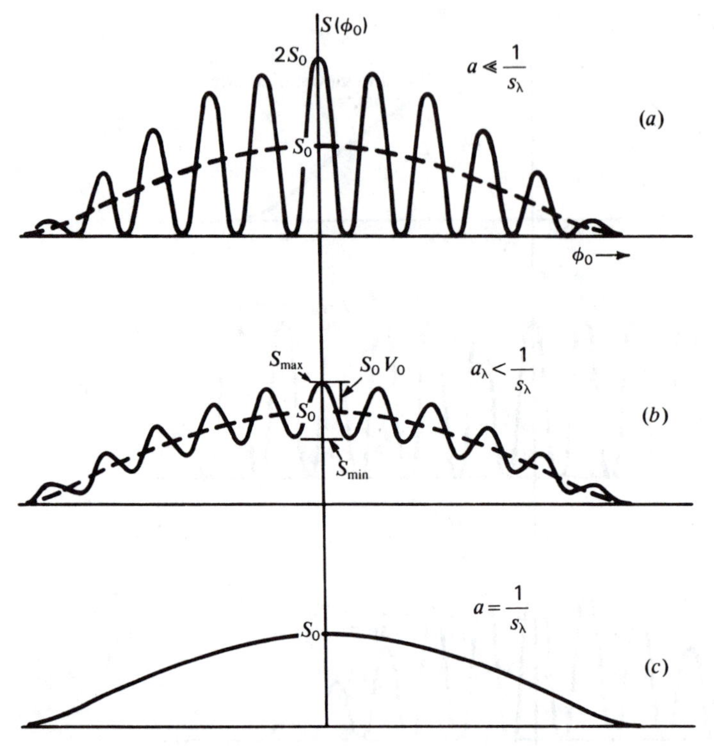

Figure 11-72 Interferometer pattern (*a*) for point source, (*b*) for a uniform extended source of angle $\alpha < 1/s_\lambda$ and (*c*) for a uniform extended source of angle $\alpha = 1/s_\lambda$.

where $S(\phi_0, s_\lambda)$ = observed flux-density distribution, W m^{-2} Hz^{-1}

$B(\phi)$ = true source (brightness) distribution, W m^{-2} Hz^{-1} sr^{-1}

ϕ_0 = displacement angle (=hour angle), rad

α = source extent, rad

$s_\lambda = s/\lambda$ (where s = interferometer element spacing)

S_0 = flux density of source, W m^{-2} Hz^{-1}

The observed distribution as a function of displacement or scan angle is shown in Fig. 11-72 for 3 cases: Fig. 11-72*a*, source extent very small compared with the lobe spacing ($\alpha \ll 1/s_\lambda$), the same as in Fig. 11-71; Fig. 11-72*b*, source extent comparable to, but smaller than, the lobe spacing ($\alpha < 1/s_\lambda$); and Fig. 11-72*c*, source distribution uniform and equal in extent to the lobe spacing ($\alpha = 1/s_\lambda$).

Assuming that the observations are made broadside to the array or that the source is tracked by the individual array elements, so that $|E_n(\phi)|^2 = 1$, (7) becomes

$$S(\phi_0, s_\lambda) = S_0 + \int_{-\alpha/2}^{+\alpha/2} B(\phi) \cos [2\pi s_\lambda \sin (\phi_0 - \phi)] \, d\phi \tag{8}$$

If the source is small, so that $\phi_0 - \phi \ll \pi$, we may write

$$S(\phi_0, s_\lambda) = S_0 + \cos 2\pi s_\lambda \phi_0 \int_{-\alpha/2}^{+\alpha/2} B(\phi) \cos 2\pi s_\lambda \phi \, d\phi$$

$$+ \sin 2\pi s_\lambda \phi_0 \int_{-\alpha/2}^{+\alpha/2} B(\phi) \sin 2\pi s_\lambda \phi \, d\phi \tag{9}$$

$S(\phi_0, s_\lambda)$ may also be expressed as the sum of a constant term and a variable term (sum of 2 terms). Thus,

$$S(\phi_0, s_\lambda) = S_0[1 + V(\phi_0, s_\lambda)] \tag{10}$$

where

$$V(\phi_0, s_\lambda) = \frac{1}{S_0} \cos 2\pi s_\lambda \phi_0 \int_{-\alpha/2}^{+\alpha/2} B(\phi) \cos 2\pi s_\lambda \phi \, d\phi$$

$$+ \frac{1}{S_0} \sin 2\pi s_\lambda \phi_0 \int_{-\alpha/2}^{+\alpha/2} B(\phi) \sin 2\pi s_\lambda \phi \, d\phi \tag{11}$$

The variable term may also be expressed as a cosine function with a displacement $\Delta\phi_0$. Thus,

$$V(\phi_0, s_\lambda) = V_0(s_\lambda) \cos [2\pi s_\lambda(\phi_0 - \Delta\phi_0)] \tag{12}$$

or $$V(\phi_0, s_\lambda) = V_0(s_\lambda)(\cos 2\pi s_\lambda \phi_0 \cos 2\pi s_\lambda \Delta\phi_0 + \sin 2\pi s_\lambda \phi_0 \sin 2\pi s_\lambda \Delta\phi_0) \tag{13}$$

The quantity $V_0(s_\lambda)$ represents the amplitude of the observed lobe pattern, i.e., the *fringe amplitude.* It is also called the *fringe visibility* or simply the *visibility.* As a function of s_λ, it may be referred to as the *visibility function.* The angle $\Delta\phi_0$ represents the fringe displacement from the position with a point source. Hence,

$$V_0(s_\lambda) \cos 2\pi s_\lambda \Delta\phi_0 = \frac{1}{S_0} \int_{-\alpha/2}^{+\alpha/2} B(\phi) \cos 2\pi s_\lambda \phi \, d\phi \tag{14}$$

and $$V_0(s_\lambda) \sin 2\pi s_\lambda \Delta\phi_0 = \frac{1}{S_0} \int_{-\alpha/2}^{+\alpha/2} B(\phi) \sin 2\pi s_\lambda \phi \, d\phi \tag{15}$$

It follows that

$$V_0(s_\lambda) e^{j2\pi s_\lambda \Delta\phi_0} = \frac{1}{S_0} \int_{-\alpha/2}^{+\alpha/2} B(\phi) e^{j2\pi s_\lambda \phi} \, d\phi \tag{16}$$

The quantity $V_0(s_\lambda)e^{j2\pi s_\lambda \Delta\phi_0}$ is called the *complex visibility function.* If the source is contained within a small angle, the limits can be extended to infinity without appreciable error, giving

$$V_0(s_\lambda) e^{j2\pi s_\lambda \Delta\phi_0} = \frac{1}{S_0} \int_{-\infty}^{+\infty} B(\phi) e^{j2\pi s_\lambda \phi} \, d\phi \tag{17}$$

According to (17), the complex visibility function is equal to the Fourier transform of the source brightness distribution (times $1/S_0$). By the inverse Fourier transform we obtain

$$B(\phi_0) = S_0 \int_{-\infty}^{+\infty} V_0(s_\lambda) e^{j2\pi s_\lambda \Delta\phi_0} e^{-j2\pi s_\lambda \phi_0}\, ds_\lambda \tag{18}$$

or

$$B(\phi_0) = S_0 \int_{-\infty}^{+\infty} V_0(s_\lambda) e^{-j2\pi s_\lambda(\phi_0 - \Delta\phi_0)}\, ds_\lambda \tag{19}$$

According to (18) and (19), the true brightness distribution of a source may be obtained, in principle, as the Fourier transform of the complex visibility function (an observable quantity).

To do this in practice requires observations at suitable intervals out to sufficiently large spacings, a high source signal-to-noise ratio and no other (confusing) sources of significant power in the individual-element response pattern. Thus, there are practical limits to the detail with which the source distribution can be determined. According to Bracewell,[1] the spacing interval need be no smaller than $1/\alpha$, where α is the full source extent.

Referring to Fig. 11-72*b*, the visibility may be read from the observed record as

$$V_0(s_\lambda) = \frac{S_{\max} - S_{\min}}{S_{\max} + S_{\min}} = \text{visibility} \tag{20}$$

where

$$V_0(s_\lambda) = \text{visibility } [0 \le V_0(s_\lambda) \le 1]$$

Referring to Fig. 11-73*a*, the value of the integral in (8) is proportional to the net shaded area, the areas above the ϕ axis being positive and the areas below negative. This integral (times $1/S_0$) is the variable quantity $V(\phi_0, s_\lambda)$, and its variation with respect to ϕ_0 for a fixed s_λ is a cosine function, as suggested by the solid curves in Fig. 11-73*b*, one for a point source ($\alpha \to 0$) and the other for an extended source. For symmetrical source distributions (even functions) the fringe displacement is zero or $\frac{1}{2}$-fringe ($\Delta\phi_0 = \frac{1}{2}s_\lambda$). For unsymmetrical sources, such as the one shown by the dashed lines in Fig. 11-73*a*, the fringes will have a displacement $\Delta\phi_0$, as suggested in Fig. 11-73*b*.

For symmetrical sources the visibility is, from (14),

$$V_0(s_\lambda) = \pm \frac{1}{S_0} \int_{-\alpha/2}^{+\alpha/2} B(\phi) \cos 2\pi s_\lambda \phi\, d\phi \tag{21}$$

For a uniform source [$B(\phi)$ = constant] and noting that $\alpha B(\phi) = S_0$, (21) reduces to

$$V_0(s_\lambda) = \pm \frac{\sin 2\pi s_\lambda(\alpha/2)}{2\pi s_\lambda(\alpha/2)} \tag{22}$$

---

[1] R. N. Bracewell, "Radio Interferometry of Discrete Sources," *Proc. IRE*, **46**, 97–105, January 1958.

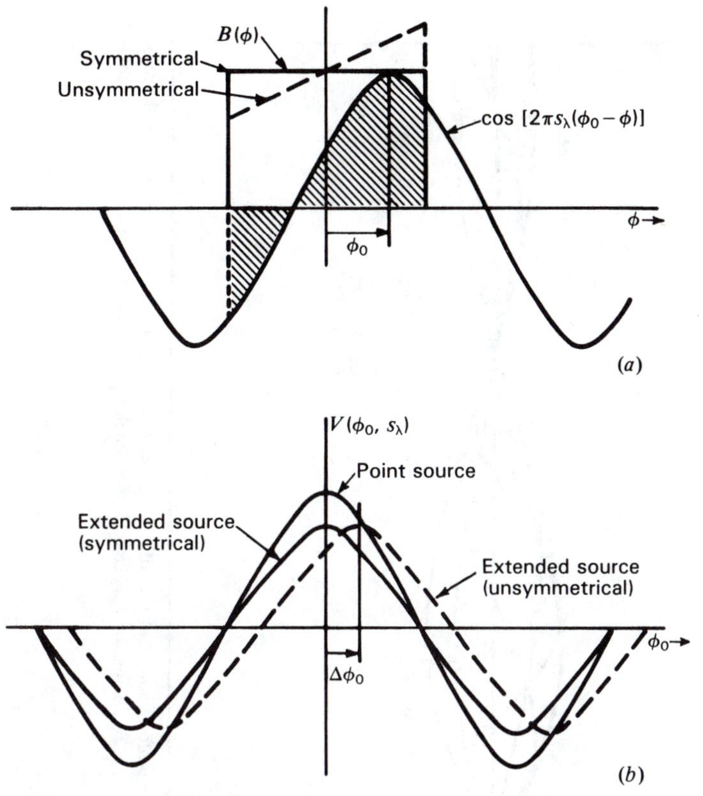

Figure 11-73 Interferometer patterns for symmetrical and unsymmetrical source distributions.

A graph of the visibility $V_0(s_\lambda)$ as a function of s_λ is presented in Fig. 11-74 for the case where the source is uniform and 1° in width. As the source extent becomes small compared to the fringe spacing ($\alpha \ll 1/s_\lambda$), the visibility $V_0(s_\lambda)$ approaches unity, but as the fringe spacing becomes very small compared to the source extent ($\alpha \gg 1/s_\lambda$), $V_0(s_\lambda)$ tends to zero. The visibility is also zero (for a uniform source) when the source extent is equal to the fringe spacing ($1/s_\lambda$) or integral multiples thereof. For symmetrical sources we have from (19) that the source power distribution is given by the Fourier cosine transform of the visibility function, or

$$B(\phi_0) = S_0 \int_{-\infty}^{+\infty} V_0(s_\lambda) \cos 2\pi s_\lambda \phi_0 \, ds_\lambda \tag{23}$$

Also, from (13),

$$V(\phi_0, s_\lambda) = V_0(s_\lambda) \cos 2\pi s_\lambda \phi_0 \tag{24}$$

so that another form for (23) is

$$B(\phi_0) = S_0 \int_{-\infty}^{+\infty} V(\phi_0, s_\lambda) \, ds_\lambda = 2S_0 \int_{0}^{+\infty} V(\phi_0, s_\lambda) \, ds_\lambda \tag{25}$$

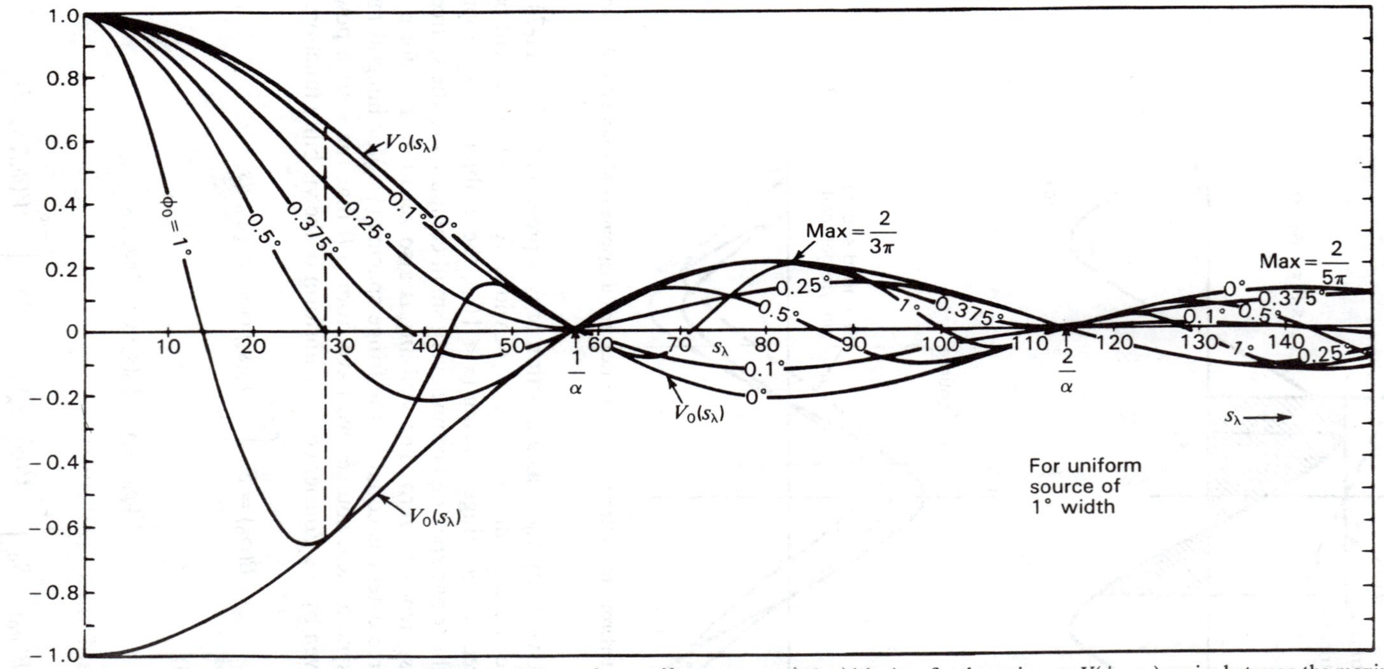

Figure 11-74 Visibility functions versus interferometer spacing s_λ for a uniform source of 1° width. At a fixed spacing s_λ, $V(\phi_0, s_\lambda)$ varies between the maximum and minimum curves $V_0(s_\lambda)$ as a function of the fringe drift or displacement ϕ_0. Thus, for $s_\lambda = 28.6$ (dashed line), $V(\phi_0, s_\lambda)$ is about 0.65 for $\phi_0 = 0°$, about 0.25 for $\phi_0 = 0.375°$, 0.0 for $\phi_0 = 0.5°$ and about −0.65 for $\phi_0 = 1°$.

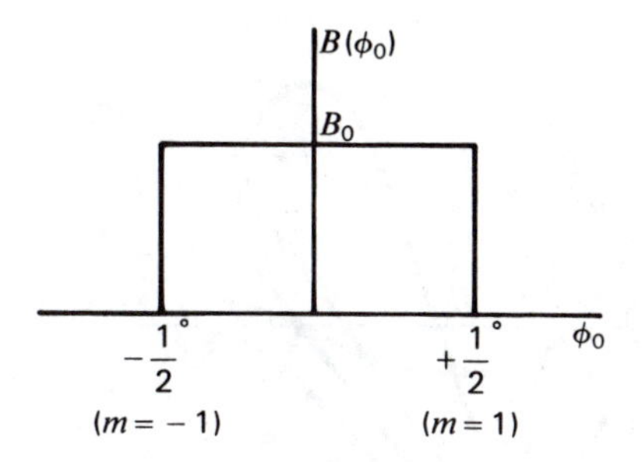

Figure 11-75 Source distribution reconstructed from visibility function.

Curves of $V(\phi_0, s_\lambda)$ as a function of s_λ for several values of ϕ_0 are also shown in Fig. 11-74 for the uniform 1° source.

Example. A uniform source of unknown extent is observed with a simple 2-element interferometer at spacings up to 100λ for determining the source visibility function. Find the source extent.

Solution. Substituting (22) in (23),

$$B(\phi_0) = 2S_0 \int_0^\infty \cos 2\pi s_\lambda \phi_0 \frac{\sin 2\pi s_\lambda(\alpha/2)}{2\pi s_\lambda(\alpha/2)} ds_\lambda \tag{26}$$

This equation has the form

$$B(\phi_0) = \frac{2S_0}{\alpha\pi} \int_0^\infty \cos mx \frac{\sin x}{x} dx \tag{27}$$

where $x = 2\pi s_\lambda \alpha/2$
$m = 2\phi_0/\alpha$

The definite integral (27) is well known and yields

$$B(\phi_0) = \frac{2S_0}{\alpha\pi}\frac{\pi}{2} = \frac{S_0}{\alpha} \qquad (\text{W m}^{-2}\text{ Hz}^{-1}\text{ sr}^{-1})$$

for $-1 < m < +1$ and $B(\phi_0) = 0$ for $m < -1$ and $m > +1$.

Suppose that the visibility function is like the one in Fig. 11-74 with the visibility going through zero at an interferometer spacing of 57.3λ ($s_\lambda = 57.3$) or

$$\alpha \text{ (source width)} = \frac{1 \text{ rad}}{s_\lambda} = \frac{57.3°}{57.3} = 1°$$

with the resulting source brightness

$$B_0 = \frac{S_0}{\alpha} = 57.3S_0 \qquad (\text{W m}^{-2}\text{ Hz}^{-1}\text{ rad}^{-1})$$

where S_0 = source power density

The source has this brightness in the range $-\frac{1}{2}° < \phi_0 < +\frac{1}{2}°$ $(-1 < m < +1)$ and is zero outside, as indicated in Fig. 11-75.

By scanning a source with an interferometer over a range of spacings the visibility function was obtained. Then by taking the Fourier transform of the visibility a source brightness distribution was reconstructed (Fig. 11-75) which is the same

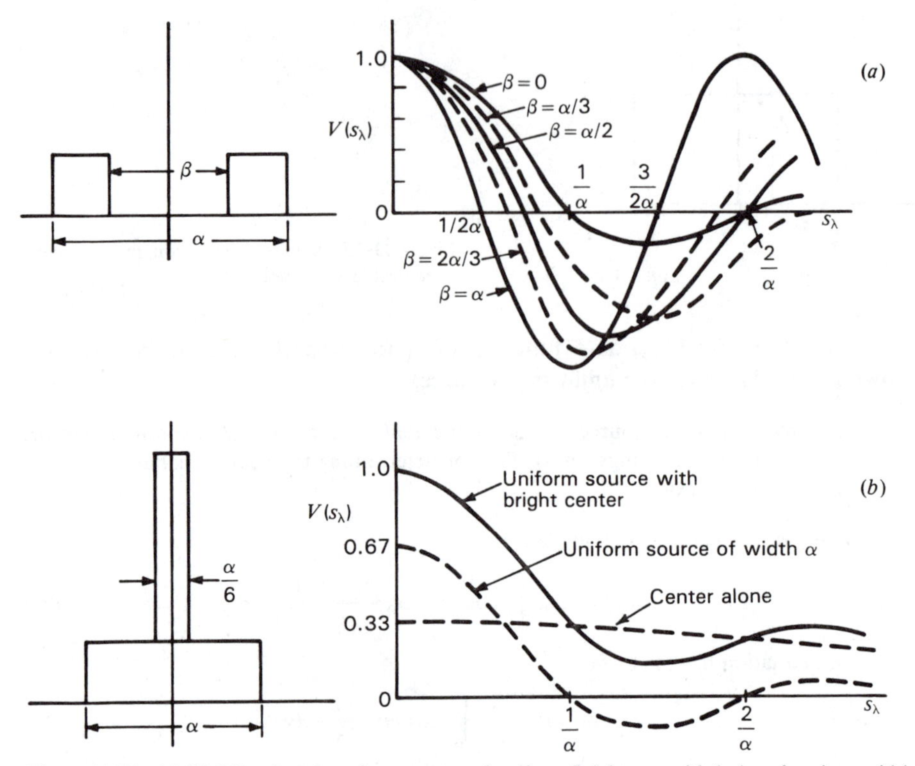

Figure 11-76 (*a*) Visibility functions for a source of uniform brightness with holes of various widths β; (*b*) visibility function for a uniform source with a bright center.

one discussed earlier in our analysis. Thus, in this hypothetical example we have gone full circle.

Examples of visibility functions for other source distributions are given in Fig. 11-76. The case of a uniform source of extent α with a symmetrical hole of extent β is shown in Fig. 11-76*a* for several cases of hole width. When β approaches α, the distribution approaches that of 2 point sources with a separation α. When $\beta = 0$, the source distribution is uniform. In Fig. 11-76*b* the visibility function is presented for a uniform source with a bright center of 4 times the side brightness.

It is to be noted that the visibility functions of Fig. 11-76*a* can be obtained as the visibility for a uniform source distribution of width α minus the visibility of a uniform source of width β, while in Fig. 11-76*b* the result can be obtained as the sum of 2 uniform distributions of widths α and β.

Comparing (17) with (11-22-1) and (11-22-3), it is apparent that the complex visibility function is related to the source brightness distribution in the same manner that the far-field pattern of an antenna is related to the antenna aperture distribution. Accordingly, the graphs of Fig. 11-66 may also be interpreted as giving the visibility functions for various source distributions and the graphs of Fig. 11-76 the field patterns for various aperture distributions. The restriction holds that the source extent is small and the aperture extent large.

For simplicity only 1-dimensional distributions have been considered. This case is of considerable practical importance. The principles may be extended also to the more general 2-dimensional case. A thorough treatment of the 2-dimensional problem is given by Bracewell.[1]

In the above discussion monochromatic radiation at a single frequency is assumed. If the antennas and receiver respond uniformly over a bandwidth $f_0 \pm \Delta f/2$ with the radiation considered to be made up of mutually incoherent monochromatic components, the result for a point source is similar in form to the one above for a uniform source of width α, but with $\alpha/2$ replaced by $\Delta f/2$. A result of too wide a bandwidth is that the higher-order fringes may be obliterated.

11-25 APERTURE SYNTHESIS AND MULTI-APERTURE ARRAYS. The example of the preceding section is illustrative of 1-dimensional aperture synthesis. To obtain resolution of the 1° uniform source of the example to a Rayleigh resolution (=BWFN/2) of 0.1° requires a single aperture 573λ in length [$=114.6°/(2 \times 0.1°)$]. An interferometer, on the other hand, can produce the same result with two small (say, 5λ) apertures operated at various spacings provided that the source is strong enough to give a satisfactory signal-to-noise (S/N) ratio with the small apertures. Thus, the interferometer can synthesize a large continuous aperture—hence the term *aperture synthesis.*

Instead of using only 2 apertures and moving one or both of them, a number of apertures can be employed with unequal spacings in order to provide data points on the visibility curve. Many types of interferometers including phase-switched (multiplying), multi-element, grating, compound and cross arrays and the use of phase-closure and self-calibration techniques are discussed by Kraus.[2]

In radio astronomy observations, the baseline separation of the apertures and the angle of the baseline changes with the earth's rotation providing more visibility data, making it possible to do 2-dimensional aperture synthesis and produce 2-dimensional maps of the source distribution. Spreading the apertures over a plane (instead of all in-line) also improves the data. Typically, a celestial object is tracked (as the earth rotates) by a number of radio telescope antennas with each antenna-pair combination producing a *complex visibility function* (amplitude *and* phase) as a function of displacement (hour) angle. A map or image

---

[1] R. N. Bracewell, "Radio Interferometry of Discrete Sources," *Proc. IRE*, **46**, 97–105, January 1958. See also A. R. Thompson, J. M. Moran and G. W. Swenson, Jr., *Interferometry and Synthesis in Radio Astronomy*, Wiley-Interscience, 1986.

[2] J. D. Kraus, *Radio Astronomy*, 2nd ed., Cygnus-Quasar, 1986, chap. 6.

Figure 11-77 Antennas of the Very Large Array (VLA) of the National Radio Astronomy Observatory in compact configuration. The array, on the Plains of San Augustin, New Mexico, has 27 steerable Cassegrain-type reflector antennas 25 m in diameter mounted on 3 rail tracks (radials at 120°), each 21 km long for deployment in a variety of aperture spacings. (*Courtesy NRAO/AUI.*)

of the object or sky region is then constructed as a Fourier transform of the complex visibility.

The largest and most elaborate aperture synthesis array is presently the VLA (Very Large Array) of the National Radio Astronomy Observatory built at a cost of $78M (1975 $) on the Plains of San Augustin near Socorro, New Mexico. This array consists of 27 Cassegrain-type reflector antennas (apertures), shown in Fig. 11-77, each with a diameter of 25 m. Each dish can be moved along one of three radial railroad tracks 21 km long so that the dishes can be deployed in a Y-shaped configuration with spacings of 0.6 to 36 km in order to maximize the visibility data with the earth's rotation. In the photograph 9 dishes are shown in their most compact, close-in configuration along one track. Synthesized sky maps of 10^7 pixels (picture elements) can be produced with sensitivities of 1 mJy or better. At $\lambda = 6$ cm the resolution is about $\frac{1}{2}$ arcsecond.[1]

Radio telescope antennas separated by intercontinental distances have also been operated as interferometers at even higher resolution. With antenna apertures separated by 10000 km, milliarcsecond resolution is possible at $\lambda = 6$ cm. In this Very Long Baseline Interferometry (VLBI), signals are downconverted

[1] P. J. Napier, A. R. Thompson and R. D. Ekers, "The Very Large Array: Design and Performance of a Modern Synthesis Radio Telescope," *Proc. IEEE*, **71**, 1295–1320, November 1983 (includes 155 references).

and the IF signals taped and sent to a central location. By synchronizing the tapes using atomic standards, a real-time comparison can be simulated. Such observations require a high degree of coordination between participating observatories and being time-consuming have only been performed on an intermittent basis. A dedicated full-time Very Long Baseline Array (VLBA) has been proposed which has at least 10 antennas (apertures) with locations coast-to-coast in the continental United States and in Puerto Rico and Hawaii.[1] The cost is estimated at $75M (1985 $) with completion scheduled for 1995.

A logical next step is to add an orbiting antenna to the array, increasing the resolution and visibility data resulting in maps with better detail and dynamic range. The rapid change of baseline distances for an orbiting antenna also reduces the mapping time. Ultimately with 2 or more orbiting antennas, all of the interferometry could be done above the earth's atmosphere obtaining higher phase stability. High elliptical or circular orbits of 10 000 to 60 000 km apogee or radius are contemplated.[2]

Aperture-synthesis arrays were pioneered by Martin Ryle of Cambridge University, England, and used by him and his students for mapping celestial radio sources.[3] Currently, the most complete and definitive work on aperture synthesis is the book by Thompson, Moran and Swenson.[4]

11-26 GRATING LOBES. If a uniform linear array of n elements has a spacing d_λ between elements exceeding unity, sidelobes appear which are equal in amplitude to the main (center) lobe. These so-called *grating lobes* have a spacing from the main lobe of

$$\phi_G = \sin^{-1}\frac{m}{d_\lambda}\ \text{(rad)}, \qquad \text{where } m = 1, 2, 3, \ldots \tag{1}$$

If $d_\lambda \gg 1$, this reduces approximately to

$$\phi_G = \frac{m}{d_\lambda} \qquad \text{(rad)} \tag{2}$$

---

[1] K. I. Kellermann and A. R. Thompson, "The Very Long Baseline Array," *Science*, **229**, 123–130, 1985.

[2] B. F. Burke, "Orbiting VLBI: A Survey," in R. Fanti, K. Kellermann and G. Setti (eds.), *VLBI and Compact Radio Sources*, Reidel, 1984.
R. A. Preston, B. F. Burke, R. Doxsey, J. F. Jordan, S. H. Morgan, D. H. Roberts and I. I. Shapiro, "The Future of VLBI Observations in Space," in F. Biraud (ed.), *Very Long Baseline Interferometry Techniques*, Cepadues, 1983.

[3] See, for example, M. Ryle, A. Hewish and J. R. Shakeshaft, "The Synthesis of Large Radio Telescopes by the Use of Radio Interferometers," *IRE Trans. Ants. Prop.*, **AP-7**, S120–124, December 1959.

[4] A. R. Thompson, J. M. Moran and G. W. Swenson, Jr., *Interferometry and Synthesis in Radio Astronomy*, Wiley-Interscience, 1986.

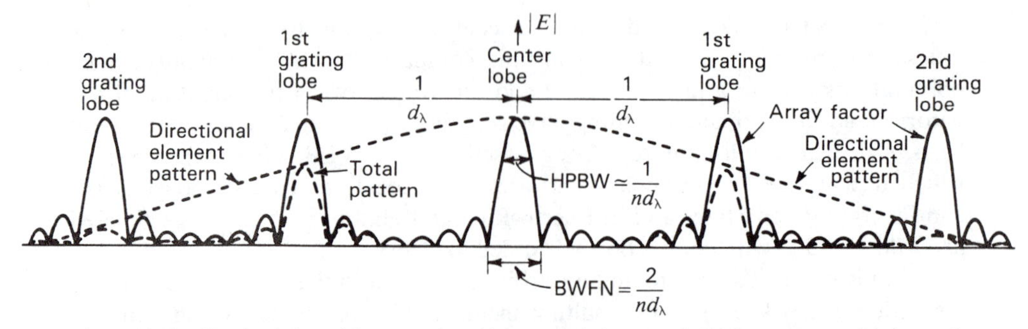

Figure 11-78 Grating lobes with array of n widely spaced elements. Solid line: pattern with isotropic elements (array factor). Dashed line: total pattern with directional elements, each with pattern of dotted line.

with the first sidelobe ($m = 1$) at $1/d_\lambda$ as given by the solid line (array factor) in Fig. 11-78 [individual array elements are nondirectional (isotropic)]. To suppress all grating lobes including the first requires directional elements, each with an aperture of approximately d_λ. This puts the first null of the individual element pattern on the first grating lobe, but the array is now equivalent to a continuous aperture. With less directivity (smaller aperture) elements, with pattern as suggested by the upper dashed line in Fig. 11-78, the resultant (total) pattern is as suggested by the lower dashed line with some grating lobe suppression but not elimination.

ADDITIONAL REFERENCES

Allen, J. L.: "Array Antennas: New Applications for an Old Technique," *IEEE Spectrum*, **1**, 115–130, November 1964. Excellent introductory article with 50 references.

Barton, P.: "Digital Beamforming for Radar," *IEE Proc.*, **127**, August 1980.

Fenn, A. J.: "Maximizing Jammer Effectiveness for Evaluating the Performance of Adaptive Nulling Array Antennas," *IEEE Trans. Ants. Prop.*, **AP-33**, 1131–1142, October 1985.

Gabriel, W. F.: "Spectral Analysis and Adaptive Array Superresolution," *Proc. IEEE*, **68**, June 1980.

Howell, J. M.: "Phased Arrays for Microwave Landing Systems," *Microwave J.*, **30**, 129–137, January 1987.

Hudson, J. E.: *Adaptive Array Principles*, Peregrinus, 1981.

Kinzel, J. A.: "GaAs Technology for Millimeter-Wave Phased Arrays," *IEEE Ants. Prop. Soc. Newsletter*, **29**, 12–14, February 1987.

Kominami, M., D. M. Pozar and D. H. Schaubert: "Dipole and Slot Elements and Arrays on Semi-infinite Substrates," *IEEE Trans. Ants. Prop.*, **AP-33**, 600–607, June 1985.

Lindell, I. V., E. Alanen and K. Mannersalo: "Exact Image Method for Impedance Computation of Antennas above the Ground," *IEEE Trans. Ants. Prop.*, **AP-33**, 937–945, September 1985.

Mailloux, R. J.: "Antennas and Radar," *Microwave J.*, **30**, 26–30, January 1987.

Moynihan, R. L.: "Phased Arrays for Airborne ECM," *Microwave J.*, **30**, 34–53, January 1987.

Pozar, D. M.: "General Relations for a Phased Array of Printed Antennas Derived from Infinite Current Sheets," *IEEE Trans. Ants. Prop.*, **AP-33**, 498–504, May 1985.

Rhodes, D. R.: "On a Fundamental Principle in the Theory of Planar Antennas," *Proc. IEEE*, **52**, 1013–1021, September 1964.

Rhodes, D. R.: "On the Stored Energy of Planar Apertures," *IEEE Trans. Ants. Prop.*, **AP-14**, 676–683, November 1966.

Rhodes, D. R.: "On the Aperture and Pattern Space Factors for Rectangular and Circular Apertures," *IEEE Trans. Ants. Prop.*, **AP-19**, 763–770, November 1971.

Rhodes, D. R.: "Observable Stored Energies of Electromagnetic Systems," *J. Franklin Inst.*, **302**, 225–237, September 1976.

Sato, M., M. Iguchi and R. Sato: "Transient Response of Coupled Linear Dipole Antennas," *IEEE Trans. Ants. Prop.*, **AP-32**, 133–140, February 1984.

Tai, C.-T.: "The Optimum Directivity of Uniformly Spaced Broadside Arrays of Dipoles," *IEEE Trans. Ants. Prop.*, **AP-12**, 447–454, 1964.

Taylor, T. T.: "Design of Line-Source Antennas for Narrow Beamwidth and Low Side Lobes," *IRE Trans. Ants. Prop.*, **AP-3**, 16–28, 1955.

Woodward, P. M., and J. D. Lawson: "The Theoretical Precision with Which an Arbitrary Radiation Pattern May Be Obtained from a Source of Finite Size," *J. Inst. Elec. Engng.*, pt. 3, **95**, 363–370, September 1948.

PROBLEMS[1]

★11-1 Two λ/2-element broadside array.

(*a*) Calculate and plot the gain of a broadside array of 2 side-by-side $\lambda/2$ elements in free space as a function of the spacing d for values of d from 0 to 2λ. Express the gain with respect to a single $\lambda/2$ element. Assume all elements are 100 percent efficient.

(*b*) What spacing results in the largest gain?

(*c*) Calculate and plot the radiation field patterns for $\lambda/2$ spacing. Show also the patterns of the $\lambda/2$ reference antenna to the proper relative scale.

11-2 Two λ/2-element end-fire array. A 2-element end-fire array in free space consists of 2 vertical side-by-side $\lambda/2$ elements with equal out-of-phase currents. At what angles in the horizontal plane is the gain equal to unity:

(*a*) When the spacing is $\lambda/2$?

(*b*) When the spacing is $\lambda/4$?

11-3 Two-element VP array. Calculate and plot the field and phase patterns of the far field for an array of 2 vertical side-by-side $\lambda/2$ elements in free space with $\lambda/4$ spacing when the elements are:

(*a*) In phase and

(*b*) 180° out of phase.

For the in-phase case also include on the graph the patterns in both the yz or vertical plane and xy or horizontal plane of Fig. 11-2*a*. For the out-of-phase case do the same for the patterns in both the xz or vertical plane and xy or horizontal plane of Fig. 11-7*a*.

11-4 W8JK array. Calculate the vertical and horizontal plane free-space field patterns of a W8JK antenna consisting of two horizontal out-of-phase $\lambda/2$ elements spaced $\lambda/8$. Assume a loss resistance of 1 Ω and show the relative patterns of a $\lambda/2$ reference antenna with the same power input.

11-5 Sixteen-element and W8JK array gain equations. Confirm (11-6-6) and (11-7-11).

11-6 Two-element array with unequal currents.

(*a*) Consider two $\lambda/2$ side-by-side vertical elements spaced a distance d with currents related by $I_2 = aI_1 \underline{/\delta}$. Develop the gain expression in a plane parallel to the elements and the gain expression in a plane normal to the elements, taking

[1] Answers to starred (★) problems are given in App. D.

a vertical $\lambda/2$ element with the same power input as reference ($0 \le a \le 1$). Check that these reduce to (11-4-15) and (11-4-13) when $a = 1$.

(*b*) Plot the field patterns in both planes and also show the field pattern of the reference antenna in proper relative proportion for the case where $d = \lambda/4$, $a = \frac{1}{2}$ and $\delta = 120°$.

***11-7 Impedance of D-T array.**

(*a*) Calculate the driving-point impedance at the center of each element of an in-phase broadside array of 6 side-by-side $\lambda/2$ elements spaced $\lambda/2$ apart. The currents have a Dolph-Tchebyscheff distribution such that the minor lobes have $\frac{1}{5}$ the field intensity of the major lobe.

(*b*) Design a feed system for the array.

11-8 Stacked λ/2 elements and W8JK array above ground.

(*a*) Develop (11-7-12).

(*b*) Calculate and plot from (11-7-12) the gain in field intensity for an array of 2 in-phase horizontal $\lambda/2$ elements stacked $\lambda/2$ apart (as in Fig. 11-31) over a $\lambda/2$ antenna in free space with the same power input as a function of h up to $h = 1.5\lambda$ for an elevation angle $\alpha = 10°$. Also calculate and plot for comparison on the same graph the gains at $\alpha = 10°$ for a 2-element horizontal W8JK antenna over a single horizontal $\lambda/2$ antenna as a function of the height above ground from $h = 0$ to $h = 1.5\lambda$. Note differences of these curves and those for $\alpha = 20°$ in Fig. 11-32.

11-9 Two-tower BC array. A broadcast-station antenna array consists of two vertical $\lambda/4$ towers spaced $\lambda/4$ apart. The currents are equal in magnitude and in phase quadrature. Assume a perfectly conducting ground and zero loss resistance. Calculate and plot the azimuthal field pattern in millivolts (rms) per meter at 1.6 km with 1 kW input for vertical elevation angles $\alpha = 0$, 20, 40, 60 and 80°. The towers are series fed at the base. Assume that the towers are infinitesimally thin.

11-10 Two-tower BC array. Calculate and plot the relative field pattern in the vertical plane through the axis of the 2-tower broadcast array fulfilling the requirements of Prob. 4-19 if the towers are $\lambda/4$ high and are series fed at the base. Assume that the towers are infinitesimally thin and that the ground is perfectly conducting.

11-11 Three-tower BC array. Calculate and plot the relative field pattern in the vertical plane through the axis of the 3-tower broadcast array fulfilling the requirements of Prob. 4-20 if the towers are $3\lambda/8$ high and are series fed at the base. Assume that the towers are infinitesimally thin and that the ground is perfectly conducting.

11-12 BC array with null at $\alpha = 30°$. Design a broadcast-station antenna array of 2 vertical base-fed towers $\lambda/4$ high and spaced $3\lambda/8$ which produces a broad maximum of field intensity to the north in the horizontal plane and a null at an elevation angle $\alpha = 30°$ and azimuth angle $\phi = 135°$ measured ccw from north to reduce interference via ionospheric bounce. Assume that the towers are infinitesimally thin, that the ground is perfectly conducting and that the base currents of the two towers are equal. Specify the orientation and phasing of the towers. Calculate and plot the azimuthal field pattern at $\alpha = 0°$ and $\alpha = 30°$ and also the pattern in the vertical plane through $\phi = 135°$. The suggested procedure is as follows. Solve (11-8-4) for ϕ' at the null. Then set ϕ in the pattern factor in (11-4-13) equal to ϕ' and solve for the value of δ which makes the pattern factor zero. The relative field intensity at any angle (ϕ, α) is then given by (11-4-14) where $\sin\theta = \cos\phi' = \cos\alpha\cos\phi$ in the first pattern factor and $\theta = 90° - \alpha$ in the second pattern factor.

11-13 BC array with null to west at all α. Design a broadcast-station array of 2 vertical base-fed towers $\lambda/4$ high that produces a broad maximum of field intensity to the north in the horizontal plane and a null at all vertical angles to the west. Assume that the towers are infinitesimally thin and that the ground is perfectly conducting. Specify the spacing, orientation and phasing of the towers. Calculate and plot the azimuthal relative field patterns at elevation angles of $\alpha = 0$, 30 and 60°.

★11-14 Impedance and gain of 2-element array. Two thin center-fed $\lambda/2$ antennas are driven in phase opposition. Assume that the current distributions are sinusoidal. If the antennas are parallel and spaced 0.2λ,

(*a*) Calculate the mutual impedance of the antennas.

(*b*) Calculate the gain of the array in free space over one of the antennas alone.

11-15 Triangle array. Three isotropic point sources of equal amplitude are arranged at the corners of an equilateral triangle, as in Fig. P11-15. If all sources are in phase, determine and plot the far-field pattern.

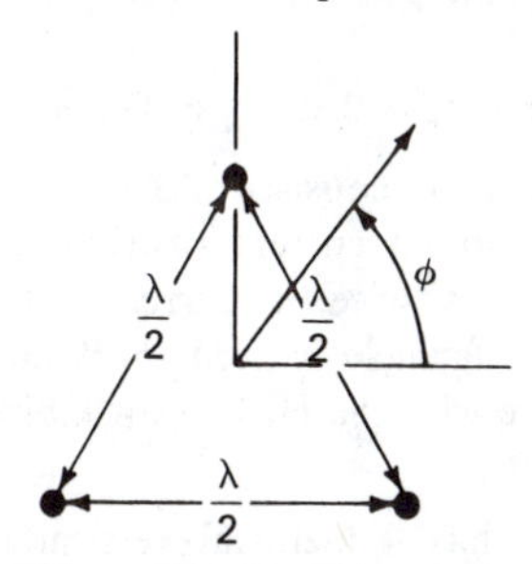

Figure P11-15 Triangle array.

11-16 Square array. Four isotropic point sources of equal amplitude are arranged at the corners of a square, as in Fig. P11-16. If the phases are as indicated by the arrows, determine and plot the far-field pattern.

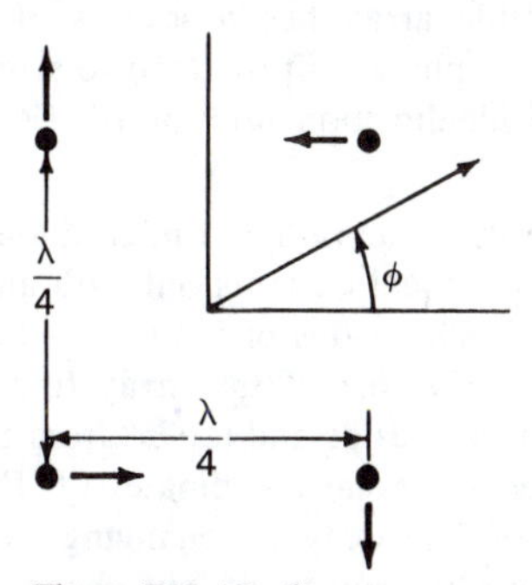

Figure P11-16 Square array.

★11-17 Terminated V. Traveling wave.

(*a*) Calculate and plot the far-field pattern of a terminated-V antenna with 5λ legs and 45° included angle.

(*b*) What is the HPBW?

★11-18 Seven short dipoles. 4-dB angle. A linear broadside (in-phase) array of 7 short dipoles has a separation of 0.35λ between dipoles. Find the angle from the maximum field for which the field is 4 dB down (to nearest 0.1°).

11-19 Square array. Four identical short dipoles (perpendicular to page) are arranged at the corners of a square $\lambda/2$ on a side. The upper left and lower right dipoles are in the same phase while the 2 dipoles at the other corners are in the opposite phase. If the direction to the right (x direction) corresponds to $\phi = 0°$, find the angles ϕ for all maxima and minima of the field pattern in the plane of the page.

11-20 Pencil-beam patterns. For symmetrical circular aperture pencil-beam patterns (function only of θ) show that the main beam solid angle Ω_M is given by

$1.13\ \theta_{HP}^2$ for a Gaussian pattern
$0.988\ \theta_{HP}^2$ for a $(\sin x)/x$ pattern
$1.008\ \theta_{HP}^2$ for a Bessel pattern

where θ_{HP} is the half-power beam width. Also show that Ω_M is given by

$1.036\ \theta_{HP}\,\phi_{HP}$ for a $(\sin u)/u$ (square aperture) pattern

where $u = x = y$. Note that for a Gaussian pattern $\theta_{HP}^2 = 0.88\ \Omega_A$ so $\Omega_M = \Omega_A$.

11-21 Collinear array of three λ/2 dipoles. An antenna array consists of 3 in-phase collinear thin $\lambda/2$ dipole antennas, each with sinusoidal current distribution and spaced $\lambda/2$ apart. The current in the center dipole is twice the current in the end dipoles (binomial array). (*a*) Calculate and plot the far-field pattern. (*b*) What is the HPBW? (*c*) How does this HPBW value compare with the HPBW for a binomial array of 3 isotropic point sources spaced $\lambda/2$ apart?

*****11-22 Four-tower broadcast array.** A broadcast array has 4 identical vertical towers arranged in an east-west line with a spacing d and progressive phase shift δ. Find (*a*) d and (*b*) δ so that there is a maximum field at $\phi = 45°$ (northeast) and a null at $\phi = 90°$ (north). There can be other nulls and maxima, but no maximum can exceed the one at 45°. The distance d must be less than $\lambda/2$.

11-23 Eight-source scanning array. A linear broadside array has 8 sources of equal amplitude and $\lambda/2$ spacing. Find the progressive phase shift required to swing the beam (*a*) 5°, (*b*) 10° and (*c*) 15° from the broadside direction. (*d*) Find BWFN when all sources are in phase.

11-24 A 24-dipole scanning array. A linear array consists of an in-line configuration of 24 $\lambda/2$ dipoles spaced $\lambda/2$. The dipoles are fed with equal currents but with an arbitrary progressive phase shift δ between dipoles. What value of δ is required to put the main-lobe maximum (*a*) perpendicular to the line of the array (broadside condition), (*b*) 25° from broadside, (*c*) 50° from broadside and (*d*) 75° from broadside? (*e*) Calculate and plot the four field patterns in polar coordinates. (*f*) Discuss the feasibility of this arrangement for a scanning array by changing feed-line lengths to change δ or by keeping the array physically fixed but changing the frequency. What practical limits occur in both cases?

11-25 Three-helix scanning array. Three 4-turn right-handed monofilar axial mode helical antennas spaced 1.5λ apart are arranged in a broadside array as in Fig. 11-46. The pitch angle $\alpha = 12.5°$ and the circumference $C = \lambda$. (*a*) If the outer two helices rotate on their axes in opposite directions while the center helix is fixed, determine the angle ϕ of the main lobe with respect to the broadside direction and describe how ϕ varies as the helices rotate. (*b*) What is the maximum scan angle ϕ? (*c*) What is the main-lobe HPBW as a function of ϕ?

★**11-26 E-type rhombic.** Design a maximum E-type rhombic antenna for an elevation angle $\alpha = 17.5°$.

11-27 Alignment rhombic. Design an alignment-type rhombic antenna for an elevation angle $\alpha = 17.5°$.

★**11-28 Compromise rhombic.** Design a compromise-type rhombic antenna for an elevation angle $\alpha = 17.5°$ at a height above ground of $\lambda/2$.

11-29 Compromise rhombic. Design a compromise-type rhombic antenna for an elevation angle $\alpha = 17.5°$ with a leg length of 3λ.

★**11-30 Compromise rhombic.** Design a compromise-type rhombic antenna for an elevation angle $\alpha = 17.5°$ at a height above ground of $\lambda/2$ and a leg length of 3λ.

11-31 Rhombic patterns. Calculate the relative vertical plane patterns in the axial direction for the rhombics of Probs. 11-26, 11-27, 11-28, 11-29 and 11-30. Compare the patterns with the main lobes adjusted to the same maximum value.

11-32 Rhombic equation. Derive (11-16-1) for the relative field intensity of a horizontal rhombic antenna above a perfectly conducting ground.

11-33 Alignment rhombic equation. Verify (11-16-3), (11-16-8) and (11-16-9) for the alignment design rhombic antenna.

★**11-34 Sixteen source broadside array.** A uniform linear array has 16 isotropic in-phase point sources with a spacing of $\lambda/2$. Calculate exactly (*a*) the half-power beam width, (*b*) the level of the first sidelobe, (*c*) the beam solid angle, (*d*) the beam efficiency, (*e*) the directivity and (*f*) the effective aperture.

11-35 Four aperture distributions. For the following aperture distributions show that the far-field patterns are as given:

(*a*) L_λ Stepped $\quad E(\phi) = \dfrac{2}{3}\dfrac{\sin\psi}{\psi} + \dfrac{1}{3}\dfrac{\sin\psi/2}{\psi/2} \quad$ where $\psi = \pi L_\lambda \sin\phi$

(*b*) Circular $\quad E(\phi) = \dfrac{2J_1(\psi)}{\psi} \quad$ where $\psi = 2\pi L_\lambda \sin\phi$

(*c*) Triangular $\quad E(\phi) = \dfrac{\sin^2\psi}{\psi^2} \quad$ where $\psi = \dfrac{\pi}{2} L_\lambda \sin\phi$

(*d*) Triangular asymmetric $\quad E(\phi) = \dfrac{\sin\psi}{\psi} + j\left(\dfrac{\cos\psi}{\psi} - \dfrac{\sin\psi}{\psi^2}\right) \quad$ where $\psi = \pi L_\lambda \sin\phi$

11-36 Fourier transform. Apply the Fourier transform method to obtain the far-field pattern of an array of 2 equal in-phase isotropic point sources with a separation d. Reduce the expression to its simplest trigonometric form.

11-37 Interferometer. Pattern multiplication. An interferometer antenna consists of 2 square broadside in-phase apertures with uniform field distribution.

(*a*) If the apertures are 10λ square and are separated 60λ on centers, calculate and plot the far-field pattern to the first null of the single aperture pattern.

(*b*) How many lobes are contained between first nulls of the aperture pattern?

(*c*) What is the effective aperture?

(*d*) What is the HPBW of the central interferometer lobe?

(*e*) How does this HPBW compare with the HPBW for the central lobe of two isotropic in-phase point sources separated 60λ?

11-38 Visibility function. Show that the visibility function observed with a simple interferometer of spacing s_λ for a uniform source of width α with a symmetrical uniform bright center of width $\beta = \alpha/6$ is

$$V(s_\lambda) = \frac{\sin(\pi s_\lambda \alpha) + 3\sin(\pi s_\lambda \alpha/6)}{3\pi s_\lambda \alpha/2}$$

if the center brightness is 4 times the side brightness (see Fig. 11-76*b*).

11-39 Visibility function. Show that the visibility function observed with a simple interferometer of spacing s_λ for 2 equal uniform sources of width $\alpha/6$ spaced between centers by $5\alpha/6$ is

$$V(s_\lambda) = \frac{\sin \pi s_\lambda \alpha}{\pi s_\lambda \alpha} - \frac{2}{3}\frac{\sin(2\pi s_\lambda \alpha/3)}{2\pi s_\lambda \alpha/3}$$

★**11-40 Pattern smoothing.** An idealized antenna pattern-brightness distribution is illustrated by the 1-dimensional diagram in Fig. P11-40. The brightness distribution consists of a point source of flux density S and a uniform source 2° wide, also of flux density S. The point source is 2° from the center of the 2° source. The antenna pattern is triangular (symmetrical) with a 2° beam width between zero points and with zero response beyond.

(*a*) Draw an accurate graph of the observed flux density as a function of angle from the center of the 2° source.

(*b*) What is the maximum ratio of the observed to the actual total flux density ($2S$)?

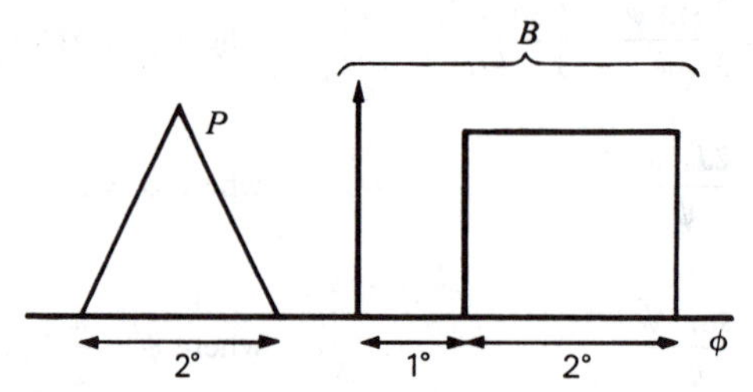

Figure P11-40 Pattern smoothing.

11-41 Interferometer output. Show that the output of a simple interferometer of spacing s and bandwidth $f_0 \pm \Delta f/2$ for a point source is given by

$$S_0 \Delta f \left(1 + \frac{\sin \dfrac{2\pi s}{c}\dfrac{\Delta f}{2}\phi}{\dfrac{2\pi s}{c}\dfrac{\Delta f}{2}\phi} \cos \frac{2\pi s}{c} f_0 \phi\right)$$

where c = velocity of light.

11-42 Interferometer bandwidth. Show that for an interferometer with bandwidth $f_0 \pm \Delta f/2$ (Prob. 11-41), the condition $\Delta f/f_0 \ll 1/n$, where n = fringe order, must hold in order that the fringe amplitude not be decreased.

11-43 Number of elements. In Fig. 11-78 how many elements n have been assumed?

CHAPTER 12

REFLECTOR ANTENNAS AND THEIR FEED SYSTEMS

12-1 INTRODUCTION. Reflectors are widely used to modify the radiation pattern of a radiating element. For example, the backward radiation from an antenna may be eliminated with a plane sheet reflector of large enough dimensions. In the more general case, a beam of predetermined characteristics may be produced by means of a large, suitably shaped, and illuminated reflector surface. The characteristics of antennas with sheet reflectors or their equivalent are considered in this chapter.

Several reflector types are illustrated in Fig. 12-1. The arrangement in Fig. 12-1*a* has a large, flat sheet reflector near a linear dipole antenna to reduce the backward radiation (to the left in the figure). With small spacings between the antenna and sheet this arrangement also yields a substantial gain in the forward radiation. This case was discussed in Sec. 11-7a with the ground acting as the flat sheet reflector. The desirable properties of the sheet reflector may be largely preserved with the reflector reduced in size as in Fig. 12-1*b* and even in the limiting case of Fig. 12-1*c*. Here the sheet has degenerated into a thin reflector element. Whereas the properties of the large sheet are relatively insensitive to small frequency changes, the thin reflector element is highly sensitive to frequency changes. The case of a $\lambda/2$ antenna with parasitic reflector element was treated in Sec. 11-9a.

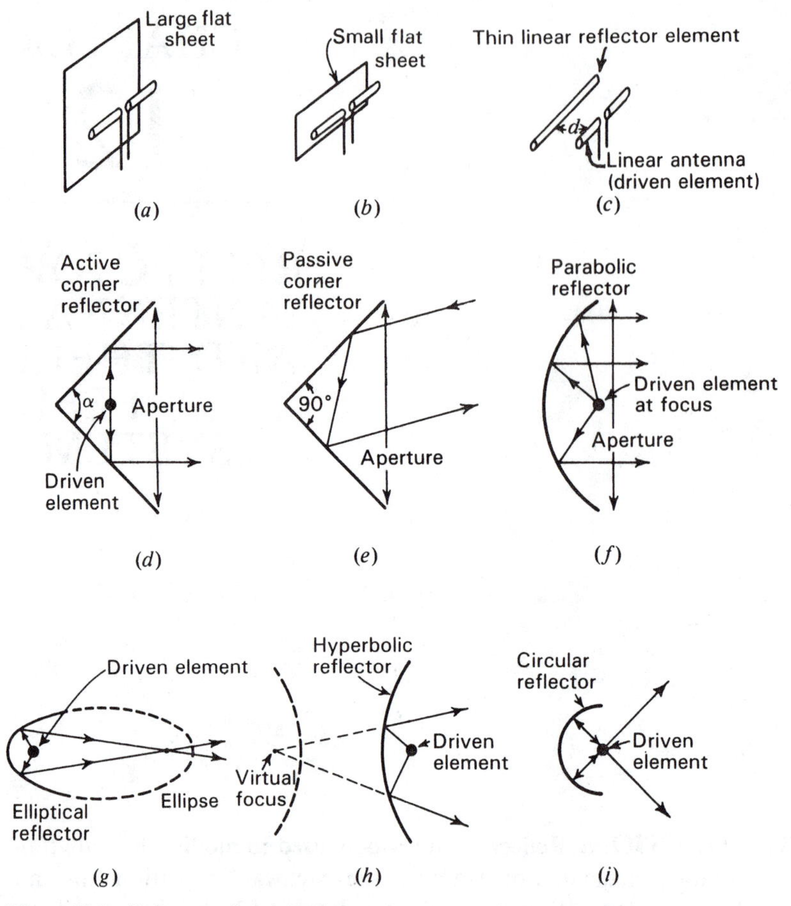

Figure 12-1 Reflectors of various shapes.

With two flat sheets intersecting at an angle α ($<180^\circ$) as in Fig. 12-1*d*, a sharper radiation pattern than from a flat sheet reflector ($\alpha = 180^\circ$) can be obtained. This arrangement, called an *active corner reflector antenna*, is most practical where apertures of 1 or 2λ are of convenient size. A corner reflector without an exciting antenna can be used as a *passive reflector* or target for radar waves. In this application the aperture may be many wavelengths, and the corner angle is *always 90°*. Reflectors with this angle have the property that an incident wave is reflected back toward its source as in Fig. 12-1*e*, the corner acting as a *retroreflector*.

When it is feasible to build antennas with apertures of many wavelengths, parabolic reflectors can be used to provide highly directional antennas. A parabolic reflector antenna is shown in Fig. 12-1*f*. The parabola reflects the waves originating from a source at the focus into a parallel beam, the parabola transforming the curved wave front from the feed antenna at the focus into a plane wave front. Many other shapes of reflectors can be employed for special applications. For instance, with an antenna at one focus, the elliptical reflector (Fig. 12-1*g*) produces a diverging beam with all reflected waves passing through the second focus of the ellipse. Examples of reflectors of other shapes are the hyperbolic[1] and the circular reflectors[2] shown in Fig. 12-1*h* and *i*.

The plane sheet reflector, the corner reflector, the parabolic reflector and other reflectors are discussed in more detail in the following sections. In addition, feed systems, aperture blockage, aperture efficiency, diffraction, surface irregularities, gain and frequency-selective surfaces are considered.

12-2 PLANE SHEET REFLECTORS AND DIFFRACTION.

The problem of an antenna at a distance S from a perfectly conducting plane sheet reflector of infinite extent is readily handled by the method of images.[3] In this method the reflector is replaced by an image of the antenna at a distance $2S$ from the antenna, as in Fig. 12-2. This situation is identical with the one considered in Sec. 11-7, of a horizontal antenna above ground. If the antenna is a $\lambda/2$ dipole this in turn reduces to the problem of the W8JK antenna discussed in Sec. 11-5. Assuming zero reflector losses, the gain in field intensity of a $\lambda/2$ dipole antenna at a distance S from an infinite plane reflector is, from (11-5-8),

$$G_f(\phi) = 2\sqrt{\frac{R_{11} + R_L}{R_{11} + R_L - R_{12}}}\,|\sin(S_r \cos\phi)| \tag{1}$$

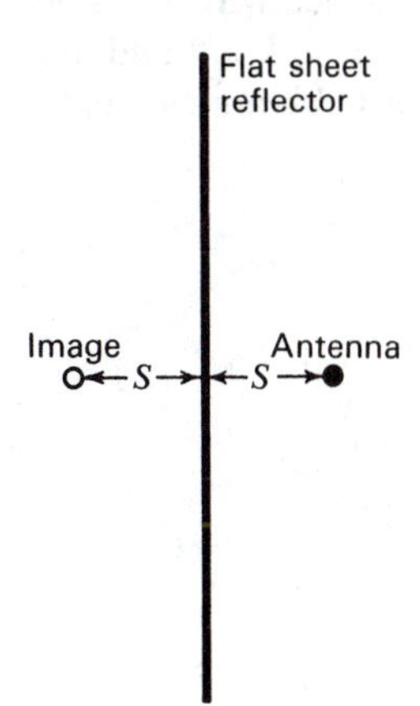

Figure 12-2 Antenna with flat sheet reflector.

[1] G. Stavis and A. Dorne, in *Very High Frequency Techniques*, Radio Research Laboratory Staff, McGraw-Hill, New York, 1947, chap. 6.

[2] J. Ashmead and A. B. Pippard, "The Use of Spherical Reflectors as Microwave Scanning Aerials," *JIEE* (Lond.), **93**, pt. IIIA, no. 4, 627–632, 1946.

[3] See, for example, G. H. Brown, "Directional Antannas," *Proc. IRE*, **25**, 122, January 1937.

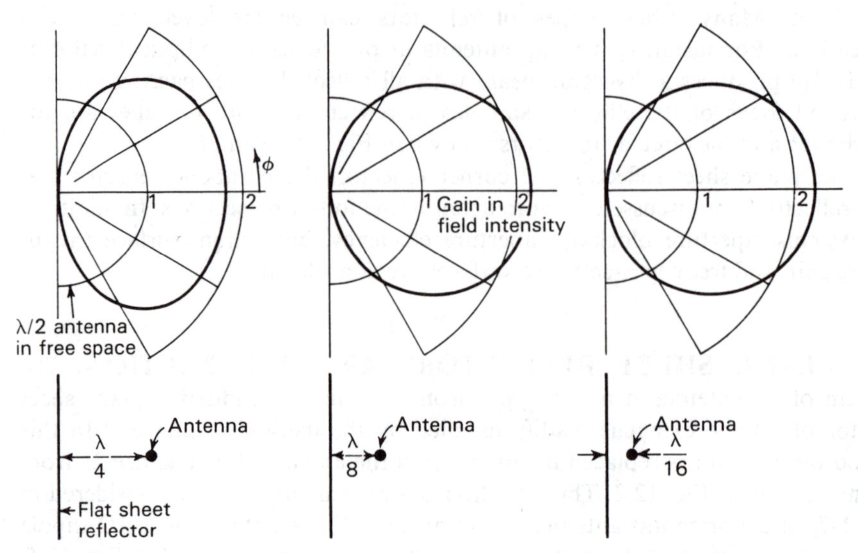

Figure 12-3 Field patterns of a $\lambda/2$ antenna at spacings of $\frac{1}{4}$, $\frac{1}{8}$ and $\frac{1}{16}\lambda$ from an infinite flat sheet reflector. Patterns give gain in field intensity over a $\lambda/2$ antenna in free space with same power input. For $\lambda/8$ spacing, the gain is 2.2(=6.7 dB = 8.9 dBi).

where $S_r = 2\pi \dfrac{S}{\lambda}$

The gain in (1) is expressed relative to a $\lambda/2$ antenna in free space with the same power input. The field patterns of $\lambda/2$ antennas at distances $S = \lambda/4$, $\lambda/8$ and $\lambda/16$ from the flat sheet reflector are shown in Fig. 12-3. These patterns are calculated from (1) for the case where $R_L = 0$.

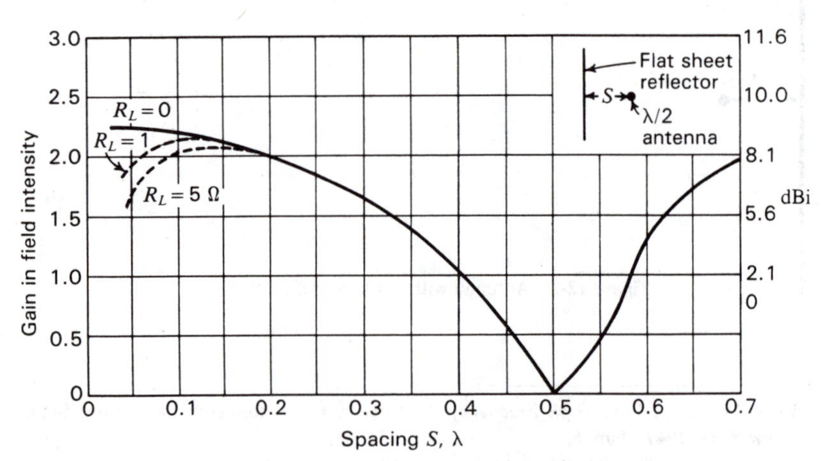

Figure 12-4 Gain in field intensity of $\lambda/2$ dipole antenna at distance S from flat sheet reflector. Gain is relative to $\lambda/2$ dipole antenna in free space with the same power input. Gain in dBi is also shown. Gain is in direction $\phi = 0$ and is shown for assumed loss resistances $R_L = 0, 1$ and $5\ \Omega$.

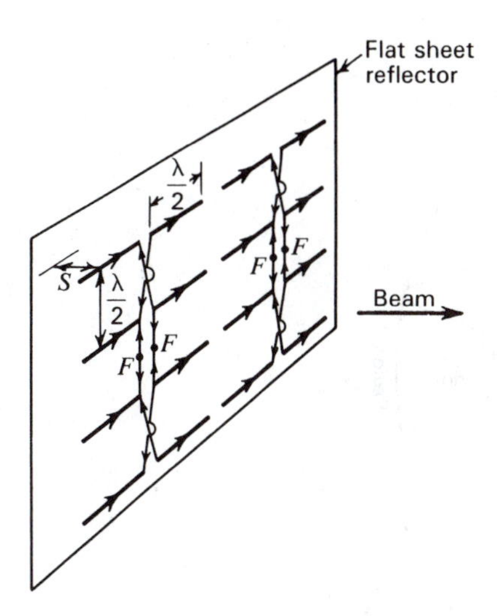

Figure 12-5 Array of $\lambda/2$ elements with flat sheet reflector (billboard antenna).

The gain as a function of the spacing S is presented in Fig. 12-4 for assumed antenna loss resistances $R_L = 0$, 1 and 5 Ω. These curves are calculated from (1) for $\phi = 0$. It is apparent that very small spacings can be used effectively provided that losses are small. However, the bandwidth is narrow for small spacings, as discussed in Sec. 11-5. With wide spacings the gain is less, but the bandwidth is larger. Assuming an antenna loss resistance of 1 Ω, a spacing of 0.125λ yields the maximum gain.

A large flat sheet reflector can convert a bidirectional antenna array into a unidirectional system. An example is shown in Fig. 12-5. Here a broadside array of 16 in-phase $\lambda/2$ elements spaced $\lambda/2$ apart is backed up by a large sheet reflector so that a unidirectional beam is produced. The feed system for the array is indicated, equal in-phase voltages being applied at the 2 pairs of terminals *F*-*F*. If the edges of the sheet extend some distance beyond the array, the assumption that the flat sheet is infinite in extent is a good first approximation. The choice of the spacing S between the array and the sheet usually involves a compromise between gain and bandwidth. If a spacing of $\lambda/8$ is used, the radiation resistance of the elements of a large array remains about the same as with no reflector present.[1] This spacing also has the advantage over wider spacings of reduced interaction between elements. On the other hand, a spacing such as $\lambda/4$ provides a greater bandwidth, and the precise value of S is less critical in its effect on the element impedance.

When the reflecting sheet is reduced in size, the analysis is less simple. The situation is shown in Fig. 12-6*a*. There are 3 principal angular regions:

[1] H. A. Wheeler, "The Radiation Resistance of an Antenna in an Infinite Array or Waveguide," *Proc. IRE*, **36**, 478–487, April 1948.

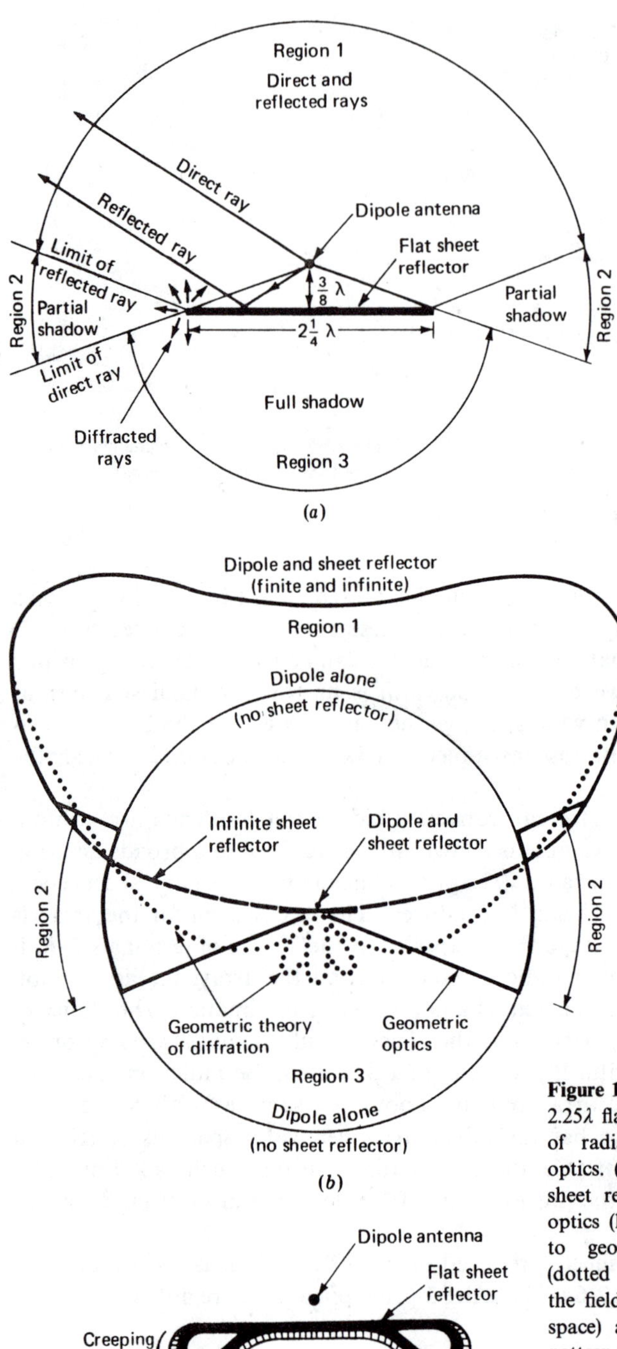

Figure 12-6 (*a*) Dipole antenna with 2.25λ flat sheet reflector with 3 regions of radiation according to geometric optics. (*b*) Field pattern of dipole and sheet reflector according to geometric optics (heavy solid line) and according to geometric theory of diffraction (dotted line). The solid circle indicates the field from the dipole alone (in free space) and the dashed line gives the pattern for dipole with an infinite sheet reflector. (*c*) Modification of edges of sheet to reduce diffracted back-side radiation (in region 3).

Region 1 (above or in front of the sheet). In this region the radiated field is given by the resultant of the *direct field* of the dipole and the *reflected field* from the sheet.

Region 2 (above and below at the sides of the sheet). In this region there is only the direct field from the dipole. This region is in the shadow of the reflected field.

Region 3 (below or behind the sheet). In this region the sheet acts as a shield, producing a full shadow (no direct or reflected fields, only diffracted fields).

If the sheet is 1 or 2λ in width and the dipole is close to it, image theory accounts adequately for the radiation pattern in region 1. In region 2, the distant field is dominated by the direct ray from the dipole. In the full shadow behind the sheet (region 3) the Geometrical Theory of Diffraction (GTD) must be used.[1] The pattern in this region is effectively that of 2 weak line sources, one along each edge. The fields in the 3 regions are shown in Fig. 12-6*b* for the case of the sheet width dimension $D = 2.25\lambda$ and the dipole spacing $d = 3\lambda/8$. It is assumed that the sheet is very long perpendicular to the page ($\gg D$).

Narrower reflecting sheets result in more radiation into region 3 but this diffracted radiation can be minimized by using a rolled edge (radius of curvature $> \lambda/4$) and absorbing material, as suggested in Fig. 12-6*c*.

12-3 CORNER REFLECTORS

12-3a Active (Kraus) Corner Reflector. In 1938 while analyzing the radiation from a dipole parallel to and closely spaced from a flat reflecting sheet, I realized that when the sheet is replaced by its image, the dipole and its image form a W8JK array. When the flat sheet (180° included angle) is folded into a square (90°) corner the theory calls for 3 images, and my calculations showed correspondingly higher gain. Thus, the corner reflector developed as an extension of my analysis of the W8JK array. I immediately constructed several corner reflectors to obtain experimental confirmation. I tried parallel-wire grid reflectors, modifying both the spacing and length of the reflector wires to determine the limiting dimensions required. Figure 12-7 shows my first corner reflector, a 90° corner for $\lambda = 5$ m operation built in 1938. Subsequent tests were done with smaller corners of various angles at $\lambda = 1$ m with patterns measured by rotating the antennas on a turntable.

---

[1] Diffraction is also discussed in Secs. 2-18, 4-15, 4-16, 13-3, 17-5 and 18-3d. For more details on the principles of physical and geometrical diffraction theories see J. D. Kraus, *Electromagnetics*, 3rd ed., McGraw-Hill, 1984, pp. 524–531 (2nd ed., 1972, pp. 464–479), and J. D. Kraus, *Radio Astronomy*, 2nd ed., Cygnus-Quasar, 1986, pp. 642–646.

Figure 12-7 The author with the first corner reflector which he built and tested in 1938 for $\lambda = 5$ m operation.

The following presentation is based mainly on my publications on the corner reflector in the *Proceedings of the Institute of Radio Engineers* (November 1940), in *Radio* (March 1939) and in a patent application filed Jan. 31, 1940.[1]

Two flat reflecting sheets intersecting at an angle or corner as in Fig. 12-8 form an effective directional antenna. When the corner angle $\alpha = 90°$, the sheets intersect at right angles, forming a square-corner reflector. Corner angles both greater or less than 90° can be used although there are practical disadvantages to angles much less than 90°. A corner reflector with $\alpha = 180°$ is equivalent to a flat sheet reflector and may be considered as a limiting case of the corner reflector. This case has been treated in Sec. 12-2.

Assuming perfectly conducting reflecting sheets of infinite extent, the *method of images* can be applied to analyze the corner reflector antenna for angles $\alpha = 180°/n$, where n is any positive integer. This method of handling corners is well known in electrostatics.[2] Corner angles of 180° (flat sheet), 90°, 60°, etc., can be treated in this way. Corner reflectors of intermediate angle cannot be determined by this method but can be interpolated approximately from the others.

In the analysis of the 90° corner reflector there are 3 image elements, 2, 3 and 4, located as shown in Fig. 12-9*a*. The driven antenna 1 and the 3 images have currents of equal magnitude. The phase of the currents in 1 and 4 is the same. The phase of the currents in 2 and 3 is the same but 180° out of phase with respect to the currents in 1 and 4. All elements are assumed to be $\lambda/2$ long.

At the point P at a large distance D from the antenna, the field intensity is

$$E(\phi) = 2kI_1 \left| [\cos (S_r \cos \phi) - \cos (S_r \sin \phi)] \right| \tag{1}$$

where I_1 = current in each element

S_r = spacing of each element from the corner, rad

$= 2\pi(S/\lambda)$

k = constant involving the distance D, etc.

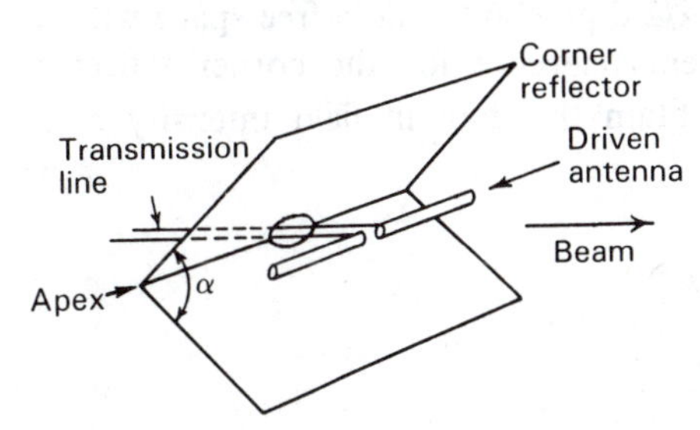

Figure 12-8 Corner reflector antenna.

[1] J. D. Kraus, "The Corner Reflector Antenna," *Proc. IRE*, **28**, 513–519, November 1940.
J. D. Kraus, "The Square-Corner Reflector," *Radio*, no. 237, 19–24, March 1939.
J. D. Kraus, "Corner Reflector Antenna," U.S. Patent 2,270,314, granted Jan. 20, 1942.

[2] See, for example, Sir James Jeans, *Mathematical Theory of Electricity and Magnetism*, 5th ed., Cambridge University Press, London, p. 188. For arbitrary corner angles, analysis involves integrations of cylindrical functions which can be approximated by infinite sums as shown by R. W. Klopfenstein, "Corner Reflector Antennas with Arbitrary Dipole Orientation and Apex Angle," *IRE Trans. Ants. Prop.*, **AP-5**, 297–305, July 1957.

The emf V_1 at the terminals at the center of the driven element is

$$V_1 = I_1 Z_{11} + I_1 R_{1L} + I_1 Z_{14} - 2I_1 Z_{12} \tag{2}$$

where Z_{11} = self-impedance of driven element
R_{1L} = equivalent loss resistance of driven element
Z_{12} = mutual impedance of elements 1 and 2
Z_{14} = mutual impedance of elements 1 and 4

Similar expressions can be written for the emf's at the terminals of each of the images. Then if P is the power delivered to the driven element (power to each image element is also P), we have from symmetry that

$$I_1 = \sqrt{\frac{P}{R_{11} + R_{1L} + R_{14} - 2R_{12}}} \tag{3}$$

Substituting (3) in (1) yields

$$E(\phi) = 2k \sqrt{\frac{P}{R_{11} + R_{1L} + R_{14} - 2R_{12}}} \, |[\cos (S_r \cos \phi) - \cos (S_r \sin \phi)]| \tag{4}$$

The field intensity at the point P at a distance D from the driven $\lambda/2$ element with the reflector removed is

$$E_{\text{HW}}(\phi) = k \sqrt{\frac{P}{R_{11} + R_{1L}}} \tag{5}$$

where k = the same constant as in (1) and (4)

This is the relation for field intensity of a $\lambda/2$ dipole antenna in free space with a power input P and provides a convenient reference for the corner reflector antenna. Thus, dividing (4) by (5), we obtain the gain in field intensity of a

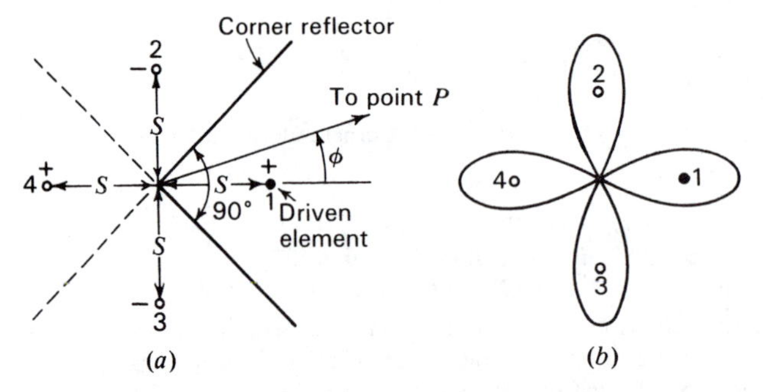

Figure 12-9 Square-corner reflector with images used in analysis (*a*) and 4-lobed pattern of driven element and images (*b*).

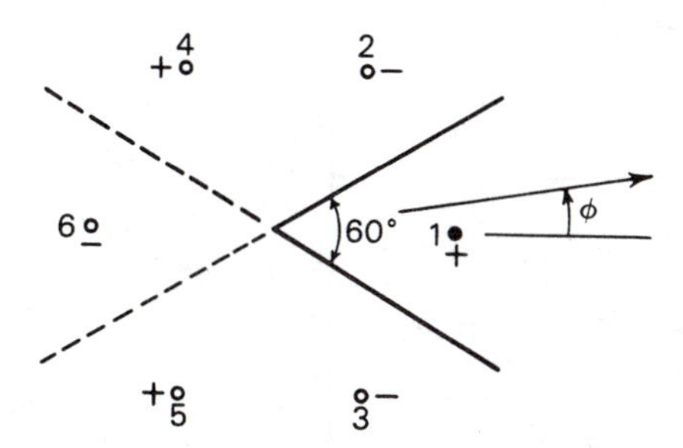

Figure 12-10 A 60° corner reflector with images used in analysis.

square-corner reflector antenna over a single $\lambda/2$ antenna in free space with the same power input, or

$$G_f(\phi) = \frac{E(\phi)}{E_{HW}(\phi)}$$

$$= 2\sqrt{\frac{R_{11} + R_{1L}}{R_{11} + R_{1L} + R_{14} - 2R_{12}}}\,|[\cos(S_r \cos\phi) - \cos(S_r \sin\phi)]| \quad (6)$$

where the expression in brackets is the pattern factor and the expression included under the radical sign is the coupling factor. The pattern shape is a function of both the angle ϕ and the antenna-to-corner spacing S. The pattern calculated by (6) has 4 lobes as shown in Fig. 12-9*b*. However, only one of the lobes is real.

Expressions for the gain in field intensity of corner reflectors with corner angles of 60, 45°, etc., can be obtained in a similar manner. The driven element is a $\lambda/2$ dipole. For the 60° corner the analysis requires a total of 6 elements, 1 actual antenna and 5 images as in Fig. 12-10. Gain-pattern expressions for corner reflectors of 90 and 60° are listed in Table 12-1. The expression for a 180° "corner" or flat sheet is also included.

Table 12-1 Gain-pattern formulas for corner reflector antennas

Corner angle, deg	Number of elements in analysis	Gain in field intensity over λ/2 antenna in free space with same power input
180	2	$2\sqrt{\frac{R_{11}+R_{1L}}{R_{11}+R_{1L}-R_{12}}}\sin(S_r \cos\phi)$
90	4	$2\sqrt{\frac{R_{11}+R_{1L}}{R_{11}+R_{1L}+R_{14}-2R_{12}}}\,|[\cos(S_r\cos\phi)-\cos(S_r\sin\phi)]|$
60	6	$2\sqrt{\frac{R_{11}+R_{1L}}{R_{11}+R_{1L}+2R_{14}-2R_{12}-R_{16}}}$ $\times\,|\{\sin(S_r\cos\phi)-\sin[S_r\cos(60°-\phi)]-\sin[S_r\cos(60°+\phi)]\}|$

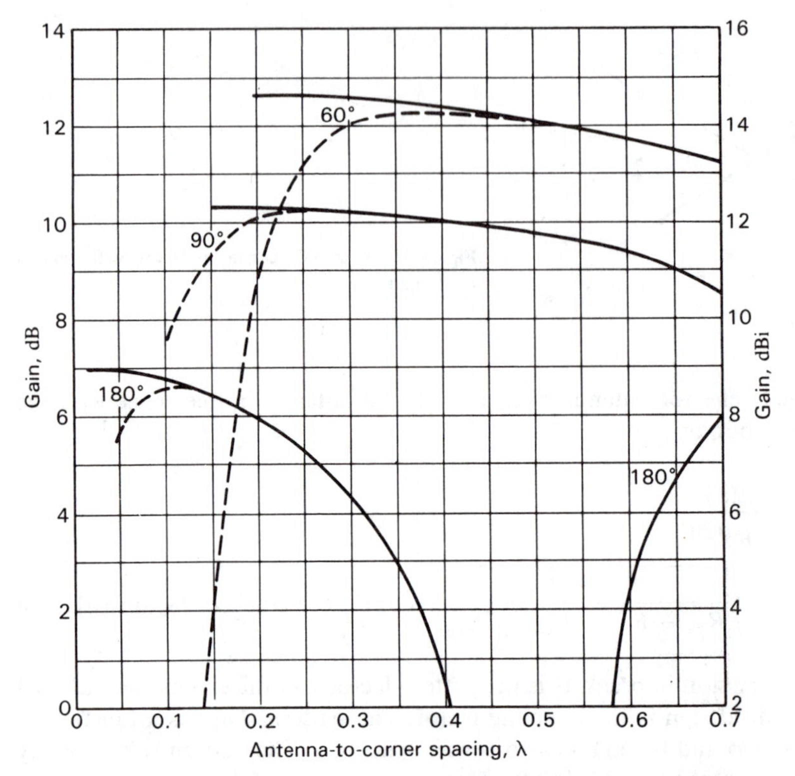

Figure 12-11 Gain of corner reflector antennas over a $\lambda/2$ dipole antenna in free space with the same power input as a function of the antenna-to-corner spacing. Gain is in the direction $\phi = 0$ and is shown for zero loss resistance (solid curves) and for an assumed loss resistance of 1 Ω ($R_{1L} = 1$ Ω) (dashed curves). (*After Kraus.*)

In the formulas of Table 12-1 it is assumed that the reflector sheets are perfectly conducting and of infinite extent. Curves of gain versus spacing calculated from these relations are presented in Fig. 12-11. The gain given is in the direction $\phi = 0$. Two curves are shown for each corner angle. The solid curve in each case is computed for zero losses ($R_{1L} = 0$), while the dashed curve is for an assumed loss resistance $R_{1L} = 1$ Ω. It is apparent that for efficient operation too small a spacing should be avoided. A small spacing is also objectionable because of narrow bandwidth. On the other hand, too large a spacing results in less gain.[1]

The calculated pattern of a 90° corner reflector with antenna-to-corner spacing $S = 0.5\lambda$ is shown in Fig. 12-12*a*. The gain is nearly 10 dB over a reference $\lambda/2$ antenna or 12 dBi. This pattern is typical if the spacing S is not too large. If S exceeds a certain value, a multilobed pattern may be obtained. For

[1] Displacing the driven dipole from the bisector of the corner angle shifts (squints) the beam direction to the other side of the bisector (see Prob. 16-20).

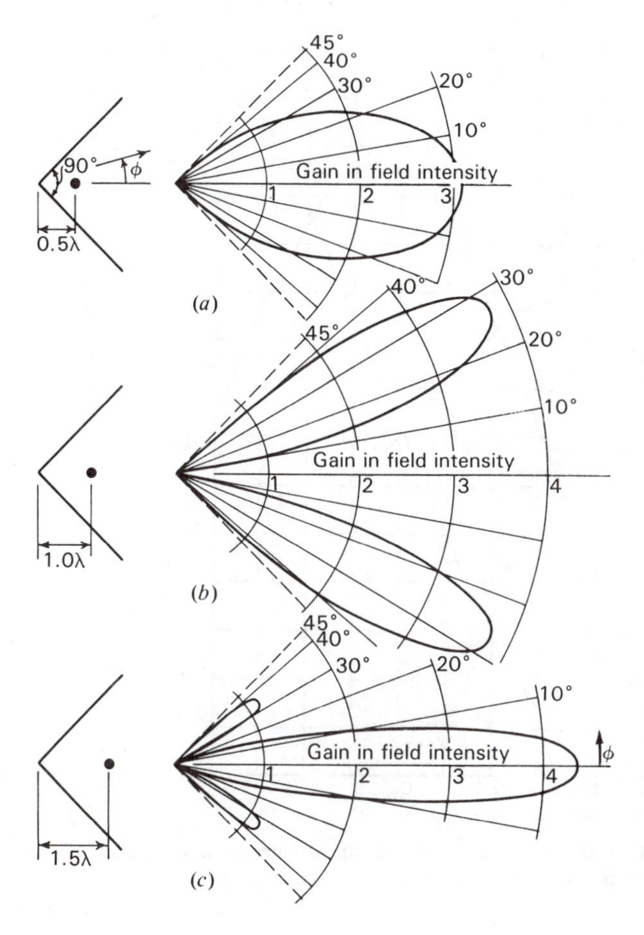

Figure 12-12 Calculated patterns of square-corner reflector antennas with antenna-to-corner spacings of (*a*) 0.5λ, (*b*) 1.0λ and (*c*) 1.5λ. Patterns give gain relative to the $\lambda/2$ dipole antenna in free space with the same power input.

example, a square-corner reflector with $S = 1.0\lambda$ has a 2-lobed pattern as in Fig. 12-12*b*. If the spacing is increased to 1.5, the pattern shown in Fig. 12-12*c* is obtained with the major lobe in the $\phi = 0$ direction but with minor lobes present. This pattern may be considered as belonging to a higher-order radiation mode of the antenna. The gain over a single $\lambda/2$ dipole antenna is 12.9 dB ($\simeq 15$ dBi).

Restricting patterns to the lower-order radiation mode (no minor lobes), it is generally desirable that S lie between the following limits:

α	S
90°	0.25–0.7λ
180° (flat sheet)	0.1–0.3λ

The terminal resistance R_T of the driven antenna is obtained by dividing (2) by I_1 and taking the real parts of the impedances. Thus,

$$R_T = R_{11} + R_{1L} + R_{14} - 2R_{12} \tag{7}$$

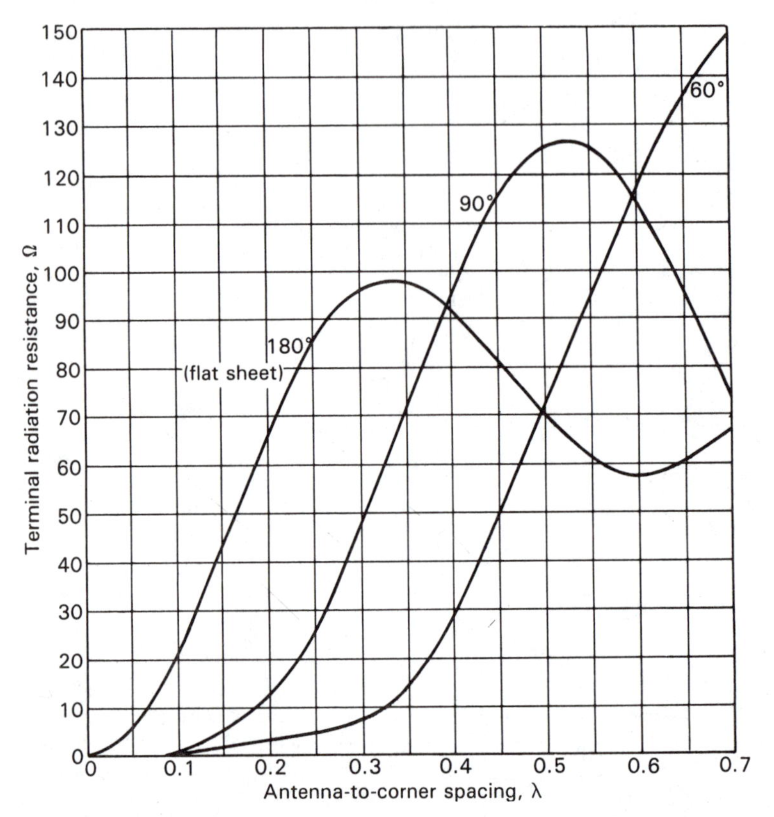

Figure 12-13a Terminal radiation resistance of driven $\lambda/2$ dipole antenna as a function of the antenna-to-corner spacing for corner reflectors of 3 corner angles. (*After Kraus.*)

If $R_{1L} = 0$, the terminal resistance is all radiation resistance. The variation of the terminal radiation resistance of the driven element is presented in Fig. 12-13a as a function of the spacing S for corner angles $\alpha = 180$, 90 and 60°. We note that for $\alpha = 90°$ and $S = 0.35\lambda$, the resistance of the driven $\lambda/2$ dipole is the same as for a $\lambda/2$ dipole in free space.

In the above analysis it is assumed that the reflectors are perfectly conducting and of infinite extent, with the exception that the gains with a finitely conducting reflector may be approximated with a proper choice of R_{1L}. The analysis provides a good first approximation to the gain-pattern characteristics of actual corner reflectors with finite sides provided that the sides are not too small. Neglecting edge effects, a suitable value for the length of sides may be arrived at by the following line of reasoning. An essential region of the reflector is that near the point at which a wave from the driven antenna is reflected parallel to the axis. For example, this is the point A of the square-corner reflector of Fig. 12-13b. This point is at a distance of $1.41S$ from the corner C, where S is the antenna-to-corner spacing. If the reflector ends at the point B at a distance $L = 2S$ from the

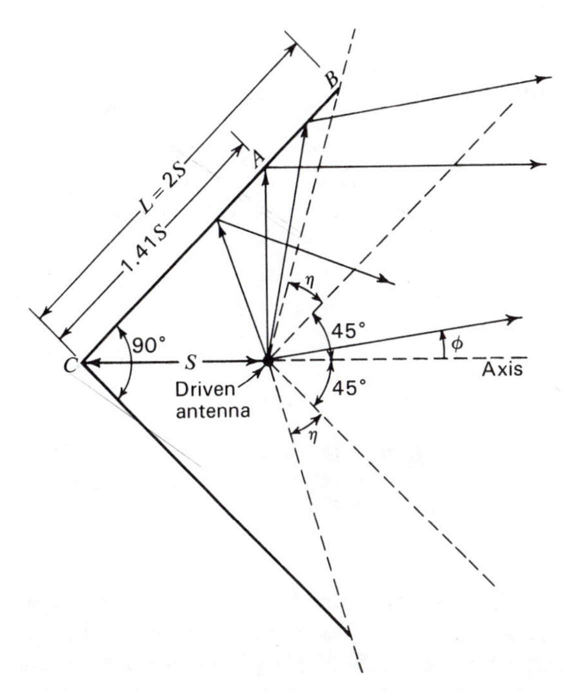

Figure 12-13b Square-corner reflector with sides of length L equal to twice the antenna-to-corner spacing S.

corner, as in Fig. 12-13b, the reflector extends approximately $0.6S$ beyond A. With the reflector ending at B, it is to be noted that the only waves reflected from infinite sides, but not from finite sides, are those radiated in the sectors η. Furthermore, these waves are reflected with infinite sides into a direction that is at a considerable angle ϕ with respect to the axis. Hence, the absence of the reflector beyond B should not have a large effect. It should also have relatively little effect on the driving-point impedance. The most noticeable effect with finite sides is that the measured pattern is appreciably broader than that calculated for infinite sides and a null does not occur at $\phi = 45°$ but at a somewhat larger angle. If this is not objectionable, a side length of twice the antenna-to-corner spacing ($L = 2S$) is a practical minimum value for square-corner reflectors.

Although the gain of a corner reflector with infinite sides can be increased by reducing the corner angle, it does not follow that the gain of a corner reflector with finite sides of fixed length will increase as the corner angle is decreased. To maintain a given efficiency with a smaller corner angle requires that S be increased. Also on a 60° reflector, for example, the point at which a wave is reflected parallel to the axis is at a distance of $1.73S$ from the corner as compared to $1.41S$ for the square-corner type. Hence, to realize the increase in gain requires that the length of the reflector sides be much larger than for a square-corner reflector designed for the same frequency. Usually this is a practical disadvantage in view of the relatively small increase to be expected in gain.

To reduce the wind resistance offered by a solid reflector, a grid of parallel wires or conductors can be used as in Fig. 12-14. The supporting member joining

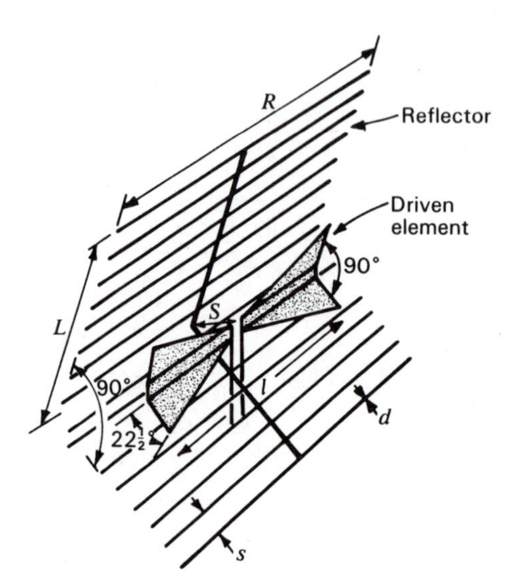

Figure 12-14 Square-corner (grid) reflector with bow-tie dipole for wideband operation (see Table 12-2).

the midpoints of the reflector conductors may be either a conductor or an insulator. In general the spacing s between reflector conductors should be equal to or less than $\lambda/8$. With a $\lambda/2$ driven element the length R of the reflector conductors should be equal to or greater than 0.7λ. If the length R is reduced to values of less than 0.6λ, radiation to the sides and rear tends to increase and the gain decreases. When R is decreased to as little as 0.3λ, the strongest radiation is no longer forward and the "reflector" acts as a director.

Table 12-2 Design data for wideband 90° corner reflector with bow-tie dipole (see Fig. 12-14)

Dipole-to-corner spacing, $S = \lambda/4$ at lowest frequency
Length of reflector, $L = 3\lambda/4$ at lowest frequency
Reflector rod length, $R = 4\lambda/5$ at lowest frequency
Reflector rod spacing, $s = \lambda/8$ at highest frequency
Reflector rod diameter, $d = \lambda/50$ at highest frequency
Bow-tie dipole length, $l = 4\lambda/5$ at mid frequency

	Lowest frequency f_1	Mid frequency f_2	Highest frequency f_3	Units
Dipole-to-corner spacing, S	0.27	0.40	0.54	λ
Length of reflector, L	0.75	1.13	1.50	λ
Reflector rod length, R	0.81	1.20	1.62	λ
Reflector rod spacing, s	0.061	0.092	0.122	λ
Reflector rod diameter, d	0.01	0.015	0.02	λ
Bow-tie dipole length, l	0.53	0.80	1.06	λ
Gain	11.0	13.0	14.0	dBi

The square-corner reflector is a simple, practical, inherently wideband antenna producing substantial gains (11 to 14 dBi). Typical design data for a 90° (square) corner reflector with bow-tie dipole for wideband (2 to 1 frequency range) operation are given in Table 12-2.

The driven element is a 45° Brown-Woodward (bow-tie) dipole bent at 90° as suggested in the figure, so that its flat sides are parallel to the reflector. The dipole can be fed by a 300 or 400 Ω twin line with low VSWR over the 2 to 1 frequency range. None of the corner reflector dimensions are critical. A moderate increase or decrease in the reflector dimensions L and R from the values in Table 12-2 results in only small changes in gain. Thus, a 10 percent increase in L and R can increase the gain by a decibel or less while a 10 percent decrease in L and R can decrease the gain by a decibel or more. Also the (center-to-center) spacing s between reflector rods can be increased with only a small gain reduction. However, an increase in the rod diameter d can compensate for the larger spacing, keeping the gain essentially constant.

For operation at a single frequency the dimensions for f_1, f_2 or f_3 may be used depending on the gain desired. However, for f_1 or f_2 values of S, L, R and l, f_3 values may be used for s and d, resulting in fewer reflector elements.

A closed-sleeve dipole, as shown in Fig. 12-15, may be used as the driven element in place of the Brown-Woodward dipole (see Fig. 16-13c for details of the

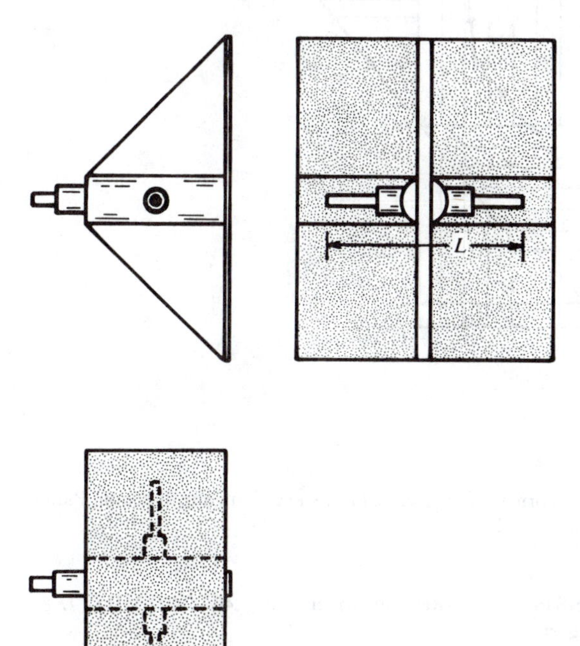

Figure 12-15 Square-corner (solid sheet) reflector with sleeve dipole for wideband operation in side, front and top (or bottom) views. $L = \lambda$ at the highest frequency. Drawing is to scale. (See Fig. 16-13c for dipole details.)

sleeve dipole). Wong and King[1] describe an open-sleeve dipole as the driven element. See also the corner–Yagi-Uda antenna of Sec. 11-9c and the Yagi-Uda–corner–log-periodic antenna of Sec. 15-6.

If a 3-dimensional square corner (see Fig. 12-16) is *driven* by a $\lambda/2$ to $3\lambda/4$ *monopole* spaced 0.9λ from the corner with $d = 2\lambda$, a beam is obtained in a direction making approximately equal angles with the 3 coordinate axes, and Inagaki[2] reports a gain of 17 dBi with higher gains for smaller corner angles. By placing a cylindrical surface of constant radius from the apex of a corner reflector (of 2 or 3 dimensions), Elkamchouchi[3] reports improved performance.

To distinguish the corner reflector with dipole (driven element) discussed in this section from corners without dipoles (or monopoles), the ones with the driven element may be called *active* (Kraus) corners and the others *passive* (retro) corners. While the active corners may have any included angle, with 90° the most common, the passive corners are *always* square (90°). These corners are discussed further in the next section.

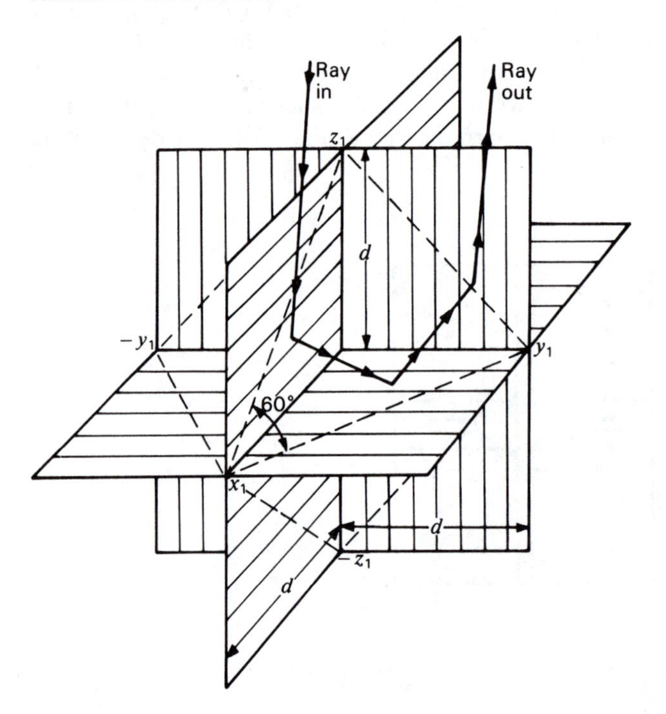

Figure 12-16 Retroreflector of 8 square corners for reflecting back waves from any direction. Path of ray returning via triple bounce is shown.

---

[1] J. L. Wong and H. E. King, "A Wide-Band Corner-Reflector Antenna for 240 to 400 MHz," *IEEE Trans. Ants. Prop.*, **AP-33**, 891–892, August 1985.

[2] N. Inagaki, "Three Dimensional Corner Reflector Antennas," *IEEE Trans. Ants. Prop.*, **AP-22**, 580, July 1974.

[3] H. M. Elkamchouchi, "Cylindrical and Three-Dimensional Corner Reflector Antennas," *IEEE Trans. Ants. Prop.*, **AP-31**, 451–455, May 1983.

12-3b Passive (Retro) Corner Reflector. A passive (retro) corner consisting of 2 flat reflecting sheets is shown in Fig. 12-1*e*.

With 3 mutually perpendicular reflecting sheets, as in Fig. 12-16, each sheet extending a distance d ($= \pm x_1 = \pm y_1 = \pm z_1$) from the origin, we have a cluster of eight *3-dimensional square-corner retroreflectors.* Each square corner occupies one octant ($4\pi/8$ sr $= \pi/2$ sr $= 5157$ square deg). Any ray or wave incident within this solid angle is reflected back in the same direction, as suggested in the figure. Together the cluster of 8 square corners form a retroreflector for waves from *any direction* within a full sphere solid angle ($=4\pi$ sr $= 41\,253$ square deg) with the equivalent (normal incidence) flat-sheet reflecting area being a function of the angle of incidence. Its maximum value is $\sqrt{3}\,d^2$ except that in the 6 directions, exactly on the 3 axes (x, y and z), the area is $4d^2$. Just off these directions the area approaches zero.

The enhanced reflection of radar signals from such passive corner clusters makes them useful for many applications. For example, small water-craft commonly carry one (usually with reflecting surfaces of wire mesh) on a tall mast to make the craft's presence more visible on radar screens. To be most effective, the reflector dimensions should be many λ and the periphery of the mesh hole $<\lambda/2$. The surface should also be flat to better than $\lambda/16$ and, to increase the probability that the radar echo will be noticed, the reflector can be rotated to avoid a persistent null condition.

Truncating the sides of the reflecting sheets along the diagonal lines (dashed lines in Fig. 12-16) results in 8 truncated corners, each bounded by an equilateral triangle (included angle 60°) with an aperture area of $\frac{3}{4}d^2$. If the surface of a sphere is divided into triangles (in the manner of a Buckminster Fuller geodesic dome) and a truncated corner inserted in each triangular area, it is possible to array dozens of corner reflectors over the sphere and obtain a more uniform echo area as a function of angle of incidence over 4π sr, but at the expense of a smaller maximum value.

Retroreflectors can also be constructed in other ways. Thus, a Luneburg lens (Sec. 14-10) with a reflecting cap over, say π sr, acts as a retroreflector over this angle. The Van Atta array (Sec. 11-12) is another example.

12-4 THE PARABOLA. GENERAL PROPERTIES.

Suppose that we have a point source and that we wish to produce a plane-wave front over a large aperture by means of a sheet reflector. Referring to Fig. 12-17*a*, it is then required that the distance from the source to the plane-wave front via path 1 and 2 be equal or[1]

$$2L = R(1 + \cos\theta) \tag{1}$$

and

$$R = \frac{2L}{1 + \cos\theta} \tag{2}$$

[1] This is an application of the *principle of equality of path length* (*Fermat's principle*) to the special case where all paths are in the same medium. For the more general situation involving more than one medium see Chap. 14.

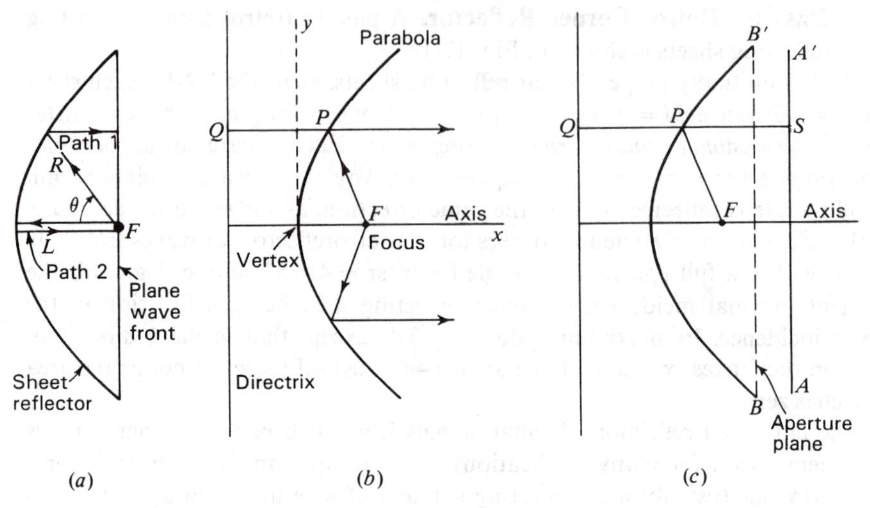

Figure 12-17 Parabolic reflectors.

This is the equation for the required surface contour. It is the equation of a parabola with the focus at F.

Referring to Fig. 12-17b, the parabolic curve may be defined as follows. The distance from any point P on a parabolic curve to a fixed point F, called the *focus*, is equal to the perpendicular distance to a fixed line called the *directrix*. Thus, in Fig. 12-17b, $PF = PQ$. Referring now to Fig. 12-17c, let AA' be a line normal to the axis at an arbitrary distance QS from the directrix. Since $PS = QS - PQ$ and $PF = PQ$, it follows that the distance from the focus to S is

$$PF + PS = PF + QS - PQ = QS \tag{3}$$

Thus, a property of a parabolic reflector is that all waves from an isotropic source at the focus that are reflected from the parabola arrive at a line AA' with equal phase. The "image" of the focus is the directrix, and the reflected field along the line AA' appears as though it originated at the directrix as a plane wave. The plane BB' (Fig. 12-17c) at which a reflector is cut off is called the *aperture plane*.

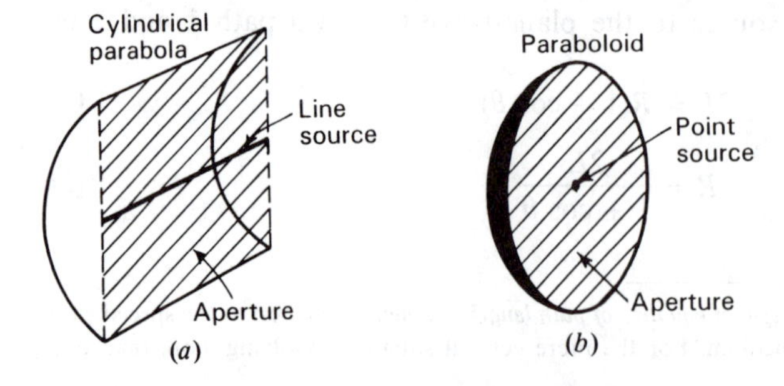

Figure 12-18 Line source and cylindrical parabolic reflector (a) and point source and paraboloidal reflector (b).

A cylindrical parabola converts a cylindrical wave radiated by an in-phase line source at the focus, as in Fig. 12-18*a*, into a plane wave at the aperture, or a paraboloid-of-revolution converts a spherical wave from an isotropic source at the focus, as in Fig. 12-18*b*, into a uniform plane wave at the aperture. Confining our attention to a single ray or wave path, the paraboloid has the property of directing or collimating radiation from the focus into a beam parallel to the axis (see Fig. 12-17).

12-5 A COMPARISON BETWEEN PARABOLIC AND CORNER REFLECTORS. Referring to Fig. 12-17, any radiation from the primary source or feed antenna at the focus of the parabola which is not directed into the parabola is not collimated but is distributed by direct paths over a large solid angle. This is not only inefficient but the distributed radiation can degrade the pattern of the radiation from the parabola. Thus, it is essential that a parabolic reflector have a directional feed which radiates all or most of the energy into the parabola. A corner reflector, on the other hand, does not require a directional feed since the direct and reflected waves are properly combined (image theory). Furthermore, a corner reflector has no specific focal point.

Practical aperture dimensions for a square-corner reflector are 1 to 2λ. For larger apertures parabolic reflectors should be used, and for a large parabola of many λ aperture, a practical choice for the feed can be a corner reflector with a corner angle of 90 to 180° depending on the F/D ratio of the parabola (see Sec. 12-6). Corner angles of about 120° have the advantage that the beam widths are approximately equal in both principal planes.

Although the corner reflector differs in principle from the parabolic reflector, there are situations in which the two may be nearly equivalent. This may be illustrated with the aid of Fig. 12-19. Let a linear antenna be located at the focus

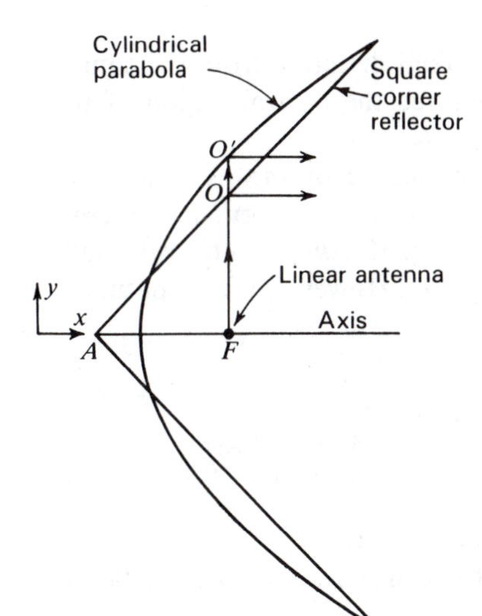

Figure 12-19 Cylindrical parabolic reflector compared with square-corner reflector.

F of a cylindrical parabolic reflector, and let this arrangement be compared with a square-corner reflector of the same aperture and with an antenna-to-corner spacing AF. The parabolic and corner reflectors are superimposed for comparison in Fig. 12-19. A wave radiated in the positive y direction from F is reflected at 0 by the corner reflector and at 0′ by the cylindrical parabolic reflector. Hence, this wave travels a shorter distance in the corner reflector by an amount 00′. If $AF = 2\lambda$, the electrical length of 00′ is about 180° so that a marked difference would be expected in the field patterns of the two reflectors. However, if $AF = 0.35\lambda$ the electrical length of 00′ is only about 30°, and this will cause only a slight difference in the field patterns. It follows that if AF is small in terms of the wavelength the exact shape of the reflector is not of great importance. The practical advantage of the corner reflector is the simplicity and ease of construction of the flat sides.

12-6 THE PARABOLOIDAL REFLECTOR.[1]

The surface generated by the revolution of a parabola around its axis is called a *paraboloid* or a *parabola of revolution.* If an isotropic source is placed at the focus of a paraboloidal reflector as in Fig. 12-20*a*, the portion A of the source radiation that is intercepted by the paraboloid is reflected as a plane wave of circular cross section provided that the reflector surface deviates from a true parabolic surface by no more than a small fraction of a wavelength.

If the distance L between the focus and vertex of the paraboloid is an even number of $\lambda/4$, the direct radiation in the axial direction from the source will be in opposite phase and will tend to cancel the central region of the reflected wave. However, if

$$L = \frac{n\lambda}{4} \tag{1}$$

where $n = 1, 3, 5, \ldots$, the direct radiation in the axial direction from the source will be in the same phase and will tend to reinforce the central region of the reflected wave.

Direct radiation from the source can be eliminated by means of a directional source or primary antenna[2] as in Fig. 12-20*b* and *c*. A primary antenna with the idealized hemispherical pattern shown in Fig. 12-20*b* (solid curve) results in a wave of uniform phase over the reflector aperture. However, the amplitude is

---

[1] S. Silver (ed.), *Microwave Antenna Theory and Design*, McGraw-Hill, New York, 1949.
H. T. Friis and W. D. Lewis, "Radar Antennas," *Bell System Tech. J.*, **26**, 219–317, April, 1947.
C. C. Cutler, "Parabolic Antenna Design for Microwaves," *Proc. IRE*, **37**, 1284–1294, November 1947.
J. C. Slater, *Microwave Transmission*, McGraw-Hill, New York, 1942, pp. 272–276.

[2] It is convenient to refer to the pattern of the source or primary antenna as the *primary pattern* and the pattern of the entire antenna as the *secondary pattern*.

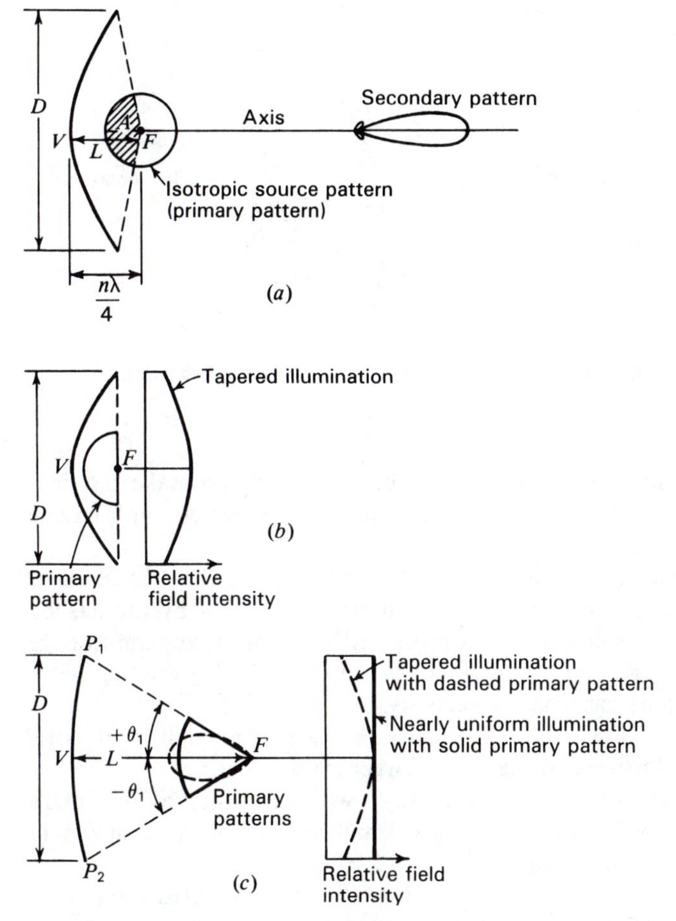

Figure 12-20 Parabolic reflectors of different focal lengths (L) and with sources of different patterns.

tapered as indicated. To obtain a more uniform aperture field distribution or illumination, it is necessary to make θ_1 small, as suggested in Fig. 12-20*c*, by increasing the focal length L while keeping the reflector diameter D constant.[1] If the source pattern is uniform between the angles $\pm\theta_1$ (solid curve), the aperture illumination is more nearly uniform (solid curve) but not entirely so. The path length from F to the edges (at P_1 and P_2) is greater than from F to V (at the vertex). Although ray paths via the edges (at P_1 and P_2) and via the vertex V are of equal total length to the plane wave front (see Fig. 12-17*a*), giving phase equality, there is more path length to the edges in a spherical ($1/r$) attenuating wave.

[1] That is, by increasing the ratio of L to D. This ratio is called the *F ratio*, the *F/D ratio* or the *focal ratio* ($= L/D = FV/D$).

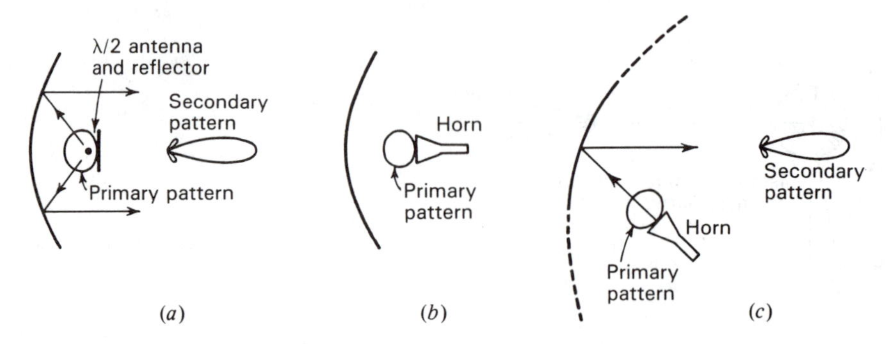

Figure 12-21 Full parabolic reflectors (*a* and *b*) and partial reflector with offset feed (*c*).

Thus, the field at the edges is weaker, as will be calculated. To make the field completely uniform across the aperture would require a feed pattern with *inverse* taper.

A typical pattern for a directional source as indicated by the dashed curve at (*c*) (left) gives a more tapered aperture distribution as shown by the dashed curve at (*c*) (right). The greater amount of taper with resultant reduction in edge illumination may be desirable in order to reduce the minor-lobe level, this being achieved, however, at some sacrifice in directivity.

The arrangement of Fig. 12-20*b* illustrates the case of a small focal ratio. The arrangement at (*c*) illustrates the case of a larger ratio.

Suitable directional patterns may be obtained with various types of primary antennas. As examples, a $\lambda/2$ antenna with a small ground plane is shown in Fig. 12-21*a*, and a small horn antenna in Fig. 12-21*b*.

The presence of the primary antenna in the path of the reflected wave, as in the above examples, has 2 principal disadvantages. These are, first, that waves reflected from the parabola back to the primary antenna produce interaction and mismatching.[1] Second, the primary antenna acts as an obstruction, blocking out the central portion of the aperture and increasing the minor lobes. To avoid both effects, a portion of the paraboloid can be used and the primary antenna displaced as in Fig. 12-21*c*. This is called an *offset feed*.

Let us next develop an expression for the field distribution across the aperture of a parabolic reflector. Since the development is simpler for a cylindrical parabola, this case is treated first, as an introduction to the case for a paraboloid. Consider a *cylindrical parabolic reflector* with line source as in Fig. 12-22*a*. The line source is isotropic in a plane perpendicular to its axis (plane of page). For a

[1] This may be greatly reduced by using a circularly polarized primary antenna, such as an axial-mode helix. If the primary antenna radiation is right-circularly polarized, the wave reflected from the parabola is mostly left-circularly polarized and the primary antenna is insensitive to this polarization.

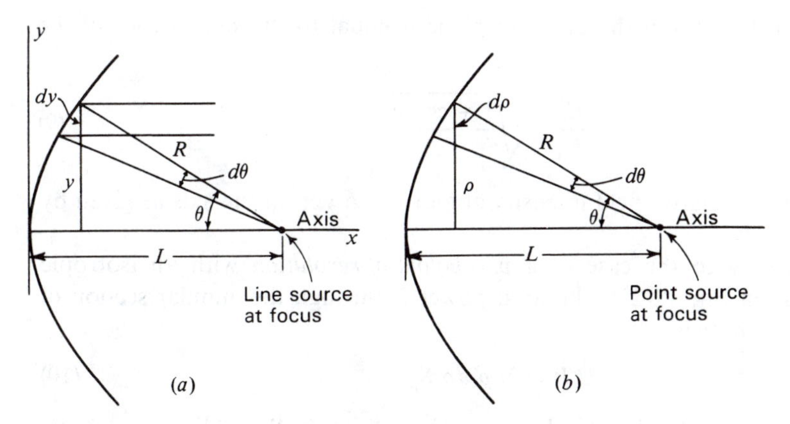

Figure 12-22 Cross sections of cylindrical parabola (*a*) and of paraboloid-of-revolution (*b*).

unit distance in the z direction (normal to page in Fig. 12-22*a*) the power P in a strip of width dy is

$$P = dy\ S_y \tag{2}$$

where S_y = the power density at y, W m^{-2}

However, we also have that

$$P = d\theta\ U' \tag{3}$$

where U' = the power per unit angle per unit length in the z direction

Thus,

$$S_y\ dy = U'\ d\theta \tag{4}$$

and

$$\frac{S_y}{U'} = \frac{1}{(d/d\theta)(R \sin \theta)} \tag{5}$$

where

$$R = \frac{2L}{1 + \cos \theta} \tag{6}$$

This yields

$$S_y = \frac{1 + \cos \theta}{2L} U' \tag{7}$$

The ratio of the power density S_θ at θ to the power density S_0 at $\theta = 0$ is then given by the ratio of (7) when $\theta = \theta$ to (7) when $\theta = 0$, or

$$\frac{S_\theta}{S_0} = \frac{1 + \cos \theta}{2} \tag{8}$$

The field-intensity ratio in the aperture plane is equal to the square root of the power ratio or

$$\frac{E_\theta}{E_0} = \sqrt{\frac{1 + \cos\theta}{2}} \tag{9}$$

where E_θ/E_0 is the relative field intensity at a distance y from the axis as given by $y = R \sin\theta$.

Turning now to the case of a *paraboloid-of-revolution* with an isotropic point source as in Fig. 12-22*b*, the total power P through the annular section of radius ρ and width $d\rho$ is

$$P = 2\pi\rho \; d\rho \; S_\rho \tag{10}$$

where S_ρ = the power density at a distance ρ from the axis, W m^{-2}

This power must be equal to the power radiated by the isotropic source over the solid angle $2\pi \sin\theta \; d\theta$. Thus,

$$P = 2\pi \sin\theta \; d\theta \; U \tag{11}$$

where U = the radiation intensity, W sr^{-1}

Then

$$\rho \; d\rho \; S_\rho = \sin\theta \; d\theta \; U \tag{12}$$

or

$$\frac{S_\rho}{U} = \frac{\sin\theta}{\rho(d\rho/d\theta)} \tag{13}$$

where $\rho = R \sin\theta = \dfrac{2L \sin\theta}{1 + \cos\theta}$

This yields

$$S_\rho = \frac{(1 + \cos\theta)^2}{4L^2} U \tag{14}$$

The ratio of the power density S_θ at the angle θ to the power density S_0 at $\theta = 0$ is then

$$\frac{S_\theta}{S_0} = \frac{(1 + \cos\theta)^2}{4} \tag{15}$$

The field-intensity ratio in the aperture plane is equal to the square root of the power ratio or

$$\frac{E_\theta}{E_0} = \frac{1 + \cos\theta}{2} \tag{16}$$

where E_θ/E_0 is the relative field intensity at a radius ρ from the axis as given by $\rho = R \sin\theta$.

12-7 PATTERNS OF LARGE CIRCULAR APERTURES WITH UNIFORM ILLUMINATION. The radiation from a large paraboloid with uniformly illuminated aperture is essentially equivalent to that from a circular aperture of the same diameter D in an infinite metal plate with a uniform plane wave incident on the plate as in Fig. 12-23. The radiation-field pattern for such a uniformly illuminated aperture can be calculated[1] by applying Huygens' principle in a similar way to that for a rectangular aperture in Chap. 4. The normalized field pattern $E(\phi)$ as a function of ϕ and D is

$$E(\phi) = \frac{2\lambda}{\pi D} \frac{J_1[(\pi D/\lambda) \sin \phi]}{\sin \phi} \tag{1}$$

where D = diameter of aperture, m
λ = free-space wavelength, m
ϕ = angle with respect to the normal to the aperture (Fig. 12-23)
J_1 = first-order Bessel function

The angle ϕ_0 to the first nulls of the radiation pattern are given by

$$\frac{\pi D}{\lambda} \sin \phi_0 = 3.83 \tag{2}$$

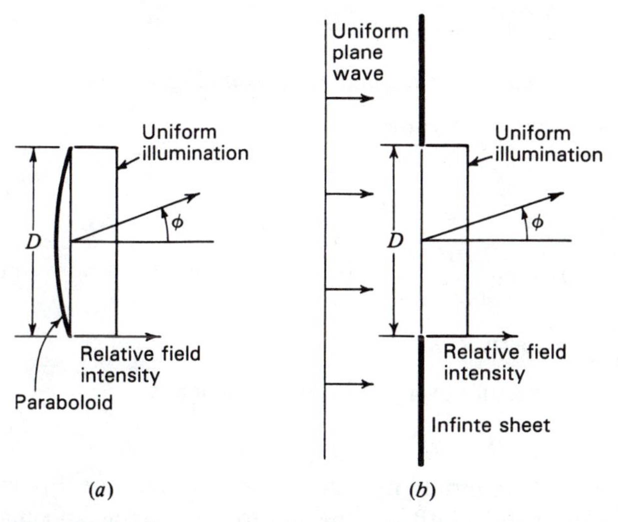

Figure 12-23 Large paraboloid with uniformly illuminated aperture (*a*) and equivalent uniformly illuminated aperture of same diameter D in infinite flat sheet (*b*).

[1] See, for example, J. C. Slater and N. H. Frank, *Introduction to Theoretical Physics*, McGraw-Hill, New York, 1933, p. 325. Also see S. Silver (ed.), *Microwave Antenna Theory and Design*, McGraw-Hill, New York, 1949, p. 194.

since $J_1(x) = 0$ when $x = 3.83$. Thus,

$$\phi_0 = \arcsin \frac{3.83\lambda}{\pi D} = \arcsin \frac{1.22\lambda}{D} \tag{3}$$

When ϕ_0 is very small (aperture large)

$$\phi_0 \simeq \frac{1.22}{D_\lambda} \text{ (rad)} = \frac{70}{D_\lambda} \text{ (deg)} \tag{4}$$

where $D_\lambda = D/\lambda$ = diameter of aperture, λ

The beam width between first nulls is twice this. Hence for large *circular apertures*, the *beam width between first nulls*

$$\text{BWFN} = \frac{140}{D_\lambda} \qquad \text{(deg)} \tag{5}$$

By way of comparison, the beam width between first nulls for a large uniformly illuminated *rectangular aperture* or a long linear array is

$$\text{BWFN} = \frac{115}{L_\lambda} \qquad \text{(deg)} \tag{6}$$

where L_λ = length of aperture, λ

The *beam width between half-power points* for a large circular aperture is[1]

$$\text{HPBW} = \frac{58}{D_\lambda} \qquad \text{(deg)} \tag{7}$$

The directivity D of a large *uniformly illuminated aperture* is given by

$$D = 4\pi \frac{\text{area}}{\lambda^2} \tag{8}$$

For a *circular aperture*

$$D = 4\pi \frac{\pi D^2}{4\lambda^2} = 9.87 D_\lambda^2 \tag{9}$$

where D_λ = the diameter of the aperture, λ

The power gain G of a circular aperture over a $\lambda/2$ dipole antenna is

$$G = 6D_\lambda^2 \tag{10}$$

For example, an antenna with a uniformly illuminated circular aperture 10λ in diameter has a gain of 600 or nearly 28 dB with respect to a $\lambda/2$ dipole antenna ($\simeq$30 dBi).

---

[1] S. Silver (ed.), *Microwave Antenna Theory and Design*, McGraw-Hill, New York, 1949, p. 194.

For a *square aperture*, the directivity is

$$D = 4\pi \frac{L^2}{\lambda^2} = 12.6L_\lambda^2 \tag{11}$$

and the power gain over a $\lambda/2$ dipole is

$$G = 7.7L_\lambda^2 \tag{12}$$

where L_λ = the length of a side, λ

For example, an antenna with a square aperture 10λ on a side has a gain of 770 or nearly 29 dB over a $\lambda/2$ dipole (=31 dBi).

The above directivity and gain relations are for uniformly illuminated apertures at least several wavelengths across. If the illumination is tapered, the directivity and gain are less.

The patterns for a square aperture 10λ on a side and for a circular aperture 10λ in diameter are compared in Fig. 12-24. In both cases the field is assumed to be uniform in both magnitude and phase across the aperture. The patterns are given as a function of ϕ in the xy plane. The patterns in the xz plane are identical to those in the xy plane. Although the beam width for the circular aperture is greater than for the square aperture, the sidelobe level for the circular aperture is smaller. A similar effect could be produced with the square aperture by tapering the illumination.

Beam widths, directivities and gains are summarized in Table 12-3. Beam widths are compared with horn antennas in Table 13-1.

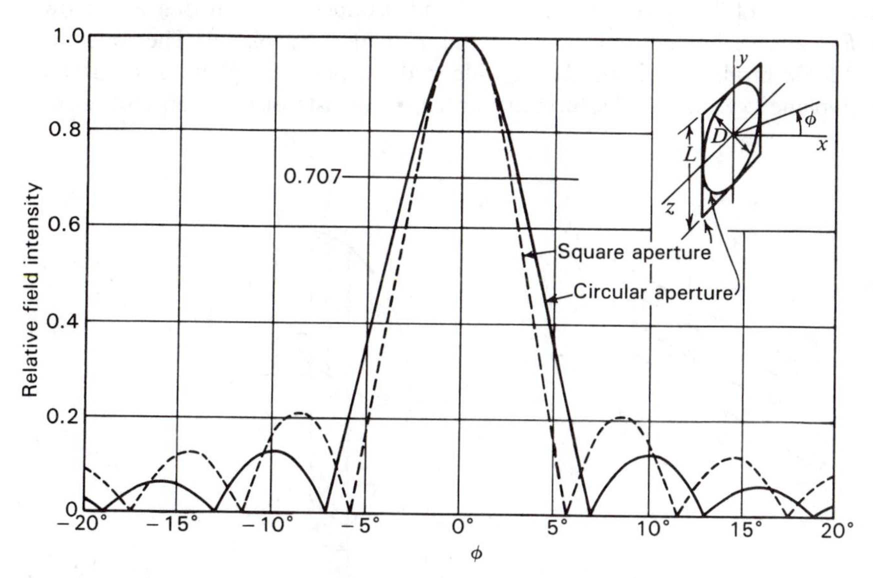

Figure 12-24 Relative radiation patterns of circular aperture of diameter $D = 10\lambda$ and of square aperture of side length $L = 10\lambda$.

Table 12-3 Beam widths, directivities and gains of circular and rectangular apertures with uniform aperture distributions†

	Aperture	
	Circular	**Rectangular**
Half-power beam width	$\frac{58^\circ}{D_\lambda}$	$\frac{51^\circ}{L_\lambda}$
Beam width between first nulls	$\frac{140^\circ}{D_\lambda}$	$\frac{115^\circ}{L_\lambda}$
Directivity (gain over isotropic source)	$9.9D_\lambda^2$	$12.6L_\lambda L'_\lambda$
Gain over $\lambda/2$ dipole	$6D_\lambda^2$	$7.7L_\lambda L'_\lambda$

where D_λ = diameter, λ
L_λ = side length, λ
L'_λ = length of other side (if aperture is square, $L'_\lambda = L_\lambda$)

† Apertures are assumed to be large compared to λ. With tapered distributions beam widths are larger, and directivities, gains and minor lobes are less.

12-8 THE CYLINDRICAL PARABOLIC REFLECTOR. The cylindrical parabolic reflector is used with a line-source type of primary antenna. Two types are illustrated in Fig. 12-25. Both produce fan beams, i.e., a field pattern that is wide in one plane and narrow in the other. The antenna in Fig. 12-25*a* has a line source of 8 in-phase $\lambda/2$ antennas and produces a beam that is narrow in the E plane (xz plane) and wide in the H plane (xy plane). The antenna in Fig. 12-25*b* produces a beam that is wide in the E plane (xz plane) and narrow in the H plane (xy plane). The primary antenna consists of a driven stub element

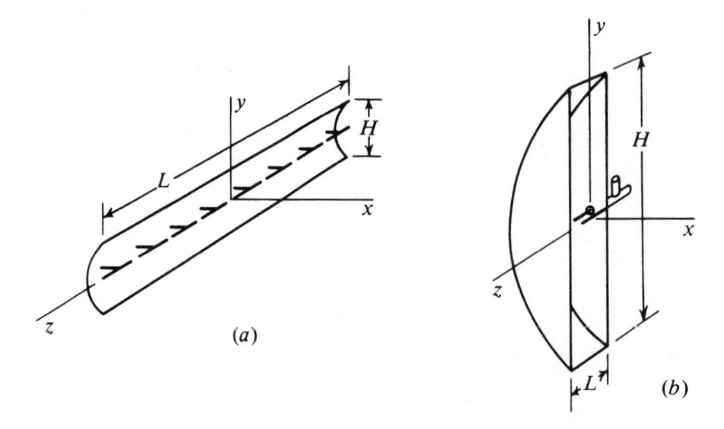

Figure 12-25 Parabolic reflector with linear array of 8 in-phase $\lambda/2$ dipole antennas (a) and "pillbox" or "cheese" antenna (b).

with a reflector element. The driven element is fed by a coaxial line. The side plates act as a parallel-plane waveguide. They guide the radiation from the primary antenna to the parabolic reflector. This type of antenna is called a "pillbox" or "cheese" antenna. If $L < \lambda/2$, propagation between the planes is restricted to the principal or TEM mode. In this case the source may be a stub antenna of length less than $\lambda/2$ (as in Fig. 12-25*b*), or the source may be an open-ended waveguide or small horn. Neglecting edge effects, the patterns of the antennas of Fig. 12-25 are those of rectangular apertures of side dimensions L by H. If the illumination is substantially uniform over the aperture, the relations developed for rectangular apertures in Chap. 4 can be applied to calculate the patterns, provided that the side length is large compared to the wavelength.

12-9 APERTURE DISTRIBUTIONS AND EFFICIENCIES.[1]

As an introduction to aperture distributions and efficiencies, a few basic concepts discussed in earlier chapters are reviewed briefly. Then a number of criteria useful in antenna design are developed.

Let a plane wave of power density S (W m^{-2}) be incident on an antenna, as in Fig. 12-26. The power P delivered by the antenna to the receiver is then

$$P = SA_e \tag{1}$$

where A_e = *effective aperture* of the antenna

or

$$A_e = \frac{P}{S} \tag{2}$$

If ohmic losses are not negligible, as assumed above, we may distinguish

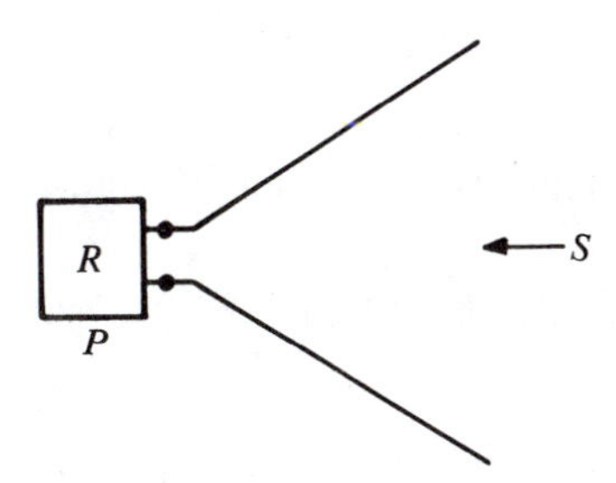

Figure 12-26 Wave of flux density S incident on antenna delivers a power P to the receiver R.

[1] This section (12-9) and the following sections (12-10 and 12-11) are from J. D. Kraus, *Radio Astronomy*, 2nd ed., Cygnus-Quasar, 1986.

between the actual effective aperture (including the effect of ohmic losses) and an effective aperture based entirely on the pattern (losses neglected). Thus, we may write

$$A_e = k_o A_{ep} \tag{3}$$

where A_e = actual effective aperture, m^2
k_o = ohmic-loss factor, dimensionless ($0 \leq k_o \leq 1$)
A_{ep} = effective aperture as determined entirely by *pattern*, m^2

Using these symbols, the directivity of an antenna is given by

$$D = \frac{4\pi}{\Omega_A} = \frac{4\pi}{\lambda^2} A_{ep} \tag{4}$$

from which[1]

$$A_{ep} \Omega_A = \lambda^2 \tag{5}$$

and

$$A_e \Omega_A = k_o \lambda^2 \tag{6}$$

where Ω_A = antenna beam solid angle, sr
λ = wavelength, m

The *aperture efficiency* is defined as the ratio of the effective aperture to the physical aperture, or

$$\varepsilon_{\mathrm{ap}} = \frac{A_e}{A_p} \tag{7}$$

so that the ratio of the aperture and beam efficiencies is

$$\frac{\varepsilon_{\mathrm{ap}}}{\varepsilon_M} = \frac{A_e \Omega_A}{A_p \Omega_M} = \frac{k_o \lambda^2}{A_p \Omega_M} \tag{8}$$

where $\varepsilon_M = \Omega_M / \Omega_A$ = beam efficiency, dimensionless
Ω_M = main beam area, sr
Ω_A = total beam area, sr

Although the physical aperture A_p may not have a unique meaning on some apertures, its value tends to be readily defined on apertures that are large in terms of wavelength.

The directivity of an antenna depends only on the radiation pattern, so that the *directivity D* is given by

$$D = \frac{4\pi}{\lambda^2} A_{ep} \tag{9}$$

---

[1] In (2-22-5) this is given as $A_{em} \Omega_A = \lambda^2$ where A_{em} as defined in Chap. 2 is equal to A_{ep} as defined here.

The *gain G* is then

$$G = Dk_o = k_o \frac{4\pi}{\lambda^2} A_{ep} \tag{10}$$

The *maximum directivity* D_m will be defined as the directivity obtainable from an antenna (assumed to be large in terms of λ) if the field is uniform over the aperture, i.e., the physical aperture. Hence,

$$D_m = \frac{4\pi}{\lambda^2} A_p \tag{11}$$

In designing an antenna, one may have a certain *design directivity* D_d one wishes to achieve. In general, this will be less than D_m, since, to reduce sidelobes, some taper will probably be introduced into the aperture distribution. Thus we may write

$$D_d = \frac{4\pi}{\lambda^2} A_p k_u = D_m k_u \tag{12}$$

where D_d = design directivity

The factor k_u is called the *utilization factor*. It is the ratio of the directivity chosen by design to that obtainable with a uniform aperture distribution, or

$$k_u = \frac{D_d}{D_m} \qquad 0 \le k_u \le 1 \tag{13}$$

After designing and building the antenna and measuring its performance, it will probably be found, however, that the actual directivity D is less than the design directivity D_d. It is possible, but unlikely, that D exceeds D_d. The actual directivity may then be expressed as

$$D = D_d k_a = D_m k_u k_a \tag{14}$$

where k_a is the *achievement factor*, which is a measure of how well the objective has been achieved. Thus,

$$k_a = \frac{D}{D_d} \qquad 0 \le k_a \le 1 \text{ (usually)} \tag{15}$$

The gain G of the antenna can now be written as

$$G = D_m k_o k_u k_a = \frac{4\pi}{\lambda^2} A_p k_o k_u k_a = \frac{4\pi}{\lambda^2} A_p \varepsilon_{\text{ap}} \tag{16}$$

where A_p = physical aperture, m^2
k_o = ohmic efficiency factor, dimensionless
k_u = utilization factor, dimensionless
k_a = achievement factor, dimensionless
ε_{ap} = aperture efficiency, dimensionless

We may also write

$$A_{ep} = k_u k_e A_p \tag{17}$$

$$A_e = k_o k_u k_a A_p \tag{18}$$

$$\varepsilon_{\text{ap}} = \frac{A_e}{A_p} = k_o k_u k_a \tag{19}$$

where A_{ep} = effective aperture (as determined entirely by pattern)
A_e = actual effective aperture

A basic definition for the directivity of an antenna is that the directivity is equal to the ratio of the maximum to the average radiation intensity from the antenna (assumed transmitting). By reciprocity the directivity will be the same in the receiving case. Hence,

$$D = \frac{U_m}{U_{\text{av}}} \tag{20}$$

where U_m = maximum radiation intensity, W sr^{-1}
U_{av} = average radiation intensity, W sr^{-1}

The average value may be expressed as the integral of the radiation intensity $U(\theta, \phi)$ over a solid angle of 4π divided by 4π. Thus

$$D = \frac{U_m}{\dfrac{1}{4\pi}\displaystyle\iint_{4\pi} U(\theta, \phi)\, d\Omega} = \frac{4\pi U_m}{P} \tag{21}$$

where P is the total power radiated. From (21) the effective power (power which would need to be radiated if the antenna were isotropic) is

$$DP = 4\pi U_m \tag{22}$$

Let us now consider an aperture under two conditions. Under Condition 1 the field distribution is assumed to be uniform, so that the directivity is D_m. The power radiated is P'. Under Condition 2 the field has its actual distribution, and the directivity and power radiated are D and P respectively. If the effective powers are equal in the two conditions,

$$DP = D_m P' \tag{23}$$

and

$$D = D_m \frac{P'}{P} = \frac{4\pi}{\lambda^2} A_p \frac{\dfrac{E_{\text{av}} E_{\text{av}}^*}{Z} A_p}{\displaystyle\iint_{A_p} \frac{E(x, y)E^*(x, y)}{Z}\, dx\, dy} \tag{24}$$

In (24) the surface of integration has been collapsed over the antenna, with part of the surface coinciding with the aperture. Further, it is assumed that all the power radiated flows out through the aperture. In (24) Z is the intrinsic impedance of the medium (Ω per square) and E_{av} is the average field across the aperture, as given by

$$E_{av} = \frac{1}{A_p} \iint_{A_p} E(x, y)\, dx\, dy \tag{25}$$

where $E(x, y)$ is the field at any point (x, y) of the aperture. Rearranging (24), we have for the actual directivity

$$D = \frac{4\pi}{\lambda^2} A_p \frac{1}{\dfrac{1}{A_p} \displaystyle\iint_{A_p} \left[\frac{E(x, y)}{E_{av}}\right]\left[\frac{E(x, y)}{E_{av}}\right]^* dx\, dy} \tag{26}$$

This relation was developed originally by Ronald N. Bracewell.[1] Following Bracewell's discussion and also elaborating on it by introducing the utilization factor, we can write for the design directivity

$$D_d = \frac{4\pi}{\lambda^2} A_p \frac{1}{\dfrac{1}{A_p} \displaystyle\iint_{A_p} \left[\frac{E'(x, y)}{E'_{av}}\right]\left[\frac{E'(x, y)}{E'_{av}}\right]^* dx\, dy} \tag{27}$$

where the primes indicate the design field values. The right-hand factor in (27) may be recognized as the *utilization factor* k_u. Multiplying and dividing (26) by k_u, as given in (27), yields

$$D = \frac{4\pi}{\lambda^2} A_p \frac{1}{\dfrac{1}{A_p} \displaystyle\iint_{A_p} \left[\frac{E'(x, y)}{E'_{av}}\right]\left[\frac{E'(x, y)}{E'_{av}}\right]^* dx\, dy} \times \frac{\dfrac{1}{A_p} \displaystyle\iint_{A_p} \left[\frac{E'(x, y)}{E'_{av}}\right]\left[\frac{E'(x, y)}{E'_{av}}\right]^* dx\, dy}{\dfrac{1}{A_p} \displaystyle\iint_{A_p} \left[\frac{E(x, y)}{E_{av}}\right]\left[\frac{E(x, y)}{E_{av}}\right]^* dx\, dy} \tag{28}$$

We also have, from (14),

$$D = D_m k_u k_a \tag{29}$$

[1] R. N. Bracewell, "Tolerance Theory of Large Antennas," *IRE Trans. Ants. Prop.*, **AP-9**, 49–58, January 1961.

Thus, the last factor in (28) is the *achievement factor*, a result given by Bracewell. Further, as done by Bracewell, let

$$E(x, y) = E_{av} + \delta E_{av} \tag{30}$$

or

$$\frac{E(x, y)}{E_{av}} = 1 + \delta \tag{31}$$

where δ is the *complex deviation* (*factor*) of the field from its average value. Thus, the denominator of the last factor in (28) may be written as

$$\frac{1}{A_p} \iint_{A_p} (1 + \delta)(1 + \delta)^* \, dx \, dy \tag{32}$$

Since the average of δ or δ^* over the aperture is zero, (32) simplifies to

$$1 + \frac{1}{A_p} \iint_{A_p} \delta\delta^* \, dx \, dy = 1 + \text{var } \delta \tag{33}$$

where var δ is used to signify the *variance* of δ or average of $\delta\delta^*$ ($= |\delta|^2$) over the aperture. Thus, (28) may be written more concisely as

$$D = \frac{4\pi}{\lambda^2} A_p \frac{1}{1 + \text{var } \delta'} \frac{1 + \text{var } \delta'}{1 + \text{var } \delta} \tag{34}$$

where δ' = design value
δ = actual value

We also have that the *utilization factor*

$$k_u = \frac{1}{1 + \text{var } \delta'} \tag{35}$$

and the *achievement factor*

$$k_a = \frac{1 + \text{var } \delta'}{1 + \text{var } \delta} \tag{36}$$

Turning now to the *beam efficiency* ε_M, or the ratio of the solid angle of the main beam Ω_M to the total beam solid angle Ω_A, we have

$$\varepsilon_M = \frac{\Omega_M}{\Omega_A} = \frac{\displaystyle\iint_{\substack{\text{main}\\\text{lobe}}} P(\theta, \phi) \, d\Omega}{\displaystyle\iint_{4\pi} P(\theta, \phi) \, d\Omega} \qquad 0 \le \varepsilon_M \le 1 \tag{37}$$

where $P(\theta, \phi)$ = antenna power pattern ($= EE^* = |E|^2$)

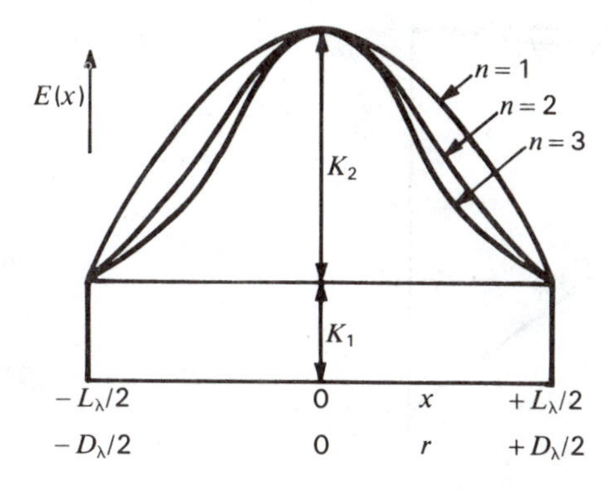

Figure 12-27 Various shapes of aperture distribution.

Let us consider next the effect of the aperture field distribution on the beam and aperture efficiencies. We take first the simplest case of a 1-dimensional distribution, i.e., a rectangular aperture (L_λ by L'_λ) with a uniform distribution in the y direction (aperture L'_λ) and a distribution in the x direction (aperture L_λ) as in Fig. 12-27, given by

$$E(x) = K_1 + K_2\left[1 - \left(\frac{2x}{L_\lambda}\right)^2\right]^n \tag{38}$$

where $E(x)$ = field distribution
K_1 = constant (see Fig. 12-27)
K_2 = constant (see Fig. 12-27)
$L_\lambda = L/\lambda$ = aperture width, λ
n = integer (= 1, 2, 3, ...)

If $K_2 = 0$, the distribution is uniform. If $K_1 = 0$, the distribution is parabolic for $n = 1$ and more severely tapered toward the edges for larger values of n, as indicated in Fig. 12-27. The beam and aperture efficiency of a 1-dimensional aperture with $n = 1$ has been calculated by Nash[1] as a function of the ratio $K_1/(K_1 + K_2)$, with the results shown in Fig. 12-28. For $K_1 = 0$, the distribution tapers to zero at the edge (maximum taper). As the abscissa value in Fig. 12-28 [ratio $K_1/(K_1 + K_2)$] increases, the taper decreases until at an abscissa value of 1 ($K_2 = 0$), there is no taper; i.e., the distribution is uniform.[2] The curves of Fig. 12-28 show that the beam efficiency tends to increase with an increase in taper but the aperture efficiency decreases. *Maximum aperture efficiency occurs for*

[1] R. T. Nash, "Beam Efficiency Limitations of Large Antennas," *IEEE Trans. Ants. Prop.*, **AP-12**, 918–923, December 1964.

[2] The step function or pedestal K_1 is an idealization which is not physically realizable. The actual field cutoff must be more gradual. Rhodes shows that the electric field component perpendicular to the edge of a planar structure must vanish as the first power of the distance from the edge and that the component parallel to the edge must vanish as the second power. See D. R. Rhodes, "On a New Condition for Physical Realizability of Planar Antennas," *IEEE Trans. Ants. Prop.*, **AP-19**, 162–166, March 1971.

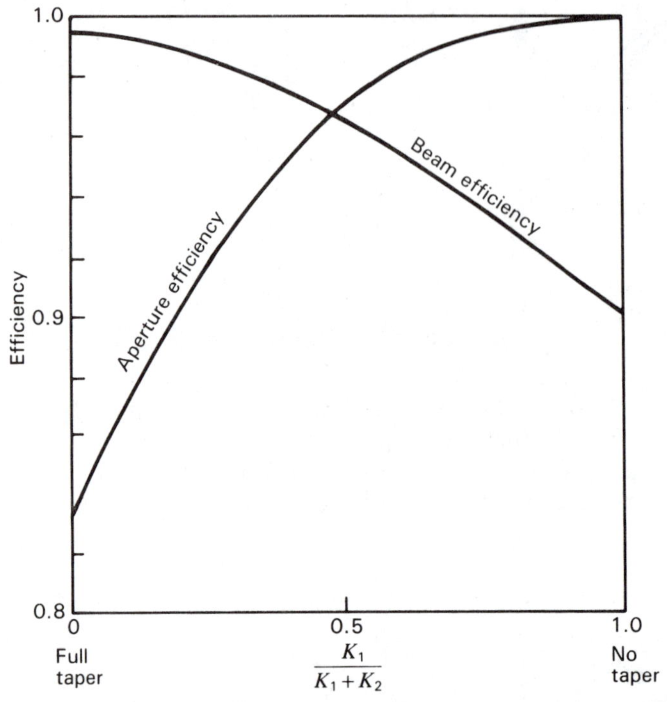

Figure 12-28 Beam and aperture efficiencies for a 1-dimensional aperture as a function of taper. The aperture efficiency is a maximum with no taper, while the beam efficiency is a maximum with full taper. A parabolic distribution for K_2 is assumed (see $n = 1$ in Fig. 12-27). (*After R. T. Nash, "Beam Efficiency Limitations for Large Antennas,"* IEEE Trans. Ants Prop., **AP-12**, *918–923, December 1964.*)

a uniform aperture distribution, but maximum beam efficiency occurs for a highly tapered distribution. In most cases a taper is used that is intermediate between the two extremes of Fig. 12-28 $[K_1/(K_1 + K_2) = 0 \text{ or } 1]$, and a compromise is reached between large beam and aperture efficiencies.

For a 2-dimensional aperture distribution, i.e., a rectangular aperture (L_λ by L'_λ) with the same type of distribution in both the x and y directions, the beam and aperture efficiencies as a function of taper have been calculated by Nash.[1] A field distribution as given by (38) with $n = 1$ is assumed (K_2 part of distribution parabolic). Thus, for this case

$$E(x, y) = \left\{K_1 + K_2\left[1 - \left(\frac{2x}{L_\lambda}\right)^2\right]\right\}\left\{K_1 + K_2\left[1 - \left(\frac{2y}{L'_\lambda}\right)^2\right]\right\} \tag{39}$$

[1] R. T. Nash, "Beam Efficiency Limitations of Large Antennas," *IEEE Trans. Ants. Prop.*, **AP-12**, 918–923, December 1964.

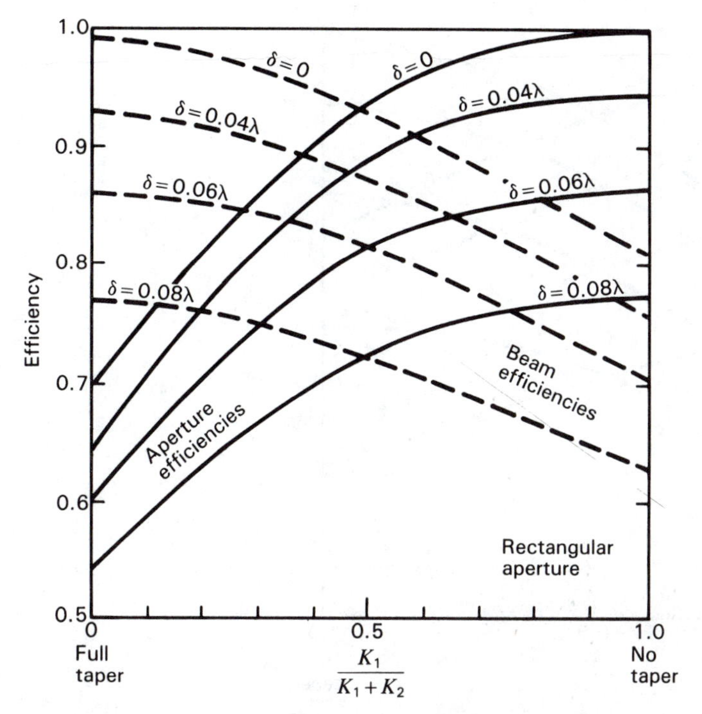

Figure 12-29a Aperture efficiency (solid) and beam efficiency (dashed) of a rectangular aperture as a function of taper and phase error. A parabolic distribution for K_2 is assumed (see $n = 1$ in Fig. 12-27). (*After R. T. Nash, "Beam Efficiency Limitations of Large Antennas,"* IEEE Trans. Ants. Prop., **AP-12**, *918–923, December 1964*).

Nash's data, which also show the effect of random phase errors across the aperture, are presented in Fig. 12-29a. There is a family of 4 curves for the aperture efficiency (solid) and another family of 4 curves for the beam efficiency (dashed) for random displacements (or errors) of 0, 0.04, 0.06 and 0.08λ rms deviation. To obtain these curves the aperture and beam efficiencies calculated for the smooth distribution (39) were multiplied by the gain-degradation (or gain-loss) factor of Ruze,[1] given by

$$k_g = e^{-(2\pi\delta/\lambda)^2} \tag{40}$$

where δ is the rms phase front displacement from planar over the aperture. It is assumed that the correlation intervals of the deviations are greater than the wavelength. The curves of Fig. 12-29a indicate that the controlling effect of the taper on the efficiencies (beam and aperture) tends to decrease as the phase error increases. The efficiencies are also reduced by the presence of the phase error,

[1] J. Ruze, "Physical Limitations of Antennas," MIT Res. Lab. Electron. Tech. Rept. 248, 1952.

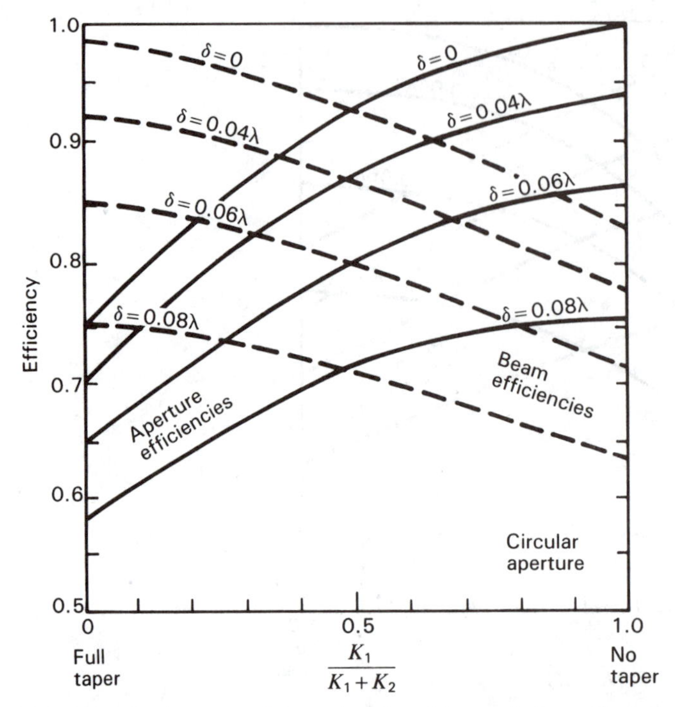

Figure 12-29b Aperture efficiency (solid) and beam efficiency (dashed) of a circular aperture as a function of taper and phase error. A parabolic distribution for K_2 is assumed (see $n = 1$ in Fig. 12-27). (*After R. T. Nash, "Beam Efficiency Limitations of Large Antennas,"* IEEE Trans. Ants. Prop., **AP-12,** *918–923, December 1964*).

since such errors tend to scatter radiation into the sidelobe regions. Thus, the phase errors constitute a primary limitation on the antenna efficiency.

The gain-loss formula (40) of Ruze assumes specifically that the deviations from the best-fit paraboloid are random and distributed in a Gaussian manner, that the errors are uniformly distributed, that the region over which the errors are substantially constant is large compared to the wavelength and that the number of such uncorrelated regions is large.

A circular aperture of diameter D_λ with a distribution as given by (38), where x is replaced by r and L_λ by D_λ, has also been investigated by Nash, with the results shown in Fig. 12-29b. Two families of 4 curves each are given for the beam and aperture efficiencies for 4 conditions of random phase error. It is assumed that $n = 1$ (K_2 part of distribution parabolic). The curves of Fig. 12-29b (circular aperture) are seen to be very similar to those of Fig. 12-29a for the rectangular aperture. For reflector antennas with an rms surface deviation δ' it should be noted that the phase front deviation δ in (40) will be approximately twice as large; that is, $\delta = 2\delta'$.

Another problem to be considered with reflector antennas is the efficiency with which the primary, or feed, antenna illuminates the reflector. This may be defined as the *feed efficiency* ε_f, where

$$\varepsilon_f = \frac{\iint_{\Omega_R} P_f(\theta, \phi)\, d\Omega}{\iint_{4\pi} P_f(\theta, \phi)\, d\Omega} \tag{41}$$

where $P_f(\theta, \phi)$ = power pattern of feed
Ω_R = solid angle subtended by reflector as viewed from feed point

If the first null of the feed antenna pattern coincides with the edge of the reflector, the feed efficiency given by (41) is identical with the beam efficiency of the feed.[1]

In the general situation with a reflector antenna, the beam efficiency of the system may be expressed as

$$\varepsilon_M = e^{-(4\pi\delta'/\lambda)^2} \frac{\iint_{\text{main lobe}} P(\theta, \phi)\, d\Omega \iint_{\Omega_R} P_f(\theta, \phi)\, d\Omega}{\iint_{4\pi} P(\theta, \phi)\, d\Omega \iint_{4\pi} P_f(\theta, \phi)\, d\Omega} \tag{42}$$

where δ' = rms surface error of reflector
$P(\theta, \phi)$ = power pattern of reflector due to aperture distribution produced by the feed assuming no phase error
$P_f(\theta, \phi)$ = power pattern of feed

Surface leakage is neglected. If it is appreciable, another factor is required.

An experimental procedure for determining the *beam efficiency* ε_M of a large antenna using a *celestial source* is as follows. The main-beam solid angle Ω_M is evaluated from

$$\Omega_M = k_p\, \theta_{\text{HP}}\, \phi_{\text{HP}} \tag{43}$$

where k_p = factor between about 1.0 for a uniform aperture distribution and 1.13 for a Gaussian power pattern
θ_{HP} = half-power beam width in θ plane, sr
ϕ_{HP} = half-power beam width in ϕ plane, sr

The half-power beam widths are measured, while k_p may be calculated or estimated from the pattern shape. The half-power beam width in right ascension in degrees is given by the observed half-power beam width of a drift profile in

[1] Common practice is to taper the feed illumination by 10 dB or so at the edges of the reflector.

minutes of time multiplied by cos $\delta/4$, where δ is the declination of the observed point source; that is,

$$\text{HPBW(deg)} = \frac{\text{HPBW(min)} \cos \delta}{4} \tag{44}$$

The total beam solid angle Ω_A is obtained from the relation

$$\Omega_A = \frac{k_0 \lambda^2}{A_e} \tag{45}$$

where the effective aperture A_e is determined by observing a celestial point source of known flux density;[1] that is,

$$A_e = \frac{2kT_A}{S} \tag{46}$$

Combining relations, the *beam efficiency* is then given by

$$\varepsilon_M = \frac{\Omega_M}{\Omega_A} = \frac{k_p \theta_{HP} \phi_{HP} 2kT_A}{k_o \lambda^2 S} \tag{47}$$

where k = Boltzmann's constant ($=1.38 \times 10^{-23}$ J K^{-1})
T_A = antenna temperature due to radio source (measured value corrected for cable loss), K
S = flux density of radio source, W m^{-2} Hz^{-1}
λ = wavelength, m
k_o = antenna ohmic-loss factor, dimensionless

and where k_p, θ_{HP} and ϕ_{HP} are as defined in (43).

A procedure for determining the *aperture efficiency* ε_{ap} of a large antenna is to observe a *celestial source* of known flux density to find A_e by (46), from which

$$\varepsilon_{ap} = \frac{A_e}{A_p} \tag{48}$$

where A_p = physical aperture

It is often of interest to the antenna designer, however, to break ε_{ap} up into a number of factors, as in (19), in order to account as completely as possible for all the causes of efficiency degradation. Thus, from knowledge of the antenna structure and its conductivity the ohmic-loss factor k_o may be determined. The utilization factor k_u can be calculated from design considerations. The achievement factor k_a could be calculated from the relation in (28) if the actual fields across the aperture were accurately known; however, these are rarely measured.[2] As an alternative the achievement factor can be separated into many factors

[1] See the table of celestial sources, App. A, Sec. A-8. See also Sec. 18-6e.

[2] See the discussion on p. 615 regarding a *holographic measurement* of the aperture distribution.

involving the random surface error, feed efficiency, aperture blocking, feed displacement normal to axis (squint or coma), feed displacement parallel to axis (astigmatism), etc., of the (parabolic) reflector antenna. It is assumed that each of these subfactors can be independently calculated or estimated.

The aperture efficiency ε_{ap} may then be expressed as

$$\varepsilon_{\text{ap}} = \frac{A_e}{A_p} = k_o k_u k_a = k_o k_u k_1 k_2 \cdots k_N$$

$$= k_o k_u \prod_{n=1}^{N} k_n \tag{49}$$

where k_o = ohmic-loss factor
k_u = utilization factor (by design)
k_a = achievement factor
$k_1 = e^{-(4\pi\delta'/\lambda)^2}$ = random reflector-surface-error (gain-loss) factor = k_g
$k_2 = \varepsilon_f$ = feed-efficiency factor
k_3 = aperture-blocking factor
k_4 = squint factor
k_5 = astigmatism factor
k_6 = surface-leakage factor
etc.

A close agreement between ε_{ap} as calculated by (49) and as measured by (48) does not necessarily mean that the designer has taken all factors properly into account (some could have been overestimated and others underestimated), but it does provide some confidence in understanding the factors involved. However, if there is significant disagreement between the two methods, the designer knows the analysis is incorrect or incomplete in one or more respects. This comparative method has been used by Nash[1] in analyzing the performance characteristics of a 110-m radio telescope (Big Ear). A discussion of tolerances in large antennas is given by Bracewell.[2]

For in-phase fields over a lossless aperture, the aperture efficiency is given from (26) by

$$\varepsilon_{\text{ap}} = \frac{E_{\text{av}}^2}{(E^2)_{\text{av}}} \tag{50}$$

where E = field at any point in aperture
E_{av} = average of E over aperture
$(E^2)_{\text{av}}$ = average of E^2 over aperture

---

[1] R. T. Nash, "A Multi-reflector Meridian Transit Radio Telescope Antenna for the Observation of Waves of Extra-terrestrial Origin," Ph.D. thesis, Ohio State University, 1961.

[2] R. N. Bracewell, "Tolerance Theory of Large Antennas," *IRE Trans. Ants. Prop.*, **AP-9**, 49–58, January 1961.

Example. Determine the aperture efficiency and the beam efficiency $\varepsilon_M = \Omega_M/\Omega_A$ for a 50λ diameter circular aperture with aperture distribution of the form $(1 - r^2)^n$, where $r = 0$ at the center and $r = 1$ at the rim for various values of n. Show the results graphically.

Solution. The aperture efficiency is obtained from (50). The beam efficiency is obtained by integrating the pattern over 4π for Ω_A and over the main beam for Ω_M, using numerical methods. The results are shown in Fig. 12-30 with efficiency as ordinate and sidelobe level as abscissa, with aperture distribution shape as given by n also indicated. For an $n = 2$ distribution the aperture efficiency is about 56 percent and the first sidelobe 31 dB down.

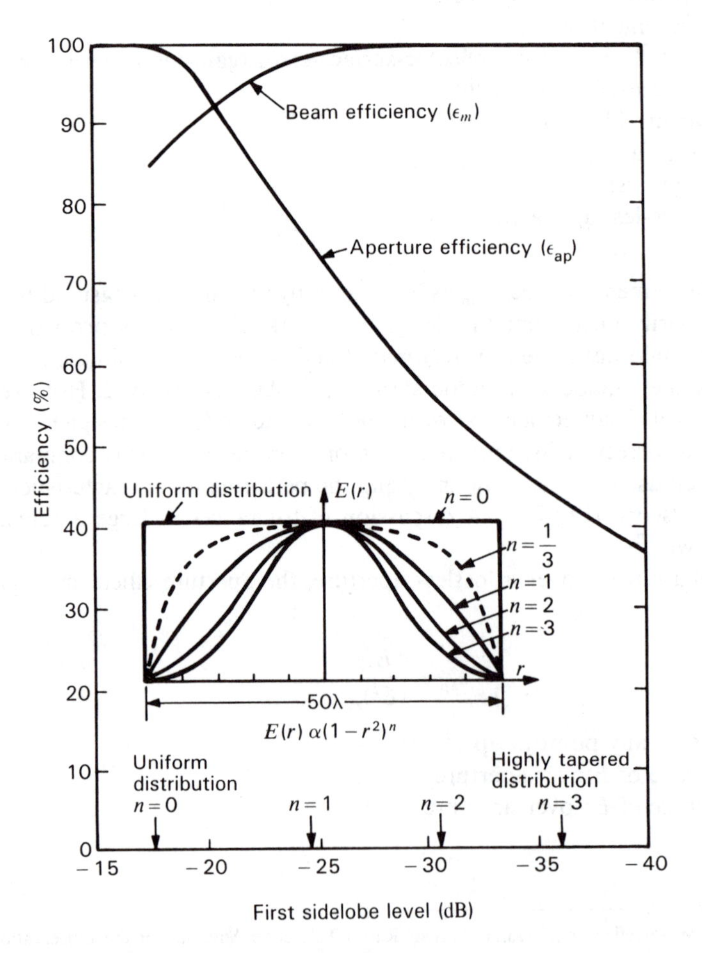

Figure 12-30 Aperture and beam efficiencies for various aperture distributions and sidelobe levels from worked example. Uniform phase is assumed. With phase variation, aperture efficiency decreases and sidelobes increase.

For continuous apertures which are large ($\gg \lambda$) some conclusions are:

1. A uniform amplitude distribution yields the maximum directivity (nonuniform edge-enhanced distributions for supergain being considered impractical).
2. Tapering the amplitude from a maximum at the center to a smaller value at the edges reduces the sidelobe level but results in larger (main-lobe) beam width and less directivity.
3. A distribution with an inverse taper (amplitude depression at center) results in smaller (main-lobe) beam width but in increased sidelobe level and less directivity. (The amplitude depression at the center might be produced inadvertently by a primary (feed) antenna blocking the center of the aperture.)
4. Maximum aperture efficiency occurs for a uniform aperture distribution, but maximum beam efficiency occurs for a highly tapered distribution.
5. Aperture phase errors are a primary limitation on antenna efficiency.
6. Depending on aperture size (in λ) and phase error, there is a frequency (or λ) for which the gain peaks, rolling off to smaller values as the frequency is raised (see Fig. 12-34.)

12-10 SURFACE IRREGULARITIES AND GAIN LOSS.[1]

Referring to Fig. 12-31, consider the idealized case of a reflecting surface with irregularities which depart a distance δ' above and below the ideal surface. Plane waves reflected from the irregular surface will be advanced or retarded with respect to waves reflected from the uniform surface by

$$2\pi \frac{\delta'}{\lambda} \times 2 = 2\pi \frac{\delta}{\lambda} \qquad \text{(rad)} \tag{1}$$

where δ' = surface deviation (surface error)
δ = twice surface deviation = $2\delta'$

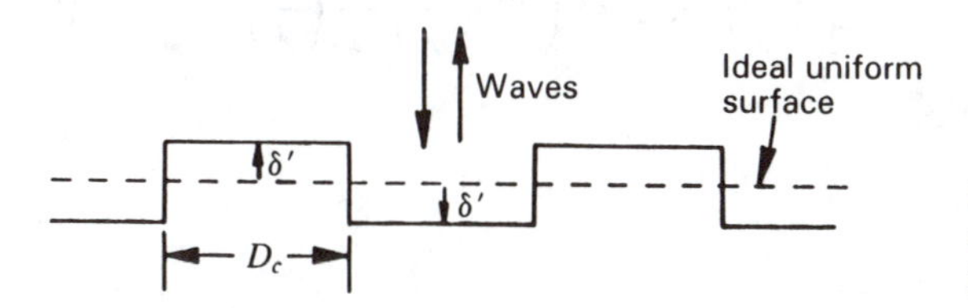

Figure 12-31 Geometry for determining the effect of surface irregularities.

[1] W. N. Christiansen and J. A. Hogbom (*Radiotelescopes*, Cambridge University Press, 1985) give an excellent, lucid discussion of the effect of surface irregularities on gain which served as an inspiration in preparing parts of this section.

Referring to the phase diagram of Fig. 12-32, the resultant field

$$E = E_0 \cos \Delta\phi \tag{2}$$

where $\Delta\phi = 4\pi(\delta'/\lambda)$ rad $= 720° \, (\delta'/\lambda) = 360° \, (\delta/\lambda) =$ phase error

Assuming that the separation (correlation distance D_c) of the irregularities is at least as large as the wavelength and that there are as many positive as negative irregularities, the (normalized) surface gain-loss factor is

$$k_g = \left(\frac{E_0 \cos \Delta\phi}{E_0}\right)^2 = \cos^2 \Delta\phi = \cos^2\left(720° \, \frac{\delta'}{\lambda}\right) \tag{3}$$

where k_g = gain-loss factor ($0 \leq k_g \leq 1$), dimensionless

The variation of the gain-loss factor as a function of rms surface deviation in wavelengths and degrees as calculated from (3) is given in Fig. 12-33. The gain-loss factor as calculated from the Ruze relation (12-9-40) ($=\exp[-(4\pi\delta'/\lambda)^2]$) is also shown. Correlation distances of at least a wavelength are assumed.

Example. A surface has an rms deviation $\delta' = \lambda/20$. Find the reduction in gain.

Solution. From (3),

$$\Delta\phi = 4\pi \frac{\delta'}{\lambda} \text{ (rad)} = 720° \, \frac{\delta'}{\lambda} = 720° \, \frac{1}{20} = 36° \tag{4}$$

and

$$k_g = \cos^2 36° = 0.65 \tag{5}$$

or a gain reduction of 1.8 dB (as may be noted in Fig. 12-33).

Assuming 50 percent aperture efficiency ($A_e = \frac{1}{2}A_p$) and a circular dish diameter D,

$$A_e = \frac{1}{2}\pi\left(\frac{D}{2}\right)^2$$

Introducing the gain-loss factor for surface irregularities from (3),

$$G = k_g \frac{4\pi}{\lambda^2} \frac{\pi}{2}\left(\frac{D}{2}\right)^2 = k_g \frac{\pi^2}{2}\left(\frac{D}{\lambda}\right)^2 = \cos^2\left(720° \, \frac{\delta'}{\lambda}\right)\frac{\pi^2}{2}\left(\frac{D}{\lambda}\right)^2 \tag{6}$$

where δ' = rms surface deviation

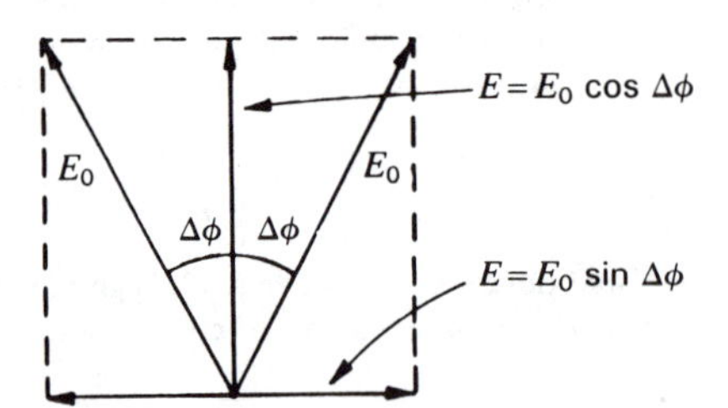

Figure 12-32 Phase variation due to surface irregularities.

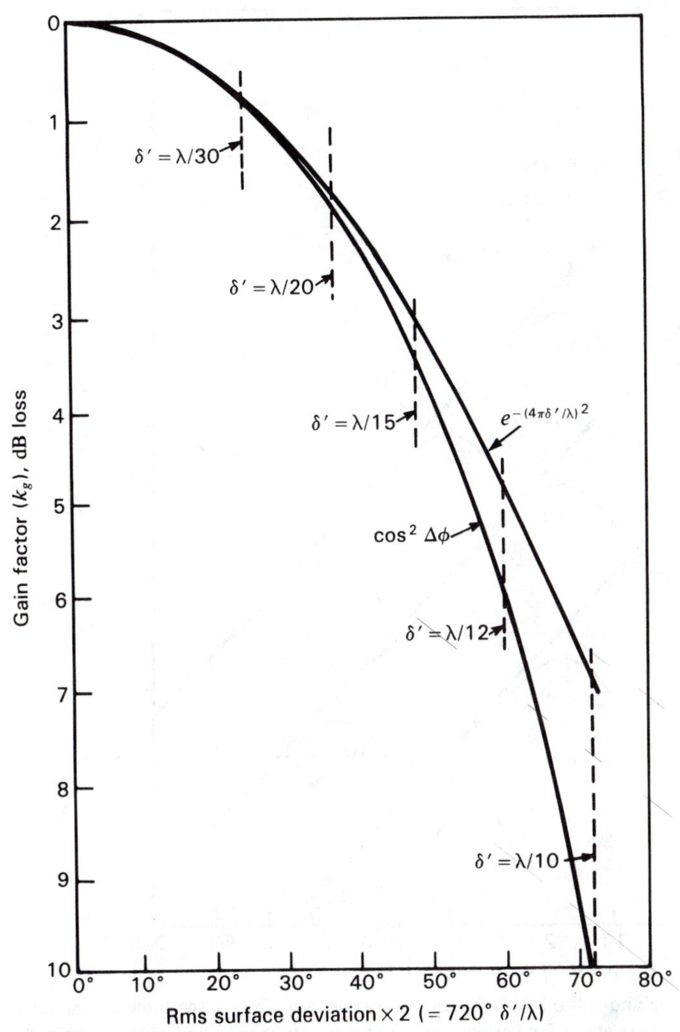

Figure 12-33 Gain-loss factor of reflector antenna as a function of twice the rms surface deviation expressed in degrees of phase angle (of the reflected wave) as calculated from (3) and (12-9-40). The vertical dashed lines correspond to phase angles for rms surface deviations expressed in fractions of a wavelength. See text for assumptions involved.

Values of the gain calculated from (6) are shown in Fig. 12-34 for various surface deviations δ' and circular dish diameters $D = 4$, 20 and 100 m. Measured gains of three radio telescope dishes (dashed curves) are shown for comparison. It may be inferred from Fig. 12-34 that the equivalent rms surface deviations of the Onsala and Nobeyama dishes are about 150 μm and of the Bonn dish about 500 μm.

Referring to Fig. 12-32, the quadrature field components are given by

$$E_q = E_0 \sin \Delta\phi \tag{7}$$

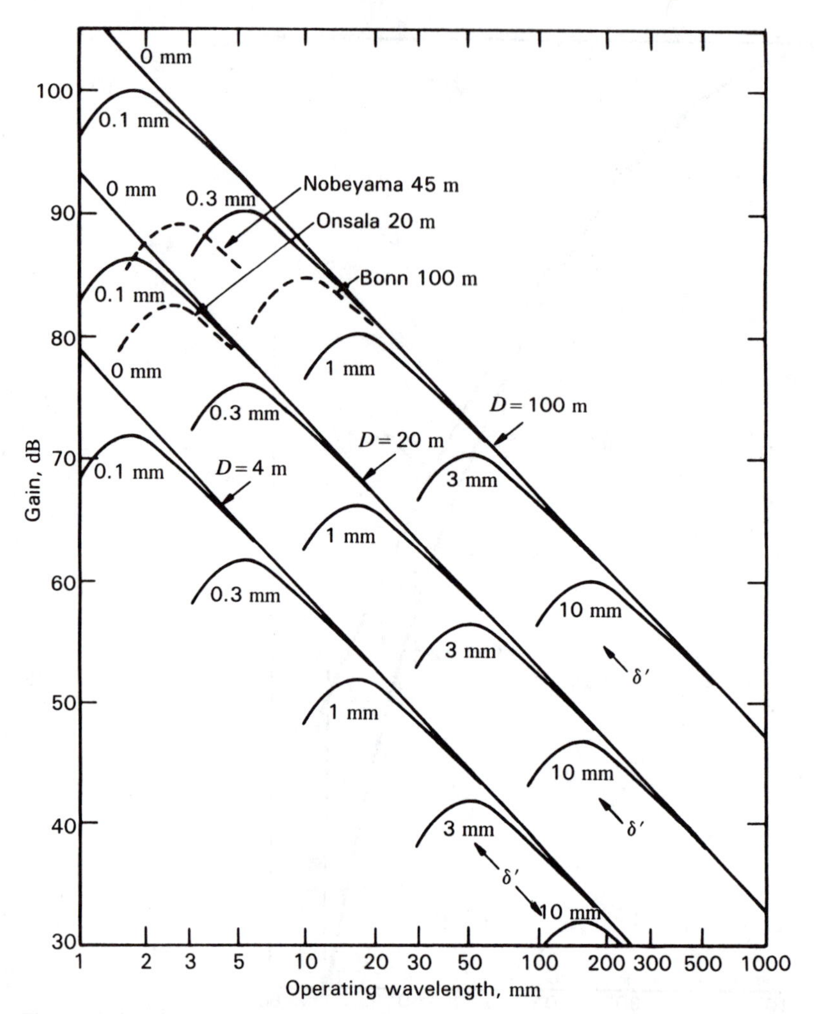

Figure 12-34 Gain peaking and roll-off with decreasing wavelength. The curves show antenna gain as a function of operating wavelength for various rms surface deviations (δ') and antenna diameters (D) equal to 4, 20 and 100 m as calculated from (6). Since scales are logarithmic, other values of D and δ' can be readily interpolated. Curves for Onsala, Nobeyama and Bonn dishes are shown dashed. See text for assumptions involved. Gain is in dBi.

For equal positive and negative irregularities these total zero. Suppose, however, that in some other direction they all *add in phase*. The power level of this minor lobe relative to the main lobe is then

$$\left(\frac{E_{\text{minor}}}{E_{\text{major}}}\right)^2 = \left(\frac{NE_0 \sin \Delta\phi}{NE_0 \cos \Delta\phi}\right)^2 = \tan^2 \Delta\phi \tag{8}$$

where N = number of irregularities

Equation (8) gives the maximum level a minor lobe could have due to these irregularities. Consider, on the other hand, that the quadrature fields *add randomly*. Their total is then $\sqrt{N}\,E_0 \sin \Delta\phi$ and the power level of the resulting minor lobe relative to the main lobe is

$$\left(\frac{E_{\text{minor}}}{E_{\text{major}}}\right)^2 = \left(\frac{\sqrt{N}\,E_0 \sin \Delta\phi}{NE_0 \cos \Delta\phi}\right)^2 = \frac{1}{N}\tan^2 \Delta\phi \tag{9}$$

Example. If the rms phase error is 36° and the number of irregularities $N = 100$, find the maximum and random sidelobe levels.

Solution. From (8),

$$\text{Maximum sidelobe level} = \tan^2 36° = 0.527$$

or down 2.8 dB from the main lobe. From (9),

$$\text{Sidelobe level (random case)} = \frac{1}{N}\tan^2 36° = \frac{0.527}{100} = 0.00527$$

or down 22.8 dB from the main lobe.

Although the gain-loss factor (3) is not a function of the number N of irregularities, the sidelobe level (9) for the random case does depend on N and decreases as N increases. Small-scale irregularities with correlation distances which are small compared to the wavelength result in fields which smooth out at large distances from the reflecting surface and, accordingly, these small-scale irregularities do not affect the gain or sidelobe level as discussed above.

For a reflector antenna with perfect surface the main lobe and near sidelobes are determined principally by the aperture distribution and taper, while the minor lobes at larger angles to 90° or so are influenced by scatter from the feed support structure. Back-radiation is a function of spillover and diffraction around the edge of the reflector. For a circular dish a substantial back lobe may occur on axis (180° from the main lobe) due to all diffracted waves adding in phase (see Fig. 12-35). To reduce diffraction the reflector should have a rolled edge with radius of curvature at least $\lambda/4$ at the longest wavelength of operation.[1]

Due to the force of gravity, a ground-based steerable parabolic reflector deforms as the reflector is tilted so that for a given wavelength there is a maximum diameter which cannot be exceeded by adding metal to the backup structure. However, this limit can be exceeded by a *homologous design* in which one paraboloid deforms into another. The next limit is then imposed by thermal deformation.[2]

---

[1] W. D. Burnside, M. C. Gilreath, B. M. Kent and G. C. Clerici, "Curved Edge Modification of Compact Range Reflector," *IEEE Trans. Ants. Prop.*, **AP-35**, 176–182, Feburary 1987.

[2] S. Von Hoerner, "The Design and Improvement of Tiltable Radio Telescopes," *Vistas in Astron.*, **20**, pt. 4, 411–444, 1977.

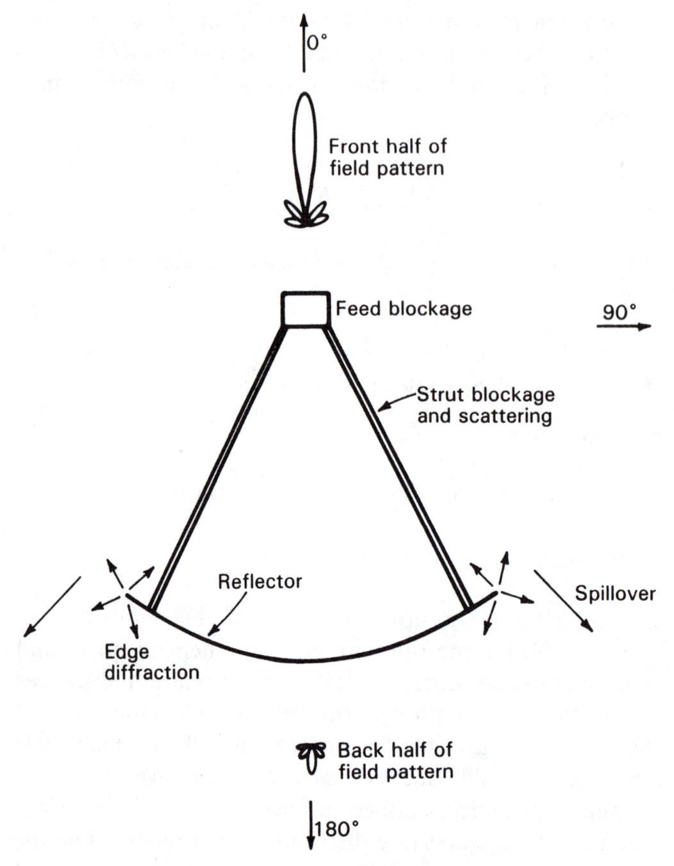

Figure 12-35 Typical parabolic reflector with feed (or subreflector). Front part of field pattern (main lobe and sidelobes) is determined by aperture distribution, surface irregularities and scattering from feed and struts. Back half of pattern (axial back lobe and near lobes) is determined by spillover and diffraction around dish.

12-11 OFF-AXIS OPERATION OF PARABOLIC REFLECTORS.

When the feed of a parabolic reflector antenna is displaced laterally from the focal point the beam is shifted (squinted) off-axis in the opposite direction to the feed displacement. Such squinting is accompanied by gain loss, beam broadening and the appearance of a coma sidelobe. The amount of such degradation in performance is a function of the F/D ratio and the feed displacement, as illustrated in Fig. 12-36.[1]

[1] D. E. Baker, "The Focal Regions of Paraboloidal Reflectors of Arbitrary F/D Ratio," Ph.D. thesis, Ohio State University, 1974.

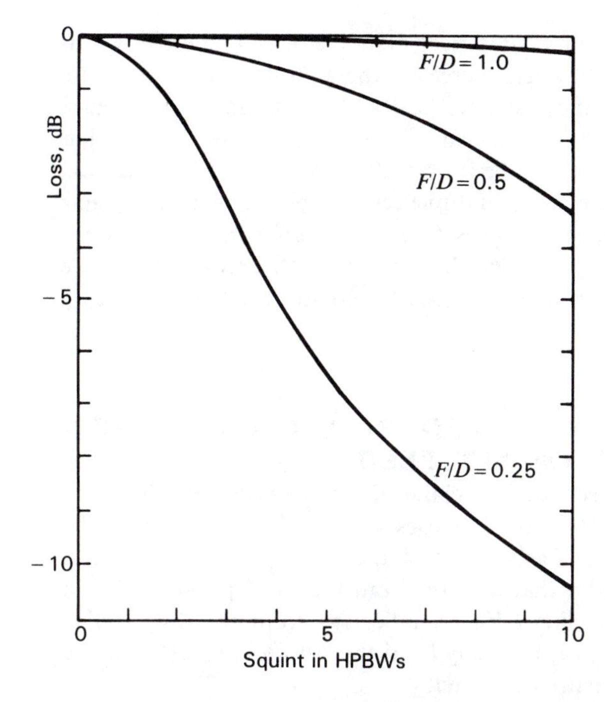

Figure 12-36 Universal squint diagram for parabolic antennas of any size showing the gain loss as a function of the squint displacement in HPBWs for 3 values of F/D ratio. The loss for other F/D ratios may be interpolated. (*After D. E. Baker, "The Focal Regions of Paraboloidal Reflectors of Arbitrary* F/D *Ratio," Ph.D. thesis, Ohio State University, 1974*).

With $F/D = 1.0$, the gain loss is 0.3 dB for a feed displacement of 10 HPBWs but with $F/D = 0.25$ the loss is 10 dB for the same displacement. Thus, a long focal length parabola is more tolerant of feed displacement than a short focal length parabola.

Calculated patterns for a 530λ diameter parabola with $F/D = 1.2$ are shown in Fig. 12-37 for 5 values of squint angle as calculated by Baker.[1] The aperture

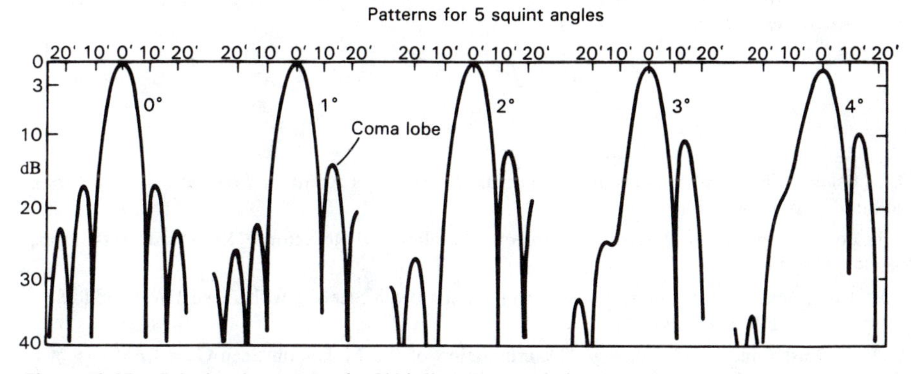

Figure 12-37 Calculated patterns of a 530λ diameter parabola with $F/D = 1.2$ for 5 squint angles. At 4° squint the gain is 1 dB down, the HPBW has increased from 8′ to 8.5′ (arcminutes) and the coma lobe is up 7.1 dB. (*After D. E. Baker, 1977.*)

[1] D. E. Baker, personal communication, 1977.

distribution is uniform, resulting in symmetrical first sidelobes 17.6 dB down as shown for 0° squint. As the squint angle increases, the gain decreases, the beam width increases, the pattern becomes asymmetrical and a first sidelobe increases (coma lobe). Thus, at a squint of 4° the gain is down 1 dB, the HPBW has increased from 8′ to 8.5′ and the coma lobe is only 10.5 dB down (up 7.1 dB from the 0° squint condition). Squinting with multiple feeds deployed near the prime focus (or by a movable feed for tracking) is frequently useful in spite of some degradation in performance. More detailed discussions of off-axis operation are given by Baker[1], Lo,[2] Rudge and Withers,[3] Rusch and Ludwig,[4] Sandler[5] and von Gniss and Ries.[6]

12-12 CASSEGRAIN FEED, SHAPED REFLECTORS, SPHERICAL REFLECTORS AND OFFSET FEED. Sub-reflectors offer flexibility of design for reflecting telescopes. The classical arrangement introduced by N. Cassegrain of France over 300 years ago uses a sub-reflector of hyperbolic shape which surrounds the prime focus point of the main parabolic reflector. Referring to Fig. 12-38, we require that all rays from the focal point F form a spherical wave front (circle of radius CF′) on reflection from the (hyperbolic) sub-reflector (as though radiating isotropically from the parabola focus F′) or by Fermat's principle of equality of path length that

$$C'A' + FA' = CA + FA \tag{1}$$

Noting that $CA = CF' - AF'$ and that $FA - AF' = 2\,OA$ we obtain

$$FA' - A'F' = 2\,OA = BA \tag{2}$$

which is the relation for an hyperbola with standard form

$$\frac{x^2}{a^2} - \frac{y^2}{f^2 - a^2} = 1 \tag{3}$$

[1] D. E. Baker, "The Focal Regions of Paraboloidal Reflectors of Arbitrary *F/D* Ratio," Ph.D. thesis, Ohio State University, 1974.

[2] Y. T. Lo, "On the Beam Deviation Factor of a Paraboloidal Reflector," *IRE Trans. Ants. Prop.*, **AP-8**, 347, May 1960.

[3] A. W. Rudge and M. J. Withers, "New Technique for Beam Steering with Fixed Parabolic Reflectors," *Proc. IEE*, **118**, 857, July 1971.

[4] W. V. T. Rusch and A. C. Ludwig, "Determination of the Maximum Scan-Gain Contours of a Beam-Scanning Paraboloid and Their Relation to the Petzval Surface," *IEEE Trans. Ants. Prop.*, **AP-21**, 141, March 1973.

[5] S. S. Sandler, "Paraboloidal Reflector Patterns for Off-Axis Feed," *IRE Trans. Ants. Prop.*, **AP-8**, 368, July 1960.

[6] H. von Gniss and G. Ries, "Feld Bild um den Brennpunkt von Parabolreflektoren mit Kleinen *F/D* Verhaltnis," *Archiv der Elek. Ubertrag.*, **23**, 481, October 1969.

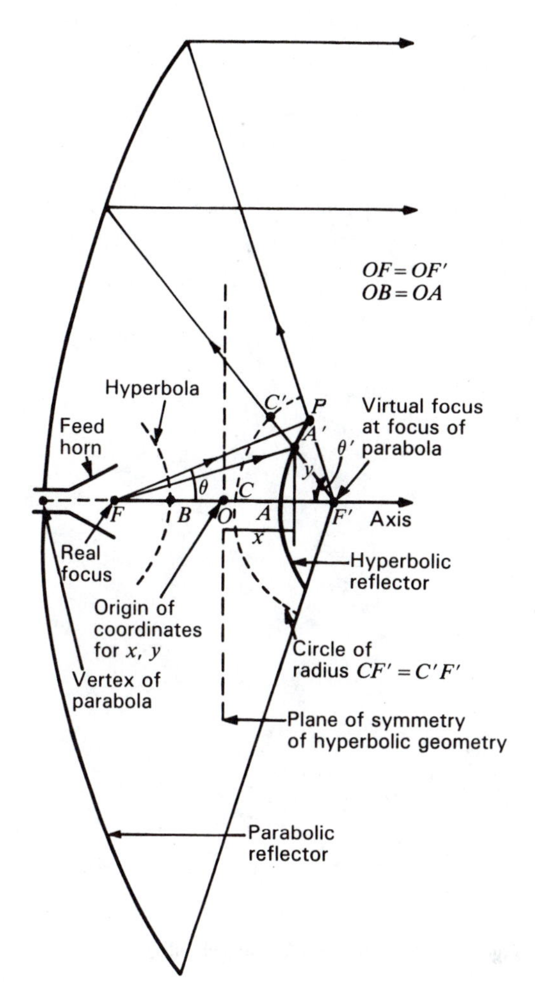

Figure 12-38 Geometry for Cassegrain reflector.

where $a = OA = OB, f = OF' = OF$, and x and y are as shown in Fig. 12-38. Or

$$y^2 = \left(\frac{x^2}{a^2} - 1\right)(f^2 - a^2) \tag{4}$$

The hyperbolic sub-reflector is then truncated at point P for which a ray reflected from the hyperbola hits the edge of the parabolic reflector. The hyperbolic reflector then subtends an angle θ from the feed location at the focal point F while the (main) parabolic reflector subtends an angle θ' from the focal point F' of the parabola. Thus, the feed horn beam angle is increased in the ratio θ'/θ to fill the parabola aperture.

An advantage of the Cassegrain design is its compactness, with feed and amplifier near the vertex of the parabola. Higher aperture efficiency may also be realized by shaping the sub-reflector (or modifying the hyperbolic contour) to

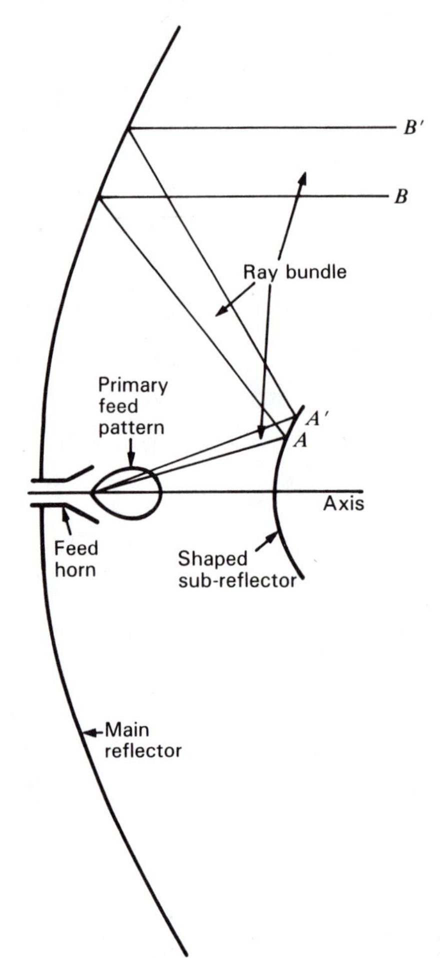

Figure 12-39 Geometry for shaped reflectors.

correct the primary pattern of the feed and produce a more uniform field distribution across the parabola aperture.[1]

Referring to Fig. 12-39, the surface at AA' is deformed to enlarge or restrict the incremental ray bundle BB', thereby decreasing or increasing the watts per steradian in the bundle and finally the watts per square meter in the aperture plane of the parabola. This *shaping technique* may be extended over the entire sub-reflector and often both sub-reflector and parabola are shaped. As a result a more uniform aperture distribution and higher aperture efficiency can be

[1] If the rays are allowed to pass through the focus of the parabola instead of being intercepted by the hyperbolic (convex) Cassegrain surface, as in Fig. 12-38, they can be reflected by a concave ellipsoidal surface beyond (to the right of) the focus. This type of sub-reflector is called *Gregorian* after James Gregory of England who devised it about 1660.

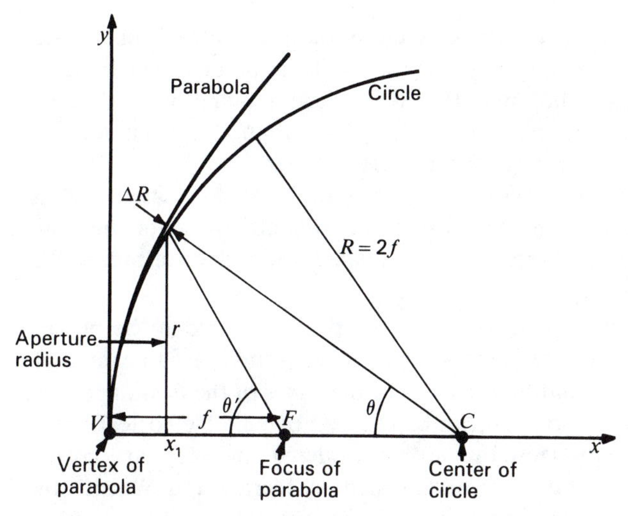

Figure 12-40 Circle and parabola compared, with radius of circle equal to twice the focal length of the parabola.

achieved but with higher first sidelobes and also more rapid gain loss as the feed is moved off-axis to squint the beam.

A constraint on the Cassegrain arrangement is that to minimize blockage the sub-reflector should be small compared to the parabola, yet the sub-reflector must be large compared to the wavelength. Details on Cassegrain reflectors are given by Love.[1]

In Fig. 12-40 a *parabola* is given by

$$y^2 = 4fx \tag{5}$$

where f = focal distance = VF

This parabola is compared with a *circle* of radius $R = VC$. It may be shown that for small values of x, the circle is of nearly the same form as the parabola when

$$R = 2f \tag{6}$$

Over an angle θ and *aperture radius*

$$r = R \sin \theta \tag{7}$$

the circle differs from the parabola by less than ΔR. If $\Delta R \ll \lambda$ (or specifically $<\lambda/16$) the field radiated from a point source at F within an angle θ' and reflected from the circle will be within 45° (=2 × 360°/16) of the phase of a field radiated from F and reflected from the parabola.

[1] A. W. Love, "Spherical Reflecting Antennas with Corrected Line Sources," *IRE Trans. Ants. Prop.*, **AP-10**, 529–537, September 1962.

Let Fig. 12-40 represent the cross section of a reflecting sphere and paraboloid-of-revolution. Then a feed antenna at the focal point F which illuminates the sphere only within the angle θ' will produce a plane wave over the aperture of diameter $2r$ having a phase deviation of less than 45°, this amount of deviation occurring only near the edge of the aperture.

It is apparent that with a *spherical reflector* of 2π sr solid angle, which remains fixed in position, a beam can be scanned over almost all of this solid angle by moving a suitable feed antenna along a spherical surface of radius $R/2$, as suggested in Fig. 12-41.

Although the spherical reflector is fixed and does not need to be moved, only a fraction of its total aperture (πR^2) is used.[1] If the primary feed beam angle θ' is increased, the phase degradation increases at the edges of the beam aperture. The effects of this spherical aberration can be reduced by suitable corrections of the feed distribution. Ashmead and Pippard[2] considered the necessary corrections when the spherical aberration is small. Spencer, Sletten and Walsh[3] and Love[4] describe a phase-correcting line source. Diffraction theory of large spherical reflectors is given by Schell[5] and means of correcting for spherical aberration are discussed by Rumsey.[6]

The most famous example of a large spherical reflector is the 305-m diameter dish of 265-m radius of curvature suspended in a mountain valley at Arecibo, Puerto Rico (see Sec. 12-14 and Figs. 12-47, 12-48). The feeds for this spherical reflector antenna are slotted-waveguide line sources which can be swung through an angle of 20° from vertical, allowing beam scanning over a solid angle around the zenith of 6 percent of the sky above the horizon.[7] At 430 MHz, aperture efficiency is about 53 percent with sidelobes 16 dB down. A discussion of the deficiencies of the Arecibo line-source feeds for circular polarization is given by Kildal[8] with 8 conclusions and recommendations for improvements.

The loss in aperture due to feed antenna blockage with attendant scattering and minor lobes can be avoided by the use of an *offset feed* as in Fig. 12-21*c*.

---

[1] However, with multiple feeds multiple simultaneous beams could be produced.

[2] J. Ashmead and A. B. Pippard, "The Use of Spherical Reflectors as Microwave Scanning Aerials," *JIEE*, **93**, 627–632, 1946.

[3] R. C. Spencer, C. J. Sletten and J. E. Walsh, "Correction of Spherical Aberration by a Phased Line Source," *Proc. Natl. Elect. Conf.*, **5**, 320, 1949.

[4] A. W. Love, "Spherical Reflecting Antennas with Corrected Line Sources," *IRE Trans. Ants. Prop.*, **AP-10**, 529–537, September 1962.

[5] A. C. Schell, "The Diffraction Theory of Large Aperture Spherical Reflector Antennas," *IEEE Trans. Ants. Prop.*, **AP-11**, 428–432, July 1963.

[6] V. H. Rumsey, "On the Design and Performance of Feeds for Correcting Spherical Aberration," *IEEE Trans. Ants. Prop.*, **AP-18**, 343–351, May 1970.

[7] A. W. Love, "Scale Model Development of a High Efficiency Dual Polarized Line Feed for the Arecibo Spherical Reflector," *IEEE Trans. Ants. Prop.*, **AP-21**, 628–639, September 1973.

[8] P-S. Kildal, "Study of Element Patterns and Excitations of the Line Feeds of the Spherical Reflector at Arecibo," *IEEE Trans. Ants. Prop.*, **AP-34**, 197–207, February 1986.

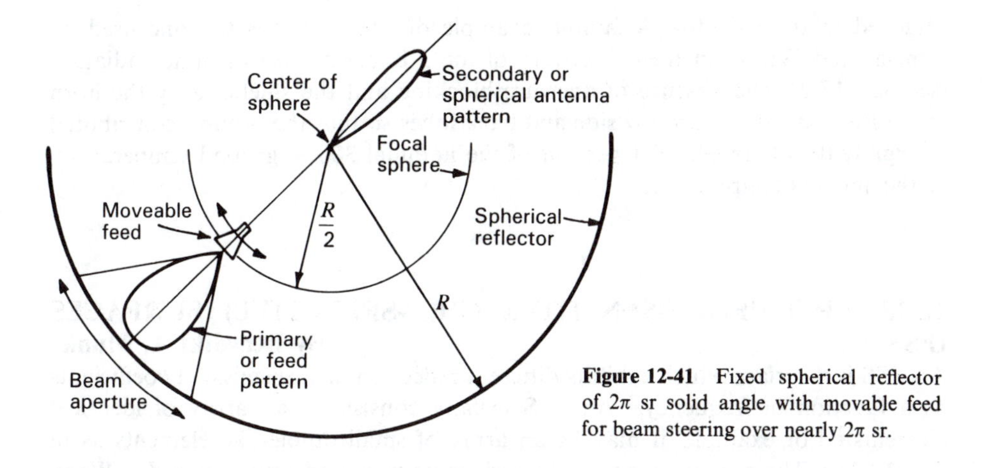

Figure 12-41 Fixed spherical reflector of 2π sr solid angle with movable feed for beam steering over nearly 2π sr.

Although the asymmetry of the offset feed makes full mechanical steerability more awkward for ground-based antennas, the asymmetry is less of a problem under the weightless conditions of space. However, with the standing-curved-reflector tiltable-flat-reflector (Kraus-type) radio telescope, aperture blockage can be made zero or nearly so (see Sec. 12-14 and Fig. 12-51). A fully steerable offset-fed paraboloid is also described in Sec. 12-14 (see Fig. 12-49).

An extreme example of an offset feed is provided by the horn-reflector of Fig. 12-42 in which the energy from the feed point is guided inside a horn

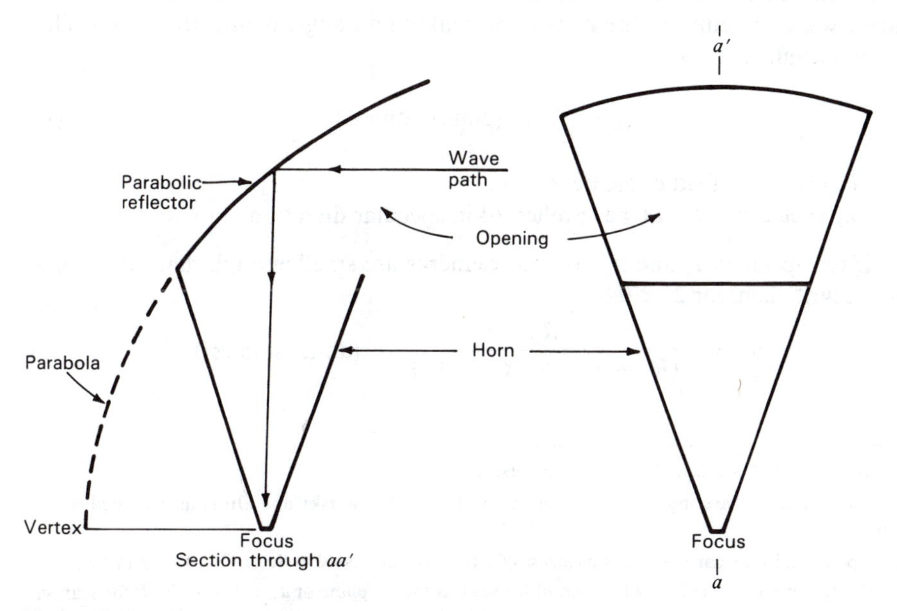

Figure 12-42 "Sugar scoop" low-sidelobe horn-reflector antenna in cross section and front view.

attached to the reflector. A famous example of this design is the one used by Penzias and Wilson in their discovery of the 3 K cosmic background radiation (see Sec. 17-2). The absence of aperture blocking and the shielding by the horn structure resulted in very low side and back lobes so that the ground contributed negligibly (less than $\frac{3}{100}$ of 1 percent of the nominal 300 K ground temperature) to the antenna temperature.

12-13 FREQUENCY-SENSITIVE (OR -SELECTIVE) SURFACES (FSS)

By Benedikt A. Munk[1]

An FSS is a surface which exhibits different reflection or transmission coefficients as a function of frequency.[2] An FSS usually consists of an array of identical elements.[2] For example, it may be an array of small dipole-like elements as in Fig. 12-43*a*. These may be either shorted or have a load impedance Z_L. When the elements resonate and $R_L = \text{Re } Z_L = 0$ the array can provide *complete reflection.* Distinguish between R_L for *load* resistance as used here and R_L for *loss* resistance as used in some earlier sections.

Alternatively, an array of slots in a conducting surface, as in Fig. 12-43*b*, can make the surface *completely transparent.*[3] Frequency-sensitive or -selective surfaces are used extensively for hybrid radomes (a mixture of FSS and dielectric slabs), dichroic (dual-frequency) surfaces, analog absorbers, etc.

Effect of Element Spacings d_x and d_z. Let us consider the reflection or transmission characteristics with frequency of the FSS array in Fig. 12-43*a* for an incident wave traveling in the $\hat{\mathbf{r}}$ direction making an angle θ with the y axis. The reflection coefficient

$$\rho_v = \frac{E_r}{E_i} \qquad \text{(dimensionless)} \qquad (1)$$

where E_i = electric field of incident wave
E_r = electric field of wave reflected in specular direction

If the spacings d_x and d_z between elements are small enough that no grating lobes occur,[4] then, for $2l < 0.6\lambda$,

$$\rho_v = -\frac{R_A}{(R_A + R_L) + j(X_A + X_L)} \qquad \text{(dimensionless)} \qquad (2)$$

[1] ElectroScience Laboratory, Ohio State University.

[2] References to the pioneering work of Kieburtz and Ishimaru, Twersky and Ott *et al.* are listed at the end of the chapter.

[3] The dipoles and slots constitute *complementary* surfaces as discussed in Secs. 13-4, 13-5 and 13-6.

[4] No grating lobes occur if $d_x < \lambda/(1 + \sin\theta)$ for scan in the xy plane or $d_z < \lambda/(1 + \sin\theta)$ for scan in the yz plane.

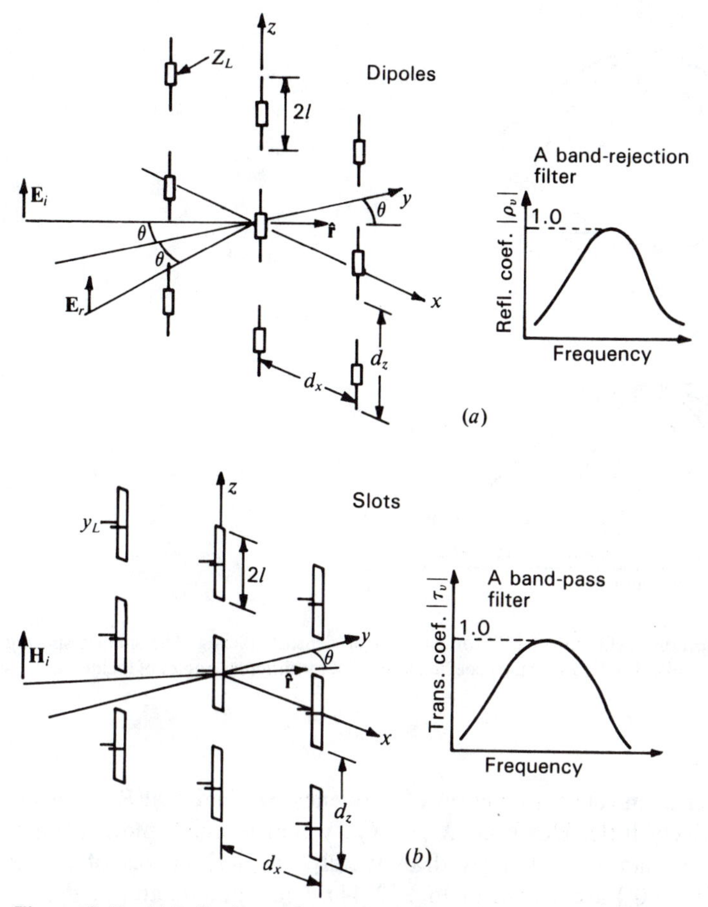

Figure 12-43 (*a*) Reflecting surface of dipole-like elements acts as a *band-rejection* filter. The response curve is shown at right. (*b*) Transparent surface of slots acts as a *bandpass* filter. The response curve is shown at right. The transmission coefficient $\tau_v = 1 + \rho_v$.

where $$R_A = \frac{Z_0}{2d_x\, d_z}\frac{p^2}{\cos\theta},\ \Omega \text{ (for scan in } xy \text{ plane)} \tag{3a}$$

$$R_A = \frac{Z_0\, p^2}{2d_x\, d_z},\ \Omega \text{ (for scan in } yz \text{ plane)} \tag{3b}$$

$$Z_0 = 377\ \Omega$$

$$p = \frac{1}{I_0}\int_{-l}^{l} I(z) \exp(-j\beta\hat{\mathbf{r}}\cdot\mathbf{l})\, dz = \text{pattern function,}^1 \text{ m} \tag{4}$$

I_0 = element terminal current, A

[1] For short elements, p varies little with angle of incidence and frequency.

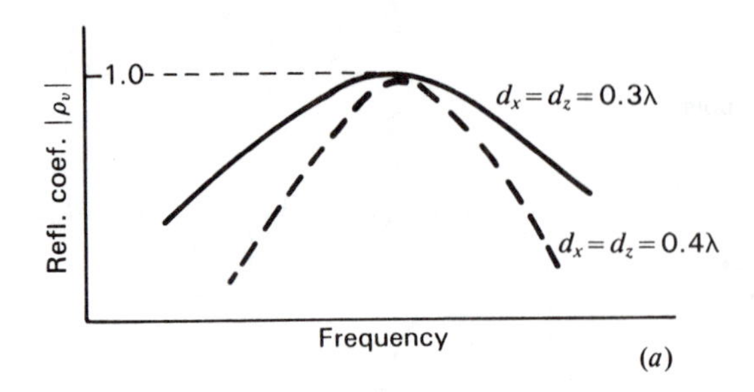

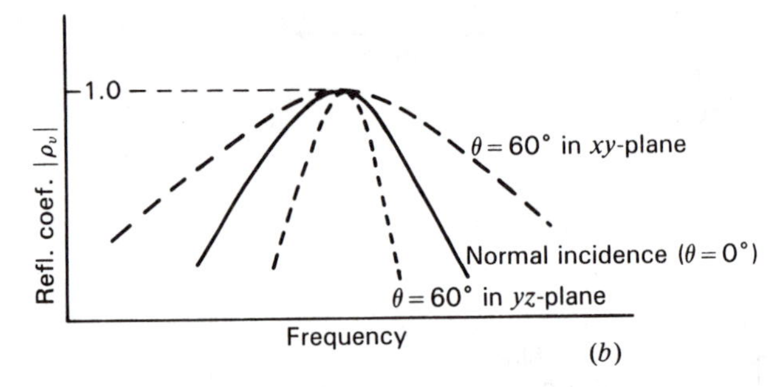

Figure 12-44 (*a*) Reflection coefficient of FSS for 2 values of element spacing. The smaller spacing results in the broader bandwidth. (*b*) Reflection coefficient of FSS for different angles of incidence θ.

If we consider ρ_v in (2) as a function of frequency, we find that R_A as given by (3) varies relatively little. However, $X_A + X_L$, which is much more complicated to determine than R_A, changes dramatically. Typical curves of ρ_v for normal incidence ($\theta = 0°$) are shown in Fig. 12-44*a*. Changing d_x and/or d_z, $X_A + X_L$ remains essentially constant, provided d_x and $d_z < \lambda/2$. However, R_A is inversely proportional to both of these, i.e., becomes larger for smaller d_x and d_z. Inspection of (2) shows that smaller spacings make ρ_v more broadbanded, as illustrated in Fig. 12-44*a*. However, the maximum magnitude of $|\rho_v|$ will always be unity provided $R_L = 0$ and no grating lobes are present.

Effect of Angle of Incidence θ. Let us next consider the case where the spacings d_x and d_z are fixed ($< \sim 0.45\lambda$) and determine how the reflection coefficient ρ_v varies with angle of incidence θ. We again find that $X_A + X_L$ changes relatively little with angle of incidence. However, by inspection of (3), we note that R_A increases with angle of incidence for scan in the *xy* plane and decreases for scan in the *yz* plane. Further, we observe that the bandwidth of ρ_v becomes larger and smaller respectively, as indicated in Fig. 12-44*b*. This change of bandwidth with angle of incidence is quite large in all FSS regardless of the types of elements used. It can be compensated for in a number of ways. The most promising

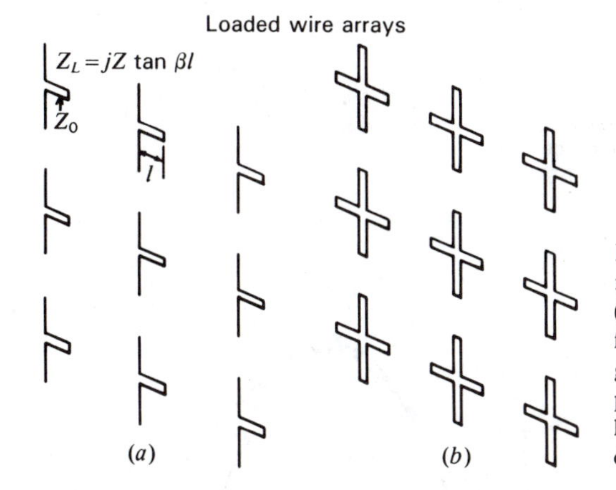

Figure 12-45-1 (*a*) Single loaded-dipole reflecting surface for vertical polarization. (*b*) Double loaded-dipole reflecting surface for arbitrary polarization. This form is generally preferable because of the cross-polarization component in the single loaded-dipole type at high angle of incidence in the vertical (*yz*) plane (Fig. 12-43).

appears to be by using dielectric slabs mounted on both sides of the FSS as done by Munk[1] and by Munk and Kornbau.[2]

Control of Bandwidth. It was shown above that the bandwidth could be broadened by decreasing d_x and/or d_z, or vice versa, for smaller element spacings (this feature is quite universally independent of element type). Another way to affect the bandwidth is to make the load reactance vary faster with frequency. This makes ρ_v more narrowbanded as shown by Munk[3] and by Munk, Kouyoumjian and Peters.[4] A practical type of load may typically consist of a 2-wire transmission line stub as shown in Fig. 12-45-1*a*. A variation using two transmission lines in parallel is shown in Fig. 12-45-1*b*. It has the advantage of being capable of handling arbitrary polarizations.

Finally, the diameter of the wire elements can be reduced, leading to an even narrower bandwidth of ρ_v. However, compared to the other two methods above, this is much less effective since it changes only logarithmically with wire diameter.

Cascading or Stacking More FSS. A single FSS, as shown above, has only one frequency where $|\rho_v| = 1$. However, by cascading or stacking two or more surfaces the reflection coefficient curve can be flattened at the top and can roll off faster as indicated in Fig. 12-45-2. This approach may be used for the reflecting (dipole) as well as the transparent (slot) surfaces. However, while a *single dipole*

[1] B. A. Munk, "Space Filter," U.S. Patent 4,125,841, November 1978.

[2] B. A. Munk and T. W. Kornbau, "On Stabilization of the Bandwidth of a Dichroic Surface by Use of Dielectric Slabs," *Electromag.*, **5**, 349–373, 1985.

[3] B. A. Munk, "Periodic Surface for Large Scan Angle," U.S. Patent 3,789,404, Jan. 29, 1974.

[4] B. A. Munk, R. G. Kouyoumjian and L. Peters, Jr., "Reflection Properties of Periodic Surfaces of Loaded Dipoles," *IEEE Trans. Ants. Prop.*, **AP-19**, 612–617, September 1971.

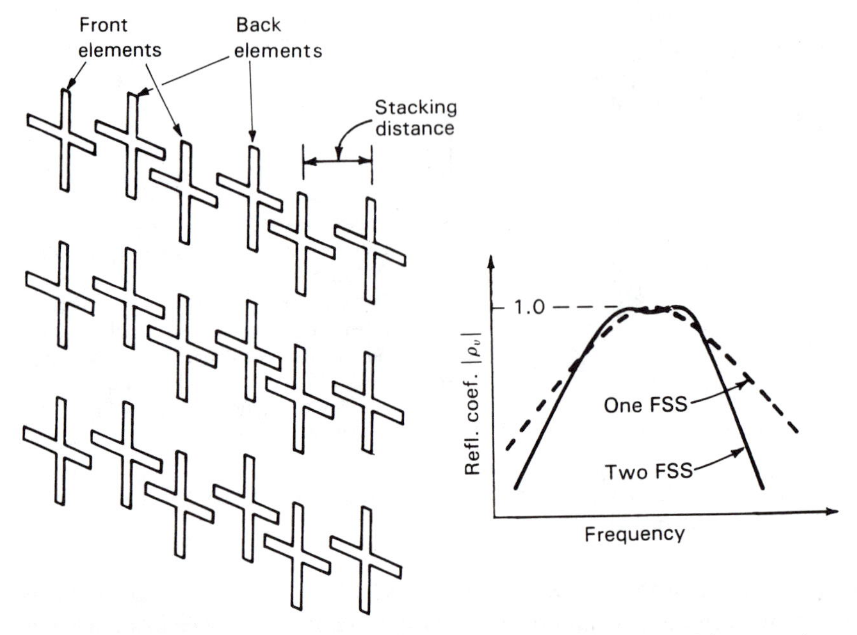

Figure 12-45-2 Cascading or stacking 2 or more FSS (in y direction) as at left can produce a flatter-topped reflection coefficient curve with faster roll-off as shown at right. Note that the figure shows two 3×3 (=9) element arrays stacked one in front of (or to the left of) the other.

surface has the same reflection curve as its complementary surface's transmission curve (Babinet's principle), this is not the case when surfaces are stacked or when dielectric is added to a single surface.

Element Types. Nearly any type of element will resonate at some frequency. However, some types are much better than others. First, the element should be sufficiently small that the element spacing d_x and d_z can remain small, meeting the grating lobe conditions. Consequently, a straight element as shown in Fig. 12-45-3*a* is rarely used for high-performance FSS, being too long ($\sim\lambda/2$). Much better performance can be obtained with inductively loaded elements as shown in Fig. 12-45-3*b*, *c* and *d* (also Fig. 12-45-1*a* and *b*). The latter can also be constructed in 3-legged loaded form as shown in Fig. 12-45-3*e* and in the unloaded 3-legged form of Fig. 12-45-3*f* (Pelton and Munk[1]). Four-legged unloaded elements of 2 crossed wires (or slots) are used but are not recommended since they will, in general, exhibit a double resonance in one scan plane but not in the other, a feature considered undesirable for most applications, as pointed out

[1] E. L. Pelton and B. A. Munk, "Periodic Antenna Surface of Tripole Slot Elements," AF invention 10,270, Aug. 17, 1976.

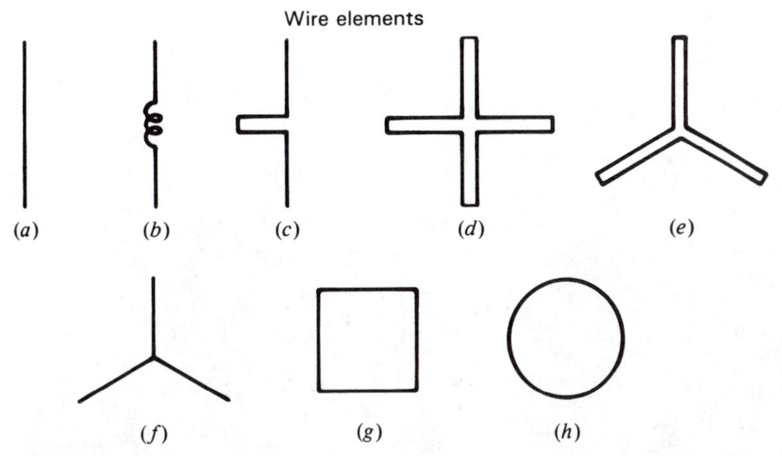

Figure 12-45-3 Various types of wire elements for use with FSS.

by Pelton and Munk.[1] All of the elements described are characterized by having medium or narrow bandwidth (4 to 25 percent). In contrast, the two loop elements shown in Fig. 12-45-3*g* and *h* can be made to operate over more than an octave bandwidth.

12-14 SOME EXAMPLES OF REFLECTOR ANTENNAS.

A number of reflector antennas illustrative of different designs are given in this section.

Bonn. The world's largest fully steerable reflector telescope is shown in Fig. 12-46. Standing 40 stories tall in a valley of the Eifel mountains near Bonn, West Germany, the 3200-tonne 100-m diameter parabolic reflector antenna is used for radio astronomy and space research. The dish can be elevated 20° per minute and rotated in azimuth on a circular track 40° per minute. It can operate at wavelengths as short as 1 cm. Feeds may be either prime focus or Cassegrain. See the gain versus wavelength curve in Fig. 12-34.

Arecibo. The fixed 305-m diameter spherical reflector at Arecibo, discussed in the preceding section, is shown in cross section in Fig. 12-47 and from above in Fig. 12-48.

Bell Telephone Laboratories. The 7-m diameter parabolic reflector with offset feed of the Bell Telephone Laboratories is shown in Fig. 12-49. This antenna,

[1] E. L. Pelton and B. A. Munk, "Scattering from Periodic Arrays of Crossed Dipoles," *IEEE Trans. Ants. Prop.*, **AP-27**, 323–330, May 1979.

Figure 12-46 Fully steerable 100-m diameter antenna near Bonn, West Germany, standing nearly 40 stories tall. The moving parts weigh 3200 tonnes. Note the massive backup structure required to provide rigidity to the dish. (*Courtesy Dr. R. Wielebinski, Max Planck Institute for Radioastronomy.*)

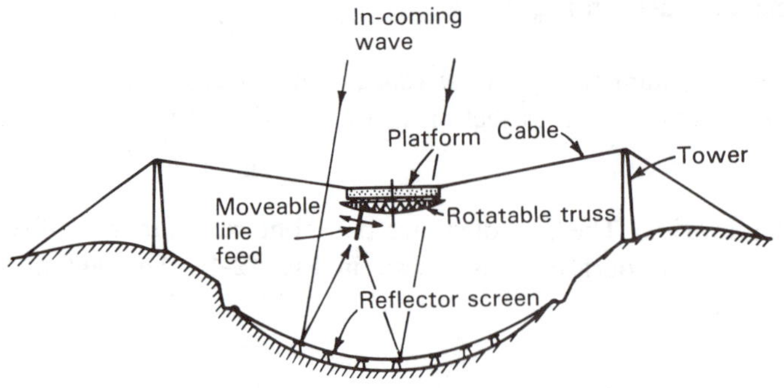

Figure 12-47 Elevation cross section of 305-m diameter fixed spherical reflector suspended in mountain valley at Arecibo, Puerto Rico.

Figure 12-48 Air view of 305-m Arecibo dish. Feed structure is supported by cables from 3 towers. Beam steering by moving the line feed allows observations at angles up to 20° from the zenith. Away from the zenith, only part of the reflector can be used with resulting decrease in aperture efficiency. (*Cornell University.*)

located at Crawford Hill, New Jersey, has a 100-μm rms surface accuracy permitting operation at wavelengths as short as 3 mm (100 GHz) with 59 percent aperture efficiency, a gain of 77 dBi and HPBW of 2 arcminutes.[1] At 1 cm (30 GHz)

[1] T-S Chu, R. W. Wilson, R. W. England, D. A. Gray and W. E. Legg, "The Crawford Hill 7-m Millimeter Wave Antenna," *BSTJ*, **57**, 1257–1288, May–June 1978.

Figure 12-49 The 7-m offset Cassegrain-fed low-sidelobe millimeter-wave antenna of the Bell Telephone Laboratories at Crawford Hill, New Jersey. Key structural parts are covered with insulation to reduce thermal effects. (*Courtesy Dr. T-S Chu, Bell Telephone Laboratories.*)

the aperture efficiency is 69 percent and the gain 66.5 dBi. With a 15-dB feed taper and no aperture blockage the sidelobe level is over 40 dB down 1° off-axis at 30 GHz. The antenna also has very low cross-polarization (over 40 dB down across the main beam). These low levels are achieved using a quasi-optical feed system and conical corrugated horn with hybrid-mode launcher (see Fig. 13-31).

Whereas the atmosphere contributes a temperature of only 2.3 K at 4 GHz (at the zenith), the dry atmosphere temperature at 100 GHz is about 100 K so that the temperature contribution due to spillover (12 K) and mylar waveguide window loss (1 K) add little of significance to the system temperature at 100 GHz.[1]

In radio astronomy observations, receiver noise reflected back into prime focus or Cassegrain feeds can produce spurious spectral lines. This effect is eliminated with an offset feed as in the Crawford Hill antenna.

Nobeyama. The world's largest millimeter-wave dish, shown in Fig. 12-50, is

Figure 12-50 The 45-m fully steerable parabolic reflector antenna with gain of nearly 90 dBi at wavelengths of 2 to 3 mm at the Nobeyama Radio Observatory in the Yatsugatake Mountains, Japan. (*NRO photo.*)

[1] T-S Chu, personal communication, 1985.

located at the Nobeyama Radio Observatory at a high elevation in the central mountains of Honshu Island, Japan. It has a 200-μm rms overall equivalent surface deviation and a pointing accuracy of 2 arcseconds.[1] Maximum gain is almost 90 dBi at 2 to 3 mm wavelengths. See the gain curve in Fig. 12-34. Thermal effects are minimized by covering the dish structure with insulated panels and circulating air inside the enclosure. Each of the 600 rigid panels of the dish have a surface accuracy of 60 μm rms.

Ohio State University. The first radio telescope antenna I designed and built at the Ohio State University in 1951 consisted of an array of 96 helical antennas mounted on a tiltable flat panel 50 m long for operation at 200 to 300 MHz. The array is shown in Fig. 7-4.

A few years later I began designing a larger dual-reflector antenna on which construction began in 1956.[2] Work was done mostly by part-time university students and took over 10 years to complete. The antenna, often called a Kraus-type reflector, consists of a fixed-standing-curved reflector and a tiltable-flat reflector. The one we built, called "Big Ear," is shown in Fig. 12-51. The standing-curved reflector is a rectangular section of a sphere or paraboloid-of-revolution with dimensions of 110 by 21 m. The tiltable-flat reflector dimensions are 104 by 31 m and the two reflectors are joined by a flat conducting ground plane. The flat reflector is shown in Fig. 12-52.

The basic design consideration of the antenna was that of obtaining the maximum aperture per unit of cost. It also has other advantages such as less than $\frac{1}{2}$ of 1 percent effective aperture blocking, reduced susceptibility to terrestrial interference because of the low profile of the feed (right at ground level) and shielding by the large reflectors, and a spacious underground receiver laboratory directly below the prime focus where weight restrictions are not a consideration.

Tilting the flat reflector allows observations between declinations of -36 and $+64°$ (a range of 100°) which gives a coverage of 90 percent of the sky observable from the site. Movement of the feed car, shown in Fig. 12-53, permits beam steering of 15° in azimuth (or tracking of sources for an hour or more in

---

[1] K. Akabane, M. Morimoto, N. Kaifu and M. Ishigro, "The Nobeyama Radio Observatory," *Sky and Tel.*, 495, December 1983.

[2] J. D. Kraus, "The Ohio State Radio Telescope," *Sky and Tel.*, **12**, 157–159, April 1953.

J. D. Kraus, "Radio Telescopes," *Sci. Am.*, **192**, 33–43, March 1955.

J. D. Kraus, "Radio Telescopes of Large Aperture," *Proc. IRE*, **46**, 92–97, January 1958.

J. D. Kraus, R. T. Nash and H. C. Ko, "Some Characteristics of the OSU 360-ft Radio Telescope," *IRE Trans. Ants. Prop.*, **AP-9**, 4–8, January 1961.

J. D. Kraus, "The Large Radio Telescope of the Ohio State University," *Sky and Tel.*, **26**, 12–16, July 1963.

J. D. Kraus, *Big Ear*, Cygnus-Quasar, 1976.

Figure 12-51 "Big Ear," the 110-m Kraus-type telescope antenna at the Ohio State University. The low-profile design features a large aperture per unit cost, negligible aperture blocking, reduced susceptibility to terrestrial interference, convenient feed location, long focal length and extended tracking capability. This telescope was used for the Ohio Sky Survey, in which 20 000 radio sources were cataloged and mapped at 1415 MHz and many unique sources discovered, including the most distant known objects in the universe. (*Tom Root photo.*)

Figure 12-52 The tiltable flat reflector of Big Ear. For scale, note the man halfway up the ladder on the far side.

Figure 12-53 Horn feeds for 1.4 to 1.7 GHz on car of Big Ear for beam steering or tracking sources in azimuth (right ascension). (See also Fig. 13-29.)

right ascension). The telescope has a very long focal length with $F/D = 1.17$ in azimuth (right ascension) and $F/D = 6.0$ in elevation (declination) so that it is possible to deploy many feed systems efficiently in the focal region for simultaneous operation. It is noteworthy that because of the long focal length the curved reflector can be described as either a parabola or a sphere since both are almost identical except at the extreme E-W edges where they differ by only a few millimeters. The antenna has been operated routinely at frequencies as low as 20 MHz (15 m) and as high as 3 GHz (10 cm). With installation of a finer mesh screen, efficient operation could extend to even higher frequencies.

The principle of operation is indicated in the elevation cross section of Fig. 12-54. Incoming waves are deflected by the flat reflector into the parabola, which brings the waves to a focus at ground level near the base of the flat reflector. By moving the flat reflector through 50° the antenna beam is tilted through a 100° range in elevation.

The antenna may be operated in two modes. In one mode, illustrated in Fig. 12-55*a*, the feed horn axis is aligned with the center of the parabola and the ground plane is incidental. In the second mode (Fig. 12-55*c*), the horn axis is coincident with the conducting ground plane, which joins the parabola and flat reflector. The ground plane serves as a guiding boundary surface. In this mode polarization must be vertical, and the feed horn required is $\frac{1}{4}$ the height and $\frac{1}{2}$ the length of the horn required in the first mode. This difference in horn size may be

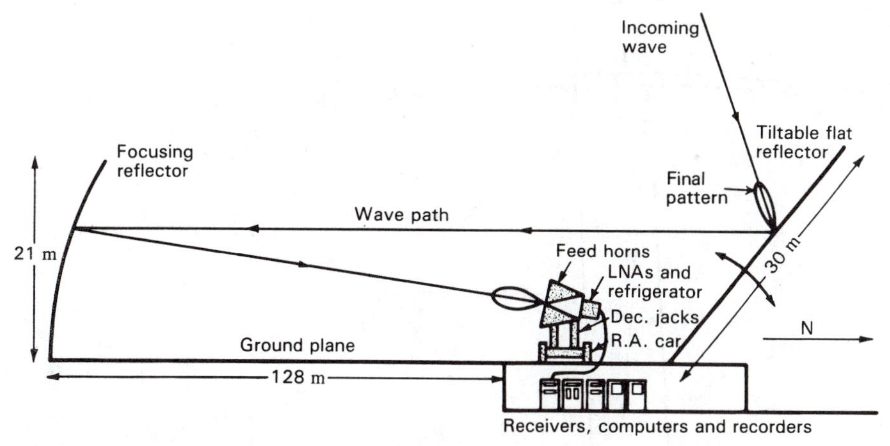

Figure 12-54 Elevation cross section through Big Ear.

inferred with the aid of the 3 diagrams in Fig. 12-55. If a horn of the first mode (Fig. 12-55*a*) is placed with its axis coincident with the ground plane, the lower half is an image and may be discarded, as in Fig. 12-55*b*. However, the beam width of the horn is too narrow (by a factor of 2), so that its dimensions must be halved, as in Fig. 12-55*c*. Although the ground plane serves no primary function in the first mode of operation, its presence reduces the antenna temperature significantly by shielding minor lobes from direct ground pickup.

Another application of Big Ear is as a compact measuring range for frequencies below 3 GHz, as discussed in Sec. 18-3d.

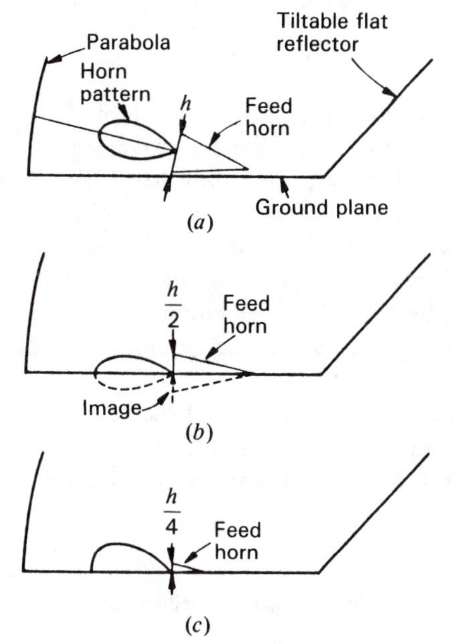

Figure 12-55 Arrangement for feeding antenna without ground plane (*a*) requires 4 times horn height of arrangement at (*c*) using ground plane. Diagram (*b*) shows that half the horn in (*a*) produces too sharp a pattern when used with the ground plane and must be reduced in size, as at (*c*).

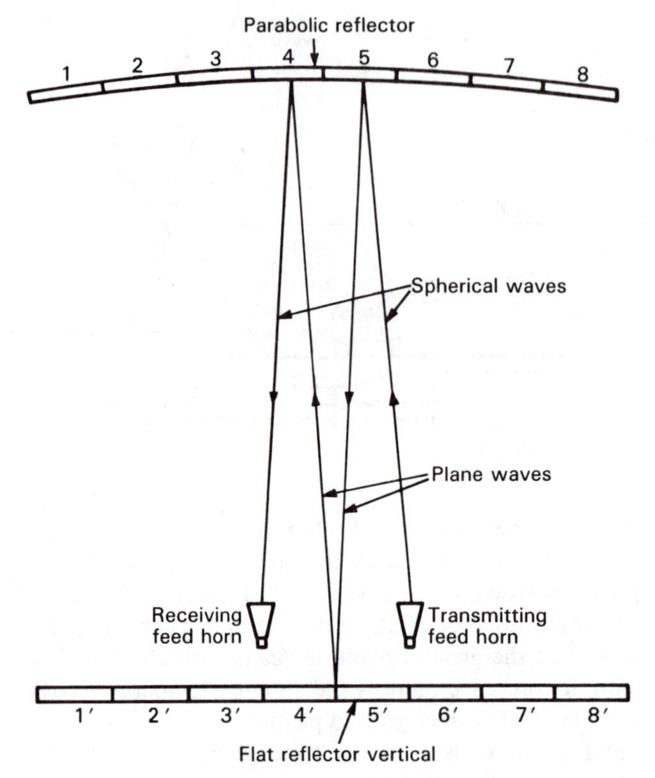

Figure 12-56 Arrangement of standing-parabolic and tiltable-flat reflectors of 25-m millimeter-wave antenna at Gorki, USSR, used by Albert Kislyakov at a wavelength of 1 mm, with flat reflector turned vertically, for self-calibration.

Gorki. Albert Kislyakov has constructed a standing-parabola tiltable-flat-reflector antenna near Gorki, USSR, for operation at millimeter wavelengths. Although the basic principle is the same as for Big Ear, Kislyakov has added a unique feature of *self-calibration* by providing that the flat reflector can be set vertically.[1] With flat reflector vertical and with feed displaced laterally off-axis (and transmitting), radiation is brought to a focus on the opposite side of the axis where the signal can be received as suggested by Fig. 12-56. By moving the receiving horn sideways or up and down, the *far-field pattern* of the antenna can be measured with an arbitrarily high S/N ratio.

With the same configuration (flat reflector vertical), panels of the reflectors can be adjusted to maximize the gain. For example, with panels 1, 2, 3, 5, 6 and 7 covered with absorbing material, only panels 4, 4′, 8 and 8′ are operational, and,

[1] A. G. Kislyakov, "Radioastronomical Investigations at Millimetric and Submillimetric Wavelengths," *UFN*, **101**, 607, 1970.

Figure 12-57 Front and back views of the Five College Radio Astronomy Observatory 14-m radome-enclosed radio telescope.

with 4 and 4′ as reference, panels 8 and 8′ can be adjusted for maximum signal or gain. The same procedure is then repeated with other pairs of panels until all panels are adjusted.

Five College Observatory. The 14-m dish of the Five College Radio Astronomy Observatory is an example of a millimeter-wave antenna enclosed in a Buckminster Fuller triangular-panel geodesic radome. Front and back views of the dish are shown in Fig. 12-57. The reflector is constructed of 72 panels contoured to a section of a paraboloid with a 65-μm rms surface accuracy, or $\lambda/20$ at 1.3 mm. To set the panels accurately a *holographic technique* was used in which the amplitude *and* phase of a 38-GHz satellite beacon signal was measured over an angular extent much larger than the main beam of the antenna. The amplitude and phase of the antenna aperture field distribution was then obtained by a 2-dimensional Fourier transform.[1] This indicated a 90-μm rms residual panel

[1] A complete description of the procedure is given by P. F. Goldsmith and N. R. Erickson in sec. 6-22 of J. D. Kraus, *Radio Astronomy*, 2nd ed., Cygnus-Quasar, 1986. For additional information on holographic techniques see Bennet *et al.*, Godwin, Whitaker and Anderson, Mayer *et al.* and Rahmat-Samii in the references listed at the end of this chapter.

Figure 12-58 The 5.5 by 2.4 m *Offsat* uplink-downlink antenna with full offset feed for C and Ku bands. (*Comtech Antenna Corporation.*)

position error with the antenna at a 45° elevation angle. At 38 GHz ($\lambda = 7.9$ mm) the gain-loss factor from (12-10-3) is

$$k_g = \cos^2\left(720° \frac{0.09 \text{ mm}}{7.9 \text{ mm}}\right) = 0.98$$

or only 0.09 dB loss from surface error.

At 230 GHz ($\lambda = 1.3$ mm) the loss is larger and irregular lens behavior of the radome fabric becomes significant.

Offsat. A parabolic one-piece offset-feed fiberglass 5.5 by 2.4 m antenna is shown in Fig. 12-58. This *Offsat* antenna was developed by Comtech Antenna Corporation to meet the FCC-ITU 2° spacing requirement for satellites in the Clarke orbit. The full-offset prime-focus feed eliminates aperture blocking. The gain at 4 GHz is 40 dBi, at 6 GHz is 46 dBi, at 12 GHz is 51 dBi and at 14 GHz is 52 dBi. The dish weighs 1 tonne, has a noise temperature of 17 K at 40° elevation angle with the first sidelobe 24 dB down. (See Sec. 16-16.)

12-15 LOW-SIDELOBE CONSIDERATIONS. Referring to Fig. 12-35, it is to be noted that the sidelobes in the forward direction (0 to 90°) of a large

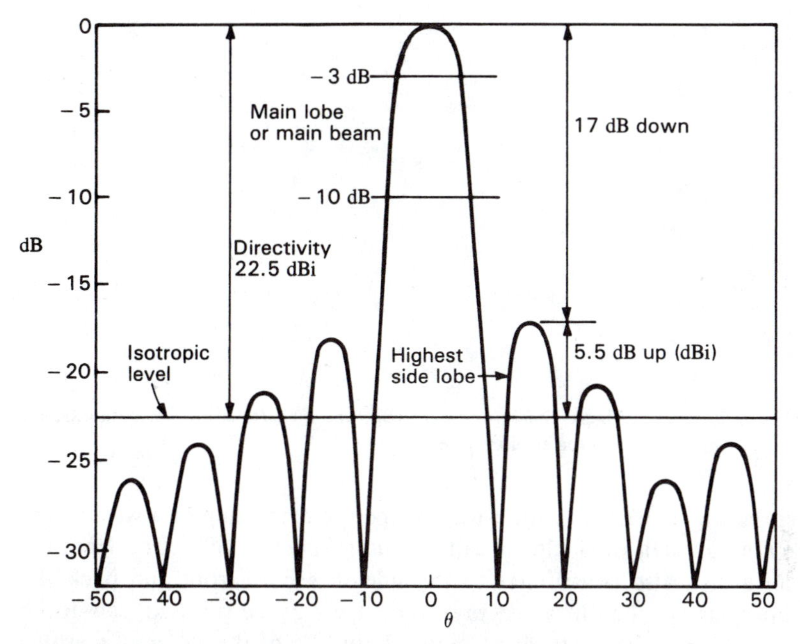

Figure 12-59 Main beam and sidelobes with respect to the isotropic level.

parabolic dish reflector are determined (1) by the aperture distribution, (2) by the irregularities of the dish surface, (3) by scattering or diffraction from the feed structure and support struts and (4) by diffraction from the edge of the dish. Effect (3) is absent with offset feeds or with the horn-reflector antenna. The side-lobes to the rear (90 to 180°) are determined (1) by the spillover and (2) by diffraction around the edges of the dish. Schrank[1] gives a thorough summary of low-sidelobe reflector antennas.

For an isotropic source the directivity $D = 1$ (0 dBi) so the *isotropic level* and the relative power level (uniform in all directions) are the same. As D increases, the main lobe rises above the isotropic level in proportion to D. In the hypothetical (but typical) pattern shown in Fig. 12-59, the directivity $D = 22.5$ dBi (=lossless gain) and the highest sidelobe is 17 dB *below* (the main-lobe maximum) or 5.5 dB *above* the isotropic level (5.5 dBi). Sidelobe levels are usually referred to the main lobe but sometimes to the isotropic level.

As shown in Fig. 11-66, a triangular or cosine tapered aperture distribution drops to zero field at the edge of the aperture, yet it results in front sidelobes. Back sidelobes should, in principle, be absent. However, the cosine squared or Gaussian distributions have no sidelobes but the HPBWs are greater.

---

[1] H. E. Schrank, "Low Sidelobe Reflector Antennas," *IEEE Ant. Prop. Soc. Newsletter*, **27**, 5–16, April 1985.

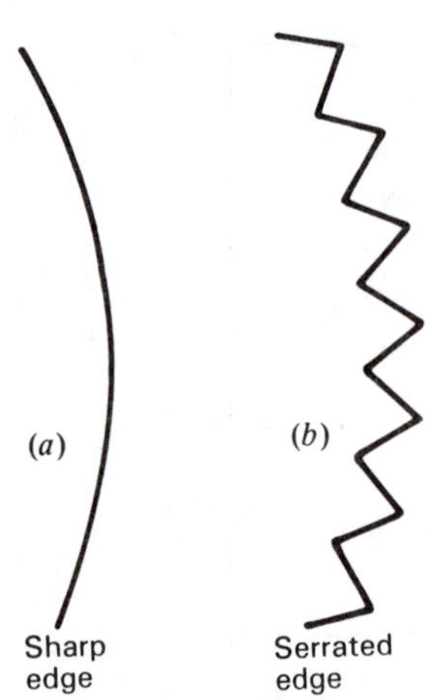

Figure 12-60 (*a*) Sharp edge and (*b*) serrated or sawtooth edge for reduced sidelobe level.

Dish surface irregularities with a regular (periodic) spacing of a wavelength or more are apt to result in sidelobes called *grating lobes*.[1] Diffraction from the sharp edge of a dish also contributes to the sidelobes both front and back. To randomize the phase of the diffracted rays, the edge may be serrated (saw-tooth edge) as in Fig. 12-60. The tooth dimensions should be of the order of a wavelength or more. The straight-versus-diagonal or sawtooth effect may also be noted with a square dish or ground plane. Thus, as suggested in Fig. 12-61, the sidelobes in the plane of the diagonal tend to be less than in the plane of the sides.

Edge diffraction may be reduced by means of a rolled edge with (or without) absorbing material, as illustrated in Fig. 12-6*c*. An oversize parabolic dish with reduced-edge illumination might also be used to reduce edge diffraction but a problem with this approach is that the underilluminated edge and outer regions of the parabola (with its irregularities) still contribute to a diffracted field. However, if the outer region of the dish is curved away from a parabolic contour

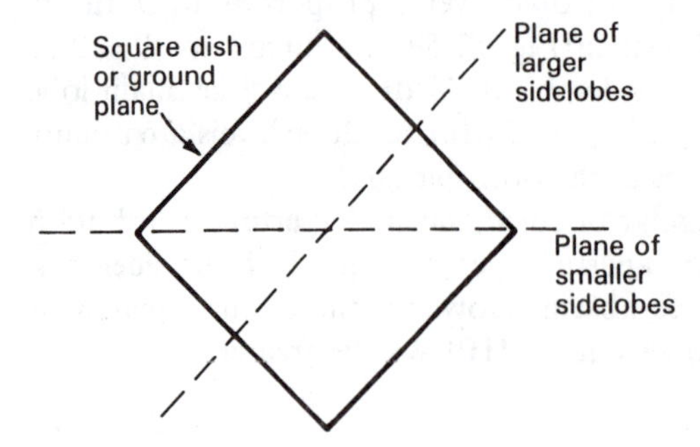

Figure 12-61 Square dish or ground plane with lower sidelobes in plane of diagonal.

[1] Grating lobes are typical for arrays with interelement spacings of λ or more (see Sec. 11-26).

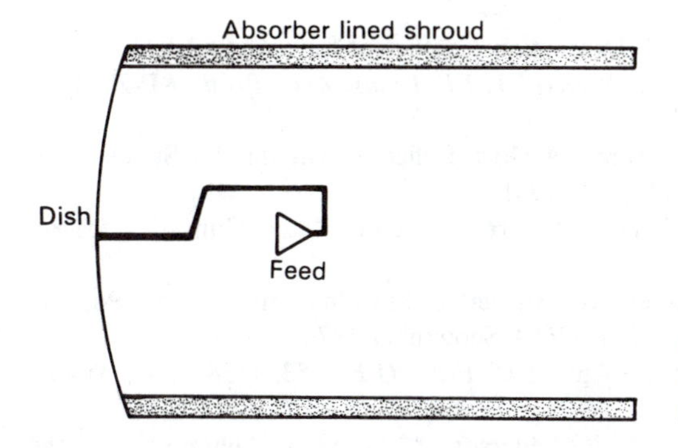

Figure 12-62 Reflector antenna with cylindrical shroud of absorber for reducing far-out sidelobes.

and blended into a rolled or curved edge, diffraction effects are much reduced[1] (see Sec. 18-3d).

Although the horn-reflector antenna (Fig. 12-42) has very low wide-angle sidelobes, a sidelobe tends to appear at $\theta = 90°$, which may be objectionable. Thomas[2] has used a serrated edge (blinder) to reduce it.

Another side and back lobe suppression technique involves the addition of a cylindrical absorbing shroud attached to the edge of the dish as in Fig. 12-62. The outer surface of the shroud may be metal or dielectric. Dybdal and King[3] found that with a shroud twice as long as the dish diameter the far-out sidelobes were 60 to 75 dB down, although the on-axis back lobe ($\theta = 180°$) was only 50 dB down.

ADDITIONAL REFERENCES

Bach, H., and H-H. Viskum: "The SNFGTD (Spherical Near-Field Geometrical Theory of Diffraction) Method and Its Accuracy," *IEEE Trans. Ants. Prop.*, **AP-35**, 169–175, February 1987.

Bennet, J. C., A. P. Anderson, P. A. McInnes and A. J. T. Whitaker: "Microwave Holographic Metrology of Large Reflector Antennas," *IEEE Trans. Ants. Prop.*, **AP-24**, 295–303, 1976.

Burnside, W. D., M. C. Gilreath and B. M. Kent: "Rolled Edge Modification of Compact Range Reflector," *AMTA Symp.*, San Diego, 1984.

Chu, T-S: "An Imaging Beam Waveguide Feed," *IEEE Trans. Ants. Prop.*, **AP-31**, 614–619, July 1983.

Chu, T-S, and R. H. Turrin: "Depolarization Properties of Offset Reflector Antennas," *IEEE Trans. Ants. Prop.*, **AP-21**, 339–345, May 1973.

[1] W. D. Burnside, M. C. Gilreath, B. M. Kent and G. C. Clerici, "Curved Edge Modification of Compact Range Reflector," *IEEE Trans. Ants. Prop.*, **AP-35**, 176–182, Feburary 1987.

[2] D. T. Thomas, "Design of Multiple-Edge Blinders for Large Horn Reflector Antennas," *IEEE Trans. Ants. Prop.*, **AP-21**, 153–158, March 1973.

[3] R. B. Dybdal and H. E. King, "Performance of Reflector Antennas with Absorber-Lined Tunnels," *IEEE Symp. Ants. Prop.*, Seattle, June 1979.

Chuang, C. W., and W. D. Burnside: "A Diffraction Coefficient for a Cylindrically Truncated Planar Surface (Rolled Edge Effect)," *IEEE Trans. Ants. Prop.*, **AP-28**, 177–182, March 1980.

Crawford, A. B., D. C. Hogg and L. E. Hunt: "A Horn-Reflector Antenna for Space Communication," *BSTJ*, **40**, 1045–1116, July 1961.

Cross, D. C., D. D. Howard and J. W. Titus: "Mirror-Antenna Radar Concept," *Microwave J.*, **29**, 323–335, May 1986.

Dragone, C.: "Offset Multireflector Antennas with Perfect Pattern Symmetry and Polarization Discrimination," *BSTJ*, **57**, 2663–2684, September 1978.

Ehrenspeck, H. W.: "The Short Backfire Antenna," *Proc. IEEE*, **53**, 1138–1140, August 1965.

Godwin, M. P., A. J. T. Whitaker and A. P. Anderson: "Microwave Diagnostics of the Chilbolton 25 m Antenna Using the OTS Satellite," *Proc. Inst. Elec. Eng. Int. Conf.*, York, England, pp. 232–236, 1981.

James, G. L.: *Geometrical Theory of Diffraction for Electromagnetic Waves*, Peregrinus, 1980.

Kaplan, P. D.: "Predicting Antenna Sidelobe Performance (Sidelobe Pop-Up Probability)," *Microwave J.*, **29**, 201–206, September 1986.

Keller, J. B.: "Geometrical Theory of Diffraction," *J. Opt. Soc. Am.*, **52**, 116–130, February 1962.

Kieburtz, R. B., and A. Ishimaru: "Scattering by a Periodically Apertured Conducting Screen," *IRE Trans. Ants. Prop.*, **AP-9**, 506–514, November 1961.

Kieburtz, R. B., and A. Ishimaru: "Aperture Fields of an Array of Rectangular Apertures," *IRE Trans. Ants. Prop.*, **AP-10**, 663–671, November 1962.

Koffman, I.: "Feed Polarization for Parallel Currents in Reflectors Generated by Conic Sections," *IEEE Trans. Ants. Prop.*, **AP-14**, 37–40, January 1966.

Kouyoumjian, R. G., and P. H. Pathak: "A Uniform Geometrical Theory of Diffraction for an Edge in a Perfectly Conducting Surface," *Proc. IEEE*, **62**, 1448–1461, November 1974.

Lamb, J. W.: "Analysis of a Cassegrain Antenna Having a Gaussian Illumination Pattern," Helsinki Univ. Tech. Rept. S147, 1983.

Ludwig, A. C.: "Low Sidelobe Aperture Distribution for Blocked and Unblocked Circular Apertures," *IEEE Trans. Ants. Prop.*, **30**, 933–946, September 1982.

Mayer, C. E., J. H. Davis, W. L. Peters and W. J. Vogel: "A Holographic Surface Measurement of the Texas 4.9-meter Antenna at 86 GHz," *IEEE Trans. Instrum. Meas.*, **IM-32**, 102–109, 1983.

Nash, R. T.: "Stepped Amplitude Distributions," *IEEE Trans. Ants. Prop.*, **AP-12**, July 1964.

Ott, R. H., R. G. Kouyoumjian and L. Peters, Jr.: "The Scattering by a Two-Dimensional Periodic Narrow Array of Plates," *Radio Sci.*, November 1967.

Ott, R. H., and J. H. Richmond: "A Linear Equation Solution for Scattering by a Periodic Plane Array of Thin Conducting Plates," ElectroSci. Lab. Rept. 1774-9, 1965.

Peters, L., Jr., and C. E. Ryan, Jr.: "Empirical Formulas for E-Plane Creeping Waves on General Smooth Conducting Bodies," *IEEE Trans. Ants. Prop.*, **AP-18**, 432–434, May 1970.

Rahmat-Samii, Y.: "Surface Diagnosis of Large Reflector Antennas Using Microwave Holographic Metrology—An Iterative Approach," *Radio Sci.*, **13**, 1205–1217, September–October 1984.

Rahmat-Samii, Y.: "Microwave Holography of Large Reflector Antennas: Simulation Algorithms," *IEEE Trans. Ants. Prop.*, **AP-33**, 1194–1203, November 1985.

Ratnasiri, P. A. J., R. G. Kouyoumjian and P. H. Pathak: "The Wide Angle Side Lobes of Reflector Antennas," OSU Elect. Sci. Lab. Rept. 5635-4, 1970.

Rudge, A. W., and N. A. Adatia: "Offset-Parabolic-Antennas: A Review," *Proc. IEEE*, **66**, 1592–1618, December 1978.

Rudge, A. W., and M. J. Withers: "Design of Flared-Horn Primary Feeds for Parabolic Reflector Antennas," *Proc. IEE*, **117**, 1741–1747, September 1970.

Rudge, A. W., and M. J. Withers: "New Technique for Beam Steering with Fixed Parabolic Reflectors," *Proc. IEE*, **118**, 857–863, July 1971.

Rusch, W. V. T., Y. Rahmat-Samii and R. A. Shore: "The Equivalent Paraboloid of an Optimized Off-Set Cassegrain Antenna," *Dig. IEEE Int. Symp. Ants. Prop.*, 1986.

Scott, P. F., and M. Ryle: "A Rapid Method for Measuring the Figure of a Radio Telescope Reflector," *Royal Ast. Soc. Mon. Notices*, **178**, 539–545, 1977.

Shore, R. A.: "A Simple Derivation of the Basic Design Equation for Offset Dual Reflector Antennas with Rotational Symmetry and Zero Cross Polarization," *IEEE Trans. Ants. Prop.*, **AP-33**, 114–116, January 1985.

Sletten, C. J., W. Payne and W. Shillue: "Offset Dual Reflector Antennas for Very Low Sidelobes," *Microwave J.*, **29**, 221–240, May 1986.

Tanaka, H., and M. Mizuasawa: "Elimination of Cross-Polarization in Offset Dual-Reflector Antennas," *Elec. Commun. (Japan)*, **58-B**, 71–78, 1975.

Twersky, V.: "Multiple Scattering of Waves and Optical Phenomena," *J. Opt. Soc. Am.*, **52**, 145–171, 1962.

PROBLEMS[1]

12-1 Flat sheet reflector. Calculate and plot the radiation pattern of a $\lambda/2$ dipole antenna spaced 0.15λ from an infinite flat sheet for assumed antenna loss resistances $R_L = 0$ and $10\ \Omega$. Express the patterns in gain over a $\lambda/2$ dipole antenna in free space with the same power input (and zero loss resistance).

12-2 Square-corner reflector. A square-corner reflector has a driven $\lambda/2$ dipole antenna spaced $\lambda/2$ from the corner. Assume perfectly conducting sheet reflectors of infinite extent (ideal reflector). Calculate and plot the radiation pattern in a plane at right angles to the driven element.

12-3 Square-corner reflector. Calculate and plot the pattern of an ideal square-corner reflector with $\lambda/2$ driven antenna spaced $\lambda/2$ from the corner but with the antenna displaced 20° from the bisector of the corner angle. The pattern to be calculated is in a plane perpendicular to the antenna and to the reflecting sides.

12-4 Paraboloidal reflector. Calculate and plot the radiation patterns of a paraboloidal reflector with uniformly illuminated aperture when the diameter is 8λ and when the diameter is 16λ.

12-5 Cylindrical parabolic reflector. Calculate the radiation pattern of a cylindrical parabolic reflector of square aperture 16λ on a side when the illumination is uniform over the aperture and when the field intensity across the aperture follows a cosine variation with maximum intensity at the center and zero intensity at the edges. Compare the two cases by plotting the normalized curves on the same graph.

---

[1] Answers to starred (*) problems are given in App. D.

★12-6 Square-corner reflector.

(*a*) Calculate and plot the pattern of a 90° corner reflector with a thin center-fed $\lambda/2$ driven antenna spaced 0.35λ from the corner. Assume that the corner reflector is of infinite extent.

(*b*) Calculate the radiation resistance of the driven antenna.

(*c*) Calculate the gain of the antenna and corner reflector over the antenna alone. Assume that losses are negligible.

12-7 Square-corner reflector versus array of its image elements. Assume that the corner reflector of Prob. 12-6 is removed and that in its place the three images used in the analysis are present physically, resulting in a 4-element driven array.

(*a*) Calculate and plot the pattern of this array.

(*b*) Calculate the radiation resistance at the center of one of the antennas.

(*c*) Calculate the gain of the array over one of the antennas alone.

★12-8 Square-corner reflector array. Four 90° corner-reflector antennas are arranged in line as a broadside array. The corner edges are parallel and side-by-side as in Fig. P12-8. The spacing between corners is 1λ. The driven antenna in each corner is a $\lambda/2$ element spaced 0.4λ from the corner. All antennas are energized in phase and have equal current amplitude. Assuming that the properties of each corner are the same as if its sides were of infinite extent, what is (*a*) the gain of the array over a single $\lambda/2$ antenna and (*b*) the half-power beam width in the *H* plane?

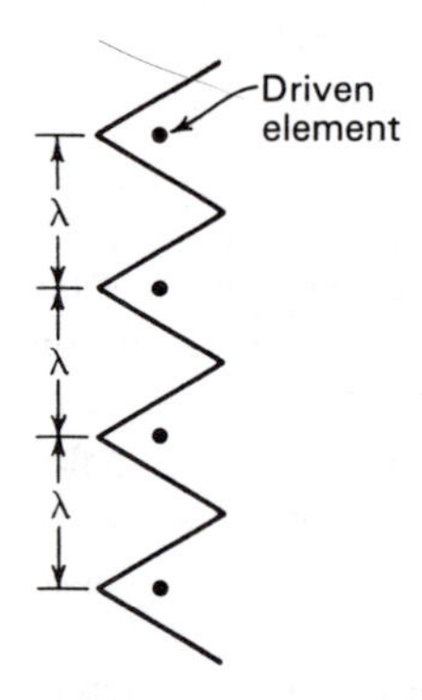

Figure P12-8 Square-corner reflector array.

12-9 Paraboloidal reflector aperture distribution.

(*a*) Show that the variation of field across the aperture of paraboloidal reflector with an isotropic source is proportional to $1/[1 + (\rho/2L)^2]$ where ρ is the radial distance from the axis of the paraboloid. Show that this relation is equivalent to $(1 + \cos\theta)/2$.

(*b*) If the parabola extends to the focal plane and the feed is isotropic over the hemisphere subtended by the parabola, calculate the aperture efficiency.

12-10 Square-corner reflector.

(*a*) Show that the relative field pattern in the plane of the driven $\lambda/2$ element of a square-corner reflector is given by

$$E = [1 - \cos(S_r \sin\theta)] \frac{\cos(90^\circ \cos\theta)}{\sin\theta}$$

where θ is the angle with respect to the element axis. Assume that the corner-reflector sheets are perfectly conducting and of infinite extent.

(*b*) Calculate and plot the field pattern in the plane of the driven element for a spacing of $\lambda/2$ to the corner. Compare with the pattern at right angles (Prob. 12-2).

12-11 Corner reflector. λ/4 to the driven element. A square-corner reflector has a spacing of $\lambda/4$ between the driven $\lambda/2$ element and the corner. Show that the directivity $D = 12.8$ dBi.

12-12 Corner reflector. λ/2 to the driven element. A square-corner reflector has a driven $\lambda/2$ element $\lambda/2$ from the corner.

(*a*) Calculate and plot the far-field pattern in both principal planes.

(*b*) What are the HPBWs in the two principal planes?

(*c*) What is the terminal impedance of the driven element?

(*d*) Calculate the directivity in two ways: (1) from impedances of driven and image dipoles and (2) from HPBWs, and compare. Assume perfectly conducting sheet reflectors of infinite extent.

***12-13 Parabolic reflector with missing sector.** A circular parabolic dish antenna has an effective aperture of 100 m^2. If one 45° sector of the parabola is removed, find the new effective aperture. The rest of the antenna, including the feed, is unchanged.

***12-14 Efficiency of rectangular aperture with partial taper.** Calculate the aperture efficiency and directivity of an antenna with rectangular aperture $x_1 y_1$ with a uniform field distribution in the y direction and a cosine field distribution in the x direction (zero at edges, maximum at center) if $x_1 = 20\lambda$ and $y_1 = 10\lambda$.

***12-15 Efficiency of rectangular aperture with full taper.** Repeat Prob. 12-14 for the case where the aperture field has a cosine distribution in both the x and y directions.

12-16 Efficiency of aperture with phase ripple. A square unidirectional aperture $(x_1 y_1)$ is 10λ on a side and has a design distribution for the electric field which is uniform in the x direction but triangular in the y direction with maximum at the center and zero at the edges. Design phase is constant across the aperture. However, in the actual aperture distribution there is a plus-and-minus-30° sinusoidal phase variation in the x direction with a phase cycle per wavelength. Calculate (*a*) the design directivity, (*b*) the utilization factor, (*c*) the actual directivity, (*d*) the achievement factor, (*e*) the effective aperture and (*f*) the aperture efficiency.

***12-17 Rectangular aperture. Cosine taper.** An antenna with rectangular aperture $x_1 y_1$ has a uniform field in the y direction and a cosine field distribution in the x direction (zero at edges, maximum at center). If $x_1 = 16\lambda$ and $y_1 = 8\lambda$, calculate (*a*) the aperture efficiency and (*b*) the directivity.

12-18 Rectangular aperture. Cosine tapers. Repeat Prob. 12-17 for the case where the aperture field has a cosine distribution in both the x and y directions.

***12-19 A 20λ line source. Cosine-squared taper.**

(*a*) Calculate and plot the far-field pattern of a continuous in-phase line source 20λ long with cosine-squared field distribution.

(*b*) What is the HPBW?

CHAPTER 13

SLOT, HORN AND COMPLEMENTARY ANTENNAS

13-1 INTRODUCTION. Sections 13-2 and 13-3 of this chapter deal with slot antennas and their patterns. These antennas are useful in many applications, especially where low-profile or flush installations are required as, for example, on high-speed aircraft. Sections 13-4, 13-5 and 13-6 discuss the relation of slots to their complementary dipole forms. Any slot has its complementary form in wires or strips, so that pattern and impedance data of these forms can be used to predict the patterns and impedances of the corresponding slots. The discussion is based largely on a generalization and extension of Babinet's (Ba-bi-naý's) principle by Henry Booker. A slotted cylinder antenna is discussed in Sec. 13-7.[1] The next four sections of the chapter describe horn antennas, both rectangular and conical. The remaining sections involve ridge, septum, corrugated and matched horns.

13-2 SLOT ANTENNAS. The antenna shown in Fig. 13-1*a*, consisting of two resonant $\lambda/4$ stubs connected to a 2-wire transmission line, forms an inefficient radiator. The long wires are closely spaced ($w \ll \lambda$) and carry currents of opposite phase so that their fields tend to cancel. The end wires carry currents in

[1] Patch or microstrip antennas, which may be regarded as derived from slot antennas, are discussed in Sec. 16-12.

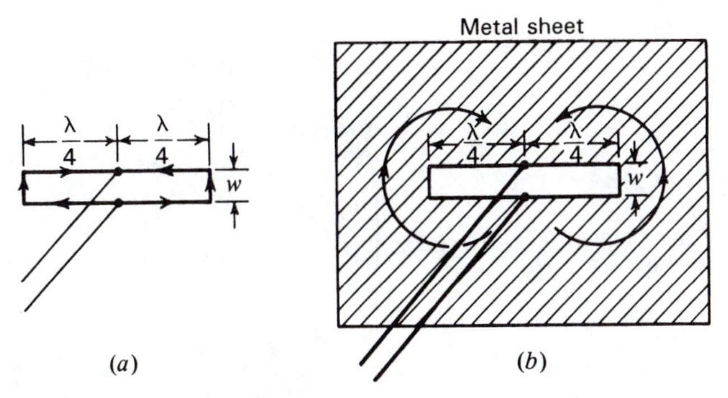

Figure 13-1 Parallel connected $\lambda/4$ stubs (a) and simple slot antenna (b).

the same phase, but they are too short to radiate efficiently. Hence, enormous currents may be required to radiate appreciable amounts of power.

The antenna in Fig. 13-1b, on the other hand, is a very efficient radiator. In this arrangement a $\lambda/2$ slot is cut in a flat metal sheet. Although the width of the slot is small ($w \ll \lambda$), the currents are not confined to the edges of the slot but spread out over the sheet. This is a simple type of slot antenna. Radiation occurs equally from both sides of the sheet. If the slot is horizontal, as shown, the radiation normal to the sheet is vertically polarized.

A slot antenna may be conveniently energized with a coaxial transmission line as in Fig. 13-2a. The outer conductor of the cable is bonded to the metal sheet. Since the terminal resistance at the center of a resonant $\lambda/2$ slot in a large sheet is about 500 Ω and the characteristic impedance of coaxial transmission lines is usually much less, an off-center feed such as shown in Fig. 13-2b may be used to provide a better impedance match. For a 50-Ω coaxial cable the distance s should be about $\lambda/20$. Slot antennas fed by a coaxial line in this manner are illustrated in Fig. 13-2c and d. The radiation normal to the sheet with the horizontal slot (Fig. 13-2c) is vertically polarized while radiation normal to the sheet with the vertical slot (Fig. 13-2d) is horizontally polarized. The slot may be $\lambda/2$ long, as shown, or more.

A flat sheet with a $\lambda/2$ slot radiates equally on both sides of the sheet. However, if the sheet is very large (ideally infinite) and boxed in as in Fig. 13-3a, radiation occurs only from one side. If the depth d of the box is appropriate ($d \sim \lambda/4$ for a thin slot), no appreciable shunt susceptance appears across the terminals. With such a zero susceptance box, the terminal impedance of the resonant $\lambda/2$ slot is nonreactive and approximately twice its value without the box, or about 1000 Ω.

The boxed-in slot antenna might be applied even at relatively long wavelengths by using the ground as the flat conducting sheet and excavating a trench $\lambda/2$ long by $\lambda/4$ deep as suggested in Fig. 13-3b. The absence of any structure above ground level might make this type of antenna attractive, for example, in applications near airports. To improve the ground conductivity, the walls of the trench and the ground surrounding the slot can be covered with copper sheet or

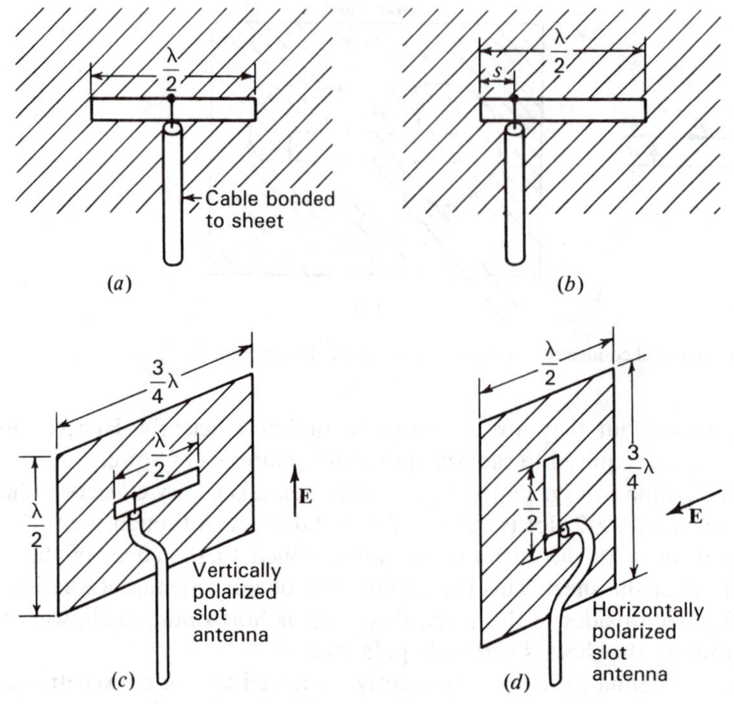

Figure 13-2 Slot antennas fed by coaxial transmission lines.

screen. Radiation is maximum in all directions at right angles to the slot and is zero along the ground in the directions of the ends of the slot, as suggested in Fig. 13-3*b*. The radiation along the ground is vertically polarized.

Radiation from only one side of a large flat sheet may also be achieved by a slot fed with a waveguide as in Fig. 13-4*a*. With transmission in the guide in the TE_{10} mode, the direction of the electric field **E** is as shown. The width L of the guide must be more than $\lambda/2$ to transmit energy, but it should be less than 1λ to suppress higher-order transmission modes. With the slot horizontal, as shown,

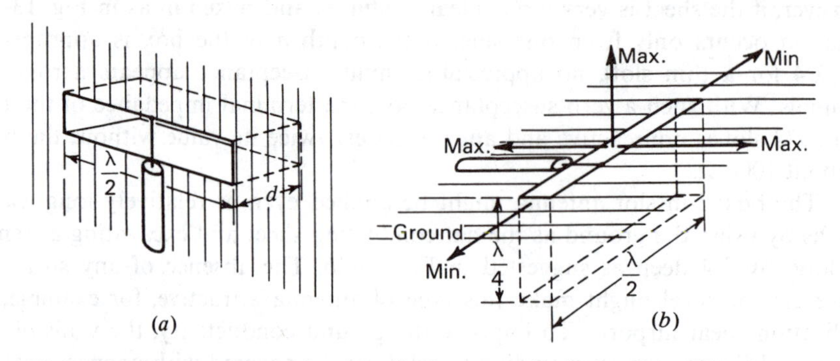

Figure 13-3 Boxed-in slot antenna (*a*) and application to provide flush radiator (*b*).

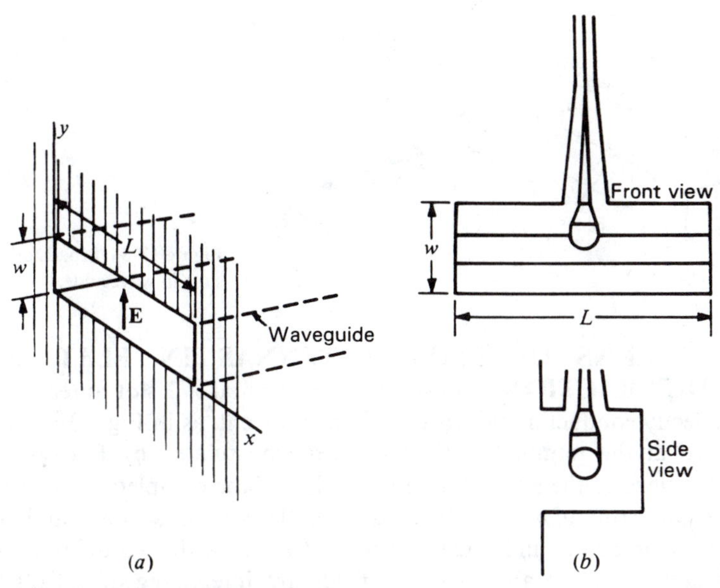

Figure 13-4 Waveguide-fed slot (*a*) and T-fed slot (*b*).

the radiation normal to the sheet is vertically polarized. The slot opening constitutes an abrupt termination to the waveguide. It has been found[1] that the resulting impedance mismatch is least over a wide frequency band if the ratio L/w is less than 3.

A compact wideband method for feeding a boxed-in slot is illustrated in Fig. 13-4*b*. In this *T*-fed arrangement[1] the bar compensates the impedance characteristics so as to provide a VSWR on a 50-Ω feed line of less than 2 over a frequency range of nearly 2 to 1. The ratio L/w of the length to width of the slot is about 3.

Dispensing with the flat sheet altogether, an array of slots may be cut in the waveguide as in Fig. 13-5 so as to produce a directional radiation pattern.[2] With transmission in the guide in the TE_{10} mode, the instantaneous direction of the electric field **E** inside the guide is as indicated by the dashed arrows. By cutting inclined slots as shown at intervals of $\lambda_g/2$ (where λ_g is the wavelength in the guide), the slots are energized in phase and produce a directional pattern with maximum radiation broadside to the guide. If the guide is horizontal and **E** inside the guide is vertical, the radiated field is horizontally polarized as suggested in Fig. 13-5.

---

[1] A. Dorne and D. Lazarus, in *Very High Frequency Techniques*, Radio Research Laboratory Staff, McGraw-Hill, New York, 1947, chap. 7.

[2] W. H. Watson, *The Physical Principles of Wave Guide Transmission and Antenna Systems*, Oxford University Press, London, 1947.

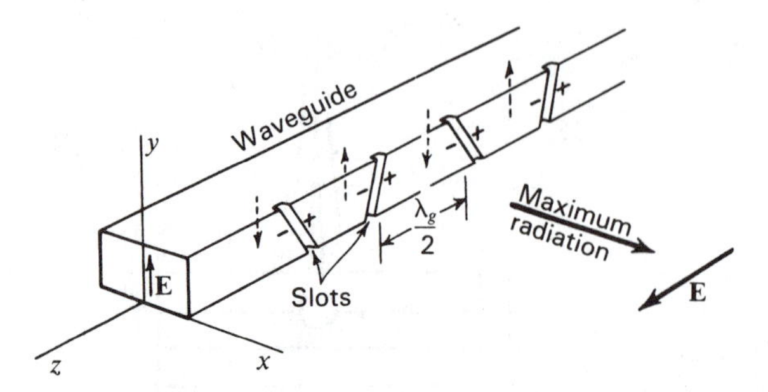

Figure 13-5 Broadside array of slots in waveguide.

13-3 PATTERNS OF SLOT ANTENNAS IN FLAT SHEETS. EDGE DIFFRACTION. Consider the horizontal $\lambda/2$ slot antenna of width w in a perfectly conducting flat sheet of infinite extent, as in Fig. 13-6*a*. The sheet is energized at the terminals FF. It has been postulated by Booker[1] that the radiation pattern of the slot is the same as that of the complementary horizontal $\lambda/2$ dipole consisting of a perfectly conducting flat strip of width w and energized at the terminals FF, as indicated in Fig. 13-6*b*, but with two differences. These are (1) that the electric and magnetic fields are interchanged and (2) that the component of the electric field of the slot normal to the sheet is discontinuous from one side of the sheet to the other, the direction of the field reversing. The tangential component of the magnetic field is, likewise, discontinuous.

The patterns of the $\lambda/2$ slot and the complementary dipole are compared in Fig. 13-7. The infinite flat sheet is coincident with the xz plane, and the long dimension of the slot is in the x direction (Fig. 13-7*a*). The complementary dipole is coincident with the x axis (Fig. 13-7*b*). The radiation-field patterns have the same doughnut shape, as indicated, but the directions of **E** and **H** are interchanged. The solid arrows indicate the direction of the electric field **E** and the dashed arrows the direction of the magnetic field **H**.

If the xy plane is horizontal and the z axis vertical as in Fig. 13-7*a*, the radiation from the horizontal slot is vertically polarized everywhere in the xy plane. Turning the slot to a vertical position (coincident with the z axis) rotates

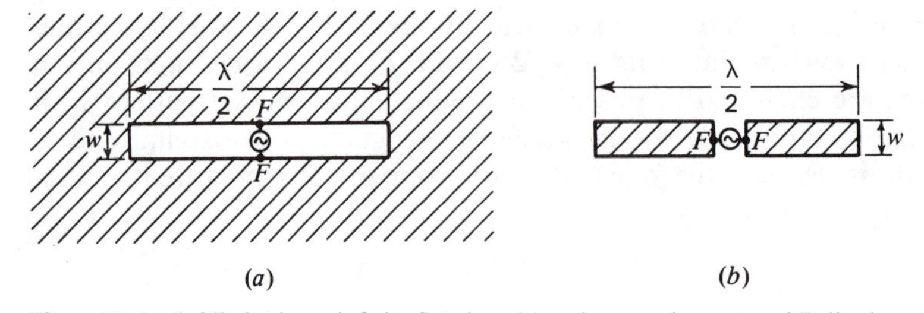

Figure 13-6 A $\lambda/2$ slot in an infinite flat sheet (*a*) and a complementary $\lambda/2$ dipole antenna (*b*).

[1] H. G. Booker, "Slot Aerials and Their Relation to Complementary Wire Aerials," *JIEE* (*Lond.*), **93**, pt. IIIA, no. 4, 1946.

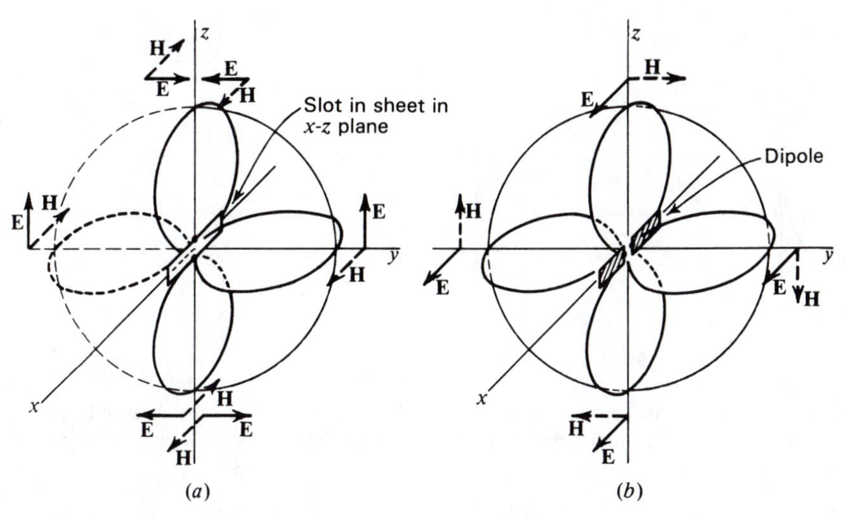

Figure 13-7 Radiation-field patterns of slot in an infinite sheet (*a*) and of complementary dipole antenna (*b*). The patterns have the same shape but with **E** and **H** interchanged.

the radiation pattern through 90° to the position shown in Fig. 13-8. The radiation in this case is everywhere horizontally polarized; i.e., the electric field has only an E_ϕ component. If the slot is very thin ($w \ll \lambda$) and $\lambda/2$ long ($L = \lambda/2$), the variation of E_ϕ as a function of θ is, from (5-5-12), given by

$$E_\phi(\theta) = \frac{\cos\,[(\pi/2)\cos\theta]}{\sin\theta} \tag{1}$$

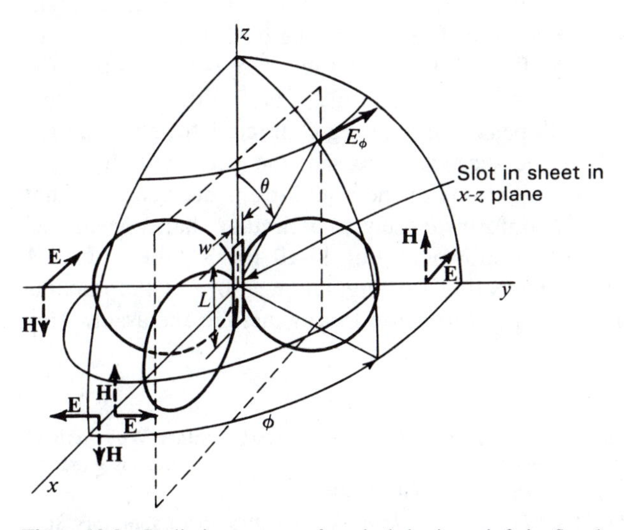

Figure 13-8 Radiation pattern of vertical slot in an infinite flat sheet.

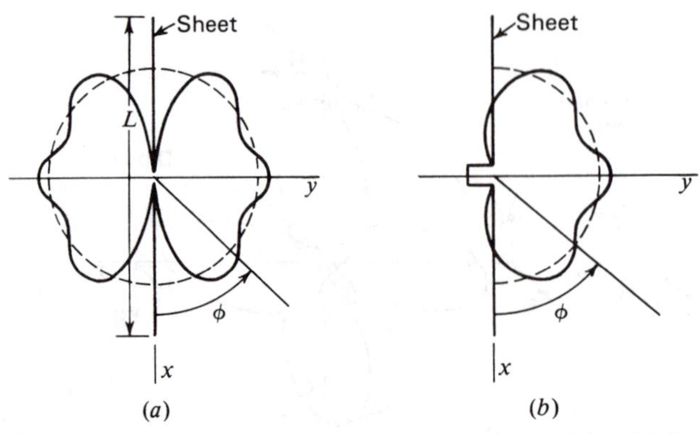

Figure 13-9 Solid curves show patterns in xy plane of Fig. 13-8 for slot in finite sheet of length L. Slot is open on both sides in (a) and closed on left side in (b). Dashed curves show pattern for infinite sheet. All patterns idealized.

Assuming that the sheet is perfectly conducting and infinite in extent, the magnitude of the field component E_ϕ remains constant as a function of ϕ for any value of θ. Thus,

$$E_\phi(\phi) = \text{constant} \tag{2}$$

Consider now the situation where the slot is cut in a sheet of finite extent as suggested by the dashed lines in Fig. 13-8. This change produces relatively little effect on the $E_\phi(\theta)$ pattern given by (1). However, there must be a drastic change in the $E_\phi(\phi)$ pattern since in the x direction, for example, the fields radiated from the two sides of the sheet are equal in magnitude but opposite in phase so that they cancel. Hence, there is a null in all directions in the plane of the sheet. For a sheet of given length L in the x direction, the field pattern in the xy plane might then be as indicated by the solid curve in Fig. 13-9a. The dashed curve is for an infinite sheet ($L = \infty$). If one side of the slot is boxed in, there is radiation in the plane of the sheet as suggested by the pattern in Fig. 13-9b.[1] With a finite sheet the pattern usually exhibits a scalloped or undulating characteristic, as suggested in Fig. 13-9. As the length L of the sheet is increased, the pattern undulations become more numerous but the magnitude of the undulations decreases, so that for a very large sheet the pattern conforms closely to a circular shape. Measured patterns[2] illustrating this effect are shown in Fig. 13-10 for 3 values of L. A method due to Andrew Alford for locating the angular positions of the maxima and minima is described by Dorne and Lazarus.[2] In this method the assumption

[1] According to H. G. Booker, "Slot Aerials and Their Relation to Complementary Wire Aerials," *JIEE* (*Lond.*), **93**, pt. IIIA, no. 4, 1946, the energy density in the $\phi = 0$ or 180° directions is $\frac{1}{2}$ that for an infinite sheet, or the field intensity is 0.707 that for an infinite sheet.

[2] A. Dorne and D. Lazarus, in *Very High Frequency Techniques*, Radio Research Laboratory Staff, McGraw-Hill, New York, 1947, chap. 7 (see sec. 7-3).

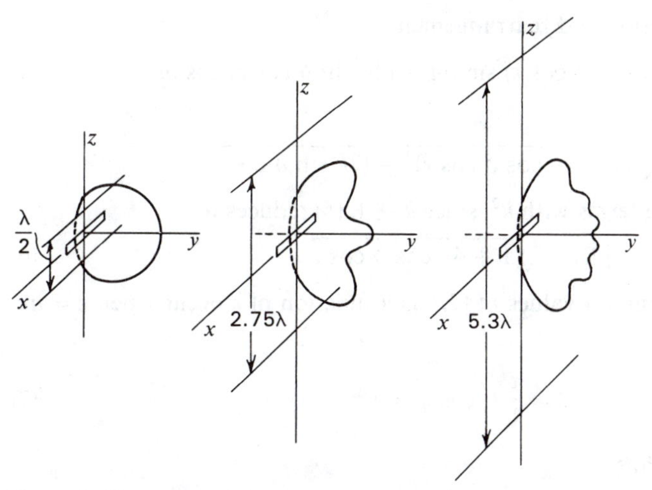

Figure 13-10 Measured ϕ-plane patterns of $\lambda/2$ boxed-in slot antennas in finite sheets of three lengths $L = 0.5$, 2.75 and 5.3λ. The width of the slots is 0.1λ. (*After A. Dorne and D. Lazarus, in* Very High Frequency Techniques, *Radio Research Laboratory Staff, McGraw-Hill, New York, 1947.*)

is made that the far field is produced by three sources (see Fig. 13-11), one (1) at the slot of strength 1 sin ωt and two (2 and 3) at the edges of the sheet (edge diffraction effect) with a strength $k \sin(\omega t - \delta)$, where $k \ll 1$ and δ gives the phase difference of the edge sources with respect to the source (1) at the slot. At the point P at a large distance in the direction ϕ, the relative field intensity is then

$$E = \sin \omega t + k \sin(\omega t - \delta - \varepsilon) + k \sin(\omega t - \delta + \varepsilon) \tag{3}$$

where $\varepsilon = (\pi/\lambda)L \cos \phi$

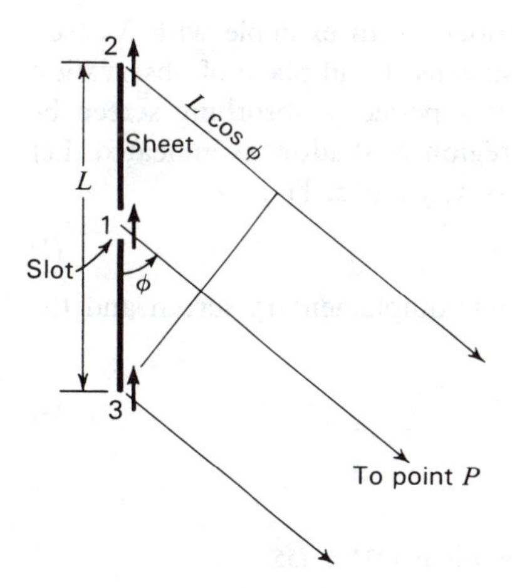

Figure 13-11 Construction for locating maxima and minima of ϕ pattern for slot in a finite sheet.

By trigonometric expansion and rearrangement,

$$E = (1 + 2k \cos \delta \cos \varepsilon) \sin \omega t - (2k \sin \delta \cos \varepsilon) \cos \omega t \tag{4}$$

and the modulus of E is

$$|E| = \sqrt{(1 + 2k \cos \delta \cos \varepsilon)^2 + (2k \sin \delta \cos \varepsilon)^2} \tag{5}$$

Squaring and neglecting terms with k^2, since $k \ll 1$, (5) reduces to

$$|E| = \sqrt{1 + 4k \cos \delta \cos \varepsilon} \tag{6}$$

The maximum and minimum values of $|E|$ as a function of ε occur when $\varepsilon = n\pi$, so that

$$\varepsilon = \frac{\pi}{\lambda} L \cos \phi = n\pi \tag{7}$$

where n is an integer. Thus

$$\cos \phi = \frac{n\lambda}{L} \qquad \text{and} \qquad \phi = \arccos \frac{n\lambda}{L} \tag{8}$$

The values of ϕ for maxima and minima in the ϕ pattern are given by (8). These locations are independent of k and δ. If $\cos \delta$ is positive, then the maxima correspond to even values of n and the minima to odd values of n.

13-4 BABINET'S PRINCIPLE AND COMPLEMENTARY ANTENNAS.

By means of Babinet's (Ba-bi-naý's) principle many of the problems of slot antennas can be reduced to situations involving complementary linear antennas for which solutions have already been obtained. In optics Babinet's principle[1] may be stated as follows:

> *The field at any point behind a plane having a screen, if added to the field at the same point when the complementary screen is substituted, is equal to the field at the point when no screen is present.*

The principle may be illustrated by considering an example with 3 cases. Let a source and 2 imaginary planes, plane of screens A and plane of observation B, be arranged as in Fig. 13-12. As Case 1, let a perfectly absorbing screen be placed in plane A. Then in plane B there is a region of shadow as indicated. Let the field behind this screen be some function f_1 of x, y and z. Thus,

$$F_s = f_1(x, y, z) \tag{1}$$

As Case 2 let the first screen be replaced by its complementary screen and the field behind it be given by

$$F_{cs} = f_2(x, y, z) \tag{2}$$

[1] See, for example, Max Born, *Optik*, Verlag Julius Springer, Berlin, 1933, p. 155.

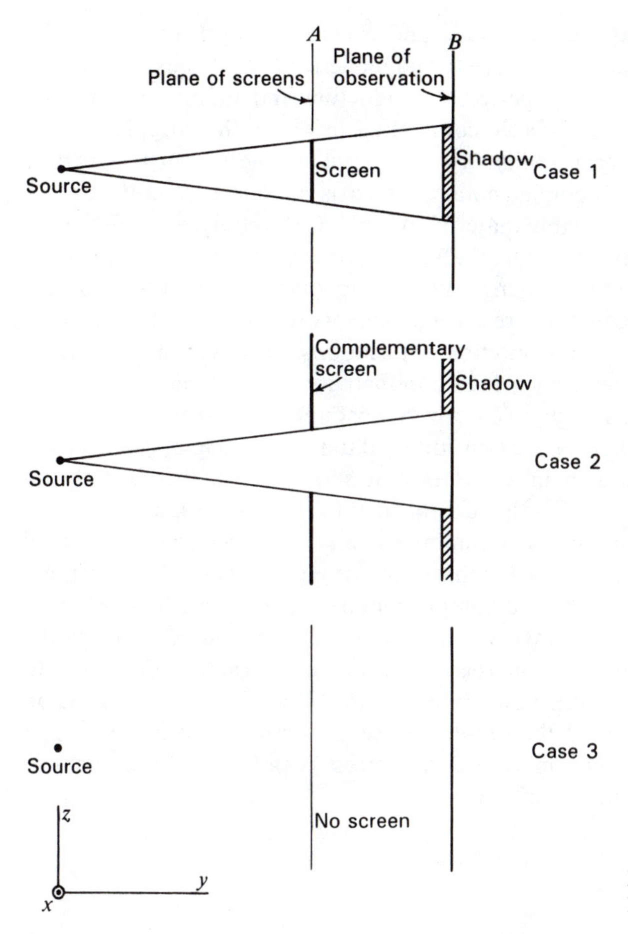

Figure 13-12 Illustration of Babinet's principle.

As Case 3 with no screen present the field is

$$F_0 = f_3(x, y, z) \tag{3}$$

Then Babinet's principle asserts that at the same point x_1, y_1, z_1,

$$F_s + F_{cs} = F_0 \tag{4}$$

The source may be a point as in the above example or a distribution of sources. The principle applies not only to points in the plane of observation B as suggested in Fig. 13-12 but also to any point behind screen A. Although the principle is obvious enough for the simple shadow case above, it also applies where diffraction is considered.

Babinet's principle has been extended and generalized by Booker[1] to take into account the vector nature of the electromagnetic field. In this extension it is assumed that the screen is plane, perfectly conducting and infinitesimally thin. Furthermore, if one screen is perfectly conducting ($\sigma = \infty$), the complementary screen must have infinite permeability ($\mu = \infty$). Thus, if one screen is a perfect conductor of electricity, the complementary screen is a perfect "conductor" of magnetism. No infinitely permeable material exists, but the equivalent effect may be obtained by making both the original and complementary screens of perfectly conducting material and interchanging electric and magnetic quantities everywhere. The only perfect conductors are *superconductors* which soon may be available at ordinary temperatures for antenna applications. However, many metals, such as silver and copper, have such high conductivity that we may assume the conductivity is infinite with a negligible error in most applications.

As an illustration of Booker's extension of Babinet's principle, consider the cases in Fig. 13-13. The source in all cases is a short dipole, theoretically an infinitesimal dipole. In Case 1 the dipole is horizontal and the original screen is an infinite, perfectly conducting, plane, infinitesimally thin sheet with a vertical slot cut out as indicated. At a point P behind the screen the field is E_1. In Case 2 the original screen is replaced by the complementary screen consisting of a perfectly conducting, plane, infinitesimally thin strip of the same dimensions as the slot in the original screen. In addition, the dipole source is turned vertical so as to interchange $\mathbf{E}$ and $\mathbf{H}$. At the same point P behind the screen the field is E_2. As an alternative situation for Case 2 the dipole source is horizontal and the strip is also turned horizontal. Finally, in Case 3 no screen is present and the field at point P is E_0. Then, by Babinet's principle,

$$E_1 + E_2 = E_0 \tag{5}$$

or

$$\frac{E_1}{E_0} + \frac{E_2}{E_0} = 1 \tag{6}$$

Babinet's principle may also be applied to points in front of the screens. In the situation of Case 1 (Fig. 13-13) a large amount of energy may be transmitted through the slot so that the field E_1 may be about equal to the field E_0 with no intermediate screen (Case 3). In such a situation the complementary dipole acts like a reflector and E_2 is very small. (Recall Sec. 12-13 on frequency-sensitive surfaces.) Since a metal sheet with a $\lambda/2$ slot or, in general, an orifice of at least 1λ perimeter may transmit considerable energy, slots or orifices of this size should be assiduously avoided in sheet reflectors such as described in Chap. 12 when $\mathbf{E}$ is not parallel to the slot.

---

[1] H. G. Booker, "Slot Aerials and Their Relation to Complementary Wire Aerials," *JIEE* (*Lond.*), **93**, pt. IIIA, no. 4, 1946.

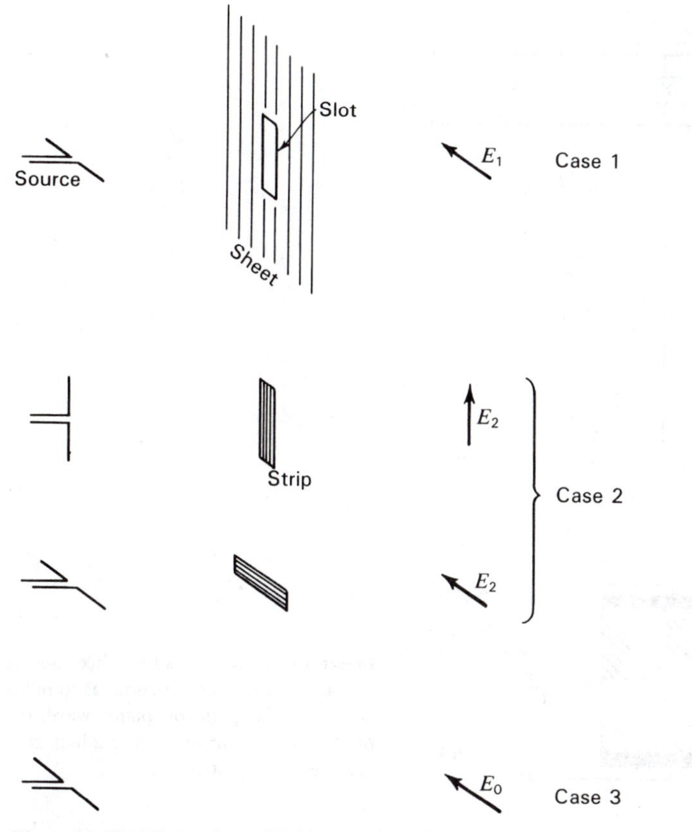

Figure 13-13 Illustration of Babinet's principle applied to a slot in an infinite metal sheet and the complementary metal strip.

13-5 THE IMPEDANCE OF COMPLEMENTARY SCREENS.

In this section Babinet's principle is applied with the aid of a transmission-line analogy to finding the relation between the surface impedance Z_1 of a screen and the surface impedance Z_2 of the complementary metal screen.[1]

Consider the infinite transmission line shown in Fig. 13-14*a* of characteristic impedance Z_0 or characteristic admittance $Y_0 = 1/Z_0$. Let a shunt admittance Y_1 be placed across the line. An incident wave traveling to the right of voltage V_i is partly reflected at Y_1 as a wave of voltage V_r and partly transmitted beyond Y_1 as a wave of voltage V_t. The voltages are measured across the line.

[1] The treatment follows that given by H. G. Booker. See "Slot Aerials and Their Relation to Complementary Wire Aerials," *JIEE* (*Lond.*), **93**, pt. IIIA, no. 4, 1946.

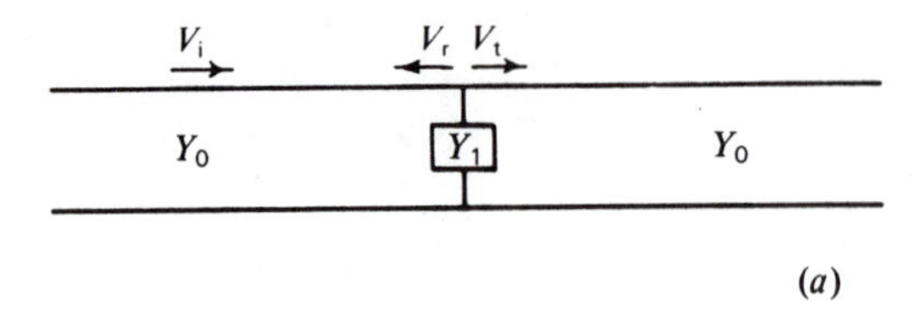

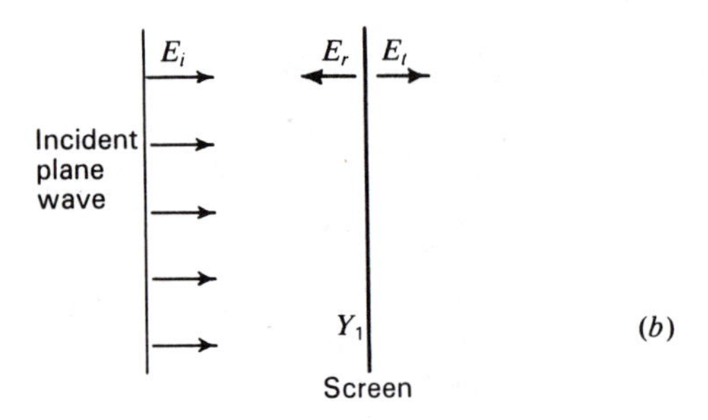

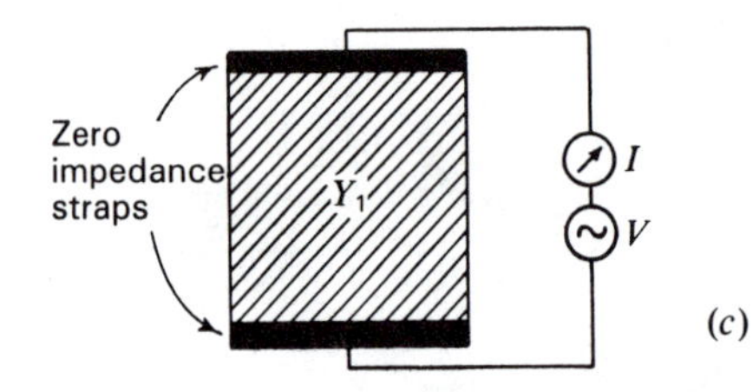

Figure 13-14 Shunt admittance across transmission line (*a*) is analogous to infinite screen in path of plane wave (*b*). Method of measuring surface admittance of screen is suggested in (*c*).

This situation is analogous to a plane wave of field intensity E_i incident normally on a plane screen of infinite extent with a surface admittance, or admittance per square, of Y_1; that is, the admittance measured between the opposite edges of any square section of the sheet as in Fig. 13-14*c* is Y_1. Neglecting the impedance of the leads the admittance

$$Y_1 = \frac{I}{V} \qquad (\mho \text{ per square}) \tag{1}$$

The value of Y is the same for *any square section* of the sheet. Thus, the section may be 1 cm square or 1 meter square. Hence, (1) has the dimensions of admittance rather than of admittance per length squared and is called a *surface admittance*, or *admittance per square*. The field intensities of the waves reflected and transmitted normally to the screen are E_r and E_t. Let the medium surrounding the screen be free space. It has a characteristic admittance Y_0 which is a pure conductance G_0. Thus,

$$Y_0 = \frac{1}{Z_0} = \frac{1}{377} = G_0 \qquad (\mho) \tag{2}$$

The ratio of the magnetic to the electric field intensity of any plane traveling wave in free space has this value. Hence,

$$Y_0 = \frac{H_i}{E_i} = -\frac{H_r}{E_r} = \frac{H_t}{E_t} \qquad (\mho) \tag{3}$$

where H_i, H_r and H_t are the magnetic field intensities of the incident, reflected and transmitted waves respectively.

The transmission coefficient for volatge τ_v of the transmission line[1] (Fig. 13-14*a*) is

$$\tau_v = \frac{V_t}{V_i} = \frac{2Y_0}{2Y_0 + Y_1} \tag{4}$$

By analogy the transmission coefficient for the electric field (Fig. 13-14*b*) is

$$\tau_E = \frac{E_t}{E_i} = \frac{2Y_0}{2Y_0 + Y_1} \tag{5}$$

If now the original screen is replaced by its complementary screen with an admittance per square of Y_2, the new transmission coefficient is the ratio of the new transmitted field E_t' to the incident field. Thus,

$$\tau_E' = \frac{E_t'}{E_i} = \frac{2Y_0}{2Y_0 + Y_2} \tag{6}$$

Applying Babinet's principle, we have from (13-4-6) that

$$\frac{E_t}{E_i} + \frac{E_t'}{E_i} = 1 \tag{7}$$

or

$$\tau_E + \tau_E' = 1 \tag{8}$$

Therefore,

$$\frac{2Y_0}{2Y_0 + Y_1} + \frac{2Y_0}{2Y_0 + Y_2} = 1 \tag{9}$$

and we obtain Booker's result that

$$Y_1 Y_2 = 4Y_0^2 \tag{10}$$

Since $Y_1 = 1/Z_1$, $Y_2 = 1/Z_2$ and $Y_0 = 1/Z_0$,

$$Z_1 Z_2 = \frac{Z_0^2}{4} \qquad \text{or} \qquad \sqrt{Z_1 Z_2} = \frac{Z_0}{2} \tag{11}$$

[1] See, for example, S. A. Schelkunoff, *Electromagnetic Waves*, Van Nostrand, New York, 1943, p. 212.

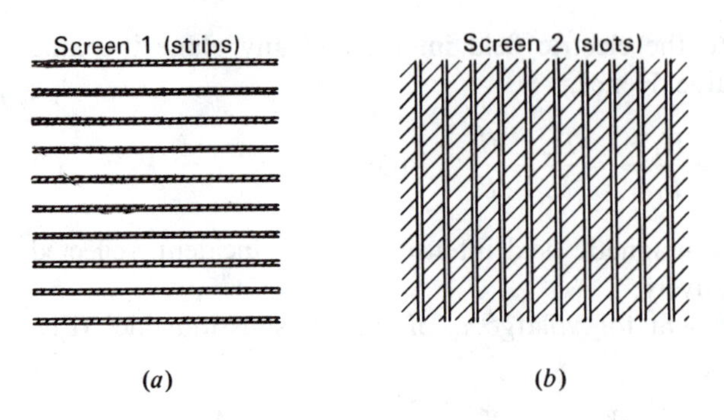

Figure 13-15 Screen of parallel strips (*a*) and complementary screen of slots (*b*).

Thus, the geometric mean of the impedances of the two screens equals $\frac{1}{2}$ the intrinsic impedance of the surrounding medium. Since, for free space, $Z_0 = 376.7\ \Omega$,

$$Z_1 = \frac{35\,476}{Z_2} \qquad (\Omega) \tag{12}$$

If screen 1 is an infinite grating of narrow parallel metal strips as in Fig. 13-15*a*, then the complementary screen (screen 2) is an infinite grating of narrow slots as shown in Fig. 13-15*b*. Suppose that a low-frequency plane wave is incident normally on screen 1 with the electric field parallel to the strips. Then the grating acts as a perfectly reflecting screen and zero field penetrates to the rear. Thus $Z_1 = 0$ and, from (12), $Z_2 = \infty$, so that the complementary screen of slots (screen 2) offers no impediment to the passage of the wave. If the frequency is increased sufficiently, screen 1 begins to transmit part of the incident wave. If at the frequency F_0 screen 1 has a surface impedance $Z_1 = j188\ \Omega$ per square, the impedance Z_2 of screen 2 is $-j188\ \Omega$ per square, so that both screens transmit equally well. If screen 1 becomes more transparent (Z_1 larger) as the frequency is further increased, screen 2 will become more opaque (Z_2 smaller). At any frequency the sum of the fields transmitted through screen 1 and through screen 2 is a constant and equal to the field without any screen present.

The dipoles and slots of Sec. 12-13 are more specialized examples of complementary surfaces or screens.

13-6 THE IMPEDANCE OF SLOT ANTENNAS.

In this section a relation is developed for the impedance Z_s of a slot antenna in terms of the impedance Z_d of the complementary dipole antenna.[1] Knowing Z_d for the dipole, the impedance Z_s of the slot can then be determined.

[1] The treatment follows that given by H. G. Booker, "Slot Aerials and Their Relation to Complementary Wire Aerials," *JIEE* (*Lond.*), **93**, pt. IIIA, no. 4, 1946, with minor embellishments suggested by V. H. Rumsey. See also Sec. 15-2.

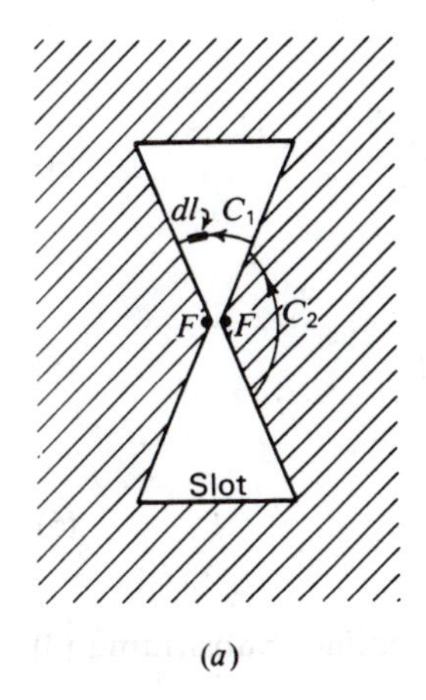

(a)

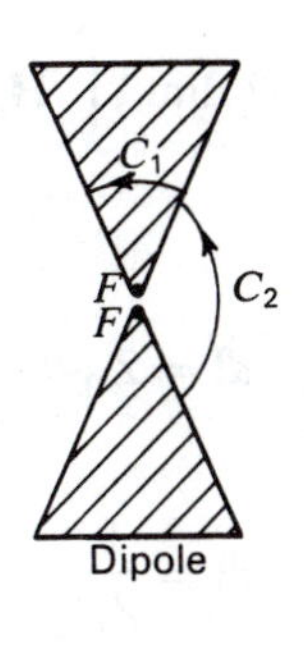

(b)

Figure 13-16 Slot antenna (*a*) and complementary dipole antenna (*b*).

Consider the slot antenna shown in Fig. 13-16*a* and the complementary dipole antenna shown in Fig. 13-16*b*. The terminals of each antenna are indicated by FF, and it is assumed that they are separated by an infinitesimal distance. It is assumed that the dipole and slot are cut from an infinitesimally thin, plane, perfectly conducting sheet.

Let a generator be connected to the terminals of the slot. The driving-point impedance Z_s at the terminals is the ratio of the terminal voltage V_s to the terminal current I_s. Let $\mathbf{E}_s$ and $\mathbf{H}_s$ be the electric and magnetic fields of the slot at any point P. Then the voltage V_s at the terminals FF of the slot is given by the line integral of $\mathbf{E}_s$ over the path C_1 (Fig. 13-16*a*) as C_1 approaches zero. Thus,

$$V_s = \lim_{C_1 \to 0} \int_{C_1} \mathbf{E}_s \cdot d\mathbf{l} \tag{1}$$

where $d\mathbf{l}$ = infinitesimal vector element of length $d\mathbf{l}$ along the contour or path C_1

The current I_s at the terminals of the slot is

$$I_s = 2 \lim_{C_2 \to 0} \int_{C_2} \mathbf{H}_s \cdot d\mathbf{l} \tag{2}$$

The path C_2 is just outside the metal sheet and parallel to its surface. The factor 2 enters because only $\frac{1}{2}$ of the closed line integral is taken, the line integral over the other side of the sheet being equal by symmetry.

Turning our attention to the complementary dipole antenna, let a generator be connected to the terminals of the dipole. The driving-point impedance Z_d at the terminals is the ratio of the terminal voltage V_d to the terminal current I_d. Let $\mathbf{E}_d$ and $\mathbf{H}_d$ be the electric and magnetic fields of the dipole at any point P. Then the voltage at the dipole terminals is

$$V_d = \lim_{C_2 \to 0} \int_{C_2} \mathbf{E}_d \cdot d\mathbf{l} \tag{3}$$

and the current is

$$I_d = 2 \lim_{C_1 \to 0} \int_{C_1} \mathbf{H}_d \cdot d\mathbf{l} \tag{4}$$

However,

$$\lim_{C_2 \to 0} \int_{C_2} \mathbf{E}_d \cdot d\mathbf{l} = Z_0 \lim_{C_2 \to 0} \int_{C_2} \mathbf{H}_s \cdot d\mathbf{l} \tag{5}$$

and

$$\lim_{C_1 \to 0} \int_{C_1} \mathbf{H}_d \cdot d\mathbf{l} = \frac{1}{Z_0} \lim_{C_1 \to 0} \int_{C_1} \mathbf{E}_s \cdot d\mathbf{l} \tag{6}$$

where Z_0 is the intrinsic impedance of the surrounding medium. Subsituting (3) and (2) in (5) yields

$$V_d = \frac{Z_0}{2} I_s \tag{7}$$

Substituting (4) and (1) in (6) gives

$$V_s = \frac{Z_0}{2} I_d \tag{8}$$

Multiplying (7) and (8) we have

$$\frac{V_s}{I_s}\frac{V_d}{I_d} = \frac{Z_0^2}{4} \tag{9}$$

or

$$Z_s Z_d = \frac{Z_0^2}{4} \qquad \text{or} \qquad Z_s = \frac{Z_0^2}{4Z_d} \tag{10}$$

Thus, we obtain Booker's result that the terminal impedance Z_s of a slot antenna is equal to $\frac{1}{4}$ of the square of the intrinsic impedance of the surrounding medium divided by the terminal impedance Z_d of the complementary dipole antenna. For free space $Z_0 = 376.7\ \Omega$, so[1]

$$Z_s = \frac{Z_0^2}{4Z_d} = \frac{35\,476}{Z_d} \qquad (\Omega) \tag{11}$$

---

[1] If the intrinsic impedance Z_0 of free space were unknown, (11) provides a means of determining it by measurements of the impedance Z_s of a slot antenna and the impedance Z_d of the complementary dipole antenna. The impedance Z_0 is twice the geometric means of Z_s and Z_d or

$$Z_0 = 2\sqrt{Z_s Z_d} \tag{12}$$

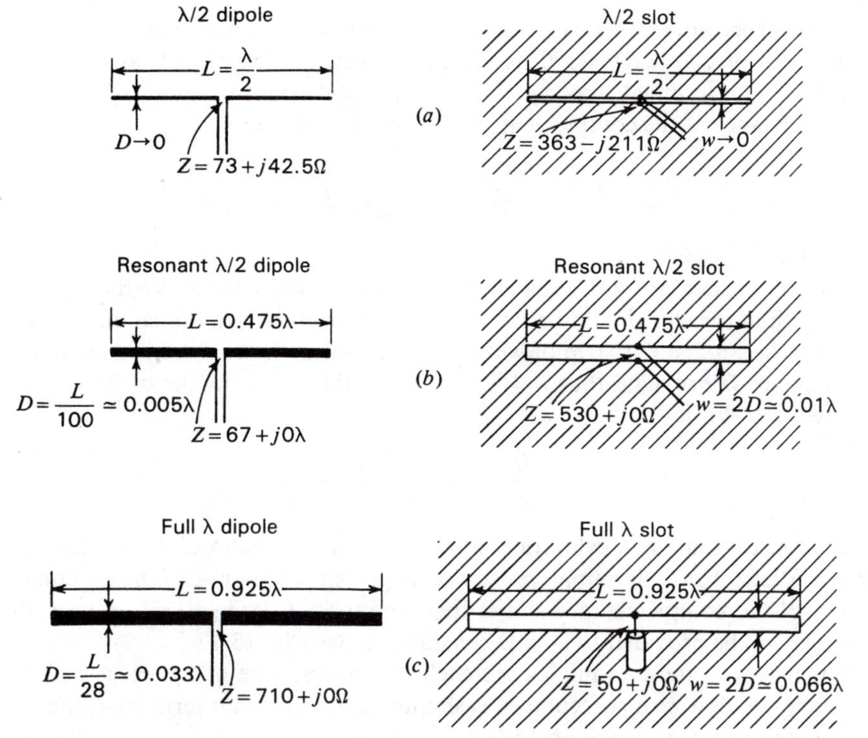

Figure 13-17 Comparison of impedances of cylindrical dipole antennas with complementary slot antennas. The slot in (*c*) matches directly to the 50 Ω coaxial line.

The impedance of the slot is proportional to the admittance of the dipole, or vice versa. Since, in general, Z_d may be complex, we may write

$$Z_s = \frac{35\,476}{R_d + jX_d} = \frac{35\,476}{R_d^2 + X_d^2}(R_d - jX_d) \tag{13}$$

where R_d and X_d are the resistive and reactive components of the dipole terminal impedance Z_d. Thus, if the dipole antenna is inductive, the slot is capacitative, and vice versa. Lengthening a $\lambda/2$ dipole makes it more inductive, but lengthening a $\lambda/2$ slot makes it more capacitative.

Let us now consider some numerical examples, proceeding from known dipole types to the complementary slot types. The impedance of an infinitesimally thin $\lambda/2$ antenna ($L = 0.5\lambda$ and $L/D = \infty$) is $73 + j42.5\ \Omega$ (see Chap. 10). Therefore, the terminal impedance of an infinitesimally thin $\lambda/2$ slot antenna ($L = 0.5\lambda$ and $L/w = \infty$) is

$$Z_1 = \frac{35\,476}{73 + j42.5} = 363 - j211\ \Omega \tag{14}$$

See Fig. 13-17*a*.

As another more practical example, a cylindrical antenna with a length-diameter ratio of 100 ($L/D = 100$) is resonant when the length is about 0.475λ ($L = 0.475\lambda$). The terminal impedance is resistive and equal to about 67 Ω. The terminal resistance of the complementary slot antenna is then

$$Z_1 = \frac{35\,476}{67} = 530 + j0\ \Omega \tag{15}$$

See Fig. 13-17*b*.

The complementary slot has a length $L = 0.475\lambda$, the same as for the dipole, but the width of the slot should be twice the diameter of the cylindrical dipole. As indicated in Sec. 9-7, a flat strip of width w is equivalent to a cylindrical conductor of diameter D provided that $w = 2D$. Thus, in this example, the width of the complementary slot is

$$w = 2D = \frac{2L}{100} = \frac{2 \times 0.475\lambda}{100} \simeq 0.01\lambda \tag{16}$$

As a third example, a cylindrical dipole with an L/D ratio of 28 and length of about 0.925λ has a terminal resistance of about $710 + j0\ \Omega$. The terminal resistance of the complementary slot is then about $50 + j0\ \Omega$ so that an impedance match will be provided to a 50 Ω coaxial line. See Fig. 13-17*c*.

If the slots in these examples are enclosed on one side of the sheet with a box of such size that zero susceptance is shunted across the slot terminals, due to the box, the impedances are doubled.

The bandwidth or selectivity characteristics of a slot antenna are the same as for the complementary dipole. Thus, widening a slot (smaller L/w ratio) increases the bandwidth of the slot antenna, the same as increasing the thickness of a dipole antenna (smaller L/D ratio) increases its bandwidth.

The above discussion of this section applies to slots in sheets of infinite extent. If the sheet is finite, the impedance values are substantially the same provided that the edge of the sheet is at least a wavelength from the slot. However, the measured slot impedance is sensitive to the nature of the terminal connections.

13-7 SLOTTED CYLINDER ANTENNAS.[1] A slotted sheet antenna is shown in Fig. 13-18*a*. By bending the sheet into a U-shape as in (*b*) and finally into a cylinder as in (*c*), we arrive at a slotted cylinder antenna. The impedance of the path around the circumference of the cylinder may be sufficiently low so that

[1] A. Alford, "Long Slot Antennas," *Proc. Natl. Electronics Conf.*, 1946, p. 143.
A. Alford, "Antenna for F-M Station WGHF," *Communications*, 26, 22, February 1946.
E. C. Jordan and W. E. Miller, "Slotted Cylinder Antennas," *Electronics*, **20**, 90–93, February 1947.
George Sinclair, "The Patterns of Slotted Cylinder Antennas," *Proc. IRE*, **36**, 1487–1492, December 1948.

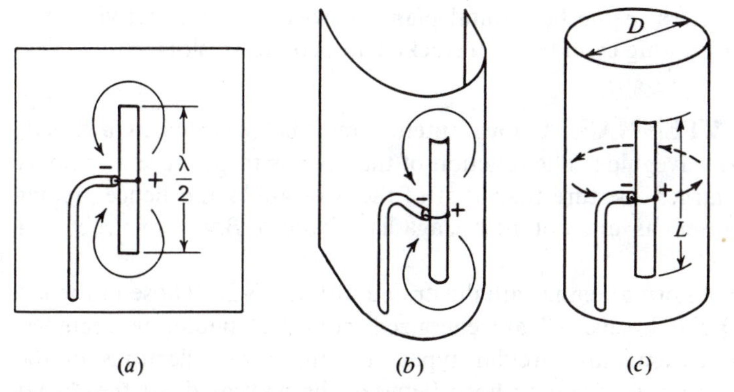

Figure 13-18 Evolution of slotted cylinder from slotted sheet.

most of the current tends to flow in horizontal loops around the cylinder as suggested. If the diameter D of the cylinder is a sufficiently small fraction of a wavelength, say, less than $\lambda/8$, the vertical slotted cylinder radiates a horizontally polarized field with a pattern in the horizontal plane which is nearly circular. As the diameter of the cylinder is increased, the pattern in the horizontal plane tends to become more unidirectional with the maximum radiation from the side of the cylinder with the slot. For resonance, the length L of the slot is greater than $\lambda/2$. This may be explained as follows. Referring to Fig. 13-19*a*, the 2-wire transmission line is resonant when it is $\lambda/2$ long. However, if this line is loaded with a series of loops of diameter D as at (*b*), the phase velocity of wave transmission on the line can be increased, so that the resonant frequency is raised. With a sufficient number of shunt loops the arrangement of (*b*) becomes equivalent to a slotted cylinder of diameter D. Typical slotted cylinder dimensions for resonance are $D = 0.125\lambda$, $L = 0.75\lambda$ and the slot width about 0.02λ.

This type of antenna, pioneered by Andrew Alford, has found considerable application for broadcasting a horizontally polarized wave with an omnidirec-

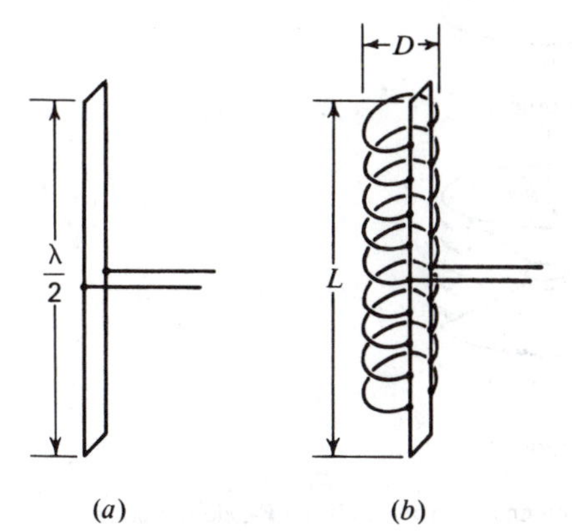

Figure 13-19 Slotted cylinder as a loop-loaded transmission line.

tional or circular pattern in the horizontal plane. Vertical-plane directivity may be increased by using a long cylinder with stacked, i.e., collinear, slots.

13-8 HORN ANTENNAS. A horn antenna may be regarded as a flared-out (or opened-out) waveguide. The function of the horn is to produce a uniform phase front with a larger aperture than that of the waveguide and hence greater directivity. Horn antennas are not new. Jagadis Chandra Bose constructed a pyramidal horn in 1897.

Several types of horn antennas are illustrated in Fig. 13-20. Those in the left column are rectangular horns. All are energized from rectangular waveguides. Those in the right column are circular types. To minimize reflections of the guided wave, the transition region or horn between the waveguide at the throat

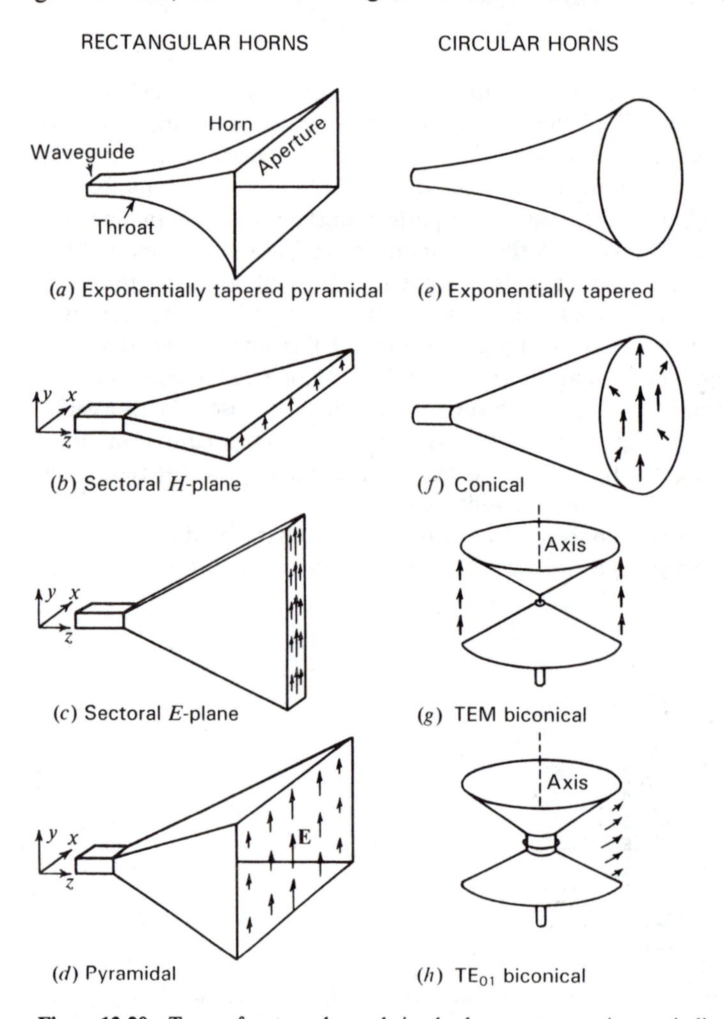

Figure 13-20 Types of rectangular and circular horn antennas. Arrows indicate **E**-field directions.

and free space at the aperture could be given a gradual exponential taper as in Fig. 13-20*a* or *e*. However, it is the general practice to make horns with straight flares as suggested by the other types in Fig. 13-20.[1] The types in Fig. 13-20*b* and *c* are sectoral horns. They are rectangular types with a flare in only one dimension. Assuming that the rectangular waveguide is energized with a TE_{10} mode wave electric field (**E** in the y direction), the horn in Fig. 13-20*b* is flared out in a plane perpendicular to **E**. This is the plane of the magnetic field **H**. Hence, this type of horn is called a sectoral horn flared in the H plane or simply *an H-plane sectoral horn*. The horn in Fig. 13-20*c* is flared out in the plane of the electric field **E**, and, hence, is called an *E-plane sectoral horn*. A rectangular horn with flare in both planes, as in Fig. 13-20*d*, is called a *pyramidal horn*. With a TE_{10} wave in the waveguide the magnitude of the electric field is quite uniform in the y direction across the apertures of the horns of Fig. 13-20*b*, *c* and *d* but tapers to zero in the x direction across the apertures. This variation is suggested by the arrows at the apertures in Fig. 13-20*b*, *c* and *d*. The arrows indicate the direction of the electric field **E**, and their length gives an approximate indication of the magnitude of the field intensity. For small flare angles the field variation across the aperture of the rectangular horns is similar to the sinusoidal distribution of the TE_{10} mode across the waveguide.

The horn shown in Fig. 13-20*f* is a *conical type*. When excited with a circular guide carrying a TE_{11} mode wave, the electric field distribution at the aperture is as shown by the arrows. The horns in Fig. 13-20*g* and *h* are *biconical* types. The one in Fig. 13-20*g* is excited in the TEM mode by a vertical radiator while the one in Fig. 13-20*h* is excited in the TE_{01} mode by a small horizontal loop antenna. These biconical horns are nondirectional in the horizontal plane. The biconical horn of Fig. 13-20*g* is like the one shown in Fig. 2-28*c*.

Neglecting edge effects, the radiation pattern of a horn antenna can be determined if the aperture dimensions and aperture field distribution are known. For a given aperture the directivity is maximum for a uniform distribution. Variations in the magnitude or phase of the field across the aperture decrease the directivity. Since the H-plane sectoral horn (Fig. 13-20*b*) has a field distribution over the x dimension which tapers to zero at the edges of the aperture, one would expect a pattern in the xz plane relatively free of minor lobes as compared to the yz plane pattern of an E-plane sectoral horn (Fig. 13-20*c*) for which the magnitude of **E** is quite constant over the y dimension of the aperture. This is borne out experimentally.

The *principle of equality of path length* (*Fermat's principle*) is applicable to the horn design but with a different emphasis. Instead of requiring a constant phase across the horn mouth, the requirement is relaxed to one where the phase may deviate, but by less than a specified amount δ, equal to the path length difference between a ray traveling along the side and along the axis of the horn.

[1] Horns with a straight flare tend to have a *constant phase center* while those with a taper do not.

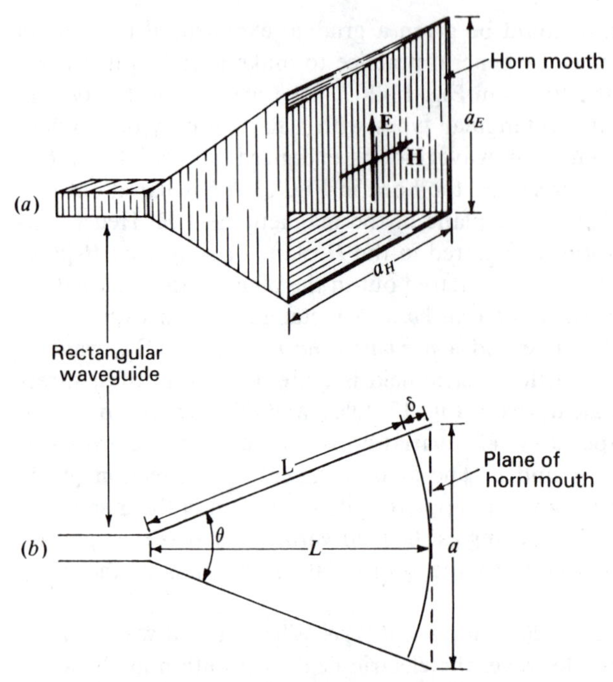

Figure 13-21 (*a*) Pyramidal horn antenna. (*b*) Cross section with dimensions used in analysis. The diagram can be for either *E*-plane or *H*-plane cross sections. For the *E* plane the flare angle is θ_E and aperture a_E. For the *H* plane the flare angle is θ_H and the aperture a_H. See Fig. 13-22.

From Fig. 13-21,

$$\cos\frac{\theta}{2} = \frac{L}{L+\delta} \tag{1}$$

$$\sin\frac{\theta}{2} = \frac{a}{2(L+\delta)} \tag{2}$$

$$\tan\frac{\theta}{2} = \frac{a}{2L} \tag{3}$$

where θ = flare angle (θ_E for *E* plane, θ_H for *H* plane)
a = aperture (a_E for *E* plane, a_H for *H* plane)
L = horn length

From the geometry we have also that

$$L = \frac{a^2}{8\delta} \qquad (\delta \ll L) \tag{4}$$

and

$$\theta = 2\tan^{-1}\frac{a}{2L} = 2\cos^{-1}\frac{L}{L+\delta} \tag{5}$$

In the E plane of the horn, δ is usually held to 0.25λ or less. However, in the H plane, δ can be larger, or about 0.4λ, since $\mathbf{E}$ goes to zero at the horn edges (boundary condition, $E_t = 0$ satisfied).

To obtain as uniform an aperture distribution as possible, a very long horn with a small flare angle is required. However, from the standpoint of practical convenience the horn should be as short as possible. An *optimum horn* is between these extremes and has the minimum beam width without excessive sidelobe level (or most gain) for a given length.

If δ is a sufficiently small fraction of a wavelength, the field has nearly uniform phase over the entire aperture. For a constant length L, the directivity of the horn increases (beam width decreases) as the aperture a and flare angle θ are increased. However, if the aperture and flare angle become so large that δ is equivalent to 180 electrical degrees, the field at the edge of the aperture is in phase opposition to the field on the axis. For all but very large flare angles the ratio $L/(L + \delta)$ is so nearly unity that the effect of the additional path length δ on the distribution of the field magnitude can be neglected. However, when $\delta = 180°$, the phase reversal at the edges of the aperture reduces the directivity (increases sidelobes). It follows that the maximum directivity occurs at the largest flare angle for which δ does not exceed a certain value (δ_0). Thus, from (1) the *optimum horn dimensions* can be related by

$$\delta_0 = \frac{L}{\cos(\theta/2)} - L \tag{6}$$

or

$$L = \frac{\delta_0 \cos(\theta/2)}{1 - \cos(\theta/2)} \tag{7}$$

It turns out that the value of δ_0 must usually be in the range of 0.1 to 0.4 free-space wavelength.[1] Suppose that for an optimum horn $\delta_0 = 0.25\lambda$ and that the axial length $L = 10\lambda$. Then from (5), $\theta = 25°$. This flare angle then results in the maximum directivity for a 10λ horn.

The path length, or δ effect, discussed above is an inherent limitation of all horn antennas of the conventional type.[2] The relations of (1) through (7) can be applied to all the horns of Fig. 13-20 to determine the optimum dimensions. However, the appropriate value of δ_0 may differ as discussed in the following sections. Another limitation of horn antennas is that for the most uniform aperture illumination higher modes of transmission in the horn must be suppressed. It

[1] At a given frequency the wavelength in the horn λ_h is always equal to or greater than the free-space wavelength λ. Since λ_h depends on the horn dimensions, it is more convenient to express δ_0 in free-space wavelengths λ.

[2] In the lens-compensated type of horn antenna (see Fig. 14-18*b* and Fig. 13-30) the velocity of the wave is increased near the edge of the horn with respect to the velocity at the axis in order to equalize the phase over the aperture.

follows that the width of the waveguide at the throat of the horn must be between $\lambda/2$ and 1λ, or if the excitation system is symmetrical, so that even modes are not energized, the width must be between $\lambda/2$ and $3\lambda/2$.

13-9 THE RECTANGULAR HORN ANTENNA.[1] Provided that the aperture in both planes of a rectangular horn exceeds 1λ, the pattern in one plane is substantially independent of the aperture in the other plane. Hence, in general, the H-plane pattern of an H-plane sectoral horn is the same as the H-plane pattern of a pyramidal horn with the same H-plane cross section. Likewise, the E-plane pattern of an E-plane sectoral horn is the same as the E-plane pattern of a pyramidal horn with the same E-plane cross section. Referring to Fig. 13-22, the total flare angle in the E plane is θ_E and the total flare angle in the H plane is

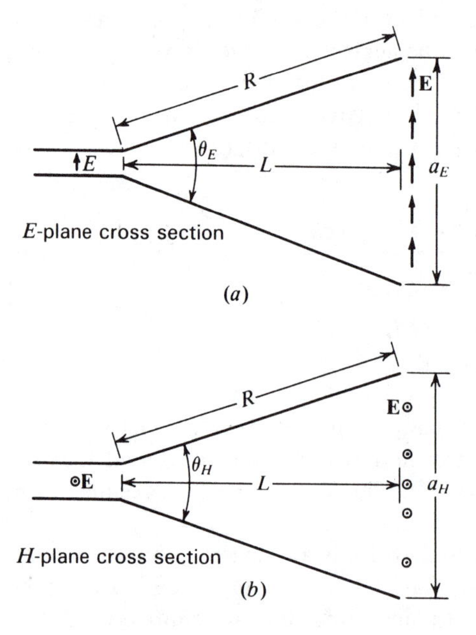

Figure 13-22 E-plane and H-plane cross sections.

[1] W. L. Barrow and F. D. Lewis, "The Sectoral Electromagnetic Horn," *Proc. IRE*, **27**, 41–50, January 1939.
W. L. Barrow and L. J. Chu, "Theory of the Electromagnetic Horn," *Proc. IRE*, **27**, 51–64, January 1939.
L. J. Chu and W. L. Barrow, "Electromagnetic Horn Design," *Trans. AIEE*, **58**, 333–337, July 1939.
F. E. Terman, *Radio Engineers' Handbook*, McGraw-Hill, New York, 1943, pp. 824–837 (this reference includes a summary of design data on horns).
J. R. Risser, in S. Silver (ed.), *Microwave Antenna Theory and Design*, McGraw-Hill, New York, 1949, chap. 10, pp. 349–365.
G. Stavis and A. Dorne, in *Very High Frequency Techniques*, Radio Research Laboratory Staff, McGraw-Hill, New York, 1947, chap. 6.

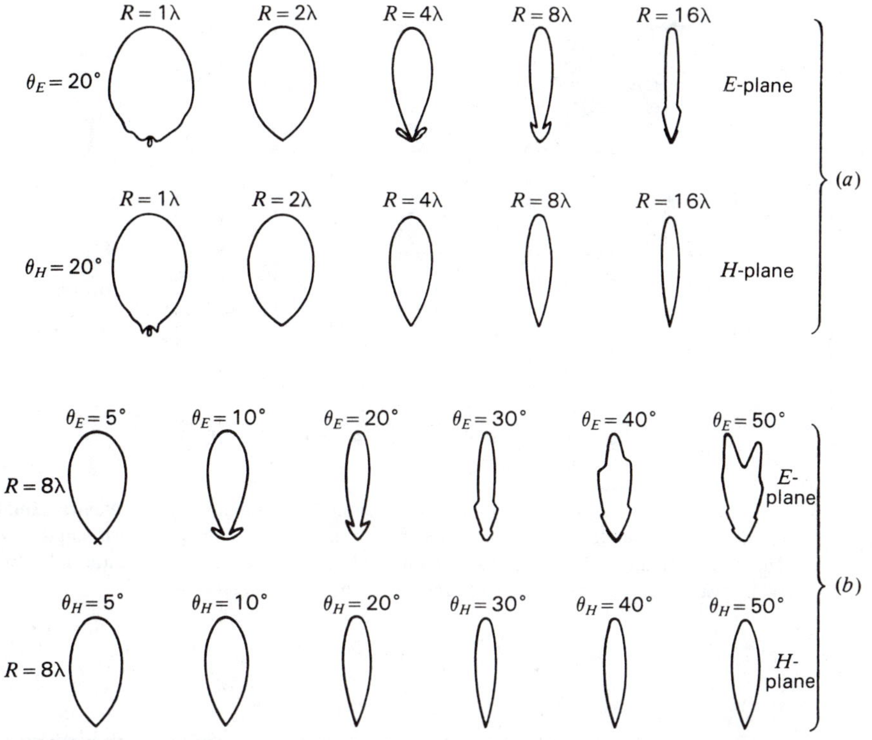

Figure 13-23 Measured E- and H-plane field patterns of rectangular horns as a function of flare angle and horn length. (*After D. R. Rhodes, "An Experimental Investigation of the Radiation Patterns of Electromagnetic Horn Antennas,"* Proc. IRE, **36**, *1101–1105, September 1948.*)

θ_H. The axial length of the horn from throat to aperture is L and the radial length is R. Patterns measured by Donald Rhodes[1] are shown in Fig. 13-23. In (a) the patterns in the E plane and H plane are compared as a function of R. Both sets are for a flare angle of 20°. The E-plane patterns have minor lobes whereas the H-plane patterns have practically none. In (b) measured patterns for horns with $R = 8\lambda$ are compared as a function of flare angle. In the upper row E-plane patterns are given as a function of the E-plane flare angle θ_E and in the lower row H-plane patterns are shown as a function of the H-plane flare angle θ_H. For a flare angle $\theta_E = 50°$ the E-plane pattern is split, whereas for $\theta_H = 50°$ the H-plane pattern is not. This is because a given phase shift at the aperture in the E-plane horn has more effect on the pattern than the same phase shift in the H-plane horn. In the H-plane horn the field goes to zero at the edges of the aperture, so

[1] D. R. Rhodes, "An Experimental Investigation of the Radiation Patterns of Electromagnetic Horn Antennas," *Proc. IRE*, **36**, 1101–1105, September 1948.

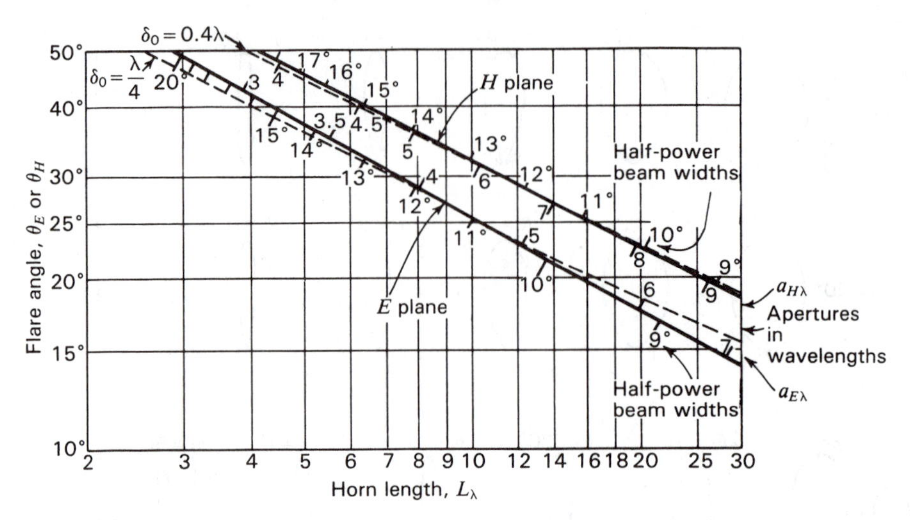

Figure 13-24 Experimentally determined optimum dimensions for rectangular horn antennas. Solid curves give relation of flare angle θ_E in E plane and flare angle θ_H in H plane to horn length (see Fig. 13-22). The corresponding half-power beam widths and apertures in wavelengths are indicated along the curves. Dashed curves show calculated dimensions for $\delta_0 = 0.25$ and 0.4λ.

the phase near the edge is relatively less important. Accordingly, we should expect the value of δ_0 for the H plane to be larger than for the E plane. This is illustrated in Fig. 13-24 and discussed in the next paragraph.

From Rhodes's experimental patterns, optimum dimensions were selected for both E- and H-plane flare as a function of flare angle and horn length L. These optimum dimensions are indicated by the solid lines in Fig. 13-24. The corresponding half-power beam widths and apertures in wavelengths are also indicated. The dashed curves show the calculated dimensions for a path length $\delta_0 = 0.25\lambda$ and $\delta_0 = 0.4\lambda$. The value of 0.25λ gives a curve close to the experimental curve for E-plane flare, while the value of 0.4λ gives a curve close to the experimental one for H-plane flare over a considerable range of horn length. Thus, the tolerance in path length is greater for H-plane flare than for E-plane flare, as indicated above.

Suppose we wish to construct an optimum horn with length $L = 10\lambda$. From Fig. 13-24 we note that for this length the HPBW (E plane) = 11° and the HPBW (H plane) = 13°, the E-plane aperture $a_E = 4.5\lambda$ and the H-plane aperture $a_H = 5.8\lambda$. Thus, although the E-plane aperture is not so large as the H-plane aperture, the beam width is less (but minor lobes larger) because the E-plane aperture distribution is more uniform. For horn operation over a frequency band it is desirable to determine the optimum dimensions for the highest frequency to be used, since δ as measured in wavelengths is largest at this highest frequency.

The directivity (or gain, assuming no loss) of a horn antenna can be expressed in terms of its effective aperture. Thus,

$$D = \frac{4\pi A_e}{\lambda^2} = \frac{4\pi \varepsilon_{ap} A_p}{\lambda^2} \tag{1}$$

where A_e = effective aperture, m^2
A_p = physical aperture, m^2
ε_{ap} = aperture efficiency = A_e/A_p
λ = wavelength, m

For a rectangular horn $A_p = a_E a_H$ and for a conical horn $A_p = \pi r^2$, where r = aperture radius. It is assumed that a_E, a_H or r are all at least 1λ. Taking $\varepsilon_{ap} \simeq 0.6$, (1) becomes

$$D \simeq \frac{7.5 A_p}{\lambda^2} \tag{2}$$

or

$$D \simeq 10 \log \left(\frac{7.5 A_p}{\lambda^2} \right) \quad \text{(dBi)} \tag{3}$$

For a pyramidal (rectangular) horn (3) can also be expressed as

$$D \simeq 10 \log (7.5 a_{E\lambda} a_{H\lambda}) \tag{4}$$

where $a_{E\lambda}$ = E-plane aperture in λ
$a_{H\lambda}$ = H-plane aperture in λ

Example. (*a*) Determine the length *L*, *H*-plane aperture and flare angles θ_E and θ_H (in the *E* and *H* planes respectively) of a pyramidal horn as in Fig. 13-20*d* for which the *E*-plane aperture $a_E = 10\lambda$. The horn is fed by a rectangular waveguide with TE_{10} mode. Let $\delta = 0.2\lambda$ in the *E* plane and 0.375λ in the *H* plane. (*b*) What are the beam widths? (*c*) What is the directivity?

Solution. Taking $\delta = \lambda/5$ in the *E* plane, we have from (13-8-4) that the required horn length

$$L = \frac{a^2}{8\delta} = \frac{100\lambda}{8/5} = 62.5\lambda \tag{5}$$

and from (13-8-5) that the flare angle in the *E* plane

$$\theta_E = 2 \tan^{-1} \frac{a}{2L} = 2 \tan^{-1} \frac{10}{125} = 9.1^\circ \tag{6}$$

Taking $\delta = 3\lambda/8$ in the *H* plane we have from (13-8-5) that the flare angle in the *H* plane

$$\theta_H = 2 \cos^{-1} \frac{L}{L + \delta} = 2 \cos^{-1} \frac{62.5}{62.5 + 0.375} = 12.52^\circ \tag{7}$$

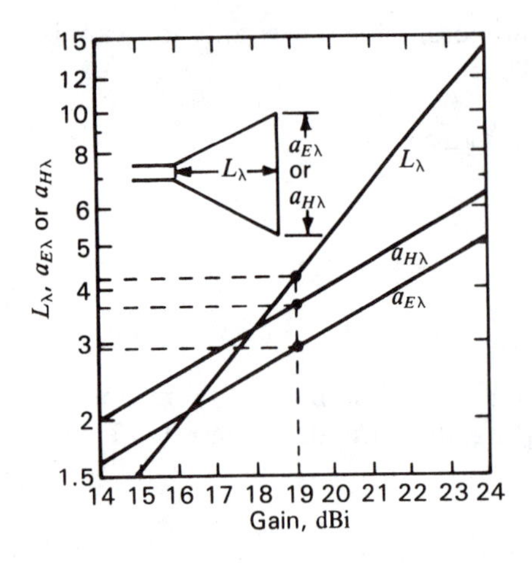

Figure 13-25a Dimensions of rectangular (pyramidal) horns (in wavelengths) versus directivity (or gain, if no loss). Thus, noting the dashed lines, a gain of 19 dBi requires a horn length $L_\lambda = 4.25$, an H-plane aperture $a_{H\lambda} = 3.7$ and an E-plane aperture $a_{E\lambda} = 2.9$. These are inside dimensions. It is assumed that δ (E plane) $= 0.25\lambda$ and δ (H plane) $= 0.4\lambda$, making the dimensions close to optimum. It is also assumed that $\varepsilon_{ap} = 0.6$.

and from (13-8-5) that the H-plane aperture

$$a_H = 2L \tan \frac{\theta_H}{2} = 2 \times 62.5\lambda \tan 6.26° = 13.7\lambda \tag{8}$$

From Table 13-1,

$$\text{HPBW } (E \text{ plane}) = \frac{56°}{a_{E\lambda}} = \frac{56°}{10} = 5.6° \tag{9a}$$

$$\text{HPBW } (H \text{ plane}) = \frac{67°}{a_{H\lambda}} = \frac{67°}{13.7} = 4.9° \tag{9b}$$

From (3),

$$D \simeq 10 \log \left(\frac{7.5A_p}{\lambda^2}\right) = 10 \log (7.5 \times 10 \times 13.7) = 30.1 \text{ dBi} \tag{10}$$

The δ values used in this example are conservative. For an optimum horn, the δ values are larger, resulting in a considerably shorter horn but at the expense of slightly less gain (because fields are less uniform across the aperture of an optimum horn).

Figure 13-25a shows the optimum dimensions for pyramidal (rectangular) horns versus gain (or directivity, if no loss).[1] For a given desired gain, the graph gives the dimensions for the length L_λ, E-plane aperture $a_{E\lambda}$ and H-plane aperture $a_{H\lambda}$, all in wavelengths. For a given length, the graph also gives the appropriate apertures and the gain. The dimensions are close to optimum. Such dimensions are only of importance on large horns (many wavelengths long)

[1] A similar but somewhat different version has been given by H. Schrank ("Optimum Horns," *IEEE Ant. Prop. Soc. Newsletter*, **27**, 13–14, February 1985), who credits T. A. Milligan for sending him the data.

Table 13-1†

Type of aperture	Beam width, deg	
	Between first nulls	Between half-power points
Uniformly illuminated rectangular aperture or linear array	$\frac{115}{L_\lambda}$	$\frac{51}{L_\lambda}$
Uniformly illuminated circular aperture	$\frac{140}{D_\lambda}$	$\frac{58}{D_\lambda}$
Optimum *E*-plane rectangular horn	$\frac{115}{a_{E\lambda}}$	$\frac{56}{a_{E\lambda}}$
Optimum *H*-plane rectangular horn	$\frac{172}{a_{H\lambda}}$	$\frac{67}{a_{H\lambda}}$

† L_λ = length of rectangular aperture or linear array in free-space wavelengths
D_λ = diameter of circular aperture in free-space wavelengths
$a_{E\lambda}$ = aperture in *E* plane in free-space wavelengths
$a_{H\lambda}$ = aperture in *H* plane in free-space wavelengths

where it is desired that the length be a minimum. For small (short) horns, optimization is usually unwarranted.

13-10 BEAM-WIDTH COMPARISON. It is interesting to compare the beam width between first nulls and between half-power points for uniformly illuminated rectangular and circular apertures obtained in previous chapters with those for optimum rectangular horn antennas (sectoral or pyramidal). This is done in Table 13-1. In general, the relations apply to apertures that are at least several wavelengths long. The beam widths between nulls for the horns are calculated and the half-power beam widths are empirical.[1]

13-11 CONICAL HORN ANTENNAS. The conical horn[2] (Fig. 13-20*f*) can be directly excited from a circular waveguide. Dimensions can be determined from (13-8-5), (13-8-6) and (13-8-7) by taking $\delta_0 = 0.32\lambda$.

[1] G. Stavis and A. Dorne, in *Very High Frequency Techniques*, Radio Research Laboratory Staff, McGraw-Hill, New York, 1947, chap. 6.

[2] G. C. Southworth and A. P. King, "Metal Horns as Directive Receivers of Ultrashort Waves," *Proc. IRE*, **27**, 95–102, February 1939.

A. P. King, "The Radiation Characteristics of Conical Horn Antennas," *Proc. IRE*, **38**, 249–251, March 1950. For optimum conical horns King gives half-power beam widths of $60/a_{E\lambda}$ in the *E* plane and $70/a_{H\lambda}$ in the *H* plane. These are about 6 per cent more than the values for a rectangular horn as given in Table 13-1.

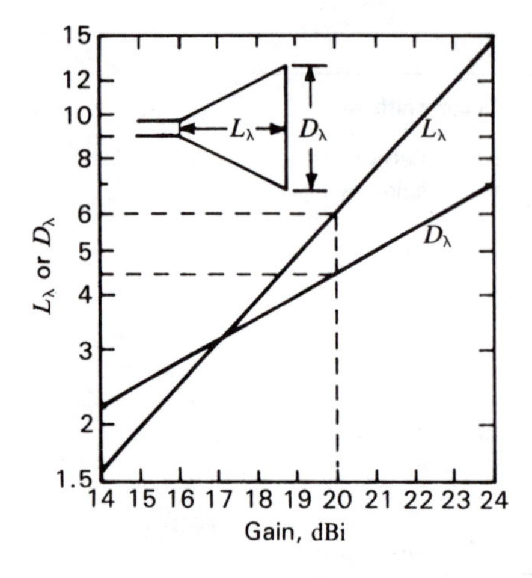

Figure 13-25b Dimensions of conical horn (in wavelengths) versus directivity (or gain, if no loss). Thus, noting the dashed lines, a gain of 20 dBi requires a horn length $L_\lambda = 6.0$ and a diameter $D_\lambda = 4.3$. These (inside) dimensions are close to optimum.

The biconical horns[1] (Fig. 13-20*g* and *h*) have patterns that are nondirectional in the horizontal plane (axis of horns vertical). These horns may be regarded as modified pyramidal horns with a 360° flare angle in the horizontal plane. The optimum vertical-plane flare angle is about the same as for a sectoral horn of the same cross section excited in the same mode.

Figure 13-25b shows optimum dimensions for conical horns versus directivity (or gain if no loss) as adapted from King.[2] For a given desired gain, the graph gives the length L_λ and diameter D_λ, or for a given length, the graph gives the appropriate aperture and gain.

13-12 RIDGE HORNS. A central ridge loads a waveguide and increases its useful bandwidth by lowering the cutoff frequency of the dominant mode.[3] A

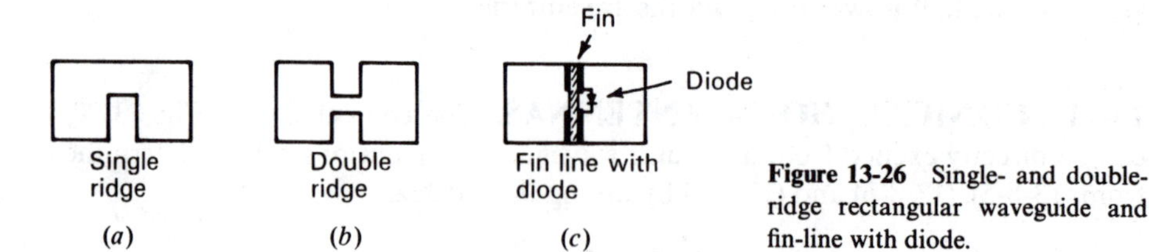

Figure 13-26 Single- and double-ridge rectangular waveguide and fin-line with diode.

[1] W. L. Barrow, L. J. Chu and J. J. Jansen, "Biconical Electromagnetic Horns," *Proc. IRE*, **27**, 769–779, December 1939.

[2] A. P. King, "The Radiation Characteristics of Conical Horn Antennas," *Proc. IRE*, **38**, 249–251, 1950.

[3] S. B. Cohn, "Properties of the Ridge Wave Guide," *Proc. IRE*, **35**, 783–789, August 1947.
T-S Chen, "Calculation of the Parameters of Ridge Waveguides," *IRE Trans. Microwave Theory Tech.*, **MTT-26**, 726–732, October 1978.

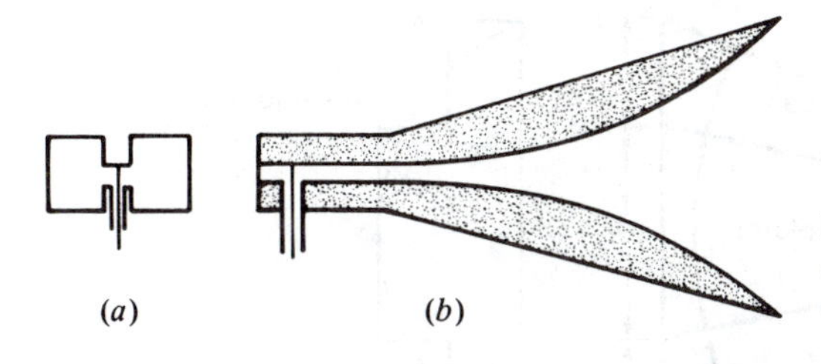

Figure 13-27 Double-ridge horn with coaxial feed. The view at (*a*) is a cross section at the feed point.

rectangular guide with single ridge is shown in Fig. 13-26*a* and with a double ridge in Fig. 13-26*b*. A very thin ridge or fin is also effective in producing the loading of a central ridge. It may consist of a metal-clad ceramic sheet which facilitates the installation of shunt circuit elements as suggested in Fig. 13-26*c*.[1]

Of course, the cutoff frequency can be lowered by placing dielectric material in the waveguide, but this does not increase the bandwidth and it may increase losses.

By continuing a double-ridge structure from a waveguide into a pyramidal horn as suggested in Fig. 13-27, the useful bandwidth of the horn can be increased manyfold[2] (see also Fig. 15-1). A quadruple-ridge horn connected to a dual-fed quadruple-ridge square waveguide can provide dual orthogonal linear polarizations over bandwidths of more than 6 to 1.

13-13 SEPTUM HORNS.

Although the electric field in the *H* plane of a pyramidal horn tends to zero at the edges, resulting in a tapered distribution and reduced sidelobes, the electric field in the *E* plane may be close to uniform in amplitude to the edges, resulting in significant sidelobes. By introducing septum plates bonded to the horn walls, a stepped-amplitude distribution can be achieved in the *E* plane with a reduction in *E*-plane sidelobes. Typically, the first sidelobes of a uniform amplitude distribution are down about 13 dB. With a 2-septum horn Peace and Swartz[3] were able to achieve a sidelobe level more than 30 dB down, which is lower than the sidelobe level in the *H* plane.

A cosine field distribution is approximated with a 1 : 2 : 1 stepped amplitude distribution with apertures also in the ratio 1 : 2 : 1 as suggested in Fig. 13-28. To achieve this distribution the septums must be appropriately spaced at the throat of the horn.

Figure 13-29 shows a stepped-amplitude septum horn with metal-plate lens for feeding "Big Ear," the Ohio State University 110-m radio telescope. The dual-

[1] P. J. Meier, "Integrated Fin-Line Millimeter Components," *IEEE Trans. Microwave Theory Tech.*, **MTT-22**, 1209–1216, December 1974.

[2] K. L. Walton and V. C. Sundberg, "Broadband Ridged Horn Design," *Microwave J.*, **7**, 96–101, March 1964.

[3] G. M. Peace and E. E. Swartz, "Amplitude Compensated Horn Antenna," *Microwave J.*, **7**, 66–68, February 1964.

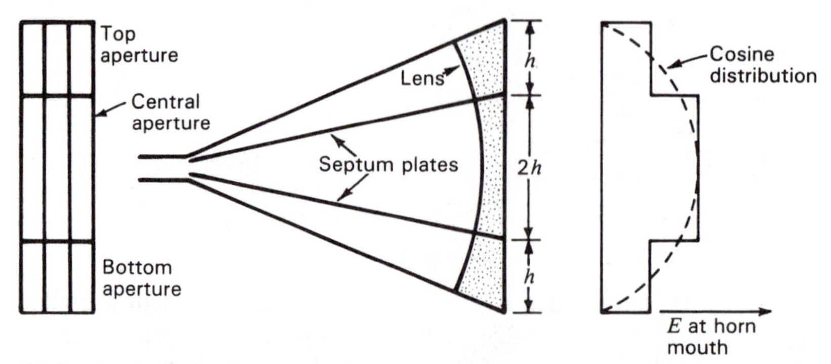

Figure 13-28 Two-septum horn with 1 : 2 : 1 stepped amplitude distribution in field intensity at mouth of horn (approximating a cosine distribution).

feed stacked twin horn with metal-plate E-plane lens, as in Fig. 13-28, provides a compact arrangement $\frac{1}{4}$ the length of a single horn. This twin horn is one of a pair used for radio astronomy observations at 1 to 2 GHz.[1] Both the aperture and field distributions have the ratios 1 : 2 : 1 (binomial series ratio).

Figure 13-29 Stacked septum horns with E-plane metal-plate lenses for feeding 110-m Ohio State University radio telescope. The aperture height ($=4h$ in Fig. 13-28) is 3 m. (See also Fig. 12-53.)

[1] R. T. Nash, "Stepped-Amplitude Distributions," *IEEE Trans. Ants. Prop.*, **AP-12**, July 1964.
R. T. Nash, "Beam Efficiency Limitations of Large Antennas," *IEEE Trans. Ants. Prop.*, **AP-12**, 918–923, December 1964.
J. D. Kraus, *Radio Astronomy*, 2nd ed., Cygnus-Quasar, 1986.

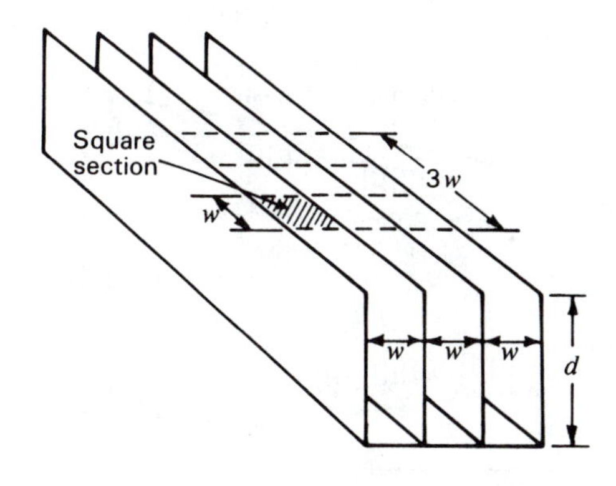

Figure 13-30 Corrugations of width w and depth d.

13-14 CORRUGATED HORNS.

Corrugated horns can provide reduced edge diffraction, improved pattern symmetry and reduced cross-polarization (less **E** field in the **H** plane).

Corrugations on the horn walls acting as $\lambda/4$ chokes are used to reduce **E** to very low values at all horn edges for all polarizations. These prevent waves from diffracting around the edges of the horn (or surface currents flowing around the edge and over the outside).[1]

Consider the corrugations of width w and depth d shown in Fig. 13-30. A square cross section (w by w), as indicated in the figure, constitutes the open end of a short-circuited field-cell transmission line of length d with a characteristic impedance $Z = 377\ \Omega$.[2] The reactance at the open end is given approximately by

$$X \simeq 377 \tan\left(\frac{2\pi d}{\lambda}\right) \qquad (\Omega) \tag{1}$$

where d = depth, λ

The reactance for any square area of the corrugations (such as $3w \times 3w$) is also as given by (1). Thus, this is the surface reactance in ohms per square. It is assumed that the corrugations are air-filled and that the wall thickness is small enough to be neglected.

When $d = \lambda/4$, X becomes infinite, while when $d = \lambda/2$, $X = 0$ and, assuming no radiation or loss, $R = 0$ and, hence, $Z = 0$.

As an example, a circular waveguide-fed corrugated horn with a corrugated transition is shown in Fig. 13-31. In the transition section the corrugation depth changes from $\lambda/2$, where the corrugations act like a conducting surface, to $\lambda/4$, where the corrugations present a high impedance. The corrugation spacing or

---

[1] A. F. Kay, "The Scalar Feed," AFCRL Rept. 64–347, March 1964.

[2] J. D. Kraus, *Electromagnetics*, 3rd ed., McGraw-Hill, 1984, sec. 10-5.

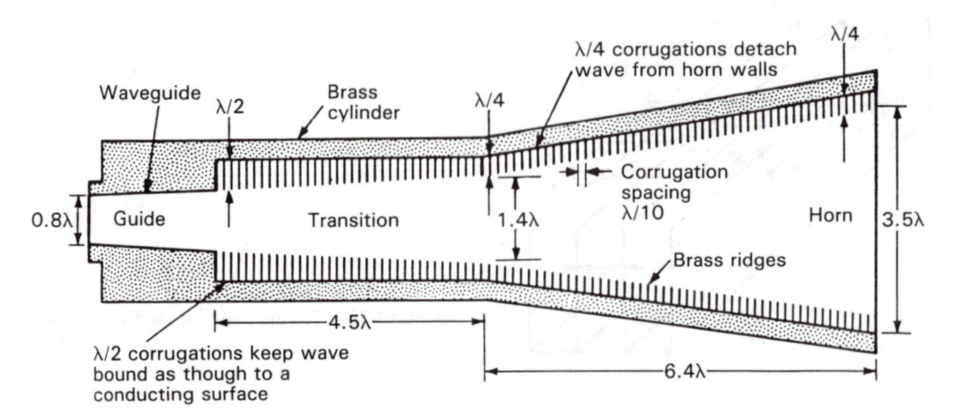

Figure 13-31 Cross section of circular waveguide-fed corrugated horn with corrugated transition. Corrugations with depth of $\lambda/2$ at waveguide act like a conducting surface while corrugations with $\lambda/4$ depth in horn present a high impedance. (*After T. S. Chu* et al., *"Crawford Hill 7-m Millimeter Wave Antenna,"* Bell Sys. Tech. J., **57**, *1257–1288, May–June 1978.*)

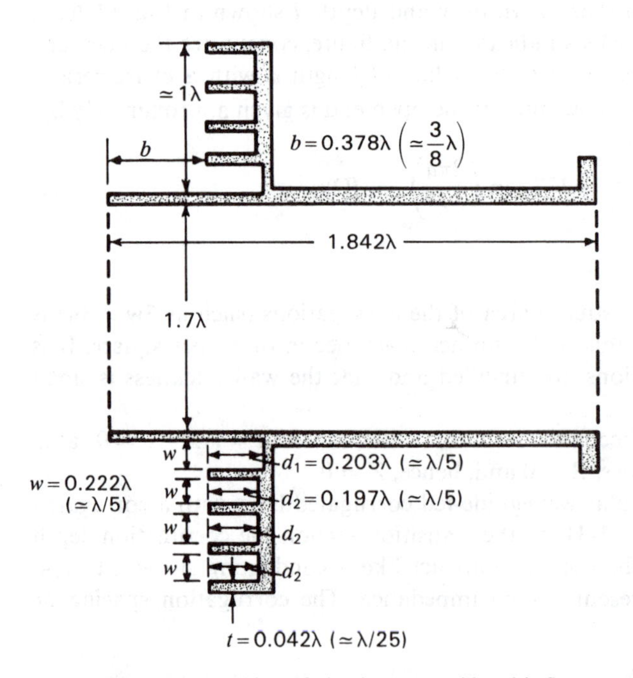

Figure 13-32 Cross section of circular waveguide with flange and 4 chokes for wide-beam-width high-efficiency feed of low F/D parabolic reflectors. (*After R. Wohlleben, H. Mattes and* O. *Lochner, "Simple, Small, Primary Feed for Large Opening Angles and High Aperture Efficiency,"* Electronics Letters, **8**, *19, September 1972.*)

width $w = \lambda/10$. The corrugations are air-filled. This corrugated horn was developed by Chu *et al.*[1] for feeding a 7-m millimeter-wave reflector antenna of the Bell Telephone Laboratories.

A simpler feed with choke corrugations was developed for deep dishes with F/D ratios of less than 0.35 by Wohlleben, Mattes and Lochner[2] for use on the Bonn 100-m radiotelescope. It consists of a circular waveguide equipped with a disc (or flange) projecting 1λ beyond the guide and 4 chokes $\lambda/5$ deep as shown in Fig. 13-32. The location of the disc with chokes $3\lambda/8$ behind the waveguide opening gives a broad 130° 10-dB beam width with a steep edge taper. This results in high aperture efficiency.

13-15 APERTURE-MATCHED HORN. By attaching a smooth curved (or rolled) surface section to the outside of the aperture edge of a horn, Burnside and Chuang[3] have achieved a significant improvement in the pattern, impedance and bandwidth characteristics. This arrangement, shown in Fig. 13-33 is an attractive alternative to a corrugated horn. The shape of the rolled edge is not critical but its radius of curvature should be at least $\lambda/4$.

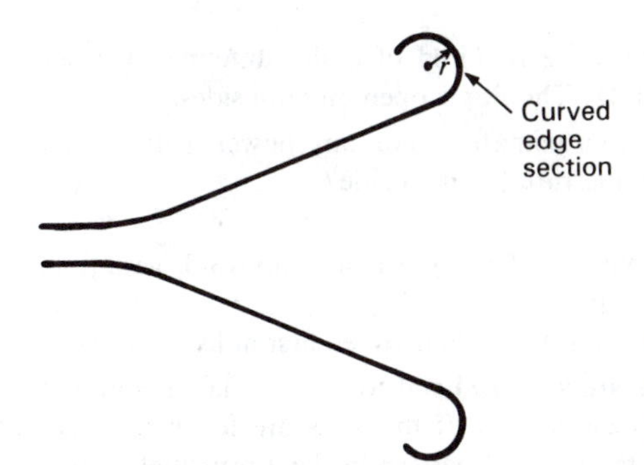

Figure 13-33 Cross section of Burnside and Chuang's aperture-matched horn. The radius of curvature r of the rolled edge should be at least $\lambda/4$.

ADDITIONAL REFERENCES

Ando, M., K. Sakurai and N. Goto: "Characteristics of a Radial Line Slot Antenna for 12 GHz Band Satellite TV Reception," *IEEE Trans. Ants. Prop.*, **AP-34**, 1269–1272, October 1986.

Clarricoats, P. J. B., and A. D. Olver: *Corrugated Horns for Microwave Antennas*, IEE Electromagnetic Wave Series, 1984.

Dragone, C.: "A Rectangular Horn of Four Corrugated Plates," *IEEE Trans. Ants. Prop.*, **AP-33**, 160–164, February 1985.

[1] T. S. Chu, R. W. Wilson, R. W. England, D. A. Gray and W. E. Legg, "Crawford Hill 7-m Millimeter Wave Antenna," *Bell Sys. Tech. J.*, **57**, 1257–1288, May–June 1978.

[2] R. Wohlleben, H. Mattes and O. Lochner, "Simple, Small, Primary Feed for Large Opening Angles and High Aperture Efficiency," *Electronics Letters*, **8**, 19, September 1972.

[3] W. D. Burnside and C. W. Chuang, "An Aperture-Matched Horn Design," *IEEE Trans. Ants. Prop.*, **AP-30**, 790–796, July 1982.

Geyer, H.: "Runder Hornstrahler mit Ringförmigen Sperrtöpfen zur Gleichzeitigen Übertragen zweier Polarizationsentkoppelter Wellen," *Frequenz*, **20**, 22–28, 1966.

Kay, A. F.: "The Scalar Feed," AFCRL Rept. 64-347, 1964.

Kisliuk, M., and A. Axelrod: "A Broadband Double-Slot Waveguide Antenna," *Microwave J.*, **30**, 223–226, September 1987.

Lagrone, A. H., and G. F. Roberts: "Minor Lobe Suppression in a Rectangular Horn Antenna through the Utilization of a High Impedance Choke Flange," *IEEE Trans. Ants. Prop.*, **AP-14**, 102–104, 1966.

Love, A. W.: *Electromagnetic Horn Antennas*, IEEE Press, New York, 1976.

Mentzer, C. A., and L. Peters, Jr.: "Pattern Analysis of Corrugated Horn Antennas," *IEEE Trans. Ants. Prop.*, **AP-24**, 304–309, May 1976.

Rumsey, V. H.: "Horn Antennas with Uniform Power Patterns around Their Axes," *IEEE Trans. Ants. Prop.*, **AP-14**, 656–658, September 1966.

Yngvesson, K. S., J. J. Johansson and E. K. Kollberg: "A New Integrated Slot Element Feed Array for Multibeam Systems," *IEEE Trans. Ants. Prop.*, **AP-34**, 1372–1376, November 1986.

PROBLEMS[1]

★13-1 Boxed-slot impedance. What is the terminal impedance of a slot antenna boxed to radiate only in one half-space whose complementary dipole antenna has a driving-point impedance of $Z = 100 + j0\ \Omega$? The box adds no shunt susceptance across the terminals.

13-2 Open-slot impedance. What dimensions are required of a slot antenna in order that its terminal impedance be $75 + j0\ \Omega$? The slot is open on both sides.

★13-3 Optimum horn gain. What is the approximate maximum power gain of an optimum horn antenna with a square aperture 10λ on a side?

13-4 Horn pattern.

(*a*) Calculate and plot the E-plane pattern of the horn of Prob. 13-3, assuming uniform illumination over the aperture.

(*b*) What is the half-power beamwidth and the angle between first nulls?

13-5 Two $\lambda/2$ slots. Two $\lambda/2$-slot antennas are arranged end-to-end in a large conducting sheet with a spacing of 1λ between centers. If the slots are fed with equal in-phase voltages, calculate and plot the far-field pattern in the 2 principal planes. Note that the H plane coincides with the line of the slots.

★13-6 Boxed slot. The complementary dipole of a slot antenna has a terminal impedance $Z = 90 + j10\ \Omega$. If the slot antenna is boxed so that it radiates only in one half-space, what is the terminal impedance of the slot antenna? The box adds no shunt susceptance at the terminals.

13-7 Pyramidal horn.

(*a*) Determine the length L, aperture a_H and half-angles in E and H planes for a pyramidal electromagnetic horn for which the aperture $a_E = 8\lambda$. The horn is fed with a rectangular waveguide with TE_{10} mode. Take $\delta = \lambda/10$ in the E plane and $\delta = \lambda/4$ in the H plane.

(*b*) What are the HPBWs in both E and H planes?

(*c*) What is the directivity?

(*d*) What is the aperture efficiency?

---

[1] Answers to starred (★) problems are given in App. D.

CHAPTER 14

LENS ANTENNAS

14-1 INTRODUCTION. Lens antennas may be divided into two distinct types: (1) delay lenses, in which the electrical path length is increased by the lens medium, and (2) fast lenses, in which the electrical path length is decreased by the lens medium. In delay lenses the wave is retarded by the lens medium. Dielectric lenses and H-plane metal-plate lenses are of the delay type. E-plane metal-plate lenses are of the fast type. The actions of a dielectric lens and an E-plane metal-plate lens are compared in Fig. 14-1.

The dielectric lenses may be divided into two groups:

1. Lenses constructed of nonmetallic dielectrics, such as lucite or polystyrene
2. Lenses constructed of metallic or artificial dielectrics

These types are considered in the next two sections (14-2 and 14-3). E-plane metal-plate lenses are discussed in Sec. 14-4, tolerances in Sec. 14-5 and the H-plane metal-plate lens in Sec. 14-6. A reflector lens is presented in Sec. 14-7.

All lens antennas of the delay type may be regarded basically as end-fire antennas with the polyrod and monofilar axial-mode helical antennas as rudimentary forms as suggested in Fig. 14-2. Likewise, the director structure of a many-element Yagi-Uda antenna is a rudimentary lens. Polyrods are covered in Sec. 14-8, monofilar axial-mode helical antennas having already been discussed in Chap. 7. Yagi-Uda antennas are considered in Chap. 11. Lenses of multiple helices are also described in Sec. 14-9. The last section of this chapter (14-10) discusses two spherically symmetric lenses of special type, the Luneburg and Einstein lenses, the latter utilizing a large mass as the focusing device.

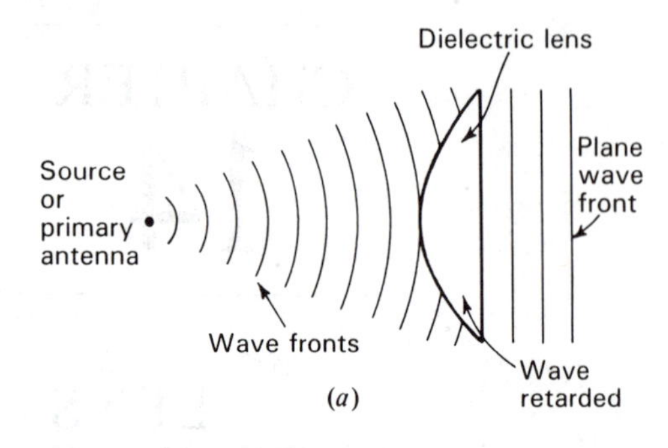

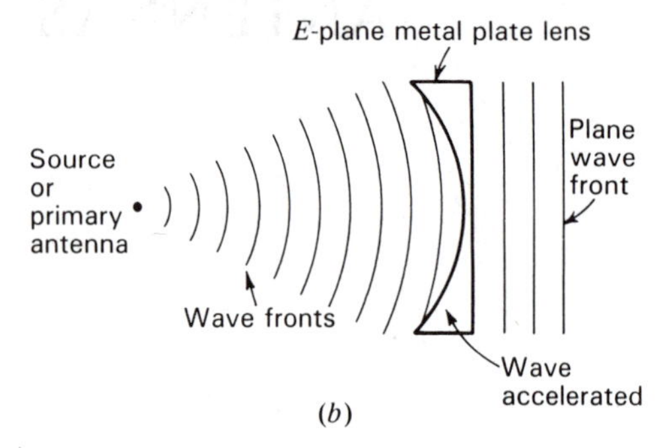

Figure 14-1 Comparison of dielectric (delay) lens and *E*-plane metal-plate (fast) lens actions.

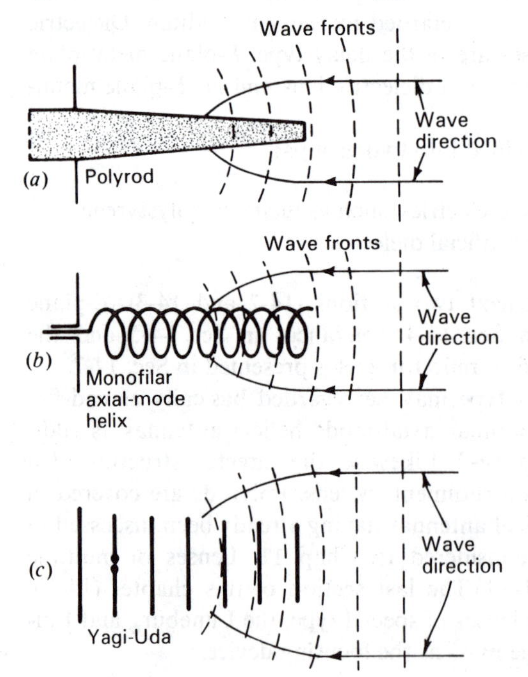

Figure 14-2 Three forms of rudimentary lens antennas.

At millimeter wavelengths low-loss dielectric lens antennas are competitive in weight and performance with reflector antennas.[1]

14-2 NONMETALLIC DIELECTRIC LENS ANTENNAS.[2] FERMAT'S PRINCIPLE. This type is similar to the optical lens. It may be designed by the ray analysis methods of geometrical optics. As an example, let us determine the shape of the plano-convex lens of Fig. 14-1*a* for transforming the spherical wave front from an isotropic point source or primary antenna into a plane wave front.[3] The field over the plane surface can be made everywhere in phase by shaping the lens so that all paths from the source to the plane are of equal electrical length. This is the principle of equality of electrical (or optical) path length (Fermat's principle). Thus, in Fig. 14-3, the electrical length of the path OPP' must equal the electrical length of the path $OQQ'Q''$, or more simply OP must equal OQ'. Let $OQ = L$ and $OP = R$, and let the medium surrounding the lens be air or vacuum. Then

$$\frac{R}{\lambda_0} = \frac{L}{\lambda_0} + \frac{R\cos\theta - L}{\lambda_d} \tag{1}$$

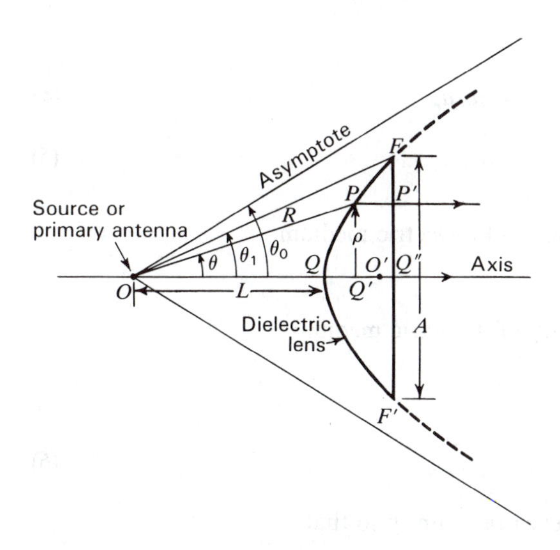

Figure 14-3 Path lengths in dielectric lens.

---

[1] P. F. Goldsmith and E. L. Moore, "Gaussian Optics Lens Antennas (GOLAs)," *Microwave J.*, **26**, 153–156, July 1984.

P. F. Goldsmith, "Quasioptical Techniques at Millimeter and Submillimeter Wavelengths," in K. Button (ed.), *Infrared and Millimeter Waves*, vol. 6, Academic Press, 1982, pp. 277–343.

P. F. Goldsmith and G. J. Gill, "Dielectric Wedge Conical Scanned Gaussian Optics Lens Antenna," *Microwave J.*, **29**, 207–212, September 1986.

[2] A detailed discussion is given by J. R. Risser, in S. Silver (ed.), *Microwave Antenna Theory and Design*, McGraw-Hill, New York, 1949, chap. 11.

[3] A wave front is defined as a surface on all points of which the field is in the same phase.

where λ_0 = wavelength in free space (air or vacuum)
λ_d = wavelength in the lens

Multiplying (1) by λ_0,

$$R = L + n(R \cos \theta - L) \tag{2}$$

where $n = \lambda_0/\lambda_d$ = index of refraction

In general,

$$n = \frac{\lambda_0}{\lambda_d} = \frac{f\lambda_0}{f\lambda_d} = \frac{v_0}{v_d} = \frac{\sqrt{\mu\varepsilon}}{\sqrt{\mu_0 \varepsilon_0}} \tag{3}$$

where f = frequency, Hz
v_0 = velocity in free space, m s^{-1}
v_d = velocity in dielectric, m s^{-1}
μ = permeability of the dielectric medium, H m^{-1}
ε = permittivity of the dielectric medium, F m^{-1}
μ_0 = permeability of free space = $4\pi \times 10^{-7}$, H m^{-1}
ε_0 = permittivity of free space = 8.85×10^{-12}, F m^{-1}

However,

$$\mu = \mu_0 \mu_r \tag{4}$$

and

$$\varepsilon = \varepsilon_0 \varepsilon_r \tag{5}$$

where $\mu_r = \dfrac{\mu}{\mu_0}$ = relative permeability of dielectric medium

$\varepsilon_r = \dfrac{\varepsilon}{\varepsilon_0}$ = relative permittivity of dielectric medium

Thus, from (3),

$$n = \sqrt{\mu_r \varepsilon_r} \tag{6}$$

For nonmagnetic materials μ_r is very nearly unity so that

$$n = \sqrt{\varepsilon_r}$$

The index of refraction of dielectric substances is always greater than unity. For vacuum, $\varepsilon_r = 1$ by definition. For air at atmospheric pressure, $\varepsilon_r = 1.0006$, but in most applications it is sufficiently accurate to take $\varepsilon_r = 1$ for air. The relative permittivity (or dielectric constant), index of refraction and power factor for a number of lens materials are listed in Table 14-1 in order of increasing ε_r. Although the *permittivity* of materials may vary with frequency (ε_r for water is 81 at radio frequencies and about 1.8 at optical frequencies), the table values are

Table 14-1

Material	Relative permittivity ε_r	Index of refraction n	Power factor§
Paraffin	2.1	1.4	0.0003
Polyethylene	2.2	1.5	0.0003
Lucite or Plexiglas (methacrylic resin)	2.6	1.6	0.01
Polystyrene	2.5	1.6	0.0004
Flint glass	7	2.5	0.004
Polyglas (TiO_2 or titanate fillers)	4–16†	2–4	0.003
Rutile (TiO_2)	85–170‡	9–13	0.0006

† Depends on composition.
‡ Depends on orientation of crystal with respect to field.
§ Equals cosine of angle between conduction and total currents.

appropriate at radio wavelengths down to the order of 1 cm. The power factor is also a function of frequency. The values listed merely indicate the order of magnitude at radio frequencies.

Returning now to Eq. (2) and solving for R, we have

$$R = \frac{(n-1)L}{n\cos\theta - 1} \tag{7}$$

This equation gives the required shape of the lens. It is the equation of a hyperbola. Referring to Fig. 14-3, the distance L is the focal length of the lens.[1] The asymptotes of the hyperbola are at an angle θ_0 with respect to the axis. The angle θ_0 may be determined from (7) by letting $R = \infty$. Thus,

$$\theta_0 = \arccos\frac{1}{n} \tag{8}$$

The point O is at one focus of the hyperbola. The other focus is at O'. For a point source at the focus, the 3-dimensional lens surface is a spherical hyperbola obtained by rotating the hyperbola on its axis. For an in-phase line source normal to the page (Fig. 14-3) as the primary antenna, the lens surface is a cylindrical hyperbola obtained by translating the hyperbola parallel to the line source.

[1] The F number of a lens is the ratio of the focal distance to the diameter A of the lens aperture. Thus, $F = L/A$.

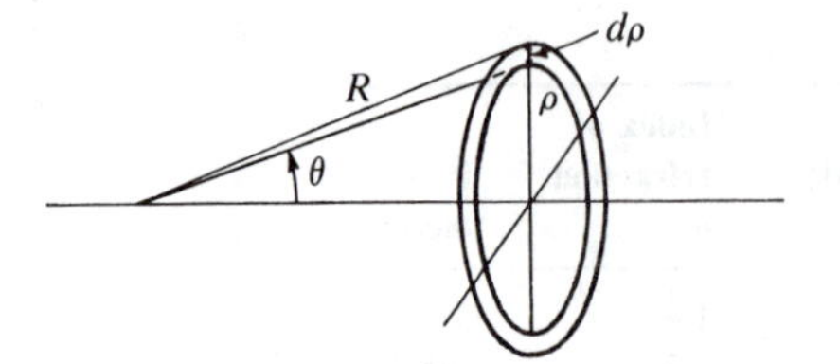

Figure 14-4 Annular zone.

Although Eq. (7) for the lens surface was derived without using Snell's laws of refraction,[1] these laws are satisfied by the lens boundary as given by (7).

The plane wave emerging from the right side of the lens produces a secondary pattern with maximum radiation in the direction of the axis. The shape of the secondary pattern is a function of both the aperture A and the type of illumination. This aperture-pattern relation has been discussed in previous chapters.

For an isotropic point-source primary antenna and a given focal distance L, the field at the edge of the lens ($\theta = \theta_1$) is less than at the center ($\theta = 0$), the effects of reflections at the lens surfaces and losses in the lens material being neglected. The variation of field intensity in the aperture plane of the spherical lens can be determined by calculating the power per unit area passing through an annular section of the aperture as a function of the radius ρ.[2] Referring to Fig. 14-4, the total power P through the annular section of radius ρ and width $d\rho$ is given by

$$P = 2\pi\rho \, d\rho \, S_\rho \tag{9}$$

where S_ρ = power density or Poynting vector (power per unit area) at radius ρ

This power must be equal to that radiated by the isotropic source over the solid angle $2\pi \sin\theta \, d\theta$. Thus,

$$P = 2\pi \sin\theta \, d\theta \, U \tag{10}$$

where U = radiation intensity of the isotropic source (power per unit solid angle)

Equating (9) and (10),

$$\rho \, d\rho \, S_\rho = \sin\theta \, d\theta \, U \tag{11}$$

and

$$\frac{S_\rho}{U} = \frac{\sin\theta}{\rho(d\rho/d\theta)} \tag{12}$$

---

[1] Snell's laws of refraction are (1) that the incident ray, the refracted ray and the normal to the surface lie in a plane and (2) that the ratio of the sine of the angle of incidence to the sine of the angle of refraction equals a constant for any two media. If the medium of the incident wave is air, the constant is the index of refraction n of the medium with the refracted ray. Thus, $\sin\alpha/\sin\beta = n$, where α is the angle between the incident ray in air and the normal to the surface and β is the angle between the refracted ray in the dielectric medium and the normal to the surface.

[2] J. R. Risser, in S. Silver (ed.), *Microwave Antenna Theory and Design*, McGraw-Hill, New York, 1949, chap. 11.

However, $\rho = R \sin \theta$, and introducing the value of R from (7),

$$S_\rho = \frac{(n \cos \theta - 1)^3}{(n-1)^2(n - \cos \theta)L^2} U \tag{13}$$

The ratio of the power density S_θ at the angle θ to the power density S_0 at the axis ($\theta = 0$) is given by the ratio of (13) when $\theta = \theta$, to (13) when $\theta = 0$. Thus,

$$\frac{S_\theta}{S_0} = \frac{(n \cos \theta - 1)^3}{(n-1)^2(n - \cos \theta)} \tag{14}$$

In the aperture plane the field-intensity ratio is equal to the square root of (14), or[1]

$$\frac{E_\theta}{E_0} = \sqrt{\frac{S_\theta}{S_0}} = \frac{1}{n-1}\sqrt{\frac{(n \cos \theta - 1)^3}{n - \cos \theta}} \tag{15a}$$

The ratio E_θ/E_0 is the relative field intensity at a radius ρ given by $\rho = R \sin \theta$. For $n = 1.5$,

$$\frac{E_\theta}{E_0} = 0.7 \text{ at } \theta = 20^\circ$$

and

$$\frac{E_\theta}{E_0} = 0.14 \text{ at } \theta = 40^\circ$$

Hence, for nearly uniform aperture illumination an angle θ_1 to the edge of the lens even less than 20° is essential unless the pattern of the primary antenna is an inverted type, i.e., one with less intensity in the axial direction ($\theta = 0$) than in directions off the axis.

Instead of uniform aperture illumination, a tapered illumination may be desired in order to suppress minor lobes. Thus, in the above example with $\theta_1 =$ 40°, the field at the edge of the lens is 0.14 its value at the center. The disadvantage of this method of producing a taper is that the lens is bulky (Fig. 14-5*a*). An alternative arrangement, shown in Fig. 14-5*b*, has a lens of smaller θ_1 value with the desired taper obtained with a directional primary antenna at a larger focal distance (relative to the aperture). The lens in this case is less bulky, but the focal distance is larger (F number $= L/A$ larger).

For compactness and mechanical lightness it would be desirable to combine the short focal distance of the lens at (*a*) with the light weight of the lens at (*b*). This combination may be largely achieved with the short focal distance zoned

[1] Equation (15*a*) is for a spherical lens. Attenuation in the lens is neglected. For a cylindrical lens the field-intensity ratio is

$$\frac{E_\theta}{E_0} = \frac{n \cos \theta - 1}{\sqrt{(n-1)(n - \cos \theta)}} \tag{15b}$$

where E_θ/E_0 is the relative field intensity at a distance y from the axis given by $y = R \sin \theta$.

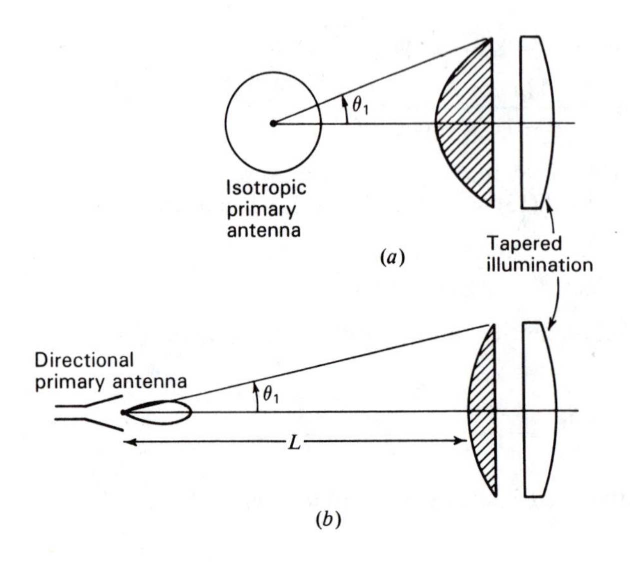

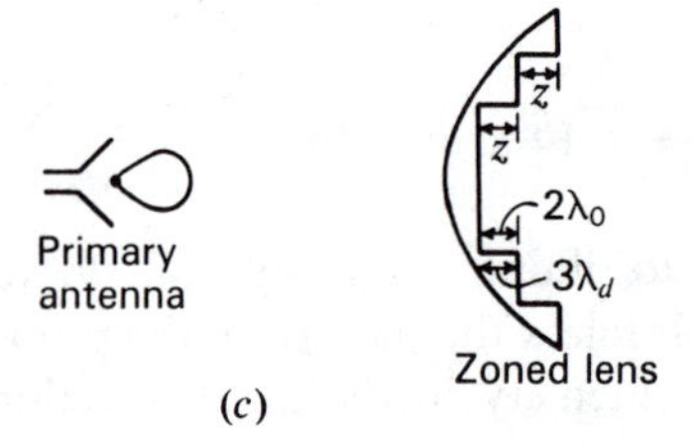

Figure 14-5 Short-focus lens (*a*), long-focus lens (*b*) and zoned lens (*c*).

lens of Fig. 14-5*c*. The weight of this lens is reduced by the removal of sections or zones, the geometry of the zones being such that the lens performance is substantially unaffected at the design frequency. Whereas the unzoned lens is not frequency sensitive, the zoned lens is, and this may be a disadvantage. The thickness z of a zone step is such that the electrical length of z in the dielectric is an integral number of wavelengths longer (usually unity) than the electrical length of z in air. Thus, for a 1λ difference,

$$\frac{z}{\lambda_d} - \frac{z}{\lambda_0} = 1 \tag{16}$$

or

$$z = \frac{\lambda_0}{n - 1} \tag{17}$$

For a dielectric with index of refraction $n = 1.5$,

$$z = 2\lambda_0$$

that is, each zone step is twice the free-space wavelength. Since $n = \lambda_0/\lambda_d$,

$$z = 3\lambda_d$$

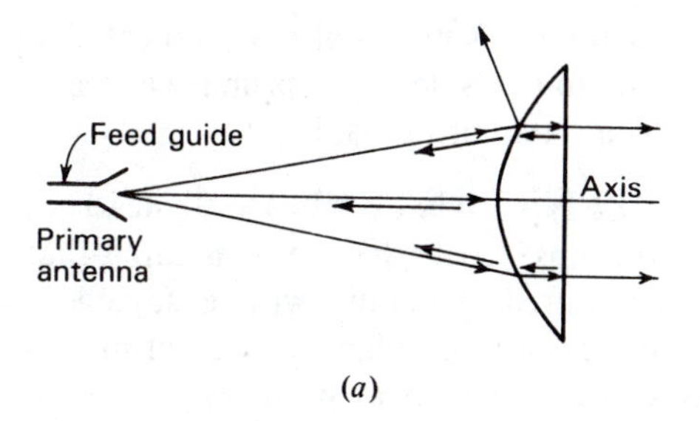

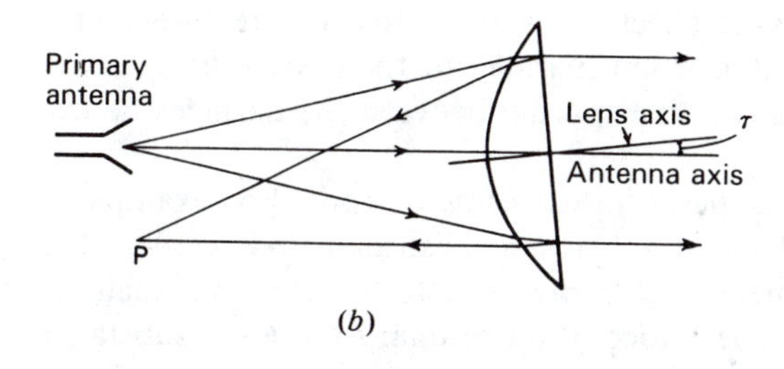

Figure 14-6 Reflected waves entering primary antenna (*a*) and refocused to one side of primary antenna (*b*).

Thus, in this case, the electrical length of z in the dielectric is $3\,\lambda$, while the electrical length of z in air is $2\,\lambda$ (see Fig. 14-5*c*).

In lens antennas the primary antenna does not interfere with the plane wave leaving the aperture as it does in a symmetrical parabolic reflector antenna (with prime focus feed) (see Chap. 12). However, the energy reflected from the lens surfaces may be sufficient to cause a mismatch of the primary antenna to its feed line or guide. In the lens of Fig. 14-6*a* reflections from the convex surface of the lens do not return to the source except from points at or near the axis. This is not serious, but the wave reflected internally from the plane lens surface is refocused at the primary antenna and may be disturbing. In this case, the wave is reflected at normal incidence, and the reflection coefficient is

$$\rho = \frac{Z_0 - Z}{Z_0 + Z} \tag{18}$$

where Z_0 = intrinsic impedance of free space $= \sqrt{\mu_0/\varepsilon_0}$

Z = intrinsic impedance of dielectric lens material $= \sqrt{\mu/\varepsilon}$

Thus,

$$\rho = \frac{(Z_0/Z) - 1}{(Z_0/Z) + 1} = \frac{n - 1}{n + 1} \tag{19}$$

where n = the index of refraction of the dielectric lens material

For $n = 1.5$, $\rho = 0.2$, while for $n = 4$, $\rho = 0.6$. Hence, for a small reflection a low index of refraction is desirable. The reflection can also be minimized by other methods. For example, a $\lambda/4$ plate can be applied to the plane lens surface with the refractive index of the plate made equal to $\sqrt{n}$, where n is the refractive index of the lens proper.[1] Another method is to tilt the lens slightly as indicated in Fig. 14-6*b* so that the reflected wave refocuses to one side of the primary antenna. Even though the lens is tilted, the antenna beam remains on axis.

14-3 ARTIFICIAL DIELECTRIC LENS ANTENNAS. Instead of using ordinary, nonmetallic dielectrics for the lens, Kock[2] has demonstrated that artificial or metallic dielectrics can be substituted, generally with a saving in weight. Whereas the ordinary dielectric consists of molecular particles of microscopic size, the artificial dielectric consists of discrete metal particles of macroscopic size. The size of the metal particles should be small compared to the design wavelength to avoid resonance effects. It is found that this requirement is satisfied if the maximum particle dimension (parallel to the electric field) is less than $\lambda/4$. A second requirement is that the spacing between the particles be less than λ to avoid diffraction effects.

The particles may be metal spheres, discs, strips or rods. For example, a plano-convex lens constructed of metal spheres is illustrated in Fig. 14-7. The spheres are arranged in a 3-dimensional array or lattice structure. Such an arrangement simulates the crystalline lattice of an ordinary dielectric substance but on a much larger scale. The radio waves from the source or primary antenna cause oscillating currents to flow on the spheres. The spheres are, thus, analogous to the oscillating molecular dipoles of an ordinary dielectric.

An artificial dielectric lens can be designed in the same manner as an ordinary dielectric lens (Sec. 14-2). To do this, it is necessary to know the effective index of refraction of the artificial dielectric. This can be measured experimentally with a slab of the material, or it can be calculated approximately by the following method of analysis.[2] Metal discs or strips are generally preferable to spheres because they are lighter in weight. The strips may be continuous in a direction

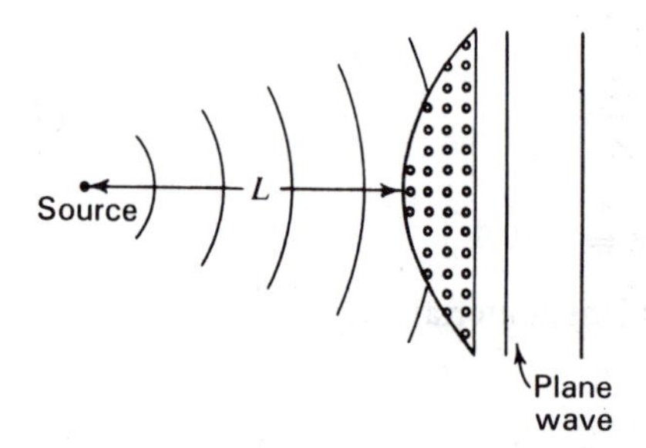

Figure 14-7 Artificial dielectric lens of metal spheres.

[1] In general the refractive index of a $\lambda/4$ matching plate between two media should be equal to the geometric mean of the indices of the two media. This is equivalent to saying that the intrinsic impedance Z_p of the plate material is made equal to the geometric mean of the intrinsic impedances Z_1 and Z_2 of the two media. Thus, $Z_p = \sqrt{Z_1 Z_2}$.

[2] W. E. Kock, "Metallic Delay Lens," *Bell System Tech. J.*, **27**, 58–82, January 1948.

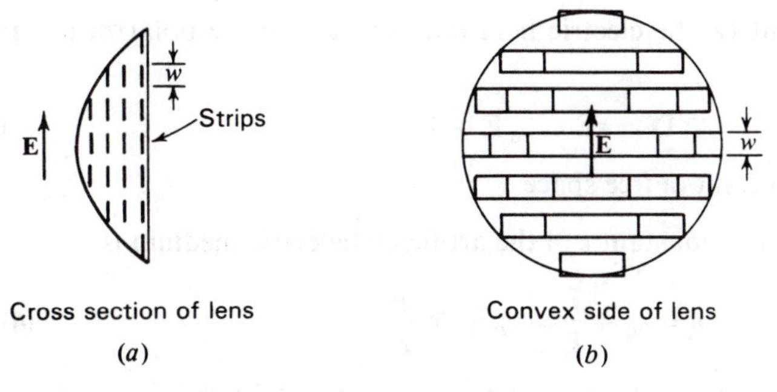

Figure 14-8 Artificial dielectric lens of flat metal strips of width w.

perpendicular to the electric field as indicated in Fig. 14-8. Since, however, the sphere is more readily analyzed, the method will be illustrated for the case of the sphere.

Let an uncharged conducting sphere be placed in an electric field **E** as in Fig. 14-9*a*. The field induces positive and negative charges as indicated. At a distance the effect of these charges may be represented by point charges $+q$ and $-q$ separated by a distance l as in Fig. 14-9*b*. Such a configuration is an *electric dipole of dipole moment ql*. At a distance $r \gg l$ the potential due to the dipole is given by

$$V = \frac{ql \cos \theta}{4\pi\varepsilon_0 r^2} \tag{1}$$

The polarization **P** of the artificial dielectric is given by

$$\mathbf{P} = Nq\mathbf{l} \tag{2}$$

where N = number of spheres per cubic meter
$\mathbf{l}$ = vector of length l joining the charges q

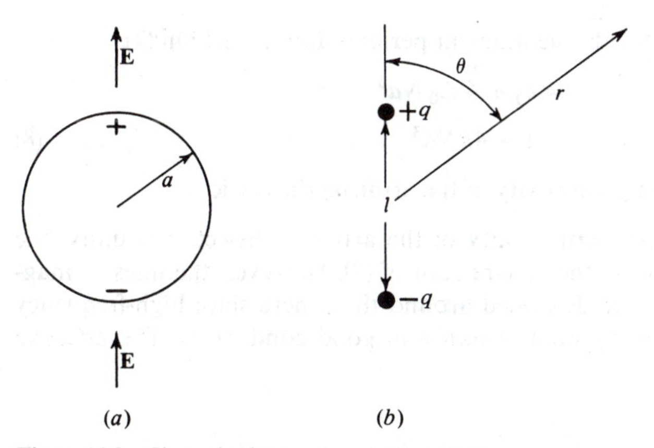

Figure 14-9 Charged sphere and equivalent dipole.

The electric displacement **D**, the electric field intensity **E** and the polarization **P** are related by

$$\mathbf{D} = \varepsilon \mathbf{E} = \varepsilon_0 \mathbf{E} + \mathbf{P} \tag{3}$$

where ε_0 = dielectric constant of free space

Thus, the effective dielectric constant ε of the artificial dielectric medium is

$$\varepsilon = \varepsilon_0 + \frac{\mathbf{P}}{\mathbf{E}} = \varepsilon_0 + N\frac{q\mathbf{l}}{\mathbf{E}} \tag{4}$$

Hence, if the number of spheres per unit volume and the dipole moment of one sphere per unit applied field are known, the effective dielectric constant can be determined. Let us now determine the dipole moment per unit applied field.

We have $\mathbf{E} = -\nabla V$. Then in a uniform field the potential

$$V = -\int_0^r E \cos\theta \, dr = -Er\cos\theta \tag{5}$$

where θ is the angle between the radius vector and the field (see Fig. 14-9*b*). The potential V_0 outside the sphere placed in an originally uniform field is the sum of (1) and (5), or

$$V_0 = -Er\cos\theta + \frac{ql\cos\theta}{4\pi\varepsilon_0 r^2} \tag{6}$$

At the sphere (radius a) (6) becomes[1]

$$0 = -Ea\cos\theta + \frac{ql\cos\theta}{4\pi\varepsilon_0 a^2}$$

and solving for ql/E we obtain

$$\frac{ql}{E} = 4\pi\varepsilon_0 a^3 \tag{7}$$

Introducing this value for the dipole moment per unit applied field in (4),

$$\varepsilon = \varepsilon_0 + 4\pi\varepsilon_0 Na^3$$

or

$$\varepsilon_r = 1 + 4\pi Na^3 \tag{8}$$

where ε_r = effective relative permittivity of the artificial dielectric

If the effective relative permeability of the artificial dielectric is unity, the index of refraction is given by the square root of (8). However, the lines of magnetic field of a radio wave are deformed around the sphere since high-frequency fields penetrate to only a very small distance in good conductors. The effective

[1] The potential of the sphere is zero since there is as much positive as negative charge on its surface.

Table 14-2 Artificial dielectric materials†

Type of particle	Relative permittivity ε_r	Relative permeability μ_r	Index of refraction n
Sphere	$1 + 4\pi N a^3$	$1 - 2\pi N a^3$	$\sqrt{(1 + 4\pi N a^3)(1 - 2\pi N a^3)}$
Disc	$1 + 5.33 N a^3$	~ 1	$\sqrt{1 + 5.33 N a^3}$
Strip	$1 + 7.85 N' w^2$	~ 1	$\sqrt{1 + 7.85 N' w^2}$

† N = number of spheres or discs per cubic meter
a = radius of sphere or disc, m
N' = number of strips per square meter in lens cross section (see Fig. 14-8*a*)
w = width of strips, m (see Fig. 14-8)

relative permeability of an artificial dielectric of conducting spheres is

$$\mu_r = 1 - 2\pi N a^3 \tag{9}$$

The effective index of refraction of the artificial dielectric of conducting spheres is then given by

$$n = \sqrt{\varepsilon_r \mu_r} = \sqrt{(1 + 4\pi N a^3)(1 - 2\pi N a^3)} \tag{10}$$

Equation (10) gives a smaller n than obtained by the square root of (8) alone. According to (10) the index of refraction of an artificial dielectric of conducting spheres can be calculated if the radius a of the sphere (in meters) and the number N of spheres per cubic meter are known. The relative permeability of disc or strip-type artificial dielectrics is more nearly unity so that one can take $\sqrt{\varepsilon_r}$ as their index of refraction. Theoretical values of ε_r, μ_r and n for artificial dielectrics made of conducting spheres, discs and strips are listed in Table 14-2.[1] According to Kock the table values are reliable only for $\varepsilon_r \leq 1.5$, and only approximate for larger ε_r. For $\varepsilon_r > 1.5$, N becomes sufficiently large that the particles interact because of their close spacing. This effect is neglected by the formulas.

14-4 *E*-PLANE METAL-PLATE LENS ANTENNAS.[2]

Whereas the ordinary and artificial dielectric lens depend for their action on a retardation of the wave in the lens, the *E*-plane metal-plate type of lens depends for its action on an *acceleration* of the wave by the lens. In this type of lens the metal plates are parallel to the *E* plane (or plane of the electric field). Referring to Fig. 14-10, the velocity v of propagation of a TE_{10} wave (**E** as indicated) in the x direction

---

[1] From W. E. Kock, "Metallic Delay Lens," *Bell System Tech. J.*, **27**, 58–82, January 1948.

[2] W. E. Kock, "Metal Lens Antennas," *Proc. IRE*, **34**, 828–836, November 1946.

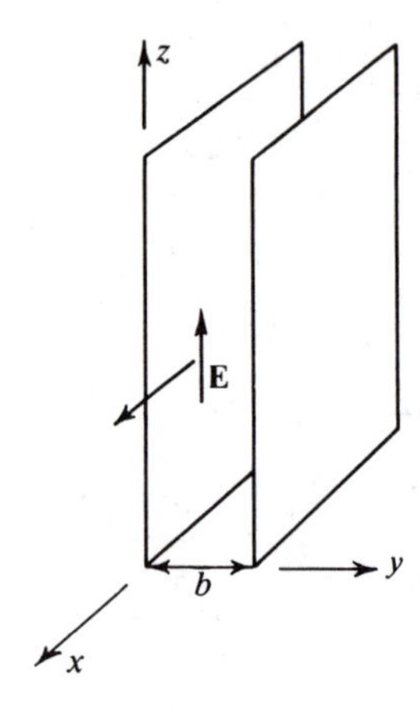

Figure 14-10 Wave between plates in *E*-plane type of metal-plate lens.

between two parallel conducting plates of large extent is given by[1]

$$v = \frac{v_0}{\sqrt{1 - (\lambda_0/2b)^2}} \tag{1}$$

where v_0 = velocity in free space
λ_0 = wavelength in free space
b = spacing of plates or sheets

The plates act as a guide, transmitting the wave for values of $b \geq \lambda_0/2$. The spacing $b = \lambda_0/2$ is the critical spacing since for smaller values of b the guide is opaque and the wave is not transmitted.[2] The variation of the velocity for a fixed wavelength as a function of the plate spacing b is illustrated in Fig. 14-11. The velocity of the wave between the plates is always greater than the free-space velocity v_0. It approaches infinity as b approaches $0.5\lambda_0$, and it approaches v_0 as b becomes infinite.

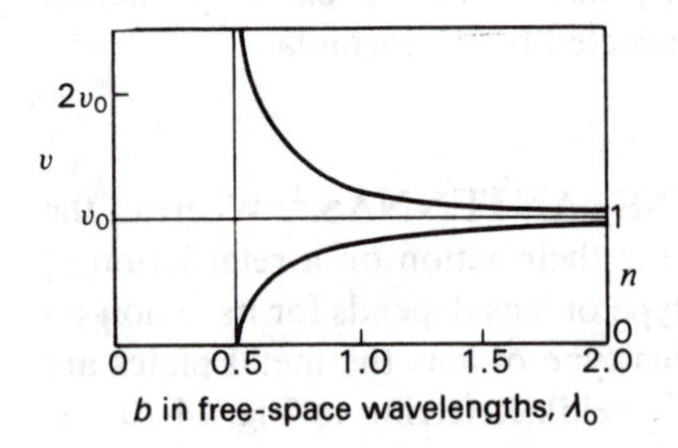

Figure 14-11 Velocity v of wave between parallel plates and equivalent index of refraction n as a function of spacing b between plates.

[1] L. J. Chu and W. L. Barrow, "Electromagnetic Waves in Hollow Metal Tubes of Rectangular Cross Section," *Proc. IRE*, **26**, 1520–1555, December 1938.

[2] However, for typical spacings of $b \simeq 3\lambda/4$ at normal incidence the transmission coefficient is nearly unity.

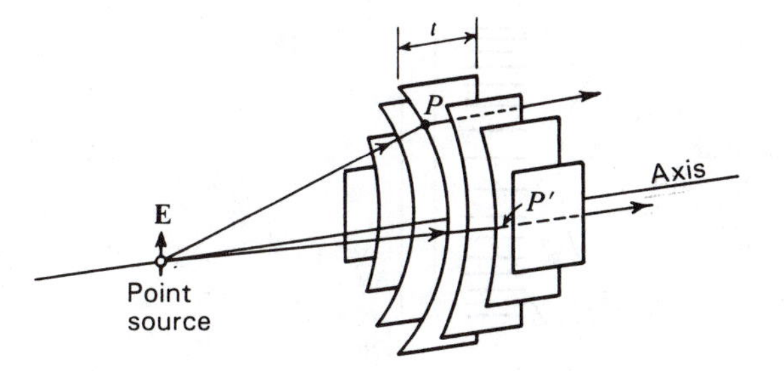

Figure 14-12 *E*-plane type of metal-plate lens.

The equivalent index of refraction of a medium constructed of many such parallel plates with a spacing b is

$$n = \frac{v_0}{v} = \sqrt{1 - \left(\frac{\lambda_0}{2b}\right)^2} \tag{2}$$

The index is always *less* than unity, as shown in Fig. 14-11.

The acceleration of waves between plates has been applied[1] in a metal-plate lens for focusing radio waves. For instance, a metal lens equivalent to the plano-convex dielectric lens of Fig. 14-1*a* or Fig. 14-3 is a plano-concave type as illustrated in Fig. 14-12. The plates are cut from flat sheets, the thickness t at any point being such as to transform the spherical wave from the source into a plane wave on the plane side of the lens. The electric field is parallel to the plates.

The lens plate on the axis of the lens in Fig. 14-12 is shown in Fig. 14-13. The shape of the plate can be determined by the principle of equality of electrical path length (Fermat's principle). Thus, in Fig. 14-13, OPP' must be equal to OQQ' in electrical length, or

$$\frac{L}{\lambda_0} = \frac{R}{\lambda_0} + \frac{L - R\cos\theta}{\lambda_g} \tag{3}$$

where λ_0 = wavelength in free space
λ_g = wavelength in lens

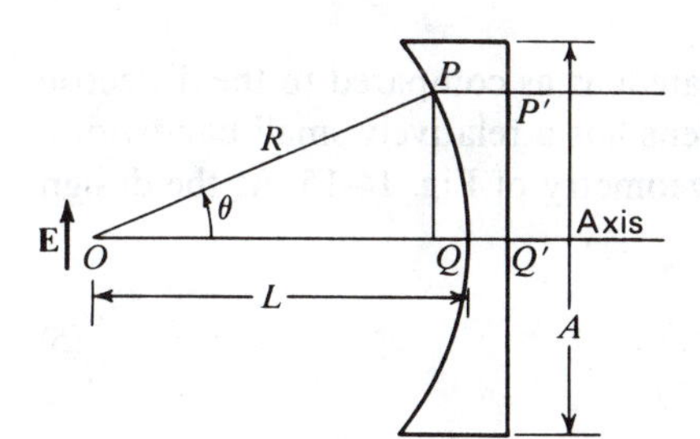

Figure 14-13 Geometry for *E*-plane type of metal-plate lens.

[1] W. E. Kock, "Metal Lens Antennas," *Proc. IRE*, **34**, 828–836, November 1946.

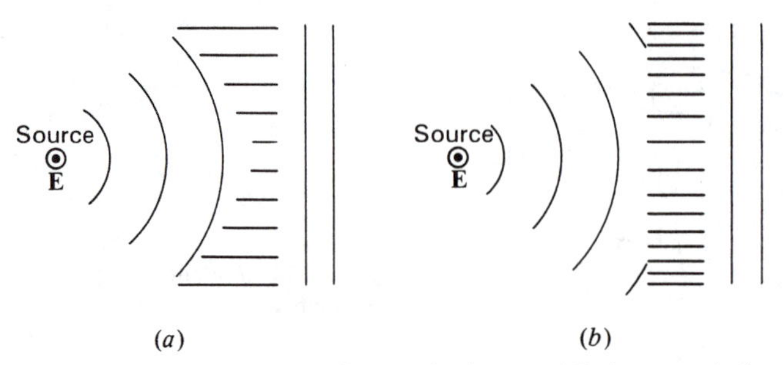

Figure 14-14 Cross sections of constrained types of *E*-plane metal-plate lenses.

Then

$$R = \frac{(1 - n)L}{1 - n \cos \theta} \tag{4}$$

This relation is identical with (14-2-7). However, to keep both numerator and denominator positive (since $n < 1$ in the present case), the numerator and denominator of (14-2-7) should be multiplied by minus 1. With $n < 1$, (4) is the equation of an ellipse. The 3-dimensional concave surface of the lens in Fig. 14-12 would be generated by rotating the contour for the center plate, as given by (4), on the axis. If the primary antenna were a line source perpendicular to the page in Fig. 14-13, all the plates would be identical and the lens surface would be in the form of an elliptical cylinder.

Waves entering the lens of Fig. 14-12 at the point P obey Snell's laws of refraction. However, this is not necessarily the case for waves entering at P' where the metal plates constrain the wave to travel between them. E-plane metal-plate lenses may be constructed that have only such constrained refraction. Two types are illustrated in cross section in Fig. 14-14. Both have a line source normal to the page. The electric field **E** is parallel to the source. All lens cross sections perpendicular to the line sources are the same as the ones shown in the figure. In the lens at (*a*) the spacing between plates is uniform, but the width varies from plate to plate. In the lens at (*b*) all plates have the same width, but the spacing varies.

A disadvantage of the E-plane metal-plate lens as compared to the dielectric type is that it is frequency-sensitive, i.e., the lens has a relatively small bandwidth. To determine the bandwidth,[1] consider the geometry of Fig. 14-15. At the design frequency f we have from (3) that

$$\frac{L}{\lambda_0} = \frac{R}{\lambda_0} + \frac{t}{\lambda_g} \tag{5}$$

[1] J. R. Risser, in S. Silver (ed.), *Microwave Antenna Theory and Design*, McGraw-Hill, New York, 1949, chap. 11.

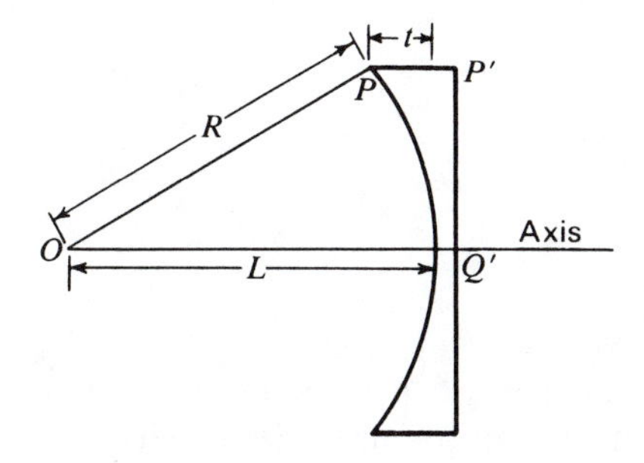

Figure 14-15 Geometry for bandwidth considerations.

or

$$L = R + nt \tag{6}$$

where n = index of refraction at the design frequency f

At some other frequency f',

$$L + \delta = R + n't \tag{7}$$

where δ = the difference in electrical path length of OQ and OPP'
n' = index of refraction at the frequency f'

Subtracting (6) from (7),

$$\delta = \Delta n\, t \tag{8}$$

where $\Delta n = n' - n$

For a small wavelength difference $\Delta\lambda_0$,

$$\Delta n = \frac{\partial n}{\partial \lambda_0} \Delta\lambda_0 \tag{9}$$

Introducing n from (2) into (9), differentiating and substituting this value of Δn in (8) yields

$$\delta = \frac{n^2 - 1}{n} \frac{\Delta\lambda_0}{\lambda_0} t \tag{10}$$

or

$$\left| \frac{\Delta\lambda_0}{\lambda_0} \right| = \frac{n\delta}{(1 - n^2)t} \tag{11}$$

The total bandwidth B is twice (11) so

$$B = \frac{2n\delta}{(1 - n^2)t} = \frac{2n}{1 - n^2} \frac{\delta_\lambda}{t_\lambda} \tag{12}$$

where δ_λ = maximum tolerable path difference in free-space wavelengths
t_λ = thickness of lens plate at edge of lens in free-space wavelengths

If we arbitrarily take $\delta = 0.25\lambda$,

$$B = \frac{50n}{(1 - n^2)t_\lambda} \qquad (\%) \tag{13}$$

For $n = 0.5$ and $t = 6\lambda$, the bandwidth

$$B = 5.5\%$$

Thus, the usable frequency band for this antenna is 5.5 percent of the design frequency.[1] Although zoning a dielectric lens introduces frequency sensitivity, the effect of zoning an E-plane metal-plate lens is to decrease the frequency sensitivity. Hence, zoning is desirable with E-plane metal-plate lens, both to save weight and to increase the bandwidth. An E-plane metal-plate lens 40λ square with nine zones is illustrated in Fig. 14-16. The patterns of this lens, fed with a short primary horn antenna, are shown in Fig. 14-17.

The bandwidth of a zoned E-plane metal-plate lens is given approximately by

$$B = \frac{50n}{1 + Kn} \qquad (\%) \tag{14}$$

where n = index of refraction at the design frequency
K = number of zones, the zone on the axis of the lens being counted as the first zone

A zoned lens comparable to the unzoned lens of $n = 0.5$, $t = 6\lambda_0$ and $B = 5.5$ percent has $n = 0.5$ and $K = 3$ since with $n = 0.5$, $K \sim t_\lambda/2$. The bandwidth B of this zoned lens is 10 percent, or nearly double the bandwidth of the unzoned lens.

The aperture efficiency to be expected of large lens antennas is about 0.6 so that the directivity is about the same as for optimum horns of the same size aperture (see Chap. 12).

Referring to Fig. 14-18a, the thickness z of a zone step is given by

$$\frac{z}{\lambda_0} - \frac{z}{\lambda_g} = 1$$

or

$$z = \frac{\lambda_0}{1 - n} \tag{15}$$

---

[1] $\dfrac{2\Delta\lambda}{\lambda} = \dfrac{\lambda_2 - \lambda_1}{\lambda} = \dfrac{(1/f_2) - (1/f_1)}{(1/f)} = \dfrac{f(f_1 - f_2)}{f_1 f_2} \simeq \dfrac{f_1 - f_2}{f} = \dfrac{2\Delta f}{f}$ (for $\Delta f \ll f$)

where λ = design wavelength
f = design frequency
λ_1 = short wavelength limit of band
λ_2 = long wavelength limit of band
f_1 = high-frequency limit of band
f_2 = low-frequency limit of band

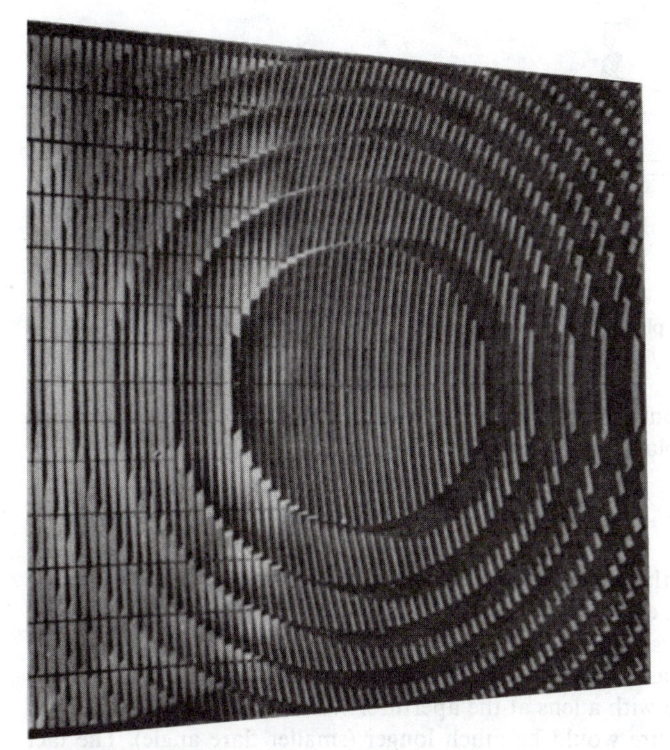

Figure 14-16 Zoned type of E-plane metal-plate lens with a square aperture 40λ on a side. (*Courtesy W. E. Kock, Bell Telephone Laboratories.*)

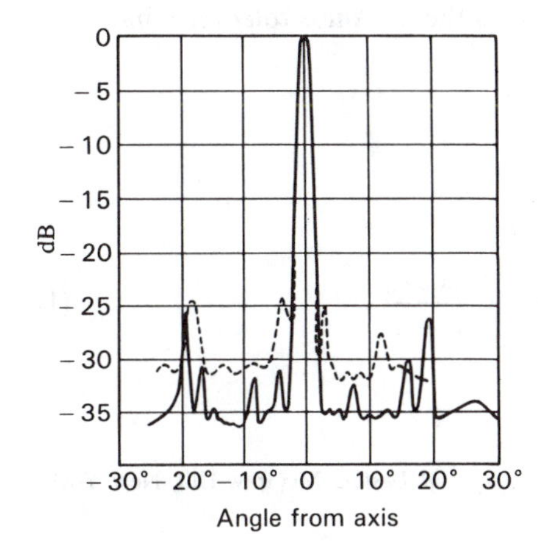

Figure 14-17 E-plane pattern (solid) and H-plane pattern (dashed) of 40λ square zoned E-plane lens of Fig. 14-16. (*After W. E. Kock.*)

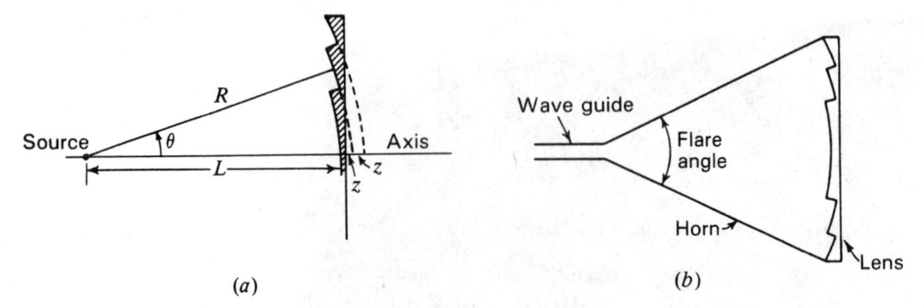

Figure 14-18 (*a*) Zoned lens plate. (*b*) Horn with lens.

The equation for the contour of the zoned lens is the same as (4) for the unzoned lens except that L is replaced by L_k, where

$$L_k = L + \frac{(K-1)\lambda_0}{1-n} = L + (K-1)z \tag{16}$$

For the first zone (on the axis) $L_k = L$, for the second zone $L_k = L + z$, for the third zone $L_k = L + 2z$, etc.

To shield against stray radiation from the source side of a lens, a metallic enclosure may be used as in Fig. 14-18*b*. This enclosure forms an electromagnetic horn of wide flare angle with a lens at the aperture. Without the lens an optimum horn of the same aperture would be much longer (smaller flare angle). The fact that the lens permits a much shorter structure for the same size aperture is, perhaps, the principal advantage of a lens or lens-horn combination over a simple horn antenna.

14-5 TOLERANCES ON LENS ANTENNAS.[1]

In a *dielectric lens*, differences in the path length may be caused by deviations in thickness from the ideal contour and by variations in the index of refraction. Assigning an allowable variation of $\lambda_0/32$ rms to either cause, we have as the *thickness tolerance* that

$$\frac{\Delta t}{\lambda_d} - \frac{\Delta t}{\lambda_0} = \frac{1}{32}$$

or

$$\Delta t = \frac{\lambda_0}{32(n-1)}$$

or

$$\Delta t = \frac{\lambda_0}{32(n-1)} = \frac{0.03\lambda_0}{n-1} \tag{1}$$

[1] J. R. Risser, in S. Silver (ed.), *Microwave Antenna Theory and Design*, McGraw-Hill, New York, 1949, chap. 11.

For $n = 1.5$,

$$\Delta t = 0.06\lambda_0$$

For the *tolerance on n*,

$$\Delta n \, t = \frac{\lambda_0}{32}$$

or

$$\Delta n = \frac{0.03}{t_\lambda} \tag{2}$$

where t_λ = thickness of lens in free-space wavelengths

Dividing (2) by n,

$$\frac{\Delta n}{n} = \frac{3}{n t_\lambda} \qquad (\%) \tag{3}$$

If $n = 1.5$ and $t = 4\lambda_0$, $\Delta n/n = \frac{1}{2}\%$.

In an *E-plane metal-plate lens* the path length may be affected by both the thickness of the lens and by the spacing b between lens plates. Assigning $\lambda_0/32$ to each cause, we have as the *thickness tolerance* that

$$\Delta t = \frac{\lambda_0}{32(1-n)} = \frac{0.03\lambda_0}{1-n} \tag{4}$$

and for the *tolerance on the spacing b between plates*

$$\frac{\Delta b}{b} = \frac{3n}{(1-n^2)t_\lambda} \qquad (\%) \tag{5}$$

It is interesting to compare these tolerances with the surface contour requirement of a *parabolic reflector*. A displacement Δx normal to the surface of the reflector at the vertex (i.e., a displacement in the axial direction) results in a change in wave path of $2\Delta x$. Thus, for the same effect as either Δt or Δn, Δx must be $\frac{1}{2}$ as large.

This is a severe requirement for a large reflector at a small wavelength, placing a strict limitation on the allowable warping or twisting of the reflector. In contrast to this, *the thickness tolerance on a lens refers only to the thickness dimension*. It does not imply that the lens contour be maintained to this accuracy. With a lens, a relatively large amount of warping or twisting can be tolerated, and this is an important advantage of this type of antenna. Furthermore, the lens axis can

Table 14-3 Tolerances on lens and reflector antennas

Type of antenna	Type of tolerance	Amount of tolerance (rms)
Parabolic reflector	Surface contour	$0.016\lambda_0$
Dielectric lens† (unzoned)	Thickness	$\dfrac{0.03\lambda_0}{n-1}$
	Index of refraction	$\dfrac{3}{nt_\lambda}$ %
Dielectric lens† (zoned)	Thickness	3%
	Index of refraction	$\dfrac{3(n-1)}{n}$ %
E-plane metal-plate lens‡ (unzoned)	Thickness	$\dfrac{0.03\lambda_0}{1-n}$
	Plate spacing	$\dfrac{3n}{(1-n^2)t_\lambda}$ %
E-plane metal-plate lens‡ (zoned)	Thickness	3%
	Plate spacing	$\dfrac{3n}{1+n}$ %

n = index of refraction
t = lens thickness
t_λ = lens thickness in free-space wavelengths
† $n > 1$.
‡ $n < 1$.

be tilted a considerable angle τ with respect to the axis through the primary antenna and center of the lens (see Fig. 14-6*b*) without serious effects.[1]

The above-mentioned tolerances are summarized in Table 14-3. Tolerances for zoned lenses are also listed. These are derived from the unzoned lens tolerances by taking the dielectric lens thickness as nearly equal to $\lambda_0/(n-1)$ and the metal-plate lens thickness as nearly equal to $\lambda_0/(1-n)$. All tolerances in the table assume $\lambda_0/32$ rms for the individual lens variations and $\lambda_0/64$ rms for the reflector variation resulting, in each case, in a gain-loss factor $k_g = 0.16$ dB as calculated from (12-10-3), provided the correlation distance $>\lambda$ and other conditions of (12-10-3) are met. For a combination of random variations (as index *and*

[1] Little difference in radiation-field patterns of an *E*-plane metal-plate lens antenna is revealed for a tilt angle τ as large as 30° according to patterns presented by Friis and Lewis. See H. T. Friis and W. D. Lewis, "Radar Antennas," *Bell System Tech. J.*, **26**, 270, April 1947.

thickness) the net effective variation is given by the quadrature sum of the individual variations.

14-6 *H*-PLANE METAL-PLATE LENS ANTENNAS.[1] A wave entering a stack of metal plates oriented parallel to the *H* plane (perpendicular to the *E* plane) as in Fig. 14-19*a* is little affected in its velocity. However, the wave is constrained to pass between the plates so that, once inside, the path length can be increased if the plates are deformed, as suggested in Fig. 14-19*b*. An increase in path length can also be produced by slanting the plates as at (*c*). The increase of path length is $S - T$. Using the slant-plate method of increasing the path length, an *H*-plane metal-plate lens can be designed by applying the principle of equality of electrical path length. This type of lens is called an *H*-plane type since the plates are parallel to the magnetic field (perpendicular to the *E* plane).

Referring to Fig. 14-19*d*, the condition for equality of electrical path length requires that

$$R = L + \frac{R\cos\theta - L}{\cos\xi} \tag{1}$$

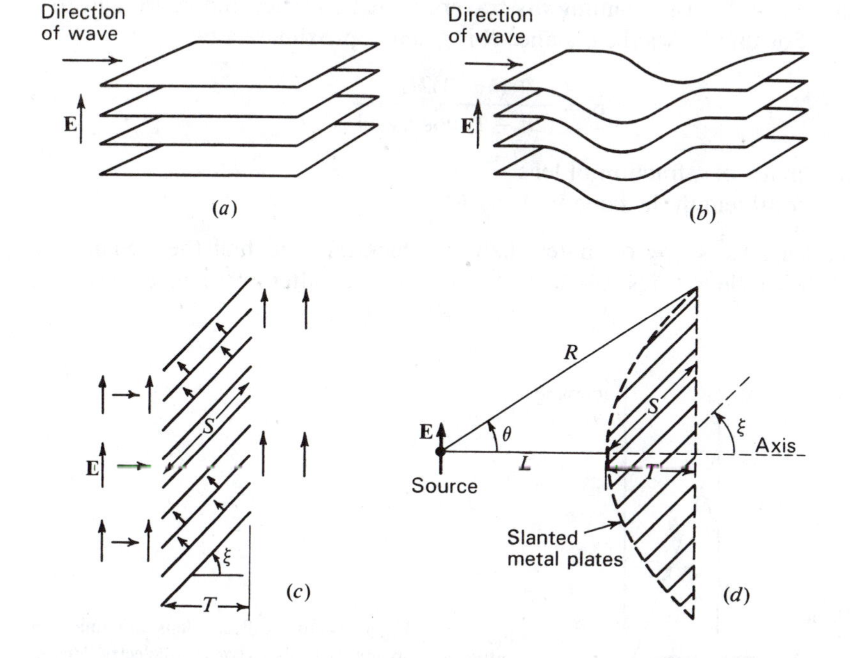

Figure 14-19 (*a*) *H*-plane stack of flat metal plates. (*b*) *H*-plane stack with increased path length. (*c*) Slanted *H*-plane plates. (*d*) *H*-plane metal-plate lens using slanted plate construction.

---

[1] W. E. Kock, "Path Length Microwave Lenses," *Proc. IRE*, **37**, 852–855, August 1949.

or

$$R = \frac{(n-1)L}{n \cos\theta - 1} \tag{2}$$

where $n = 1/\cos\xi$ = effective index of refraction of the slant-plate lens medium

In this case the index of refraction is equal to or greater than unity so that (2) is identical with (14-2-7) for a dielectric lens. The index n depends only on the plate slant angle ξ and is not a function of the frequency as in the E-plane type of metal-plate lens. The most critical dimension is the path length S in the lens. This may be affected by a change in T or in ξ. Assuming a maximum allowable variation $\delta = \lambda_0/8$ in electrical path length, the tolerance in S is given by

$$\Delta S = \pm 0.06\lambda_0 \tag{3}$$

A disadvantage of the H-plane metal-plate lens is that this type of construction tends to produce unsymmetrical aperture illumination in the E plane.

14-7 REFLECTOR-LENS ANTENNA. About 1960 I filled a notebook with many unique reflector designs and at long last in 1982 published several of them.[1] One of these combines a dielectric lens with a flat reflecting sheet as suggested in Fig. 14-20. An incoming ray traverses the lens twice and is brought to a focus at F. For thin lenses the distance R is given approximately by

$$R = \frac{(n-1)2L}{(2n-1)\cos\theta - 1} \tag{1}$$

where n = index of refraction of lens
L = focal length (in same units as R)

This reflector lens is approximately half the thickness and half the weight of a simple dielectric lens (Figs. 14-1a or 14-3) and is an alternative to a parabolic

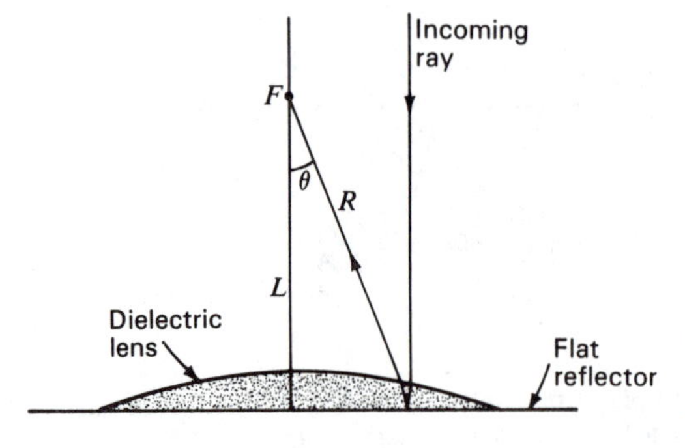

Figure 14-20 Reflector-lens antenna consisting of a plano-convex dielectric lens on a flat reflecting sheet.

[1] J. D. Kraus, "Some Unique Reflector Antennas," *IEEE Ant. Prop. Soc. Newsletter*, **24**, 10, April 1982.

reflector. An interesting, specialized application would be to use a pool of conducting liquid for the flat surface with a long focal length reflector lens situated above as in Fig. 14-20. Beam squinting by horizontal displacements of the feed would allow observations of a region near the zenith. Only the thickness of the lens is critical since the flatness of the reflecting surface is maintained automatically, thanks to gravity. The lens need not be in contact with the flat surface. A meridian-transit millimeter-wave radio astronomy application at low cost is envisioned.

14-8 POLYRODS. A dielectric rod or wire can act as a guide for electromagnetic waves.[1] The guiding action, however, is imperfect since considerable power may escape through the wall of the rod and be radiated. This tendency to radiate is turned to advantage in the *polyrod antenna*,[2] so called because the dielectric rod is usually made of polystyrene. A 6λ long polyrod antenna is shown in cross section in Fig. 14-21*a*. The rod is fed by a short section of cylindrical

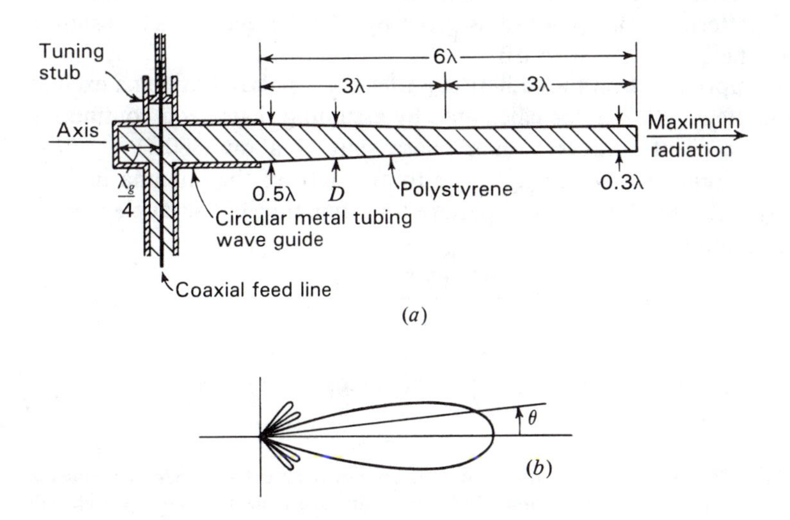

Figure 14-21 (*a*) Cross section of cylindrical polystyrene antenna 6λ long. (*b*) Radiation pattern. (*After G. E. Mueller and W. A. Tyrrell, "Polyrod Antennas,"* Bell System Tech. J., **26**, *837–851, October 1947.*)

[1] D. Hondros and P. Debye, "Elektromagnetische Wellen an dielektrischen Drahten," *Ann. Physik*, **32**, 465–476, 1910.

S. A. Schelkunoff, *Electromagnetic Waves*, Van Nostrand, New York, 1943, pp. 425–428.

R. M. Whitmer, "Fields in Non-metallic Guides," *Proc. IRE*, **36**, 1105–1109, September 1948.

[2] G. E. Mueller and W. A. Tyrrell, "Polyrod Antennas," *Bell System Tech. J.*, **26**, 837–851, October 1947.

waveguide which, in turn, is energized by a coaxial transmission line. This type of polyrod acts as an end-fire antenna.[1]

The phase velocity of wave propagation in the rod and also the ratio of the power guided outside the rod to the power guided inside are both functions of the rod diameter D in wavelengths and the dielectric constant[2] of the rod material.[3] For polystyrene rods with $D < \lambda/4$, the rod possesses little guiding effect on the wave, and only a small fraction of the power is confined to the inside of the rod. The phase velocity in the rod is also close to that for the surrounding medium (free space). For diameters of the order of a wavelength, however, most of the power is confined to the rod, and the phase velocity in the rod is nearly the same as in an unbounded medium of polystyrene. For increased directivity operation the diameter D_λ in free-space wavelengths of a uniform rod (length $L_\lambda > 2$ and $2 < \varepsilon_r < 5$) is

$$D_\lambda \simeq \frac{3}{\varepsilon_r^{3/2}\sqrt{1 + 2L_\lambda}} + 0.2 \tag{1}$$

In practice, polystyrene rod diameters in the range 0.5 to 0.3λ are used.[4] The rod may be uniform or to reduce minor lobes can be tapered as in Fig. 14-21*a*. This polyrod is tapered halfway and is uniform in cross section the remainder of its length. The diameter D is 0.5λ at the butt end and 0.3λ at the far end. The radiation-field pattern for this polyrod as given by Mueller and Tyrrell is shown in Fig. 14-21*b*. The gain is about 16 dBi.

To a first approximation the radiation pattern of a polyrod antenna excited uniformly along its length may be calculated by assuming that it is a continuous array of isotropic point sources with a phase shift of about 360 $(1 + 1/2L_\lambda)$ deg/wavelength of antenna, where L_λ is the total length of the antenna in free-space wavelengths.[5] The relative field pattern as a function of the angle θ from the axis is then given by

$$E(\theta) = \frac{\sin(\psi'/2)}{\psi'/2} \tag{2}$$

where $\psi' = 2\pi L_\lambda \cos\theta - 2\pi L_\lambda\left(1 + \frac{1}{2L_\lambda}\right) = 2\pi\left[L_\lambda(\cos\theta - 1) - \frac{1}{2}\right]$

[1] An end-fire polyrod antenna may be regarded as a degenerate or rudimentary form of lens antenna with an effective lens cross section of the order of a wavelength. See Gilbert Wilkes, "Wavelength Lens," *Proc. IRE*, **36**, 206–212, February 1948.

[2] The relative permittivity $\varepsilon_r = 2.5$ for polystyrene. See Table 14-1.

[3] G. E. Mueller and W. A. Tyrrell, "Polyrod Antennas," *Bell System Tech. J.*, **26**, 837–851, October 1947.

[4] To transmit the lowest (TE_{11}) mode in a circular waveguide, the diameter D of the guide must be at least $0.58\lambda/\sqrt{\varepsilon_r}$, where λ is the free-space wavelength and ε_r is the relative permittivity of the guide. Thus, for a rod of polystyrene ($\varepsilon_r = 2.5$) fed from a circular waveguide as in Fig. 14-21*a*, the guide diameter must be at least 0.37λ to allow transmission in the metal tube.

[5] This is the Hansen and Woodyard condition for increased directivity of an end-fire array. See Sec. 4-6d.

The radiation field could be calculated exactly by applying Schelkunoff's equivalence principle, provided the fields on the surface were known.[1] By this principle the fields at the rod surfaces are replaced by equivalent electric and fictitious magnetic current sheets, and the radiation field is calculated from these currents. An approximate calculation has been made by assuming a field distribution.[2]

The directivity D of a polyrod antenna is given approximately by[3]

$$D \simeq 8L_\lambda \tag{3}$$

and the half-power beam width by

$$\text{HPBW} \simeq \frac{60^\circ}{\sqrt{L_\lambda}} \tag{4}$$

where L_λ = length of polyrod in free-space wavelengths

Polyrod antennas may also be of square or rectangular cross section. Another possibility is to use a dielectric sleeve of circular or square cross section, the interior of the sleeve being air-filled. In this case the appropriate diameter of the sleeve may be of the order of 1λ.[4]

The rods and cones of the retina of a human eye are similar to polyrod antennas but with unity front-to-back ratio and diameters of 1 to 2λ as compared to the high front-to-back ratio and $\lambda/2$ diameter or less of the polyrod of Fig. 14-21. The retina has an array of more than 100 million "polyrods." Curiously, to an antenna engineer, all of the feed system and connections (axons and dendrites) are *in front* of the "polyrods." However, this is all right because the "wiring" is transparent.[5]

14-9 MULTIPLE-HELIX LENSES. As mentioned earlier, an axial-mode monofilar helical antenna is a rudimentary form of lens antenna (see Fig. 14-2*b*). A multiplicity of parasitic axial-mode monofilar helices can also be arranged as in Fig. 14-22 to form a *phase-controlled lens.* Helices of opposite hand are mounted back-to-back on a spherical shell with the center at the focus. The helices on the outer side of the shell receive the incoming wave which is retransmitted to the focal point by the helices on the inner side of the shell. The

---

[1] S. A. Schelkunoff, "Equivalence Theorems in Electromagnetics," *Bell System Tech. J.*, **15**, 92–112, 1936.

[2] R. B. Watson and C. W. Horton, "The Radiation Patterns of Dielectric Rods—Experiment and Theory," *J. Appl. Phys.*, **19**, 661–670, July 1948.

[3] G. E. Mueller and W. A. Tyrrell, "Polyrod Antennas," *Bell System Tech. J.*, **26**, 837–851, October 1947.

[4] D. G. Kiely, *Dielectric Aerials*, Methuen, 1953.

[5] For more detail see, for example, J. D. Kraus, *Electromagnetics*, 3rd ed., McGraw-Hill, 1984, pp. 596–597.

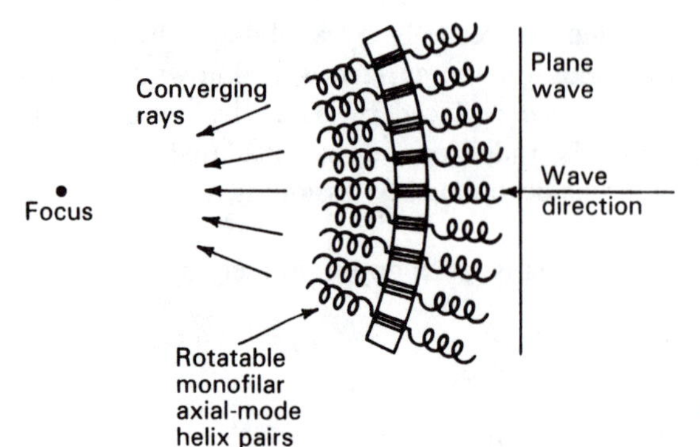

Figure 14-22 Helix lens. Focusing is accomplished by rotating helix pairs.

lens is focused by rotating each helix pair with respect to a reference pair (at the center) to compensate for path-length differences. Rotation of a helix pair through an angle θ shifts the phase by 2θ. The number of turns n of each helix is given by

$$n = \frac{4\pi A_p}{12 S_\lambda \lambda^2 N} \tag{1}$$

where A_p = lens area

S_λ = helix turn spacing

N = number of helix pairs

To avoid grating lobes the spacing between helix pairs should be less than λ. The helices and coaxial interconnections should, of course, be matched. A helix lens 1.3 m in diameter with 1213 helix pairs has been constructed for operation at 8.15 GHz ($\lambda = 3.7$ cm).[1] At this wavelength the lens diameter is 35λ producing a half-power beam width of about 2°.

14-10 LUNEBURG AND EINSTEIN LENSES.

The *Luneburg lens*[2] shown in Fig. 14-23 is a spherically symmetric delay-type lens formed of a dielectric with index of refraction n which varies as a function of radius, as given by

$$n = \sqrt{2 - \left(\frac{r}{R}\right)^2} \tag{1}$$

where r = radial distance from center of sphere
R = radius of sphere

[1] D. T. Nakatami and T. S. Ajioka, "Lens Designs Using Rotatable Phasing Elements," *IEEE APS Int. Symp. Dig.*, 357–360, June 1977.

[2] R. K. Luneburg, *Mathematical Theory of Optics*, Brown University Press, 1944. (Luneburg has sometimes been misspelled as Luneberg.)

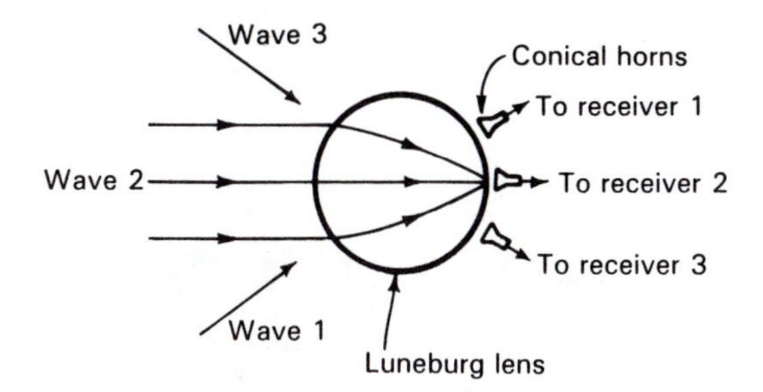

Figure 14-23 Luneburg lens. A plane wave incident on one side is brought to a focus on the opposite side.

When $r = R$, $n = 1$ while at the center of the sphere ($r = 0$), $n = 2$ (its maximum value). The lens has the property that an incident plane wave is brought to a focus on the opposite side of the sphere as suggested for wave 2 in Fig. 14-23. Simultaneously, waves from other directions will be brought to a focus at a point on the opposite side of the sphere, as suggested in Fig. 14-23 for waves 1 and 3. Thus, signals can be received simultaneously with a Luneburg lens from as many directions as there is space available on the sphere to place feed horns or other receiving devices. For steering a single beam the receiver (or transmitter) can be switched to different feed horns, or a single movable feed horn can be used. The variable index required can be obtained with an artificial dielectric material (Sec. 14-3) or with concentric shells of dielectric of different indices of refraction.

If a Luneburg sphere is cut in half and a reflecting sheet placed on the flat side, a Luneburg reflector-lens antenna results with incoming wave at an angle of incidence θ_i brought to a focus at the corresponding angle of reflection $\theta_r = \theta_i$.

The full spherical Luneburg lens provides beam steering in both polar coordinates (θ and ϕ). For steering in only one coordinate (ϕ), a plane (parallel-sided) section through the center of the sphere can be used. However, the beam is no longer the same in both coordinates due to vingetting in the θ direction. Kelleher gives a good review on the Luneburg lens and its variants.[1]

Although the idea had been mentioned earlier, Albert Einstein's brief note in *Science*[2] in 1936 was the first analysis of lens action by the gravitational field of a star such as the sun.[3] Incident electromagnetic waves passing around a star are deflected through an angle which is proportional to the mass of the star and brought to a focus on the far side, as suggested by the *Einstein gravity lens* of Fig. 14-24. There is no single focal point, rather a *focal line* extending from a minimum focal distance to infinity. The gain of the lens is proportional to the mass of the star and inversely proportional to the wavelength. At $\lambda = 1$ mm a solar lens can, in principle, give a gain of more than 80 dB. If the spacecraft on

---

[1] K. S. Kelleher, "Electromechanical Scanning Antennas," in R. C. Johnson and H. Jasik (eds.), *Antennas Engineering Handbook*, 2nd ed., McGraw-Hill, 1984, chap. 18.

[2] A. Einstein, "Lens-like Action of a Star by the Deviation of Light in the Gravitational Field," *Science*, **84**, 506, 1936.

[3] Actually any large mass—the sun, Jupiter, the earth, a neutron star or a black hole—would do.

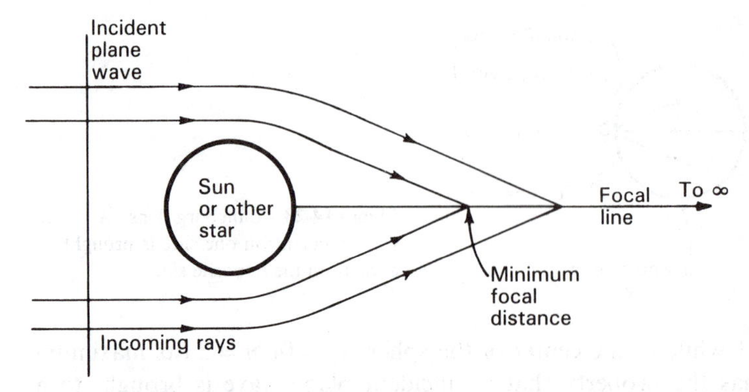

Figure 14-24 Einstein gravity lens. A plane wave incident on a large mass (such as the sun) is brought to a focus along a line extending from a minimum distance to infinity. Large gains are possible.

the focal line has an antenna with 80 dB gain, the total system gain is 160 dB, equivalent to the gain of an array of 100 million 80 dB-gain antennas.

The minimum focal distance in the case of the sun is about a dozen times the distance of Pluto so that we must wait until it is possible to send a properly equipped spacecraft to that distance before the sun can be put to use as a gravity lens.

I give more details on the Einstein lens with a worked example in my book *Radio Astronomy*,[1] based on the very extensive treatment of Von R. Eshleman.[2]

ADDITIONAL REFERENCE

Goldsmith, P. F., and G. J. Gill: "Dielectric Wedge Conical Scanned Gaussian Optics Lens Antenna," *Microwave J.*, **29**, 207–212, September 1986.

PROBLEMS[3]

14-1 Dielectric lens.

(*a*) Design a plano-convex dielectric lens for 5 GHz with a diameter of 10λ. The lens material is to be paraffin and the F number is to be unity. Draw the lens cross section.

(*b*) What type of primary antenna pattern is required to produce a uniform aperture distribution?

14-2 Artificial dielectric. Design an artificial dielectric with relative permittivity of 1.4 for use at 3 GHz when the artificial dielectric consists of (*a*) copper spheres, (*b*) copper discs, (*c*) copper strips.

[1] J. D. Kraus, *Radio Astronomy*, 2nd ed., Cygnus-Quasar, 1986.

[2] V. R. Eshleman, "Gravitational Lens of the Sun: Its Potential for Observations and Communications over Interstellar Distances," *Science*, **205**, 1133, 1979.

[3] Answers to starred (★) problems are given in App. D.

★**14-3 Unzoned metal-plate lens.** Design an unzoned plano-concave E-plane type of metal-plate lens of the unconstrained type with an aperture 10λ square for use with a 3-GHz line source 10λ long. The source is to be 20λ from the lens ($F = 2$). Make the index of refraction 0.6.

(*a*) What should the spacing between the plates be?

(*b*) Draw the shape of the lens and give dimensions.

(*c*) What is the bandwidth of the lens if the maximum tolerable path difference is $\lambda/4$?

14-4 Helix lens. Confirm (14-9-1).

14-5 Cylindrical lens. Prove (14-2-15*b*).

CHAPTER

15

BROADBAND AND FREQUENCY-INDEPENDENT ANTENNAS

15-1 BROADBAND ANTENNAS. Many antennas are highly resonant, operating over bandwidths of only a few percent. Such "tuned," narrow-bandwidth antennas may be entirely satisfactory or even desirable for single-frequency or narrowband applications. In many situations, however, wider bandwidths may be required.

The volcano smoke unipole antenna of Fig. 15-1*a* and the twin Alpine horn antenna of Fig. 15-1*b* are examples of basic wide-bandwidth antennas[1] (see earlier discussion of the antennas in Sec. 2-32). The gradual, smooth transition from coaxial or twin line to a radiating structure can provide an almost constant input impedance over very wide bandwidths. The high-frequency limit of the Alpine horn antenna may be said to occur when the transmission-line spacing $d > \lambda/10$ and the low-frequency limit when the open end spacing $D < \lambda/2$. Thus, if $D = 1000d$, the antenna has a theoretical 200 to 1 bandwidth.

A compact version of the twin Alpine horn, shown in Fig. 15-1*c* and *d*, has a double-ridge waveguide as the launcher on an exponentially flaring 2-conductor balanced transmission line. The design in Fig. 15-1*c* and *d* incorporates features

[1] I built and tested a volcano smoke antenna at the Harvard University Radio Research Laboratory in 1945 at the suggestion of Andrew Alford. The impedance bandwidth was very broad as anticipated. Drawings of both volcano smoke and twin Alpine horn antennas appeared in the first edition of this book (1950).

used by Kerr[1] and by Baker and Van der Neut.[2] The exponential taper is of the form

$$y = k_1 e^{k_2 x} \tag{1}$$

where k_1 and k_2 are constants. The exact curvature is not critical provided it is gradual.

The fields are bound sufficiently close to the ridges that the horn beyond the launcher may be omitted. The version shown is a compromise with the top and bottom of the horn present but solid sides replaced by a grid of conductors with a spacing of about $\lambda/10$ at the lowest frequency. The grid reduces the pattern width in the H-plane, increasing the low-frequency gain. The Chuang-Burnside[3] cylindrical end sections on the ridges reduce the back radiation and VSWR. Absorber on the top and bottom of the ridges (or horn) also reduces back-radiation and VSWR.

Depending on the ratio of the open end dimension D to the spacing d of the ridges at the feed end, almost arbitrarily large bandwidths are possible with gain increasing with frequency. Thus, for the antenna of Fig. 15-1c and d, the feed dimension $d = 1.5$ mm, so that the shortest wavelength $\lambda = 15$ mm ($\lambda/10 = 1.5$ mm) and the open-end dimension $D = 128$ mm so that the longest wavelength $\lambda = 256$ mm ($\lambda/2 = 128$ mm) for a bandwidth of 17 to 1 ($=256/15$). For a similar antenna without end cylinders, Kerr reports bandwidths of 17 to 1 with gains up to 14 dBi.

The design of Fig. 15-1c and d is linearly polarized (vertical). With two orthogonal sets of ridges forming a quadruply ridged waveguide-fed horn, either vertical or horizontal or circular polarization can be obtained.

The 120° wide-cone-angle biconical antenna of Fig. 15-1e may be regarded as an evolved form of the volcano smoke antenna of Fig. 15-1a. The properties of the biconical antenna have been discussed in Chap. 8. An example of a 120° wide-cone biconical antenna is shown in Fig. 2-28c and, even though not fitted with end caps or absorber, has a nearly constant 50-Ω input impedance over a 6 to 1 bandwidth.

Although the parabolic reflector is also a broadband device, its useful bandwidth is often restricted by the bandwidth of its feed. Other examples of broadband antennas are the monofilar axial-mode helical antenna (Chap. 7) and the corner reflector (Chap. 12), both with bandwidths of 2 to 1 or more. Tapered helices or clustered helices have reported bandwidths of 5 to 1 or more (see

---

[1] J. L. Kerr, "Short Axial Length Broad Band Horns," *IEEE Trans. Ants. Prop.*, **AP-21**, 710–714, September 1973.

[2] D. E. Baker and C. A. Van der Neut, "A Compact, Broadband, Balanced Transmission Line Antenna from Double-Ridged Waveguide," *IEEE Ant. Prop. Soc. Symp.*, Albuquerque, 1982, pp. 568–570.

[3] C. W. Chuang and W. D. Burnside, "A Diffraction Coefficient for a Cylindrically Truncated Planar Surface," *IEEE Trans. Ants. Prop.*, **AP-28**, 177–182, March 1980.

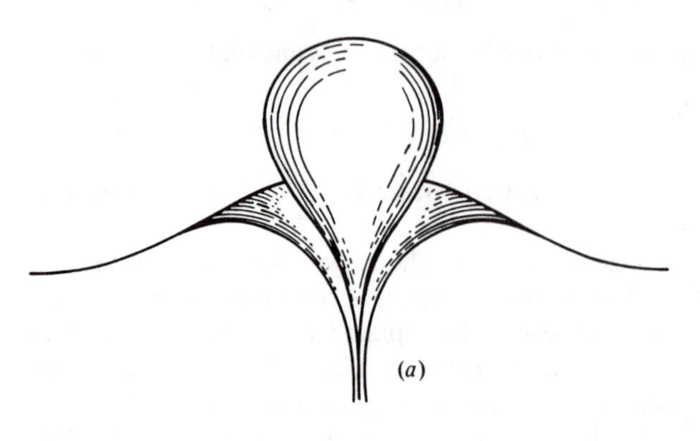
(a)

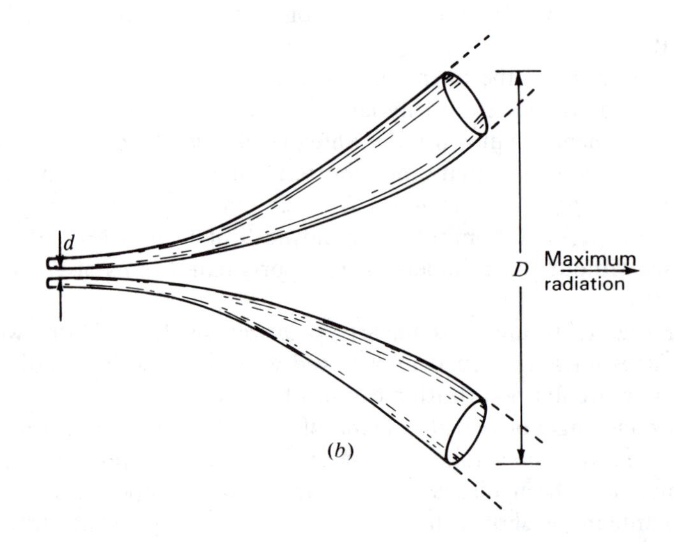
d
D
Maximum radiation
(b)

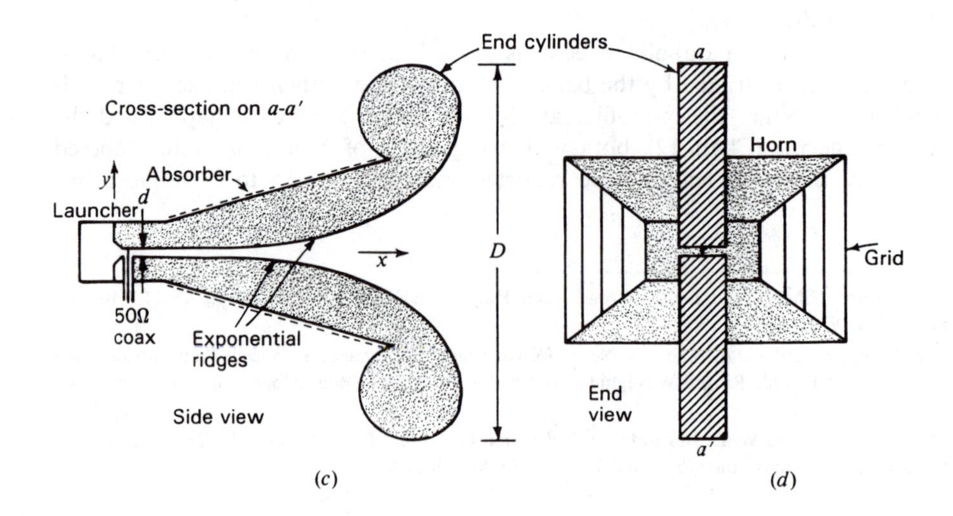
End cylinders
a
Cross-section on a-a′
Horn
y
Absorber
d
Launcher
x
D
Grid
50Ω coax
Exponential ridges
Side view
End view
a′
(c)

(d)

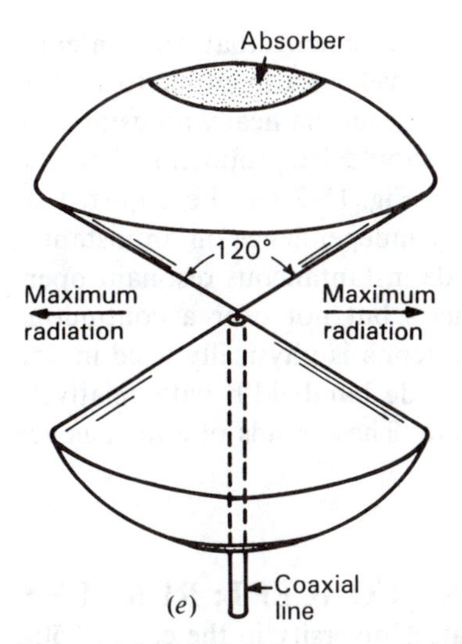

Figure 15-1 Wideband antennas. Volcano smoke (*a*) in cutaway view and twin Alpine horn (*b*) are basic types. Volcano smoke is omnidirectional in azimuth. The twin Alpine horn is unidirectional. Practical design derived from the twin Alpine horn with 17 to 1 design bandwidth is shown in side view (*c*) and end view (*d*). The 120° wide-cone-angle biconical antenna in (*e*) has end caps with absorber.

Secs. 7-16, 7-17 and 15-4). The rhombic antenna (Sec. 11-16b) and the discone (Sec. 16-5) are also broadband types.

All of the above antennas may have relatively constant input impedance and satisfactory pattern and gain over wide bandwidths with beam widths tending to become smaller and gains larger with increasing frequency. Although the increased gain may be highly desirable and useful, these antennas are not frequency independent in the sense that *all* parameters (impedance, pattern, polarization, gain) are constant or nearly so as a function of frequency.

Consider, for example, the arrangement of Fig. 15-2, which consists of an adjustable $\lambda/2$ dipole made of two drum-type (roll-up) pocket rulers. If L is adjusted to approximately $\lambda/2$ at the frequency of operation, the impedance and pattern remain the same. Strictly speaking, the element thickness or width w and the size of the drum housing should also be adjusted, but if these dimensions remain small compared to λ, this effect is small and for many purposes may be

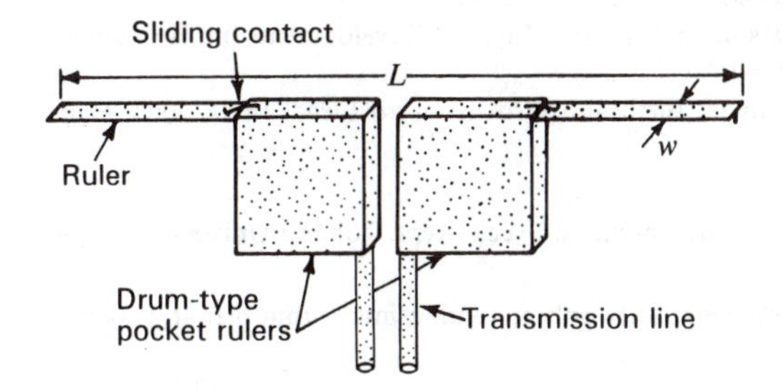

Figure 15-2 Adjustable $\lambda/2$ dipole of 2 drum-type rulers illustrates the requirement that to be frequency independent an antenna must expand or contract in proportion to the wavelength.

negligible. This simple antenna illustrates the requirement that the antenna should expand or contract in proportion to the wavelength in order to be frequency independent, or if the antenna structure is not mechanically adjustable as above, the size of the active or radiating region should be proportional to the wavelength. Although the roll-up pocket ruler of Fig. 15-2 can be adjusted to different frequencies, it does not provide frequency independence on an instantaneous basis. A dipole with tuned traps can provide instantaneous resonant operation at a number of widely separated frequencies but not over a continuous wide bandwidth. A true frequency-independent antenna is physically fixed in size and operates on an instantaneous basis over a wide bandwidth with relatively constant impedance, pattern, polarization and gain. These kinds of antennas are discussed in subsequent sections.

15-2 THE FREQUENCY-INDEPENDENT CONCEPT: RUMSEY'S PRINCIPLE. Beginning while at the Ohio State University in the early 1950s, continuing from 1954 to 1957 at the University of Illinois, and later at the University of California, first at Berkeley and subsequently at San Diego, Victor H. Rumsey developed and introduced a new way of looking at antennas and their operation as a function of the frequency.[1]

Rumsey was intrigued with Mushiake's observation[2] in 1949 that *self-complementary antennas*[3] have a constant impedance of $Z_0/2$, or half the intrinsic impedance of space, at all frequencies. This is remarkable since there is an infinity of self-complementary shapes. A self-complementary planar antenna has a metal area congruent to the open area, i.e., the two areas can be brought into coincidence by a rigid motion. Three examples of self-complementary antennas are shown in Fig. 15-3. The metal and open areas are congruent since a rotation of either brings both into coincidence.

The slot and complementary dipole antennas of Chap. 13 are similarly related but usually require a translation for coincidence. Mushiake's $Z_0/2$ result comes directly from Booker's relation for complementary slots and dipoles as given by (13-6-10).

Rumsey's principle is that the impedance and pattern properties of an antenna

---

[1] V. H. Rumsey, *Frequency Independent Antennas*, Academic Press, 1966.

E. C. Jordan, G. A. Deschamps, J. D. Dyson and P. E. Mayes, "Developments in Broadband Antennas," *IEEE Spectrum*, **1**, 58–71, April 1964.

P. E. Mayes, "Frequency-Independent Antennas: Birth and Growth of an Idea," *IEEE Ants. Prop. Soc. Newsletter*, **24**, 5–8, August 1982.

[2] Y. Mushiake, *J. IEE Japan*, **69**, 1949.

S. Uda and Y. Mushiake, "Input Impedance of Slit Antennas," Tech. Rept. Tohoku University 14, pp. 46–59, 1949.

[3] The concept of complementary antennas applies strictly only to infinitesimally thin planar, perfectly conducting shapes of infinite extent.

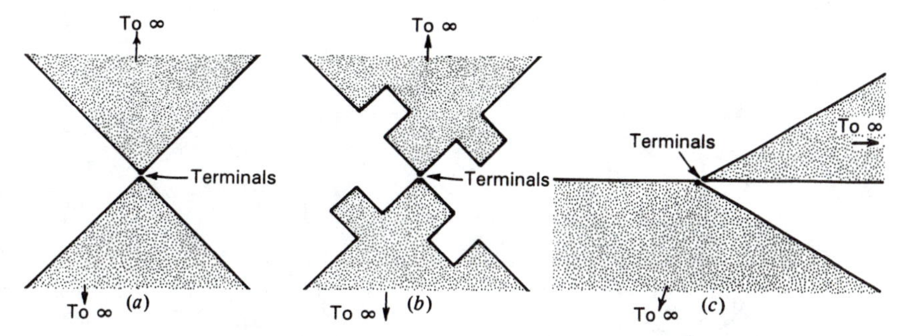

Figure 15-3 Three self-complementary planar antennas. Theoretical terminal impedance is 188 Ω.

will be frequency independent if the antenna shape is specified only in terms of angles. Thus, an infinite logarithmic spiral should meet the requirement.

The biconical antenna of Chap. 8 is an example of an antenna that can be specified only in terms of the included cone angle, but it is frequency independent only if it is infinitely long. When truncated (without a matched termination) there is a reflected wave from the ends of the cones which results in modified impedance and pattern characteristics.

To meet the frequency-independent requirement in a finite structure requires that the current attenuate along the structure and be negligible at the point of truncation. For radiation and attenuation to occur, as stated in Sec. 2-37, charge must be accelerated (or decelerated) and this happens when a conductor is curved or bent normally to the direction in which the charge is traveling. Thus, the curvature of a spiral results in radiation and attenuation so that, even when truncated, the spiral provides frequency-independent operation over a wide bandwidth.

Rumsey's principle was implemented experimentally by John D. Dyson at the University of Illinois, who constructed the first practical frequency-independent spiral antennas in 1958, first the bidirectional planar spiral and then the unidirectional conical spiral. These two types are described in the next two sections.

15-3 THE FREQUENCY-INDEPENDENT PLANAR LOG-SPIRAL ANTENNA.

The equation for a *logarithmic* (*or log*) *spiral* is given by

$$r = a^{\theta} \tag{1}$$

or

$$\ln r = \theta \ln a \tag{2}$$

where, referring to Fig. 15-4,

r = radial distance to point P on spiral
θ = angle with respect to x axis
a = a constant

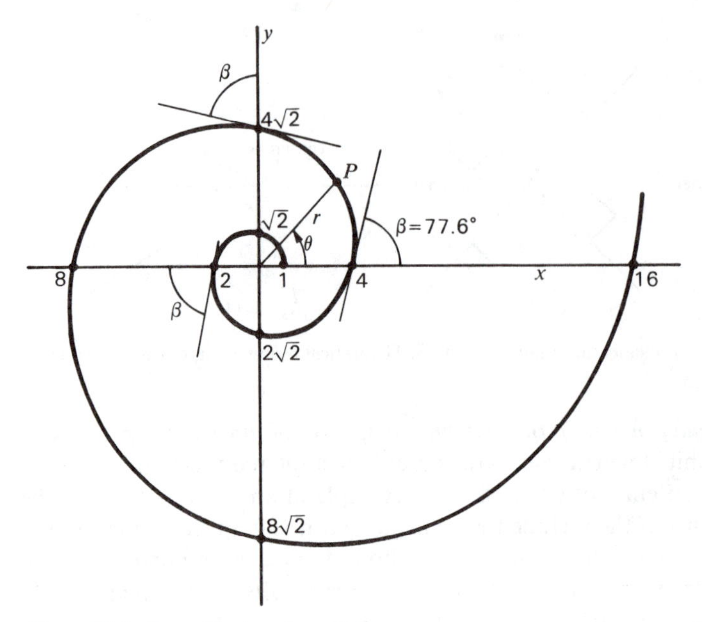

Figure 15-4 Logarithmic or log spiral.

From (1), the rate of change of radius with angle is

$$\frac{dr}{d\theta} = a^{\theta} \ln a = r \ln a \tag{3}$$

The constant a in (3) is related to the angle β between the spiral and a radial line from the origin as given by

$$\ln a = \frac{dr}{r\,d\theta} = \frac{1}{\tan \beta} \tag{4}$$

Thus, from (4) and (2),

$$\theta = \tan \beta \ln r \tag{5}$$

The log spiral in Fig. 15-4 was constructed so as to make $r = 1$ at $\theta = 0$ and $r = 2$ at $\theta = \pi$. These conditions determine the value of the constants a and β. Thus, from (4) and (5), $\beta = 77.6°$ and $a = 1.247$. Thus, the shape of the spiral is determined by the angle β which is the same for all points on the spiral.[1]

[1] Although it is a broadband antenna, the *Archimedes spiral* is not regarded as completely frequency independent. The angle β for an Archimedes spiral as given by $r = a\theta$ is not a constant but is a function of position along the spiral. However, remote from the origin on a tight Archimedes spiral, β approaches a nearly constant angle and the Archimedes spiral becomes a close approximation of a tightly wound log spiral. See R. Bawer and J. J. Wolfe, "A Printed Circuit Balun for Use with Spiral Antennas," *IRE Trans. Microwave Th. Tech.*, **MTT-8**, 319–325, May 1960; and P. E. Mayes and J. D. Dyson, "A Note on the Difference between Equiangular and Archimedes Spiral Antennas," *IRE Trans. Microwave Th. Tech.*, **MTT-9**, 203–205, March 1961.

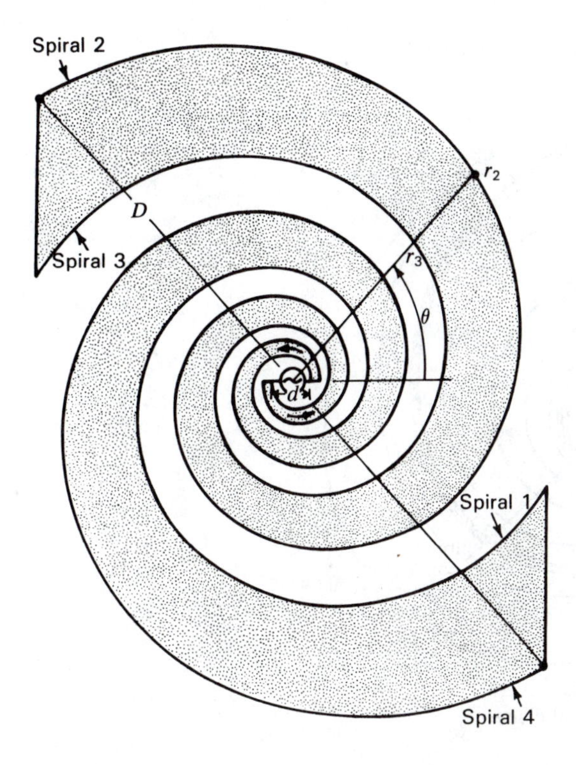

Figure 15-5 Frequency-independent planar spiral antenna.

Let a second log spiral, identical in form to the one in Fig. 15-4, be generated by an angular rotation δ so that (1) becomes

$$r_2 = a^{\theta - \delta} \tag{6}$$

and a third and fourth spiral given by

$$r_3 = a^{\theta - \pi} \tag{7}$$

and

$$r_4 = a^{\theta - \pi - \delta} \tag{8}$$

Then, for a rotation $\delta = \pi/2$ we have 4 spirals at 90° angles. Metalizing the areas between spirals 1 and 4 and 2 and 3, with the other areas open, self-complementary and congruence conditions are satisfied. Connecting a generator or receiver across the inner terminals, we obtain Dyson's frequency-independent planar spiral antenna of Fig. 15-5.[1]

The arrows indicate the direction of the outgoing waves traveling along the conductors resulting in right-circularly polarized (RCP) radiation (IEEE definition) outward from the page and left-circularly polarized radiation into the

[1] J. D. Dyson, "The Equiangular Spiral Antenna," *IRE Trans. Ants. Prop.*, **AP-7**, 181–187, April 1959.

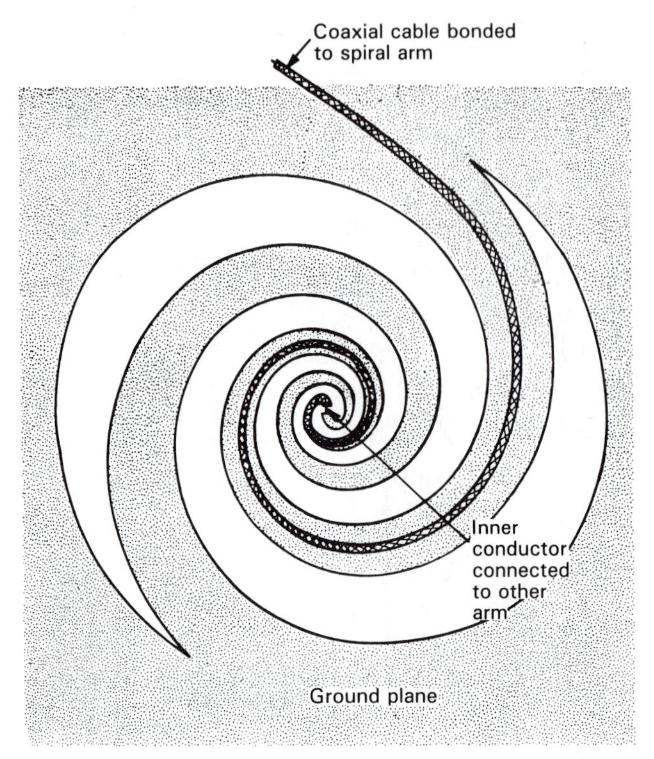

Figure 15-6 Frequency-independent planar spiral antenna cut from large ground plane.

page. The high-frequency limit of operation is determined by the spacing d of the input terminal and the low-frequency limit by the overall diameter D. The ratio D/d for the antenna of Fig. 15-5 is about 25 to 1. If we take $d = \lambda/10$ at the high-frequency limit and $D = \lambda/2$ at the low-frequency limit, the antenna bandwidth is 5 to 1. The spiral should be continued to a smaller radius but, for clarity, the terminal separation shown in Fig. 15-5 is larger than it should be. Halving it doubles the bandwidth.

In practice, it is more convenient to cut the slots for the antenna from a large ground plane, as done by Dyson, and feed the antenna with a coaxial cable bonded to one of the spiral arms as in Fig. 15-6, the spiral acting as a balun.[1] A dummy cable may be bonded to the other arm for symmetry but is not shown.

Radiation for the antennas of Figs. 15-5 and 15-6 is bidirectional broadside to the plane of the spiral. The patterns in both directions have a single broad lobe so that the gain is only a few dBi. The input impedance depends on the parameters δ and a and the terminal separation. According to Dyson, typical values are in the range 50 to 100 Ω, or considerably less than the theoretical

[1] Balance-to-unbalance transformer.

188 Ω ($=Z_0/2$). The smaller measured values are apparently due to the finite thickness of spirals.

Referring to Fig. 15-5, the ratio K of the radii across any arm, such as between spirals 2 and 3, is given by the ratio of (7) to (6), or

$$K = \frac{r_3}{r_2} = a^{-\pi+\delta} \tag{9}$$

For the antenna of Fig. 15-5, $\delta = \pi/2$ so

$$K = \frac{r_3}{r_2} = a^{-\pi/2} = 0.707 \; (=1/\sqrt{2}) \tag{10}$$

This is seen to be the ratio of the radial distances to the spiral of Fig. 15-4 at successive 90° intervals.

15-4 THE FREQUENCY-INDEPENDENT CONICAL-SPIRAL ANTENNA. A tapered helix is a conical-spiral antenna and these were described and investigated extensively in the years following 1947. In my first article on the helical antenna published in 1947,[1] I describe a tapered helix which I constructed and measured. Figure 15-7*a* and *b* shows tapered helical or conical spiral antennas in which the pitch angle is constant with diameter and turn spacing variable. These figures appeared in the first edition of this book (1950) and also appear as Fig. 7-59*a* and *b* of the present edition. These tapered helix or conical spirals were investigated by Springer (1950),[2] by Chatterjee (1953, 1955)[3]

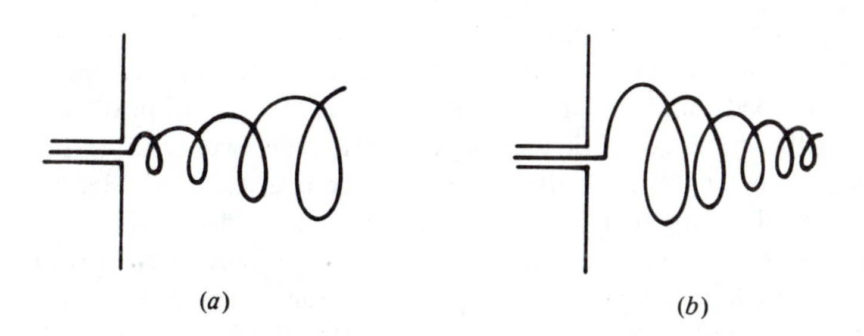

Figure 15-7 Tapered helical or conical-spiral (forward-fire) CP antennas.

[1] J. D. Kraus, "Helical Beam Antenna," *Electronics*, **20**, 109–111, April 1947.

[2] P. S. Springer, "End-Loaded and Expanding Helices as Broad-Band Circularly Polarized Radiators," Electronic Subdiv. Tech. Rept. 6104, Wright-Patterson AFB, 1950.

[3] J. S. Chatterjee, "Radiation Field of a Conical Helix," *J. Appl. Phys.*, **24**, 550–559, May 1953; "Radiation Characteristics of a Conical Helix of Low Pitch Angle," *J. Appl. Phys.*, **26**, 331–335, March 1955.

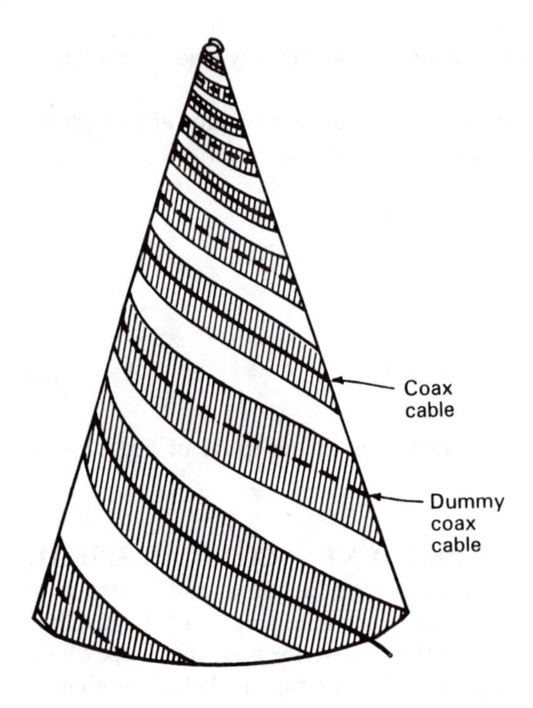

Figure 15-8 Dyson 2-arm balanced conical-spiral (backward-fire) antenna. Polarization is RCP. Inner conductor of coax connects to dummy at apex.

and others, and more recently by Nakano, Mikawa and Yamauchi,[1] and found capable of bandwidths of 5 to 1 or more (see Sec. 7-17). Chatterjee also described a planar spiral antenna.

However, it was not until 1958 that John D. Dyson[2] at the University of Illinois made the tapered helix or conical spiral fully frequency independent by wrapping or projecting multiple planar spirals onto a conical surface.

A typical balanced 2-arm Dyson conical spiral is shown in Fig. 15-8. The conical spiral retains the frequency-independent properties of the planar spiral while providing broad-lobed unidirectional circularly polarized radiation off the small end or apex of the cone. As with the planar spiral, the two arms of the conical spiral are fed at the centerpoint or apex from a coaxial cable bonded to one of the arms, the spiral acting as a balun. For symmetry a dummy cable may be bonded to the other arm, as suggested in Fig. 15-8. In some models the metal straps are dispensed with and the cables alone used as the spiral conductors. According to Dyson, the input impedance is between 100 and 150 Ω for a pitch angle $\alpha = 17°$ and full cone angles of 20 to 60°. The smaller cone angles (30° or less) have higher front-to-back ratios of radiation. The bandwidth, as with the

[1] H. Nakano, T. Mikawa and J. Yamauchi, "Numerical Analysis of Monofilar Conical Helix," *IEEE AP-S Int. Symp.*, **1**, 177–180, 1984.

[2] J. D. Dyson, "The Unidirectional Spiral Antenna," *IRE Trans. Ants. Prop.*, **AP-7**, 329–334, October 1959.

planar spiral, depends on the ratio of the base diameter ($\sim\lambda/2$ at the lowest frequency) to the truncated apex diameter ($\sim\lambda/4$ at the highest frequency). This ratio may be made arbitrarily large.

Conical and planar spirals with more than 2 arms are also possible and have been investigated by Dyson and Mayes[1] and by Deschamps,[2] all at the University of Illinois, and also by Atia and Mei.[3]

15-5 THE LOG-PERIODIC ANTENNA.

While the planar and conical spirals were being developed, Raymond DuHamel and Dwight Isbell,[4] also at the University of Illinois, created a new type of frequency-independent antenna with a self-complementary toothed structure as suggested in Fig. 15-9. In an alternative version, the metal and slot areas of Fig. 15-9 are interchanged. Since

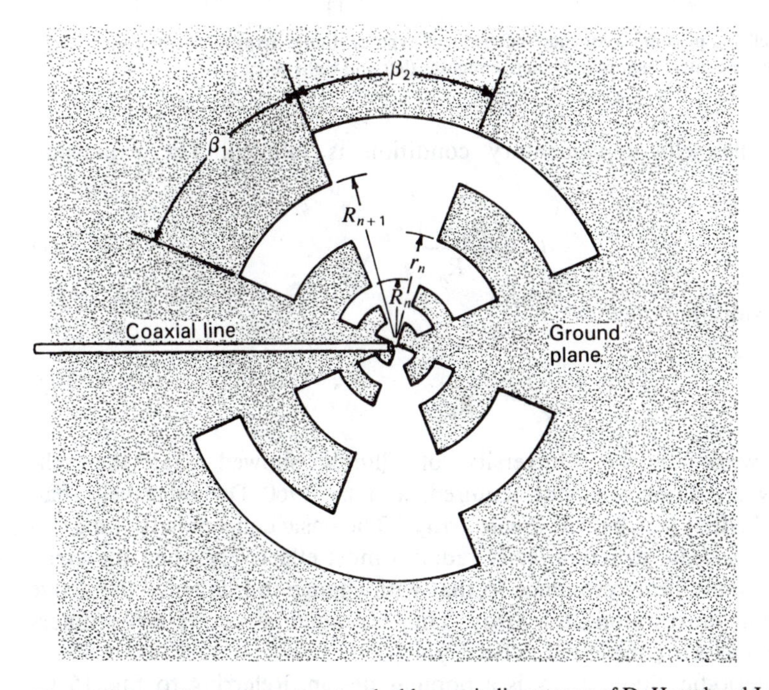

Figure 15-9 Self-complementary toothed log-periodic antenna of DuHamel and Isbell.

---

[1] J. D. Dyson and P. E. Mayes, "New Circularly-Polarized Frequency-Independent Antennas with Conical Beam or Omnidirectional Patterns," *IRE Trans. Ants. Prop.*, **AP-9**, 334–342, July 1961.

[2] G. A. Deschamps, "Impedance Properties of Complementary Multiterminal Planar Structures," *IRE Trans. Ants. Prop.*, **AP-7**, S371–378, December 1959.

[3] A. E. Atia and K. K. Mei, "Analysis of Multiple Arm Conical Log-Spiral Antennas," *IEEE Trans. Ants. Prop.*, **AP-19**, 320–331, May 1971.

[4] R. H. DuHamel and D. E. Isbell, "Broadband Logarithmically Periodic Antenna Structures," *IRE Natl. Conv. Rec.*, pt. 1, 119–128, 1957.

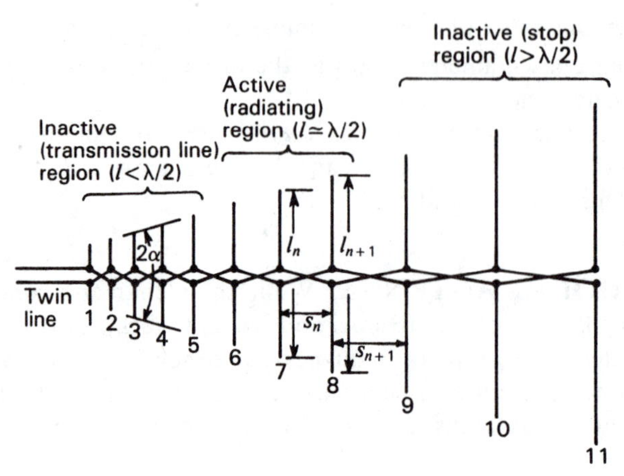

Figure 15-10 Isbell log-periodic frequency-independent type of dipole array of 7 dBi gain with 11 dipoles showing active central region and inactive regions (left and right ends).

$\beta_1 + \beta_2 = 90°$ the self-complementary condition is fulfilled. The expansion parameter

$$k_1 = \frac{R_n + 1}{R_n} \tag{1}$$

and the tooth-width parameter

$$k_2 = \frac{r_n}{R_n} \tag{2}$$

Further work at the University of Illinois showed that the self-complementary condition was not required, and by 1960 Dwight Isbell[1] had demonstrated the first *log-periodic dipole array.*[2] The basic concept is that *a gradually expanding periodic structure array* radiates most effectively when the array elements (dipoles) are near resonance so that with change in frequency the active (radiating) region moves along the array. This *expanding* structure array differs from the *uniform* arrays considered in Sec. 7-11.

The log-periodic dipole array is a popular design. Referring to Fig. 15-10, the dipole lengths increase along the antenna so that the included angle α is a constant, and the lengths l and spacings s of adjacent elements are scaled so that

$$\frac{l_{n+1}}{l_n} = \frac{s_{n+1}}{s_n} = k \tag{3}$$

---

[1] D. E. Isbell, "Log Periodic Dipole Arrays," *IRE Trans. Ants. Prop.*, **AP-8**, 260–267, May 1960.

[2] So called because the structure repeats periodically with the logarithm of the frequency. Put another way, the structure doubles for each doubling of the wavelength [see (8)].

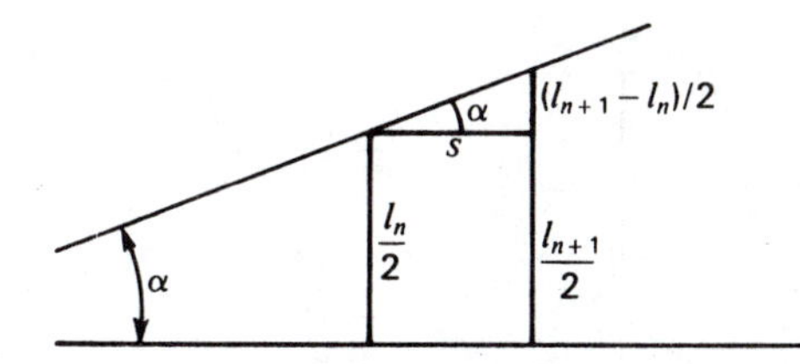

Figure 15-11 Log-periodic array geometry for determining the relation of parameters.

where k is a constant. At a wavelength near the middle of the operating range, radiation occurs primarily from the central region of the antenna, as suggested in Fig. 15-10. The elements in this active region are about $\lambda/2$ long.

Elements 9, 10 and 11 are in the neighborhood of 1λ long and carry only small currents (they present a large inductive reactance to the line). The small currents in elements 9, 10 and 11 mean that the antenna is effectively truncated at the right of the active region. Any small fields from elements 9, 10 and 11 also tend to cancel in both forward and backward directions. However, some radiation may occur broadside since the currents are approximately in phase. The elements at the left (1, 2, 3, etc.) are less than $\lambda/2$ long and present a large capacitative reactance to the line. Hence, currents in these elements are small and radiation is small.

Thus, at a wavelength λ, radiation occurs from the middle portion where the dipole elements are $\sim\lambda/2$ long. When the wavelength is increased the radiation zone moves to the right and when the wavelength is decreased it moves to the left with maximum radiation toward the apex or feed point of the array.

At any given frequency only a fraction of the antenna is used (where the dipoles are about $\lambda/2$ long). At the short-wavelength limit of the bandwidth only 15 percent of the length may be used, while at the long-wavelength limit a larger fraction is used but still less than 50 percent.

From the geometry of Fig. 15-11 for a section of the array, we have

$$\tan \alpha = \frac{(l_{n+1} - l_n)/2}{s} \tag{4}$$

or from (3),

$$\tan \alpha = \frac{[1 - (1/k)](l_{n+1}/2)}{s} \tag{5}$$

Taking $l_{n+1} = \lambda/2$ (when active) we have

$$\tan \alpha = \frac{1 - (1/k)}{4s_\lambda} \tag{6}$$

where α = apex angle
k = scale factor
s_λ = spacing in wavelengths shortward of $\lambda/2$ element

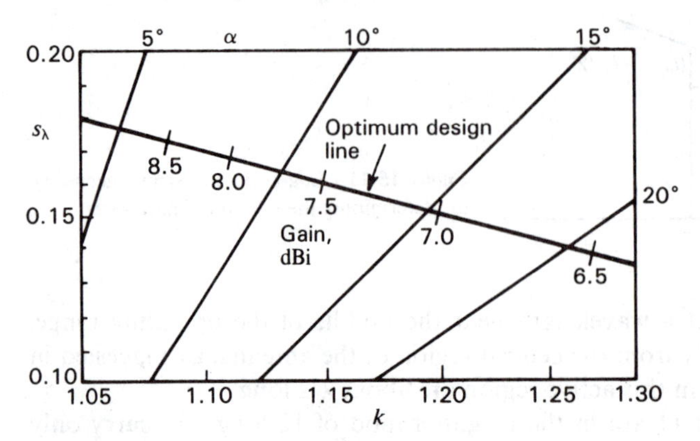

Figure 15-12 Relation of log-periodic array parameters of apex angle α, scale factor k and spacing s_λ [from (6)] with optimum design line and gain values according to Carrel and others (see text for details).

Specifying any 2 of the 3 parameters α, k and s_λ determines the third. The relationship of the 3 parameters is displayed in Fig. 15-12 with the optimum design line (maximum gain for a given value of scale factor k) and gain along this line from calculations of Carrel,[1] Cheong and King,[2] De Vito and Stracca[3] and Butson and Thomson.[4]

The length l (and spacing s) for any element $n + 1$ is k^n greater than for element 1, or

$$\frac{l_{n+1}}{l_1} = k^n = F \tag{7}$$

where F = frequency ratio or bandwidth

Thus, if $k = 1.19$ and $n = 4$, $F = k^4 = 1.19^4 = 2$ and element 5 ($= n + 1$) is twice the length l_1 of element 1. Thus, with 5 elements and $k = 1.19$, the frequency ratio is 2 to 1.

Example. Design a log-periodic dipole array with 7 dBi gain and a 4 to 1 bandwidth. Specify apex angle α, scale constant k and number of elements.

[1] R. L. Carrel, "The Design of Log-Periodic Dipole Antennas," *IRE Int. Conv. Rec.*, **1**, 61–75, 1961.

[2] W. M. Cheong and R. W. P. King, "Log Periodic Dipole Antenna," *Radio Sci.*, **2**, 1315–1325, November 1967.

[3] G. De Vito and G. B. Stracca, "Comments on the Design of Log-Periodic Dipole Antennas," *IEEE Trans. Ants. Prop.*, **AP-21**, 303–308, May 1973; **AP-22**, 714–718, September 1974.

[4] P. C. Butson and G. T. Thomson, "A Note on the Calculation of the Gain of Log-Periodic Dipole Antennas," *IEEE Trans. Ants. Prop.*, **AP-24**, 105–106, January 1976.

Solution. From Fig. 15-12, the 7 dBi point on the maximum gain line corresponds to the apex angle $\alpha = 15°$ and $k = 1.2$. (We also note that $s_\lambda = 0.15$.) From (7),

$$k^n = F \qquad \text{or} \qquad n \ln k = \ln F \tag{8}$$

and

$$n = \frac{\ln F}{\ln k} = \frac{\ln 4}{\ln 1.2} = \frac{1.386}{0.182} = 7.6 \tag{9}$$

Taking $n = 8$, $n + 1 = 9$. Adding 2 more elements for a conservative design brings the total to 11.[1]

The array in Fig. 15-10 corresponds to the parameters of the above example. The E-plane HPBW $\simeq 60°$. The H-plane beam width is a function of the gain as given by

$$\text{HPBW } (H \text{ plane}) \le \frac{41\,000}{D \times 60°} \qquad \text{(deg)} \tag{10}$$

For the antenna of the example $D = 5$, since $\log_{10} 5 = 7$ dBi, so

$$\text{HPBW } (H \text{ plane}) \le \frac{41\,000}{5 \times 60°} = 137° \tag{11}$$

Details of construction and feeding are shown in Fig. 15-13. The arrangement in (*a*) is fed with coaxial cable, the one at (*b*) with twin line.

To obtain more gain than with a single log-periodic dipole array, 2 arrays may be stacked. However, for frequency-independent operation, Rumsey's principle requires that the locations of all elements be specified by angles rather than distances. This means that both log-periodic arrays must have a common apex, and, accordingly, the beams of the 2 arrays point in different directions. For a stacking angle of 60°, the situation is as suggested in Fig. 15-14*a* for dipole arrays of the type shown in Figs. 15-10 and 15-13. The array in Fig. 15-14*b* is a skeleton-tooth or edge-fed trapezoidal type. Wires supported by a central boom replace the teeth of the antenna of Fig. 15-9.

For very wide bandwidths the log-periodic array must be correspondingly long. To shorten the structure, Paul Mayes and Robert Carrel,[2] of the University of Illinois, developed a more compact V-dipole array which can operate in several modes. In the lowest mode, with the central region dipoles $\sim\lambda/2$ long, operation is as already described. However, as the frequency is increased to the point where the shortest elements are too long to give $\lambda/2$ resonance, the longest elements become active at $3\lambda/2$ resonance. As the frequency is increased further,

---

[1] The number of extra elements needed depends, for example, on the design gain. Thus, for a high-gain design the active region requires more elements than for a low-gain design so that the bandwidth is less than given by k^n.

[2] P. E. Mayes and R. L. Carrel, "Log Periodic Resonant-V Arrays," San Francisco Wescon Conf., August 1961.

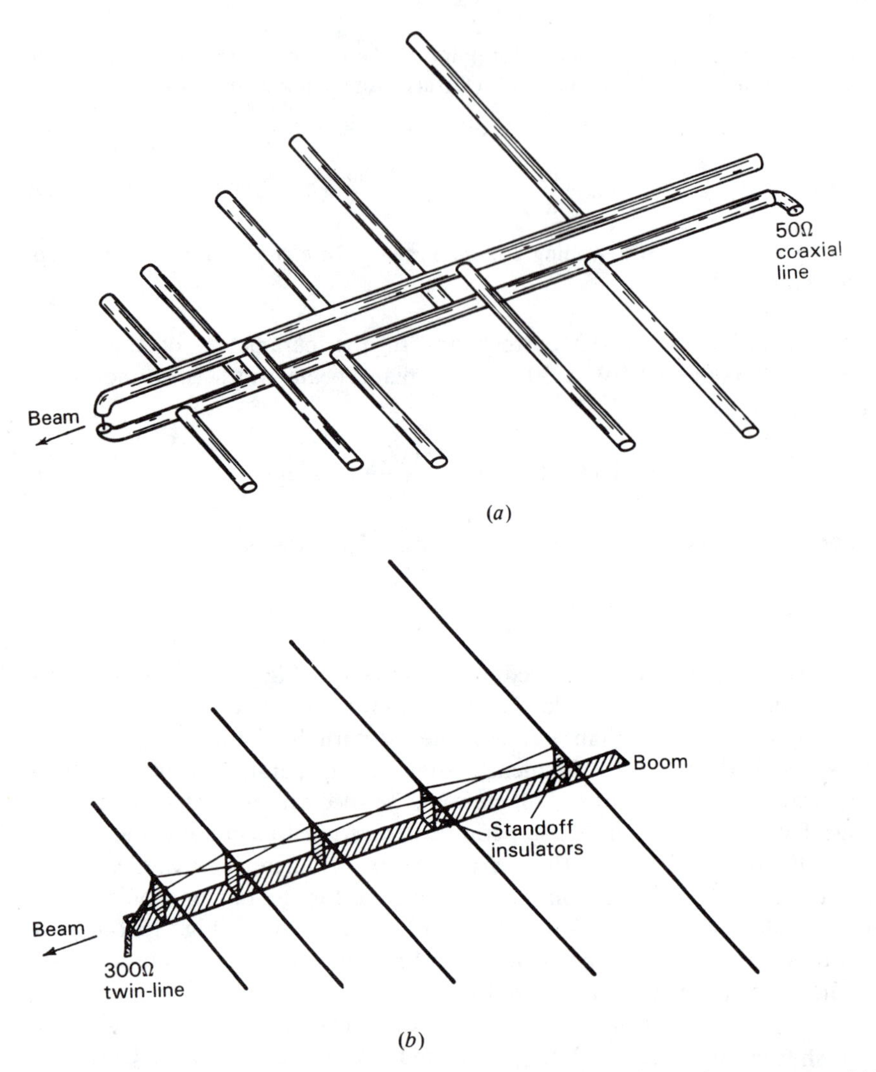

Figure 15-13 Construction and feed details of log-periodic dipole array. Arrangement at (*a*) has a 50- or 75-Ω coaxial feed. The one at (*b*) has criss-crossed open-wire line for 300-Ω twin-line feed.

the active region moves to the small end in the $3\lambda/2$ mode. With still further increase in frequency the large end becomes active in still higher-order modes. The forward tilt of the V-dipoles has little effect on the $\lambda/2$ mode but in the higher modes provides essential forward beaming. An example of a V-dipole array is shown in Fig. 15-15.

15-6 THE YAGI-UDA–CORNER–LOG-PERIODIC (YUCOLP) ARRAY. For ultimate compactness and gain, to cover the 54 to 890 MHz U.S. TV and FM bands, a hybrid YUCOLP (Yagi-Uda–Corner–Log-Periodic) array

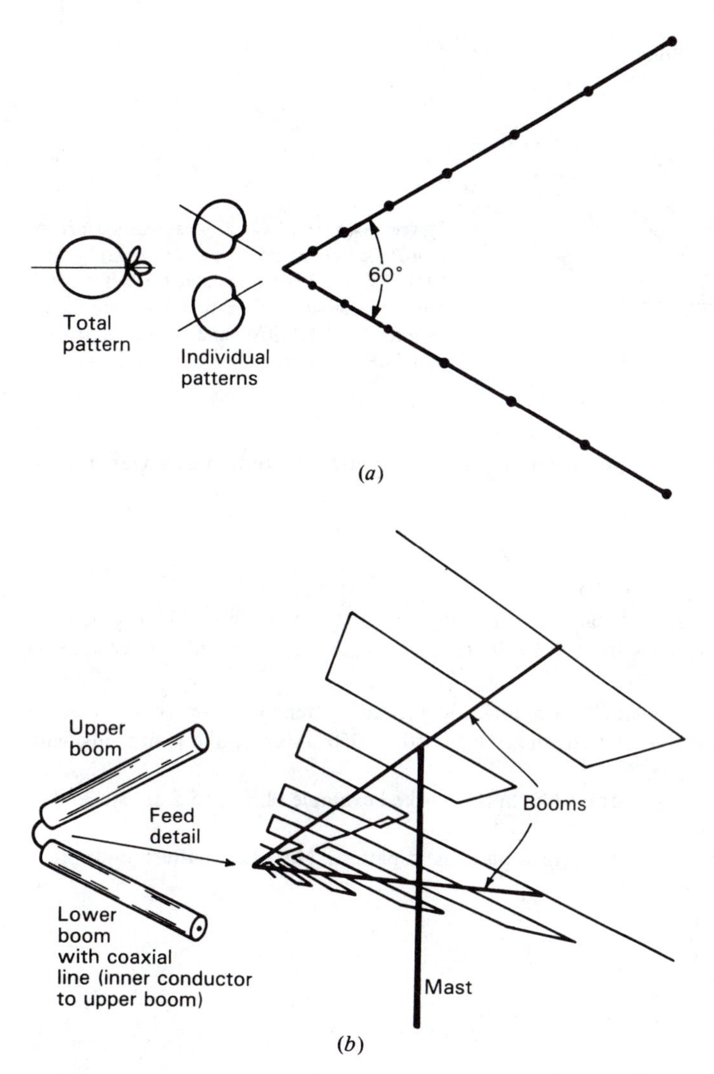

Figure 15-14 Stacked log-periodic arrays, with dipole type, as in Figs. 15-10 and 15-11, at (*a*) and trapezoidal or edge-fed type at (*b*).

is a popular design. A typical model, shown in Fig. 15-15, has an $\alpha = 43°$, $k = 1.3$ LP array of 5 V-dipoles to cover the 54 to 108 MHz TV and FM bands with a 6 dBi gain in the $\lambda/2$ mode, the 174 to 216 MHz band with 8 or 9 dBi gain in the $3\lambda/2$ mode and a square-corner–YU array to cover the 470 to 890 UHF TV band with a 7 to 10 dBi gain. The total included angle of the V-dipoles is 120°. The corner–YU array is similar in design to the one in Fig. 11-42. As frequency increases, the active region moves from the large to the small end of the LP array in the $\lambda/2$ mode, then from the large to the small end in the $3\lambda/2$ mode, next to the corner reflector and finally to the YU array. The corner–YU array provides

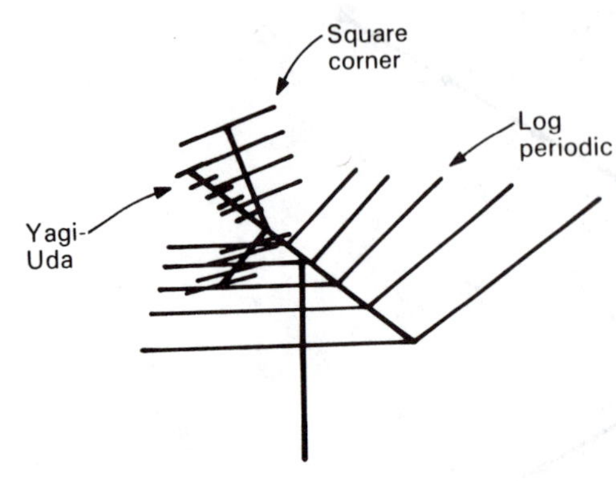

Figure 15-15 YUCOLP (Yagi-Uda–Corner–Log-Periodic) hybrid array for covering U.S. VHF TV and FM bands and UHF TV band. The YU–corner combination provides higher gain for the UHF TV band than an extension of the LP dipole array.

more gain for the UHF band than possible with a high-frequency extension of the LP array.

PROBLEMS

15-1 Log spiral. Design a planar log-spiral antenna of the type shown in Fig. 15-5 to operate at frequencies from 1 to 10 GHz. Make a drawing with dimensions in millimeters.

15-2 Log periodic. Design an "optimum" log-periodic antenna of the type shown in Fig. 15-10 to operate at frequencies from 50 to 250 MHz. Make a drawing with dimensions in meters.

15-3 Stacked LPs. Two LP arrays like in the worked example of Sec. 15-5 are stacked as in Fig. 15-14*a*.

(*a*) Calculate and plot the vertical plane field pattern. Note that pattern multiplication cannot be applied.

(*b*) What is the gain?

CHAPTER 16

ANTENNAS FOR SPECIAL APPLICATIONS: FEEDING CONSIDERATIONS

16-1 INTRODUCTION. Previous chapters cover the properties of many basic types of antennas which are the mainstream of antenna technology. In this chapter, more specialized, application-oriented antennas are discussed. These include antennas that are electrically small (but sometimes physically very large) and ones that are physically small (but sometimes electrically large), siting of antennas and the effect of typical ground, feeding and matching arrangements, and antennas for specialized applications from vehicular to satellite communication.

16-2 ELECTRICALLY SMALL ANTENNAS. The *radiation efficiency* of an antenna is given by

$$\text{RE} = \frac{R_r}{R_r + R_L} \qquad \text{(dimensionless)} \tag{1}$$

where R_r = radiation resistance, Ω
R_L = loss resistance, Ω

From (5-3-15), radiation resistance of a short dipole is given by

$$R_r = 200\, l_\lambda^2 \qquad (\Omega) \tag{2}$$

where l_λ = length of dipole in wavelengths, dimensionless

If $l_\lambda = 0.1$, $R_r = 2\ \Omega$; if $l_\lambda = 0.01$, $R_r = 0.02\ \Omega$. With such low values of radiation resistance it is apparent that even small values of loss resistance can result in low radiation efficiency.

Example. Referring to Fig. 16-1*a*, a 100-kHz transmitter feeds a 150-m vertical antenna against ground. The loss resistance is 2 Ω. Find (*a*) the effective height, (*b*) the radiation resistance and (*c*) the radiation efficiency.

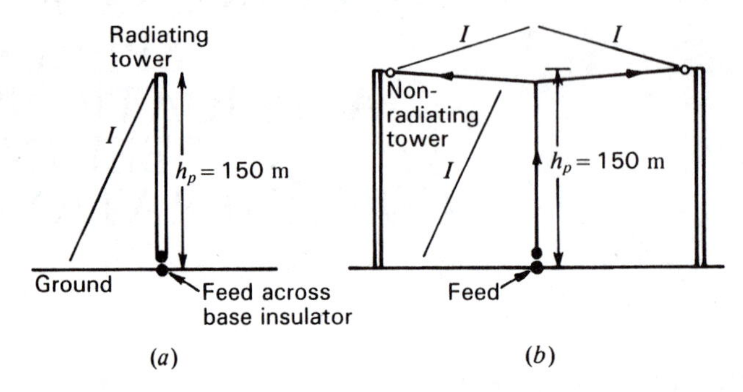

Figure 16-1 Electrically small antennas. (*a*) 150-m tower for operation at $\lambda_0 = 3000$ m. (*b*) Two non-radiating towers supporting flat top for increasing effective height.

Solution

(*a*) At 100 kHz ($\lambda = 3000$ m), the antenna (physical) height $h_p = 0.05\lambda$ ($=150/3000$) and with a triangular current distribution (current zero at top increasing linearly to a maximum value I_0 at the base) we have, from (2-19-3), that the effective height $h_e = h_p/2 = 0.025\lambda$.

(*b*) The radiation resistance for a short vertical monopole is given by

$$R_r = 400\left(\frac{h_p}{\lambda}\right)^2 = 1600\left(\frac{h_e}{\lambda}\right)^2 \quad (\Omega) \tag{3}$$[1]

where h_p = physical height, m
h_e = effective height, m

and for the 150-m vertical radiator, we have

$$R_r = 400 \times 0.05^2 = 1\ \Omega$$

(*c*) From (1) the radiation efficiency

$$\text{RE} = \frac{R_r}{R_r + R_L} = \frac{1}{2+1} = 0.33 \text{ or } 33 \text{ percent}$$

To increase the radiation efficiency requires an increase in the radiation resistance R_r or a decrease in the loss resistance R_L or both. The loss resistance

[1] It is assumed in (2) and (3) that the average antenna current is $\frac{1}{2}$ the terminal current.

involves losses in the ground system, antenna insulators, tuning coil, conductors (including the tower itself) and corona. These may be reduced but only at a certain cost. On the other hand, doubling the physical height of the antenna quadruples the radiation resistance and increases the radiation efficiency. However, to double the tower height might cost 8 times as much (cost proportional approximately to the cube of the height). Another alternative would be to use 2 towers 150 m tall supporting a vertical wire with flat top as suggested in Fig. 16-1*b*, making the current distribution along the vertical conductor more uniform ($I_{av} > \frac{1}{2}I_0$) and the effective height more nearly equal to the physical height. This will not quadruple R_r but could increase it by some factor less than 4. The currents on the flat top are in phase opposition and, being separated by a small fraction of a wavelength, have only a small effect on the radiation. What steps, if any, are taken to increase the radiation efficiency involve trade-offs with cost considerations a determining factor.

According to Harold Wheeler,[1] a figure-of-merit for an electrically small antenna is its *radiation power factor:*

$$\text{PF} = \frac{\text{radiated power}}{\text{reactive power}} = \frac{R_r}{X} \ll 1 \tag{4}$$

where R_r = radiation resistance
X = reactance

This is not the same as $1/Q$ since

$$Q = \frac{\text{energy stored per unit time}}{\text{energy lost per unit time}} = \frac{X}{R_r + R_L} = \frac{\text{center frequency}}{\text{bandwidth}} \tag{5}$$

Thus, increasing either R_r or R_L, or both, reduces Q and broadens the bandwidth. However, only an increase in R_r increases Wheeler's radiation power factor.

Harold Wheeler defines an electrically small antenna as one which occupies a volume of less than a *radian sphere*, i.e., a sphere of radius $r = \lambda/2\pi$ ($=0.16\ \lambda$). The significance of this definition is that stored energy dominates inside the radian sphere while radiated energy is important outside. Wheeler shows further that the radiation power factor for a small antenna is equal to the ratio of the antenna volume to a radian sphere or

$$\text{PF} = \frac{\text{antenna volume}}{\text{radian sphere}} = \frac{\frac{4}{3}\pi r^3}{\frac{4}{3}\pi(\lambda/2\pi)^3} = \left(\frac{2\pi r}{\lambda}\right)^3 \tag{6}$$

[1] H. A. Wheeler, "Fundamental Limitations of Small Antennas," *Proc. IRE*, **35**, 1479–1484, December 1947.
H. A. Wheeler, "Small Antennas," *IEEE Trans. Ants. Prop.*, **AP-23**, 462–469, July 1975.
See also R. C. Hansen, "Fundamental Limitations of Antennas," *Proc. IEEE*, **69**, 170–182, February 1981.

where r = radius of antenna volume, m
λ = wavelength, m

According to (6), the PF or figure-of-merit of a small antenna varies as the cube of its dimension. If $R_L = 0$ in (5), we have from (4) and (6) that

$$Q = \frac{1}{(2\pi r_\lambda)^3} \tag{7}$$

Thus, an electrically small lossless antenna ($2\pi r_\lambda \ll 1$) is inherently high Q and narrow bandwidth. We note from (5) that losses decrease Q and increase bandwidth.

To reduce the attenuation through salt water, very low frequencies (10 to 50 kHz) are used for transmitting to submerged submarines. At a frequency of 10 kHz ($\lambda = 30$ km) antennas may have flat tops covering many square kilometers and rank as the world's largest antennas (physically), yet they are electrically small ($h_e < 0.01\lambda$). (See Prob. 16-21.)

16-3 PHYSICALLY SMALL ANTENNAS. Heinrich Hertz and other radio pioneers used meter and centimeter wavelengths, but the demonstration of long-distance communication with kilometer wavelengths by Guglielmo Marconi and others soon moved interest from the short to the long wavelengths. However, over the intervening years this trend has been reversed and now active radio technology extends down to millimeter and submillimeter wavelengths. One of the bonuses of the short wavelengths is the spectrum space available for many wide-bandwidth video and high-data-rate channels.

Antennas for these short wavelengths include printed and patch antennas (Sec. 16-12) and ones with built-in (integrated) active elements (amplifiers and detectors) (Sec. 16-13). These active elements can compensate for the increased transmission-line losses at millimeter wavelengths.

The twin Alpine exponential horn described in Secs. 2-2, 2-28 and 15-1 is well adapted for printed-circuit fabrication, as suggested in Fig. 16-2. Here the

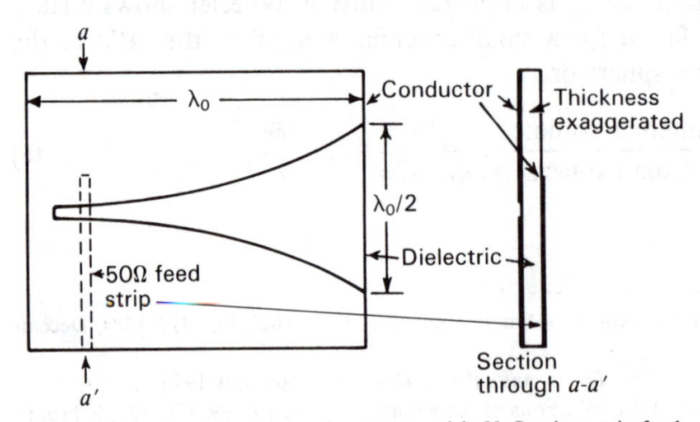

Figure 16-2 Exponential notch antenna with 50-Ω microstrip feed.

horn takes the form of an exponential notch in the conducting surface of a circuit board with coupling from a 50-Ω strip line on the other surface of the board. Bandwidths of 5 to 1 are possible but with only small gain.[1] However, many such printed notch radiators can be stacked to form highly directional phased arrays.[2]

Visible wavelengths (blue to red) extend from 400 to 700 nm. Infrared wavelengths extend from 700 nm into the micrometer range where they merge with radio. Whether these micrometer wavelengths are called infrared or radio is arbitrary. With printed-circuit technologies now able to fabricate structures in the micrometer and nanometer range,[3] it may be possible to construct antennas for visible wavelengths.

An array of $\lambda/2$ dipoles connected to detector (rectifier) elements has been described by Marks[4] for the direct, high-efficiency conversion of light to dc power. An important application is power generation from solar radiation.

In the design proposed by Marks, a broadside array of $\lambda/2$ dipoles is used to collect the incident radiation and convert it to direct current via a 2-conductor RF transmission line which terminates in a full-wave bridge rectifier. The dc output of the rectifier feeds the conductors of a dc bus. However, my variant, shown in Fig. 16-3, consists of an array of λ dipoles (2 colinear $\lambda/2$ dipoles), reducing the number of dipoles required to $\frac{1}{2}$.

From the table of Sec. 2-24 the effective aperture of a resonant $\lambda/2$ dipole is $0.13\lambda^2$. If the array is backed by a suitable reflector (solid or grid) the effective aperture of each dipole can be doubled (approximately) to $0.26\lambda^2$, which is about the same as an area $\lambda/2$ square, as shown shaded for one $\lambda/2$ dipole in Fig. 16-3. It is apparent that the aperture efficiency of the array should be about 100 percent.[5]

This efficiency implies linear polarization. However, solar radiation is *unpolarized* (equivalent to 2 orthogonal polarizations[6] in random phase), so that the aperture efficiency for sunlight is only about 50 percent. To capture the cross-polarized radiation requires an additional orthogonal array (with grid reflector) placed above the other array with all dc outputs combined. Phasing between the

---

[1] S. N. Prasad and S. Mahapatra, "A New MIC Slot-Line Aerial," *IEEE Trans. Ants. Prop.*, **AP-31**, 525–527, May 1983.

[2] J. L. Armitage, "Electronic Warfare Solid-State Phased Arrays," *Microwave J.*, **29**, 109–122, February 1986.

[3] A. L. Robinson, "Cornell Submicron Facility Dedicated," *Science*, **214**, 777–778, 1981.

[4] A. M. Marks, "Device for Conversion of Light Power to Electric Power," U.S. Patent 4,445,050, Apr. 24, 1984.

[5] Note that proper phasing for broadside operation requires that adjacent dipoles (whether $\lambda/2$ long as in Marks' design or λ long as in Fig. 16-3) be cross-connected. An alternative is to space the dipoles by 1λ and have 2 such identical arrays, each with its own transmission line and rectifiers, one array above (or below) and staggered $\lambda/2$ with respect to the other array. The reflector could be spaced $\lambda/4$ and $3\lambda/4$ respectively from the 2 arrays.

[6] Two orthogonal linear polarizations or left- and right-circular polarizations. Note that an *unpolarized* wave is not the same as a single circularly polarized wave.

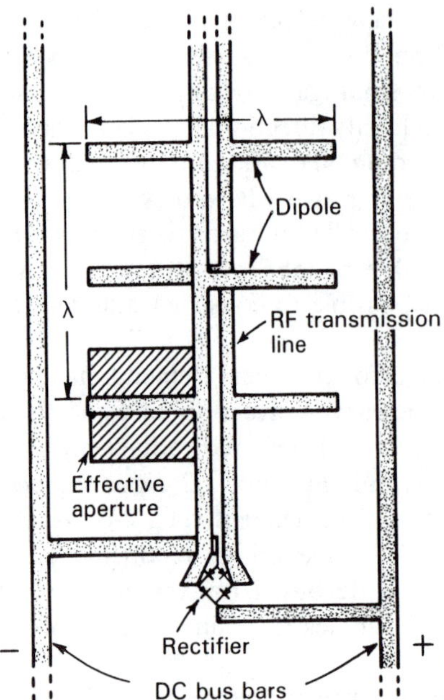

Figure 16-3 Dipole array for direct conversion of light to dc power. The wavelength $\lambda \simeq 600$ nm.

arrays is of no concern. With such a combination array, 100 percent aperture efficiency is possible, in principle. Although transmission line and rectifier loss could reduce the overall light-to-dc efficiency to 75 or 80 percent, this is several times the efficiency of present photocell devices. Since the array may be imbedded in a dielectric, the dipole dimensions for resonance will be less than in air or less than about 500 nm for the center-fed full-λ dipole.

Although not operating in a strict sense as an antenna, the rods and cones of the retina of the human eye are very similar in design to a polyrod antenna (see Sec. 14-8), with conversion of photon to dc impulses transmitted via bipolar cells, dendrites and axons (transmission lines) to the brain for signal processing and image recognition.[1] The retina contains an array of over 100 million rods and cones. A typical cone is 50 μm long with diameter tapering from about 3 μm at one end to 1 μm at the other. At a light wavelength of 500 nm ($\frac{1}{2}$ μm) these dimensions correspond to a length of 100λ and diameters of 6 to 2λ. It is interesting that the index of refraction of the cone medium is nearly the same as employed in typical commercial optical fibers ($n \simeq 1.45$).

16-4 ANTENNA SITING AND THE EFFECT OF TYPICAL (IMPERFECT) GROUND.

In Secs. 11-7 and 11-8 the vertical plane patterns for horizontal and vertical antennas were calculated assuming that the

---

[1] J. D. Kraus, *Electromagnetics*, 3rd ed., McGraw-Hill, 1984, p. 596.
J. M. Enoch and F. L. Tobey (eds.), *Vertebrate Photoreceptor Optics*, Springer Verlag, 1981.

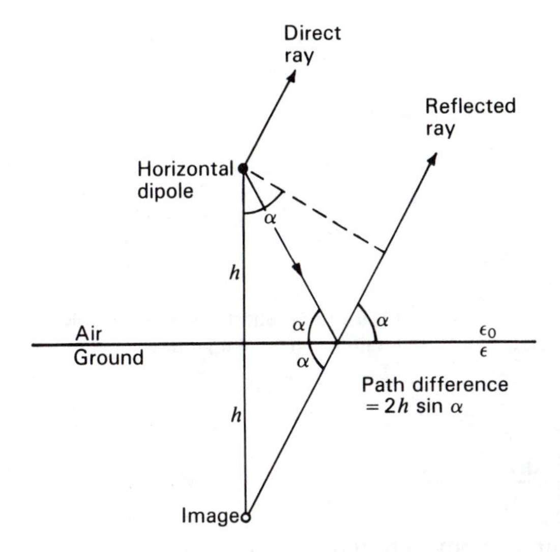

Figure 16-4 Geometry for horizontal dipole at height h above a flat earth or ground.

ground was perfectly conducting ($\sigma = \infty$). Let us consider now the more general situation of a flat ground which is not perfectly conducting, considering first the case for *horizontal polarization.* Referring to Fig. 16-4, the electric field from the horizontal dipole is perpendicular ($\perp$) to the page or plane of incidence. Assuming the ground is nonmagnetic ($\mu = \mu_0$), it may be shown that the reflection coefficient ($\rho_\perp$) is given by[1]

$$\rho_\perp = \frac{\sin\alpha - \sqrt{\varepsilon_r - \cos^2\alpha}}{\sin\alpha + \sqrt{\varepsilon_r - \cos^2\alpha}} \tag{1}$$

where ε_r = relative permittivity of the ground = $\varepsilon/\varepsilon_0$ where
ε = ground permittivity
ε_0 = air (or vacuum) permittivity
α = elevation angle (=90° minus angle of incidence measured from the zenith)

If $\varepsilon_r \gg 1$, (1) reduces to

$$\rho_\perp = \frac{\sin\alpha - n}{\sin\alpha + n} \tag{2}$$

where $n = \sqrt{\varepsilon_r}$ = index of refraction

The relative electric field at a large distance is the resultant of the direct ray from the dipole and the ray reflected from the ground, as suggested in Fig. 16-4, as

[1] See, for example, J. D. Kraus, *Electromagnetics*, 3rd ed., McGraw-Hill, 1984, p. 515.

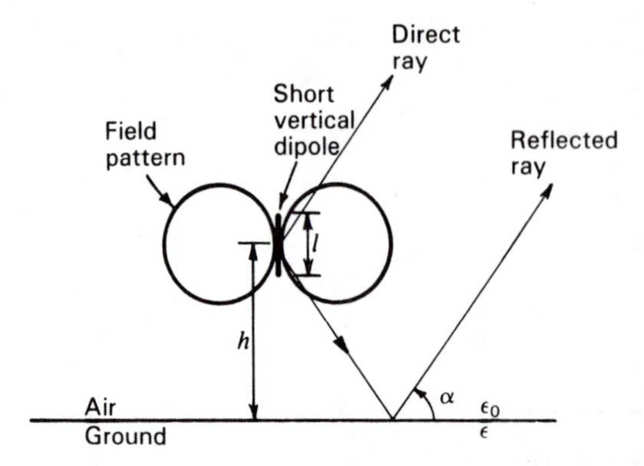

Figure 16-5 Short vertical dipole at height h above a flat ground.

given by

$$E_\perp = 1 + p_\perp \underline{/2\beta h \sin \alpha} \tag{3}$$

where $\beta = 2\pi/\lambda$
h = height of horizontal dipole above ground
α = elevation angle
$\rho_\perp$ = reflection coefficient
$2\beta h \sin \alpha$ = path length difference of direct and reflected rays, rad

In its complex form

$$\varepsilon_r = \varepsilon_r' - j\varepsilon_r'' = \varepsilon_r' - j\frac{\sigma}{\omega\varepsilon_0} \tag{4}$$

From (2) we note that if the ground is: (1) perfectly conducting ($\sigma = \infty$) or (2) a lossless dielectric ($\sigma = 0$) of large relative permittivity ($\varepsilon_r' \gg 1$),

$$\rho_\perp = -1 \tag{5}$$

and the patterns for the two cases are the same.

Consider now the situation for *vertical polarization.* Referring to Fig. 16-5, the electric field from the short vertical dipole is parallel (‖) to the page or plane of incidence. Assuming that the ground is nonmagnetic ($\mu = \mu_0$), it may be shown that the reflection coefficient ($\rho_\parallel$) is given by[1]

$$\rho_\parallel = \frac{\varepsilon_r \sin \alpha - \sqrt{\varepsilon_r - \cos^2 \alpha}}{\varepsilon_r \sin \alpha + \sqrt{\varepsilon_r - \cos^2 \alpha}} \tag{6}$$

If $\varepsilon_r \gg 1$, (6) reduces to

$$\rho_\parallel = \frac{n \sin \alpha - 1}{n \sin \alpha + 1} \tag{7}$$

where $n = \sqrt{\varepsilon_r}$ = index of refraction

[1] See, for example, J. D. Kraus, *Electromagnetics*, 3rd ed., McGraw-Hill, 1984, p. 518.

As with the horizontal dipole, the field at a large distance is the resultant of the direct ray from the vertical dipole and the ray reflected from the ground, as suggested in Fig. 16-5, as given by

$$E_{\parallel} = \cos\alpha[1 + \rho_{\parallel}\,\underline{/2\beta h \sin\alpha}] \tag{8}$$

where $\beta = 2\pi/\lambda$
h = height of center of short vertical dipole above ground
α = elevation angle
$\rho_{\parallel}$ = reflection coefficient
$2\beta h \sin\alpha$ = path length difference of direct and reflected rays, rad

From (7) we note that if the ground is: (1) perfectly conducting ($\sigma = \infty$) or (2) a lossless dielectric ($\sigma = 0$) of large relative permittivity ($\varepsilon_r' \gg 1$),

$$\rho_{\parallel} = +1 \tag{9}$$

and the patterns for the two cases are similar, except at small values of $\sin\alpha$.

To illustrate the significance of the above relations let us consider several important cases.[1]

Example 1. A horizontal $\lambda/2$ dipole is situated $\lambda/2$ above a homogeneous flat ground of rich, clay soil typical of many midwestern U.S. states ($\varepsilon_r' = 16$, $\sigma = 10^{-2}\ \mho\ \text{m}^{-1}$). Calculate and plot the vertical-plane electrical field pattern broadside to the dipole at (*a*) 1 MHz and (*b*) 100 MHz.

Solution. From (4), the loss term of the relative permittivity at 1 MHz is equal to

$$\varepsilon_r'' = \frac{\sigma}{\omega\varepsilon_0} = \frac{10^{-2}}{2\pi \times 10^6 \times 8.85 \times 10^{-12}} = 180$$

which is large compared to the dielectric term $\varepsilon_r' = 16$. However, at 100 MHz, $\varepsilon_r'' = 1.8$ and is small compared to $\varepsilon_r' = 16$.

(*a*) *1 MHz case.* Putting $\varepsilon_r = 16 - j180$ in (2) for $\rho_{\perp}$ as a function of the elevation angle α and then in (3) with $h = \lambda/2$, the electric field pattern as a function of α is given by the solid curve in Fig. 16-6. The loss permittivity ε_r'' is sufficiently large at 1 MHz that the pattern is essentially the same as for a perfectly conducting ground.

(*b*) *100 MHz case.* Putting $\varepsilon_r = 16 - j1.8$ in (2) for $\rho_{\perp}$ as a function of α and then in (3) with $h = \lambda/2$, the pattern is as shown by the dashed curve in Fig. 16-6.

Comparing the 2 cases, we note that at 1 MHz the pattern has a vertical null (at $\alpha = 90°$) which is filled in at 100 MHz and also that the gain is greater at the lower frequency (up 1 dB at $\alpha = 30°$).

---

[1] For table of ground and water constants, see App. A, Sec. A-6.

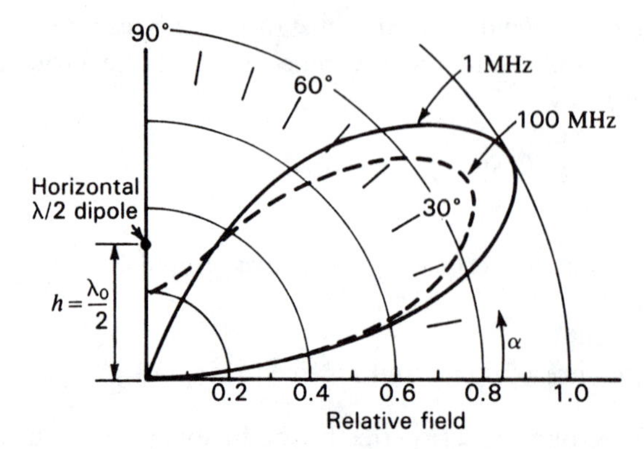

Figure 16-6 Vertical plane patterns for Example 1 of horizontal $\lambda/2$ dipole $\lambda/2$ above a flat earth with $\varepsilon_r' = 16$ and $\sigma = 10^{-2}$ ℧ m^{-1} at 1 MHz (solid curve) and at 100 MHz (dashed curve). The pattern for perfectly conducting ground ($\sigma = \infty$) is essentially the same as the solid curve. Patterns to left ($90° < \alpha < 180°$) are mirror images.

Example 2. A short vertical dipole ($l \leq \lambda/10$) [$E(\theta) = \cos\alpha$] is located $\lambda/2$ above a ground with the same constants as for Example 1. Calculate and plot the electric field pattern at (*a*) 1 MHz and (*b*) 100 MHz.

Solution

(*a*) *1 MHz case.* Putting $\varepsilon_r = 16 - j180$ in (7) for $\rho_{\parallel}$ as a function of α and then in (8) with $h = \lambda/2$, the pattern is given by the dashed curve in Fig. 16-7.

(*b*) *100 MHz case.* Putting $\varepsilon_r = 16 - j1.8$ in (7) for $\rho_{\parallel}$ and then in (8) with $h = \lambda/2$, the pattern is as shown by the dotted curve in Fig. 16-7. The solid pattern is for perfectly conducting ground ($\sigma = \infty$) (same at all frequencies).

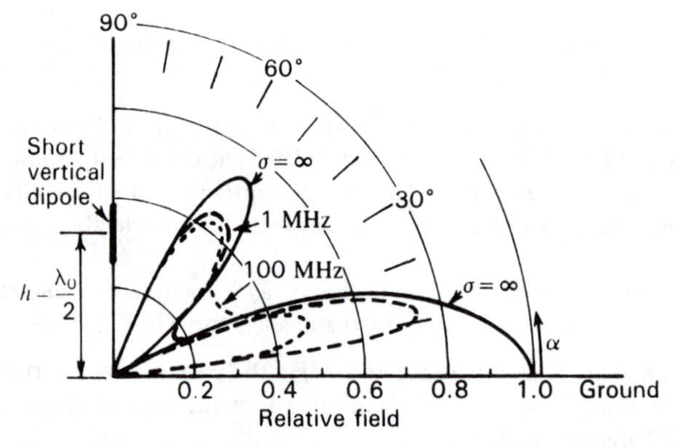

Figure 16-7 Vertical plane patterns for Example 2 of short vertical dipole $\lambda/2$ above a flat earth with $\varepsilon_r' = 16$ and $\sigma = 10^{-2}$ ℧ m^{-1} at 1 MHz (dashed curve) and at 100 MHz (dotted curve). The pattern for perfectly conducting ground ($\sigma = \infty$), as shown by the solid curve, is the same at all frequencies. Patterns to left ($90° < \alpha < 180°$) are mirror images.

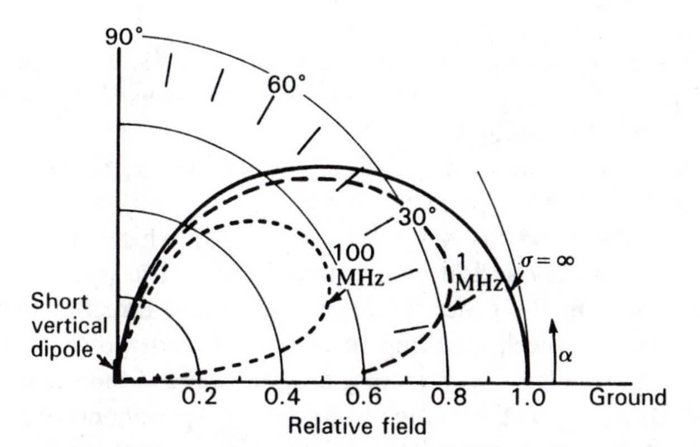

Figure 16-8 Vertical plane patterns for Example 3 of short vertical dipole at the surface of a flat earth with $\varepsilon_r' = 16$ and $\sigma = 10^{-2}$ ℧ m^{-1} at 1 MHz (dashed curve) and at 100 MHz (dotted curve). The pattern for perfectly conducting ground (same at all frequencies) is shown by the solid curve. Patterns to left ($90° < \alpha < 180°$) are mirror images.

Example 3. Repeat Example 2 for the case where the vertical dipole is at the ground (vertical monopole) so that we may set $h = 0$.

Solution. Since $h = 0$, (8) reduces to

$$E = \cos \alpha (1 + \rho_{\parallel}) \tag{10}$$

and introducing (7),

$$E = \cos \alpha \frac{2n \sin \alpha}{n \sin \alpha + 1} \tag{11}$$

Evaluating (11) for the 2 frequencies results in the curves shown in Fig. 16-8. The pattern for 1 MHz is given by the dashed curve and at 100 MHz by the dotted curve. The solid pattern is for perfectly conducting ground ($\sigma = \infty$).

At 100 MHz ($\varepsilon_r'' = 1.8$), the field along the ground ($\alpha = 0°$) is zero and the maximum field is down almost 5 dB from the field for perfectly conducting ground ($\sigma = \infty$). At 1 MHz ($\varepsilon_r'' = 180$), the situation is intermediate with the maximum field down about 2 dB from the perfectly conducting case. For higher ground conductivity the pattern approaches the $\sigma = \infty$ curve.

The approximate solutions in the above examples involve the assumption that $\varepsilon_r \gg 1$. More rigorous, and necessarily more complex, studies have been conducted, beginning with Arnold Sommerfeld's classic solution of 1909.[1] Nevertheless, our examples illustrate some of the principle changes caused by typical ground as compared to perfectly conducting ground.

Let us summarize some of the principal differences of nonperfectly versus perfectly conducting ground. With a perfectly conducting ground the reflected-ray amplitude is equal to the direct-ray amplitude (Fig. 16-4), which means that in some directions the 2 fields may add in phase, doubling the field (quadrupling the power) for a 6-dB gain. On the other hand, in some other directions the 2 rays may be out of phase or cancel, resulting in zero field (zero power) for an infinite dB loss. Thus, with a perfectly conducting ground, there is the possibility of anything from a $+6$ dB to a $-\infty$ dB change from the free-space condition.

With nonperfectly conducting ground the reflected-ray amplitude tends to be *less than* the direct-ray amplitude. Thus, in directions for which the fields add in phase, the maximum gain is less than 6 dB but in directions for which the fields are out of phase there tends not to be complete signal cancellation (a null) except at low frequencies. Referring to Figs. 16-6, 16-7 and 16-8, the above noted trends are apparent, with less gain and filled nulls in Figs. 16-6 and 16-7 and also much reduced field strength along the ground ($\alpha = 0°$) for the vertical antennas in Figs. 16-7 and 16-8.

Comparing Figs. 16-6 and 16-7, we note that for perfectly conducting ground the maximum for horizontal polarization is at an elevation angle of 30° with a null at 0°, while for vertical polarization the null is at 30° with the maximum at 0°. If both the vertical and horizontal antennas are short dipoles, $\lambda/2$ above perfectly conducting ground, and are connected together as a George Brown turnstile to transmit circular polarization, a circularly polarized antenna will receive a constant signal as a function of elevation between 0 and 30° (no nulls, no maxima) but 6 dB below the vertical or horizontal maxima.

In the above discussion it is assumed that the direct and reflected rays are parallel (distance very large). However, if the receiving antenna is closer so that the direct and reflected rays are *not* parallel, the received signal can fluctuate many times between maxima and nulls as a function of height, linear polarization being assumed. In general, avoiding a null with linear polarization may require either raising or *lowering* the receiving antenna (see Prob. 16-14, the accompanying figure and solution). Thus, although the siting procedure of Fig. 16-9 is unorthodox, the result has some credibility.

---

[1] A. Sommerfeld, "Über die Ausbreitung der Wellen in der drahtlosen Telegraphie," *Ann. Phys.*, **28**, 665, 1909.

F. E. Terman, *Radio Engineering Handbook*, 1st ed., McGraw-Hill, 1943, sec. 10, 28 refs.

R. W. P. King, *Theory of Linear Antennas*, Harvard, 1956, chap. 7, 75 refs.

I. V. L. Lindell, E. Alanen and K. Mannersalo, "Exact Image Method for Impedance Computation of Antennas above the Ground," Helsinki University Radio Lab. Rept. S 161, 1984, 23 refs.

R. W. P. King, "Electromagnetic Surface Waves; New Formulas and Applications," *IEEE Trans. Ants. Prop.*, **AP-33**, 1204–1212, November 1985.

Figure 16-9 Siting the antenna. (*Reprinted with special permission of King Features Syndicate, Inc.*)

Another siting effect occurs when, for example, a Yagi-Uda array is located less than $\lambda/2$ above ground, the proximity detuning the elements and reducing the gain.

16-5 GROUND-PLANE ANTENNAS. Several types of ground-plane or related antennas are shown in Fig. 16-10. The type at (*a*) has a vertical $\lambda/4$ stub with a circular-sheet ground plane about $\lambda/2$ in diameter. The antenna is fed by a coaxial transmission line with the inner conductor connected to the $\lambda/4$ stub and the outer conductor terminating in the ground plane. In (*b*) the ground plane has been modified to a skirt or cone. By replacing the $\lambda/4$ stub with a disc as in (*c*), a *Kandoian discone* antenna[1] is obtained. The dimensions given are appropriate for the center frequency of operation. In Fig. 16-10*d* the solid-sheet ground plane is replaced by 4 radial conductors. A modification of this antenna is shown at (*e*) in which a short-circuited $\lambda/4$ section of coaxial line is connected in parallel with the antenna terminals.[2] This widens the impedance band width and also places the stub antenna at dc ground potential. This is desirable to protect the transmission line from lightning surges.

With reference to solid-sheet ground-plane antennas, it should be noted that the radiation pattern of a vertical $\lambda/4$ stub on a *finite* ground sheet differs appreciably from the pattern with an infinite sheet. This is illustrated by Fig. 16-11. The solid curve is the calculated pattern with a ground sheet of infinite extent. The dashed curve is for a sheet several wavelengths in diameter and the

---

[1] A. G. Kandoian, "Three New Antenna Types and Their Applications," *Proc. IRE*, **34**, 70W–75W, February 1946.
A. G. Kandoian, W. Sichak and G. A. Felsenheld, "High Gain with Discone Antennas," *Proc. Natl. Electronics Conf.*, **3**, 318–328, 1947.

[2] These radial conductor ground-plane antennas were invented in 1938 by George H. Brown, Jess Epstein and Robert Lewis.

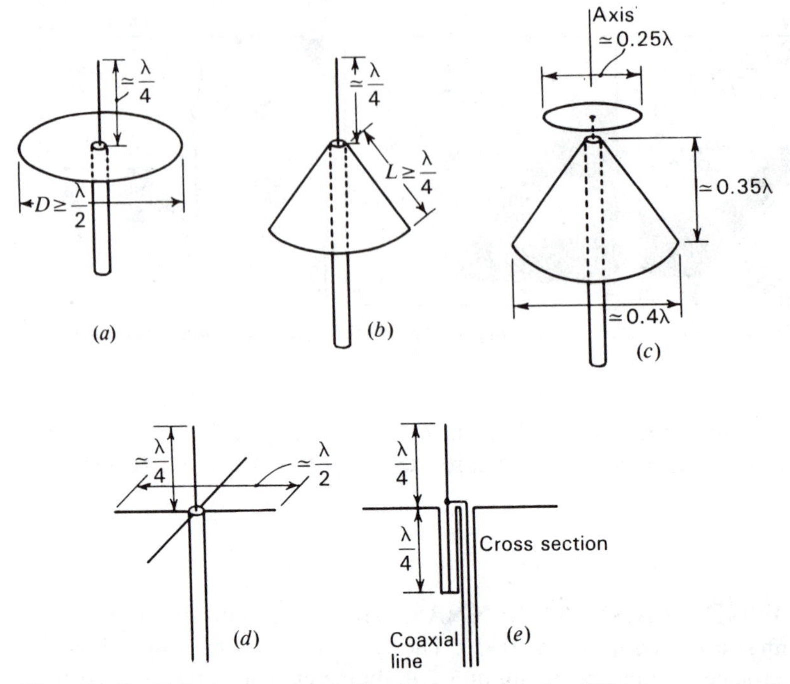

Figure 16-10 (*a*) Stub antenna with flat circular ground plane, (*b*) same antenna with ground plane modified to skirt or cone, (*c*) Kandoian discone antenna, (*d*) stub antenna with 4 radial conductors to simulate ground plane and (*e*) a method of feeding ground-plane antenna.

dotted curve for a sheet of the order of 1λ in diameter. With finite solid-sheet ground planes the maximum radiation is generally not in the direction of the ground plane but at an angle α above it. In order that maximum radiation be in the horizontal plane, the ground plane may be modified as in Fig. 16-10*b* or *c*. The maximum radiation from the Kandoian discone antenna is nearly horizontal (normal to axis) over a considerable bandwidth.

By top-loading a vertical stub antenna, it may be modified through the successive stages of Fig. 16-12 to the form in Fig. 16-12*d*. This antenna consists of a circular disc with an annular slot between it and the ground plane. The ground

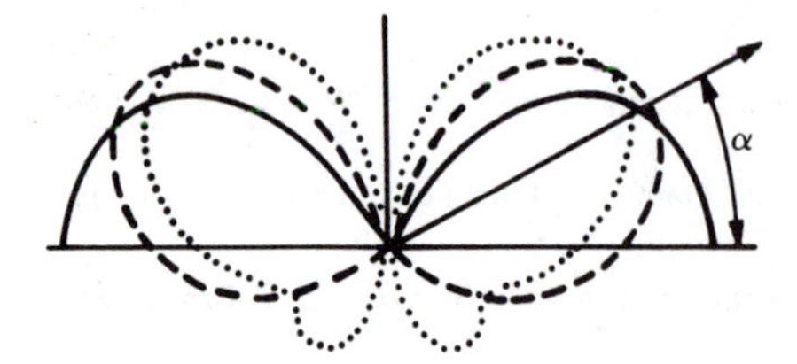

Figure 16-11 Vertical-plane patterns of $\lambda/4$ stub antenna on infinite ground plane (solid), and on finite ground planes several wavelengths in diameter (dashed) and about 1λ in diameter (dotted).

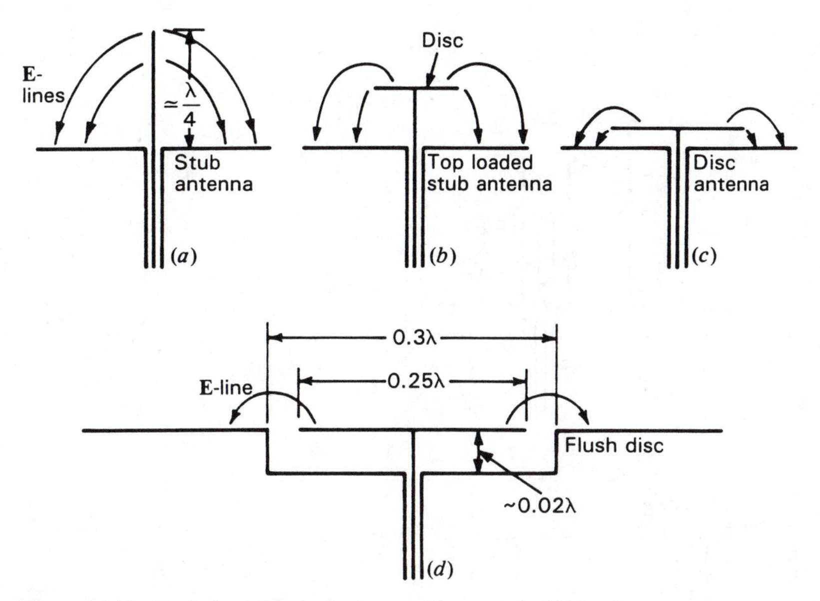

Figure 16-12 Evolution of flush-disc antenna from vertical $\lambda/4$ stub antenna.

plane is depressed below the disc forming a shallow cavity.[1,2] The radiation pattern of the antenna at (*d*) is quite similar to the pattern for the vertical stub at (*a*).[2]

16-6 SLEEVE ANTENNAS.

Carrying the ground-plane modification of Fig. 16-10*b* a step further results in the vertical $\lambda/2$ sleeve antenna of Fig. 16-13*a*. Here the ground plane has degenerated into a sleeve or cylinder $\lambda/4$ long. Maximum radiation is normal to the axis of this antenna.

Another variety of sleeve antenna is illustrated in Fig. 16-13*b*.[3] The antenna is similar to a stub antenna with ground plane but with the feed point moved to approximately the center of the stub. This is accomplished by enclosing the lower end of the stub in a cylindrical sleeve. By varying the characteristic impedance of this $\lambda/8$ section, some control is afforded over the impedance presented to the coaxial line at the ground plane.

A balanced-sleeve dipole antenna corresponding to the sleeve stub type of Fig. 16-13*b* is illustrated in Fig. 16-13*c*. It is shown with a coaxial line feed and balance-to-unbalance transformer or balun.[3] This antenna may be operated over

---

[1] A. A. Pistolkors, "Theory of Circular Diffraction Antenna," *Proc. IRE*, **36**, 56–60, January 1948.

[2] D. R. Rhodes, "Flush-Mounted Antenna for Mobile Application," *Electronics*, **22**, 115–117, March 1949.

[3] E. L. Bock, J. A. Nelson and A. Dorne, in *Very High Frequency Techniques*, Radio Research Laboratory Staff, McGraw-Hill, New York, 1947, chap. 5.

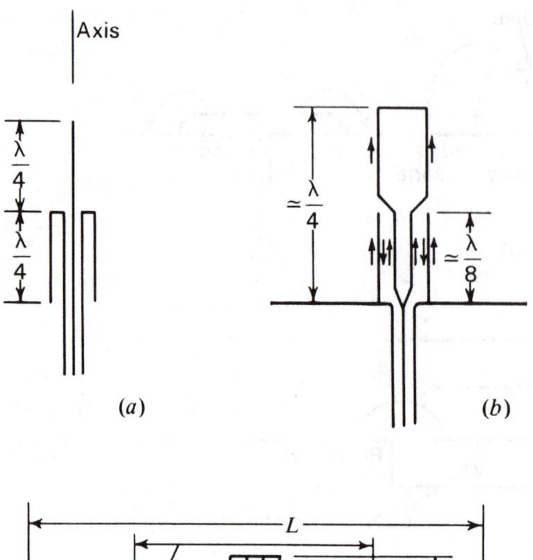

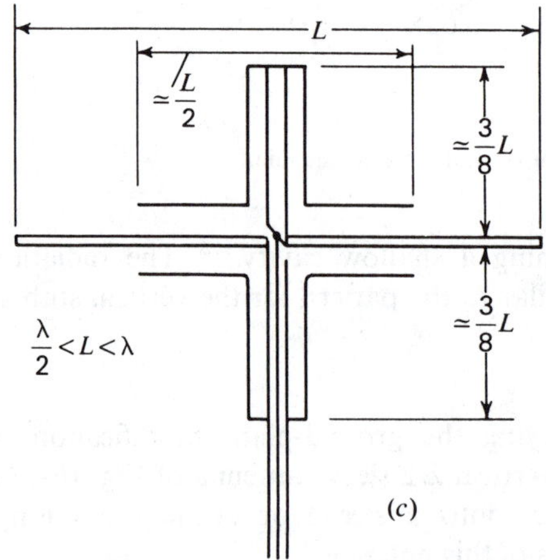

Figure 16-13 (*a*) $\lambda/2$ sleeve antenna, (*b*) sleeve antenna above ground plane and (*c*) balanced-sleeve antenna.

a frequency range of about 2 to 1 such that L is in the range from about $\frac{1}{2}$ to 1λ. A corner reflector with this type of sleeve is shown in Fig. 12-15.

16-7 TURNSTILE ANTENNA. Consider 2 crossed infinitesimal dipoles energized with currents of equal magnitude but in phase quadrature. This arrangement, shown in plan view in Fig. 16-14*a*, produces a circular pattern in the θ plane since the field pattern E as a function of θ and time is given by

$$E = \sin\theta \cos\omega t + \cos\theta \sin\omega t \tag{1}$$

which reduces to

$$E = \sin(\theta + \omega t) \tag{2}$$

At any value of θ the maximum amplitude of E is unity at some instant during each cycle. Hence, the rms field pattern is circular, as shown by the circle in Fig. 16-14*b*. At any instant of time the pattern is a figure-of-eight of the same

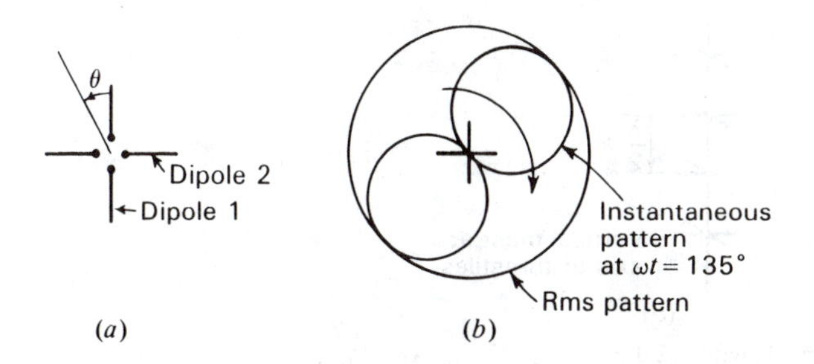

Figure 16-14 George Brown turnstile with short (infinitesimal) dipoles.

shape as for a single infinitesimal dipole. An instantaneous pattern is shown in Fig. 16-14*b* for $\omega t = 135°$. As a function of time this pattern rotates, completing 1 revolution per cycle. In the case being considered in Fig. 16-14, the pattern rotates clockwise. Thus, the phase of the field as a function of θ is given by $\theta + \omega t =$ constant and, if the constant is zero, by

$$\omega t = -\theta \tag{3}$$

If the field is a maximum in the direction $\theta = 0$ at a given instant, then according to (3) the field is a maximum in the $\theta = -45°$ direction $\frac{1}{8}$-period later.

The above discussion concerns the field in the θ plane (plane of the crossed dipoles). The field in the axial direction (normal to the crossed infinitesimal dipoles) has a constant magnitude given by

$$|E| = \sqrt{\cos^2 \omega t + \sin^2 \omega t} = 1 \tag{4}$$

Thus, the field normal to the infinitesimal dipoles is circularly polarized. In the case being considered in Fig. 16-14 the field rotates in a clockwise direction.

Replacing the infinitesimal dipoles by $\lambda/2$ dipoles results in a practical type of antenna with approximately the same pattern characteristics. This kind of antenna is a *George Brown turnstile antenna.*[1] Since the pattern of a $\lambda/2$ element is slightly sharper than for an infinitesimal dipole, the θ-plane pattern of the turnstile with $\lambda/2$ elements is not quite circular but departs from a circle by about ± 5 percent. The relative pattern is shown in Fig. 16-15*a*. The relative field as a function of θ and time is expressed by

$$E = \frac{\cos(90° \cos\theta)}{\sin\theta} \cos\omega t + \frac{\cos(90° \sin\theta)}{\cos\theta} \sin\omega t \tag{5}$$

Although the θ-plane pattern with $\lambda/2$ elements differs from the pattern with infinitesimal dipoles, the radiation is circularly polarized in the axial direction from the $\lambda/2$ elements provided that the currents are equal in magnitude and in phase quadrature.

A turnstile antenna may be conveniently mounted on a vertical mast. The mast is coincident with the axis of the turnstile. To increase the vertical plane

[1] G. H. Brown, "The Turnstile Antenna," *Electronics*, **9**, 15, April 1936.

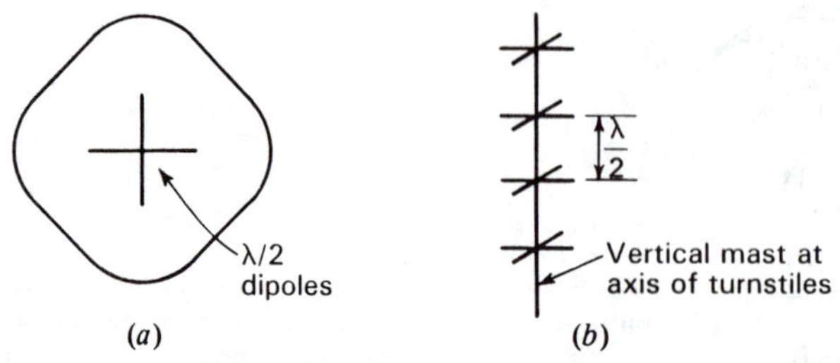

Figure 16-15 George Brown turnstile with $\lambda/2$ dipoles.

directivity, several turnstile units can be stacked at about $\lambda/2$ intervals as in Fig. 16-15*b*. The arrangement at (*b*) is called a "4-bay" turnstile. It requires 2 bays to obtain a field intensity approximately equal to the maximum field from a single $\lambda/2$ dipole with the same power input.

In order that the currents on the $\lambda/2$ dipoles be in phase quadrature, the dipoles may be connected to separate nonresonant lines of unequal length. Suppose, for example, that the terminal impedance of each dipole in a single-bay turnstile antenna is $70 + j0\ \Omega$. Then by connecting 70-Ω lines (dual coaxial type), as in the schematic diagram of Fig. 16-16*a*, with the length of one line 90 electrical degrees longer than the other, the dipoles will be driven with currents of equal magnitude and in phase quadrature. By connecting a 35-Ω line between the junction point P of the two 70-Ω lines and the transmitter, the entire transmission-line system is matched.

Another method of obtaining quadrature currents is by introducing reactance in series with one of the dipoles.[1] Suppose, for example, that the length and diameter of the dipoles in Fig. 16-16*b* result in a terminal impedance of $70 - j70\ \Omega$. By introducing a series reactance (inductive) of $+j70\ \Omega$ at each ter-

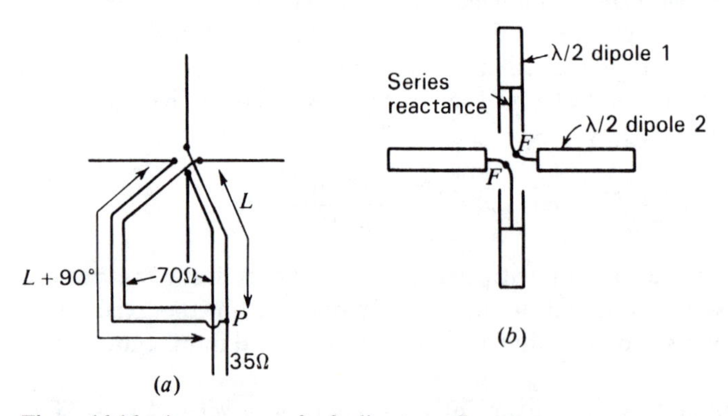

Figure 16-16 Arrangements for feeding turnstile antennas.

[1] G. H. Brown and J. Epstein, "A Pretuned Turnstile Antenna," *Electronics*, **18**, 102–107, June 1945.

minal of dipole 1 as in Fig. 16-16*b*, the terminal impedance of this dipole becomes $70 + j70\ \Omega$. With the 2 dipoles connected in parallel, the currents are

$$I_1 = \frac{V}{70 + j70}$$

and
$$I_2 = \frac{V}{70 - j70} \tag{6}$$

where V = impressed emf
I_1 = current at terminals of dipole 1
I_2 = current at terminals of dipole 2

Thus,

$$I_1 = \frac{V}{99} \angle -45^\circ$$

and
$$I_2 = \frac{V}{99} \angle +45^\circ \tag{7}$$

so that I_1 and I_2 are equal in magnitude but I_2 leads I_1 by 90°. The 2 impedances in parallel yield

$$Z = \frac{1}{Y} = \frac{1}{[1/(70 + j70)] + [1/(70 - j70)]} = 70 + j0 \qquad (\Omega) \tag{8}$$

so that a 70-Ω (dual coaxial) line will be properly matched when connected to the terminals FF.

16-8 SUPERTURNSTILE ANTENNA. In order to obtain a very low VSWR over a considerable bandwidth, the turnstile described above has been modified by Masters[1] to the form shown in the photograph of Fig. 16-17. In this arrangement, or *Masters superturnstile*, the simple dipole elements are replaced by flat sheets or their equivalent.

Each "dipole" is equivalent to a slotted sheet about 0.7 by 0.5λ as in Fig. 16-18*a*. The terminals are at FF. As in the slotted cylinder antenna, the length of the slot for resonance is more than $\lambda/2$ (about 0.7λ). The dipole can be mounted on a mast as in Fig. 16-18*b*. To reduce wind resistance, the solid sheet is replaced by a grid of conductors. Typical dimensions for the center frequency of operation are shown. This arrangement gives a VSWR of about 1.1 or less over about a 30 percent bandwidth, which makes it convenient as a mast-mounted television transmitting antenna for frequencies as low as about 50 MHz. Unlike the simple turnstile there is relatively little radiation in the axial direction (along the mast), and only one bay is required to obtain a field intensity approximately

[1] R. W. Masters, "The Super-turnstile Antenna," *Broadcast News*, **42**, January 1946.

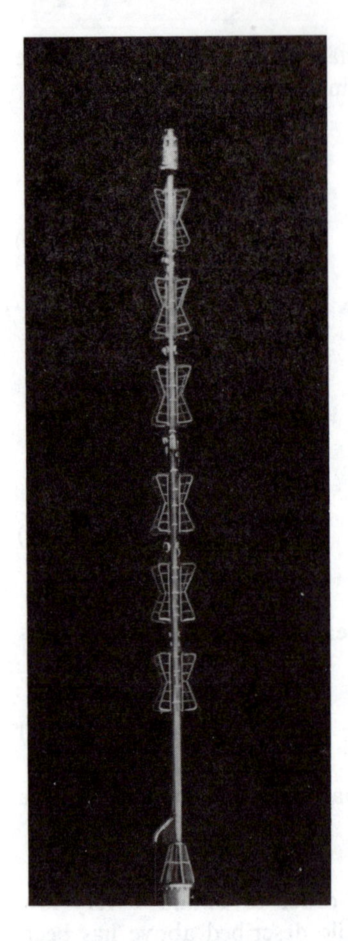

Figure 16-17 Six-bay Masters superturnstile antenna. (*Courtesy RCA.*)

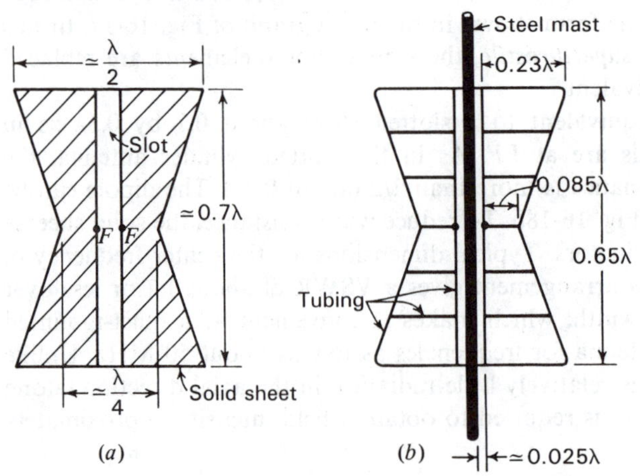

Figure 16-18 Single dipole element of Masters superturnstile antenna. (*a*) Solid-sheet construction, (*b*) tubing construction showing method of mounting on mast.

equal to the maximum field from a single $\lambda/2$ dipole with the same power input. For decreased beam width in the vertical plane the superturnstile bays are stacked at intervals of about 1λ between centers. Impedance matching is discussed by Sato *et al.*[1]

16-9 OTHER OMNIDIRECTIONAL ANTENNAS.

The radiation patterns of the slotted-cylinder and turnstile antennas are nearly circular in the horizontal plane. Such antennas are sometimes referred to as omnidirectional types, it being understood that "omnidirectional" refers only to the horizontal plane.

As shown in Chap. 6, a circular loop with a uniform current radiates a maximum in the plane of the loop provided that the diameter D is less than about 0.58λ. The pattern is doughnut shaped with a null in the axial direction as suggested by the vertical plane cross section in Fig. 16-19*a*.

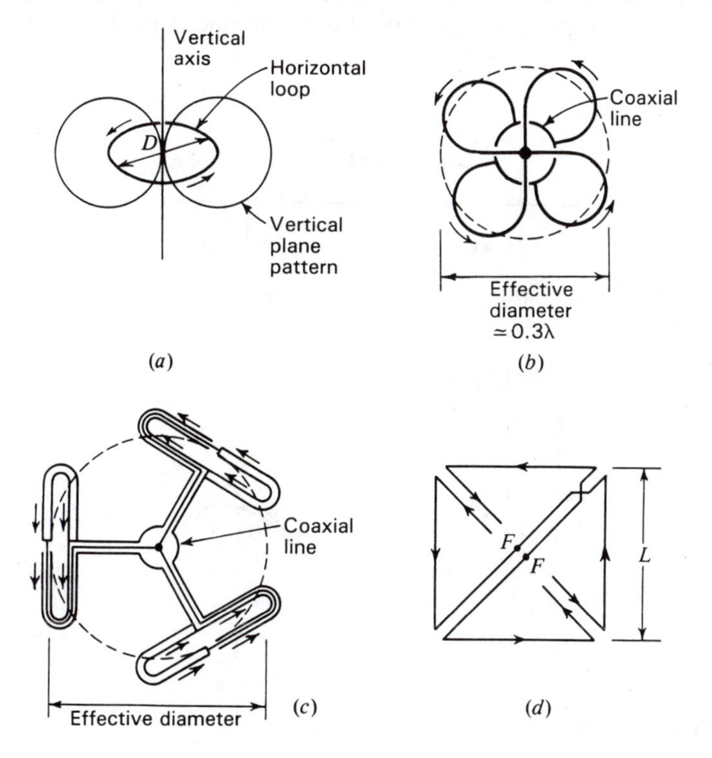

Figure 16-19 (*a*) Circular loop antenna and (*b*) approximately equivalent arrangements of "cloverleaf" type, (*c*) "triangular-loop" type and (*d*) square or Alford loop.

[1] G. Sato, H. Kawakami, H. Sato and R. W. Masters, "Design Method for Fine Impedance Matching of Superturnstile Antenna and Characteristics of the Modified Batwing Antenna," *Trans. IECE (Japan)*, **E65**, 271–278, May 1982.

One method of simulating the uniform loop is illustrated in Fig. 16-19*b*. Here 4 smaller loops are connected in parallel across a coaxial line. This arrangement is called a "cloverleaf" antenna.[1] Another method is shown in Fig. 16-19*c*, 3 folded dipoles being connected in parallel across a coaxial line.[2] A third method utilizing a square loop is illustrated in Fig. 16-19*d*.[3] The terminals are at FF. The side length L may be of the order of $\lambda/4$. A single equivalent loop or bay of any of these types produces approximately the same field intensity as the maximum field from a single $\lambda/2$ dipole with the same power input. For increased directivity in the vertical plane, several loops may be stacked, forming a multibay arrangement.

16-10 CIRCULARLY POLARIZED ANTENNAS. Circularly polarized radiation may be produced with various antennas. The monofilar axial-mode helical antenna (Fig. 16-20*a*) is a simple, effective type of antenna for generating circular polarization. Circular polarization may also be produced in the axial

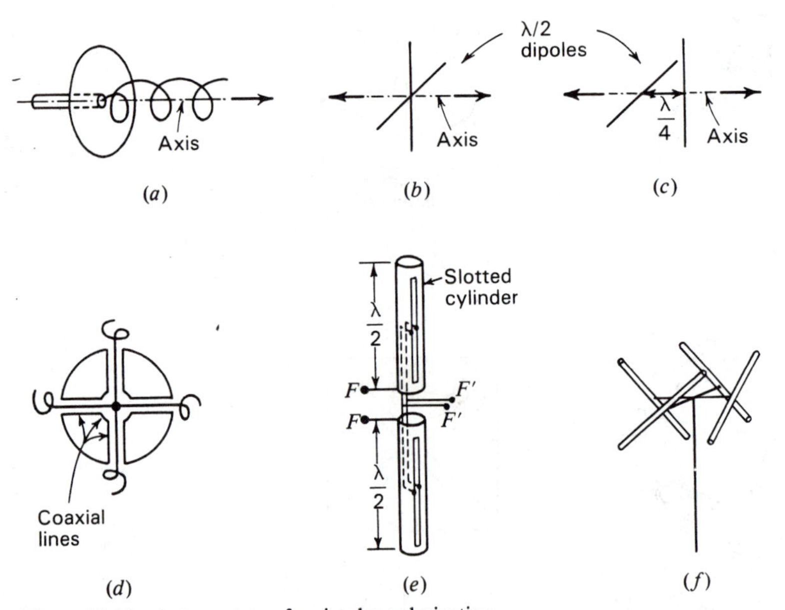

Figure 16-20 Antenna types for circular polarization.

[1] P. H. Smith, "Cloverleaf Antenna for FM Broadcasting," *Proc. IRE*, **35**, 1556–1563, December 1947.

[2] A. G. Kandoian and R. A. Felsenheld, "Triangular High-Band TV Loop Antenna System," *Communications*, **29**, 16–18, August 1949.

[3] A. Alford and A. G. Kandoian, "Ultra-high Frequency Loop Antennas," *Trans. AIEE*, **59**, 843–848, 1940.

direction from a pair of crossed $\lambda/2$ dipoles with equal currents in phase quadrature (Fig. 16-20*b*) as in a turnstile antenna. If radiation in one axial direction is right-circularly polarized, it is left-circularly polarized in the opposite axial direction.

A third type of circularly polarized antenna consists of 2 in-phase crossed dipoles separated in space by $\lambda/4$ as in Fig. 16-20*c*. With this arrangement the type of circular polarization is the same in both axial directions.

Any of these 3 arrangements can serve as a primary antenna that illuminates a parabolic reflector or they can be placed within a circular waveguide so as to generate a circularly polarized TE_{11} mode wave. By flaring the guide out into a conical horn, a circularly polarized beam can be produced.

Another technique by which a circularly polarized beam may be obtained with a parabolic reflector of large focal length with respect to the diameter is with the aid of a metal grid or grating of parallel wires spaced $\lambda/8$ from the reflector and oriented at 45° with respect to the plane of polarization of the wave from a linearly polarized primary antenna.

Three arrangements for producing an omnidirectional pattern of circularly polarized radiation are illustrated by Fig. 16-20*d*, *e* and *f*. At (*d*) 4 short axial-mode helices of the same type are disposed around a metal cylinder with axis vertical and fed in phase from a central coaxial line.[1] In the system at (*e*) vertically polarized omnidirectional radiation is obtained from two vertical $\lambda/2$ cylinders when fed at FF and horizontally polarized omnidirectional radiation is obtained from the slots fed at $F'F'$. By adjusting the power and phasing to the 2 sets of terminals so that the vertically polarized and horizontally polarized fields are equal in magnitude and in phase quadrature, a circularly polarized omnidirectional pattern is produced.[2] At (*f*) 4 in-phase $\lambda/2$ dipoles are mounted around the circumference of an imaginary circle about $\lambda/3$ in diameter.[3] Each dipole is inclined to the horizontal plane as suggested in the figure.

In general, any linearly polarized wave can be transformed to an elliptically or circularly polarized wave, or vice versa, by means of a wave polarizer.[4] For example, assume that a linearly polarized wave is traveling in the negative z direction in Fig. 16-21 and that the plane of polarization is at a 45° angle with respect to the positive x axis. Suppose that this wave is incident on a large grating of many dielectric slabs of depth L with air spaces between. A section of

---

[1] J. D. Kraus, "Helical Beam Antenna for Wide-Band Applications," *Proc. IRE*, **36**, 1236–1242, October 1948.

[2] C. E. Smith and R. A. Fouty, "Circular Polarization in F-M Broadcasting," *Electronics*, **21**, 103–107, September 1948.

[3] G. H. Brown and O. M. Woodward, Jr., "Circularly-Polarized Omnidirectional Antenna," *RCA Rev.*, **8**, 259–269, June 1947.

[4] F. Braun, "Elektrische Schwingungen und drahtlose Telegraphie," *Jahrbuch der drahtlosen Telegraphie und Telephonie*, **4**, no. 1, 17, 1910.

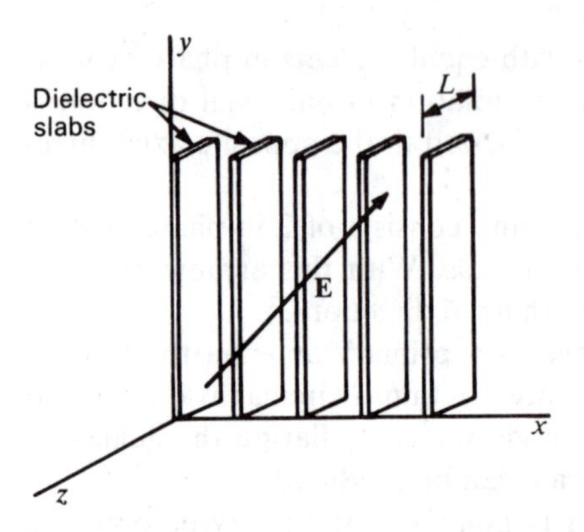

Figure 16-21 Wave polarizer.

this grating is shown in Fig. 16-21. The slab spacing (in the x direction) is assumed to be a small part of a wavelength.

The incident electric field **E** can be resolved into two components, one parallel to the x axis (E_x) and the other parallel to the y axis (E_y); that is, $\mathbf{E} = \hat{\mathbf{x}}E_x + \hat{\mathbf{y}}E_y$. The x component (E_x) will be relatively unaffected by the slabs. However, E_y will be retarded (velocity reduced). If the depth L of the slabs is just sufficient to retard E_y by 90° in time phase behind E_x, the wave emerging from the back side of the slabs will be circularly polarized if $|E_x| = |E_y|$. Viewing the approaching wave from a point on the negative z axis, the **E** vector rotates clockwise.

If the depth of the slabs is increased to $2L$, the wave emerging from the back side will again be linearly polarized since E_x and E_y are in opposite phase, but **E** is at a negative angle of 45° with respect to the positive x axis. Increasing the slab depth to $3L$ makes the emerging wave circularly polarized but this time with a counterclockwise rotation direction for **E** (as viewed from a point on the negative z axis). Finally, if the slab depth is increased to $4L$, the emerging wave is linearly polarized at a slant angle of 45°, the same as the incident wave. The dielectric grating in this example behaves in a similar way to the atomic planes of a uniaxial crystal, such as calcite or rutile, to the propagation of light. For such crystals the velocity of propagation of light, linearly polarized parallel to the optic axis, is different from the velocity for light, linearly polarized perpendicular to the optic axis.

16-11 MATCHING ARRANGEMENTS, BALUNS AND TRAPS.

Impedance matching between a transmission line and antenna may be accomplished in various ways.[1] As illustrations, several methods for matching a trans-

[1] Only arrangements with transmission-line elements will be described. These are convenient at high frequencies. However, at low or medium frequencies the length of the required transmission-line sections may be inconveniently large so that it is the usual practice to use matching circuits with lumped elements. Radio-frequency transformers and π, T and L sections are employed in this application.

mission line to a simple $\lambda/2$ dipole will be considered. Suppose that the antenna is a cylindrical dipole with a length-diameter ratio of 60 ($L/D = 60$) and that the measured terminal impedances at 5 frequencies are as follows:

Frequency	Antenna length, λ	Terminal impedance, Ω
$1.15F_0$	$L = 0.53$	$110 + j90$
$1.07F_0$	$L = 0.49$	$80 + j40$
F_0 = center frequency	$L = 0.46$	$65 + j0$
$0.93F_0$	$L = 0.43$	$52 - j40$
$0.85F_0$	$L = 0.39$	$40 - j100$

The center frequency F_0 corresponds to the resonant frequency of the antenna. At this frequency the terminal impedance is $65 + j0\ \Omega$.

The most direct arrangement for obtaining an impedance match is to feed the dipole with a dual coaxial transmission line of 65 Ω characteristic impedance as in Fig. 16-22*a*. The variation of the antenna impedance referred to 65 Ω is shown by the solid curve in the Smith chart[1] of Fig. 16-23. The normalized impedances plotted on the chart are obtained by dividing the antenna terminal impedances by 65. The VSWR on the 65-Ω line as a function of frequency and antenna length is shown by the solid curve in Fig. 16-24.

The dipole antenna may also be energized with a 2-wire open type of transmission line. Since the characteristic impedance of convenient sizes of open 2-wire line is in the range of 200 to 600 Ω, an impedance transformer is required between the line and the antenna. A suitable transformer design may be deduced as follows. Referring to Fig. 16-22*b*, the impedance Z_B at the terminals of a lossless transmission line terminated in an impedance Z_A is

$$Z_B = Z_0 \frac{Z_A + jZ_0 \tan \beta x}{Z_0 + jZ_A \tan \beta x} \tag{1}$$

where $\beta x = (2\pi/\lambda)x$ = length of line, rad
Z_0 = characteristic impedance of the transmission line (since the line is assumed to be lossless, Z_0 is a pure resistance), Ω

Equation (1) may be reexpressed as

$$Z_B = Z_0 \frac{(Z_A/\tan \beta x) + jZ_0}{(Z_0/\tan \beta x) + jZ_A} \tag{2}$$

[1] P. H. Smith, "An Improved Transmission Line Calculator," *Electronics*, **17**, 130, January 1944. See J. D. Kraus, "Electromagnetics" 3rd ed., McGraw-Hill, 1984, pp. 408–411, for Smith chart solutions of single and double stub matching.

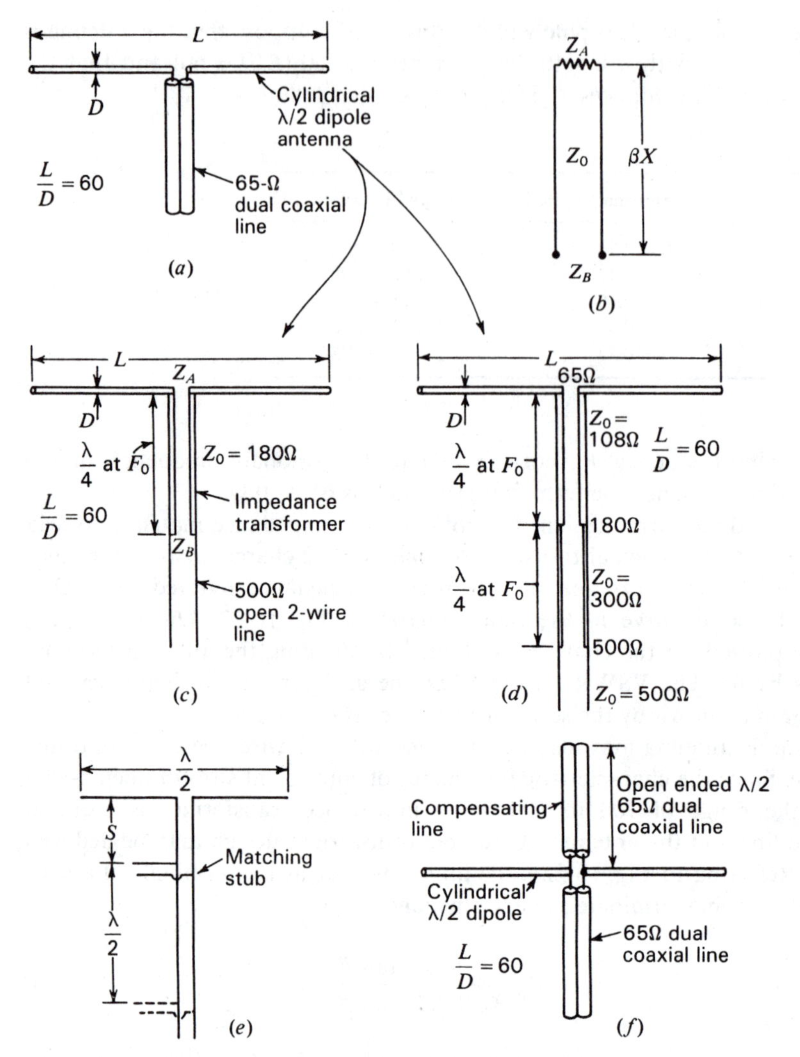

Figure 16-22 Matching arrangements for cylindrical $\lambda/2$ dipole antenna.

When the line is $\lambda/4$ long ($\beta x = 90°$), (2) reduces to

$$Z_B = \frac{Z_0^2}{Z_A} \tag{3}$$

or

$$Z_0^2 = Z_A Z_B \tag{4}$$

or

$$Z_0 = \sqrt{Z_A Z_B} \tag{5}$$

If Z_A is the antenna terminal impedance and Z_B is the characteristic impedance of the transmission line we wish to use, the two can be matched with a $\lambda/4$ section having a characteristic impedance Z_0 given by (5). The arrangement is

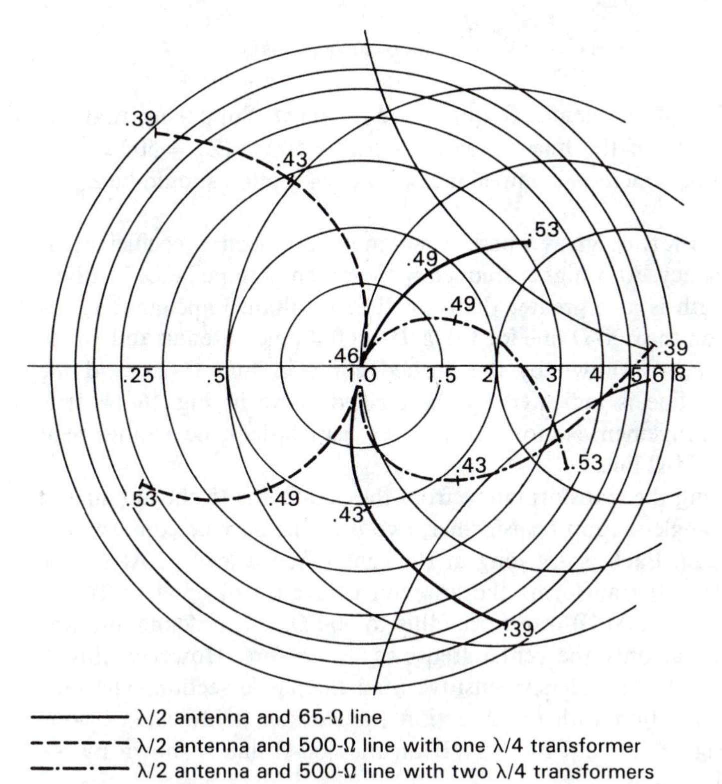

Figure 16-23 Normalized impedance variation for cylindrical $\lambda/2$ dipole antenna ($L/D = 60$) fed directly by a 65-Ω line (solid), by a 500-Ω line with one $\lambda/4$ transformer (dashed) and by a 500-Ω line with two $\lambda/4$ transformers in series (dash-dot).

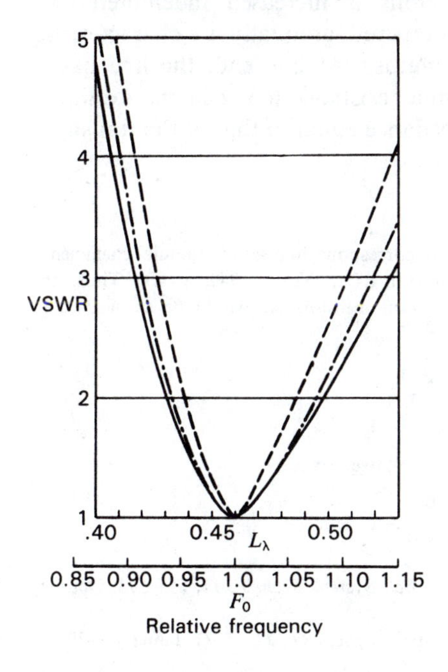

Figure 16-24 VSWR as a function of antenna length L in wavelengths and as a function of the frequency (the resonant frequency F_0 is taken as unity). The VSWR curves are for the same 3 cases of Fig. 16-23.

shown in Fig. 16-22*c*. At the center frequency, $Z_A = 65\ \Omega$. Supposing that the characteristic impedance of the line we wish to use is 500 Ω ($Z_B = 500\ \Omega$), we have from (5) that the characteristic impedance of the $\lambda/4$ section should be $Z_0 = 180\ \Omega$.

This type of transformer gives a perfect match (zero reflection coefficient) at only the center frequency. At a higher frequency the antenna impedance is different, and the line length is also greater than $\lambda/4$. The resultant impedance variation with frequency on the 500-Ω line for the $L/D = 60$ dipole antenna and 180-Ω transformer ($\lambda/4$ at F_0) is shown by the dashed curve in Fig. 16-23, and the VSWR on the 500-Ω line is indicated by the dashed curve in Fig. 16-24. It is apparent that this arrangement is more frequency sensitive than the arrangement with the dual coaxial 65-Ω line.

Instead of making the transformation from the 500-Ω line to the antenna in a single step with a single-section transformer, two sections may be connected in series as in Fig. 16-22*d*. Each is $\lambda/4$ long at the center frequency F_0. At F_0 the first section ($Z_0 = 108\ \Omega$) transforms the antenna resistance of 65 Ω to 180 Ω. The second section ($Z_0 = 300\ \Omega$) transforms this to 500 Ω. The antenna and line are perfectly matched at only the center frequency, as before. However, this 2-section arrangement is less frequency sensitive than the single section. The normalized impedance variation with the 2-section transformer is indicated by the dash-dot curve in Fig. 16-23, and the VSWR on the 500-Ω line is shown by the dash-dot curve in Fig. 16-24.

If the number of sections in the transformer is increased further, it should be possible to approach closer to the frequency sensitivity with the direct connected 65-Ω line.[1] As the number of sections is increased indefinitely, we approach in the limit a transmission line tapered gradually in characteristic impedance over a distance of many wavelengths.[2] At one end, the line has a characteristic impedance equal to the antenna resistance (65 Ω in the example) and at the other end has a characteristic impedance equal to that of the transmission line we wish to use (500 Ω in the example).

---

[1] The logarithms of the impedance ratios may be made to correspond to a set of binomial coefficients. (See J. C. Slater, *Microwave Transmission*, McGraw-Hill, New York, 1942, p. 60.) Thus, the logarithms of the impedance ratios for 2-, 3- and 4-section transformers would be as in Pascal's triangle:

2-sections: 1 : 2 : 1
3-sections: 1 : 3 : 3 : 1
4-sections: 1 : 4 : 6 : 4 : 1

In the 2-section transformer of Fig. 16-22*d* these ratios are followed since

$$\log \frac{108}{65} : \log \frac{300}{108} : \log \frac{500}{300} \simeq 1 : 2 : 1$$

[2] C. R. Burrows, "The Exponential Transmission Line," *Bell System Tech. J.*, **17**, 555–573, October 1938.
H. A. Wheeler, "Transmission Lines with Exponential Taper," *Proc. IRE*, **27**, 65–71, January 1939.

Another more frequency-sensitive method of matching a 500-Ω line to a $\lambda/2$ dipole is with a stub,[1] as shown in Fig. 16-22*e*. The line between the stub and the transmitter may be nonresonant or perfectly matched to the antenna at one frequency with the stub as shown. The stub may also be placed $\lambda/2$ farther from the antenna, as shown by the dashed lines.[2] In this case, however, the resonant line between the stub and antenna is longer, and this arrangement is more frequency sensitive than with the stub closer to the antenna. In general, it is desirable to place matching or compensating networks as close to the antenna as possible if frequency sensitivity is to be a minimum.

With the single stub as in Fig. 16-22*e* both the length of the stub and its distance S from the antenna are adjustable. The stub may be open or short-circuited at the end remote from the line, the stub length being $\lambda/4$ different for the 2 cases. To adapt this arrangement to a coaxial line requires that a line stretcher be inserted between the stub and the antenna. An alternative arrangement is a double-stub tuner which has 2 stubs at fixed distances from the antenna but with the lengths of both stubs adjustable.

The frequency sensitivity[3] of a dipole antenna may be made less than for the $L/D = 60$ dipole with a direct-connected 65-Ω line, as above, in several ways. A large-diameter dipole can be used (smaller L/D ratio) since, as shown in Chap. 9, the impedance variation with frequency is inherently less for thick dipoles as compared to thin dipoles. The thick dipole is desirable for very wide-band applications. If such a dipole is inconvenient, the impedance variation can often be reduced over a moderate bandwidth by means of a compensating network. For example, the frequency sensitivity of the $L/D = 60$ dipole with a direct-connected 65-Ω line can be reduced over a considerable bandwidth by connecting a compensating line in parallel with the antenna terminals as in Fig. 16-22*f*. If this line or stub has an electrical length of $\lambda/2$ at the center frequency and has a 65-Ω characteristic impedance, the same as the transmission line, the variation of normalized antenna terminal impedance with frequency, as referred to 65 Ω, is shown by the dash-dot curve in Fig. 16-25*a*. The variation without compensation (antenna of Fig. 16-22*a*) is given by the solid curve (same curve as in Fig. 16-23). The VSWRs on a 65-Ω line are compared in Fig. 16-25*b* for the antenna without compensation (solid curve) and with the compensating stub (dash-dot curve). The frequency sensitivity of the compensated arrangement is appreciably less over the frequency range shown. For instance, the bandwidth for VSWR ≤ 2 is about 14 percent for the uncompensated dipole but is about 18 percent for the compensated dipole.

The action of the parallel-connected compensating line or stub is as follows.

---

[1] F. E. Terman, *Radio Engineers' Handbook*, McGraw-Hill, New York, 1943, pp. 187–191. Gives design charts for open stub, closed stub and reentrant matching arrangements.

[2] In general, the distance of the stub from the antenna can be increased by $n\lambda/2$ where n is an integer.

[3] Frequency sensitivity as used here refers only to impedance, not antenna pattern.

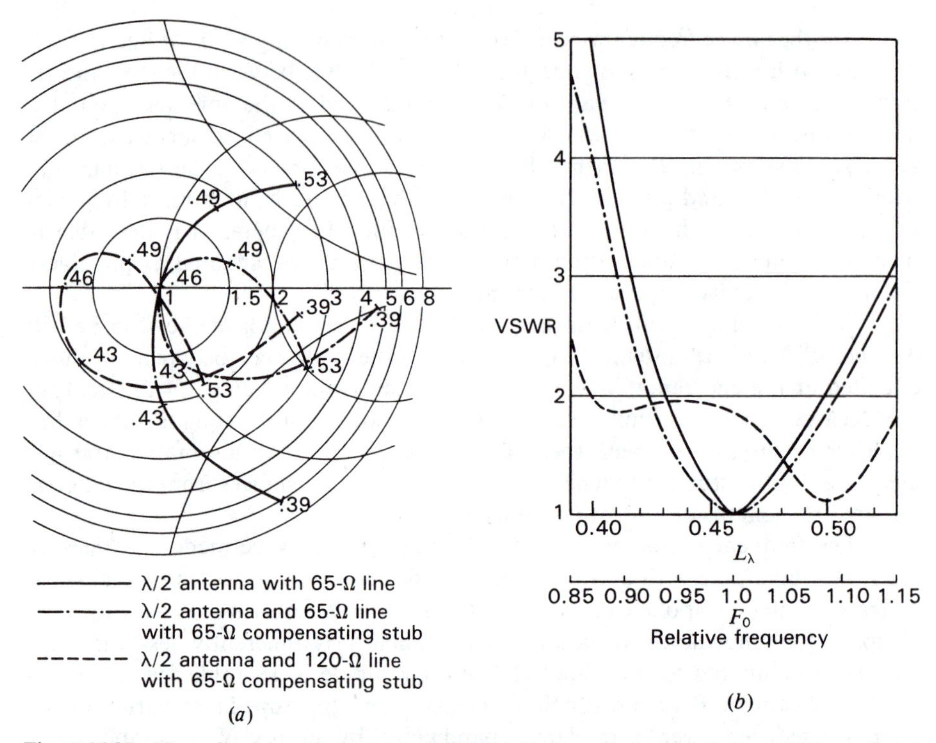

Figure 16-25 Normalized impedance (*a*) and VSWR (*b*) for cylindrical $\lambda/2$ dipole ($L/D = 60$) fed directly with a 65-Ω line as in Fig. 16-22 (solid curves); with a 65-Ω line and 65-Ω $\lambda/2$ compensating stub as in Fig. 16-22*f* (dash-dot curves); and with a 120-Ω line and 65-Ω $\lambda/2$ compensating stub (dashed curves).

At the center frequency F_0 it is 180° in length. Since it is open ended, it places an infinite impedance across the antenna terminals and has no effect. At a frequency slightly above F_0 the line becomes capacitative. Hence, it places a positive susceptance in parallel with the antenna admittance which at this frequency has a negative susceptance.[1] Admittances in parallel are additive so this tends to reduce the total susceptance at the antenna terminals and, therefore, the VSWR on the line. At a frequency slightly below F_0 the result is similar, but in this case the stub is inductive and the antenna has capacitative reactance.

The above matching arrangements provide for a perfect impedance match (VSWR = 1) at the resonant frequency of the antenna. Sometimes a perfect impedance match is not required at any frequency, and it is sufficient to make the

[1] The antenna impedance at this frequency has a positive reactance. Hence,

$$Y = \frac{1}{Z} = \frac{1}{R + jX} = G - jB$$

where G is the conductance component and B the susceptance component of the admittance Y.

VSWR less than a certain value over as wide a frequency band as possible. For example, the VSWR for the $\lambda/2$ dipole ($L/D = 60$) may be made less than 2 over nearly the entire frequency band under consideration if the antenna with a 65-Ω compensating stub is fed with a 120-Ω line instead of a 65-Ω line. The impedance and VSWR curves for this case are shown by the dashed lines in Fig. 16-25*a* and *b*.

Although the above discussion deals specifically with matching arrangements between a $\lambda/2$ dipole and a 2-conductor transmission line, the principles are general and can be applied to other types of antennas and to coaxial lines.

Antenna impedance characteristics may also be compensated by series reactances or by combinations of series and parallel reactances.[1] Many of the techniques of impedance compensation are discussed with examples by J. A. Nelson and G. Stavis.[2]

It is often convenient to use a single coaxial cable to feed a balanced antenna. This may be accomplished with the aid of a balance-to-unbalance transformer or balun.[3] One type of balun suitable for operation over a wide frequency band is illustrated in Fig. 16-13*c*. Another more compact type is shown in Fig. 16-26*a*. The gap spacing at the center of the dipole is made small to minimize unbalance. The length L may be about $\lambda/4$ at the center frequency with operation over a frequency range of 2 to 1 or more. With this arrangement a reactive impedance $Z = jZ_0 \tan \beta L$ appears in parallel with the antenna impedance at the gap, Z_0 being the characteristic impedance of the 2-conductor line of length L. Yet another form of balun is shown in Fig. 16-26*b*. This form provides a balanced transformation only when L is $\lambda/4$ and, accordingly, is suitable only for operation over a few percent bandwidth.

If a $\lambda/2$ dipole antenna is fed by a single coaxial line as in Fig. 16-27*a*, current can flow back along the outside of the coaxial cable making the current distribution on the dipole unbalanced, as suggested in the figure. To feed the dipole with a single coaxial line in a balanced manner, a dummy $\lambda/4$ cable can be placed parallel to the active coaxial line as in Fig. 16-28*a* (see also Fig. 16-26*a*). However, this arrangement is narrowband and may be cumbersome in some applications. A wideband, more-compact arrangement is to wind the coaxial cables around a ferrite core, as suggested in Fig. 16-28*b*, and locate this balun at the center of the dipole, as indicated in Fig. 16-27*b*. The high decoupling impedance of the coils on the ferrite core affects mainly the exterior fields of the coaxial

---

[1] F. D. Bennett, P. D. Coleman and A. S. Meier, "The Design of Broadband Aircraft-Antenna Systems," *Proc. IRE*, **33**, 671–700, October 1945.
H. J. Rowland, "The Series Reactance in Coaxial Lines," *Proc. IRE*, **36**, 65–69, January 1948.
J. R. Whinnery, H. W. Jamieson and T. E. Robbins, "Coaxial-Line Discontinuities," *Proc. IRE*, **32**, 695–709, November 1944.

[2] J. A. Nelson and G. Stavis, in *Very High Frequency Techniques*, Radio Research Laboratory Staff, McGraw-Hill, New York, 1947, chap. 3, pp. 53–92.

[3] Hu Shuhao, "The Balun Family," *Microwave J.*, **30**, 227–229, September 1987.

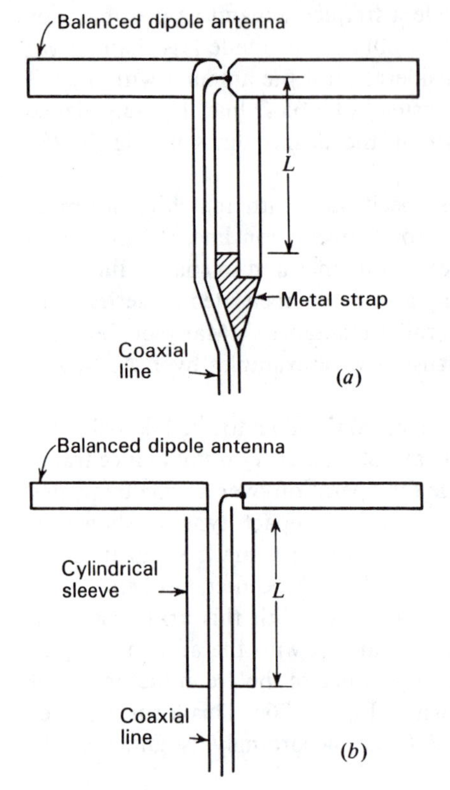

Figure 16-26 Methods of feeding a balanced antenna with a single coaxial line.

cables rather than their interior so that the coaxial line impedance (50 or 75 Ω) is unchanged.

Another method of feeding a balanced dipole or a driven dipole array (W8JK type) is shown in Fig. 16-29. A 2-conductor transmission line of aluminum tubing is terminated in a short-circuited telescoping section of slightly larger diameter tubing which can slide trombone-fashion in order to adjust the antenna and line to resonance. The coaxial line tap point at a distance d above the shorting bar is then adjusted by a second trombone for minimum VSWR on the coax. Since the antenna is resonated, its terminal impedance is immaterial, making the arrangement effective for matching a coaxial line to a wide variety of balanced

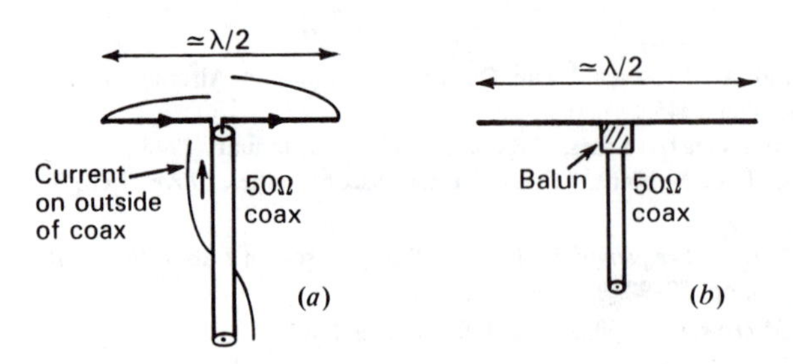

Figure 16-27 (*a*) Unbalanced $\lambda/2$ dipole fed from a 50-Ω coaxial line with currents on outside of the coaxial line. (*b*) Balanced $\lambda/2$ dipole balun-fed from coaxial line.

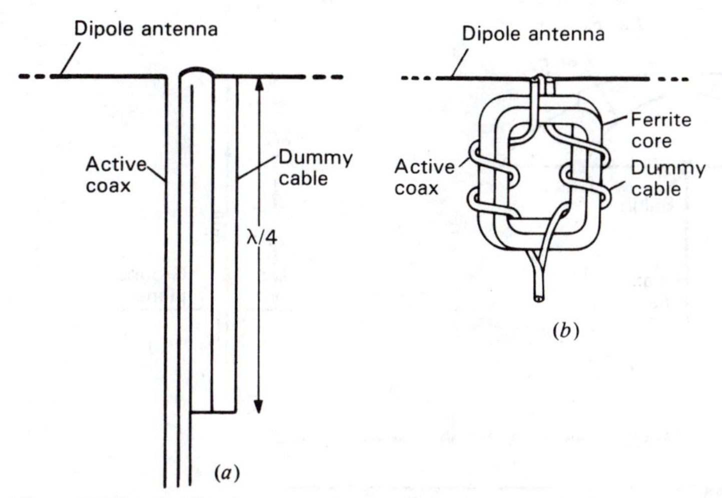

Figure 16-28 (*a*) $\lambda/4$ balance-to-unbalance transformer (balun). (*b*) More compact balun with cables wound on a ferrite core.

antennas with high efficiency since the 2-conductor aluminum line has low losses. By sliding the trombones, the antenna may be resonated and matched at any frequency over a range of several octaves, with the bandwidth at a particular setting dependent on the Q of the antenna.

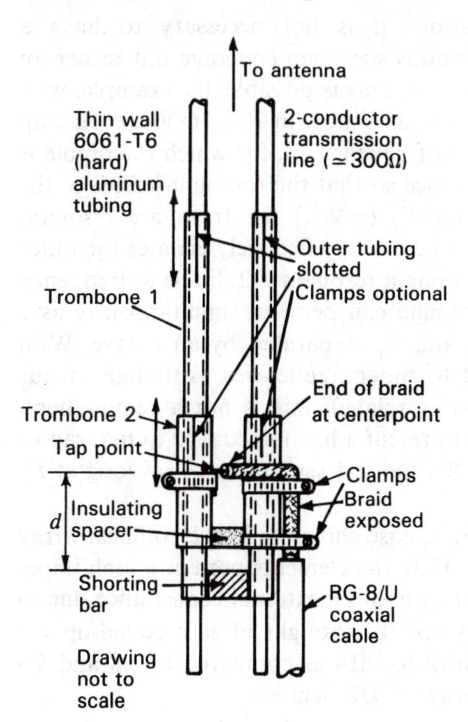

Figure 16-29 Trombone tuners for resonant antenna system with balun.

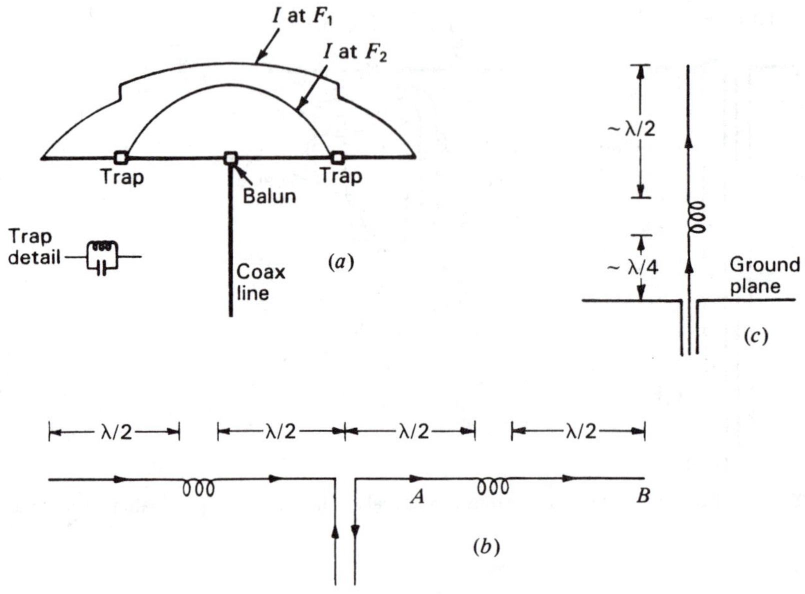

Figure 16-30 (*a*) Dipole with traps for operation at 2 frequencies separated by an octave ($F_2 = 2F_1$). (*b*) Four in-phase $\lambda/2$ elements with phase-reversing coils. (*c*) Vertical omnidirectional in-phase $3\lambda/4$ monopole.

In many wide-bandwidth applications it is not necessary to have a frequency-independent antenna for continuous spectrum coverage but rather an antenna which can operate at *spot frequencies.* This is possible, for example, with a center-fed dipole by means of tuned traps, as shown in Fig. 16-30*a*, each trap consisting of a parallel-tuned *LC* circuit. At frequency F_1, for which the dipole is $\lambda/2$ long, the traps introduce some inductance so that the resonant length of the dipole is reduced. At twice the frequency F_2 $(=2F_1)$ the traps are resonant $(\omega L = 1/\omega C)$ and the high impedance they introduce effectively isolates the outer sections of the dipole, the inner part becoming a resonant $\lambda/2$ dipole at frequency F_2. Thus, in this simple example, the antenna can perform simultaneously as a matched $\lambda/2$ dipole at 2 frequencies, F_1 and F_2, separated by an octave. With more traps, operation may be extended to other frequencies. Although in our example the 2 frequencies are harmonically related, this is not a requirement. Note, however, that the end segment must present a low impedance to the trap to be isolated (i.e., mismatched). In Fig. 16-30*a* the end segment is $\sim\lambda/4$ long at F_2 so this requirement is met.

A coil (or trap) can also act as a 180° phase shifter as in the collinear array of 4 in-phase $\lambda/2$ elements in Fig. 16-30*b*. Here the elements present a high impedance to the coil which may be resonated without an external capacitance due to its distributed capacitance. The coil may also be thought of as a coiled-up $\lambda/2$ element. This 4-element array has a gain of 6.4 dBi as compared to 3.8 and 5.3 dBi gain for 2- and 3-element collinear arrays of $\lambda/2$ elements.

Cutting the antenna of Fig. 16-30*b* at point *A* and turning the section *AB* vertical above a ground plane results in the in-phase $3\lambda/4$ omnidirectional monopole antenna of Fig. 16-30*c* with 8.3 dBi gain. To match the approximately 150-Ω terminal resistance to a 50- to 75-Ω coaxial line, a capacitance-inductance L network can be used.

16-12 PATCH OR MICROSTRIP ANTENNAS. These antennas are popular for low-profile applications at frequencies above 100 MHz ($\lambda_0 < 3$ m). They commonly consist of a rectangular metal patch on a dielectric-coated ground plane (circuit board). A microstrip or patch antenna is shown in Fig. 16-31 with the dielectric substrate material having a typical relative permittivity $\varepsilon_r \simeq 2$ and thickness $t \simeq \lambda_0/100$. They differ in size and method of feeding from the flush-disc antenna of Fig. 16-12*d*.

Typical patch dimensions of length *L*, width *W* and thickness *t* are indicated in Fig. 16-31 with feed from a coaxial line at the center of the left edge. The horizontal components of the electric fields at the left and right edges are in the same direction, giving in-phase linearly polarized radiation with a maximum broadside to the patch (up in Fig. 16-31*b*).

The patch acts as a resonant $\lambda/2$ parallel-plate microstrip transmission line with characteristic impedance equal to the reciprocal of the number *n* of parallel

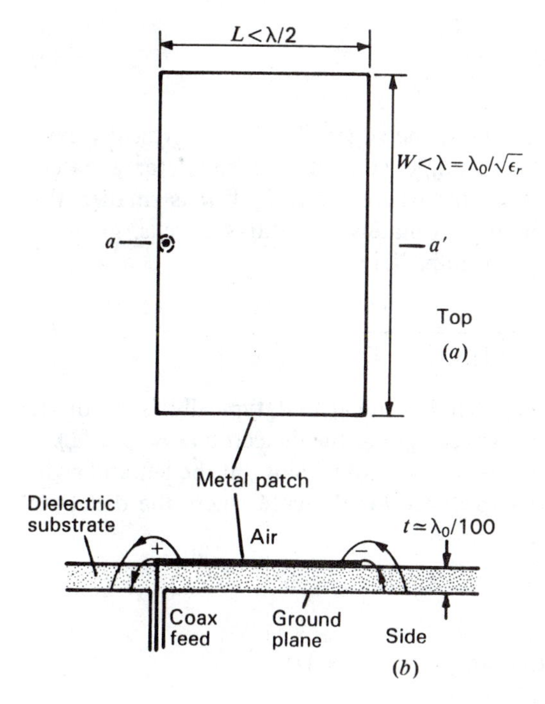

Figure 16-31 Top view (*a*) and side (cross-section) view (*b*) of patch antenna fed by coaxial line at left edge.

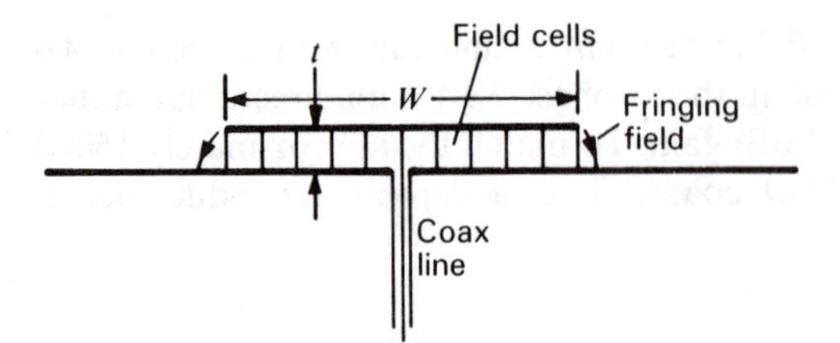

Figure 16-32 View from coax feed side of patch of Fig. 16-31 with gap (or slot) divided into (square) field cells.

field-cell transmission lines.[1] Each field-cell transmission line has a characteristic impedance equal to the intrinsic impedance Z_i of the medium where

$$Z_i = \sqrt{\frac{\mu}{\varepsilon}} = Z_0\sqrt{\frac{\mu_r}{\varepsilon_r}} \quad (\Omega) \tag{1}$$

For air, $\mu_r = \varepsilon_r = 1$ and $Z_i = Z_0 = 377\ \Omega$.

An end view of the patch (from the left side) is shown in Fig. 16-32. The cross section (as drawn) has 10 parallel field-cell transmission lines so that for $\varepsilon_r = 2$ the microstrip characteristic impedance Z_c is given by

$$Z_c = \frac{Z_0}{n\sqrt{\varepsilon_r}} = \frac{377}{10\sqrt{2}} = 26.7\ \Omega \tag{2}$$

Since $n = W/t$, a more general relation is

$$Z_c = \frac{Z_0 t}{W\sqrt{\varepsilon_r}} \tag{3}$$

This relation neglects fringing of the field at the edges. Since W is typically even much larger than the t value in this example, this effect is small for a patch. However, for a microstrip transmission line where the ratio W/t is smaller, the fringing effect can be accounted for approximately by adding 2 cells, giving a more accurate formula for microstrip line impedance as

$$Z_c = \frac{Z_0}{\sqrt{\varepsilon_r}\,[(W/t) + 2]} \tag{4}$$

The resonant length L of the patch is critical and typically is a couple of percent less than $\lambda/2$, where λ is the wavelength in the dielectric ($\lambda = \lambda_0/\sqrt{\varepsilon_r}$).

Radiation from the patch occurs as though from 2 slots, at the left and right in Fig. 16-31. Let us calculate their impedance for the case where the dielectric substrate is air ($\varepsilon_r = 1$).

[1] J. D. Kraus, *Electromagnetics*, 3rd ed., McGraw-Hill, 1984, pp. 388, 397.

From (13-6-10) the impedance Z_s of a slot antenna is given by

$$Z_s = \frac{Z_0^2}{4Z_d} = \frac{35\,476}{Z_d} \quad (\Omega) \tag{5}$$

where Z_0 = intrinsic impedance of empty space = 377 Ω
Z_d = impedance of complementary dipole, Ω

For a slot λ long and a few $\lambda/100$ wide, the complementary dipole impedance is approximately $700 + j0\Omega$. Thus, the approximate slot impedance is

$$Z_s \simeq \frac{35\,476}{700} \simeq 50\ \Omega \tag{6}$$

This is the impedance of a slot in a conducting sheet open on both sides.

The $\lambda/2$ microstrip is equivalent to a $\lambda/2$ section of transmission line shunted with a radiation (and loss) resistance at each end. The centerpoint of the line is at low potential and can be grounded with little effect on operation. Thus, the slot at each end is effectively boxed-in, doubling its impedance. However, the 2 patch "slots" act in parallel, which halves the impedance resulting in an input resistance

$$R_{\text{in}} \simeq 50\ \Omega$$

for $W \simeq \lambda_0$, which is a typical value even with $\varepsilon_r > 1$. If W is smaller, R_{in} increases proportionately.

The radiation pattern of the patch is broad. Typically, the beam area Ω_A is $\frac{1}{2}$ of a half-space or about π sr. The resulting patch directivity D is then given by

$$D = \frac{4\pi}{\Omega_A} = \frac{4\pi}{\pi} = 4 \text{ (or 6 dBi)} \tag{7}$$

The impedance bandwidth of a patch antenna is usually much narrower than its pattern bandwidth. The impedance bandwidth is proportional to the thickness t of the dielectric substrate. Since t is small, the bandwidth is small, typically a few percent.

Although the concept of effective height h_e may not be appropriate when applied to a patch, it is interesting to evaluate h_e for a typical patch antenna.

From (2-19-6) the effective height h_e of an antenna is given by

$$h_e = \sqrt{\frac{2R_r A_e}{Z_0}} \tag{8}$$

where R_r = radiation resistance, Ω
A_e = effective aperture, λ^2
Z_0 = intrinsic impedance of space, Ω

If we take $D = 4$ and $R_r = 50\ \Omega$ for a typical patch we have as its effective aperture

$$A_e = \frac{D\lambda_0^2}{4\pi} = \frac{\lambda_0^2}{\pi} \tag{9}$$

and as its effective height

$$h_e = \sqrt{\frac{2 \times 50\lambda_0^2}{377\pi}} \simeq 0.3\lambda_0 \tag{10}$$

It is interesting that an antenna which extends only $\lambda_0/100$ above a flat ground plane has an "effective height" 30 times as much.

The dimensions of a patch are not electrically small and at, say, 2 GHz ($\lambda = \lambda_0/\sqrt{2} = 106$ mm) a patch is also not physically very small.

Although the above discussion is considerably simplified, it outlines some of the important properties of patch antennas. Many other shapes and configurations are possible. For example, for matching purposes the feed point can be moved in from the edge. Patches also lend themselves to microstrip arrays in which the patches are fed by microstrip transmission lines. The configuration of a 4-patch array with $\varepsilon_r = 4$ is shown in Fig. 16-33. See also Fig. 11-54 showing an array of 896 microstrip elements.

The literature on patch or microstrip antennas is extensive. Some basic references are those of Munson,[1] Derneryd,[2] Carver and Mink[3] and Pozar.[4]

Monolithic Microwave Integrated Circuits (MMICs) combine microstrip antennas and associated circuitry in very compact form with applications at frequencies from 50 MHz to 100 GHz.[5]

Printed-circuit technology is suitable for printing a variety of antenna elements as, for example, dipoles.[6] To feed these balanced center-fed elements

---

[1] R. E. Munson, "Conformal Microstrip Antennas and Microstrip Phased Arrays," *IEEE Trans. Ants. Prop.*, **AP-22**, 74–78, January 1974.
R. E. Munson, "Microstrip Antennas," *Antenna Engineering Handbook*, McGraw-Hill, 1984, chap. 7.

[2] A. Derneryd, "A Theoretical Investigation of the Rectangular Microstrip Antenna Element," *IEEE Trans. Ants. Prop.*, **AP-26**, 532–535, July 1978.

[3] K. R. Carver and J. W. Mink, "Microstrip Antenna Technology," *IEEE Trans. Ants. Prop.*, **AP-29**, 25–38, January 1981.

[4] D. M. Pozar, "An Update on Microstrip Antenna Theory and Design Including Some Novel Feeding Techniques," *IEEE Ants. Prop. Soc. Newsletter*, **28**, 5–9, October 1986.

[5] E. Brookner, "Array Radars: An Update," *Microwave J.*, **30**, 134–138, February 1987, and 167–174, March 1987.

[6] D. M. Pozar, "Considerations for Millimeter Wave Printed Antennas," *IEEE Trans. Ants. Prop.*, **31**, 740–747, September 1983.
E. Levine, J. Ashenasy and D. Treves, "Printed Dipole Arrays on a Cylinder," *Microwave J.*, **30**, 85–92, March 1987.

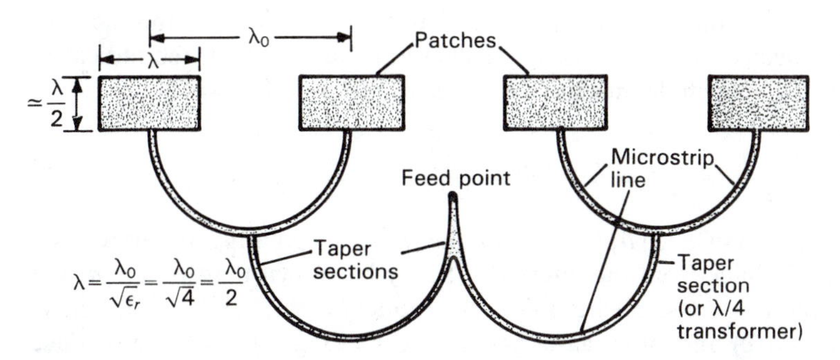

Figure 16-33 Microstrip array of 4 patches with corporate feed by microstrip transmission lines.

requires 2-conductor microstrip transmission lines, which may be a disadvantage. However, a dipole element occupies less area than a patch. It is interesting that the input resistance variation with dielectric thickness t of a center-fed 1λ dipole is similar to that of a patch whose slot is complementary to the dipole.

16-13 THE HIGH-GAIN OMNI. A high-gain (high-directivity) antenna which is also omnidirectional involves a contradiction. The directivity D of an antenna is given by

$$D = \frac{4\pi}{\Omega_A} \tag{1}$$

where Ω_A = beam area

A hypothetical isotropic point source has a beam area $\Omega_A = 4\pi$, making it completely omnidirectional. From (1), its directivity $D = 1$, which is the lowest possible directivity. For the directivity (or gain of a lossless radiator) to be more than unity requires that the beam area be less than 4π so that the antenna is no longer omnidirectional. Thus, for a simple antenna, high directivity is incompatible with an omnidirectional (4π sr) pattern. The combination of both is a theoretical and a practical impossibility. However, in the digital signal-processing domain of a large phased array the combination is theoretically possible, giving an arbitrarily large number of simultaneous high-directivity beams but, due to inherent losses, not necessarily ones of high gain (see Sec. 11-13).

16-14 SUBMERGED ANTENNAS. In (16-4-1) for the reflection coefficient $\rho_\perp$, it is assumed that the wave originates in the less-dense medium (air) and is incident on the earth or ground of relative permittivity ε_r. If the wave originates in the denser medium and travels from it into air (16-4-1) becomes

$$\rho_\perp = \frac{\sin\alpha - \sqrt{(1/\varepsilon_r) - \cos^2\alpha}}{\sin\alpha + \sqrt{(1/\varepsilon_r) - \cos^2\alpha}} \tag{1}$$

where ε_r = relative permittivity of denser medium

If $1/\varepsilon_r \leq \cos^2 \alpha$, $\rho_\perp$ is complex and $|\rho_\perp| = 1$. Under this condition, the incident wave is totally reflected back into the more-dense medium.[1] When the radical in (1) is zero, $\rho_\perp = 1\underline{/0^\circ}$, which defines the *critical angle* (see Fig. 16-34)

$$\alpha_c = \cos^{-1}\sqrt{\frac{1}{\varepsilon_r}} \tag{2}$$

For all angles less than α_c, $|\rho_\perp| = 1$, and the wave originating in the denser medium is reflected back from the interface. It may be shown that for $\alpha < \alpha_c$ the electric field in the less-dense medium decays exponentially away from the interface (evanescent wave) and propagates without loss along the interface (surface wave).[2]

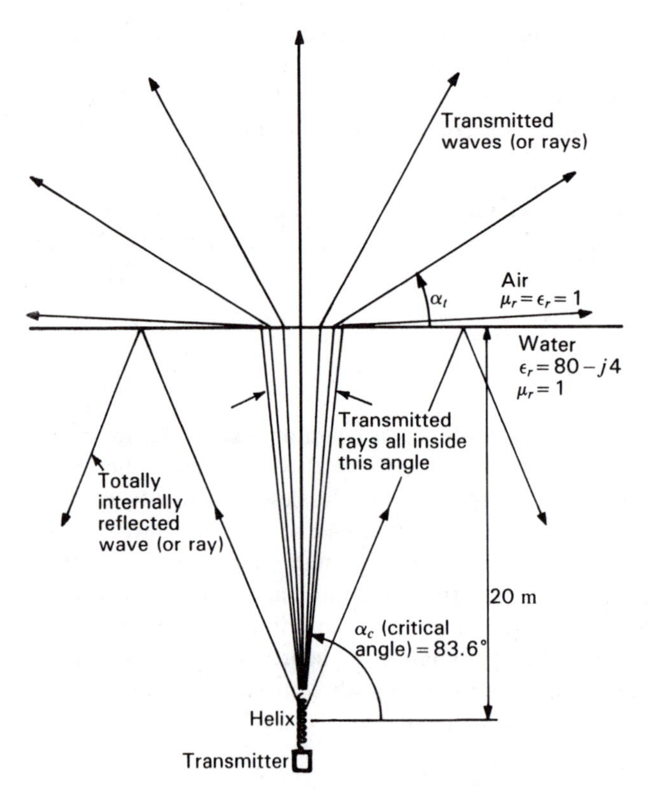

Figure 16-34 Rays from submerged antenna at angles α below 83.6° are totally internally reflected. Rays between 83.6 and 90° are transmitted through the surface into air above and spread out over almost 180°.

[1] This is the case for either perpendicular or parallel polarization.

[2] J. D. Kraus, *Electromagnetics*, 3rd ed., McGraw-Hill, 1984, p. 515. For a detailed exposition of antennas in material media, see R. W. P. King and G. S. Smith (with M. Owens and T. T. Wu), *Antennas in Matter*, MIT Press, 1980. See also P. Delogne, *Leaky Feeders and Subsurface Radio Communication*, IEE, London, 1982.

Example. A 60-MHz ($\lambda_0 = 5$ m) radio transmitter feeding a monofilar axial-mode helical antenna is situated 20 m below the surface of a freshwater lake with constants $\mu_r = 1$, $\varepsilon_r' = 80$ and $\sigma = 1.33 \times 10^{-2}$ ℧ m^{-1}. The helix axis is vertical. See Fig. 16-34. (*a*) For a given transmitter power how many turns should the helix have to maximize the signal at an elevation angle of 10° above the surface? (*b*) What is the relative power density radiated above the surface as a function of the elevation angle above the surface? (*c*) If the transmitter power is 100 W, what is the signal level at an elevation angle of 10° at a distance of 1 km from the submerged antenna site? The receiving antenna is a George Brown turnstile with reflector elements.

Solution

(*a*) At 60 MHz the loss term of the relative permittivity

$$\varepsilon_r'' = \frac{\sigma}{\varepsilon_0 \omega} = \frac{1.33 \times 10^{-2}}{8.85 \times 10^{-12} \times 2\pi \times 60 \times 10^6} = 4 \tag{3}$$

and the complex relative permittivity

$$\varepsilon_r = \varepsilon_r' - j\varepsilon_r'' = 80 - j4 \tag{4}$$

The water is not lossless but $\varepsilon_r' \gg \varepsilon_r''$, so neglecting ε_r'' we have from (2) that the *critical angle* is given by

$$\alpha_c = \cos^{-1} \sqrt{\frac{1}{80}} = 83.58° \tag{5}$$

Thus, all rays from the helix at elevation angles less than 83.58° are reflected back into the water, and only rays at elevation angles greater than 83.58° emerge into the air, as suggested in Fig. 16-34. The hole through which the rays emerge is 12.84° wide [$= 2\,(90° - 83.58°)$] centered on the zenith ($\alpha = 90°$).

From Snell's law we have approximately that

$$\cos \alpha_t \simeq \sqrt{\varepsilon_r'} \cos \alpha = 8.94 \cos \alpha \tag{6}$$

where α_t = elevation angle of ray transmitted above the surface

For $\alpha_t = 10°$, we have from (6) that $\alpha = 83.68°$ or just 0.1° ($= 83.68° - 83.58°$) above the critical angle.

For a monofilar axial-mode helical antenna with number of turns $n > 3$, pitch angle $12° < \alpha < 14°$ and circumference $0.8 < C_\lambda < 1.15$, the beam width and directivity from (7-4-4) and (7-4-7) are

$$\text{HPBW} \simeq \frac{52°}{C_\lambda \sqrt{nS_\lambda}} \tag{7}$$

and

$$D \simeq 12 C_\lambda^2 n S_\lambda \tag{8}$$

Taking $C_\lambda = 1$, $\alpha = 12.5°$ and $n = 12$, the HPBW = 32.0° and $D = 31.7$.[1] The half-power angle off axis is given by HPBW/2 = 16.0°, as compared to 6.32°

[1] Do not confuse α here for *helix pitch angle* with α in (6) for *elevation angle*.

($=90° - 83.68°$) for the ray which emerges at $\alpha_t = 10°$. Thus, with 12 turns the beam is broad enough that at or close to the critical angle the signal level is down from the maximum (on axis) only about 0.3 dB. Doubling the length of the helix ($n = 24$) increases the directivity from (8) by about 3 dB, but narrows the beam width sufficiently (by $1/\sqrt{2}$) that at the critical angle the signal level is down about another 0.5 dB. Thus, doubling n results in a net increase in the level of the radiation emerging at an elevation angle $\alpha_t = 10°$ above the surface. However, the narrower beam requires that the helix axis be set within 1° of vertical. Thus, a helix with about 20 turns is a reasonable compromise.

(*b*) The relative power density S of a ray reflected back into the water is given by

$$S_r = \frac{(E_\perp |\rho_\perp|)^2 + (E_\parallel |\rho_\parallel|)^2}{E_\perp^2 + E_\parallel^2} \tag{9}$$

where $E_\perp$ = electric field perpendicular to the plane of incidence
$E_\parallel$ = electric field parallel to the plane of incidence
$\rho_\perp$ = reflection coefficient for perpendicular component [Eq. (16-4-1) with ε_r inverted]
$\rho_\parallel$ = reflection coefficient for parallel component [Eq. (16-4-6) with ε_r inverted]

Assuming circular polarization ($E_\perp = E_\parallel$) from the helix within the critical-angle hole, (9) reduces to

$$S_r = \tfrac{1}{2}(|\rho_\perp|^2 + |\rho_\parallel|^2) \tag{10}$$

The relative power density S_t transmitted through the surface and emerging above it is then given by

$$S_t = 1 - S_r \tag{11}$$

Evaluating (10) and (11) for $\varepsilon_r = 80$ as a function of the elevation angle α_t yields the curve of Fig. 16-35*a* and the pattern of Fig. 16-35*b*. Both graphs give the power radiated relative to its value at the zenith ($\alpha_t = 90°$).

(*c*) Note that the dimensions of the helix are reduced to $1/\sqrt{\varepsilon_r} = 0.11$ or 11 percent of the dimensions in air. Thus, the submerged helix diameter $D = \lambda_0/(\sqrt{\varepsilon_r}\,\pi) = 5\text{ m}/(\sqrt{80}\,\pi) = 0.178$ m.

The loss component of the permittivity $\varepsilon_r'' = 4$ and since $\varepsilon_r'' \ll \varepsilon_r'$ the *attenuation constant* for the water is given by[1]

$$\alpha = \frac{\pi}{\lambda_0}\frac{\varepsilon_r''}{\sqrt{\varepsilon_r'}} = \frac{\pi}{5}\frac{4}{\sqrt{80}} = 0.281 \text{ Np m}^{-1}$$

and the attenuation in the water path d by

$$\text{Attenuation (water)} = 20 \log e^{-\alpha d} = 20 \log e^{-0.281 \times 20} = 49 \text{ dB}$$

The loss at the interface is given from (11) for elevation angle α (water) = 83.68° as 10 log 0.51 = 3 dB.

[1] Do not confuse α here as the *attenuation constant* (=real part of propagation constant γ) with α for *helix pitch angle* or α for *elevation angle*. Unfortunately, the English and Greek alphabets do not have enough letters to go around.

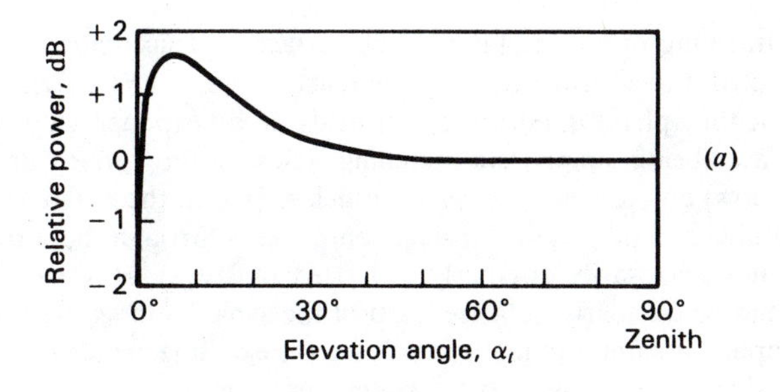

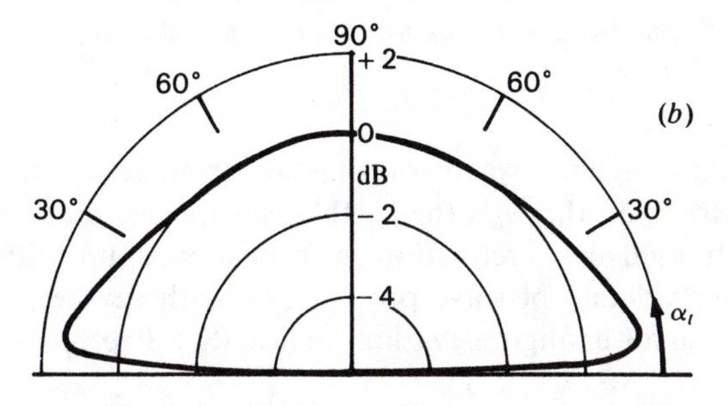

Figure 16-35 (*a*) Ratio of power density at elevation angle α_t to power density at zenith (90°) for waves transmitted above surface of lake from submerged transmitter. (*b*) Field pattern of radiation transmitted above surface in polar coordinates.

The directivity of the 20-turn helix is given by

$$D \simeq 12C_\lambda^2 nS_\lambda = 12 \times 20 \times 0.22 = 52.8$$

and its effective aperture by

$$A_e = \frac{D\lambda_0^2}{4\pi} = \frac{52.8 \times 5^2}{4\pi} = 105 \text{ m}^2$$

A George Brown turnstile (Sec. 16-7) with reflector elements has a directivity $D \simeq 3$ and effective aperture at $\lambda_0 = 5$ m of 6 m^2. Taking the distance between transmitter and receiver as 1 km we have from the Friis transmission formula that the received power less water attenuation and surface (interface) loss is

$$P_r = \frac{P_t A_{et} A_{er}}{r^2 \lambda_0^2} = \frac{100 \times 105 \times 3}{1000^2 \times 5^2} = 1.25 \text{ mW}$$

The water attenuation and surface (interface) loss total 52 dB (=49 + 3) so the net (actual) received power is given by

$$P_r = 1.25 \times 10^{-3}/\text{antilog } 5.2 \simeq 8 \times 10^{-9} \text{ W} = 8 \text{ nW}$$

or a bit less than 1 mV on a 100-Ω transmission line.

The transmitted energy-carrying field emerging above the surface goes to zero at the surface (elevation angle $\alpha_t = 0°$), as indicated in Fig. 16-35*a* and *b*. There is also a reactive wave traveling out along (above) the surface that accompanies the totally internally reflected wave. However, it is a reactive wave and carries no energy (**E** and **H** are in time-phase quadrature). Its fields decay exponentially with height above the surface (being simply the matching fields at the surface of the reflected-wave fields below) and are called *evanescent* fields. Due to the difference in transmission of the parallel and perpendicular field components through the surface, the received wave is not necessarily circularly polarized (AR $\neq 1$). However, the receiving antenna should be circularly polarized and of the same hand as the transmitting antenna, an implicit assumption made in solving the example problem.

It is also assumed in the example that the water surface is smooth. If it is not smooth, as under windy conditions, the situation is different and fluctuations (noise) will occur in the received signal.

The problem of electromagnetic wave transmission from water into air is analogous to that of transmission through the earth's ionosphere to an extraterrestrial point above it. In each case, refraction is from a medium with lower velocity to one with higher velocity of wave propagation with ray-bending and total internal reflection at small enough elevation angles. (See Probs. 17-18 and 17-19.)

However, there are some fundamental differences in that the ionosphere is inhomogeneous and anisotropic (a magnetized plasma) with polarization changes by Faraday rotation. These effects make circularly polarized antennas essential.

We note that the radio situation differs from the optical one in that $\varepsilon_r \simeq 80$ for water at radio wavelengths but $\varepsilon_r \simeq 1.75$ for water at optical wavelengths. Thus, for light waves the critical angle $\alpha_c = 41°$ as compared to 84° for radio waves.

16-15 SURFACE-WAVE AND LEAKY-WAVE ANTENNAS.

Traveling-wave antennas discussed in previous chapters include the monofilar axial-mode helix, long-wire and polyrod antennas. *Surface-wave* and *leaky-wave* antennas are also *traveling-wave* antennas but are ones which are adapted to flush or low-profile installations as on the skin of high-speed aircraft. Typically, their bandwidth is narrow (10 percent) and their gain is moderate ($\sim$15 dBi).

Consider the plane boundary between air and a perfect conductor as shown in Fig. 16-36*a* with a vertically polarized plane TEM wave traveling to the right along the boundary. From the boundary condition that the tangential component of the electric field vanishes at the surface of a perfect conductor, the electric field of a TEM wave traveling parallel to the boundary must be exactly normal to the boundary, or vertical, as in Fig. 16-36*a*. However, if the conductivity σ of the conductor is not infinite, there will be a tangential electric field E_x at the boundary, as well as the vertical component E_y, so that the total field will have a *forward tilt* as in Fig. 16-36*b*.

The direction and magnitude of the power flow per unit area are given by the Poynting vector with average value

$$\mathbf{S}_{av} = \tfrac{1}{2}\,\text{Re}\,\mathbf{E} \times \mathbf{H}^* \qquad (\text{W m}^{-2}) \tag{1}$$

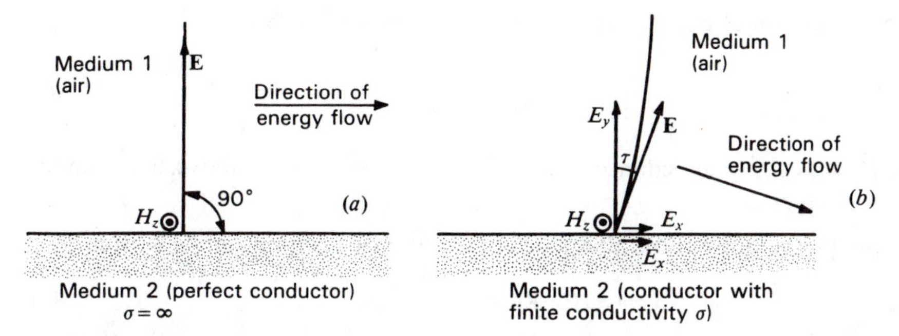

Figure 16-36 Vertically polarized wave traveling to right (*a*) along surface of a perfectly conducting medium and (*b*) along surface of medium with finite conductivity.

At the conducting surface the average power into the conductor ($-y$ direction) is

$$S_y = -\tfrac{1}{2} \operatorname{Re} E_x H_z^* \tag{2}$$

This is power dissipated as heat in the conductor.

The space relation of E_x, H_z (or H_z^*) and S_y is shown in Fig. 16-37*a*. Since

$$\frac{E_x}{H_z} = Z_c \tag{3}$$

where Z_c = intrinsic impedance of the conductor

then from (2) the power flow into the conductor can be written as

$$S_y = -\tfrac{1}{2} H_z H_z^* \operatorname{Re} Z_c = -\tfrac{1}{2} H_{z0}^2 \operatorname{Re} Z_c \tag{4}$$

where $H_z = H_{z0}\, e^{-j\xi - \gamma x}$
$H_z^* = H_{z0}\, e^{j\xi + \gamma x}$ = complex conjugate of H_z
ξ = phase lag of H_z with respect to E_x
γ = propagation constant = $\alpha + j\beta$

The power flow parallel to the surface (x direction) is

$$S_x = \tfrac{1}{2} \operatorname{Re} E_y H_z^* \qquad (\text{W m}^{-2}) \tag{5}$$

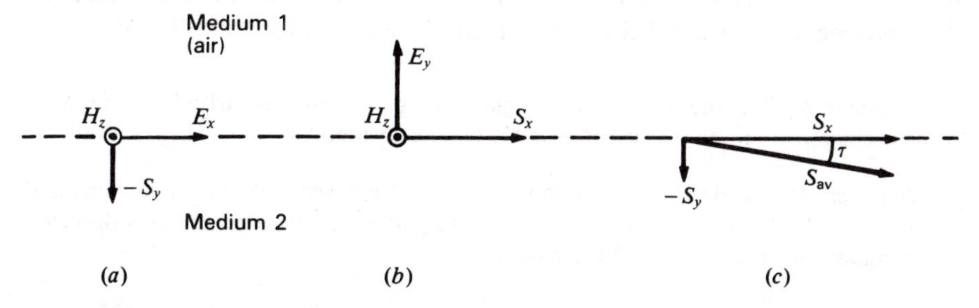

Figure 16-37 Fields and Poynting vector at the surface of a conducting medium with a vertically polarized wave traveling parallel to the surface.

The space relation of E_y, H_z and S_x is shown in Fig. 16-37*b*. Since

$$\frac{E_y}{H_z} = Z_d \tag{6}$$

where Z_d = intrinsic impedance of the dielectric medium (air) above the conductor

then from (5) the power flow along the conductor (x-direction) can be written

$$S_x = \tfrac{1}{2}H_{z0}^2 \text{ Re } Z_d \qquad (\text{W m}^{-2}) \tag{7}$$

The total average Poynting vector is then

$$\mathbf{S}_{\text{av}} = \hat{\mathbf{x}}S_x + \hat{\mathbf{y}}S_y = \frac{H_{z0}^2}{2}(\hat{\mathbf{x}} \text{ Re } Z_d - \hat{\mathbf{y}} \text{ Re } Z_c) \tag{8}$$

Figure 16-37*c* shows the relation of $\mathbf{S}_{\text{av}}$ to its x and y components. It is apparent that the average power flow is not parallel to the surface but downward at an angle τ. This angle is the same as the forward tilt angle of the electric field (see Fig. 16-36*b*). If the conductivity of the conductor is infinite ($\sigma = \infty$), the tilt is zero ($\tau = 0$).

From (8) the tilt angle

$$\tau = \tan^{-1} \frac{\text{Re } Z_c}{\text{Re } Z_d} \tag{9}$$

Example 1. Find the tilt angle τ for a vertically polarized 3-GHz wave traveling in air along a flat copper sheet.

Solution. At 3 GHz, we have for copper that Re Z_c = 14.4 mΩ. The intrinsic impedance of air is real and independent of frequency (=377 Ω). From (9) we have

$$\tau = \tan^{-1} \frac{14.4 \times 10^{-3}}{377} = 0.0022^\circ \tag{10}$$

Although τ is very small it is not zero, indicating some power flow into the copper sheet. If σ is small or if the conductor is replaced by a dielectric, τ may amount to a few degrees. In the Beverage antenna (Sec. 11-16c) the horizontal field component (E_x) is parallel to the antenna wire and induces emf's in it.

Example 2. Find the forward tilt angle τ for a vertically polarized 3-GHz wave traveling in air along the smooth surface of a freshwater lake.

Solution. At 3 GHz the conduction current of fresh water is negligible compared with the displacement current ($\varepsilon_r'' \ll \varepsilon_r'$) so the water may be regarded as a dielectric medium with $\varepsilon_r \simeq \varepsilon_r' = 80$. Thus, from (9),

$$\tau = \tan^{-1} \frac{1}{\sqrt{80}} = 6.4^\circ \tag{11}$$

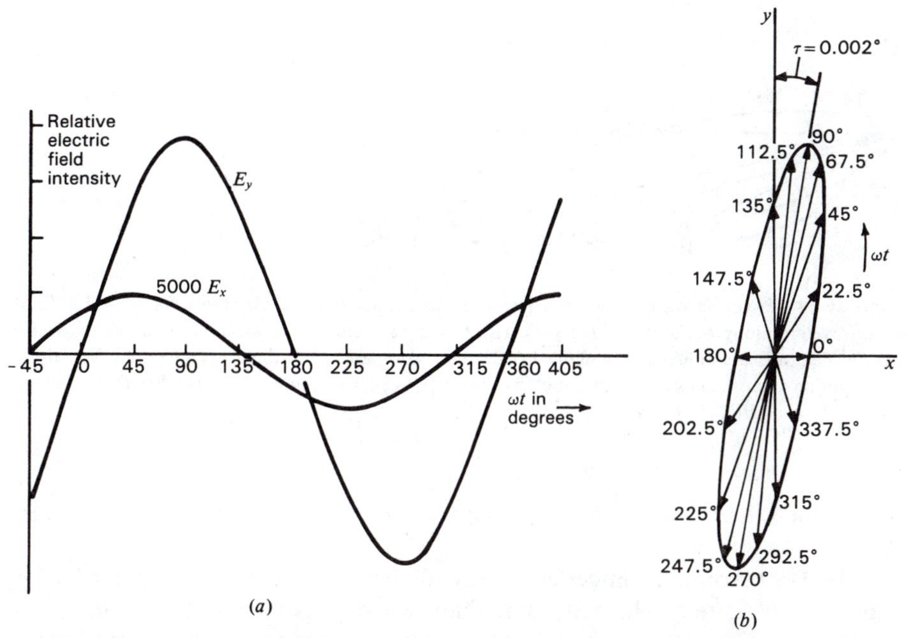

Figure 16-38 (*a*) Magnitude variation with time of E_y and E_x components of **E** in air at the surface of a copper region for a 3-GHz TEM wave traveling parallel to the surface. (*b*) Resultant values of **E** (space vector) at 22.5° intervals over one cycle, illustrating elliptical cross-field at the surface of the copper region. The wave is traveling to the right. Note that although **E** has x and y components (cross-field) it is linearly polarized as seen from the x direction. (*From J. D. Kraus,* Electromagnetics, *3rd ed., McGraw-Hill, 1984.*)

The tilt angle τ calculated above is an average value. In general, its instantaneous direction varies with time. For the wave traveling along the copper sheet, E_y and E_x are in phase octature (45° phase difference), so that at one instant the total field **E** may be in the x direction and $\frac{1}{8}$-period later in the y direction. For the 3-GHz wave of Example 1, E_x and E_y vary with time as in Fig. 16-38*a*. The locus of the tip of **E** describes a cross-field ellipse as in Fig. 16-38*b*, with positions shown as a function of time (ωt). The ellipse is not to scale, the E_x values being magnified by 5000. The variation of the instantaneous Poynting vector is shown in Fig. 16-39 with ordinate values magnified by 5000. It is of note that the tip of the Poynting vector travels around the ellipse *twice* per cycle.

Whereas copper has a complex intrinsic impedance, fresh water at 3 GHz (as in Example 2) has a real intrinsic impedance so that E_x and E_y are essentially in phase and the cross-field ellipse collapses to a straight line with a forward tilt of 6.4°.

The forward tilt of **E** or downward tilt of **S** in the above examples tend to make the wave energy hug the surface, resulting in a *bound wave* or *surface wave.* The phase velocity v of this wave is less than the velocity c of light, i.e., it is a *slow wave* ($v < c$). To initiate a wave along the surface, a launching device, such as a

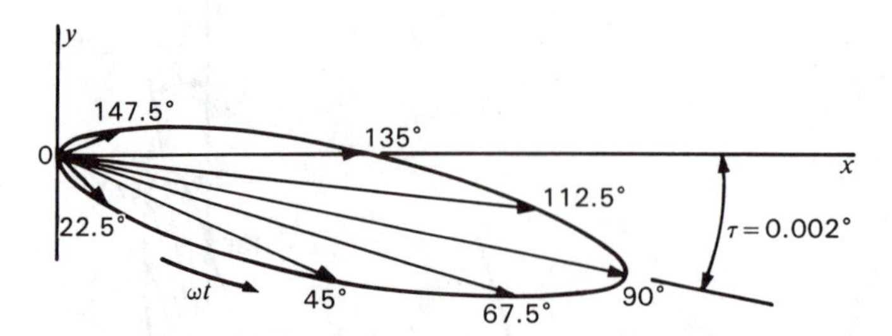

Figure 16-39 Poynting vector in air at a point on the surface of a copper region for a 3-GHz TEM wave traveling along the surface (to right). The Poynting vector is shown at 22.5° intervals over a $\frac{1}{2}$-cycle. The ordinate values are magnified 5000 times compared with the abscissa values. The tip of the Poynting vector travels around the ellipse *twice* per cycle. (*From J. D. Kraus,* Electromagnetics, *3rd ed., McGraw-Hill, 1984.*)

horn, with a height of several wavelengths can be used as in Figs. 16-40 and 16-41.

In 1899 Arnold Sommerfeld[1] showed that a wave could be guided by a round wire of finite conductivity. Jonathan Zenneck[2] pointed out that for similar reasons a wave traveling along the earth's surface would tend to be guided by it.

The guiding action of a flat conducting surface can be enhanced by adding metal corrugations. These and the horn launcher in Fig. 16-40 form a surface-wave antenna. The corrugations have many teeth per wavelength ($s < \lambda/5$). The slots between the teeth support a TEM wave involving E_x, which travels up and down the depth of the slot (a standing wave). Each slot acts like a short-circuited section of a parallel-plane 2-conductor transmission line of depth d. Assuming lossless materials, the impedance Z presented to a wave traveling down the slots is a pure reactance given by

$$Z = jZ_d \tan \frac{2\pi\sqrt{\varepsilon_r}\, d}{\lambda_0} \quad (\Omega) \tag{12}$$

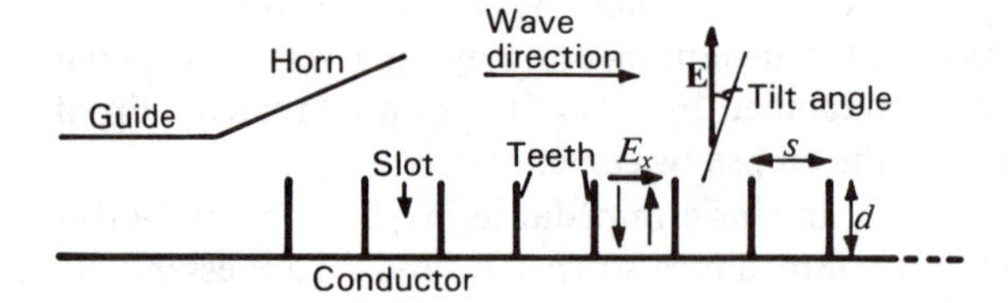

Figure 16-40 Corrugated surface-wave antenna with horn wave-launcher.

[1] A. Sommerfeld, "Fortpflanzung elektrodynamischer Wellen an einem zylindrischen Leiter," *Ann. Phys. u. Chem.*, **67**, 233, December 1899.

[2] J. Zenneck, "Über die Fortpflanzung ebener elektromagnetischer Wellen längs einer ebenen Leiterfläche und ihre Beziehung zur drahtlosen Telegraphie," *Ann. Phys.*, **23**, 846–866, 1907.

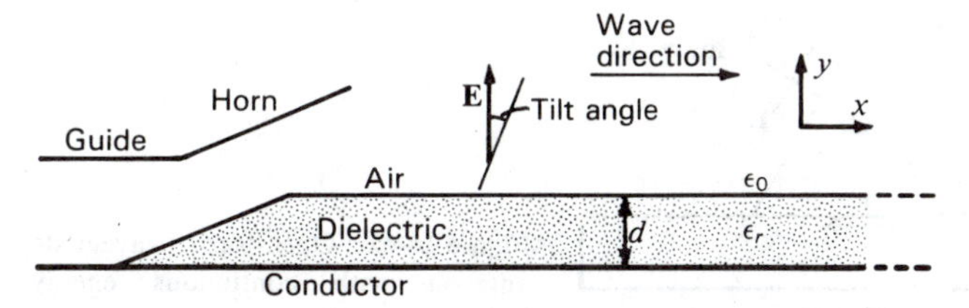

Figure 16-41 Dielectric-slab surface-wave antenna with horn wave-launcher.

where Z_d = intrinsic impedance of medium filling the slots, Ω
ε_r = relative permittivity of the medium, dimensionless
λ_0 = free-space wavelength, m
d = depth of slots, m

For air-filled slots ($Z_d = 377\ \Omega$ and $\varepsilon_r = 1$), (12) reduces to

$$Z \simeq j120\pi \tan \frac{2\pi d}{\lambda_0} \qquad (\Omega) \tag{13}$$

The slots store energy from the passing wave. When $d < \lambda_0/4$, the plane along the top of the teeth is inductively reactive. When $d = \lambda_0/4$, $Z = \infty$ and the plane along the top is like an open circuit (nothing below), while when $d = \lambda_0/2$, $Z = 0$ and the plane appears like the conducting sheet below it (a short circuit).

The guiding action of a flat conducting surface can also be enhanced by adding a dielectric coating or slab of thickness d. With a launcher, as in Fig. 16-41, the combination forms another type of surface-wave antenna. The electric field is vertically polarized but has a small forward tilt at the dielectric surface. For a sufficient thickness d, the fields attenuate perpendicular to the surface (y direction) as $e^{-\alpha y}$, where

$$\alpha = \frac{2\pi}{\lambda_0} \sqrt{\varepsilon_r - 1} \qquad (\text{Np m}^{-1}) \tag{14}$$

For $\varepsilon_r = 2$, the attenuation is over 50 dB λ^{-1}. Thus, the fields are confined close to the surface.

Corrugated and dielectric slab surface-wave antennas, as in Figs. 16-40 and 16-41, with a length of several λ and a width of 1λ or more (into page), produce end-fire beams with gain proportional to their length and width. It is assumed that the conducting surface (ground plane) extends beyond the end of the corrugations or slab. If not, the beam direction tends to be off end-fire (elevated). For optimum patterns, the depth of the slots or thickness of the slab may be tapered at both ends.

Although *surface-wave antennas* take many forms, the ones described above are typical. They are traveling-wave antennas carrying a bound wave with the energy flowing *above* the guiding surface and with velocity $v < c$ (*slow wave*). **E** would be perpendicular to the surface except for its forward tilt.

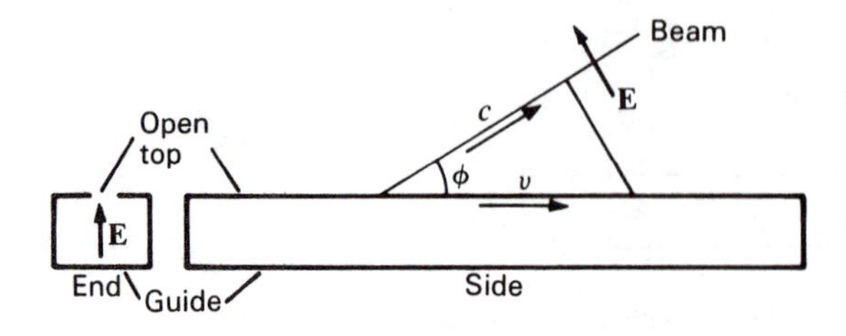

Figure 16-42 Open-top waveguide antenna with continuous energy leakage.

Leaky-wave antennas are also traveling-wave types but ones in which radiating energy leaks continuously or periodically from along the length of the guiding structure, with most of the energy flow *within* the structure. Typically, but not necessarily, the structure carries a *fast wave* ($v > c$). A hollow metal waveguide is an example. With one wall removed, energy can leak out continuously all along the guide.

A leaky-waveguide antenna of this type is shown in Fig. 16-42. Since the wave velocity v in the guide is faster than the velocity c of light, the radiation forms a beam inclined at an angle ϕ with the guide as given by

$$\phi = \cos^{-1} \frac{c}{v} \tag{15}$$

For $v = 1.5c$, $\phi = 48°$. Since v is a function of the frequency, the beam angle ϕ may be scanned by a change in frequency.

A leaky-wave antenna may also be constructed by cutting holes or slots at a regular spacing along the waveguide wall as in Fig. 16-43 (see also the slotted waveguide of Fig. 13-5). Leakage may be controlled by the slot or hole size. For a slot or hole perimeter of $\sim 1\lambda$, leakage is large but decreases rapidly with a decrease in perimeter. This periodic structure radiates at a beam angle ϕ given from (7-11-8) by

$$\phi = \cos^{-1} \left[\frac{\lambda_0}{\lambda_g} + \frac{m}{s/\lambda_0} \right] \tag{16}$$

where s = hole or slot spacing, m
λ_0 = free-space wavelength, m
λ_g = wavelength in guide, m
m = mode number, $0, \pm\frac{1}{2}, \pm 1, \ldots$

Example. Find the beam angle ϕ for $\lambda_g = 1.5\lambda_0$, $s = \lambda_0$ and $m = -1$.

Solution. From (16),

$$\phi = \cos^{-1} \left[\frac{1}{1.5} - \frac{1}{1} \right] = 109.5° \qquad \text{(in back-fire direction)}$$

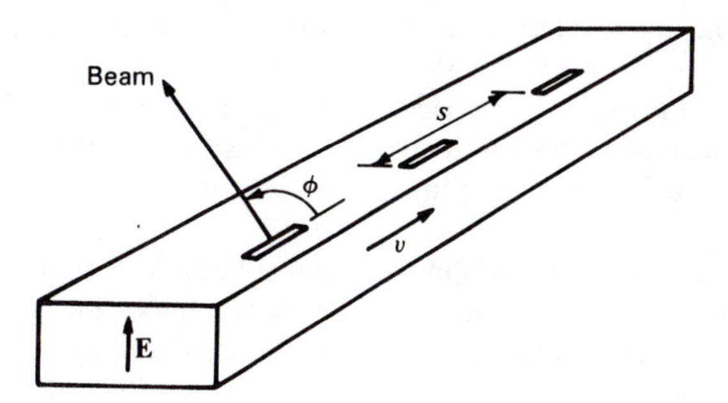

Figure 16-43 Slotted-waveguide antenna with leakage at periodic (discrete) intervals.

If the slots are spaced $\lambda/2$ instead of 1λ, proper phasing requires that alternate ones be placed on opposite sides of the centerline. For this case $m = -\frac{1}{2}$ in (16). The beam angle ϕ may be scanned by a change in frequency. Although described here as a leaky-wave antenna, the antenna of Fig. 16-43 may also be considered as simply an array of waveguide slots.

Consider now a dielectric slab waveguide of thickness d with wave injected at small enough α (below the critical angle) so that the wave is totally internally reflected and propagates by multiple reflections inside the slab. The situation is similar to that for the submerged radio transmitter in Sec. 16-14, except that here the denser medium has upper *and* lower boundaries and a thickness of the order of $1\lambda_0$. Although the energy is transmitted inside the slab, fields exist above and below. These, however, are evanescent and decay exponentially away from the slab. They convey no energy. (Recall the evanescent wave above the water surface in Sec. 16-14.)

It might be supposed that any wave injected into the slab at an angle below the critical value will propagate, but, because of multiple-reflection interference, waves will only propagate at certain angles (eigenvalues).[1] Figure 16-44 is an end view of a dielectric slab waveguide carrying a wave with electric field parallel to

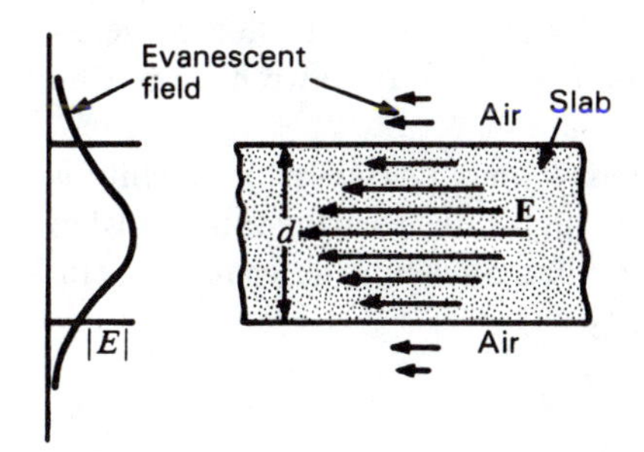

Figure 16-44 End view of dielectric slab waveguide with **E** parallel to the slab surfaces. Propagation is into the page. Arrows and graph (at left) indicate variation in magnitude of **E** across the thickness d of the slab. The field outside the slab is evanescent and transmits no energy.

[1] J. D. Kraus, *Electromagnetics*, 3rd ed., McGraw-Hill, 1984, p. 590.
D. Marcuse, *Theory of Dielectric Optical Waveguides*, Academic Press, 1974.

the surfaces (perpendicular to the plane of incidence). In the notation of Sec. 16-14 this is an $\mathbf{E}_{\perp}$ field. Propagation is into the page. The evanescent field above and below the slab matches the field inside but carries no energy. However, discontinuities on the slab surface can cause energy leakage from inside and radiation. Thus, a dielectric slab with surface discontinuities can act as a leaky-wave antenna.

A dielectric cylinder can also serve as a guide in the same manner as a slab. At or near optical wavelengths the cylinder diameter can be physically small or threadlike. Such guides, or *optical fibers*, consist typically of a transparent glass *core* surrounded by a glass *sheath* or *cladding* of slightly lower index of refraction, with the *sheath* enclosed in an opaque protective jacket. A typical core fiber 25 μm in diameter is as fine as a human hair and can carry one thousand 2-way voice communication channels with an attenuation as small as 1 dB km^{-1} at light or infrared wavelengths ($\frac{1}{2}$ to 1 μm).

The literature on surface- and leaky-wave antennas is extensive. Summary treatments are given by Zucker and by Mittra.[1]

16-16 ANTENNA DESIGN CONSIDERATIONS FOR SATELLITE COMMUNICATION. In 1945, while a radar officer in the Royal Air Force, Arthur C. Clarke[2] published an article in *Wireless World* in which he proposed the use of artificial satellites and, in particular, those in a geostationary orbit, as a solution to the world's communication problem. The idea then seemed far fetched to many but 12 years later Sputnik went up and only 6 years after that, in 1963, the first successful geostationary satellite, Syncom 2, was put into orbit and Clarke's proposal became a reality. Now there are hundreds of satellites[3] in the geostationary or Clarke orbit with more being added at frequent intervals. The satellites move with the earth as though attached to it so that from the earth each appears to remain stationary above a fixed point on the equator. These satellites form a ring around the earth at a height of 36 000 km above the equator, putting the earth in the company of other ringed planets, Saturn and Uranus, with the difference that the earth's ring is man-made and not natural. Arthur Clarke has observed that the ancient superstition that our destinies are controlled by celestial bodies has at last come true, except that the bodies are ones we have put up there ourselves.

Having a transmitting (transponder) antenna on a Clarke-orbit satellite is like having it on an invisible tower 36 000 km high. Since earth-station dishes look upward at the sky, ground reflections are eliminated but the ground is still

---

[1] F. J. Zucker (chap. 12) and R. Mittra (chap. 10), in *Antenna Engineering Handbook*, McGraw-Hill, 1984.

[2] A. C. Clarke, "Extra-terrestrial Relays," *Wireless World*, **51**, 305–308, October 1945.

[3] Nearly 500 at the end of 1986.

involved through its effect on the antenna noise temperature via side and back lobes.

The spacing of satellites in the Clarke orbit is closely connected with the design of both satellite and earth-station antennas. Thus, if beam widths and sidelobe levels are reduced, the spacing can be reduced and more satellites accommodated in orbit.

The full orbit utilization problem involves not only satellite placement and spacing but also the available spectrum. Thus, satellites may be parked closer together than beam widths would allow if they operate in different frequency bands. However, both *orbit space* and *spectrum space* are limited resources which must be allocated in an optimum manner to meet the tremendous international demand for satellite communications. The problem of maximizing access to the Clarke orbit is under study and planning by the satellite orbit-use (ORB) sessions of the World Administrative Radio Conference (WARC), with its first session (ORB-1) in 1985 and a second session (ORB-2) scheduled for 1988.[1]

To permit closer spacings than otherwise, North American satellites sharing the same frequencies use an alternate linear-polarization technique. For example, the even-numbered channels of a satellite may be polarized parallel to the orbit with odd-numbered channels perpendicular to it, while the adjacent satellites on either side have the polarizations reversed (odd-numbered parallel and even-numbered perpendicular to the orbit). Some satellites serving other parts of the world use circular polarizations of opposite hand to do the same thing. Thus, the response level of the earth-station antenna to the opposite state of polarization is a factor.

It is necessary that the earth-station antenna have gain and pattern properties capable of providing a satisfactory S/N ratio with respect to noise and interference sources (as from adjacent satellites). The satellite transponder power, antenna gain and pattern are also factors. At the Clarke-orbit height of 36 000 km, a satellite spacing of 1° amounts to a physical separation of 630 km. Although each satellite wanders about its assigned orbital location (usually in a systematic manner), this "station-keeping" motion is required by regulation to be less than 60 km (0.1°) so that spacing even closer than 1° is possible without danger of satellites colliding.[2]

Some of the antenna design aspects may be illustrated by an example.

Example. Determine the required parabolic dish diameter of a 4-GHz (C-band) earth-station antenna if its system temperature = 100 K for an S/N ratio of 20 dB

[1] H. J. Weiss, "Maximizing Access to the Geostationary-Satellite Orbit," *ITU J.*, **53**, 469–477, August 1986.

[2] The satellite antenna pointing accuracy in roll, pitch and yaw are also usually of the order of 0.1° or less. Roll, pitch and yaw are of primary concern in the satellite antenna design but of secondary importance in the earth-station antenna design. The Astra satellites have specifications of 0.06° roll, 0.07° pitch, 0.22° yaw and ±0.05° station keeping.

and bandwidth = 30 MHz with satellite transponder power = 5 W, satellite parabolic dish diameter = 2 m and spacing between satellites = 2°.

Solution. From the Friis transmission formula and the Nyquist noise-power equation, we have from (17-3-9) that the S/N (actually C/N or carrier-to-noise) ratio for isotropic antennas and a transmitter power of 1 W is given by

$$\frac{S}{N} = \frac{\lambda^2}{16\pi^2 r^2 k T_{sys} B} \tag{1}$$

where λ = wavelength
r = downlink distance, 36 000 km
k = Boltzmann's constant = 1.38×10^{-23} J K^{-1}
T_{sys} = system temperature, K
B = bandwidth, Hz

Introducing the indicated values in (1) we obtain

$$\frac{S}{N} = -61.8 \text{ dB} \tag{2}$$

for the downlink at 1 W isotropic.

For the $D = 2$ m diameter satellite dish at 50 percent aperture efficiency, the antenna gain G_s is given by

$$G_s = \frac{4\pi A_e}{\lambda^2} = 35.5 \text{ dB} \tag{3}$$

where $A_e = \frac{1}{2}\pi\left(\frac{D^2}{4}\right) = 1.6 \text{ m}^2$
$\lambda = 0.075$ m

The transponder power of 5 W gives an additional 7 dB (=log 5) gain. The required earth-station antenna gain G_E must then be sufficient to make the system S/N ratio ≥ 20 dB, or

$$G_E = 20 + 61.8 - 35.5 - 7 = 39.3 \text{ dB} \tag{4}$$

Thus, the required earth-station effective aperture

$$A_e = \frac{G_E \lambda^2}{4\pi} = 3.8 \text{ m}^2 \tag{5}$$

At an assumed aperture efficiency of 50 percent the required physical aperture A_p is twice this, or 7.6 m^2, making the required diameter of the earth-station antenna

$$D_E = 2\sqrt{\frac{A_p}{\pi}} = 3.1 \text{ m} \tag{6}$$

This diameter meets the S/N ratio requirement. To determine if it also meets the 2° spacing requirement we should know the illumination taper across the dish aper-

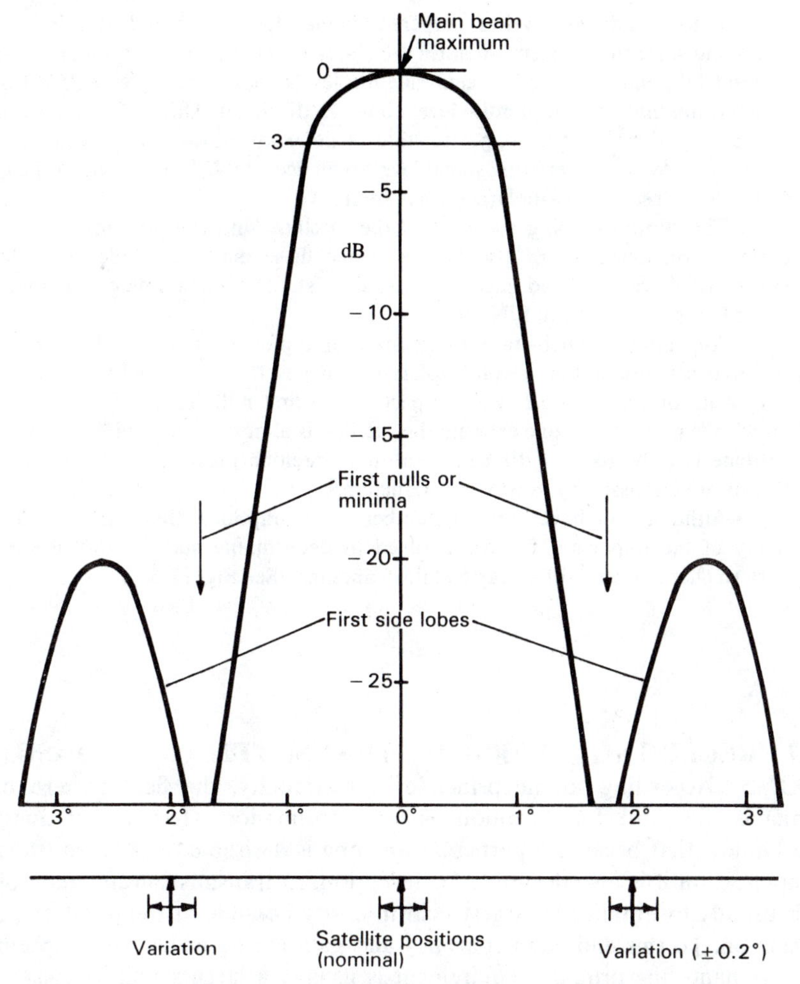

Figure 16-45 Pattern of C-band earth-station dish for worked example. Pattern is shown with respect to satellite positions at 2° spacing in the Clarke orbit.

ture. This is not given. However, we assumed 50 percent aperture efficiency which implies substantial taper. Accordingly, the beam widths may be estimated as

$$\text{HPBW} \simeq \frac{65^\circ}{D_\lambda} = \frac{65^\circ}{3.1 \text{ m}/0.075 \text{ m}} = 1.6^\circ \tag{7}$$

and

$$\text{BWFN} \simeq \frac{145^\circ}{D_\lambda} = 3.5^\circ \tag{8}$$

Thus, first nulls or minima are approximately 1.75° either side of the satellite or 0.25° less than the 2° spacing of the adjacent satellites, as indicated in Fig. 16-45.

The first sidelobe level of a tapered circular aperture should be at least 20 dB down and since the adjacent satellites are closer to the first nulls or minima than the first sidelobe maxima, the adjacent satellite levels should be at least 25 dB down, which could make the total noise level about 18 dB down. Although this may not be satisfactory, the use of an opposite state of polarization on the adjacent satellites should reduce their interfering signal level to an acceptable level provided the earth-station response to cross-polarization is small.

The station-keeping accuracy of the satellites and the pointing error of the earth-station antenna are also factors. Since these may vary independently, the effects will be random and must be assessed by statistical methods giving upper and lower limits to the overall S/N ratio.

Assuming an earth-station antenna pointing error of $\pm 0.1^\circ$, the same as the required satellite station-keeping tolerance, means that the satellite position may vary a maximum of $\pm 0.2^\circ$ with respect to the first null. This is indicated in Fig. 16-45. We note that at one extreme the satellite is almost at the null but at the other extreme is only about 4 dB below the first sidelobe maximum. These and other factors are discussed by Jansky and Jeruchim.[1]

Although we have made a number of assumptions, the example illustrates many of the important factors involved in determining suitable dimensions and pattern characteristics of an earth-station antenna. (See Fig. 12-58.)

16-17 RECEIVING VERSUS TRANSMITTING CONSIDERATIONS. According to the principle of reciprocity, the field pattern of an antenna is the same for reception as for transmission. However, it does not always follow that because a particular antenna is desirable for a given transmitting application it is also desirable for reception. In transmission the main objective is usually to obtain the largest field intensity possible at the point or points of reception. To this end, high efficiency and gain are desirable. In reception, on the other hand, the primary requirement is usually a large signal-to-noise ratio. Thus, although high efficiency and also gain may be desirable, they are important only insofar as they improve the signal-to-noise ratio. As an example, a receiving antenna with the pattern of Fig. 16-46*a* may be preferable to a higher-gain antenna with the more directional pattern of Fig. 16-46*b*, if there is an interfering signal or noise arriving from the back direction as indicated. Although the gain of the antenna with the pattern at (*a*) is less, it may provide a higher signal-to-noise ratio since its pattern has a null directed toward the source of the noise or interference (see Sec. 11-13 on adaptive arrays).

However, by way of contrast, suppose that circuit noise in the receiver is the

---

[1] D. M. Jansky and M. C. Jeruchim, *Communication Satellites in the Geostationary Orbit*, Artech House, 1983.

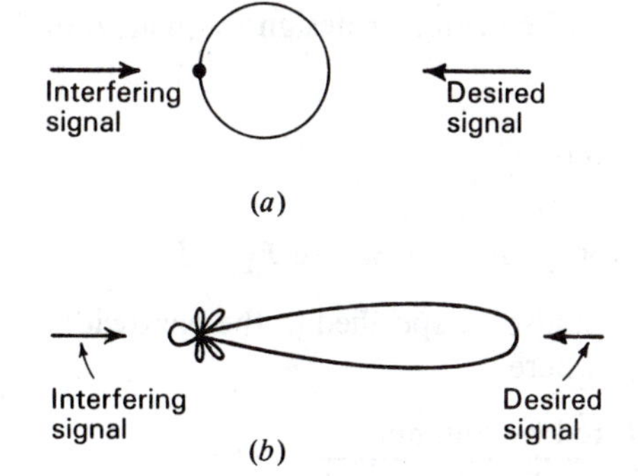

Figure 16-46 Low-gain antenna with high front-to-back ratio at (*a*) is better than higher-gain antenna with lower front-to-back ratio at (*b*) for reception in the presence of a back-side interfering signal. However, for transmission the higher-gain antenna at (*b*) is superior.

limiting factor. Then high antenna gain and efficiency would be important in order to raise the signal-to-noise ratio.

In direction-finding antennas, the directional characteristic of the antenna is employed to determine the direction of arrival of the radio wave. If the signal-to-noise ratio is high, a null in the field pattern may be used to find the direction of arrival. With a low signal-to-noise ratio, however, the maximum of the main lobe may provide a more satisfactory indication.[1]

16-18 BANDWIDTH CONSIDERATIONS. The useful bandwidth of an antenna depends, in general, on *both its pattern and impedance* characteristics.[2] In thin dipole antennas the bandwidth is usually determined by the impedance variation since the pattern changes less rapidly.[3] However, with very thick cylindrical antennas or biconical antennas of considerable cone angle, the impedance characteristics may be satisfactory over so wide a bandwidth that the pattern variation determines one or both of the frequency limits. The pattern may also determine the useful bandwidth of horn antennas, metal-plate lens antennas or zoned lens antennas.

If the acceptable bandwidth for pattern exceeds that for impedance, the bandwidth can be arbitrarily specified by the frequency limits F_1 and F_2 at which the VSWR on the transmission line exceeds an acceptable value. What is acceptable varies widely depending on the application. In some cases the VSWR must be close to unity. In others it may be as high as 10 to 1 or higher. The bandwidth

[1] A. Alford, J. D. Kraus and E. C. Barkofsky, in *Very High Frequency Techniques*, Radio Research Laboratory Staff, McGraw-Hill, New York, 1947, chap. 9.

[2] The pattern is understood to include *polarization* characteristics.

[3] A dipole $\lambda/2$ long has a half-power beam width of 78°. If the frequency is reduced so that the dipole length approaches an infinitesimal fraction of a wavelength, the beam width only increases from 78 to 90°, while if the frequency is doubled so that the dipole is 1λ long the beam width decreases from 78 to about 47°.

can be specified as the ratio of $F_2 - F_1$ to F_0 (the center or design frequency) or in percent as

$$\frac{F_2 - F_1}{F_0} \times 100$$

Another definition is the simple ratio F_2/F_1 (or F_2/F_1 to 1) where $F_2 > F_1$.

The bandwidth due to the impedance can also be specified (if the bandwidth is small) in terms of its reciprocal or Q at F_0, where

$$Q = 2\pi \frac{\text{total energy stored by antenna}}{\text{energy dissipated or radiated per cycle}}$$

16-19 GRAVITY-WAVE ANTENNAS. ROTATING BOOM FOR TRANSMITTING AND WEBER BAR FOR RECEIVING. Gravity, which we frequently think of as a force which holds us to our seat in a chair, is a manifestation of Albert Einstein's theory of general relativity in which space, time and matter are related in a geometric concept.[1] Although accelerating masses may, in theory, radiate gravity waves (analogous to the electromagnetic waves radiated by accelerated charges), gravitational radiation is a smaller-order effect and tends to be weak.[2]

Consider, for example, the gravitational radiation from a uniform solid steel bar rotating end-over-end as a gravity-wave transmitting antenna. According to Misner, Thorne and Wheeler[3] in their classic book *Gravitation*, the bar will radiate a power

$$P = 1.5 \times 10^{-54} M^2 L^4 \omega^6 \qquad \text{(W)} \tag{1}$$

where M = mass of bar, kg
L = length of bar, m
ω = angular velocity, rad s^{-1}

If the bar's mass is 500 tonnes ($M = 5 \times 10^5$ kg), is 20 m long ($L = 20$ m) and rotates end-over-end 270 times per minute, which is about as fast as the tensile strength of the bar will permit ($\omega = 2\pi \times 270/60$ rad s^{-1}), we have, introducing these values in (1), that the very small power

$$P = 2.2 \times 10^{-29} \text{ W} \tag{2}$$

is radiated in gravity waves.

---

[1] A. Einstein, "Zur allgemeinen Relativitätstheorie," *Preuss. Akad. Wiss.*, **47**, 778–786, 799–801, 1915.

[2] Electric charges, being of two signs (positive and negative), can form dipoles but masses, being of only one sign, cannot. Thus, gravitational dipole radiation is not possible but gravitational quadrupolar radiation is.

[3] C. W. Misner, K. S. Thorne and J. A. Wheeler, *Gravitation*, Freeman, 1973.

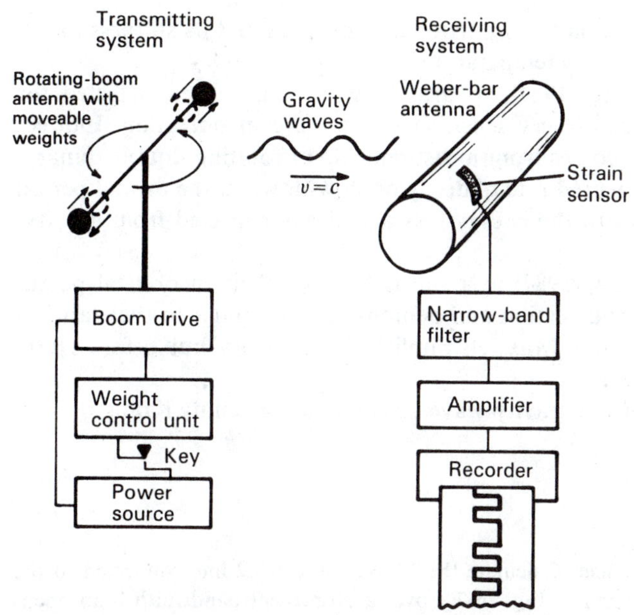

Figure 16-47 Proposed gravity-wave communication link with rotating-boom transmitting antenna and Weber-bar receiving antenna. The Weber bar should be suspended in a cushioned, evacuated, refrigerated tank (not shown).

Although the above example shows that gravity-wave radiation is very inefficient and weak, let us design a hypothetical gravity-wave communication link. Instead of a uniform solid bar as the transmitting antenna let us use a pair of moveable weights on a rotating boom as in Fig. 16-47. The rotating boom idles with the weights at the ends. With weights drawn in, the boom speeds up and sends out waves of a higher frequency to which a Weber-bar receiving antenna is tuned. With the position of the weights under the control of a telegraph key, we have a frequency-shift CW system with the transmitted gravity-wave frequency equal to twice the boom rotation rate. The velocity of propagation of the gravity waves is presumed to be equal to the velocity of light.

Turning next to the receiving system, let us use an antenna similar to the one developed by the pioneer gravity-wave scientist Professor Joseph Weber of the University of Maryland. It consists of a several-tonne aluminum bar, or Weber-bar antenna, suspended at its midpoint and isolated from sound and vibration by enclosing it in a refrigerated, cushioned, evacuated tank (Fig. 16-47). A passing gravity wave should make the bar vibrate as though tapped by a tiny hammer. The vibration of the bar generates electrical signals in strain sensors attached to the bar. These signals are then amplified and recorded.

For optimum performance the bar's natural frequency should be the same as the frequency of the gravity waves from the rotating transmitting antenna. The Weber-bar antenna should also be broadside to the wave direction. The sensitivity of the Weber-bar antenna is proportional to its mass and its stiffness (or Q) and inversely proportional to its temperature.

As of the present date (1987) no gravity waves have been detected for certain although a number of very sensitive systems are in operation. Indirect evidence of their existence comes from measurements of rotating double (binary) star systems where the measured rate of decay or slow-down of the orbital period is found to agree closely with the energy loss and decay expected from gravity-wave radiation.

Gravity waves are now (1987) in a status between their theoretical postulation by Einstein and their detection and demonstration, similar to the status of radio waves in the years after Maxwell published his theory but before Hertz produced and measured them.

Additional discussions of gravity-wave detection are given by Kraus.[1]

PROBLEMS[2]

16-1 VSWR for dipole antenna. Calculate the VSWR on a 65-Ω line connected to the $L/D = 60$ dipole antenna of Fig. 16-22*f* over a 30 percent bandwidth if an open-ended line of 40 Ω characteristic impedance is connected in parallel with the antenna terminals. The line is 180° long at the center frequency F_0.

★16-2 Stub impedance.

(*a*) What is the terminal impedance of a ground-plane mounted stub antenna fed with a 50-Ω air-filled coaxial line if the VSWR on the line is 2.5 and the first voltage minimum is 0.17λ from the terminals?

(*b*) Design a transformer so that the VSWR = 1.

16-3 Square loop. Calculate and plot the far-field pattern in the plane of a loop antenna consisting of four $\lambda/2$ center-fed dipoles with sinusoidal current distribution arranged to form a square $\lambda/2$ on a side. The dipoles are all in phase around the square.

16-4 Triangular loop. Calculate and plot the far-field pattern in the plane of a loop antenna consisting of three $\lambda/2$ center-fed dipoles with sinusoidal current distribution arranged to form an equilateral triangle $\lambda/2$ on a side. The dipoles are all in phase around the triangle.

16-5 Microstrip line. For a polystyrene substrate ($\varepsilon_r = 2.7$) what width–substrate thickness ratio results in a 50-Ω microstrip transmission line?

★16-6 Surface-wave powers. A 100-MHz wave is traveling parallel to a copper sheet ($|Z_c| = 3.7 \times 10^{-3}$ Ω) with **E** (=100 V m^{-1} rms) perpendicular to the sheet. Find (*a*) the Poynting vector (watts per square meter) parallel to sheet and (*b*) the Poynting vector into the sheet.

[1] J. D. Kraus, *Radio Astronomy*, 2nd ed., Cygnus-Quasar, 1986, pp. 9–29, 9–33; J. D. Kraus, *Our Cosmic Universe*, Cygnus-Quasar, 1980, pp. 211–214, 225.

[2] Answers to starred (★) problems are given in App. D.

16-7 Surface-wave powers. A 100-MHz wave is traveling parallel to a conducting sheet for which $|Z_c| = 0.02\ \Omega$. If **E** is perpendicular to the sheet and equal to 150 V m^{-1} (rms), find (*a*) watts per square meter traveling parallel to the sheet and (*b*) watts per square meter into the sheet.

★16-8 Surface-wave power. A plane 3-GHz wave in air is traveling parallel to the boundary of a conducting medium with **H** parallel to the boundary. The constants for the conducting medium are $\sigma = 10^7$ ℧ m^{-1} and $\mu_r = \varepsilon_r = 1$. If the traveling-wave rms electric field $E = 75$ mV m^{-1}, find the average power per unit area lost in the conducting medium.

16-9 Surface-wave current sheet. A TEM wave is traveling in air parallel to the plane boundary of a conducting medium. Show that if $K = \rho_s v$, where K is the sheet-current density in amperes per meter, ρ_s is the surface charge density in coulombs per square meter and v the velocity of the wave in meters per second, it follows that $K = H$, where H is the magnitude of the **H** field of the wave.

16-10 Coated-surface wave power. Show that for a dielectric-coated conductor, as in Fig. 16-41, the ratio of the power transmitted in the dielectric P_d to the power transmitted in the air P_a is given by

$$\frac{P_d}{P_a} = \frac{\cos \beta d}{\sin^3 \beta d}(\sin 2\beta d - 2\beta d)$$

where d is the thickness of the dielectric coating.

★16-11 Coated-surface wave cutoff. A perfectly conducting flat sheet of large extent has a dielectric coating ($\varepsilon_r = 3$) of thickness $d = 5$ mm. Find the cutoff frequency for the TM_0 (dominant) mode and its attenuation per unit distance.

16-12 Lunar communication by surface wave. Discuss the possibilities of using dielectric-slab surface-wave modes for radio communication around the moon over long distances (1000 km or more). Note that the moon has no ionosphere. See, for example, W. W. Salisbury and D. L. Fernald, "Postocculation Reception of Lunar Ship Endeavour Radio Transmission," *Nature*, **234,** 95, Nov. 12, 1971; also A. F. Wickersham, Jr., "Generation, Detection and Propagation on the Earth of HF and VHF Radio Surface Waves," *Nature*, **230**, 125–130, Apr. 5, 1971.

16-13 Horizontal dipole above ground. A thin $\lambda/2$ dipole is parallel to a flat, perfectly conducting ground at a height h above it. (*a*) Calculate and plot the gain of the dipole in the zenith direction as a function of height h for heights from zero to λ. Express the gain with respect to a $\lambda/2$ dipole in free space. Assume zero losses. (*b*) Repeat (*a*) for dipole loss resistance $R_L = 1\ \Omega$.

★16-14 Overland TV for HP, VP and CP.

(*a*) A typical overland microwave communication circuit for AM, FM or TV between a transmitter on a tall building and a distant receiver involves 2 paths of transmission, one a direct path (length r_0) and one an indirect path with ground reflection (length $r_1 + r_2$), as suggested in Fig. P16-14. Let $h_1 = 300$ m and $d = 5$ km. For a frequency of 100 MHz calculate the ratio of the power received per unit area to the transmitted power as a function of the height h_2 of the receiving antenna. Plot these results in decibels as abscissa versus h_2 as ordinate for 3 cases with transmitting and receiving antennas both (1) vertically polarized, (2) horizontally polarized and (3) right-circularly polarized for h_2 values from 0 to 100 m. Assume that the transmitting antenna is isotropic and

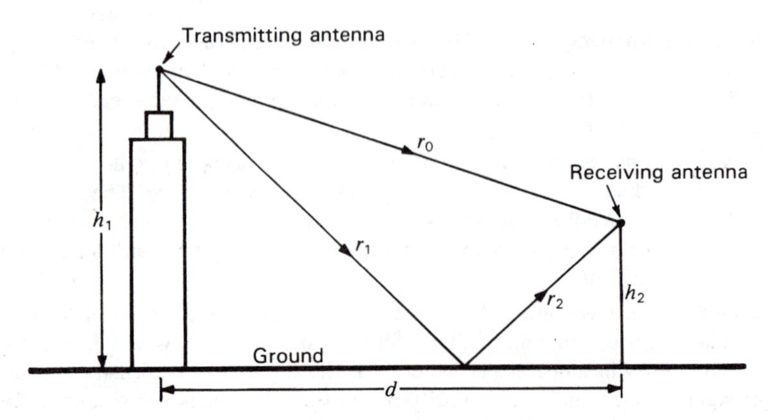

Figure P16-14 Overland microwave communication circuit.

that the receiving antennas are also isotropic (all have the same effective aperture). Consider that the ground is flat and perfectly conducting.

(*b*) Compare the results for the 3 types of polarization and show that circular polarization is best from the standpoint of both the noncriticalness of the height h_2 and of the absence of echo or ghost signals. Thus, for horizontal or vertical polarization the direct and ground-reflected waves may cancel at certain heights while at other heights, where they reinforce, the images on the TV screen may be objectionable because the time difference via the 2 paths produces a double image (a direct image and its ghost).

(*c*) Extend the comparison of (*b*) to consider the effect of other buildings or structures which may produce additional paths of transmission.

Note that direct satellite-to-earth TV downlinks are substantially free of these reflection and ghost image effects.

16-15 Horizontal dipole above imperfect ground. Calculate the vertical plane field pattern broadside to a horizontal $\lambda/2$ dipole antenna $\lambda/4$ above actual homogeneous ground with constants $\epsilon_r' = 12$ and $\sigma = 2 \times 10^{-3}$ ℧ m^{-1} at (*a*) 100 kHz and (*b*) 100 MHz.

16-16 Short vertical dipole above imperfect ground. The center of a short vertical dipole ($l < \lambda/10$) is located $\lambda/4$ above actual homogeneous ground with the same constants as in Prob. 16-15. Calculate the vertical plane field pattern at (*a*) 1 MHz and (*b*) 100 MHz.

***16-17 DF and monopulse.** Many direction-finder (DF) antennas consist of small (in terms of λ) loops giving a figure-of-eight pattern as in Fig. P16-17*a*. Although the null is sharp the bearing (direction of transmitter signal) may have considerable uncertainty unless the S/N ratio is large. To resolve the 180° ambiguity of the loop pattern, an auxiliary antenna may be used with the loop to give a cardiod pattern with broad maximum in the signal direction and null in the opposite direction.

The maximum of a beam antenna pattern, as in Fig. P16-17*b*, can be employed to obtain a bearing with the advantage of a higher S/N ratio but with reduced pattern change per unit angle. However, if 2 receivers and 2 displaced beams are used, as in Fig. P16-17*c*, a large power-pattern change can be combined with a high S/N ratio. An arrangement of this kind for receiving radar echo signals

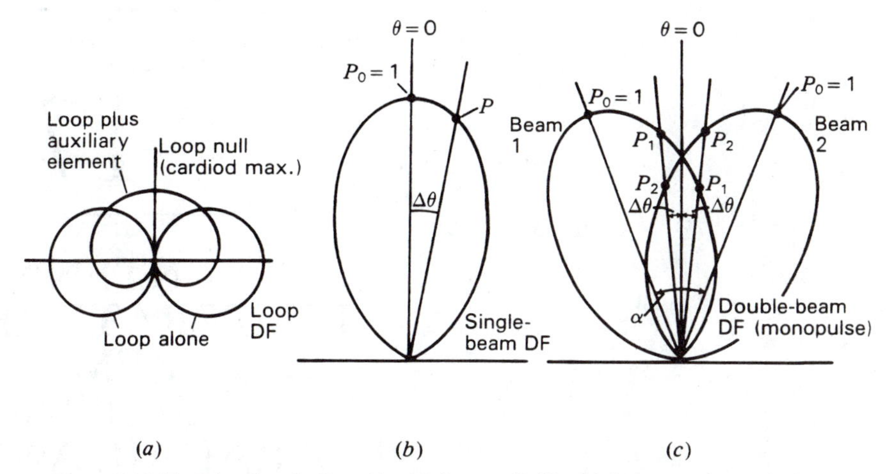

Figure P16-17 Direction finding: (*a*) with loop null, (*b*) with beam maximum and (*c*) with double beam (monopulse).

can give bearing information on a single pulse (*monopulse radar*). If the power received on beam 1 is P_1 and on beam 2 is P_2, then if $P_2 > P_1$ the bearing is to the right. If $P_1 > P_2$ the bearing is to the left and if $P_1 = P_2$ the bearing is on axis (boresight). (With 4 antennas, bearing information left-right *and* up-down can be obtained.)

(*a*) If the power pattern is proportional to $\cos^4 \theta$, as in Fig. P16-17*c*, determine P_2/P_1 if the interbeam (squint) angle $\alpha = 40°$ for $\Delta\theta = 5$ and $10°$.

(*b*) Repeat for $\alpha = 50°$.

(*c*) Determine the P_0/P_1 of the single power pattern of Fig. P16-17*b* for $\Delta\theta = 5$ and $10°$ if the power pattern is also proportional to $\cos^4 \theta$.

(*d*) Tabulate the results for comparison and indicate any improvement of the double over the single beam.

16-18 Path difference on overland radio link. If $h_1 = h_2$ and $d \gg h_1$ in Prob. 16-14 (Fig. P16-14), show that the path difference of direct and reflected rays is $2h_1^2/d$.

16-19 Antennas over imperfect ground. Write and run computer programs for Probs. 16-15 and 16-16 with a menu for height, frequency and ground constants ε_r' and σ.

16-20 Square-corner monopulse radar antenna. Design a square-corner reflector antenna with 2 off-axis feeds to operate as a monopulse radar. See footnote preceding equation (12-3-7).

★16-21 Signalling to submerged submarines. Calculate the depths at which a 1 μV m^{-1} field will be obtained with E at the surface equal to 1 V m^{-1} at frequencies of 1, 10, 100 and 1000 kHz. What combination of frequency and antennas is most suitable?

CHAPTER

17

ANTENNA TEMPERATURE, REMOTE SENSING, RADAR AND SCATTERING

17-1 INTRODUCTION. The concepts of *antenna temperature* and *system temperature*,[1] mentioned in earlier chapters, are developed here in more detail and their relation shown to the *signal-to-noise ratio*. The application of the temperature concept to *remote sensing* is then discussed. The next sections cover *radar* and the *radar equation*, leading to considerations of *scattering* and the *radar or scattering cross section.*

17-2 ANTENNA TEMPERATURE, INCREMENTAL AND TOTAL. Referring to Fig. 17-1*a*, the *noise power per unit bandwidth* available at the terminals of a resistor of resistance R and temperature T is given by the Nyquist[2] relation as

$$p = kT \qquad (\text{W Hz}^{-1}) \tag{1}$$

[1] For more detailed treatment see J. D. Kraus, *Radio Astronomy*, 2nd ed., Cygnus-Quasar, 1986.

[2] H. Nyquist, "Thermal Agitation of Electric Charge in Conductors," *Phys. Rev.*, **32**, 110–113, 1928.

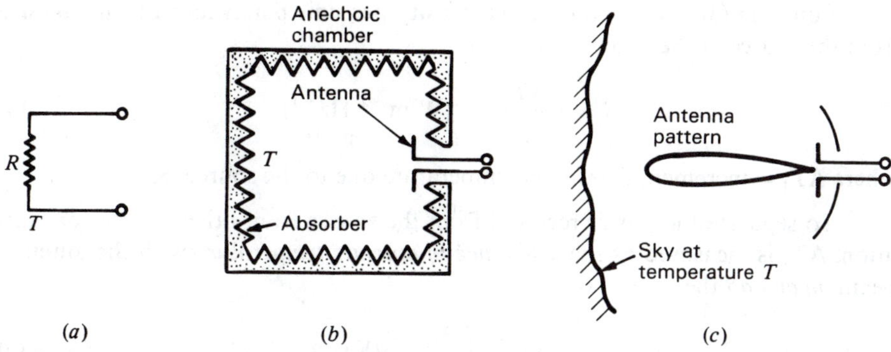

Figure 17-1 (*a*) Resistor at temperature T. (*b*) Antenna in an anechoic chamber at temperature T. (*c*) Antenna observing sky at temperature T. The same noise power per unit bandwidth is available at the terminals in all 3 cases.

where p = power per unit bandwidth, W Hz^{-1}
k = Boltzmann's constant $= 1.38 \times 10^{-23}$ J K^{-1}
T = absolute or Kelvin temperature, K

If the resistor R is replaced by a lossless antenna of radiation resistance R in an anechoic chamber, as in Fig. 17-1*b*, at temperature T the noise power per unit bandwidth available at the terminals is unchanged.

Now if the antenna is removed from the chamber and pointed at a sky of temperature T, the noise power at the terminals is the same as for the two previous cases.[1]

If p is independent of frequency, the total power P is given by p multiplied by the bandwidth Δf, or

$$P = kT\,\Delta f \qquad \text{(W)} \tag{2}$$

where P = power, W
Δf = bandwidth, Hz

Suppose that the antenna of Fig. 17-1*c* has an effective aperture A_e and that its beam is directed at a source of radiation which produces a power density per unit bandwidth or flux density S at the antenna. The power received from the source is then given by

$$P = SA_e\,\Delta f \qquad \text{(W)} \tag{3}$$

where S = power density per unit bandwidth, W m^{-2} Hz^{-1}
A_e = effective aperture, m^2
Δf = bandwidth, Hz

[1] It is assumed that the entire antenna pattern "sees" the sky of temperature T.

Equating (3) and (2), the power density per unit bandwidth or *flux density* from the source at the antenna is

$$S = \frac{k\,\Delta T_A}{A_e} \qquad (\text{W m}^{-2}\ \text{Hz}^{-1}) \tag{4}$$

where ΔT_A = incremental antenna temperature due to the source, K

To separate the power received from the source from other sources of radiation, ΔT_A is measured as the *difference in antenna temperature* with the antenna beam *on and off* the source, or

$$\Delta T_A = \frac{SA_e}{k} \qquad (\text{K}) \tag{5}$$

This *incremental antenna temperature* ΔT_A is equal to the *change in temperature* of the resistor R (substituted for the antenna) required to produce an equal power P as given by (2).

In practice it is not necessary to make the resistor temperature change equal to ΔT_A in order to measure ΔT_A. Thus, the receiver may be switched alternately to antenna and resistor. Assuming that the receiver output indicator has a linear power scale, let the deflection with the antenna be d_A and with the resistor d_R. If the resistor temperature change is ΔT_R, then the incremental antenna temperature ΔT_A is given by

$$\Delta T_A = \frac{d_A}{d_R}\,\Delta T_R \qquad (\text{K}) \tag{6}$$

where ΔT_A = incremental antenna temperature, K
d_A = deflection with antenna, arbitrary units
d_R = deflection with resistor, same units as for d_A
ΔT_R = resistor temperature change, K

It is assumed that the antenna and resistor are both matched to the transmission line and receiver.

If the antenna has no side or back lobes and its *beam is narrower* than the source, the incremental antenna temperature ΔT_A is equal to the source temperature T_s. Thus, the antenna and receiver act as a *passive remote sensing device* which can determine the temperature of near or distant regions within the antenna beam. It is as though the antenna's radiation resistance R is coupled by the beam to the remote regions and acquires a temperature ΔT_A equal to the remote region (or source) temperature T_s.

Let us consider now a different situation, i.e., one in which the antenna *beam is much wider* than the source extent, as in Fig. 17-2. Then, obviously, ΔT_A will be less than T_s. However, if the source solid angle Ω_s and the antenna beam solid angle Ω_A are known, the source temperature is given very simply by

$$T_s = \frac{\Omega_A}{\Omega_s}\,\Delta T_A \tag{7}$$

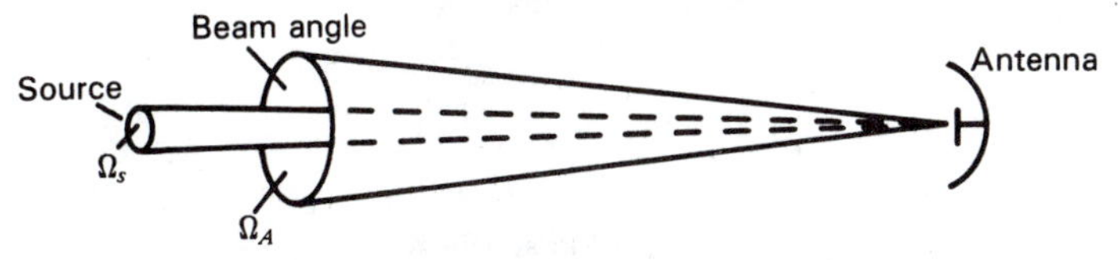

Figure 17-2 Situation where source extent Ω_s is smaller than the beam area Ω_A.

where T_s = source temperature, K
ΔT_A = incremental antenna (noise) temperature, K
Ω_s = source solid angle (see Fig. 17-2), sr
Ω_A = antenna beam solid angle (see Fig. 17-2), sr

It is important to note that the antenna temperature has nothing to do with the physical temperature of the antenna provided the antenna is lossless.

Example 1 Mars temperature. The incremental antenna temperature for the planet Mars measured with the U.S. Naval Research Laboratory[1] 15-m radio telescope antenna at 31.5 mm wavelength was 0.24 K. Mars subtended an angle of 0.005° at the time of the measurement. The antenna HPBW = 0.116°. Find the average temperature of Mars at 31.5 mm wavelength.

Solution. Assuming that Ω_A is given by the solid angle within the HPBW, we have from (7) that the Mars temperature

$$T_s = \frac{\Omega_A}{\Omega_s} \Delta T_A \simeq \frac{0.116^2}{\pi(0.005^2/4)} 0.24 = 164 \text{ K}$$

This temperature is less than the infrared temperature measured for the sunlit side (250 K), implying that the 31.5-mm radiation may originate further below the Martian surface than the infrared radiation. This is an example of *remote-sensing* the surface of another planet from the earth.

The source temperature in the above discussion and example is an *equivalent temperature.* It may represent the physical temperature of a planetary surface, as in the example, but, on the other hand, a celestial plasma cloud with oscillating electrons which is at a physical temperature close to absolute zero may generate radiation with an equivalent temperature of thousands of kelvins. The temperatures we are discussing are thermal (noise) temperatures like those of a perfect emitting-absorbing object called a *blackbody.* A hot object filling the beam of a receiving antenna will ideally produce an antenna temperature equal to its thermometer-measured temperature.[2] However, the oscillating currents of a transmitting antenna can produce an equivalent temperature of millions of degrees (K) even though the antenna structure is at normal outdoor temperature.

---

[1] C. H. Mayer, T. P. McCullough and R. M. Sloanaker, "Observations of Mars and Jupiter at a Wavelength of 3.15 cm," *Astrophys, J.*, **127**, 11–16, January 1958.

[2] Assuming the object's intrinsic impedance = 377 Ω.

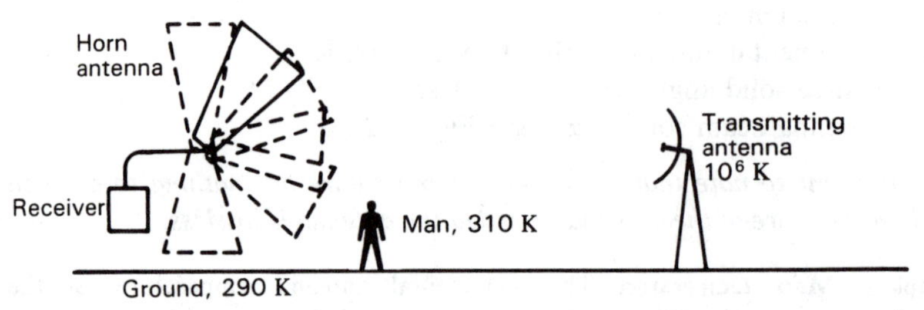

Figure 17-3 Horn antenna directed at various objects senses different temperatures as suggested.

It may be said that the antenna (and its currents) have an equivalent blackbody (or noise) temperature of millions of degrees.

All objects not at absolute zero produce radiation which, in principle, may be detected with a radio antenna-receiver. A few objects are shown in Fig. 17-3 with the equivalent temperatures measured when the horn antenna is pointed at them. Thus, the temperature of a distant quasar is over 10^6 K, of Mars 164 K, of a transmitter on the earth 10^6 K, of a man 310 K, of the ground 290 K,[1] while the empty sky at the zenith is 3 K. This temperature, called the 3 K sky background, is the residual temperature of the primordial fireball which created the universe and is the *minimum possible temperature* of any antenna looking at the sky.

An assumption was made in the above discussion which requires comment. It was assumed that the antenna and source polarizations were matched (same polarization states on the Poincaré sphere; see Sec. 2-36). Although this is possible for the transmitting antenna in Fig. 17-3, it is not possible for the other sources because their radiation is *unpolarized*[2] and *any* antenna, whether linearly or circularly polarized, receives only half of the available power. Hence, for such sources the flux density of the source at the antenna is given by twice (4) or

$$S = \frac{2k\,\Delta T_A}{A_e} \qquad (\mathrm{W\ m^{-2}\ Hz^{-1}}) \tag{8}$$

[1] Due to reflection, more realistic values for a man and the ground might be less.

[2] See J. D. Kraus, *Radio Astronomy*, 2nd ed., Cygnus-Quasar, 1986, Secs. 4-4, 4-5 and 4-6.

We have considered two extreme cases, one where the source extent is much broader than the antenna beam width and one where the source extent is much less than the beam width. Let us consider now the general situation for any source beam width–size relation. For this general situation the *total antenna temperature* is

$$T_A = \frac{1}{\Omega_A} \int_0^{\pi} \int_0^{2\pi} T_s(\theta, \phi)\, P_n(\theta, \phi)\, d\Omega \qquad \text{(K)} \tag{9}$$

where T_A = total antenna temperature (not ΔT_A), K
$T_s(\theta, \phi)$ = brightness temperature of source or sources as a function of angle, K
$P_n(\theta, \phi)$ = normalized antenna power pattern, dimensionless
Ω_A = antenna beam solid angle, sr
$d\Omega = \sin\theta \, d\theta \, d\phi$ = infinitesimal element of solid angle, sr

Note that T_A in (9) is the *total antenna temperature* including not only contributions from a particular source in the main beam but from sources of radiation in all directions in proportion to the pattern response. Note also that the temperatures are in *kelvins*, K (=Celsius degrees above absolute zero).

Example 2 Antenna temperature. A circular reflector antenna of 500 m^2 effective aperture operating at $\lambda = 20$ cm is directed at the zenith. What is the total antenna temperature assuming the sky temperature is uniform and equal to 10 K? Take the ground temperature equal to 300 K and assume that half the minor-lobe beam area is in the back direction (toward the ground).[1] The beam efficiency is 0.7 ($=\Omega_M/\Omega_A$).

Solution. Assuming that the antenna aperture efficiency is 50 percent, its physical aperture is 1000 m^2 and its diameter 35.7 m ($=2\sqrt{1000/\pi}$ m). At $\lambda = 0.2$ m the diameter is 179λ, implying that the HPBW $\simeq 0.4°$ ($=70°/179$). Thus, the antenna is highly directional with the main beam directed entirely at the sky (close to the zenith).

Since the (main) beam efficiency is 0.7, 70 percent of the beam area Ω_A is directed at the 10 K sky, half of the remainder or 15 percent at the sky and the other half of the remainder or 15 percent at the 300 K ground. Thus, integrating (9) in 3 steps, we have

$$\text{Sky contribution} = \frac{1}{\Omega_A}(10 \times 0.7\,\Omega_A) = 7 \text{ K}$$

$$\text{Sidelobe contribution} = \frac{1}{\Omega_A}(10 \times \tfrac{1}{2} \times 0.3\,\Omega_A) = 1.5 \text{ K}$$

$$\text{Back-lobe contribution} = \frac{1}{\Omega_A}(300 \times \tfrac{1}{2} \times 0.3\,\Omega_A) = 45 \text{ K}$$

[1] Since much sky radiation reaches the antenna via reflection from the ground, a more realistic ground temperature might be less than 300 K.

and $$T_A = 7 + 1.5 + 45 = 53.5 \text{ K}$$

Note that 45 of the 53.5 K, or 84 percent, of the total antenna temperature results from the back-lobe pickup from the ground. With no back lobes the antenna temperature could ideally be only 10 K, so that in this example the back lobes are very detrimental to the system sensitivity (see Sec. 17-3). It is for this reason that radio telescope and space communication antennas are usually designed to reduce back- and sidelobe response to a minimum.

The information given regarding aperture and wavelength is relevant to the problem only to the extent that it indicates that the main beam is directed entirely at the sky.

In contrast to the above example, let us recall the temperature measurements made by Arno Penzias and Robert Wilson[1] in 1965 at 4 GHz on their 6.2-m horn-reflector antenna which resulted in their discovery of the 3-K sky background. When directed at regions of "empty" sky near the zenith, Penzias and Wilson measured a total antenna temperature $T_A = 6.7$ K.

Contributions to this temperature were measured as[2]

2.3 ± 0.3 K due to the atmosphere
0.8 ± 0.4 K due to ohmic losses
<0.1 K due to back lobes into the ground

3.2 ± 0.5

The difference, $6.7 - 3.2 = 3.5$ K, they attributed to the sky background. Theirs was the first measurement of the residual temperature of the primordial (Big Bang) fireball which created the universe, and sets a lower limit to the temperature of any antenna looking at the sky.

The 0.1-K ground pickup by the antenna of Penzias and Wilson is one of the smallest values ever measured for an antenna. Note also that, in their analysis, they attributed 0.8 K to *ohmic losses* in the antenna and rotary joint.

The antenna noise temperature from the sky as a function of frequency (and wavelength) is presented in Fig. 17-4. A beam angle (HPBW) of less than a few degrees and 100 percent (main) beam efficiency are assumed. Curves are given for beam angles from the zenith (complementary to elevation angles). At lower frequencies the temperature is dominated by radiation from the galaxy. At higher frequencies the atmosphere introduces noise due to absorption. Above the earth's atmosphere (in space) this noise is avoided, but there is a universal photon or quantum noise temperature limit at still higher frequencies given by the photon energy hf divided by Boltzmann's constant, or

---

[1] A. A. Penzias and R. W. Wilson, "A Measurement of Excess Antenna Temperature at 4080 MHz," *Astrophys. J.*, **142**, 419–421, 1965.

[2] A discussion of the errors is given in Sec. 18-12.

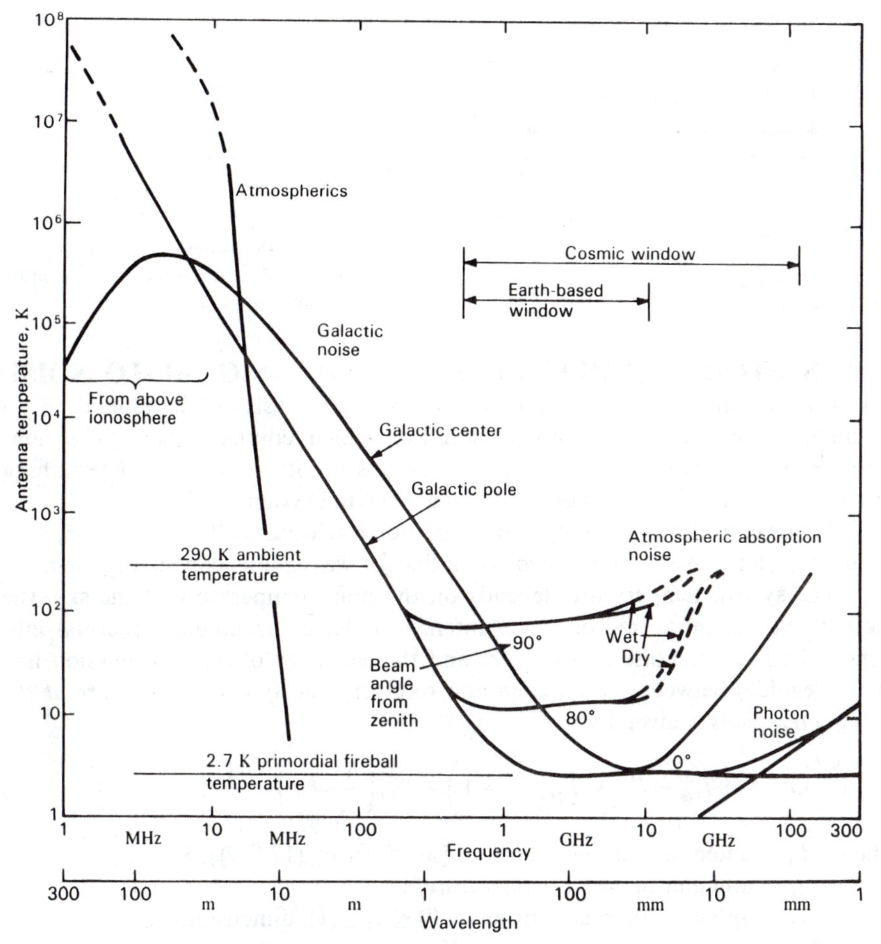

Figure 17-4 Antenna (noise) temperature from the sky as a function of frequency. See text for explanation. (*From J. D. Kraus,* Radio Astronomy, *2nd ed., Cygnus-Quasar, 1986.*)

$$T = \frac{hf}{k} \quad \text{(K)} \tag{10}$$

where h = Planck's constant = 6.63×10^{-34} J s
f = frequency, Hz
k = Boltzmann's constant = 1.38×10^{-23} J K^{-1}

Across the spectrum between these sources of noise there is the noise background or floor of 3 K (or more precisely 2.7 K) due to radiation from the primordial fireball. The low-noise region between galactic radiation and atmospheric absorption defines an *earth-based radio window*, while the region between galactic radiation and quantum limit establishes a *cosmic radio window*.

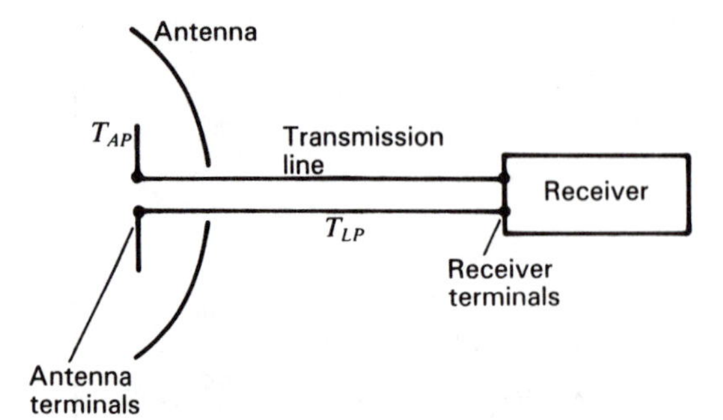

Figure 17-5 Antenna, transmission line and receiver for system temperature determination.

17-3 SYSTEM TEMPERATURE AND SIGNAL-TO-NOISE RATIO. An antenna is part of a receiving system consisting, in general, of an antenna, a receiver and a transmission line which connects them. The temperature of the system, or *system temperature*, is a critical factor in determining the sensitivity and signal-to-noise ratio of a receiving system.

Let us consider a receiving system as shown schematically in Fig. 17-5, with an antenna, a receiver and a transmission line (or waveguide) connecting them.

The system temperature depends on the noise temperature of the sky, the ground and antenna environs, the antenna pattern, the antenna thermal efficiency, the receiver noise temperature and the efficiency of the transmission line (or waveguide) between the antenna and receiver. The *system temperature at the antenna terminals* is given by

$$T_{\text{sys}} = T_A + T_{AP}\left(\frac{1}{\varepsilon_1} - 1\right) + T_{LP}\left(\frac{1}{\varepsilon_2} - 1\right) + \frac{1}{\varepsilon_2} T_R \tag{1}$$

where T_A = antenna noise temperature [as given by (17-2-9)], K
T_{AP} = antenna physical temperature, K
ε_1 = antenna (thermal) efficiency ($0 \leq \varepsilon_1 \leq 1$), dimensionless
T_{LP} = line physical temperature, K
ε_2 = line efficiency ($0 \leq \varepsilon_2 \leq 1$),[1] dimensionless
T_R = receiver noise temperature (see next paragraph), K

The *receiver noise temperature* is given by

$$T_R = T_1 + \frac{T_2}{G_1} + \frac{T_3}{G_1 G_2} + \cdots \tag{2}$$

where T_1 = noise temperature of first stage of receiver, K
T_2 = noise temperature of second stage, K
T_3 = noise temperature of third stage, K
G_1 = power gain of first stage
G_2 = power gain of second stage

[1] $\varepsilon_2 = e^{-\alpha l}$, where α = attenuation constant (Np m^{-1}) and l = length of line (m).

Terms for additional stages may be required if the temperatures are sufficiently high and the gains sufficiently low.

Example 1 System temperature. A receiving system has an antenna with a total noise temperature of 50 K, a physical temperature of 300 K and an efficiency of 99 percent, a transmission line at a physical temperature 300 K and an efficiency of 90 percent, and a receiver with the first 3 stages all of 80 K noise temperature and 13 dB gain. Find the system temperature.

Solution. From (2) the receiver noise temperature is

$$T_R = 80 + \frac{80}{20} + \frac{80}{20^2} = 80 + 4 + 0.2 = 84.2 \text{ K}$$

From (1) the system temperature is

$$T_{sys} = 50 + 300\left(\frac{1}{0.99} - 1\right) + 300\left(\frac{1}{0.9} - 1\right) + \frac{1}{0.9}\,84.2$$

$$= 50 + 3 + 33.3 + 93.6 \simeq 180 \text{ K}$$

Note that due to losses in the antenna, its physical temperature contributes 3 K. The line contributes about 33 K and the receiver about 94 K.

The sensitivity, or *minimum detectable temperature*, ΔT_{min}, of a receiving system is equal to the rms noise temperature ΔT_{rms} of the system, as given by

$$\Delta T_{min} = \frac{k' T_{sys}}{\sqrt{\Delta f\, t}} = \Delta T_{rms} \tag{3}$$

where k' = system constant (order of unity), dimensionless
T_{sys} = system temperature [sum of antenna, line and receiver temperatures as given by (1)], K
T_{rms} = rms noise temperature = ΔT_{min}, K
Δf = predetection bandwidth of receiver, Hz
t = postdetection time constant, s

The *criterion of detectability* is that the incremental antenna temperature ΔT_A due to a radio source be equal to or exceed ΔT_{min}, that is,

$$\Delta T_A \geq \Delta T_{min} \tag{4}$$

and the *signal-to-noise* (S/N) *ratio* is then[1]

$$\frac{S}{N} = \frac{\Delta T_A}{\Delta T_{\text{min}}} \qquad (5)^2$$

Many space communication systems, radio telescopes and remote sensing systems operate at such high sensitivity (low signal levels) that a low system temperature is essential.

Example 2 Minimum detectable flux density. The Ohio State University 110 by 21 m radio telescope antenna (see Fig. 12-51) has a physical aperture of 2208 m^2, and at 1415 MHz an aperture efficiency of 54 percent and a system temperature of 50 K. The rf bandwidth is 100 MHz, the output time constant is 10 s and the system constant is 2.2. Find the minimum detectable flux density.

Solution. From (3) the minimum detectable temperature is

$$\Delta T_{\text{min}} = \frac{k' T_{\text{sys}}}{\sqrt{\Delta f\, t}} = \frac{2.2 \times 50}{\sqrt{100 \times 10^6 \times 10}} = 0.0035 \text{ K}$$

The effective aperture $A_e = A_p \varepsilon_{\text{ap}} = 2208 \times 0.545 = 1203 \text{ m}^2$. From (17-2-8) the minimum detectable flux density is

$$\Delta S_{\text{min}} = \frac{2k\, \Delta T_{\text{min}}}{A_e} = \frac{2 \times 1.38 \times 10^{-23} \times 0.0035}{1203}$$

$$= 8.1 \times 10^{-29} \text{ W m}^{-2} \text{ Hz}^{-1} \simeq 8 \text{ mJy}^3$$

By repeating observations and averaging, the minimum can be further reduced ($\propto \sqrt{1/n}$, where n = number of observations).

In a large sky survey at 1415 MHz, about 20 000 radio sources were detected and cataloged at flux densities above 180 mJy. Thus, the signal-to-noise ratio for these cataloged sources is

$$\frac{S}{N} = \frac{180 \text{ mJy}}{8.0 \text{ mJy}} = \frac{\Delta T_A}{\Delta T_{\text{min}}} = \frac{0.0785 \text{ K}}{0.0035 \text{ K}} \simeq 22.5$$

Let us consider next the signal-to-noise ratio for a receiving system which is part of a *communication* link. If a transmitter radiates a power P_t isotropically and uniformly over a bandwidth Δf_t, it produces a flux density at a distance r of $P_t/(4\pi r^2\, \Delta f_t)$. A receiving antenna of effective aperture A_{er} at a distance r can

[1] Distinguish between S here for *signal* and S elsewhere for *flux density* or *Poynting vector*.

[2] The *minimum detectable signal* is given by $S = N$ or $S/N = 1$. The ratio is sometimes expressed $(S + N)/N$ and for the minimum detectable signal case this ratio equals 2.

[3] 8 millijansky, where 1 Jy = 10^{-26} W m^{-2} Hz^{-1}.

collect a power

$$P_r = \frac{P_t A_{er} \Delta f_r}{4\pi r^2 \Delta f_t} \qquad \text{(W)} \tag{6}$$

where P_t = radiated transmitter power, W
A_{er} = effective aperture of receiving antenna, m^2
Δf_r = receiver bandwidth, Hz
Δf_t = transmitter bandwidth, Hz
r = distance between transmitter and receiver, m

It is assumed that $\Delta f_r \leq \Delta f_t$.

With a transmitting antenna of directivity $D = 4\pi A_{et}/\lambda^2$, the received power becomes

$$P_r = \frac{P_t A_{er} A_{et}}{r^2 \lambda^2} \frac{\Delta f_r}{\Delta f_t} \qquad \text{(W)} \tag{7}$$

where λ = wavelength, m
A_{et} = transmitter effective aperture, m^2

For $\Delta f_r = \Delta f_t$ (bandwidths matched), (7) is the Friis transmission formula (2-25-5).

Whether this amount of received power is useful for communication depends on the signal-to-noise ratio S/N, where the *signal power* is given by (7) and the *noise power* by the Nyquist relation (17-2-1) as

$$P_n = kT_{\text{sys}} \Delta f_r = N \tag{8}$$

where k = Boltzmann's constant = 1.38×10^{-23} J K^{-1}
T_{sys} = system temperature, K

For matched bandwidths ($\Delta f_r = \Delta f_t$), the ratio of (7) to (8) gives the *signal-to-noise ratio* as

$$\frac{S}{N} = \frac{P_r}{P_n} = \frac{P_t A_{er} A_{et}}{kT_{\text{sys}} r^2 \lambda^2 \Delta f_r} \qquad \text{(dimensionless)} \tag{9}$$

Example 3 Down-link signal-to-noise ratio. A Clarke-orbit-satellite-to-earth FM-TV downlink, as in Fig. 17-6, has a transmitter (transponder) power $P_t = 5$ W, transmitter antenna physical aperture $A_{pt} = 6$ m^2, earth-station receiving antenna physical aperture $A_{pr} = 12$ m^2, path length $r = 40\,000$ km and frequency 4 GHz. The earth-station receiving antenna has an antenna temperature contribution from the main beam directed at the sky region of the Clarke orbit when the galactic plane is crossing (worst condition) (see Fig. 17-4) of 7 K. Beam efficiency is 80 percent with half of the minor lobes above the horizon "seeing" an average of 10 K while the remaining minor lobes "see" the earth at 290 K. The receiver has a 3-stage front-end amplifier with the same characteristics as the receiver in Example 1. The receiving antenna has a 99 percent thermal efficiency and short transmission line (amplifier close to antenna terminals) with 98 percent efficiency. The transmitting antenna on the Clarke-orbit satellite and the earth-station receiving antenna both

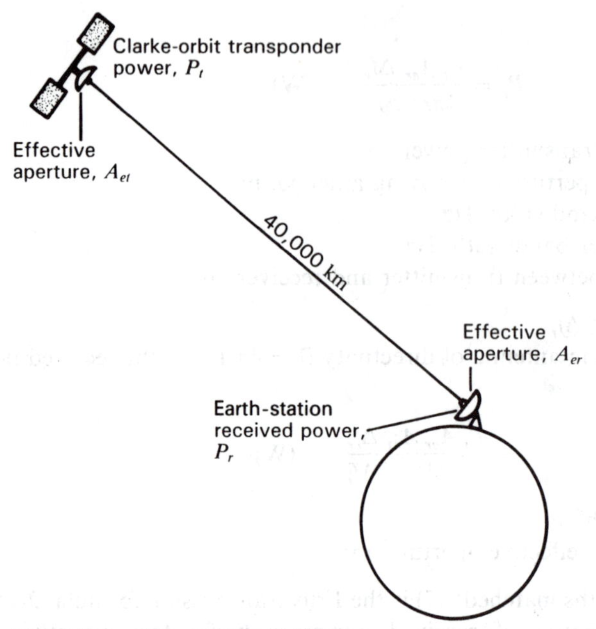

Figure 17-6 Clarke-orbit downlink.

have an aperture efficiency ε_{ap} of 60 percent. Find the signal-to-noise ratio for a 30-MHz bandwidth. Both receiver and transmission line are at a physical (ambient) temperature of 300 K. It is assumed that beams are aligned and that polarizations and bandwidths are matched.

Solution. At 60 percent aperture efficiency, the effective apertures of the two antennas are

$$A_{et} = \varepsilon_{ap} A_{pt} = 0.6 \times 6 = 3.6 \text{ m}^2$$

$$A_{er} = \varepsilon_{ap} A_{pr} = 0.6 \times 12 = 7.2 \text{ m}^2$$

The receiving antenna temperature from (17-2-9) is

$$T_A = 0.8 \times 7 + 0.1 \times 10 + 0.1 \times 290 \simeq 36 \text{ K}$$

The receiving system temperature from (1) is

$$T_{sys} \simeq 36 + 300 \times 0.01 + 300 \times 0.02 + 1.02 \times 84$$
$$= 36 + 3 + 6 + 86 = 131 \text{ K}$$

At 4 GHz, $\lambda = 0.075$ m. The signal-to-noise ratio from (9) is

$$\frac{S}{N} = \frac{5 \times 7.2 \times 3.6}{1.38 \times 10^{-23} \times 131 \times 4^2 \times 10^{14} \times 0.075^2 \times 3 \times 10^7}$$
$$= 265 \text{ or } 24 \text{ dB}$$

which would be satisfactory for most applications.

Examples 2 and 3 dealt with receiving systems operating at 1.4 and 4 GHz. Referring to Fig. 17-4, we note that at these frequencies and at large elevation angles (small zenith angles) the sky temperature is less than 10 K. However, at angles near the horizon (zenith angle $>80°$) the temperature may approach 100 K due to atmospheric absorption. At frequencies below 1 GHz the noise from our galaxy becomes important. At 50 MHz the galactic noise ranges from about 2000 K when the antenna is looking at the galactic poles to 20 000 K when it is looking at the galactic center. Under these conditions reducing the noise temperature of a receiver from, say, 200 to 100 K would make but a small difference on the system temperature and the signal-to-noise ratio. Thus, if $T_A = 5000$ K, $T_R = 200$ K and neglecting other contributions, the system temperature $T_{sys} = 5200$ K as compared to 5100 K for $T_R = 100$ K. The improvement in T_{sys} and minimum noise level is only 0.08 dB [$=10 \log (52/51)$].

Sometimes the parameter *noise figure* is used instead of the *noise temperature*. They are related as follows:

$$T = (F - 1)T_0 \tag{10}$$

where T = noise temperature, K
$T_0 = 290$ K
F = noise figure, dimensionless

Thus,

$$F = \frac{T + T_0}{T_0} \tag{11}$$

or

$$F(\text{dB}) = 10 \log F \tag{12}$$

where $F(\text{dB})$ = noise figure in decibels. The relation of the noise figure F and its value in decibels to the noise temperature T are shown in Fig. 17-7.

17-4 PASSIVE REMOTE SENSING.

A radio telescope is a remote sensing device whether it is earth-based and pointed at the sky for observing celestial objects[1] or on an aircraft or satellite and pointed at the earth. In this section we consider the case where the radiation detected or sensed by the telescope originates in the objects being observed, making for a *passive remote sensing* system in distinction to *radar or active remote sensing* where signals are transmitted and their reflections observed and analyzed. The active case is discussed in the next section.

---

[1] See the Mars temperature example, Example 1 in Sec. 17-2, and the minimum detectable flux density example, Example 2 in Sec. 17-3.

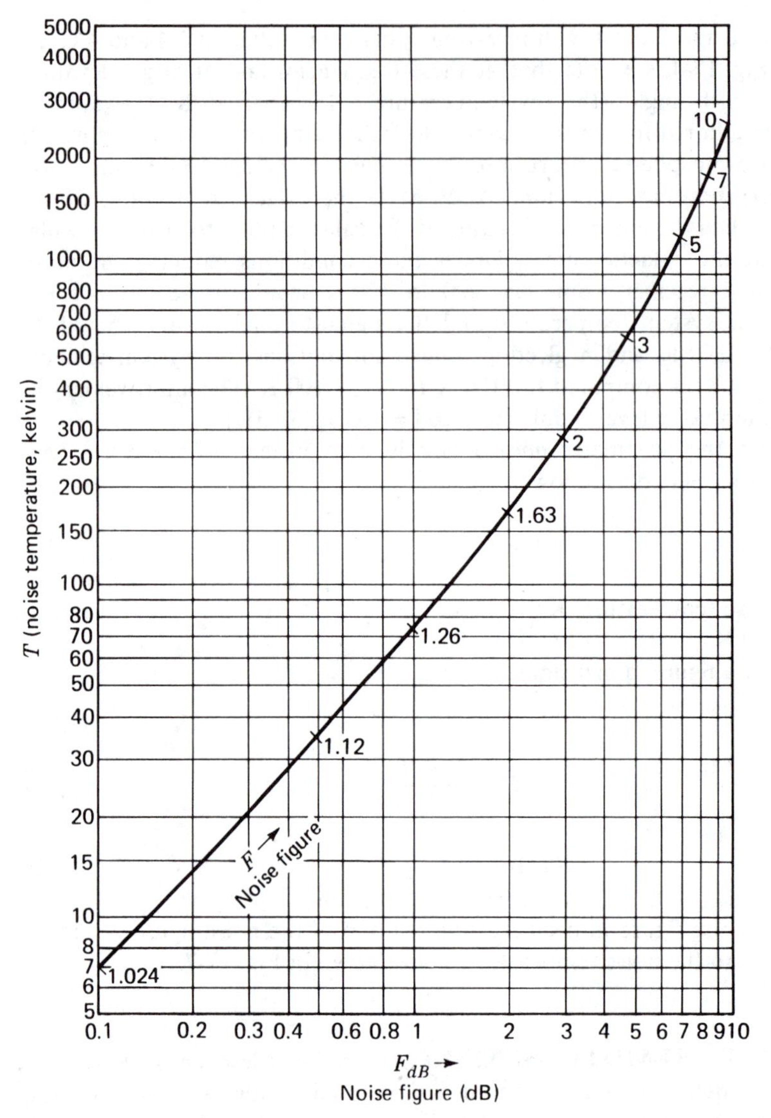

Figure 17-7 Noise-temperature–noise-figure chart.

Consider the situation of Fig. 17-8*a*, in which the earth-based radio telescope antenna beam is completely subtended by a celestial source of temperature T_s with an intervening absorbing-emitting cloud of temperature T_c. *With no cloud present*, the incremental antenna temperature $\Delta T_A = T_s$, but *with the cloud* it may be shown that the observed antenna temperature

$$\Delta T_A = T_c(1 - e^{-\tau_c}) + T_s e^{-\tau_c} \qquad \text{(K)} \tag{1}$$

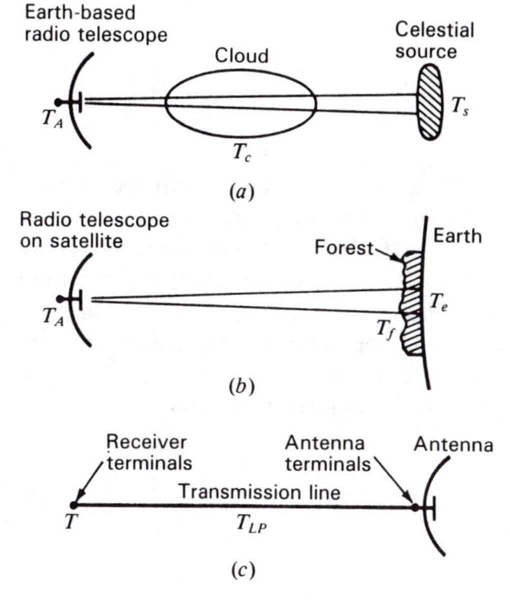

Figure 17-8 (*a*) Earth-based radio telescope remote-sensing celestial source through intervening interstellar cloud. (*b*) Radio telescope on satellite remote-sensing the earth through forest. (*c*) Receiver detecting antenna output through transmission line. The cloud, forest and transmission line have analogous emitting-absorbing properties.

where τ_c = absorption coefficient of the cloud[1] (=0 for no absorption and $=\infty$ for infinite absorption). Thus, knowing T_s and τ_c, the cloud's equivalent blackbody temperature T_c can be determined.

Now, referring to Fig. 17-8*b*, let us reverse the situation and put the radio telescope on an orbiting satellite for observing the surface of the earth at temperature T_e with the antenna beam completely subtended by a large forest at a temperature T_f. The incremental satellite antenna temperature is then

$$\Delta T_A = T_f(1 - e^{-\tau_f}) + T_e e^{-\tau_f} \qquad \text{(K)} \tag{2}$$

where τ_f = absorption coefficient of the forest. Knowing T_e and τ_f, the temperature of the forest can be determined, or knowing T_e and T_f, the absorption coefficient can be deduced. It is by such a technique that the whole earth can be surveyed and much information obtained about the temperatures of land and water areas, and from absorption coefficients about the nature of the surface cover.

Example Forest temperature by remote sensing. The remote-sensing antenna of a 3-GHz orbiting satellite measures a temperature $\Delta T_A = 300$ K when directed at a tropical forest region having an absorption coefficient $\tau_f = 0.693$ at vertical incidence. If the earth temperature $T_e = 305$ K, find the temperature of the forest.

---

[1] Astronomers call τ_c the "optical depth." The quantity $e^{-\tau_c}$ is equivalent to the efficiency factor ε in (17-3-1).

Solution. Since $\tau_f = 0.693$, $e^{-\tau_f} = 0.5$, so from (2),

$$T_f = \frac{\Delta T_A - T_e e^{-\tau_f}}{1 - e^{-\tau_f}} = \frac{300 - 305 \times 0.5}{1 - 0.5} = 295 \text{ K}$$

If the antenna-transmission line-receiver system is viewed from the receiver terminals as in Fig. 17-8*c* (instead of from the antenna terminals as in Fig. 17-5), we note that the analogy between the remote-sensing situations discussed above extends here to the transmission line. Thus, the emitting-absorbing transmission line of Fig. 17-8*c* is like the emitting-absorbing cloud of Fig. 17-8*a* and like the emitting-absorbing forest of Fig. 17-8*b*. The analogy may be emphasized by comparing (1) and (2) with the temperature as seen from the receiver terminals, so that the equations for the 3 situations have identical form as follows:

Antenna looking at celestial source (see Fig. 17-8*a*):

$$\Delta T_A = T_c(1 - e^{-\tau_c}) + T_s e^{-\tau_c} \quad \text{(K)} \tag{3}$$

Antenna looking at earth from satellite (see Fig. 17-8*b*):

$$\Delta T_A = T_f(1 - e^{-\tau_f}) + T_e e^{-\tau_f} \quad \text{(K)} \tag{4}$$

Receiver looking at antenna (see Fig. 17-8*c*):

$$T = T_{LP}(1 - e^{-\alpha l}) + T_A e^{-\alpha l} \quad \text{(K)} \tag{5}$$

where ΔT_A = incremental antenna temperature, K
T_c = cloud temperature, K
τ_c = cloud absorption coefficient (optical depth), dimensionless
T_s = celestial source temperature, K
T_f = forest temperature, K
τ_f = forest absorption coefficient, dimensionless
T_e = temperature of earth, K
T_{LP} = transmission line physical temperature, K
α = transmission line attenuation constant, Np m^{-1}
l = length of transmission line, m

Note that the *system temperature* should be referred to the antenna terminals as in (17-3-1) and *not* to the receiver terminals as in (5). Thus, if the line is completely lossy ($e^{-\alpha l} = \varepsilon_2 = 0$), (17-3-1) gives an infinite system temperature which is correct, meaning that the system has no sensitivity whatever. However, with this condition, a "system temperature" viewed from the receiver terminals, as in (5), would equal the temperature T_{LP} of the line plus the receiver temperature, a completely misleading result since it indicates that the system still has sensitivity.

17-5 RADAR,[1] SCATTERING AND ACTIVE REMOTE SENSING

By Robert G. Kouyoumjian[2]

Consider an object in the far zone of a radar antenna as shown in Fig. 17-9. Typically, in pulse radar the antenna is connected to the transmitter for the pulse transmission and then switched to the receiver in time to receive the pulse echo. In practice, the antenna may be connected continuously to both the transmitter and receiver but with the receiver blanked during transmission of the pulse.

During transmission the power density incident at the object (radar target) is

$$S_{\text{inc}} = \frac{P_t G}{4\pi r^2} \qquad (\text{W m}^{-2}) \tag{1}$$

where P_t = transmitted power, W
G = antenna gain, dimensionless
r = distance between antenna and radar target, m

The radar target scatters the incident power in all directions. The total amount of this scattered power is P_{ts} as given by

$$P_{ts} = S_{\text{inc}}\,\sigma_t \qquad (\text{W}) \tag{2}$$

where σ_t = total scattering cross section, m^2

The *total scattering cross section* σ_t may be regarded as an effective aperture which "collects" a power P_{ts} from the incident wave and reradiates it in all directions (4π sr).

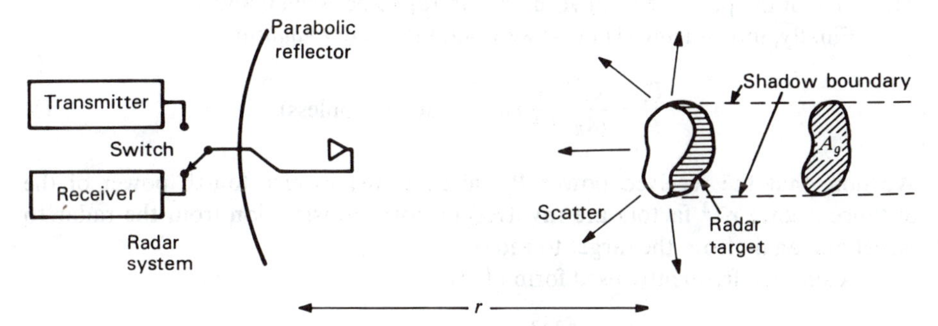

Figure 17-9 Radar system and scattering object (radar target).

[1] RADAR is an acronym for RAdio Direction And Range.

[2] Department of Electrical Engineering, The Ohio State University.

The scattered power incident back on the antenna (backscattered power) S_r is given by

$$S_r = \frac{D_s P_{ts}}{4\pi r^2} = \frac{S_{\text{inc}} D_s \sigma_t}{4\pi r^2} \qquad (\text{W m}^{-2}) \tag{3}$$

where D_s = radar target (backscattering) directivity and the antenna is in the far zone of the radar target.

The power P_r reaching the radar receiver is then

$$P_r = S_r A_e = \frac{S_{\text{inc}} A_e \sigma}{4\pi r^2} \qquad (\text{W}) \tag{4}$$

where $\sigma = D_s \sigma_t =$ *radar cross section*, m^2

$A_e = A'_e pq =$ effective aperture of the radar antenna,[1] m^2

$A'_e = \dfrac{\lambda^2}{4\pi} G =$ maximum effective aperture of radar antenna, m^2

$p =$ polarization mismatch factor $= \cos^2 (MM_a/2)$, dimensionless

$MM_a =$ angular distance between polarization states M and M_a

$M =$ polarization state of wave (see Sec. 2-36 on Poincaré sphere).

$M_a =$ polarization state of antenna

$q =$ impedance mismatch factor $= \dfrac{4R_a R_r}{(R_a + R_r)^2 + (X_a + X_r)^2}$, dimensionless

$R_a =$ antenna resistance, Ω

$R_r =$ receiver resistance, Ω

$X_a =$ antenna reactance, Ω

$X_r =$ receiver reactance, Ω

The value of the product pq in A_e in (4) can range between 0 and 1.

Finally, introducing (1) in (4) we obtain the *radar equation*

$$\frac{P_r}{P_t} = \frac{G^2 \lambda^2 \sigma}{(4\pi)^3 r^4} pq \qquad (\text{dimensionless}) \tag{5}$$

We note that the received power P_r varies as the inverse fourth power of the distance r since r^{-2} factors are involved in both transmission from the radar to target and again from the target to radar.

A simpler, frequently used form of (5) is

$$\frac{P_r}{P_t} = \frac{G^2 \lambda^2 \sigma}{(4\pi)^3 r^4} \qquad (\text{dimensionless}) \tag{6}$$

---

[1] C-T Tai, "On the Definition of the Effective Aperture of Antennas," *IRE Trans. Ants. Prop.*, **AP-9**, 224–225, March 1961.

where it is assumed that there is no cross-polarized component of the backscattered wave ($p = 1$)[1] and also that the antenna and receiver are matched ($q = 1$).

The power received by a radar is proportional to the *radar cross section*

$$\sigma = D_s \sigma_t = \frac{4\pi r^2 S_r}{S_{\text{inc}}} \qquad (\text{m}^2) \tag{7}$$

The *radar cross section* σ is the effective area intercepting the incident power density which, if scattered isotropically, would result in the backscattered power density S_r. The *total scattering cross section* σ_t $(=\sigma/D_s)$ is also of interest, being important in calculating the transmission of radio and light waves *through* clouds of particles (interstellar dust, rain, smoke or clouds of insects).

The radar cross section depends on:

1. Target shape
2. Target material
3. Frequency of the incident wave
4. Polarization of the incident wave
5. Aspect (or orientation) of the target

The geometrical (or physical) cross section of a sphere is

$$A_g = A_p = \pi a^2 \qquad (\text{m}^2) \tag{8}$$

where $A_g = A_p$ = geometric (or physical) cross section. For complex shapes A_g is the cross-sectional area of the shadow which is cast by the target when illuminated by a plane wave as from a searchlight (see Fig. 17-9).

It is convenient to divide our discussion of σ into three regions:

1. The high-frequency (optical) region where the target is large compared to the wavelength
2. The low-frequency (Rayleigh) region where the target is small compared to the wavelength
3. The resonance region between the high- and low-frequency regions

These regions are evident in Fig. 17-10 for a sphere and a flat disc showing the radar backscattering efficiency as a function of the sphere or disc circumference C_λ $(=2\pi a/\lambda)$, where the *radar backscattering efficiency* is given by

$$\varepsilon_s = \frac{\sigma}{A_g} \qquad (\text{dimensionless}) \tag{9}$$

---

[1] We note that if the target is a large, flat, perfectly conducting sheet with wave incident normally, $p = 1$ for linear polarization but $p = 0$ for circular polarization.

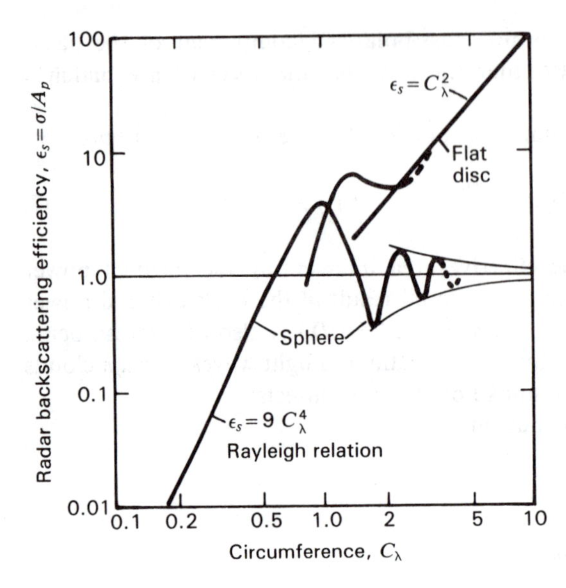

Figure 17-10 Broadside radar backscattering efficiency of flat disc and backscatter efficiency of sphere as a function of disc or sphere circumference (in wavelengths). (*Disc after Kouyoumjian, sphere after G. Mie.*)

A resonance peak for the sphere occurs for $C_\lambda = 1$ followed by oscillating values which converge to unity at large values of C_λ. At small C_λ ($<\frac{1}{2}$) the backscatter efficiency follows the *Rayleigh relation*[1]

$$\varepsilon_s = 9\ C_\lambda^4 \qquad \text{(dimensionless)} \tag{10}$$

The sphere efficiency at larger C_λ is from Mie.[2]

The broadside radar (backscattering) cross sections of thin wires and of wire loops are shown in Figs. 17-11 and 17-12.[3]

Figure 17-11 illustrates the variation of the radar cross section in square wavelengths for straight wires as a function of the wire length in wavelengths for 2 wire thicknesses. Both wires resonate at lengths somewhat less than $\lambda/2$ with the thinner wire resonating closer to $\lambda/2$. These lengths do not represent optimum lengths when the wire is used as a reflector element in an array because the mutual impedances cannot be neglected. The wires for Fig. 17-11 are remote from other objects.

The radar cross section in square wavelengths for loops is presented in Fig. 17-12 as a function of the loop circumference in wavelengths for 2 wire thicknesses. Both loops resonate at circumferences of approximately 1λ, with the

---

[1] Lord Rayleigh, "On the Incidence of Aerial and Electric Waves upon Small Obstacles in the Form of Ellipsoids or Elliptic Cylinders," *Phil. Mag.*, **44**, 28–52, 1897.
H. C. Van de Hulst, *Light Scattering by Small Particles*, Wiley, 1957.

[2] G. Mie, "Beiträge zur Optik trüber Medien, Speziell Kolloidal Metaläsungen," *Ann. Phys.*, **25**, 377–446, 1908.

[3] R. G. Kouyoumjian, "The Backscattering from a Circular Loop," *Appl. Sci. Res.*, **6B**, 165–179, 1957.
R. G. Kouyoumjian, "The Calculation of the Echo Areas of Perfectly Conducting Objects by the Variational Method," Ph.D. dissertation, Ohio State University, 1953.

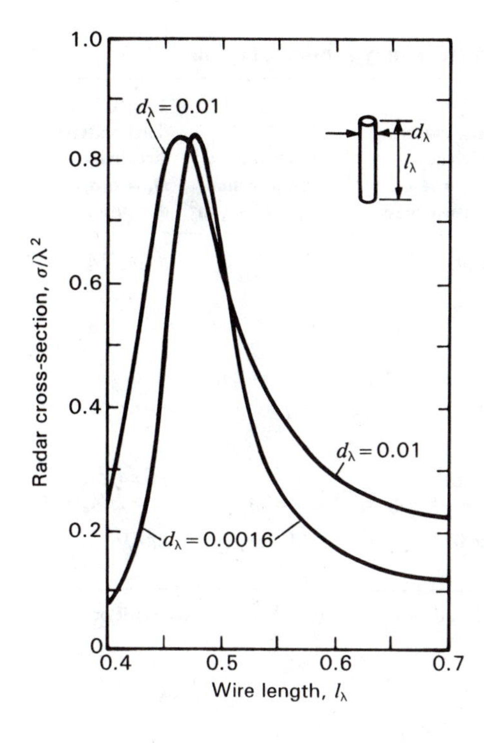

Figure 17-11 Broadside radar (backscattering) cross section (in λ^2) of straight wires with **E** parallel to the wires as a function of wire length l_λ for wire diameters $d_\lambda = 0.0016$ and 0.01. (*After R. G. Kouyoumjian.*)

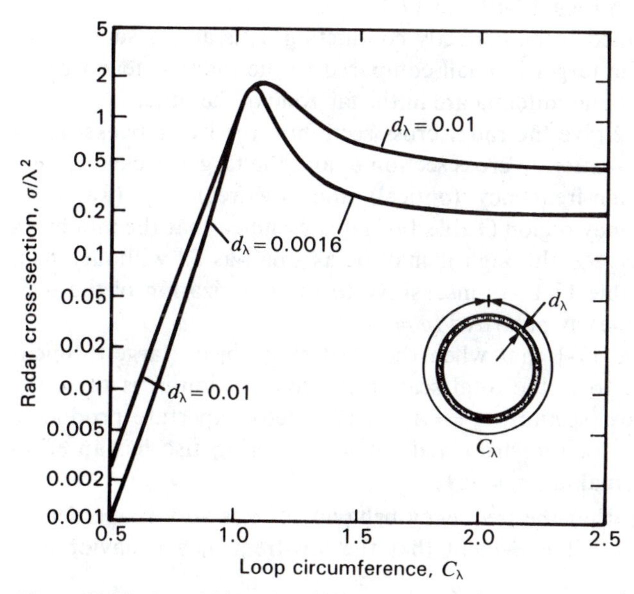

Figure 17-12 Broadside radar (backscattering) cross section (in λ^2) of loops as a function of loop circumference C_λ for loop conductor diameters $d_\lambda = 0.0016$ and 0.01. (*After R. G. Kouyoumjian.*)

Table 17-1 Radar quantities for the high-frequency (optical) region†

Target	Radar cross section $\sigma = D_s \sigma_t$, m²	Radar backscattering efficiency $\varepsilon_s = \sigma/A_g$, dimensionless	Total scattering cross section $\sigma_t = \sigma/D_s$, m²	Target (backscattering) directivity $D_s = \sigma/\sigma_t$, dimensionless
Object bounded by a smooth curved surface with principal radii a_1 and a_2 at the specular point	$\pi a_1 a_2$	$\pi a_1 a_2/A_g$	$2A_g$	$\pi a_1 a_2/2A_g$
Sphere, radius a	$\pi a^2 = A_g$	1	$2\pi a^2 = 2A_g$	$\frac{1}{2}$
Cylinder of radius a and length L, normal incidence	$2\pi a_\lambda L^2 = C_\lambda L^2$	πL_λ	$4aL$	$(\pi/2)L_\lambda$
Flat disc of radius a, normal incidence	$(2\pi a_\lambda)^2 \pi a^2 = 4\pi A_g^2/\lambda^2$	$(2\pi a_\lambda)^2 = C_\lambda^2 = 4\pi A_g/\lambda^2$	$2\pi a^2 = 2A_g$	$\frac{1}{2}(2\pi a_\lambda)^2 = 2\pi A_g/\lambda^2$
Flat plate of area A $(=A_g)$, normal incidence	$4\pi A^2/\lambda^2$	$4\pi A/\lambda^2$	$2A$	$2\pi A/\lambda^2$

† $a \gg \lambda$, $L \gg \lambda$, $a_\lambda = a/\lambda$, $L_\lambda = L/\lambda$, C = circumference $= 2\pi a$, $C_\lambda = C/\lambda = 2\pi a/\lambda = 2\pi a_\lambda$, A_g = geometric or physical cross section $= A_p$.

Targets are assumed to be perfectly conducting.

thinner loop resonating closer to this value. Also it is worth noting that σ of the loops at first resonance is larger than σ at the first resonance for the sphere, disc and straight wires shown in Figs. 17-10 and 17-11.

All targets are assumed to be perfectly conducting. It is also assumed that the angle subtended by the target is small compared to the radar antenna beam width and that both target and antenna are in the far zone of the other.

Tables 17-1 and 17-2 give the radar cross section σ, the radar backscattering efficiency ε_s, the total scattering cross section σ_t and the target backscattering directivity D_s for the high-frequency (optical) and low-frequency (Rayleigh) regions. In the high-frequency region (Table 17-1) it is assumed that the minimum dimension of the target is $\gg\lambda$ although it may be as small as 3λ without much error. The relations of Table 17-1 are insensitive to the polarization of the incident wave provided it is linearly polarized ($p = 1$).

We note from Table 17-1 that when the scattering object (target) dimensions are large compared to λ, the total scattering cross section σ_t is *twice* the geometric or physical cross section A_g $(=A_p)$. The effective aperture producing scattering equals A_g while the forward (small angle) scattering also has an effective aperture equal to A_g, making $\sigma_t = 2A_g$.[1]

In Table 17-1 it is seen that the frequency behavior of σ is very shape dependent whereas in Table 17-2 it is evident that the low-frequency behavior of σ

[1] H. C. Van de Hulst, *Light Scattering by Small Particles*, Wiley, 1957, p. 107.

Table 17-2 Radar quantities for the low-frequency (Rayleigh) region†

Target	Radar cross section $\sigma = D_s \sigma_t$, m²	Radar backscattering efficiency $\varepsilon_s = \sigma / A_g$, dimensionless	Total scattering cross section $\sigma_t = \sigma / D_s$, m²	Target (backscattering) directivity $D_s = \sigma/\sigma_t$, dimensionless
Sphere, radius a	$14\,026 a_\lambda^4 \pi a^2 = 9 C_\lambda^4 A_g$	$14\,026 a_\lambda^4 = 9 C_\lambda^4$	$5195 a_\lambda^4 \pi a^2 = 3.33 C_\lambda^4 A_g$	2.7
Flat disc, radius a, normal incidence	$1123 a_\lambda^4 \pi a^2 = 114 A_g^3/\lambda^4$	$1123 a_\lambda^4 = 114 A_g^2/\lambda^4$	$749 a_\lambda^4 \pi a^2 = 75.9 A_g^3/\lambda^4$	1.5
Square plate, side length L, normal incidence with **E** parallel to edge	$132 L_\lambda^4 L^2 = 132 A_g^3/\lambda^4$	$132 L_\lambda^4 = 132 A_g^2/\lambda^4$	$88 L_\lambda^4 L^2 = 88 A_g^3/\lambda^4$	1.5
Thin wire loop, radius a, wire radius b, normal incidence	$\dfrac{15\,382}{[\ln(8a/b) - 2]^2} a_\lambda^4 \pi a^2 = 9.87 C_\lambda^4 A_g / [\ln(8a/b) - 2]^2$		$\dfrac{10\,255}{[\ln(8a/b) - 2]^2} a_\lambda^4 \pi a^2 = 6.58 C_\lambda^4 A_g / [\ln(8a/b) - 2]^2$	1.5
Thin straight wire, radius b, length L, normal incidence with **E** parallel to wire	$\dfrac{34}{[\ln(L/2b) - 1]^2} L_\lambda^4 L^2$		$\dfrac{22.7}{[\ln(L/2b) - 1]^2} L_\lambda^4 L^2$	1.5

† $a < \lambda/10$, $L < \lambda/10$, $a_\lambda = a/\lambda$, $L_\lambda = L/\lambda$, C = circumference = $2\pi a$, $C_\lambda = C/\lambda = 2\pi a/\lambda = 2\pi a_\lambda$, A_g = geometric or physical cross section = A_p.
Targets are assumed to be perfectly conducting.

varies as $1/\lambda^4$ for all targets, which is characteristic of scattering in the Rayleigh region. Additional information on radar cross-section calculations and measurements is given in the references at the end of the chapter. See especially W. E. Blore and the *IEEE Proceedings* special issues for August 1965 and February 1985.

Whereas passive remote sensing can provide information on temperature, polarization and absorption and also on velocity (for line emitting objects), active (radar) remote sensing has the capability of providing much additional information such as distance, shape and composition. A returning pulse carries a characteristic signature by which different objects may be recognized, the pulse response being the inverse Fourier transform of the object's frequency response. Both passive and active remote-sensing techniques can be used for mapping or imaging of extended objects.

ADDITIONAL REFERENCES

Arvas, E., R. F. Harrington and J. R. Mautz: "Radiation and Scattering from Electrically Small Conducting Bodies of Arbitrary Shape," *IEEE Trans. Ants. Prop.*, **AP-34**, 66–77, January 1986.

Blore, W. E.: "The Radar Cross-Section of Ogives, Double-Backed Cones, Double-Rounded Cones and Cone Spheres," *IEEE Trans. Ants. Prop.*, **AP-12**, 582–590, September 1964.

Bowman, J. J., T. B. A. Senior and P. L. E. Uslenghi: *Electromagnetic and Acoustic Scattering by Simple Shapes*, North Holland, Amsterdam, 1969.

Chuang, C. W., and D. L. Moffatt: "Natural Resonances of Radar Targets via Prony's Method and Target Discrimination," *IEEE Trans. Aerospace and Elect. Sys.*, **AES-12**, September 1976.

Clarke, J. (ed.): *Advances in Radar Techniques*, Peregrinus, 1985.

Dalle Mees, E., M. Mancianti, L. Verrazzani and A. Cantoni: "Target Identification by Means of Radar," *Microwave J.*, **27**, 85–102, December 1984.

Davidovitz, M., and W.-M. Boerner: "Extension of Kennaugh's Optimal Polarization Concept to the Asymmetric Scattering Matrix Case," *IEEE Trans. Ants. Prop.*, **AP-34**, 569–574, April 1986.

Deschamps, G. A.: "High Frequency Diffraction by Wedges," *IEEE Trans. Ants. Prop.*, **AP-33**, 357, April 1985 (56 references).

Eftimiu, C.: "Scattering by Rough Surfaces: A Simple Model," *IEEE Trans. Ants. Prop.*, **AP-34**, 626–630, May 1986.

Evans, J. V., and T. Hagfors: *Radar Astronomy*, McGraw-Hill, 1968.

Hall, R. C., and R. Mittra: "Scattering from a Periodic Array of Resistive Strips," *IEEE Trans. Ants. Prop.*, **AP-33**, 1009–1011, September 1985.

Hansen, R. C. (ed.): *Geometric Theory of Diffraction*, IEEE Press, 1981.

Keller, J. B.: "Backscattering from a Finite Cone," *IRE Trans. Ants. Prop.*, **9**, 411–412, 1961.

Kennaugh, E. M.: "The K-Pulse Concept," *IEEE Trans. Ants. Prop.*, **AP-29**, 327–331, March 1981.

Kennaugh, E. M., and D. L. Moffatt: "Transient and Impulse Response Approximation," *Proc. IEEE*, **53**, August 1965.

Kim, H. T., N. Wang and D. L. Moffatt: "K-Pulse for a Thin Circular Loop," *IEEE Trans. Ants. Prop.*, **AP-33**, 1403–1407, December 1985.

Kouyoumjian, R. G., and P. H. Pathak: "A Uniform Geometrical Theory of Diffraction for an Edge in a Perfectly Conducting Surface," *Proc. IEEE*, **62**, 1448–1461, November 1974.

Medgyesi-Mitschang, L. N., and J. M. Putnam: "Electromagnetic Scattering from Extended Wires and 2- and 3-Dimensional Surfaces," *IEEE Trans. Ants. Prop.*, **AP-33**, 1090–1100, October 1985.

Nakano, H., A. Yoshizawa and J. Yamauchi: "Characteristics of a Crossed-Wire Scatterer without a Junction Point for an Incident Wave of Circular Polarization," *IEEE Trans. Ants. Prop.*, **AP-33**, 409–415, April 1985.

Newman, E. H.: "TM and TE Scattering by a Dielectric/Ferrite Cylinder in the Presence of a Half Plane," *IEEE Trans. Ants. Prop.*, **AP-34**, 804–813, June 1986.

Rao, S. M., T. K. Sakar and S. A. Dianat: "A Novel Technique to the Solution of Transient Electromagnetic Scattering from Thin Wires," *IEEE Trans. Ants. Prop.*, **AP-34**, 630–634, May 1986.

Richmond, J. H.: "Digital Computer Solutions of the Rigorous Equations for Scattering Problems," *Proc. IEEE*, **53**, 796–804, August 1965.

Richmond, J. H.: "On the Edge Mode in the Theory of TM Scattering by a Strip or Strip Grating," *IEEE Trans. Ants. Prop.*, **AP-28**, 883–887, November 1980.

Richmond, J. H. (see also other Richmond references at the end of Chap. 9).

Rusch, W. V. T., and R. J. Poyorzelski: "A Mixed-Field Solution for Scattering from Composite Bodies," *IEEE Trans. Ants. Prop.*, **AP-34**, 955–958, July 1986.

Schuerman, D. W.: *Light Scattering by Irregularly Shaped Particles*, Plenum Press, 1980.

Senior, T. B. A., and J. L. Volakis: "Scattering by an Imperfect Right-Angled Wedge," *IEEE Trans. Ants. Prop.*, **AP-34**, 681–689, May 1986.

Shaffer, J. F.: "EM Scattering from Bodies of Revolution with Attached Wires," *IEEE Trans. Ants. Prop.*, **AP-30**, 426–431, May 1982.

Siegel, K. M.: "Far Field Scattering from Bodies of Revolution," *Appl. Sci. Res.*, **7B**, 293–328, 1959.

Skolnik, M. I. (ed.): *Radar Handbook*, McGraw-Hill, 1980.

Ulaby, F. T., R. K. Moore and A. K. Fung: *Microwave Remote Sensing, Active and Passive*, Addison-Wesley, 1981.

Umashankar, K., A. Taflove and S. M. Rao: "Electromagnetic Scattering by Arbitrary Shaped 3-Dimensional Homogeneous Lossy Dielectric Objects," *IEEE Trans. Ants. Prop.*, **AP-34**, 758–766, June 1986.

Volakis, J. L., W. D. Burnside and L. Peters, Jr.: "Electromagnetic Scattering from Appendages on a Smooth Surface," *IEEE Trans. Ants. Prop.*, **AP-33**, 736–743, July 1985.
Wang, D. S., and L. N. Medgyesi-Mitschang: "Electromagnetic Scattering from Finite Circular and Elliptic Cones," *IEEE Trans. Ants. Prop.*, **AP-33**, 488–497, May 1985.
Wang, N.: "Electromagnetic Scattering from a Dielectric-Coated Circular Cylinder," *IEEE Trans. Ants. Prop.*, **AP-33**, 960–963, September 1985.
Yaghjian, A. D., and R. V. McGahan: "Broadside Radar Cross Section of the Perfectly Conducting Cube," *IEEE Trans. Ants. Prop.*, **AP-33**, 321–329, March 1985.

Proceedings of the IEEE, Special Issue on Radar Reflectivity, **53**, August 1965.
Proceedings of the IEEE, Special Issue on Radar, **73**, February 1985.

See also backscatter references at the end of Chap. 9.

PROBLEMS[1]

★17-1 Satellite TV downlink. A transmitter (transponder) on a Clarke orbit satellite produces an effective radiated power (ERP) at an earth station of 35 dB over 1 W isotropic.

(*a*) Determine the S/N ratio (dB) if the earth station antenna diameter is 3 m, the antenna temperature 25 K, the receiver temperature 75 K and the bandwidth 30 MHz. Take the satellite distance as 36 000 km. Assume the antenna is a parabolic reflector (dish-type) of 50 percent efficiency.

(*b*) If a 10-dB S/N ratio is acceptable, what is the required diameter of the earth station antenna?

Note: For FM-modulated video signals, as employed by the Clarke-orbit satellites, the S/N ratio as used above is actually a carrier-to-noise (C/N) ratio, the ultimate video signal-to-noise ratio for typical North American domestic system satellites being almost 40 dB higher. This is an advantage of FM modulation. If the C/N exceeds a few decibels, then, in principle, a perfect picture results. However, a C/N $\geq$ 10 dB is desirable to allow for misalignment of the earth-station antenna, attenuation due to water or snow in the dish, a decrease in transponder power, etc.

★17-2 Antenna temperature. An end-fire array is directed at the zenith. The array is located over flat nonreflecting ground. If $0.9\ \Omega_A$ is within 45° of the zenith and $0.08\ \Omega_A$ between 45° and the horizon calculate the antenna temperature. The sky brightness temperature is 5 K between the zenith and 45° from the zenith, 50 K between 45° from the zenith and the horizon and 300 K for the ground (below the horizon). The antenna is 99 percent efficient and is at a physical temperature of 300 K.

17-3 Earth-station antenna temperature. An earth-station dish of 100 m^2 effective aperture is directed at the zenith. Calculate the antenna temperature assuming that the sky temperature is uniform and equal to 6 K. Take the ground temperature equal to 300 K and assume that $\frac{1}{3}$ of the minor-lobe beam area is in the back direction. The wavelength is 75 mm and the beam efficiency is 0.8.

[1] Answers to starred (★) problems are given in App. D.

***17-4** **System temperature.** The digital output of a 1.4-GHz radio telescope gives the following values (arbitrary units) as a function of the sidereal time while scanning a uniform brightness region. The integration time is 14 s, with 1 s idle time for print-out. The output units are proportional to power.

Time	Output	Time	Output
31^m30^s	234	32^m45^s	229
31 45	235	33 00	236
32 00	224	33 15	233
32 15	226	33 30	230
32 30	239	33 45	226

If the temperature calibration gives 170 units for 2.9 K applied, find (*a*) the rms noise at the receiver, (*b*) the minimum detectable temperature, (*c*) the system temperature and (*d*) the minimum detectable flux density. The calibration signal is introduced at the receiver. The transmission line from the antenna to the receiver has 0.5 dB attenuation. The antenna effective aperture is 500 m^2. The receiver bandwidth is 7 MHz. The receiver constant $k' = 2$.

17-5 **System temperature.** Find the system temperature of a receiving system with 15 K antenna temperature, 0.95 transmission-line efficiency, 300 K transmission-line temperature, 75 K receiver first-stage temperature, 100 K receiver second-stage temperature and 200 K receiver third-stage temperature. Each receiver stage has 16 dB gain.

***17-6** **Minimum detectable temperature.** A radio telescope has the following characteristics: antenna noise temperature 50 K, receiver noise temperature 50 K, transmission-line between antenna and receiver 1 dB loss and 270 K physical temperature, receiver bandwidth 5 MHz, receiver integration time 5 s, receiver (system) constant $k' = \pi/\sqrt{2}$ and antenna effective aperture 500 m^2. If two records are averaged, find (*a*) the minimum detectable temperature and (*b*) the minimum detectable flux density.

17-7 **Minimum detectable temperature.** A radio telescope operates at 2650 MHz with the following parameters: system temperature 150 K, predetection bandwidth 100 MHz, postdetection time constant 5 s, system constant $k' = 2.2$ and effective aperture of antenna 800 m^2. Find (*a*) the minimum detectable temperature and (*b*) the minimum detectable flux density. (*c*) If four records are averaged, what change results in (*a*) and (*b*)?

***17-8** **Antenna temperature with absorbing cloud.** A radio source is occulted by an intervening emitting and absorbing cloud of unity optical depth and brightness temperature 100 K. The source has a uniform brightness distribution of 200 K and a solid angle of 1 square degree. The radio telescope has an effective aperture of 50 m^2. If the wavelength is 50 cm, find the antenna temperature when the radio telescope is directed at the source. The cloud is of uniform thickness and has an angular extent of 5 square degrees. Assume that the antenna has uniform response over the source and cloud.

17-9 **Passive remote-sensing antenna.** Design a 30-GHz antenna for an earth-resource passive remote-sensing satellite to measure earth-surface temperatures with 1 km^2 resolution from a 300-km orbital height.

17-10 Forest absorption. An earth-resource satellite passive remote-sensing antenna directed at the Amazon River Basin measures a night-time temperature $T_A = 21°C$. If the earth temperature $T_e = 27°C$ and the Amazon forest temperature $T_f = 15°C$, find the forest absorption coefficient τ_f.

★17-11 Solar interference to earth station.

(*a*) Twice a year the sun passes through the apparent declination of the geostationary Clarke-orbit satellites, causing solar-noise interference to earth stations. If the equivalent temperature of the sun at 4 GHz is 50 000 K, find the sun's signal-to-noise ratio (in decibels) for an earth station with a 3-m parabolic dish antenna at 4 GHz. Take the sun's diameter as 0.5° and the earth-station system temperature as 100 K.

(*b*) Compare this result with that for the carrier-to-noise ratio calculated in Prob. 17-1 for a typical Clarke-orbit TV transponder.

(*c*) How long does the interference last?

Note that the relation $\Omega_A = \lambda^2/A_e$ gives the solid beam angle in steradians and not in square degrees.

17-12 Radar detection. A radar receiver has a sensitivity of 10^{-12} W. If the radar antenna effective aperture is 1 m^2 and the wavelength is 10 cm, find the transmitter power required to detect an object with a 5-m^2 radar cross section at a distance of 1 km.

17-13 Venus and moon radar.

(*a*) Design an earth-based radar system capable of delivering 10^{-15} W of peak echo power from Venus to a receiver. The radar is to operate at 2 GHz and the same antenna is to be used for both transmitting and receiving. Specify the effective aperture of the antenna and the peak transmitter power. Take the earth-Venus distance as 3 light-minutes, the diameter of Venus as 12.6 Mm and the radar cross section of Venus as 10 percent of the physical cross section.

(*b*) If the system of (*a*) is used to observe the moon, what will the received power be? Take the moon diameter as 3.5 Mm and the moon radar cross section as 10 percent of the physical cross section.

17-14 Thompson scatter. The alternating electric field of a passing electromagnetic wave causes an electron (initially at rest) to oscillate. This oscillation of the electron makes it equivalent to a dipole radiator. Show that the ratio of the power scattered per steradian to the incident Poynting vector is given by $(\mu_0 e^2 \sin\theta/4\pi m)^2$, where e and m are the charge and mass of the electron and θ is the angle of the scattered radiation with respect to the direction of the electric field $\mathbf{E}$ of the incident wave. This ratio times 4π is the radar cross section of the electron. Such reradiation is called *Thompson scatter*.

17-15 Thompson-scatter radar. A ground-based vertical-looking radar can be used to determine electron densities in the earth's ionosphere by means of Thompson scatter (see Prob. 17-14). The scattered-power radar return is proportional to the electron density. If a short pulse is transmitted by the radar, the backscattered power as a function of time is a measure of the electron density as a function of height. Design a Thompson-scatter radar operating at 430 MHz capable of measuring ionospheric electron densities with 1 km resolution in height and horizontal position to heights of 1 Mm. The radar should also be capable of detecting a minimum of 100 electrons at a height of 1 Mm. The design should specify radar peak power, pulse length, antenna size and receiver sensitivity. See W. E. Gordon,

"Radar Backscatter from the Earth's Ionosphere," *IEEE Trans. Ants. Prop.*, **AP-12**, 873–876, December 1964.

★**17-16 Jupiter signals.** Flux densities of 10^{-20} W m^{-2} Hz^{-1} are commonly received from Jupiter at 20 MHz. What is the power per unit bandwidth radiated at the source? Take the earth-Jupiter distance as 40 light-minutes and assume that the source radiates isotropically.

17-17 Red shifts. Powers. Some radio sources have been identified with optical objects and the Doppler or red shift z ($=\Delta\lambda/\lambda$) measured from an optical spectrum. Assume that the objects with larger red shift are more distant, according to the Hubble relation

$$R = \frac{v}{H_0} = \frac{m-1}{m+1}\frac{c}{H_0}$$

where R = distance in megaparsecs (1 megaparsec = 1 Mpc = 3.26×10^6 light-years), v = velocity of recession of object in m s^{-1}, $m = (z+1)^2$, c = velocity of light and H_0 = Hubble's constant = 75 km s^{-1} Mpc^{-1}. Determine the distance R in light-years to the following radio sources: (*a*) Cygnus A (prototype radio galaxy), $z = 0.06$; (*b*) 3C273 (quasistellar radio source, or quasar), $z = 0.16$; and (*c*) OQ172 (distant quasar), $z = 3.53$. The above sources have flux densities as follows at 3 GHz: Cygnus A, 600 Jy; 3C273, 30 Jy; OQ172, 2 Jy (1 Jy = 10^{-26} W m^{-2} Hz^{-1}). (*d*) Determine the radio power per unit bandwidth radiated by each source. Assume that the source radiates isotropically.

★**17-18 Critical frequency. MUF.** Layers may be said to exist in the earth's ionosphere where the ionization gradient is sufficient to refract radio waves back to the earth. [Although the wave actually may be bent gradually along a curved path in an ionized region of considerable thickness, a useful simplification for some situations is to assume that the wave is reflected as though from a horizontal perfectly conducting surface situated at a (*virtual*) *height h*.] The highest frequency at which this layer reflects a vertically incident wave back to the earth is called the *critical frequency* f_0. Higher frequencies at vertical incidence pass through. For waves at oblique incidence ($\phi > 0$ in Fig. P17-18) the *maximum usable frequency* (MUF) for point-to-point communication on the earth is given by MUF $= f_0/\cos\phi$, where ϕ = angle of incidence. The critical frequency $f_0 = 9\sqrt{N}$, where N = electron density (number m^{-3}). N is a function of solar irradiation and other factors. Both f_0 and h vary with time of day, season, latitude and phase of the 11-year sunspot cycle. Find the MUF for (*a*) a distance d = 1.3 Mm by F_2-layer (h = 325 km)

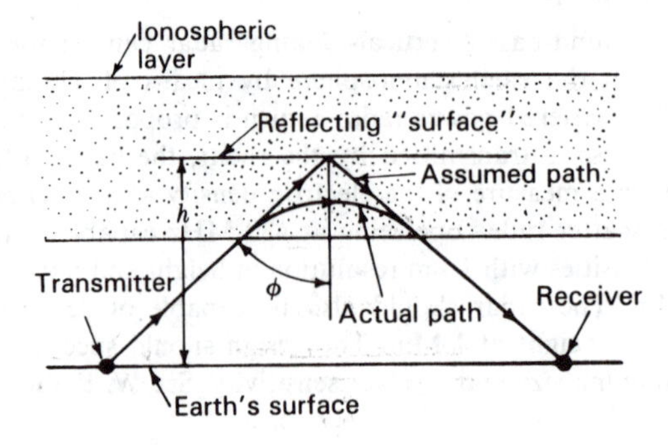

Figure P17-18 Communication path via reflection from ionospheric layer.

reflection with F_2-layer electron density $N = 6 \times 10^{11}$ m^{-3}; (*b*) a distance $d = 1.5$ Mm by F_2-layer ($h = 275$ km) reflection with $N = 10^{12}$ m^{-3}; and (*c*) a distance $d = 1$ Mm by sporadic E-layer ($h = 100$ km) reflection with $N = 8 \times 10^{11}$ m^{-3}. Neglect earth curvature.

17-19 mUF for Clarke-orbit satellites. Stationary communication (relay) satellites are placed in the Clarke orbit at heights of about 36 Mm. This is far above the ionosphere, so that the transmission path passes completely through the ionosphere twice, as in Fig. P17-19. Since frequencies of 2 GHz and above are usually used, the ionosphere has little effect. The high frequency also permits wide bandwidths. If the ionosphere consists of a layer 200 km thick between heights of 200 and 400 km with a uniform electron density $N = 10^{12}$ m^{-3}, find the lowest frequency (or *minimum usable frequency*, mUF) which can be used with a communication satellite (*a*) for vertical incidence and (*b*) for paths 30° from the zenith. (*c*) For an earth station on the equator, what is the mUF for a satellite 15° above the eastern or western horizon?

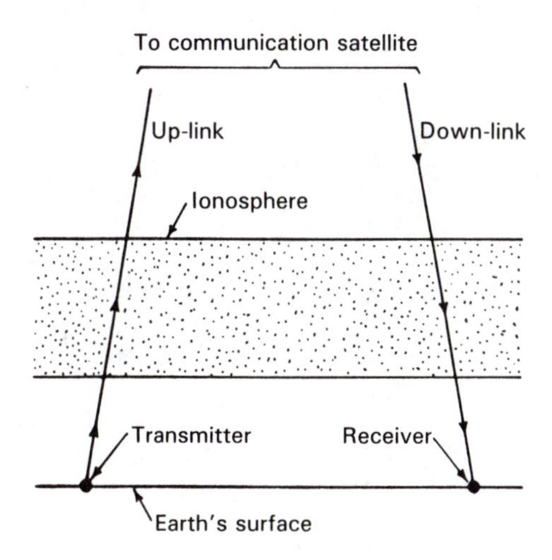

Figure P17-19 Communication path via geostationary Clarke-orbit relay satellite.

17-20 S/N ratio. Show that the S/N ratio for a radio link with 1 W transmitter and isotropic antennas is

$$\frac{S}{N} = \frac{\lambda^2}{16\pi^2 r^2 k T_{\text{sys}}\,\Delta f}$$

where symbols are as given in (17-3-6) and (17-3-8).

17-21 Disc backscatter. Show that the backscattering efficiency of a large broadside flat disc is twice its backscattering directivity or $C_\lambda^2 = 2D_{\text{obj}}$.

17-22 Large sphere. Show that the total scattering cross section of a large sphere (radius $r \gg \lambda$) is twice its geometric or physical cross section.

17-23 Effect of resonance on radar cross section of short dipoles.

(*a*) Calculate the radar cross section of a lossless resonant dipole ($Z_L = -jX_A$) with length $= \lambda/10$ and diameter $= \lambda/100$. (See Secs. 2-14, 2-20 and 9-17).

(*b*) Calculate the radar cross section of the same dipole from Table 17-2.

(*c*) Compare both values with the maximum radar cross section of Kouyoumjian shown in Fig. 17-11. Comment on the results.

17-24 Loop, wire, sphere and disc radar cross sections. Confirm Kouyoumjian's statement that the loop's radar cross section at first resonance is more than the radar cross sections of a wire, sphere or disc at their first resonances. Why is the loop's σ the highest?

CHAPTER
18

ANTENNA MEASUREMENTS

18-1 INTRODUCTION. The understanding of physical phenomena involves a balance of theory and experiment. Since theoretical analyses usually treat idealizations or simplifications of actual situations, theory may only approximate the real world. So while theory is essential to our understanding, experimental measurements determine the actual performance, but only if the measurements are done properly.

In this chapter methods and techniques are discussed for experimental measurements on antennas. There are sections on the measurement of pattern, gain, current distribution, impedance and polarization. According to the reciprocity relation, the same pattern will be measured whether the antenna is transmitting or receiving. Reciprocity also applies to certain other characteristics,[1] so that it is convenient in some cases to regard the antenna as a radiator and in other situations as a receiver.

18-2 PATTERNS. The far-field pattern of an antenna is one of its most important characteristics. The complete field pattern is a 3-dimensional or space pattern and its complete description requires field intensity measurements in all directions (over 4π sr).

Consider that the antenna under test is situated at the origin of the coordinates of Fig. 18-1 with the z axis vertical. Then patterns of the θ and ϕ com-

[1] But not to current distributions.

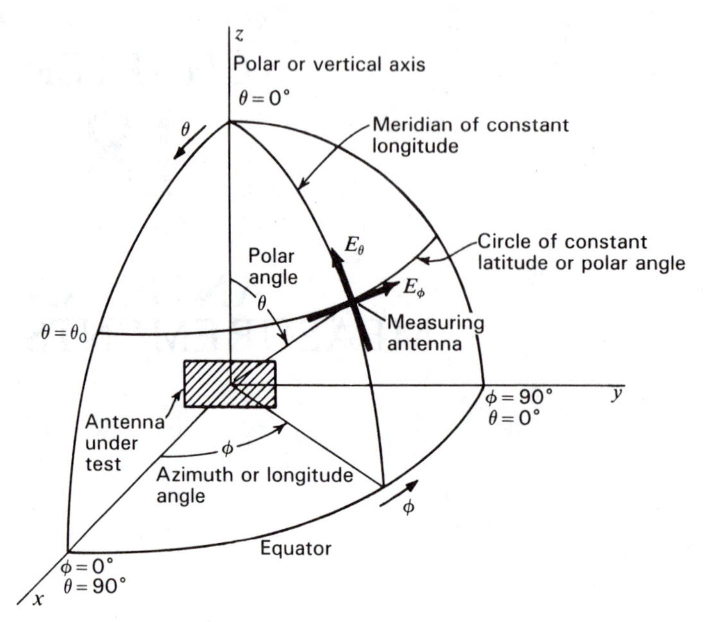

Figure 18-1 Antenna and coordinates for pattern measurements.

ponents of the electric field (E_θ and E_ϕ) are measured as a function of ϕ along constant θ circles, where ϕ is the longitude or azimuth angle and θ the zenith angle (complement of the latitude or elevation angle). These patterns may be determined by moving the measuring antenna with the antenna under test fixed or by rotating the antenna under test on its vertical (z) axis with the measuring antenna fixed. For complete information, the phase angle δ between E_θ and E_ϕ is also required to establish the polarization, although the polarization ellipse may also be determined by rotation of a linearly polarized measuring antenna (see Sec. 18-9). With sufficient data, 3-dimensional intensity diagrams of E_θ, E_ϕ and δ can be produced. An alternative is to make a 3-dimensional contour map of the power (proportional to $E_\theta^2 + E_\phi^2$) and superimpose polarization ellipse axes as suggested in Fig. 18-2.

Although detailed pattern measurements as above are sometimes required, fewer patterns are frequently sufficient. Thus, suppose that the antenna is a directional type with a main beam in the x direction, as suggested in Fig. 18-3. Then two patterns, called *principal plane patterns,* bisecting the main beam may suffice. If the antenna is horizontally polarized, then xz and xy plane patterns of E_ϕ, as indicated in Fig. 18-3*a*, are measured. If the antenna is vertically polarized then xz and xy plane patterns of E_θ, as indicated in Fig. 18-3*b*, are measured. If the antenna is elliptically or circularly polarized, both sets of measurements (4 patterns) plus axial ratio data are required. Even if the antenna is believed to be linearly polarized, measurement of the 4 patterns plus axial ratios may be desirable to establish polarization purity.

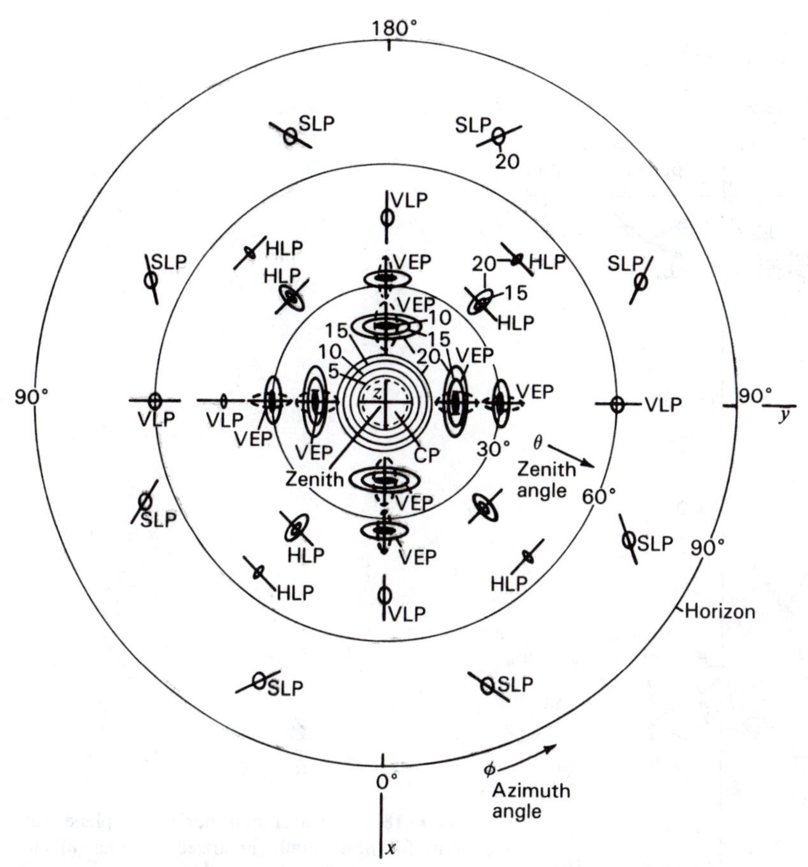

Figure 18-2 Three-dimensional power pattern in 5-dB increments for circularly polarized antenna with main beam vertical (in direction of z axis or out of the page). Near sidelobes are 10 dB down, others 15 and 20 dB down. The polarization ellipses (for circular and elliptical polarization) are shown dotted with the solid orthogonal lines indicating major and minor axes. For essentially pure linear polarization (vertical, horizontal or slant) the ellipse collapses to the major axis line. The solid contours represent signal strength. The main beam is circularly polarized (CP), while the near sidelobes are elliptically polarized with the major axis vertical (VEP). Other minor lobes are linearly polarized vertically (VLP), horizontally (HLP) or at a slant angle (SLP).

To summarize, the 4 patterns are:

$E_\phi(\theta = 90°, \phi)$ = pattern of ϕ component of electric field as a function of ϕ in xy plane ($\theta = 90°$)

$E_\phi(\theta, \phi = 0°)$ = pattern of ϕ component as a function of θ in xz plane ($\phi = 0°$)

$E_\theta(\theta = 90°, \phi)$ = pattern of θ component as a function of ϕ in xy plane ($\theta = 90°$)

$E_\theta(\theta, \phi = 0°)$ = pattern of θ component as a function of θ in xz plane ($\phi = 0°$)

18-3 PATTERN MEASUREMENT ARRANGEMENTS. Consider the arrangement in Fig. 18-4, with the antenna under test acting as a receiving

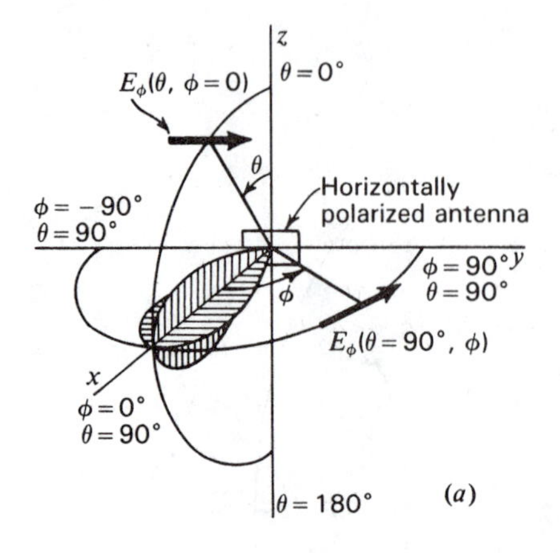

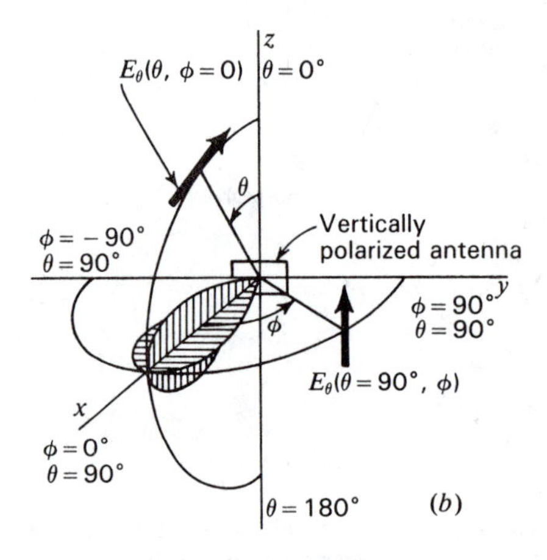

Figure 18-3 Vertical and horizontal plane patterns for horizontally polarized antenna (*a*) and vertically polarized antenna (*b*).

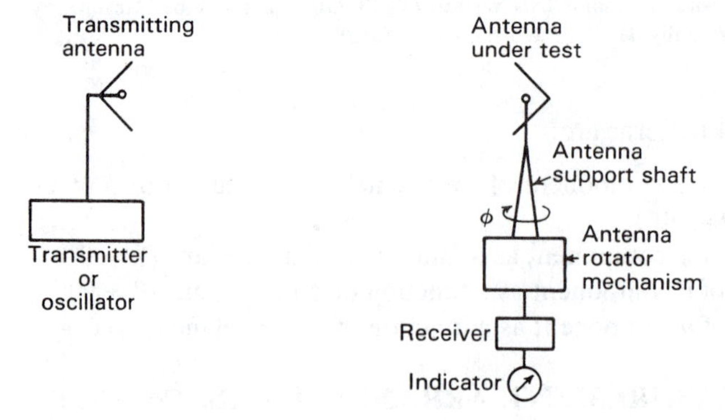

Figure 18-4 Antenna pattern-measuring arrangement.

antenna situated under suitable "illumination" from a transmitting antenna as suggested. The transmitting antenna is fixed in position and the antenna under test is rotated on a vertical axis by the antenna support shaft. Assuming that both antennas are linearly polarized, the E_ϕ ($\theta = 90°$, ϕ) pattern is measured by rotating the antenna support shaft with both antennas horizontal as in Fig. 18-4. To measure the E_ϕ (θ, $\phi = 0$) pattern, the antenna support shaft is rotated with both antennas vertical.

Indication may be on a direct-reading meter calibrated in field intensity or the meter may always be adjusted to a constant value by means of a calibrated attenuator. Where many pattern measurements are involved, work is facilitated with an automatic pattern recorder.

18-3a Distance Requirement for Uniform Phase. For an accurate far-field or Fraunhofer pattern of an antenna a first requirement is that the measurements be made at a sufficiently large distance that the field at the antenna under test approximates a uniform plane wave. Suppose that the antenna to be measured is a broadside array consisting of a number of in-phase linear elements as suggested in Fig. 18-5. The width or physical size of the array is a. At an infinite distance normal to the center of the array, the fields from all parts of the array will arrive in the same phase. However, at any finite distance r, as in Fig. 18-5, the field from the edge of the array must travel a distance $r + \delta$ and, hence, is retarded in phase by $360°\delta/\lambda$ with respect to the field from the center of the array. If δ is a large enough fraction of a wavelength, the measured pattern will depart appreciably from the true far-field pattern.[1] Referring to Fig. 18-5,

$$r^2 + 2r\delta + \delta^2 = r^2 + \frac{a^2}{4} \tag{1}$$

If $\delta \ll a$ and $\delta \ll r$,

$$r \simeq \frac{a^2}{8\delta} \tag{2}$$

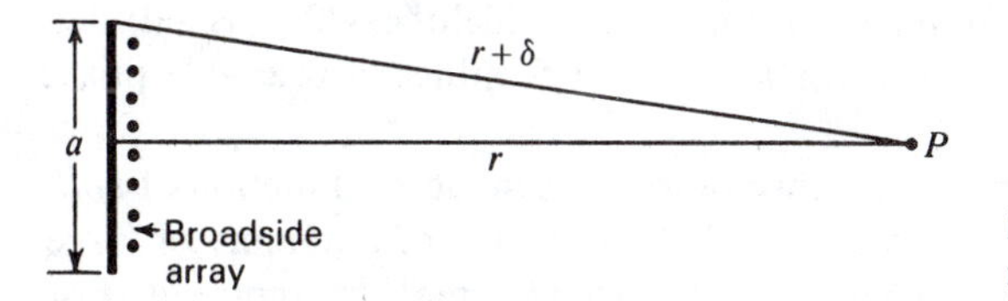

Figure 18-5 Geometry for distance requirement.

[1] If the distance is insufficient, the near field or Fresnel pattern is measured. In general, this pattern is a function of the distance at which it is measured. However, since the far-field pattern is the *Fourier transform* of the aperture field distribution (see Sec. 11-22), it is possible to deduce a *far-field pattern from near-field measurements*. See, for example, R. C. Johnson, H. A. Ecker and J. S. Hollis, "Determination of Far Field Patterns from Near Field Measurements," *Proc. IEEE*, **61**, 1668–1694, December 1973.

Table 18-1 Tolerable phase difference data

Maximum tolerable phase difference, deg	δ	k
5	$\frac{\lambda}{72}$	9
10	$\frac{\lambda}{36}$	4.5
22.5	$\frac{\lambda}{16}$	2
30	$\frac{\lambda}{12}$	1.5
45	$\frac{\lambda}{8}$	1

Thus, the minimum distance r depends on the maximum value of δ which can be tolerated. Some workers[1] recommended that δ be equal to or less than $\lambda/16$. Then

$$r \geq 2\frac{a^2}{\lambda} \tag{3}$$

In general, the constant factor [equal to 2 in (3)] may be represented by k. Thus,

$$r \geq k\frac{a^2}{\lambda} \tag{4}$$

The phase difference equals $360°\delta/\lambda$, which for $\delta = \lambda/16$ is 22.5°. In some special cases phase differences of more than 22.5° can be tolerated and in other cases less. Hacker and Schrank[2] indicate that on very low sidelobe antennas (-30 to -40 dB), phase differences of 5° or less may be required to resolve near-in sidelobes but for measurements of the far-out sidelobes 22.5° is satisfactory. Table 18-1 gives the constant factor k in (4) for 5 values of tolerable phase difference.

According to (4) the minimum distance of measurement is a function of *both* the antenna aperture a and the wavelength λ. In the case of antennas of large physical aperture and small wavelength, large distances may be required. For

[1] C. C. Cutler, A. P. King and W. E. Kock, "Microwave Antenna Measurements," *Proc. IRE*, **35**, 1462–1471, December 1947.

[2] P. S. Hacker and H. E. Schrank, "Range Distance Requirements for Measuring Low and Ultralow Sidelobe Antenna Patterns," *IEEE Trans. Ants. Prop.*, **AP-30**, 956–966, September 1982.

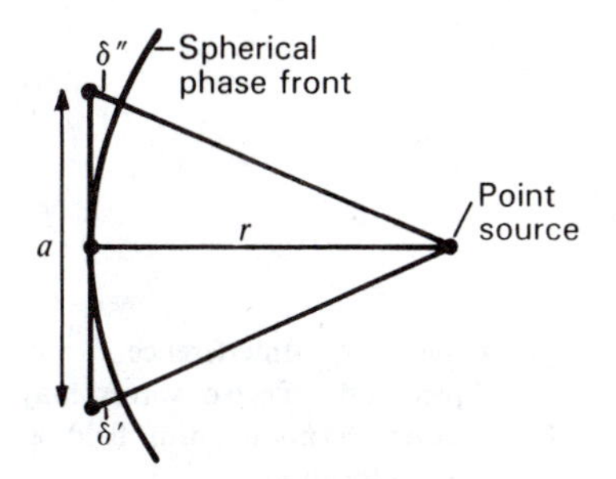

Figure 18-6 Passive distance determination by measurement of phase differences δ' and δ''.

example, consider a 3-GHz broadside beam antenna with a physical size of 20 m. Taking $k = 2$, we obtain for the minimum distance $r = 4$ km.

The constant phase front of a wave radiated by a point source is a sphere. At a distance r the phase front departs by distances δ' and δ'' from the ends of a nominally perpendicular line of length a as indicated in Fig. 18-6. By comparing phases at the ends of the line with the phase at the center, δ' and δ'' can be determined and knowing the length a, the distance r of the source is given by

$$r = \frac{a^2}{4(\delta' + \delta'')} \tag{5}$$

Thus, changing the distance requirement of (3) around, we use a phase measurement to determine the distance. This is a *passive distance measuring method.* It has been proposed that with radio telescopes in space on a very long baseline, the distance to all radio-emitting objects in the universe can be measured by this method.[1]

18-3b Uniform Field Amplitude Requirement. Further requirements are that the field in the test or target zone (volume containing antenna or scattering object under test) have small amplitude taper, small amplitude ripple and small cross-polarization (or high polarization purity). The polarization requirement means that for linear vertical or horizontal polarization the polarization state be on or close to the equator of the Poincaré sphere and for circular polarization on or close to one of the poles.

On outdoor measuring ranges, field variations can be produced by interference of the direct wave with waves reflected from the ground, as in Fig. 18-7, or from other objects. The effect of the ground reflection may be reduced by using a directional transmitting antenna and placing both antennas on towers as in Fig. 18-8*a* or near the edges of adjacent buildings as in Fig. 18-8*b*. With such arrangements the amplitude of the reflected wave is reduced since the groundward radiation from the transmitting antenna is less and also since the path

---

[1] N. Kardashev *et al.*, Acad. Sci. USSR, Space Res. Inst., Rept. PR-373, 1977.

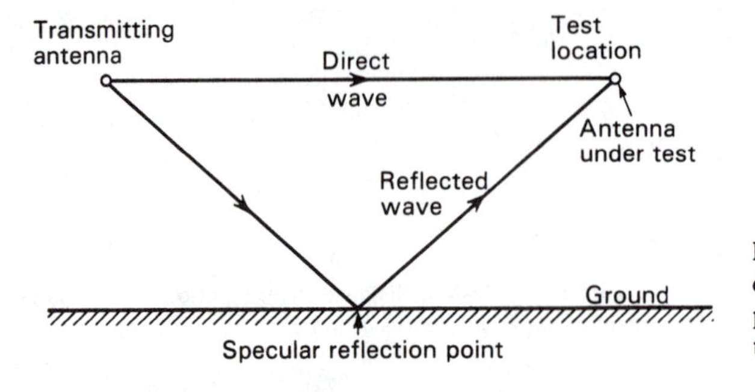

Figure 18-7 Interference of direct and reflected waves may produce a nonuniform field at the test location.

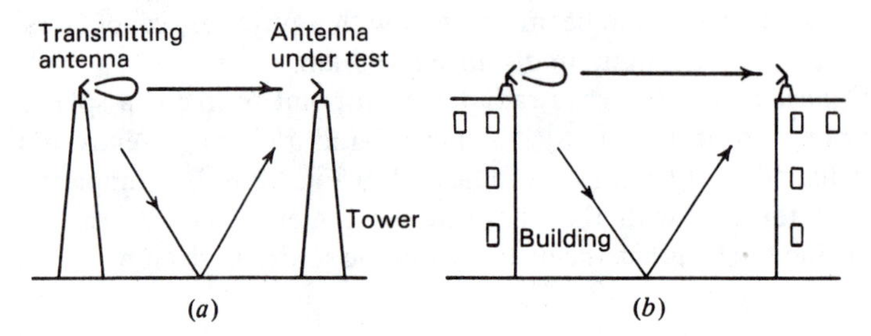

Figure 18-8 Antenna test arrangements.

length of the reflected wave is considerably greater than the path length of the direct wave. In a typical case, the variation in field intensity as a function of height at the test location may be as indicated by the solid curve in Fig. 18-9. The transmitting antenna is directional and is at a fixed height h. There is a considerable target zone region near the height h with but small ripple. If the transmitting antenna is nondirectional the ripple is much greater, as suggested by the dashed curve in Fig. 18-9.

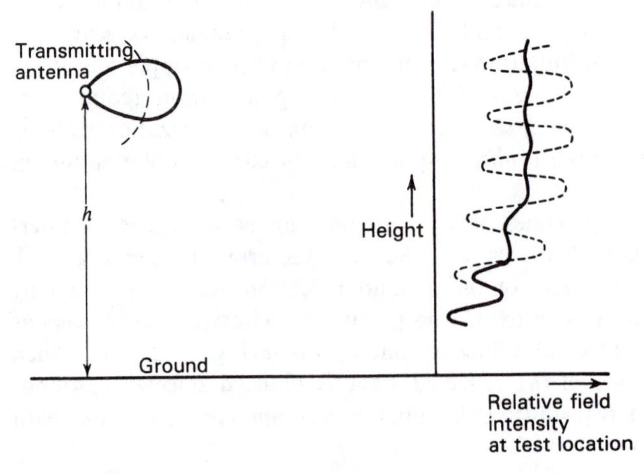

Figure 18-9 Variation or ripple in field intensity with height at the test location with a directional transmitting antenna relatively close (solid curve). If the transmitting antenna is nondirectional in the vertical plane, the ripple is greater (dashed curve).

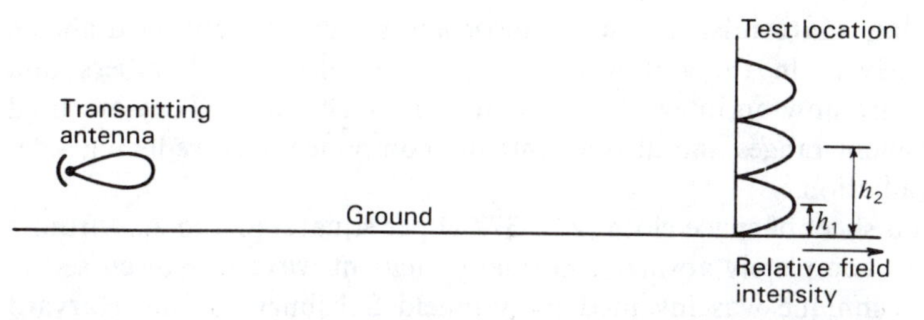

Figure 18-10 Variation of field intensity with height at the test location with a directional transmitting antenna at a large distance.

Sometimes the distance requirement of (4) is so large that the tower height needed may be impractical. In this case, the test antenna can be situated in a region of maximum field intensity such as at the heights h_1 or h_2 in Fig. 18-10. (See also Prob. 16-14.) This arrangement has the limitation that the height of the test antenna may need to be adjusted for each change in frequency.

To reduce ground reflection, the ground can be covered with absorbing material around the specular reflection point or one or more conducting fences can be installed. However, the edge of a fence can diffract waves into the test area and generate ripple.

Ranges constructed indoors require appropriately placed wave-absorbing material. In the simple, inexpensive range shown in Fig. 18-11, absorbing panels are placed at the specular reflection points on the walls, ceiling and floor as suggested. With small aperture antennas the distance requirement may be satisfied while with a directional transmitting antenna and the absorbing panels the field uniformity may be adequate.

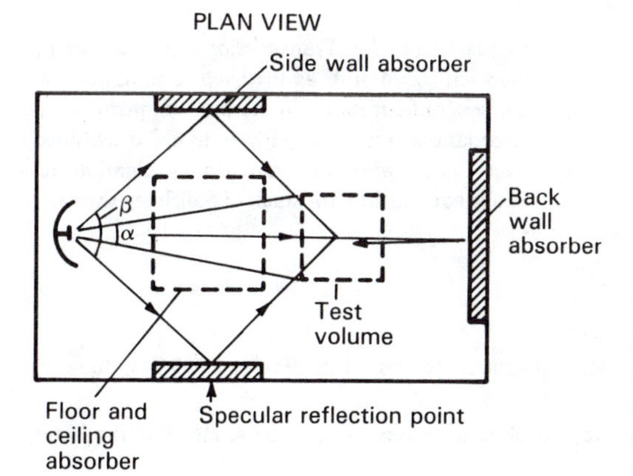

Figure 18-11 Simple low-cost indoor range in plan view with wave-absorbing panels only at specular reflection points. For minimum field taper across the test volume, the transmitting antenna beam width should be considerably greater than the test volume angle (HPBW $\gg \alpha$) but considerably smaller than the angle of specular-reflection (HPBW $\ll \beta$).

18-3c Absorbing Materials. The wave-absorbing materials mentioned above, and more extensively in the following sections on anechoic chambers and compact ranges, are now an integral part of antenna technology. They are used both in measurement ranges and also as antenna components for reducing side- and back-lobe radiation.

The use of a sheet of space cloth ($Z = 377\ \Omega$ per square) placed $\lambda_0/4$ from a reflecting plate to completely absorb a normally incident wave was discussed in Sec. 2-18. This technique was invented by Winfield Salisbury[1] at the Harvard Radio Research Laboratory during World War II and the resistive (carbon-impregnated) cloth sheets he used are called *Salisbury screens.* At normal incidence the arrangement gives a 1.3 to 1 bandwidth for a reflected wave at least 20 dB down.[2]

The transmission line equivalent is shown in Fig. 18-12*a* with the characteristic impedance of the transmission line equal to 377 Ω. For simplicity let us

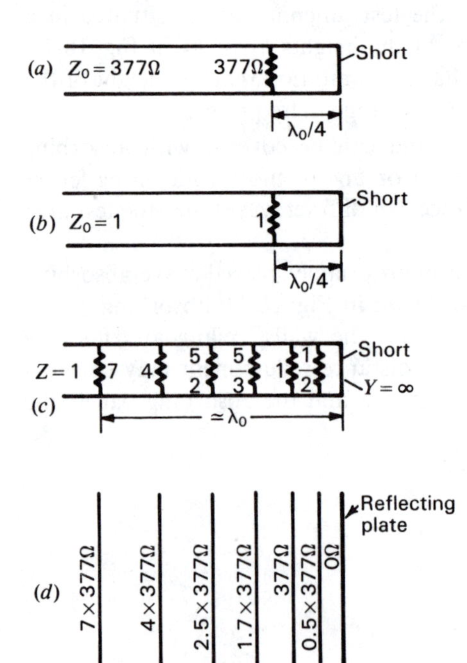

Figure 18-12 (*a*) Transmission line and single load with $\lambda_0/4$ stub as matched termination. (*b*) Same configuration in terms of normalized impedances. (*c*) Line with 6 loads distributed over $1\lambda_0$ as wideband matched termination. (*d*) Space equivalent with stack of Salisbury sheets.

---

[1] W. W. Salisbury, "Absorbent Body for Electromagnetic Waves," U.S. Patent 2,599,944, June 10, 1952.

[2] The wave absorption resonates also for sheet-to-plate distances of $3\lambda/4$, $5\lambda/4$, etc., but the bandwidth is narrower.

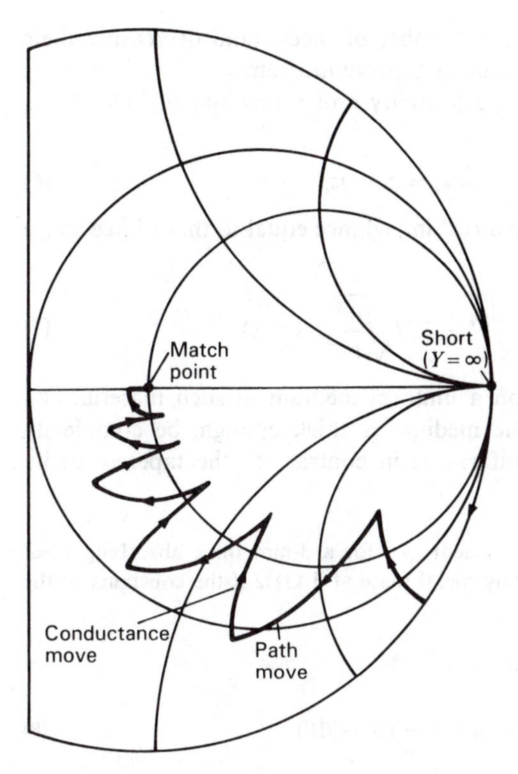

Figure 18-13 Path of normalized admittance Y from short via incremental steps to match point at center of Smith chart for the line with 6 distributed loads of Fig. 18-12*c* (5 to 1 bandwidth for reflected wave 20 dB down).

divide by 377 Ω, obtaining *normalized* (*dimensionless*) *impedances* as in Fig. 18-12*b*.

Consider now the situation shown in Fig. 18-12*c* with a number of resistances shunting the line over a distance of about 1λ. The (normalized) resistances range from small to large values with distance from the short. The spacings also increase with distance from the short.[1] As shown in Fig. 18-13, the path of the normalized line admittance Y moves from the $Y = \infty$ position at the short via incremental steps of distance and conductance to the center of the chart (match point). The advantage of this incrementally tapered termination is that it provides a low reflection coefficient over wider bandwidths than the single-resistor termination of Fig. 18-12*a* and *b*. The space equivalent of Fig. 18-12*c* is shown in Fig. 18-12*d* consisting of a λ_0 stack of Salisbury sheets with impedances per square as indicated and backed by a reflecting plate. Stacks of this kind with sheets sandwiched between layers of plastic (dielectric) were developed in Germany during World War II by J. Jaumann. An historical summary of the development of wave-absorbing material and its application is given by Emerson.[2]

[1] Both resistance and spacing increase in an approximately exponential manner.

[2] W. H. Emerson, "Electromagnetic Wave Absorbers and Anechoic Chambers through the Years," *IEEE Trans. Ants. Prop.*, **AP-21**, 484–489, July 1973 (49 references).

Increasing both the resistance and number of sheets (and decreasing their spacing) results in the limit in a continuously tapered medium.

If both the permeability μ and permittivity ε of a medium include a loss term but with

$$\mu_r = \mu_r' - j\mu_r'' = \varepsilon_r = \varepsilon_r' - j\varepsilon_r'' \tag{6}$$

the medium, although lossy, will have a real impedance equal to that of free space as given by

$$Z = \sqrt{\frac{\mu}{\varepsilon}} = \sqrt{\frac{\mu_0}{\varepsilon_0}}\sqrt{\frac{\mu_r}{\varepsilon_r}} = 377\sqrt{\frac{\mu_r}{\varepsilon_r}} = 377\ \Omega \tag{7}$$

In principle, a wave incident on a uniform medium of such material can enter it without reflection and, if the medium is thick enough, be completely absorbed. Such a medium, being uniform, is in contrast to the tapered media discussed above.

Example 1. Find the reflection coefficient $|\rho_v|$ for a 3-mm thick absorbing sheet backed by a flat perfectly conducting metal plate at 3 GHz if the constants of the sheet are $\sigma = 0$, $\mu_r = \varepsilon_r = 10 - j10$.

Solution. The propagation constant

$$\gamma = \alpha + j\beta = j\frac{2\pi}{\lambda_0}(10 - j10) \tag{8}$$

and $$\lambda_0 = \frac{c}{f} = \frac{3 \times 10^8}{3 \times 10^9} = 0.1 \text{ m}$$

Therefore,

$$\alpha = 2\pi \times 100 = 628 \text{ Np m}^{-1}$$

and the relative field intensity of the wave emerging from the sheet after reflection from the metal plate is given by

$$\frac{E}{E_0} = |\rho_v| = e^{-2\alpha x} = e^{-3.77} = 0.023$$

or down 33 dB from the incident wave.

Although lossy media with $\mu_r = \varepsilon_r$ appear attractive in principle, the parameters of the more popular types of absorber are typically: $\mu_r = 1$ and $\varepsilon_r \simeq 2 - j1$. Popular shapes are in the form of pyramids and wedges as illustrated in Fig. 18-14*a* and *b*. The pyramids behave like a tapered transition (as discussed above) for normal (nose-on) incidence. However, DeWitt and Burnside[1] find that pyramid absorbers tend to scatter as a random rough surface with large reflec-

[1] B. T. DeWitt and W. D. Burnside, "Electromagnetic Scattering by Pyramidal and Wedge Absorber," *IEEE Trans. Ants. Prop.*, 1988.

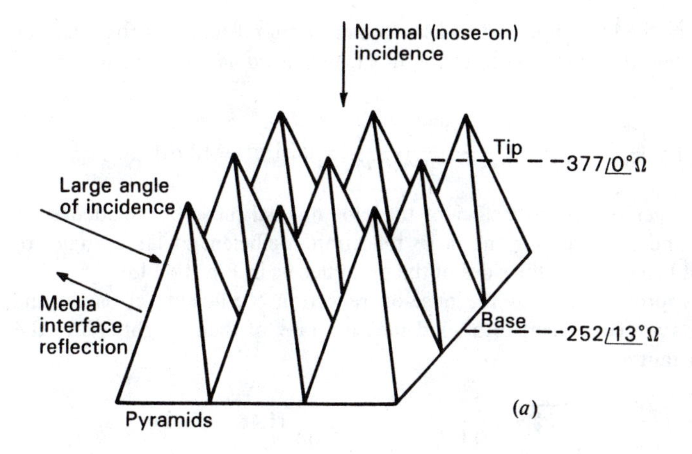

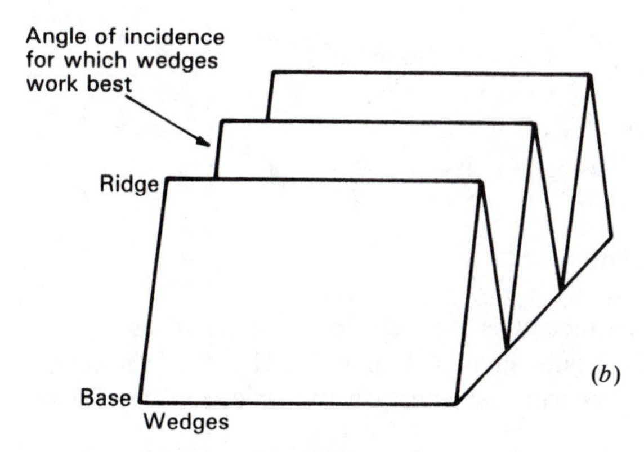

Figure 18-14 (*a*) Pyramid and (*b*) wedge forms of wave absorbers.

tion coefficient at large angles of incidence, wedges being far superior at these angles provided the wave direction is nearly *parallel* to the ridge of the wedge. At large incidence angles the wave direction is almost broadside to the side faces of the pyramids, resulting in reflection due to media mismatch.

Example 2. (*a*) Find the normal (nose-on) reflection coefficient at 3 GHz ($\lambda = 100$ mm) for an array of pyramids 30 cm from tip to base with $\sigma = 0$, $\mu_r = 1$ and $\varepsilon_r = 2 - j1$. (*b*) Find the reflection coefficient at 10 GHz ($\lambda = 30$ mm) assuming that σ, μ_r and ε_r are the same as at 3 GHz.

Solution. Referring to Fig. 18-14*a*, the effective impedance presented by the pyramid array to a normally incident wave is essentially 377 $\underline{/0^\circ}$ at the tip, increasing gradually (over 3λ at 3 GHz and 10λ at 10 GHz) to an impedance of

$$Z = \frac{377}{\sqrt{\varepsilon_r}} = \frac{377}{\sqrt{2 - j1}} = 252 \underline{/13.3^\circ}\ \Omega$$

at the base. Due to the long taper, let us assume negligible reflection at the interface of the media (air and pyramid). Without the taper, the *media mismatch* would give a reflection coefficient of

$$|\rho_v| = \left|\frac{Z_L - Z_0}{Z_L + Z_0}\right| = \left|\frac{252 \,\underline{/13^\circ} - 377}{252 \,\underline{/13^\circ} + 377}\right| = 0.24 \text{ or } -12 \text{ dB}$$

This is a much larger reflection coefficient than for normal (nose-on) incidence (as calculated next) and accounts for the large reflection coefficient at large angles of incidence (almost broadside to the sides of the pyramids as in Fig. 18-14*a*).

As a first approximation for the nose-on reflection coefficient, let us assume that the pyramids are equivalent to a solid medium of $\frac{1}{3}$ of their height. Thus, the propagation constant

$$\gamma = j\frac{2\pi}{\lambda_0}\sqrt{\varepsilon_r' - j\varepsilon_r''} = j\frac{2\pi}{0.1}\sqrt{2 - j1} = j\frac{2\pi}{0.1}(1.46 - j0.35)$$

and the attenuation constant

$$\alpha = \frac{2\pi}{0.1} \times 0.35 = 22 \text{ Np m}^{-1}$$

The reflection coefficient is then

$$|\rho_v| = e^{-2\alpha x} = e^{-2 \times 22 \times 0.1} = 0.0123$$

(*a*) At 3 GHz this is −38 dB.
(*b*) At 10 GHz it is about −125 dB.

In practice, the reflection coefficient is unlikely to be as small as this at 10 GHz although it may be substantially smaller than at 3 GHz. The inhomogeneity of some commercial absorbers can also increase the reflection coefficient and backscatter.[1]

18-3d The Anechoic Chamber Compact Range. The transition from the simple indoor range of Fig. 18-11 to what might be called an *anechoic* (*no echo*) *chamber* is accomplished by completely covering all room surfaces with absorbing material. Thus, the side walls, ceiling and floor are covered by wedge absorbers with a ridge direction parallel to the path from the transmitter to the test site, while the back wall and wall behind the transmitter are covered with pyramids. The philosophy is to provide a nonreflecting environment like in outer space except that the walls are at ambient (~300 K) temperature instead of 3 K (or somewhat more), but with the distinct advantage that the room provides shielding from all of the external electromagnetic noise and interference (natural

[1] B. T. DeWitt and W. D. Burnside, "Electromagnetic Scattering by Pyramidal and Wedge Absorber," *IEEE Trans. Ants. Prop.*, 1988.

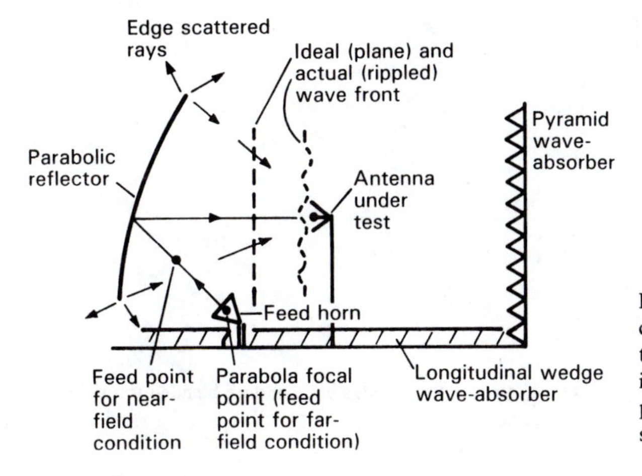

Figure 18-15 Compact range configuration. Radiation scattered from edges of the parabolic reflector degrades desired plane (far-field) wavefront at test site into one with ripples.

and man-made). However, simple indoor ranges are usually limited by the distance requirement.

A parabolic reflector is a spherical-to-plane-wave transformer and may be used to produce a plane wave in the test zone of an indoor range as suggested in Fig. 18-15. In effect, the parabolic reflector moves the far-field region in very close, making this foreshortened or *compact range* equivalent to a much larger conventional range.[1]

A conventional parabolic dish reflector will diffract significant radiation from its edges into the test zone of a compact range, limiting the usable test volume as suggested in Fig. 18-15. Serrated edges and absorber material are sometimes used to reduce this scattering. Burnside *et al.*[2] have shown that a rolled edge is effective in reducing the edge scattering effect. A minimum radius of curvature $r > \lambda/4$ at the lowest frequency is required. Furthermore, it is shown by Pistorius and Burnside[3] that diffraction effects can be made still smaller with a very smooth or blended transition of the reflector surface from the parabola to the rolled edge, as suggested in Fig. 18-16. The compact range with 5.8 m high elliptic rolled-edge parabolic reflector at the Electro-Science Laboratory at the Ohio State University is shown in Fig. 18-17.[4]

---

[1] R. C. Johnson, H. A. Ecker and R. A. Moore, "Compact Range Techniques and Measurements," *IEEE Trans. Ants. Prop.*, **AP-17**, 568–576, September 1969.

[2] W. D. Burnside, M. C. Gilreath, B. M. Kent and G. C. Clerici, "Curved Edge Modification of Compact Range Reflector," *IEEE Trans. Ants. Prop.*, **AP-35**, 176–182, February 1987.
W. D. Burnside, M. C. Gilreath and B. M. Kent, "Rolled-Edge Modification of Compact Range Reflector," *AMTA Symp.*, San Diego, 1984.

[3] C. W. I. Pistorius and W. D. Burnside, "An Improved Main Reflector Design for Compact Range Applications," *IEEE Trans. Ants. Prop.*, **AP-35**, 342–346, March 1987.

[4] E. K. Walton and J. D. Young, "The Ohio State University Compact Radar Cross-Section Measurement Range," *IEEE Trans. Ants. Prop.*, **AP-32**, 1218–1223, November 1984.

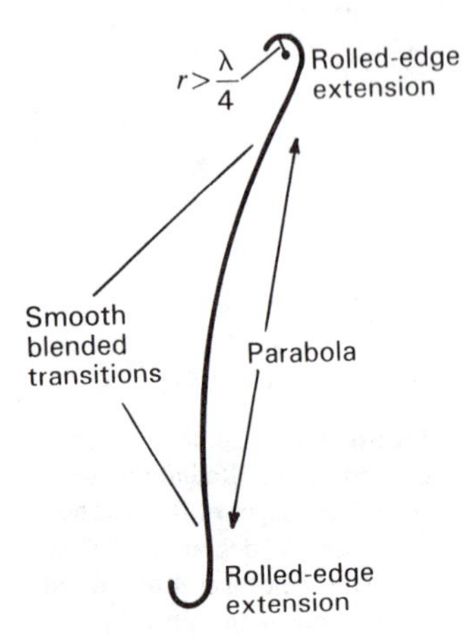

Figure 18-16 Rolled edge with smooth blended transition to parabola.

Pistorius and Burnside also indicate that combining a 4-point parabola as in Fig. 18-18 with a blended rolled edge leads to even greater improvement, the 4-point design being superior to circular or rectangular shapes.

The ultimate in compact range design, as proposed by Pistorius, Clerici and

Figure 18-17 Compact range at the Electro-Science Laboratory of the Ohio State University with 5.8-m high blended-rolled-edge parabolic dish reflector. The test object pedestal is in the foreground and the transmitter feed close to the floor between it and the dish. Wedge absorbers line the floor, ceiling and side walls with pyramid absorbers on the wall behind the dish and the back wall (behind our point-of-view in the photo). (*Courtesy W. D. Burnside, Electro-Science Laboratory.*)

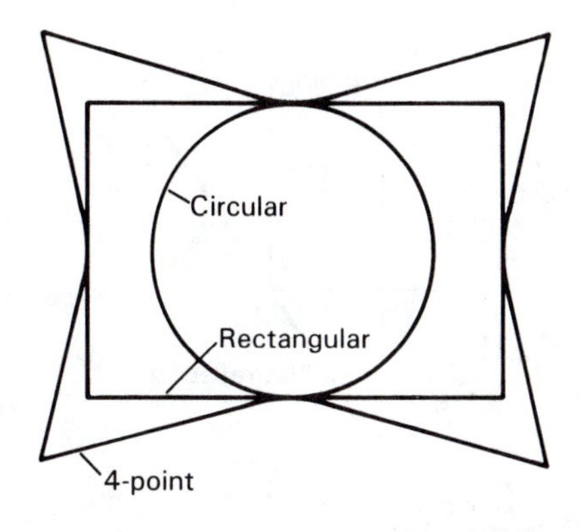

Figure 18-18 Four-point parabola compared with circular and rectangular designs. (*After C. W. I. Pistorius and W. D. Burnside, "An Improved Main Reflector Design for Compact Range Applications,"* IEEE Trans. Ants. Prop., **AP-35**, *342–346, March 1987.*)

Burnside,[1] is a dual-chamber Gregorian-fed system as shown in Fig. 18-19. This arrangement can provide very small taper, ripple and cross-polarization in the test zone.

Gating (or turning the receiver on and off) in the interval of pulse return from the target can further improve performance and the ability to measure the echo from objects with very small radar cross section.

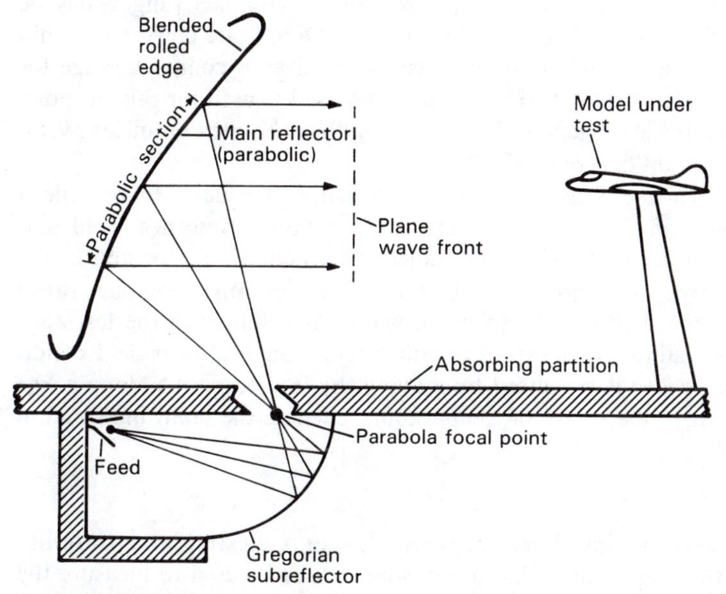

Figure 18-19 Dual-chamber Gregorian-fed Compact Range. (*After C. W. I. Pistorius, G. Clerici and W. D. Burnside, "A Dual-Chamber Gregorian Subreflector System for Compact Range Applications,"* IEEE Trans. Ant. Prop., **AP-36**, *1988.*)

---

[1] C. W. Pistorius, G. Clerici and W. D. Burnside, "A Dual-Chamber Gregorian Subreflector System for Compact Range Applications," *IEEE Trans. Ants. Prop.*, **AP-36**, 1988.

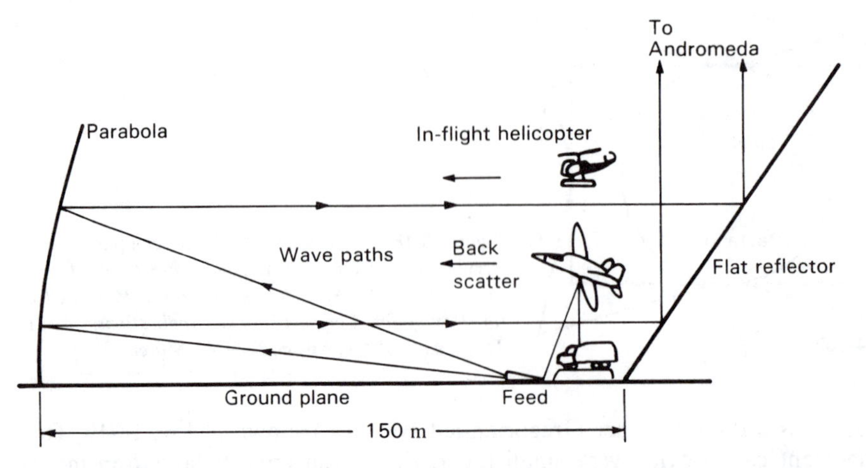

Figure 18-20 Big Ear radio telescope of the Ohio State University as a compact range for scattering measurements of large objects. (See Fig. 12-51 for photograph.) Actual full-size vehicles, aircraft and in-flight helicopters can be accommodated. The flat reflector deflects waves into the empty sky, replacing absorbing materials used in standard compact ranges.

For lower-frequency operation, even large indoor compact ranges may be too small. Although not initially designed as one, the Ohio State University radio telescope (photo in Fig. 12-51) is almost ideally suited as a compact range for frequencies from 3 GHz to 100 MHz or even lower, and is used for this purpose. Figure 18-20 shows the arrangement for measuring the echo area of full size vehicles, aircraft and in-flight helicopters.[1]

Although a prime objective of a compact range is usually to provide a far-field environment in the test zone, there are situations where near-field patterns are desired, such as where the test object may actually be located in the near field. It is shown by Rudduck *et al.*[2] that by defocusing a compact range reflector it is possible to produce a spherical-wave environment in the test zone with an adjustable radius of curvature to suit a wide range of near-field conditions. The spherical wave is produced by moving the feed to points along a line from the focal point to the center of the parabola, such as the point indicated in Fig. 18-15.

18-3e Pattern and Squint Measurements Using Celestial and Satellite Radio Sources.

Celestial and satellite radio sources can be used to measure the

---

[1] J. D. Kraus, "Ohio State's Big Ear Detects Edge of the Universe and Doubles as a Compact Range," *IEEE Ants. Prop. Soc. Newsletter*, **27**, 5–10, January 1985.

[2] R. C. Rudduck, M. C. Liang, W. D. Burnside and J. S. Yu, "Feasibility of Compact Ranges for Near-Zone Measurements," *IEEE Trans. Ants. Prop.*, **AP-35**, 280–286, March 1987.

far-field patterns and squint of antennas, especially ones with large apertures in terms of wavelengths where the distance to the far field is large (10s or 100s of kilometers). For pattern measurements, the radio source should have a small angular extent (much less than the antenna HPBW), be strong and isolated from nearby sources. For squint measurements, the position of the source should be accurately known. The table in Sec. A-8 (App. A) lists a few celestial radio sources which meet most or all of the above requirements. Most of these sources are at distances of more than 1 billion light-years (1 light-year = 10^{13} kilometers). For more details see Kraus.[1] Celestial radio sources can also be used for antenna gain, beam efficiency and aperture efficiency measurements (see Sec. 18-6e).

18-4 PHASE MEASUREMENTS. The preceding sections on pattern measurements deal only with the magnitude of the field intensity. To measure the phase variation of the field, an arrangement such as shown in Fig. 18-21 can be used. The antenna under test is operated as a transmitting antenna. The output of a receiving antenna is combined with the signal conveyed by cable from the oscillator. The receiving antenna is then moved so as to maintain either a minimum or a maximum indication. The path traced out in this way is a line of constant phase.

In another type of measurement the receiving antenna is moved along a reference line. A calibrated line stretcher or phase shifter is then adjusted to

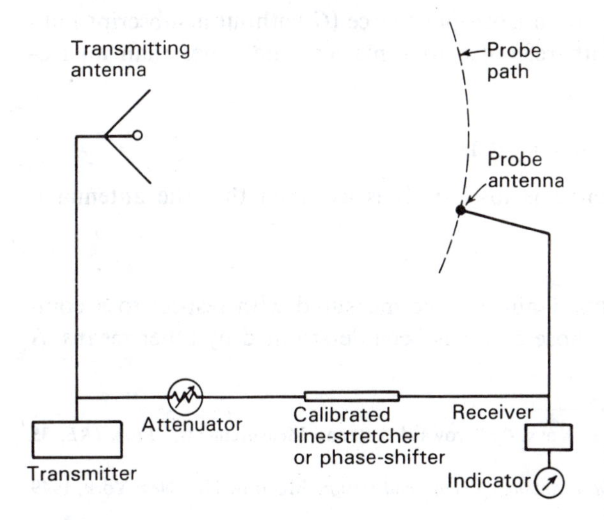

Figure 18-21 Arrangement for phase measurements.

[1] J. D. Kraus, *Radio Astronomy*, 2nd ed., Cygnus-Quasar, 1986, chap. 6.

maintain a maximum or minimum indication. The measured phase shift can then be plotted as a function of position along the reference line.[1]

18-5 DIRECTIVITY. The directivity of an antenna can be determined from the measured field pattern. Thus, as defined in Chap. 2, the directivity of an antenna is

$$D = \frac{4\pi}{\iint f(\theta, \phi) \sin \theta \, d\theta \, d\phi} \tag{1}$$

where $f(\theta, \phi)$ = relative radiation intensity (power per square radian) as a function of the space angles θ and ϕ (see Fig. 18-1)

Since the radiation intensity is proportional to the square of the field intensity, the directivity expression (1) can be written as

$$D = \frac{4\pi}{\iint P_n(\theta, \phi) \sin \theta \, d\theta \, d\phi} \tag{2}$$

where $P_n(\theta, \phi)$ = normalized power pattern
$= (E_\theta^2 + E_\phi^2)/(E_\theta^2 + E_\phi^2)_{\max}$

The directivity is determined from the shape of the field pattern by integration and is independent of antenna loss or mismatch.

18-6 GAIN. The gain of an antenna over an isotropic source is defined in Chap. 2 as

$$G_0 = k_0 D \tag{1}$$

where G_0 = gain with respect to an isotropic source (G without a subscript indicates the gain with reference to some antenna other than an isotropic source)
D = directivity
k_0 = ohmic loss factor ($0 \le k_0 \le 1$)

The factor $k_0 = 1$ if the antenna is lossless. It is assumed that the antenna is matched.

18-6a Gain by Comparison. Gain may be measured with respect to a comparison or reference antenna whose gain has been determined by other means. A

[1] C. C. Cutler, A. P. King and W. E. Kock, "Microwave Antenna Measurements," *Proc. IRE*, **35**, 1462–1471, December 1947.
H. Krutter, in S. Silver (ed.), *Microwave Antenna Theory and Design*, McGraw-Hill, New York, 1949, chap. 15, p. 543.
Harley Iams, "Phase Plotter for Centimeter Waves," *RCA Rev.*, **8**, 270–275, June 1947. Describes automatic device for plotting phase fronts near antennas.

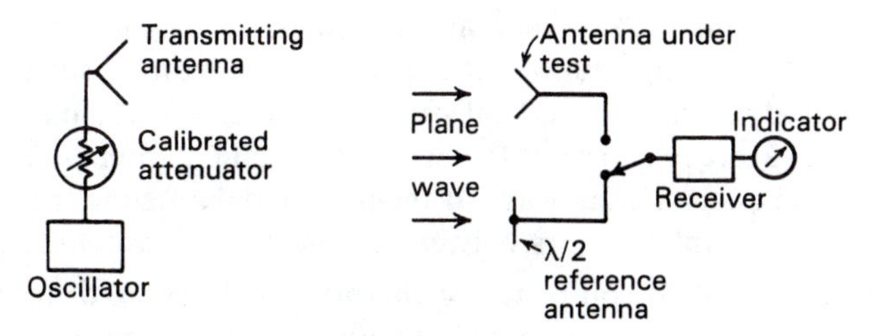

Figure 18-22 Gain measurement by comparison.

$\lambda/2$ dipole antenna or a horn antenna are commonly used as references. The gain G is then given by

$$G = \frac{P_1}{P_2} = \left(\frac{V_1}{V_2}\right)^2 \qquad (2)$$

where P_1 = power received with antenna under test, W
P_2 = power received with reference antenna, W
V_1 = voltage received with antenna under test, V
V_2 = voltage received with reference antenna, V

It is assumed that both antennas are properly matched. If both are also lossless and the reference is a $\lambda/2$ dipole, the gain G_0 over a lossless isotropic source is

$$G_0 = 1.64\ G$$

$$= 10 \log (1.64\ G) \qquad \text{(dBi)} \qquad (3)$$

The comparison should be made with both antennas in a suitable location where the wave from a distant source is substantially plane and of constant amplitude. The requirements of Secs. 18-3a and 18-3b should be fulfilled.

Both antennas may be mounted side by side as in Fig. 18-22 and the comparison made by switching the receiver from one antenna to the other. The ratio V_1/V_2 is observed on an output indicator calibrated in relative voltage. An alternative method is to adjust the power radiated by the transmitting antenna with a calibrated attenuator so that the received indication is the same for both antennas. The ratio P_1/P_2 is then obtained from the attenuator settings.

Mounting both antennas side by side as in Fig. 18-22 but in too close proximity may affect the measurements because of coupling between the antennas. To avoid such coupling, a direct substitution may be made with the idle antenna removed to some distance. If the antennas are of unequal gain, it is more important that the high-gain antenna be thus removed.

If the gain of the antenna under test is large, it is often more convenient to use a reference antenna of higher gain than that of a $\lambda/2$ dipole. At microwave frequencies electromagnetic horns are frequently employed for this purpose.[1]

[1] H. Krutter, in S. Silver (ed.), *Microwave Antennas*, McGraw-Hill, New York, 1949, chap. 15, p. 543.

Short-wave directional antenna arrays, such as used in transoceanic communication, are situated at a fixed height above the ground. The gain of such antennas is customarily referred to either a vertical or a horizontal $\lambda/2$ antenna placed at a height equal to the average height of the array. This gain comparison is at the elevation angle α of the downcoming wave. If the directional antenna is a high-gain type and any mutual coupling exists between it and the $\lambda/2$ antenna, the directional antenna can be rendered completely inoperative by lowering it to the ground or sectionalizing its elements when receiving with the $\lambda/2$ antenna.

In the above discussion it has been assumed that the antennas are perfectly matched. It is not always practical to provide such matching. This is particularly true with wideband receiving antennas that are only approximately matched to the transmission line. In general, another mismatch may occur between the transmission line and the receiver. In such cases the measured gain is a function of the receiver input impedance and the length of the transmission line.[1] To determine the range of fluctuation of gain of such wideband antennas with a given receiver as a function of the frequency and line length, the length of the line can be adjusted at each frequency to a length giving maximum gain and then to a length giving minimum gain. The average of this maximum and minimum may be called the average gain.

18-6b Absolute Gain of Identical Antennas. The gain can also be measured by a so-called *absolute method*[2] in which two identical antennas are arranged in free space as in Fig. 18-23*a*. One antenna acts as a transmitter and the other as a receiver. By the Friis transmission formula

$$\frac{P_r}{P_t} = \frac{A_{er} A_{et}}{\lambda^2 r^2} \quad \text{(dimensionless)} \tag{4}$$

where P_r = received power, W
P_t = transmitted power, W
A_{er} = effective aperture of receiving antenna, m^2
A_{et} = effective aperture of transmitting antenna, m^2
λ = wavelength, m
r = distance between antennas, m

The distance requirement of Sec. 18-3a should be fulfilled. If r is large compared

[1] J. D. Kraus, H. K. Clark, E. C. Barkofsky and G. Stavis, in *Very High Frequency Techniques*, Radio Research Laboratory Staff, McGraw-Hill, New York, 1947, chap. 10, pp. 232 and 271.

[2] C. C. Cutler, A. P. King and W. E. Kock, "Microwave Antenna Measurements," *Proc. IRE*, **35**, 1462–1471, December 1947; also H. Krutter, in S. Silver (ed.), *Microwave Antennas*, McGraw-Hill, New York, 1949, chap. 15, p. 543. The gain is *absolute* in the sense that it depends on distance and power measurements which are independent of the antenna itself or the gain of other antennas.

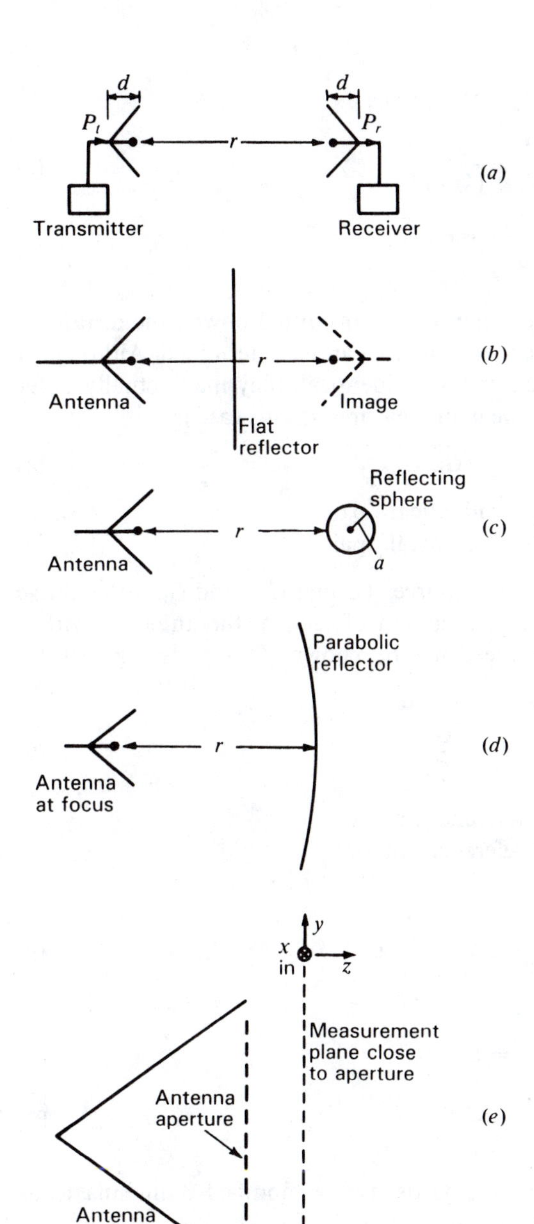

Figure 18-23 Arrangements for absolute gain measurements.

to the depth d of the antenna, the precise points on the antennas between which r is measured will not be critical. Since

$$A_{er} = G_0 \frac{\lambda^2}{4\pi} \tag{5}$$

where G_0 = gain of antenna over an isotropic source

and since it is assumed that $A_{er} = A_{et}$, (4) becomes

$$\frac{P_r}{P_t} = \frac{G_0^2 \lambda^2}{(4\pi)^2 r^2} \tag{6}$$

and

$$G_0 = \frac{4\pi r}{\lambda} \sqrt{\frac{P_r}{P_t}} \tag{7}$$

Thus, by measuring the ratio of the received to transmitted power, the distance r and the wavelength λ, the gain of either antenna can be determined. Although it may have been intended that the antennas be identical, they may actually differ in gain by an appreciable amount. The gain measured in this case is

$$G_0 = \sqrt{G_{01} G_{02}} \tag{8}$$

where G_{01} = gain of antenna 1 of the "identical" pair
G_{02} = gain of antenna 2 of the "identical" pair

with both gains referred to an isotropic source. To find G_{01} and G_{02}, the above measurement is supplemented by a comparison of each of the antennas with a third reference antenna whose gain need not be known. This gives a gain ratio between "identical" antennas of

$$G' = \frac{G_1}{G_2} \tag{9}$$

where G_1 = gain of antenna 1 over reference antenna
G_2 = gain of antenna 2 over reference antenna

Then since

$$G' = \frac{G_1}{G_2} = \frac{G_{01}}{G_{02}} \tag{10}$$

we have

$$G_{01} = G_0 \sqrt{G'} \tag{11}$$

and

$$G_{02} = \frac{G_0}{\sqrt{G'}} \tag{12}$$

The U.S. National Bureau of Standards uses a modified 3-antenna technique for accurate antenna gain measurements.[1]

18-6c Absolute Gain of Single Antenna. Using radar techniques the method of the preceding section involving 2 or 3 antennas can be extended in several ways to measuring the absolute gain of a single antenna.

---

[1] A. C. Newell, C. F. Stubenrauch and R. C. Baird, "Calibration of Microwave Antenna Gain Standards," *Proc. IEEE*, **74**, 129–132, January 1986.

1. **By flat sheet reflector.** By replacing the second antenna of Fig. 18-23*a* with a sufficiently large, flat, perfectly reflecting sheet, as in Fig. 18-23*b*, the gain of the single (transmitting-receiving) antenna is given by (7) where r now equals the distance from the antenna to its image behind the reflector. This distance must meet the far-field requirement and this may require a very large flat sheet reflector.
2. **By reflecting sphere.** As discussed in Sec. 17-5, the radar cross section σ of a perfectly reflecting sphere is equal to its physical cross section (πa^2) when its radius $a \gg \lambda$. With a sphere as the radar target, as in Fig. 18-23*c*, we have from the radar equation (17-5-6) that the antenna gain

$$G = \frac{8\pi r^2}{\lambda a}\sqrt{\frac{P_r}{P_t}} \tag{13}$$

where r = distance from antenna to sphere, m
a = radius of sphere, m

The distance r must meet the far-field requirement while the sphere radius requirement is that $a \gg \lambda$.

3. **By parabolic reflector.** A more compact configuration involves the use of a parabolic reflector as in Fig. 18-23*d* with the antenna at the focus of the parabola. For this configuration the gain

$$G = 4\pi r_\lambda \sqrt{\frac{P_r}{P_t}} \tag{14}$$

where r_λ = focal distance of parabola in wavelengths, dimensionless

Equation (14) is identical to (7).

18-6d Gain by Near-Field Measurements. Referring to Fig. 18-23*e*, measurements of the near field of a large antenna with a probe can be used to obtain the gain from Bracewell's relation (12-9-26) as

$$G = \frac{4\pi A_p}{\lambda^2} \frac{1}{\dfrac{1}{A_p}\displaystyle\iint_{A_p} \left[\frac{E(x, y)}{E_{\text{av}}}\right]\left[\frac{E(x, y)}{E_{\text{av}}}\right]^* dx\, dy} \tag{15}$$

where $E(x, y)$ = electric field at any point x, y in the aperture, V m^{-1}

$E_{\text{av}} = \dfrac{1}{A_p}\displaystyle\iint_{A_p} E(x, y)\, dx\, dy$ = average electric field over the aperture, V m^{-1}

A_p = area of (aperture) plane over which measurements are made, m^2

$*$ = complex conjugate

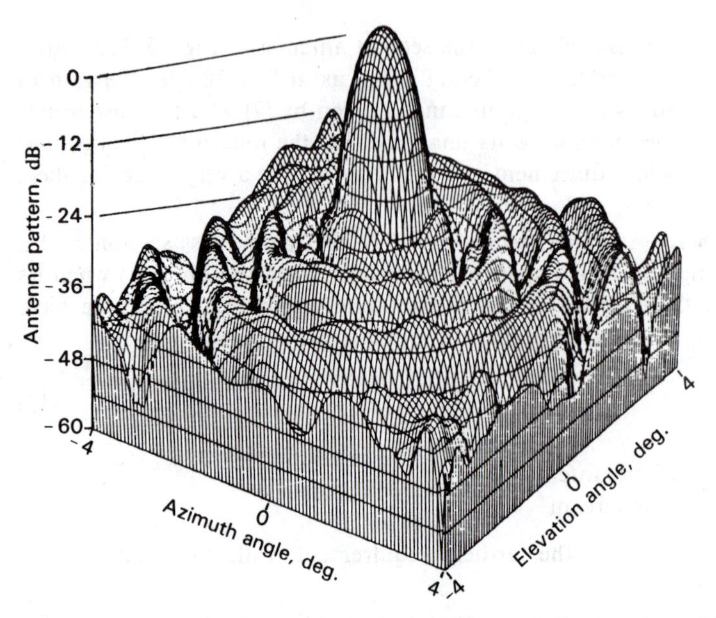

Figure 18-24 Far-field pattern in dB of high-gain paraboloidal reflector antenna over a 64 square-degree area centered on the beam axis as obtained from near-field measurements by Newell, Stubenrauch and Baird of the U.S. National Bureau of Standards. (*From A. C. Newell, C. F. Stubenrauch and R. C. Baird, "Calibration of Microwave Antenna Gain Standards,"* Proc. IEEE, **74**, *129–132, January 1986.*)

It is assumed that all of the radiated power flows through A_p.

This general method[1] is employed by the U.S. National Bureau of Standards (Boulder, Colorado) for gain measurements to an overall accuracy of the order of ± 0.2 dB. In addition, far-field patterns are obtained using the Fourier transform. An example of the pattern of a high-gain antenna obtained in this way is shown in Fig. 18-24.[2]

18-6e Gain and Aperture Efficiency from Celestial Source Measurements.[3] For gain measurements using a celestial radio source, an accurate flux density of the source is required and, generally, the source should be essentially unpolarized (less than 1 or 2 percent). Most of the sources in the table of Sec. A-8 meet these requirements, but since flux densities are given at only discrete frequencies it may be necessary to interpolate the fluxes at other frequencies.

---

[1] W. H. Kummer and E. S. Gillespie, "Antenna Measurements," *Proc. IEEE*, **66**, 483–507, April 1978.

[2] A. C. Newell, C. F. Stubenrauch and R. C. Baird, "Calibration of Microwave Antenna Gain Standards," *Proc. IEEE*, **74**, 129–132, January 1986.

[3] See also Secs. 18-3e and 12-9.

From (17-2-8) the effective aperture A_e of an antenna is related to the known flux density S and measured incremental antenna temperature ΔT_A as given by

$$A_e = \frac{2k\,\Delta T_A}{S} \tag{16}$$

from which the gain is

$$G = \frac{4\pi A_e}{\lambda^2} = \frac{8\pi k\,\Delta T_A}{S\lambda^2} \tag{17}$$

where k = Boltzmann's constant = 1.38×10^{-23} J K^{-1}
ΔT_A = measured source temperature, K
S = source flux density, W m^{-2} Hz^{-1}
λ = wavelength, m

Thus, knowing S (from the table of Sec. A-8) and λ, a measurement of ΔT_A determines the gain. This measurement includes the effect of any (ohmic) loss in the antenna and any mismatch.

Example. Find the gain and aperture efficiency of the Ohio State University 110-m radio telescope antenna at 1.4 GHz if the measured increase in antenna temperature from Cygnus A = 687 K. The physical aperture is 2208 m^2.

Solution. From the table of Sec. A-8, the flux density of Cygnus A at 1.4 GHz is 1590 Jy. From (16) the effective aperture

$$A_e = \frac{2k\,\Delta T_A}{S} = \frac{2 \times 1.38 \times 10^{-23} \times 687}{1590 \times 10^{-26}} = 1193 \text{ m}^2$$

and from (17) the *gain*

$$G = \frac{4\pi A_e}{\lambda^2} = \frac{4\pi \times 1193}{0.214^2} = 3.27 \times 10^5 = 55 \text{ dBi}$$

Since the physical aperture A_p of the antenna is 2208 m^2, the *aperture efficiency*

$$\varepsilon_{ap} = \frac{A_e}{A_p} = \frac{1193}{2208} = 0.54 \text{ or } 54 \text{ percent}$$

The U.S. National Bureau of Standards offers a service for measuring antenna gain using celestial radio sources.[1]

See discussion of (12-9-43) to (12-9-47) for *beam efficiency* measurements with a celestial source.

[1] R. C. Baird, "Microwave Antenna Measurement Services at the National Bureau of Standards," *Ant. Meas. Symp.* (*Ant. Meas. Tech. Assoc.*), October 1981.

18-7 TERMINAL IMPEDANCE MEASUREMENTS. In general, any antenna impedance Z_A terminating a transmission line will produce a reflected wave with reflection coefficient ρ_v and a voltage standing wave ratio (VSWR) related as follows:

$$\rho_v = \frac{|\text{reflected voltage}|}{|\text{incident voltage}|} = \frac{|V_r|}{|V_i|} = \frac{\text{VSWR} - 1}{\text{VSWR} + 1} \tag{1}$$

where the VSWR is the ratio of the maximum to minimum voltage on the line.

The reflection coefficient is a complex quantity with magnitude $|\rho_v|$ and phase angle θ_v. Thus,

$$\rho_v = |\rho_v| \underline{/\theta_v} \tag{2}$$

The VSWR can be measured by moving a voltage probe along a slotted measuring line (Fig. 18-25). The value of $|\rho_v|$ is then given from (1). Using the probe to locate the voltage minimum point on the line, the reflection coefficient phase angle θ_v is found from

$$\theta_v = 720°\left(\frac{x_{vm}}{\lambda} - \frac{1}{4}\right) \tag{3}$$

where x_{vm} = distance of voltage minimum from antenna terminals, m
λ = wavelength along line, m

Knowing both the magnitude and phase angle of the reflection coefficient, the antenna impedance Z_A is given by

$$Z_A = Z_l \frac{1 + |\rho_v| \underline{/\theta_v}}{1 - |\rho_v| \underline{/\theta_v}} \quad (\Omega) \tag{4}$$

where Z_l = transmission line impedance, Ω

In a slotted (coaxial) transmission line a probe is introduced through a longitudinal slot in the outer conductor as indicated in Fig. 18-25. Since currents

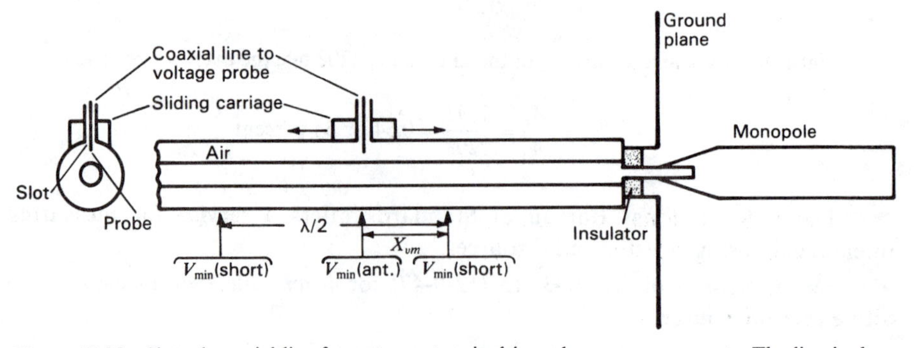

Figure 18-25 Slotted coaxial line for antenna terminal impedance measurements. The line is shown in longitudinal and transverse cross sections with movable probe for measuring voltage as a function of distance along the line.

flow parallel to the slot, little field escapes the line. With the probe connected to a voltage indicator, the voltage variation along the line can be determined, giving both VSWR and x_{vm}. Replacing the antenna with a short circuit, two successive minima $V_{\min}$ (short) are located on the line. Their separation is equal to $\lambda/2$. With the antenna connected, the distance between a voltage minimum point $V_{\min}$ (ant) and $V_{\min}$ (short) is equal to x_{vm}, as shown in Fig. 18-25. It is usually preferable to measure x_{vm} with respect to the first $V_{\min}$ (short) rather than to the terminals, due to uncertainties from end effects, as from an insulator, which modify the electrical distance.

Instead of a slotted line, *directional couplers* can be used which give outputs proportional to the reflected and incident voltages V_r and V_i from which $|\rho_v|$ can be determined as in (1) and the VSWR from

$$\text{VSWR} = \frac{1 + |\rho_v|}{1 - |\rho_v|} \tag{5}$$

With directional couplers and a sweep frequency generator, the VSWR can be monitored over a bandwidth and displayed continuously on a cathode ray tube (CRT) while adjustments are made on the antenna. If the phase difference of V_r and V_i is also monitored, the antenna impedance can be displayed over a bandwidth on a CRT Smith chart.[1]

Although the impedance measuring arrangement shown in Fig. 18-25 is appropriate for monopoles (or other unbalanced antennas), it can also be used to measure the terminal impedance of a center-fed dipole (or other balanced antenna) by measuring $\frac{1}{2}$ of the antenna and multiplying the measured impedance by 2. Thus, instead of measuring a balanced $\lambda/2$ dipole with a 2-wire transmission line, measurements are made on $\frac{1}{2}$ of the dipole or $\lambda/4$ monopole (or stub) antenna with a large ground plane (Fig. 18-25). Ideally the ground plane should be perfectly conducting and infinite in extent to produce a perfect image of the stub antenna. The ground plane of finite extent used in practice should, therefore, be as large as possible. Even though the ground plane is several wavelengths in diameter, the measured impedance of a stub antenna varies appreciably as a function of the ground-plane diameter.[2] This variation is reduced as the ground-plane diameter is increased. Meier and Summers[2] found that a large square ground plane results in about half the variation of impedance observed with a circular ground plane of approximately the same size. The antennas were mounted symmetrically on both ground planes. The reduced variation with the square ground plane is presumably due to the partial cancellation of waves reflected to the antenna terminals from the edge of the ground plane. These waves

[1] For more details and procedures for using Smith charts see, for example, J. D. Kraus, *Electromagnetics*, 3rd ed., McGraw-Hill, 1984, secs. 10-7 and 10-8.

[2] A. S. Meier and W. P. Summers, "Measured Impedance of Vertical Antennas over Finite Ground Planes," *Proc. IRE*, 37, 609–616, June 1949.

travel different distances on a square ground plane, and, hence, all cannot arrive in the same phase. The ratio of the longest to the shortest distance is the ratio of the diagonal of a square to the length of one side (1.41). With a circular ground plane and symmetrically located antenna, all the waves reflected from the edge return in the same phase.

The ground (image) plane technique can also be used to advantage in measuring the terminal impedance of slot antennas. A sheet with a half-slot (equal in length to the full-slot but of $\frac{1}{2}$ the width) is butted against an image plane placed perpendicular to the slot plane. The half-slot is energized by a coaxial line with the inner conductor connected to the terminal of the slot and the outer conductor terminated in the image plane. The terminal impedance of the full-slot is twice the impedance of the half-slot. The impedance $Z_{s/2}$ of the half-slot is related to the impedance $Z_{d/2}$ of the complementary stub antenna or half-dipole by $Z_{s/2} = 8869/Z_{d/2}$.

With horn or slot antennas that are fed with a waveguide, measurements of the field in the guide can be made with a slotted waveguide and probe arrangement. In this way measurements of the VSWR, reflection coefficient and equivalent load impedance may be obtained in a manner analogous to that used with a coaxial line.

18-8 CURRENT DISTRIBUTION MEASUREMENTS. In many cases it is important to know the current distribution along an antenna. For example, if both the magnitude and phase of the current is known at all points along an antenna, the far field of the antenna can be calculated.

The current can be sampled by a small pickup loop placed close to the antenna conductor. Loop and indicator may be combined into a single unit. However, at short wavelengths the indicating instrument may be too large to place near the antenna without disturbing the field. To remove the indicator from the antenna field the arrangement of Fig. 18-26 can be used. Here the loop projects through a longitudinal slot in the hollow antenna conductor. The output cable from the loop is confined within the antenna conductor and is brought out through the end of a grounded stub as shown. The arrangement in Fig. 18-26 permits both amplitude and phase measurements. The phase is measured by comparison with a reference current as suggested by the dashed connections in the figure. The signal picked up by the loop is mixed with a signal of approximately equal amplitude extracted by a probe on a matched slotted line. With the antenna sampling loop fixed, the line probe is moved to give a minimum indication. When the antenna sampling loop is displaced to a new location, the line probe is moved so as to maintain a minimum indication. The phase shift between the line-probe positions then equals the phase shift between the two antenna sampling-loop locations. The phase shift is a linear function of distance on a line with matched termination. Assuming the phase velocity equals that of light in free space, the phase shift θ along the line in degrees per unit length is given by $360°/\lambda_0$, where λ_0 is the free-space wavelength of the applied signal. The phase

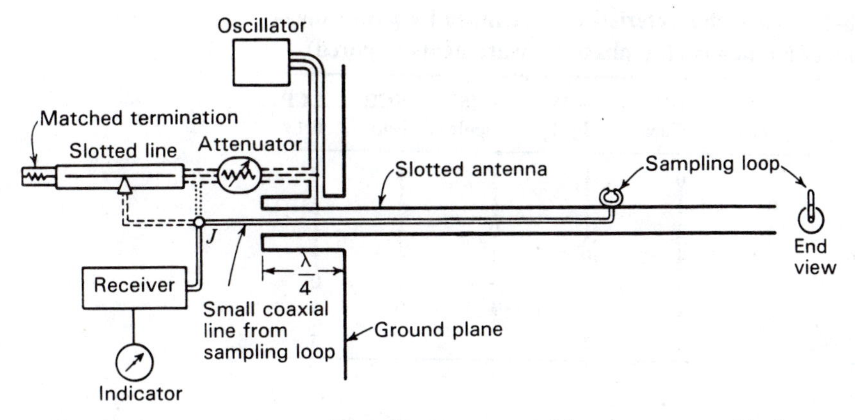

Figure 18-26 Slotted antenna and sampling loop arranged for measurement of both current amplitude and phase.

change between two points on the line is then the distance between the points multiplied by θ.

In Chap. 9 on cylindrical antennas Fig. 9-18 shows the current distribution on a 5λ cylindrical monopole antenna 0.2λ in diameter measured using an arrangement similar to that in Fig. 18-26.

18-9 POLARIZATION MEASUREMENTS.

Four methods for polarization measurements are:

1. *Polarization-pattern method.* A linearly polarized antenna is used to measure a polarization pattern and two circularly polarized antennas are used to determine the hand of rotation.
2. *Linear-component method.* Two perpendicular linearly polarized antennas are used to measure the linearly polarized components of the wave and also their phase difference.
3. *Circular-component method.* Two circularly polarized antennas are used to measure the circularly polarized components of the wave of opposite hand and the phase angle between them.
4. *Power measurement (without phase) method.* Some waves may consist of the superposition of a large number of statistically independent waves of a variety of polarizations. The resultant wave is said to be *randomly polarized* or *unpolarized.* Thus, in general, waves may be *partially polarized* and *partially unpolarized.* In ordinary communications the waves are usually completely polarized but in radio astronomy the waves from celestial sources are, in general, partially polarized and in many cases completely unpolarized. To deal with the most general situation it is convenient to use Stokes' parameters. A detailed discussion of polarization parameters, Stokes' parameters and

Table 18-2 Wave characteristics determined by power measurements of 6 antennas (no phase measurements required)

Wave	VP dipole	HP dipole	+45° dipole	−45° dipole	RCP helix	LCP helix
VP	1	0	$\frac{1}{2}$	$\frac{1}{2}$	$\frac{1}{2}$	$\frac{1}{2}$
HP	0	1	$\frac{1}{2}$	$\frac{1}{2}$	$\frac{1}{2}$	$\frac{1}{2}$
+45° LP	$\frac{1}{2}$	$\frac{1}{2}$	1	0	$\frac{1}{2}$	$\frac{1}{2}$
−45° LP	$\frac{1}{2}$	$\frac{1}{2}$	0	1	$\frac{1}{2}$	$\frac{1}{2}$
RCP	$\frac{1}{2}$	$\frac{1}{2}$	$\frac{1}{2}$	$\frac{1}{2}$	1	0
LCP	$\frac{1}{2}$	$\frac{1}{2}$	$\frac{1}{2}$	$\frac{1}{2}$	0	1
Unpolarized	$\frac{1}{2}$	$\frac{1}{2}$	$\frac{1}{2}$	$\frac{1}{2}$	$\frac{1}{2}$	$\frac{1}{2}$

polarization measurements is given by Kraus.[1] Of interest here is that the polarization characteristics of a wave (including any unpolarized components) may be completely determined *without any phase measurements* by noting the power response of 6 antennas: 1 vertically polarized (VP), 1 horizontally polarized (HP), 1 linearly polarized (LP) at a slant angle of +45°, 1 linearly polarized (LP) at a slant angle of −45° and 2 circularly polarized (CP) antennas, one right-circularly polarized (RCP) and the other left-circularly polarized (LCP). The linearly polarized antennas may be dipoles and the circularly polarized antennas monofilar axial-mode helices, one wound right-handed and the other left-handed. For a completely polarized wave only 3 independent measurements are necessary so there is some redundancy.

An example of the responses of the 6 antennas to a wave of unit incident power density is shown in Table 18-2. The power response of all 6 antennas is normalized to unity for a wave of unit incident power density of the same polarization. We note that each type of wave polarization produces a different set of power responses.

18-9a Polarization-Pattern Method. In this method a rotatable linearly polarized antenna, such as the $\lambda/2$ dipole antenna in Fig. 18-27*a*, is connected to a receiver calibrated to read relative voltage.[2] Let the wave be approaching (out of page). Then as the antenna is rotated in the plane of the page, the voltage observed at each position is proportional to the maximum component of **E** in the direction of the antenna. Such measurements of the incident wave with a rotatable linearly polarized antenna do not yield the *polarization ellipse* of the wave but rather its *polarization pattern* (Fig. 18-27*b*). Thus, if the tip of the electric vector **E** describes the *polarization ellipse* shown in Fig. 18-27*b* (dashed curve),

[1] J. D. Kraus, *Radio Astronomy*, 2nd ed., Cygnus-Quasar, 1986, chap. 4.

[2] In practice a linearly polarized antenna of considerable directivity is preferable to a simple $\lambda/2$ dipole.

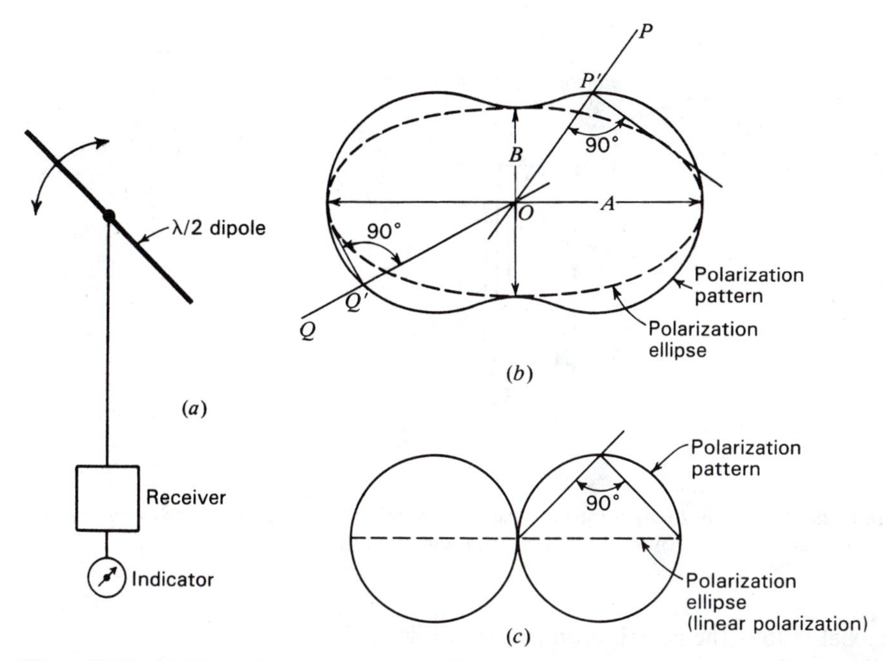

Figure 18-27 (*a*) Schematic arrangement for measuring wave polarization by the *polarization-pattern method.* (*b*) Measured polarization pattern and polarization ellipse for elliptical polarization. (*c*) Measured polarization pattern for linear polarization with polarization ellipse collapsed to a line.

the variation measured with a linearly polarized receiving antenna is given by the *polarization pattern* in Fig. 18-27*b* (solid line). For a given orientation OP of the linearly polarized antenna, the response is proportional to the greatest ellipse dimension measured normally to OP. As shown in Fig. 18-27*b*, this is the length OP'. If the linearly polarized antenna orientation is OQ, the response is proportional to the length OQ'. For the case of linear polarization, the polarization ellipse degenerates to a straight line and the corresponding polarization pattern is a figure-of-eight, as indicated in Fig. 18-27*c*. By graphical construction as in Fig. 18-27*b* and *c*, the polarization ellipse can be constructed if the polarization pattern is known, or vice versa. To determine the direction of rotation of **E** an auxiliary measurement is necessary. For example, the output of 2 circularly polarized antennas could be compared, one responsive to right- and the other to left-circular polarization. The rotation direction of **E** then corresponds to the polarization of the antenna with the larger response.

Thus, by this method the polarization ellipse can be drawn and the rotation direction indicated. Although such a diagram completely describes the polarization characteristics of a wave, it is simpler to measure merely the maximum amplitude $A/2$ and the minimum amplitude $B/2$ and take the ratio of the two amplitudes which is the *axial ratio of the polarization ellipse* or simply the *axial ratio* (AR). The axial ratio is expressed so that it is equal to or greater than unity.

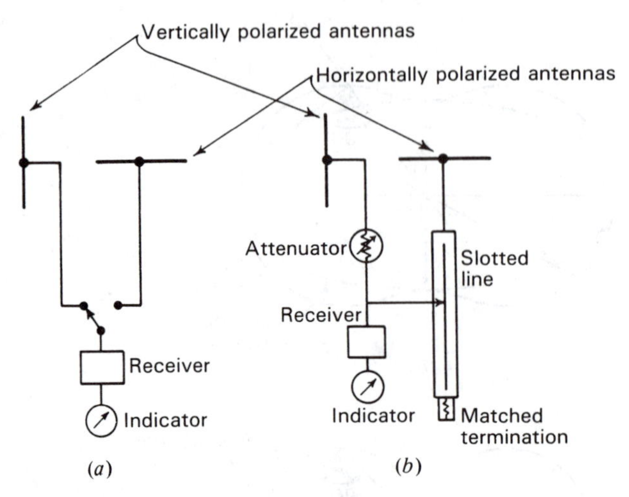

Figure 18-28 Schematic arrangement for measuring polarization by the *linear-component method* with vertical and horizontal components given by (*a*) and phase by (*b*).

The axial ratio of the polarization ellipse of Fig. 18-27*b* is

$$\text{AR} = \frac{A}{B}$$

Thus, by specifying AR, the tilt angle and the rotation direction of **E** the polarization characteristics are completely described. (See Secs. 2-34, 2-35 and 2-36.)

18-9b Linear-Component Method. In this method 2 fixed linearly polarized antennas can be mounted at right angles, like the two $\lambda/2$ antennas in Fig. 18-28*a*. The wave is approaching normally out of the page. By connecting the receiver first to the terminals of one antenna and then the other, the ratio E_2/E_1 can be measured. Then, by connecting both antennas to a phase comparator, the angle δ can be measured. This may be done as in Fig. 18-28*b*, using a matched slotted line. From a knowledge of E_1, E_2 and δ the polarization ellipse can be calculated and the direction of rotation **E** determined.

18-9c Circular-Component Method. In this method 2 circularly polarized antennas of opposite hand are connected successively to the receiver and the amplitudes E_L and E_R of the circularly polarized component waves measured. The antennas can very conveniently consist of 2 long monofilar axial-mode helical antennas, one wound left-handed and the other wound right-handed as in Fig. 18-29. The left-handed helix responds to left-circular polarization and the right-handed helix to right-circular polarization (IEEE definition). The left-circular component E_L of the wave is measured with the switch to the left as in Fig. 18-29 so that the receiver is connected to the left-handed helix. The right-circular component E_R of the wave is measured with the switch thrown to the

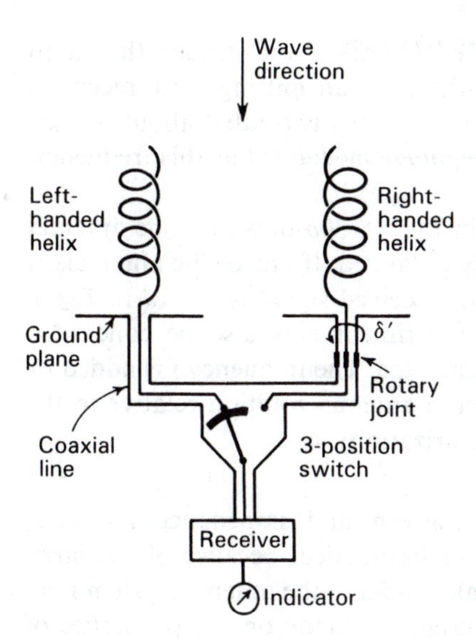

Figure 18-29 Arrangement for measuring polarization by the *circular-component method.* Left- and right-handed components are measured by individual helices and phase angle by rotating one helix with both connected.

right so that the receiver is connected to the right-handed helix. The axial ratio (AR) of the received wave is then given by

$$\text{AR} = \frac{E_R + E_L}{E_R - E_L} \tag{1}$$

According to (1) the axial ratio may have values between $+1$ and $+\infty$ and between -1 and $-\infty$. For positive values of AR the wave is right-elliptical and for negative values it is left-elliptical. The tilt angle τ of the polarization ellipse may be measured by finding the direction of maximum **E** with a rotatable linearly polarized antenna, or τ may be determined with the helical antennas of Fig. 18-29 by rotating one helix on its axis with both helices connected in parallel to the receiver (switch segment up in Fig. 18-29). Assuming that the axes of the helices are in a horizontal plane, let the helix rotation angle be δ' and let its reference point ($\delta' = 0$) be taken when the receiver output is a minimum for a horizontally polarized incident wave. Then for any type of polarization with the polarization ellipse at a tilt angle τ to the horizontal, $\tau = \delta'/2$. Thus, 3 measurements, E_L, E_R and δ', with the helical antennas determine the polarization characteristics of the received wave completely.

The circular-component method using helical antennas is suitable for measurements over a considerable frequency range. The accuracy depends on the circularity of polarization of the helices. This is improved (AR nearer unity) by making the helices long since

$$\text{AR} = \frac{2n + 1}{2n} \tag{2}$$

where n = number of turns of the helix

18-10 ANTENNA ROTATION EXPERIMENTS. Consider the radio circuit shown in Fig. 18-30*a* in which both the transmitting and receiving antennas are linearly polarized. If either of the antennas is rotated about its axis at a frequency f (r/s), the received signal is *amplitude modulated* at this frequency. The direction of rotation is immaterial.

Consider next the radio circuit shown in Fig. 18-30*b* in which one antenna is circularly polarized and the other is linearly polarized. If one of the antennas is rotated about its axis at a frequency f (r/s), the received signal is *shifted* to $F \pm f$, where F is the transmitter frequency. This experiment may also be conducted with 2 circularly polarized antennas of the same type. The frequency f is added or subtracted from F depending on the direction of antenna rotation relative to the rotation direction of **E** (or hand of circular polarization).

18-11 MODEL MEASUREMENTS. Pattern and impedance measurements of actual antennas are often difficult or impractical because of the large size of the antenna system. In such cases a scale model of the antenna system may be built to a convenient size and then measurements made on the properties of the model.[1] This technique is especially useful in measuring patterns of antennas mounted on aircraft. Although the antenna proper may be small, it may excite currents over much of the airplane surface so that the entire airplane becomes part of the antenna system, and, hence, the measurements must be made of the antenna *with* airplane. Another advantage is that the patterns of antennas on aircraft in flight (remote from the ground) can be easily simulated by the model technique by placing the model on a suitable tower. To measure such patterns on actual aircraft is both tedious and expensive.

Let the scale factor for the model be p. Then any length dimension L_m on the model is related to the corresponding dimension L on the actual antenna by

$$L_m = \frac{L}{p} \tag{1}$$

Then the frequency f_m used to measure the model must be related to the frequency f used with the actual antenna by

$$f_m = pf \tag{2}$$

A further requirement of an accurate model for pattern and impedance measurements is that the conductivity of the antenna metal be scaled according to the relation

$$\sigma_m = p\sigma \tag{3}$$

[1] George Sinclair, "Theory of Models of Electromagnetic Systems," *Proc. IRE*, **36**, 1364–1370, November 1948.
G. H. Brown and Ronold King, "High-Frequency Models in Antenna Investigations," *Proc. IRE*, **22**, 457–480, April 1934.

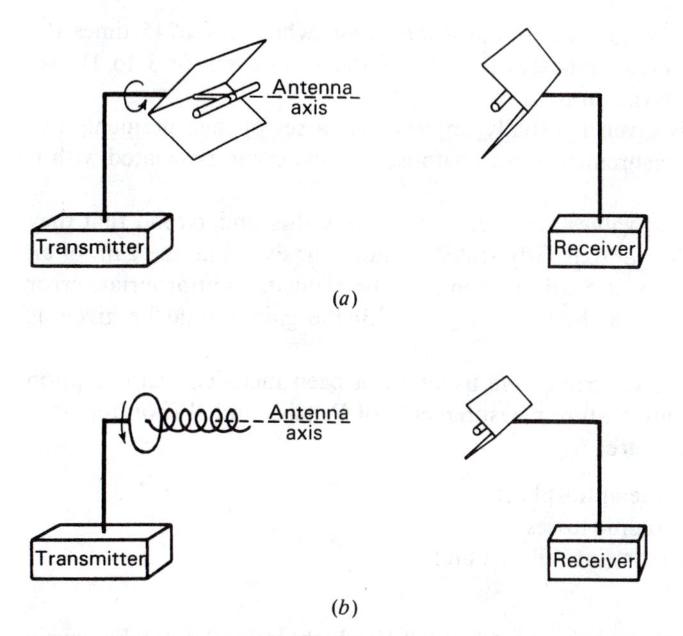

Figure 18-30 Arrangements for antenna rotation experiments. (*a*) Antenna rotation produces amplitude modulation. (*b*) Rotating the monofilar axial-mode helix increases or decreases the signal frequency by the rotation rate.

where σ_m = conductivity of metal in model
σ = conductivity of metal in actual antenna

However, if σ is large enough, the metal can be considered to be a "perfect conductor" ($\sigma = \infty$) and the conductivity need not be modeled. Thus, actual antennas of copper can usually be modeled in copper. It is assumed that ferromagnetic materials are excluded from both actual antenna and model and that the model is measured in air.

18-12 MEASUREMENT ERROR. All measured quantities involve error. Thus, the measured area of a sphere might be given as 2.76 ± 0.03 m^2, indicating an error or uncertainty of ± 0.03 m^2. If this is the *root mean square* (rms) or *standard deviation*, it means that the chances are roughly 2 to 1 that the true value is between the limits, i.e., greater than 2.73 and less than 2.79 m^2. For a finite number n of readings the

$$\text{rms deviation} = \sqrt{\frac{d_1^2 + d_2^2 + \cdots + d_n^2}{n-1}} \tag{1}$$

where d_1, d_2, etc., are the measured deviations from the mean of a set of n observations.

Sometimes the error given is the *probable error*, which is 0.6745 times the rms error. A probable error indicates that the chances are even (or 1 to 1) that the true value is between the limits cited.

When an error is given it usually implies that a set of measurements has been made. A single measurement is anomalous, and any error associated with it must be an estimate.

Wherever measured values are cited in this book it is understood that they are subject to error whether explicitly stated or not. For example, the gain of an antenna may be quoted as 36.5 dBi. However, to be explicit, an appropriate error should be included. Thus, if the error is ± 0.5 dBi the gain should be given as 36.5 ± 0.5 dBi.

However, for brevity, errors have usually not been included. An exception involves the antenna temperature measurements of Penzias and Wilson reported in Sec. 17-2. Their results are

2.3 ± 0.3 K due to the atmosphere
0.8 ± 0.4 K due to ohmic losses
<0.1 K due to back lobes into the ground
3.2 ± 0.5 K total[1]

Their measured sky temperature was 6.7 ± 0.8 K which, less 3.2 ± 0.5 K, gave a residual of 3.5 ± 1.0 K.[1]

Penzias and Wilson's attention to the errors led to their discovery of the 3-K sky background for which they subsequently received a Nobel prize.

ADDITIONAL REFERENCES

Burnside, W. D., and R. W. Burgener: "High Frequency Scattering by a Thin Lossless Dielectric Slab," *IEEE Trans. Ants. Prop.*, **AP-31**, 104–110, January 1983.

Collington, G., Y. Michael, F. Robin and J. C. Bolomey: "Quick Microwave Field Mapping for Large Antennas," *Microwave J.*, **25**, 129–132, December 1982.

IEEE Standard Test Procedures for Antennas, IEEE Standard 149, 1979.

Johnson, R. C.: "Some Design Parameters for Point-Source Compact Ranges," *IEEE Trans. Ants. Prop.*, **AP-34**, 845–847, June 1986.

Kummer, W. H., and E. S. Gillespie: "Antenna Measurements," *Proc. IEEE*, **66**, 483–507, April 1978.

Mostafavi, M., J. C. Bolomey and D. Picard: "Far-Field Accuracy Investigation Using Modulated Scattering Technique for Fast Near-Field Measurements," *IEEE Trans. Ants. Prop.*, **AP-33**, 279–285, March 1985.

Nortier, J. R., C. A. Van der Neut and D. E. Baker: "Tables for the Design of Jaumann Microwave Absorber," *Microwave J.*, **30**, 219–222, September 1987.

Rhodes, D. R.: "On Minimum Range for Radiation Patterns," *Proc. IRE*, **42**, 1409–1410, September 1954.

[1] Note that these total errors are rss (root sum square) (quantities unrelated) not rms (root mean square) (single quantity error). Thus, 0.3 and 0.4 (in above tabulation) are rms but the total 0.5 is rss.

References on Radiation Hazards

Mumford, W. W.: "Some Technical Aspects of Microwave Radiation Hazards," *Proc. IRE*, **49**, 427–447, February 1961.

Tell, R. A., and F. Harlen: "A Review of Selected Biological Effects and Dosimetric Data Useful for Development of Radio Frequency Safety Standards for Human Exposure," *J. Microwave Power*, **14**, 405–424, 1979.

PROBLEMS[1]

★18-1 Absorbing material. $1/e$ depths. A medium has constants $\sigma = 10^2$ ℧ m^{-1}, $\mu_r = 2$ and $\varepsilon_r = 3$. If the constants do not change with frequency, find the $1/e$ and 1 percent depths of penetration at (*a*) 60 Hz, (*b*) 2 MHz and (*c*) 3 GHz.

18-2 Lossy medium. At 200 MHz a solid ferrite-titanate medium has constants $\sigma = 0$, $\mu_r = 15(1 - j3)$ and $\varepsilon_r = 50(1 - j1)$. Find (*a*) Z/Z_0, (*b*) λ/λ_0, (*c*) v/v_0, (*d*) $1/e$ depth, (*e*) the dB attenuation for a 5-mm thickness and (*f*) the reflection coefficient ρ_v for a wave in air incident normally on the flat surface of the medium. The zero subscripts refer to parameters for air (or vacuum).

★18-3 Lossy medium. Complex constants. A medium has constants $\sigma = 3.34$ ℧ m^{-1} and $\mu_r = \varepsilon_r = 5 + j2$. Find the $1/e$ depth of penetration at 30 GHz.

18-4 Attenuation by lossy slab. A nonconducting slab 200 mm thick has constants $\mu_r = \varepsilon_r = 2 - j2$. Find the dB attenuation of the slab to a 600-MHz wave.

★18-5 Attenuation of lossy sheet. A nonconducting sheet 6 mm thick has constants $\mu_r = \varepsilon_r = 5 - j5$. Find the dB attenuation of the sheet to a 300-MHz wave.

18-6 Attenuation of conducting sheet. A nonmagnetic conducting sheet 2 mm thick has a conductivity $\sigma = 10^3$ ℧ m^{-1}. Find the dB attenuation of the sheet to an 800-MHz wave.

★18-7 Attenuation in lossy medium. A medium has constants $\sigma = 1.112 \times 10^{-2}$ ℧ m^{-1}, $\mu_r = 5 - j4$ and $\varepsilon_r = 5 - j2$. At 100 MHz find (*a*) the impedance of the medium and (*b*) the distance required to attenuate a wave by 20 dB after entering the medium.

★18-8 Absorbing sheet. A large flat sheet of nonconducting material is backed by aluminum foil. At 500 MHz the constants of the ferrous dielectric medium are $\mu_r = \varepsilon_r = 6 - j6$. How thick must the sheet be for a 500-MHz wave (in air) incident on the sheet to be reduced upon reflection by 30 dB if the wave is incident normally?

★18-9 Reflection from dielectric medium. A plane 3-GHz wave is incident normally from air onto a half-space of dielectric with constants $\sigma = 0$, $\mu_r = 1$ and $\varepsilon_r = 2 - j2$. Find the dB value of the reflected power.

18-10 CP wave reflection and transmission. A circularly polarized 200-MHz wave in air with $E = 2$ V m^{-1} (rms) is incident normally on a half-space of nonconducting medium with $\mu_r = \varepsilon_r = 3 - j3$. Find the Poynting vector (*a*) for the reflected wave and (*b*) for the transmitted wave at a depth of 200 mm in the medium.

18-11 Radar cross section. The radar return from an object on a compact range is 8 dB more than for a 10λ diameter sphere when substituted for the object. What is the object's radar cross section?

[1] Answers to starred (★) problems are given in App. D. Permittivity relations are given in Sec. A-7.

★**18-12 Aperture efficiency from Cygnus A.** At 2.7 GHz the increase in antenna temperature from Cygnus A with a 20-m dish antenna is 51 K. What is the aperture efficiency of the dish?

★**18-13 Jaumann sandwich absorber.** Three absorber sheets of 250, 625 and 1563 Ω per square are stacked at $\lambda/4$ intervals in an expanded PVC ($\varepsilon_r = 1.1$) sandwich. Find the bandwidth for which the reflected wave is at least 20 dB down. (See Nortier, Van der Neut and Baker reference.)

APPENDIX A

TABLES FOR REFERENCE

A-1 TABLE OF ANTENNA AND ANTENNA SYSTEM RELATIONS[1]

Aperture efficiency, $\varepsilon_{ap} = \frac{A_e}{A_p} = \frac{E_{av}^2}{(E^2)_{av}}$ (dimensionless)

Aperture, effective, $A_e = \frac{\lambda^2}{\Omega_A}$ (m^2)

Array factor (n sources of equal amplitude and spacing),

$$E_n = \frac{\sin (n\psi/2)}{n \sin (\psi/2)} \quad \text{(dimensionless)}$$

where $\psi = \beta\, d \cos \theta + \delta$ (rad or deg)

Beam efficiency, $\varepsilon_M = \frac{\Omega_M}{\Omega_A}$ (dimensionless)

Beam solid angle, $\Omega_A = \iint P_n(\theta, \phi)\, d\Omega$ (sr)

[1] See index for page references giving more details on these relations. Also see index for tables (list of) for other relations.

Beam solid angle (approx.), $\Omega_A = \theta_{HP}\,\phi_{HP}$ (sr) $= \theta^\circ_{HP}\,\phi^\circ_{HP}$ (deg²)

Charge-current continuity, $\dot{I}l = q\dot{v}$

Circular aperture, HPBW $= \dfrac{58^\circ}{D_\lambda}$ (D_λ = diameter in λ)
(uniform distribution)

Circular aperture, BWFN $= \dfrac{140^\circ}{D_\lambda}$
(uniform distribution)

Circular aperture, directivity $= 9.9D_\lambda^2$
(uniform distribution)

Circular aperture, gain over $\lambda/2$ dipole $= 6D_\lambda^2$
(uniform distribution)

Directivity, $D = \dfrac{4\pi A_e}{\lambda^2} = \dfrac{4\pi}{\Omega_A} = \dfrac{P(\theta, \phi)_{\max}}{P_{av}}$ (dimensionless)

Directivity (approx.), $D \simeq \dfrac{4\pi(\text{sr})}{\theta_{HP}\,\phi_{HP}(\text{sr})} \simeq \dfrac{41\,000\ (\text{deg}^2)}{\theta^\circ_{HP}\,\phi^\circ_{HP}}$

Directivity (better approx.), $D \simeq \dfrac{41\,000\ \varepsilon_M}{k_P\,\theta_{HP}\,\phi_{HP}}$

Dipole (short), directivity, $D = 1.5$ (= 1.76 dBi)

Dipole (short), radiation resistance, $R = 80\pi^2\left(\dfrac{l}{\lambda}\right)^2\left(\dfrac{I_{av}}{I_0}\right)^2$ (Ω)

Dipole ($\lambda/2$), directivity, $D = 1.64$ (= 2.15 dBi)

Dipole ($\lambda/2$), self-impedance $Z = R_r + jX = 73 + j42.5\ \Omega$

Friis formula, $\dfrac{P_r}{P_t} = \dfrac{A_{er}A_{et}}{r^2\lambda^2}$ (dimensionless)

Flux density, $S = \dfrac{2kT_A}{A_e}$ (W m⁻² Hz⁻¹)

Flux density, minimum detectable, $\Delta S_{\min} = \dfrac{2k\,\Delta T_{\min}}{A_e}$ (W m⁻² Hz⁻¹)

Gain, $G = kD$ (dimensionless)

Height, effective, $h_e = \dfrac{V}{E} = \dfrac{I_{av}}{I_0}\,h_p = \dfrac{1}{I_0}\displaystyle\int_0^{h_p} I(z)\,dz$

$$= \sqrt{\frac{2R_r A_e}{Z_0}} \quad \text{(m)}$$

Helical antenna, monofilar axial-mode, directivity, $D = 12\left(\frac{C}{\lambda}\right)^2 \frac{nS}{\lambda}$

Linear array (long, uniform, in phase), $\text{HPBW} = \frac{1}{L_\lambda}\ (\text{rad}) = \frac{57.3°}{L_\lambda}$

Loop (single turn) radiation resistance, $R_r = 197C_\lambda^4 \quad (\Omega)$

Near-field–far-field boundary, $R = \frac{2L^2}{\lambda} \quad (\text{m})$

Noise power, receiver, $N = kT_{\text{sys}}\,\Delta f \quad (\text{W})$

Nyquist power, $w = kT \quad (\text{W Hz}^{-1})$

Radar equation, $\frac{P_r}{P_t} = \frac{A^2\sigma}{4\pi\lambda^2 r^4} \quad (\text{dimensionless})$

Radiation power, $P = \frac{\mu^2 q^2 \dot{v}^2}{6\pi Z} \quad (\text{W})$

Radiation resistance, $R_r = \frac{S(\theta, \phi)_{\max} r^2\, \Omega_A}{I^2} \quad (\Omega)$

Rectangular aperture, $\text{HPBW} = \frac{51°}{L_\lambda}$
(uniform distribution)

Rectangular aperture, $\text{BWFN} = \frac{115°}{L_\lambda}$

Rectangular aperture, directivity $= 12.6 L_\lambda L'_\lambda$

Rectangular aperture, gain over $\lambda/2$ dipole $= 7.7 L_\lambda L'_\lambda$

Resolution angle $\simeq \frac{\text{BWFN}}{2}$

Signal-to-noise ratio, $\frac{S}{N} = \frac{P_r}{P_n} = \frac{P_t A_{et} A_{er}}{r^2 \lambda^2 k T_{\text{sys}}\,\Delta f}$

Signal-to-noise ratio, (above 1 W isotropic) $\frac{S}{N} = \frac{\lambda^2}{16\pi^2 r^2 k T_{\text{sys}}\,\Delta f}$

System temperature, $T_{\text{sys}} = T_A + T_{\text{LP}}\left(\frac{1}{\varepsilon} - 1\right) + \frac{T_R}{\varepsilon} \quad (\text{K})$

Temperature, antenna (through emitting-absorbing cloud),

$$T_A = T_c(1 - e^{-\tau_c}) + T_s e^{-\tau_c} \quad (\text{K})$$

Temperature, minimum detectable, $\Delta T_{\text{min}} = \dfrac{k' T_{\text{sys}}}{\sqrt{t\,\Delta f}} = \Delta T_{\text{rms}}$ (K)

Wave power (average, elliptically polarized),

$$S_{\text{av}} = \frac{1}{2}\,\hat{\mathbf{z}}\,\frac{E_1^2 + E_2^2}{Z_0} \qquad (\text{W m}^{-2})$$

Wavelength and frequency, $\lambda = \dfrac{c}{f}$ (m)
(air or vacuum)

A-2 FORMULAS FOR INPUT IMPEDANCE OF TERMINATED TRANSMISSION LINES.

Formulas for the input impedance Z_x appearing at a distance x from a load or terminating impedance Z_L on a transmission line of characteristic impedance Z_0 as shown in Fig. A-1 are listed in the table for 3 load conditions: (1) any value of impedance Z_L, (2) $Z_L = 0$ or short-circuited line and (3) $Z_L = \infty$ or open-circuited line. For each load condition there are columns for 2 cases: (1) the general case in which attenuation is present on the line ($\alpha \neq 0$) and (2) the lossless case where the line losses are negligible ($\alpha = 0$).

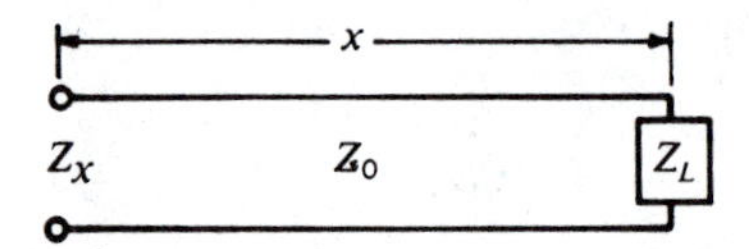

Figure A-1 Transmission line of characteristic impedance Z_0, length x and load Z_L.

Load condition	General case ($\alpha \neq 0$)	Lossless case ($\alpha = 0$)
Any value Z_L	$Z_x = Z_0 \dfrac{Z_L + Z_0 \tanh \gamma x}{Z_0 + Z_L \tanh \gamma x}$	$Z_x = Z_0 \dfrac{Z_L + jZ_0 \tan \beta x}{Z_0 + jZ_L \tan \beta x}$ $Z_x = Z_0^2/Z_L$†
$Z_L = 0$ Short-circuited line	$Z_x = Z_0 \tanh \gamma x$ $= Z_0 \dfrac{\tanh \alpha x + j \tan \beta x}{1 + j \tanh \alpha x \tan \beta x}$ $Z_x = Z_0 \coth \alpha x$†	$Z_x = jZ_0 \tan \beta x$
$Z_L = \infty$ Open-circuited line	$Z_x = Z_0 \coth \gamma x$ $= Z_0 \dfrac{1 + j \tanh \alpha x \tan \beta x}{\tanh \alpha x + j \tan \beta x}$ $Z_x = Z_0 \tanh \alpha x$†	$Z_x = -jZ_0 \cot \beta x$

† When $\beta x = n\pi/2$ where $n = 1, 3, 5, \ldots$.
In the table $\gamma = \alpha + j\beta$ where α = attenuation constant and $\beta = 2\pi/\lambda$.

A-3 REFLECTION AND TRANSMISSION COEFFICIENTS AND VSWR.

For a transmission line of characteristic impedance Z_0 terminated in a load impedance Z_L, the reflection coefficient for voltage ρ_v, the reflection coefficient for current ρ_i, the transmission coefficient for voltage or relative voltage at the load τ_v, the transmission coefficient for current or relative current at the load τ_i and the VSWR are given by

Reflection coefficient for voltage
$$\rho_v = \frac{Z_L - Z_0}{Z_L + Z_0}$$

Reflection coefficient for current
$$\rho_i = \frac{Z_0 - Z_L}{Z_0 + Z_L} = -\rho_v$$

Transmission coefficient for voltage
$$\tau_v = \frac{2Z_L}{Z_0 + Z_L} = 1 + \rho_v$$

Transmission coefficient for current
$$\tau_i = \frac{2Z_0}{Z_0 + Z_L} = 1 + \rho_i$$

Voltage standing-wave ratio (VSWR)
$$\frac{1 + |\rho_v|}{1 - |\rho_v|} = \frac{1 + |\rho_i|}{1 - |\rho_i|}$$

Magnitude of reflection coefficient
$$|\rho_v| = |\rho_i| = \frac{\text{VSWR} - 1}{\text{VSWR} + 1}$$

A-4 CHARACTERISTIC IMPEDANCE OF COAXIAL, 2-WIRE AND MICROSTRIP TRANSMISSION LINES

Type of line	Characteristic impedance, Ω
Coaxial (filled with medium of relative permittivity ε_r)	$Z_0 = \frac{138}{\sqrt{\varepsilon_r}} \log \frac{b}{a}$
Coaxial (air-filled)	$Z_0 = 138 \log \frac{b}{a}$
Two-wire (in medium of relative permittivity ε_r) ($D \gg a$)	$Z_0 = \frac{276}{\sqrt{\varepsilon_r}} \log \frac{D}{a}$
Two-wire (in air) ($D \gg a$)	$Z_0 = 276 \log \frac{D}{a}$
Microstrip ($w \geq 2h$)	$Z_0 \simeq \frac{377}{\sqrt{\varepsilon_r}[(w/h) + 2]}$

where b = inside radius of outer conductor
a = radius of inner conductor or wire
D = spacing between centers of wires
w = width of strip
h = height or thickness of substrate

It is assumed the lines are lossless (or $R \ll \omega L$ and $G \ll \omega C$) and also that the currents are confined to the conductor surfaces to which the radii refer. This condition is approximated at high frequencies owing to the small depth of penetration. It is also assumed that the lines are operating in the TEM mode.

The microstrip relation approaches exactness as the ratio w/h becomes very large. For strips of width w less than $2h$, the formula for a single wire above a ground plane can be used. Thus,

$$Z_0 \simeq 138 \log \frac{D}{a} = 138 \log \frac{8}{w/h} \qquad (\Omega)$$

where D = spacing between wire and its image = $2h$

a = wire radius = $w/4$

The flat strip is considered equivalent to a circular conductor of radius $\frac{1}{4}$ of the strip width.

A-5 CHARACTERISTIC IMPEDANCE OF TRANSMISSION LINES IN TERMS OF DISTRIBUTED PARAMETERS. In the following table the characteristic impedance Z_0 of a transmission line is given for 3 cases: (1) general case where losses are present, (2) special case where losses are small and (3) lossless case. In the table

Z_0 = characteristic impedance, Ω
R_0 = characteristic resistance, Ω
Z = series impedance, Ω m^{-1}
R = series resistance, Ω m^{-1}
L = series inductance, H m^{-1}
Y = shunt admittance, ℧ m^{-1}
G = shunt conductance, ℧ m^{-1}
C = shunt capacitance, F m^{-1}
$Z = R + j\omega L$
$Y = G + j\omega C$

General case	$Z_0 = \sqrt{\frac{Z}{Y}} = \sqrt{\frac{R + j\omega L}{G + j\omega C}}$
Small losses	$Z_0 = \sqrt{\frac{L}{C}}\left[1 + j\left(\frac{G}{2\omega C} - \frac{R}{2\omega L}\right)\right]$
Lossless case† $R = 0, G = 0$	$Z_0 = \sqrt{\frac{L}{C}} = R_0$

† Also holds approximately for the case where losses are not zero but $\omega L \gg R$ and $\omega C \gg G$.

A-6 MATERIAL CONSTANTS (PERMITTIVITY, CONDUCTIVITY AND DIELECTRIC STRENGTH)

Material	Relative permittivity		Conductivity	Dielectric strength,
	ε_r'	ε_r''	σ, ℧ m^{-1}	MV m^{-1}
Air (atmospheric pressure)	1.0006	0	0	3
Aluminum	1	0	3.5×10^7	
Bakelite	5	0.05	10^{-14}	25
Carbon			3×10^4	
Copper	1	0	5.8×10^7	
Glass (plate)	6	0.03	10^{-13}	30
Graphite			10^5	
Mica	6	0.2	10^{-15}	200
Oil, mineral	2.2	0.0002	10^{-14}	15
Paper (impregnated)	3	0.1		50
Paraffin	2.1	0.0004	$\sim 10^{-15}$	20
Plexiglas	3.4			
Polyfoam	$\sim$1.05			
Polystyrene	2.7	0.0002	10^{-16}	20
Polyvinyl chloride (PVC)	2.7			
Porcelain	5	0.004		
PVC (expanded)	$\sim$1.1			
Rubber, neoprene	5	0.02	10^{-13}	25
Quartz	5	0.001	10^{-17}	35
Rutile (titanium dioxide)	100	0.02		
Snow, fresh	1.5	0.5–0.0003		
Soil, clay	14		5×10^{-3}	
Soil, sandy	10		2×10^{-3}	
Stone (limestone)			10^{-2}	
Stone, slate	7			
Styrofoam	1.03			
Urban ground	4		2×10^{-4}	
Vacuum	1†	0	0	
Vaseline	2.2	0.0003		
Teflon	2.1	0.005	10^{-15}	60
Water, distilled	80		10^{-4}	
Water, fresh	80		10^{-2} to 10^{-3}	
Water, sea	80		4 to 5	
Wood, fir plywood	2	0.04		

† By definition.

Note: Both ε_r' and ε_r'' are, in general, a function of frequency. Values given are typical of the kilohertz to gigahertz range. The permittivity is also a function of the temperature. Values given are typical of temperatures near 25°C except for snow.

A-7 PERMITTIVITY RELATIONS

$$\varepsilon_r = \varepsilon_r' - j\varepsilon_r''$$

$$\varepsilon_r = \varepsilon_r' - j\,\frac{\sigma'}{\omega\varepsilon_0}$$

$$\varepsilon_r = \varepsilon_r' - j\left(\frac{\sigma + \omega\varepsilon''}{\omega\varepsilon_0}\right)$$

$$\varepsilon_r = \varepsilon_r' - j\left(\frac{\varepsilon''}{\varepsilon_0} + \frac{\sigma}{\omega\varepsilon_0}\right)$$

$$\varepsilon_r = \varepsilon_r' - j(\varepsilon_{rh}'' + \varepsilon_{rc}'')$$

$\varepsilon_r \simeq \varepsilon_r' - j\varepsilon_r'\text{PF}$, for small PF

Power loss $= \sigma E^2 + \omega\varepsilon'' E^2$ (W m^{-3})

where ε_r = relative permittivity = $\varepsilon/\varepsilon_0$, dimensionless
ε = permittivity, F m^{-1}
ε_0 = permittivity of vacuum = 8.85×10^{-12} F m^{-1}
ε_r' = relative permittivity related to displacement current
ε_r'' = relative permittivity related to equivalent conduction current
σ = dc conductivity, ℧ m^{-1}
σ' = equivalent conductivity, ℧ m^{-1}
ε_{rh}'' = relative permittivity related to hysteresis effects
ε_{rc}'' = relative permittivity related to dc conductivity
PF = power factor = $\sigma'/\omega\varepsilon$ for $\sigma' \ll \omega\varepsilon$
E = electric field, V m^{-1}
σE^2 = power loss due to dc conductivity, W m^{-3}
$\omega\varepsilon'' E^2$ = power loss due to hysteresis effects, W m^{-3}

A-8 CELESTIAL RADIO SOURCES FOR PATTERN, SQUINT, GAIN AND APERTURE EFFICIENCY MEASUREMENTS

	Position (epoch 1950.0)																	
	Right ascension				Declination						Flux density, Jy‡							
Source	Hours	Min.	Sec.	Sec. error	Deg.	Arc min.	Arc sec.	Arc sec. error	Size, arc min.	Isolation, deg.	38 MHz	178 MHz	750 MHz	1.4 GHz	2.7 GHz	5 GHz	Polarization, percent	Distance, light-years (and red shift)§
3C20	00	40	19.6	0.4	51	47	09	2	1	1	112	43	17	11.3	6.4	4.2	1.8	
3C196	08	09	59.4	0.2	48	22	07	2	0.25	1	166	68	23	13.9	7.7	4.4	1.2	8×10^9 ($z = 0.9$)
3C273	12	26	32.9	0.1	02	19	39	2	0.35	1	140	63	45	40	42	45	2	3×10^9 ($z = 0.16$)
3C295	14	09	33.6	0.3	52	26	14	2	0.3	0.5	94	83.5	35	22.4	12	6.5	0.4	7.5×10^9 ($z = 0.85$)
3C348	16	48	40.1	0.6	05	04	28	5	2	1	1690	351	84	44.5	22.4	11.9	1.5	3×10^9 ($z = 0.15$)
3C380	18	28	13.4	0.2	48	42	40	2	0.25	0.5	211	59	22	14.4	10	7.5	1.2	7×10^9 ($z = 0.7$)
Cygnus A	19	57	44.5	0.5	40	35	02		1.6	†	22 000	8700	2980	1590	785	371	0.2	10^9
3C123	04	33	55.2	0.2	29	34	14	2	0.1		57	189	72	45.8	27	16	1.2	
3C286	13	28	49.7	0.2	30	46	02	2	0.04	0.5	32	24	18	14.6	10	7.5	9	
Cas A	23	21	07		58	33	48		5	†	37 200	11 600	3880	2410	1470	910	—	11 000

† In complex galactic plane region.
‡ Jy = 10^{-26} W m^{-2} Hz^{-1}.
§ 1 light year = 9.46×10^{12} km.

For pattern measurements, the source should have a small angular extent (considerably less than the antenna HPBW), be relatively strong and be well isolated from nearby sources. For gain measurements, accurate flux densities are required and generally the source should be essentially unpolarized (less than 1 or 2 percent). For squint measurements, the position of the sources should be accurately known.

The upper part of the table lists a few selected radio sources which meet most or all of the above requirements, i.e., they are relatively strong, $\frac{1}{2}°$ to 1° from the nearest neighboring sources, of small angular extent, are essentially unpolarized, have accurate positions and also accurate flux densities over a wide frequency range. Three more sources (listed in the lower part of the table) do not meet all of the above requirements but nevertheless may be useful.

For more information see J. D. Kraus, *Radio Astronomy*, 2nd ed., Cygnus-Quasar, 1986.

A-9 MAXWELL'S EQUATIONS[1]

Maxwell's equations in differential form

	From Ampère	From Faraday	From Gauss	From Gauss
Case / Dimensions:	Electric current / Area	Electric potential / Area	Electric flux / Volume	Magnetic flux / Volume
General	$\nabla \times \mathbf{H} = \mathbf{J} + \frac{\partial \mathbf{D}}{\partial t}$	$\nabla \times \mathbf{E} = -\frac{\partial \mathbf{B}}{\partial t}$	$\nabla \cdot \mathbf{D} = \rho$	$\nabla \cdot \mathbf{B} = 0$
Free space	$\nabla \times \mathbf{H} = \frac{\partial \mathbf{D}}{\partial t}$	$\nabla \times \mathbf{E} = -\frac{\partial \mathbf{B}}{\partial t}$	$\nabla \cdot \mathbf{D} = 0$	$\nabla \cdot \mathbf{B} = 0$
Harmonic variation	$\nabla \times \mathbf{H} = (\sigma + j\omega\varepsilon)\mathbf{E}$	$\nabla \times \mathbf{E} = -j\omega\mu\mathbf{H}$	$\nabla \cdot \mathbf{D} = \rho$	$\nabla \cdot \mathbf{B} = 0$
Steady	$\nabla \times \mathbf{H} = \mathbf{J}$	$\nabla \times \mathbf{E} = 0$	$\nabla \cdot \mathbf{D} = \rho$	$\nabla \cdot \mathbf{B} = 0$
Static	$\nabla \times \mathbf{H} = 0$	$\nabla \times \mathbf{E} = 0$	$\nabla \cdot \mathbf{D} = \rho$	$\nabla \cdot \mathbf{B} = 0$

[1] The first table gives Maxwell's equations in differential form and the second table in integral form. The equations are stated for the general case, free-space case, harmonic-variation case, steady case (static fields but with conduction currents) and static case (static fields with no currents). In the table giving the integral form, the equivalence is also indicated between the various equations and the electric potential or emf V, the magnetic potential or mmf U, the electric current I, the electric flux ψ and the magnetic flux ψ_m.

Maxwell's equations in integral form

	From Ampère	From Faraday	From Gauss	From Gauss
Dimensions:	**Magnetic potential**	**Electric potential**	**Electric flux**	**Magnetic flux**
Mks units:	**amperes**	**volts**	**coulombs**	**webers**
Case				
General	$U = \oint \mathbf{H} \cdot d\mathbf{l} = \iint \left(\mathbf{J} + \frac{\partial \mathbf{D}}{\partial t}\right) \cdot d\mathbf{s} = I_{\text{total}}$	$V = \oint \mathbf{E} \cdot d\mathbf{l} = -\iint \frac{\partial \mathbf{B}}{\partial t} \cdot d\mathbf{s}$	$\psi = \iint \mathbf{D} \cdot d\mathbf{s} = \iiint \rho \, d\tau$	$\psi_m = \iint \mathbf{B} \cdot d\mathbf{s} = 0$
Free space	$U = \oint \mathbf{H} \cdot d\mathbf{l} = \iint \frac{\partial \mathbf{D}}{\partial t} \cdot d\mathbf{s} = I_{\text{disp}}$	$V = \oint \mathbf{E} \cdot d\mathbf{l} = -\iint \frac{\partial \mathbf{B}}{\partial t} \cdot d\mathbf{s}$	$\psi = \iint \mathbf{D} \cdot d\mathbf{s} = 0$	$\psi_m = \iint \mathbf{B} \cdot d\mathbf{s} = 0$
Harmonic variation	$U = \oint \mathbf{H} \cdot d\mathbf{l} = (\sigma + j\omega\varepsilon) \iint \mathbf{E} \cdot d\mathbf{s} = I_{\text{total}}$	$V = \oint \mathbf{E} \cdot d\mathbf{l} = -j\omega\mu \iint \mathbf{H} \cdot d\mathbf{s}$	$\psi = \iint \mathbf{D} \cdot d\mathbf{s} = \iiint \rho \, d\tau$	$\psi_m = \iint \mathbf{B} \cdot d\mathbf{s} = 0$
Steady	$U = \oint \mathbf{H} \cdot d\mathbf{l} = \iint \mathbf{J} \cdot d\mathbf{s} = I_{\text{cond}}$	$V = \oint \mathbf{E} \cdot d\mathbf{l} = 0$	$\psi = \iint \mathbf{D} \cdot d\mathbf{s} = \iiint \rho \, d\tau$	$\psi_m = \iint \mathbf{B} \cdot d\mathbf{s} = 0$
Static	$U = \oint \mathbf{H} \cdot d\mathbf{l} = 0$	$V = \oint \mathbf{E} \cdot d\mathbf{l} = 0$	$\psi = \iint \mathbf{D} \cdot d\mathbf{s} = \iiint \rho \, d\tau$	$\psi_m = \iint \mathbf{B} \cdot d\mathbf{s} = 0$

A-10 BEAM WIDTH AND SIDELOBE LEVEL FOR RECTANGULAR AND CIRCULAR APERTURE DISTRIBUTIONS†

Aperture field distribution		Half-power beam width	Level of first sidelobe, dB
Rectangular or linear apertures	$E(x)$		
Tapered to $\frac{1}{3}$ at edge (~10 dB down) $E(x) = 1 - 2x^2/3$	−1 0 x +1	$\frac{59°}{L_\lambda}$	−19
Tapered to zero at edge $E(x) = 1 - x^2 \simeq \cos(\pi x/2)$	L_λ	$\frac{66°}{L_\lambda}$	−21
Tapered to zero at edge $E(x) = \cos^2(\pi x/2)$		$\frac{83°}{L_\lambda}$	−32
Circular apertures	$E(r)$		
Uniform	D_λ	$\frac{58°}{D_\lambda}$	−18
Tapered to $\frac{1}{3}$ at edge (~10 dB down) $E(r) = 1 - 2r^2/3$	1 0 r 1	$\frac{66°}{D_\lambda}$	−23
Tapered to zero at edge $E(r) = 1 - r^2$		$\frac{73°}{D_\lambda}$	−25
Tapered to zero at edge $E(r) = (1 - r^2)^2$		$\frac{84°}{D_\lambda}$	−31

† $L_\lambda = L/\lambda$, $D_\lambda = D/\lambda$. It is assumed that $L_\lambda \gg 1$ and $D_\lambda \gg 1$. For a uniform rectangular or linear aperture HPBW $\simeq 51°/L_\lambda$ with first sidelobe −13 dB. See also Tables 12-3 and 13-1.

APPENDIX B

COMPUTER PROGRAMS (CODES)

1. **Horizontal dipole arrays over imperfect ground.** Computer programs for pattern and gain calculations of HF horizontal dipole arrays over imperfect ground were developed in 1986 by the International Radio Consultive Committee (CCIR). These programs in BASIC are available on disks from the International Telecommunication Union, General Secretariat (Sales Section), Place de Nations, CH-1211 Geneva 20, Switzerland. The programs have the code name HFMULSLW–HFDUASLW.
2. **Three-*d* pattern plots.** A computer program that plots antenna patterns in a 3-dimensional form (with hidden lines not plotted) was written in 1982 by W. A. Sandrin in FORTRAN IV. It is available as ASIS-NAPS Document NAPS-04053 in photocopy or microfiche from NAPS c/o Microfiche Publications, PO Box 3513, Grand Central Station, New York, NY 10163.
10163.
3. **The Numerical/Electromagnetics Code (NEC).**[1] This program or code has been under development for many years. It is a hybrid code which uses an Electric Field Integral Equation (EFIE) to model wire-like objects and a Magnetic Field Integral Equation (MFIE) to model surface-like objects with time-

[1] J. K. Breakall, G. J. Burke and E. K. Miller, "The Numerical/Electromagnetics Code," Lawrence Livermore Natl. Lab., Document UCRL-90560, 1984.

harmonic excitation. A 3-term sinusoidal spline basis is used for the wire current and a pulse basis for the surface current. Antennas and scatterers can be modeled in their environments. *Users of this program and also of almost all computer models should be aware of the uncertainty of solution accuracy and validity.* Therefore, are the number of wire segments or surface patches adequate? Adding more can be expensive and may actually be ambiguous if convergence is not monotonic.

An 8000-line program by G. J. Burke, A. J. Poggio and E. K. Miller of Lawrence Livermore National Laboratory is available for radiation or scattering from a wire structure in free space or over a ground plane (date 1980).

4. **Wire Antenna Program.** A 5000-line FORTRAN IV program by R. J. Marhefka of the ElectroScience Lab (OSU) is available from W. D. Burnside of the laboratory for radiation, gain and scattering of wire antennas near conducting structures. Incorporates NEC (date 1978).
5. **Wire and Plate Program.** A 4000-line FORTRAN program by E. H. Newman and D. M. Pozar of the ElectroScience Lab (OSU) is available from E. H. Newman for impedance, currents, radiation and scattering of 3-dimensional objects consisting of wires and/or plates. See E. H. Newman and D. M. Pozar, "Electromagnetic Modeling of Composite Wire and Surface Geometries," *IEEE Trans. Ants. Prop.*, **AP-26**, November 1978.

Other codes (programs) are listed at the end of Chap. 9.

B-1 ADDITIONAL COMPUTER PROGRAM REFERENCES

Baum, C. E.: "Emerging Technology for Transient and Broadband Analysis and Synthesis of Antennas and Scatterers," *Proc. IEEE*, **64**, November 1976.

Brittingham, J. N., E. K. Miller and J. L. Willows: "Pole Extraction from Real-Frequency Information," *Proc. IEEE*, **68**, 263–273, February 1980.

Burke, G. J., and E. K. Miller: "Modeling Antennas Near to and Penetrating a Lossy Surface," Lawrence Livermore Lab. Rept. 89838, 1983.

Burke, G. J., E. K. Miller, J. N. Brittingham, D. L. Lager, R. J. Lytle and J. T. Okada: "Computer Modeling of Antennas near the Ground," *Electromag.*, **1**, 29–49, 1981. Includes current distributions and elevation patterns of Beverage antennas.

Harrington, R. F., and J. R. Mantz: "Theory of Characteristic Modes for Conducting Bodies," *IEEE Trans. Ants. Prop.*, **AP-19**, 622–629, 1971.

McDonald, B. H., and A. Wexler: "Finite-Element Solution of Unbounded Field Problems," *IEEE Trans. Microwave Theory Tech.*, **MTT-20**, 841–847, December 1972.

Miller, E. K.: "Numerical Modeling Techniques," Lawrence Livermore Lab. Rept. 89613, 1983.

Miller, E. K., G. J. Burke and E. S. Selden: "Accuracy Modeling Guidelines for Integral-Equation Evaluation of Thin-Wire Structures," *IEEE Trans. Ants. Prop.*, **AP-19**, 1971.

Miller, E. K., and F. J. Deadrick: "Some Computational Aspects of Thin-Wire Modeling," in R. Mittra (ed.), *Numerical and Asymptotic Techniques in Electromagnetics*, Springer-Verlag, 1975, chap. 4.

Miller, E. K., F. J. Deadrick and G. J. Burke: "Computer Graphics Applications in Electromagnetic Computer Modeling," *Electromag.*, **1**, 135–153, 1981.

Rautio, J. C.: "Reflection Coefficient Analysis of the Effect of Ground on Antenna Patterns," *IEEE Ants. Prop. Soc. Newsletter*, **29**, 5–11, February 1987.

Wilton, D. R., S. S. M. Rao and A. W. Glisson, "Triangular Patch Modeling of Arbitrary Bodies," *IEEE Ants. Prop. Soc. Symp.*, University of Washington, 1979.

See also references at the end of Chap. 9.

B-2 BASIC PHASED-ARRAY ANTENNA PATTERN PROGRAMS.[1] These BASIC programs supplement those of Chap. 4 for linear arrays of isotropic sources of equal amplitude and spacing. These programs, based on (4-6-9) and (4-6-14), provide normalized field and power patterns in polar and rectangular coordinates for a variety of conditions including broadside, end-fire, angle-fire and interferometer cases for N sources with spacing D (λ), phasing S (rad) and multiplying (or scale) factor MF. These programs are Applesoft BASIC. To change to IBM GW-BASIC see note on page 862.

General program (*field pattern, polar plot*):

```
10 N = ?: D = ?: S = ?: MF = ?
20 HOME: CD = 6.28*D
30 HGR
40 HCOLOR = 3
50 FOR A = .01 TO 6.27 STEP .01
60 CA = COS(A): PF = CD*CA + S
70 R = SIN(N*PF/2)/SIN(PF/2)
80 R = MF*ABS(R)
90 HPLOT 138 + R*CA, 79 + R*SIN(A)
100 NEXT A
```

Program 1. Broadside array of 4 sources with $\lambda/2$ spacing. First or menu line should read:

```
10 N = 4: D = .5: S = 0: MF = 16.8
```

Program 2. Ordinary end-fire array of 8 sources with $\lambda/4$ spacing. First line should read:

```
10 N = 8: D = .25: S = -1.57: MF = 8.4
```

Program 3. End-fire array with increased-directivity of 8 sources with $\lambda/4$ spacing. First line should read:

```
10 N = 8: D = .25: S = -1.96: MF = 13.1
```

Program 4. End-fire array with increased-directivity of 24 sources with $\lambda/4$ spacing. First line should read:

```
10 N = 24: D = .25: S = -1.70: MF = 4.38
```

---

[1] The assistance of Marc Abel in preparing these Apple compatible programs is gratefully acknowledged.

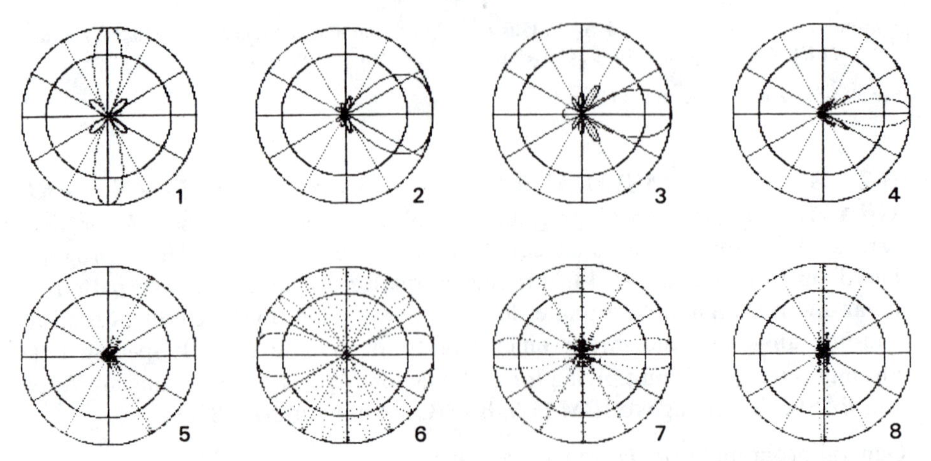

Figure B-1 Polar field patterns for the 8 computer program examples of linear phased arrays of N isotropic sources along the horizontal axis. The 3-dimensional patterns are figures-of-revolution around the horizontal axis. Thus, in the broadside patterns 1 and 8 the mainlobe is a disc and in the end-fire patterns 2, 3 and 4 the main lobe is like a balloon or zeppelin. The main lobe in 5 is conical, as are the lobes between broadside and end-fire in 6 and 7. The inner circle is at half-power.

Program 5. Angle-fire array (beam at 30° from broadside) of 16 sources with $\lambda/2$ spacing. First line should read:

```
10 N = 16: D = .5: S = -1.57: MF = 4.2
```

Program 6. Interferometer 2-source array with 4λ spacing. First line should read:

```
10 N = 2: D = 4: S = 0: MF = 33.5
```

Program 7. Broadside array of 8 sources with 2λ spacing and grating lobes. First line should read:

```
10 N = 8: D = 2: S = 0: MF = 8.4
```

Program 8. Broadside array 12 sources with $\lambda/2$ spacing. First line should read:

```
10 N = 12: D = .5: S = 0: MF = 5.6
```

Polar field patterns for the above 8 programs are shown in Fig. B-1. Compare the grating lobes of (7) with Fig. 11-78.

For different numbers of sources (N), spacings (D) and phasings (S) an unlimited variety of patterns are possible. Combinations of broadside and end-fire arrays are left as an exercise. For example, a broadside array (as in Program 1 but with $D = 1.5$) of 4 end-fire arrays (as in Program 3) results in a 32-source area array of high gain and small sidelobes in the plane of the array.

Although the magnification factor MF is arbitrary, the product of N and MF should be a constant for all pattern maxima to be equal (or normalized) when sources add in phase. Thus, in the above examples N*MF = 67 or

MF = 67/*N*. For increased directivity, however, all sources do not add in phase at the pattern maximum so make MF = 67*SIN(1.57/N).

For *power pattern, polar plot* change line 80 to:

```
80 R = MF*ABS(R)*ABS(R)/N
```

For adding polar plot coordinate lines to programs 1 through 8 continue from line 80 as follows:

```
90 HPLOT 138+R*.83*CA, 79+R*SIN(A)
100 NEXT A
110 FOR X = -58 TO 58 STEP 1
120 Y = 0
130 HPLOT 138+X, 79+Y
140 NEXT X
150 FOR Y = -67 TO 67 STEP 1
160 X = 0
170 HPLOT 138+X, 79+Y
180 NEXT Y
190 FOR X = -50 TO 50 STEP 1
200 Y = 1.2*.577*X
210 HPLOT 138+X, 79+Y
220 HPLOT 138-X, 79-Y
230 HPLOT 138+X, 79-Y
240 HPLOT 138-X, 79+Y
250 NEXT X
260 FOR X = -29 TO 29 STEP 1
270 Y = 2.08*X
280 HPLOT 138+X, 79+Y
290 HPLOT 138-X, 79-Y
300 HPLOT 138+X, 79-Y
310 HPLOT 138-X, 79+Y
320 NEXT X
330 FOR A = .01 TO 6.27 STEP .01
340 HPLOT 138+56*COS(A), 79+67*SIN(A)
350 HPLOT 138+40*COS(A), 79+47*SIN(A)
360 NEXT A
RUN
```

By entering the above 36-line program and storing it on a disc, an infinity of array patterns can be calculated and drawn by simply entering a new line 10 with the desired parameters as in the 8 example programs.

The factor *.83 in line 90 equalizes the *X* and *Y* scales on the printer used. Equivalent equalizing factors are written into the *X* and *Y* instructions for the coordinates. These factors may be modified or omitted.

For *field pattern, rectangular plot* change lines 80 and 90 to:

```
80 R = MF*R
90 HPLOT A*30, 75-R
```

For *power pattern, rectangular plot* change lines 80 and 90 to:

```
80 R = MF*ABS(R)*ABS(R)/N
90 HPLOT A*30, 75-R
```

To run the above programs as **IBM GW-BASIC** programs:

Change HOME to CLS.
Change HGR to SCREEN 2.
Delete line 40.
Change HPLOT to PSET with parentheses enclosing the rest of the line.

APPENDIX C

BOOKS AND VIDEO TAPES

C-1 BOOKS

Abraham, M., and R. Becker: *Electricity and Magnetism*, Stechert, 1932.
Aharoni, J.: *Antennae*, Oxford, 1946.
Bahl, I. J., and P. Bhartia: *Microstrip Antennas*, Artech House, 1980.
Balanis, C. A.: *Antenna Theory: Analysis and Design*, Harper and Row, 1982.
Biraud, F. (ed.): *Very Long Baseline Interferometry Techniques*, Cepadues, 1983.
Blake, L. V.: *Antennas*, Artech House, 1984.
Born, M.: *Optik*, Springer, 1933.
Bowman, J. J., T. B. A. Senior and P. L. E. Uslenghi: *Electromagnetic and Acoustic Scattering by Simple Shapes*, North Holland, Amsterdam, 1969.
Bracewell, R. N.: *The Fourier Transform and Its Applications*, McGraw-Hill, 1965.
Bracewell, R. N.: *The Hartley Transform*, Clarendon Press, Oxford, 1986.
Brillouin, L.: *Wave Propagation in Periodic Structures*, McGraw-Hill, 1946.
Brown, George H.: *And Part of Which I Was*, Angus Cupar (117 Hunt Drive, Princeton, NJ 08540), 1982.
Brückmann, H.: *Antennen, ihre Theorie und Technik*, Hirzel, 1939.
Burrows, M. L.: *ELF Communications Antennas*, Peregrinus, 1978.
Cady, W. M., M. B. Karelitz and L. A. Turner: *Radar Scanners and Radomes*, McGraw-Hill, 1948.
Christiansen, W. N., and J. A. Hogbom: *Radio Telescopes*, Cambridge, 1969, 1985.
Clarke, J. (ed.): *Advances in Radar Techniques*, Peregrinus, 1985.
Clarricoats, P. J. B., and A. D. Olver: *Corrugated Horns for Microwave Antennas*, Peregrinus, 1984.
Collin, R. E.: *Antennas and Radiowave Propagation*, McGraw-Hill, 1985.
Collin, R. E., and F. J. Zucker (eds.): *Antenna Theory*, McGraw-Hill, 1969.
Cornbleet, S.: *Microwave Optics: The Optics of Microwave Antenna Design*, Academic Press, 1984.
Delogne, P.: *Leaky Feeders and Subsurface Radio Communication*, IEE, London, 1982.
Elliott, R. S.: *Antenna Theory and Design*, Prentice-Hall, 1981.
Evans, J. V., and T. Hagfors (eds.): *Radar Astronomy*, McGraw-Hill, 1968.

Fanti, R., K. Kellermann and G. Setti (eds.): *VLBI and Compact Radio Sources*, Reidel, 1984.
Faraday, M.: *Experimental Researches in Electricity*, Quaritch, 1839, 1855.
Galejs, J.: *Antennas in Inhomogeneous Media*, Pergamon, 1969.
Hall, G. L. (ed.): *ARRL Antenna Book*, American Radio Relay League, 1984.
Hallén, E.: *Teoretisk Electricitetslära*, Skrivbyran Standard, 1945.
Hansen, R. C.: *Microwave Scanning Antennas*, vols. 1, 2, 3, Academic Press, 1966.
Harper, A. E.: *Rhombic Antenna Design*, Van Nostrand, 1941.
Harrington, R. F.: *Field Computation by Moment Methods*, Macmillan, 1968.
Hertz, Heinrich, R.: *Electric Waves*, Macmillan, 1893; Dover, 1962.
Hertz, Heinrich: *Memoirs, Letters, Diaries*, San Francisco Press, 1977.
Hord, R. M.: *Remote Sensing, Methods and Applications*, Wiley, 1986.
Hudson, J. E.: *Adaptive Array Principles*, Peregrinus, 1981.
Huygens, C.: *Traite de la Luminière*, Leyden, 1690.
James, G. L.: *Geometrical Theory of Diffraction for Electromagnetic Waves*, Peregrinus, 1980.
James, J. R., P. S. Hall and C. Wood: *Microstrip Antenna Theory and Design*, Peregrinus, 1982.
Jansky, D. M., and M. C. Jeruchim: *Communication Satellites in the Geostationary Orbit*, Artech House, 1983.
Johnson, R. C., and H. Jasik (eds.): *Antenna Engineering Handbook*, McGraw-Hill, 1984.
Johnson, R. C., and H. Jasik (eds.): *Antenna Applications Reference Guide*, McGraw-Hill, 1987. (Selected chapters from *Antenna Engineering Handbook*.)
Jordan, E. C., and K. G. Balmain: *Electromagnetic Waves and Radiating Systems*, Prentice-Hall, 1968.
Jull, E. V.: *Antennas and Diffraction Theory*, Peregrinus, 1981.
Kiely, D. G.: *Dielectric Aerials*, Methuen, 1953.
King, R. W. P.: *The Theory of Linear Antennas*, Harvard, 1956.
King, R. W. P., R. B. Mack and S. S. Sandler: *Arrays of Cylindrical Dipole Antennas*, Cambridge, 1968.
King, R. W. P., H. R. Mimno and A. H. Wing: *Transmission Lines, Antennas and Wave Guides*, McGraw-Hill, 1945.
King, R. W. P., and G. S. Smith: *Antennas in Matter*, MIT Press, 1980.
Kraus, J. D.: *Electromagnetics*, McGraw-Hill, 1953, 1973, 1984.
Kraus, J. D.: *Radio Astronomy*, Cygnus-Quasar (PO Box 85, Powell, OH 43065), 1966, 1986.
Kraus, J. D.: *Big Ear*, Cygnus-Quasar, 1976.
Kuzmin, A. D., and A. S. Solomonovich: *Radio Astronomical Methods for the Measurement of Antenna Parameters*, Soviet Radio, 1974.
Landau, L., and E. Lifshitz: *The Classical Theory of Fields*, Addison-Wesley, 1951.
Laport, E. A.: *Radio Antenna Engineering*, McGraw-Hill, 1952.
Law, P. E., Jr.: *Shipboard Antennas*, Artech House, 1986.
Leonov, A. I., and K. I. Fomichev: *Monopulse Radar*, Artech House, 1986.
Lo, Y. T. (ed.): *Handbook of Antenna Theory and Design*, Van Nostrand Reinhold, 1987.
Love, A. W.: *Electromagnetic Horn Antennas*, IEEE Press, 1976.
Love, A. W.: *Reflector Antennas*, IEEE Press, 1978.
Luneburg, R. K.: *Mathematical Theory of Optics*, Brown University Press, 1944.
Ma, M. T.: *Theory and Application of Antenna Arrays*, Wiley, 1974.
Mar, J. W., and H. Liebowitz (eds.): *Structures Technology for Large Radar and Radio Telescope Systems*, MIT, 1969.
Marconi, Degna: *My Father Marconi*, McGraw-Hill, 1962.
Marcuse, D.: *Theory of Dielectric Optical Waveguides*, Academic Press, 1974.
Maxwell, J. C.: *A Treatise on Electricity and Magnetism*, Oxford, 1873.
Monzingo, R. A., and T. W. Miller: *Introduction to Adaptive Arrays*, Wiley, 1980.
Moullin, E. B.: *Radio Aerials*, Oxford, 1949.
Poincaré, H.: *Theorie Mathematique de la Luminiere*, Carre, 1892.
Popovic, B. D., M. B. Dragovic and A. R. Djordjevic: *Analysis and Synthesis of Wire Antennas*, Wiley, 1982.
Pozar, D.: *Antenna Design Using Personal Computers*, Artech House, 1985.

Rayleigh, Lord: *The Theory of Sound*, Macmillan, 1877, 1878, 1929, 1937.
Reich, H. J. (ed.): *Very High Frequency Techniques*, McGraw-Hill, 1947.
Reintjes, J. F. (ed.): *Principles of Radar*, McGraw-Hill, 1946.
Rhodes, D. R.: *Introduction to Monopulse*, McGraw-Hill, 1959.
Rhodes, D. R.: *Synthesis of Planar Antenna Sources*, Clarendon Press, Oxford, 1974.
Rudge, A. W., K. Milne, A. D. Olver and P. Knight (eds.): *Handbook of Antenna Design*, Peregrinus, 1983.
Rumsey, V. H.: *Frequency Independent Antennas*, Academic Press, 1966.
Rusch, W. V. T., and P. D. Potter: *Analysis of Reflector Antennas*, Academic Press, 1970.
Schelkunoff, S. A.: *Electromagnetic Waves*, Van Nostrand, 1948.
Schelkunoff, S. A.: *Advanced Antenna Theory*, Wiley, 1952.
Schelkunoff, S. A., and H. T. Friis: *Antennas: Theory and Practice*, Wiley, 1952.
Sherman, S. M.: *Monopulse Principles and Techniques*, Artech House, 1984.
Silver, S.: *Microwave Antenna Theory and Design*, McGraw-Hill, 1949.
Skolnik, M. I.: *Introduction to Radar Systems*, McGraw-Hill, 1980.
Slater, J. C.: *Microwave Transmission*, McGraw-Hill, 1942.
Slater, J. C., and N. H. Frank: *Introduction to Theoretical Physics*, McGraw-Hill, 1933.
Smith, C. E.: *Directional Antennas*, Cleveland Institute of Radio Electronics, 1946.
Smith, C. E.: *Theory and Design of Directional Antenna Systems*, National Association of Broadcasters, 1949.
Steinberg, B. D.: *Principles of Aperture and Array System Design*, Wiley, 1976.
Stutzman, W. L., and G. A. Thiele: *Antenna Theory and Design*, Wiley, 1981.
Tai, C-T: *Dyadic Green's Functions in Electromagnetic Theory*, Intext, 1971.
Terman, F. E.: *Radio Engineers' Handbook*, McGraw-Hill, 1943.
Thompson, A. R., J. M. Moran and G. W. Swenson, Jr.: *Interferometry and Synthesis in Radio Astronomy*, Wiley–Interscience, 1986.
Tseitlin, N. M.: *Practical Methods of Radioastronomical and Antenna Techniques*, Soviet Radio, 1966.
Uchida, H.: *Fundamentals of Coupled Lines and Multiwire Antennas*, Sasaki (Sendai), 1967.
Uda, Shintaro: *On the Wireless Beam of Short Electric Waves*. Series of 11 articles on his wave canal (Yagi-Uda) antenna published in *J. IEE Japan*, between March 1926 and July 1929, plus earlier and later articles on meter wavelength experiments. Privately published bound volume.
Uda, S.: *Short Wave Projector*, Tohoku University, 1974.
Uda, S., and Y. Mushiake: *Yagi-Uda Antenna*, Tohoku University, 1954.
Ulaby, F. T., R. K. Moore and A. K. Fung: *Microwave Remote Sensing, Active and Passive*, Addison-Wesley, 1981.
Wait, J. R.: *Antennas and Propagation*, Peregrinus, 1986.
Walter, C. H.: *Traveling Wave Antennas*, Dover, 1972.
Watson, W. H.: *The Physical Principals of Wave Guide Transmission and Antenna Systems*, Oxford, 1947.
Weeks, W. L.: *Antenna Engineering*, McGraw-Hill, 1968.
Wood, P. J.: *Reflector Antenna Analysis and Design*, Peregrinus, 1980.
Zenneck, J.: *Wireless Telegraphy*, McGraw-Hill, 1915.

C-2 VIDEO TAPES

Kraus, J. D.: "Antennas and Radiation," Lecture-demonstration, excellent teaching supplement, 70 min. color, VHS. Cygnus-Quasar, P.O. Box 85, Powell, OH 43065.
Landt, J. A., and E. K. Miller: "Computer Graphics of Transient Radiation and Scattering Phenomena on Antennas and Wire Structures," Fields and currents in slow motion, 15 min, color, VHS. Cygnus-Quasar, P.O. Box 85, Powell, OH 43065.

APPENDIX D

ANSWERS TO STARRED PROBLEMS

CHAPTER 2

2-4. $71.6\lambda^2$
2-6. ~ 11 kW
2-9. $152\ m^2$ RCP
2-11. (*a*) AR = 1.5, (*b*) $\tau = 90°$, (*c*) CW
2-13. (*b*) AR = 1.38, (*c*) $\tau = 45°$
2-15. Straight line with $\tau = 45°$
2-18. (*a*) AR = -2.33 (RH), (*b*) $\tau = -45°$, (*c*) RH
2-20. (*a*) AR = -5, (*b*) RH, (*c*) 34 mW m^{-2}
2-22. (*a*) AR = 3.0, (*b*) $\tau = -22.5°$, (*c*) CW, (*d*) LH
2-25. 14.9 kW

CHAPTER 3

3-1. (*a*) 5.1, 6, 7.07; (*b*) 3.8, 4.6, 6.1
3-3. (*a*) 1539 W m^{-2}, (*b*) 4.29×10^{26} W, (*c*) 762 V m^{-1}

CHAPTER 4

4-1. (*c*) Max. at 0°, ±90°, 180°
Nulls at ±30°, ±150°
Half-power at ±14.5°, ±165.5°, ±48.6°, ±131.4°

(*d*) Max. at 0°, 180°, ±41.8°, ±138.2°
Nulls at ±19.4°, ±90°, ±160.6°
Half-power at ±9.6°, ±170.4°, ±30°, ±150°, ±56.5°, ±123.5°

(*e*) Max. at 0°, 180°, ±14.5°, ±165.5°, ±30°, ±150°, ±49°, ±131°, ±90°
Nulls at ±7°, ±173°, ±22°, ±158°, ±39°, ±141°, ±61°, ±119°
Half-power at ±3.6°, ±176.5°, ±11°, ±169°, ±18.5°, ±161.5°, ±26°, ±154° ±34.5°, ±145.5°, ±43.5°, ±136.5°, ±54.5°, ±125.5°, ±70°, ±110°

(*f*) Max. at 0°, 180°
Half-power at ±90°

4-4. (*a*) $E(\phi) = \frac{1}{4}\frac{\sin(\frac{5}{2}\pi \sin\phi)}{\sin(\frac{5}{8}\pi \sin\phi)} \Big/ \frac{15}{8}\pi \sin\phi$

(*c*) $E(\phi) = \cos^3(\frac{5}{8}\pi \sin\phi)$ or $\frac{3}{4}\cos(\frac{5}{8}\pi \sin\phi) + \frac{1}{4}\cos(\frac{15}{8}\pi \sin\phi)$

4-6. (*a*) 0.52, 0.82, 1.00, 0.82, 0.52
(*b*) Max. at ±39°, ±141°, ±90°
Nulls at ±30°, ±54°, ±126°, ±150°
(*d*) 24°

4-10. (*a*) 2

4-15. $R = 5$: 0.93, 0.84, 1.00, 1.00, 0.84, 0.93
$R = 7$: 0.69, 0.80, 1.00, 1.00, 0.80, 0.69
$R = 10$; 0.53, 0.78, 1.00, 1.00, 0.78, 0.53

4-18. (*a*) $E = \frac{\sin \frac{5}{2}\psi}{\sin \psi/2}$ where $\psi = d_r \cos\phi + \delta$
(*b*) $\delta = 0$, (1): 1, 1, 1, 1, 1, (2): 1, 4, 6, 4, 1, (3): 1, 0, 0, 0, 1

4-22. (*a*) and (*b*) 1 major and 5 minor lobes, (*c*) ordinary $D \sim 7$; inc. dir.: $D \sim 12$

4-23. 2500

4-25. (*b*) 6.6, (*c*) 6.3

4-34. 0.61

4-38. Max. 0°, 180°, ±60°, ±90°, ±120°
Min. ±41.4°, ±75.5°, ±104.5, ±138.6

4-44. (*a*) 44°, (*b*) −13.3 dB, (*c*) 0.17π sr, (*d*) 0.89, (*e*) 24, (*f*) $1.9\lambda^2$

CHAPTER 5

5-1. (*b*) $E_r = \frac{Ql\cos\theta}{2\pi\varepsilon r^3}$, $E_\theta = \frac{Ql\sin\theta}{4\pi\varepsilon r^3}$, $E_\phi = 0$

5-3. (*a*) $E = \tan\theta \sin\left(\frac{\pi}{2}\cos\theta\right)$, (*b*) 168 Ω, (*c*) 168, 73, 197 Ω

5-6. (*a*) $E = \frac{\sin\theta}{1 - p\cos\theta}\left[\sin \pi\left(\frac{1}{p} - \cos\theta\right)\right]$; 4-lobed patterns
(*b*) 40-lobed pattern

5-7. 3.33 Ω

5-9. (*a*) π sr, (*b*) 4

5-11. 354 Ω

5-13. (*a*) 8.16, (*b*) 653 Ω

CHAPTER 6

6-2. 4-lobed pattern

6-4. 1890 Ω

6-7. (1) 180 Ω, 1.5; (2) 1550 Ω, 1.2; (3) 4100 Ω, 3.6

CHAPTER 7

7-1. (1) 0.802, (2) 0.763

7-6. (*a*) $D_\lambda = \sqrt{2H_\lambda/\pi}$, (*b*) $E = \sin\theta$

CHAPTER 8

8-3. $270 + j350$ Ω

CHAPTER 10

10-1. $121 + j46$ Ω

10-5. $R_a = 50\ \Omega$, $R_b = 20\ \Omega$, $R_c = 50\ \Omega$

CHAPTER 11

11-1. (*b*) $\sim 0.67\lambda$

11-7. (*a*) 1 and 6: $63 + j29$; 2 and 5: $46 - j2$; 3 and 4: $53 + j10$ Ω

11-14. (*a*) $52 - j21\Omega$; (*b*) $G_f\,(\max) = 1.55$

11-17. (*b*) $\sim 17°$

11-18. 11.7°

11-22. (*a*) 0.354λ, (*b*) $-\pi/2$

11-26. $H_\lambda = 0.83$, $\phi = 72.5°$, $L_\lambda = 5.5$

11-28. $\phi = 72.5°$, $L_\lambda = 5.14$

11-30. $\phi = 60°$

11-34. (*a*) 6°22′, (*b*) −13.15 dB, (*c*) $\pi/4$ sr, (*d*) 0.89, (*e*) 16, (*f*) $1.27\lambda^2$

11-40. $\frac{1}{2}$

CHAPTER 12

12-6. (*b*) 73 Ω, (*c*) 10 dB

12-8. (*a*) 16 dBi, (*b*) 12.6°

12-13. 76.6 m^2

12-14. 0.81, 2036

12-15. 0.66, 1651

12-17. 81.1%, 1304 or 31.2 dBi

12-19. (*b*) 4.2°

CHAPTER 13

13-1. 710 Ω

13-3. 750

13-6. $779 - j87$ Ω

CHAPTER 14

14-3. (*a*) 62.5 mm, (*c*) 28%

CHAPTER 16

16-2. $56 - j50$ Ω

16-6. (*a*) 26.5 W m^{-2}, (*b*) 184 μW m^{-2}

16-8. 1.35 nW m^{-2}

16-11. $f_c = 0$, $8.89/\lambda_0$ Np m^{-1}

16-14. See App. E and Fig. E-1.

16-17. (*a*) 5°, 1.1 dB; 10°, 2.2 dB
(*b*) 5°, 1.4 dB; 10°, 2.9 dB
(*c*) 5°, 0.06 dB; 10°, 0.26 dB

16-21. 106, 35, 11 and 3.5 m

CHAPTER 17

17-1. (*a*) 12.2 dB, (*b*) 2.3 m

17-2. 14.5 K

17-4. (*a*) 0.08 K, (*b*) 0.09 K, (*c*) 445 K, (*d*) 500 mJy

17-6. (*a*) 0.06 K, (*b*) 320 mJy

17-8. 23.6 K

17-11. (*a*) 12.7 dB, (*b*) 0.5 dB, (*c*) ~13 min

17-16. 65.1 kW Hz^{-1}

17-18. (*a*) 15.6 MHz, (*b*) 26.1 MHz, (*c*) 41.0 MHz

CHAPTER 18

18-1. (*a*) 4.60 m, 21.2 m; (*b*) 25.2 mm, 116 mm; (*c*) 650 μm, 3.00 mm

18-3. 398 μm

18-5. 1.64 dB

18-7. (*a*) 377 Ω, (*b*) 275 mm

18-8. 27.5 mm

18-9. −9.9 dB

18-12. 57%

18-13. ~3:1

APPENDIX E

PROBLEM SUPPLEMENT

16-14 (*b*) Note that the level with the CP antennas is the same as would be obtained with either the VP or HP antennas if the signal was received by the direct path only (no ground reflection). However, with CP antennas the signal level is essentially independent of the

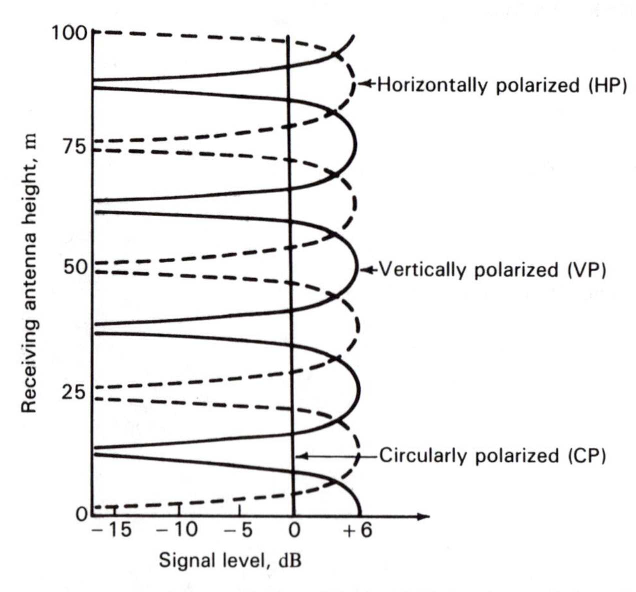

Figure E-1 Solution to part (*a*) of Prob. 16-14 showing variation of vertical, horizontal and circular polarization signals with height above ground.

height, while with the HP and VP antennas the level varies from 6 dB more to no signal at all.

Another important factor for TV reception is that with CP antennas the signal is received (ideally) over only the direct path while with the HP and VP antennas the signal is received via both direct and reflected paths. At maximum, the direct and reflected path signals are essentially equal in level but arrive at different times. If the time difference is of the order of a microsecond, objectionable ghost images will occur, degrading the picture quality.

NAME INDEX†

† n after a page number signifies *footnote*.

SUBJECT INDEX†

† n after a page number signifies *footnote.*

CONSTANTS AND CONVERSIONS

Quantity	Symbol or abbreviation	Nominal value	More accurate value†
Astronomical unit	AU	1.5×10^8 km	1.496×10^8
Boltzmann's constant	k	1.38×10^{-23} J K^{-1}	1.38062×10^{-23}
Earth mass		6.0×10^{24} kg	5.98×10^{24}
Earth radius (average)		6.37 Mm	
Electron charge	e	-1.60×10^{-19} C	-1.602×10^{-19}
Electron rest mass	m	9.11×10^{-31} kg	9.10956×10^{-31}
Electron charge-mass ratio	e/m	1.76×10^{11} C kg^{-1}	1.758803×10^{11}
Flux density (power)	Jy	10^{-26} W m^{-2} Hz^{-1}	10^{-26} (by definition)
Hydrogen atom (mass)		1.673×10^{-27} kg	
Hydrogen line rest frequency		1420.405 MHz	
Light-second		300 Mm	
Light, velocity of	c	300 Mm s^{-1}	299.7925
Light-year	LY	9.46×10^{12} km	9.4605×10^{12}
$\log x = \log_{10} x$ (common logarithm)			
$\ln x = \log_e x$ (natural logarithm)			
Logarithm, base	e	2.72	2.718282
reciprocal of base	$1/e$	0.368	0.36788
Logarithm conversion		$\ln x = 2.3 \log x$	$\ln x = 2.3026 \log x$
		$\log x = 0.43 \ln x$	$\log x = 0.4343 \ln x$
Moon distance (average)		380 Mm	
Moon mass		6.7×10^{22} kg	
Moon radius (average)		1.738 Mm	
Parsec	pc	3.1×10^{13} km	3.0856×10^{13}
Parsec	pc	3.26 Ly	3.2615
Parsec	pc	2.06×10^5 AU	2.06265
Permeability of vacuum†	μ_0	1260 nH m^{-1}	400π (exact value)
Permittivity of vacuum†	ε_0	8.85 pF m^{-1}	$8.854185 = 1/\mu_0 c^2$
Pi	π	3.14	3.1415927
Planck's constant	h	6.63×10^{-34} J s	6.62620×10^{-34}
Proton rest mass		1.67×10^{-27} kg	1.67261×10^{-27}
Radian	rad	57.3°	57.2958°
Space, impedance of†	Z	376.7 ($\approx 120\pi$) Ω	$376.7304 = \mu_0 c$
Sphere, solid angle		12.6 sr	$4\pi = 12.5664$
Sphere, solid angle		41253 deg^2	41252.96
Square degree	deg^2	3.05×10^{-4} sr	3.04617×10^{-4}
Stefan-Boltzmann constant		5.67×10^{-8} Wm^{-2} K^{-4}	5.6692×10^{-8}
Steradian (= square radian)	sr	3283 deg^2	$(180/\pi)^2 = 3282.806$
Sun, distance	AU	1.5×10^8 km	1.496×10^8
Sun mass	$M_\odot$	2.0×10^{30} kg	1.99×10^{30}
Sun radius (average)	$R_\odot$	700 Mm	695.3
Year (tropical)		365.24 days = 3.1556925×10^7 s	

† Same units as nominal value. Regarding permittivity ε_0 and space impedance Z, note that the values for these quantities are determined by the exact (definition) value of μ_0 and the measured value of c (velocity of light).